Rational Function

$$f(x) = \frac{N(x)}{D(x)} = \frac{a_n x^n + \cdots + a_1 x + a_0}{b_m x^m + \cdots + b_1 x + b_0}$$

$N(x)$ and $D(x)$ are polynomials.

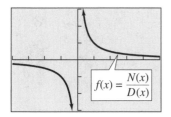

Domain: all real numbers, $D \neq 0$

Range: $(-\infty, 0), (0, \infty)$

y-intercept: $(0, f(x))$, if $f(0)$ exists

x-intercept(s): zeros of N

Vertical asymptotes: zeros of D

Horizontal asymptote:

$\quad y = 0$ if $n < m$

$\quad y = a_n/b_m$ if $n = m$

Exponential Function

$f(x) = a^x, \ a > 0$

$f(x) = a^{-x}, \ a > 0$

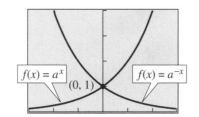

Domain: $(-\infty, \infty)$

Range: $(0, \infty)$

y-intercept: $(0, 1)$

Asymptote: $y = 0$

Logarithmic Function

$f(x) = \log_a x, \ a > 1$

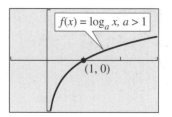
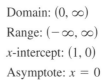

Domain: $(0, \infty)$

Range: $(-\infty, \infty)$

x-intercept: $(1, 0)$

Asymptote: $x = 0$

Sine Function

$f(x) = \sin x$

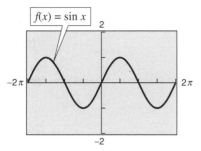

Domain: all real numbers

Range: $[-1, 1]$

Period: 2π

Cosine Function

$f(x) = \cos x$

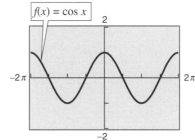

Domain: all real numbers

Range: $[-1, 1]$

Period: 2π

Tangent Function

$f(x) = \tan x$

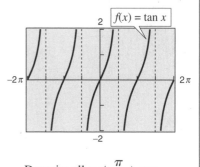

Domain: all $x \neq \dfrac{\pi}{2} + n\pi$

Range: $(-\infty, \infty)$

Period: π

Vertical asymptotes: $x = \dfrac{\pi}{2} + n\pi$

Cosecant Function

$f(x) = \csc x$

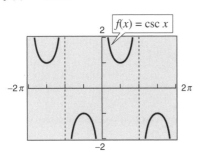

Domain: all $x \neq n\pi$

Range: $(-\infty, -1]$ and $[1, \infty)$
Period: 2π

Vertical asymptotes: $x = n\pi$

Secant Function

$f(x) = \sec x$

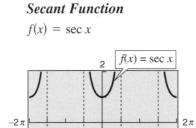

Domain: all $x \neq \dfrac{\pi}{2} + n\pi$

Range: $(-\infty, -1]$ and $[1, \infty)$
Period: 2π

Vertical asymptotes: $x = \dfrac{\pi}{2} + n\pi$

Cotangent Function

$f(x) = \cot x$

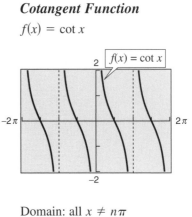

Domain: all $x \neq n\pi$

Range: $(-\infty, \infty)$
Period: π

Vertical asymptotes: $x = n\pi$

Inverse Sine Function

$f(x) = \arcsin x$

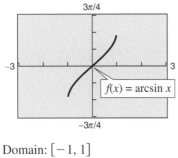

Domain: $[-1, 1]$

Range: $\left[-\dfrac{\pi}{2}, \dfrac{\pi}{2} \right]$

Inverse Cosine Function

$f(x) = \arccos x$

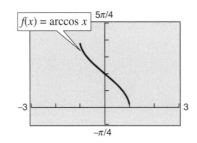

Domain: $[-1, 1]$

Range: $[0, \pi]$

Inverse Tangent Function

$f(x) = \arctan x$

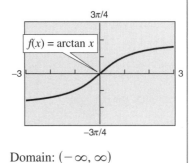

Domain: $(-\infty, \infty)$

Range: $\left(-\dfrac{\pi}{2}, \dfrac{\pi}{2} \right)$

Precalculus Functions and Graphs
A Graphing Approach

Third Edition

Ron Larson
Robert P. Hostetler
The Pennsylvania State University
The Behrend College

Bruce H. Edwards
University of Florida

With the assistance of David C. Falvo
The Pennsylvania State University
The Behrend College

Houghton Mifflin Company   Boston   New York

Editor-in-Chief: Jack Shira
Managing Editor: Cathy Cantin
Senior Associate Editor: Maureen Ross
Associate Editor: Laura Wheel
Assistant Editor: Carolyn Johnson
Supervising Editor: Karen Carter
Project Editor: Patty Bergin
Editorial Assistant: Kate Hartke
Art Supervisor: Gary Crespo
Senior Manufacturing Coordinator: Sally Culler
Marketing Manager: Michael Busnach
Freelance Development Editor: Michael Richards
Freelance Project Editor: Kathleen Deselle
Composition and Art: Meridian Creative Group
Cover Images: Meridian Creative Group
Cover Design: Gary Crespo

We have included examples and exercises that use real-life data as well as technology output from a variety of software. This would not have been possible without the help of many people and organizations. Our wholehearted thanks go to all for their time and effort.

Printed in the U.S.A.

Library of Congress Catalog Card Number: 00-133806

ISBN: 0-618-05290-9

3456789–DOW–04 03 02 01

Contents

A Word from the Authors

Welcome to *Precalculus Functions and Graphs: A Graphing Approach*, Third Edition. In this revision we have focused on student success, accessibility, and flexibility.

Accessibility: Over the years we have taken care to write a text for the student. We have paid careful attention to the presentation, using precise mathematical language and clear writing to create an effective learning tool. We believe that every student can learn mathematics and we are committed to providing a text that makes the mathematics within it accessible to all students.

In the Third Edition, we have revised and improved upon many text features designed for this purpose. Our pedagogical approach includes presenting solutions to examples from multiple perspectives—algebraic, graphic, and numeric. The side-by-side format allows students to see that a problem can be solved in more than one way, and to compare the accuracy of the solution methods.

Technology has been fully integrated into the text presentation. Also, the *Exploration* and *Study Tip* features have been expanded. *Chapter Tests*, which give students an opportunity for self-assessment, now follow every chapter in the Third Edition. The exercise sets now contain both *Synthesis* exercises, which check students' conceptual understanding, and *Review* exercises, which reinforce skills learned in previous sections and chapters. Students also have access to several media resources—videotapes, *Interactive Precalculus Functions and Graphs: A Graphing Approach* CD-ROM, and a *Precalculus Functions and Graphs: A Graphing Approach* website—that provide additional text-specific support.

Student Success: During our past 30 years of teaching and writing, we have learned many things about the teaching and learning of mathematics. We have found that students are most successful when they know what they are expected to learn and why it is important to learn it. With that in mind, we have restructured the Third Edition to include a thematic study thread in every chapter.

Each chapter begins with a study guide called *How to Study This Chapter*, which includes a comprehensive overview of the chapter concepts (*The Big Picture*), a list of *Important Vocabulary* that is integral to learning *The Big Picture* concepts, a list of study resources, and a general study tip. The study guide allows students to get organized and prepare for the chapter.

An old pedagogical recipe goes something like this, "First I'm going to tell you what I'm going to teach you, then I will teach it to you, and finally I will go over what I taught you." Following this recipe, we have included a set of learning objectives in every section that outlines what students are expected to learn, followed by an interesting real-life application that illustrates why it is important to learn the concepts in that section. Finally, the chapter summary (*What did you learn?*), which reinforces the section objectives, and the chapter *Review Exercises*, which are correlated to the chapter summary, provide additional study support at the conclusion of each chapter.

Our new *Student Success Organizer* supplement takes this study thread one step further, providing a content-based study aid.

Flexibility: From the time we first began writing in the early 1970s, we have always viewed part of our authoring role as that of providing instructors with flexible teaching programs. The optional features within the text allow instructors with different pedagogical approaches to design their courses to meet both their instructional needs and the needs of their students. In addition, we provide several print and media resources to support instructors, including a new *Instructor Success Organizer*.

We hope you enjoy the Third Edition.

Ron Larson

Robert P. Hostetler

Bruce H. Edwards

Acknowledgments

We would like to thank the many people who have helped us prepare the text and the supplements package. Their encouragement, criticisms, and suggestions have been invaluable to us.

Third Edition Reviewers

Jamie Whitehead Ashby, Texarkana College; Teresa Barton, Western New England College; Diane Burleson, Central Piedmont Community College; Alexander Burstein, University of Rhode Island; Victor M. Cornell, Mesa Community College; Marcia Drost, Texas A & M University; Kenny Fister, Murray State University; Susan C. Fleming, Virginia Highlands Community College; Nicholas E. Geller, Collin County Community College; Betty Givan, Eastern Kentucky University; John Kendall, Shelby State Community College; Donna M. Krawczyk, University of Arizona; JoAnn Lewin, Edison Community College; David E. Meel, Bowling Green University; Beverly Michael, University of Pittsburgh; Jon Odell, Richland Community College; Laura Reger, Milwaukee Area Technical College; Craig M. Steenberg, Lewis-Clark State College; Mary Jane Sterling, Bradley University; Ellen Vilas, York Technical College. In addition, we would like to thank all the algebra & trigonometry instructors who took the time to respond to our survey.

We would like to extend a special thanks to Ellen Vilas for her contributions to this revision.

We would like to thank the staff of Larson Texts, Inc., and the staff of Meridian Creative Group, who assisted in proofreading the manuscript, preparing and proofreading the art package, and typesetting the supplements.

On a personal level, we are grateful to our wives, Deanna Gilbert Larson, Eloise Hostetler, and Consuelo Edwards for their love, patience, and support. Also, a special thanks goes to R. Scott O'Neil.

If you have suggestions for improving this text, please feel free to write to us. Over the past two decades we have received many useful comments from both instructors and students, and we value these very much.

Ron Larson
Robert P. Hostetler
Bruce H. Edwards

Features Highlights

Student Success Tools

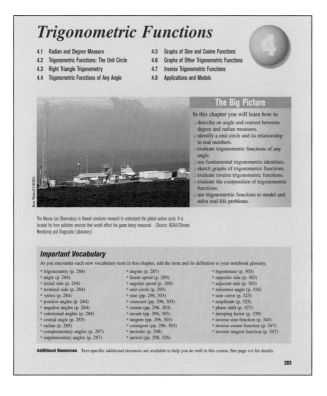

Trigonometric Functions

4.1 Radian and Degree Measure
4.2 Trigonometric Functions: The Unit Circle
4.3 Right Triangle Trigonometry
4.4 Trigonometric Functions of Any Angle
4.5 Graphs of Sine and Cosine Functions
4.6 Graphs of Other Trigonometric Functions
4.7 Inverse Trigonometric Functions
4.8 Applications and Models

The Big Picture

In this chapter you will learn how to

❏ describe an angle and convert between degree and radian measures.
❏ identify a unit circle and its relationship to real numbers.
❏ evaluate trigonometric functions of any angle.
❏ use fundamental trigonometric identities.
❏ sketch graphs of trigonometric functions.
❏ evaluate inverse trigonometric functions.
❏ evaluate the composition of trigonometric functions.
❏ use trigonometric functions to model and solve real-life problems.

The Mauna Loa Observatory in Hawaii conducts research to understand the global carbon cycle. It is located far from pollution sources that would affect the gases being measured. (Source: NOAA/Climate Monitoring and Diagnostic Laboratory)

Important Vocabulary

As you encounter each new vocabulary term in this chapter, add the term and its definition to your notebook glossary.

* trigonometry (p. 284)
* angle (p. 284)
* initial side (p. 284)
* terminal side (p. 284)
* vertex (p. 284)
* positive angles (p. 284)
* negative angles (p. 284)
* coterminal angles (p. 284)
* central angle (p. 285)
* radian (p. 285)
* complementary angles (p. 287)
* supplementary angles (p. 287)

* degree (p. 287)
* linear speed (p. 289)
* angular speed (p. 289)
* unit circle (p. 295)
* sine (pp. 296, 303)
* cosecant (pp. 296, 303)
* cosine (pp. 296, 303)
* secant (pp. 296, 303)
* tangent (pp. 296, 303)
* cotangent (pp. 296, 303)
* periodic (p. 298)
* period (pp. 298, 326)

* hypotenuse (p. 303)
* opposite side (p. 303)
* adjacent side (p. 303)
* reference angle (p. 316)
* sine curve (p. 323)
* amplitude (p. 325)
* phase shift (p. 327)
* damping factor (p. 339)
* inverse sine function (p. 345)
* inverse cosine function (p. 347)
* inverse tangent function (p. 347)

Additional Resources Text-specific additional resources are available to help you do well in this course. See page xvi for details.

283

New Chapter Openers include:

▶ **The Big Picture**

An objective-based overview of the main concepts of the chapter.

▶ **Important Vocabulary**

Mathematical terms integral to learning *The Big Picture* concepts.

New Section Openers include:

▶ **"What you should learn"**

Objectives outline the main concepts and help keep students focused on *The Big Picture*.

▶ **"Why you should learn it"**

A real-life application or a reference to other branches of mathematics illustrates the relevance of the section's content.

284 *Chapter 4 • Trigonometric Functions*

4.1 Radian and Degree Measure

Angles

As derived from the Greek language, the word **trigonometry** means "*measurement of triangles.*" Initially, trigonometry dealt with relationships among the sides and angles of triangles and was used in the development of astronomy, navigation, and surveying. With the development of calculus and the physical sciences in the 17th century, a different perspective arose—one that viewed the classic trigonometric relationships as *functions* with the set of real numbers as their domains. Consequently, the applications of trigonometry expanded to include a vast number of physical phenomena involving rotations and vibrations, including sound waves, light rays, planetary orbits, vibrating strings, pendulums, and orbits of atomic particles. The approach in this text incorporates *both* perspectives, starting with angles and their measure.

What You Should Learn:
• How to describe angles
• How to use radian measure
• How to use degree measure
• How to use angles to model and solve real-life problems

Why You Should Learn It:
You can use radian measures of angles to model and solve real-life problems. For instance, in Exercise 93 on page 293 you are asked to determine measures of angles through which figure skaters jump.

(a) (b)
Figure 4.1

An **angle** is determined by rotating a ray (half-line) about its endpoint. The starting position of the ray is the **initial side** of the angle, and the position after rotation is the **terminal side**, as shown in Figure 4.1(a). The endpoint of the ray is the **vertex** of the angle. This perception of an angle fits a coordinate system in which the origin is the vertex and the initial side coincides with the positive *x*-axis. Such an angle is in **standard position**, as shown in Figure 4.1(b). **Positive angles** are generated by counterclockwise rotation, and **negative angles** by clockwise rotation, as shown in Figure 4.2. Angles are labeled with Greek letters α (alpha), β (beta), and θ (theta), as well as uppercase letters *A*, *B*, and *C*. In Figure 4.3, note that angles α and β have the same initial and terminal sides. Such angles are **coterminal**.

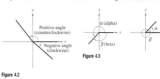

Figure 4.2

α (alpha)
β (beta)

Figure 4.3

"What did you learn?" Chapter Summary

This chapter summary provides a concise, section-by-section review of the section objectives. These objectives are correlated to the chapter Review Exercises.

Flexibility and Accessibility

▶ Algebraic, Graphical, and Numerical Approach

- Many examples present solutions from multiple approaches—algebraic, graphical, and numerical.
- Solutions are displayed side-by-side.
- The multiple-approach format shows students different solution methods that can be used to reach the same answer.
- The solution format helps students expand their problem-solving abilities.

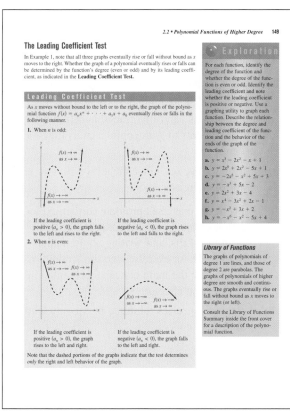

The Leading Coefficient Test

In Example 1, note that all three graphs eventually rise or fall without bound as x moves to the right. Whether the graph of a polynomial eventually rises or falls can be determined by the function's degree (even or odd) and by its leading coefficient, as indicated in the **Leading Coefficient Test.**

Leading Coefficient Test

As x moves without bound to the left or to the right, the graph of the polynomial function $f(x) = a_n x^n + \cdots + a_1 x + a_0$ eventually rises or falls in the following manner.

1. When n is odd:

If the leading coefficient is positive $(a_n > 0)$, the graph falls to the left and rises to the right.

If the leading coefficient is negative $(a_n < 0)$, the graph rises to the left and falls to the right.

2. When n is even:

If the leading coefficient is positive $(a_n > 0)$, the graph rises to the left and right.

If the leading coefficient is negative $(a_n < 0)$, the graph falls to the left and right.

Note that the dashed portions of the graphs indicate that the test determines *only* the right and left behavior of the graph.

Exploration

For each function, identify the degree of the function and whether the degree of the function is even or odd. Identify the leading coefficient and note whether the leading coefficient is positive or negative. Use a graphing utility to graph each function. Describe the relationship between the degree and leading coefficient of the function and the behavior of the ends of the graph of the function.

a. $y = x^3 - 2x^2 - x + 1$
b. $y = 2x^5 + 2x^2 - 5x + 1$
c. $y = -2x^5 - x^2 + 5x + 3$
d. $y = -x^3 + 5x - 2$
e. $y = 2x^3 + 3x - 4$
f. $y = x^4 - 3x^2 + 2x - 1$
g. $y = -x^2 + 3x + 2$
h. $y = -x^6 - x^2 - 5x + 4$

Library of Functions

The graphs of polynomials of degree 1 are lines, and those of degree 2 are parabolas. The graphs of polynomials of higher degree are smooth and continuous. The graphs eventually rise or fall without bound as x moves to the right (or left).

Consult the Library of Functions Summary inside the front cover for a description of the polynomial function.

Exploration

- Before introducing selected topics, *Exploration* engages students in active discovery of mathematical concepts and relationships, often through the power of technology.

- *Exploration* strengthens students' critical thinking skills and helps them develop an intuitive understanding of theoretical concepts.

- *Exploration* is an optional feature and can be omitted without loss of continuity in coverage.

▶ **Examples**

- Each example was carefully chosen to illustrate a particular mathematical concept or problem-solving skill.

- Every example contains step-by-step solutions, most with side-by-side explanations that lead students through the solution process.

- Many examples provide side-by-side solutions utilizing two separate approaches.

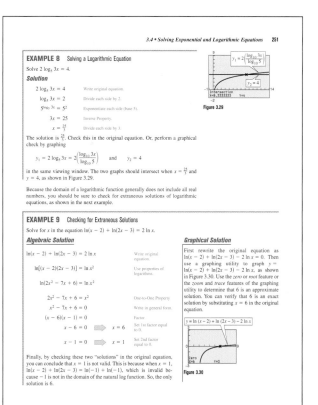

EXAMPLE 8 Solving a Logarithmic Equation

Solve $2 \log_5 3x = 4$.

Solution

$2 \log_5 3x = 4$	Write original equation.
$\log_5 3x = 2$	Divide each side by 2.
$5^{\log_5 3x} = 5^2$	Exponentiate each side (base 5).
$3x = 25$	Inverse Property.
$x = \frac{25}{3}$	Divide each side by 3.

The solution is $\frac{25}{3}$. Check this in the original equation. Or, perform a graphical check by graphing

$$y_1 = 2 \log_5 3x = 2\left(\frac{\log_{10} 3x}{\log_{10} 5}\right) \quad \text{and} \quad y_2 = 4$$

in the same viewing window. The two graphs should intersect when $x = \frac{25}{3}$ and $y = 4$, as shown in Figure 3.29.

Figure 3.29

Because the domain of a logarithmic function generally does not include all real numbers, you should be sure to check for extraneous solutions of logarithmic equations, as shown in the next example.

EXAMPLE 9 Checking for Extraneous Solutions

Solve for x in the equation $\ln(x - 2) + \ln(2x - 3) = 2 \ln x$.

Algebraic Solution

$\ln(x - 2) + \ln(2x - 3) = 2 \ln x$	Write original equation.
$\ln[(x - 2)(2x - 3)] = \ln x^2$	Use properties of logarithms.
$\ln(2x^2 - 7x + 6) = \ln x^2$	
$2x^2 - 7x + 6 = x^2$	One-to-One Property
$x^2 - 7x + 6 = 0$	Write in general form.
$(x - 6)(x - 1) = 0$	Factor
$x - 6 = 0 \Rightarrow x = 6$	Set 1st factor equal to 0.
$x - 1 = 0 \Rightarrow x = 1$	Set 2nd factor equal to 0.

Finally, by checking these two "solutions" in the original equation, you can conclude that $x = 1$ is not valid. This is because when $x = 1$, $\ln(x - 2) + \ln(2x - 3) = \ln(-1) + \ln(-1)$, which is invalid because -1 is not in the domain of the natural log function. So, the only solution is 6.

Graphical Solution

First rewrite the original equation as $\ln(x - 2) + \ln(2x - 3) - 2 \ln x = 0$. Then use a graphing utility to graph $y = \ln(x - 2) + \ln(2x - 3) - 2 \ln x$, as shown in Figure 3.30. Use the *zero* or *root* feature or the *zoom* and *trace* features of the graphing utility to determine that 6 is an approximate solution. You can verify that 6 is an exact solution by substituting $x = 6$ in the original equation.

Figure 3.30

Revised Exercises and Applications

400   *Chapter 5 • Analytic Trigonometry*

5.3   E x e r c i s e s

In Exercises 1–6, verify that the x-values are solutions of the equation.

1. $2\cos x - 1 = 0$

   (a) $x = \dfrac{\pi}{3}$ (b) $x = \dfrac{5\pi}{3}$

2. $\csc x - 2 = 0$

   (a) $x = \dfrac{\pi}{6}$ (b) $x = \dfrac{5\pi}{6}$

3. $3\tan^2 2x - 1 = 0$

   (a) $x = \dfrac{\pi}{12}$ (b) $x = \dfrac{5\pi}{12}$

4. $4\cos^2 2x - 2 = 0$

   (a) $x = \dfrac{\pi}{8}$ (b) $x = \dfrac{7\pi}{8}$

5. $2\cos^2 x + 3\cos x + 1 = 0$

   (a) $x = \dfrac{4\pi}{3}$ (b) $x = \pi$

6. $\sec^4 x - 3\sec^2 x - 4 = 0$

   (a) $x = \dfrac{2\pi}{3}$ (b) $x = \dfrac{5\pi}{3}$

In Exercises 7–10, find the x-intercepts of the graph. Use a graphing utility to check your solutions.

7. $y = \sin\dfrac{\pi x}{2} + 1$ 8. $y = \sin \pi x + \cos \pi x$

9. $y = \tan^2\!\left(\dfrac{\pi x}{6}\right) - 3$ 10. $y = \sec^4\!\left(\dfrac{\pi x}{8}\right) - 4$

In Exercises 11–24, solve the equation for $0 \le x < 2\pi$.

11. $2\cos x + 1 = 0$ 12. $\sqrt{2}\sin x + 1 = 0$
13. $\sqrt{3}\sec x - 2 = 0$ 14. $\cot x + 1 = 0$
15. $3\csc^2 x - 4 = 0$ 16. $\csc^2 x - 2 = 0$
17. $2\sin^2 2x = 1$ 18. $\tan^2 3x = 3$
19. $4\cos^2 x - 3 = 0$ 20. $\cos x(\cos x - 1) = 0$
21. $\sin^2 x = 3\cos^2 x$ 22. $\tan 3x(\tan x - 1) = 0$
23. $(3\tan^2 x - 1)(\tan^2 x - 3) = 0$
24. $\cos 2x(2\cos x + 1) = 0$

In Exercises 25–42, find all solutions of the equation in the interval $[0, 2\pi)$ algebraically. Use the *table* feature of a graphing utility to confirm your answers numerically.

25. $\cos^3 x = \cos x$ 26. $\tan^2 x - 1 = 0$
27. $3\tan^3 x = \tan x$ 28. $2\sin^2 x = 2 + \cos x$
29. $\sec^2 x - \sec x = 2$ 30. $\sec x \csc x = 2\csc x$
31. $2\sin x + \csc x = 0$ 32. $\sin 2x = -\dfrac{\sqrt{3}}{2}$
33. $\csc x + \cot x = 1$ 34. $\tan 3x = 1$
35. $\cos\dfrac{x}{2} = \dfrac{\sqrt{2}}{2}$ 36. $\sec 4x = 2$
37. $\dfrac{1 + \cos x}{1 - \cos x} = 0$
38. $2\sin^2 x + 3\sin x + 1 = 0$
39. $2\sec^2 x + \tan^2 x - 3 = 0$
40. $\cos x + \sin x \tan x = 2$
41. $\sec^2 x + \tan x = 3$
42. $\csc^2 x - 4\cot x = -2$

In Exercises 43–48, use a graphing utility to approximate the solutions of the equation in the interval $[0, 2\pi)$ by setting the equation equal to 0, graphing the new equation, and using the *zero* or *root* feature to approximate the x-intercepts of the graph.

43. $9\cos x + 2 = 3$ 44. $4\sin x = \cos x - 2$
45. $4\sin^2 x = 2\cos x + 1$ 46. $\csc^2 x = 3\csc x + 4$
47. $4\sin^3 x + 2\sin^2 x = 2\sin x + 1$
48. $\dfrac{1 + \sin x}{\cos x} + \dfrac{\cos x}{1 + \sin x} = 4$

▶ Exercises

- Exercise sets consist of a variety of computational, conceptual, and applied problems.

- Exercise sets carefully graded in difficulty allow students to gain confidence as they progress.

- Each exercise set now concludes with two new types of exercises:

  - *Synthesis* exercises promote further exploration of mathematical concepts, critical thinking skills, and writing about mathematics. These exercises require students to synthesize the main concepts presented in the section and chapter.

  - *Review* exercises reinforce previously learned skills and concepts.

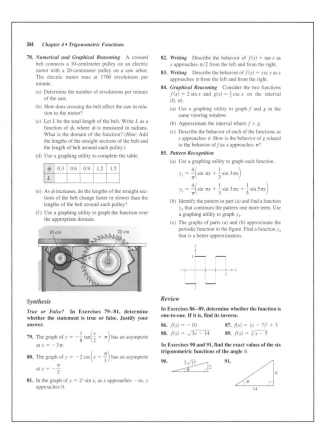

344   *Chapter 4 • Trigonometric Functions*

78. *Numerical and Graphical Reasoning*   A crossed belt connects a 10-centimeter pulley on an electric motor with a 20-centimeter pulley on a saw arbor. The electric motor runs at 1700 revolutions per minute.

   (a) Determine the number of revolutions per minute of the saw.

   (b) How does crossing the belt affect the saw in relation to the motor?

   (c) Let L be the total length of the belt. Write L as a function of ϕ, where ϕ is measured in radians. What is the domain of the function? (*Hint:* Add the lengths of the straight sections of the belt and the length of belt around each pulley.)

   (d) Use a graphing utility to complete the table.

ϕ	0.3	0.6	0.9	1.2	1.5
L					

   (e) As ϕ increases, do the lengths of the straight sections of the belt change faster or slower than the lengths of the belt around each pulley?

   (f) Use a graphing utility to graph the function over the appropriate domain.

82. *Writing*   Describe the behavior of $f(x) = \tan x$ as x approaches $\pi/2$ from the left and from the right.

83. *Writing*   Describe the behavior of $f(x) = \csc x$ as x approaches π from the left and from the right.

84. *Graphical Reasoning*   Consider the two functions $f(x) = 2\sin x$ and $g(x) = \frac{1}{2}\csc x$ on the interval $(0, \pi)$.

   (a) Use a graphing utility to graph f and g in the same viewing window.

   (b) Approximate the interval where $f > g$.

   (c) Describe the behavior of each of the functions as x approaches π. How is the behavior of g related to the behavior of f as x approaches π?

85. *Pattern Recognition*

   (a) Use a graphing utility to graph each function.

   $$y_1 = \frac{4}{\pi}\left(\sin \pi x + \frac{1}{3}\sin 3\pi x\right)$$

   $$y_2 = \frac{4}{\pi}\left(\sin \pi x + \frac{1}{3}\sin 3\pi x + \frac{1}{5}\sin 5\pi x\right)$$

   (b) Identify the pattern in part (a) and find a function y_3 that continues the pattern one more term. Use a graphing utility to graph y_3.

   (c) The graphs of parts (a) and (b) approximate the periodic function in the figure. Find a function y_4 that is a better approximation.

Synthesis

True or False?   In Exercises 79–81, determine whether the statement is true or false. Justify your answer.

79. The graph of $y = -\frac{1}{8}\tan\left(\frac{x}{2} + \pi\right)$ has an asymptote at $x = -3\pi$.

80. The graph of $y = -2\csc\left(x + \frac{\pi}{3}\right)$ has an asymptote at $x = -\frac{\pi}{3}$.

81. In the graph of $y = 2^x \sin x$, as x approaches $-\infty$, y approaches 0.

Review

In Exercises 86–89, determine whether the function is one-to-one. If it is, find its inverse.

86. $f(x) = -10$ 87. $f(x) = (x - 7)^2 + 3$
88. $f(x) = \sqrt{3x - 14}$ 89. $f(x) = \sqrt[3]{x - 5}$

In Exercises 90 and 91, find the exact values of the six trigonometric functions of the angle θ.

90.   91.

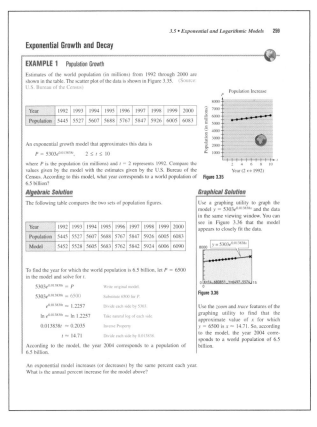

3.5 • *Exponential and Logarithmic Models* **259**

▶ Real-Life Applications

- A wide variety of real-life applications, many using current, real data, are integrated throughout the examples and exercises.

- The icon 🌐 indicates an example that involves a real-life application.

▶ Algebra of Calculus

- Special emphasis is given to the algebraic techniques used in calculus.
- Algebra of Calculus examples and exercises are integrated throughout the text.
- The symbol 🔵 indicates an example or exercise in which the Algebra of Calculus is featured.

▶ Additional Features

Carefully crafted learning tools designed to create a rich learning environment can be found throughout the text. These learning tools include a Library of Functions, Study Tips, Historical Notes, Writing About Math, Chapter Projects, Chapter Review Exercises, Chapter Tests, Cumulative Tests, and an extensive art program.

Supplements

Resources

Website (*college.hmco.com*)
Many additional text-specific study and interactive features for students and instructors can be found at the Houghton Mifflin website.

For the Student

Student Success Organizer

Study and Solutions Guide by Bruce H. Edwards (University of Florida)

Graphing Technology Guide by Benjamin N. Levy and Laurel Technical Services

Instructional Videotapes by Dana Mosely

Instructional Videotapes for Graphing Calculators by Dana Mosely

For the Instructor

Instructor's Annotated Edition

Instructor Success Organizer

Complete Solutions Guide by Bruce H. Edwards (University of Florida)

Test Item File

Problem Solving, Modeling, and Data Analysis Labs by Wendy Metzger (Palomar College)

Computerized Testing (Windows, Macintosh)

HMClassPrep Instructor's CD-ROM

An Introduction to Graphing Utilities

Graphing utilities such as graphing calculators and computers with graphing software are very valuable tools for visualizing mathematical principles, verifying solutions to equations, exploring mathematical ideas, and developing mathematical models. Although graphing utilities are extremely helpful in learning mathematics, their use does not mean that learning algebra is any less important. In fact, the combination of knowledge of mathematics and the use of graphing utilities allows you to explore mathematics more easily and to a greater depth. If you are using a graphing utility in this course, it is up to you to learn its capabilities and to practice using this tool to enhance your mathematical learning.

In this text there are many opportunities to use a graphing utility, some of which are described below.

Some Uses of a Graphing Utility

A graphing utility can be used to

- check or validate answers to problems obtained using algebraic methods.
- discover and explore algebraic properties, rules, and concepts.
- graph functions, and approximate solutions to equations involving functions.
- efficiently perform complicated mathematical procedures such as those found in many real-life applications.
- find mathematical models for sets of data.

In this introduction, the features of graphing utilities are discussed from a generic perspective. To learn how to use the features of a specific graphing utility, consult your user's manual or the website for this text found at *college.hmco.com.* Additionally, keystroke guides are available for most graphing utilities, and your college library may have a videotape on how to use your graphing utility.

The Equation Editor

Many graphing utilities are designed to act as "function graphers." In this course, you will study functions and their graphs in detail. You may recall from previous courses that a function can be thought of as a rule that describes the relationship between two variables. These rules are frequently written in terms of x and y. For example, the equation $y = 3x + 5$ represents y as a function of x.

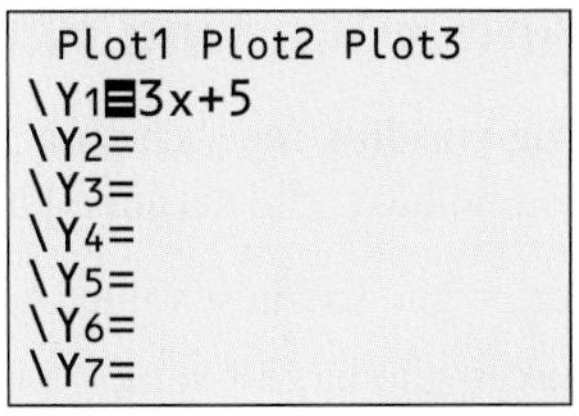

```
Plot1  Plot2  Plot3
\Y1■3x+5
\Y2=
\Y3=
\Y4=
\Y5=
\Y6=
\Y7=
```

Figure 1

Many graphing utilities have an equation editor that requires an equation to be written in "$y =$" form in order to be entered, as shown in Figure 1. (You should note that your equation editor screen may not look like the screen shown in Figure 1.) To determine exactly how to enter an equation into your graphing utility, consult your user's manual.

The Table Feature

Most graphing utilities are capable of displaying a table of values with x-values and one or more corresponding y-values. These tables can be used to check solutions of an equation and to generate ordered pairs to assist in graphing an equation.

To use the *table* feature, enter an equation into the equation editor in "$y =$" form. The table may have a setup screen, which allows you to select the starting x-value and the table step or x-increment. You may then have the option of automatically generating values for x and y or building your own table using the *ask* mode. In the *ask* mode, you enter a value for x and the graphing utility displays the y-value.

For example, enter the equation

$$y = \frac{3x}{x + 2}$$

into the equation editor, as shown in Figure 2. In the table setup screen, set the table to start at $x = -4$ and set the table step to 1. When you view the table, notice that the first x-value is -4 and each value after it increases by 1. Also notice that the Y_1 column gives the resulting y-value for each x-value, as shown in Figure 3. The table shows that the y-value when $x = -2$ is ERROR. This means that the variable x may not take on the value -2 in this equation.

With the same equation in the equation editor, set the table to *ask* mode. In this mode you do not need to set the starting x-value or the table step, because you are entering any value you choose for x. You may enter any real value for x—integers, fractions, decimals, irrational numbers, and so forth. If you enter $x = 1 + \sqrt{3}$, the graphing utility may rewrite the number as a decimal approximation, as shown in Figure 4. You can continue to build your own table by entering additional x-values in order to generate y-values.

If you have several equations in the equation editor, the table may generate y-values for each equation.

Figure 2

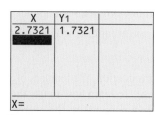

Figure 3

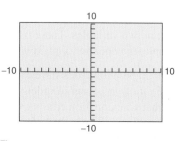

Figure 4

Creating a Viewing Window

A **viewing window** for a graph is a rectangular portion of the coordinate plane. A viewing window is determined by the following six values.

  Xmin = the smallest value of x

  Xmax = the largest value of x

  Xscl = the number of units per tick mark on the x-axis

  Ymin = the smallest value of y

  Ymax = the largest value of y

  Yscl = the number of units per tick mark on the y-axis

When you enter these six values into a graphing utility, you are setting the viewing window. Some graphing utilities have a standard viewing window, as shown in Figure 5.

Figure 5

By choosing different viewing windows for a graph, it is possible to obtain very different impressions of the graph's shape. For instance, Figure 6 shows four different viewing windows for the graph of

$$y = 0.1x^4 - x^3 + 2x^2.$$

Of these, the view shown in part (a) is the most complete.

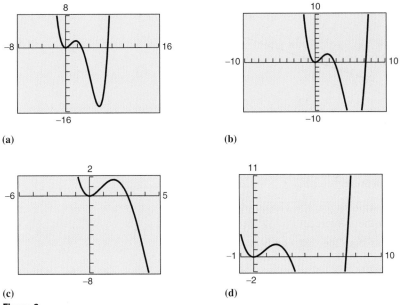

Figure 6

On most graphing utilities, the display screen is two-thirds as high as it is wide. On such screens, you can obtain a graph with a true geometric perspective by using **a square setting**—one in which

$$\frac{\text{Ymax} - \text{Ymin}}{\text{Xmax} - \text{Xmin}} = \frac{2}{3}.$$

One such setting is shown in Figure 7. Notice that the x and y tick marks are equally spaced on a square setting, but not on a standard setting.

To see how the viewing window affects the geometric perspective, graph the semicircles $y_1 = \sqrt{9 - x^2}$ and $y_2 = -\sqrt{9 - x^2}$ in a standard viewing window. Then graph y_1 and y_2 in a square window. Note the difference in the shapes of the circles.

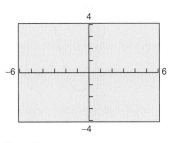

Figure 7

Zoom and Trace Features

When you graph an equation, you can move from point to point along its graph using the *trace* feature. As you trace the graph, the coordinates of each point are displayed, as shown in Figure 8. The *trace* feature combined with the *zoom* feature allows you to obtain better and better approximations of desired points on a graph. For instance, you can use the *zoom* feature of a graphing utility to approximate the x-intercept(s) of a graph [the point(s) where the graph crosses the x-axis]. Suppose you want to approximate the x-intercept(s) of the graph of $y = 2x^3 - 3x + 2$.

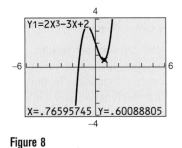

Figure 8

Begin by graphing the equation, as shown in Figure 9(a). From the viewing window shown, the graph appears to have only one x-intercept. This intercept lies between -2 and -1. By zooming in on the intercept, you can improve the approximation, as shown in Figure 9(b). To three decimal places, the solution is $x \approx -1.476$.

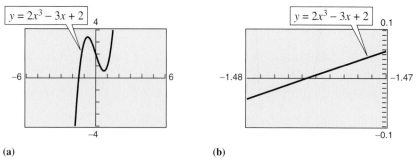

(a)   (b)

Figure 9

Here are some suggestions for using the *zoom* feature.

1. With each successive zoom-in, adjust the x-scale so that the viewing window shows at least one tick mark on each side of the x-intercept.

2. The error in your approximation will be less than the distance between two scale marks.

3. The *trace* feature can usually be used to add one more decimal place of accuracy without changing the viewing window.

Figure 10(a) shows the graph of $y = x^2 - 5x + 3$. Figures 10(b) and 10(c) show "zoom-in views" of the two x-intercepts. From these views, you can approximate the x-intercepts to be $x \approx 0.697$ and $x \approx 4.303$.

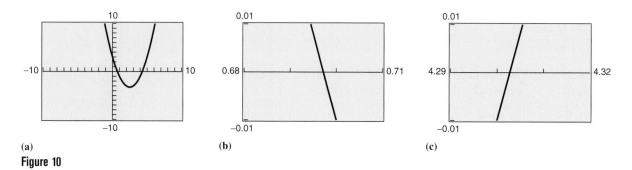

(a)   (b)   (c)

Figure 10

Zero or Root Feature

Using the *zero* or *root* feature, you can find the real zeros of functions of the various types studied in this text—polynomial, exponential, logarithmic, and trigonometric functions. To find the zeros of a function such as $f(x) = \frac{3}{4}x - 2$, first enter the function as $y_1 = \frac{3}{4}x - 2$. Then use the *zero* or *root* feature, which may require entering lower and upper bound estimates of the root, as shown in Figure 11.

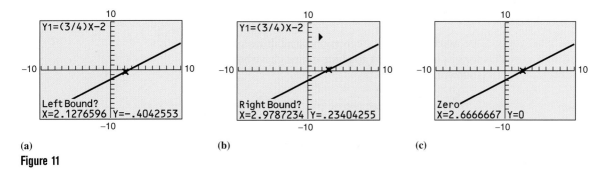

(a)   **(b)**   **(c)**

Figure 11

In Figure 11(c), you can see that the zero is $x = 2.6666667 \approx 2\frac{2}{3}$.

Intersect Feature

To find the points of intersection of two graphs, you can use the *intersect* feature. For instance, to find the points of intersection of the graphs of $y_1 = -x + 2$ and $y_2 = x + 4$, enter these two functions and use the *intersect* feature, as shown in Figure 12.

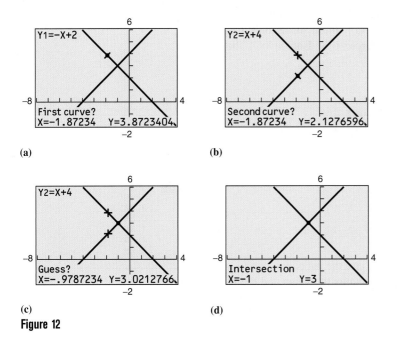

(a)   **(b)**

(c)   **(d)**

Figure 12

From Figure 12(d), you can see that the point of intersection is $(-1, 3)$.

Regression Capabilities

Throughout the text, you will be asked to use the regression capabilities of a graphing utility to find models for sets of data. Most graphing utilities have built-in regression programs for the following.

Regression	*Form of Model*
Linear	$y = ax + b$
Quadratic	$y = ax^2 + bx + c$
Cubic	$y = ax^3 + bx^2 + cx + d$
Quartic	$y = ax^4 + bx^3 + cx^2 + dx + e$
Logarithmic	$y = a + b \ln(x)$
Exponential	$y = ab^x$
Power	$y = ax^b$
Logistic	$y = \dfrac{c}{1 + ae^{-bx}}$
Sine	$y = a \sin(bx + c) + d$

For instance, you can find the linear regression model for the average hourly wages y (in dollars per hour) of production workers in manufacturing industries from 1987 through 1997 shown in the table. (Source: U.S. Bureau of Labor Statistics)

Year	1987	1988	1989	1990	1991	1992
y	9.91	10.19	10.48	10.83	11.18	11.46

Year	1993	1994	1995	1996	1997
y	11.74	12.06	12.37	12.78	13.17

First, let $x = 0$ correspond to 1990 and enter the data into the list editor, as shown in Figure 13. Note that the list in the first column contains the years and the list in the second column contains the hourly wages that correspond to the years. Run your graphing utility's built-in linear regression program to obtain the coefficients a and b for the model $y = ax + b$, as shown in Figure 14. So, a linear model for the data is

$$y \approx 0.321x + 10.83.$$

When you run some regression programs, you may obtain an "r-value," which gives a measure of how well the model fits the data. The closer the value of $|r|$ is to 1, the better the fit. For the data in the table above, $r \approx 0.999$, which implies that the model is a very good fit.

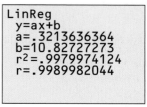

Figure 13

Figure 14

Prerequisites

P.1   Graphical Representation of Data

P.2   Graphs of Equations

P.3   Lines in the Plane

P.4   Solving Equations Algebraically and Graphically

P.5   Solving Inequalities Algebraically and Graphically

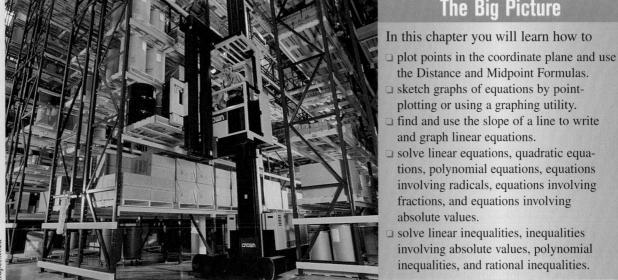

The Big Picture

In this chapter you will learn how to

- ❑ plot points in the coordinate plane and use the Distance and Midpoint Formulas.
- ❑ sketch graphs of equations by point-plotting or using a graphing utility.
- ❑ find and use the slope of a line to write and graph linear equations.
- ❑ solve linear equations, quadratic equations, polynomial equations, equations involving radicals, equations involving fractions, and equations involving absolute values.
- ❑ solve linear inequalities, inequalities involving absolute values, polynomial inequalities, and rational inequalities.

In 1997, over $24 billion worth of corrugated and solid fiber boxes were produced to create shipping containers. (Source: U.S. Bureau of the Census)

Important Vocabulary

As you encounter each new vocabulary term in this chapter, add the term and its definition to your notebook glossary.

- rectangular coordinate system (p. 2)
- Cartesian plane (p. 2)
- Distance Formula (p. 5)
- Midpoint Formula (p. 7)
- circle of radius r (p. 8)
- standard form of the equation of a circle (p. 8)
- solution point (p. 14)
- graph of an equation (p. 14)
- intercepts (p. 15)
- slope (p. 25)

- point-slope form (p. 27)
- slope-intercept form (p. 29)
- general form (p. 30)
- parallel (p. 31)
- perpendicular (p. 31)
- equation (p. 38)
- solution (p. 38)
- linear equation in one variable x (p. 38)
- extraneous (p. 39)
- x-intercept (p. 39)

- y-intercept (p. 39)
- point of intersection (p. 43)
- solutions of an inequality (p. 54)
- graph of an inequality (p. 54)
- properties of inequalities (p. 54)
- equivalent inequalities (p. 54)
- linear inequality (p. 55)
- double inequality (p. 56)
- critical numbers (p. 58)
- test intervals (p. 58)

Additional Resources   Text-specific additional resources are available to help you do well in this course. See page xvi for details.

1

P.1 Graphical Representation of Data

The Cartesian Plane

Just as you can represent real numbers by points on a real number line, you can represent ordered pairs of real numbers by points in a plane called the **rectangular coordinate system,** or the **Cartesian plane,** after the French mathematician René Descartes (1596–1650).

The Cartesian plane is formed by using two real number lines intersecting at right angles, as shown in Figure P.1. The horizontal real number line is usually called the **x-axis,** and the vertical real number line is usually called the **y-axis.** The point of intersection of these two axes is the **origin,** and the two axes divide the plane into four parts called **quadrants.**

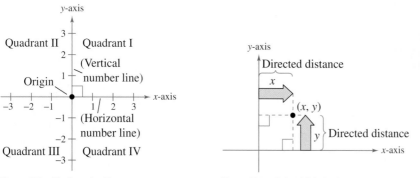

Figure P.1 *The Cartesian Plane*

Figure P.2 *Ordered Pair (x, y)*

Each point in the plane corresponds to an **ordered pair** (x, y) of real numbers x and y, called **coordinates** of the point. The **x-coordinate** represents the directed distance from the y-axis to the point, and the **y-coordinate** represents the directed distance from the x-axis to the point, as shown in Figure P.2.

$$\text{Directed distance} \quad (x, y) \quad \text{Directed distance}$$
$$\text{from } y\text{-axis} \quad\quad\quad\quad \text{from } x\text{-axis}$$

The notation (x, y) denotes both a point in the plane and an open interval on the real number line. The context will tell you which meaning is intended.

EXAMPLE 1 Plotting Points in the Cartesian Plane

Plot the points $(-1, 2)$, $(3, 4)$, $(0, 0)$, $(3, 0)$, and $(-2, -3)$.

Solution
To plot the point $(-1, 2)$, imagine a vertical line through -1 on the x-axis and a horizontal line through 2 on the y-axis. The intersection of these two lines is the point $(-1, 2)$. This point is 1 unit to the left of the y-axis and 2 units up from the x-axis. The other four points can be plotted in a similar way (see Figure P.3).

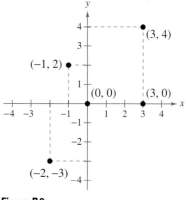

Figure P.3

EXAMPLE 2 Translating Points in the Plane

The triangle in Figure P.4(a) has vertices at the points $(-1, 2)$, $(1, -4)$, and $(2, 3)$. Shift the triangle three units to the right and two units up and find the vertices of the shifted triangle, as shown in Figure P.4(b).

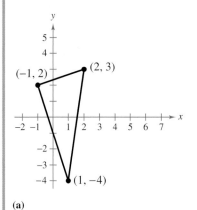

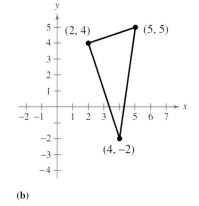

(a)   **(b)**

Figure P.4

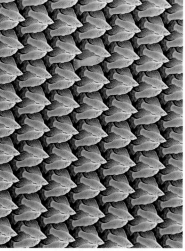

Much of computer graphics, including this computer-generated goldfish tessellation, consists of transformations of points in a coordinate plane. One type of transformation, a translation, is illustrated in Example 2. Other types include reflections, rotations, and stretches.

A computer animation of this example appears in the *Interactive* CD-ROM and *Internet* versions of this text.

Solution

To shift the vertices three units to the right, add 3 to each of the x-coordinates. To shift the vertices two units up, add 2 to each of the y-coordinates.

Original Point	Translated Point
$(-1, 2)$	$(-1 + 3, 2 + 2) = (2, 4)$
$(1, -4)$	$(1 + 3, -4 + 2) = (4, -2)$
$(2, 3)$	$(2 + 3, 3 + 2) = (5, 5)$

Plotting the translated points and sketching the line segments between them produces the shifted triangle shown in Figure P.4(b).

Example 2 shows how to translate points in a coordinate plane. The following transformed points are related to the original points as follows.

Original Point	Transformed Point	
(x, y)	$(-x, y)$	$(-x, y)$ is a reflection of the original point in the y-axis.
(x, y)	$(x, -y)$	$(x, -y)$ is a reflection of the original point in the x-axis.
(x, y)	$(-x, -y)$	$(-x, -y)$ is a reflection of the original point through the origin.

Representing Data Graphically

The beauty of a rectangular coordinate system is that it allows you to see relationships between two variables. It would be difficult to overestimate the importance of Descartes's introduction of coordinates to the plane. Today, his ideas are in common use in virtually every scientific and business-related field.

EXAMPLE 3   Sketching a Scatter Plot, Bar Graph, and Line Graph

From 1988 through 1997, the amount *A* (in millions of dollars) spent on archery equipment in the United States is given in the table, where *t* represents the year. (a) Sketch a scatter plot of the data. (b) Sketch a bar graph of the data. (c) Sketch a line graph of the data.   (Source: National Sporting Goods Association)

A computer animation of this example appears in the *Interactive* CD-ROM and *Internet* versions of this text.

t	1988	1989	1990	1991	1992	1993	1994	1995	1996	1997
A	235	261	265	270	334	285	306	287	272	273

Solution
a. To sketch a *scatter plot* of the data given in the table, you simply represent each pair of values by an ordered pair (t, A) and plot the resulting points, as shown in Figure P.5. For instance, the first pair of values is represented by the ordered pair (1988, 235). Note that the break in the *t*-axis indicates that the numbers between 0 and 1988 have been omitted.

b. To create a *bar graph*, begin by drawing a vertical axis to represent the amount (in millions of dollars) and a horizontal axis to represent the year. Then for each value of *t* in the table, draw a vertical bar whose height is the corresponding value *A*. The bar graph is shown in Figure P.6.

c. To draw a *line graph*, begin by drawing a vertical axis to represent the amount (in millions of dollars). Then label the horizontal axis with years and plot the points given in the table. Finally, connect the points with line segments, as shown in Figure P.7.

> ### STUDY T!P
>
> You can use a graphing utility to sketch each of the graphs in Example 3. First, enter the data into the graphing utility. You can then use the *statistical plotting* feature to sketch a scatter plot, a bar graph, and a line graph.

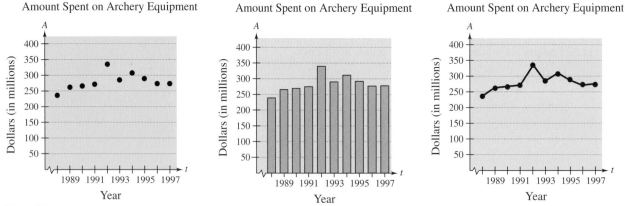

Figure P.5

Figure P.6

Figure P.7

In Example 3, you could have let $t = 1$ represent the year 1988. In that case, the horizontal axis of each graph would not have been broken, and the tick marks would have been labeled 1 through 10 (instead of 1988 through 1997).

EXAMPLE 4 Interpreting a Population Model

The population (in millions) of North Carolina from 1990 to 1997 can be modeled by $y = 0.11x + 6.63$, where x is the time in years, with $x = 0$ corresponding to 1990. During which year did the population exceed 7 million?

Graphical Solution

You can use the *statistical plotting* feature of a graphing utility to create a bar graph of the population values obtained using the model. On the same viewing window, graph the line $y = 7$.

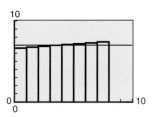

Figure P.8

Numerical Solution

You can use the *table* feature of a graphing utility to evaluate the model $y = 0.11x + 6.63$ for different years. Enter the model into the equation editor. Then create a table such as the one below.

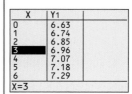

X	Y1
0	6.63
1	6.74
2	6.85
3	6.96
4	7.07
5	7.18
6	7.29

X=3

Figure P.9

From the graph in Figure P.8 you can see that the bars that correspond to $x = 0, 1, 2$, and 3 are less than 7. The bar that corresponds to $x = 4$ is the first bar greater than 7, so the population exceeded 7 million sometime during 1993.

From the table in Figure P.9, you can see that the value of y when $x = 3$ is 6.96, and the value of y when $x = 4$ is 7.07. So, the population exceeded 7 million when x is greater than 3 and less than 4, or sometime during 1993.

The Distance Formula

Recall from the Pythagorean Theorem that for a right triangle with hypotenuse of length c and sides of lengths a and b, you have $a^2 + b^2 = c^2$ as shown in Figure P.10. (The converse is also true. That is, if $a^2 + b^2 = c^2$, the triangle is a right triangle.)

   Suppose you want to determine the distance d between two points (x_1, y_1) and (x_2, y_2) in the plane. With these two points, a right triangle can be formed, as shown in Figure P.11. The length of the vertical side of the triangle is $|y_2 - y_1|$, and the length of the horizontal side is $|x_2 - x_1|$. By the Pythagorean Theorem,

$$d^2 = |x_2 - x_1|^2 + |y_2 - y_1|^2$$

$$d = \sqrt{|x_2 - x_1|^2 + |y_2 - y_1|^2}$$

$$d = \sqrt{(x_2 - x_1)^2 + (y_2 - y_1)^2}.$$

This result is the **Distance Formula.**

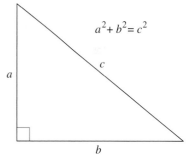

Figure P.10

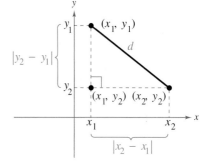

Figure P.11

The Distance Formula

The distance d between the points (x_1, y_1) and (x_2, y_2) in the plane is

$$d = \sqrt{(x_2 - x_1)^2 + (y_2 - y_1)^2}.$$

EXAMPLE 5   Finding a Distance

Find the distance between the points $(-2, 1)$ and $(3, 4)$.

Algebraic Solution

Let $(x_1, y_1) = (-2, 1)$ and $(x_2, y_2) = (3, 4)$. Then apply the Distance Formula as follows.

$$d = \sqrt{(x_2 - x_1)^2 + (y_2 - y_1)^2} \qquad \text{Distance Formula}$$

$$= \sqrt{[3 - (-2)]^2 + (4 - 1)^2} \qquad \begin{array}{l}\text{Substitute for} \\ x_1, y_1, x_2, \text{ and } y_2.\end{array}$$

$$= \sqrt{(5)^2 + (3)^2} \qquad \text{Simplify.}$$

$$= \sqrt{34}$$

$$\approx 5.83 \qquad \text{Use a calculator.}$$

So, the distance between the points is about 5.83 units.

You can use the Pythagorean Theorem to check that the distance is correct.

$$d^2 \stackrel{?}{=} 3^2 + 5^2 \qquad \text{Pythagorean Theorem}$$

$$\left(\sqrt{34}\right)^2 \stackrel{?}{=} 3^2 + 5^2 \qquad \text{Substitute for } d.$$

$$34 = 34 \qquad \text{Distance checks.} \ \checkmark$$

Graphical Solution

Use centimeter graph paper to plot the points A and B. Carefully sketch the line segment from A to B. Then use a centimeter ruler to measure the length of the segment.

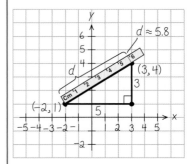

Figure P.12

The line segment measures about 5.8 centimeters as shown in Figure P.12. So, the distance between the points is about 5.8 units.

EXAMPLE 6   Verifying a Right Triangle

Show that the points $(2, 1)$, $(4, 0)$, and $(5, 7)$ are vertices of a right triangle.

Solution

The three points are plotted in Figure P.13. Using the Distance Formula, you can find the lengths of the three sides as follows.

$$d_1 = \sqrt{(5 - 2)^2 + (7 - 1)^2} = \sqrt{9 + 36} = \sqrt{45}$$

$$d_2 = \sqrt{(4 - 2)^2 + (0 - 1)^2} = \sqrt{4 + 1} = \sqrt{5}$$

$$d_3 = \sqrt{(5 - 4)^2 + (7 - 0)^2} = \sqrt{1 + 49} = \sqrt{50}$$

Because

$$d_1{}^2 + d_2{}^2 = 45 + 5 = 50 = d_3{}^2,$$

you can conclude that the triangle must be a right triangle.

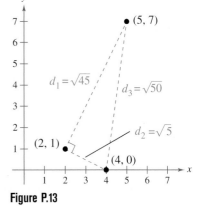

Figure P.13

The figure provided with Example 6 was not really essential to the solution. Nevertheless, it is strongly recommended that you develop the habit of including sketches with your solutions—even if they are not required.

The Midpoint Formula

To find the **midpoint** of the line segment that joins two points in a coordinate plane, you can simply find the average values of the respective coordinates of the two endpoints using the **Midpoint Formula.** (See Appendix A for a proof of the Midpoint Formula.)

> ### The Midpoint Formula
>
> The midpoint of the segment joining the points (x_1, y_1) and (x_2, y_2) is given by the Midpoint Formula
>
> $$\text{Midpoint} = \left(\frac{x_1 + x_2}{2}, \frac{y_1 + y_2}{2}\right).$$

EXAMPLE 7   Finding a Line Segment's Midpoint

Find the midpoint of the line segment joining the points $(-5, -3)$ and $(9, 3)$, as shown in Figure P.14.

Solution

Let $(x_1, y_1) = (-5, -3)$ and $(x_2, y_2) = (9, 3)$.

$$\text{Midpoint} = \left(\frac{x_1 + x_2}{2}, \frac{y_1 + y_2}{2}\right) \qquad \text{Midpoint Formula}$$

$$= \left(\frac{-5 + 9}{2}, \frac{-3 + 3}{2}\right) \qquad \text{Substitute for } x_1, y_1, x_2, \text{ and } y_2.$$

$$= (2, 0) \qquad \text{Simplify.}$$

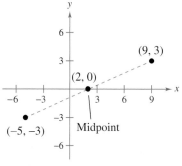

Figure P.14

EXAMPLE 8   Estimating Annual Sales

Winn-Dixie Stores had annual sales of $1.30 billion in 1996 and $1.36 billion in 1998. Without knowing any additional information, what would you estimate the 1997 sales to have been?   (Source: Winn-Dixie Stores, Inc.)

Solution

One solution to the problem is to assume that sales followed a linear pattern. With this assumption, you can estimate the 1997 sales by finding the midpoint of the segment connecting the points $(1996, 1.30)$ and $(1998, 1.36)$.

$$\text{Midpoint} = \left(\frac{1996 + 1998}{2}, \frac{1.30 + 1.36}{2}\right)$$

$$= (1997, 1.33)$$

So, you would estimate the 1997 sales to have been about $1.33 billion, as shown in Figure P.15 (the actual 1997 sales were $1.32 billion).

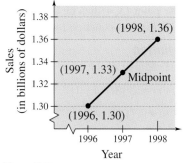

Figure P.15

The Equation of a Circle

The Distance Formula provides a convenient way to define circles. A **circle of radius** *r* with center at the point (h, k) is shown in Figure P.16. The point (x, y) is on this circle if and only if its distance from the center (h, k) is *r*. This means that a **circle** in the plane consists of all points (x, y) that are a given positive distance *r* from a fixed point (h, k). Using the Distance Formula, you can express this relationship by saying that the point (x, y) lies on the circle if and only if

$$\sqrt{(x - h)^2 + (y - k)^2} = r.$$

By squaring both sides of this equation, you can obtain the **standard form of the equation of a circle.**

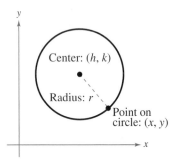

Figure P.16

EXAMPLE 9   Finding an Equation of a Circle

The point $(3, 4)$ lies on a circle whose center is at $(-1, 2)$, as shown in Figure P.17. Find an equation for the circle.

Solution

The radius *r* of the circle is the distance between $(-1, 2)$ and $(3, 4)$.

$$r = \sqrt{[3 - (-1)]^2 + (4 - 2)^2}$$

$$= \sqrt{16 + 4} = \sqrt{20}$$

So, the center of the circle is $(h, k) = (-1, 2)$ and the radius is $r = \sqrt{20}$, and you can write the standard form of the equation of the circle as follows.

$$(x - h)^2 + (y - k)^2 = r^2 \qquad \text{Standard form}$$

$$[x - (-1)]^2 + (y - 2)^2 = \left(\sqrt{20}\right)^2 \qquad \text{Substitute for } h, k, \text{ and } r.$$

$$(x + 1)^2 + (y - 2)^2 = 20 \qquad \text{Equation of circle}$$

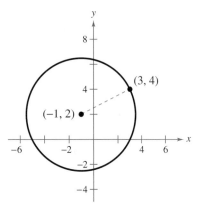

Figure P.17

P.1 Exercises

In Exercises 1–4, sketch the polygon with the indicated vertices.

1. Triangle: $(-1, 1), (2, -1), (3, 4)$

2. Triangle: $(0, 3), (-1, -2), (4, 8)$

3. Square: $(2, 4), (5, 1), (2, -2), (-1, 1)$

4. Parallelogram: $(5, 2), (7, 0), (1, -2), (-1, 0)$

In Exercises 5–8, approximate the coordinates of the points.

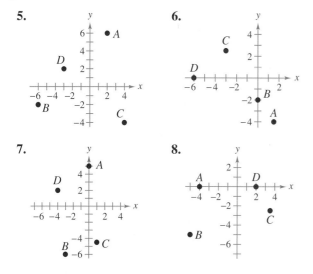

5.

6.

7.

8.

In Exercises 9–12, find the coordinates of the point.

9. The point is located three units to the left of the y-axis and four units above the x-axis.

10. The point is located eight units below the x-axis and four units to the right of the y-axis.

11. The point is located five units below the x-axis and the coordinates of the point are equal.

12. The point is on the x-axis and twelve units to the left of the y-axis.

In Exercises 13–22, determine the quadrant(s) in which (x, y) is located so that the condition(s) is (are) satisfied.

13. $x > 0$ and $y < 0$ **14.** $x < 0$ and $y < 0$

15. $x = -4$ and $y > 0$ **16.** $x > 2$ and $y = 3$

17. $y < -5$ **18.** $x > 4$

19. $(x, -y)$ is in the second quadrant.

20. $(-x, y)$ is in the fourth quadrant.

21. $xy > 0$ **22.** $xy < 0$

In Exercises 23 and 24, the polygon is shifted to a new position in the plane. Find the coordinates of the vertices of the polygon in its new position.

23. Shift: 5 units up, 2 units to the right

24. Shift: 3 units down, 6 units to the right

In Exercises 25–30, find the distance between the points algebraically and verify graphically by using centimeter graph paper and a centimeter ruler.

25. $(6, -3), (6, 5)$ **26.** $(1, 4), (8, 4)$

27. $(-3, -1), (2, -1)$ **28.** $(-3, -4), (-3, 6)$

29. $(-2, 6), (3, -6)$ **30.** $(8, 5), (0, 20)$

In Exercises 31–42, (a) plot the points, (b) find the distance between the points, and (c) find the midpoint of the line segment joining the points.

31. $(1, 1), (9, 7)$ **32.** $(1, 12), (6, 0)$

33. $(-4, 10), (4, -5)$ **34.** $(-7, -4), (2, 8)$

35. $(-1, 2), (5, 4)$ **36.** $(2, 10), (10, 2)$

37. $\left(\frac{1}{2}, 1\right), \left(-\frac{5}{2}, \frac{4}{3}\right)$ **38.** $\left(-\frac{1}{3}, -\frac{1}{3}\right), \left(-\frac{1}{6}, -\frac{1}{2}\right)$

39. $(6.2, 5.4), (-3.7, 1.8)$

40. $(-16.8, 12.3), (5.6, 4.9)$

41. $(-36, -18), (48, -72)$

42. $(1.451, 3.051), (5.906, 11.360)$

In Exercises 43–46, (a) find the length of each side of a right triangle and (b) show that these lengths satisfy the Pythagorean Theorem.

43.   **44.**

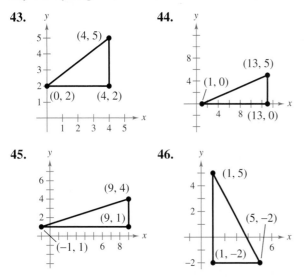

45.   **46.**

Business   In Exercises 47 and 48, estimate the sales of a company in 1998, given the sales in 1996 and 2000. Assume the sales followed a linear pattern.

47.

Year	1996	2000
Sales	$520,000	$740,000

48.

Year	1996	2000
Sales	$4,200,000	$5,650,000

In Exercises 49–52, show that the points form the vertices of the polygon.

49. Right triangle: $(4, 0), (2, 1), (-1, -5)$

50. Isosceles triangle: $(1, -3), (3, 2), (-2, 4)$

51. Parallelogram: $(2, 5), (0, 9), (-2, 0), (0, -4)$

52. Parallelogram: $(0, 1), (3, 7), (4, 4), (1, -2)$

53. *Exploration*   A line segment has (x_1, y_1) as one endpoint and (x_m, y_m) as its midpoint. Find the other endpoint (x_2, y_2) of the line segment in terms of $x_1, y_1, x_m,$ and y_m. Use the result to find the coordinates of the endpoint of a line segment if the coordinates of the other endpoint and midpoint are, respectively,

(a) $(1, -2), (4, -1)$ (b) $(-5, 11), (2, 4)$.

54. *Exploration*   Use the Midpoint Formula three times to find the three points that divide the line segment joining (x_1, y_1) and (x_2, y_2) into four parts. Use the result to find the points that divide the line segment joining the given points into four equal parts.

(a) $(1, -2), (4, -1)$ (b) $(-2, -3), (0, 0)$

In Exercises 55–62, find the standard form of the equation of the specified circle.

55. Center: $(0, 0)$; radius: 3

56. Center: $(0, 0)$; radius: 5

57. Center: $(2, -1)$; radius: 4

58. Center: $\left(0, \frac{1}{3}\right)$; radius: $\frac{1}{3}$

59. Center: $(-1, 2)$; solution point: $(0, 0)$

60. Center: $(3, -2)$; solution point: $(-1, 1)$

61. Endpoints of a diameter: $(0, 0), (6, 8)$

62. Endpoints of a diameter: $(-4, -1), (4, 1)$

In Exercises 63–68, find the center and radius, and sketch the circle.

63. $x^2 + y^2 = 4$

64. $x^2 + y^2 = 16$

65. $(x - 1)^2 + (y + 3)^2 = 4$

66. $x^2 + (y - 1)^2 = 4$

67. $\left(x - \frac{1}{2}\right)^2 + \left(y - \frac{1}{2}\right)^2 = \frac{9}{4}$

68. $\left(x - \frac{2}{3}\right)^2 + \left(y + \frac{1}{4}\right)^2 = \frac{25}{9}$

In Exercises 69 and 70, sketch a scatter plot of the data given in the table.

69. *Meteorology*   The table shows the lowest temperature of record y (in degrees Fahrenheit) in Duluth, Minnesota, for each month x, where $x = 1$ represents January.   (Source: NOAA)

x	1	2	3	4	5	6
y	-39	-33	-29	-5	17	27

x	7	8	9	10	11	12
y	35	32	22	8	-23	-34

70. *Business* The table shows the number y of Wal-Mart stores for each year x from 1992 through 1999. (Source: Wal-Mart Stores, Inc.)

x	1992	1993	1994	1995
y	2136	2440	2759	2943

x	1996	1997	1998	1999
y	3054	3406	3630	3815

Milk Prices **In Exercises 71 and 72, use the graph below, which shows the average retail price of one-half gallon of milk from 1992 to 1997.** (Source: U.S. Bureau of Labor Statistics)

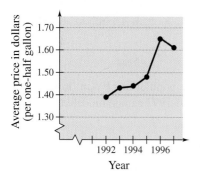

71. Approximate the highest price of one-half gallon of milk shown in the graph. When did this occur?

72. Approximate the difference in the price of milk from the highest price shown in the graph to the price in 1992.

Advertising **In Exercises 73 and 74, use the graph below, which shows the cost of a 30-second television spot (in thousands of dollars) during the Super Bowl from 1987 to 1999.** (Source: USA Today Research)

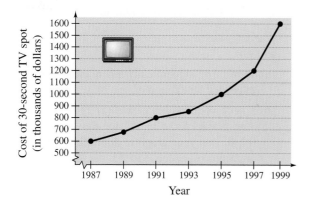

73. Approximate the percent increase in cost of a 30-second spot from Super Bowl XXI in 1987 to Super Bowl XXXIII in 1999.

74. Estimate the increase in cost of a 30-second spot (a) from Super Bowl XXI to Super Bowl XXVII, and (b) from Super Bowl XXVII to Super Bowl XXXIII.

Analyzing Data **In Exercises 75 and 76, refer to the scatter plot, which shows the mathematics entrance test scores x and the final examination scores y in an algebra course for a sample of 10 students.**

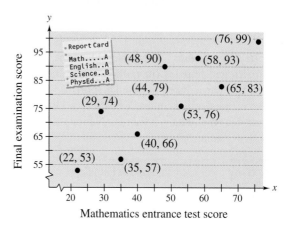

75. Find the entrance exam score of any student with a final exam score in the 80s.

76. Does a higher entrance exam score necessarily imply a higher final exam score? Explain.

77. *Food Production* The double bar graph shows the production and exports (in millions of metric tons) of corn, soybeans, and wheat for the year 1997. Approximate the percent of each product that is exported. (Source: U.S. Department of Agriculture)

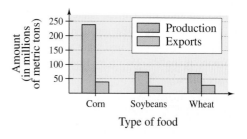

78. *Fruit Crops* The table shows farmers' cash receipts (in millions of dollars) from fruit crops in 1996. Construct a bar graph for the data. (Source: U.S. Department of Agriculture)

Fruit	Receipts	Fruit	Receipts
Apples	1846	Oranges	1798
Cherries	264	Peaches	380
Cranberries	246	Pears	292
Grapes	2334	Plums and Prunes	295
Lemons	228	Strawberries	770

79. *Sports Participants* The table shows the number of males and females (in millions) over the age of seven that participated in popular sports activities in 1996 in the United States. Construct a double bar graph for the data. (Source: National Sporting Goods Association)

Activity	Male	Female
Aerobic exercising	5.3	18.8
Basketball	22.4	10.9
Bicycling	28.6	24.7
Bowling	22.6	20.3
Camping	24.1	20.6
Exercise walking	26.7	46.6
Running	12.3	9.9
Swimming	29.1	31.1

80. *Oil Imports* The table shows the amount of crude oil imported into the United States (in millions of barrels) for the years 1988 through 1997. Construct a line graph for the data and state what information the graph reveals. (Source: Energy Information Administration)

Year	1988	1989	1990	1991	1992
Imports	1864	2133	2151	2110	2220

Year	1993	1994	1995	1996	1997
Imports	2477	2578	2643	2748	2918

81. *Personal Savings* The line graph shows the percent of disposable income saved in the United States in selected years from 1980 to 1997. (Source: U.S. Bureau of Economic Analysis)

(a) Determine the percent decrease in the rate of saving from 1980 to 1997.

(b) Is the trend shown in the line graph good for the country? Explain your reasoning.

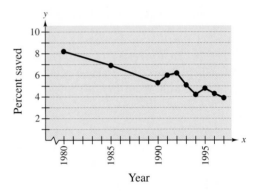

82. *Population* The population P (in millions) of Texas from 1990 to 1997 can be modeled by

$$P = 0.35x + 16.99$$

where x is the time in years, with $x = 0$ corresponding to 1990. (Source: U.S. Bureau of the Census)

(a) Use your graphing utility to create a bar graph of the model for the years 1990 to 1997. Use the graph to determine when the population of Texas exceeded 19 million.

(b) Determine algebraically when the population of Texas exceeded 19 million. Use your graph from part (a) to check your answer.

83. *Health* The average patient cost C to a community hospital per day in the United States from 1989 to 1996 can be modeled by

$$C = -2.37t^2 + 66.44t + 696.39$$

where t is the time in years, with $t = 0$ corresponding to 1990. (Source: American Hospital Association)

(a) Determine algebraically when the average cost exceeded $900 per day.

(b) Check your answer in part (a) by constructing a line graph of the model for the years 1989 to 1996. Use your graph to approximate when the average cost exceeded $900 per day.

84. *Sports*   In a football game, a quarterback throws a pass from the 15-yard line, 10 yards from the sideline as shown in the figure. The pass is caught on the 40-yard line, 45 yards from the same sideline. How long is the pass?

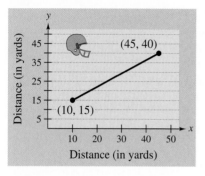

Distance (in yards)

85. *Flying Distance*   A plane flies in a straight line to a city that is 100 kilometers east and 150 kilometers north of the point of departure. How far does it fly?

86. *Make a Conjecture*   Plot the points $(2, 1)$, $(-3, 5)$, and $(7, -3)$ on a rectangular coordinate system. Then change the sign of the indicated coordinate(s) of each point and plot the three new points on the same rectangular coordinate system. Make a conjecture about the location of a point when

(a) the sign of the *x*-coordinate is changed;

(b) the sign of the *y*-coordinate is changed;

(c) the sign of both the *x*- and *y*-coordinates are changed.

87. *Rock and Roll Hall of Fame*   The graph below shows the number of recording artists who were elected to the Rock and Roll Hall of Fame from 1986 to 1999.

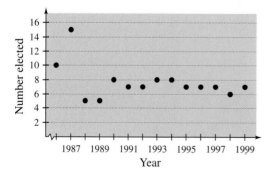

(a) Describe any trends in the data. From these trends, estimate the number of artists that will be elected in 2001.

(b) Why do you think the numbers elected in 1986 and 1987 were greater than in other years?

88. *Business*   Starbucks Corporation had annual sales of \$696.5 million in 1996 and \$1308.7 million in 1998. Use the Midpoint Formula to estimate the 1997 sales.   (Source: Starbucks Corporation)

89. *Business*   Lands' End, Inc., had annual sales of \$1118.7 million in 1996 and \$1371.4 million in 1998. Use the Midpoint Formula to estimate the 1997 sales.   (Source: Lands' End, Inc.)

Synthesis

True or False?   **In Exercises 90–92, determine whether the statement is true or false. Justify your answer.**

90. In order to divide a line segment into 16 equal parts, you would have to use the Midpoint Formula 16 times.

91. The points $(-8, 4)$, $(2, 11)$, and $(-5, 1)$ represent the vertices of an isosceles triangle.

92. If four points represent the vertices of a polygon, and the four sides are equal, then the polygon must be a square.

93. *Think About It*   What is the *y*-coordinate of any point on the *x*-axis? What is the *x*-coordinate of any point on the *y*-axis?

94. *Think About It*   When plotting points on the rectangular coordinate system, is it true that the scales on the *x*- and *y*-axes must be the same? Explain.

P.2   Graphs of Equations

The Graph of an Equation

News magazines often show graphs comparing the rate of inflation, the federal deficit, wholesale prices, or the unemployment rate to the time of year. Industrial firms and businesses use graphs to report their monthly production and sales statistics. Such graphs provide geometric pictures of the way one quantity changes with respect to another. Frequently, the relationship between two quantities is expressed as an equation. This section introduces the basic procedure for determining the geometric picture associated with an equation.

For an equation in variables x and y, a point (a, b) is a **solution point** if the substitution of $x = a$ and $y = b$ satisfies the equation. Most equations have *infinitely* many solution points. For example, the equation

$$3x + y = 5$$

has solution points $(0, 5)$, $(1, 2)$, $(2, -1)$, $(3, -4)$, and so on. The set of all solution points of an equation is the **graph of the equation.**

EXAMPLE 1   Determining Solution Points

Determine whether each point lies on the graph of $y = 10x - 7$.

a. $(2, 13)$ **b.** $(-1, -3)$

Solution

a. The point $(2, 13)$ lies on the graph of $y = 10x - 7$ because it is a solution point of the equation.

$$y = 10x - 7 \qquad \text{Write original equation.}$$
$$13 \overset{?}{=} 10(2) - 7 \qquad \text{Substitute 2 for } x \text{ and 13 for } y.$$
$$13 = 13 \qquad (2, 13) \text{ is a solution.} \checkmark$$

b. The point $(-1, -3)$ does not lie on the graph of $y = 10x - 7$ because it is not a solution point of the equation.

$$y = 10x - 7 \qquad \text{Write original equation.}$$
$$-3 \overset{?}{=} 10(-1) - 7 \qquad \text{Substitute } -1 \text{ for } x \text{ and } -3 \text{ for } y.$$
$$-3 \neq -17 \qquad (-1, -3) \text{ is not a solution.}$$

How to Sketch the Graph of an Equation by Point Plotting

1. If possible, rewrite the equation so that one of the variables is isolated on one side of the equation.
2. Make a table of several solution points.
3. Plot these points in the coordinate plane.
4. Connect the points with a smooth curve.

What You Should Learn:

- How to sketch graphs of equations by point plotting
- How to sketch graphs of equations using a graphing utility
- How to use graphs of equations in real-life problems

Why You Should Learn It:

The graph of an equation can help you see relationships between real-life quantities. For example, Exercise 75 on page 24 shows how a graph can be used to understand the relationship between life expectancy and the year a child is born.

Bruce Ayres/Tony Stone Images

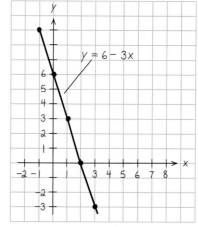

Figure P.18

EXAMPLE 2   Sketching a Graph by Point Plotting

Use point plotting and graph paper to sketch the graph of

$$3x + y = 6.$$

Solution

In this case you can isolate the variable y.

$$y = 6 - 3x \qquad \text{Solve equation for } y.$$

Using negative, zero, and positive values for x, you can obtain the following table of values (solution points).

x	-1	0	1	2	3
$y = 6 - 3x$	9	6	3	0	-3

Next, plot these points and connect them, as shown in Figure P.18. It appears that the graph is a straight line. You will study lines extensively in Section P.3.

The points at which a graph touches or crosses an axis are the **intercepts** of the graph. For instance, in Example 2 the point $(0, 6)$ is the y-intercept of the graph because the graph crosses the y-axis at that point. The point $(2, 0)$ is the x-intercept of the graph because the graph crosses the x-axis at that point.

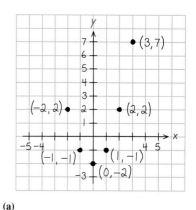

(a)

EXAMPLE 3   Sketching a Graph by Point Plotting

Use point plotting and graph paper to sketch the graph of $y = x^2 - 2$.

Solution

First, make a table of values by choosing several convenient values of x and calculating the corresponding values of y.

x	-2	-1	0	1	2	3
$y = x^2 - 2$	2	-1	-2	-1	2	7

Next, plot the corresponding solution points, as shown in Figure P.19(a). Finally, connect the points with a smooth curve, as shown in Figure P.19(b). This graph is called a *parabola*. You will study parabolas in Section 2.1.

In this text, you will study two basic ways to create graphs: *by hand* and *using a graphing utility*. For instance, the graphs in Figures P.18 and P.19 were sketched by hand and the graph in Figure P.21 was sketched using a graphing utility.

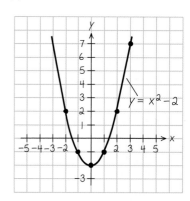

(b)
Figure P.19

A computer animation of this example appears in the *Interactive* CD-ROM and *Internet* versions of this text.

Using a Graphing Utility

One of the disadvantages of the point-plotting method is that to get a good idea about the shape of a graph you need to plot *many* points. With only a few points, you could badly misrepresent the graph. For instance, consider the equation

$$y = \frac{1}{30}x(x^4 - 10x^2 + 39).$$

Suppose you plotted only five points: $(-3, -3), (-1, -1), (0, 0), (1, 1)$, and $(3, 3)$, as shown in Figure P.20(a). From these five points, you might assume that the graph of the equation is a straight line. That, however, is not correct. By plotting several more points, you can see that the actual graph is not straight at all, as shown in Figure P.20(b).

STUDY T!P

This section presents a brief overview of how to use a graphing utility to graph an equation. For more extensive coverage on this topic, see "An Introduction to Graphing Utilities" on pages xvii–xxii, and the *Graphing Technology Guide*.

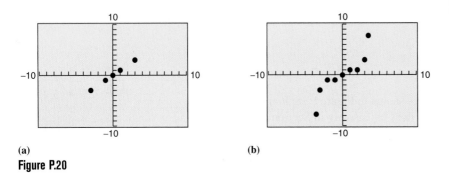

(a)   (b)

Figure P.20

From this, you can see that the point-plotting method leaves you with a dilemma. On the one hand, the method can be very inaccurate if only a few points are plotted. But on the other hand, it is very time-consuming to plot a dozen (or more) points. Technology can help solve this dilemma. Plotting several (even several hundred) points on a rectangular coordinate system is something that a computer or calculator can do easily.

The point-plotting method is the method used by *all* graphing utilities. Each computer or calculator screen is made up of a grid of hundreds or thousands of small areas called *pixels*. Screens that have many pixels per square inch are said to have a higher *resolution* than screens with fewer pixels.

Using a Graphing Utility to Graph an Equation

To graph an equation involving *x* and *y* on a graphing utility, use the following procedure.

1. Rewrite the equation so that *y* is isolated on the left side.
2. Enter the equation into a graphing utility.
3. Determine a *viewing window* that shows all important features of the graph.
4. Graph the equation.

EXAMPLE 4   Using a Graphing Utility to Graph an Equation

Use a graphing utility to graph $2y + x^3 = 4x$.

Solution

To begin, solve the equation for y in terms of x.

$$2y + x^3 = 4x \qquad \text{Write original equation.}$$

$$2y = -x^3 + 4x \qquad \text{Subtract } x^3 \text{ from each side.}$$

$$y = -\frac{1}{2}x^3 + 2x \qquad \text{Divide each side by 2.}$$

Now, by entering this equation into a graphing utility (using a standard viewing window), you can obtain the graph shown in Figure P.21.

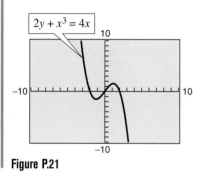

Figure P.21

STUDY T!P

Many graphing utilities have the capability of creating a table, such as the following table, which shows some points of the graph in Figure P.21.

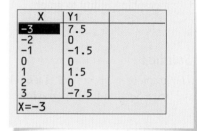

X	Y1	
-3	7.5	
-2	0	
-1	-1.5	
0	0	
1	1.5	
2	0	
3	-7.5	
X=-3		

By choosing different viewing windows for a graph, it is possible to obtain very different impressions of the graph's shape. For instance, Figure P.22 shows four different viewing windows for the graph of the equation in Example 4. None of these views shows *all* of the important features of the graph as Figure P.21 does.

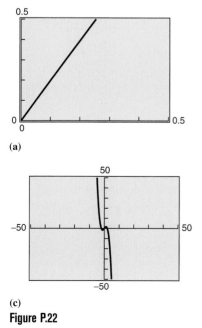

(a)

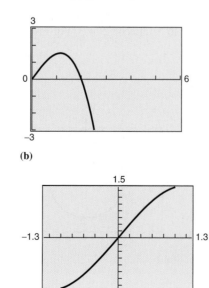

(b)

(c)   (d)

Figure P.22

The standard viewing window on many graphing utilities does not give a true geometric perspective. That is, perpendicular lines will not appear to be perpendicular and circles will not appear to be circular. To overcome this, you can use a *square setting*, as demonstrated in Example 5.

EXAMPLE 5   Sketching a Circle with a Graphing Utility

Use a graphing utility to graph

$$x^2 + y^2 = 9.$$

Solution

The graph of $x^2 + y^2 = 9$ is a circle whose center is the origin and whose radius is 3. (See Section P.1.) To graph the equation, begin by solving the equation for y.

$$x^2 + y^2 = 9 \qquad \text{Write original equation.}$$

$$y^2 = 9 - x^2 \qquad \text{Subtract } x^2 \text{ from each side.}$$

$$y = \pm\sqrt{9 - x^2} \qquad \text{Take square root of each side.}$$

The graph of

$$y = \sqrt{9 - x^2} \qquad \text{Upper semicircle}$$

is the upper semicircle. The graph of

$$y = -\sqrt{9 - x^2} \qquad \text{Lower semicircle}$$

is the lower semicircle. Enter *both* equations in your graphing utility and generate the resulting graphs. In Figure P.23(a), note that if you use a standard viewing window, the two graphs do not appear to form a circle. You can overcome this problem by using a *square setting*, in which the horizontal and vertical tick marks have equal spacing, as shown in Figure P.23(b). On many graphing utilities, a square setting can be obtained by using a y to x ratio of 2 to 3. For instance, in Figure 1.6(b), the y to x ratio is

$$\frac{Y_{max} - Y_{min}}{X_{max} - X_{min}} = \frac{4 - (-4)}{6 - (-6)} = \frac{8}{12} = \frac{2}{3}.$$

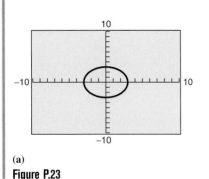

(a)

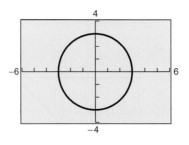

(b)

Figure P.23

The *Interactive* CD-ROM and *Internet* versions of this text show every example with its solution; clicking on the *Try It!* button brings up similar problems. Guided Examples and Integrated Examples show step-by-step solutions to additional examples. Integrated Examples are related to several concepts in the section.

Throughout this course, you will learn that there are many ways to approach a problem. Two of the three common approaches are illustrated in Example 6.

A Numerical Approach: Construct and use a table.

A Graphical Approach: Draw and use a graph.

An Algebraic Approach: Use the rules of algebra.

We recommend that you habitually use at least two approaches with every problem to help build your intuition and check that your answer is reasonable.

Applications

The following two applications show how to develop mathematical models to represent real-world situations. You will see that both a graphing utility and algebra can be used to understand and solve the problems posed.

> **STUDY T!P**
>
> In applications, it is convenient to use variable names that suggest real-life quantities: *d* for distance, *t* for time, and so on. Most graphing utilities, however, require the variable names to be *x* and *y*.

EXAMPLE 6   Running a Marathon

A runner runs at a constant rate of 4.9 miles per hour. The verbal model and algebraic equation relating distance run and elapsed time are as follows.

Verbal Model: Distance = Rate · Time   *Equation:* $d = 4.9t$

a. Determine how far the runner can run in 3.1 hours.

b. Determine how long it will take to run a 26.2-mile marathon.

Graphical Solution

a. To begin, use a graphing utility to graph the equation $d = 4.9t$. (Represent *d* by *y* and *t* by *x*.) Be sure to use a viewing window that shows the graph when $x = 3.1$. Then use the *value* feature or *zoom* and *trace* features of the graphing utility to estimate that when $x = 3.1$, the distance is $y \approx 15.2$ miles, as shown in Figure P.24(a).

b. Adjust the viewing window so that it shows the graph when $y = 26.2$. Use the *value* feature or *zoom* and *trace* features to estimate that when $y = 26.2$, the time is $x \approx 5.4$ hours, as shown in Figure P.24(b).

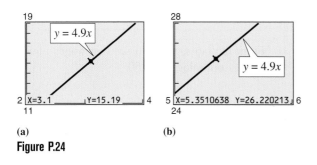

(a)   (b)

Figure P.24

Note that the viewing window on your graphing utility may differ slightly from those shown in Figure P.24.

Algebraic Solution

a. To find how far the runner can run in 3.1 hours, substitute 3.1 for *t* in the equation.

$$d = 4.9t \qquad \text{Write original equation.}$$
$$= 4.9(3.1) \qquad \text{Substitute 3.1 for } t.$$
$$\approx 15.2 \qquad \text{Use a calculator.}$$

So, a runner can run about 15.2 miles in 3.1 hours. Use estimation to check your answer. Because 4.9 is about 5 and 3.1 is about 3, the distance is about $5(3) = 15$. So 15.2 is reasonable.

b. You can find how long it will take to run a 26.2-mile marathon as follows. (For help with solving linear equations see Appendix C.)

$$d = 4.9t \qquad \text{Write original equation.}$$
$$\frac{d}{4.9} = t \qquad \text{Divide each side by 4.9.}$$
$$\frac{26.2}{4.9} = t \qquad \text{Substitute 26.2 for } d.$$
$$5.3 \approx t \qquad \text{Use a calculator.}$$

So, it will take about 5.3 hours to run 26.2 miles.

EXAMPLE 7   Monthly Wages

You receive a monthly salary of $2000 plus a commission of 10% of sales. The verbal model and algebraic equations relating the wages, the salary, and the commission are as follows.

Verbal Model:	Wages $=$ Salary $+$ Commission on Sales	*Equation:* $y = 2000 + 0.1x$

a. If sales are $x = 1480$ in August, what are your wages for that month?

b. If you receive $2225 for September, what are your sales for that month?

Graphical Solution

a. You can use a graphing utility to graph $y = 2000 + 0.1x$ and then estimate the wages when $x = 1480$. Because $x \geq 0$ and the monthly wages are at least $2000, a reasonable viewing window is the one shown in Figure P.25(a). Using the *value* feature or *zoom* and *trace* features near $x = 1480$ shows that the wages are about $2148.

b. Use the graphing utility to find the value along the x-axis (sales) that corresponds to a y-value of 2225 (wages). Beginning with Figure P.25(b) and using the *value* feature or *zoom* and *trace* features, you can estimate the sales to be about $2250.

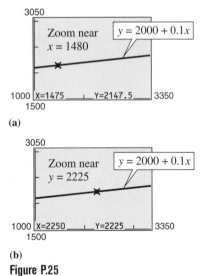

(a)

(b)

Figure P.25

Numerical Solution

a. To find the wages in August, evaluate the equation when $x = 1480$.

$$y = 2000 + 0.1x \qquad \text{Write original equation.}$$
$$= 2000 + 0.1(1480) \qquad \text{Substitute 1480 for } x.$$
$$= 2148 \qquad \text{Simplify.}$$

So, the wages in August are $2148.

b. The table shows the wages for different amounts of sales.

Sales x	2000	2100	2200	2300	2400
Wages y	2200	2210	2220	2230	2240

From the table, you can see that wages of $2225 result from sales between $2200 and $2300. You can improve this estimate by making a table similar to the one below.

Sales x	2210	2220	2230	2240	2250
Wages y	2221	2222	2223	2224	2225

Sales x	2260	2270	2280	2290
Wages y	2226	2227	2228	2229

From the table, you can see that wages of $2225 result from sales of $2250.

Writing About Math   *Comparison of Wages*

Your employer offers you a choice of wage scales: a monthly salary of $3000 plus commission of 7% of sales or a salary of $3400 plus a 5% commission. Write a short paragraph discussing how you would choose your option. At what sales level would the options yield the same salary? Be sure to take advantage of your graphing utility to make your decision.

P.2 Exercises

In Exercises 1–8, determine whether the points lie on the graph of the equation.

Equation	Points
1. $y = \sqrt{x + 4}$	(a) $(0, 2)$ (b) $(5, 3)$
2. $y = x^2 - 3x + 2$	(a) $(2, 0)$ (b) $(-2, 8)$
3. $y = 4 - |x - 2|$	(a) $(1, 5)$ (b) $(1.2, 3.2)$
4. $y = \dfrac{1}{x^2 + 1}$	(a) $(0, 0)$ (b) $(3, 0.1)$
5. $2x - y - 3 = 0$	(a) $(1, 2)$ (b) $(1, -1)$
6. $x^2 + y^2 = 20$	(a) $(3, -2)$ (b) $(-4, 2)$
7. $x^2y - x^2 + 4y = 0$	(a) $\left(1, \frac{1}{5}\right)$ (b) $\left(2, \frac{1}{2}\right)$
8. $y = \frac{1}{3}x^3 - 2x^2$	(a) $\left(2, -\frac{16}{3}\right)$ (b) $(-3, 9)$

In Exercises 9–14, complete the table. Use the resulting solution points to sketch the graph of the equation. Use a graphing utility to verify the graph.

9. $y = -2x + 3$

x	-1	0	1	$\frac{3}{2}$	2
y					

10. $y = \frac{3}{2}x - 1$

x	-2	0	$\frac{2}{3}$	1	2
y					

11. $y = x^2 - 2x$

x	-1	0	1	2	3
y					

12. $y = 4 - x^2$

x	-2	-1	0	1	2
y					

13. $y = 3 - |x - 2|$

x	0	1	2	3	4
y					

14. $y = \sqrt{x - 1}$

x	1	2	5	10	17
y					

15. *Exploration*

(a) Complete the table for the equation $y = \frac{1}{4}x - 3$.

x	-2	-1	0	1	2
y					

(b) Use the resulting solution points to sketch its graph. Then use a graphing utility to verify the graph.

(c) Repeat parts (a) and (b) for the equation $y = -\frac{1}{4}x - 3$. Use the result to describe any differences between the graphs.

16. *Exploration*

(a) Complete the table for the equation $y = \dfrac{6x}{x^{-2} + 1}$.

x	-2	-1	0	1	2
y					

(b) Use the resulting solution points to sketch its graph. Then use a graphing utility to verify the graph.

(c) Continue the table in part (a) for x-values of 5, 10, 20, and 40. What is the value of y approaching? Can y be negative for positive values of x? Explain.

In Exercises 17–22, match the equation with its graph. [The graphs are labeled (a), (b), (c), (d), (e), and (f).]

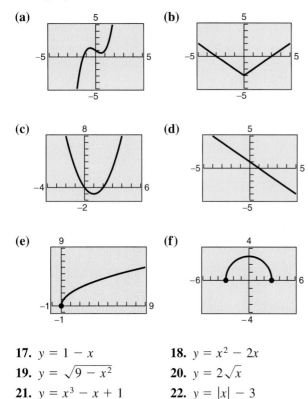

(a) (b)

(c) (d)

(e) (f)

17. $y = 1 - x$ **18.** $y = x^2 - 2x$

19. $y = \sqrt{9 - x^2}$ **20.** $y = 2\sqrt{x}$

21. $y = x^3 - x + 1$ **22.** $y = |x| - 3$

In Exercises 23–36, sketch the graph of the equation.

23. $y = -3x + 2$ **24.** $y = 2x - 3$

25. $y = 1 - x^2$ **26.** $y = x^2 - 1$

27. $y = x^2 - 3x$ **28.** $y = -x^2 - 4x$

29. $y = x^3 + 2$ **30.** $y = x^3 - 1$

31. $y = \sqrt{x - 3}$ **32.** $y = \sqrt{1 - x}$

33. $y = |x - 2|$ **34.** $y = 4 - |x|$

35. $x = y^2 - 1$ **36.** $x = y^2 - 4$

In Exercises 37–50, use a graphing utility to graph the equation. Use a standard viewing window. Approximate any x- or y-intercepts of the graph.

37. $y = x - 5$ **38.** $y = (x + 1)(x - 3)$

39. $y = 3 - \frac{1}{2}x$ **40.** $y = \frac{2}{3}x - 1$

41. $y = x^2 - 4x + 3$ **42.** $y = \frac{1}{2}(x + 4)(x - 2)$

43. $y = x(x - 2)^2$ **44.** $y = \dfrac{4}{x^2 + 1}$

45. $y = \dfrac{2x}{x - 1}$ **46.** $y = \dfrac{4}{x}$

47. $y = x\sqrt{x + 6}$ **48.** $y = (6 - x)\sqrt{x}$

49. $y = \sqrt[3]{x}$ **50.** $y = \sqrt[3]{x + 1}$

In Exercises 51–54, use a graphing utility to sketch the graph of the equation. Begin by using a standard viewing window. Then graph the equation a second time using the specified viewing window. Which viewing window is better? Explain.

51. $y = \frac{5}{2}x + 5$ **52.** $y = -3x + 50$

Xmin = 0
Xmax = 6
Xscl = 1
Ymin = 0
Ymax = 10
Yscl = 1

Xmin = -1
Xmax = 4
Xscl = 1
Ymin = -5
Ymax = 60
Yscl = 5

53. $y = -x^2 + 10x - 5$ **54.** $y = 4(x + 5)\sqrt{4 - x}$

Xmin = -1
Xmax = 11
Xscl = 1
Ymin = -5
Ymax = 25
Yscl = 2

Xmin = -6
Xmax = 6
Xscl = 1
Ymin = -5
Ymax = 50
Yscl = 4

In Exercises 55–58, describe the viewing window of the graph.

55. $y = 4x^2 - 25$ **56.** $y = x^3 - 3x^2 + 4$

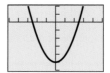 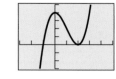

57. $y = |x| + |x - 10|$ **58.** $y = 8\sqrt[3]{x - 6}$

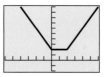 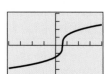

In Exercises 59–62, solve for y and use a graphing utility to graph each of the resulting equations in the same viewing window. Adjust the viewing window so that a circle really does appear circular.

59. $x^2 + y^2 = 64$ **60.** $(x - 1)^2 + (y - 2)^2 = 16$

61. $x^2 + y^2 = 49$ **62.** $(x - 3)^2 + (y - 1)^2 = 25$

In Exercises 63–66, explain how to use a graphing utility to verify that $y_1 = y_2$. Identify the rule of algebra that is illustrated.

63. $y_1 = \frac{1}{4}(x^2 - 8)$ **64.** $y_1 = \frac{1}{2}x + (x + 1)$

$\quad y_2 = \frac{1}{4}x^2 - 2$ $\quad y_2 = \frac{3}{2}x + 1$

65. $y_1 = \frac{1}{5}[10(x^2 - 1)]$ **66.** $y_1 = (x - 3) \cdot \dfrac{1}{x - 3}$

$\quad y_2 = 2(x^2 - 1)$ $\quad y_2 = 1$

In Exercises 67–70, use a graphing utility to graph the equation. Move the cursor along the curve to approximate the unknown coordinate of each solution point accurate to two decimal places. (*Hint:* You may need to use the *zoom* feature of the graphing utility to obtain the required accuracy.)

67. $y = \sqrt{5 - x}$ **68.** $y = x^3(x - 3)$

(a) $(2, y)$ (a) $(2.25, y)$

(b) $(x, 3)$ (b) $(x, 20)$

69. $y = x^5 - 5x$ **70.** $y = |x^2 - 6x + 5|$

(a) $(-0.5, y)$ (a) $(2, y)$

(b) $(x, -4)$ (b) $(x, 1.5)$

71. *Business* A manufacturing plant purchases a new molding machine for $225,000. The depreciated value y after t years is

$$y = 225,000 - 20,000t, \quad 0 \le t \le 8.$$

(a) Use the constraints of the model to determine an appropriate viewing window.

(b) Use a graphing utility to graph the equation.

(c) Use the *value* feature or *zoom* and *trace* features of your graphing utility to determine the value of y when $t = 5.8$. Verify your answer algebraically.

(d) Use the *value* feature or *zoom* and *trace* features of your graphing utility to determine the value of y when $t = 2.35$. Verify your answer algebraically.

72. *Consumerism* You purchase a personal watercraft for $8100. The depreciated value y after t years is

$$y = 8100 - 929t, \quad 0 \le t \le 6.$$

(a) Use the constraints of the model to determine an appropriate viewing window.

(b) Use a graphing utility to graph the equation.

(c) Use the *value* feature or *zoom* and *trace* features of your graphing utility to determine the value of t when $y = 5545.25$. Verify your answer algebraically.

(d) Use the *value* feature or *zoom* and *trace* features of your graphing utility to determine the value of y when $t = 5.5$. Verify your answer algebraically.

73. *Geometry* A rectangle of length x and width w has a perimeter of 12 meters.

(a) Draw a diagram to represent the rectangle. Use the specified variables to label its sides.

(b) Show that $w = 6 - x$ is the width of the rectangle and that $A = x(6 - x)$ is its area.

(c) Use a graphing utility to graph the area equation.

(d) Use the *zoom* and *trace* features of your graphing utility to determine the value of A when $w = 4.9$ meters. Verify your answer algebraically.

(e) From the graph in part (c), estimate the dimensions of the rectangle that yield a maximum area.

74. *Federal Debt* The table shows the per capita U.S. federal debt for several years. (Sources: U.S. Treasury Department; U.S. Bureau of the Census)

Year	1950	1960	1970	1980
Per Capita Debt	$1688	$1572	$1807	$3981

Year	1990	1994	1997	1998
Per Capita Debt	$12,848	$15,750	$20,063	$20,513

A model for the per capita debt during this period is

$$y = 0.223t^3 - 0.733t^2 - 78.255t + 1837.433$$

where y represents the per capita debt and t is the time in years, with $t = 0$ corresponding to 1950.

(a) Use the *value* feature or *zoom* and *trace* features of your graphing utility to find the per capita federal debt in 1975 and 1992. Verify your answers algebraically.

(b) Use your graphing utility to determine during which year the per capita federal debt exceeded $10,200.

(c) Use the model to estimate the per capita federal debt in 2002 and 2004.

75. *Population Statistics* The table gives the life expectancy of a child (at birth) in the United States for selected years from 1920 to 2000. (Source: U.S. National Center for Health Statistics)

Year	1920	1930	1940	1950
Life Expectancy	54.1	59.7	62.9	68.2

Year	1960	1970	1980	1990	2000
Life Expectancy	69.7	70.8	73.7	75.4	76.4

A model for the life expectancy during this period is

$$y = \frac{66.93 + t}{1 + 0.01t}$$

where y represents the life expectancy and t is the time in years, with $t = 0$ corresponding to 1950.

(a) What does the y-intercept of the graph of the model represent?

(b) Use your graphing utility to determine the year when the life expectancy was 73.2. Verify your answer algebraically.

(c) Determine the life expectancy in 1948 both graphically and algebraically.

(d) Use the model to estimate the life expectancy of a child born in 2005.

76. *Finance* The dividends declared per share of Procter & Gamble Company from 1992 to 1998 can be approximated by the model

$$y = 0.464 + 0.091t, \qquad 0 \le t \le 6$$

where y is the dividend (in dollars) and t is the time (in years), with $t = 0$ corresponding to 1992. (Source: Procter & Gamble Company)

(a) Create a table showing the dividends y for the years 1992 through 1998. From your table, determine the year during which the dividend was $0.65.

(b) Use your graphing utility to graph the model.

(c) Verify your answer to part (a) by using the *zoom* and *trace* features of your graphing utility to determine the year when $y = 0.65$.

(d) Determine the value of y in 1997 algebraically.

(e) Use the model to estimate the value of y in 2002.

77. *Copper Wire* The resistance y in ohms of 1000 feet of solid copper wire at $77°$F can be approximated by the mathematical model

$$y = \frac{10{,}770}{x^2} - 0.37, \qquad 5 \le x \le 100$$

where x is the diameter of the wire in mils (0.001 in.). (Source: American Wire Gage)

(a) Complete the table.

x	10	20	30	40	50	60	70	80	90	100
y										

(b) Use your table to approximate the value of x when the resistance is 4.8 ohms. Then determine the answer algebraically.

(c) Use the *value* feature or *zoom* and *trace* features of your graphing utility to determine the resistance when $x = 85.5$.

(d) What can you conclude in general about the relationship between the diameter of the copper wire and the resistance?

Synthesis

True or False? **In Exercises 78 and 79, determine whether the statement is true or false. Justify your answer.**

78. A parabola can have only one x-intercept.

79. The graph of a linear equation can have either no x-intercepts or only one x-intercept.

80. *Think About It* Find a and b if the x-intercepts of the graph of $y = (x - a)(x - b)$ are $(-2, 0)$ and $(5, 0)$.

81. *Writing* Explain how to find an appropriate viewing window for the graph of an equation.

P.3   Lines in the Plane

The Slope of a Line

In this section, you will study lines and their equations. The **slope** of a nonvertical line represents the number of units a line rises or falls vertically for each unit of horizontal change from left to right. For instance, consider the two points (x_1, y_1) and (x_2, y_2) on the line shown in Figure P.26. As you move from left to right along this line, a change of $(y_2 - y_1)$ units in the vertical direction corresponds to a change of $(x_2 - x_1)$ units in the horizontal direction. That is,

$$y_2 - y_1 = \text{the change in } y$$

and

$$x_2 - x_1 = \text{the change in } x.$$

The slope of the line is given by the ratio of these two changes.

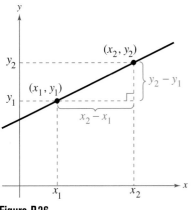

Figure P.26

Allsport

Definition of the Slope of a Line

The **slope** m of the nonvertical line through (x_1, y_1) and (x_2, y_2) is

$$m = \frac{y_2 - y_1}{x_2 - x_1} = \frac{\text{change in } y}{\text{change in } x}$$

where $x_1 \neq x_2$.

When this formula is used, the *order of subtraction* is important. Given two points on a line, you are free to label either one of them as (x_1, y_1) and the other as (x_2, y_2). However, once this has been done, you must form the numerator and denominator using the same order of subtraction.

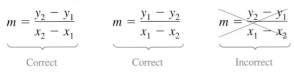

$$m = \frac{y_2 - y_1}{x_2 - x_1} \qquad m = \frac{y_1 - y_2}{x_1 - x_2} \qquad m = \frac{y_2 - y_1}{x_1 - x_2}$$

Correct   Correct   Incorrect

Throughout this text, the term *line* always means a *straight* line.

EXAMPLE 1 Finding the Slope of a Line

Find the slope of the line passing through each pair of points.

a. $(-2, 0)$ and $(3, 1)$ **b.** $(-1, 2)$ and $(2, 2)$ **c.** $(0, 4)$ and $(1, -1)$

Solution

Difference in *y*-values

a. $m = \dfrac{y_2 - y_1}{x_2 - x_1} = \dfrac{1 - 0}{3 - (-2)} = \dfrac{1}{3 + 2} = \dfrac{1}{5}$

Difference in *x*-values

b. $m = \dfrac{2 - 2}{2 - (-1)} = \dfrac{0}{3} = 0$

c. $m = \dfrac{-1 - 4}{1 - 0} = \dfrac{-5}{1} = -5$

The graphs of the three lines are shown in Figure P.27. Note that the square setting gives the correct "steepness" of the lines.

Exploration

Use a graphing utility to compare the slopes of the lines $y = 0.5x$, $y = x$, $y = 2x$, and $y = 4x$. What do you observe about these lines? Compare the slopes of the lines $y = -0.5x$, $y = -x$, $y = -2x$, and $y = -4x$. What do you observe about these lines? (*Hint:* Use a square setting to guarantee a true geometric perspective.)

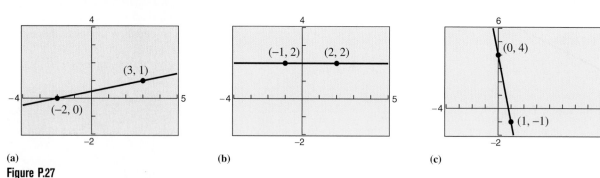

(a)   (b)   (c)

Figure P.27

The definition of slope does not apply to vertical lines. For instance, consider the points $(3, 4)$ and $(3, 1)$ on the vertical line shown in Figure P.28. Applying the formula for slope, you obtain

$$m = \frac{4 - 1}{3 - 3} = \frac{3}{0}. \qquad \text{Undefined}$$

Because division by zero is undefined, the slope of a vertical line is undefined.

From the slopes of the lines shown in Figures P.27 and P.28, you can make the following generalizations about the slope of a line.

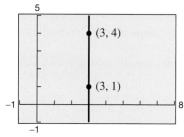

Figure P.28

The Slope of a Line

1. A line with positive slope ($m > 0$) *rises* from left to right.
2. A line with negative slope ($m < 0$) *falls* from left to right.
3. A line with zero slope ($m = 0$) is *horizontal*.
4. A line with undefined slope is *vertical*.

The Point-Slope Form of the Equation of a Line

If you know the slope of a line *and* you also know the coordinates of one point on the line, you can find an equation for the line. For instance, in Figure P.29, let (x_1, y_1) be a given point on the line whose slope is m. If (x, y) is any *other* point on the line, it follows that

$$\frac{y - y_1}{x - x_1} = m.$$

This equation in the variables x and y can be rewritten in the **point-slope form** of the equation of a line.

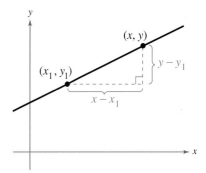

Figure P.29

Point-Slope Form of the Equation of a Line

The **point-slope form** of the equation of the line that passes through the point (x_1, y_1) and has a slope of m is

$$y - y_1 = m(x - x_1).$$

EXAMPLE 2   The Point-Slope Form of the Equation of a Line

Find an equation of the line that passes through the point $(1, -2)$ and has a slope of 3.

Solution

$$
\begin{aligned}
y - y_1 &= m(x - x_1) && \text{Point-slope form} \\
y - (-2) &= 3(x - 1) && \text{Substitute for } y_1, m, \text{ and } x_1. \\
y + 2 &= 3x - 3 \\
y &= 3x - 5
\end{aligned}
$$

This line is shown in Figure P.30.

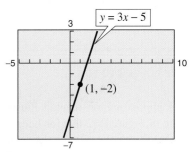

Figure P.30

The point-slope form can be used to find an equation of a nonvertical line passing through two points (x_1, y_1) and (x_2, y_2). First, use the formula for the slope of the line passing through two points.

$$m = \frac{y_2 - y_1}{x_2 - x_1}$$

Then, once you know the slope, use the point-slope form to obtain the equation

$$
\begin{aligned}
y - y_1 &= m(x - x_1) \\
&= \frac{y_2 - y_1}{x_2 - x_1}(x - x_1).
\end{aligned}
$$

This is sometimes called the **two-point form** of the equation of a line.

EXAMPLE 3   A Linear Model for Sales Prediction

During 1997, Barnes & Noble's net sales were $2.8 billion, and in 1998 net sales
were $3.0 billion.   (Source: Barnes & Noble, Inc.)

a. Write a linear equation giving the net sales y in terms of the year x.

b. Use the equation to estimate the net sales during 2000.

Solution

a. Let $x = 7$ represent 1997. In Figure P.31, let $(7, 2.8)$ and $(8, 3.0)$ be two points
on the line representing the net sales. The slope of the line passing through
these two points is

$$m = \frac{3.0 - 2.8}{8 - 7} = 0.2. \qquad m = \frac{y_2 - y_1}{x_2 - x_1}$$

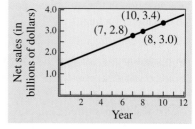

Figure P.31

By the point-slope form, the equation of the line is as follows.

$$y - y_1 = m(x - x_1) \qquad \text{Point-slope form}$$

$$y - 2.8 = 0.2(x - 7) \qquad \text{Substitute for } y_1, m, \text{ and } x_1.$$

$$y = 0.2x - 1.4 + 2.8$$

$$y = 0.2x + 1.4 \qquad \text{Simplify.}$$

b. Using the equation from part (a), estimate the 2000 net sales $(x = 10)$ to be

$$y = 0.2(10) + 1.4$$

$$= 2 + 1.4$$

$$= \$3.4 \text{ billion.}$$

The approximation method illustrated in Example 3 is **linear extrapolation.**
Note in Figure P.32 that for linear extrapolation, the estimated point lies outside
of the given points. When the estimated point lies *between* two given points, the
procedure is called **linear interpolation.**

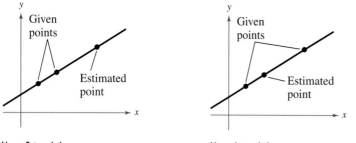

Linear Extrapolation
Figure P.32

Linear Interpolation

Library of Functions

In Section 1.1, you will be introduced to the precise meaning of
the term *function*. The simplest
type of function is a linear function and has the form

$$y = mx + b.$$

As its name implies, the graph
of a linear function is a line that
has a slope of m and a y-intercept
at $(0, b)$.

Consult the Library of Functions
Summary inside the front cover
for a description of the linear
function.

Sketching Graphs of Lines

Many problems in coordinate geometry can be classified as follows.

1. Given a graph (or parts of it), find its equation.

2. Given an equation, sketch its graph.

For lines, the first problem is solved easily by using the point-slope form. This formula, however, is not particularly useful for solving the second type of problem. The form that is better suited to graphing linear equations is the **slope-intercept form,** $y = mx + b$, of the equation of a line.

Slope-Intercept Form of the Equation of a Line

The graph of the equation

$$y = mx + b$$

is a line whose slope is m and whose y-intercept is $(0, b)$.

A computer simulation of this concept appears in the *Interactive* CD-ROM and *Internet* versions of this text.

EXAMPLE 4   Using the Slope-Intercept Form

Determine the slope and y-intercept of each linear equation. Then describe its graph.

a. $x + y = 2$ **b.** $y = 2$

Algebraic Solution

a. Begin by writing the equation in slope-intercept form.

$$x + y = 2 \qquad \text{Write original equation.}$$

$$y = 2 - x \qquad \text{Subtract } x \text{ from each side.}$$

$$y = -x + 2 \qquad \text{Slope-intercept form}$$

From the slope-intercept form of the equation, the slope is -1 and the y-intercept is $(0, 2)$. Because the slope is negative, you know that the graph of the equation is a line that falls one unit for every unit it moves to the right.

b. By writing the equation $y = 2$ in slope-intercept form

$$y = (0)x + 2$$

you can see that the slope is 0 and the y-intercept is $(0, 2)$. A zero slope implies that the line is horizontal.

Graphical Solution

a. Solve the equation for y to obtain $y = 2 - x$. Enter this equation in your graphing utility. Use a decimal viewing window to graph the equation as shown in Figure P.33(a).

To find the y-intercept, use the *value* or *trace* feature. When $x = 0$, $y = 2$. So, the y-intercept is $(0, 2)$. To find the slope, continue to use the *trace* feature. Move the cursor along the line until $x = 1$. At this point, $y = 1$. So the graph falls 1 unit for every unit it moves to the right, and the slope is -1.

b. Enter the equation $y = 2$ in your graphing utility and graph the equation as shown in Figure P.33(b). Use the *trace* feature to verify the y-intercept $(0, 2)$ and to see that the value of y is the same for all values of x. So, the slope of the horizontal line is 0.

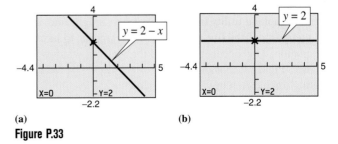

(a)   **(b)**

Figure P.33

M

From the slope-intercept form of the equation of a line, you can see that a horizontal line ($m = 0$) has an equation of the form $y = b$. This is consistent with the fact that each point on a horizontal line through $(0, b)$ has a y-coordinate of b.

Similarly, each point on a vertical line through $(a, 0)$ has an x-coordinate of a. So, a vertical line has an equation of the form $x = a$. This equation cannot be written in the slope-intercept form, because the slope of a vertical line is undefined. However, *every* line has an equation that can be written in the **general form**

$$Ax + By + C = 0 \qquad \text{General form of the equation of a line}$$

where A and B are not *both* zero.

Summary of Equations of Lines

1. **General form:** $Ax + By + C = 0$
2. **Vertical line:** $x = a$
3. **Horizontal line:** $y = b$
4. **Slope-intercept form:** $y = mx + b$
5. **Point-slope form:** $y - y_1 = m(x - x_1)$

Exploration

Graph the lines $y = 2x + 1$, $y = \frac{1}{2}x + 1$, and $y = -2x + 1$ in the same viewing window. What do you observe?

Graph the lines $y = 2x + 1$, $y = 2x$, and $y = 2x - 1$ in the same viewing window. What do

EXAMPLE 5   Different Viewing Windows

The graphs of the two lines

$$y = -x - 1 \qquad \text{and} \qquad y = -10x - 1$$

are shown in Figure P.34. Even though the slopes of these lines are quite different (-1 and -10, respectively), the graphs seem misleadingly similar because the viewing windows are different.

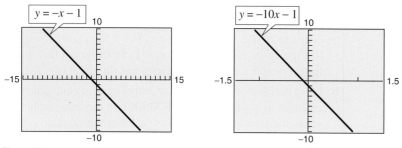

Figure P.34

When a graphing utility is used to sketch a straight line, it is important to realize that the graph of the line may not visually appear to have the slope indicated by its equation. This occurs because of the viewing window used for the graph. For instance, Figure P.35 shows graphs of $y = 2x + 1$ produced on a graphing utility using three different viewing windows.

Notice that the slopes in Figure P.35(a) and (b) do not visually appear to be equal to 2. However, if you use the *square* viewing window, as in Figure P.35(c), the slope visually appears to be 2. In general, two graphs of the same equation can appear to be quite different depending on the viewing window selected.

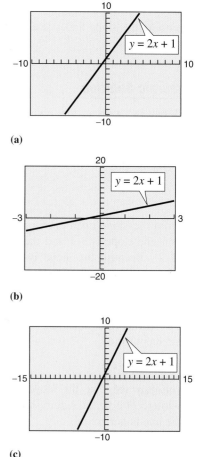

(a)

(b)

(c)
Figure P.35

Parallel and Perpendicular Lines

The slope of a line is a convenient tool for determining whether two lines are parallel or perpendicular.

Parallel Lines

Two distinct nonvertical lines are **parallel** if and only if their slopes are equal.

EXAMPLE 6   Equations of Parallel Lines

Find the slope-intercept form of the equation of the line that passes through the point $(2, -1)$ and is parallel to the line $2x - 3y = 5$.

Solution

Begin by finding the slope of the given line.

$$2x - 3y = 5 \qquad \text{Write original equation.}$$

$$-2x + 3y = -5 \qquad \text{Multiply by } -1.$$

$$3y = 2x - 5 \qquad \text{Add } 2x \text{ to each side.}$$

$$y = \frac{2}{3}x - \frac{5}{3} \qquad \text{Write in slope-intercept form.}$$

Therefore, the given line has a slope of $m = \frac{2}{3}$. Because any line parallel to the given line must also have a slope of $\frac{2}{3}$, the required line through $(2, -1)$ has the following equation.

$$y - (-1) = \frac{2}{3}(x - 2) \qquad \text{Substitute for } m, x_1, \text{ and } y_1 \text{ in point-slope form.}$$

$$y + 1 = \frac{2}{3}x - \frac{4}{3} \qquad \text{Simplify.}$$

$$y = \frac{2}{3}x - \frac{4}{3} - 1 \qquad \text{Subtract 1 from each side.}$$

$$y = \frac{2}{3}x - \frac{7}{3} \qquad \text{Write in slope-intercept form.}$$

Notice the similarity between the slope-intercept form of the original equation and the slope-intercept form of the parallel equation. The graphs of both equations are shown in Figure P.36.

Perpendicular Lines

Two nonvertical lines are **perpendicular** if and only if their slopes are negative reciprocals of each other. That is,

$$m_1 = -\frac{1}{m_2}.$$

STUDY T!P

Be careful when you graph equations such as $y = \frac{2}{3}x - \frac{7}{3}$ on your graphing utility. A common mistake is to type it in as

$$Y_1 = 2/3X - 7/3,$$

which may not be interpreted as the original equation by your graphing utility. You should use one of the following formulas.

$$Y_1 = 2X/3 - 7/3$$

$$Y_1 = (2/3)X - 7/3$$

Do you see why?

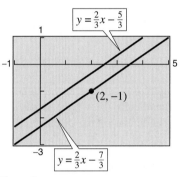

Figure P.36

EXAMPLE 7   Equations of Perpendicular Lines

Find an equation of the line that passes through the point $(2, -1)$ and is perpendicular to the line $2x - 3y = 5$.

Solution

By writing the given line in the form $y = \frac{2}{3}x - \frac{5}{3}$, you can see that the line has a slope of $\frac{2}{3}$. So, any line that is perpendicular to this line must have a slope of $-\frac{3}{2}$ (because $-\frac{3}{2}$ is the negative reciprocal of $\frac{2}{3}$). Therefore, the required line through the point $(2, -1)$ has the following equation.

$$y - (-1) = -\frac{3}{2}(x - 2) \qquad \text{Substitute for } m, x_1, \text{ and } y_1 \text{ in point-slope form.}$$

$$y + 1 = -\frac{3}{2}x + 3 \qquad \text{Simplify.}$$

$$y = -\frac{3}{2}x + 3 - 1 \qquad \text{Subtract 1 from each side.}$$

$$y = -\frac{3}{2}x + 2 \qquad \text{Slope-intercept form}$$

The graphs of both equations are shown in Figure P.37.

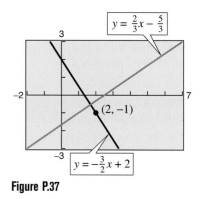

Figure P.37

EXAMPLE 8   Graphs of Perpendicular Lines

Use a graphing utility to graph the lines

$$y = x + 1$$

and

$$y = -x + 3.$$

Display *both* graphs in the same viewing window. The lines are supposed to be perpendicular (they have slopes of $m_1 = 1$ and $m_2 = -1$). Do they appear to be perpendicular on the display?

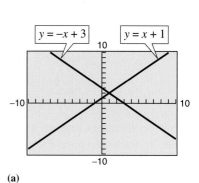

(a)

Solution

If the viewing window is nonsquare, as in Figure P.38(a), the two lines will not appear perpendicular. If, however, the viewing window is square, as in Figure P.38(b), the lines will appear perpendicular.

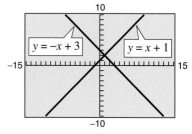

(b)

Figure P.38

Writing About Math   *An Application of Slope*

In 1990, a college had an enrollment of 5500 students. By 2000, the enrollment had increased to 7000 students.

a. What was the average annual change in enrollment from 1990 to 2000?

b. Use the average annual change in enrollment to estimate the enrollments in 1993, 1997, and 1999.

c. Write the equation of the line that represents the data in part (b). What is its slope? Interpret the slope in the context of the problem.

d. Write a short paragraph discussing the concepts of *slope* and *average rate of change*.

P.3 E X E R C I S E S

In Exercises 1 and 2, identify the line that has the specified slope.

1. (a) $m = \frac{2}{3}$ (b) m is undefined.   (c) $m = -2$

2. (a) $m = 0$ (b) $m = -\frac{3}{4}$ (c) $m = 1$

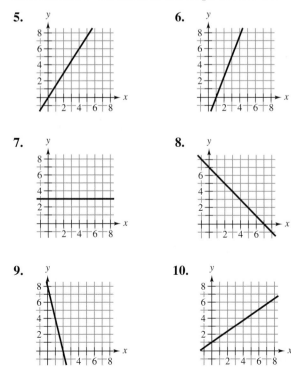

FIGURE FOR 1   FIGURE FOR 2

In Exercises 3 and 4, sketch the line through the point with each indicated slope on the same set of coordinate axes.

Point	Slopes
3. $(2, 3)$	(a) 0 (b) 1 (c) 2 (d) -3
4. $(-4, 1)$	(a) 3 (b) -3 (c) $\frac{1}{2}$ (d) Undefined

In Exercises 5–10, estimate the slope of the line.

5.

6.

7.

8.

9.

10.

In Exercises 11–14, plot the points and find the slope of the line passing through the points. Verify the slope using the *draw* feature on your graphing utility to graph the line segment connecting the two points. (Use a square setting.)

11. $(0, -10), (-4, 0)$ **12.** $(2, 4), (4, -4)$

13. $(-6, -1), (-6, 4)$ **14.** $(-3, -2), (1, 6)$

In Exercises 15–20, you are given the slope of the line and a point on the line. Find three additional points through which the line passes. (There are many correct answers.)

Point	Slope
15. $(2, 1)$	$m = 0$
16. $(-4, 1)$	m is undefined.
17. $(-5, 4)$	$m = 2$
18. $(0, -9)$	$m = -2$
19. $(7, -2)$	$m = \frac{1}{2}$
20. $(-1, -6)$	$m = -\frac{1}{2}$

In Exercises 21–24, determine whether the lines L_1 and L_2 passing through the pairs of points are parallel, perpendicular, or neither. Use a graphing utility to graph the line segments connecting the pairs of points on the respective lines. (Use a square setting.)

21. L_1: $(0, -1), (5, 9)$ **22.** L_1: $(-2, -1), (1, 5)$
   L_2: $(0, 3), (4, 1)$ L_2: $(1, 3), (5, -5)$

23. L_1: $(3, 6), (-6, 0)$ **24.** L_1: $(4, 8), (-4, 2)$
   L_2: $(0, -1), \left(5, \frac{7}{3}\right)$ L_2: $(3, -5), \left(-1, \frac{1}{3}\right)$

In Exercises 25–32, (a) find the slope and y-intercept (if possible) of the equation of the line algebraically, (b) sketch the line by hand, and (c) use a graphing utility to verify your answers to parts (a) and (b).

25. $5x - y + 3 = 0$ **26.** $2x + 3y - 9 = 0$

27. $5x - 2 = 0$ **28.** $3x + 7 = 0$

29. $3y + 5 = 0$ **30.** $-11 - 8y = 0$

31. $7x + 6y - 30 = 0$ **32.** $x - y - 10 = 0$

In Exercises 33–42, find the general form of the equation of the line that passes through the given point and has the indicated slope. Sketch the line by hand. Use a graphing utility to verify your sketch.

	Point	*Slope*
33.	$(0, -2)$	$m = 3$
34.	$(0, 10)$	$m = -1$
35.	$(-3, 6)$	$m = -2$
36.	$(0, 0)$	$m = 4$
37.	$(4, 0)$	$m = -\frac{1}{3}$
38.	$(-2, -5)$	$m = \frac{3}{4}$
39.	$(6, -1)$	m is undefined.
40.	$(-10, 4)$	$m = 0$
41.	$\left(-\frac{1}{2}, \frac{3}{2}\right)$	$m = -3$
42.	$(2.3, -8.5)$	$m = -\frac{5}{2}$

In Exercises 43–52, find the general form of the equation of the line that passes through the points. Use a graphing utility to sketch the line.

43. $(5, -1), (-5, 5)$ **44.** $(4, 3), (-4, -4)$

45. $(-8, 1), (-8, 7)$ **46.** $(-1, 4), (6, 4)$

47. $\left(2, \frac{1}{2}\right), \left(\frac{1}{2}, \frac{5}{4}\right)$ **48.** $\left(1, 1\right), \left(6, -\frac{2}{3}\right)$

49. $\left(-\frac{1}{10}, -\frac{3}{5}\right), \left(\frac{9}{10}, -\frac{9}{5}\right)$ **50.** $\left(\frac{3}{4}, \frac{3}{2}\right), \left(-\frac{4}{3}, \frac{7}{4}\right)$

51. $(1, 0.6), (-2, -0.6)$ **52.** $(-8, 0.6), (2, -2.4)$

Exploration   **In Exercises 53 and 54, use the values of** *a* **and** *b* **and a graphing utility to graph the equation of the line**

$$\frac{x}{a} + \frac{y}{b} = 1, \qquad a \neq 0, b \neq 0.$$

Use the graphs to make a conjecture about what *a* **and** *b* **represent. Verify your conjecture.**

53. $a = 5, \quad b = -3$ **54.** $a = -6, \quad b = 2$

In Exercises 55–58, use the results of Exercises 53 and 54 to write an equation of the line that passes through the points.

55. x-intercept: $(2, 0)$ **56.** x-intercept: $(-5, 0)$
     y-intercept: $(0, 3)$ y-intercept: $(0, -4)$

57. x-intercept: $\left(-\frac{1}{6}, 0\right)$ **58.** x-intercept: $\left(\frac{3}{4}, 0\right)$
     y-intercept: $\left(0, -\frac{2}{3}\right)$ y-intercept: $\left(0, \frac{4}{5}\right)$

In Exercises 59 and 60, use a graphing utility to graph the equation using each of the suggested viewing windows. Describe the difference between the two graphs.

59. $y = 0.5x - 3$

Xmin = -5	Xmin = -2
Xmax = 10	Xmax = 10
Xscl = 1	Xscl = 1
Ymin = -1	Ymin = -4
Ymax = 10	Ymax = 1
Yscl = 1	Yscl = 1

60. $y = -8x + 5$

Xmin = -5	Xmin = -5
Xmax = 5	Xmax = 10
Xscl = 1	Xscl = 1
Ymin = -10	Ymin = -80
Ymax = 10	Ymax = 80
Yscl = 1	Yscl = 20

Graphical Analysis   **In Exercises 61–64, use a graphing utility to graph the three equations in the same viewing window. Adjust the viewing window so that the slope appears visually correct. Identify any relationships that exist among the lines. Use the slope of the lines to verify your results.**

61. (a) $y = 2x$ (b) $y = -2x$ (c) $y = \frac{1}{2}x$

62. (a) $y = \frac{2}{3}x$ (b) $y = -\frac{3}{2}x$ (c) $y = \frac{2}{3}x + 2$

63. (a) $y = -\frac{1}{2}x$ (b) $y = -\frac{1}{2}x + 3$ (c) $y = 2x - 4$

64. (a) $y = x - 8$ (b) $y = x + 1$ (c) $y = -x + 3$

In Exercises 65–70, write equations of the lines through the given point (a) parallel to the given line and (b) perpendicular to the given line.

	Point	*Line*
65.	$(2, 1)$	$4x - 2y = 3$
66.	$(-3, 2)$	$x + y = 7$
67.	$\left(-\frac{2}{3}, \frac{7}{8}\right)$	$3x + 4y = 7$
68.	$\left(\frac{7}{8}, \frac{3}{4}\right)$	$5x + 3y = 0$
69.	$(2.5, 6.8)$	$x - y = 4$
70.	$(-3.9, -1.4)$	$6x + 2y = 9$

In Exercises 71 and 72, find a relationship between *x* **and** *y* **such that** (x, y) **is equidistant from the two points.**

71. $(4, -1), (-2, 3)$ **72.** $(3, -2), (-7, 1)$

73. Business The slopes are the slopes of lines representing annual sales y in terms of time x in years. Use each slope to interpret any change in annual sales for a 1-year increase in time.

(a) The line has a slope of $m = 135$.

(b) The line has a slope of $m = 0$.

(c) The line has a slope of $m = -40$.

74. Business The slopes are the slopes of lines representing daily revenues y in terms of time x in days. Use each slope to interpret any change in daily revenues for a 1-day increase in time.

(a) The line has a slope of $m = 400$.

(b) The line has a slope of $m = 100$.

(c) The line has a slope of $m = 0$.

75. Business The graph shows the earnings per share of stock for the Kellogg Company for the years 1988 through 1998. (Source: Kellogg Company)

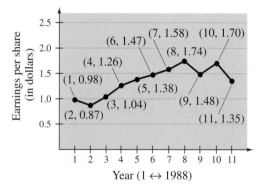

(a) Use the slopes to determine the year(s) when the earnings per share showed the greatest increase and decrease.

(b) Find the equation of the line between the years 1988 and 1998.

(c) Interpret the meaning of the slope in the equation from part (b) in the context of the problem.

(d) Use the equation from part (b) to estimate the earnings per share of stock for the year 2001. Do you think this is an accurate estimation? Explain.

76. Business The graph shows the dividends declared per share of stock for the Colgate-Palmolive Company for the years 1988 through 1998. (Source: Colgate-Palmolive Company)

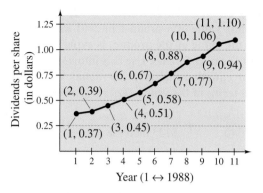

(a) Use the slopes to determine the years when the dividends declared per share showed the greatest increase and the smallest increase.

(b) Find the equation of the line between the years 1988 and 1998.

(c) Interpret the meaning of the slope in the equation from part (b) in the context of the problem.

(d) Use the equation from part (b) to estimate the dividends declared per share for the year 2001. Do you think this is an accurate estimation? Explain.

77. Driving When driving down a mountain road, you notice warning signs indicating that it is a "12% grade." This means that the slope of the road is $-\frac{12}{100}$. Approximate the amount of horizontal change in your position if you note from elevation markers that you have descended 2000 feet vertically.

78. Attic Height The "rise to run" ratio of the roof of a house determines the steepness of the roof. Suppose the rise to run ratio of a roof is 3 to 4. Determine the maximum height in the attic of the house if the house is 32 feet wide.

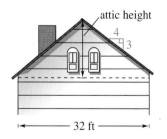

Rate of Change **In Exercises 79–82, you are given the dollar value of a product in 2001 and the rate at which the value of the product is expected to change during the next 5 years. Use this information to write a linear equation that gives the dollar value V of the item in terms of the year t. (Let $t = 1$ represent 2001.)**

	2001 Value	*Rate*
79.	$2540	$125 increase per year
80.	$156	$4.50 increase per year
81.	$20,400	$2000 decrease per year
82.	$245,000	$5600 decrease per year

Graphical Interpretation **In Exercises 83–86, match the description with its graph. Also determine the slope and how it is interpreted in the situation. [The graphs are labeled (a), (b), (c), and (d).]**

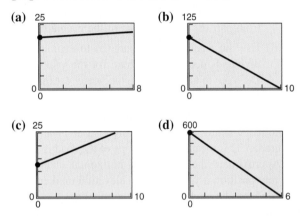

83. A person is paying $10 per week to a friend to repay a $100 loan.

84. An employee is paid $12.50 per hour plus $1.50 for each unit produced per hour.

85. A sales representative receives $20 per day for food plus $0.25 for each mile traveled.

86. A word processor that was purchased for $600 depreciates $100 per year.

87. *Temperature*   Find the equation of the line that shows the relationship between the temperature in degrees Celsius C and degrees Fahrenheit F. Remember that water freezes at $0°C$ ($32°F$) and boils at $100°C$ ($212°F$).

88. *Temperature*   Use the result of Exercise 87 to complete the table.

C		$-10°$	$10°$			$177°$
F	$0°$			$68°$	$90°$	

89. *Finance*   Your salary was $28,500 in 1998 and $32,900 in 2000. If your salary follows a linear growth pattern, what will your salary be in 2003?

90. *College Enrollment*   A small college had 2546 students in 1998 and 2702 students in 2000. If the enrollment follows a linear growth pattern, how many students will the college have in 2004?

91. *Business*   A small business purchases a fax machine for $875. After 5 years, the fax machine will be outdated and have no value.

  (a) Write a linear equation giving the value V of the fax machine during the 5 years it will be used.

  (b) Use a graphing utility to graph the linear equation representing the depreciation of the fax machine, and use the *value* or *trace* feature to complete the table.

t	0	1	2	3	4	5
V						

  (c) Verify your answers in part (b) algebraically by using the equation you found in part (a).

92. *Business*   A small business purchases a computer network system for $25,000. After 10 years, the system will have to be replaced. Its value at that time is expected to be $2000.

  (a) Write a linear equation giving the value V of the system during the 10 years it will be used.

  (b) Use a graphing utility to graph the linear equation representing the depreciation of the system, and use the *value* or *trace* feature to complete the table.

t	0	1	2	3	4	5	6	7	8	9	10
V											

  (c) Verify your answers in part (b) algebraically by using the equation you found in part (a).

93. *Business*   A contractor purchases a bulldozer for $36,500. The bulldozer requires an average expenditure of $5.25 per hour for fuel and maintenance, and the operator is paid $11.50 per hour.

  (a) Write a linear equation giving the total cost C of operating the bulldozer for t hours. (Include the purchase cost of the bulldozer.)

  (b) Assuming that customers are charged $27 per hour of bulldozer use, write an equation for the revenue R derived from t hours of use.

(c) Use the formula for profit ($P = R - C$) to write an equation for the profit derived from t hours of use.

(d) Use the result of part (c) to find the break-even point (the number of hours the bulldozer must be used to yield a profit of 0 dollars).

94. *Real Estate Purchase* A real estate office handles an apartment complex with 50 units. When the rent per unit is $580 per month, all 50 units are occupied. However, when the rent is $625 per month, the average number of occupied units drops to 47. Assume that the relationship between the monthly rent p and the demand x is linear.

(a) Write the equation of the line giving the demand x in terms of the rent p.

(b) Use a graphing utility to graph the demand equation and use the *trace* feature to estimate the number of units occupied if the rent is raised to $655. Verify your answer algebraically.

(c) Use the demand equation to estimate the number of units occupied if the rent is lowered to $595. Verify your answer graphically.

95. *Sports* The average annual salaries of Major League Baseball players (in thousands of dollars) from 1988 to 1998 are shown in the scatter plot. Let y represent the average salary and let t represent the year, with $t = 0$ corresponding to 1988. (Source: Major League Baseball Player Relations Committee)

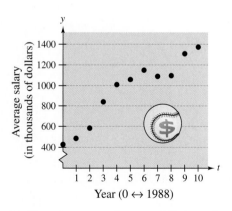

(a) Use the regression capabilities of a graphing utility to find the line that best fits the data.

(b) Use the regression line to estimate the average salary in the year 2000.

(c) Interpret the meaning of the slope of the regression line.

Synthesis

True or False? **In Exercises 96 and 97, determine whether the statement is true or false. Justify your answer.**

96. The line through $(-8, 2)$ and $(-1, 4)$ and the line through $(0, -4)$ and $(-7, 7)$ are parallel.

97. If the points $(10, -3)$ and $(2, -9)$ lie on the same line, then the point $\left(-12, -\frac{37}{2}\right)$ also lies on that line.

98. *Writing* Explain how you could show that the points $A(2, 3)$, $B(2, 9)$, and $C(7, 3)$ are the vertices of a right triangle.

99. *Think About It* The slopes of two lines are -4 and $\frac{5}{2}$. Which is steeper?

100. *Writing* Write a brief paragraph explaining whether or not any pair of points on a line can be used to calculate the slope of the line.

101. *Think About It* Is it possible for two lines with positive slopes to be perpendicular? Explain.

P.4 Solving Equations Algebraically and Graphically

Equations and Solutions of Equations

An **equation** is a statement that two algebraic expressions are equal. To **solve** an equation in x means to find all values of x for which the equation is true. Such values are **solutions.** For instance, $x = 4$ is a solution of the equation $3x - 5 = 7$, because $3(4) - 5 = 7$ is a true statement.

The solutions of an equation depend upon the kinds of numbers being considered. For instance, in the set of rational numbers, $x^2 = 10$ has no solution because there is no rational number whose square is 10. However, in the set of real numbers the equation has the two solutions $\sqrt{10}$ and $-\sqrt{10}$.

An equation that is true for every real number in the domain of the variable is called an **identity.** For example, $x^2 - 9 = (x + 3)(x - 3)$ is an identity because it is a true statement for any real value of x.

An equation that is true for just *some* (or even none) of the real numbers in the domain of the variable is called a **conditional equation.** For example, the equation $x^2 - 9 = 0$ is conditional because $x = 3$ and $x = -3$ are the only values in the domain that satisfy the equation. The equation $2x + 1 = 2x - 3$ is also conditional because it is not true for any values of x. A **linear equation in one variable** x is an equation that can be written in the standard form $ax + b = 0$, where a and b are real numbers with $a \neq 0$. For a review of solving one- and two-step equations, see Appendix C.

To solve an equation involving fractional expressions, find the least common denominator (LCD) of all terms in the equation and multiply every term by this LCD. This procedure clears the equation of fractions.

What You Should Learn:

- How to solve linear equations
- How to find x- and y-intercepts of graphs of equations
- How to find solutions of equations graphically
- How to find the points of intersection of two graphs
- How to solve polynomial equations
- How to solve equations involving radicals, fractions, or absolute values

Why You Should Learn It:

Knowing how to solve equations graphically and algebraically can help you solve real-life problems. For instance, in Exercise 161 on page 53, you can find the number of copies of a book that will by sold at a certain price by solving a demand equation involving a radical.

Robert Holmes/CORBIS

EXAMPLE 1   Solving an Equation Involving Fractions

$$\frac{x}{3} + \frac{3x}{4} = 2 \qquad \text{Original equation}$$

$$(12)\frac{x}{3} + (12)\frac{3x}{4} = (12)2 \qquad \text{Multiply by the LCD.}$$

$$4x + 9x = 24 \qquad \text{Simplify.}$$

$$13x = 24 \qquad \text{Combine like terms.}$$

$$x = \frac{24}{13} \qquad \text{Divide each side by 13.}$$

Check

After solving an equation, check the solution in the original equation.

$$\frac{\frac{24}{13}}{3} + \frac{3\left(\frac{24}{13}\right)}{4} \overset{?}{=} 2 \qquad \text{Substitute } \tfrac{24}{13} \text{ for } x.$$

$$2 = 2 \qquad \text{Solution checks. } \checkmark$$

When multiplying or dividing an equation by a *variable* expression, it is possible to introduce an **extraneous** solution—one that does not satisfy the original equation. The next example demonstrates the importance of checking your solution when you have multiplied or divided by a variable expression.

EXAMPLE 2   An Equation with an Extraneous Solution

Solve for x: $\dfrac{1}{x-2} = \dfrac{3}{x+2} - \dfrac{6x}{x^2-4}$

Algebraic Solution

In this case, the LCD is

$$x^2 - 4 = (x+2)(x-2).$$

Multiplying each term by the LCD and simplifying produces the following.

$$\frac{1}{x-2}(x+2)(x-2)$$

$$= \frac{3}{x+2}(x+2)(x-2) - \frac{6x}{x^2-4}(x+2)(x-2)$$

$$x + 2 = 3(x-2) - 6x, \qquad x \neq \pm 2$$

$$x + 2 = 3x - 6 - 6x$$

$$4x = -8$$

$$x = -2$$

A check of $x = -2$ in the original equation shows that it yields a denominator of zero. So, $x = -2$ is extraneous, and the equation has *no solution*.

Graphical Solution

Use a graphing utility to graph the left and right sides of the equation

$$y_1 = \frac{1}{x-2} \quad \text{and} \quad y_2 = \frac{3}{x+2} - \frac{6x}{x^2-4}$$

in the same viewing window as shown in Figure P.39. The graphs of the equations do not appear to intersect. This means that there is no point for which the left side of the equation $1/(x-2)$ is equal to the right side of the equation

$$\frac{3}{x+2} - \frac{6x}{x^2-4}.$$

So, the equation appears to have *no solution*.

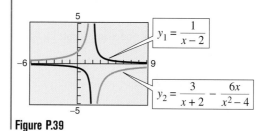

Figure P.39

Intercepts and Solutions

In Section P.2, you learned that the intercepts of a graph are the points at which the graph intersects the x- or y-axis.

Definition of Intercepts

1. The point $(a, 0)$ is called an **x-intercept** of the graph of an equation if it is a solution point of the equation. To find the x-intercept(s), let $y = 0$ and solve the equation for x.
2. The point $(0, b)$ is called a **y-intercept** of the graph of an equation if it is a solution point of the equation. To find the y-intercept(s), let $x = 0$ and solve the equation for y.

An ancient Egyptian papyrus, discovered in 1858, contains one of the earliest examples of mathematical writing in existence. The papyrus itself dates back to around 1650 B.C., but it is actually a copy of writings from two centuries earlier. The algebraic equations on the papyrus were written in words. Diophantus, a Greek who lived around A.D. 250, is often called the Father of Algebra. He was the first to use abbreviated word forms in equations.

Sometimes it is convenient to denote the *x*-intercept as simply the *x*-coordinate of the point $(a, 0)$ rather than the point itself. Unless it is necessary to make a distinction, "intercept" will be used to mean either the point or the coordinate.

It is possible that a particular graph will have no intercepts or several intercepts. For instance, consider the three graphs in Figure P.40.

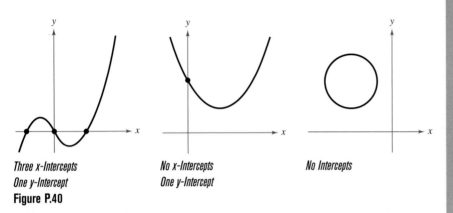

Three x-Intercepts
One y-Intercept
Figure P.40

No x-Intercepts
One y-Intercept

No Intercepts

Exploration

Use a graphing utility to graph each equation. How many times does each of the graphs intersect the *x*-axis and how many times does each intersect the *y*-axis?

$$y_1 = x^4 - 5x^2 + 4$$

$$y_2 = 3 + x^2$$

$$y_3 = x^2 - 3$$

$$y_4 = 2 \pm \sqrt{4x - x^2 - 3}$$

In general, how many times can an equation's graph intersect the *x*-axis? How many times can it intersect the *y*-axis?

EXAMPLE 3   Finding *x*- and *y*-Intercepts

Find the *x*- and *y*-intercepts of the graph of $2x + 3y = 5$.

Solution

To find the *x*-intercept, let $y = 0$. This produces

$$2x = 5 \quad \Longrightarrow \quad x = \frac{5}{2}$$

which implies that the graph has one *x*-intercept: $\left(\frac{5}{2}, 0\right)$. To find the *y*-intercept, let $x = 0$. This produces

$$3y = 5 \quad \Longrightarrow \quad y = \frac{5}{3}$$

which implies that the graph has one *y*-intercept: $\left(0, \frac{5}{3}\right)$. See Figure P.41.

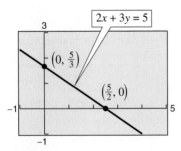

Figure P.41

The *Interactive* CD-ROM and *Internet* versions of this text offer a built-in graphing calculator, which can be used with the Examples, Explorations, and Exercises.

The concepts of *x*-intercepts and solutions of equations are closely related. In fact, the following statements are equivalent.

1. The point $(a, 0)$ is an *x-intercept* of the graph of an equation.

2. The number a is a *solution* of the equation $y = 0$.

This close connection among x-intercepts and solutions described on page 40 is crucial to our study of algebra, and you can take advantage of this connection in two basic ways. You can use your algebraic "equation-solving skills" to find the x-intercepts of a graph, and you can use your "graphing skills" to approximate the solutions of an equation.

Finding Solutions Graphically

Polynomial equations of degree 1 or 2 can be solved in relatively straightforward ways. Polynomial equations of higher degrees can, however, be quite difficult to solve, especially if you rely only on algebraic techniques. For such equations, a graphing utility can be very helpful.

Graphical Approximations of Solutions of an Equation

1. Write the equation in *general form, $y = 0$,* with the nonzero terms on one side of the equation and zero on the other side.

2. Use a graphing utility to graph the equation. Be sure the viewing window shows all the relevant features of the graph.

3. Use the *zero* or *root* feature or the *zoom* and *trace* features of the graphing utility to approximate each of the x-intercepts of the graph. Remember that a graph can have more than one x-intercept, so you may need to change the viewing window a few times.

In Chapter 2 you will learn techniques for determining the number of solutions of a polynomial equation. For now, you should know that a polynomial equation of degree n cannot have more than n different solutions.

EXAMPLE 4   Finding Solutions of an Equation Graphically

Use a graphing utility to approximate the solutions of $2x^3 - 3x + 2 = 0$.

Solution
Begin by graphing the equation $y = 2x^3 - 3x + 2$, as shown in Figure P.42. You can see from the graph that there is only one x-intercept. It lies between -1 and -2 and is approximately -1.5. By using the *zero* or *root* feature of a graphing utility you can improve the approximation. To three-decimal-place accuracy, the solution is $x \approx -1.476$. Check this approximation on your calculator. You will find that the value of y is $y = 2(-1.476)^3 - 3(-1.476) + 2 \approx -0.003$.

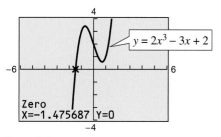

Figure P.42

Exploration

In Chapter 2 you will learn that a cubic equation such as

$$24x^3 - 36x + 17 = 0$$

can have up to 3 real solutions. Use a graphing utility to graph

$$y = 24x^3 - 36x + 17.$$

Describe a viewing window that allows you to determine the number of real solutions of the equation

$$24x^3 - 36x + 17 = 0.$$

Use the same technique to determine the number of real solutions of

$$97x^3 - 102x^2 - 200x - 63 = 0.$$

You can also use a graphing calculator's *zoom* and *trace* features to approximate the solution of an equation. Here are some suggestions for using the *zoom-in* feature of a graphing utility.

1. With each successive zoom-in, adjust the *x*-scale (if necessary) so that the resulting viewing window shows at least the two scale marks between which the solution lies.

2. The accuracy of the approximation will always be such that the error is less than the distance between two scale marks.

3. If you have a *trace* feature on your graphing utility, you can generally add one more decimal place of accuracy without changing the viewing window.

Unless stated otherwise, this book will approximate all real solutions with an error of *at most* 0.01.

STUDY T!P

Remember that the more decimal places in the solution, the more accurate it is. You can reach the desired accuracy when zooming in as follows.

- To approximate the zero to the nearest hundredth, set the *x*-scale to 0.01.
- To approximate the zero to the nearest thousandth, set the *x*-scale to 0.001.

EXAMPLE 5   Approximating Solutions of an Equation Graphically

Use a graphing utility to approximate the solutions of $x^2 + 3 = 5x$.

Solution

In general form, this equation is

$$x^2 - 5x + 3 = 0. \qquad \text{Equation in general form}$$

So, you can begin by graphing

$$y = x^2 - 5x + 3 \qquad \text{Equation to be graphed}$$

as shown in Figure P.43(a). This graph has two *x*-intercepts, and by using the *zoom* and *trace* features you can approximate the corresponding solutions to be $x \approx 0.70$ and $x \approx 4.30$, as shown in Figures P.43(b) and P.43(c).

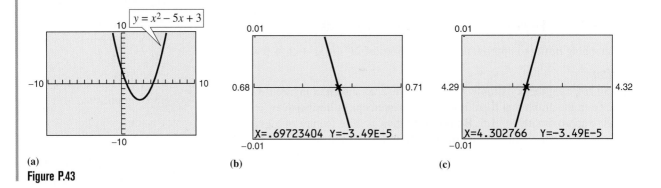

(a)   **(b)**   **(c)**

Figure P.43

The built-in *zero* or *root* programs of a graphing utility will approximate solutions of equations or approximate *x*-intercepts of graphs. If your graphing utility has such features, try using them to approximate the solutions in Example 5.

Points of Intersection of Two Graphs

An ordered pair that is a solution of two different equations is called a **point of intersection** of the graphs of the two equations. For instance, in Figure P.44 you can see that the graphs of the following equations have two points of intersection.

$$y = x + 2 \qquad \text{Equation 1}$$

$$y = x^2 - 2x - 2 \qquad \text{Equation 2}$$

The point $(-1, 1)$ is a solution of both equations, and the point $(4, 6)$ is a solution of both equations. To check this algebraically, substitute -1 and 4 into each equation.

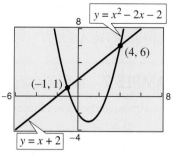

Figure P.44

Check that $(-1, 1)$ is a solution.

Equation 1: $y = -1 + 2 = 1$ Solution checks. ✓

Equation 2: $y = (-1)^2 - 2(-1) - 2$

$$= 1 \qquad \text{Solution checks. ✓}$$

Check that $(4, 6)$ is a solution.

Equation 1: $y = 4 + 2$

$$= 6 \qquad \text{Solution checks. ✓}$$

Equation 2: $y = (4)^2 - 2(4) - 2$

$$= 6 \qquad \text{Solution checks. ✓}$$

To find the points of intersection of the graphs of two equations, solve each equation for y (or x) and set the two results equal to each other. The resulting equation will be an equation in one variable, which can be solved using standard procedures, as shown in Example 6.

EXAMPLE 6   Finding Points of Intersection

Find the points of intersection of the graphs of $2x - 3y = -2$ and $4x - y = 6$.

Algebraic Solution

To begin, solve each equation for y to obtain

$$y = \frac{2}{3}x + \frac{2}{3} \quad \text{and} \quad y = 4x - 6.$$

Next, set the two expressions for y equal to each other and solve the resulting equation for x, as follows.

$$\frac{2}{3}x + \frac{2}{3} = 4x - 6 \qquad \text{Equate expressions for } y.$$

$$2x + 2 = 12x - 18 \qquad \text{Multiply each side by 3.}$$

$$-10x = -20 \qquad \text{Subtract } 12x \text{ and 2 from each side.}$$

$$x = 2 \qquad \text{Divide each side by } -10.$$

When $x = 2$, the y-value of each of the given equations is 2. So, the point of intersection is $(2, 2)$.

Graphical Solution

To begin, solve each equation for y to obtain $y_1 = \frac{2}{3}x + \frac{2}{3}$ and $y_2 = 4x - 6$. Then use a graphing utility to graph both equations in the same viewing window. In Figure P.45, the graphs appear to have one point of intersection. Use the *intersect* feature of the graphing utility to approximate the point of intersection to be $(2, 2)$.

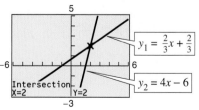

Figure P.45

Another way to approximate points of intersection of two graphs is to graph both equations and use the *zoom* and *trace* features to find the point or points at which the two graphs intersect.

EXAMPLE 7   Approximating Points of Intersection Graphically

Approximate the point(s) of intersection of the graphs of the following equations.

$$y = x^2 - 3x - 4 \qquad\qquad \text{Equation 1 (quadratic equation)}$$

$$y = x^3 + 3x^2 - 2x - 1 \qquad\qquad \text{Equation 2 (cubic equation)}$$

Solution

Begin by using a graphing utility to graph both equations, as shown in Figure P.46. From this display, you can see that the two graphs have only one point of intersection. Then, using the *zoom* and *trace* features, approximate the point of intersection to be $(-2.17, 7.25)$. To test the reasonableness of this approximation, you can evaluate both equations when $x = -2.17$.

Quadratic Equation:

$$y = (-2.17)^2 - 3(-2.17) - 4$$

$$\approx 7.22$$

Cubic Equation:

$$y = (-2.17)^3 + 3(-2.17)^2 - 2(-2.17) - 1$$

$$\approx 7.25$$

Because both equations yield approximately the same y-value, you can conclude that the approximate coordinates of the point of intersection are $x \approx -2.17$ and $y \approx 7.25$.

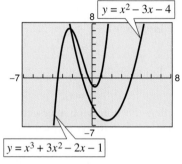

Figure P.46

The method shown in Example 7 gives a nice graphical picture of the points of intersection of two graphs. However, for actual approximation purposes, it is better to use the algebraic procedure described in Example 6. That is, the point of intersection of $y = x^2 - 3x - 4$ and $y = x^3 + 3x^2 - 2x - 1$ coincides with the solution of the equation

$$x^3 + 3x^2 - 2x - 1 = x^2 - 3x - 4 \qquad \text{Equate } y\text{-values.}$$

$$x^3 + 2x^2 + x + 3 = 0. \qquad\qquad \text{Write in general form.}$$

By graphing $y = x^3 + 2x^2 + x + 3$ on a graphing utility and using the *zoom* and *trace* features (or the *zero* or *root* feature), you can approximate the solution of this equation to be $x \approx -2.17$. The corresponding y-value for *both* of the equations given in Example 7 is $y \approx 7.25$.

Solving Polynomial Equations Algebraically

Polynomial equations can be classified by their degree.

Degree	Name	Example
First degree	Linear equation	$6x + 2 = 4$
Second degree	Quadratic equation	$2x^2 - 5x + 3 = 0$
Third degree	Cubic equation	$x^3 - x = 0$
Fourth degree	Quartic equation	$x^4 - 3x^2 + 2 = 0$
Fifth degree	Quintic equation	$x^5 - 12x^2 + 7x + 4 = 0$

In general, the higher the degree, the more difficult it is to solve the equation—either algebraically or graphically.

You should be familiar with the following four methods for solving quadratic equations *algebraically*.

Solving a Quadratic Equation

Method

Factoring: If $ab = 0$, then $a = 0$ or $b = 0$.

Square Root Principle: If $u^2 = c$, where $c > 0$, then $u = \pm\sqrt{c}$.

Completing the Square: If $x^2 + bx = c$, then

$$x^2 + bx + \left(\frac{b}{2}\right)^2 = c + \left(\frac{b}{2}\right)^2$$

$$\left(x + \frac{b}{2}\right)^2 = c + \frac{b^2}{4}.$$

Quadratic Formula: If $ax^2 + bx + c = 0$, then

$$x = \frac{-b \pm \sqrt{b^2 - 4ac}}{2a}.$$

Example

$$x^2 - x - 6 = 0$$
$$(x - 3)(x + 2) = 0$$
$$x - 3 = 0 \implies x = 3$$
$$x + 2 = 0 \implies x = -2$$

$$(x + 3)^2 = 16$$
$$x + 3 = \pm 4$$
$$x = -3 \pm 4$$
$$x = 1 \quad \text{or} \quad x = -7$$

$$x^2 + 6x = 5$$
$$x^2 + 6x + 3^2 = 5 + 3^2$$
$$(x + 3)^2 = 14$$
$$x + 3 = \pm\sqrt{14}$$
$$x = -3 \pm \sqrt{14}$$

$$2x^2 + 3x - 1 = 0$$
$$x = \frac{-3 \pm \sqrt{3^2 - 4(2)(-1)}}{2(2)}$$
$$= \frac{-3 \pm \sqrt{17}}{4}$$

The methods used to solve quadratic equations can sometimes be extended to polynomial equations of higher degree, as shown in the next two examples.

EXAMPLE 8   Solving an Equation of Quadratic Type

Solve the equation $x^4 - 3x^2 + 2 = 0$.

Solution

The expression $x^4 - 3x^2 + 2$ is said to be in *quadratic form* because it is written in the form $au^2 + bu + c$, where u is any expression in x, namely x^2. You can use factoring to solve the equation as follows.

$$x^4 - 3x^2 + 2 = 0 \qquad \text{Write original equation.}$$
$$(x^2)^2 - 3(x^2) + 2 = 0 \qquad \text{Write in quadratic form in } x^2.$$
$$(x^2 - 1)(x^2 - 2) = 0 \qquad \text{Partially factor.}$$
$$(x + 1)(x - 1)(x^2 - 2) = 0 \qquad \text{Factor.}$$
$$x + 1 = 0 \quad \Longrightarrow \quad x = -1$$
$$x - 1 = 0 \quad \Longrightarrow \quad x = 1$$
$$x^2 - 2 = 0 \quad \Longrightarrow \quad x = \pm\sqrt{2}$$

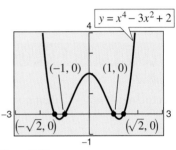

The equation has four solutions: -1, 1, $\sqrt{2}$, and $-\sqrt{2}$. Check these solutions in the original equation. The graph of $y = x^4 - 3x^2 + 2$, as shown in Figure P.47, verifies the solutions graphically.

Figure P.47

EXAMPLE 9   Solving a Polynomial Equation by Factoring

Solve the equation $2x^3 - 6x^2 - 6x + 18 = 0$.

Solution

This equation has a common factor of 2. To simplify things, first divide each side of the equation by 2.

$$2x^3 - 6x^2 - 6x + 18 = 0 \qquad \text{Write original equation.}$$
$$x^3 - 3x^2 - 3x + 9 = 0 \qquad \text{Divide each side by 2.}$$
$$x^2(x - 3) - 3(x - 3) = 0 \qquad \text{Factor out common monomial factors.}$$
$$(x - 3)(x^2 - 3) = 0 \qquad \text{Factor by grouping.}$$
$$x - 3 = 0 \quad \Longrightarrow \quad x = 3$$
$$x^2 - 3 = 0 \quad \Longrightarrow \quad x = \pm\sqrt{3}$$

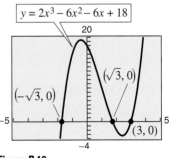

The equation has three solutions: 3, $\sqrt{3}$, and $-\sqrt{3}$. Check these solutions in the original equation. The graph of $y = 2x^3 - 6x^2 - 6x + 18$, as shown in Figure P.48, verifies the solutions graphically.

Figure P.48

Other Types of Equations

An equation involving a radical expression can often be cleared of radicals by raising both sides of the equation to an appropriate power. When using this procedure, remember that it can introduce extraneous solutions—so be sure to check each solution in the original equation.

EXAMPLE 10 Solving an Equation Involving a Radical

Solve $\sqrt{2x + 7} - x = 2$.

Algebraic Solution

$\sqrt{2x + 7} - x = 2$	Write original equation.
$\sqrt{2x + 7} = x + 2$	Isolate radical.
$2x + 7 = x^2 + 4x + 4$	Square each side.
$x^2 + 2x - 3 = 0$	Write in general form.
$(x + 3)(x - 1) = 0$	Factor.
$x + 3 = 0 \implies x = -3$	Set 1st factor equal to 0.
$x - 1 = 0 \implies x = 1$	Set 2nd factor equal to 0.

By substituting into the original equation, you can determine that -3 is extraneous, whereas 1 is valid. So, the equation has only one real solution: $x = 1$.

Graphical Solution

First rewrite the equation as $\sqrt{2x + 7} - x - 2 = 0$. Then use a graphing utility to graph $y = \sqrt{2x + 7} - x - 2$, as shown in Figure P.49(a). Notice that the domain is $x \geq -\frac{7}{2}$ because the expression under the radical cannot be negative. There appears to be one solution near $x = 1$. Use the *zoom* and *trace* features, as shown in Figure P.49(b), to approximate the only solution to be $x = 1$.

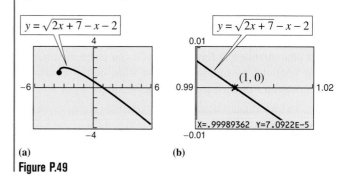

(a) **(b)**
Figure P.49

EXAMPLE 11 Solving an Equation Involving Two Radicals

Solve the equation $\sqrt{2x + 6} - \sqrt{x + 4} = 1$.

Solution

$\sqrt{2x + 6} - \sqrt{x + 4} = 1$	Write original equation.
$2x + 6 = 1 + 2\sqrt{x + 4} + (x + 4)$	Isolate radical and square each side.
$x + 1 = 2\sqrt{x + 4}$	Isolate radical.
$x^2 + 2x + 1 = 4(x + 4)$	Square each side.
$x^2 - 2x - 15 = 0$	Write in general form.
$(x - 5)(x + 3) = 0$	Factor.
$x - 5 = 0 \implies x = 5$	Set 1st factor equal to 0.
$x + 3 = 0 \implies x = -3$	Set 2nd factor equal to 0.

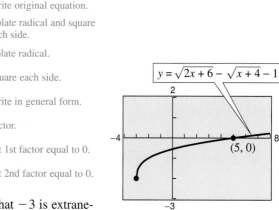

Figure P.50

By substituting into the original equation, you can determine that -3 is extraneous, whereas 5 is valid. Figure P.50 verifies that $x = 5$ is the only solution.

Sometimes radicals in equations are represented by fractional exponents.

EXAMPLE 12   Solving an Equation with Rational Exponents

Solve $(x + 1)^{2/3} = 4$.

Algebraic Solution

$(x + 1)^{2/3} = 4$	Write original equation.
$\sqrt[3]{(x + 1)^2} = 4$	Rewrite with radical sign.
$(x + 1)^2 = 64$	Cube each side.
$x + 1 = \pm 8$	Take square root of each side.
$x = -9, x = 7$	Subtract 1 from each side.

Substitute $x = -9$ and $x = 7$ into the original equation to determine that both are valid solutions.

Graphical Solution

Use a graphing utility to graph $y_1 = \sqrt[3]{(x + 1)^2}$ and $y_2 = 4$ in the same viewing window, as shown in Figure P.51. Use the *intersect* feature of the graphing utility to approximate the solutions to be $x = -9$ and $x = 7$.

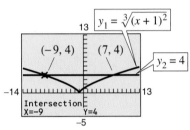

Figure P.51

As demonstrated in Example 1, you can algebraically solve an equation involving fractions by multiplying both sides of the equation by the least common denominator of each term in the equation to "clear an equation of fractions."

EXAMPLE 13   Solving an Equation Involving Fractions

Solve $\dfrac{2}{x} = \dfrac{3}{x - 2} - 1$.

Solution

For this equation, the least common denominator of the three terms is $x(x - 2)$, so you can begin by multiplying each term in the equation by this expression.

$$\frac{2}{x} = \frac{3}{x - 2} - 1$$

$$x(x - 2)\frac{2}{x} = x(x - 2)\frac{3}{x - 2} - x(x - 2)(1)$$

$$2(x - 2) = 3x - x(x - 2), \qquad x \neq 0, 2$$

$$2x - 4 = -x^2 + 5x$$

$$x^2 - 3x - 4 = 0$$

$$(x - 4)(x + 1) = 0$$

$$x - 4 = 0 \quad \Longrightarrow \quad x = 4$$

$$x + 1 = 0 \quad \Longrightarrow \quad x = -1$$

The equation has two solutions: 4 and -1. Check these solutions in the original equation. Use a graphing utility to verify these solutions graphically.

Exploration

Using *dot* mode, graph the two equations

$$y_1 = \frac{2}{x}$$

$$y_2 = \frac{3}{x - 2} - 1$$

in the same viewing window. How many times do the graphs of the equations intersect? What does this tell you about the solution to Example 13?

STUDY T!P

Graphs of equations involving variable denominators can be tricky because of the way graphing utilities skip over points where the denominator is zero. You will study graphs of such equations in Sections 2.6 and 2.7.

EXAMPLE 14   Solving an Equation Involving Absolute Value

Solve $|x^2 - 3x| = -4x + 6$.

Solution

Begin by writing the equation as $|x^2 - 3x| + 4x - 6 = 0$. From the graph of $y = |x^2 - 3x| + 4x - 6$ in Figure P.52, you can estimate the solutions to be -3 and 1. These can be verified by substitution into the equation. To solve an equation involving an absolute value *algebraically*, you must consider the fact that the expression inside the absolute value symbols can be positive or negative. This consideration results in *two* separate equations, each of which must be solved.

First Equation:

$$x^2 - 3x = -4x + 6 \qquad \text{Use positive expression.}$$

$$x^2 + x - 6 = 0 \qquad \text{Write in general form.}$$

$$(x + 3)(x - 2) = 0 \qquad \text{Factor.}$$

$$x + 3 = 0 \implies x = -3 \qquad \text{Set 1st factor equal to 0.}$$

$$x - 2 = 0 \implies x = 2 \qquad \text{Set 2nd factor equal to 0.}$$

Second Equation:

$$-(x^2 - 3x) = -4x + 6 \qquad \text{Use negative expression.}$$

$$x^2 - 7x + 6 = 0 \qquad \text{Write in general form.}$$

$$(x - 1)(x - 6) = 0 \qquad \text{Factor.}$$

$$x - 1 = 0 \implies x = 1 \qquad \text{Set 1st factor equal to 0.}$$

$$x - 6 = 0 \implies x = 6 \qquad \text{Set 2nd factor equal to 0.}$$

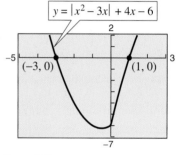

Figure P.52

Check

$$|(-3)^2 - 3(-3)| \overset{?}{=} -4(-3) + 6 \qquad \text{Substitute } -3 \text{ for } x.$$

$$18 = 18 \qquad -3 \text{ checks. } \checkmark$$

$$|2^2 - 3(2)| \overset{?}{=} -4(2) + 6 \qquad \text{Substitute 2 for } x.$$

$$2 \neq -2 \qquad 2 \text{ does not check.}$$

$$|1^2 - 3(1)| \overset{?}{=} -4(1) + 6 \qquad \text{Substitute 1 for } x.$$

$$2 = 2 \qquad 1 \text{ checks. } \checkmark$$

$$|6^2 - 3(6)| \overset{?}{=} -4(6) + 6 \qquad \text{Substitute 6 for } x.$$

$$18 \neq -18 \qquad 6 \text{ does not check.}$$

The equation has only two solutions: -3 and 1, just as you obtained by graphing.

In Figure P.52, the graph of $y = |x^2 - 3x| + 4x - 6$ appears to be a straight line to the right of the *y*-axis. Is it? Explain how you decided.

P.4 E x e r c i s e s

In Exercises 1–6, determine whether the given values of *x* are solutions of the equation.

Equation	*Values*

1. $\dfrac{5}{2x} - \dfrac{4}{x} = 3$ (a) $x = -\frac{1}{2}$ (b) $x = 4$

                         (c) $x = 0$ (d) $x = \frac{1}{4}$

2. $\dfrac{x}{2} + \dfrac{6x}{7} = \dfrac{19}{14}$ (a) $x = -2$ (b) $x = 1$

                         (c) $x = \dfrac{1}{2}$ (d) $x = 7$

3. $3 + \dfrac{1}{x + 2} = 4$ (a) $x = -1$ (b) $x = -2$

                         (c) $x = 0$ (d) $x = 5$

4. $\dfrac{(x + 5)(x - 3)}{2} = 24$ (a) $x = -3$ (b) $x = -2$

                         (c) $x = 7$ (d) $x = 9$

5. $\dfrac{\sqrt{x + 4}}{6} + 3 = 4$ (a) $x = -3$ (b) $x = 0$

                         (c) $x = 21$ (d) $x = 32$

6. $\dfrac{\sqrt[3]{x - 8}}{3} = -\dfrac{2}{3}$ (a) $x = -16$ (b) $x = 0$

                         (c) $x = 9$ (d) $x = 16$

In Exercises 7–12, determine whether the equation is an identity or a conditional equation.

7. $2(x - 1) = 2x - 2$

8. $-7(x - 3) + 4x = 3(7 - x)$

9. $x^2 - 8x + 5 = (x - 4)^2 - 11$

10. $x^2 + 2(3x - 2) = x^2 + 6x - 4$

11. $3 + \dfrac{1}{x + 1} = \dfrac{4x}{x + 1}$ **12.** $\dfrac{5}{x} + \dfrac{3}{x} = 24$

In Exercises 13 and 14, solve the equation in two ways. Then explain which way was easier for you.

13. $\dfrac{3x}{8} - \dfrac{4x}{3} = 4$ **14.** $\dfrac{3z}{8} - \dfrac{z}{10} = 6$

In Exercises 15–30, solve the equation (if possible). Then use a graphing utility to verify your solution.

15. $\dfrac{5x}{4} + \dfrac{1}{2} = x - \dfrac{1}{2}$ **16.** $\dfrac{x}{5} - \dfrac{x}{2} = 3$

17. $\frac{3}{2}(z + 5) - \frac{1}{4}(z + 24) = 0$

18. $\dfrac{3x}{2} + \dfrac{1}{4}(x - 2) = 10$

19. $\dfrac{100 - 4u}{3} = \dfrac{5u + 6}{4} + 6$

20. $\dfrac{17 + y}{y} + \dfrac{32 + y}{y} = 100$

21. $\dfrac{5x - 4}{5x + 4} = \dfrac{2}{3}$ **22.** $\dfrac{15}{x} - 4 = \dfrac{6}{x} + 3$

23. $\dfrac{1}{x - 3} + \dfrac{1}{x + 3} = \dfrac{10}{x^2 - 9}$

24. $\dfrac{1}{x - 2} + \dfrac{3}{x + 3} = \dfrac{4}{x^2 + x - 6}$

25. $\dfrac{7}{2x + 1} - \dfrac{8x}{2x - 1} = -4$

26. $\dfrac{4}{u - 1} + \dfrac{6}{3u + 1} = \dfrac{15}{3u + 1}$

27. $\dfrac{1}{x} + \dfrac{2}{x - 5} = 0$ **28.** $\dfrac{6}{x} - \dfrac{2}{x + 3} = \dfrac{3(x + 5)}{x(x + 3)}$

29. $\dfrac{3}{x(x - 3)} + \dfrac{4}{x} = \dfrac{1}{x - 3}$

30. $3 = 2 + \dfrac{2}{z + 2}$

In Exercises 31–40, find the *x*- and *y*-intercepts of the graph of the equation.

31. $y = x - 5$ **32.** $y = -\frac{3}{4}x - 3$

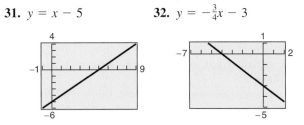

33. $y = x^2 + x - 2$ **34.** $y = 4 - x^2$

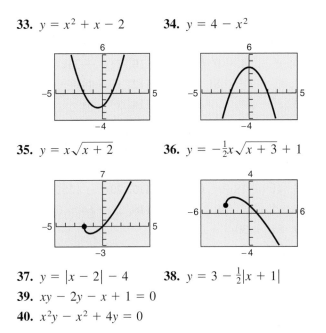

35. $y = x\sqrt{x + 2}$ **36.** $y = -\frac{1}{2}x\sqrt{x + 3} + 1$

37. $y = |x - 2| - 4$ **38.** $y = 3 - \frac{1}{2}|x + 1|$

39. $xy - 2y - x + 1 = 0$

40. $x^2y - x^2 + 4y = 0$

In Exercises 41–46, the solution(s) of the equation are given. Verify the solution(s) both algebraically and graphically.

	Equation	*Solution(s)*
41.	$y = 12 - 4x$	$x = 3$
42.	$y = 3(x - 5) + 9$	$x = 2$
43.	$y = x^2 - 2.5x - 6$	$x = -1.5, 4$
44.	$y = x^3 - 9x^2 + 18x$	$x = 0, 3, 6$
45.	$y = \dfrac{x + 2}{3} - \dfrac{x - 1}{5} - 1$	$x = 1$
46.	$y = x - 3 - \dfrac{10}{x}$	$x = -2, 5$

Graphical Analysis   In Exercises 47–50, use a graphing utility to graph the equation and approximate any x-intercepts. Set $y = 0$ and solve the resulting equation. Compare the results with the x-intercepts of the graph.

47. $y = 2(x - 1) - 4$ **48.** $y = \frac{4}{3}x + 2$

49. $y = 20 - (3x - 10)$ **50.** $y = 10 + 2(x - 2)$

In Exercises 51–70, solve the equation algebraically. Then write the equation in the form $y = 0$ and use a graphing utility to verify the algebraic solution.

51. $27 - 4x = 12$ **52.** $3.5x - 8 = 0.5x$

53. $25(x - 3) = 12(x + 2) - 10$

54. $1200 = 300 + 2(x - 500)$

55. $\dfrac{3x}{2} + \dfrac{1}{4}(x - 2) = 10$

56. $0.60x + 0.40(100 - x) = 50$

57. $\dfrac{2x}{3} = 10 - \dfrac{24}{x}$ **58.** $\dfrac{x - 3}{25} = \dfrac{x - 5}{12}$

59. $\dfrac{3}{x + 2} - \dfrac{4}{x - 2} = 5$ **60.** $\dfrac{6}{x} + \dfrac{8}{x + 5} = 3$

61. $3(x + 3) = 5(1 - x) - 1$

62. $(x + 1)^2 + 2(x - 2) = (x + 1)(x - 2)$

63. $2x^3 - x^2 - 18x + 9 = 0$

64. $4x^3 + 12x^2 - 26x - 24 = 0$

65. $x^4 = 2x^3 + 1$ **66.** $x^5 = 3 + 2x^3$

67. $\dfrac{2}{x + 2} = 3$ **68.** $\dfrac{5}{x} = 1 + \dfrac{3}{x + 2}$

69. $|x - 3| = 4$ **70.** $\sqrt{x - 2} = 3$

In Exercises 71–74, determine any point(s) of intersection algebraically. Then verify your result numerically by creating a table of values for each equation.

71. $y = 2 - x$ **72.** $2x + y = 6$
     $y = 2x - 1$ $-x + y = 0$

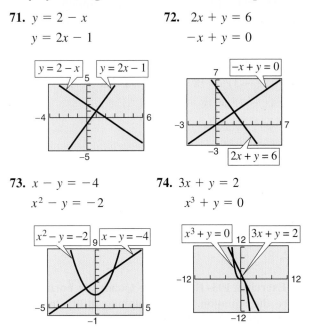

73. $x - y = -4$ **74.** $3x + y = 2$
     $x^2 - y = -2$ $x^3 + y = 0$

In Exercises 75–80, use a graphing utility to approximate any points of intersection (accurate to three decimal places) of the graphs of equations. Verify your results algebraically.

75. $y = 9 - 2x$
$y = x - 3$

76. $y = \frac{1}{3}x + 2$
$y = \frac{5}{2}x - 11$

77. $y = 4 - x^2$
$y = 2x - 1$

78. $y = -x$
$y = 2x - x^2$

79. $y = 2x^2$
$y = x^4 - 2x^2$

80. $y = x^3 - 3$
$y = 5 - 2x$

In Exercises 81–88, solve the quadratic equation by factoring. Check your solutions in the original equation.

81. $6x^2 + 3x = 0$

82. $9x^2 - 1 = 0$

83. $x^2 - 2x - 8 = 0$

84. $x^2 - 10x + 9 = 0$

85. $3 + 5x - 2x^2 = 0$

86. $2x^2 = 19x + 33$

87. $x^2 + 4x = 12$

88. $-x^2 + 8x = 12$

In Exercises 89–96, solve the equation by extracting square roots. List both the exact solution and the decimal solution rounded to two decimal places.

89. $x^2 = 7$

90. $9x^2 = 25$

91. $(x - 12)^2 = 18$

92. $(x - 5)^2 = 20$

93. $(2x - 1)^2 = 18$

94. $(4x + 7)^2 = 44$

95. $(x - 7)^2 = (x + 3)^2$

96. $(x + 5)^2 = (x + 4)^2$

In Exercises 97–102, solve the quadratic equation by completing the square. Verify your answer graphically.

97. $x^2 + 4x - 32 = 0$

98. $x^2 - 2x - 3 = 0$

99. $x^2 + 6x + 2 = 0$

100. $x^2 + 8x + 14 = 0$

101. $9x^2 - 18x + 3 = 0$

102. $9x^2 - 12x - 14 = 0$

In Exercises 103–108, use the Quadratic Formula to solve the equation. Use a graphing utility to verify your solutions graphically.

103. $2 + 2x - x^2 = 0$

104. $x^2 - 10x + 22 = 0$

105. $x^2 + 8x - 4 = 0$

106. $4x^2 - 4x - 4 = 0$

107. $4x^2 + 16x + 15 = 0$

108. $9x^2 - 6x - 35 = 0$

In Exercises 109–114, solve the equation by any convenient method.

109. $x^2 - 2x - 1 = 0$

110. $11x^2 + 33x = 0$

111. $(x + 3)^2 = 81$

112. $x^2 - 14x + 49 = 0$

113. $x^2 - x - \frac{11}{4} = 0$

114. $x^2 + 3x - \frac{3}{4} = 0$

In Exercises 115–130, find all solutions of the equation. Use a graphing utility to verify the solutions graphically.

115. $4x^4 - 18x^2 = 0$

116. $20x^3 - 125x = 0$

117. $x^4 - 81 = 0$

118. $x^6 - 64 = 0$

119. $5x^3 + 30x^2 + 45x = 0$

120. $9x^4 - 24x^3 + 16x^2 = 0$

121. $x^3 - 3x^2 - x + 3 = 0$

122. $x^4 + 2x^3 - 8x - 16 = 0$

123. $x^4 - 4x^2 + 3 = 0$

124. $x^4 + 5x^2 - 36 = 0$

125. $4x^4 - 65x^2 + 16 = 0$

126. $36t^4 + 29t^2 - 7 = 0$

127. $\frac{1}{t^2} + \frac{8}{t} + 15 = 0$

128. $6\left(\frac{s}{s+1}\right)^2 + 5\left(\frac{s}{s+1}\right) - 6 = 0$

129. $2x + 9\sqrt{x} - 5 = 0$

130. $3x^{1/3} + 2x^{2/3} = 5$

In Exercises 131–150, find all solutions of the equation algebraically. Check your solutions both algebraically and graphically.

131. $\sqrt{x - 10} - 4 = 0$

132. $\sqrt[3]{2x + 5} + 3 = 0$

133. $\sqrt{x + 1} - 3x = 1$

134. $\sqrt{x + 5} = \sqrt{x - 5}$

135. $\sqrt{x} - \sqrt{x - 5} = 1$

136. $\sqrt{x} + \sqrt{x - 20} = 10$

137. $(x - 5)^{2/3} = 16$

138. $(x^2 - x - 22)^{4/3} = 16$

139. $3x(x - 1)^{1/2} + 2(x - 1)^{3/2} = 0$

140. $4x^2(x - 1)^{1/3} + 6x(x - 1)^{4/3} = 0$

141. $\frac{20 - x}{x} = x$

142. $\frac{4}{x} - \frac{5}{3} = \frac{x}{6}$

143. $\frac{1}{x} - \frac{1}{x + 1} = 3$

144. $\frac{x}{x^2 - 4} + \frac{1}{x + 2} = 3$

145. $x = \frac{3}{x} + \frac{1}{2}$

146. $4x + 1 = \frac{3}{x}$

147. $|2x - 1| = 5$

148. $|3x + 2| = 7$

149. $|x| = x^2 + x - 3$

150. $|x - 10| = x^2 - 10x$

Graphical Analysis **In Exercises 151–158, (a) use a graphing utility to graph the equation, (b) use the graph to approximate any *x*-intercepts of the graph, (c) set *y* = 0 and solve the resulting equation, and (d) compare the result with the *x*-intercepts of the graph.**

151. $y = x^3 - 2x^2 - 3x$ **152.** $y = x^4 - 10x^2 + 9$

153. $y = \sqrt{11x - 30} - x$ **154.** $y = 2x - \sqrt{15 - 4x}$

155. $y = \dfrac{1}{x} - \dfrac{4}{x - 1} - 1$ **156.** $y = x + \dfrac{9}{x + 1} - 5$

157. $y = |x + 1| - 2$ **158.** $y = |x - 2| - 3$

159. *Per Capita Utilization*   The per capita utilization (in pounds) of nectarines and peaches *N* and cucumbers *C* from 1991 through 1996 can be modeled by

$$N = -0.37t + 6.88$$

$$C = 0.27t + 4.42$$

where $t = 1$ represents 1991.   (Source: U.S. Department of Agriculture)

(a) What does the intersection of the graphs of these equations represent?

(b) Find the point of intersection of the graphs algebraically.

(c) Verify your answer to part (b) using the *zoom* and *trace* features of a graphing utility.

160. *Saturated Steam*   The temperature *T* (in degrees Fahrenheit) of saturated steam increases as pressure increases. This relationship is approximated by

$$T = 75.82 - 2.11x + 43.51\sqrt{x}, \quad 5 \le x \le 40$$

where *x* is the absolute pressure in pounds per square inch.

(a) Use a graphing utility to graph the temperature equation over the specified domain.

(b) The temperature of steam at sea level ($x = 14.696$) is 212°F. Evaluate the model at this pressure and verify the result graphically.

(c) Use the model to approximate the pressure for a steam temperature of 240°F.

161. *Demand Equation*   The marketing department at a publisher is asked to determine the price of a book. The department determines that the demand for the book depends on the price of the book according to the formula

$$p = 40 - \sqrt{0.0001x + 1}, \quad x \ge 0$$

where *p* is the price per book in dollars and *x* is the number of books sold at the given price. For instance, if the price were $39, then (according to the model) no one would be willing to buy the book. On the other hand, if the price were $17.60, 5 million copies could be sold.

(a) Use a graphing utility to graph the demand equation over the specified domain.

(b) If the publisher set the price at $12.95, how many copies would be sold?

Synthesis

True or False?   **In Exercises 162 and 163, determine whether the statement is true or false. Justify your answer.**

162. An equation can never have more than one extraneous solution.

163. Two linear equations can have either one point of intersection or no points of intersection.

164. *Exploration*   Given that *a* and *b* are nonzero real numbers, determine the solutions of the equations.

(a) $ax^2 + bx = 0$ (b) $ax^2 - ax = 0$

P.5 Solving Inequalities Algebraically and Graphically

Properties of Inequalities

The inequality symbols $<, \leq, >,$ and $\geq$ are used to compare two numbers and to denote subsets of real numbers. For instance, the simple inequality $x \geq 3$ denotes all real numbers x that are greater than or equal to 3. In this section you will study inequalities that contain more involved statements such as

$$5x - 7 > 3x + 9 \qquad \text{and} \qquad -3 \leq 6x - 1 < 3.$$

As with an equation, you **solve an inequality** in the variable x by finding all values of x for which the inequality is true. These values are **solutions** of the inequality and are said to **satisfy** the inequality. For instance, the number 9 is a solution of the first inequality listed above because

$$5(9) - 7 > 3(9) + 9$$

$$38 > 36.$$

On the other hand, the number 7 is not a solution because

$$5(7) - 7 \not> 3(7) + 9$$

$$28 \not> 30.$$

The set of all real numbers that are solutions of an inequality is the **solution set** of the inequality.

The set of all points on the real number line that represent the solution set is the **graph of the inequality.** Graphs of many types of inequalities consist of intervals on the real number line.

The procedures for solving linear inequalities in one variable are much like those for solving linear equations. To isolate the variable you can make use of the **properties of inequalities.** These properties are similar to the properties of equality, but there are two important exceptions. When both sides of an inequality are multiplied or divided by a negative number, *the direction of the inequality symbol must be reversed.* Here is an example.

$$-2 < 5 \qquad \text{Write original inequality.}$$

$$(-3)(-2) > (-3)(5) \qquad \text{Multiply each side by } -3 \text{ and reverse inequality.}$$

$$6 > -15 \qquad \text{New inequality}$$

Two inequalities that have the same solution set are **equivalent inequalities.** The properties listed at the top of the next page describe operations that can be used to create equivalent inequalities.

Properties of Inequalities

Let a, b, c, and d be real numbers.

1. *Transitive Property*

$$a < b \text{ and } b < c \quad \Longrightarrow \quad a < c$$

2. *Addition of Inequalities*

$$a < b \text{ and } c < d \quad \Longrightarrow \quad a + c < b + d$$

3. *Addition of a Constant*

$$a < b \quad \Longrightarrow \quad a + c < b + c$$

4. *Multiplying by a Constant*

For $c > 0$, $a < b \quad \Longrightarrow \quad ac < bc$

For $c < 0$, $a < b \quad \Longrightarrow \quad ac > bc$

Each of the properties above is true if the symbol $<$ is replaced by $\le$ and $>$ is replaced by $\ge$. For instance, another form of Property 3 would be as follows.

$$a \le b \quad \Longrightarrow \quad a + c \le b + c$$

Solving a Linear Inequality

The simplest type of inequality to solve is a **linear inequality** in a single variable, such as $2x + 3 > 4$. (See Appendix C for help with solving one-step linear inequalities.)

EXAMPLE 1   Solving a Linear Inequality

Solve the inequality

$$5x - 7 > 3x + 9.$$

Solution

$5x - 7 > 3x + 9$	Write original inequality.
$5x > 3x + 16$	Add 7 to each side.
$5x - 3x > 16$	Subtract $3x$ from each side.
$2x > 16$	Combine like terms.
$x > 8$	Divide each side by 2.

So, the solution set consists of all real numbers that are greater than 8. The interval notation for this solution set is $(8, \infty)$. The number line graph of this solution set is shown in Figure P.53.

Note that the five inequalities forming the solution steps of Example 1 are all *equivalent* in the sense that each has the same solution set.

> ## Exploration
>
> Use a graphing utility to graph $y_1 = 5x - 7$ and $y_2 = 3x + 9$ in the same viewing window. (Use $-1 \le x \le 15$ and $-5 \le y \le 50$.) For which values of x does the graph of y_1 lie above the graph of y_2? Explain how the answer to this question can be used to solve the inequality in Example 1.

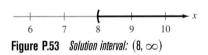

Figure P.53 *Solution interval:* $(8, \infty)$

> ## STUDY T!P
>
> Checking the solution set of an inequality is not as simple as checking the solution of an equation because there are simply too many x-values to substitute into the original inequality. However, you can get an indication of the validity of the solution set by substituting a few convenient values of x. For instance, in Example 1 try substituting $x = 5$ and $x = 10$ into the original inequality.

EXAMPLE 2   Solving an Inequality

Solve $1 - \dfrac{3x}{2} \geq x - 4$.

Algebraic Solution

$1 - \dfrac{3x}{2} \geq x - 4$ Write original inequality.

$2 - 3x \geq 2x - 8$ Multiply each side by the LCD.

$-3x \geq 2x - 10$ Subtract 2 from each side.

$-5x \geq -10$ Subtract $2x$ from each side.

$x \leq 2$ Divide each side by -5 and reverse inequality.

The solution set consists of all real numbers that are less than or equal to 2. The interval notation for this solution set is $(-\infty, 2]$. The number line graph of this solution set is shown in Figure P.54.

Figure P.54   *Solution interval:* $(-\infty, 2]$

Graphical Solution

Use a graphing utility to graph $y_1 = 1 - (3x)/2$ and $y_2 = x - 4$ in the same viewing window. In Figure P.55, you can see that the graphs appear to intersect at the point $(2, -2)$. Use the *intersect* feature of the graphing utility to confirm this. The graph of y_1 lies above the graph of y_2 to the left of their point of intersection, which implies that $y_1 \geq y_2$ for all $x \leq 2$.

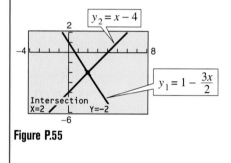

Figure P.55

Sometimes it is possible to write two inequalities as a **double inequality.** For instance, you can write the two inequalities $-4 \leq 5x - 2$ and $5x - 2 < 7$ more simply as $-4 \leq 5x - 2 < 7$, so you can solve the two inequalities together.

EXAMPLE 3   Solving a Double Inequality

Solve $-3 \leq 6x - 1 < 3$.

Algebraic Solution

$-3 \leq 6x - 1 < 3$ Write original inequality.

$-2 \leq 6x < 4$ Add 1 to all three parts.

$-\dfrac{1}{3} \leq x < \dfrac{2}{3}$ Divide by 6 and simplify.

The solution set consists of all real numbers that are greater than or equal to $-\frac{1}{3}$ *and* less than $\frac{2}{3}$. The interval notation for this solution set is $\left[-\frac{1}{3}, \frac{2}{3}\right)$. The number line graph of this solution set is shown in Figure P.56.

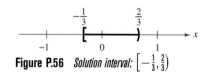

Figure P.56   *Solution interval:* $\left[-\frac{1}{3}, \frac{2}{3}\right)$

Graphical Solution

Use a graphing utility to graph $y_1 = 6x - 1$, $y_2 = -3$, and $y_3 = 3$ in the same viewing window. In Figure P.57, you can see that the graphs appear to intersect at the points $\left(-\frac{1}{3}, -3\right)$ and $\left(\frac{2}{3}, 3\right)$. Use the *intersect* feature of the graphing utility to confirm this. The graph of y_1 lies above the graph of y_2 to the right of $\left(-\frac{1}{3}, -3\right)$ *and* the graph of y_1 lies below the graph of y_3 to the left of $\left(\frac{2}{3}, 3\right)$. This implies that $y_2 \leq y_1 < y_3$ when $-\frac{1}{3} \leq x < \frac{2}{3}$.

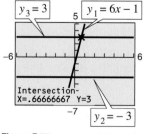

Figure P.57

Inequalities Involving Absolute Value

Solving an Absolute Value Inequality

Let x be a variable or an algebraic expression and let a be a real number such that $a \geq 0$.

1. The solutions of $|x| < a$ are all values of x that lie between $-a$ and a.

$$|x| < a \quad \text{if and only if} \quad -a < x < a.$$

2. The solutions of $|x| > a$ are all values of x that are less than $-a$ or greater than a.

$$|x| > a \quad \text{if and only if} \quad x < -a \quad \text{or} \quad x > a.$$

These rules are also valid if $<$ is replaced by $\leq$ and $>$ is replaced by $\geq$.

EXAMPLE 4   Solving Absolute Value Inequalities

Solve each inequality.

a. $|x - 5| < 2$ **b.** $|x - 5| > 2$

Algebraic Solution

a.

$|x - 5| < 2$	Write original inequality.
$-2 < x - 5 < 2$	Equivalent inequalities
$-2 + 5 < x - 5 + 5 < 2 + 5$	Add 5 to all three parts.
$3 < x < 7$	Simplify.

The solution set is all real numbers that are greater than 3 *and* less than 7. The interval notation for this solution set is $(3, 7)$. The number line graph of this solution set is shown in Figure P.58.

b. The absolute value inequality $|x - 5| > 2$ is equivalent to the following compound inequality.

$$x - 5 < -2 \quad or \quad x - 5 > 2$$

Solve first inequality: $x - 5 < -2$	Write first inequality.
$x < 3$	Add 5 to each side.
Solve second inequality: $x - 5 > 2$	Write second inequality.
$x > 7$	Add 5 to each side.

The solution is all real numbers that are less than -3 *or* greater than 7. The interval notation for this solution set is $(-\infty, 3) \cup (7, \infty)$. The symbol $\cup$ is called a *union* symbol and is used to denote combining of two sets. The number line graph of this solution set is shown in Figure P.59.

Graphical Solution

a. Use a graphing utility to graph $y_1 = |x - 5|$ and $y_2 = 2$ in the same viewing window. In Figure P.60, you can see that the graphs appear to intersect at the points $(3, 2)$ and $(7, 2)$. Use the *intersect* feature of the graphing utility to confirm this. The graph of y_1 lies below the graph of y_2 when $3 < x < 7$. So, you can approximate the solution set to be all real numbers greater than 3 *and* less than 7.

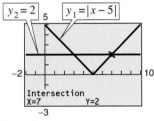

Figure P.60

b. In Figure P.60, you can see that the graph of y_1 lies above the graph of y_2 when $x < 3$ or when $x > 7$. So, you can approximate the solution set to be all real numbers that are less than 3 *or* greater than 7.

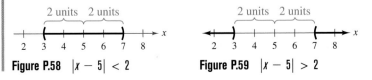

Figure P.58 $|x - 5| < 2$ **Figure P.59** $|x - 5| > 2$

Polynomial Inequalities

To solve a polynomial inequality such as $x^2 - 2x - 3 < 0$, use the fact that a polynomial can change signs only at its zeros (the x-values that make the polynomial equal to zero). Between two consecutive zeros a polynomial must be entirely positive or entirely negative. This means that when the real zeros of a polynomial are put in order, they divide the real number line into intervals in which the polynomial has no sign changes. These zeros are the **critical numbers** of the inequality, and the resulting open intervals are the **test intervals** for the inequality. For instance, the polynomial

$$x^2 - 2x - 3 = (x + 1)(x - 3)$$

has two zeros, $x = -1$ and $x = 3$, which divide the real number line into three test intervals: $(-\infty, -1)$, $(-1, 3)$, and $(3, \infty)$. To solve the inequality $x^2 - 2x - 3 < 0$, you only need to test one value from each test interval.

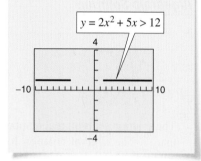

STUDY T!P

Some graphing utilities will produce graphs of inequalities. For instance, you can graph $2x^2 + 5x > 12$ by setting the graphing utility to *dot* mode and entering $y = 2x^2 + 5x > 12$. Using $-10 \le x \le 10$ and $-4 \le y \le 4$, your graph should look like the graph shown below. Solve the problem algebraically to verify that the solution is $(-\infty, -4) \cup \left(\frac{3}{2}, \infty\right)$.

$y = 2x^2 + 5x > 12$

Finding Test Intervals for a Polynomial

To determine the intervals on which the values of a polynomial are entirely negative or entirely positive, use the following steps.

1. Find all real zeros of the polynomial, and arrange the zeros in increasing order. The zeros of a polynomial are its **critical numbers.**

2. Use the critical numbers of the polynomial to determine its **test intervals.**

3. Choose one representative x-value in each test interval and evaluate the polynomial at that value. If the value of the polynomial is negative, the polynomial will have negative values for *every* x-value in the interval. If the value of the polynomial is positive, the polynomial will have positive values for *every* x-value in the interval.

EXAMPLE 5   Investigating Polynomial Behavior

To determine the intervals on which $x^2 - x - 6$ is entirely negative and those on which it is entirely positive, factor the quadratic as

$$x^2 - x - 6 = (x + 2)(x - 3).$$

The critical numbers occur at $x = -2$ and $x = 3$. So, the test intervals for the quadratic are $(-\infty, -2)$, $(-2, 3)$, and $(3, \infty)$. In each test interval, choose a representative x-value and evaluate the polynomial, as shown in the table.

Interval	x-Value	Value of Polynomial	Sign of Polynomial
$(-\infty, -2)$	$x = -3$	$(-3)^2 - (-3) - 6 = 6$	Positive
$(-2, 3)$	$x = 0$	$(0)^2 - (0) - 6 = -6$	Negative
$(3, \infty)$	$x = 5$	$(5)^2 - (5) - 6 = 14$	Positive

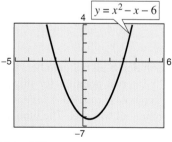

$y = x^2 - x - 6$

Figure P.61

The polynomial has negative values for every x in the interval $(-2, 3)$ and positive values for every x in the intervals $(-\infty, -2)$ and $(3, \infty)$. This result is shown graphically in Figure P.61.

To determine the test intervals for a polynomial inequality, the inequality must first be written in general form with the polynomial on one side.

EXAMPLE 6 Solving a Polynomial Inequality

Solve $2x^2 + 5x > 12$.

Algebraic Solution

$$2x^2 + 5x > 12 \qquad \text{Write original inequality.}$$

$$2x^2 + 5x - 12 > 0 \qquad \text{Write in general form.}$$

$$(x + 4)(2x - 3) > 0 \qquad \text{Factor.}$$

Critical Numbers: $x = -4, x = \frac{3}{2}$

Test Intervals: $(-\infty, -4), \left(-4, \frac{3}{2}\right), \left(\frac{3}{2}, \infty\right)$

Test: Is $(x + 4)(2x - 3) > 0$?

After testing these intervals, you can see that the polynomial $2x^2 + 5x - 12$ is positive in the open intervals $(-\infty, -4)$ and $\left(\frac{3}{2}, \infty\right)$. Therefore, the solution set of the inequality is

$$(-\infty, -4) \cup \left(\tfrac{3}{2}, \infty\right).$$

Graphical Solution

First write the polynomial inequality $2x^2 + 5x > 12$ as $2x^2 + 5x - 12 > 0$. Then use a graphing utility to graph $y = 2x^2 + 5x - 12$. In Figure P.62, you can see that the graph is *above* the x-axis when x is less than -4 *or* when x is greater than $\frac{3}{2}$. So, you can graphically approximate the solution set to be $(-\infty, -4) \cup \left(\frac{3}{2}, \infty\right)$.

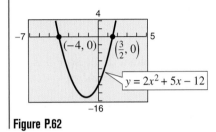

Figure P.62

EXAMPLE 7 Solving a Polynomial Inequality

Solve $2x^3 - 3x^2 - 32x > -48$.

Solution

Begin by writing the inequality in general form.

$$2x^3 - 3x^2 - 32x > -48 \qquad \text{Write original inequality.}$$

$$2x^3 - 3x^2 - 32x + 48 > 0 \qquad \text{Write in general form.}$$

$$x^2(2x - 3) - 16(2x - 3) > 0 \qquad \text{Factor by grouping.}$$

$$(x^2 - 16)(2x - 3) > 0 \qquad \text{Distributive Property}$$

$$(x - 4)(x + 4)(2x - 3) > 0 \qquad \text{Factor difference of two squares.}$$

The critical numbers are $x = -4, x = \frac{3}{2}$, and $x = 4$; and the test intervals are $(-\infty, -4), \left(-4, \frac{3}{2}\right), \left(\frac{3}{2}, 4\right)$, and $(4, \infty)$.

Interval	x-Value	Polynomial Value	Conclusion
$(-\infty, -4)$	$x = -5$	$2(-5)^3 - 3(-5)^2 - 32(-5) + 48 = -117$	Negative
$\left(-4, \frac{3}{2}\right)$	$x = 0$	$2(0)^3 - 3(0)^2 - 32(0) + 48 = 48$	Positive
$\left(\frac{3}{2}, 4\right)$	$x = 2$	$2(2)^3 - 3(2)^2 - 32(2) + 48 = -12$	Negative
$(4, \infty)$	$x = 5$	$2(5)^3 - 3(5)^2 - 32(5) + 48 = 63$	Positive

From this you can conclude that the polynomial $2x^3 - 3x^2 - 32x + 48$ is positive on the open intervals $\left(-4, \frac{3}{2}\right)$ and $(4, \infty)$. Therefore, the solution set consists of all real numbers in the intervals $\left(-4, \frac{3}{2}\right)$ and $(4, \infty)$.

STUDY T!P

When solving a quadratic inequality, be sure you have accounted for the particular type of inequality symbol given in the inequality. For instance, in Example 7, note that the original inequality contained a "greater than" symbol and the solution consisted of two open intervals. If the original inequality had been

$$2x^3 - 3x^2 + 32x \geq -48,$$

the solution would have consisted of the closed interval $\left[-4, \frac{3}{2}\right]$ and the interval $[4, \infty)$.

EXAMPLE 8   Unusual Solution Sets

a. The solution set of

$$x^2 + 2x + 4 > 0$$

consists of the entire set of real numbers, $(-\infty, \infty)$. In other words, the quadratic $x^2 + 2x + 4$ is positive for every real value of x, as indicated in Figure P.63(a). (Note that this quadratic inequality has *no* critical numbers. In such a case, there is only one test interval—the entire real number line.)

b. The solution set of

$$x^2 + 2x + 1 \le 0$$

consists of the single real number -1, because the graph touches the x-axis only at -1, as shown in Figure P.63(b).

c. The solution set of

$$x^2 + 3x + 5 < 0$$

is empty. In other words, the quadratic $x^2 + 3x + 5$ is not less than zero for any value of x, as indicated in Figure P.63(c).

d. The solution set of

$$x^2 - 4x + 4 > 0$$

consists of all real numbers *except* the number 2. In interval notation, this solution set can be written as $(-\infty, 2) \cup (2, \infty)$. The graph of $x^2 - 4x + 4$ lies above the x-axis except at $x = 2$, where it touches it, as indicated in Figure P.63(d).

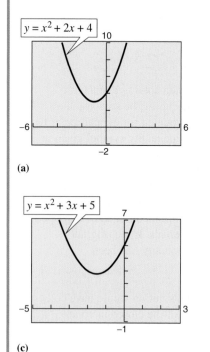

$y = x^2 + 2x + 4$

(a)

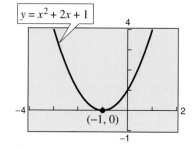

$y = x^2 + 2x + 1$

$(-1, 0)$

(b)

$y = x^2 + 3x + 5$

(c)

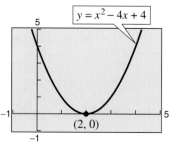

$y = x^2 - 4x + 4$

$(2, 0)$

(d)

Figure P.63

Rational Inequalities

The concepts of critical numbers and test intervals can be extended to inequalities involving rational expressions. To do this, use the fact that the value of a rational expression can change sign only at its *zeros* (the *x*-values for which its numerator is zero) and its *undefined values* (the *x*-values for which its denominator is zero). These two types of numbers make up the *critical numbers* of a rational inequality.

EXAMPLE 9   Solving a Rational Inequality

Solve $\dfrac{2x - 7}{x - 5} \le 3$.

Algebraic Solution

$$\frac{2x - 7}{x - 5} \le 3 \qquad \text{Write original inequality.}$$

$$\frac{2x - 7}{x - 5} - 3 \le 0 \qquad \text{Write in standard form.}$$

$$\frac{2x - 7 - 3x + 15}{x - 5} \le 0 \qquad \text{Write as single fraction.}$$

$$\frac{-x + 8}{x - 5} \le 0 \qquad \text{Simplify.}$$

Now, in standard form you can see that the critical numbers are 5 and 8, and you can proceed as follows.

Critical Numbers: $x = 5, x = 8$

Test Intervals: $(-\infty, 5), (5, 8), (8, \infty)$

Test: Is $\dfrac{-x + 8}{x - 5} \le 0$?

Interval	x-Value	Polynomial Value	Conclusion
$(-\infty, 5)$	$x = 0$	$\dfrac{-0 + 8}{0 - 5} = -\dfrac{8}{5}$	Negative
$(5, 8)$	$x = 6$	$\dfrac{-6 + 8}{6 - 5} = 2$	Positive
$(8, \infty)$	$x = 9$	$\dfrac{-9 + 8}{9 - 5} = -\dfrac{1}{4}$	Negative

By testing these intervals, you can determine that the rational expression $(-x + 8)/(x - 5)$ is negative in the open intervals $(-\infty, 5)$ and $(8, \infty)$. Moreover, because $(-x + 8)/(x - 5) = 0$ when $x = 8$, you can conclude that the solution set of the inequality is $(-\infty, 5) \cup [8, \infty)$.

Graphical Solution

Use a graphing utility to graph

$$y_1 = \frac{2x - 7}{x - 5} \quad \text{and} \quad y_2 = 3$$

in the same viewing window. In Figure P.64, you can see that the graphs appear to intersect at the point $(8, 3)$. Use the *intersect* feature of the graphing utility to confirm this. The graph of y_1 lies below the graph of y_2 in the intervals $(-\infty, 5)$ and $[8, \infty)$. So, you can graphically approximate the solution set to be all real numbers less than 5 *or* all real numbers greater than or equal to 8.

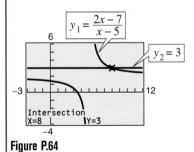

Figure P.64

Note in Example 9 that $x = 5$ is not included in the solution set because the inequality is undefined when $x = 5$.

Applications

The *implied domain* of a function is the set of all *x*-values for which a function is defined. A common type of implied domain is that used to avoid even roots of negative numbers, as shown in Example 10.

EXAMPLE 10   Finding the Domain of an Expression

Find the domain of

$$\sqrt{64 - 4x^2}.$$

Solution

Because $\sqrt{64 - 4x^2}$ is defined only if $64 - 4x^2$ is nonnegative, the domain is given by $64 - 4x^2 \geq 0$.

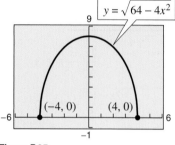

$$64 - 4x^2 \geq 0 \qquad \text{Write in general form.}$$

$$16 - x^2 \geq 0 \qquad \text{Divide each side by 4.}$$

$$(4 - x)(4 + x) \geq 0 \qquad \text{Factor.}$$

The inequality has two critical numbers: -4 and 4. A test shows that $64 - 4x^2 \geq 0$ in the *closed interval* $[-4, 4]$. The graph of $y = \sqrt{64 - 4x^2}$, shown in Figure P.65, confirms that the domain is $[-4, 4]$.

Figure P.65

EXAMPLE 11   The Height of a Projectile

A projectile is fired straight upward from ground level with an initial velocity of 384 feet per second. During what time period will its height exceed 2000 feet?

Solution

The position of an object moving vertically can be modeled by the *position equation*

$$s = -16t^2 + v_0 t + s_0,$$

where s is the height in feet and t is the time in seconds. In this case, $s_0 = 0$ and $v_0 = 384$. So, you need to solve the inequality $-16t^2 + 384t > 2000$. Using a graphing utility, graph $s = -16t^2 + 384t$ and $s = 2000$, as shown in Figure P.66. From the graph, you can determine that $-16t^2 + 384t > 2000$ for t between approximately 7.6 and 16.4. You can verify this result algebraically as follows.

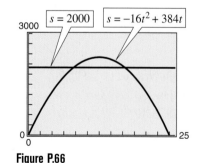

Figure P.66

$$-16t^2 + 384t > 2000 \qquad \text{Write original inequality.}$$

$$t^2 - 24t < -125 \qquad \text{Divide by } -16 \text{ and reverse inequality.}$$

$$t^2 - 24t + 125 < 0 \qquad \text{Write in general form.}$$

By the Quadratic Formula the critical numbers are $12 - \sqrt{19}$ and $12 + \sqrt{19}$, or approximately 7.64 and 16.36. A test will verify that the height of the projectile will exceed 2000 feet when $7.64 < t < 16.36$, that is, during the time interval $(7.64, 16.36)$ seconds.

P.5 E x e r c i s e s

In Exercises 1–4, match the inequality with its graph. [The graphs are labeled (a), (b), (c), and (d).]

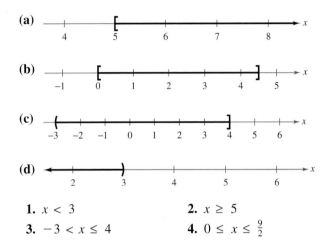

1. $x < 3$

2. $x \geq 5$

3. $-3 < x \leq 4$

4. $0 \leq x \leq \frac{9}{2}$

In Exercises 5–8, determine whether each given value of x is a solution of the inequality.

Inequality	Values
5. $5x - 12 > 0$	(a) $x = 3$ (b) $x = -3$
	(c) $x = \frac{5}{2}$ (d) $x = \frac{3}{2}$
6. $-5 < 2x - 1 \leq 1$	(a) $x = -\frac{1}{2}$ (b) $x = -\frac{5}{2}$
	(c) $x = \frac{4}{3}$ (d) $x = 0$
7. $-1 < \dfrac{3 - x}{2} \leq 1$	(a) $x = 0$ (b) $x = \sqrt{5}$
	(c) $x = 1$ (d) $x = 5$
8. $\lvert x - 10 \rvert \geq 3$	(a) $x = 13$ (b) $x = -1$
	(c) $x = 14$ (d) $x = 9$

In Exercises 9–18, solve the inequality and sketch the solution on the real number line. Use a graphing utility to verify your solution graphically.

9. $-10x < 40$

10. $2x > 3$

11. $4(x + 1) < 2x + 3$

12. $2x + 7 < 3$

13. $1 < 2x + 3 < 9$

14. $-2 < 3x + 1 < 10$

15. $-8 \leq 1 - 3(x - 2) < 13$

16. $0 \leq 2(x + 4) < 20$

17. $-4 < \dfrac{2x - 3}{3} < 4$

18. $0 \leq \dfrac{x + 3}{2} < 5$

Graphical Analysis In Exercises 19–24, use a graphing utility to approximate the solution.

19. $6x > 12$

20. $3x - 1 \leq 5$

21. $5 - 2x \geq 1$

22. $3(x + 1) < x + 7$

23. $-9 < 6x - 1 < 1$

24. $-10 < 4(x - 3) \leq 8$

In Exercises 25–28, use a graphing utility to graph the equation and graphically approximate the values of x that satisfy the specified inequalities. Then solve each inequality algebraically.

	Equation	Inequalities	
25.	$y = 2x - 3$	(a) $y \geq 1$	(b) $y \leq 0$
26.	$y = \frac{2}{3}x + 1$	(a) $y \leq 5$	(b) $y \geq 0$
27.	$y = -\frac{1}{2}x + 2$	(a) $0 \leq y \leq 3$	(b) $y \geq 0$
28.	$y = -3x + 8$	(a) $-1 \leq y \leq 3$	(b) $y \leq 0$

In Exercises 29–36, solve the inequality and sketch the solution on the real number line.

29. $\lvert 5x \rvert > 10$

30. $\lvert x - 20 \rvert \leq 4$

31. $\lvert x - 7 \rvert < 6$

32. $\lvert x - 20 \rvert \geq 4$

33. $\lvert x + 14 \rvert + 3 > 17$

34. $\left\lvert \dfrac{x - 3}{2} \right\rvert \geq 5$

35. $\lvert 1 - 2x \rvert < 5$

36. $3\lvert 4 - 5x \rvert \leq 9$

In Exercises 37 and 38, use a graphing utility to graph the equation and graphically approximate the values of x that satisfy the specified inequalities. Then solve each inequality algebraically.

	Equation	Inequalities	
37.	$y = \lvert x - 3 \rvert$	(a) $y \leq 2$	(b) $y \geq 4$
38.	$y = \left\lvert \frac{1}{2}x + 1 \right\rvert$	(a) $y \leq 4$	(b) $y \geq 1$

In Exercises 39–44, use absolute value notation to define each interval (or pair of intervals) on the real number line.

39.

40.

41.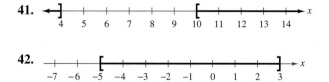

42.

43. All real numbers within 10 units of 12

44. All real numbers whose distances from -3 are more than 5

In Exercises 45–50, solve the inequality and graph the solution on the real number line. Use a graphing utility to verify your solution graphically.

45. $(x + 2)^2 < 25$ **46.** $(x + 6)^2 \leq 8$

47. $x^2 + 4x + 4 \geq 9$ **48.** $x^2 - 6x + 9 < 16$

49. $x^3 - 4x \geq 0$ **50.** $x^4(x - 3) \leq 0$

In Exercises 51–54, use a graphing utility to graph the equation and graphically approximate the values of x that satisfy the specified inequalities. Then solve each inequality algebraically.

	Equation	*Inequalities*
51.	$y = -x^2 + 2x + 3$	(a) $y \leq 0$ (b) $y \geq 3$
52.	$y = \frac{1}{2}x^2 - 2x + 1$	(a) $y \leq 1$ (b) $y \geq 7$
53.	$y = \frac{1}{8}x^3 - \frac{1}{2}x$	(a) $y \geq 0$ (b) $y \leq 6$
54.	$y = x^3 - x^2 - 16x + 16$	(a) $y \leq 0$ (b) $y \geq 36$

In Exercises 55–58, solve the inequality and graph the solution on the real number line. Use a graphing utility to verify your solution graphically.

55. $\dfrac{1}{x} - x > 0$ **56.** $\dfrac{1}{x} - 4 < 0$

57. $\dfrac{x + 6}{x + 1} - 2 < 0$ **58.** $\dfrac{x + 12}{x + 2} - 3 \geq 0$

In Exercises 59–62, use a graphing utility to graph the equation and graphically approximate the values of x that satisfy the specified inequalities. Then solve each inequality algebraically.

	Equation	*Inequalities*
59.	$y = \dfrac{3x}{x - 2}$	(a) $y \leq 0$ (b) $y \geq 6$
60.	$y = \dfrac{2(x - 2)}{x + 1}$	(a) $y \leq 0$ (b) $y \geq 8$
61.	$y = \dfrac{2x^2}{x^2 + 4}$	(a) $y \geq 1$ (b) $y \leq 2$

62. $y = \dfrac{5x}{x^2 + 4}$ (a) $y \geq 1$ (b) $y \leq 0$

In Exercises 63–68, find the domain of x in the expression.

63. $\sqrt{x - 5}$ **64.** $\sqrt{x^2 - 4}$

65. $\sqrt[3]{6 - x}$ **66.** $\sqrt[3]{2x^2 - 8}$

67. $\sqrt[4]{6x + 15}$ **68.** $\sqrt[4]{4 - x^2}$

69. *Data Analysis*   You want to determine whether there is a relationship between an athlete's weight x (in pounds) and the athlete's maximum bench-press weight y (in pounds). The table shows a sample of 12 athletes.

x	165	184	150	210	196	240
y	170	185	200	255	205	295

x	202	170	185	190	230	160
y	190	175	195	185	250	155

(a) Use a graphing utility to plot the data.

(b) A model for this data is $y = 1.266x - 35.766$. Use a graphing utility to graph the equation on the same display used in part (a).

(c) Use the graph to estimate the values of x that estimate a maximum bench-press weight of at least 200 pounds.

(d) Use the graph to write a statement about the accuracy of the model. If you think the graph indicates that an athlete's weight is not a particularly good indicator of the athlete's maximum bench-press weight, list other factors that may influence an individual's maximum bench-press weight.

70. *Educational Degrees*   The number D (in thousands) of earned degrees conferred annually in the United States from 1950 to 1995 is approximated by the model

$$D = -0.0977t^2 + 47.1174t + 291.5651$$

where $t = 0$ represents 1950. (Source: U.S. National Center for Education Statistics)

(a) Use a graphing utility to graph the model.

(b) According to this model, estimate when the number of degrees will exceed 2,500,000.

71. ***Height***   The height h of two-thirds of the members of a certain population satisfies the inequality

$$\left|\frac{h - 68.5}{2.7}\right| \le 1$$

where h is measured in inches. Determine the interval on the real number line in which these heights lie.

72. ***Meteorology***   A certain electronic device is to be operated in an environment with relative humidity h in the interval defined by

$$|h - 50| \le 30.$$

What are the minimum and maximum relative humidities for the operation of this device?

73. ***Music***   Michael Kasha of Florida State University used physics and mathematics to design a new classical guitar. He used the model for the frequency of the vibrations on a circular plate

$$v = \frac{2.6t}{d^2}\sqrt{\frac{E}{\rho}}$$

where v is the frequency, t is the plate thickness, d is the diameter, E is the elasticity of the plate material, and ρ is the density of the plate material. For fixed values of d, E, and ρ, the graph of the equation is a line, as shown below.

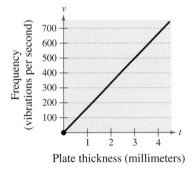

(a) Estimate the frequency if the plate thickness is 2 millimeters.

(b) Estimate the plate thickness if the frequency is 600 vibrations per second.

(c) Approximate the interval for the plate thickness if the frequency is between 200 and 400 vibrations per second.

(d) Approximate the interval for the frequency if the plate thickness is less than 3 millimeters.

Synthesis

True or False?   **In Exercises 74 and 75, determine whether the statement is true or false. Justify your answer.**

74. If $a, b,$ and c are real numbers, and $a \le b$, then $ac \le bc$.

75. If $-10 \le x \le 8$, then $-10 \ge -x$ and $-x \ge -8$.

76. Identify the solution of the inequality $|x - a| \ge 2$.

77. Consider the polynomial $(x - a)(x - b)$ and the real number line.

(a) Identify the points on the line where the polynomial is zero.

(b) In each of the three subintervals of the line, write the sign of each factor and the sign of the product.

(c) For which x-values does a polynomial possibly change signs?

P   Chapter Summary

What did you learn?

P   Review Exercises

P.1 **In Exercises 1–4, plot the point and determine the quadrant in which it is located.**

1. $(8, -3)$
2. $(-4, -9)$
3. $\left(-\frac{5}{2}, 10\right)$
4. $(-6.5, -0.5)$

In Exercises 5 and 6, determine the quadrant(s) in which (x, y) is located so that the conditions are satisfied.

5. $x > 0$ and $y = -2$
6. (x, y), $xy = 4$

7. *Patents*   The number of patents P (in thousands) issued in the United States from 1988 through 1996 is shown in the table. (Source: U.S. Patent and Trademark Office)

Year	1988	1989	1990	1991	1992
P	84.4	102.7	99.2	106.8	107.4

Year	1993	1994	1995	1996
P	109.7	113.6	113.8	121.7

(a) Sketch a scatter plot of the data.

(b) What statement can be made about the number of patents issued in the United States?

8. *Business*   The net profits (in millions of dollars) for the Progressive Corporation for the years 1994 through 1998 are shown in the table. Create a bar graph and a line graph for the data. (Source: Progressive Corporation)

Year	1994	1995	1996	1997	1998
Profits	228.1	250.5	316.6	400.0	456.7

In Exercises 9 and 10, plot the points and find the distance between the points.

9. $(-3, 8), (1, 5)$
10. $(5.6, 0), (0, 8.2)$

Geometry   **In Exercises 11 and 12, plot the points and verify that the points form the polygon.**

11. *Right Triangle:* $(2, 3), (13, 11), (5, 22)$
12. *Parallelogram:* $(1, 2), (8, 3), (9, 6), (2, 5)$

In Exercises 13 and 14, plot the points and find the midpoint of the line segment joining the points.

13. $(-12, 5), (4, -7)$
14. $(1.8, 7.4), (-0.6, -14.5)$

15. *Business*   The Sbarro restaurant chain had revenues of \$329.5 million in 1996 and \$375.2 million in 1998. (Source: Sbarro, Inc.)

(a) Without any additional information, what would you estimate the 1997 revenues to have been?

(b) The actual revenue for 1997 was \$349.4 million. How accurate is your estimate?

In Exercises 16 and 17, find the standard form of the equation of the specified circle.

16. Center: $(3, -1)$; Solution point: $(-5, 1)$
17. End points of a diameter: $(-4, 6), (10, -2)$

P.2 **In Exercises 18 and 19, complete the table. Use the resulting solution points to sketch the graph of the equation.**

18. $y = -\frac{1}{2}x + 2$

x	-2	0	2	3	4
y					

19. $y = x^2 - 3x$

x	-1	0	1	2	3
y					

In Exercises 20–29, sketch the graph of the equation by hand.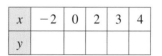

20. $y - 2x - 3 = 0$
21. $3x + 2y + 6 = 0$
22. $x - 5 = 0$
23. $y = 8 - |x|$
24. $y = \sqrt{5 - x}$
25. $y = \sqrt{x + 2}$
26. $y + 2x^2 = 0$
27. $y = x^2 - 4x$
28. $x + y^2 = 9$
29. $x^2 + y^2 = 10$

In Exercises 30–37, use a graphing utility to graph the equation. Approximate any intercepts.

30. $y = \frac{1}{4}(x + 1)^3$ **31.** $y = 4 - (x - 4)^2$

32. $y = \frac{1}{4}x^4 - 2x^2$ **33.** $y = \frac{1}{4}x^3 - 3x$

34. $y = x\sqrt{9 - x^2}$ **35.** $y = x\sqrt{x + 3}$

36. $y = |x - 4| - 4$ **37.** $y = |x + 2| + |3 - x|$

38. *Data Analysis*   The average expenditures y (in dollars) for automobile insurance per insured vehicle from 1992 through 1996 are shown in the table. (Source: National Association of Insurance Commissioners)

x	1992	1993	1994	1995	1996
y	616	638	651	667	685

(a) Use a graphing utility to plot the data.

(b) Use the regression capabilities of a graphing utility to find the best-fitting linear model (let $t = 2$ correspond to 1992).

(c) Graph the model in the same viewing window with the data.

(d) Use the model to estimate the values of y for the years 2000 and 2002.

P.3 **In Exercises 39–44, plot the two points and find the slope of the line that passes through the points.**

39. $(-3, 2), (8, 2)$ **40.** $(7, -1), (7, 12)$

41. $\left(\frac{3}{2}, 1\right), \left(5, \frac{5}{2}\right)$ **42.** $\left(-\frac{3}{4}, \frac{5}{6}\right), \left(\frac{1}{2}, -\frac{5}{2}\right)$

43. $(-4.5, 6), (2.1, 3)$

44. $(-2.7, -6.3), (-1, -1.2)$

In Exercises 45–48, use the concept of slope to find t such that the three points are collinear.

45. $(-2, 5), (0, t), (1, 1)$ **46.** $(-6, 1), (1, t), (10, 5)$

47. $(1, -4), (t, 3), (5, 10)$ **48.** $(-3, 3), (t, -1), (8, 6)$

In Exercises 49–58, (a) find an equation of the line that passes through the given point and has the specified slope, and (b) find three additional points through which the line passes.

Point	Slope
49. $(2, -1)$	$m = \frac{1}{4}$
50. $(-3, 5)$	$m = -\frac{3}{2}$
51. $(0, -5)$	$m = \frac{3}{2}$
52. $(3, 0)$	$m = -\frac{2}{3}$
53. $\left(\frac{1}{5}, -5\right)$	$m = -1$
54. $\left(0, \frac{7}{8}\right)$	$m = -4$
55. $(-2, 6)$	$m = 0$
56. $(-8, 8)$	$m = 0$
57. $(10, -6)$	m is undefined.
58. $(5, 4)$	m is undefined.

In Exercises 59–64, (a) find an equation of the line (in slope-intercept form) that passes through the points and (b) sketch the graph of the equation.

59. $(2, -1), (4, -1)$ **60.** $(0, 0), (0, 10)$

61. $(2, 1), (14, 6)$ **62.** $(-2, 2), (3, -10)$

63. $(-1, 0), (6, 2)$ **64.** $(1, 6), (4, 2)$

In Exercises 65–68, write equations of the lines through the point (a) parallel to the given line and (b) perpendicular to the given line. Verify your result with a graphing utility (use a square setting).

Point	Line
65. $(3, -2)$	$5x - 4y = 8$
66. $(-8, 3)$	$2x + 3y = 5$
67. $(-6, 2)$	$x = 4$
68. $(3, -4)$	$y = 2$

P.4 **In Exercises 69–72, solve the equation (if possible) and use a graphing utility to verify your solution.**

69. $14 + \dfrac{2}{x - 1} = 10$ **70.** $6 - \dfrac{11}{x} = 3 + \dfrac{7}{x}$

71. $\dfrac{9x}{3x - 1} - \dfrac{4}{3x + 1} = 3$

72. $\dfrac{5}{x - 5} + \dfrac{1}{x + 5} = \dfrac{2}{x^2 - 25}$

In Exercises 73–76, determine the x- and y-intercepts of the graph of the equation algebraically. Use a graphing utility to verify your answer.

73. $-x + y = 3$ **74.** $x - 5y = 20$

75. $y = x^2 - 9x + 8$ **76.** $y = 25 - x^2$

In Exercises 77 and 78, use a graphing utility to graph the equation and approximate the x- and y-intercepts.

77. $y = -|x + 5| - 2$ **78.** $y = 6 - 2|x - 3|$

In Exercises 79–84, use a graphing utility to approximate any solutions (accurate to three decimal places) of the equation.

79. $5(x - 2) - 1 = 0$ **80.** $12 - 5(x - 7) = 0$

81. $3x^3 - 2x + 4 = 0$ **82.** $\frac{1}{3}x^3 - x + 4 = 0$

83. $x^4 - 3x + 1 = 0$ **84.** $6 - \frac{1}{2}x^2 + \frac{5}{6}x^4 = 0$

In Exercises 85–88, determine algebraically any points of intersection of the graphs of the equations. Use a graphing utility to verify your answer(s).

85. $3x + 5y = -7$ **86.** $x - y = 3$

$\quad -x - 2y = 3$ $\qquad 2x + y = 12$

87. $x^2 + 2y = 14$ **88.** $y = -x + 7$

$\quad 3x + 4y = 1$ $\qquad y = 2x^3 - x + 9$

In Exercises 89–98, use any method to solve the equation. Use a graphing utility to verify your solution(s).

89. $6x = 3x^2$

90. $15 + x - 2x^2 = 0$

91. $(x + 4)^2 = 18$

92. $16x^2 = 25$

93. $x^2 - 12x + 30 = 0$

94. $x^2 + 6x - 3 = 0$

95. $2x^2 + 9x - 5 = 0$

96. $-x^2 - x + 15 = 0$

97. $x^2 - 4x - 10 = 0$

98. $-2x^2 - 13x = 0$

In Exercises 99–116, solve the equation (if possible) and use a graphing utility to verify your solution.

99. $3x^3 - 26x^2 + 16x = 0$ **100.** $216x^4 - x = 0$

101. $5x^4 - 12x^3 = 0$ **102.** $4x^3 - 6x^2 = 0$

103. $\sqrt{x + 4} = 3$ **104.** $\sqrt{x - 2} - 8 = 0$

105. $\sqrt{2x + 3} + \sqrt{x - 2} = 2$

106. $5\sqrt{x} - \sqrt{x - 1} = 6$

107. $(x - 1)^{2/3} - 25 = 0$ **108.** $(x + 2)^{3/4} = 27$

109. $3\left(1 - \dfrac{1}{5t}\right) = 0$ **110.** $\dfrac{1}{x - 2} = 3$

111. $\dfrac{4}{(x - 4)^2} = 1$ **112.** $\dfrac{1}{(t + 1)^2} = 1$

113. $|x - 5| = 10$ **114.** $|2x + 3| = 7$

115. $|x^2 - 3| = 2x$ **116.** $|x^2 - 6| = x$

P.5 In Exercises 117–134, solve the inequality and graph the solution on the real number line. Use a graphing utility to verify your solution.

117. $8x - 3 < 6x + 15$

118. $\frac{1}{2}(3 - x) > \frac{1}{3}(2 - 3x)$

119. $-2 < -x + 7 \le 10$

120. $-6 \le 3 - 2(x - 5) < 14$

121. $|x - 2| < 1$ **122.** $|x| \le 4$

123. $\left|x - \frac{3}{2}\right| \ge \frac{3}{2}$ **124.** $|x - 3| > 4$

125. $4|3 - 2x| \le 16$ **126.** $|x + 9| + 7 > 19$

127. $x^2 - 2x \ge 3$ **128.** $4x^2 - 23x \le 6$

129. $x^3 - 16x \ge 0$ **130.** $12x^3 - 20x^2 < 0$

131. $\dfrac{x - 5}{3 - x} < 0$ **132.** $\dfrac{2}{x + 1} \le \dfrac{3}{x - 1}$

133. $\dfrac{3x + 8}{x - 3} \le 4$ **134.** $\dfrac{x + 8}{x + 5} - 2 < 0$

135. *Accuracy of Measurement* The side of a square is measured as 20.8 inches with a possible error of $\frac{1}{16}$ inch. Using these measurements, determine the interval containing the area of the square.

Synthesis

True or False? **In Exercises 136 and 137, determine whether the statement is true or false. Justify your answer.**

136. If $ab = 0$, then the point (a, b) lies on the x-axis or on the y-axis.

137. The graph of an equation may have two distinct y-intercepts.

138. In your own words, explain the difference between an identity and a conditional equation.

139. In your own words, explain what is meant by equivalent equations. Describe the steps used to transform an equation into an equivalent equation.

Chapter Project *Modeling the Volume of a Box*

Many mathematical results are discovered experimentally by calculating examples and looking for patterns. Prior to the 1950s, this mode of discovery was very time-consuming because the calculations had to be done by hand. The introduction of computer and calculator technology has removed much of this drudgery. In the following project, you are asked to model a real-life situation and solve a problem by looking for patterns in the corresponding data.

Consider a rectangular box with a square base and a surface area of 216 square inches. Let x represent the length (in inches) of each side of the base and let h represent the height (in inches) of the box, as shown at the right. Your goal is to answer this question: "Of all rectangular boxes with square bases and surface area of 216 square inches, which has the greatest volume?"

a. Express the areas of the base, top, and sides in terms of x and h.

b. Find an expression in terms of x and h for the surface area of the box.

c. Use the fact that the surface area is 216 square inches to express the variable h in terms of x.

d. Find an expression for the volume of the box in terms of x alone.

e. Use the expression in part (d) and a graphing utility to complete the table. Then use the results to decide which box has the greatest volume.

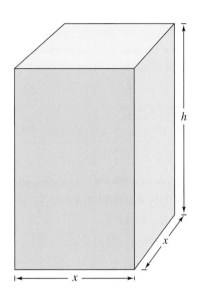

Base, x	Height, h	Surface Area	Volume
1.0	53.5	216.0	53.5
1.5	35.3	216.0	79.3
2.0	26.0	216.0	104.0
⋮	⋮	⋮	⋮
10.0	0.4	216.0	40.0

A computer simulation to accompany this project appears in the *Interactive* CD-ROM and *Internet* versions of this text.

Questions for Further Exploration

1. What happens to the height of the box as x gets closer and closer to 0? Of all boxes with square bases and a surface area of 216 square inches, is there a tallest? Explain your reasoning.

2. What is the maximum value of x? What happens to the height of the box as x gets closer and closer to this maximum value? Is there a shortest box that has a square base and a surface area of 216 square inches? Explain your reasoning.

3. Complete the table. Does it lend further support to your answer to part (e)? Explain.

x	5.9	5.99	5.999	6.001	6.01	6.1
V						

4. Of all rectangular boxes with surface area of 216 square inches and a base that is x inches by $2x$ inches, which has the maximum volume? Explain your reasoning.

P Chapter Test

Take this test as you would take a test in class. After you are done, check your work against the answers in the back of the book.

The *Interactive* CD-ROM and *Internet* versions of this text provide answers to the Chapter Tests and Cumulative Tests. They also offer Chapter Pre-Tests (which test key skills and concepts covered in previous chapters) and Chapter Post-Tests, both of which have randomly generated exercises with diagnostic capabilities.

1. Plot the points $(-2, 5)$ and $(6, 0)$. Find the coordinates of the midpoint of the line segment joining the points and the distance between the points.

2. The numbers (in millions) of votes cast for the Democratic candidates for president in 1980, 1984, 1988, 1992, and 1996 were 35.5, 37.6, 41.8, 44.9, and 47.4, respectively. Create a bar graph for this data. (Source: Congressional Quarterly, Inc.)

3. Find the standard form of the equation of a circle with center $(4, -1)$ and a solution point $(1, 4)$.

In Exercises 4–9, use the point-plotting method to graph the equation and identify any intercepts. Verify your results using a graphing utility.

4. $y = 4 - \frac{3}{4}|x|$ 5. $y = 4 - (x - 2)^2$ 6. $y = x - x^3$

7. $y = -x^3 + 2x - 4$ 8. $y = \sqrt{3 - x}$ 9. $y = \frac{1}{2}x\sqrt{x + 3}$

10. A line passes through the point $(3, -1)$ with slope $m = \frac{3}{2}$. Write the equation of the line. List three additional points on the line. Then sketch the line.

11. Find an equation of the line that passes through the point $(0, 4)$ and is perpendicular to the line $5x + 2y = 3$.

12. Solve $\dfrac{12}{x} - 7 = -\dfrac{27}{x} + 6$ and use a graphing utility to verify your solution.

In Exercises 13–15, use a graphing utility to graph the equation. Determine the number of x-intercepts of the graph.

13. $y = 3x^2 + 1$ 14. $y = x^3 + x$ 15. $y = x^3 - 4x^2 + 5x$

In Exercises 16–21, find all solutions of the equation. Check your solution(s) algebraically and graphically.

16. $x^2 - 10x + 9 = 0$ 17. $4x^2 - 81 = 0$

18. $3x^3 - 4x^2 - 12x + 16 = 0$ 19. $x + \sqrt{22 - 3x} = 6$

20. $(x^2 + 6)^{2/3} = 16$ 21. $|8x - 1| = 21$

In Exercises 22–24, solve the inequality and sketch the solution on the real number line. Use a graphing utility to verify your solution.

22. $-\frac{5}{6} < x - 2 < \frac{1}{8}$ 23. $2|x - 8| < 10$ 24. $\dfrac{3 - 5x}{2 + 3x} < -2$

25. The table at the right shows the number of local telephone calls C (in billions) in the United States from 1989 through 1996, where $t = 0$ represents 1990. Use the regression capabilities of a graphing utility to find a linear model for the data. Determine algebraically and graphically when the annual number of local calls will be about 600 billion. (Source: U.S. Federal Communications Commission)

t	C
-1	389
0	402
1	416
2	434
3	447
4	465
5	484
6	504

LIBRARY OF FUNCTIONS

In Chapter 1, you will be introduced to the concept of a *function*. As you proceed through the text, you will see that functions play a primary role in modeling real-life situations.

Over the past few hundred years, many different types of functions have been introduced and studied. Those that have proven to be most important in modeling real life have come to be known as *elementary functions*. There are three basic types of elementary functions: algebraic functions, exponential and logarithmic functions, and trigonometric and inverse trigonometric functions.

You will also encounter other types of functions in this text, such as functions defined by real-life data and piecewise-defined functions.

Library of Functions

Each time a new type of function is studied in detail in this text, it will be highlighted in a box like this one. For instance, a linear function is highlighted in the Library of Functions box in Section 1.2 because lines are discussed in that section.

In addition, there is a Library of Functions Summary inside the front cover which describes the functions listed below.

Functions and Their Graphs

1.1 Functions

1.2 Graphs of Functions

1.3 Shifting, Reflecting, and Stretching Graphs

1.4 Combinations of Functions

1.5 Inverse Functions

The Big Picture

In this chapter you will learn how to

❑ evaluate functions and find their domains.

❑ analyze graphs of functions.

❑ identify and graph shifts, reflections, and nonrigid transformations of functions.

❑ find arithmetic combinations and compositions of functions.

❑ find inverses of functions graphically and algebraically.

Superstock

One and one-half percent of all households own a horse as a pet. Some owners board their horses on farms. (Source: American Veterinary Medical Association)

Important Vocabulary

As you encounter each new vocabulary term in this chapter, add the term and its definition to your notebook glossary.

- function (p. 74)
- domain (p. 74)
- range (p. 74)
- independent variable (p. 76)
- dependent variable (p. 76)
- function notation (p. 76)
- implied domain (p. 78)
- graph of a function (p. 88)
- Vertical Line Test (p. 89)

- increasing (p. 90)
- decreasing (p. 90)
- constant (p. 90)
- relative minimum (p. 91)
- relative maximum (p. 91)
- greatest integer function (p. 93)
- step function (p. 93)
- even function (p. 94)
- odd function (p. 94)

- vertical and horizontal shifts (p. 101)
- reflection (p. 103)
- rigid transformations (p. 105)
- nonrigid transformations (p. 105)
- arithmetic combination (p. 109)
- composition (p. 111)
- inverse function (p. 120)
- one-to-one (p. 124)
- Horizontal Line Test (p. 124)

Additional Resources Text-specific additional resources are available to help you do well in this course. See page xvi for details.

1.1   Functions

Introduction to Functions

Many everyday phenomena involve pairs of quantities that are related to each other by some rule of correspondence. The mathematical term for such a rule of correspondence is a **relation.** Here are two examples.

1. The simple interest I earned on an investment of \$1000 for 1 year is related to the annual interest rate r by the formula $I = 1000r$.

2. The area A of a circle is related to its radius r by the formula $A = \pi r^2$.

Not all relations have simple mathematical formulas. For instance, people commonly match up NFL starting quarterbacks with touchdown passes, and hours of the day with temperature. In each of these cases, however, there is some relation that matches each item from one set with exactly one item from a different set. Such a relation is called a **function.**

Definition of a Function

A **function** f from a set A to a set B is a relation that assigns to each element x in the set A exactly one element y in the set B. The set A is the **domain** (or set of inputs) of the function f, and the set B contains the **range** (or set of outputs).

To help understand this definition, look at the function in Figure 1.1.

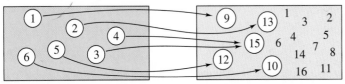

Time of day   Temperature (in degrees C)

Set A is the domain.   Set B contains the range.
Inputs: 1, 2, 3, 4, 5, 6   Outputs: 9, 10, 12, 13, 15

Figure 1.1

This function can be represented by the ordered pairs $\{(1, 9°), (2, 13°), (3, 15°), (4, 15°), (5, 12°), (6, 10°)\}$. In each ordered pair, the first coordinate is the **input** and the second coordinate is the **output.** In this example, note the following characteristics of a function.

1. Each element in A must be matched with an element of B.

2. Some elements in B may not be matched with any element in A.

3. Two or more elements of A may be matched with the same element of B.

The converse of the third statement is not true. That is, an element of A (the domain) cannot be matched with two different elements of B. In other words, each element of A must be matched with only one element of B.

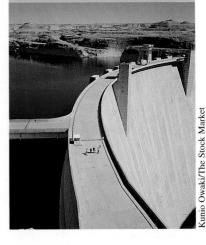

Kunio Owaki/The Stock Market

Functions are commonly represented in four ways.

1. *Verbally* by a sentence that describes how the input variable is related to the output variable
2. *Numerically* by a table or a list of ordered pairs that matches input values with output values
3. *Graphically* by points on a graph in a coordinate plane in which the input values are represented by the horizontal axis and the output values are represented by the vertical axis
4. *Algebraically* by an equation in two variables

In the following example, you are asked to decide whether the given relation is a function. To do this, you must decide whether each input value is matched with exactly one output value. If any input value is matched with two or more output values, the relation is not a function.

> ### Library of Functions
> Many functions do not have simple mathematical formulas but are defined by real-life data. Such functions arise when you are using collections of data to model real-life applications. You will see that it is often convenient to approximate the data using a mathematical model or formula.
>
> Consult the Library of Functions Summary inside the front cover for a description of functions used to model data.

EXAMPLE 1   Testing for Functions

Decide whether the description represents y as a function of x.

a. The input value x is the number of representatives from a state and the output value y is the number of senators.

b.

Input x	2	2	3	4	5
Output y	11	10	8	5	1

c.

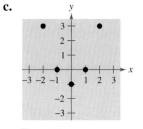

Figure 1.2

> ### STUDY T!P
> Be sure you see that the *range* of a function is not the same as the use of *range* relating to the viewing window.

Solution

a. This verbal description does describe y as a function of x. Regardless of the value of x, the value of y is always 2. Such functions are called *constant functions*.

b. This table does not describe y as a function of x. The input value 2 is mapped to two different y-values.

c. The graph in Figure 1.2 does describe y as a function of x. No input value is mapped to two output values.

Representing functions by sets of ordered pairs is common in *discrete mathematics*. For instance, a function that relates the number of units produced or sold in business to the profit could be represented by sets of ordered pairs. In algebra, however, it is more common to assume that the variables can be any real value and represent functions by equations or formulas involving two variables.

For instance, the equation

$$y = x^2 \qquad \text{\textit{y} is a function of \textit{x}.}$$

represents the variable y as a function of the variable x. In this equation, x is the **independent variable** and y is the **dependent variable.** The domain of the function is the set of all values taken on by the independent variable x, and the range of the function is the set of all values taken on by the dependent variable y.

EXAMPLE 2   Testing for Functions Represented Algebraically

Which of the equations represents y as a function of x?

a. $x^2 + y = 1$ **b.** $-x + y^2 = 1$

Solution

To determine whether y is a function of x, try to solve for y in terms of x.

a. Solving for y yields

$$x^2 + y = 1 \qquad \text{Write original equation.}$$

$$y = 1 - x^2. \qquad \text{Solve for \textit{y}.}$$

To each value of x there corresponds exactly one value of y. So, y *is* a function of x.

b. Solving for y yields

$$-x + y^2 = 1 \qquad \text{Write original equation.}$$

$$y^2 = 1 + x \qquad \text{Add \textit{x} to each side.}$$

$$y = \pm\sqrt{1 + x}. \qquad \text{Solve for \textit{y}.}$$

The $\pm$ indicates that for a given value of x there correspond two values of y. So, y *is not* a function of x.

Function Notation

When an equation is used to represent a function, it is convenient to name the function so that it can be referenced easily. For example, you know that the equation $y = 1 - x^2$ describes y as a function of x. Suppose you give this function the name "f." Then you can use the following **function notation.**

Input	*Output*	*Equation*
x	$f(x)$	$f(x) = 1 - x^2$

The symbol $f(x)$ is read as the *value of f at x* or simply *f of x.* The symbol $f(x)$ corresponds to the *y*-value for a given x. So, you can write $y = f(x)$. Keep in mind that f is the *name* of the function, whereas $f(x)$ is the *output value* of the function at the *input value x*. In function notation, the *input* is the independent variable and the *output* is the dependent variable. For instance, the function $f(x) = 3 - 2x$ has *function values* denoted by $f(-1), f(0), f(2)$, and so on. To find these values, substitute the specified input values into the given equation.

For $x = -1$, $f(-1) = 3 - 2(-1) = 3 + 2 = 5.$

For $x = 0$, $f(0) = 3 - 2(0) = 3 - 0 = 3.$

For $x = 2$, $f(2) = 3 - 2(2) = 3 - 4 = -1.$

Exploration

Use a graphing utility to graph $x^2 + y = 1$. Then use the graph to write a convincing argument that each x-value has at most one y-value.

Use a graphing utility to graph $-x + y^2 = 1$. (*Hint:* You will need to use two equations.) Then use the graph to find an x-value that corresponds to two y-values. Does the graph represent y as a function of x? Explain.

The *Interactive* CD-ROM and *Internet* versions of this text offer a built-in graphing calculator, which can be used with the Examples, Explorations, and Exercises.

STUDY T!P

You can use a graphing utility to evaluate a function. On the website *college.hmco.com*, you will find the graphing utility program EVALUATE. The program will prompt you for a value of x, and then evaluate the expression in the equation editor for that value of x. Try using the program to evaluate several different functions of x.

Although *f* is often used as a convenient function name and *x* is often used as the independent variable, you can use other letters. For instance,

$$f(x) = x^2 - 4x + 7, \quad f(t) = t^2 - 4t + 7, \quad \text{and} \quad g(s) = s^2 - 4s + 7$$

all define the same function. In fact, the role of the independent variable is that of a "placeholder." Consequently, the function could be described by

$$f(\quad) = (\quad)^2 - 4(\quad) + 7.$$

Library of Functions

The function in Example 4 is a *piecewise-defined* function. This means that the function is defined by two or more equations over a specified domain. In Example 4, you use the top equation for all *x*-values less than 0, and the bottom equation for all *x*-values greater than or equal to 0.

Consult the Library of Functions Summary inside the front cover for a description of the piecewise-defined function.

EXAMPLE 3   Evaluating a Function

Let $g(x) = -x^2 + 4x + 1$ and find

a. $g(2)$ **b.** $g(t)$ **c.** $g(x + 2)$.

Solution

a. Replacing *x* with 2 in $g(x) = -x^2 + 4x + 1$ yields the following.

$$g(2) = -(2)^2 + 4(2) + 1 = -4 + 8 + 1 = 5$$

b. Replacing *x* with *t* yields the following.

$$g(t) = -(t)^2 + 4(t) + 1 = -t^2 + 4t + 1$$

c. Replacing *x* with $x + 2$ yields the following.

$$
\begin{aligned}
g(x + 2) &= -(x + 2)^2 + 4(x + 2) + 1 && \text{Substitute } x + 2 \text{ for } x. \\
&= -(x^2 + 4x + 4) + 4x + 8 + 1 && \text{Multiply.} \\
&= -x^2 - 4x - 4 + 4x + 8 + 1 && \text{Distributive Property} \\
&= -x^2 + 5 && \text{Simplify.}
\end{aligned}
$$

In Example 3, note that $g(x + 2)$ is not equal to $g(x) + g(2)$. In general, $g(u + v) \neq g(u) + g(v)$.

Sometimes a function is defined by more than one equation. An illustration of this is given in Example 4.

EXAMPLE 4   A Piecewise–Defined Function

Evaluate the function when $x = -1, 0$, and 1.

$$f(x) = \begin{cases} x^2 + 1, & x < 0 \\ x - 1, & x \geq 0 \end{cases}$$

Solution

Because $x = -1$ is less than 0, use $f(x) = x^2 + 1$ to obtain

$$f(-1) = (-1)^2 + 1 = 2.$$

For $x = 0$, use $f(x) = x - 1$ to obtain

$$f(0) = (0) - 1 = -1.$$

For $x = 1$, use $f(x) = x - 1$ to obtain

$$f(1) = (1) - 1 = 0.$$

STUDY T!P

Most graphing utilities can graph functions that are defined piecewise. For instructions on how to enter a piecewise-defined function into your graphing utility, consult your user's manual. You may find it helpful to set your graphing utility to *dot mode* before graphing.

The Domain of a Function

The domain of a function is the set of all values of the independent variable for which the function is defined. If x is in the domain of f, f is said to be *defined* at x. If x is not in the domain of f, f is said to be *undefined* at x.

The domain of a function can be described explicitly or it can be *implied* by the expression used to define the function. The **implied domain** is the set of all real numbers for which the expression is defined. For instance, the function

$$f(x) = \frac{1}{x^2 - 4}$$ Domain excludes x-values that result in division by zero.

has an implied domain that consists of all real x other than $x = \pm 2$. These two values are excluded from the domain because division by zero is undefined. Another common type of implied domain is that used to avoid even roots of negative numbers. For example, the function

$$f(x) = \sqrt{x}$$ Domain excludes x-values that result in even roots of negative numbers.

is defined only for $x \geq 0$. So, its implied domain is the interval $[0, \infty)$. In general, the domain of a function *excludes* values that would cause division by zero *or* result in the even root of a negative number.

EXAMPLE 5   Finding the Domain of a Function

Find the domain of each function.

a. $f: \{(-3, 0), (-1, 4), (0, 2), (2, 2), (4, -1)\}$

b. $g(x) = -3x^2 + 4x + 5$ **c.** $h(x) = \dfrac{1}{x + 5}$

d. Volume of a sphere: $V = \frac{4}{3}\pi r^3$ **e.** $k(x) = \sqrt{4 - 3x}$

Solution

a. The domain of f consists of all first coordinates in the set of ordered pairs.

Domain $= \{-3, -1, 0, 2, 4\}$.

b. The domain of g is the set of all real numbers.

c. Excluding x-values that yield zero in the denominator, the domain of h is the set of all real numbers $x \neq -5$.

d. Because this function represents the volume of a sphere, the values of the radius r must be positive. So, the domain is the set of all real numbers r such that $r > 0$.

e. This function is defined only for x-values for which

$$4 - 3x \geq 0.$$

The domain of k is all real numbers that are less than or equal to $\frac{4}{3}$.

In Example 5(d), note that the *domain of a function may be implied by the physical context*. For instance, from the equation $V = \frac{4}{3}\pi r^3$, you would have no reason to restrict r to positive values, but the physical context implies that a sphere cannot have a negative or zero radius.

Library of Functions

The *square root* function $f(x) = \sqrt{x}$ is not defined for $x < 0$. This means that you must be careful when analyzing the domain of complicated functions involving the square root symbol.

Consult the Library of Functions Summary inside the front cover for a description of the square root or radical function.

The *Interactive* CD-ROM and *Internet* versions of this text show every example with its solution; clicking on the *Try It!* button brings up similar problems. Guided Examples and Integrated Examples show step-by-step solutions to additional examples. Integrated Examples are related to several concepts in the section.

Applications

$\dfrac{h}{r} = 4$

$|\leftarrow r \rightarrow|$

EXAMPLE 6   The Dimensions of a Container

You work in the marketing department of a soft-drink company and are experimenting with a new soft-drink can that is slightly narrower and taller than a standard can. For your experimental can, the ratio of the height to the radius is 4, as shown in Figure 1.3.

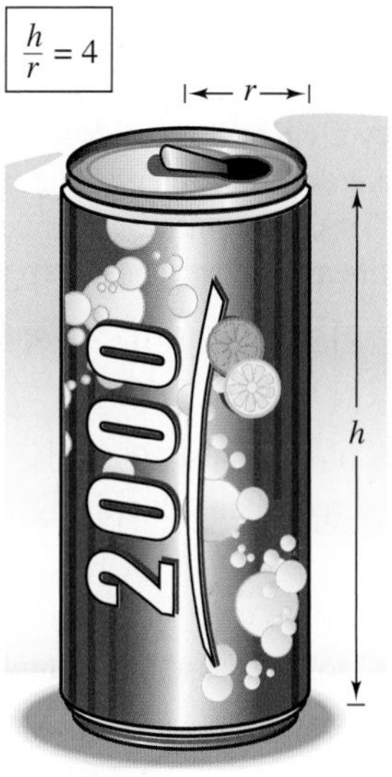

a. Express the volume of the can as a function of the radius r.

b. Express the volume of the can as a function of the height h.

Solution

The volume of a right circular cylinder is $V = \pi r^2 h$.

a. To write the volume as a function of the radius, use the fact that $h = 4r$.

$$V(r) = \pi r^2 h = \pi r^2 (4r) = 4\pi r^3 \qquad \text{Write } V \text{ as a function of } r.$$

b. To write the volume as a function of the height, use the fact that $r = h/4$.

$$V(h) = \pi r^2 h = \pi \left(\frac{h}{4}\right)^2 h = \frac{\pi h^3}{16} \qquad \text{Write } V \text{ as a function of } h.$$

Figure 1.3

EXAMPLE 7   The Path of a Baseball

A baseball is hit at a point 3 feet above ground at a velocity of 100 feet per second and an angle of 45°. The path of the baseball is given by the function

$$f(x) = -0.0032x^2 + x + 3$$

where y and x are measured in feet. Will the baseball clear a 10-foot fence located 300 feet from home plate?

Algebraic Solution

The height of the baseball is a function of the horizontal distance from home plate. When $x = 300$, you can find the height of the baseball as follows.

$$f(x) = -0.0032x^2 + x + 3 \qquad \text{Write original equation.}$$

$$f(300) = -0.0032(300)^2 + 300 + 3 \qquad \text{Substitute 300 for } x.$$

$$= 15 \qquad \text{Simplify.}$$

When $x = 300$, the height of the baseball is 15 feet, so the baseball will clear a 10-foot fence.

Graphical Solution

Use a graphing utility to graph the function $y = -0.0032x^2 + x + 3$, as shown in Figure 1.4.

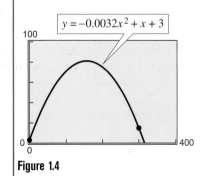

$y = -0.0032x^2 + x + 3$

Figure 1.4

Use the *value* feature or *zoom* and *trace* features of the graphing utility to estimate that $y = 15$ when $x = 300$. So, the ball will clear a 10-foot fence.

EXAMPLE 8 Direct Mail Advertising

The money C (in billions of dollars) spent for direct mail advertising in the United States increased in a linear pattern from 1990 to 1992, as shown in Figure 1.5. Then, in 1993, the money spent took a jump and, until 1996, increased in a *different* linear pattern. These two patterns can be approximated by the function

$$C(t) = \begin{cases} 23.40 + 1.01t & 0 \le t \le 2 \\ 19.85 + 2.50t & 3 \le t \le 6 \end{cases}$$

where $t = 0$ represents 1990. Use this function to approximate the total amount spent for direct mail advertising between 1990 and 1996. (Source: McCann-Erickson)

Solution
From 1990 to 1992, use the formula $C(t) = 23.40 + 1.01t$.

$\underbrace{\$23.40,}_{1990}$ $\underbrace{\$24.41,}_{1991}$ $\underbrace{\$25.42}_{1992}$

From 1993 to 1996, use the formula $C(t) = 19.85 + 2.50t$.

$\underbrace{\$27.35,}_{1993}$ $\underbrace{\$29.85,}_{1994}$ $\underbrace{\$32.35,}_{1995}$ $\underbrace{\$34.85}_{1996}$

The total of these seven amounts is $197.63, which implies that the total amount spent was approximately $197,630,000,000.

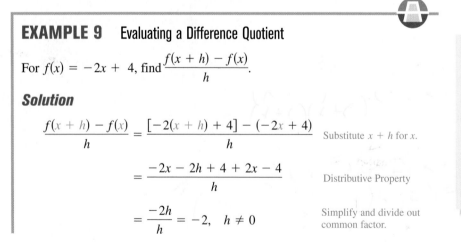

One of the basic definitions in calculus employs the ratio

$$\frac{f(x + h) - f(x)}{h}, \qquad h \ne 0.$$

This ratio is called a **difference quotient,** as illustrated in Examples 9 and 10.

EXAMPLE 9 Evaluating a Difference Quotient

For $f(x) = -2x + 4$, find $\dfrac{f(x + h) - f(x)}{h}$.

Solution

$$\begin{aligned} \frac{f(x + h) - f(x)}{h} &= \frac{[-2(x + h) + 4] - (-2x + 4)}{h} \qquad \text{Substitute } x + h \text{ for } x. \\ &= \frac{-2x - 2h + 4 + 2x - 4}{h} \qquad \text{Distributive Property} \\ &= \frac{-2h}{h} = -2, \quad h \ne 0 \qquad \text{Simplify and divide out common factor.} \end{aligned}$$

> **STUDY TIP**
>
> Notice in Example 9 that h cannot be zero in the original expression. So, you must restrict the domain of the simplified expression by adding $h \ne 0$ so that the simplified expression is equivalent to the original expression.

The symbol indicates an example or exercise that highlights algebraic techniques specifically used in calculus.

Direct-Mail Advertising

Figure 1.5

EXAMPLE 10   Evaluating a Difference Quotient

For $f(x) = x^2 - 4x + 7$, find $\dfrac{f(x + h) - f(x)}{h}$.

Solution

$$\frac{f(x + h) - f(x)}{h} = \frac{[(x + h)^2 - 4(x + h) + 7] - (x^2 - 4x + 7)}{h}$$

$$= \frac{x^2 + 2xh + h^2 - 4x - 4h + 7 - x^2 + 4x - 7}{h}$$

$$= \frac{2xh + h^2 - 4h}{h}$$

$$= \frac{h(2x + h - 4)}{h} = 2x + h - 4, \quad h \neq 0$$

Summary of Function Terminology

Function: A **function** is a relationship between two variables such that to each value of the independent variable there corresponds exactly one value of the dependent variable.

Function Notation: $y = f(x)$
   f is the *name* of the function.
   y is the **dependent variable,** or output value.
   x is the **independent variable,** or input value.
   $f(x)$ is the *value of the function at x.*

Domain: The **domain** of a function is the set of all values (inputs) of the independent variable for which the function is defined.

Range: The **range** of a function is the set of all values (outputs) assumed by the dependent variable (that is, the set of all function values).

Implied Domain: If f is defined by an algebraic expression and the domain is not specified, the **implied domain** consists of all real numbers for which the expression is defined.

Leonhard Euler (1707–1783), a Swiss mathematician, is considered to have been the most prolific and productive mathematician in history. One of his greatest influences on mathematics was his use of symbols, or notation. The function notation $y = f(x)$ was introduced by Euler.

The Granger Collection

Writing About Math   *Modeling with Piecewise-Defined Functions*

x	y
1	5.2
2	5.6
3	6.6
4	8.3
5	11.5
6	15.8

x	y
7	12.8
8	10.1
9	8.6
10	6.9
11	4.5
12	2.7

The table at the left shows the monthly revenue y (in thousands of dollars) for one year of a landscaping business, with $x = 1$ representing January.

A mathematical model that represents this data is

$$f(x) = \begin{cases} -1.97x + 26.33 \\ 0.51x^2 - 1.47x + 6.31. \end{cases}$$

What is the domain of each part of the piecewise-defined function? How can you tell? Explain your reasoning.

Find $f(5)$ and $f(11)$, and interpret your results in the context of the problem. How do these model values compare with the actual data values?

1.1   E x e r c i s e s

In Exercises 1–4, is the relationship a function?

1. *Domain   Range*   **2.** *Domain   Range*

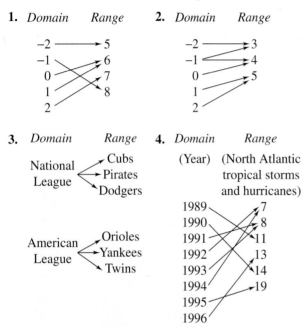

In Exercises 5–8, does the table describe a function? Explain your reasoning.

5.

Input Value	-2	-1	0	1	2
Output Value	-8	-1	0	1	8

6.

Input Value	0	1	2	1	0
Output Value	-4	-2	0	2	4

7.

Input Value	10	7	4	7	10
Output Value	3	6	9	12	15

8.

Input Value	0	3	9	12	15
Output Value	3	3	3	3	3

In Exercises 9 and 10, which sets of ordered pairs represent function(s) from A to B? Explain.

9. $A = \{0, 1, 2, 3\}$ and $B = \{-2, -1, 0, 1, 2\}$

(a) $\{(0, 1), (1, -2), (2, 0), (3, 2)\}$

(b) $\{(0, -1), (2, 2), (1, -2), (3, 0), (1, 1)\}$

(c) $\{(0, 0), (1, 0), (2, 0), (3, 0)\}$

(d) $\{(0, 2), (3, 0) (1, 1)\}$

10. $A = \{a, b, c\}$ and $B = \{0, 1, 2, 3\}$

(a) $\{(a, 1), (c, 2), (c, 3), (b, 3)\}$

(b) $\{(a, 1), (b, 2), (c, 3)\}$

(c) $\{(1, a), (0, a), (2, c), (3, b)\}$

(d) $\{(c, 0), (b, 0), (a, 3)\}$

Circulation of Newspapers   **In Exercises 11 and 12, use the graph, which shows the circulation (in millions) of daily newspapers in the United States.** (Source: Editor & Publisher Company)

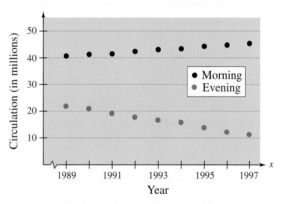

11. Is the circulation of morning newspapers a function of the year? Is the circulation of evening newspapers a function of the year? Explain.

12. Let $f(x)$ represent the circulation of evening newspapers in year x. Find $f(1994)$.

In Exercises 13–24, determine if the equation represents y as a function of x.

13. $x^2 + y^2 = 4$ **14.** $x = y^2$

15. $x^2 + y = 4$ **16.** $x + y^2 = 4$

17. $2x + 3y = 4$ **18.** $x = -y + 5$

19. $y^2 = x^2 - 1$ **20.** $y = \sqrt{x + 5}$

21. $y = |4 - x|$ **22.** $|y| = 4 - x$

23. $x = 4$ **24.** $y = -2$

In Exercises 25 and 26, fill in the blanks using the specified function and the given values of the independent variable. Simplify the result.

25. $f(x) = \dfrac{1}{x+1}$

    (a) $f(4) = \dfrac{1}{(\rule{1cm}{0.3cm})+1}$

    (b) $f(0) = \dfrac{1}{(\rule{1cm}{0.3cm})+1}$

    (c) $f(4t) = \dfrac{1}{(\rule{1cm}{0.3cm})+1}$

    (d) $f(x+c) = \dfrac{1}{(\rule{1cm}{0.3cm})+1}$

26. $g(x) = x^2 - 2x$

    (a) $g(2) = (\rule{1cm}{0.3cm})^2 - 2(\rule{1cm}{0.3cm})$

    (b) $g(-3) = (\rule{1cm}{0.3cm})^2 - 2(\rule{1cm}{0.3cm})$

    (c) $g(t+1) = (\rule{1cm}{0.3cm})^2 - 2(\rule{1cm}{0.3cm})$

    (d) $g(x+c) = (\rule{1cm}{0.3cm})^2 - 2(\rule{1cm}{0.3cm})$

In Exercises 27–38, evaluate the function at each specified value of the independent variable and simplify.

27. $f(x) = 2x - 3$

    (a) $f(1)$ (b) $f(-3)$ (c) $f(x-1)$

28. $g(y) = 7 - 3y$

    (a) $g(0)$ (b) $g\left(\frac{7}{3}\right)$ (c) $g(s+2)$

29. $h(t) = t^2 - 2t$

    (a) $h(2)$ (b) $h(1.5)$ (c) $h(x+2)$

30. $V(r) = \frac{4}{3}\pi r^3$

    (a) $V(3)$ (b) $V\left(\frac{3}{2}\right)$ (c) $V(2r)$

31. $f(y) = 3 - \sqrt{y}$

    (a) $f(4)$ (b) $f(0.25)$ (c) $f(4x^2)$

32. $f(x) = \sqrt{x+8} + 2$

    (a) $f(-8)$ (b) $f(1)$ (c) $f(x-8)$

33. $q(x) = \dfrac{1}{x^2 - 9}$

    (a) $q(0)$ (b) $q(3)$ (c) $q(y+3)$

34. $q(t) = \dfrac{2t^2 + 3}{t^2}$

    (a) $q(2)$ (b) $q(0)$ (c) $q(-x)$

35. $f(x) = \dfrac{|x|}{x}$

    (a) $f(2)$ (b) $f(-2)$ (c) $f(x^2)$

36. $f(x) = |x| + 4$

    (a) $f(2)$ (b) $f(-2)$ (c) $f(x^2)$

37. $f(x) = \begin{cases} 2x + 1, & x < 0 \\ 2x + 2, & x \ge 0 \end{cases}$

    (a) $f(-1)$ (b) $f(0)$ (c) $f(2)$

38. $f(x) = \begin{cases} x^2 + 2, & x \le 1 \\ 2x^2 + 2, & x > 1 \end{cases}$

    (a) $f(-2)$ (b) $f(1)$ (c) $f(2)$

In Exercises 39–44, complete the table.

39. $f(x) = x^2 - 3$

x	-2	-1	0	1	2
$f(x)$					

40. $g(x) = \sqrt{x - 3}$

x	3	4	5	6	7
$g(x)$					

41. $h(t) = \frac{1}{2}|t + 3|$

t	-5	-4	-3	-2	-1
$h(t)$					

42. $f(s) = \dfrac{|s - 2|}{s - 2}$

s	0	1	$\frac{3}{2}$	$\frac{5}{2}$	4
$f(s)$					

43. $f(x) = \begin{cases} -\frac{1}{2}x + 4, & x \le 0 \\ (x - 2)^2, & x > 0 \end{cases}$

x	-2	-1	0	1	2
$f(x)$					

44. $h(x) = \begin{cases} 9 - x^2, & x < 3 \\ x - 3, & x \ge 3 \end{cases}$

x	1	2	3	4	5
$h(x)$					

In Exercises 45–52, find all real values of x such that $f(x) = 0$.

45. $f(x) = 15 - 3x$

46. $f(x) = 5x + 1$

47. $f(x) = \dfrac{3x - 4}{5}$

48. $f(x) = \dfrac{12 - x^2}{5}$

49. $f(x) = x^2 - 9$

50. $f(x) = x^3 - x$

51. $f(x) = \sqrt{x^2 - 16}$

52. $f(x) = \sqrt{4x^2 - x}$

In Exercises 53–56, find the value(s) of x for which $f(x) = g(x)$.

53. $f(x) = x^2, \quad g(x) = x + 2$

54. $f(x) = x^2 + 2x + 1, \quad g(x) = 3x + 3$

55. $f(x) = \sqrt{3x} + 1, \quad g(x) = x + 1$

56. $f(x) = x^4 - 2x^2, \quad g(x) = 2x^2$

In Exercises 57–70, determine the domain of the function.

57. $f(x) = 5x^2 + 2x - 1$

58. $g(x) = 1 - 2x^2$

59. $h(t) = \dfrac{4}{t}$

60. $s(y) = \dfrac{3y}{y + 5}$

61. $g(y) = \sqrt{y - 10}$

62. $f(t) = \sqrt[3]{t + 4}$

63. $f(x) = \sqrt[4]{1 - x^2}$

64. $f(x) = \sqrt[4]{x^2 + 3x}$

65. $g(x) = \dfrac{1}{x} - \dfrac{3}{x + 2}$

66. $h(x) = \dfrac{10}{x^2 - 2x}$

67. $f(s) = \dfrac{\sqrt{s - 1}}{s - 4}$

68. $f(x) = \dfrac{\sqrt{x + 6}}{6 + x}$

69. $f(x) = \dfrac{\sqrt[3]{x - 4}}{x}$

70. $f(x) = \dfrac{x - 5}{\sqrt[4]{x^2 - 9}}$

In Exercises 71–74, assume that the domain of f is the set $A = \{-2, -1, 0, 1, 2\}$. Determine the set of ordered pairs representing the function f.

71. $f(x) = x^2$

72. $f(x) = \dfrac{2x}{x^2 + 1}$

73. $f(x) = \sqrt{x + 2}$

74. $f(x) = |x + 1|$

Exploration **In Exercises 75–78, select a function from $f(x) = cx$, $g(x) = cx^2$, $h(x) = c\sqrt{|x|}$, or $r(x) = c/x$ and determine the value of the constant c such that the function fits the data given in the table.**

75.

x	-4	-1	0	1	4
y	-32	-2	0	-2	-32

76.

x	-4	-1	0	1	4
y	-1	$-\frac{1}{4}$	0	$\frac{1}{4}$	1

77.

x	-4	-1	0	1	4
y	-8	-32	Undef.	32	8

78.

x	-4	-1	0	1	4
y	6	3	0	3	6

In Exercises 79–86, find the difference quotient and simplify your answer.

79. $f(x) = 2x, \quad \dfrac{f(x + c) - f(x)}{c}, \quad c \neq 0$

80. $g(x) = 3x - 1, \quad \dfrac{g(x + h) - g(x)}{h}, \quad h \neq 0$

81. $f(x) = x^2 - x + 1, \quad \dfrac{f(2 + h) - f(2)}{h}, \quad h \neq 0$

82. $f(x) = 5x - x^2, \quad \dfrac{f(5 + h) - f(5)}{h}, \quad h \neq 0$

83. $f(x) = x^3, \quad \dfrac{f(x + c) - f(x)}{c}, \quad c \neq 0$

84. $f(x) = x^3 + x, \quad \dfrac{f(x + h) - f(x)}{h}, \quad h \neq 0$

85. $f(t) = \dfrac{1}{t}, \quad \dfrac{f(t) - f(1)}{t - 1}, \quad t \neq 1$

86. $f(x) = \dfrac{4}{x + 1}, \quad \dfrac{f(x) - f(7)}{x - 7}, \quad x \neq 7$

87. *Geometry* Express the area A of a circle as a function of its circumference C.

88. *Geometry* Express the area A of an equilateral triangle as a function of the length s of its sides.

89. *Geometry* Express the area A of an isosceles right triangle as a function of the length s of one of its two equal sides.

90. *Geometry* Express the area A of an equilateral triangle as a function of the height of the triangle.

The symbol ⊕ indicates an example or exercise that highlights algebraic techniques specifically used in calculus.

91. Exploration An open box of maximum volume is to be made from a square piece of material, 24 centimeters on a side, by cutting equal squares from the corners and turning up the sides.

(a) Use the *table* feature of a graphing utility to complete six rows of the table. Use the result to guess the maximum volume.

Height, x	Width	Volume, V
1	$24 - 2(1)$	$1[24 - 2(1)]^2 = 484$
2	$24 - 2(2)$	$2[24 - 2(2)]^2 = 800$

(b) Write the volume V as a function of x, and determine its domain. Use a graphing utility to graph the function.

(c) Use the *value* feature or *zoom* and *trace* features of the graphing utility to approximate V when $x = 9$ and when $x = 10$.

(d) Verify your answers to part (c) algebraically.

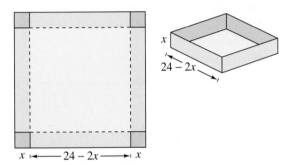

92. Exploration The cost per unit of a certain radio model is $60. The manufacturer charges $90 per unit for orders of 100 or less. To encourage large orders, the manufacturer reduces the charge by $0.15 per radio for each unit ordered in excess of 100 (i.e., there would be a charge of $87 per radio for an order size of 120).

(a) Use the *table* feature of a graphing utility to complete six rows of the table. Use the result to estimate the maximum profit.

Units, x	Price, p	Profit, P
110	$90 - 10(0.15)$	$xp - 110(60)$
120	$90 - 20(0.15)$	$xp - 120(60)$

(b) Write the profit P as a function of x, and determine its domain. Use a graphing utility to graph the function.

(c) Use the *value* feature or *zoom* and *trace* features of the graphing utility to approximate P when $x = 120, 130,$ and 140.

(d) Verify your answers to part (c) algebraically.

93. Geometry A right triangle is formed in the first quadrant by the x- and y-axes and a line through the point $(2, 1)$. Write the area of the triangle as a function of x, and determine the domain of the function.

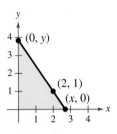

94. Geometry A rectangle is bounded by the x-axis and the semicircle $y = \sqrt{36 - x^2}$. Write the area of the rectangle as a function of x, and determine the domain of the function.

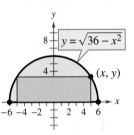

95. Geometry A rectangular package to be sent by the U.S. Postal Service can have a maximum combined length and girth (perimeter of a cross section) of 108 inches.

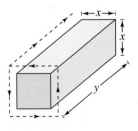

(a) Write the volume of the package as a function of x.

(b) What is the domain of the function?

(c) Use a graphing utility to graph the function. Be sure to use the appropriate viewing window.

(d) What dimensions will maximize the volume of the package? Explain your answer.

96. *Mobile Home Prices* The average price p (in thousands of dollars) of a new mobile home in the United States from 1974 to 1997 can be approximated by the model

$$p(t) = \begin{cases} 17.27 + 1.036t, & -6 \le t \le 11 \\ -4.807 + 2.882t + 0.011t^2, & 12 \le t \le 17 \end{cases}$$

where $t = 0$ represents 1980. Use a graphing utility to graph this model. Then use the *value* feature or *zoom* and *trace* features to find the average price of a mobile home in 1978, 1988, 1993, and 1997. (Source: U.S. Bureau of the Census, *Construction Reports*)

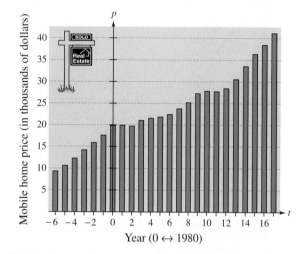

Year (0 ↔ 1980)

97. *Business* A company produces a toy for which the variable cost is $12.30 per unit and the fixed costs are $98,000. The toy sells for $17.98. Let x be the number of units produced and sold.

(a) Write the total cost C as a function of the number of units produced.

(b) Write the revenue R as a function of the number of units sold.

(c) Write the profit P as a function of the number of units sold. (*Note: P = R − C.*)

98. *Business* The inventor of a new game believes that the variable cost for producing the game is $0.95 per unit and the fixed costs are $6000. The inventor sells each game for $1.69. Let x be the number of games sold.

(a) Write the total cost C as a function of the number of games sold.

(b) Write the average cost per unit $\overline{C} = C/x$ as a function of x.

99. *Charter Bus Fares* For groups of 80 or more people, a charter bus company determines the rate per person according to the formula

$$\text{Rate} = 8 - 0.05(n - 80), \quad n \ge 80$$

where the rate is given in dollars and n is the number of people.

(a) Express the revenue R for the bus company as a function of n.

(b) Use the function from part (a) to complete the table. What can you conclude?

n	90	100	110	120	130	140	150
$R(n)$							

(c) Use a graphing utility to graph R and determine the number of people that will produce a maximum revenue. Compare the result with your conclusion from part (b).

100. *Physics* The force F (in tons) of water against the face of a dam is a function $F(y) = 149.76\sqrt{10}y^{5/2}$, where y is the depth of the water in feet. Complete the table.

y	5	10	20	30	40
$F(y)$					

(a) What can you conclude from the table?

(b) Use a graphing utility to graph the function. Describe your viewing window.

(c) Use the table to approximate the depth at which the force against the dam is 1,000,000 tons. How could you find a better estimate?

(d) Verify your answer in part (c) graphically.

101. *Height of a Balloon* A balloon carrying a transmitter ascends vertically from a point 3000 feet from the receiving station.

(a) Draw a diagram to represent the problem. Let h represent the height of the balloon and let d represent the distance between the balloon and the receiving station.

(b) Express the height of the balloon as a function of d. What is the domain of the function?

(c) Use a graphing utility to graph the function in part (b). Describe your viewing window.

(d) Graphically find the height of the balloon when $d = 10,000$ feet. Verify your answer algebraically.

102. ***Biology*** The graph below shows the lynx population from 1988 through 1995 in a 350-square-kilometer region of the Yukon territory in Canada. Let $f(t)$ represent the number of lynx in year t. (Source: Kluane Boreal Forest Ecosystem Project)

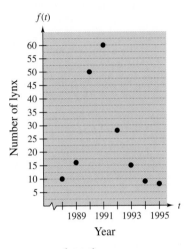

(a) Find $f(1992)$.

(b) Find $\dfrac{f(1994) - f(1991)}{1994 - 1991}$

and interpret the result in the context of the problem.

(c) An approximate model for the function is

$$N(t) = \frac{434t + 4387}{45t^2 - 55t + 100}$$

where N is the number of lynx and t is the time in years, with $t = 0$ corresponding to 1990. Complete the table and compare the result with the data. Use a graphing utility to graph the model and data in the same viewing window. Comment on the validity of the model.

t	1988	1989	1990	1991
N				

t	1992	1993	1994	1995
N				

Synthesis

True or False? **In Exercises 103 and 104, determine whether the statement is true or false. Justify your answer.**

103. The domain of the function $f(x) = x^4 - 1$ is $(-\infty, \infty)$, and the range of $f(x)$ is $(0, \infty)$.

104. The set of ordered pairs $\{(-8, -2),\ (-6, 0),\ (-4, 0),\ (-2, 2),\ (0, 4), (2, -2)\}$ represents a function.

105. ***Think About It*** Does the relationship shown in the figure represent a function from set A to set B? Explain.

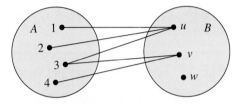

106. ***Writing*** In your own words, explain the meaning of domain and range.

107. ***Think About It*** Describe an advantage of function notation.

Review

In Exercises 108–111, sketch the graph of the equation using the point-plotting method.

108. $y = -2x - 5$

109. $y = -x^2 - 3x$

110. $y = \sqrt{x - 6}$

111. $y = x^3 - 2$

In Exercises 112–115, find the standard form of the equation of the specified circle.

112. Center: $(0, 0)$; radius: 9

113. Center: $(-8, -5)$; radius: $\frac{3}{4}$

114. Center: $(10, 1)$; solution point: $(-2, -4)$

115. Endpoints of a diameter: $(6, -5), (-2, 7)$

1.2 Graphs of Functions

The Graph of a Function

In Section 1.1 functions were represented graphically by points on a graph in a coordinate plane in which the input values are represented by the horizontal axis and the output values are represented by the vertical axis. The **graph of a function** f is the collection of ordered pairs $(x, f(x))$ such that x is in the domain of f. As you study this section, remember the following geometrical interpretation of x and $f(x)$.

$x =$ the directed distance from the y-axis

$f(x) =$ the directed distance from the x-axis

Example 1 shows how to use the graph of a function to find the domain and range of the function.

EXAMPLE 1 Finding the Domain and Range of a Function

Use the graph of the function f shown in Figure 1.6 to find (a) the domain of f, (b) the function values $f(-1)$ and $f(2)$, and (c) the range of f.

Solution

a. The closed dot (on the left) indicates that $x = -1$ is in the domain of f, whereas the open dot (on the right) indicates that $x = 4$ is not in the domain. So, the domain of f is all x in the interval $[-1, 4)$.

b. Because $(-1, -5)$ is a point on the graph of f, it follows that

$$f(-1) = -5.$$

Similarly, because $(2, 4)$ is a point on the graph of f, it follows that

$$f(2) = 4.$$

c. Because the graph does not extend below $f(-1) = -5$ or above $f(2) = 4$, the range of f is the interval $[-5, 4]$.

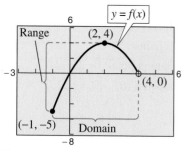

Figure 1.6

What You Should Learn:

- How to find the domains and ranges of functions
- How to use the Vertical Line Test for functions
- How to determine intervals on which functions are increasing or decreasing
- How to determine relative maximum and relative minimum values of functions
- How to identify and graph step functions and other piecewise-defined functions
- How to identify even and odd functions

Why You Should Learn It:

Graphs of functions provide a visual relationship between two variables. Exercise 87 on page 98 shows how the graph of a step function can represent the cost of a telephone call.

The use of dots (open or closed) at the extreme left and right points of a graph indicates that the graph does not extend beyond these points. If no such dots are shown, assume that the graph extends beyond these points.

EXAMPLE 2   Finding the Domain and Range of a Function

Find the domain and range of $f(x) = \sqrt{x - 4}$.

Algebraic Solution

Because the expression under a radical cannot be negative, the domain of $f(x) = \sqrt{x - 4}$ is the set of all real numbers such that $x - 4 \geq 0$. Solve this linear inequality for x as follows. (For help with solving linear inequalities, see Appendix C.)

$$x - 4 \geq 0 \qquad \text{Write original inequality.}$$

$$x \geq 4 \qquad \text{Add 4 to each side.}$$

So, the domain is the set of all real numbers greater than or equal to 4. Because the value of a radical expression is never negative, the range of $f(x) = \sqrt{x - 4}$ is the set of all nonnegative real numbers.

Graphical Solution

Use a graphing utility to graph the equation $y = \sqrt{x - 4}$, as shown in Figure 1.7. Use the *trace* feature to determine that the x-coordinates of points on the graph extend from 4 to the right. When x is greater than or equal to 4, the expression under the radical is nonnegative. So, you can conclude that the domain is the set of all real numbers greater than or equal to 4. From the graph, you can see that the y-coordinates of points on the graph extend from 0 upwards. So you can estimate the range to be the set of all nonnegative real numbers.

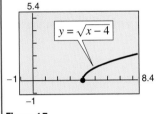

Figure 1.7

By the definition of a function, at most one y-value corresponds to a given x-value. It follows, then, that a vertical line can intersect the graph of a function at most once. This leads to the **Vertical Line Test** for functions.

Vertical Line Test for Functions

A set of points in a coordinate plane is the graph of y as a function of x if and only if no vertical line intersects the graph at more than one point.

EXAMPLE 3   Vertical Line Test for Functions

Which of the graphs in Figure 1.8 represent y as a function of x?

Solution

a. This *is not* a graph of y as a function of x because you can find a vertical line that intersects the graph twice.

b. This *is* a graph of y as a function of x because every vertical line intersects the graph at most once.

c. This *is* a graph of y as a function of x. (Note that if a vertical line does not intersect the graph, it simply means that the function is undefined for that particular value of x.)

Most graphing utilities are designed to graph functions of x more easily than other types of equations. For instance, the graph shown in Figure 1.8(a) represents the equation $x - (y - 1)^2 = 0$. To use a graphing utility to duplicate this graph you must first solve the equation for y to obtain $y = 1 \pm \sqrt{x}$, and then graph the two equations $y_1 = 1 + \sqrt{x}$ and $y_2 = 1 - \sqrt{x}$ in the same viewing window.

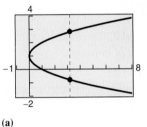

(a)

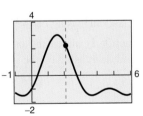

(b)

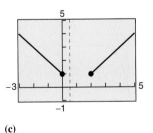

(c)
Figure 1.8

Increasing and Decreasing Functions

The more you know about the graph of a function, the more you know about the function itself. Consider the graph shown in Figure 1.9. Moving from *left to right,* this graph falls from $x = -2$ to $x = 0$, is constant from $x = 0$ to $x = 2$, and rises from $x = 2$ to $x = 4$.

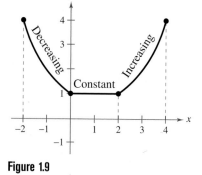

Figure 1.9

Increasing, Decreasing, and Constant Functions

A function f is **increasing** on an interval if, for any x_1 and x_2 in the interval,

$$x_1 < x_2 \text{ implies } f(x_1) < f(x_2).$$

A function f is **decreasing** on an interval if, for any x_1 and x_2 in the interval,

$$x_1 < x_2 \text{ implies } f(x_1) > f(x_2).$$

A function f is **constant** on an interval if, for any x_1 and x_2 in the interval,

$$f(x_1) = f(x_2).$$

EXAMPLE 4 Increasing and Decreasing Functions

In Figure 1.10, determine the open intervals on which each function is increasing, decreasing, or constant.

Solution

a. Although it might appear that there is an interval in which this function is constant, you can see that if $x_1 < x_2$, then $(x_1)^3 < (x_2)^3$, which implies that $f(x_1) < f(x_2)$. So, the function is increasing over the entire real line.

b. This function is increasing on the interval $(-\infty, -1)$, decreasing on the interval $(-1, 1)$, and increasing on the interval $(1, \infty)$.

c. This function is increasing on the interval $(-\infty, 0)$, constant on the interval $(0, 2)$, and decreasing on the interval $(2, \infty)$.

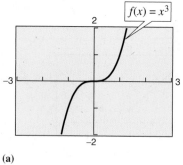

(a)

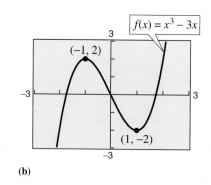

(b)

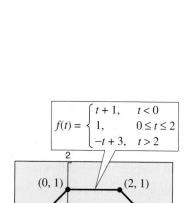

(c)

Figure 1.10

Relative Minimum and Maximum Values

The points at which a function changes its increasing, decreasing, or constant behavior are helpful in determining the relative maximum or relative minimum values of the function.

> ### Definition of Relative Minimum and Relative Maximum
>
> A function value $f(a)$ is called a **relative minimum** of f if there exists an interval (x_1, x_2) that contains a such that
>
> $$x_1 < x < x_2 \quad \text{implies} \quad f(a) \le f(x).$$
>
> A function value $f(a)$ is called a **relative maximum** of f if there exists an interval (x_1, x_2) that contains a such that
>
> $$x_1 < x < x_2 \quad \text{implies} \quad f(a) \ge f(x).$$

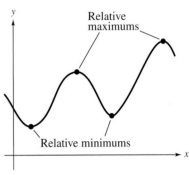

Figure 1.11

Figure 1.11 shows several different examples of relative minimums and relative maximums. In Section 2.1, you will study a technique for finding the *exact points* at which a second-degree polynomial function has a relative minimum or relative maximum. For the time being, however, you can use a graphing utility to find reasonable approximations of these points.

EXAMPLE 5   Approximating a Relative Minimum

Use a graphing utility to approximate the relative minimum of the function $f(x) = 3x^2 - 4x - 2$.

Solution

The graph of f is shown in Figure 1.12. By using the *zoom* and *trace* features of a graphing utility, you can estimate that the function has a relative minimum at the point

$(0.67, -3.33).$ Approximate relative minimum

Later, in Section 2.1, you will be able to determine that the exact point at which the relative minimum occurs is $\left(\frac{2}{3}, -\frac{10}{3}\right)$.

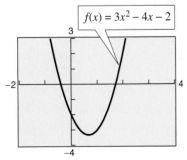

Figure 1.12

> ### STUDY T!P
>
> When you use a graphing utility to estimate x- and y-values of a relative minimum or relative maximum, the automatic *zoom* feature will often produce graphs that are nearly flat. To overcome this problem, you can manually change the vertical setting of the viewing window. The graph will vertically stretch if the values of Ymin and Ymax are closer together.

EXAMPLE 6   Approximating Relative Minimums and Maximums

Use a graphing utility to approximate the relative minimum and relative maximum of the function $f(x) = -x^3 + x$.

Solution

A sketch of the graph of f is shown in Figure 1.13. By using the *zoom* and *trace* features of the graphing utility, you can estimate that the function has a relative minimum at the point

$(-0.58, -0.38)$ Approximate relative minimum

and a relative maximum at the point

$(0.58, 0.38)$.   Approximate relative maximum

If you go on to take a course in calculus, you will learn a technique for finding the exact points at which this function has a relative minimum and a relative maximum.

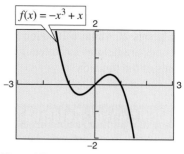

Figure 1.13

STUDY T!P

Some graphing utilities have built-in programs that will find minimum or maximum values. If your graphing utility has such features, try using them to rework Example 6.

EXAMPLE 7   Bowling Equipment Sales

During the 1990s, the sales of bowling equipment in the United States increased and then decreased according to the model

$$C = 0.165t^3 - 7.16t^2 + 100.6t - 303.1, \qquad 8 \le t \le 16$$

where C is the sales of bowling equipment (in millions of dollars) and t represents the year, with $t = 8$ corresponding to 1988. According to this model, during which years were bowling equipment sales increasing? During which years were bowling equipment sales decreasing? Approximate the maximum amount of bowling equipment sales between 1988 and 1996.   (Source: National Sporting Goods Association)

Solution

To solve this problem graph the function, as shown in Figure 1.14. From the graph, you can see that the bowling equipment sales increased from 1988 until 1992. Then, from 1992 to 1996, the sales decreased. The maximum amount of bowling equipment sales during the 8-year period was approximately $158 million.

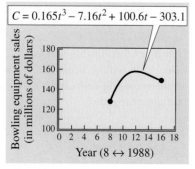

Figure 1.14

Graphing Step Functions and Piecewise-Defined Functions

The **greatest integer function** is denoted by $[\![x]\!]$ and is defined by

$$f(x) = [\![x]\!] = \text{the greatest integer less than or equal to } x.$$

The graph of this function is shown in Figure 1.15.

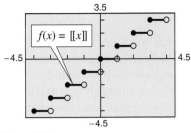

$f(x) = [\![x]\!]$

Figure 1.15

The graph of the greatest integer function jumps vertically one unit at each integer and is constant (a horizontal line segment) between each pair of consecutive integers. Because of these jumps, the greatest integer function is an example of a **step function** whose graph resembles a set of stair steps. Some values of the greatest integer function are as follows.

$$[\![-2]\!] = -2 \qquad [\![-1.5]\!] = -2 \qquad [\![-1]\!] = -1$$
$$[\![-0.5]\!] = -1 \qquad [\![0]\!] = 0 \qquad [\![0.5]\!] = 0$$
$$[\![1]\!] = 1 \qquad [\![1.5]\!] = 1 \qquad [\![2]\!] = 2$$

The range of the greatest integer function is the set of all integers.

In Section 1.1, you learned that a piecewise-defined function is a function that is defined by two or more equations over a specified domain. To sketch the graph of a piecewise-defined function, you need to sketch the graph of each equation on the appropriate portion of the domain.

EXAMPLE 8   Graphing a Piecewise-Defined Function

Sketch the graph of $f(x) = \begin{cases} 2x + 3, & x \le 1 \\ -x + 4, & x > 1 \end{cases}$.

Solution

This piecewise-defined function is composed of two linear functions. To the left of $x = 1$, the graph is the line given by $y = 2x + 3$. To the right of $x = 1$, the graph is the line given by $y = -x + 4$ (see Figure 1.16). Notice that the point $(1, 5)$ is a solid dot and the point $(1, 3)$ is an open dot. This is because $f(1) = 5$.

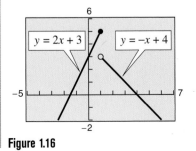

$y = 2x + 3$ $y = -x + 4$

Figure 1.16

Library of Functions

The *greatest integer* function has an infinite number of breaks or steps—one at each integer value in its domain. Could you describe the greatest integer function using a piecewise-defined function? How does the graph of the greatest integer function differ from the graph of a line with zero slope?

Consult the Library of Functions Summary inside the front cover for a description of the greatest integer function.

STUDY T!P

Most graphing utilities display graphs in *connected mode*, which means that the graph has no breaks. When you are sketching graphs that do have breaks, it is better to use *dot mode*. Graph the greatest integer function [often called Int (x)] in connected and dot modes, and compare the two results.

Even and Odd Functions

A graph has *symmetry with respect to the y-axis* if whenever (x, y) is on the graph, so is the point $(-x, y)$. A graph has *symmetry with respect to the origin* if whenever (x, y) is on the graph, so is the point $(-x, -y)$. A graph has *symmetry with respect to the x-axis* if whenever (x, y) is on the graph, so is the point $(x, -y)$. A function whose graph is symmetric with respect to the y-axis is an **even** function. A function whose graph is symmetric with respect to the origin is an **odd** function. The graph of a (nonzero) function cannot be symmetric with respect to the x-axis. These three types of symmetry are illustrated in Figure 1.17.

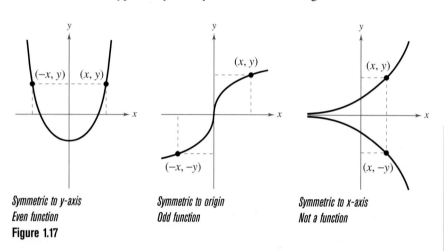

Symmetric to y-axis
Even function

Symmetric to origin
Odd function

Symmetric to x-axis
Not a function

Figure 1.17

STUDY T!P

The $\vee$-shaped graph in Example 9 is a graph of the *absolute value function*. It was produced using the *abs* key on a graphing utility and the standard viewing window.

Library of Functions

The *absolute value* function can be expressed as a piecewise-defined function.

$$|x| = \begin{cases} -x, & x < 0 \\ x, & x \geq 0 \end{cases}$$

Use this definition to show that $|-5| = 5$ and $|0| = 0$.

Consult the Library of Functions Summary inside the front cover for a description of the absolute value function.

Test for Even and Odd Functions

A function f is **even** if, for each x in the domain of f, $f(-x) = f(x)$.

A function f is **odd** if, for each x in the domain of f, $f(-x) = -f(x)$.

EXAMPLE 9 Testing for Evenness and Oddness

Is the function $f(x) = |x|$ even, odd, or neither?

Algebraic Solution

This function is even because

$$f(-x) = |-x|$$
$$= |x|$$
$$= f(x).$$

Graphical Solution

Use a graphing utility to graph $y = |x|$, as shown in Figure 1.18. You can see that the graph appears to be symmetric about the y-axis. So, the function is even.

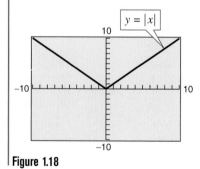

Figure 1.18

EXAMPLE 10   Even and Odd Functions

Determine whether each function is even, odd, or neither.

a. $g(x) = x^3 - x$ **b.** $h(x) = x^2 + 1$ **c.** $f(x) = x^3 - 1$

Algebraic Solution

a. This function is odd because

$$g(-x) = (-x)^3 - (-x)$$
$$= -x^3 + x$$
$$= -(x^3 - x)$$
$$= -g(x).$$

b. This function is even because

$$h(-x) = (-x)^2 + 1$$
$$= x^2 + 1$$
$$= h(x).$$

c. Substituting $-x$ for x produces

$$f(-x) = (-x)^3 - 1$$
$$= -x^3 - 1.$$

Because $f(x) = x^3 - 1$ and $-f(x) = -x^3 + 1$, you can conclude that $f(-x) \neq f(x)$ and $f(-x) \neq -f(x)$. So, the function is neither even nor odd.

Graphical Solution

a. In Figure 1.19(a), the graph is symmetric with respect to the origin. So, this function is odd.

b. In Figure 1.19(b), the graph is symmetric with respect to the y-axis. So, this function is even.

c. In Figure 1.19(c), the graph is neither symmetric to the origin nor to the y-axis. So, this function is neither even nor odd.

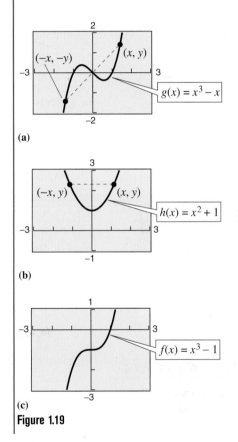

Figure 1.19

Writing About Math   *Increasing and Decreasing Functions*

Write a short paragraph describing three different functions that represent quantities between 1985 and 2000. Describe one that decreased during this time, one that increased, and one that was constant. For instance, the value of the dollar decreased, the cost of first-class postage increased, and the land size of the United States remained constant. Present your results graphically.

1.2 Exercises

In Exercises 1–6, find the domain and range of the function.

1. $f(x) = 1 - x^2$

2. $f(x) = x^3 - 3x + 2$

3. $f(x) = \sqrt{x^2 - 1}$

4. $h(x) = \sqrt{16 - x^2}$

5. $f(x) = \frac{1}{2}|x - 2|$

6. $g(x) = -|x - 1|$

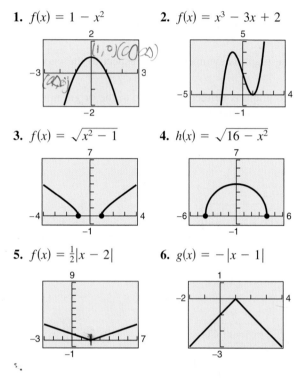

In Exercises 7–12, use a graphing utility to graph the function and estimate its domain and range. Then find the domain and range algebraically.

7. $f(x) = 2x^2 + 3$

8. $f(x) = -x^2 - 1$

9. $f(x) = \sqrt{x - 1}$

10. $h(t) = \sqrt{4 - t^2}$

11. $f(x) = |x + 3|$

12. $f(x) = -\frac{1}{4}|x - 5|$

In Exercises 13–18, use the Vertical Line Test to determine whether y is a function of x. Describe how you can use a graphing utility to produce the given graph.

13. $y = \frac{1}{2}x^2$

14. $y = \frac{1}{4}x^3$

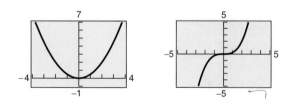

15. $x - y^2 = 1$

16. $x^2 + y^2 = 25$

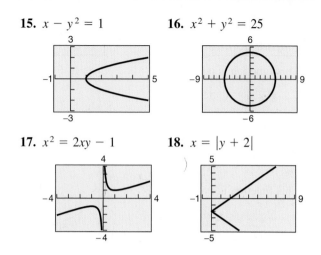

17. $x^2 = 2xy - 1$

18. $x = |y + 2|$

In Exercises 19–22, (a) determine the intervals over which the function is increasing, decreasing, or constant and (b) determine whether the function is even, odd, or neither.

19. $f(x) = \frac{3}{2}x$

20. $f(x) = x^2 - 4x$

21. $f(x) = x^3 - 3x^2 + 2$

22. $f(x) = \sqrt{x^2 - 1}$

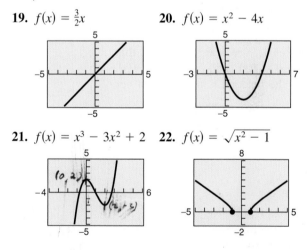

In Exercises 23–30, (a) use a graphing utility to graph the function; (b) determine the intervals over which the function is increasing, decreasing, or constant; and (c) determine whether the function is even, odd, or neither.

23. $f(x) = 3x^4 - 6x^2$

24. $f(x) = -x^6 - 2x^4$

25. $f(x) = x^{2/3}$

26. $f(x) = -x^{3/4}$

27. $f(x) = x\sqrt{x + 3}$ **28.** $f(x) = x(x^2 + 1)^{1/2}$
29. $f(x) = |x + 1| + |x - 1|$
30. $f(x) = -|x + 4| - |x + 1|$

In Exercises 31–36, use a graphing utility to approximate (to two decimal places) any relative minimum or maximum values of the function.

31. $f(x) = x^2 - 6x$ **32.** $f(x) = 3x^2 - 2x - 5$
33. $y = 2x^3 + 3x^2 - 12x$ **34.** $y = x^3 - 6x^2 + 15$
35. $h(x) = (x - 1)\sqrt{x}$ **36.** $g(x) = x\sqrt{4 - x}$

In Exercises 37–42, (a) approximate the relative minimum or maximum values of the function by sketching its graph using the point-plotting method, (b) use a graphing utility to approximate (to two decimal places) any relative minimum or maximum values, and (c) compare your answers from parts (a) and (b).

37. $f(x) = x^2 - 4x - 5$ **38.** $f(x) = 3x^2 - 12$
39. $f(x) = x^3 - 8x$ **40.** $f(x) = -x^3 + 7x$
41. $f(x) = (x - 4)^{2/3}$ **42.** $f(x) = \sqrt{4x^2 + 1}$

In Exercises 43–46, use a graphing utility to graph the piecewise-defined function.

43. $f(x) = \begin{cases} 2x + 3, & x < 0 \\ 3 - x, & x \geq 0 \end{cases}$

44. $f(x) = \begin{cases} x^2 + 5, & x \leq 1 \\ -x^2 + 4x + 3, & x > 1 \end{cases}$

45. $f(x) = \begin{cases} \sqrt{4 + x}, & x < 0 \\ \sqrt{4 - x}, & x \geq 0 \end{cases}$

46. $f(x) = \begin{cases} 1 - (x - 1)^2, & x \leq 2 \\ \sqrt{x - 2}, & x > 2 \end{cases}$

In Exercises 47–54, algebraically determine whether the function is even, odd, or neither. Verify your answer using a graphing utility.

47. $f(t) = t^2 + 2t - 3$ **48.** $f(x) = x^6 - 2x^2 + 3$
49. $g(x) = x^3 - 5x$ **50.** $h(x) = x^3 - 5$
51. $f(x) = x\sqrt{1 - x^2}$ **52.** $f(x) = x\sqrt{x + 5}$
53. $g(s) = 4s^{2/3}$ **54.** $f(s) = 4s^{3/2}$

Think About It **In Exercises 55–60, find the coordinates of a second point on the graph of a function f if the given point is on the graph and the function is (a) even and (b) odd.**

55. $\left(-\frac{3}{2}, 4\right)$ **56.** $\left(-\frac{5}{3}, -7\right)$

57. $(4, 9)$ **58.** $(5, -1)$
59. $(x, -y)$ **60.** $(2a, 2c)$

In Exercises 61–72, use a graphing utility to graph the function and determine whether it is even, odd, or neither. Verify your answer algebraically.

61. $f(x) = 5$ **62.** $f(x) = -9$
63. $f(x) = 3x - 2$ **64.** $f(x) = 5 - 3x$
65. $h(x) = x^2 - 4$ **66.** $f(x) = -x^2 - 8$
67. $f(x) = \sqrt{1 - x}$ **68.** $g(t) = \sqrt[3]{t - 1}$
69. $f(x) = |x + 2|$ **70.** $f(x) = -|x - 5|$

71. $f(x) = \begin{cases} x + 3, & x \leq 0 \\ 3, & 0 < x \leq 2 \\ 2x - 1, & x > 2 \end{cases}$

72. $f(x) = \begin{cases} 2x + 1, & x \leq -1 \\ x^2 - 2, & x > -1 \end{cases}$

In Exercises 73–82, graph the function and determine the interval(s) (if any) on the real axis for which $f(x) \geq 0$. Use a graphing utility to verify your results.

73. $f(x) = 4 - x$ **74.** $f(x) = 4x + 2$
75. $f(x) = x^2 - 9$ **76.** $f(x) = x^2 - 4x$
77. $f(x) = 1 - x^4$ **78.** $f(x) = x^2 + 1$
79. $f(x) = \sqrt{x + 2}$ **80.** $f(x) = -2\sqrt{x - 3}$
81. $f(x) = -\left(1 + |x|\right)$ **82.** $f(x) = \frac{1}{2}\left(2 + |x|\right)$

In Exercises 83 and 84, use a graphing utility to graph the function. State the domain and range of the function. Describe the pattern of the graph.

83. $s(x) = 2\left(\frac{1}{4}x - \left[\!\left[\frac{1}{4}x\right]\!\right]\right)$ **84.** $g(x) = 2\left(\frac{1}{4}x - \left[\!\left[\frac{1}{4}x\right]\!\right]\right)^2$

85. ***Geometry*** The perimeter of a rectangle is 100 meters.

(a) Show that the area of the rectangle is $A = x(50 - x)$, where x is its length.

(b) Use a graphing utility to graph the area function.

(c) Use a graphing utility to approximate the maximum area of the rectangle and the dimensions that yield the maximum area.

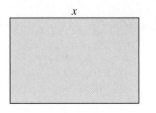

86. *Business* The marketing department of a company estimates that the demand for a product is

$$p = 100 - 0.0001x$$

where p is the price per unit and x is the number of units. The cost of producing x units is

$$C = 350,000 + 30x$$

and the profit for producing and selling x units is

$$P = R - C = xp - C.$$

Use a graphing utility to graph the profit function and estimate the number of units that would produce a maximum profit.

87. *Communications* The cost of using a telephone calling card is $1.05 for the first minute and $0.38 for each additional minute.

(a) Which of the following is the appropriate model for the cost C of a telephone call lasting t minutes? Explain.

$$C_1(t) = 1.05 + 0.38[\![t - 1]\!]$$

$$C_2(t) = 1.05 - 0.38[\![-(t - 1)]\!]$$

(b) Use a graphing utility to graph the appropriate model. Use the *value* feature or *zoom* and *trace* features to estimate the cost of a call lasting 18 minutes and 45 seconds.

88. *Delivery Service* Suppose that the cost of sending an overnight package from New York to Atlanta is $9.80 for under one pound and $2.50 for each additional pound. Use the greatest integer function to create a model for the cost C of overnight delivery of a package weighing x pounds where $x > 0$. Sketch the graph of the function.

In Exercises 89–92, write the height h of the rectangle as a function of x.

89.   **90.**

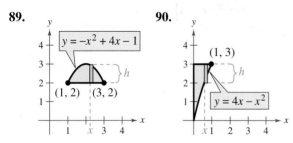

91.   **92.**

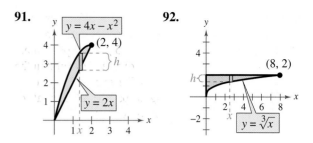

In Exercises 93 and 94, write the length L of the rectangle as a function of y.

93.   **94.**

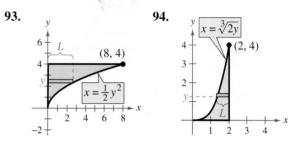

95. *Data Analysis* The table shows the amount y (in billions of dollars) of the merchandise trade balance of the United States for the years 1990 through 1997. (Source: U.S. International Trade Administration)

Year	1990	1991	1992	1993
y	-101.7	-66.7	-84.5	-115.6

Year	1994	1995	1996	1997
y	-150.6	-158.7	-170.2	-181.5

(a) Use the regression capabilities of a graphing utility to find a cubic model for the data. Let x be the time (in years), with $x = 0$ corresponding to 1990.

(b) What is the domain of the model?

(c) Use a graphing utility to graph the data and the model on the same viewing window.

(d) For which year does the model most accurately estimate the actual data? During which year is it least accurate?

(e) If this model remained valid in the future, would the economy show improvement? Explain your answer.

96. Fluid Flow The intake pipe of a 100-gallon tank has a flow rate of 10 gallons per minute, and two drain pipes have a flow rate of 5 gallons per minute each. The graph shows the volume V of fluid in the tank as a function of time t. Determine the pipes in which the fluid is flowing in specific subintervals of the 1 hour of time shown on the graph. (There are many correct answers.)

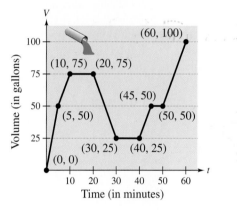

Synthesis

True or False? **In Exercises 97 and 98, determine whether the statement is true or false. Justify your answer.**

97. A function with a square root cannot have a domain that is the set of all real numbers.

98. It is possible for an odd function to have the interval $[0, \infty)$ as its domain.

99. Prove that a function of the following form is odd.

$$y = a_{2n+1}x^{2n+1} + a_{2n-1}x^{2n-1} + \cdots + a_3x^3 + a_1x$$

100. Prove that a function of the following form is even.

$$y = a_{2n}x^{2n} + a_{2n-2}x^{2n-2} + \cdots + a_2x^2 + a_0$$

101. If f is an even function, determine if g is even, odd, or neither. Explain.

  (a) $g(x) = -f(x)$ (b) $g(x) = f(-x)$

  (c) $g(x) = f(x) - 2$ (d) $g(x) = -f(x - 2)$

102. Think About It Does the graph of $x - y^2 = 1$ represent x as a function of y? Explain.

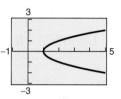

103. Think About It Does the graph of $x^2 + y^2 = 25$ represent x as a function of y? Explain.

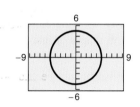

Review

In Exercises 104–107, find (a) the distance between the two points and (b) the midpoint of the segment joining the points.

104. $(-2, 7), (6, 3)$ 105. $(-5, 0), (3, 6)$

106. $\left(\frac{5}{2}, -1\right), \left(-\frac{3}{2}, 4\right)$ 107. $\left(-6, \frac{2}{3}\right), \left(\frac{3}{4}, \frac{1}{6}\right)$

In Exercises 108–111, evaluate the function at each specified value of the independent variable and simplify.

108. $f(x) = 5x - 1$

  (a) $f(6)$ (b) $f(-1)$ (c) $f(x - 3)$

109. $f(x) = -x^2 - x + 3$

  (a) $f(4)$ (b) $f(-2)$ (c) $f(x - 2)$

110. $f(x) = x\sqrt{x - 3}$

  (a) $f(3)$ (b) $f(12)$ (c) $f(6)$

111. $f(x) = -\frac{1}{2}x|x + 1|$

  (a) $f(-4)$ (b) $f(10)$ (c) $f\left(-\frac{2}{3}\right)$

In Exercises 112 and 113, find the difference quotient.

112. $f(x) = x^2 - 2x + 9, \dfrac{f(3 + h) - f(3)}{h}, \; h \neq 0$

113. $f(x) = 5 + 6x - x^2, \dfrac{f(6 + h) - f(6)}{h}, \; h \neq 0$

1.3 Shifting, Reflecting, and Stretching Graphs

Summary of Graphs of Common Functions

One of the goals of this text is to enable you to build your intuition for the basic shapes of the graphs of different types of functions. For instance, from your study of lines in Section P.3, you can determine the basic shape of the graph of the linear function $f(x) = mx + b$. Specifically, you know that the graph of this function is a line whose slope is m and whose y-intercept is b.

The six graphs shown in Figure 1.20 represent the most commonly used functions in algebra. Familiarity with the basic characteristics of these simple graphs will help you analyze the shapes of more complicated graphs.

What You Should Learn:

- How to recognize graphs of common functions
- How to use vertical and horizontal shifts and reflections to sketch graphs of functions
- How to use nonrigid transformations to sketch graphs of functions

Why You Should Learn It:

Knowing the graphs of common functions and knowing how to shift, reflect, and stretch graphs of functions can help you sketch a wide variety of simple functions by hand. This skill is useful in sketching graphs of functions that model real-life data, such as in Exercise 77 on page 108, where you are asked to sketch a function that models the amount of fuel used by trucks from 1980 through 1996.

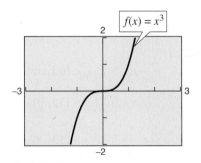

Index Stock

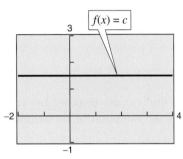

(a) Constant Function

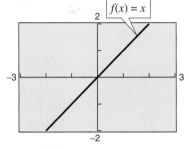

(b) Identity Function

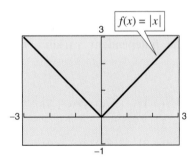

(c) Absolute Value Function

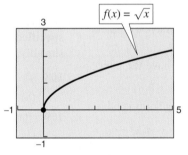

(d) Square Root Function

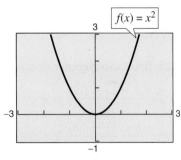

(e) Quadratic Function

(f) Cubic Function

Figure 1.20

MJL

Vertical and Horizontal Shifts

Many functions have graphs that are simple transformations of the common graphs summarized in Figure 1.20. For example, you can obtain the graph of

$$h(x) = x^2 + 2$$

by shifting the graph of $f(x) = x^2$ *up* two units, as shown in Figure 1.21. In function notation, h and f are related as follows.

$$h(x) = x^2 + 2$$
$$ = f(x) + 2 \qquad \text{Upward shift of 2}$$

Similarly, you can obtain the graph of

$$g(x) = (x - 2)^2$$

by shifting the graph of $f(x) = x^2$ to the *right* two units, as shown in Figure 1.22. In this case, the functions g and f have the following relationship.

$$g(x) = (x - 2)^2$$
$$ = f(x - 2) \qquad \text{Right shift of 2}$$

A computer animation of this concept appears in the *Interactive* CD-ROM and *Internet* versions of this text.

Exploration

Use a graphing utility to display the graphs of $y = x^2 + c$, where $c = -4, -2, 0, 2,$ and 4. Use the result to describe the effect that c has on the graph. Use a graphing utility to display the graphs of $y = (x + c)^2$, where $c = -2, 0, 2,$ and 4. Use the result to describe the effect that c has on the graph.

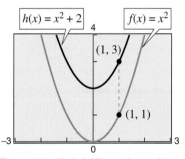

Figure 1.21   *Vertical shift upward: two units*

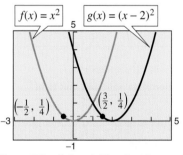

Figure 1.22   *Horizontal shift to the right: two units*

The following list summarizes this discussion about horizontal and vertical shifts.

Vertical and Horizontal Shifts

Let c be a positive real number. **Vertical and horizontal shifts** in the graph of $y = f(x)$ are represented as follows.

1. Vertical shift c units *upward:* $h(x) = f(x) + c$
2. Vertical shift c units *downward:* $h(x) = f(x) - c$
3. Horizontal shift c units to the *right:* $h(x) = f(x - c)$
4. Horizontal shift c units to the *left:* $h(x) = f(x + c)$

In items 3 and 4, be sure you see that $h(x) = f(x - c)$ corresponds to a *right* shift and $h(x) = f(x + c)$ corresponds to a *left* shift for $c > 0$.

EXAMPLE 1   Shifts in the Graph of a Function

Compare the graph of each function with the graph of $f(x) = x^3$.

a. $g(x) = x^3 - 1$ **b.** $h(x) = (x - 1)^3$ **c.** $k(x) = (x + 2)^3 + 1$

Solution

a. Graph $f(x) = x^3$ and $g(x) = x^3 - 1$ [see Figure 1.23(a)]. You can see that you can obtain the graph of g by shifting the graph of f one unit down.

b. Graph $f(x) = x^3$ and $h(x) = (x - 1)^3$ [see Figure 1.23(b)]. You can obtain the graph of h by shifting the graph of f one unit to the right.

c. Graph $f(x) = x^3$ and $k(x) = (x + 2)^3 + 1$ [see Figure 1.23(c)]. You can see that you can obtain the graph of k by shifting the graph of f two units to the left and then one unit up.

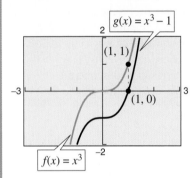 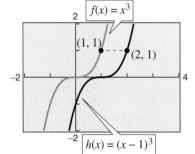 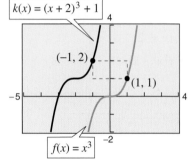

(a) **Vertical shift: one unit down**   (b) **Horizontal shift: one unit right**   (c) **Two units left and one unit up**

Figure 1.23

EXAMPLE 2   Finding Equations from Graphs

The graphs shown in Figures 1.24 (b) and (c) are shifts of the graph of $f(x) = x^2$ shown in Figure 1.24 (a). Find equations for g and h.

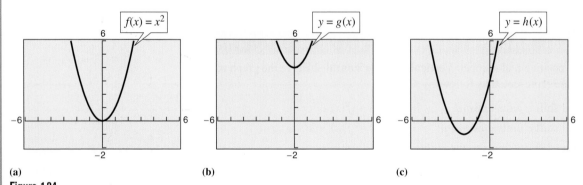

(a)   (b)   (c)

Figure 1.24

Solution

a. The graph of g is a vertical shift of four units upward of the graph of $f(x) = x^2$. So, the equation for g is $g(x) = x^2 + 4$.

b. The graph of h is a horizontal shift of two units to the left of the graph of $f(x) = x^2$ and one unit downward. So, the equation for h is $h(x) = (x + 2)^2 - 1$.

Reflecting Graphs

The second common type of transformation is called a **reflection.** For instance, if you consider the x-axis to be a mirror, the graph of $h(x) = -x^2$ is the mirror image (or reflection) of the graph of $f(x) = x^2$ (see Figure 1.25).

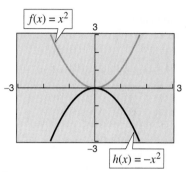

Figure 1.25

 Exploration

Compare the graph of each function with the graph of $f(x) = x^2$ by using a graphing utility to graph the function and f in the same viewing window. Describe the transformation.

a. $g(x) = -x^2$

b. $h(x) = (-x)^2$

Reflections in the Coordinate Axes

Reflections in the coordinate axes of the graph of $y = f(x)$ are represented as follows.

1. Reflection in the x-axis: $h(x) = -f(x)$
2. Reflection in the y-axis: $h(x) = f(-x)$

A computer animation of this concept appears in the *Interactive* CD-ROM and *Internet* versions of this text.

EXAMPLE 3   Finding Equations from Graphs

Each of the graphs shown in Figure 1.27 is a transformation of the graph of $f(x) = x^4$ (see Figure 1.26). Find an equation of each function.

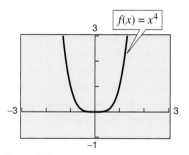

Figure 1.26

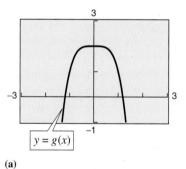

(a)

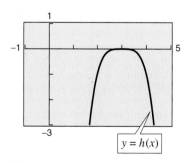

(b)

Figure 1.27

Solution

a. The graph of g is a reflection in the x-axis followed by an upward shift of two units of the graph of $f(x) = x^4$. So, the equation for g is $g(x) = -x^4 + 2$.

b. The graph of h is a horizontal shift of three units to the right followed by a reflection in the x-axis of the graph of $f(x) = x^4$. So, the equation for h is $h(x) = -(x - 3)^4$.

EXAMPLE 4   Reflections and Shifts

Compare the graph of each function with the graph of $f(x) = \sqrt{x}$.

a. $g(x) = -\sqrt{x}$ **b.** $h(x) = \sqrt{-x}$ **c.** $k(x) = -\sqrt{x+2}$

Algebraic Solution

a. Relative to the graph of $f(x) = \sqrt{x}$, the graph of g is a reflection in the x-axis because

$$g(x) = -\sqrt{x}$$
$$= -f(x).$$

b. The graph of h is a reflection of the graph of $f(x) = \sqrt{x}$ in the y-axis because

$$h(x) = \sqrt{-x}$$
$$= f(-x).$$

c. From the equation

$$k(x) = -\sqrt{x+2}$$
$$= -f(x+2)$$

you can conclude that the graph of k is a left shift of two units, followed by a reflection in the x-axis.

Graphical Solution

a. Use a graphing utility to graph f and g in the same viewing window. From the graph in Figure 1.28(a), you can see that the graph of g is a reflection of the graph of f in the x-axis.

b. Use a graphing utility to graph f and h in the same viewing window. From the graph in Figure 1.28(b), you can see that the graph of h is a reflection of the graph of f in the y-axis.

c. Use a graphing utility to graph f and k in the same viewing window. From the graph in Figure 1.28(c), you can see that the graph of k is a left shift of the graph of f of two units, followed by a reflection in the x-axis.

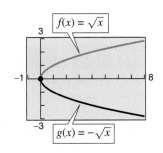

(a)

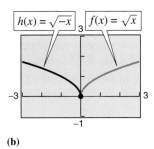

(b)

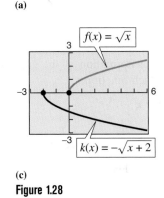

(c)

Figure 1.28

When sketching the graphs of functions involving square roots, remember that the domain must be restricted to exclude negative numbers inside the radical. For instance, here are the domains of the functions in Example 4.

Domain of $g(x) = -\sqrt{x}$: $x \geq 0$

Domain of $h(x) = \sqrt{-x}$: $x \leq 0$

Domain of $k(x) = -\sqrt{x+2}$: $x \geq -2$

Nonrigid Transformations

Horizontal shifts, vertical shifts, and reflections are called **rigid transformations** because the basic shape of the graph is unchanged. These transformations change only the *position* of the graph in the *xy*-plane. **Nonrigid transformations** are those that cause a *distortion*—a change in the shape of the original graph. For instance, a nonrigid transformation of the graph of $y = f(x)$ is represented by $y = cf(x)$, where the transformation is a **vertical stretch** if $c > 1$ and a **vertical shrink** if $0 < c < 1$.

EXAMPLE 5 Nonrigid Transformations

Compare the graph of each function with the graph of $f(x) = |x|$.

a. $h(x) = 3|x|$ **b.** $g(x) = \dfrac{1}{3}|x|$

Solution
a. Relative to the graph of $f(x) = |x|$, the graph of

$$h(x) = 3|x|$$
$$= 3f(x)$$

is a vertical stretch (multiply each *y*-value by 3) of the graph of f.

b. Similarly, the equation

$$g(x) = \frac{1}{3}|x|$$

$$= \frac{1}{3}f(x)$$

indicates that the graph of g is a vertical shrink $\left(\text{multiply each } y\text{-value by } \tfrac{1}{3}\right)$ of the graph of f.

The graphs of all three functions are shown in Figure 1.29.

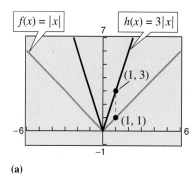

(a)

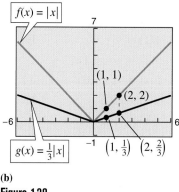

(b)

Figure 1.29

EXAMPLE 6 Sequences of Nonrigid Transformations

Use a graphing utility to graph the two functions

$$g(x) = 5(x^2 - 2) \quad \text{and} \quad h(x) = 5x^2 - 2$$

in the same viewing window. Describe how each function was obtained from $f(x) = x^2$ as a sequence of shifts and stretches.

Solution
Notice that $g(x)$ and $h(x)$ in Figure 1.30 are different. The graph of g is a downward shift of f of two units followed by a vertical stretch, whereas h is a vertical stretch of f followed by a downward shift of two units. So, the order of applying the transformations that include nonrigid transformations is important.

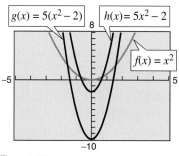

Figure 1.30

1.3 Exercises

In Exercises 1–12, sketch the graphs of the three functions by hand on the same rectangular coordinate system. Verify your result with a graphing utility.

1. $f(x) = x$
$g(x) = x - 4$
$h(x) = 3x$

2. $f(x) = \frac{1}{2}x$
$g(x) = \frac{1}{2}x + 2$
$h(x) = \frac{1}{2}(x - 2)$

3. $f(x) = x^2$
$g(x) = x^2 + 2$
$h(x) = (x - 2)^2$

4. $f(x) = x^2$
$g(x) = x^2 - 4$
$h(x) = (x + 2)^2 + 1$

5. $f(x) = -x^2$
$g(x) = -x^2 + 1$
$h(x) = -(x - 2)^2$

6. $f(x) = (x - 2)^2$
$g(x) = (x - 2)^2 + 2$
$h(x) = -(x - 2)^2 + 4$

7. $f(x) = x^2$
$g(x) = \frac{1}{2}x^2$
$h(x) = 2x^2$

8. $f(x) = x^2$
$g(x) = \frac{1}{4}x^2 + 2$
$h(x) = -\frac{1}{4}x^2$

9. $f(x) = |x|$
$g(x) = |x| - 1$
$h(x) = |x - 3|$

10. $f(x) = |x|$
$g(x) = 2|x|$
$h(x) = -2|x + 2| - 1$

11. $f(x) = \sqrt{x}$
$g(x) = \sqrt{x + 1}$
$h(x) = \sqrt{x - 2} + 1$

12. $f(x) = \sqrt{x}$
$g(x) = \frac{1}{2}\sqrt{x}$
$h(x) = -\frac{1}{2}\sqrt{x + 4}$

13. Use the graph of f to sketch each graph.
(a) $y = f(x) + 2$
(b) $y = -f(x)$
(c) $y = f(x - 2)$
(d) $y = f(x + 3)$
(e) $y = 2f(x)$
(f) $y = f(-x)$

14. Use the graph of f to sketch each graph.
(a) $y = f(x) - 1$
(b) $y = f(x + 1)$
(c) $y = f(x - 1)$
(d) $y = -f(x - 2)$
(e) $y = f(-x)$
(f) $y = \frac{1}{2}f(x)$

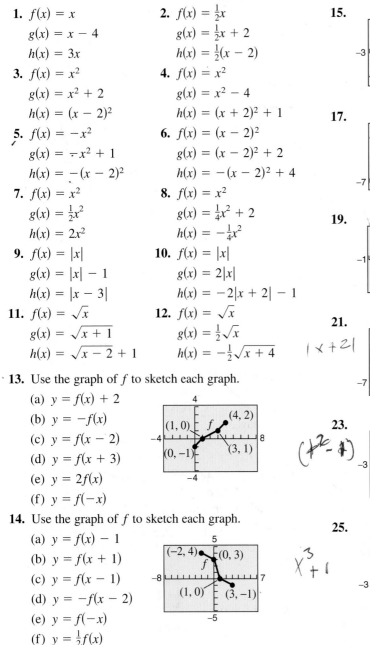

In Exercises 15–26, identify the common function and the transformation shown in the graph. Write the formula for the graphed function.

15.

16.

17.

18.

19.

20.

21.

22.

23.

24.

25.

26.

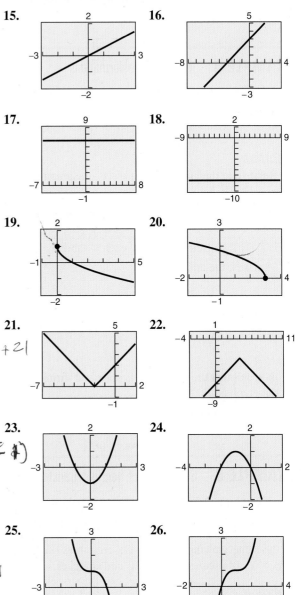

In Exercises 27–32, compare the graph of the function with the graph of $f(x) = \sqrt{x}$.

27. $y = \sqrt{x} + 2$ **28.** $y = -\sqrt{x} - 1$

29. $y = \sqrt{x - 2}$ **30.** $y = \sqrt{x + 3}$

31. $y = 2\sqrt{x}$ **32.** $y = \sqrt{-x + 3}$

In Exercises 33–38, compare the graph of the function with the graph of $f(x) = |x|$.

33. $y = |x + 2|$ **34.** $y = |x| - 3$

35. $y = -|x|$ **36.** $y = |-x|$

37. $y = \frac{1}{3}|x|$ **38.** $y = \frac{1}{2}|x|$

In Exercises 39–44, specify the sequence of transformations that will yield the graph of the given function from the graph of the function $f(x) = x^3$.

39. $g(x) = 4 - x^3$ **40.** $g(x) = -(x - 4)^3$

41. $h(x) = \frac{1}{4}(x + 2)^3$ **42.** $h(x) = -2(x - 1)^3 + 3$

43. $p(x) = \frac{1}{3}x^3 + 2$ **44.** $p(x) = [3(x - 2)]^3$

In Exercises 45–48, use a graphing utility to graph the three functions in the same viewing window. Describe the graphs of g and h relative to the graph of f.

45. $f(x) = x^3 - 3x^2$ **46.** $f(x) = x^3 - 3x^2 + 2$

$\quad g(x) = f(x + 2)$ $\quad g(x) = f(x - 1)$

$\quad h(x) = \frac{1}{2}f(x)$ $\quad h(x) = 2f(x)$

47. $f(x) = x^3 - 3x^2$ **48.** $f(x) = x^3 - 3x^2 + 2$

$\quad g(x) = -\frac{1}{3}f(x)$ $\quad g(x) = -f(x)$

$\quad h(x) = f(-x)$ $\quad h(x) = f(-x)$

In Exercises 49 and 50, use the graph of $f(x) = x^3 - 3x^2$ (see Exercise 45) to write a formula for the function g shown in the graph.

49.   **50.**

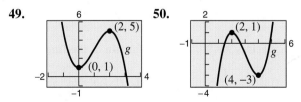

In Exercises 51–74, function g is related to one of the six common functions on page 100. (a) Identify the common function f that is related to g. (b) Describe the sequence of transformations from f to g. (c) Sketch the graph of g. (d) Use function notation to write g in terms of the common function f.

51. $g(x) = 12 - x^2$ **52.** $g(x) = (x - 8)^2$

53. $g(x) = 2 - (x + 5)^2$ **54.** $g(x) = -(x + 10)^2 + 5$

55. $g(x) = 3 + 2(x - 4)^2$

56. $g(x) = -\frac{1}{4}(x + 2)^2 - 2$

57. $g(x) = x^3 + 7$ **58.** $g(x) = -x^3 - 1$

59. $g(x) = (x - 1)^3 + 2$ **60.** $g(x) = -(x + 3)^3 - 10$

61. $g(x) = 3(x - 2)^3$ **62.** $g(x) = -\frac{1}{2}(x + 1)^3 - 5$

63. $g(x) = -|x| - 2$ **64.** $g(x) = 6 - |x + 5|$

65. $g(x) = -|x + 4| + 8$ **66.** $g(x) = |-x + 3| + 9$

67. $g(x) = -2|x - 1|$ **68.** $g(x) = \frac{1}{2}|x - 2| - 3$

69. $g(x) = \sqrt{x - 9}$ **70.** $g(x) = \sqrt{x + 4} + 8$

71. $g(x) = \sqrt{7 - x} - 2$ **72.** $g(x) = -\sqrt{x + 1} - 6$

73. $g(x) = 4\sqrt{x - 1}$

74. $g(x) = -\frac{1}{2}\sqrt{x + 3} - 1$

75. *Profit*   The profit P per week on a certain product is given by the model

$$P(x) = 80 + 20x - 0.5x^2, \qquad 0 \le x \le 20$$

where x is the amount spent on advertising. In this model, x and P are both measured in hundreds of dollars.

(a) Use a graphing utility to graph the profit function.

(b) The business estimates that taxes and operating costs will increase by an average of \$2500 per week during the next year. Rewrite the profit equation to reflect this expected decrease in profits. Identify the type of transformation applied to the graph of the equation.

(c) Rewrite the profit equation so that x measures advertising expenditures in dollars. [Find $P\left(\frac{x}{100}\right)$.] Identify the type of transformation applied to the graph of the profit function.

76. *Automobile Aerodynamics*   The number of horsepower H required to overcome wind drag on a certain automobile is approximated by

$$H(x) = 0.002x^2 + 0.005x - 0.029, \quad 10 \le x \le 100$$

where x is the speed of the car in miles per hour.

(a) Use a graphing utility to graph the power function.

(b) Rewrite the power function so that x represents the speed in kilometers per hour. [Find $H(x/1.6)$.] Identify the type of transformation applied to the graph of the power function.

77. *Energy* The amount of fuel F (in billions of gallons) used by trucks from 1980 through 1996 can be approximated by the function

$$F(t) = 20.46 + 0.04t^2$$

where $t = 0$ represents 1980.   (Source: U.S. Federal Highway Administration)

(a) Describe how F can be obtained from the common function $f(x) = x^2$. Then sketch the graph over the interval $0 \leq t \leq 16$.

(b) Rewrite the function so that $t = 0$ represents 1990. Explain how you got your answer.

78. *Finance* The amount of mortgage debt outstanding M (in billions of dollars) in the United States from 1985 through 1997 can be approximated by the function

$$M(t) = 1.5\sqrt{t} - 1.25$$

where $t = 5$ represents 1985.   (Source: Board of Governors of the Federal Reserve System)

(a) Describe how M can be obtained from the common function $f(x) = \sqrt{x}$. Then sketch the graph over the interval $5 \leq t \leq 17$.

(b) Rewrite the function so that $t = 5$ represents 1995. Explain how you got your answer.

79. *Graphical Reasoning* An electronically controlled thermostat in a home is programmed to automatically lower the temperature at night (see figure). The temperature in the house T, in degrees Fahrenheit, is given in terms of t, the time (in hours) on a 24-hour clock.

(a) Explain why T is a function of t.

(b) Approximate $T(4)$ and $T(15)$.

(c) Suppose the thermostat were reprogrammed to produce a temperature H where $H(t) = T(t - 1)$. How would this change the temperature in the house? Explain.

(d) Suppose the thermostat were reprogrammed to produce a temperature H where $H(t) = T(t) - 1$. How would this change the temperature in the house? Explain.

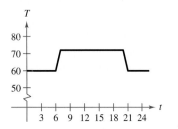

Synthesis

True or False?   **In Exercises 80 and 81, determine whether the statement is true or false. Justify your answer.**

80. The graphs of $f(x) = |x| - 5$ and $g(x) = |-x| - 5$ are identical.

81. If the graph of the common function $f(x) = x^2$ is moved 6 units to the right, 3 units up, and reflected in the x-axis, then the point $(-1, 28)$ will lie on the graph of the transformation.

82. *Exploration* Use a graphing utility to graph each function. Describe any similarities and differences you observe among the graphs.

(a) $y = x$ (b) $y = x^2$ (c) $y = x^3$

(d) $y = x^4$ (e) $y = x^5$ (f) $y = x^6$

83. *Conjecture* Use the results of Exercise 82 to make a conjecture about the shapes of the graphs of $y = x^7$ and $y = x^8$. Use a graphing utility to verify your conjecture.

84. Use the results of Exercise 82 to sketch the graph of $y = (x - 3)^3$ by hand. Use a graphing utility to verify your graph.

85. Use the results of Exercise 82 to sketch the graph of $y = (x + 1)^2$ by hand. Use a graphing utility to verify your graph.

Conjecture   **In Exercises 86–89, use the results of Exercise 82 to make a conjecture about the shape of the graph of the function. Use a graphing utility to verify your conjecture.**

86. $f(x) = x^2(x - 6)^2$ **87.** $f(x) = x^3(x - 6)^2$

88. $f(x) = x^2(x - 6)^3$ **89.** $f(x) = x^3(x - 6)^3$

Review

In Exercises 90–93, find the domain of the function.

90. $f(x) = \dfrac{4}{9 - x}$ **91.** $f(x) = \dfrac{\sqrt{x - 5}}{x - 7}$

92. $f(x) = \sqrt{100 - x^2}$ **93.** $f(x) = \sqrt[3]{16 - x^2}$

In Exercises 94–99, sketch a graph of the equation. Use a graphing utility to verify your graph.

94. $y = -x - 7$ **95.** $y = 9 - 4x$

96. $y = x^2 - 4x$ **97.** $y = -x^3 - 3$

98. $y = \sqrt{9 - x}$ **99.** $y = 5 - 2|3x|$

1.4 Combinations of Functions

Arithmetic Combinations of Functions

Just as two real numbers can be combined by the operations of addition, subtraction, multiplication, and division to form other real numbers, two *functions* can be combined to create new functions. If $f(x) = 2x - 3$ and $g(x) = x^2 - 1$ you can form the sum, difference, product, and quotient of f and g as follows.

$$f(x) + g(x) = (2x - 3) + (x^2 - 1)$$
$$= x^2 + 2x - 4 \qquad \text{Sum}$$

$$f(x) - g(x) = (2x - 3) - (x^2 - 1)$$
$$= -x^2 + 2x - 2 \qquad \text{Difference}$$

$$f(x) \cdot g(x) = (2x - 3)(x^2 - 1)$$
$$= 2x^3 - 3x^2 - 2x + 3 \qquad \text{Product}$$

$$\frac{f(x)}{g(x)} = \frac{2x - 3}{x^2 - 1}, \qquad x \neq \pm 1 \qquad \text{Quotient}$$

The domain of an **arithmetic combination** of functions f and g consists of all real numbers that are common to the domains of f and g. In the case of the quotient $f(x)/g(x)$, there is the further restriction that $g(x) \neq 0$.

Sum, Difference, Product, and Quotient of Functions

Let f and g be two functions with overlapping domains. Then, for all x common to both domains, the sum, difference, product, and quotient of f and g are defined as follows.

1. **Sum:** $(f + g)(x) = f(x) + g(x)$
2. **Difference:** $(f - g)(x) = f(x) - g(x)$
3. **Product:** $(fg)(x) = f(x) \cdot g(x)$
4. **Quotient:** $\left(\dfrac{f}{g}\right)(x) = \dfrac{f(x)}{g(x)}, \quad g(x) \neq 0$

What You Should Learn:

- How to add, subtract, multiply, and divide functions
- How to find compositions of one function with another function
- How to use combinations of functions to model and solve real-life problems

Why You Should Learn It:

Combining functions can sometimes help you better understand the bigger picture. For instance, Exercises 81 and 82 on page 118 show how to use combinations of functions to analyze U.S. health care expenditures.

Superstock

A computer animation of this concept appears in the *Interactive* CD-ROM and *Internet* versions of this text.

EXAMPLE 1 Finding the Sum of Two Functions

Find $(f + g)(x)$ for the functions $f(x) = 2x + 1$ and $g(x) = x^2 + 2x - 1$. Then evaluate the sum when $x = 2$.

Solution

$$(f + g)(x) = f(x) + g(x)$$
$$= (2x + 1) + (x^2 + 2x - 1)$$
$$= x^2 + 4x$$

When $x = 2$, the value of this sum is $(f + g)(2) = 2^2 + 4(2) = 12$.

EXAMPLE 2   Finding the Difference of Two Functions

Evaluate $(f - g)(x)$ for the functions

$$f(x) = 2x + 1 \quad \text{and} \quad g(x) = x^2 + 2x - 1$$

when $x = 2$.

Algebraic Solution

The difference of the functions f and g is

$$
\begin{aligned}
(f - g)(x) &= f(x) - g(x) \\
&= (2x + 1) - (x^2 + 2x - 1) \\
&= -x^2 + 2.
\end{aligned}
$$

When $x = 2$, the value of this difference is

$$
\begin{aligned}
(f - g)(2) &= -(2)^2 + 2 \\
&= -2.
\end{aligned}
$$

Note that $(f - g)(2)$ can also be evaluated as follows.

$$
\begin{aligned}
(f - g)(2) &= f(2) - g(2) \\
&= [2(2) + 1] - [2^2 + 2(2) - 1] \\
&= 5 - 7 \\
&= -2
\end{aligned}
$$

Graphical Solution

You can use a graphing utility to graph the difference of two functions. Enter the functions as follows.

$$y_1 = 2x + 1$$

$$y_2 = x^2 + 2x - 1$$

$$y_3 = y_1 - y_2$$

Graph y_3 as shown in Figure 1.31. Then use the *value* feature or *zoom* and *trace* features to estimate that the value of the difference when $x = 2$ is -2.

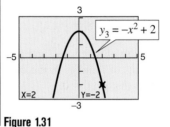

Figure 1.31

In Examples 1 and 2, both f and g have domains that consist of all real numbers. So, the domain of both $(f + g)$ and $(f - g)$ is also the set of all real numbers. Remember that any restrictions on the domains of f or g must be taken into account when forming the sum, difference, product, or quotient of f and g. For instance, the domain of $f(x) = 1/x$ is all $x \neq 0$, and the domain of $g(x) = \sqrt{x}$ is $[0, \infty)$. This implies that the domain of $f + g$ is $(0, \infty)$.

EXAMPLE 3   Finding the Product of Two Functions

Given the functions $f(x) = x^2$ and $g(x) = x - 3$, find the product of f and g. Then evaluate the product when $x = 4$.

Solution

$$
\begin{aligned}
(fg)(x) &= f(x)g(x) \\
&= (x^2)(x - 3) \\
&= x^3 - 3x^2
\end{aligned}
$$

When $x = 4$, the value of this product is

$$
\begin{aligned}
(fg)(4) &= 4^3 - 3(4)^2 \\
&= 16.
\end{aligned}
$$

EXAMPLE 4   Finding the Quotient of Two Functions

Find $\left(\dfrac{f}{g}\right)(x)$ and $\left(\dfrac{g}{f}\right)(x)$ for the functions $f(x) = \sqrt{x}$ and $g(x) = \sqrt{4 - x^2}$.
Then find the domains of f/g and g/f.

Solution
The quotient of f and g is

$$\left(\frac{f}{g}\right)(x) = \frac{f(x)}{g(x)} = \frac{\sqrt{x}}{\sqrt{4 - x^2}},$$

and the quotient of g and f is

$$\left(\frac{g}{f}\right)(x) = \frac{g(x)}{f(x)} = \frac{\sqrt{4 - x^2}}{\sqrt{x}}.$$

The domain of f is $[0, \infty)$ and the domain of g is $[-2, 2]$. The intersection of these domains is $[0, 2]$. So, the domains for f/g and g/f are as follows.

Domain of $\dfrac{f}{g}$: $[0, 2)$ Domain of $\dfrac{g}{f}$: $(0, 2]$

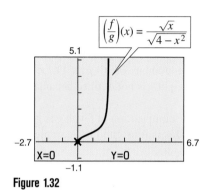

Figure 1.32

Can you see why these two domains differ slightly? You can confirm the domain of f/g in Example 4 with your graphing utility by entering the three functions

$$y_1 = \sqrt{x} \qquad y_2 = \sqrt{(4 - x^2)} \qquad y_3 = \frac{y_1}{y_2}$$

and graphing y_3, as shown in Figure 1.32.

Use the *trace* feature to determine that the x-coordinates of points on the graph extend from 0 to 2 but do not include 2. So, you can estimate the domain of f/g to be $[0, 2)$. You can confirm the domain of g/f in Example 4 by entering $y_4 = y_2/y_1$ and graphing y_4, as shown in Figure 1.33. Use the *trace* feature to determine that the x-coordinates of points on the graph extend from 0 to 2 but do not include 0. So, you can estimate the domain of g/f to be $(0, 2]$.

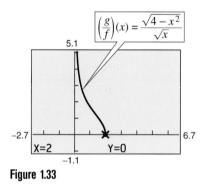

Figure 1.33

Compositions of Functions

Another way of combining two functions is to form the **composition** of one with the other. For instance, if $f(x) = x^2$ and $g(x) = x + 1$, the composition of f with g is

$$f(g(x)) = f(x + 1) = (x + 1)^2.$$

This composition is denoted as $f \circ g$.

A computer animation of this concept appears in the *Interactive* CD-ROM and *Internet* versions of this text.

Definition of Composition of Two Functions

The **composition** of the function f with g is

$$(f \circ g)(x) = f(g(x)).$$

The domain of $f \circ g$ is the set of all x in the domain of g such that $g(x)$ is in the domain of f. (See Figure 1.34.)

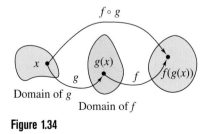

Figure 1.34

EXAMPLE 5   Forming the Composition of *f* with *g*

Find $(f \circ g)(x)$ for $f(x) = \sqrt{x}, x \geq 0$, and $g(x) = x - 1, x \geq 1$. If possible, find $(f \circ g)(2)$ and $(f \circ g)(0)$.

Solution

$$(f \circ g)(x) = f(g(x)) \qquad \text{Definition of } f \circ g$$

$$= f(x - 1) \qquad \text{Definition of } g(x)$$

$$= \sqrt{x - 1}, \quad x \geq 1 \qquad \text{Definition of } f(x)$$

The domain of $f \circ g$ is $[1, \infty)$. So, $(f \circ g)(2) = \sqrt{2 - 1} = 1$ is defined, but $(f \circ g)(0)$ is not defined because 0 is not in the domain of $f \circ g$.

The composition of *f* with *g* is generally not the same as the composition of *g* with *f*. This is illustrated in Example 6.

Exploration

Let $f(x) = x + 2$ and $g(x) = 4 - x^2$. Are the compositions $f \circ g$ and $g \circ f$ equal? You can use your graphing utility to answer this question by entering and graphing the following functions.

$$Y_1 = (4 - x^2) + 2$$
$$Y_2 = 4 - (x + 2)^2$$

What do you observe? Which function represents $f \circ g$ and which represents $g \circ f$?

EXAMPLE 6   Compositions of Functions

Given $f(x) = x + 2$ and $g(x) = 4 - x^2$, evaluate the following when $x = 0, 1, 2,$ and 3.

a. $(f \circ g)(x)$ **b.** $(g \circ f)(x)$

Algebraic Solution

a. $(f \circ g)(x) = f(g(x)) \qquad \text{Definition of } f \circ g$

$$= f(4 - x^2) \qquad \text{Definition of } g(x)$$

$$= (4 - x^2) + 2 \qquad \text{Definition of } f(x)$$

$$= -x^2 + 6$$

$$(f \circ g)(0) = -0^2 + 6 = 6$$

$$(f \circ g)(1) = -1^2 + 6 = 5$$

$$(f \circ g)(2) = -2^2 + 6 = 2$$

$$(f \circ g)(3) = -3^2 + 6 = -3$$

b. $(g \circ f)(x) = g(f(x)) \qquad \text{Definition of } g \circ f$

$$= g(x + 2) \qquad \text{Definition of } f(x)$$

$$= 4 - (x + 2)^2 \qquad \text{Definition of } g(x)$$

$$= 4 - (x^2 + 4x + 4)$$

$$= -x^2 - 4x$$

$$(g \circ f)(0) = -0^2 - 4(0) = 0$$

$$(g \circ f)(1) = -1^2 - 4(1) = -5$$

$$(g \circ f)(2) = -2^2 - 4(2) = -12$$

$$(g \circ f)(3) = -3^2 - 4(3) = -21$$

Note that $(f \circ g)(x) \neq (g \circ f)(x)$.

Numerical Solution

a. You can use a table to evaluate $f \circ g$ when $x = 0, 1, 2,$ and 3. First evaluate $g(x)$ for the values of x in the table. Then evaluate $f(g(x))$ for the values of $g(x)$ in the table.

x	0	1	2	3
$g(x)$	4	3	0	-5
$f(g(x))$	6	5	2	-3

b. To evaluate $g \circ f$ when $x = 0, 1, 2,$ and 3, first evaluate $f(x)$ for the values of x in the table. Then evaluate $g(f(x))$ for the values of $f(x)$ in the table.

x	0	1	2	3
$f(x)$	2	3	4	5
$g(f(x))$	0	-5	-12	-21

From the tables you can see that

$$(f \circ g)(x) \neq (g \circ f)(x).$$

To determine the domain of a composite function $f \circ g$, you need to restrict the outputs of g so that they are in the domain of f. For instance, to find the domain of $f \circ g$ given that $f(x) = 1/x$, and $g(x) = x + 1$, consider the outputs of g. These can be any real number. However, the domain of f is restricted to all real numbers except 0. So, the outputs of g must be restricted to all real numbers except 0. This means that $g(x) = x + 1 \neq 0$, or $x \neq -1$. So, the domain of $f \circ g$ is all real numbers except $x = -1$.

EXAMPLE 7   Finding the Domain of a Composite Function

Find the domain of the composition $(f \circ g)(x)$ for the functions

$$f(x) = x^2 - 9 \quad \text{and} \quad g(x) = \sqrt{9 - x^2}.$$

Algebraic Solution

The composition of the functions is as follows.

$$(f \circ g)(x) = f(g(x))$$
$$= f\left(\sqrt{9 - x^2}\right)$$
$$= \left(\sqrt{9 - x^2}\right)^2 - 9$$
$$= 9 - x^2 - 9$$
$$= -x^2$$

From this, it might appear that the domain of the composition is the set of all real numbers. This, however, is not true because the domain of g is $-3 \leq x \leq 3$. So, the domain of $f \circ g$ is $-3 \leq x \leq 3$.

Graphical Solution

You can use a graphing utility to graph the composition of the functions $(f \circ g)(x)$ as $y = \left(\sqrt{9 - x^2}\right)^2 - 9$. Enter the functions as follows.

$$y_1 = \sqrt{(9 - x^2)}$$
$$y_2 = y_1{}^2 - 9$$

Graph y_2 as shown in Figure 1.35. Use the *trace* feature to determine that the x-coordinates of points on the graph extend from -3 to 3. So, you can graphically estimate the domain of $(f \circ g)(x)$ to be $-3 \leq x \leq 3$.

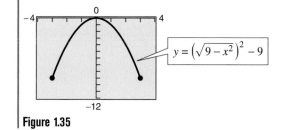

$$y = \left(\sqrt{9 - x^2}\right)^2 - 9$$

Figure 1.35

EXAMPLE 8   A Case in Which $f \circ g = g \circ f$

Given $f(x) = 2x + 3$ and $g(x) = \frac{1}{2}(x - 3)$, find the following.

a. $(f \circ g)(x)$ **b.** $(g \circ f)(x)$

Solution

a. $(f \circ g)(x) = f(g(x))$

$$= f\left(\frac{1}{2}(x - 3)\right)$$
$$= 2\left(\frac{1}{2}(x - 3)\right) + 3$$
$$= x - 3 + 3$$
$$= x$$

b. $(g \circ f)(x) = g(f(x))$

$$= g(2x + 3)$$
$$= \frac{1}{2}((2x + 3) - 3)$$
$$= \frac{1}{2}(2x)$$
$$= x$$

STUDY TIP

In Example 8, note that the two composite functions $f \circ g$ and $g \circ f$ are equal, and both represent the identity function. That is,

$$(f \circ g)(x) = x$$
$$(g \circ f)(x) = x.$$

You will study this special case in the next section.

In Examples 5, 6, 7, and 8 you formed the composition of two given functions. In calculus, it is also important to be able to identify two functions that make up a given composite function. For instance, the function h

$$h(x) = (3x - 5)^3$$

is the composition of f with g, where $f(x) = x^3$ and $g(x) = 3x - 5$. That is,

$$h(x) = (3x - 5)^3 = [g(x)]^3 = f(g(x)).$$

Basically, to "decompose" a composite function, look for an "inner" and an "outer" function. In the function h above, $g(x) = 3x - 5$ is the inner function and $f(x) = x^3$ is the outer function.

EXAMPLE 9   Identifying a Composite Function

Express the function

$$h(x) = \frac{1}{(x - 2)^2}$$

as a composition of two functions.

Solution

One way to write h as a composition of two functions is to take the inner function to be $g(x) = x - 2$ and the outer function to be

$$f(x) = \frac{1}{x^2} = x^{-2}.$$

Then you can write

$$h(x) = \frac{1}{(x - 2)^2} = (x - 2)^{-2} = f(x - 2) = f(g(x)).$$

Exploration

The function in Example 9 can be decomposed in other ways. For which of the following pairs of functions is $h(x)$ equal to $f(g(x))$?

a. $g(x) = \dfrac{1}{x - 2}$ and $f(x) = x^2$

b. $g(x) = x^2$ and $f(x) = \dfrac{1}{x - 2}$

c. $g(x) = \dfrac{1}{x}$ and $f(x) = (x - 2)^2$

Application

EXAMPLE 10   Bacteria Count

The number of bacteria in a refrigerated food is

$$N(T) = 20T^2 - 80T + 500, \qquad 2 \le T \le 14$$

where T is the temperature of the food in degrees Celsius. When the food is removed from refrigeration, the temperature is

$$T(t) = 4t + 2, \qquad 0 \le t \le 3$$

where t is the time (in hours). Find the following.

a. The composite $N(T(t))$. What does this function represent?

b. The number of bacteria in the food when $t = 2$ hours

c. The time when the bacterial count reaches 2000

Solution

a. $N(T(t)) = 20(4t + 2)^2 - 80(4t + 2) + 500$

$$= 20(16t^2 + 16t + 4) - 320t - 160 + 500$$

$$= 320t^2 + 320t + 80 - 320t - 160 + 500$$

$$= 320t^2 + 420$$

This composite function $N(T(t))$ represents the number of bacteria as a function of the amount of time the food has been out of refrigeration.

b. When $t = 2$, the number of bacteria is

$$N = 320(2)^2 + 420 = 1280 + 420 = 1700.$$

c. The bacterial count will reach $N = 2000$ when $320t^2 + 420 = 2000$. You can solve this equation for t algebraically as follows.

$$320t^2 + 420 = 2000$$

$$320t^2 = 1580$$

$$t^2 = \frac{1580}{320} = \frac{79}{16}$$

$$t = \frac{\sqrt{79}}{4} \approx 2.2 \text{ hours}$$

So, the count will reach 2000 when $t \approx 2.2$ hours. When you solve this equation, note that the negative value is rejected because it is not in the domain of the composite function. You can use a graphing utility to approximate the solution, as shown in Figure 1.36.

Exploration

Use a graphing utility to graph $y = 320t^2 + 420$ and $y = 2000$ in the same viewing window. (Use a viewing window in which $0 \le x \le 3$ and $420 \le y \le 4000$.) Explain how the graphs can be used to answer the question asked in Example 10(c). Compare your answer with that given in part (c). When will the bacteria count reach 3200?

Notice that the model for this bacteria count situation is valid only for a span of 3 hours. Now suppose the minimum number of bacteria in the food is reduced from 420 to 100. Will the number of bacteria still reach a level of 2000 within the 3-hour time span? Will the number of bacteria reach a level of 3200 within 3 hours?

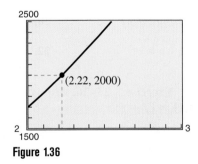

Figure 1.36

1.4 Exercises

In Exercises 1–8, find the following.

(a) $(f + g)(x)$ (b) $(f - g)(x)$

(c) $(fg)(x)$ (d) $(f/g)(x)$

(e) the domain of f/g

1. $f(x) = x + 1$, $g(x) = x - 1$

2. $f(x) = 2x - 5$, $g(x) = 1 - x$

3. $f(x) = x^2$, $g(x) = 1 - x$

4. $f(x) = 2x - 5$, $g(x) = 5$

5. $f(x) = x^2 + 5$, $g(x) = \sqrt{1 - x}$

6. $f(x) = \sqrt{x^2 - 4}$, $g(x) = \dfrac{x^2}{x^2 + 1}$

7. $f(x) = \dfrac{1}{x}$, $g(x) = \dfrac{1}{x^2}$

8. $f(x) = \dfrac{x}{x + 1}$, $g(x) = x^3$

In Exercises 9–22, evaluate the indicated function for $f(x) = x^2 + 1$ and $g(x) = x - 4$ algebraically. If possible, use a graphing utility to verify your answer.

9. $(f + g)(3)$

10. $(f - g)(-2)$

11. $(f - g)(0)$

12. $(f + g)(1)$

13. $(fg)(4)$

14. $(fg)(-6)$

15. $\left(\dfrac{f}{g}\right)(5)$

16. $\left(\dfrac{f}{g}\right)(0)$

17. $(f - g)(2t)$

18. $(f + g)(t - 4)$

19. $(fg)(-5t)$

20. $(fg)(3t^2)$

21. $\left(\dfrac{f}{g}\right)(-t)$

22. $\left(\dfrac{f}{g}\right)(t + 2)$

In Exercises 23–26, use the graphs to graph $h(x) = (f + g)(x)$.

23.

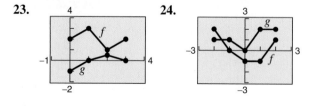

24.

25.

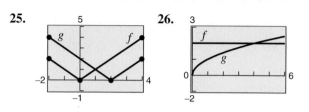

26.

In Exercises 27–30, graph the functions f, g, and $f + g$ in the same viewing window.

27. $f(x) = \tfrac{1}{2}x$, $g(x) = x - 1$

28. $f(x) = \tfrac{1}{3}x$, $g(x) = -x + 4$

29. $f(x) = x^2$, $g(x) = -2x$

30. $f(x) = 4 - x^2$, $g(x) = x$

In Exercises 31–34, use a graphing utility to sketch the graphs of f, g, and $f + g$ in the same viewing window. Which function contributes more to the magnitude of the sum when $0 \leq x \leq 2$? Which function contributes more to the magnitude of the sum when $x > 6$?

31. $f(x) = 3x$, $g(x) = -\dfrac{x^3}{10}$

32. $f(x) = \dfrac{x}{2}$, $g(x) = \sqrt{x}$

33. $f(x) = 3x + 2$, $g(x) = -\sqrt{x + 5}$

34. $f(x) = x^2 - \tfrac{1}{2}$, $g(x) = -3x^2 - 1$

In Exercises 35–38, find (a) $f \circ g$ and (b) $g \circ f$.

35. $f(x) = x^2$, $g(x) = x - 1$

36. $f(x) = \sqrt[3]{x - 1}$, $g(x) = x^3 + 1$

37. $f(x) = 3x + 5$, $g(x) = 5 - x$

38. $f(x) = x^3$, $g(x) = \dfrac{1}{x}$

In Exercises 39–44, (a) find $f \circ g$ and $g \circ f$. (b) Use a graphing utility to graph $f \circ g$ and $g \circ f$. Determine whether $f \circ g = g \circ f$.

39. $f(x) = \sqrt{x + 4}$, $g(x) = x^2$

40. $f(x) = \sqrt[3]{x + 1}$, $g(x) = x^3 - 1$

41. $f(x) = \tfrac{1}{3}x - 3$, $g(x) = 3x + 1$

42. $f(x) = \sqrt{x}$, $g(x) = \sqrt{x}$

43. $f(x) = x^{2/3}$, $g(x) = x^6$

44. $f(x) = |x|$, $g(x) = x + 6$

In Exercises 45–50, (a) find $(f \circ g)(x)$ and $(g \circ f)(x)$, (b) determine algebraically whether $(f \circ g)(x) = (g \circ f)(x)$, and (c) verify your answer to part (b) by comparing a table of values for each composition.

45. $f(x) = 5x + 4$, $g(x) = 4 - x$
46. $f(x) = \frac{1}{4}(x - 1)$, $g(x) = 4x + 1$
47. $f(x) = \sqrt{x + 6}$, $g(x) = x^2 - 5$
48. $f(x) = x^3 - 4$, $g(x) = \sqrt[3]{x + 10}$
49. $f(x) = |x + 3|$, $g(x) = 2x - 1$
50. $f(x) = \dfrac{6}{3x - 5}$, $g(x) = -x$

In Exercises 51–56, use the graphs of f and g to evaluate the functions.

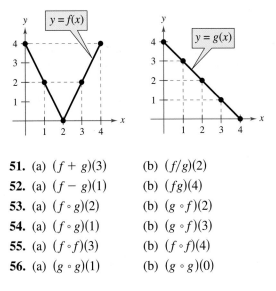

51. (a) $(f + g)(3)$ (b) $(f/g)(2)$
52. (a) $(f - g)(1)$ (b) $(fg)(4)$
53. (a) $(f \circ g)(2)$ (b) $(g \circ f)(2)$
54. (a) $(f \circ g)(1)$ (b) $(g \circ f)(3)$
55. (a) $(f \circ f)(3)$ (b) $(f \circ f)(4)$
56. (a) $(g \circ g)(1)$ (b) $(g \circ g)(0)$

In Exercises 57–64, find two functions f and g such that $(f \circ g)(x) = h(x)$. (There are many correct answers.)

57. $h(x) = (2x + 1)^2$ **58.** $h(x) = (1 - x)^3$
59. $h(x) = \sqrt[3]{x^2 - 4}$ **60.** $h(x) = \sqrt{9 - x}$
61. $h(x) = \dfrac{1}{x + 2}$ **62.** $h(x) = \dfrac{4}{(5x + 2)^2}$
63. $h(x) = (x + 4)^2 + 2(x + 4)$
64. $h(x) = (x + 3)^{3/2}$

In Exercises 65–70, determine the domains of (a) f, (b) g, and (c) $f \circ g$. Use a graphing utility to verify your answer.

65. $f(x) = \sqrt{x}$, $g(x) = x^2 + 1$

66. $f(x) = \sqrt{x + 3}$, $g(x) = \dfrac{x}{2}$
67. $f(x) = \dfrac{1}{x}$, $g(x) = x + 3$
68. $f(x) = \dfrac{1}{x}$, $g(x) = \dfrac{1}{2x}$
69. $f(x) = \dfrac{2}{|x|}$, $g(x) = x - 1$
70. $f(x) = \dfrac{3}{x^2 - 1}$, $g(x) = x + 1$

Average Rate of Change   In Exercises 71–78, find the difference quotient

$$\frac{f(x + h) - f(x)}{h}$$

and simplify your answer.

71. $f(x) = 3x - 4$ **72.** $f(x) = 5x + 1$
73. $f(x) = 1 - x^2$ **74.** $f(x) = x^2 + 4$
75. $f(x) = \dfrac{4}{x}$ **76.** $f(x) = \dfrac{2}{x^2}$
77. $f(x) = \sqrt{2x + 1}$ **78.** $f(x) = -\sqrt{4x}$

79. *Stopping Distance*   A car traveling x miles per hour stops quickly. The distance a car travels during the driver's reaction time is given by $R(x) = \frac{3}{4}x$. The distance traveled while braking is given by $B(x) = \frac{1}{15}x^2$.

(a) Find the stopping-distance function T.

(b) Use a graphing utility to graph the functions R, B, and T in the interval $0 \leq x \leq 60$.

(c) Which function contributes most to the magnitude of the sum at higher speeds? Explain.

80. *Business*   You own two restaurants. From 1995 to 2000, the sales R_1 (in thousands of dollars) for one restaurant can be modeled by

$$R_1 = 480 - 8t - 0.8t^2, \qquad t = 0, 1, 2, 3, 4, 5$$

where $t = 0$ represents 1995. During the same 6-year period, the sales R_2 (in thousands of dollars) for the other restaurant can be modeled by

$$R_2 = 254 + 0.78t, \qquad t = 0, 1, 2, 3, 4, 5.$$

(a) Write a function R_3 that represents the total sales for the two restaurants.

(b) Use a graphing utility to graph R_1, R_2, and R_3 (the total sales function) in the same viewing window.

Data Analysis **In Exercises 81 and 82, use the table, which gives the total amount spent (in billions of dollars) on health services and supplies in the United States and Puerto Rico for the years 1990 through 1996. The variables $y_1, y_2,$ and y_3 represent out-of-pocket payments, insurance premiums, and other types of payments, respectively.** (Source: U.S. Health Care Financing Administration)

Year	1990	1991	1992	1993	1994	1995	1996
y_1	144.4	151.6	159.5	163.6	164.8	166.7	171.2
y_2	238.6	259.4	282.5	303.3	315.6	326.9	337.3
y_3	21.9	24.0	25.1	27.3	29.6	31.7	32.4

81. Use a graphing utility to find a mathematical model for each of the variables. Let $t = 0$ represent 1990. Find a quadratic model $(y = at^2 + bt + c)$ for y_1 and linear models $(y = at + b)$ for y_2 and y_3.

82. Use a graphing utility to graph y_1, y_2, y_3, and $y_1 + y_2 + y_3$ in the same viewing window. Use the model to estimate the total amount spent on health services and supplies in 2000.

83. *Ripples* A pebble is dropped into a calm pond, causing ripples in the form of concentric circles. The radius (in feet) of the outer ripple is $r(t) = 0.6t$, where t is the time (in seconds) after the pebble strikes the water. The area of the circle is $A(r) = \pi r^2$. Find and interpret $(A \circ r)(t)$.

84. *Geometry* A square concrete foundation was prepared as a base for a large cylindrical gasoline tank.

    (a) Express the radius r of the tank as a function of the length x of the sides of the square.

    (b) Express the area A of the circular base of the tank as a function of the radius r.

    (c) Find and interpret $(A \circ r)(x)$.

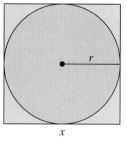

FIGURE FOR 84

85. *Business* The weekly cost of producing x units in a manufacturing process is $C(x) = 60x + 750$. The number of units produced in t hours is $x(t) = 50t$.

    (a) Find and interpret $(C \circ x)(t)$.

    (b) Use a graphing utility to graph the cost as a function of time. Use the *trace* feature to estimate (to two-decimal-place accuracy) the time that must elapse until the cost increases to $15,000.

86. *Air Traffic Control* An air traffic controller spots two planes at the same altitude flying toward each other. Their flight paths form a right angle at point P. One plane is 150 miles from point P and is moving at 450 miles per hour. The other plane is 200 miles from point P and is moving at 450 miles per hour. Write the distance s between the planes as a function of time t.

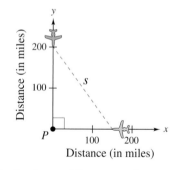

87. *Business* You are a sales representative for an automobile manufacturer. You are paid an annual salary plus a bonus of 3% of your sales over $500,000. Consider the two functions $f(x) = x - 500,000$ and $g(x) = 0.03x$. If x is greater than $500,000, which of the following represents your bonus? Explain.

    (a) $f(g(x))$ (b) $g(f(x))$

88. *Exploration* The suggested retail price of a new car is p dollars. The dealership advertised a factory rebate of \$1200 and an 8% discount.

(a) Write a function R in terms of p, giving the cost of the car after receiving the rebate from the factory.

(b) Write a function S in terms of p, giving the cost of the car after receiving the dealership discount.

(c) Form the composite functions $(R \circ S)(p)$ and $(S \circ R)(p)$, and interpret each.

(d) Find $(R \circ S)(18,400)$ and $(S \circ R)(18,400)$. Which yields the lower cost for the car? Explain.

89. *Data Analysis* The data in the table shows the circulations of morning and evening newspapers in the United States for the years 1988 through 1997. The variables y_1 and y_2 represent the circulations in millions of the morning and evening papers, respectively. (Source: Editor & Publisher Co.)

Year	1988	1989	1990	1991	1992
y_1	40.5	40.7	41.3	41.5	42.4
y_2	22.2	21.9	21.0	19.2	17.8

Year	1993	1994	1995	1996	1997
y_1	43.1	43.4	44.3	44.8	45.4
y_2	16.7	15.9	13.9	12.2	11.3

Use a graphing utility to create a scatter plot of each data set. What type of model would best fit the data? Use the regression capabilities of a graphing utility to find models for y_1 and y_2 (use $t = 8$ to represent 1988). Graph the models for y_1, y_2, and $y_1 - y_2$ in the same viewing window. What does the graph of the difference of the functions indicate about newspaper circulation in general?

Synthesis

True or False? **In Exercises 90 and 91, determine whether the statement is true or false. Justify your answer.**

90. If $f(x) = x + 1$ and $g(x) = 6x$, then

$$(f \circ g)(x) = (g \circ f)(x).$$

91. If you are given two functions $f(x)$ and $g(x)$, you can calculate $(f \circ g)(x)$ if and only if the range of g is a subset of the domain of f.

92. *Think About It* Prove that the product of two odd functions is an even function and the product of two even functions is an even function.

93. *Conjecture* Use examples to hypothesize whether the product of an odd function and an even function is even or odd. Then prove your hypothesis.

94. Given a function f, prove that $g(x)$ is even and $h(x)$ is odd where

$$g(x) = \tfrac{1}{2}[f(x) + f(-x)] \quad \text{and}$$

$$h(x) = \tfrac{1}{2}[f(x) - f(-x)].$$

95. Use the result of Exercise 94 to prove that any function can be written as a sum of even and odd functions. (*Hint:* Add the two equations in Exercise 94.)

96. Use the result of Exercise 95 to write each function as a sum of even and odd functions.

(a) $f(x) = x^2 - 2x + 1$ (b) $f(x) = \dfrac{1}{x + 1}$

Review

In Exercises 97–100, find three points that lie on the graph of the equation.

97. $y = -x^2 + x - 5$ **98.** $y = \tfrac{1}{5}x^3 - 4x^2 + 1$

99. $x^2 + y^2 = 24$ **100.** $y = \dfrac{x}{x^2 - 5}$

In Exercises 101–104, find an equation of the line that passes through the two points.

101. $(-4, -2), (-3, 8)$ **102.** $(1, 5), (-8, 2)$

103. $\left(\tfrac{3}{2}, -1\right), \left(-\tfrac{1}{3}, 4\right)$ **104.** $(0, 1.1), (-4, 3.1)$

In Exercises 105–110, use the graph of f to sketch the graph of the specified function.

105. $f(x - 4)$

106. $f(x + 2)$

107. $f(x) + 4$

108. $f(x) - 1$

109. $2f(x)$

110. $\tfrac{1}{2}f(x)$

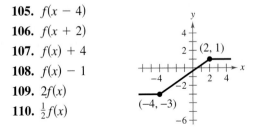

1.5 Inverse Functions

The Inverse of a Function

Recall from Section 1.1 that a function can be represented by a set of ordered pairs. For instance, the function $f(x) = x + 4$ from the set $A = \{1, 2, 3, 4\}$ to the set $B = \{5, 6, 7, 8\}$ can be written as follows.

$$f(x) = x + 4: \{(1, 5), (2, 6), (3, 7), (4, 8)\}$$

In this case, by interchanging the first and second coordinates of each of these ordered pairs, you can form the **inverse function** of f, which is denoted by f^{-1}. It is a function from the set B to the set A, and can be written as follows.

$$f^{-1}(x) = x - 4: \{(5, 1), (6, 2), (7, 3), (8, 4)\}$$

Note that the domain of f is equal to the range of f^{-1}, and vice versa, as shown in Figure 1.37. Also note that the functions f and f^{-1} have the effect of "undoing" each other. In other words, when you form the composition of f with f^{-1} or the composition of f^{-1} with f, you obtain the identity function.

$$f(f^{-1}(x)) = f(x - 4) = (x - 4) + 4 = x$$

$$f^{-1}(f(x)) = f^{-1}(x + 4) = (x + 4) - 4 = x$$

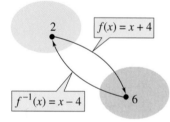

Figure 1.37

What You Should Learn:

- How to find inverse functions informally and verify that two functions are inverses of each other
- How to use graphs of functions to decide whether functions have inverses
- How to find inverse functions algebraically

Why You Should Learn It:

Inverse functions can be helpful in further exploring how two variables relate to each other. Exercise 90 on page 129 investigates the relationship between the exhaust temperature and the percent load on a diesel engine.

EXAMPLE 1 Finding Inverse Functions Informally

Find the inverse of $f(x) = 4x$. Then verify that both $f(f^{-1}(x))$ and $f^{-1}(f(x))$ are equal to the identity function.

Solution

The given function *multiplies* each input by 4. To "undo" this function, you need to *divide* each input by 4. So, the inverse function of $f(x) = 4x$ is

$$f^{-1}(x) = \frac{x}{4}.$$

You can verify that both $f(f^{-1}(x))$ and $f^{-1}(f(x))$ are equal to the identity function as follows.

$$f(f^{-1}(x)) = f\left(\frac{x}{4}\right) = 4\left(\frac{x}{4}\right) = x$$

$$f^{-1}(f(x)) = f^{-1}(4x) = \frac{4x}{4} = x$$

EXAMPLE 2   Finding Inverse Functions Informally

Find the inverse of $f(x) = x - 6$. Then verify that both $f(f^{-1}(x))$ and $f^{-1}(f(x))$ are equal to the identity function.

Solution

The given function *subtracts* 6 from each input. To "undo" this function, you need to *add* 6 to each input. So, the inverse function of $f(x) = x - 6$ is

$f^{-1}(x) = x + 6.$

You can verify that both $f(f^{-1}(x))$ and $f^{-1}(f(x))$ are equal to the identity function as follows.

$$f(f^{-1}(x)) = f(x + 6) = (x + 6) - 6 = x$$

$$f^{-1}(f(x)) = f^{-1}(x - 6) = (x - 6) + 6 = x$$

A table of values can help you understand inverse functions. For instance, the following table shows several values of the function in Example 2. Interchange the rows of this table to obtain values of the inverse function.

x	-2	-1	0	1	2
$f(x)$	-8	-7	-6	-5	-4

x	-8	-7	-6	-5	-4
$f^{-1}(x)$	-2	-1	0	1	2

In the table at the left, each output is 6 less than the input, and in the table at the right, each output is 6 more than the input.

The formal definition of the inverse of a function is as follows.

Definition of the Inverse of a Function

Let f and g be two functions such that

$f(g(x)) = x$ for every x in the domain of g

and

$g(f(x)) = x$ for every x in the domain of f.

Under these conditions, the function g is the **inverse** of the function f. The function g is denoted by f^{-1} (read "f-inverse"). So,

$f(f^{-1}(x)) = x$ and $f^{-1}(f(x)) = x.$

The domain of f must be equal to the range of f^{-1}, and the range of f must be equal to the domain of f^{-1}.

STUDY T!P

Don't be confused by the use of -1 to denote the inverse function f^{-1}. In this text, whenever f^{-1} is written, it *always* refers to the inverse of the function f and *not* to the reciprocal of $f(x)$, which is

$$\frac{1}{f(x)}.$$

If the function g is the inverse of the function f, it must also be true that the function f is the inverse of the function g. For this reason, you can say that the functions f and g are *inverses of each other*.

EXAMPLE 3 Verifying Inverse Functions Algebraically

Show that the functions are inverses of each other.

$$f(x) = 2x^3 - 1 \quad \text{and} \quad g(x) = \sqrt[3]{\frac{x + 1}{2}}$$

Solution

$$f(g(x)) = f\left(\sqrt[3]{\frac{x + 1}{2}}\right) = 2\left(\sqrt[3]{\frac{x + 1}{2}}\right)^3 - 1$$

$$= 2\left(\frac{x + 1}{2}\right) - 1$$

$$= x + 1 - 1$$

$$= x$$

$$g(f(x)) = g(2x^3 - 1) = \sqrt[3]{\frac{(2x^3 - 1) + 1}{2}}$$

$$= \sqrt[3]{\frac{2x^3}{2}}$$

$$= \sqrt[3]{x^3}$$

$$= x$$

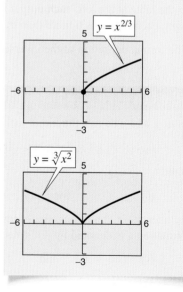

STUDY TIP

Most graphing utilities can graph $y = x^{1/3}$ in two ways:

$$y_1 = x \wedge (1/3) \quad \text{or}$$

$$y_1 = \sqrt[3]{x}.$$

However, you may not be able to obtain the complete graph of $y = x^{2/3}$ by entering $y_1 = x \wedge (2/3)$. If not, you should use

$$y_1 = (x \wedge (1/3))^2 \quad \text{or}$$

$$y_1 = \sqrt[3]{x^2}.$$

$$y = x^{2/3}$$

$$y = \sqrt[3]{x^2}$$

EXAMPLE 4 Verifying Inverse Functions Algebraically

Which of the functions is the inverse of $f(x) = \dfrac{5}{x - 2}$?

$$g(x) = \frac{x - 2}{5} \quad \text{or} \quad h(x) = \frac{5}{x} + 2$$

Solution

By forming the composition of f with g, you have

$$f(g(x)) = f\left(\frac{x - 2}{5}\right) = \frac{5}{\dfrac{x - 2}{5} - 2} = \frac{25}{x - 12} \neq x.$$

Because this composition is not equal to the identity function x, it follows that g *is not* the inverse of f. By forming the composition of f with h, you have

$$f(h(x)) = f\left(\frac{5}{x} + 2\right) = \frac{5}{\dfrac{5}{x} + 2 - 2} = \frac{5}{5/x} = x.$$

So, it appears that h is the inverse of f. You can confirm this by showing that the composition of h with f is also equal to the identity function.

The Graph of an Inverse Function

The graphs of f and f^{-1} are related to each other in the following way. If the point (a, b) lies on the graph of f, then the point (b, a) lies on the graph of f^{-1} and vice versa. This means that the graph of f^{-1} is a reflection of the graph of f in the line $y = x$, as shown in Figure 1.38.

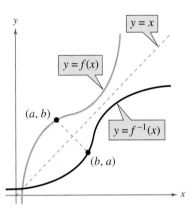

Figure 1.38

EXAMPLE 5   Verifying Inverse Functions Graphically and Numerically

Example 3 shows how to verify *algebraically* that the functions

$$f(x) = 2x^3 - 1 \quad \text{and} \quad g(x) = \sqrt[3]{\frac{x + 1}{2}}$$

are inverses of each other. Verify that f and g are inverses of each other graphically and numerically.

Graphical Solution

You can *graphically* verify that f and g are inverses of each other by using a graphing utility to graph f and g in the same viewing window. (Be sure to use a square setting.) From the graph in Figure 1.39, you can verify that the graph of g is the reflection of the graph of f in the line $y = x$.

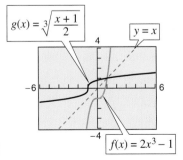

Figure 1.39

Numerical Solution

You can *numerically* verify that f and g are inverses of each other by creating two tables as shown below.

x	-2	-1	0	1	2
$f(x)$	-17	-3	-1	1	15

x	-17	-3	-1	1	15
$g(x)$	-2	-1	0	1	2

Note that the entries in the tables are the same except that their rows are interchanged. From the tables, you can verify that f and g are inverses of each other.

The Existence of an Inverse Function

A function need not have an inverse function. For instance, the function $f(x) = x^2$ has no inverse [assuming a domain of $(-\infty, \infty)$]. You can observe this numerically in the table at the right.

x	-2	-1	0	1	2
$f(x)$	4	1	0	1	4

$\downarrow$

x	4	1	0	1	4
$g(x)$	-2	-1	0	1	2

When you interchange the rows of the tables, you can find two different outputs correspond to the same input. For example, the input 4 corresponds to two different outputs, -2 and 2. So, the bottom table does not represent a function, and therefore $f(x) = x^2$ does not have an inverse.

To have an inverse, a function must be **one-to-one,** which means that no two elements in the domain of f correspond to the same element in the range of f.

Definition of a One-to-One Function

A function f is **one-to-one** if, for a and b in its domain,

$f(a) = f(b)$ implies that $a = b$.

Existence of an Inverse Function

A function f has an inverse function f^{-1} if and only if f is one-to-one.

From its graph, it is easy to tell whether a function of x is one-to-one. Simply check to see that every horizontal line intersects the graph of the function at most once. For instance, Figure 1.40 shows the graph of $y = x^4$. On the graph, you can find a horizontal line that intersects the graph twice.

Two special types of functions that pass the **Horizontal Line Test** are those that are increasing or decreasing on their entire domains.

1. If f is *increasing* on its entire domain, f is one-to-one.

2. If f is *decreasing* on its entire domain, f is one-to-one.

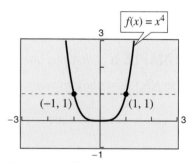

Figure 1.40 $f(x) = x^4$ *is not one-to-one.*

EXAMPLE 6   Testing for One-to-One Functions

Is the function $f(x) = \sqrt{x} + 1$ one-to-one?

Algebraic Solution

Let a and b be nonnegative real numbers with $f(a) = f(b)$.

$\sqrt{a} + 1 = \sqrt{b} + 1$ Set $f(a) = f(b)$.

$\sqrt{a} = \sqrt{b}$

$a = b$

Therefore,

$f(a) = f(b)$

implies that

$a = b$.

So, f *is* one-to-one.

Graphical Solution

Use a graphing utility to graph the function $y = \sqrt{x} + 1$. From Figure 1.41, you can see that a horizontal line will intersect the graph at most once. So, f is one-to-one.

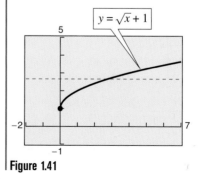

Figure 1.41

Finding Inverse Functions Algebraically

For simple functions (such as the ones in Examples 1 and 2) you can find inverse functions by inspection. For instance, the inverse of $f(x) = 8x$ is $f^{-1}(x) = x/8$. For more complicated functions, however, it is best to use the following procedure for finding the inverse of a function.

> ## Finding the Inverse of a Function
>
> To find the inverse of f, use the following steps.
>
> 1. Use the Horizontal Line Test to decide whether f has an inverse.
> 2. In the equation for $f(x)$, replace $f(x)$ by y.
> 3. Interchange the roles of x and y, and solve for y.
> 4. Replace y by $f^{-1}(x)$ in the new equation.
> 5. Verify that f and f^{-1} are inverses of each other by showing that $f(f^{-1}(x)) = x$ and $f^{-1}(f(x)) = x$.

It is important to note that in Step 1 above, the domain of f is assumed to be the entire real line. The domain of f may be restricted so that f does have an inverse. For instance, if the domain of $f(x) = x^2$ is restricted to the nonnegative real numbers, then f does have an inverse.

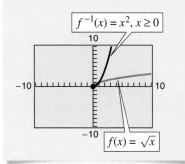

EXAMPLE 7 Finding the Inverse of a Function

Find the inverse (if it exists) of $f(x) = \dfrac{5 - 3x}{2}$.

Solution

The graph of f in Figure 1.42 passes the Horizontal Line Test, so you can see that f is one-to-one, and therefore has an inverse.

$$f(x) = \frac{5 - 3x}{2} \qquad \text{Write original equation.}$$

$$y = \frac{5 - 3x}{2} \qquad \text{Replace } f(x) \text{ by } y.$$

$$x = \frac{5 - 3y}{2} \qquad \text{Interchange } x \text{ and } y.$$

$$2x = 5 - 3y \qquad \text{Multiply each side by 2.}$$

$$3y = 5 - 2x \qquad \text{Isolate the } y\text{-term.}$$

$$y = \frac{5 - 2x}{3} \qquad \text{Solve for } y.$$

$$f^{-1}(x) = \frac{5 - 2x}{3} \qquad \text{Replace } y \text{ by } f^{-1}(x).$$

The domain and range of both f and f^{-1} consist of all real numbers. Verify that $f(f^{-1}(x)) = x$ and $f^{-1}(f(x)) = x$.

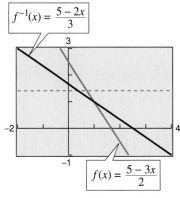

Figure 1.42

EXAMPLE 8   Finding the Inverse of a Function

Find the inverse of $f(x) = \sqrt{2x - 3}$ and sketch the graphs of f and f^{-1}.

Solution

The graph of f in Figure 1.43 passes the Horizontal Line Test, so you can see that f is one-to-one and therefore has an inverse.

$$f(x) = \sqrt{2x - 3} \qquad \text{Write original equation.}$$

$$y = \sqrt{2x - 3} \qquad \text{Replace } f(x) \text{ by } y.$$

$$x = \sqrt{2y - 3} \qquad \text{Interchange } x \text{ and } y.$$

$$x^2 = 2y - 3$$

$$2y = x^2 + 3$$

$$y = \frac{x^2 + 3}{2} \qquad \text{Solve for } y.$$

$$f^{-1}(x) = \frac{x^2 + 3}{2}, \quad x \geq 0 \qquad \text{Replace } y \text{ by } f^{-1}(x).$$

The graph of f^{-1} in Figure 1.43 is the reflection of the graph of f in the line $y = x$. Note that the range of f is the interval $[0, \infty)$, which implies that the domain of f^{-1} is the interval $[0, \infty)$. Moreover, the domain of f is the interval $\left[\frac{3}{2}, \infty\right)$, which implies that the range of f^{-1} is the interval $\left[\frac{3}{2}, \infty\right)$.

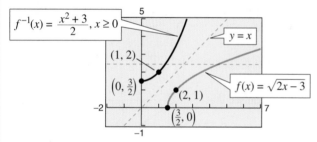

Figure 1.43

Writing About Math   *The Existence of an Inverse Function*

Write a short paragraph describing why the following functions do or do not have inverse functions. Give a numerical example for each.

a. Your hourly wage is $7.50 plus $0.90 for each unit x produced per hour. Let $f(x)$ represent your weekly wage for 40 hours of work. Does this function have an inverse?

b. Let x represent the retail price of an item (in dollars), and let $f(x)$ represent the sales tax on the item. Assume that the sales tax is 7% of the retail price *and* that the sales tax is rounded to the nearest cent. Does this function have an inverse? (*Hint:* Can you undo this function? For instance, if you know that the sales tax is $0.14, can you determine *exactly* what the retail price is?)

1.5 Exercises

In Exercises 1–4, match the graph of the function with the graph of its inverse. [The graphs of the inverse functions are labeled (a), (b), (c), and (d).]

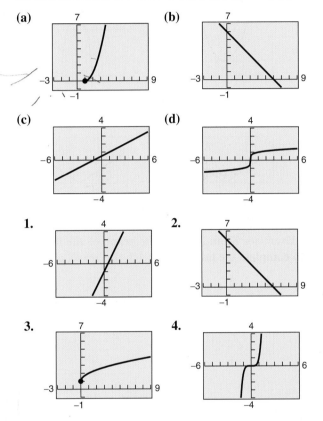

(a)

(b)

(c)

(d)

1.

2.

3.

4.

In Exercises 5–12, find the inverse of f informally. Verify that $f(f^{-1}(x)) = x$ and $f^{-1}(f(x)) = x$.

5. $f(x) = 8x$

6. $f(x) = \dfrac{1}{5}x$

7. $f(x) = x + 10$

8. $f(x) = x - 5$

9. $f(x) = 2x + 1$

10. $f(x) = \dfrac{x - 1}{4}$

11. $f(x) = \sqrt[3]{x}$

12. $f(x) = x^5$

In Exercises 13–16, show that f and g are inverse functions (a) algebraically and (b) graphically.

13. $f(x) = 2x, \quad g(x) = \dfrac{x}{2}$

14. $f(x) = x - 5, \quad g(x) = x + 5$

15. $f(x) = 5x + 1, \quad g(x) = \dfrac{x - 1}{5}$

16. $f(x) = 3 - 4x, \quad g(x) = \dfrac{3 - x}{4}$

In Exercises 17–22, show that f and g are inverse functions algebraically. Use a graphing utility to graph f and g in the same viewing window. Describe the relationship between the graphs.

17. $f(x) = x^3, \quad g(x) = \sqrt[3]{x}$

18. $f(x) = \dfrac{1}{x}, \quad g(x) = \dfrac{1}{x}$

19. $f(x) = \sqrt{x - 4}; \quad g(x) = x^2 + 4, \ x \geq 0$

20. $f(x) = 9 - x^2, \quad x \geq 0; \quad g(x) = \sqrt{9 - x}$

21. $f(x) = 1 - x^3, \quad g(x) = \sqrt[3]{1 - x}$

22. $f(x) = \dfrac{1}{1 + x}, \quad x \geq 0; \quad g(x) = \dfrac{1 - x}{x}, \ 0 < x \leq 1$

In Exercises 23–28, (a) show that f and g are inverse functions algebraically and (b) verify that f and g are inverses numerically by creating a table of values for each function.

23. $f(x) = -\dfrac{7}{2}x - 3, \quad g(x) = -\dfrac{2x + 6}{7}$

24. $f(x) = \dfrac{x + 8}{3}, \quad g(x) = 3x - 8$

25. $f(x) = x^3 + 5, \quad g(x) = \sqrt[3]{x - 5}$

26. $f(x) = \dfrac{x^3}{5}, \quad g(x) = \sqrt[3]{5x}$

27. $f(x) = -\sqrt{x - 8}; \quad g(x) = 8 + x^2, \ x \leq 0$

28. $f(x) = \sqrt[3]{3x - 10}, \quad g(x) = \dfrac{x^3 + 10}{3}$

In Exercises 29–42, use a graphing utility to graph the function and use the Horizontal Line Test to determine whether the function is one-to-one.

29. $f(x) = 3 - \tfrac{1}{2}x$

30. $g(x) = \dfrac{4 - x}{6}$

31. $h(x) = \dfrac{x^2}{x^2 + 1}$

32. $f(x) = \tfrac{1}{8}(x + 2)^2 - 1$

33. $h(x) = \sqrt{16 - x^2}$ **34.** $f(x) = -2x\sqrt{16 - x^2}$

35. $f(x) = \sqrt{x - 2}$ **36.** $f(x) = 4 - 3x^{2/3}$

37. $f(x) = 10$ **38.** $f(x) = -0.65$

39. $g(x) = (x + 5)^3$ **40.** $f(x) = x^5 - 7$

41. $h(x) = |x + 4| - |x - 4|$

42. $f(x) = -\dfrac{|x - 6|}{|x + 6|}$

In Exercises 43–54, find the inverse of the function f. Use a graphing utility to graph both f and f^{-1} in the same viewing window. Describe the relationship between the graphs.

43. $f(x) = 2x - 3$ **44.** $f(x) = 3x$

45. $f(x) = x^5$ **46.** $f(x) = x^3 + 1$

47. $f(x) = \sqrt{x}$ **48.** $f(x) = x^2, \quad x \geq 0$

49. $f(x) = \sqrt{4 - x^2}, \quad 0 \leq x \leq 2$

50. $f(x) = \sqrt{16 - x^2}, \quad -4 \leq x \leq 0$

51. $f(x) = \sqrt[3]{x - 1}$ **52.** $f(x) = x^{3/5}$

53. $f(x) = \dfrac{4}{x}$ **54.** $f(x) = \dfrac{6}{\sqrt{x}}$

In Exercises 55–68, determine algebraically whether the function is one-to-one. If it is, find its inverse. Verify your answer graphically.

55. $f(x) = x^4$ **56.** $f(x) = \dfrac{1}{x^2}$

57. $f(x) = \dfrac{3x + 4}{5}$ **58.** $f(x) = 3x + 5$

59. $f(x) = (x + 3)^2, \quad x \geq -3$

60. $q(x) = (x - 5)^2, \quad x \leq 5$

61. $h(x) = \dfrac{4}{x^2}$

62. $f(x) = |x - 2|, \quad x \leq 2$

63. $f(x) = \sqrt{2x + 3}$ **64.** $f(x) = \sqrt{x - 2}$

65. $g(x) = x^2 - x^4$ **66.** $f(x) = \dfrac{x^2}{x^2 + 1}$

67. $f(x) = ax + b, \quad a \neq 0$

68. $f(x) = c$

Think About It **In Exercises 69–72, delete part of the graph of the function so that the part that remains is one-to-one. Find the inverse of the remaining part and give the domain of the inverse. (There are many correct answers.)**

69. $f(x) = (x - 2)^2$ **70.** $f(x) = 1 - x^4$

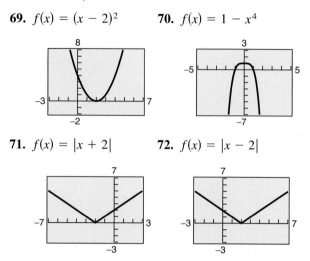

71. $f(x) = |x + 2|$ **72.** $f(x) = |x - 2|$

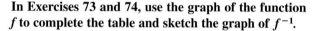

In Exercises 73 and 74, use the graph of the function f to complete the table and sketch the graph of f^{-1}.

73.

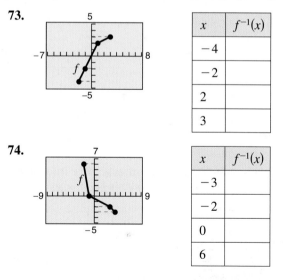

x	$f^{-1}(x)$
-4	
-2	
2	
3	

74.

x	$f^{-1}(x)$
-3	
-2	
0	
6	

Graphical Reasoning **In Exercises 75–78, (a) use a graphing utility to graph the function, (b) use the *draw inverse* feature of the graphing utility to draw the inverse of the function, and (c) determine whether the graph of the inverse relation is an inverse function, explaining your reasoning.**

75. $f(x) = x^3 + x + 1$ **76.** $h(x) = x\sqrt{4 - x^2}$

77. $g(x) = \dfrac{3x^2}{x^2 + 1}$ **78.** $f(x) = \dfrac{4x}{\sqrt{x^2 + 15}}$

In Exercises 79–84, use the functions $f(x) = \frac{1}{8}x - 3$ and $g(x) = x^3$ to find the indicated value or function.

79. $(f^{-1} \circ g^{-1})(1)$

80. $(g^{-1} \circ f^{-1})(-3)$

81. $(f^{-1} \circ f^{-1})(6)$

82. $(g^{-1} \circ g^{-1})(-4)$

83. $(f \circ g)^{-1}$

84. $g^{-1} \circ f^{-1}$

In Exercises 85–88, use the functions $f(x) = x + 4$ and $g(x) = 2x - 5$ to find the specified functions.

85. $g^{-1} \circ f^{-1}$

86. $f^{-1} \circ g^{-1}$

87. $(f \circ g)^{-1}$

88. $(g \circ f)^{-1}$

89. **Hourly Wage**   Your wage is $8.00 per hour plus $0.75 for each unit produced per hour. So, your hourly wage y in terms of the number of units produced is $y = 8 + 0.75x$.

(a) Determine the inverse of the function. What does each variable in the inverse function represent?

(b) Use a graphing utility to graph the function and its inverse.

(c) Use the *trace* feature of your graphing utility to find the hourly wage if 10 units are produced per hour.

(d) Use the *trace* feature of your graphing utility to find the number of units produced when your hourly wage is $22.25.

90. **Diesel Engine**   The function

$$y = 0.03x^2 + 254.50, \qquad 0 < x < 100$$

approximates the exhaust temperature y of a diesel engine in degrees Fahrenheit, where x is the percent load on the engine.

(a) Determine the inverse of the function. What does each variable in the inverse function represent?

(b) Use a graphing utility to graph the inverse function.

(c) Determine the percent load interval if the exhaust temperature of the engine must not exceed 500°F.

91. **Transportation**   The total value of new car sales f (in billions of dollars) in the United States from 1992 through 1997 is shown in the table. The time (in years) is given by t, with $t = 2$ corresponding to 1992. (Source: National Automobile Dealers Association)

t	2	3	4	5	6	7
$f(t)$	333.8	377.3	430.6	456.2	490.0	507.5

(a) Does f^{-1} exist?

(b) If f^{-1} exists, what does it mean in the context of the problem?

(c) If f^{-1} exists, find $f^{-1}(456.2)$.

(d) If the table above was extended to 1998 and if the total value of new car sales for that year was $430.6 billion, would f^{-1} exist? Explain.

Synthesis

True or False?   In Exercises 92 and 93, determine whether the statement is true or false. Justify your answer.

92. If f is an even function, f^{-1} exists.

93. If the inverse of f exists, the y-intercept of f is an x-intercept of f^{-1}.

94. Prove that if f and g are one-to-one functions, $(f \circ g)^{-1}(x) = (g^{-1} \circ f^{-1})(x)$.

95. Prove that if f is a one-to-one odd function, f^{-1} is an odd function.

96. ***Think About It***   The function

$$f(x) = k(2 - x - x^3)$$

is one-to-one and $f^{-1}(3) = -2$. Find k.

Review

In Exercises 97–100, evaluate the combination of functions when $f(x) = 2x^2 - 5$ and $g(x) = x - 3$ at the indicated value of x.

97. $(f + g)(-x)$ when $x = -2$

98. $(f - g)(x)$ when $x = 4$

99. $(fg)(-x)$ when $x = -3$

100. $\left(\dfrac{f}{g}\right)(x)$ when $x = \dfrac{3}{2}$

In Exercises 101–104, find the inverse of the function.

101. $y = 12x$

102. $y = 5x + 8$

103. $y = x^3 + 7$

104. $y = \sqrt[3]{x - 7}$

What did you learn?

Section 1.1	Review Exercises
☐ How to decide whether relations between two variables are functions	1–6
☐ How to use function notation and evaluate functions	7, 8
☐ How to find the domains of functions	9–14
☐ How to use functions to model and solve real-life problems	15, 16

Section 1.2	
☐ How to find the domains and ranges of functions	17–20
☐ How to use the Vertical Line Test for functions	21–24
☐ How to determine intervals on which functions are increasing or decreasing	25–28
☐ How to determine relative maximum and relative minimum values of functions	29–32
☐ How to graph step functions and other piecewise-defined functions	33, 34
☐ How to identify even and odd functions	35, 36

Section 1.3	
☐ How to recognize graphs of common functions	37, 38
☐ How to use vertical and horizontal shifts to sketch graphs of functions	39–46
☐ How to use reflections to sketch graphs of functions	47–50
☐ How to use nonrigid transformations to sketch graphs of functions	51–54

Section 1.4	
☐ How to add, subtract, multiply, and divide functions	55–58
☐ How to find compositions of one function with another function	59, 60
☐ How to use combinations of functions to model and solve real-life problems	61, 62

Section 1.5	
☐ How to find inverse functions informally and verify that two functions are inverses of each other	63–66
☐ How to use graphs of functions to decide whether functions have inverses	67–70
☐ How to find inverse functions algebraically	71–76

1 Review Exercises

1.1 **In Exercises 1 and 2, determine which of the sets of ordered pairs represents a function from A to B. Give reasons for your answers.**

1. $A = \{10, 20, 30, 40\}$ and $B = \{0, 2, 4, 6\}$
   (a) $\{(20, 4), (40, 0), (20, 6), (30, 2)\}$
   (b) $\{(10, 4), (20, 4), (30, 4), (40, 4)\}$
   (c) $\{(40, 0), (30, 2), (20, 4), (10, 6)\}$
   (d) $\{(20, 2), (10, 0), (40, 4)\}$

2. $A = \{u, v, w\}$ and $B = \{-2, -1, 0, 1, 2\}$
   (a) $\{(v, -1), (u, 2), (w, 0), (u, -2)\}$
   (b) $\{(u, -2), (v, 2), (w, 1)\}$
   (c) $\{(u, 2), (v, 2), (w, 1), (w, 1)\}$
   (d) $\{(w, -2), (v, 0), (w, 2)\}$

In Exercises 3–6, determine if the equation represents y as a function of x.

3. $16x - y^4 = 0$

4. $2x - y - 3 = 0$

5. $y = \sqrt{1 - x}$

6. $|y| = x + 2$

In Exercises 7 and 8, evaluate the function as indicated. Simplify your answers.

7. $f(x) = x^2 + 1$
   (a) $f(2)$
   (b) $f(-4)$
   (c) $f(t^2)$
   (d) $-f(x)$

8. $g(x) = x^{4/3}$
   (a) $g(8)$
   (b) $g(t + 1)$
   (c) $\dfrac{g(8) - g(1)}{8 - 1}$
   (d) $g(-x)$

In Exercises 9–14, determine the domain of the function. Verify your result with a graphing utility.

9. $f(x) = (x - 1)(x + 2)$

10. $f(x) = x^2 - 4x - 32$

11. $f(x) = \sqrt{25 - x^2}$

12. $f(x) = \sqrt{x^2 + 8x}$

13. $g(s) = \dfrac{5}{3s - 9}$

14. $f(x) = \dfrac{2}{3x + 4}$

15. ***Business*** A company produces a product for which the variable cost is $5.35 per unit and the fixed costs are $16,000. The company sells the product for $8.20 and can sell all that it produces.

(a) Find the total cost as a function of x, the number of units produced.

(b) Find the profit as a function of x.

16. ***Boating*** The retail expenditures B (in billions of dollars) on boating in the United States from 1985 to 1996 can be represented by the piecewise-defined function

$$B(t) = \begin{cases} -0.631t^2 - 2.845t + 14.160, & -5 \le t < 2 \\ 2.088t + 5.768, & 2 \le t \le 6 \end{cases}$$

where $t = 0$ represents 1990. Use a graphing utility to graph the model and find the amount spent on boating in 1985, 1990, and 1995. (Source: National Marine Manufacturers Association)

1.2 **In Exercises 17–20, find the domain and range of the function.**

17. $f(x) = 3 - 2x^2$

18. $f(x) = \sqrt{2x^2 - 1}$

19. $h(x) = \sqrt{36 - x^2}$

20. $g(x) = |x + 5|$

In Exercises 21–24, (a) use a graphing utility to graph the equation and (b) use the Vertical Line Test to determine whether y is a function of x.

21. $y = \dfrac{x^2 + 3x}{6}$

22. $y = -\frac{2}{3}|x + 5|$

23. $3x + y^2 = 2$

24. $x^2 + y^2 = 49$

In Exercises 25–28, determine the open intervals over which the function is increasing, decreasing, or constant.

25. $f(x) = x^3 - 3x$

26. $f(x) = \sqrt{x^2 - 9}$

27. $f(x) = x\sqrt{x - 6}$

28. $f(x) = \dfrac{|x + 8|}{2}$

Graphical Analysis **In Exercises 29–32, use a graphing utility to approximate (accurate to two decimal places) any relative maximum or minimum values of the function.**

29. $f(x) = (x^2 - 4)^2$

30. $f(x) = x^2 - x - 1$

31. $h(x) = 4x^3 - x^4$

32. $f(x) = x^3 - 4x^2 - 1$

In Exercises 33 and 34, sketch the graph of the piecewise-defined function by hand. Verify using a graphing utility.

33. $f(x) = \begin{cases} 3x + 5, & x < 0 \\ x - 4, & x \geq 0 \end{cases}$

34. $f(x) = \begin{cases} x^2 + 7, & x < 1 \\ x^2 - 5x + 6, & x \geq 1 \end{cases}$

In Exercises 35 and 36, determine whether the function is even, odd, or neither.

35. $f(x) = (x^2 - 8)^2$ **36.** $f(x) = 2x^3 - x^2$

1.3 **In Exercises 37 and 38, the graph is related to one of the common functions on page 100. Identify the common function and describe transformation(s) to the graph shown. Write the equation for the graphed function.**

37.   **38.**

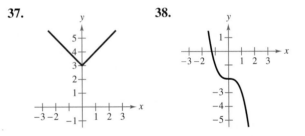

In Exercises 39–54, sketch the graph of the function by hand.

39. $f(x) = x^2 - 6$ **40.** $f(x) = (x - 3)^2 - 2$

41. $f(x) = (x - 1)^3 + 7$ **42.** $f(x) = (x + 2)^3 + 5$

43. $f(x) = \sqrt{x} - 5$ **44.** $f(x) = \sqrt{x + 5} + 4$

45. $f(x) = 7 + |x|$ **46.** $f(x) = |x + 8| - 1$

47. $f(x) = -x^2 - 3$ **48.** $f(x) = -(x + 5)^3 - 6$

49. $f(x) = -\sqrt{x + 4} - 3$ **50.** $f(x) = 1 - |-x|$

51. $f(x) = -2x^2 + 3$ **52.** $f(x) = -\frac{1}{4}(1 - x)^3 - 2$

53. $f(x) = -\frac{1}{2}|x| + 9$ **54.** $f(x) = 4\sqrt{x + 1} - 8$

1.4 **In Exercises 55–60, let** $f(x) = 3 - 2x$, $g(x) = \sqrt{x}$, **and** $h(x) = 3x^2 + 2$, **and find the indicated values.**

55. $(f - g)(4)$ **56.** $(f + h)(5)$

57. $(fh)(1)$ **58.** $\left(\dfrac{g}{h}\right)(1)$

59. $(h \circ g)(7)$ **60.** $(g \circ f)(-2)$

Data Analysis   **In Exercises 61 and 62, use the table, which shows the total value (in billions of dollars) of U.S. imports from China and Taiwan for the years 1992 through 1997. The variables y_1 and y_2 represent the total value of imports from China and Taiwan, respectively.** (Source: U.S. Bureau of the Census)

Year	1992	1993	1994	1995	1996	1997
y_1	25.7	31.5	38.8	45.5	51.5	62.6
y_2	24.6	25.1	26.7	29.0	29.9	32.6

61. Use the regression capabilities of a graphing utility to find quadratic models for each of the variables. Let $t = 2$ represent 1992.

62. Use a graphing utility to graph y_1, y_2, and $y_1 + y_2$ in the same viewing window. Use the model to estimate the total value of U.S. imports from China and Taiwan in 2002.

1.5 **In Exercises 63–66, find the inverse of f informally. Verify that $f(f^{-1}(x)) = f^{-1}(f(x)) = x$.**

63. $f(x) = 6x$ **64.** $f(x) = \frac{1}{12}x$

65. $f(x) = x - 7$ **66.** $f(x) = x + 5$

In Exercises 67–70, (a) find f^{-1}, (b) graph f and f^{-1} in the same viewing window, and (c) verify that $f^{-1}(f(x)) = x$ and $f(f^{-1}(x)) = x$.

67. $f(x) = \frac{1}{2}x - 3$ **68.** $f(x) = 5x - 7$

69. $f(x) = \sqrt{x + 1}$ **70.** $f(x) = x^3 + 2$

In Exercises 71–76, find the inverse of f algebraically.

71. $f(x) = \dfrac{x}{12}$ **72.** $f(x) = \dfrac{7x + 3}{8}$

73. $f(x) = 4x^3 - 3$ **74.** $f(x) = x^3 - 2$

75. $f(x) = \sqrt{x + 10}$ **76.** $f(x) = 4\sqrt{6 - x}$

Synthesis

True or False?   **In Exercises 77 and 78, determine whether the statement is true or false. Justify your answer.**

77. If $f(x) = x^n$ where n is odd, f^{-1} exists.

78. There exists no function f such that $f = f^{-1}$.

79. Explain why not all equations of lines are functions.

Chapter Project *Modeling the Area of a Plot*

Many real-life problems can be analyzed from a *graphical,* a *numerical,* and an *algebraic* perspective. In this project, you will use all three strategies to determine the maximum size of a rectangular plot that can be enclosed by a fixed amount of fencing.

You have 100 meters of fencing material to enclose a rectangular plot. Your goal is to determine the dimensions of the plot such that you enclose the maximum area possible.

a. Express the area $A(x)$ of the rectangular plot as a function of the length x of one side, as shown in the figure below.

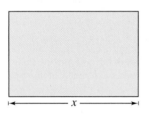

b. Analyze the problem *numerically* by completing the table.

x	0	5	10	15	20	25	30	35	40	45	50
$A(x)$											

According to this table, what do you think the dimensions of the plot should be to enclose the maximum area? Explain.

c. Use a graphing utility to graph the area function. What is the domain of the function? Solve the problem *graphically* by using the *trace* feature of your graphing utility to find the value of x that yields the maximum area.

d. Solve the problem *algebraically* by showing that the area function can be written as $A(x) = 625 - (x - 25)^2$. How does this form of the function allow you to find the dimensions that produce a maximum area?

e. Discuss the strengths and weaknesses of the three strategies used in parts (b), (c), and (d).

Questions for Further Exploration

1. Suppose you were not restricted to a rectangular plot. Would you be able to use 100 meters of fencing to enclose a greater area? Explain.

2. In the project above, you found the maximum area that can be enclosed in a rectangular plot using 100 meters of fencing. If you doubled the amount of fencing, could you enclose twice as much area? Use numerical, graphical, and algebraic approaches and explain your reasoning.

3. Suppose the rectangular plot runs along a building, so that you need to fence only three sides. What dimensions will now yield a maximum area with 100 meters of fencing?

1 Chapter Test

Take this test as you would take a test in class. After you are done, check your work against the answers in the back of the book.

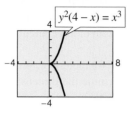

FIGURE FOR 1

1. The graph of $y^2(4 - x) = x^3$ is shown at the right. Does the graph represent y as a function of x? Explain.

In Exercises 2–5, use the function $f(x) = 10 - \sqrt{3 - x}$.

2. Evaluate: $f(-6)$

3. Simplify: $f(t - 3)$

4. Simplify: $\dfrac{f(x) - f(2)}{x - 2}$

5. Determine the domain of f.

The *Interactive* CD-ROM and *Internet* versions of this text provide answers to the Chapter Tests and Cumulative Tests. They also offer Chapter Pre-Tests (that test key skills and concepts covered in previous chapters) and Chapter Post-Tests, both of which have randomly generated exercises with diagnostic capabilities.

6. A company produces a product for which the variable cost is $5.60 and the fixed costs are $24,000. The product sells for $9.20. Write the total cost C as a function of x. Write the profit P as a function of x.

In Exercises 7–9, (a) use a graphing utility to graph the function; (b) approximate the open intervals over which the function is increasing, decreasing, or constant; and (c) determine whether the function is even, odd, or neither.

7. $f(x) = 2x^6 + 5x^4 - x^2$

8. $f(x) = 4x\sqrt{3 - x}$

9. $f(x) = |x + 5|$

10. Sketch the graph of $f(x) = \begin{cases} 3x + 7, & x \le -3 \\ 4x^2 - 1, & x > -3 \end{cases}$.

In Exercises 11 and 12, use a graphing utility to approximate (to two decimal places) any relative minimum or maximum function values.

11. $f(x) = -x^3 - 5x^2 + 12$

12. $f(x) = x^5 - x^3 + 2$

In Exercises 13–16, (a) identify the common function, (b) describe the transformation, and (c) sketch the graph of the function.

13. $f(x) = -2(x - 5)^3 + 3$

14. $f(x) = \sqrt{-x - 7} + 7$

15. $f(x) = 4|-x| - 7$

16. $f(x) = -(x - 3)^2 + 5$

17. Use the functions $f(x) = x^2$ and $g(x) = \sqrt{2 - x}$ to find the specified function and its domain.

(a) $(f - g)(x)$ (b) $\left(\dfrac{f}{g}\right)(x)$ (c) $(f \circ g)(x)$ (d) $g^{-1}(x)$

In Exercises 18–20, find the inverse function, if possible.

18. $f(x) = x^3 + 8$

19. $f(x) = |x^2 - 3| + 6$

20. $f(x) = \dfrac{3x\sqrt{x}}{8}$

21. A wire 100 inches long is to be cut into four pieces to form a rectangle whose shortest side has a length of x. Express the area A of the rectangle as a function of x. Determine the domain of the function and graph the function over that domain. Use the graph to approximate the maximum area of the rectangle.

Polynomial and
Rational Functions

The Big Picture

In this chapter you will learn how to

- ❑ sketch and analyze graphs of quadratic and polynomial functions.
- ❑ use long division and synthetic division to divide polynomials by other polynomials.
- ❑ determine the number of rational and real zeros of polynomial functions, and find them.
- ❑ perform operations with complex numbers and plot complex numbers in the complex plane.
- ❑ determine the domain, find asymptotes, and sketch the graphs of rational functions.

U.S. wheat production increased from 2183 million bushels in 1995 to 2527 million bushels in 1997, while the price per bushel dropped from $4.55 to $3.45.   (Source: U.S. Department of Agriculture)

Important Vocabulary

As you encounter each new vocabulary term in this chapter, add the term and its definition to your notebook glossary.

- polynomial function (p. 136)
- constant function (p. 136)
- linear function (p. 136)
- quadratic function (p. 136)
- parabola (p. 136)
- axis of symmetry (p. 137)
- vertex (p. 137)
- continuous (p. 147)
- Leading Coefficient Test (p. 149)
- extrema (p. 150)
- repeated zero (p. 152)
- multiplicity (p. 152)

- Intermediate Value Theorem (p. 154)
- long division of polynomials (p. 160)
- Division Algorithm (p. 161)
- synthetic division (p. 163)
- Remainder Theorem (p. 164)
- Factor Theorem (p. 164)
- Rational Zero Test (p. 166)
- upper bound (p. 168)
- lower bound (p. 168)
- imaginary unit i (p. 174)
- complex number (p. 174)
- complex conjugates (p. 177)

- complex plane (p. 178)
- imaginary axis (p. 178)
- real axis (p. 178)
- Fundamental Theorem of Algebra (p. 182)
- Linear Factorization Theorem (p. 182)
- rational function (p. 189)
- vertical asymptote (p. 190)
- horizontal asymptote (p. 190)
- slant (or oblique) asymptote (p. 202)

Additional Resources   Text-specific additional resources are available to help you do well in this course. See page xvi for details.

135

2.1 Quadratic Functions

The Graph of a Quadratic Function

In this and the next section, you will study the graphs of polynomial functions.

Definition of Polynomial Function

Let n be a nonnegative integer and let $a_n, a_{n-1}, \ldots, a_2, a_1, a_0$ be real numbers with $a_n \neq 0$. The function

$$f(x) = a_n x^n + a_{n-1} x^{n-1} + \cdots + a_2 x^2 + a_1 x + a_0$$

is called a **polynomial function of x with degree n.**

Polynomial functions are classified by degree. For instance, the polynomial function

$$f(x) = a, \quad a \neq 0 \qquad \text{Constant function}$$

has degree 0 and is called a **constant function.** In Chapter P, you learned that the graph of this type of function is a horizontal line. The polynomial function

$$f(x) = mx + b, \quad m \neq 0 \qquad \text{Linear function}$$

has degree 1 and is called a **linear function.** You also learned in Chapter P that the graph of the linear function $f(x) = mx + b$ is a line whose slope is m and whose y-intercept is $(0, b)$. In this section you will study second-degree polynomial functions, which are called **quadratic functions.**

Definition of Quadratic Function

Let a, b, and c be real numbers with $a \neq 0$. The function

$$f(x) = ax^2 + bx + c \qquad \text{Quadratic function}$$

is called a **quadratic function.**

Often real-life data can be modeled by quadratic functions. For instance, the table shows height h (in feet) of a projectile fired from a height of 6 feet with an initial velocity of 256 feet per second at any time t (in seconds). A quadratic model for the data in the table is $h(t) = -16t^2 + 256t + 6$ for $0 \leq t \leq 16$.

t	0	2	4	6	8	10	12	14	16
h	6	454	774	966	1030	966	774	454	6

The graph of a quadratic function is a special type of U-shaped curve called a **parabola.** Parabolas occur in many real-life applications—especially those involving reflective properties, such as satellite dishes or flashlight reflectors. You will study these properties in Section 10.1.

What You Should Learn:

- How to analyze graphs of quadratic functions
- How to write quadratic functions in standard form and use the results to sketch graphs of functions
- How to use quadratic functions to model and solve real-life problems

Why You Should Learn It:

Quadratic functions can be used to model data to analyze consumer behavior. For instance, Exercise 78 on page 146 shows how a quadratic function can model VCR usage in the United States.

Mary K. Kenny/PhotoEdit

All parabolas are symmetric with respect to a line called the **axis of symmetry,** or simply the **axis** of the parabola. The point where the axis intersects the parabola is the **vertex** of the parabola, as shown in Figure 2.1. If the leading coefficient a is positive, the graph of $f(x) = ax^2 + bx + c$ is a parabola that opens upward; and if the leading coefficient a is negative, the graph of $f(x) = ax^2 + bx + c$ is a parabola that opens downward.

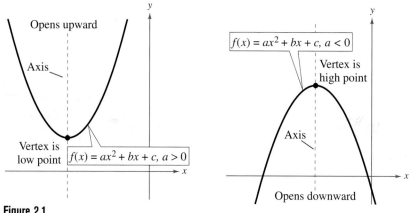

Figure 2.1

The simplest type of quadratic function is

$$f(x) = ax^2.$$

Its graph is a parabola whose vertex is $(0, 0)$. If $a > 0$, the vertex is the *minimum* point on the graph; and if $a < 0$, the vertex is the *maximum* point on the graph, as shown in Figure 2.2.

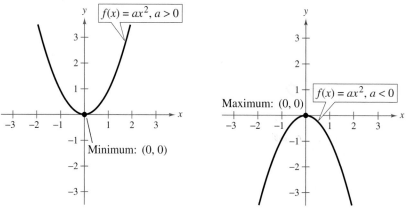

Figure 2.2

When sketching the graph of $f(x) = ax^2$, it is helpful to use the graph of $y = x^2$ as a reference, as discussed in Section 1.3. There you saw that when $a > 1$, the graph of $y = af(x)$ is a vertical stretch of the graph of $y = f(x)$. When $0 < a < 1$, the graph of $y = af(x)$ is a vertical shrink of the graph of $y = f(x)$. This is demonstrated again in Example 1.

Exploration

Use a graphing utility to graph the parabola

$$y = ax^2$$

for $a = -3, -2, -1, 1, 2,$ and 3. What can you conclude about the parabola when $a < 0$? When $a > 0$?

Library of Functions

The graph of a quadratic function is called a parabola. Graph $y_1 = x^2$ and $y_2 = |x|$ in the same viewing window. Zoom in near the origin and compare the shape of the two graphs. Which graph grows faster as x gets larger and larger? Why does the definition of a quadratic function require that $a \neq 0$?

Consult the Library of Functions Summary inside the front cover for a description of the quadratic function.

EXAMPLE 1 Graphing Simple Quadratic Functions

Describe how the graph of each function is related to the graph of $y = x^2$.

a. $f(x) = \dfrac{1}{3}x^2$ **b.** $g(x) = 2x^2$

c. $h(x) = -x^2 + 1$ **d.** $k(x) = (x + 2)^2 - 3$

Solution

a. Compared with $y = x^2$, each output of f "shrinks" by a factor of $\frac{1}{3}$. The result is a parabola that opens upward and is broader than the parabola represented by $y = x^2$, as shown in Figure 2.3(a).

b. Compared with $y = x^2$, each output of g "stretches" by a factor of 2, creating a narrower parabola, as shown in Figure 2.3(b).

c. With respect to the graph of $y = x^2$, the negative coefficient in $h(x) = -x^2 + 1$ reflects the graph *downward* and the positive constant term shifts the vertex *up* one unit. The graph of h is shown in Figure 2.3(c).

d. With respect to the graph of $y = x^2$, the graph of $k(x) = (x + 2)^2 - 3$ is obtained by a horizontal shift two units *to the left* and a vertical shift three units *down*, as shown in Figure 2.3(d).

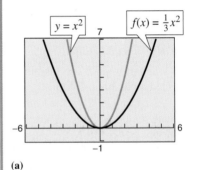

(a)

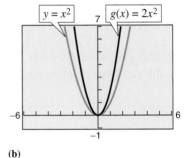

(b)

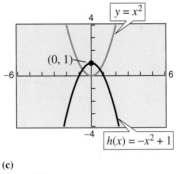

(c)

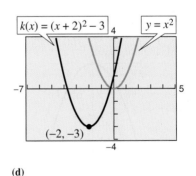

(d)

Figure 2.3

> ### STUDY T!P
>
> In Example 1, note that the coefficient a determines how widely the parabola given by $f(x) = ax^2$ opens. If $|a|$ is small, the parabola opens more widely than if $|a|$ is large.

The *Interactive* CD-ROM and *Internet* versions of this text show every example with its solution; clicking on the *Try It!* button brings up similar problems. Guided Examples and Integrated Examples show step-by-step solutions to additional examples. Integrated Examples are related to several concepts in the section.

Recall from Section 1.3 that the graphs of $y = f(x \pm c)$, $y = f(x) \pm c$, $y = -f(x)$, and $y = f(-x)$ are rigid transformations of the graph of $y = f(x)$.

$y = f(x \pm c)$	Horizontal shift	$y = -f(x)$	Reflection in x-axis
$y = f(x) \pm c$	Vertical shift	$y = f(-x)$	Reflection in y-axis

The Standard Form of a Quadratic Function

The equation in Example 1(d) is written in the **standard form**

$$f(x) = a(x - h)^2 + k.$$

This form is especially convenient for sketching a parabola because it identifies the vertex of the parabola as (h, k).

Standard Form of a Quadratic Function

The quadratic function

$$f(x) = a(x - h)^2 + k, \qquad a \neq 0$$

is said to be in **standard form.** The graph of f is a parabola whose axis is the vertical line $x = h$ and whose vertex is the point (h, k). If $a > 0$, the parabola opens upward, and if $a < 0$, the parabola opens downward.

EXAMPLE 2   Identifying the Vertex of a Quadratic Function

Describe the graph of

$$f(x) = 2x^2 + 8x + 7$$

and identify the vertex.

Solution

Write the quadratic function in standard form by completing the square. Recall that the first step is to factor out any coefficient of x^2 that is different from 1.

$f(x) = 2x^2 + 8x + 7$	Write original function.
$= 2(x^2 + 4x) + 7$	Factor 2 out of x-terms.
$= 2(x^2 + 4x + 4 - 4) + 7$	Because $b = 4$, add and subtract $(b/2)^2 = 4$ within parentheses.

$$\left(\frac{b}{2}\right)^2$$

$= 2(x^2 + 4x + 4) - 2(4) + 7$	Regroup terms.
$= 2(x + 2)^2 - 1$	Write in standard form.

From the standard form, you can see that the graph of f is a parabola that opens upward with vertex $(-2, -1)$, as shown in Figure 2.4.

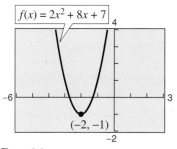

$f(x) = 2x^2 + 8x + 7$

$(-2, -1)$

Figure 2.4

To find the x-intercepts of the graph of $f(x) = ax^2 + bx + c$, solve the equation $ax^2 + bx + c = 0$. If $ax^2 + bx + c$ does not factor, you can use the Quadratic Formula to find the x-intercepts, or a graphing utility to approximate the x-intercepts. Remember, however, that a parabola may have no x-intercept.

EXAMPLE 3   Writing a Quadratic Function in Standard Form

Describe the graph of $f(x) = -x^2 + 6x - 8$. Identify any x-intercepts.

Solution

$$f(x) = -x^2 + 6x - 8 \qquad \text{Write original function.}$$

$$= -(x^2 - 6x) - 8 \qquad \text{Factor } -1 \text{ out of } x\text{-terms.}$$

$$= -(x^2 - 6x + 9 - 9) - 8 \qquad \begin{array}{l}\text{Because } b = 6, \text{ add and subtract} \\ (b/2)^2 = 9 \text{ within parentheses.}\end{array}$$

$$\left(\frac{b}{2}\right)^2$$

$$= -(x^2 - 6x + 9) - (-9) - 8 \qquad \text{Regroup terms.}$$

$$= -(x - 3)^2 + 1 \qquad \text{Write in standard form.}$$

The graph of f is a parabola that opens downward with vertex at $(3, 1)$, as shown in Figure 2.5. The x-intercepts are determined as follows.

$$-x^2 + 6x - 8 = 0$$

$$x^2 - 6x + 8 = 0$$

$$(x - 2)(x - 4) = 0$$

$$x = 2, x = 4$$

So, the x-intercepts are $(2, 0)$ and $(4, 0)$, as shown in Figure 2.5.

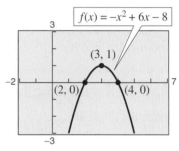

Figure 2.5

EXAMPLE 4   Finding the Equation of a Parabola in Standard Form

Find the standard form of the equation for the parabola that has its vertex at $(1, 2)$ and passes through the point $(3, -6)$, as shown in Figure 2.6.

Solution

Because the parabola has a vertex at $(h, k) = (1, 2)$, the equation has the form

$$f(x) = a(x - 1)^2 + 2. \qquad \text{Standard form}$$

Because the parabola passes through the point $(3, -6)$, it follows that $f(3) = -6$. So, you obtain

$$-6 = a(3 - 1)^2 + 2$$

$$-6 = 4a + 2$$

$$-2 = a.$$

The equation in standard form is $f(x) = -2(x - 1)^2 + 2$. Try graphing $f(x) = -2(x - 1)^2 + 2$ with a graphing utility to confirm that its vertex is $(1, 2)$ and that it passes through the point $(3, -6)$.

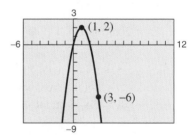

Figure 2.6

STUDY TIP

In Example 4, there are infinitely many different parabolas that have a vertex at $(1, 2)$. Of these, however, the only one that passes through the point $(3, -6)$ is the one given by

$$f(x) = -2(x - 1)^2 + 2.$$

Applications

Many applications involve finding the maximum or minimum value of a quadratic function. By writing the quadratic function $f(x) = ax^2 + bx + c$ in standard form,

$$f(x) = a\left(x + \frac{b}{2a}\right)^2 + \left(c - \frac{b^2}{4a}\right)$$

you can see that the vertex occurs at $x = -b/(2a)$, which implies the following.

1. If $a > 0$, f has a *minimum* that occurs at

$$x = -\frac{b}{2a}.$$

2. If $a < 0$, f has a *maximum* that occurs at

$$x = -\frac{b}{2a}.$$

EXAMPLE 5 The Maximum Height of a Baseball

A baseball is hit 3 feet above ground at a velocity of 100 feet per second and at an angle of 45 degrees with respect to level ground. The path of the baseball is given by the function $f(x) = -0.0032x^2 + x + 3$, where $f(x)$ is the height of the baseball (in feet) and x is the horizontal distance from home plate (in feet). What is the maximum height reached by the baseball? (See Example 7 in Section 1.1.)

A computer simulation of this example appears in the *Interactive* CD-ROM and *Internet* versions of this text.

Algebraic Solution

For this quadratic function, you have

$$f(x) = ax^2 + bx + c$$

$$= -0.0032x^2 + x + 3$$

which implies that $a = -0.0032$ and $b = 1$. Because the function has a maximum when $x = -b/(2a)$, you can conclude that the baseball reaches its maximum height when

$$x = -\frac{b}{2a}$$

$$= -\frac{1}{2(-0.0032)}$$

$$= 156.25 \text{ feet}$$

from home plate. At this distance, the maximum height is

$$f(156.25) = -0.0032(156.25)^2 + 156.25 + 3$$

$$= 81.125 \text{ feet}.$$

Graphical Solution

Use a graphing utility to graph $y = -0.0032x^2 + x + 3$ so that you can see the important features of the parabola. In Figure 2.7 the graph appears to have a maximum near $x = 150$. Use the *maximum* feature of the graphing utility or the *zoom* and *trace* features to approximate the maximum height on the graph to be $y \approx 81.125$ feet at $x \approx 156.25$. Note that when using the *zoom* and *trace* features of a graphing utility, you might have to change the y-scale in order to avoid a graph that is "too flat."

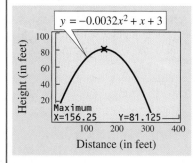

Figure 2.7

EXAMPLE 6   Charitable Contributions

According to a survey conducted by *Independent Sector*, the percent of their income that Americans give to charities is related to their household income. For families with an annual income of \$100,000 or less, the percent is approximately

$$P = 0.0014x^2 - 0.1529x + 5.855, \qquad 5 \le x \le 100$$

where P is the percent of annual income given and x is the annual income (in thousands of dollars). According to this model, what income level corresponds to the minimum percent of charitable contributions?

Algebraic Solution

Use the fact that the minimum point of the parabola occurs when $x = -b/(2a)$. For this function, you have $a = 0.0014$ and $b = -0.1529$. So,

$$x = -\frac{b}{2a}$$

$$= -\frac{-0.1529}{2(0.0014)}$$

$$\approx 54.6.$$

From this x-value, you can conclude that the minimum percent corresponds to an income level of about \$54,600.

Graphical Solution

Use a graphing utility to graph

$$y_1 = 0.0014x^2 - 0.1529x + 5.855$$

for $5 \le x \le 100$, as shown in Figure 2.8. The graph appears to have a minimum near $x = 55$. Use the *minimum* feature of the graphing utility or the *zoom* and *trace* features to approximate the minimum point of the parabola to be $x \approx 54.6$. So, the minimum point corresponds to an income level of about \$54,600.

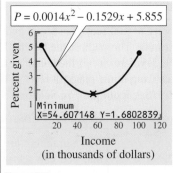

Figure 2.8

Writing About Math   *Finding an Equation for a Curve*

The parabola in the figure below has an equation of the form

$$y = ax^2 + bx - 4.$$

Find the equation for this parabola in two different ways, by hand and with technology (graphing utility or computer software). Write a paragraph describing the methods you used and comparing the results of the two methods.

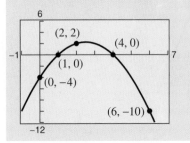

2.1   E x e r c i s e s

In Exercises 1–8, match the quadratic function with the correct graph. [The graphs are labeled (a), (b), (c), (d), (e), (f), (g), and (h).]

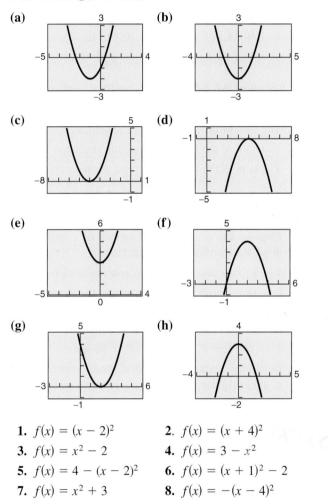

(a)

(b)

(c)

(d)

(e)

(f)

(g)

(h)

1. $f(x) = (x - 2)^2$

2. $f(x) = (x + 4)^2$

3. $f(x) = x^2 - 2$

4. $f(x) = 3 - x^2$

5. $f(x) = 4 - (x - 2)^2$

6. $f(x) = (x + 1)^2 - 2$

7. $f(x) = x^2 + 3$

8. $f(x) = -(x - 4)^2$

Exploration   **In Exercises 9–12, use a graphing utility to graph each equation. Describe how the graph of each equation is related to the graph of $y = x^2$.**

9. (a) $y = \frac{1}{2}x^2$ (b) $y = -\frac{1}{8}x^2$

     (c) $y = \frac{3}{2}x^2$ (d) $y = -3x^2$

10. (a) $y = x^2 + 1$ (b) $y = x^2 - 1$

     (c) $y = x^2 + 3$ (d) $y = x^2 - 3$

11. (a) $y = (x - 1)^2$ (b) $y = (x + 1)^2$

     (c) $y = (x - 3)^2$ (d) $y = (x + 3)^2$

12. (a) $y = -\frac{1}{2}(x - 2)^2 + 1$ (b) $y = \frac{1}{2}(x - 2)^2 + 1$

     (c) $y = -\frac{1}{2}(x + 2)^2 - 1$ (d) $y = \frac{1}{2}(x + 2)^2 - 1$

In Exercises 13–26, sketch the graph of the quadratic function. Identify the vertex and intercepts. Use a graphing utility to verify your results.

13. $f(x) = 25 - x^2$ **14.** $f(x) = x^2 - 7$

15. $f(x) = \frac{1}{2}x^2 - 4$ **16.** $f(x) = 16 - \frac{1}{4}x^2$

17. $f(x) = (x + 4)^2 - 3$ **18.** $f(x) = (x - 6)^2 + 3$

19. $h(x) = x^2 - 8x + 16$ **20.** $g(x) = x^2 + 2x + 1$

21. $f(x) = x^2 - x + \frac{5}{4}$ **22.** $f(x) = x^2 + 3x + \frac{1}{4}$

23. $f(x) = -x^2 + 2x + 5$ **24.** $f(x) = -x^2 - 4x + 1$

25. $h(x) = 4x^2 - 4x + 21$ **26.** $f(x) = 2x^2 - x + 1$

In Exercises 27–34, use a graphing utility to graph the quadratic function. Identify the vertex and intercepts. Then check your results algebraically by completing the square.

27. $f(x) = -(x^2 + 2x - 3)$

28. $f(x) = -(x^2 + x - 30)$

29. $g(x) = x^2 + 8x + 11$ **30.** $f(x) = x^2 + 10x + 14$

31. $f(x) = 2x^2 - 16x + 31$

32. $f(x) = -4x^2 + 24x - 41$

33. $g(x) = \frac{1}{2}(x^2 + 4x - 2)$ **34.** $f(x) = \frac{3}{5}(x^2 + 6x - 5)$

In Exercises 35–38, find an equation for the parabola. Use a graphing utility to graph the equation and verify your result.

35.

36.

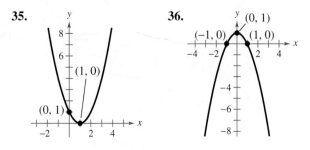

37. 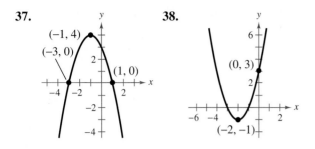 **38.**

In Exercises 39–46, find the quadratic function that has the indicated vertex and whose graph passes through the given point. Confirm your result with a graphing utility.

39. Vertex: $(-2, 5)$;   Point: $(0, 9)$

40. Vertex: $(4, -1)$;   Point: $(2, 3)$

41. Vertex: $(3, 4)$;   Point: $(1, 2)$

42. Vertex: $(2, 3)$;   Point: $(0, 2)$

43. Vertex: $(-2, -2)$;   Point: $(-1, 0)$

44. Vertex: $\left(-\frac{1}{4}, \frac{3}{2}\right)$;   Point: $(-2, 0)$

45. Vertex: $\left(\frac{5}{2}, -\frac{3}{4}\right)$;   Point: $(-2, 4)$

46. Vertex: $\left(-\frac{5}{2}, 0\right)$;   Point: $\left(-\frac{7}{2}, -\frac{16}{3}\right)$

Graphical Reasoning   In Exercises 47–50, determine the x-intercepts of the graph visually. How do the x-intercepts correspond to the solutions of the quadratic equation when $y = 0$?

47.   **48.**

49.   **50.**

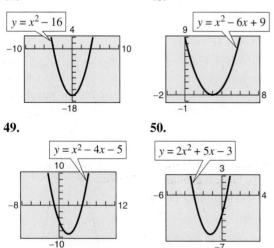

In Exercises 51–56, use a graphing utility to graph the quadratic function. Find the x-intercepts of the graph and compare them with the solutions of the corresponding quadratic equation when $y = 0$.

51. $y = x^2 - 4x$

52. $y = -2x^2 + 10x$

53. $y = 2x^2 - 7x - 30$

54. $y = 4x^2 + 25x - 21$

55. $y = -\frac{1}{2}(x^2 - 6x - 7)$

56. $y = \frac{7}{10}(x^2 + 12x - 45)$

In Exercises 57–60, find two quadratic functions, one that opens upward and one that opens downward, whose graphs have the given x-intercepts. (There are many correct answers.)

57. $(-1, 0), (3, 0)$

58. $(0, 0), (10, 0)$

59. $(-3, 0), \left(-\frac{1}{2}, 0\right)$

60. $\left(-\frac{5}{2}, 0\right), (2, 0)$

In Exercises 61–64, find two positive real numbers whose product is a maximum.

61. The sum is 110.

62. The sum is S.

63. The sum of the first and twice the second is 24.

64. The sum of the first and 3 times the second is 42.

Maximum Area   In Exercises 65 and 66, consider a rectangle of length x and perimeter P.

(a) Express the area A as a function of x and determine the domain of the function.

(b) Use a graphing utility to graph the area function.

(c) Use the graph to approximate the length and width of the rectangle of maximum area, and verify algebraically.

65. $P = 100$ feet

66. $P = 36$ meters

67. *Geometry*   An indoor physical fitness room consists of a rectangular region with a semicircle on each end. The perimeter of the room is to be a 200-meter running track.

(a) Draw a diagram to represent the problem. Let x and y represent the length and width of the rectangular region.

(b) Determine the radius of the semicircular ends of the track. Determine the distance, in terms of y, around the two semicircular parts of the track.

(c) Use the result of part (b) to write an equation, in terms of x and y, for the distance traveled in one lap around the track. Solve for y.

(d) Use the result of part (c) to write the area A of the rectangular region as a function of x.

(e) Use a graphing utility to graph the area function of part (d). Use the graph to approximate the dimensions that will produce a maximum area of the rectangle.

68. *Numerical, Graphical, and Analytical Analysis* A rancher has 200 feet of fencing to enclose two adjacent rectangular corrals. Use the following methods to determine the dimensions that will produce a maximum enclosed area.

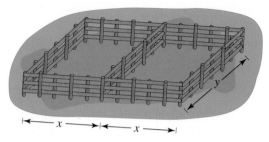

(a) Complete six rows of a table such as the one below. (The first two rows are shown.)

x	y	Area
2	$\frac{1}{3}[200 - 4(2)]$	$2xy = 256$
4	$\frac{1}{3}[200 - 4(4)]$	$2xy \approx 491$

(b) Use a graphing utility to generate additional rows of the table in part (a). Use the table to estimate the dimensions that will produce the maximum enclosed area.

(c) Write the area A as a function of x.

(d) Use a graphing utility to graph the area function. Use the graph to approximate the dimensions that will produce the maximum enclosed area.

(e) Write the area function in standard form to find algebraically the dimensions that will produce the maximum area.

(f) Compare your results from parts (b), (d), and (e).

69. *Business* A manufacturer of lighting fixtures has daily production costs of

$$C = 800 - 10x + 0.25x^2$$

where C is the total cost (in dollars) and x is the number of units produced. How many fixtures should be produced each day to yield a minimum cost?

70. *Business* A textile manufacturer has daily production costs of

$$C = 10,000 - 110x + 0.45x^2$$

where C is the total cost (in dollars) and x is the number of units produced. How many units should be produced each day to yield a minimum cost?

71. *Business* The profit P (in dollars) for a company is

$$P = -0.0002x^2 + 140x - 250,000$$

where x is the number of units sold. What sales level will yield maximum profit?

72. *Business* The profit P (in hundreds of dollars) that a company makes depends on the amount x (in hundreds of dollars) the company spends on advertising according to the model

$$P = 230 + 20x - 0.5x^2.$$

What expenditure for advertising results in the maximum profit?

73. *Trajectory of a Ball* The height y (in feet) of a ball thrown by a child is

$$y = -\frac{1}{12}x^2 + 2x + 4$$

where x is the horizontal distance (in feet) from where the ball is thrown.

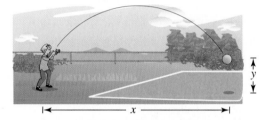

(a) Use a graphing utility to graph the path of the ball.

(b) How high is the ball when it leaves the child's hand? (*Hint:* Find y when $x = 0$.)

(c) How high is the ball when it is at its maximum height?

(d) How far from the child does the ball strike the ground?

74. *Maximum Height of a Dive* The path of a diver is

$$y = -\frac{4}{9}x^2 + \frac{24}{9}x + 12$$

where y is the height (in feet) and x is the horizontal distance (in feet) from the end of the diving board. What is the maximum height of the dive? Verify your answer using a graphing utility.

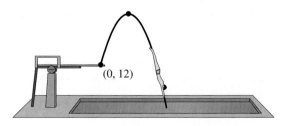

75. *Forestry* The number of board feet V in a 16-foot log is approximated by the model

$$V = 0.77x^2 - 1.32x - 9.31, \quad 5 \le x \le 40$$

where x is the diameter (in inches) of the log at the small end. (One board foot is a measure of volume equivalent to a board that is 12 inches wide, 12 inches long, and 1 inch thick.)

(a) Use a graphing utility to graph the function.

(b) Estimate the number of board feet in a 16-foot log with a diameter of 16 inches. Use a graphing utility to verify your answer.

(c) Estimate the diameter of a 16-foot log that scaled 500 board feet when the lumber was sold. Use a graphing utility to verify your answer.

76. *Automobile Aerodynamics* The number of horsepower y required to overcome wind drag on a certain automobile is approximated by

$$y = 0.002s^2 + 0.005s - 0.029, \quad 0 \le s \le 100$$

where s is the speed of the car in miles per hour.

(a) Use a graphing utility to graph the function.

(b) Graphically estimate the maximum speed of the car if the power required to overcome wind drag is not to exceed 10 horsepower. Verify your result algebraically.

77. *Graphical Analysis* For certain years from 1950 to 1990, the average annual per capita consumption C of cigarettes by Americans (18 and older) can be modeled by $C = 3248.89 + 108.64t - 2.97t^2$ for $0 \le t \le 40$, where t is the year, with $t = 0$ corresponding to 1950. (Source: U.S. Department of Agriculture)

(a) Use a graphing utility to graph the model.

(b) Use the graph of the model to approximate the maximum average annual consumption. Beginning in 1966, all cigarette packages were required by law to carry a health warning. Do you think the warning had any effect? Explain.

(c) In 1960, the U.S. population (18 and over) was 116,530,000. Of those, about 48,500,000 were smokers. What was the average annual cigarette consumption *per smoker* in 1960? What was the average daily cigarette consumption *per smoker*?

78. *Data Analysis* The number y (in millions) of VCRs in use in the United States for the years 1987 through 1996 are shown in the table. The variable t represents time (in years), with $t = 7$ corresponding to 1987.

t	7	8	9	10	11	12	13	14	15	16
y	43	51	58	63	67	69	72	74	77	79

(Source: Television Bureau of Advertising, Inc.)

(a) Use a graphing utility to sketch a scatter plot of the data.

(b) Use the regression capabilities of a graphing utility to fit a quadratic model to the data.

(c) Use a graphing utility to graph the model in the same viewing window as the scatter plot.

(d) Do you think the model can be used to estimate VCR utilization in the year 2005? Explain.

Synthesis

True or False? **In Exercises 79 and 80, determine whether the statement is true or false. Justify your answer.**

79. The function $f(x) = -12x^2 - 1$ has no x-intercepts.

80. The graphs of $f(x) = -4x^2 - 10x + 7$ and $g(x) = 12x^2 + 30x + 1$ have the same axis of symmetry.

81. *Think About It* The profits P (in millions of dollars) for a company are modeled by a quadratic function of the form $P = at^2 + bt + c$, where t represents the year. If you were president of the company, which of the models would you prefer? Explain your reasoning.

(a) a is positive and $t \ge -b/(2a)$.

(b) a is positive and $t \le -b/(2a)$.

(c) a is negative and $t \ge -b/(2a)$.

(d) a is negative and $t \le -b/(2a)$.

Review

In Exercises 82–85, determine algebraically any points of intersection of the graphs of the equations. Verify your results using the *intersect* feature of a graphing utility.

82. $x + y = 8$
$-\frac{2}{3}x + y = 6$

83. $y = 3x - 10$
$y = \frac{1}{4}x + 1$

84. $y = 9 - x^2$
$y = x + 3$

85. $y = x^3 + 2x - 1$
$y = -2x + 15$

In Exercises 86–89, decide whether the equation represents y as a function of x.

86. $3x - 4y = 12$

87. $y^2 = x^2 - 9$

88. $y = \sqrt{x + 3}$

89. $x^2 + y^2 - 6x + 8y = 0$

2.2   Polynomial Functions of Higher Degree

Graphs of Polynomial Functions

You should be able to sketch accurate graphs of polynomial functions of degrees 0, 1, and 2. The graphs of polynomial functions of degree greater than 2 are more difficult to sketch by hand. However, in this section you will learn how to recognize some of the basic features of the graphs of polynomial functions.

The graph of a polynomial function is **continuous.** Essentially, this means that the graph of a polynomial function has no breaks, holes, or gaps, as shown in Figure 2.9. Another feature of the graph of a polynomial function is that it has only smooth, rounded turns, as shown in Figure 2.10(a). It cannot have a sharp, pointed turn such as the one shown in Figure 2.10(b).

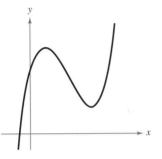

(a) Polynomial functions have continuous graphs.

Figure 2.9

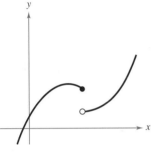

(b) Functions with graphs that are not continuous are not polynomial functions.

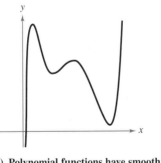

(a) Polynomial functions have smooth, rounded graphs.

Figure 2.10

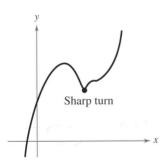

Sharp turn

(b) Graphs of polynomial functions cannot have sharp turns.

Informally, you can say that a function is continuous if its graph can be drawn with a pencil without lifting the pencil from the paper.

What You Should Learn:

- How to use transformations to sketch graphs of polynomial functions
- How to use the Leading Coefficient Test to determine the end behavior of graphs of polynomial functions
- How to find and use zeros of polynomial functions to sketch their graphs
- How to use the Intermediate Value Theorem to help locate zeros of polynomial functions

Why You Should Learn It:

You can use polynomial functions to model various aspects of nature, such as the growth of a red oak tree, as shown in Exercise 98 on page 158.

The Granger Collection

The polynomial functions that have the simplest graphs are monomials of the form $f(x) = x^n$, where n is an integer greater than zero.

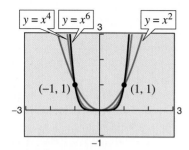

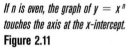

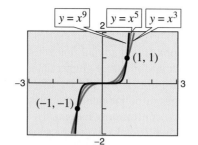

If n is even, the graph of y = xⁿ touches the axis at the x-intercept.

Figure 2.11

If n is odd, the graph of y = xⁿ crosses the axis at the x-intercept.

In Figure 2.11, you can see that when n is *even* the graph is similar to the graph of $f(x) = x^2$, and when n is *odd* the graph is similar to the graph of $f(x) = x^3$. Moreover, the greater the value of n, the flatter the graph is on the interval $(-1, 1)$.

EXAMPLE 1   Sketching Transformations of Polynomial Functions

Sketch the graph of each polynomial function.

a. $f(x) = -x^5$ **b.** $g(x) = x^4 + 1$ **c.** $h(x) = (x + 1)^4$

Solution

a. Because the degree of $f(x) = -x^5$ is odd, the graph is similar to the graph of $y = x^3$. Moreover, the negative coefficient reflects the graph in the x-axis, as shown in Figure 2.12(a).

b. The graph of $g(x) = x^4 + 1$ is an upward shift, by one unit, of the graph of $y = x^4$, as shown in Figure 2.12(b).

c. The graph of $h(x) = (x + 1)^4$ is a left shift, by one unit, of the graph of $y = x^4$, as shown in Figure 2.12(c).

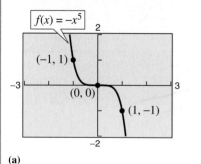

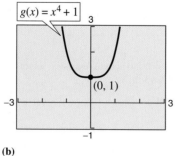

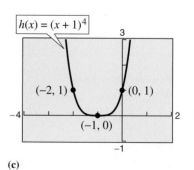

(a)

Figure 2.12

(b)

(c)

The *Interactive* CD-ROM and *Internet* versions of this text offer a built-in graphing calculator, which can be used with the Examples, Explorations, and Exercises.

The Leading Coefficient Test

In Example 1, note that all three graphs eventually rise or fall without bound as x moves to the right. Whether the graph of a polynomial eventually rises or falls can be determined by the function's degree (even or odd) and by its leading coefficient, as indicated in the **Leading Coefficient Test.**

Leading Coefficient Test

As x moves without bound to the left or to the right, the graph of the polynomial function $f(x) = a_n x^n + \cdots + a_1 x + a_0$ eventually rises or falls in the following manner.

1. When n is odd:

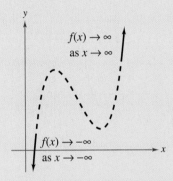 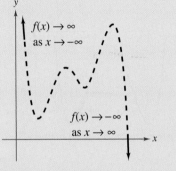

If the leading coefficient is positive ($a_n > 0$), the graph falls to the left and rises to the right.

If the leading coefficient is negative ($a_n < 0$), the graph rises to the left and falls to the right.

2. When n is even:

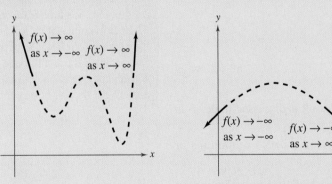

If the leading coefficient is positive ($a_n > 0$), the graph rises to the left and right.

If the leading coefficient is negative ($a_n < 0$), the graph falls to the left and right.

Note that the dashed portions of the graphs indicate that the test determines *only* the right and left behavior of the graph.

Library of Functions

The graphs of polynomials of degree 1 are lines, and those of degree 2 are parabolas. The graphs of polynomials of higher degree are smooth and continuous. The graphs eventually rise or fall without bound as x moves to the right (or left).

Consult the Library of Functions Summary inside the front cover for a description of the polynomial function.

EXAMPLE 2 Applying the Leading Coefficient Test

Use the Leading Coefficient Test to determine the left and right behavior of the graph of each polynomial function.

a. $f(x) = -x^3 + 4x$ **b.** $f(x) = x^4 - 5x^2 + 4$ **c.** $f(x) = x^5 - x$

Solution

a. Because the degree is odd and the leading coefficient is negative, the graph rises to the left and falls to the right, as shown in Figure 2.13(a).

b. Because the degree is even and the leading coefficient is positive, the graph rises to the left and right, as shown in Figure 2.13(b).

c. Because the degree is odd and the leading coefficient is positive, the graph falls to the left and rises to the right, as shown in Figure 2.13(c).

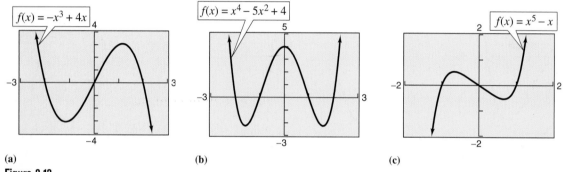

(a)   (b)   (c)

Figure 2.13

In Example 2, note that the Leading Coefficient Test only tells you whether the graph *eventually* rises or falls to the right or left. Other characteristics of the graph, such as intercepts and minimum and maximum points, must be determined by other tests.

Zeros of Polynomial Functions

It can be shown that for a polynomial function f of degree n, the following statements are true.

1. The graph of f has at most n real zeros. (This result is discussed in detail in Section 2.5.)

2. The function f has at most $n - 1$ relative **extrema** (relative **minimums** or **maximums**).

Recall that a **zero** of a function f is a number x for which $f(x) = 0$. Finding the zeros of polynomial functions is one of the most important problems in algebra. You have already seen that there is a strong interplay between graphical and algebraic approaches to this problem. Sometimes you can use information about the graph of a function to help find its zeros. In other cases you can use information about the zeros of a function to find a good viewing window.

Exploration

For each of the graphs in Figure 2.13, count the number of zeros of the polynomial function and the number of relative extrema, and compare these numbers with the degree of the polynomial. What do you observe?

Real Zeros of Polynomial Functions

If f is a polynomial function and a is a real number, the following statements are equivalent.

1. $x = a$ is a *zero* of the function f.
2. $x = a$ is a *solution* of the polynomial equation $f(x) = 0$.
3. $(x - a)$ is a *factor* of the polynomial $f(x)$.
4. $(a, 0)$ is an *x-intercept* of the graph of f.

STUDY T!P

Some graphing utilities have two features that analyze the graph of a function: (1) the *root* feature or the *zero* feature for finding zeros of a function, and (2) the *minimum* and *maximum* features for finding the relative extrema. If your graphing utility has these features, use them to confirm the results in Examples 3 and 4.

Finding zeros of polynomial functions is closely related to factoring and finding x-intercepts, as demonstrated in Examples 3, 4, and 5.

EXAMPLE 3 Finding Zeros of a Polynomial Function

Find all real zeros of $f(x) = x^3 - x^2 - 2x$.

Algebraic Solution

$$f(x) = x^3 - x^2 - 2x \qquad \text{Write original function.}$$
$$0 = x^3 - x^2 - 2x \qquad \text{Substitute 0 for } f(x).$$
$$= x(x^2 - x - 2) \qquad \text{Remove common monomial factor.}$$
$$= x(x - 2)(x + 1) \qquad \text{Factor completely.}$$

So, the real zeros are $x = 0$, $x = 2$, and $x = -1$, and the corresponding x-intercepts are $(0, 0)$, $(2, 0)$, and $(-1, 0)$.

Check

$$(0)^3 - (0)^2 - 2(0) = 0 \qquad x = 0 \text{ is a zero. } \checkmark$$
$$(2)^3 - (2)^2 - 2(2) = 0 \qquad x = 2 \text{ is a zero. } \checkmark$$
$$(-1)^3 - (-1)^2 - 2(-1) = 0 \qquad x = -1 \text{ is a zero. } \checkmark$$

Graphical Solution

Use a graphing utility to graph $y = x^3 - x^2 - 2x$. In Figure 2.14, the graph appears to have the x-intercepts $(0, 0)$, $(2, 0)$, and $(-1, 0)$. Use the *zero* or *root* feature, or the *zoom* and *trace* features, of the graphing utility to verify these intercepts. Note that this third-degree polynomial has two relative extrema.

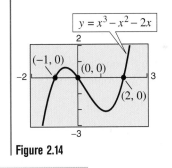

Figure 2.14

EXAMPLE 4 Analyzing a Polynomial Function

Find all real zeros and relative extrema of $f(x) = -2x^4 + 2x^2$.

Solution

$$f(x) = -2x^4 + 2x^2 \qquad \text{Write original function.}$$
$$0 = -2x^4 + 2x^2 \qquad \text{Substitute 0 for } f(x).$$
$$= -2x^2(x^2 - 1) \qquad \text{Remove common monomial factor.}$$
$$= -2x^2(x - 1)(x + 1) \qquad \text{Factor completely.}$$

So, the real zeros are $x = 0$, $x = 1$, and $x = -1$, and the corresponding x-intercepts are $(0, 0)$, $(1, 0)$, and $(-1, 0)$, as shown in Figure 2.15. Using the *minimum* and *maximum* features of a graphing utility, you can approximate the three relative extrema to be $(-0.7071, 0.5)$, $(0, 0)$, and $(0.7071, 0.5)$.

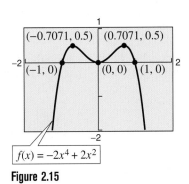

Figure 2.15

In Example 4, the real zero $x = 0$ arising from $-2x^2 = 0$ is a **repeated zero.** In general, a factor of $(x - a)^k$ yields a repeated zero $x = a$ of **multiplicity** k. If k is odd, the graph *crosses* the x-axis at $x = a$. If k is even, the graph *touches* (but does not cross) the x-axis at $x = a$, as shown in Figure 2.15.

EXAMPLE 5   Finding Zeros of a Polynomial Function

Find all real zeros of $f(x) = x^5 - 3x^3 - x^2 - 4x - 1$.

Solution

Use a graphing utility to obtain the graph shown in Figure 2.16. From the graph, you can see that there are three zeros. Using the *zero* or *root* feature, you can determine that the zeros are approximately $x \approx -1.861$, $x \approx -0.254$, and $x \approx 2.115$. It should be noted that this fifth-degree polynomial factors as

$$f(x) = x^5 - 3x^3 - x^2 - 4x - 1$$
$$= (x^2 + 1)(x^3 - 4x - 1).$$

The three zeros obtained above are the zeros of the cubic $x^3 - 4x - 1$ (the quadratic $x^2 + 1$ has no *real* zeros).

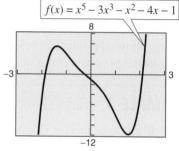

$f(x) = x^5 - 3x^3 - x^2 - 4x - 1$

Figure 2.16

EXAMPLE 6   Finding a Polynomial Function with Given Zeros

Find polynomial functions with the following zeros. (There are many correct solutions.)

a. $-2, -1, 1, 2$ **b.** $-\dfrac{1}{2}, 3, 3$ **c.** $3, 2 + \sqrt{11}, 2 - \sqrt{11}$

Solution

a. For each of the given zeros, form a corresponding factor. For instance, the zero given by $x = -2$ corresponds to the factor $(x + 2)$. So, you can write

$$f(x) = (x + 2)(x + 1)(x - 1)(x - 2)$$
$$= (x^2 - 4)(x^2 - 1) = x^4 - 5x^2 + 4.$$

b. Note that the zero $x = -\frac{1}{2}$ corresponds to either $\left(x + \frac{1}{2}\right)$ or $(2x + 1)$. To avoid fractions, choose the second factor and write

$$f(x) = (2x + 1)(x - 3)^2$$
$$= (2x + 1)(x^2 - 6x + 9) = 2x^3 - 11x^2 + 12x + 9.$$

c. For each of the given zeros, form a corresponding factor and write

$$f(x) = (x - 3)\left[x - \left(2 + \sqrt{11}\right)\right]\left[x - \left(2 - \sqrt{11}\right)\right]$$
$$= (x - 3)\left[(x - 2) - \sqrt{11}\right]\left[(x - 2) + \sqrt{11}\right]$$
$$= (x - 3)\left[(x - 2)^2 - \left(\sqrt{11}\right)^2\right]$$
$$= (x - 3)(x^2 - 4x + 4 - 11)$$
$$= (x - 3)(x^2 - 4x - 7) = x^3 - 7x^2 + 5x + 21.$$

Exploration

Use a graphing utility to graph

$$y_1 = x + 2$$
$$y_2 = (x + 2)(x - 1).$$

Predict the shape of the curve $y = (x + 2)(x - 1)(x - 3)$, and verify your answer with a graphing utility.

EXAMPLE 7   Sketching the Graph of a Polynomial Function

Sketch the graph of $f(x) = 3x^4 - 4x^3$ by hand.

Solution

1. *Apply Leading Coefficient Test.*   Because the leading coefficient is positive and the degree is even, you know that the graph eventually rises to the left and to the right [see Figure 2.17(a)].

2. *Find the Zeros of the Polynomial.*   By factoring

$$f(x) = 3x^4 - 4x^3$$
$$= x^3(3x - 4)$$

   you can see that the zeros of f are $x = 0$ (of odd multiplicity 3) and $x = \frac{4}{3}$ (of odd multiplicity 1). So, the x-intercepts occur at $(0, 0)$ and $\left(\frac{4}{3}, 0\right)$. Add these points to your graph, as shown in Figure 2.17(a).

3. *Plot a Few Additional Points.*   To sketch the graph by hand, find a few additional points, as shown in the table. Be sure to choose points between the zeros and to the left and right of the zeros. Then plot the points [see Figure 2.17(b)].

x	-1	0.5	1	1.5
$f(x)$	7	-0.3125	-1	1.6875

4. *Draw the Graph.*   Draw a continuous curve through the points, as shown in Figure 2.17(b). If you are unsure of the shape of the portion of the graph, plot some additional points.

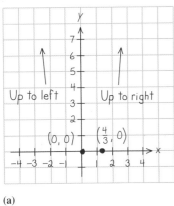

(a)

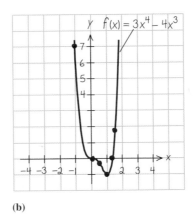

(b)

Figure 2.17

Exploration

Partner Activity   Multiply 3, 4, or 5 distinct linear factors to obtain the equation of a polynomial function that has a degree of 3, 4, or 5. Exchange equations with your partner and sketch, *by hand*, the graph of the equation that your partner wrote. When you are finished, use a graphing utility to check each other's work.

EXAMPLE 8   Sketching the Graph of a Polynomial Function

Sketch the graph of $f(x) = -2x^3 + 6x^2 - \frac{9}{2}x$.

Solution

1. *Apply Leading Coefficient Test.*   Because the leading coefficient is negative and the degree is odd, you know that the graph eventually rises to the left and falls to the right [see Figure 2.18(a)].

2. *Find the Zeros of the Polynomial.*   By factoring

$$f(x) = -2x^3 + 6x^2 - \frac{9}{2}x = -\frac{1}{2}x(4x^2 - 12x + 9) = -\frac{1}{2}x(2x - 3)^2$$

you can see that the zeros of f are

$x = 0$ (of odd multiplicity 1) and $x = \frac{3}{2}$ (of even multiplicity 2).

So, the x-intercepts occur at $(0, 0)$ and $\left(\frac{3}{2}, 0\right)$. Add these points to your graph, as shown in Figure 2.18(a).

3. *Plot a Few Additional Points.*   To sketch the graph by hand, find a few additional points, as shown in the table. Then plot the points [see Figure 2.18(b)].

x	-0.5	0.5	1	2
$f(x)$	4	-1	-0.5	-1

4. *Draw the Graph.*   Draw a continuous curve through the points, as shown in Figure 2.18(b). Notice that the graph crosses the x-axis at $(0, 0)$ and touches (but does not cross) the x-axis at $\left(\frac{3}{2}, 0\right)$.

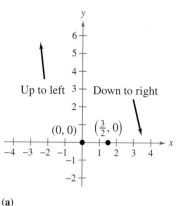

(a)

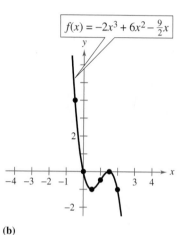

(b)

Figure 2.18

The Intermediate Value Theorem

The **Intermediate Value Theorem** concerns the existence of real zeros of polynomial functions. The theorem states that if $(a, f(a))$ and $(b, f(b))$ are two points on the graph of a polynomial function such that $f(a) \neq f(b)$, then for any number d between $f(a)$ and $f(b)$ there must be a number c between a and b such that $f(c) = d$. (See Figure 2.19.)

Intermediate Value Theorem

Let a and b be real numbers such that $a < b$. If f is a polynomial function such that $f(a) \neq f(b)$, then in the interval $[a, b]$, f takes on every value between $f(a)$ and $f(b)$.

This theorem helps locate the real zeros of a polynomial function in the following way. If you can find a value $x = a$ where a polynomial function is positive, and another $x = b$ where it is negative, you can conclude that the function has at least one real zero between these two values. For example, the function $f(x) = x^3 + x^2 + 1$ is negative when $x = -2$ and positive when $x = -1$. Therefore, it follows from the Intermediate Value Theorem that f must have a real zero somewhere between -2 and -1.

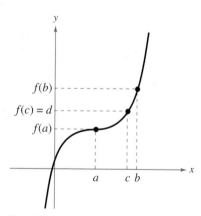

Figure 2.19

EXAMPLE 9   Approximating the Zeros of a Function

Find three intervals of length 1 in which the polynomial

$$f(x) = 12x^3 - 32x^2 + 3x + 5$$

is guaranteed to have a zero.

Graphical Solution

Use a graphing utility to graph

$$y = 12x^3 - 32x^2 + 3x + 5$$

as shown in Figure 2.20.

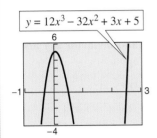

$y = 12x^3 - 32x^2 + 3x + 5$

Figure 2.20

From the figure, you can see that the graph crosses the *x*-axis three times—between -1 and 0, between 0 and 1, and another between 2 and 3. So, you can conclude that the function has zeros in the intervals $(-1, 0)$, $(0, 1)$, and $(2, 3)$.

Numerical Solution

Use the *table* feature of a graphing utility to create a table that shows the function values for $-20 \le x \le 20$ as shown in Figure 2.21.

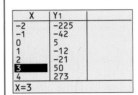

X	Y1
-2	-225
-1	-42
0	5
1	-12
2	-21
3	50
4	273

X=3

Figure 2.21

Scroll through the table looking for consecutive function values that differ in sign. For instance, from the table you can see that $f(-1)$ and $f(0)$ differ in sign. So, you can conclude from the Intermediate Value Theorem that the function has a zero between -1 and 0. Similarly, $f(0)$ and $f(1)$ differ in sign, so the function has a zero between 0 and 1. Likewise, $f(2)$ and $f(3)$ differ in sign, so the function has a zero between 2 and 3. So, you can conclude that the function has zeros in the intervals $(-1, 0)$, $(0, 1)$, and $(2, 3)$.

Writing About Math   *The Graphs of Cubic Polynomials*

The graphs of cubic polynomials can be categorized according to the four basic shapes below. Match the graph of each function with one of the basic shapes and write a short paragraph describing how you reached your conclusion. Is it possible for a polynomial of odd degree to have no real zeros? Explain.

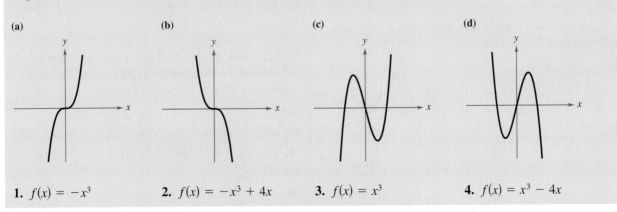

(a)

(b)

(c)

(d)

1. $f(x) = -x^3$

2. $f(x) = -x^3 + 4x$

3. $f(x) = x^3$

4. $f(x) = x^3 - 4x$

2.2   E x e r c i s e s

In Exercises 1–8, match the polynomial function with its graph. [The graphs are labeled (a) through (h).]

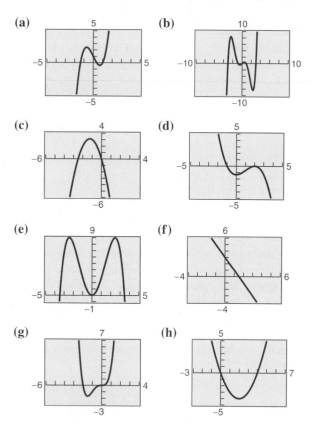

(a)

(b)

(c)

(d)

(e)

(f)

(g)

(h)

1. $f(x) = -2x + 3$ **2.** $f(x) = x^2 - 4x$

3. $f(x) = -2x^2 - 5x$ **4.** $f(x) = 2x^3 - 3x + 1$

5. $f(x) = -\frac{1}{4}x^4 + 3x^2$ **6.** $f(x) = -\frac{1}{3}x^3 + x^2 - \frac{4}{3}$

7. $f(x) = x^4 + 2x^3$ **8.** $f(x) = \frac{1}{5}x^5 - 2x^3 + \frac{9}{5}x$

In Exercises 9–12, sketch the graphs of $y = x^n$ and each specified transformation.

9. $y = x^3$

   (a) $f(x) = (x - 2)^3$ (b) $f(x) = x^3 - 2$

   (c) $f(x) = -\frac{1}{2}x^3$ (d) $f(x) = (x - 2)^3 - 2$

10. $y = x^5$

   (a) $f(x) = (x + 3)^5$ (b) $f(x) = x^5 + 3$

   (c) $f(x) = 1 - \frac{1}{2}x^5$ (d) $f(x) = -\frac{1}{2}(x + 1)^5$

11. $y = x^4$

   (a) $f(x) = (x + 5)^4$ (b) $f(x) = x^4 - 5$

   (c) $f(x) = 4 - x^4$ (d) $f(x) = \frac{1}{2}(x - 1)^4$

12. $y = x^6$

   (a) $f(x) = -\frac{1}{8}x^6$ (b) $f(x) = x^6 - 4$

   (c) $f(x) = -\frac{1}{4}x^6 + 1$ (d) $f(x) = (x + 2)^6 - 4$

Graphical Analysis   **In Exercises 13–16, use a graphing utility to graph the functions f and g in the same viewing window. Zoom out sufficiently far to show that the right-hand and left-hand behavior of f and g appear identical. Show the graphs.**

13. $f(x) = 3x^3 - 9x + 1$, $g(x) = 3x^3$

14. $f(x) = -\frac{1}{3}(x^3 - 3x + 2)$, $g(x) = -\frac{1}{3}x^3$

15. $f(x) = -(x^4 - 4x^3 + 16x)$, $g(x) = -x^4$

16. $f(x) = 3x^4 - 6x^2$, $g(x) = 3x^4$

In Exercises 17–26, use the Leading Coefficient Test to determine the right-hand and left-hand behavior of the graph of the polynomial function. Verify your result by using a graphing utility to graph the function.

17. $f(x) = 2x^2 - 3x + 1$ **18.** $f(x) = \frac{1}{3}x^3 + 5x$

19. $g(x) = 5 - \frac{7}{2}x - 3x^2$ **20.** $h(x) = 1 - x^6$

21. $f(x) = -2.1x^5 + 4x^3 - 2$

22. $f(x) = 2x^5 - 5x + 7.5$

23. $f(x) = 6 - 2x + 4x^2 - 5x^3$

24. $f(x) = \dfrac{3x^4 - 2x + 5}{4}$

25. $h(t) = -\frac{2}{3}(t^2 - 5t + 3)$

26. $f(s) = -\frac{7}{8}(s^3 + 5s^2 - 7s + 1)$

In Exercises 27–36, find all the real zeros of the polynomial function. Verify your result by using a graphing utility to graph the function.

27. $f(x) = x^2 - 25$ **28.** $f(x) = 49 - x^2$

29. $h(t) = t^2 - 6t + 9$ **30.** $f(x) = x^2 + 10x + 25$

31. $f(x) = x^2 + x - 2$ **32.** $f(x) = 2x^2 - 14x + 24$

33. $f(t) = t^3 - 4t^2 + 4t$ **34.** $f(x) = x^4 - x^3 - 20x^2$

35. $f(x) = \frac{1}{2}x^2 + \frac{5}{2}x - \frac{3}{2}$ **36.** $f(x) = \frac{5}{3}x^2 + \frac{8}{3}x - \frac{4}{3}$

Graphical Analysis   In Exercises 37–48, (a) use a graphing utility to graph the function, (b) use the graph to approximate any zeros, and (c) find the zeros algebraically.

37. $f(x) = 3x^2 - 12x + 3$

38. $g(x) = 5(x^2 - 2x - 1)$

39. $g(t) = \frac{1}{2}t^4 - \frac{1}{2}$ **40.** $y = \frac{1}{4}x^3(x^2 - 9)$

41. $f(x) = x^5 + x^3 - 6x$ **42.** $g(t) = t^5 - 6t^3 + 9t$

43. $f(x) = 2x^4 - 2x^2 - 40$

44. $f(x) = 5x^4 + 15x^2 + 10$

45. $f(x) = x^3 - 4x^2 - 25x + 100$

46. $y = 4x^3 + 4x^2 - 7x + 2$

47. $y = 4x^3 - 20x^2 + 25x$ **48.** $y = x^5 - 5x^3 + 4x$

In Exercises 49–52, use a graphing utility to graph the function and approximate (accurate to three decimal places) any relative extrema.

49. $f(x) = 2x^4 - 6x^2 + 1$

50. $f(x) = -3x^3 - 4x^2 + x - 3$

51. $f(x) = x^5 + 3x^3 - x + 6$

52. $f(x) = -\frac{3}{8}x^4 - x^3 + 2x^2 + 5$

In Exercises 53–64, find a polynomial function that has the given zeros. (There are many correct answers.)

53. $0, 12$ **54.** $0, -8$

55. $2, -6$ **56.** $-4, 5$

57. $0, -4, -3$ **58.** $0, 2, 7$

59. $4, -3, 3, 0$ **60.** $-2, -1, 0, 1, 2$

61. $1 + \sqrt{3}, 1 - \sqrt{3}$ **62.** $6 + \sqrt{3}, 6 - \sqrt{3}$

63. $2, 4 + \sqrt{5}, 4 - \sqrt{5}$ **64.** $4, 2 + \sqrt{7}, 2 - \sqrt{7}$

In Exercises 65–78, sketch the graph of the function by (a) applying the Leading Coefficient Test, (b) finding the zeros of the polynomial, (c) plotting sufficient solution points, and (d) drawing a continuous curve through the points.

65. $f(x) = x^3 - 9x$

66. $g(x) = x^4 - 4x^2$

67. $f(t) = \frac{1}{4}(t^2 - 2t + 15)$

68. $g(x) = -x^2 + 10x - 16$

69. $f(x) = x^3 - 3x^2$

70. $f(x) = 1 - x^3$

71. $f(x) = 3x^3 - 15x^2 + 18x$

72. $f(x) = -4x^3 + 4x^2 + 15x$

73. $f(x) = -x^3 - 5x^2$

74. $f(x) = 3x^4 - 48x^2$

75. $f(x) = x^2(x - 4)$

76. $h(x) = \frac{1}{3}x^3(x - 4)^2$

77. $g(t) = -\frac{1}{4}(t - 2)^2(t + 2)^2$

78. $g(x) = \frac{1}{10}(x + 1)^2(x - 3)^2$

In Exercises 79–82, (a) use the Intermediate Value Theorem and a graphing utility to find intervals of length 1 in which the polynomial function is guaranteed to have a zero, (b) use the *root* or *zero* feature of a graphing utility to approximate the zeros of the function, and (c) verify your answers in part (a) by using the *table* feature of a graphing utility.

79. $f(x) = x^3 - 3x^2 + 3$

80. $f(x) = 0.11x^3 - 2.07x^2 + 9.81x - 6.88$

81. $g(x) = 3x^4 + 4x^3 - 3$

82. $h(x) = x^4 - 10x^2 + 2$

In Exercises 83–86, use a graphing utility to graph the function. Describe a viewing window that gives a good view of the basic characteristics of the graph. (There are many correct answers.)

83. $f(x) = -\frac{3}{2}$

84. $h(x) = \frac{1}{3}x - 3$

85. $f(t) = \frac{1}{6}(t^2 - 4t + 21)$

86. $g(x) = -x^2 + 9x - 14$

In Exercises 87–94, use a graphing utility to graph the function. Identify any symmetry with respect to the *x*-axis, *y*-axis, or origin. Determine the number of *x*-intercepts of the graph.

87. $f(x) = x^2(x + 6)$

88. $h(x) = x^3(x - 4)^2$

89. $g(t) = -\frac{1}{2}(t - 4)^2(t + 4)^2$

90. $g(x) = \frac{1}{8}(x + 1)^2(x - 3)^3$

91. $f(x) = x^3 - 4x$

92. $f(x) = x^4 - 2x^2$

93. $g(x) = \frac{1}{5}(x + 1)^2(x - 3)(2x - 9)$

94. $h(x) = \frac{1}{5}(x + 2)^2(3x - 5)^2$

95. *Numerical and Graphical Analysis* An open box is to be made from a square piece of material 36 centimeters on a side by cutting equal squares from the corners and turning up the sides.

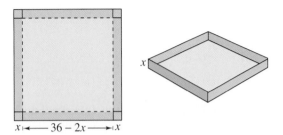

(a) Complete four rows of a table like the one below.

Height	Width	Volume
1	$36 - 2(1)$	$1[36 - 2(1)]^2 = 1156$
2	$36 - 2(2)$	$2[36 - 2(2)]^2 = 2048$

(b) Use a graphing utility to generate additional rows of the table. Use the table to estimate a range of dimensions within which the maximum volume is produced.

(c) Verify that the volume of the box is $V(x) = x(36 - 2x)^2$. Determine the domain of the function.

(d) Use a graphing utility to graph V, and use the range of dimensions from part (b) to find the x-value for which $V(x)$ is maximum.

96. *Geometry* An open box with locking tabs is to be made from a square piece of material 24 inches on a side. This is done by cutting equal squares from the corners and folding along the dashed lines, as shown in the figure.

(a) Show that the volume of the box is
$$V(x) = 8x(6 - x)(12 - x).$$

(b) Determine the domain of the function V.

(c) Sketch the graph of the function and estimate the value of x for which $V(x)$ is maximum.

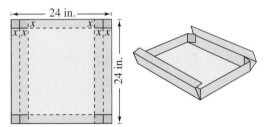

97. *Business* The total revenue R (in millions of dollars) for a company is related to its advertising expense by the function
$$R = \frac{1}{100,000}(-x^3 + 600x^2), \quad 0 \le x \le 400$$

where x is the amount spent on advertising (in tens of thousands of dollars). Use the graph of the function shown in the figure to estimate the point on the graph at which the function is increasing most rapidly. This point is called the **point of diminishing returns** because any expense above this amount will yield less return per dollar invested in advertising.

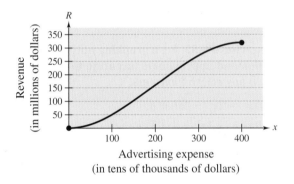

Advertising expense
(in tens of thousands of dollars)

98. *Environment* The growth of a red oak tree is approximated by the function
$$G = -0.003t^3 + 0.137t^2 + 0.458t - 0.839$$

where G is the height of the tree (in feet) and t ($2 \le t \le 34$) is its age (in years). Use a graphing utility to graph the function and estimate the age of the tree when it is growing most rapidly. This point is called the **point of diminishing returns** because the increase in growth will be less with each additional year. (*Hint:* Use a viewing window in which $-10 \le x \le 45$ and $-5 \le y \le 60$.)

99. *Data Analysis* The vertical deflection y of a 4-meter beam is measured in $\frac{1}{2}$-meter intervals x. The measurements are given by the following ordered pairs.

$(0, 0), (0.5, 0.06), (1, 0.11), (1.5, 0.15), (2, 0.16),$
$(2.5, 0.15), (3, 0.11), (3.5, 0.06), (4, 0)$

(a) Use the regression capabilities of a graphing utility to fit a quartic (fourth-degree polynomial) equation to the data.

(b) Use a graphing utility to graph the data and the regression equation in the same viewing window. How do they compare?

(c) Because $y = 0$ when $x = 0$, what should the constant term in the model be? Does your answer agree with the result of part (a)? Explain.

100. *Data Analysis* The table gives the median values of new privately owned U.S. homes for 1983 through 1997. In the table, t is the time (in years), with $t = 3$ corresponding to 1983, and y_1 and y_2 are the median prices (in thousands of dollars) in the Northeast and the South, respectively. (Sources: U.S. Bureau of the Census; U.S. Department of Housing and Urban Development)

t	3	4	5	6	7	8
y_1	82.2	88.6	103.3	125	140	149
y_2	70.9	72	75	80.2	88	92

t	9	10	11	12	13
y_1	159.6	159	155.9	169	162.6
y_2	96.4	99	100	105.5	115

t	14	15	16	17
y_1	169	180	186	190
y_2	116.9	124.5	126.2	129.6

(a) Use the regression capabilities of a graphing utility to fit a cubic model to the median prices of homes in the Northeast.

(b) Use the regression capabilities of a graphing utility to fit a cubic model to the median prices of homes in the South.

(c) Use the graphs of the models in parts (a) and (b) to write a short paragraph about the relationship between the median prices of homes in the two regions.

Synthesis

True or False? **In Exercises 101 and 102, determine whether the statement is true or false. Justify your answer.**

101. A fourth-degree polynomial can have four turning points.

102. The graph of the function

$$f(x) = 2 + x - x^2 + x^3 - x^4 + x^5 + x^6 - x^7$$

rises to the left and falls to the right.

103. *Graphical Reasoning* Use a graphing utility to graph the function $f(x) = x^4$. Explain how the graph of g differs (if it does) from the graph of f and confirm your result with a graphing utility. Determine whether g is odd, even, or neither.

(a) $g(x) = f(x) + 2$ (b) $g(x) = f(x + 2)$
(c) $g(x) = f(-x)$ (d) $g(x) = -f(x)$
(e) $g(x) = f\left(\frac{1}{2}x\right)$ (f) $g(x) = \frac{1}{2}f(x)$
(g) $g(x) = f\left(x^{3/4}\right)$ (h) $g(x) = (f \circ f)(x)$

Review

In Exercises 104–109, let $f(x) = 14x - 3$, **and** $g(x) = 8x^2$. **Find the indicated value.**

104. $(f + g)(-4)$ **105.** $(g - f)(3)$

106. $(fg)\left(-\frac{4}{7}\right)$ **107.** $\left(\frac{f}{g}\right)(-1.5)$

108. $(f \circ g)(-1)$ **109.** $(g \circ f)(0)$

In Exercises 110–113, solve the inequality and sketch the solution on the real number line. Use a graphing utility to verify your solution graphically.

110. $3(x - 5) < 4x - 7$ **111.** $2x^2 - x \geq 1$

112. $\dfrac{5x - 2}{x - 7} \leq 4$ **113.** $|x + 8| - 1 \geq 15$

In Exercises 114–117, find the quadratic function that has the indicated vertex and whose graph passes through the given point.

114. Vertex: $(3, -6)$; Point: $(-1, 2)$

115. Vertex: $(0, -8)$; Point: $(5, 9)$

116. Vertex: $(4, -4)$; Point: $(1, 10)$

117. Vertex: $(-5, -2)$; Point: $(0, 3)$

2.3   Real Zeros of Polynomial Functions

Long Division of Polynomials

Consider the graph of

$$f(x) = 6x^3 - 19x^2 + 16x - 4.$$

Notice in Figure 2.22 that $x = 2$ appears to be a zero of f. Because $f(2) = 0$, you know that $x = 2$ is a zero of the polynomial function f, and that $(x - 2)$ is a factor of $f(x)$. This means that there exists a second-degree polynomial $q(x)$ such that $f(x) = (x - 2) \cdot q(x)$. To find $q(x)$, you can use **long division of polynomials.**

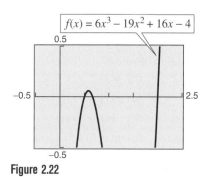

$$f(x) = 6x^3 - 19x^2 + 16x - 4$$

Figure 2.22

What You Should Learn:

- How to use long division to divide polynomials by other polynomials
- How to use synthetic division to divide polynomials by binomials of the form $(x - k)$
- How to use the Remainder Theorem and the Factor Theorem
- How to use the Rational Zero Test to determine possible rational zeros of polynomial functions
- How to determine upper and lower bounds for zeros of polynomial functions

Why You Should Learn It:

Polynomial division can help you rewrite polynomials that are used to model real-life problems. For instance, Exercise 87 on page 172 shows how polynomial division can be used to model the average monthly rates for cable television in the United States from 1988 through 1997.

Stephen Ferry/Liaison International

EXAMPLE 1   Long Division of Polynomials

Divide $f(x) = 6x^3 - 19x^2 + 16x - 4$ by $x - 2$, and use the result to factor the function f completely.

Solution

$$
\require{enclose}
\begin{array}{r}
6x^2 - 7x + 2 \\[-2pt]
x - 2 \enclose{longdiv}{6x^3 - 19x^2 + 16x - 4} \\
\end{array}
$$

$\underline{6x^3 - 12x^2}$	Multiply: $6x^2(x - 2)$.
$-7x^2 + 16x$	Subtract.
$\underline{-7x^2 + 14x}$	Multiply: $-7x(x - 2)$.
$2x - 4$	Subtract.
$\underline{2x - 4}$	Multiply: $2(x - 2)$.
0	Subtract.

Partial quotients

You can see that

$$6x^3 - 19x^2 + 16x - 4 = (x - 2)(6x^2 - 7x + 2)$$

$$= (x - 2)(2x - 1)(3x - 2).$$

Note that this factorization agrees with the graph of f (Figure 2.22) in that the three x-intercepts occur at $x = 2$, $x = \frac{1}{2}$, and $x = \frac{2}{3}$.

In Example 1, $x - 2$ is a factor of the polynomial $6x^3 - 19x^2 + 16x - 4$, and the long division process produces a remainder of zero. Often, long division will produce a nonzero remainder. For instance, if you divide $x^2 + 3x + 5$ by $x + 1$, you obtain the following.

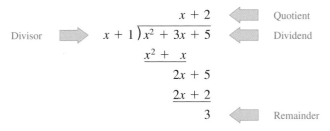

In fractional form, you can write this result as follows.

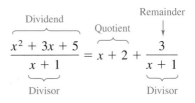

This implies that

$$x^2 + 3x + 5 = (x + 1)(x + 2) + 3$$

which illustrates the following well-known theorem called the **Division Algorithm.**

The Division Algorithm

If $f(x)$ and $d(x)$ are polynomials such that $d(x) \neq 0$, and the degree of $d(x)$ is less than or equal to the degree of $f(x)$, there exist unique polynomials $q(x)$ and $r(x)$ such that

$$f(x) = d(x)q(x) + r(x)$$

$\uparrow$ Dividend $\uparrow$ Quotient $\uparrow$
Divisor Remainder

where $r(x) = 0$ *or* the degree of $r(x)$ is less than the degree of $d(x)$. If the remainder $r(x)$ is zero, $d(x)$ *divides evenly* into $f(x)$.

The Division Algorithm can also be written as

$$\frac{f(x)}{d(x)} = q(x) + \frac{r(x)}{d(x)}.$$

In the Division Algorithm, the rational expression $f(x)/d(x)$ is **improper** because the degree of $f(x)$ is greater than or equal to the degree of $d(x)$. On the other hand, the rational expression $r(x)/d(x)$ is **proper** because the degree of $r(x)$ is less than the degree of $d(x)$.

EXAMPLE 2   Long Division of Polynomials

Divide $x^3 - 1$ by $x - 1$.

Solution

Because there is no x^2-term or x-term in the dividend, you need to line up the subtraction by using zero coefficients (or leaving spaces) for the missing terms.

$$
\begin{array}{r}
x^2 + x + 1 \\
x - 1 \overline{)\, x^3 + 0x^2 + 0x - 1} \\
\underline{x^3 - x^2} \\
x^2 \\
\underline{x^2 - x} \\
x - 1 \\
\underline{x - 1} \\
0
\end{array}
$$

So, $x - 1$ divides evenly into $x^3 - 1$, and you can write

$$
\frac{x^3 - 1}{x - 1} = x^2 + x + 1, \qquad x \neq 1.
$$

You can check the result of a division problem by multiplying. For instance, in Example 2, try checking that

$$(x - 1)(x^2 + x + 1) = x^3 - 1.$$

EXAMPLE 3   Long Division of Polynomials

Divide $2x^4 + 4x^3 - 5x^2 + 3x - 2$ by $x^2 + 2x - 3$.

Solution

$$
\begin{array}{r}
2x^2 + 1 \\
x^2 + 2x - 3 \overline{)\, 2x^4 + 4x^3 - 5x^2 + 3x - 2} \\
\underline{2x^4 + 4x^3 - 6x^2} \\
x^2 + 3x - 2 \\
\underline{x^2 + 2x - 3} \\
x + 1
\end{array}
$$

A computer animation of this example appears in the *Interactive* CD-ROM and *Internet* versions of this text.

Note that the first subtraction eliminated two terms from the dividend. When this happens, the quotient skips a term. You can write the result as

$$
\frac{2x^4 + 4x^3 - 5x^2 + 3x - 2}{x^2 + 2x - 3} = 2x^2 + 1 + \frac{x + 1}{x^2 + 2x - 3}.
$$

Synthetic Division

There is a nice shortcut for long division of polynomials by divisors of the form $x - k$. The shortcut is called **synthetic division.** The pattern for synthetic division of a cubic polynomial is summarized as follows. (The pattern for higher-degree polynomials is similar.)

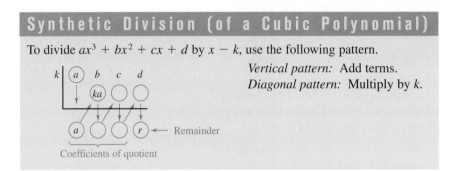

Synthetic Division (of a Cubic Polynomial)

To divide $ax^3 + bx^2 + cx + d$ by $x - k$, use the following pattern.

Vertical pattern: Add terms.
Diagonal pattern: Multiply by k.

Note that synthetic division works *only* for divisors of the form $x - k$. [Remember that $x + k = x - (-k)$.] You cannot use synthetic division to divide a polynomial by a quadratic such as $x^2 - 3$.

EXAMPLE 4   Using Synthetic Division

Use synthetic division to divide $x^4 - 10x^2 - 2x + 4$ by $x + 3$.

Solution

You should set up the array as follows. Note that a zero is included for each missing term in the dividend.

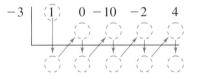

Then, use the synthetic division pattern by adding terms in columns and multiplying the results by -3.

A computer animation of this example appears in the *Interactive* CD-ROM and *Internet* versions of this text.

Divisor: $x + 3$ Dividend: $x^4 - 10x^2 - 2x + 4$

$$-3 \,\big|\, \begin{array}{rrrrr} 1 & 0 & -10 & -2 & 4 \\ & -3 & 9 & 3 & -3 \\ \hline 1 & -3 & -1 & 1 & 1 \end{array} \longleftarrow \text{Remainder: } 1$$

Quotient: $x^3 - 3x^2 - x + 1$

So, you have

$$\frac{x^4 - 10x^2 - 2x + 4}{x + 3} = x^3 - 3x^2 - x + 1 + \frac{1}{x + 3}.$$

The Remainder and Factor Theorems

The remainder obtained in the synthetic division process has an important interpretation, as described in the **Remainder Theorem.** (A proof of this theorem is given in Appendix A.)

The Remainder Theorem

If a polynomial $f(x)$ is divided by $x - k$, the remainder is

$$r = f(k).$$

The Remainder Theorem tells you that synthetic division can be used to evaluate a polynomial function. That is, to evaluate a polynomial function $f(x)$ when $x = k$, divide $f(x)$ by $x - k$. The remainder will be $f(k)$, as illustrated in Example 5.

EXAMPLE 5   Using the Remainder Theorem

Use the Remainder Theorem to evaluate the following function at $x = -2$.

$$f(x) = 3x^3 + 8x^2 + 5x - 7$$

Solution

Using synthetic division, you obtain the following.

$$
\begin{array}{r|rrrr}
-2 & 3 & 8 & 5 & -7 \\
   &   & -6 & -4 & -2 \\
\hline
   & 3 & 2 & 1 & -9
\end{array}
$$

Because the remainder is $r = -9$, you can conclude that

$$f(-2) = -9.$$

This means that $(-2, -9)$ is a point on the graph of f. You can check this by substituting $x = -2$ in the original function.

Check

$$
\begin{aligned}
f(-2) &= 3(-2)^3 + 8(-2)^2 + 5(-2) - 7 \\
      &= 3(-8) + 8(4) - 10 - 7 \\
      &= -24 + 32 - 10 - 7 \\
      &= -9
\end{aligned}
$$

Another important theorem is the **Factor Theorem,** which is stated below. This theorem states that you can test to see whether a polynomial has $(x - k)$ as a factor by evaluating the polynomial at $x = k$. If the result is 0, $(x - k)$ is a factor. For a proof of the Factor Theorem, see Appendix A.

The Factor Theorem

A polynomial $f(x)$ has a factor $(x - k)$ if and only if $f(k) = 0$.

EXAMPLE 6   Factoring a Polynomial: Repeated Division

Show that $(x - 2)$ and $(x + 3)$ are factors of

$$f(x) = 2x^4 + 7x^3 - 4x^2 - 27x - 18.$$

Then find the remaining factors of $f(x)$.

Algebraic Solution

Using synthetic division with 2 and -3 *repeatedly*, you obtain the following.

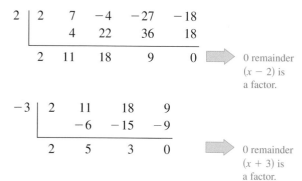

0 remainder
$(x - 2)$ is
a factor.

0 remainder
$(x + 3)$ is
a factor.

Because the resulting quadratic factors as

$$2x^2 + 5x + 3 = (2x + 3)(x + 1)$$

the complete factorization of $f(x)$ is

$$f(x) = (x - 2)(x + 3)(2x + 3)(x + 1).$$

Graphical Solution

The graph of a polynomial with factors of $(x - 2)$ and $(x + 3)$ has x-intercepts at $x = 2$ and $x = -3$. Use a graphing utility to graph

$$y = 2x^4 + 7x^3 - 4x^2 - 27x - 18.$$

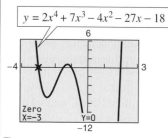

Figure 2.23

From Figure 2.23, you can see that the graph appears to cross the x-axis in two other places, near $x = -1$ and $x = -\frac{3}{2}$. Use the *zero* or *root* feature or the *zoom* and *trace* features to approximate the other two intercepts to be $x = -1$ and $x = -\frac{3}{2}$. So, the factors of f are $(x - 2)$, $(x + 3)$, $\left(x + \frac{3}{2}\right)$, and $(x + 1)$. You can rewrite the factor $\left(x + \frac{3}{2}\right)$ as $(2x + 3)$, so the complete factorization of f is $f(x) = (x - 2)(x + 3)(2x + 3)(x + 1)$.

Using the Remainder in Synthetic Division

In summary, the remainder r, obtained in the synthetic division of $f(x)$ by $x - k$, provides the following information.

1. The remainder r gives the value of f at $x = k$. That is, $r = f(k)$.

2. If $r = 0$, $(x - k)$ is a factor of $f(x)$.

3. If $r = 0$, $(k, 0)$ is an x-intercept of the graph of f.

Throughout this text, the importance of developing several problem-solving strategies has been emphasized. In the exercises for this section, try using more than one strategy to solve several of the exercises. For instance, if you find that $x - k$ divides evenly into $f(x)$, try sketching the graph of f. You should find that $(k, 0)$ is an x-intercept of the graph.

The Rational Zero Test

The **Rational Zero Test** relates the possible rational zeros of a polynomial (having integer coefficients) to the leading coefficient and to the constant term of the polynomial.

The Rational Zero Test

If the polynomial

$$f(x) = a_n x^n + a_{n-1} x^{n-1} + \cdots + a_2 x^2 + a_1 x + a_0$$

has integer coefficients, every rational zero of f has the form

$$\text{Rational zero} = \frac{p}{q}$$

where p and q have no common factors other than 1, p is a factor of the constant term a_0, and q is a factor of the leading coefficient a_n.

To use the Rational Zero Test, first list all rational numbers whose numerators are factors of the constant term and whose denominators are factors of the leading coefficient.

$$\text{Possible rational zeros} = \frac{\text{factors of constant term}}{\text{factors of leading coefficient}}$$

Now that you have formed this list of *possible rational zeros,* use a trial-and-error method to determine which, if any, are actual zeros of the polynomial. Note that when the leading coefficient is 1, the possible rational zeros are simply the factors of the constant term. This case is illustrated in Example 7.

STUDY TIP

Graph the polynomial $y = x^3 - 53x^2 + 103x - 51$ in the standard viewing window. From the graph alone, it appears that there is only one zero. From the Leading Coefficient Test, you know that because the degree of the polynomial is odd and the leading coefficient is positive, the graph falls to the left and rises to the right. So, the function must have another zero. From the Rational Zero Test, you know that ± 51 might be zeros of the function. If you zoom out several times, you will see a more complete picture of the graph. Your graph should confirm that $x = 51$ is a zero of f.

EXAMPLE 7   Rational Zero Test with Leading Coefficient of 1

Find the rational zeros of $f(x) = x^3 + x + 1$.

Solution

Because the leading coefficient is 1, the possible rational zeros are simply the factors of the constant term.

   Possible Rational Zeros: ± 1

By testing these possible zeros, you can see that neither works.

$$f(1) = (1)^3 + 1 + 1 = 3$$

$$f(-1) = (-1)^3 + (-1) + 1 = -1$$

So, the polynomial has *no* rational zeros. Note from the graph of f in Figure 2.24 that f does have one real zero (between -1 and 0). However, by the Rational Zero Test, you know that this real zero is *not* a rational number.

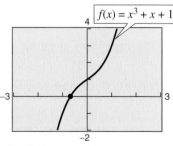

Figure 2.24

If the leading coefficient of a polynomial is not 1, the list of possible rational zeros can increase dramatically. In such cases the search can be shortened in several ways.

1. A programmable calculator can be used to speed up the calculations.

2. A graphing utility can give a good estimate of the locations of the zeros.

3. The Factor Theorem and synthetic division can be used to test the possible rational zeros.

Finding the first zero is often the most difficult part. After that, the search is simplified by working with the lower-degree polynomial obtained in synthetic division.

EXAMPLE 8   Using the Rational Zero Test

Find the rational zeros of

$$f(x) = 2x^3 + 3x^2 - 8x + 3.$$

Solution

The leading coefficient is 2 and the constant term is 3.

Possible Rational Zeros:

$$\frac{\text{Factors of 3}}{\text{Factors of 2}} = \frac{\pm 1, \pm 3}{\pm 1, \pm 2} = \pm 1, \pm 3, \pm \frac{1}{2}, \pm \frac{3}{2}$$

By synthetic division, you can determine that $x = 1$ is a zero.

$$
\begin{array}{r|rrrr}
1 & 2 & 3 & -8 & 3 \\
  &   & 2 & 5 & -3 \\
\hline
  & 2 & 5 & -3 & 0
\end{array}
$$

So, $f(x)$ factors as

$$f(x) = (x - 1)(2x^2 + 5x - 3)$$
$$= (x - 1)(2x - 1)(x + 3)$$

and you can conclude that the rational zeros of f are $x = 1$, $x = \dfrac{1}{2}$, and $x = -3$, as shown in Figure 2.25.

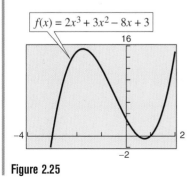

Figure 2.25

A graphing utility can help you determine which possible rational zeros to test, as demonstrated in Example 9.

EXAMPLE 9   Finding Real Zeros of a Polynomial Function

Find all the real zeros of $f(x) = 10x^3 - 15x^2 - 16x + 12$.

Solution

Because the leading coefficient is 10 and the constant term is 12, there is a long list of possible rational zeros.

Possible Rational Zeros:

$$\frac{\text{Factors of 12}}{\text{Factors of 10}} = \frac{\pm 1, \pm 2, \pm 3, \pm 4, \pm 6, \pm 12}{\pm 1, \pm 2, \pm 5, \pm 10}$$

With so many possibilities (32, in fact), it is worth your time to use a graphing utility to focus on just a few. From Figure 2.26, it looks like three reasonable choices would be $x = -\frac{6}{5}$, $x = \frac{1}{2}$, and $x = 2$. Synthetic division shows that only $x = 2$ works. (You could also use the Factor Theorem to test these choices.)

$$
\begin{array}{r|rrrr}
2 & 10 & -15 & -16 & 12 \\
  &   & 20 & 10 & -12 \\
\hline
  & 10 & 5   & -6 & 0 \\
\end{array}
$$

So, you have

$$f(x) = (x - 2)(10x^2 + 5x - 6).$$

So, $x = 2$ is one zero, and using the Quadratic Formula, you find that the two additional zeros are irrational numbers.

$$x = \frac{-5 + \sqrt{265}}{20} \approx 0.5639 \quad \text{and} \quad x = \frac{-5 - \sqrt{265}}{20} \approx -1.0639$$

Bounds for Real Zeros of Polynomial Functions

The third test for zeros of a polynomial function is related to the sign pattern in the last row of the synthetic division tableau. This test can give you an upper or lower bound of the real zeros of f, which can help you eliminate possible real zeros.

A real number b is an **upper bound** for the real zeros of f if no zeros are greater than b. Similarly, b is a **lower bound** if no real zeros of f are less than b.

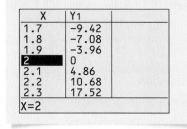

STUDY T!P

You can use the *table* feature of a graphing utility to test the possible rational zeros of the function in Example 9, as shown below. The table should be set to begin at -12, increase by increments of 0.1, and end at 12. Look through the table to determine the values of x for which y_1 is 0.

X	Y1	
1.7	-9.42	
1.8	-7.08	
1.9	-3.96	
2	**0**	
2.1	4.86	
2.2	10.68	
2.3	17.52	
X=2		

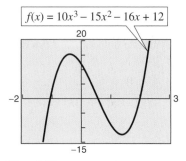

$f(x) = 10x^3 - 15x^2 - 16x + 12$

Figure 2.26

Upper and Lower Bound Rule

Let $f(x)$ be a polynomial with real coefficients and a positive leading coefficient. Suppose $f(x)$ is divided by $x - c$, using synthetic division.

1. If $c > 0$ and each number in the last row is either positive or zero, c is an **upper bound** for the real zeros of f.

2. If $c < 0$ and the numbers in the last row are alternately positive and negative (zero entries count as positive or negative), c is a **lower bound** for the real zeros of f.

EXAMPLE 10   Finding the Zeros of a Polynomial Function

Find the real zeros of $f(x) = 6x^3 - 4x^2 + 3x - 2$.

Solution

The possible real zeros are as follows.

$$\frac{\text{Factors of 2}}{\text{Factors of 6}} = \frac{\pm 1, \pm 2}{\pm 1, \pm 2, \pm 3, \pm 6} = \pm 1, \pm\frac{1}{2}, \pm\frac{1}{3}, \pm\frac{1}{6}, \pm\frac{2}{3}, \pm 2$$

Trying $x = 1$ produces the following.

$$\begin{array}{r|rrr}
1 & 6 & -4 & 3 & -2 \\
 & & 6 & 2 & 5 \\
\hline
 & 6 & 2 & 5 & 3
\end{array}$$

So, $x = 1$ is not a zero, but because the last row has all positive entries, you know that $x = 1$ is an upper bound for the real zeros. Therefore, you can restrict the search to zeros less than 1. By testing the possible rational zeros less than 1, you can determine that $x = \frac{2}{3}$ is a zero. So,

$$f(x) = \left(x - \frac{2}{3}\right)(6x^2 + 3).$$

Because $6x^2 + 3$ has no real zeros, it follows that $x = \frac{2}{3}$ is the only real zero.

Before concluding this section, two additional hints that can help you find the real zeros of a polynomial are listed below.

1. If the terms of $f(x)$ have a common monomial factor, it should be factored out before applying the tests in this section. For instance, by writing

$$f(x) = x^4 - 5x^3 + 3x^2 + x = x(x^3 - 5x^2 + 3x + 1)$$

   you can see that $x = 0$ is a zero of f and that the remaining zeros can be obtained by analyzing the cubic factor.

2. If you are able to find all but two zeros of $f(x)$, you can always use the Quadratic Formula on the remaining quadratic factor. For instance, if you succeeded in writing

$$f(x) = x^4 - 5x^3 + 3x^2 + x = x(x - 1)(x^2 - 4x - 1)$$

   you can apply the Quadratic Formula to $x^2 - 4x - 1$ to conclude that the two remaining zeros are $x = 2 + \sqrt{5}$ and $x = 2 - \sqrt{5}$.

Writing About Math   *Finding Patterns in Polynomial Division*

Complete the following polynomial divisions.

a. $\dfrac{x^2 - 1}{x - 1} = $ ▢

b. $\dfrac{x^3 - 1}{x - 1} = $ ▢

c. $\dfrac{x^4 - 1}{x - 1} = $ ▢

Write a brief description of the pattern that you obtain, and use your result to find a formula for the polynomial division $(x^n - 1)/(x - 1)$. Create a numerical example to test your formula.

2.3   E x e r c i s e s

Graphical Analysis   In Exercises 1–6, use a graphing utility to graph the two equations in the same viewing window. Use the graphs to verify that the expressions are equivalent. Verify the results algebraically.

1. $y_1 = \dfrac{4x}{x-1}$, $y_2 = 4 + \dfrac{4}{x-1}$

2. $y_1 = \dfrac{3x-5}{x-3}$, $y_2 = 3 + \dfrac{4}{x-3}$

3. $y_1 = \dfrac{x^2}{x+2}$, $y_2 = x - 2 + \dfrac{4}{x+2}$

4. $y_1 = \dfrac{x^4 - 3x^2 - 1}{x^2 + 5}$, $y_2 = x^2 - 8 + \dfrac{39}{x^2+5}$

5. $y_1 = \dfrac{x^5 - 3x^3}{x^2+1}$, $y_2 = x^3 - 4x + \dfrac{4x}{x^2+1}$

6. $y_1 = \dfrac{x^3 - 2x^2 + 5}{x^2 + x + 1}$, $y_2 = x - 3 + \dfrac{2(x+4)}{x^2+x+1}$

In Exercises 7–18, divide using long division.

7. Divide $2x^2 + 10x + 12$ by $x + 3$.

8. Divide $5x^2 - 17x - 12$ by $x - 4$.

9. Divide $4x^3 - 7x^2 - 11x + 5$ by $4x + 5$.

10. Divide $x^4 + 5x^3 + 6x^2 - x - 2$ by $x + 2$.

11. Divide $7x + 3$ by $x + 2$.

12. Divide $8x - 5$ by $2x + 1$.

13. $(6x^3 + 10x^2 + x + 8) \div (2x^2 + 1)$

14. $(x^3 - 9) \div (x^2 + 1)$

15. $\dfrac{x^4 + 3x^2 + 1}{x^2 - 2x + 3}$

16. $\dfrac{x^5 + 7}{x^3 - 1}$

17. $\dfrac{2x^3 - 4x^2 - 15x + 5}{(x-1)^2}$

18. $\dfrac{x^4}{(x-1)^3}$

In Exercises 19–28, divide using synthetic division.

19. $(3x^3 - 10x^2 + 12x - 22) \div (x - 4)$

20. $(2x^3 + 6x^2 - 14x + 9) \div (x - 1)$

21. $(6x^3 + 7x^2 - x + 26) \div (x - 3)$

22. $(2x^3 + 14x^2 - 20x + 7) \div (x + 6)$

23. $(9x^3 - 18x^2 - 16x + 32) \div (x - 2)$

24. $(5x^3 + 6x + 8) \div (x + 2)$

25. $\dfrac{x^3 + 512}{x + 8}$

26. $\dfrac{x^3 - 729}{x - 9}$

27. $\dfrac{4x^3 + 16x^2 - 23x - 15}{x + \frac{1}{2}}$

28. $\dfrac{3x^3 - 4x^2 + 5}{x - \frac{3}{2}}$

In Exercises 29–34, express the function in the form $f(x) = (x - k)q(x) + r(x)$ for the given value of k. Use a graphing utility to demonstrate that $f(k) = r$.

	Function	*Value of k*
29.	$f(x) = x^3 - x^2 - 14x + 11$	$k = 4$
30.	$f(x) = 15x^4 + 10x^3 - 6x^2 + 14$	$k = -\frac{2}{3}$
31.	$f(x) = x^3 + 3x^2 - 2x - 14$	$k = \sqrt{2}$
32.	$f(x) = x^3 + 2x^2 - 5x - 4$	$k = -\sqrt{5}$
33.	$f(x) = 4x^3 - 6x^2 - 12x - 4$	$k = 1 - \sqrt{3}$
34.	$f(x) = -3x^3 + 8x^2 + 10x - 8$	$k = 2 + \sqrt{2}$

In Exercises 35–38, use synthetic division to find each function value. Use a graphing utility to verify your result.

35. $f(x) = 4x^3 - 13x + 10$

   (a) $f(1)$ (b) $f(-2)$ (c) $f\left(\frac{1}{2}\right)$ (d) $f(8)$

36. $g(x) = x^6 - 4x^4 + 3x^2 + 2$

   (a) $g(2)$ (b) $g(-4)$ (c) $g(3)$ (d) $g(-1)$

37. $h(x) = 3x^3 + 5x^2 - 10x + 1$

   (a) $h(3)$ (b) $h\left(\frac{1}{3}\right)$ (c) $h(-2)$ (d) $h(-5)$

38. $f(x) = 0.4x^4 - 1.6x^3 + 0.7x^2 - 2$

   (a) $f(1)$ (b) $f(-2)$ (c) $f(5)$ (d) $f(-10)$

In Exercises 39–44, use synthetic division to show that x is a solution of the third-degree polynomial equation, and use the result to factor the polynomial completely. List all the real zeros of the function.

	Polynomial Equation	*Value of x*
39.	$x^3 - 7x + 6 = 0$	$x = 2$
40.	$x^3 - 28x - 48 = 0$	$x = -4$
41.	$2x^3 - 15x^2 + 27x - 10 = 0$	$x = \frac{1}{2}$
42.	$x^3 + 2x^2 - 3x - 6 = 0$	$x = -2$
43.	$x^3 + 2x^2 - 2x - 4 = 0$	$x = -2$
44.	$x^3 - x^2 - 13x - 3 = 0$	$x = -3$

In Exercises 45–50, (a) verify the given factors of the function f, (b) find the remaining factors of f, (c) use your results to write the complete factorization of f, (d) list all real zeros of f, and (e) confirm your results by using a graphing utility to graph the function.

Function	Factors
45. $f(x) = 2x^3 + x^2 - 5x + 2$	$(x + 2), (x - 1)$
46. $f(x) = 3x^3 + 2x^2 - 19x + 6$	$(x + 3), (x - 2)$
47. $f(x) = x^4 - 4x^3 - 15x^2$ $+ 58x - 40$	$(x - 5), (x + 4)$
48. $f(x) = 8x^4 - 14x^3 - 71x^2$ $- 10x + 24$	$(x + 2), (x - 4)$
49. $f(x) = 6x^3 + 41x^2 - 9x - 14$	$(2x + 1), (3x - 2)$
50. $f(x) = 2x^3 - x^2 - 10x + 5$	$(2x - 1), (x + \sqrt{5})$

In Exercises 51–54, use the Rational Zero Test to list all possible rational zeros of f. Verify that the zeros of f shown on the graph are contained in the list.

51. $f(x) = x^3 + 3x^2 - x - 3$

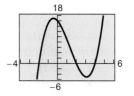

52. $f(x) = x^3 - 4x^2 - 4x + 16$

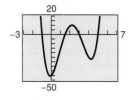

53. $f(x) = 2x^4 - 17x^3 + 35x^2 + 9x - 45$

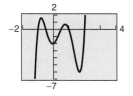

54. $f(x) = 4x^5 - 8x^4 - 5x^3 + 10x^2 + x - 2$

In Exercises 55–62, (a) list the possible rational zeros of f, (b) use a graphing utility to graph f so that some of the possible zeros in part (a) can be disregarded, and (c) determine all the real zeros of f.

55. $f(x) = x^3 + x^2 - 4x - 4$

56. $f(x) = -3x^3 + 20x^2 - 36x + 16$

57. $f(x) = -4x^3 + 15x^2 - 8x - 3$

58. $f(x) = 4x^3 - 12x^2 - x + 15$

59. $f(x) = -2x^4 + 13x^3 - 21x^2 + 2x + 8$

60. $f(x) = 4x^4 - 17x^2 + 4$

61. $f(x) = 6x^3 - x^2 - 13x + 8$

62. $f(x) = 4x^3 + 7x^2 - 11x - 18$

In Exercises 63–68, find all the real solutions of the polynomial equation.

63. $z^4 - z^3 - 2z - 4 = 0$

64. $x^4 - x^3 - 29x^2 - x - 30 = 0$

65. $x^4 - 13x^2 - 12x = 0$

66. $2y^4 + 7y^3 - 26y^2 + 23y - 6 = 0$

67. $2x^4 - 11x^3 - 6x^2 + 64x + 32 = 0$

68. $x^5 - x^4 - 3x^3 + 5x^2 - 2x = 0$

Graphical Analysis In Exercises 69–74, (a) use the *zero* or *root* feature of a graphing utility to approximate (accurate to three decimal places) the zeros of the function, (b) determine one of the exact zeros and use synthetic division to verify your result, and (c) factor the polynomial completely.

69. $f(x) = x^3 - 2x^2 - 5x + 10$

70. $g(x) = x^3 - 4x^2 - 2x + 8$

71. $h(t) = t^3 - 2t^2 - 7t + 2$

72. $f(s) = s^3 - 12s^2 + 40s - 24$

73. $h(x) = x^5 - 7x^4 + 10x^3 + 14x^2 - 24x$

74. $g(x) = 6x^4 - 11x^3 - 51x^2 + 99x - 27$

In Exercises 75–78, use synthetic division to verify the upper bound and lower bound of the zeros of f.

75. $f(x) = x^4 - 4x^3 + 15$

Upper bound: $x = 4$; Lower bound: $x = -1$

76. $f(x) = 2x^3 - 3x^2 - 12x + 8$

Upper bound: $x = 4$; Lower bound: $x = -3$

77. $f(x) = x^4 - 4x^3 + 16x - 16$

Upper bound: $x = 5$; Lower bound: $x = -3$

78. $f(x) = 2x^4 - 8x + 3$

Upper bound: $x = 3$; Lower bound: $x = -4$

In Exercises 79–82, find the rational zeros of the polynomial function.

79. $P(x) = x^4 - \frac{25}{4}x^2 + 9 = \frac{1}{4}(4x^4 - 25x^2 + 36)$

80. $f(x) = x^3 - \frac{3}{2}x^2 - \frac{23}{2}x + 6 = \frac{1}{2}(2x^3 - 3x^2 - 23x + 12)$

81. $f(x) = x^3 - \frac{1}{4}x^2 - x + \frac{1}{4} = \frac{1}{4}(4x^3 - x^2 - 4x + 1)$

82. $f(z) = z^3 + \frac{11}{6}z^2 - \frac{1}{2}z - \frac{1}{3} = \frac{1}{6}(6z^3 + 11z^2 - 3z - 2)$

In Exercises 83–86, match the cubic function with the correct number of rational and irrational zeros.

(a) **Rational zeros: 0;   Irrational zeros: 1**
(b) **Rational zeros: 3;   Irrational zeros: 0**
(c) **Rational zeros: 1;   Irrational zeros: 2**
(d) **Rational zeros: 1;   Irrational zeros: 0**

83. $f(x) = x^3 - 1$ **84.** $f(x) = x^3 - 2$

85. $f(x) = x^3 - x$ **86.** $f(x) = x^3 - 2x$

87. *Data Analysis*   The average monthly basic rates R for cable television in the United States for the years 1988 through 1997 are given in the table, where t represents the time (in years), with $t = 0$ corresponding to 1990.   (Source: Paul Kagan Associates, Inc.)

t	-2	-1	0	1	2
R	13.86	15.21	16.78	18.10	19.08

t	3	4	5	6	7
R	19.39	21.62	23.07	24.41	26.48

(a) Use a graphing utility to sketch a scatter plot of the data.

(b) Use the regression capabilities of a graphing utility to find a cubic model for the data. Then graph the model in the same viewing window as the scatter plot. Compare the model with the data.

(c) Use a graphing utility and the model to create a table of estimated values of R. Compare the estimated values with the actual data.

(d) Use the Remainder Theorem to evaluate the model for the year 2002. Even though the model is relatively accurate for estimating the given data, do you think it is accurate to predict future cable rates? Explain.

88. *Data Analysis*   The number of United States military personnel M (in thousands) on active duty for the years 1989 through 1996 is shown in the table, where t represents the time (in years), with $t = 0$ corresponding to 1990.   (Source: U.S. Department of Defense)

t	-1	0	1	2
M	2130	2044	1986	1807

t	3	4	5	6
M	1705	1611	1518	1472

(a) Use a graphing utility to sketch a scatter plot of the data.

(b) Use the regression capabilities of a graphing utility to find a cubic model for the data. Then graph the model in the same viewing window as the scatter plot. Compare the model with the data.

(c) Use a graphing utility and the model to create a table of estimated values of M. Compare the estimated values with the actual data.

(d) Use the Remainder Theorem to evaluate the model for the year 2001. Even though the model is relatively accurate for estimating the given data, would you use this model to predict the number of military personnel in the future? Explain.

89. *Geometry*   An open box is to be made from a rectangular piece of material 15 centimeters by 9 centimeters by cutting equal squares from the corners and turning up the sides.

(a) Let x represent the length of the sides of the squares. Draw a diagram showing the squares removed from the original piece of material and the resulting dimensions of the open box.

(b) Use the diagram in part (a) to write the volume V of the box as a function of x. Determine the domain of the function.

(c) Use a graphing utility to graph the function and approximate the dimensions of the box that yield maximum volume.

(d) Find values of x such that $V = 56$. Which of these values is a physical impossibility in the construction of the box? Explain.

90. *Geometry*   A rectangular package sent by a delivery service can have a maximum combined length and girth (perimeter of a cross section) of 120 inches.

(a) Show that the volume of the package is

$$V(x) = 4x^2(30 - x).$$

(b) Use a graphing utility to graph the function and approximate the dimensions of the package that yield a maximum volume.

(c) Find values of x such that $V = 13,500$. Which of these values is a physical impossibility in the construction of the package? Explain.

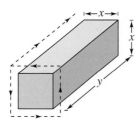

91. *Automobile Emissions*   The number of parts per million of nitric oxide emissions y from a certain car engine is approximated by the model

$$y = -5.05x^3 + 3857x - 38,411.25, \quad 13 \le x \le 18$$

where x is the air-fuel ratio.

(a) Use a graphing utility to graph the emissions function.

(b) It is observed from the graph that two air-fuel ratios produce 2400 parts per million of nitric oxide, with one being 15. Use the graph to approximate the second air-fuel ratio.

(c) Algebraically approximate the second air-fuel ratio that produces 2400 parts per million of nitric oxide. (*Hint:* Because you know that an air-fuel ratio of 15 produces the specified nitric oxide emission, you can use synthetic division.)

92. *Advertising Costs*   A company that manufactures bicycles estimates that the profit for selling a particular model is

$$P = -45x^3 + 2500x^2 - 275,000, \quad 0 \le x \le 50$$

where P is the profit (in dollars) and x is the advertising expense (in tens of thousands of dollars). According to this model, find the smaller of two advertising amounts that yield a profit of $800,000.

Synthesis

True or False?   **In Exercises 93–95, determine whether the statement is true or false. Justify your answer.**

93. If $(7x + 4)$ is a factor of some polynomial function f, then $\frac{4}{7}$ is a zero of f.

94. The rational expression $\dfrac{x^3 + 2x^2 - 13x + 10}{x^2 - 4x - 12}$ is improper.

95. $(2x - 1)$ is a factor of the polynomial

$$6x^6 + x^5 - 92x^4 + 45x^3 + 184x^2 + 4x - 48.$$

Think About It   **In Exercises 96 and 97, perform the division by assuming that n is a positive integer.**

96. $\dfrac{x^{3n} + 9x^{2n} + 27x^n + 27}{x^n + 3}$

97. $\dfrac{x^{3n} - 3x^{2n} + 5x^n - 6}{x^n - 2}$

98. *Think About It*   What does it mean for a divisor to divide evenly into a dividend?

99. *Writing*   Write a short paragraph explaining how you can check polynomial division. Give an example.

Review

In Exercises 100–103, find (a) $f \circ g$ and (b) $g \circ f$.

100. $f(x) = x^2, \quad g(x) = x + 4$

101. $f(x) = 2x - 5, \quad g(x) = x^2 + 1$

102. $f(x) = |x| + 2, \quad g(x) = x - 3$

103. $f(x) = \dfrac{1}{x}, \quad g(x) = 4x + x^2$

In Exercises 104–107, find a polynomial function that has the given zeros.

104. $0, -12$

105. $1, -3, 8$

106. $0, -1, 2, 5$

107. $2 + \sqrt{3}, 2 - \sqrt{3}$

2.4 Complex Numbers

The Imaginary Unit *i*

Some quadratic equations have no real solutions. For instance, the quadratic equation $x^2 + 1 = 0$ has no real solution because there is no real number x that can be squared to produce -1. To overcome this deficiency, mathematicians created an expanded system of numbers using the **imaginary unit *i*,** defined as

$$i = \sqrt{-1} \qquad \text{Imaginary unit}$$

where $i^2 = -1$. By adding real numbers to real multiples of this imaginary unit, you obtain the set of **complex numbers.** Each complex number can be written in the **standard form** $a + bi$. For instance, the standard form of the complex number $\sqrt{-9} - 5$ is $-5 + 3i$ because

$$\sqrt{-9} - 5 = \sqrt{3^2(-1)} - 5 = 3\sqrt{-1} - 5 = -5 + 3i.$$

In the standard form $a + bi$, the real number a is called the **real part** of the complex number and the number bi (where b is a real number) is called the **imaginary part** of the complex number.

Definition of a Complex Number

If a and b are real numbers, the number $a + bi$ is a **complex number,** and it is said to be written in **standard form.** If $b = 0$, the number $a + bi = a$ is a real number. If $b \neq 0$, the number $a + bi$ is called an **imaginary number.** A number of the form bi, where $b \neq 0$, is called a **pure imaginary number.**

The set of real numbers is a subset of the set of complex numbers, as shown in Figure 2.27. This is true because every real number a can be written as a complex number using $b = 0$. That is, for every real number a, you can write $a = a + 0i$.

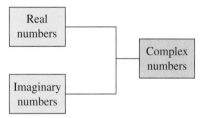

Figure 2.27

Equality of Complex Numbers

Two complex numbers $a + bi$ and $c + di$, written in standard form, are equal to each other

$$a + bi = c + di \qquad \text{Equality of two complex numbers}$$

if and only if $a = c$ and $b = d$.

What You Should Learn:

- How to use the imaginary unit *i* to write complex numbers
- How to add, subtract, and multiply complex numbers
- How to use complex conjugates to divide complex numbers
- How to plot complex numbers in the complex plane

Why You Should Learn It:

Complex numbers are used to model numerous aspects of the natural world, such as the impedance of an electrical circuit, as shown in Exercise 83 on page 181.

Phil Degginger/Tony Stone Images

Operations with Complex Numbers

To add (or subtract) two complex numbers, you add (or subtract) the real and imaginary parts of the numbers separately.

Addition and Subtraction of Complex Numbers

If $a + bi$ and $c + di$ are two complex numbers written in standard form, their sum and difference are defined as follows.

    Sum: $(a + bi) + (c + di) = (a + c) + (b + d)i$

    Difference: $(a + bi) - (c + di) = (a - c) + (b - d)i$

The **additive identity** in the complex number system is zero (the same as in the real number system). Furthermore, the **additive inverse** of the complex number $a + bi$ is

    $-(a + bi) = -a - bi.$ Additive inverse

So, you have

    $(a + bi) + (-a - bi) = 0 + 0i = 0.$

Carl Friedrich Gauss (1777–1855) proved that all the roots of any algebraic equation are "numbers" of the form $a + bi$, where a and b are real numbers and i is the square root of -1. These "numbers" are called complex.

EXAMPLE 1   Adding and Subtracting Complex Numbers

a. $(3 - i) + (2 + 3i) = 3 - i + 2 + 3i$ Remove parentheses.

               $= 3 + 2 - i + 3i$ Group like terms.

            $= (3 + 2) + (-1 + 3)i$

            $= 5 + 2i$ Write in standard form.

b. $2i + (-4 - 2i) = 2i - 4 - 2i$ Remove parentheses.

            $= -4 + 2i - 2i$ Group like terms.

            $= -4$ Write in standard form.

c. $3 - (-2 + 3i) + (-5 + i) = 3 + 2 - 3i - 5 + i$

                     $= 3 + 2 - 5 - 3i + i$

                     $= 0 - 2i$

                     $= -2i$

d. $(3 + 2i) + (4 - i) - (7 + i) = 3 + 2i + 4 - i - 7 - i$

                       $= 3 + 4 - 7 + 2i - i - i$

                       $= 0 + 0i$

                       $= 0$

In Examples 1(b) and 1(d) note that the sum of complex numbers can be a real number.

Many of the properties of real numbers are valid for complex numbers as well. Here are some examples.

Associative Property of Addition and Multiplication
Commutative Property of Addition and Multiplication
Distributive Property of Multiplication over Addition

Notice how these properties are used when two complex numbers are multiplied.

$$(a + bi)(c + di) = a(c + di) + bi(c + di) \qquad \text{Distributive Property}$$
$$= ac + (ad)i + (bc)i + (bd)i^2 \qquad \text{Distributive Property}$$
$$= ac + (ad)i + (bc)i + (bd)(-1) \qquad i^2 = -1$$
$$= ac - bd + (ad)i + (bc)i \qquad \text{Commutative Property}$$
$$= (ac - bd) + (ad + bc)i \qquad \text{Distributive Property}$$

Rather than trying to memorize this multiplication rule, you should simply remember how the distributive property is used to multiply two complex numbers. The procedure is similar to multiplying two polynomials and combining like terms (as in the FOIL Method).

EXAMPLE 2   Multiplying Complex Numbers

a. $(i)(-3i) = -3i^2$ Multiply.

$\qquad\qquad = -3(-1)$ $i^2 = -1$

$\qquad\qquad = 3$ Simplify.

b. $\sqrt{-4} \cdot \sqrt{-16} = (2i)(4i)$ Write each factor in standard form.

$\qquad\qquad = 8i^2$ Multiply.

$\qquad\qquad = 8(-1)$ $i^2 = -1$

$\qquad\qquad = -8$ Simplify.

c. $(2 - i)(4 + 3i) = 8 + 6i - 4i - 3i^2$ Product of binomials

$\qquad\qquad = 8 + 6i - 4i - 3(-1)$ $i^2 = -1$

$\qquad\qquad = 8 + 3 + 6i - 4i$ Group like terms.

$\qquad\qquad = 11 + 2i$ Write in standard form.

d. $(3 + 2i)(3 - 2i) = 9 - 6i + 6i - 4i^2$ Product of binomials

$\qquad\qquad = 9 - 4(-1)$ $i^2 = -1$

$\qquad\qquad = 9 + 4$ Simplify.

$\qquad\qquad = 13$ Write in standard form.

e. $(3 + 2i)^2 = 9 + 6i + 6i + 4i^2$ Product of binomials

$\qquad\qquad = 9 + 4(-1) + 12i$ $i^2 = -1$

$\qquad\qquad = 9 - 4 + 12i$ Simplify.

$\qquad\qquad = 5 + 12i$ Write in standard form.

Exploration

Complete the table:

$i^1 = i$	$i^7 = \boxed{}$
$i^2 = -1$	$i^8 = \boxed{}$
$i^3 = -i$	$i^9 = \boxed{}$
$i^4 = 1$	$i^{10} = \boxed{}$
$i^5 = \boxed{i}$	$i^{11} = \boxed{}$
$i^6 = \boxed{-1}$	$i^{12} = \boxed{}$

What pattern do you see? Write a brief description of how you would find i raised to any positive integer power.

STUDY TIP

Note in Example 2(b) that

$$\sqrt{-4} \cdot \sqrt{-16} \neq \sqrt{(-4)(-16)}$$
$$= \sqrt{64}$$
$$= 8.$$

Complex Conjugates and Division

Notice in Example 2(d) that the product of two complex numbers can be a real number. This occurs with pairs of complex numbers of the form $a + bi$ and $a - bi$, called **complex conjugates.**

$$(a + bi)(a - bi) = a^2 - abi + abi - b^2i^2$$

$$= a^2 - b^2(-1)$$

$$= a^2 + b^2$$

To find the quotient of $a + bi$ and $c + di$ where c and d are not both zero, multiply the numerator and denominator by the conjugate of the *denominator* to obtain

$$\frac{a + bi}{c + di} = \frac{a + bi}{c + di}\left(\frac{c - di}{c - di}\right)$$

$$= \frac{(ac + bd) + (bc - ad)i}{c^2 + d^2}.$$

EXAMPLE 3   Dividing Complex Numbers

$$\frac{1}{1 + i} = \frac{1}{1 + i}\left(\frac{1 - i}{1 - i}\right)$$ Multiply numerator and denominator by conjugate of denominator.

$$= \frac{1 - i}{1^2 - i^2}$$ Expand.

$$= \frac{1 - i}{1 - (-1)}$$ $i^2 = -1$

$$= \frac{1 - i}{2}$$ Simplify.

$$= \frac{1}{2} - \frac{1}{2}i$$ Write in standard form.

EXAMPLE 4   Dividing Complex Numbers

$$\frac{2 + 3i}{4 - 2i} = \frac{2 + 3i}{4 - 2i}\left(\frac{4 + 2i}{4 + 2i}\right)$$ Multiply numerator and denominator by conjugate of denominator.

$$= \frac{8 + 4i + 12i + 6i^2}{16 - 4i^2}$$ Expand.

$$= \frac{8 - 6 + 16i}{16 + 4}$$ $i^2 = -1$

$$= \frac{(2 + 16i)}{20}$$ Simplify.

$$= \frac{1}{10} + \frac{4}{5}i$$ Write in standard form.

STUDY T!P

Some graphing utilities can perform operations with complex numbers. For instance, on some graphing utilities, to divide $2 + 3i$ by $4 - 2i$, enter

$$\boxed{(}\ 2\ \boxed{+}\ 3\ \boxed{i}\ \boxed{)}\ \boxed{\div}$$
$$\boxed{(}\ 4\ \boxed{-}\ 2\ \boxed{i}\ \boxed{)}\ \boxed{\text{ENTER}}.$$

The display will be as follows.

$$.1 + .8i \quad \text{or} \quad \frac{1}{10} + \frac{4}{5}i$$

Consult your user's manual for specific instructions on performing operations with complex numbers.

Applications

Most applications involving complex numbers are either theoretical or very technical, and are therefore not appropriate for inclusion in this text. However, to give you some idea of how complex numbers can be used in applications, we give a general description of their use in **fractal geometry.**

To begin, consider a coordinate system called the **complex plane.** Just as every real number corresponds to a point on the real number line, every complex number corresponds to a point in the complex plane, as shown in Figure 2.28. In this figure, note that the vertical axis is the **imaginary axis** and the horizontal axis is the **real axis.** The point that corresponds to the complex number $a + bi$ is (a, b).

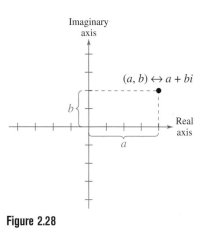

Figure 2.28

EXAMPLE 5   Plotting Complex Numbers

Plot each complex number in the complex plane.

a. $2 + 3i$ **b.** $-1 + 2i$

c. 4 **d.** $-3i$

Solution

a. To plot the complex number $2 + 3i$, move (from the origin) two units to the right on the real axis and then three units up, as shown in Figure 2.29. In other words, plotting the complex number $2 + 3i$ in the complex plane is comparable to plotting the point $(2, 3)$ in the Cartesian plane.

b. The complex number $-1 + 2i$ corresponds to the point $(-1, 2)$, as shown in Figure 2.29.

c. The complex number 4 corresponds to the point $(4, 0)$, as shown in Figure 2.29.

d. The complex number $-3i$ corresponds to the point $(0, 3)$, as shown in Figure 2.29.

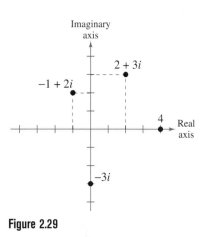

Figure 2.29

In the hands of a person who understands fractal geometry, the complex plane can become an easel on which stunning pictures, called **fractals,** can be drawn. The most famous such picture is called the **Mandelbrot Set,** named after the Polish-born mathematician Benoit Mandelbrot. To draw the Mandelbrot Set, consider the following sequence of numbers.

$$c, c^2 + c, (c^2 + c)^2 + c, [(c^2 + c)^2 + c]^2 + c, \ldots$$

The behavior of this sequence depends on the value of the complex number c. For some values of c this sequence is **bounded,** which means that all elements in the sequence are less than some fixed number N. For other values it is **unbounded,** which means that the elements in the sequence become infinitely large. If the sequence is bounded, the complex number c is in the Mandelbrot Set; if the sequence is unbounded, the complex number c is not in the Mandelbrot Set.

EXAMPLE 6   Members of the Mandelbrot Set

a. The complex number -2 is in the Mandelbrot Set because for $c = -2$, the corresponding Mandelbrot sequence is $-2, 2, 2, 2, 2, 2, \ldots$, which is bounded.

b. The complex number i is also in the Mandelbrot Set because for $c = i$, the corresponding Mandelbrot sequence is

$$i, \quad -1 + i, \quad -i, \quad -1 + i, \quad -i, \quad -1 + i, \quad \ldots$$

which is bounded.

c. The complex number $1 + i$ is not in the Mandelbrot Set because for $c = 1 + i$, the corresponding Mandelbrot sequence is

$$1 + i, \quad 1 + 3i, \quad -7 + 7i, \quad 1 - 97i, \quad -9407 - 193i,$$

$$88454401 + 3631103i, \quad \ldots$$

which is unbounded.

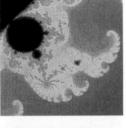

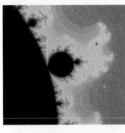

With this definition, a picture of the Mandelbrot Set would have only two colors: one color for points that are in the set (the sequence is bounded), and one for points that are outside the set (the sequence is unbounded). Figure 2.30 shows a black and yellow picture of the Mandelbrot Set. The points that are black are in the Mandelbrot Set and the points that are yellow are not.

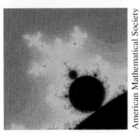

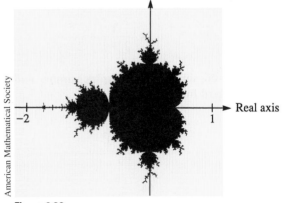

Imaginary axis

Real axis

-2 1

American Mathematical Society

Figure 2.30

American Mathematical Society

Figure 2.31

To add more interest to the picture, computer scientists discovered that the points that are not in the Mandelbrot Set can be assigned a variety of colors, depending on "how quickly" their sequences diverge. Figure 2.31 shows three different appendages of the Mandelbrot Set. (The black portions of the picture represent points that are in the Mandelbrot Set.)

Figure 2.32 shows another type of fractal. From this picture, you can see why fractals have fascinated people since their discovery (around 1980). The fractal shown was produced on a graphing calculator.

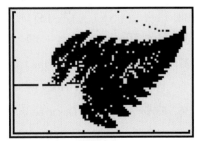

Figure 2.32   *A Fractal Fern*

2.4 Exercises

In Exercises 1–4, solve for a and b.

1. $a + bi = -9 + 4i$ **2.** $a + bi = 12 + 5i$

3. $(a - 1) + (b + 3)i = 5 + 8i$

4. $(a + 6) + 2bi = 6 - 5i$

In Exercises 5–14, write in standard form.

5. $4 + \sqrt{-25}$ **6.** $3 + \sqrt{-9}$

7. 12 **8.** 42

9. $-5i + i^2$ **10.** $-3i^2 + i$

11. $\left(\sqrt{-75}\right)^2$ **12.** $\left(\sqrt{-4}\right)^2 - 7$

13. $\sqrt{-0.09}$ **14.** $\sqrt{-0.0004}$

In Exercises 15–24, perform the addition or subtraction and write the result in standard form.

15. $(4 + i) + (7 - 2i)$ **16.** $(11 - 2i) + (-3 + 6i)$

17. $\left(-1 + \sqrt{-8}\right) + \left(8 - \sqrt{-50}\right)$

18. $\left(7 + \sqrt{-18}\right) - \left(3 + 3\sqrt{2}i\right)$

19. $13i - (14 - 7i)$ **20.** $22 + (-5 + 8i) + 10i$

21. $-\left(\frac{3}{2} + \frac{5}{2}i\right) + \left(\frac{5}{3} + \frac{11}{3}i\right)$ **22.** $-\left(\frac{3}{4} + \frac{7}{5}i\right) - \left(\frac{5}{6} - \frac{1}{6}i\right)$

23. $(1.6 + 3.2i) + (-5.8 + 4.3i)$

24. $-(-3.7 - 12.8i) - \left(6.1 - \sqrt{-24.5}\right)$

In Exercises 25–36, perform the operation and write the result in standard form.

25. $\sqrt{-6} \cdot \sqrt{-2}$ **26.** $\sqrt{-5} \cdot \sqrt{-10}$

27. $\left(\sqrt{-10}\right)^2$ **28.** $\left(\sqrt{-75}\right)^2$

29. $(1 + i)(3 - 2i)$ **30.** $(6 - 2i)(2 - 3i)$

31. $6i(5 - 2i)$ **32.** $-8i(9 + 4i)$

33. $\left(\sqrt{14} + \sqrt{10}i\right)\left(\sqrt{14} - \sqrt{10}i\right)$

34. $\left(3 + \sqrt{-5}\right)\left(7 - \sqrt{-10}\right)$

35. $(4 + 5i)^2$ **36.** $(1 - 2i)^2 - (1 + 2i)^2$

37. *Error Analysis*   Describe the error.

$$\sqrt{-6}\sqrt{-6} = \sqrt{(-6)(-6)} = \sqrt{36} = 6 \; \times$$

38. *Error Analysis*   Describe the error.

$$-i\left(\sqrt{-4} - 1\right) = -i(4i - 1) \; \times$$
$$= -4i^2 - i$$
$$= 4 - i$$

In Exercises 39–46, find the product of the number and its conjugate.

39. $4 + 3i$ **40.** $8 - 12i$

41. $-6 - \sqrt{5}i$ **42.** $-3 + \sqrt{2}i$

43. $22i$ **44.** $\sqrt{-13}$

45. $3 - \sqrt{-2}$ **46.** $1 + \sqrt{-8}$

In Exercises 47–58, perform the operation and write the result in standard form.

47. $\dfrac{6}{i}$ **48.** $-\dfrac{5}{i}$

49. $\dfrac{4}{4 - 5i}$ **50.** $\dfrac{3}{1 - i}$

51. $\dfrac{2 + i}{2 - i}$ **52.** $\dfrac{8 - 7i}{1 - 2i}$

53. $\dfrac{6 - 7i}{i}$ **54.** $\dfrac{8 + 20i}{2i}$

55. $\dfrac{1}{(4 - 5i)^2}$ **56.** $\dfrac{(2 - 3i)(5i)}{2 + 3i}$

57. $\dfrac{2}{1 + i} - \dfrac{3}{1 - i}$ **58.** $\dfrac{2i}{2 + i} + \dfrac{5}{2 - i}$

In Exercises 59–66, simplify the complex number and write it in standard form.

59. $-6i^3 + i^2$ **60.** $4i^2 - 2i^3$

61. $-5i^5$ **62.** $(-i)^3$

63. $\left(\sqrt{-75}\right)^3$ **64.** $\left(\sqrt{-2}\right)^6$

65. $\dfrac{1}{i^3}$ **66.** $\dfrac{1}{(2i)^3}$

In Exercises 67–70, determine the complex number shown in the complex plane.

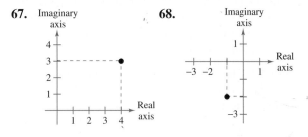

67.   **68.**

69.
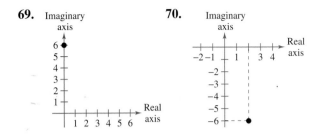

70.

(a) 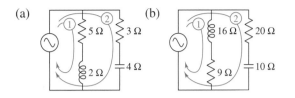 (b)

Synthesis

In Exercises 71–74, plot the complex number in the complex plane.

71. $4 - 5i$

72. $3i$

73. -6

74. $-7 + 2i$

Fractals In Exercises 75–80, find the first six terms in the following sequence.

$$c, \ c^2 + c, \ (c^2 + c)^2 + c, \ [(c^2 + c)^2 + c]^2 + c, \ \dots$$

From these terms, do you think the given complex number is in the Mandelbrot Set? Explain your reasoning.

75. $c = 0$

76. $c = 2$

77. $c = \frac{1}{2}i$

78. $c = -i$

79. $c = 1$

80. $c = -1$

81. Cube each complex number. What do you notice?

(a) 2 (b) $-1 + \sqrt{3}i$ (c) $-1 - \sqrt{3}i$

82. Raise each number to the fourth power.

(a) 2 (b) -2 (c) $2i$ (d) $-2i$

83. *Impedance* The opposition to current in an electrical circuit is called its impedance. The impedance in a parallel circuit with two pathways satisfies the equation

$$\frac{1}{z} = \frac{1}{z_1} + \frac{1}{z_2}$$

when z_1 is the impedance (in ohms) of pathway 1 and z_2 is the impedance (in ohms) of pathway 2. Use the table to determine the impedance of each parallel circuit. (*Hint:* You can find the impedance of each pathway by adding the impedance of each component in the pathway.)

	Resistor	Inductor	Capacitor
Symbol	—⩗⩗— $a\,\Omega$	—⁀⁀⁀— $b\,\Omega$	—⊣⊢— $c\,\Omega$
Impedance	a	bi	$-ci$

In Exercises 84–86, determine whether the statement is true or false. Justify your answer.

84. There is no complex number that is equal to its conjugate.

85. $-i\sqrt{6}$ is a solution of $x^4 - x^2 + 14 = 56$.

86. $i^{44} + i^{150} - i^{74} - i^{109} + i^{61} = -1$

87. Express each of the following powers of i as i, $-i$, 1, or -1.

(a) i^{40} (b) i^{25} (c) i^{50} (d) i^{67}

88. Prove that the sum of a complex number $a + bi$ and its conjugate is a real number, and that the difference of a complex number $a + bi$ and its conjugate is an imaginary number.

89. Prove that the product of a complex number $a + bi$ and its conjugate is a real number.

Review

In Exercises 90 and 91, write an equation of the line through the point (a) parallel to the given line and (b) perpendicular to the line.

90. $(3, -2)$ $5x - 4y = 8$

91. $(-8, 3)$ $2x + 3y = 5$

In Exercises 92–95, find the x- and y-intercepts of the graph of the equation.

92. $y = -x^2 + 6$

93. $y = x^2 + 2x - 8$

94. $y = |x - 4| + 1$

95. $y = |x| - 1$

96. *Mixture Problem* A 5-liter container contains a mixture with a concentration of 50%. How much of this mixture must be withdrawn and replaced by 100% concentrate to bring the mixture up to 60% concentration?

2.5 The Fundamental Theorem of Algebra

The Fundamental Theorem of Algebra

You have been using the fact that an *n*th-degree polynomial can have at most *n* real zeros. In the complex number system, this statement can be improved. That is, in the complex number system, every *n*th-degree polynomial function has *precisely n* zeros. This important result is derived from the **Fundamental Theorem of Algebra,** first proved by the famous German mathematician Carl Friedrich Gauss (1777–1855).

> ### The Fundamental Theorem of Algebra
>
> If $f(x)$ is a polynomial of degree n, where $n > 0$, f has at least one zero in the complex number system.

Using the Fundamental Theorem of Algebra and the equivalence of zeros and factors, you obtain the **Linear Factorization Theorem**.

> ### Linear Factorization Theorem
>
> If $f(x)$ is a polynomial of degree n where $n > 0$, f has precisely n linear factors
>
> $$f(x) = a_n(x - c_1)(x - c_2) \cdots (x - c_n)$$
>
> where $c_1, c_2, \ldots, c_n$ are complex numbers.

(A proof of the Linear Factorization Theorem is found in Appendix A.)

Note that neither the Fundamental Theorem of Algebra nor the Linear Factorization Theorem tells you *how* to find the zeros or factors of a polynomial. Such theorems are called *existence theorems.* To find the zeros of a polynomial function, you still must rely on other techniques.

Remember that the *n* zeros of a polynomial function can be real or complex, and they may be repeated. Examples 1 and 2 illustrate several cases.

EXAMPLE 1   Real Zeros of Polynomial Functions

a. The first-degree polynomial $f(x) = x - 2$ has exactly *one* zero: $x = 2$.

b. Counting multiplicity, the second-degree polynomial function

$$f(x) = x^2 - 6x + 9 = (x - 3)(x - 3)$$

has exactly *two* zeros: $x = 3$ and $x = 3$.

Note in Example 1 that you can use a graphing utility to verify graphically the zeros of the polynomial functions.

What You Should Learn:

- How to use the Fundamental Theorem of Algebra to determine the number of zeros of polynomial functions
- How to find all zeros of polynomial functions, including complex zeros
- How to find conjugate pairs of complex zeros
- How to find zeros of polynomials by factoring

Why You Should Learn It:

Being able to find zeros of polynomial functions is an important part of modeling real-life problems. For instance, Exercise 65 on page 188 shows how to determine whether a ball thrown with a given velocity can reach a certain height.

Jed Jacobsohn/Allsport

EXAMPLE 2   Real and Complex Zeros of Polynomial Functions

a. The third-degree polynomial function

$$f(x) = x^3 + 4x = x(x^2 + 4) = x(x - 2i)(x + 2i)$$

has exactly *three* zeros: $x = 0$, $x = 2i$, and $x = -2i$. In the graph in Figure 2.33(a), only the *real* zero $x = 0$ appears as an intercept.

b. The fourth-degree polynomial function

$$f(x) = x^4 - 1 = (x - 1)(x + 1)(x - i)(x + i)$$

has exactly *four* zeros: $x = 1$, $x = -1$, $x = i$, and $x = -i$. In the graph in Figure 2.33(b), only the *real* zeros $x = -1$ and $x = 1$ appear as x-intercepts.

Example 3 shows how to use the methods described in Sections 2.2 and 2.3 (the Rational Zero Test, synthetic division, and factoring) to find all the zeros of a polynomial function, including complex zeros.

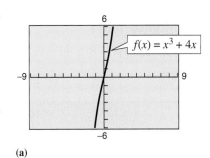

(a)

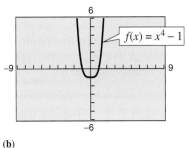

(b)

Figure 2.33

EXAMPLE 3   Finding the Zeros of a Polynomial Function

Write

$$f(x) = x^5 + x^3 + 2x^2 - 12x + 8$$

as the product of linear factors, and list all of its zeros.

Solution

The possible rational zeros are ± 1, ± 2, ± 4, and ± 8. The graph shown in Figure 2.34 indicates that 1 and -2 are good guesses, and that 1 is possibly a repeated zero because it touches the x-axis.

Using synthetic division, you can determine that -2 is a zero and 1 is a repeated zero of f. So, you have

$$f(x) = x^5 + x^3 + 2x^2 - 12x + 8$$

$$= (x - 1)(x - 1)(x + 2)(x^2 + 4).$$

By factoring $x^2 + 4$ as

$$x^2 - (-4) = \left(x - \sqrt{-4}\right)\left(x + \sqrt{-4}\right) = (x - 2i)(x + 2i)$$

you obtain

$$f(x) = (x - 1)(x - 1)(x + 2)(x - 2i)(x + 2i)$$

which gives the following five zeros of f.

$$1, \quad 1, \quad -2, \quad 2i, \quad \text{and} \quad -2i$$

Note from the graph of f shown in Figure 2.34 that the *real* zeros are the only ones that appear as x-intercepts.

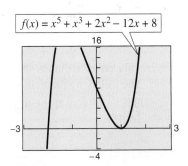

Figure 2.34

Conjugate Pairs

In Example 3, note that the two complex zeros are **conjugates.** That is, they are of the form $a + bi$ and $a - bi$.

Complex Zeros Occur in Conjugate Pairs

Let $f(x)$ be a polynomial function that has *real coefficients.* If $a + bi$, where $b \neq 0$, is a zero of the function, the conjugate $a - bi$ is also a zero of the function.

Be sure you see that this result is true only if the polynomial function has *real coefficients.* For instance, the result applies to the function $f(x) = x^2 + 1$, but not to the function $g(x) = x - i$.

Jean Le Rond d'Alembert (1717–1783) worked independently of Carl Gauss trying to prove the Fundamental Theorem of Algebra. His efforts were such that in France, the Fundamental Theorem of Algebra is frequently known as the Theorem of d'Alembert.

EXAMPLE 4   Finding a Polynomial with Given Zeros

Find a *fourth-degree* polynomial function, with real coefficients, that has -1, -1, and $3i$ as zeros.

Solution

Because $3i$ is a zero *and* the polynomial is stated to have real coefficients, you know that the conjugate $-3i$ must also be a zero. So, from the Linear Factorization Theorem, $f(x)$ can be written as

$$f(x) = a(x + 1)(x + 1)(x - 3i)(x + 3i).$$

For simplicity, let $a = 1$, to obtain

$$f(x) = (x^2 + 2x + 1)(x^2 + 9)$$

$$= x^4 + 2x^3 + 10x^2 + 18x + 9.$$

Factoring a Polynomial

The Linear Factorization Theorem shows that you can write any nth-degree polynomial as the product of n linear factors.

$$f(x) = a(x - c_1)(x - c_2)(x - c_3) \cdots (x - c_n)$$

However, this result includes the possibility that some of the values of c_i are complex. The following theorem says that even if you do not want to get involved with "complex factors," you can still write $f(x)$ as the product of linear and/or quadratic factors.

STUDY T!P

Some graphing utilities can find both real and complex zeros of a function. For instance, to find all the zeros of

$$f(x) = x^4 - 3x^3 + 6x^2$$
$$+ 2x - 60$$

use the graphing utility's *solve* feature. If your graphing utility can find complex zeros, you will obtain the following solutions.

$$x = 1 + 3i, \quad x = 1 - 3i,$$
$$x = 3, \quad x = -2$$

Factors of a Polynomial

Every polynomial of degree $n > 0$ with real coefficients can be written as the product of linear and quadratic factors with real coefficients, where the quadratic factors have no real zeros.

(For a proof of this theorem, see Appendix A.)

A quadratic factor with no real zeros is said to be **irreducible over the reals.** Be sure you see that this is not the same as being *irreducible over the rationals.* For example, the quadratic

$$x^2 + 1 = (x - i)(x + i)$$

is irreducible over the reals (and therefore over the rationals). On the other hand, the quadratic

$$x^2 - 2 = \left(x - \sqrt{2}\right)\left(x + \sqrt{2}\right)$$

is irreducible over the rationals, but *reducible* over the reals.

EXAMPLE 5   Factoring a Polynomial

Write the polynomial

$$f(x) = x^4 - x^2 - 20$$

a. as the product of factors that are irreducible over the *rationals,*

b. as the product of linear factors and quadratic factors that are irreducible over the *reals,* and

c. in completely factored form.

Solution

a. Begin by factoring the polynomial into the product of two quadratic polynomials.

$$x^4 - x^2 - 20 = (x^2 - 5)(x^2 + 4)$$

Both of these factors are irreducible over the rationals.

b. By factoring over the reals, you have

$$x^4 - x^2 - 20 = \left(x + \sqrt{5}\right)\left(x - \sqrt{5}\right)(x^2 + 4)$$

where the quadratic factor is irreducible over the reals.

c. In completely factored form, you have

$$x^4 - x^2 - 20 = \left(x + \sqrt{5}\right)\left(x - \sqrt{5}\right)(x - 2i)(x + 2i).$$

In Example 5, notice from the completely factored form that the fourth-degree polynomial has four zeros.

Throughout this chapter, the results and theorems have basically been stated in terms of zeros of polynomial functions. Be sure you see that the same results could have been stated in terms of solutions of polynomial equations. This is true because the zeros of the polynomial function

$$f(x) = a_n x^n + a_{n-1} x^{n-1} + \cdots + a_2 x^2 + a_1 x + a_0$$

are precisely the solutions of the polynomial equation

$$a_n x^n + a_{n-1} x^{n-1} + \cdots + a_2 x^2 + a_1 x + a_0 = 0.$$

EXAMPLE 6   Finding the Zeros of a Polynomial Function

Find all the zeros of

$$f(x) = x^4 - 3x^3 + 6x^2 + 2x - 60$$

given that $1 + 3i$ is a zero of f.

Algebraic Solution

Because complex zeros occur in conjugate pairs, you know that $1 - 3i$ is also a zero of f. This means that both

$$x - (1 + 3i) \quad \text{and} \quad x - (1 - 3i)$$

are factors of $f(x)$. Multiplying these two factors produces

$$[x - (1 + 3i)][x - (1 - 3i)] = [(x - 1) - 3i][(x - 1) + 3i]$$
$$= (x - 1)^2 - 9i^2$$
$$= x^2 - 2x + 10.$$

Using long division, you can divide $x^2 - 2x + 10$ into $f(x)$ to obtain the following.

$$
\begin{array}{r}
x^2 - x - 6 \\
x^2 - 2x + 10 \overline{)\,x^4 - 3x^3 + 6x^2 + 2x - 60} \\
\underline{x^4 - 2x^3 + 10x^2} \\
-x^3 - 4x^2 + 2x \\
\underline{-x^3 + 2x^2 - 10x} \\
-6x^2 + 12x - 60 \\
\underline{-6x^2 + 12x - 60} \\
0
\end{array}
$$

Therefore, you have

$$f(x) = (x^2 - 2x + 10)(x^2 - x - 6)$$
$$= (x^2 - 2x + 10)(x - 3)(x + 2)$$

and you can conclude that the zeros of f are $1 + 3i$, $1 - 3i$, 3, and -2.

Graphical Solution

Because complex zeros always occur in conjugate pairs, you know that $1 - 3i$ is also a zero of f. Because the polynomial is a fourth-degree polynomial, you know that there are at most two other zeros of the function. Use a graphing utility to graph

$$y = x^4 - 3x^3 + 6x^2 + 2x - 60$$

as shown in Figure 2.35.

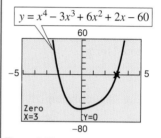

Figure 2.35

You can see that -2 and 3 appear to be x-intercepts of the graph of the function. Use the *zero* or *root* feature or the *zoom* and *trace* features of the graphing utility to confirm that $x = -2$ and $x = 3$ are x-intercepts of the graph. So, you can conclude that the zeros of f are

$$1 + 3i, \quad 1 - 3i, \quad 3, \quad \text{and} \quad -2.$$

In Example 6, if you had not been told that $1 + 3i$ is a zero of f, you could still find all zeros of the function. You could use synthetic division to find the real zeros -2 and 3. Then, you could factor the polynomial as $(x + 2)(x - 3)(x^2 - 2x + 10)$. Finally, by using the Quadratic Formula, you could determine that the zeros are $1 + 3i$, $1 - 3i$, 3, and -2.

Writing About Math   *Factoring a Polynomial*

Compile a list of all the various techniques for factoring a polynomial that have been covered so far in the text. Give an example illustrating each technique, and write a paragraph discussing when the use of each technique is appropriate.

2.5 E x e r c i s e s

In Exercises 1–8, find all the zeros of the function.

1. $f(x) = x(x - 6)^2$

2. $f(x) = x^2(x + 3)(x^2 - 1)$

3. $g(x) = (x - 2)(x + 4)^3$ **4.** $f(x) = (x + 5)(x - 8)^2$

5. $f(x) = (x + 6)(x + i)(x - i)$

6. $h(t) = (t - 3)(t - 2)(t - 3i)(t + 3i)$

7. $f(x) = (x - 2)(x + 3 - 5i)(x + 3 + 5i)$

8. $h(m) = (m - 4)^2(m - 2 + 4i)(m - 2 - 4i)$

Graphical and Analytical Analysis   **In Exercises 9–12, find all the zeros of the function. Is there a relationship between the number of real zeros and the number of x-intercepts of the graph? Explain.**

9. $f(x) = x^3 - 4x^2$ **10.** $f(x) = x^3 - 4x^2$

  $+ x - 4$ $- 4x + 16$

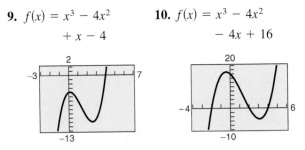

11. $f(x) = x^4 + 4x^2 + 4$ **12.** $f(x) = x^4 - 3x^2 - 4$

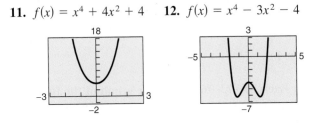

In Exercises 13–32, find all the zeros of the function and write the polynomial as a product of linear factors. Use a graphing utility to graph the function to verify your results graphically. (If your graphing utility can find complex zeros, use it to verify the complex zeros.)

13. $h(x) = x^2 - 4x + 1$ **14.** $g(x) = x^2 + 10x + 23$

15. $f(x) = x^2 - 12x + 26$ **16.** $f(x) = x^2 + 6x - 2$

17. $f(x) = x^2 + 25$ **18.** $f(x) = x^2 - x + 56$

19. $f(x) = x^4 - 81$ **20.** $f(y) = y^4 - 625$

21. $f(z) = z^2 - 2z + 2$

22. $h(x) = x^3 - 3x^2 + 4x - 2$

23. $f(t) = t^3 - 3t^2 - 15t + 125$

24. $f(x) = x^3 + 11x^2 + 39x + 29$

25. $f(x) = 16x^3 - 20x^2 - 4x + 15$

26. $f(s) = 2s^3 - 5s^2 + 12s - 5$

27. $f(x) = x^4 + 10x^2 + 9$

28. $f(x) = x^4 + 29x^2 + 100$

29. $g(x) = x^4 - 4x^3 + 8x^2 - 16x + 16$

30. $h(x) = x^4 + 6x^3 + 10x^2 + 6x + 9$

31. $f(x) = 2x^4 + 5x^3 + 4x^2 + 5x + 2$

32. $g(x) = x^5 - 8x^4 + 28x^3 - 56x^2 + 64x - 32$

In Exercises 33–40, (a) find all zeros of the function, (b) write the polynomial as a product of linear factors, (c) use your factorization to determine the x-intercepts of the graph of the function, and (d) use a graphing utility to verify that the real zeros are the only x-intercepts.

33. $f(x) = x^2 - 14x + 46$ **34.** $f(x) = x^2 - 12x + 34$

35. $f(x) = x^2 + 14x + 44$ **36.** $f(x) = x^2 - 16x + 62$

37. $f(x) = x^3 - 11x + 150$

38. $f(x) = x^3 + 10x^2 + 33x + 34$

39. $f(x) = x^4 + 25x^2 + 144$

40. $f(x) = x^4 - 8x^3 + 17x^2 - 8x + 16$

In Exercises 41–48, find a polynomial function with integer coefficients that has the given zeros. (There are many correct answers.)

41. $1, 5i, -5i$ **42.** $4, 3i, -3i$

43. $2, 4 + i, 4 - i$ **44.** $6, -5 + 2i, -5 - 2i$

45. $i, -i, 6i, -6i$ **46.** $2, 2, 2, 4i, -4i$

47. $-5, -5, 1 + \sqrt{3}i$ **48.** $0, 0, 4, 1 + \sqrt{2}i$

In Exercises 49–52, write the polynomial (a) as the product of factors that are irreducible over the *rationals*, (b) as the product of linear and quadratic factors that are irreducible over the *reals*, and (c) in completely factored form.

49. $f(x) = x^4 - 6x^2 - 7$ **50.** $f(x) = x^4 + 6x^2 - 27$

51. $f(x) = x^4 - 2x^3 - 3x^2 + 12x - 18$

  (*Hint:* One factor is $x^2 - 6$.)

52. $f(x) = x^4 - 3x^3 - x^2 - 12x - 20$

  (*Hint:* One factor is $x^2 + 4$.)

In Exercises 53–60, use the given zero to find all the zeros of the function.

Function	Zero
53. $f(x) = 2x^3 + 3x^2 + 50x + 75$	$5i$
54. $f(x) = 2x^4 - x^3 + 7x^2 - 4x - 4$	$2i$
55. $g(x) = x^3 - 7x^2 - x + 87$	$5 + 2i$
56. $g(x) = 4x^3 + 23x^2 + 34x - 10$	$-3 + i$
57. $h(x) = 3x^3 - 4x^2 + 8x + 8$	$1 - \sqrt{3}i$
58. $f(x) = x^3 + 4x^2 + 14x + 20$	$-1 - 3i$
59. $h(x) = 8x^3 - 14x^2 + 18x - 9$	$\frac{1}{2}(1 - \sqrt{5}i)$
60. $f(x) = 25x^3 - 55x^2 - 54x - 18$	$\frac{1}{5}(-2 + \sqrt{2}i)$

Graphical Analysis **In Exercises 61–64, (a) use the** *zero* **or** *root* **feature of a graphing utility to approximate the zeros of the function accurate to three decimal places, (b) determine one of the exact zeros and use synthetic division to verify your result, and (c) find the exact values of the remaining zeros.**

61. $f(x) = x^4 + 3x^3 - 5x^2 - 21x + 22$

62. $f(x) = x^3 + 4x^2 + 14x + 20$

63. $h(x) = 8x^3 - 14x^2 + 18x - 9$

64. $f(x) = 25x^3 - 55x^2 - 54x - 18$

65. *Maximum Height* A baseball is thrown upward from ground level with an initial velocity of 48 feet per second, and its height h (in feet) is

$$h = -16t^2 + 48t, \quad 0 \le t \le 3$$

where t is the time (in seconds). Suppose you are told that the ball reaches a height of 64 feet. Is this possible? Explain.

66. *Profit* The demand equation for a microwave is

$$p = 140 - 0.0001x$$

where p is the unit price (in dollars) of the microwave and x is the number of units produced and sold. The cost equation for the microwave is

$$C = 80x + 150,000$$

where C is the total cost (in dollars) and x is the number of units produced. The total profit obtained by producing and selling x units is

$$P = R - C = xp - C.$$

Suppose you are working in the marketing department that produces this microwave, and you are asked to determine a price p that would yield a profit of 9 million dollars. Is this possible? Explain.

Synthesis

True or False? **In Exercises 67 and 68, decide whether the statement is true or false. Justify your answer.**

67. It is possible for a third-degree polynomial function with integer coefficients to have no real zeros.

68. If $x = 4 + 3i$ is a zero of the function $f(x) = x^4 - 7x^3 - 13x^2 + 265x - 750$, then $x = -3i + 4$ must also be a zero of f.

69. *Exploration* Use a graphing utility to graph the function $f(x) = x^4 - 4x^2 + k$ for different values of k. Find values of k such that the zeros of f satisfy the specified characteristics. (Some parts have many correct answers.)

(a) Four real zeros

(b) Two real zeros each of multiplicity 2

(c) Two real zeros and two complex zeros

(d) Four complex zeros

70. *Think About It* Will the answers to Exercise 69 change for the function g?

(a) $g(x) = f(x - 2)$ (b) $g(x) = f(2x)$

71. Find a quadratic function f (with integer coefficients) that has $\pm \sqrt{b}i$ as zeros. Assume that b is a positive integer.

72. Find a quadratic function f (with integer coefficients) that has $a \pm bi$ as zeros. Assume that b is a positive integer.

Review

In Exercises 73–76, sketch the graph of the function. Identify the vertex and any intercepts. Use a graphing utility to verify your results.

73. $f(x) = x^2 - 7x - 8$

74. $f(x) = -x^2 + x + 6$

75. $f(x) = 6x^2 + 5x - 6$

76. $f(x) = 4x^2 + 2x - 12$

In Exercises 77–80, perform the operations and write the result in standard form.

77. $12 - (-7 + 5i) + (-2 + 3i)$

78. $-8i - i(2 + 3i)$

79. $(-3 - 8i)^2$ 80. $\dfrac{6 - i}{1 - 4i}$

2.6 Rational Functions and Asymptotes

Introduction to Rational Functions

A **rational function** can be written in the form

$$f(x) = \frac{N(x)}{D(x)}$$

where $N(x)$ and $D(x)$ are polynomials and $D(x)$ is not the zero polynomial. In this section it is assumed that $N(x)$ and $D(x)$ have no common factors.

In general, the *domain* of a rational function of x includes all real numbers except x-values that make the denominator zero. Much of the discussion of rational functions will focus on their graphical behavior near these x-values.

What You Should Learn:

• How to find domains of rational functions
• How to find horizontal and vertical asymptotes of graphs of rational functions
• How to use rational functions to model and solve real-life problems

Why You Should Learn It:

Rational functions are convenient in modeling a wide variety of real-life problems, such as environmental scenarios. For instance, Exercise 31 on page 196 shows how to determine the cost of removing pollutants from a river.

David Woodfull/Tony Stone Images

EXAMPLE 1   Finding the Domain of a Rational Function

Find the domain of $f(x) = 1/x$ and discuss the behavior of f near any excluded x-values.

Solution

Because the denominator is zero when $x = 0$, the domain of f is all real numbers except $x = 0$. To determine the behavior of f near this excluded value, evaluate $f(x)$ to the left and right of $x = 0$, as indicated in the following tables.

x	-1	-0.5	-0.1	-0.01	-0.001	$\to 0$
$f(x)$	-1	-2	-10	-100	-1000	$\to -\infty$

x	$0 \leftarrow$	0.001	0.01	0.1	0.5	1
$f(x)$	$\infty \leftarrow$	1000	100	10	2	1

Note that as x approaches 0 *from the left*, $f(x)$ decreases without bound. In contrast, as x approaches 0 *from the right*, $f(x)$ increases without bound. The graph of f is shown in Figure 2.36.

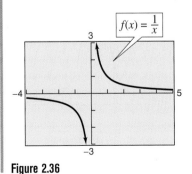

Figure 2.36

Horizontal and Vertical Asymptotes

In Example 1, the behavior of $f(x) = 1/x$ near $x = 0$ is denoted as follows.

$$f(x) \to -\infty \text{ as } x \to 0^-$$

$f(x)$ decreases without bound as x approaches 0 from the left.

$$f(x) \to \infty \text{ as } x \to 0^+$$

$f(x)$ increases without bound as x approaches 0 from the right.

The line $x = 0$ is a **vertical asymptote** of the graph of f, as shown in Figure 2.37. The graph of f also has a **horizontal asymptote**—the line $y = 0$. This means the values of $f(x) = 1/x$ approach zero as x increases or decreases without bound.

$$f(x) \to 0 \text{ as } x \to -\infty$$

$f(x)$ approaches 0 as x decreases without bound.

$$f(x) \to 0 \text{ as } x \to \infty$$

$f(x)$ approaches 0 as x increases without bound.

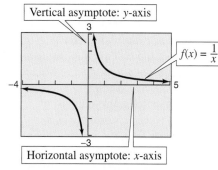

Vertical asymptote: y-axis

$f(x) = \dfrac{1}{x}$

Horizontal asymptote: x-axis

Figure 2.37

Definition of Vertical and Horizontal Asymptotes

1. The line $x = a$ is a **vertical asymptote** of the graph of f if $f(x) \to \infty$ or $f(x) \to -\infty$ as $x \to a$, either from the right or from the left.

2. The line $y = b$ is a **horizontal asymptote** of the graph of f if $f(x) \to b$ as $x \to \infty$ or $x \to -\infty$.

Eventually (as $x \to \infty$ or $x \to -\infty$), the distance between the horizontal asymptote and the points on the graph must approach zero. Figure 2.38 shows the horizontal and vertical asymptotes of the graphs of three rational functions.

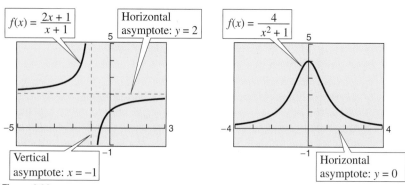

$f(x) = \dfrac{2x+1}{x+1}$

Horizontal asymptote: $y = 2$

Vertical asymptote: $x = -1$

$f(x) = \dfrac{4}{x^2+1}$

Horizontal asymptote: $y = 0$

$f(x) = \dfrac{2}{(x-1)^2}$

Vertical asymptote: $x = 1$

Horizontal asymptote: $y = 0$

Figure 2.38

Exploration

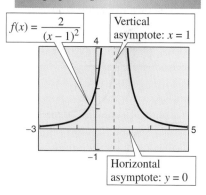

Use a graphing utility to graph $g(x) = (3x^2)/(x^2 + 2)$ in a standard viewing window. Graph $g(x)$ again, using a viewing window in which $-100 \le x \le 100$. Graph $g(x)$ a third time, using a viewing window in which $-1000 \le x \le 1000$. Describe the left and right behavior of the graphs. Complete the table.

x	10	100	1000
$g(x)$			

x	-10	-100	-1000
$g(x)$			

As the x-values become very large or very small, what do the values of $g(x)$ approach? What is the horizontal asymptote of the graph of g?

Asymptotes of a Rational Function

Let f be the rational function

$$f(x) = \frac{N(x)}{D(x)}$$

$$= \frac{a_n x^n + a_{n-1}x^{n-1} + \cdots + a_1 x + a_0}{b_m x^m + b_{m-1}x^{m-1} + \cdots + b_1 x + b_0}$$

where $N(x)$ and $D(x)$ have no common factors.

1. The graph of f has vertical asymptotes at the zeros of $D(x)$.
2. The graph of f has at most one horizontal asymptote determined by comparing the degrees of $N(x)$ and $D(x)$.

   a. If $n < m$, the line $y = 0$ (the x-axis) is a horizontal asymptote.

   b. If $n = m$, the line $y = a_n/b_m$ is a horizontal asymptote.

   c. If $n > m$, the graph of f has no horizontal asymptote.

Exploration

Use a graphing utility to compare the graphs of y_1 and y_2.

$$y_1 = \frac{3x^3 - 5x^2 + 4x - 5}{2x^2 - 6x + 7}$$

$$y_2 = \frac{3x^3}{2x^2}$$

Start with a viewing window in which $-5 \le x \le 5$ and $-10 \le y \le 10$, then zoom out. Write a convincing argument that the shape of the graph of a rational function eventually behaves like the graph of $y = a_n x^n / b_m x^m$, where $a_n x^n$ is the leading term of the numerator and $b_m x^m$ is the leading term of the denominator.

EXAMPLE 2   Finding Horizontal Asymptotes

a. The graph of

$$f(x) = \frac{2x}{3x^2 + 1}$$

has the line $y = 0$ (the x-axis) as a horizontal asymptote, as shown in Figure 2.39(a). Note that the degree of the numerator is *less than* the degree of the denominator.

b. The graph of

$$g(x) = \frac{2x^2}{3x^2 + 1}$$

has the line $y = \frac{2}{3}$ as a horizontal asymptote, as shown in Figure 2.39(b). Note that the degree of the numerator is *equal to* the degree of the denominator, and the horizontal asymptote is given by the ratio of the leading coefficients of the numerator and denominator.

c. The graph of

$$h(x) = \frac{2x^3}{3x^2 + 1}$$

has no horizontal asymptote because the degree of the numerator is *greater than* the degree of the denominator. See Figure 2.39(c).

Although the graph of the function in part (c) does not have a horizontal asymptote, it does have a *slant asymptote*—the line $y = \frac{2}{3}x$. You will study slant asymptotes in the next section.

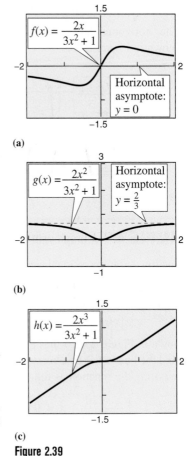

(a)

(b)

(c)

Figure 2.39

EXAMPLE 3   Finding a Function's Domain and Asymptotes

For the function f, find (a) the domain of f, (b) the vertical asymptote of f, and (c) the horizontal asymptote of f.

$$f(x) = \frac{3x^3 + 7x^2 + 2}{-4x^3 + 5}$$

Algebraic Solution

a. Because the denominator is zero when $-4x^3 + 5 = 0$, solve this equation to determine that the domain of f is all real numbers except $x = \sqrt[3]{\frac{5}{4}}$.

b. Because the denominator of f has a zero at $x = \sqrt[3]{\frac{5}{4}}$, and $\sqrt[3]{\frac{5}{4}}$ is not a zero of the numerator, the graph of f has the vertical asymptote $x = \sqrt[3]{\frac{5}{4}} \approx 1.08$, as shown in Figure 2.40.

c. Because the degrees of the numerator and denominator are the same, the horizontal asymptote is given by the ratio of the leading coefficients.

$$y = \frac{\text{leading coefficient of numerator}}{\text{leading coefficient of denominator}} = -\frac{3}{4}$$

The horizontal asymptote of f, $y = -\frac{3}{4}$, is shown in Figure 2.40.

Numerical Solution

a. Because the denominator is zero when $-4x^3 + 5 = 0$, solve this equation to determine that the domain of f is all real numbers except $x = \sqrt[3]{\frac{5}{4}}$.

b. Because the denominator of f has a zero at $x = \sqrt[3]{\frac{5}{4}}$, the graph of f has the vertical asymptote $x = \sqrt[3]{\frac{5}{4}} \approx 1.08$, as shown in Figure 2.40.

c. You can create tables like those shown in Figure 2.41 to estimate that the graph of f has a horizontal asymptote at $y = -\frac{3}{4}$ because the values of $f(x)$ become closer and closer to $-\frac{3}{4}$ as x becomes increasingly large or small.

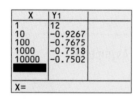

X	Y1	
1	12	
10	-0.9267	
100	-0.7675	
1000	-0.7518	
10000	-0.7502	

X=

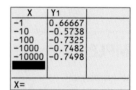

X	Y1	
-1	0.66667	
-10	-0.5738	
-100	-0.7325	
-1000	-0.7482	
-10000	-0.7498	

X=

x Increases Without Bound   *x Decreases Without Bound*

Figure 2.41

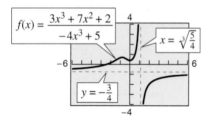

Figure 2.40

EXAMPLE 4   A Graph with Two Horizontal Asymptotes

A function that is not rational can have two horizontal asymptotes—one to the left and one to the right. For instance, the graph of

$$f(x) = \frac{x + 10}{|x| + 2}$$

is shown in Figure 2.42. It has the line $y = -1$ as a horizontal asymptote to the left and the line $y = 1$ as a horizontal asymptote to the right. You can confirm this by rewriting the function as follows.

$$f(x) = \begin{cases} \dfrac{x + 10}{-x + 2}, & x < 0 \qquad |x| = -x \text{ for } x < 0 \\[2mm] \dfrac{x + 10}{x + 2}, & x \geq 0 \qquad |x| = x \text{ for } x \geq 0 \end{cases}$$

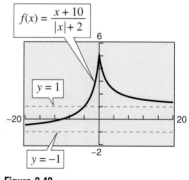

Figure 2.42

Applications

There are many examples of asymptotic behavior in real life. For instance, Example 5 shows how a vertical asymptote can be used to analyze the cost of removing pollutants from smokestack emissions.

EXAMPLE 5   Cost-Benefit Model

A utility company burns coal to generate electricity. The cost of removing a certain *percent* of the pollutants from the smokestack emissions is typically not a linear function. That is, if it costs C dollars to remove 25% of the pollutants, it would cost more than $2C$ dollars to remove 50% of the pollutants. As the percent of removed pollutants approaches 100%, the cost tends to become prohibitive. Suppose that the cost C (in dollars) of removing $p\%$ of the smokestack pollutants is

$$C = \frac{80{,}000p}{100 - p}, \quad 0 \le p < 100.$$

Sketch the graph of this function. Suppose you are a member of a state legislature that is considering a law that would require utility companies to remove 90% of the pollutants from their smokestack emissions. If the current law requires 85% removal, how much additional cost would there be to the utility company because of the new law?

Solution

The graph of this function is shown in Figure 2.43. Note that the graph has a vertical asymptote at $p = 100$. Because the current law requires 85% removal, the current cost to the utility company is

$$C = \frac{80{,}000(85)}{100 - 85} \qquad \text{Substitute 85 for } p.$$

$$\approx \$453{,}333.$$

If the new law increased the percent removal to 90%, the cost to the utility company would be

$$C = \frac{80{,}000(90)}{100 - 90} \qquad \text{Substitute 90 for } p.$$

$$= \$720{,}000.$$

So, the new law would require the utility company to spend an additional

$$720{,}000 - 453{,}333 = \$266{,}667.$$

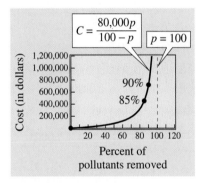

Figure 2.43

EXAMPLE 6 Average Cost of Producing a Product

A business has a cost function of $C = 0.5x + 5000$, where C is measured in dollars and x is the number of units produced. The *average cost per unit* is

$$\overline{C} = \frac{C}{x} = \frac{0.5x + 5000}{x}.$$

Find the average cost per unit when $x = 1000, 5000, 10,000,$ and $100,000$. What is the horizontal asymptote for this function, and what does it represent?

Algebraic Solution

When $x = 1000$,

$$\overline{C} = \frac{0.5(1000) + 5000}{1000} = \$5.50.$$

When $x = 5000$,

$$\overline{C} = \frac{0.5(5000) + 5000}{5000} = \$1.50.$$

When $x = 10,000$,

$$\overline{C} = \frac{0.5(10,000) + 5000}{10,000} = \$1.00.$$

When $x = 90,000$,

$$\overline{C} = \frac{0.5(90,000) + 5000}{90,000} \approx \$0.56.$$

Because the degree of the numerator and denominator are the same for

$$\overline{C} = \frac{0.5x + 5000}{x}$$

the horizontal asymptote is given by the ratio of the leading coefficients of the numerator and denominator. So, the graph has the line $\overline{C} = \$0.50$ as a horizontal asymptote. This line represents the least possible unit cost for the product.

Graphical Solution

Using a graphing utility to graph the function

$$y = \frac{0.5x + 5000}{x}$$

using a viewing window similar to the graph shown in Figure 2.44. Then use the *trace* or *value* feature to approximate the following.

When $x = 1000,\ y_1 = \$5.50.$ When $x = 5000,\ y_1 = \$1.50.$

When $x = 10,000,\ y_1 = \$1.00.$ When $x = 90,000,\ y_1 = \$0.56.$

Continue to use the *trace* or *value* feature to approximate values of $f(x)$ for larger and larger values of x. From this, you can estimate the horizontal asymptote to be $y = \$0.50$. This line represents the least possible cost for the product.

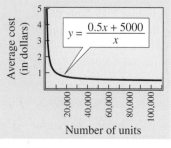

Figure 2.44

Writing About Math *Common Factors in the Numerator and Denominator*

Consider the function $f(x) = \dfrac{x(x - 1)}{x}$ which has a common factor of x in the numerator and denominator. Use a graphing utility to graph this function. Use the *zoom* and *trace* features to decide whether the function has a vertical asymptote at $x = 0$. Then write a short paragraph explaining why you should check for common factors in the numerator and denominator when graphing a rational function.

2.6   E X E R C I S E S

In Exercises 1–6, (a) complete each table, (b) determine the vertical and horizontal asymptotes of the function, and (c) find the domain of the function.

x	$f(x)$
0.5	
0.9	
0.99	
0.999	

x	$f(x)$
1.5	
1.1	
1.01	
1.001	

x	$f(x)$
5	
10	
100	
1000	

x	$f(x)$
-5	
-10	
-100	
-1000	

1. $f(x) = \dfrac{1}{x - 1}$

2. $f(x) = \dfrac{5x}{x - 1}$

3. $f(x) = \dfrac{3x}{|x - 1|}$

4. $f(x) = \dfrac{3}{|x - 1|}$

5. $f(x) = \dfrac{3x^2}{x^2 - 1}$

6. $f(x) = \dfrac{4x}{x^2 - 1}$

In Exercises 7–12, match the function with its graph. [The graphs are labeled (a), (b), (c), (d), (e), and (f).]

(a)

(b)

(c)

(d)

(e)

(f)

7. $f(x) = \dfrac{2}{x + 2}$

8. $f(x) = \dfrac{1}{x - 3}$

9. $f(x) = \dfrac{4x + 1}{x}$

10. $f(x) = \dfrac{1 - x}{x}$

11. $f(x) = \dfrac{x - 2}{x - 4}$

12. $f(x) = -\dfrac{x + 2}{x + 4}$

In Exercises 13–18, (a) find the domain of the function, (b) identify any horizontal and vertical asymptotes, and (c) verify your answer to part (a) both graphically by using a graphing utility and numerically by creating a table of values.

13. $f(x) = \dfrac{1}{x^2}$

14. $f(x) = \dfrac{3}{(x - 2)^3}$

15. $f(x) = \dfrac{3 + x}{3 - x}$

16. $f(x) = \dfrac{2 - 5x}{2 + 2x}$

17. $f(x) = \dfrac{2x^3}{x^2 - 1}$

18. $f(x) = \dfrac{3x^2 + 1}{x^2 + x + 9}$

Analytical and Numerical Explanation **In Exercises 19–22, (a) determine the domain of f and g, (b) find any vertical asymptotes of f, (c) complete the table, and (d) explain what your results suggest.**

19. $f(x) = \dfrac{x^2 - 4}{x + 2}$, $g(x) = x - 2$

x	-4	-3	-2.5	-2	-1.5	-1	0
$f(x)$							
$g(x)$							

20. $f(x) = \dfrac{x^2(x - 3)}{x^2 - 3x}$, $g(x) = x$

x	-1	0	1	2	3	3.5	4
$f(x)$							
$g(x)$							

21. $f(x) = \dfrac{x - 3}{x^2 - 3x}$, $g(x) = \dfrac{1}{x}$

x	-1	-0.5	0	0.5	2	3	4
$f(x)$							
$g(x)$							

22. $f(x) = \dfrac{2x - 8}{x^2 - 9x + 20}$, $g(x) = \dfrac{2}{x - 5}$

x	0	1	2	3	4	5	6
$f(x)$							
$g(x)$							

Exploration **In Exercises 23–26, determine the value the function f approaches as the magnitude of x increases. Is $f(x)$ greater than or less than this functional value when x is positive and large in magnitude? What about when x is negative and large in magnitude?**

23. $f(x) = 4 - \dfrac{1}{x}$

24. $f(x) = 2 + \dfrac{1}{x - 3}$

25. $f(x) = \dfrac{2x - 1}{x - 3}$

26. $f(x) = \dfrac{2x - 1}{x^2 + 1}$

In Exercises 27–30, find the zeros (if any) of the rational function. Use a graphing utility to verify your answer.

27. $g(x) = \dfrac{x^2 - 9}{x + 1}$

28. $g(x) = \dfrac{x^3 - 8}{x^2 + 4}$

29. $f(x) = 1 - \dfrac{2}{x - 5}$

30. $h(x) = 6 + \dfrac{4}{x^2 + 2}$

31. *Pollution* The cost (in millions of dollars) for removing $p\%$ of the industrial and municipal pollutants discharged into a river is

$$C = \frac{255p}{100 - p}, \quad 0 \le p < 100.$$

(a) Find the cost of removing 10% of the pollutants.

(b) Find the cost of removing 40% of the pollutants.

(c) Find the cost of removing 75% of the pollutants.

(d) Use a graphing utility to graph the cost function. Be sure to choose an appropriate viewing window. Explain why you chose the values that you used in your viewing window.

(e) According to this model, would it be possible to remove 100% of the pollutants? Explain.

32. *Recycling* In a pilot project, a rural township was given recycling bins for separating and storing recyclable products. The cost (in dollars) for supplying bins to $p\%$ of the population is

$$C = \frac{25,000p}{100 - p}, \quad 0 \le p < 100.$$

(a) Find the cost of giving bins to 15% of the population.

(b) Find the cost of giving bins to 50% of the population.

(c) Find the cost of giving bins to 90% of the population.

(d) Use a graphing utility to graph the cost function. Be sure to choose an appropriate viewing window. Explain why you chose the values that you used in your viewing window.

(e) According to this model, would it be possible to supply bins to 100% of the residents? Explain.

33. *Data Analysis*   Consider a physics laboratory experiment designed to determine an unknown mass. A flexible metal meter stick is clamped to a table with 50 centimeters overhanging the edge. Known masses M ranging from 200 grams to 2000 grams are attached to the end of the meter stick. For each mass, the meter stick is displaced vertically and then allowed to oscillate. The average time t in seconds of one oscillation for each mass is recorded in the table.

M	200	400	600	800	1000
t	0.450	0.597	0.721	0.831	0.906

M	1200	1400	1600	1800	2000
t	1.003	1.088	1.168	1.218	1.338

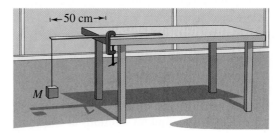

A model for the data is

$$t = \frac{38M + 16{,}965}{10(M + 5000)}.$$

(a) Use a graphing utility to create a table showing the estimated time based on the model for each of the masses shown in the table. What can you conclude?

(b) Use the model to approximate the mass of an object if the average for one oscillation is 1.056 seconds.

34. *Data Analysis*   The endpoints of the interval over which distinct vision is possible are called the *near point* and *far point* of the eye. With increasing age these points normally change. The table gives the approximate near points y in centimeters for various ages x.

x	10	20	30	40	50
y	7	10	14	22	40

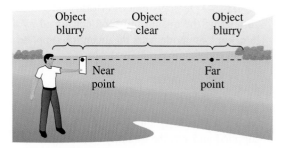

(a) Fit a rational model to the data. Take the reciprocals of the near points to generate the points $(x, 1/y)$. Use the regression capabilities of a graphing utility to fit a line to this data. The resulting line has the form

$$\frac{1}{y} = ax + b.$$

Solve for y.

(b) Use a graphing utility to create a table giving the predicted near point based on the model for each of the ages in the given table.

(c) Do you think the model can be used to predict the near point for a person who is 60 years old? Explain.

35. *Deer Population*   The game commission introduces 100 deer into newly acquired state game lands. The population of the herd is

$$N = \frac{20(5 + 3t)}{1 + 0.04t}, \quad t \geq 0$$

where t is the time (in years).

(a) Find the population when t is 5, 10, and 25.

(b) What is the limiting size of the herd as time increases? Explain your reasoning.

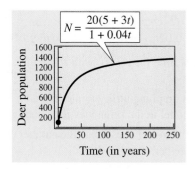

36. Insects A biology class performs an experiment comparing the quantity of food consumed by a certain kind of moth with the quantity supplied. The model for their experimental data is

$$y = \frac{1.568x - 0.001}{6.360x + 1}, \quad x > 0$$

where x is the quantity (in milligrams) of food supplied and y is the quantity (in milligrams) eaten. At what level of consumption will the moth become satiated?

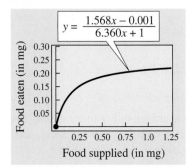

37. Military The number of United States military reserve personnel M (in thousands) for the years 1990 through 1997 is shown in the table. (Source: U.S. Department of Defense)

Year	1990	1991	1992	1993
M	1671	1786	1883	1867

Year	1994	1995	1996	1997
M	1805	1659	1550	1461

A model for the data is

$$y = \frac{1671.92 + 130.23t}{1 - 0.02t + 0.02t^2}$$

where t is time (in years), with $t = 0$ corresponding to 1990.

(a) Use a graphing utility to plot the data and graph the model in the same viewing window.

(b) Use the model to estimate the number of military reserve personnel in 2002.

(c) Would this model be useful for estimating the number of military reserve personnel for future years? Explain.

Synthesis

True or False? **In Exercises 38 and 39, determine whether the statement is true or false. Justify your answer.**

38. A rational function can have infinitely many vertical asymptotes.

39. $f(x) = x^3 - 2x^2 - 5x + 6$ is a rational function.

Think About It **In Exercises 40–43, write a rational function f having the specified characteristics. (There are many correct answers.)**

40. Vertical asymptotes: $x = -2, x = 1$

41. Vertical asymptote: None
Horizontal asymptote: $y = 0$

42. Vertical asymptote: None
Horizontal asymptote: $y = 2$

43. Vertical asymptotes: $x = 0, x = \frac{5}{2}$
Horizontal asymptote: $y = -3$

Review

In Exercises 44–49, find all solutions of the equation.

44. $225x - 50x^3 = 0$

45. $x(10 - x) = 25$

46. $2z^2 - 3z - 35 = 0$

47. $t^3 - 50t = 0$

48. $27x^3 - 147x = 0$

49. $x^4 - 225 = 0$

In Exercises 50–53, divide using synthetic division.

50. $(x^2 + 5x + 6) \div (x - 4)$

51. $(x^2 - 10x + 15) \div (x - 3)$

52. $(2x^2 + x - 11) \div (x + 5)$

53. $(4x^2 + 3x - 10) \div (x + 6)$

In Exercises 54–57, find a polynomial with integer coefficients that has the given zeros. (There are many correct answers.)

54. $8, 5i, -5i$

55. $-2, 6i, -6i$

56. $6, 3 + i, 3 - i$

57. $1, -3 + 2i, -3 - 2i$

2.7   Graphs of Rational Functions

The Graph of a Rational Function

To sketch the graph of a rational function, use the following guidelines.

Guidelines for Graphing Rational Functions

Let $f(x) = N(x)/D(x)$, where $N(x)$ and $D(x)$ are polynomials with no common factors.

1. Find and plot the y-intercept (if any) by evaluating $f(0)$.

2. Set the numerator equal to zero and solve the equation $N(x) = 0$. The real solutions represent the x-intercepts of the graph. Plot these intercepts.

3. Set the denominator equal to zero and solve the equation $D(x) = 0$. The real solutions represent the vertical asymptotes. Sketch these asymptotes using dashed vertical lines.

4. Find and sketch the horizontal asymptote of the graph using a dashed horizontal line.

5. Plot at least one point between and one point beyond each x-intercept and vertical asymptote.

6. Use smooth curves to complete the graph between and beyond the vertical asymptotes.

Testing for symmetry can be useful, especially for simple rational functions. For example, the graph of $f(x) = 1/x$ is symmetrical with respect to the origin, and the graph of $g(x) = 1/x^2$ is symmetrical with respect to the y-axis.

Graphing utilities have difficulty sketching graphs of rational functions that have vertical asymptotes. Often, the utility will connect parts of the graph that are not supposed to be connected. For instance, Figure 2.45(a) shows the graph of

$$f(x) = \frac{1}{x - 2}.$$

Notice that the graph should consist of two *unconnected* portions—one to the left of $x = 2$ and the other to the right of $x = 2$. To eliminate this problem, you can try changing the *mode* of the graphing utility to *dot mode*. The problem with this is that the graph is then represented as a collection of dots rather than as a smooth curve, as shown in Figure 2.45(b).

What You Should Learn:

- How to analyze and sketch graphs of rational functions
- How to decide whether graphs of rational functions have slant asymptotes
- How to use rational functions to model and solve real-life problems

Why You Should Learn It:

The graph of a rational function provides a good indication of the future behavior of a function. Exercise 75 on page 206 models the number of kidney transplants in the U.S., and allows you to estimate the number of transplants in the coming years.

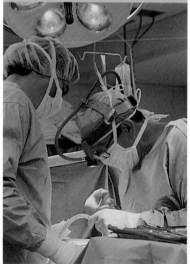

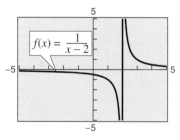

(a) **Connected mode**

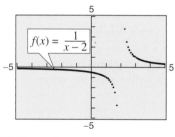

(b) **Dot mode**

Figure 2.45

EXAMPLE 1   Sketching the Graph of a Rational Function

Sketch the graph of $g(x) = \dfrac{3}{x - 2}$ by hand.

Solution

y-Intercept:	$\left(0, -\frac{3}{2}\right)$, because $g(0) = -\frac{3}{2}$.
x-Intercept:	None, because $3 \neq 0$.
Vertical Asymptote:	$x = 2$, zero of denominator
Horizontal Asymptote:	$y = 0$, degree of $N(x)$ < degree of $D(x)$

Additional Points:

x	-4	1	3	5
$g(x)$	-0.5	-3	3	1

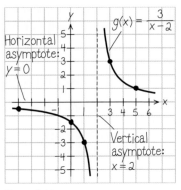

Figure 2.46

By plotting the intercepts, asymptotes, and a few additional points, you can obtain the graph shown in Figure 2.46. Confirm this with a graphing utility.

Note that the graph of g in Example 1 is a vertical stretch and a right shift of the graph of

$$f(x) = \frac{1}{x}$$

because

$$g(x) = \frac{3}{x - 2} = 3\left(\frac{1}{x - 2}\right) = 3f(x - 2).$$

EXAMPLE 2   Sketching the Graph of a Rational Function

Sketch the graph of $f(x) = \dfrac{2x - 1}{x}$ by hand.

Solution

y-Intercept:	None, because $x = 0$ is not in the domain.
x-Intercept:	$\left(\frac{1}{2}, 0\right)$, because $2x - 1 = 0$, or $x = \frac{1}{2}$.
Vertical Asymptote:	$x = 0$, zero of denominator
Horizontal Asymptote:	$y = 2$, degree of $N(x)$ = degree of $D(x)$

Additional Points:

x	-4	-1	$\frac{1}{4}$	4
$f(x)$	2.25	3	-2	1.75

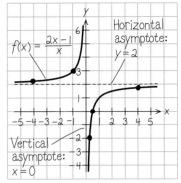

Figure 2.47

By plotting the intercepts, asymptotes, and a few additional points, you can obtain the graph shown in Figure 2.47. Confirm this with a graphing utility.

EXAMPLE 3   Sketching the Graph of a Rational Function

Sketch the graph of $f(x) = \dfrac{x}{x^2 - x - 2}$.

Solution

By factoring the denominator, you have

$$f(x) = \frac{x}{x^2 - x - 2}$$

$$= \frac{x}{(x + 1)(x - 2)}.$$

y-Intercept:	$(0, 0)$, because $f(0) = 0$.
x-Intercept:	$(0, 0)$
Vertical Asymptotes:	$x = -1$, $x = 2$, zeros of denominator
Horizontal Asymptote:	$y = 0$, degree of $N(x)$ < degree of $D(x)$
Additional Points:	

x	-3	-0.5	1	3
$f(x)$	-0.3	0.4	-0.5	0.75

The graph is shown in Figure 2.48.

EXAMPLE 4   Sketching the Graph of a Rational Function

Sketch the graph of $f(x) = \dfrac{2(x^2 - 9)}{x^2 - 4}$.

Solution

By factoring the numerator and denominator, you have

$$f(x) = \frac{2(x^2 - 9)}{x^2 - 4}$$

$$= \frac{2(x - 3)(x + 3)}{(x - 2)(x + 2)}.$$

y-Intercept:	$\left(0, \frac{9}{2}\right)$, because $f(0) = \frac{9}{2}$.
x-Intercepts:	$(-3, 0)$, $(3, 0)$
Vertical Asymptotes:	$x = -2$, $x = 2$, zeros of denominator
Horizontal Asymptote:	$y = 2$, degree of $N(x)$ = degree of $D(x)$
Symmetry:	With respect to *y*-axis, because $f(-x) = f(x)$.
Additional Points:	

x	0.5	2.5	6
$f(x)$	4.67	-2.44	1.69

The graph is shown in Figure 2.49.

 Exploration

Use a graphing utility to graph

$$f(x) = 1 + \frac{1}{x - \dfrac{1}{x}}.$$

Set the graphing utility to *dot* mode and use a decimal viewing window. Use the *trace* feature to find three "holes" or "breaks" in the graph. Do all three holes represent zeros of the denominator

$$x - \frac{1}{x}?$$

Explain.

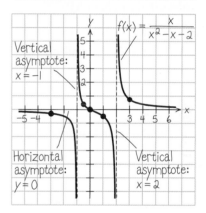

Figure 2.48

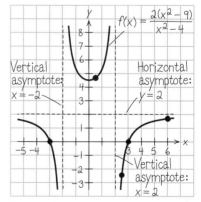

Figure 2.49

Slant Asymptotes

If the degree of the numerator of a rational function is exactly *one more* than the degree of its denominator, the graph of the function has a **slant** (or **oblique**) **asymptote.** For example, the graph of

$$f(x) = \frac{x^2 - x}{x + 1}$$

has a slant asymptote, as shown in Figure 2.50. To find the equation of a slant asymptote, use long division. For instance, by dividing $x + 1$ into $x^2 - x$, you have

$$f(x) = \frac{x^2 - x}{x + 1}$$

$$= \underbrace{x - 2}_{\substack{\text{Slant asymptote} \\ (y = x - 2)}} + \frac{2}{x + 1}. \qquad \frac{2}{x + 1} \to 0 \text{ as } x \to \pm\infty$$

In Figure 2.50, notice that the graph of f approaches the line $y = x - 2$ as x moves to the right or left.

EXAMPLE 5   A Rational Function with a Slant Asymptote

Graph the function $f(x) = \dfrac{x^2 - x - 2}{x - 1}$.

Solution

First write $f(x)$ in two different ways. Factoring the numerator

$$f(x) = \frac{x^2 - x - 2}{x - 1} = \frac{(x - 2)(x + 1)}{x - 1}$$

allows you to recognize the x-intercepts, and long division

$$f(x) = \frac{x^2 - x - 2}{x - 1}$$

$$= x - \frac{2}{x - 1} \qquad \frac{2}{x-1} \to 0 \text{ as } x \to \pm\infty$$

allows you to recognize that the line $y = x$ is a slant asymptote of the graph.

y-Intercept: $(0, 2)$, because $f(0) = 2$.
x-Intercepts: $(-1, 0), (2, 0)$
Vertical Asymptote: $x = 1$, zero of denominator
Slant Asymptote: $y = x$
Additional Points:

x	-2	0.5	1.5	3
$f(x)$	-1.33	4.5	-2.5	2

The graph is shown in Figure 2.51.

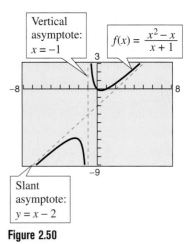

Figure 2.50

Exploration

Do you think it is possible for the graph of a rational function to cross its horizontal asymptote or its slant asymptote? Use the graphs of the following functions to investigate this question. Write a summary of your conclusion. Explain your reasoning.

$$f(x) = \frac{x}{x^2 + 1}$$

$$g(x) = \frac{2x}{3x^2 - 2x + 1}$$

$$h(x) = \frac{x^3}{x^2 + 1}$$

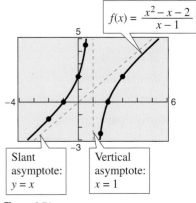

Figure 2.51

Application

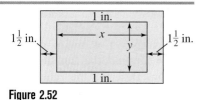

EXAMPLE 6 Finding a Minimum Area

A rectangular page is designed to contain 48 square inches of print. The margins on each side of the page are each $1\frac{1}{2}$ inches. The margins at the top and bottom are each 1 inch. What should the dimensions of the page be so that the minimum amount of paper is used?

Figure 2.52

Graphical Solution

Let *A* be the area to be minimized. From Figure 2.52, you can write

$$A = (x + 3)(y + 2).$$

The printed area inside the margins is modeled by $48 = xy$ or $y = 48/x$. To find the minimum area, rewrite the equation for *A* in terms of just one variable by substituting $48/x$ for *y*.

$$A = (x + 3)\left(\frac{48}{x} + 2\right)$$

$$= \frac{(x + 3)(48 + 2x)}{x}, \quad x > 0$$

The graph of this rational function is shown in Figure 2.53. Because *x* represents the width of the printed area, you need consider only the portion of the graph for which *x* is positive. Using a graphing utility, you can approximate the minimum value of *A* to occur when $x \approx 8.5$ inches. The corresponding value of *y* is $48/8.5 \approx 5.6$ inches. So, the dimensions should be

$$x + 3 \approx 11.5 \text{ inches by } y + 2 \approx 7.6$$
inches.

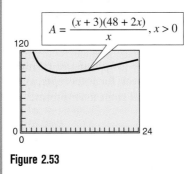

Figure 2.53

Numerical Solution

Let *A* be the area to be minimized. From Figure 2.52, you can write

$$A = (x + 3)(y + 2).$$

The printed area inside the margins is modeled by $48 = xy$ or $y = 48/x$. To find the minimum area, rewrite the equation for *A* in terms of just one variable by substituting $48/x$ for *y*.

$$A = (x + 3)\left(\frac{48}{x} + 2\right) = \frac{(x + 3)(48 + 2x)}{x}, \quad x > 0$$

Use the *table* feature of a graphing utility to create a table of values for the function $y_1 = \dfrac{(x + 3)(48 + 2x)}{x}$ beginning at $x = 1$, as shown in Figure 2.54.

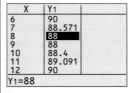

Figure 2.54

From the table, you can see that the minimum value of y_1 occurs when *x* is somewhere between 8 and 9. To approximate the minimum value of y_1 to one decimal place, change the table to begin at 8 with an increment of 0.1. The minimum value of y_1 occurs when $x \approx 8.5$, as shown in Figure 2.55. The corresponding value of *y* is $48/8.5 \approx 5.6$ inches. So, the dimensions should be $x + 3 \approx 11.5$ inches by $y + 2 \approx 7.6$ inches.

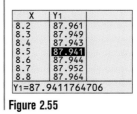

Figure 2.55

If you go on to take a course in calculus, you will learn an analytic technique for finding the exact value of *x* that produces a minimum area in Example 6. In this case, that value is $x = 6\sqrt{2} \approx 8.485$.

2.7 E x e r c i s e s

In Exercises 1–4, use a graphing utility to graph $f(x) = 2/x$ and the function g in the same viewing window. Describe the relationship between the two graphs.

1. $g(x) = f(x) + 1$ **2.** $g(x) = f(x - 1)$

3. $g(x) = -f(x)$ **4.** $g(x) = \frac{1}{2}f(x + 2)$

In Exercises 5–8, use a graphing utility to graph $f(x) = 2/x^2$ and the function g in the same viewing window. Describe the relationship between the two graphs.

5. $g(x) = f(x) - 2$ **6.** $g(x) = -f(x)$

7. $g(x) = f(x - 2)$ **8.** $g(x) = \frac{1}{4}f(x)$

In Exercises 9–12, use a graphing utility to graph $f(x) = 4/x^3$ and the function g in the same viewing window. Describe the relationship between the two graphs.

9. $g(x) = f(x + 2)$ **10.** $g(x) = f(x) + 1$

11. $g(x) = -f(x)$ **12.** $g(x) = \frac{1}{4}f(x)$

In Exercises 13–30, sketch the graph of the rational function by hand. As sketching aids, check for intercepts, symmetry, vertical asymptotes, and horizontal asymptotes. Use a graphing utility to verify your graph.

13. $f(x) = \dfrac{1}{x + 2}$ **14.** $f(x) = \dfrac{1}{x - 5}$

15. $C(x) = \dfrac{5 + 2x}{1 + x}$ **16.** $P(x) = \dfrac{1 - 3x}{1 - x}$

17. $g(x) = \dfrac{1}{x + 2} + 2$ **18.** $f(t) = \dfrac{1 - 2t}{t}$

19. $f(x) = 2 - \dfrac{3}{x^2}$ **20.** $h(x) = \dfrac{x^2}{x^2 - 9}$

21. $f(x) = \dfrac{x^2}{x^2 - 4}$ **22.** $g(x) = \dfrac{x}{x^2 - 9}$

23. $f(x) = \dfrac{x}{x^2 - 4}$ **24.** $f(x) = -\dfrac{1}{(x - 2)^2}$

25. $g(x) = \dfrac{4(x + 1)}{x(x - 4)}$ **26.** $h(x) = \dfrac{2}{x^2(x - 2)}$

27. $f(x) = \dfrac{3x}{x^2 - x - 2}$ **28.** $f(x) = \dfrac{2x}{x^2 + x - 2}$

29. $f(x) = -\dfrac{4}{\dfrac{5}{x} - 5}$ **30.** $f(x) = 2 + \dfrac{8}{3x - \dfrac{3}{x}}$

In Exercises 31–40, use a graphing utility to obtain the graph of the function. Give its domain and identify any vertical or horizontal asymptotes.

31. $f(x) = \dfrac{2 + x}{1 - x}$ **32.** $f(x) = \dfrac{3 - x}{2 - x}$

33. $f(t) = \dfrac{3t + 1}{t}$ **34.** $h(x) = \dfrac{x - 2}{x - 3}$

35. $h(t) = \dfrac{4}{t^2 + 1}$ **36.** $g(x) = -\dfrac{x}{(x - 2)^2}$

37. $f(x) = \dfrac{x + 1}{x^2 - x - 6}$ **38.** $f(x) = \dfrac{x + 4}{x^2 + x - 6}$

39. $f(x) = \dfrac{20x}{x^2 + 1} - \dfrac{1}{x}$

40. $f(x) = 5\left(\dfrac{1}{x - 4} - \dfrac{1}{x + 2}\right)$

Exploration In Exercises 41–46, use a graphing utility to obtain the graph of the function. What do you observe about its asymptotes?

41. $h(x) = \dfrac{6x}{\sqrt{x^2 + 1}}$ **42.** $f(x) = -\dfrac{x}{\sqrt{9 + x^2}}$

43. $g(x) = \dfrac{4|x - 2|}{x + 1}$ **44.** $f(x) = -\dfrac{8|3 + x|}{x - 2}$

45. $f(x) = \dfrac{4(x - 1)^2}{x^2 - 4x + 5}$ **46.** $g(x) = \dfrac{3x^4 - 5x + 3}{x^4 + 1}$

In Exercises 47–54, sketch the graph of the rational function. As sketching aids, check for intercepts, symmetry, vertical asymptotes, and slant asymptotes.

47. $f(x) = \dfrac{2x^2 + 1}{x}$ **48.** $g(x) = \dfrac{x^2 + 1}{x}$

49. $h(x) = \dfrac{x^2}{x - 1}$ **50.** $f(x) = \dfrac{x^3}{x^2 - 1}$

51. $g(x) = \dfrac{x^3}{2x^2 - 8}$ **52.** $f(x) = \dfrac{x^2 - 1}{x^2 + 4}$

53. $f(x) = \dfrac{x^3 + 2x^2 + 4}{2x^2 + 1}$ **54.** $f(x) = \dfrac{2x^2 - 5x + 5}{x - 2}$

Graphical Reasoning In Exercises 55–58, (a) use the graph to estimate any *x*-intercepts of the rational function and (b) set *y* = 0 and solve the resulting equation to confirm your result in part (a).

55. $y = \dfrac{x + 1}{x - 3}$

56. $y = \dfrac{2x}{x - 3}$

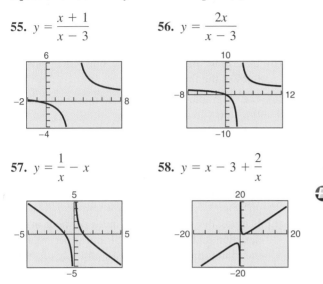

57. $y = \dfrac{1}{x} - x$

58. $y = x - 3 + \dfrac{2}{x}$

In Exercises 59–62, use a graphing utility to graph the rational function. Give the domain of the function and identify any asymptotes. Then zoom out far enough so that the graph appears as a line. Identify the line.

59. $y = \dfrac{2x^2 + x}{x + 1}$

60. $y = \dfrac{x^2 + 5x + 8}{x + 3}$

61. $y = \dfrac{1 + 3x^2 - x^3}{x^2}$

62. $y = \dfrac{12 - 2x - x^2}{2(4 + x)}$

Graphical Reasoning In Exercises 63–66, (a) use a graphing utility to graph the function and determine any *x*-intercepts, and (b) set *y* = 0 and solve the resulting equation to confirm your result in part (a).

63. $y = \dfrac{1}{x + 5} + \dfrac{4}{x}$

64. $y = 20\left(\dfrac{2}{x + 1} - \dfrac{3}{x}\right)$

65. $y = x - \dfrac{6}{x - 1}$

66. $y = x - \dfrac{9}{x}$

67. *Concentration of a Mixture* A 1000-liter tank contains 10 liters of a 25% brine solution. You add *x* liters of a 75% brine solution to the tank.

(a) Show that the concentration *C* of the final mixture is

$$C = \dfrac{3x + 10}{4(x + 10)}.$$

(b) Determine the domain of the function on the basis of the physical constraints of the problem.

(c) Use a graphing utility to graph the function. As the tank is filled, what happens to the rate at which the concentration of brine increases? How close to the horizontal asymptote is the graph of *C* when the tank is full?

68. *Geometry* A rectangular region of length *x* and width *y* has an area of 500 square meters.

(a) Express the width *y* as a function of *x*.

(b) Determine the domain of the function on the basis of the physical constraints of the problem.

(c) Sketch a graph of the function and determine the width of the rectangle if *x* = 30 meters.

69. *Page Design* A page that is *x* inches wide and *y* inches high contains 30 square inches of print. The margins at the top and bottom are 2 inches and the margins on each side are 1 inch.

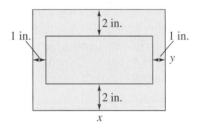

(a) Show that the total area *A* of the page is

$$A = \dfrac{2x(2x + 11)}{x - 2}.$$

(b) Determine the domain of the function on the basis of the physical constraints of the problem.

(c) Use a graphing utility to graph the area function and approximate the page size such that the minimum amount of paper will be used. Verify your answer numerically using the *table* feature of a graphing utility.

70. *Minimum Area* A right triangle is formed in the first quadrant by the *x*-axis, the *y*-axis, and a line segment through the point (3, 2).

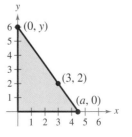

(a) Show that an equation of the line segment is

$$y = \frac{2(a - x)}{a - 3}, \quad 0 \leq x \leq a.$$

(b) Show that the area of the triangle is $A = \dfrac{a^2}{a - 3}$.

(c) Use a graphing utility to graph the area function and estimate the value of a that yields a minimum area. Estimate the minimum area. Verify your answer numerically using the *table* feature of a graphing utility.

71. *Ordering and Transportation Cost* The ordering and transportation cost C for the components used in manufacturing a certain product is

$$C = 100\left(\frac{200}{x^2} + \frac{x}{x + 30}\right), \quad x \geq 1$$

where C is measured in thousands of dollars and x is the order size (in hundreds). Use a graphing utility to obtain the graph of the cost function and, from the graph, estimate the order size that minimizes cost.

72. *Average Cost* The cost of producing x units of a product is $C = 0.2x^2 + 10x + 5$, and therefore the average cost per unit is

$$\overline{C} = \frac{C}{x} = \frac{0.2x^2 + 10x + 5}{x}, \quad x > 0.$$

Sketch the graph of the average cost function, and estimate the number of units that should be produced to minimize the average cost per unit.

73. *Medicine* The concentration of a certain chemical in the bloodstream t hours after injection into muscle tissue is

$$C = \frac{3t^2 + t}{t^3 + 50}, \quad t \geq 0.$$

(a) Determine the horizontal asymptote of the function and interpret its meaning in the context of the problem.

(b) Use a graphing utility to graph the function and approximate the time when the bloodstream concentration is greatest.

(c) Use a graphing utility to determine when the concentration is less than 0.345.

74. *Numerical and Graphical Analysis* A driver averaged 50 miles per hour on the round trip between home and a city 100 miles away. The average speeds for going and returning were x and y miles per hour, respectively.

(a) Show that $y = \dfrac{25x}{x - 25}$.

(b) Determine the vertical and horizontal asymptotes of the function.

(c) Use a graphing utility to complete the table. What do you observe?

x	30	35	40	45	50	55	60
y							

(d) Use a graphing utility to graph the function.

(e) Is it possible to average 20 miles per hour in one direction and still average 50 miles per hour on the round trip? Explain.

75. *Comparing Models* The number of kidney transplants K in the United States from 1987 through 1996 is shown in the table. (Sources: U.S. Department of Health and Human Services; United Network for Organ Sharing)

Year	1987	1988	1989	1990	1991
K	8967	9123	8890	9877	10,122

Year	1992	1993	1994	1995	1996
K	10,231	11,020	11,392	11,891	12,080

For each of the following, let t be the time (in years), with $t = 7$ corresponding to 1987.

(a) A model for the data is

$$K = \frac{8116.17 - 280t}{1 - 0.0447t}.$$

Use a graphing utility to plot the data points and graph the model in the same viewing window.

(b) Use the regression capabilities of a graphing utility to fit a line to the data.

(c) Use the regression capabilities of a graphing utility to fit a parabola to the data.

(d) Which of the three models would you recommend as an estimator of the number of kidney transplants for the years following 1996? Explain your reasoning.

76. *Comparing Models*   The number N (in thousands) of insured commercial banks in the United States for the years 1988 through 1997 is shown in the table.   (Source: U.S. Federal Deposit Insurance Corporation)

Year	1988	1989	1990	1991	1992
N	13.1	12.7	12.3	11.9	11.5

Year	1993	1994	1995	1996	1997
N	11.0	10.5	9.9	9.5	9.1

For each of the following, let t be the time (in years), with $t = 8$ corresponding to 1988.

(a) Use the regression capabilities of a graphing utility to fit a line to the data. Use a graphing utility to plot the data points and graph the model in the same viewing window.

(b) Fit a rational model to the data. Take the reciprocal of N to generate the points $(t, 1/N)$. Use the regression capabilities of a graphing utility to fit a line to this data. The resulting line has the form

$$\frac{1}{N} = at + b.$$

Solve for N. Use a graphing utility to plot the data points and graph the rational model in the same viewing window.

(c) Use a graphing utility to create a table showing the predicted number of banks based on each model for each of the years in the given table. Which model do you prefer? Why?

Synthesis

True or False?   In Exercises 77 and 78, determine whether the statement is true or false. Justify your answer.

77. If the graph of a rational function f has a vertical asymptote at $x = 5$, it is possible to sketch the graph without lifting your pencil from the paper.

78. A rational function can never cross one of its asymptotes.

Think About It   In Exercises 79 and 80, use a graphing utility to obtain the graph of the function. Explain why there is no vertical asymptote when a superficial examination of the function may indicate that there should be one.

79. $h(x) = \dfrac{6 - 2x}{3 - x}$ **80.** $g(x) = \dfrac{x^2 + x - 2}{x - 1}$

81. *Writing*   Write a paragraph discussing whether every rational function has a vertical asymptote.

Think About It   In Exercises 82–85, write a rational function satisfying the following criteria.

82. Vertical asymptote: $x = 2$
Slant asymptote: $y = x + 1$
Zero of the function: $x = -2$

83. Vertical asymptote: $x = -4$
Slant asymptote: $y = x - 2$
Zero of the function: $x = 3$

84. Vertical asymptote: $x = -1$
Horizontal asymptote: $y = 2$
Zero of the function: $x = 3$

85. Vertical asymptote: $x = 3$
Horizontal asymptote: $y = -2$
Zero of the function: $x = -6$

Review

In Exercises 86–91, sketch a graph of the equation by hand. Use a graphing utility to verify your graph.

86. $-y + 3x + 8 = 0$ **87.** $4x + 5y - 2 = 0$

88. $7x + 3 = 0$ **89.** $4y - 10 = 0$

90. $x - y - 1 = 0$ **91.** $-7 + 8x - 2y = 0$

In Exercises 92–95, use a graphing utility to graph the function and find its domain and range.

92. $f(x) = \sqrt{6 + x^2}$ **93.** $f(x) = \sqrt{121 - x^2}$

94. $f(x) = -|x + 9|$ **95.** $f(x) = -x^2 + 9$

Chapter Summary

What did you learn?

2 Review Exercises

2.1 *Graphical Reasoning*   **In Exercises 1 and 2, use a graphing utility to graph each equation in the same viewing window. Describe how each graph differs from the graph of** $y = x^2$**.**

1. (a) $y = 2x^2$ (b) $y = -2x^2$
   (c) $y = x^2 + 2$ (d) $y = (x + 2)^2$

2. (a) $y = x^2 - 4$ (b) $y = 4 - x^2$
   (c) $y = (x - 3)^2$ (d) $y = \frac{1}{2}x^2 - 1$

In Exercises 3–6, sketch the graph of the quadratic function. Identify the vertex and the intercepts.

3. $f(x) = \left(x + \frac{3}{2}\right)^2 + 1$ **4.** $f(x) = (x - 4)^2 - 4$

5. $f(x) = \frac{1}{3}(x^2 + 5x - 4)$

6. $f(x) = 3x^2 - 12x + 11$

In Exercises 7 and 8, find the quadratic function that has the indicated vertex and whose graph passes through the given point.

7. Vertex: $(1, -4)$;   Point: $(2, -3)$

8. Vertex: $(2, 3)$;   Point: $(-1, 6)$

In Exercises 9–16, find the maximum or minimum value of the quadratic function.

9. $g(x) = x^2 - 2x$ **10.** $f(x) = x^2 + 8x + 10$

11. $f(x) = 6x - x^2$ **12.** $h(x) = 3 + 4x - x^2$

13. $f(t) = -2t^2 + 4t + 1$ **14.** $h(x) = 4x^2 + 4x + 13$

15. $h(x) = x^2 + 5x - 4$ **16.** $f(x) = 4x^2 + 4x + 5$

17. *Numerical, Graphical, and Analytical Analysis*   A rectangle is inscribed in the region bounded by the x-axis, the y-axis, and the graph of $x + 2y - 8 = 0$.

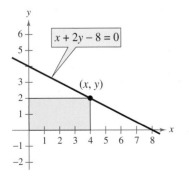

(a) Complete six rows of a table like the one below.

x	y	Area
1	$4 - \frac{1}{2}(1)$	$(1)\left[4 - \frac{1}{2}(1)\right] = \frac{7}{2}$
2	$4 - \frac{1}{2}(2)$	$(2)\left[4 - \frac{1}{2}(2)\right] = 6$

(b) Use a graphing utility to generate additional rows of the table. Use the table to estimate the dimensions that will produce the maximum area.

(c) Write the area A as a function of x. Determine the domain of the function in the context of the problem.

(d) Use a graphing utility to graph the function. Use the graph to approximate the dimensions that will produce the maximum area.

(e) Write the area function in standard form to find algebraically the dimensions that will produce the maximum area.

18. *Maximum Profit*   Let x be the amount (in hundreds of dollars) a company spends on advertising, and let P be the profit, where

$$P = 320 + 15x - \frac{1}{2}x^2.$$

(a) Use a graphing utility to graph the function.

(b) Use the *zoom* and *trace* features of a graphing utility to estimate the vertex of the function.

(c) Verify your answer in part (b) both algebraically and numerically by creating a table of values.

(d) Explain what the vertex represents.

2.2 **In Exercises 19–22, sketch the graphs of** $y = x^n$ **and the specified transformations.**

19. $y = x^4$
   (a) $f(x) = (x + 5)^4$ (b) $f(x) = x^4 - 4$
   (c) $f(x) = 3 + x^4$ (d) $f(x) = \frac{1}{4}(x - 2)^4$

20. $y = x^5$
   (a) $f(x) = (x + 4)^5$ (b) $f(x) = 6 + x^5$
   (c) $f(x) = 3 - \frac{1}{2}x^5$ (d) $f(x) = 2(x + 3)^5$

21. $y = x^6$
   (a) $f(x) = x^6 - 2$ (b) $f(x) = -\frac{1}{4}x^6$
   (c) $f(x) = -\frac{1}{2}x^6 - 5$ (d) $f(x) = -(x + 7)^6 - 5$

22. $y = x^7$

(a) $f(x) = -x^7 + 4$ (b) $f(x) = (x + 2)^7 - 1$

(c) $f(x) = -\frac{1}{3}x^7 + 1$ (d) $f(x) = -(x + 8)^7$

In Exercises 23–26, determine the right-hand and left-hand behavior of the graph of the polynomial function.

23. $f(x) = -x^2 + 6x + 9$

24. $f(x) = \frac{1}{2}x^3 + 2x$

25. $g(x) = \frac{3}{4}(x^4 + 3x^2 + 2)$

26. $h(x) = -x^5 - 7x^2 + 10x$

Graphical Analysis In Exercises 27 and 28, use a graphing utility to graph the functions f and g in the same viewing window. Zoom out far enough so that the right-hand and left-hand behavior of f and g appear identical.

27. $f(x) = \frac{1}{2}x^3 - 2x + 1, \quad g(x) = \frac{1}{2}x^3$

28. $f(x) = -x^4 + 2x^3, \quad g(x) = -x^4$

In Exercises 29–34, (a) find the zeros of the function and (b) sketch its graph.

29. $g(x) = x^4 - x^3 - 2x^2$

30. $h(x) = -2x^3 - x^2 + x$

31. $f(t) = t^3 - 3t$ **32.** $f(x) = -(x + 6)^3 - 8$

33. $f(x) = x(x + 3)^2$ **34.** $f(t) = t^4 - 4t^2$

35. *Volume* A rectangular package can have a maximum combined length and girth (perimeter of a cross section) of 216 centimeters.

(a) Write the volume V as a function of x.

(b) Use a graphing utility to graph the volume function, and then use the graph to estimate the dimensions of the package of maximum volume.

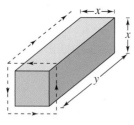

36. *Volume* Rework Exercise 35 for a cylindrical package. (The cross sections are circular.)

In Exercises 37–40, (a) use the Intermediate Value Theorem and a graphing utility to find open intervals of length 1 in which the polynomial function is guaranteed to have a zero, and (b) use the *zero* or *root* feature of a graphing utility to approximate the zeros of the function.

37. $f(x) = x^3 + 2x^2 - x - 1$

38. $f(x) = 0.24x^3 - 2.6x - 1.4$

39. $f(x) = x^4 - 6x^2 - 4$

40. $f(x) = 2x^4 + \frac{7}{2}x^3 - 2$

2.3 *Graphical Analysis* In Exercises 41 and 42, use a graphing utility to graph the two equations in the same viewing window. Use the graphs to verify that the expressions are equivalent. Verify your results algebraically.

41. $y_1 = \dfrac{x^2}{x - 2}, \quad y_2 = x + 2 + \dfrac{4}{x - 2}$

42. $y_1 = \dfrac{x^4 + 1}{x^2 + 2}, \quad y_2 = x^2 - 2 + \dfrac{5}{x^2 + 2}$

In Exercises 43–48, divide by long division.

43. $\dfrac{24x^2 - x - 8}{3x - 2}$ **44.** $\dfrac{4x + 7}{3x - 2}$

45. $\dfrac{x^4 - 3x^2 + 2}{x^2 - 1}$ **46.** $\dfrac{3x^4}{x^2 - 1}$

47. $\dfrac{x^4 + x^3 - x^2 + 2x}{x^2 + 2x}$

48. $\dfrac{6x^4 + 10x^3 + 13x^2 - 5x + 2}{2x^2 - 1}$

In Exercises 49–52, divide by synthetic division.

49. $(0.25x^4 - 4x^3) \div (x + 2)$

50. $(0.1x^3 + 0.3x^2 - 0.5) \div (x - 5)$

51. $(6x^4 - 4x^3 - 27x^2 + 18x) \div \left(x - \frac{2}{3}\right)$

52. $(2x^3 + 2x^2 - x + 2) \div \left(x - \frac{1}{2}\right)$

In Exercises 53 and 54, use synthetic division to find each function value. Use a graphing utility to verify your result.

53. $f(x) = 2x^3 + 3x^2 - 20x - 21$

(a) $f(4)$ (b) $f(-1)$

(c) $f\left(-\frac{7}{2}\right)$ (d) $f(0)$

54. $f(x) = 20x^4 + 9x^3 - 14x^2 - 3x$
  (a) $f(-1)$ (b) $f\left(\frac{3}{4}\right)$
  (c) $f(0)$ (d) $f(1)$

In Exercises 55 and 56, use synthetic division to show that x is a solution of the polynomial equation, then use the result to factor the polynomial completely. List all real zeros of the function.

55. $2x^4 + 5x^3 - 11x^2 - 20x + 12 = 0$, $x = -3$
56. $3x^3 + 2x^2 - 15x - 10 = 0$, $x = -\frac{2}{3}$

In Exercises 57–62, (a) list the possible rational zeros of f, (b) use a graphing utility to graph f so that some of the possible zeros in (a) can be disregarded, and (c) determine all the real zeros of the function.

57. $f(x) = 4x^3 - 11x^2 + 10x - 3$
58. $f(x) = 10x^3 + 21x^2 - x - 6$
59. $f(x) = 6x^3 - 5x^2 + 24x - 20$
60. $f(x) = 10x^3 - 13x^2 - 17x + 6$
61. $f(x) = 6x^4 - 25x^3 + 14x^2 + 27x - 18$
62. $f(x) = 5x^4 + 126x^2 + 25$

In Exercises 63 and 64, use synthetic division to verify the upper bound and lower bound of the zeros of f.

63. $f(x) = 4x^3 - 3x^2 + 4x - 3$
  Upper bound: $x = 1$; Lower bound: $x = -\frac{1}{4}$
64. $f(x) = 2x^3 - 5x^2 - 14x + 8$
  Upper bound: $x = 8$; Lower bound: $x = -4$

2.4 **In Exercises 65–68, write in standard form.**

65. $6 + \sqrt{-25}$ **66.** $-\sqrt{-12} + 3$
67. $-2i^2 + 7i$ **68.** $-i^2 - 4i$

In Exercises 69–80, perform the operations and write the result in standard form.

69. $(7 + 5i) + (-4 + 2i)$

70. $\left(\frac{\sqrt{2}}{2} - \frac{\sqrt{2}}{2}i\right) - \left(\frac{\sqrt{2}}{2} + \frac{\sqrt{2}}{2}i\right)$

71. $5i(13 - 8i)$ **72.** $(1 + 6i)(5 - 2i)$
73. $(10 - 8i)(2 - 3i)$ **74.** $i(6 + i)(3 - 2i)$
75. $(3 + 7i)^2 + (3 - 7i)^2$ **76.** $(4 - i)^2 - (4 + i)^2$

77. $\dfrac{6 + i}{i}$ **78.** $\dfrac{3 + 2i}{5 + i}$

79. $\dfrac{4}{-3i}$ **80.** $\dfrac{1}{(2 + i)^4}$

In Exercises 81–84, plot the complex number in the complex plane.

81. $2 - 5i$ **82.** $-1 + 4i$
83. $-6i$ **84.** $7i$

2.5 **In Exercises 85–88, use a graphing utility to (a) graph the function, (b) determine the number of real zeros of the function, and (c) approximate the real zeros of the function to the nearest hundredth.**

85. $f(x) = x^4 + 2x + 1$
86. $g(x) = x^3 - 3x^2 + 3x + 2$
87. $h(x) = x^3 - 6x^2 + 12x - 10$
88. $f(x) = x^5 + 2x^3 - 3x - 20$

In Exercises 89–94, find all the zeros of the function and write the polynomial as a product of linear factors. Use a graphing utility to verify your answer.

89. $f(x) = x^3 - 4x^2 + 6x - 4$
90. $f(x) = x^3 - 5x^2 - 7x + 51$
91. $f(x) = x^3 + 6x^2 + 11x + 12$
92. $f(x) = 2x^3 - 9x^2 + 22x - 30$
93. $f(x) = x^4 + 34x^2 + 225$
94. $f(x) = x^4 + 10x^3 + 26x^2 + 10x + 25$

In Exercises 95–98, find a polynomial function with integer coefficients that has the given zeros. (There are many correct answers.)

95. $-2, -2, -5i$ **96.** $1, -4, -3 + 5i$
97. $-\frac{2}{3}, -1, 3 + \sqrt{2}i$ **98.** $-4, -4, 1 + \sqrt{3}i$

In Exercises 99 and 100, write the polynomial (a) as the product of factors that are irreducible over the rationals, (b) as the product of linear and quadratic factors that are irreducible over the reals, and (c) in completely factored form.

99. $f(x) = x^4 - x^3 - x^2 + 5x - 20$
  (*Hint:* One factor is $x^2 - 5$.)
100. $f(x) = x^4 - 4x^3 + 3x^2 + 8x - 16$
  (*Hint:* One factor is $x^2 - x - 4$.)

2.6 In Exercises 101–104, find the domain of the function and identify any horizontal and vertical asymptotes.

101. $f(x) = \dfrac{x-8}{1-x}$

102. $f(x) = \dfrac{5x}{x+12}$

103. $f(x) = \dfrac{2}{x^2-3x-18}$

104. $f(x) = \dfrac{2x^2+3}{x^2+x+3}$

In Exercises 105–110, determine the horizontal asymptotes of the function.

105. $f(x) = \dfrac{7+x}{7-x}$

106. $f(x) = \dfrac{6x}{x^2-1}$

107. $f(x) = \dfrac{4x^2}{2x^2-3}$

108. $f(x) = \dfrac{2x^3}{x^2+2x-8}$

109. $f(x) = \dfrac{x-2}{|x|+2}$

110. $f(x) = \dfrac{2x}{|2x-1|}$

111. *Average Cost* A business has a cost of $C = 0.5x + 500$ for producing x units. The average cost per unit is

$$\overline{C} = \frac{C}{x} = \frac{0.5x+500}{x}, \quad x > 0.$$

(a) Use a graphing utility to graph the function.

(b) Find the average cost of producing $x = 50$, 100, 1000, and 10,000 units.

(c) Determine the average cost per unit as x increases without bound. (Find the horizontal asymptote.)

112. *Seizure of Illegal Drugs* The cost in millions of dollars for the U.S. government to seize $p\%$ of a certain illegal drug as it enters the country is

$$C = \frac{528p}{100-p}, \quad 0 \le p < 100.$$

(a) Find the cost of seizing 25%. (b) Find the cost of seizing 50%. (c) Find the cost of seizing 75%. (d) Use a graphing utility to graph the function. (e) According to this model, would it be possible to seize 100% of the drug?

2.7 In Exercises 113–120, sketch the graph of the rational function. As a sketching aid, check for intercepts, symmetry, vertical asymptotes, and horizontal asymptotes. Use a graphing utility to verify your graph.

113. $f(x) = \dfrac{2x-1}{x-5}$

114. $f(x) = \dfrac{x-3}{x-2}$

115. $f(x) = \dfrac{2x}{x^2+4}$

116. $f(x) = \dfrac{2x^2}{x^2-4}$

117. $f(x) = \dfrac{2}{(x+1)^2}$

118. $f(x) = \dfrac{4}{(x-1)^2}$

119. $f(x) = \dfrac{2x}{x^2+x-12}$

120. $f(x) = \dfrac{3x-1}{x^2+5x+4}$

In Exercises 121–126, sketch the graph of the rational function. As a sketching aid, check for intercepts, symmetry, vertical asymptotes, horizontal asymptotes, and slant asymptotes. Use a graphing utility to verify your graph.

121. $f(x) = \dfrac{2x^3}{x^2+1}$

122. $f(x) = \dfrac{x^3}{3x^2-6}$

123. $f(x) = \dfrac{1}{x+3}+2$

124. $f(x) = \dfrac{5x}{x^2-4}$

125. $f(x) = \dfrac{x^2-x+1}{x-3}$

126. $f(x) = \dfrac{2x^2+7x+3}{x+1}$

127. *Population of Fish* The Parks and Wildlife Commission introduces 80,000 fish into a large human-made lake. The population of the fish in thousands is

$$N = \frac{20(4+3t)}{1+0.05t}, \quad t \ge 0$$

where t is time in years.

(a) Sketch the graph of the function by hand. Use a graphing utility to verify your graph.

(b) Find the populations when t is 5, 10, and 25.

(c) What is the maximum number of fish in the lake as time increases? Explain your reasoning.

Synthesis

True or False? In Exercises 128 and 129, determine whether the statement is true or false. Justify your answer.

128. The graph of $f(x) = \dfrac{2x^3}{x+1}$ has a slant asymptote.

129. The graphs of $f(x) = (3x^2 - 10)/x^2$ and $f(x) = (3x^2 - 10)/|x^2|$ are the same, with both having a horizontal asymptote at $x = 3$.

Chapter Project   *Finding Points of Intersection*

You can use the *zoom* feature of a graphing utility to approximate the points of intersection of the graphs of equations.

In this project you will find the points of intersection of the circle and parabola given by

$$x^2 + y^2 - 3x + 5y - 11 = 0 \quad \text{and} \quad y = x^2 - 4x + 5.$$

a. Begin by writing the circle as the union of two functions. Identify the functions that represent the top half and the bottom half of the circle.

b. Use a graphing utility to graph all three functions in the same viewing window, as shown in Figure 2.56. Use the *intersect* feature of the graphing utility to estimate the points of intersection.

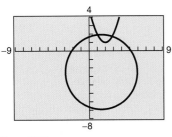

Figure 2.56

Questions for Further Exploration

1. Using a window setting of $1.05 \leq x \leq 1.06$ and $1.89 \leq y \leq 1.90$, graph the top half of the circle and the parabola in the example on the same screen. Then use the *trace* or *value* feature to approximate (accurate to three decimal places) the y-coordinate of the point of intersection that is shown on the screen.

2. Another method for finding the points of intersection is to substitute $x^2 - 4x + 5$ for y in the equation of the circle to get a fourth-degree polynomial equation. Graph this polynomial function.

   (a) Find a setting that allows you to approximate the solution $x \approx 1.055$ of the polynomial equation to two more decimal places.

   (b) Find a setting that allows you to approximate the solution $x \approx 2.841$ to two more decimal places.

3. Use a graphing utility to find the points of intersection of the circle and the parabola given by

$$x^2 + y^2 - 5x + 4y - 13 = 0$$

$$y = x^2 - 3x + 2.$$

4. *Market Equilibrium*   The *market equilibrium* of a commodity is the quantity (and corresponding price) at which the supply of the commodity and the demand for the commodity are equal. The supply and demand curves for a business dealing with wheat are

Supply: $p = 1.45 + 0.00014x^2$

Demand: $p = (2.388 - 0.007x)^2$

where p is the price (in dollars) per bushel and x is the quantity (in bushels per day). Use a graphing utility to graph the supply and demand equations and find the market equilibrium. (*Hint:* The *market equilibrium* is the point of intersection of the graphs for $x > 0$.)

2 Chapter Test

Take this test as you would take a test in class. After you are done, check your work against the answers in the back of the book.

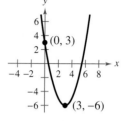

FIGURE FOR 3

1. Describe how the graph of g differs from the graph of $f(x) = x^2$.

   (a) $g(x) = 2 - x^2$ (b) $g(x) = \left(x - \frac{3}{2}\right)^2$

2. Identify the vertex and intercepts of the graph of $y = x^2 + 4x + 3$.

3. Find an equation for the parabola shown at the right.

4. The path of a ball is given by $y = -\frac{1}{20}x^2 + 3x + 5$, where y is the height in feet and x is the horizontal distance in feet.

   (a) Find the maximum height of the ball.

   (b) Which term determines the height at which the ball was thrown? Does changing this term change the coordinates of the maximum height of the ball? Explain.

5. Divide using long division: $(3x^3 + 4x - 1) \div (x^2 + 1)$.

6. Divide using synthetic division: $(2x^4 - 5x^2 - 3) \div (x - 2)$.

The *Interactive* CD-ROM and *Internet* versions of this text provide answers to the Chapter Tests and Cumulative Tests. They also offer Chapter Pre-Tests (that test key skills and concepts covered in previous chapters) and Chapter Post-Tests, both of which have randomly generated exercises with diagnostic capabilities.

In Exercises 7 and 8, list all the possible rational zeros of the function. Use a graphing utility to graph the function and find all the rational zeros.

7. $g(t) = 2t^4 - 3t^3 + 16t - 24$ 8. $h(x) = 3x^5 + 2x^4 - 3x - 2$

In Exercises 9–12, perform the operation.

9. $(-8 - 3i) + (-1 - 15i)$

10. $\left(10 + \sqrt{-20}\right) - \left(4 - \sqrt{-14}\right)$

11. $(2 + i)(6 - i)$

12. $\dfrac{8 + 5i}{6 - i}$

In Exercises 13 and 14, use the *zero* or *root* feature of a graphing utility to approximate (accurate to three decimal places) the real zeros of the function.

13. $f(x) = x^4 - x^3 - 1$ 14. $f(x) = 3x^5 + 2x^4 - 12x - 8$

In Exercises 15–17, find a polynomial function with integer coefficients that has the given zeros.

15. $0, 3, 3 + i, 3 - i$ 16. $1 - \sqrt{3}i, 2, 2$ 17. $0, -5, 1 + i$

In Exercises 18–20, sketch the graph of the rational function. As a sketching aid, check for intercepts, symmetry, vertical asymptotes, horizontal asymptotes, and slant asymptotes. Use a graphing utility to verify your graph.

18. $h(x) = \dfrac{4}{x^2} - 1$ 19. $g(x) = \dfrac{x^2 + 2}{x - 1}$ 20. $f(x) = \dfrac{2x^2 + 9}{5x^2 + 2}$

21. The average age of the groom in a wedding for a given age of the bride can be approximated by the model $y = -0.00428x^2 + 1.442x - 3.136$, $20 \le x \le 55$, where y is the age of the groom and x is the age of the bride. For what age of the bride is the average age of the groom 30? (Source: U.S. National Center for Health Statistics)

Exponential and Logarithmic Functions

3.1 Exponential Functions and Their Graphs

3.2 Logarithmic Functions and Their Graphs

3.3 Properties of Logarithms

3.4 Solving Exponential and Logarithmic Equations

3.5 Exponential and Logarithmic Models

Louise Gubb/The Image Works

The Big Picture

In this chapter you will learn how to

❏ recognize, evaluate, and graph exponential and logarithmic functions.

❏ rewrite logarithmic functions with different bases.

❏ use properties of logarithms to evaluate, rewrite, expand, or condense logarithmic expressions.

❏ solve exponential and logarithmic equations.

❏ use exponential growth models, exponential decay models, Gaussian models, logistic models, and logarithmic models to solve real-life problems.

❏ fit exponential and logarithmic models to sets of data.

Personal savings as a percent of disposable income was 3.9% in 1997 and disposable per capita income was $21,969. So, the per capita personal savings was $856.79. (Source: U.S. Bureau of Economic Analysis)

Important Vocabulary

As you encounter each new vocabulary term in this chapter, add the term and its definition to your notebook glossary.

Additional Resources   Text-specific additional resources are available to help you do well in this course. See page xvi for details.

3.1 Exponential Functions and Their Graphs

Exponential Functions

So far, this text has dealt mainly with **algebraic functions,** which include polynomial functions and rational functions. In this chapter you will study two types of nonalgebraic functions—*exponential* functions and *logarithmic* functions. These functions are examples of **transcendental functions.**

Definition of Exponential Function

The **exponential function** f with base a is denoted by

$$f(x) = a^x$$

where $a > 0$, $a \neq 1$, and x is any real number.

Note that in the definition of an exponential function, the base $a = 1$ is excluded because it yields $f(x) = 1^x = 1$. This is a constant function, not an exponential function.

You already know how to evaluate a^x for integer and rational values of x. For example, you know that $4^3 = 64$ and $4^{1/2} = 2$. However, to evaluate 4^x for any real number x, you need to interpret forms with *irrational* exponents. For the purposes of this text, it is sufficient to think of

$$a^{\sqrt{2}} \text{ (where } \sqrt{2} \approx 1.41421356)$$

as the number that has the successively closer approximations

$$a^{1.4}, a^{1.41}, a^{1.414}, a^{1.4142}, a^{1.41421}, \ldots.$$

Example 1 shows how to use a calculator to evaluate an exponential expression.

EXAMPLE 1   Evaluating Exponential Expressions

Use a calculator to evaluate each expression.

a. $2^{-3.1}$ **b.** $2^{-\pi}$ **c.** $12^{5/7}$ **d.** $(0.6)^{3/2}$

Solution

Number	Graphing Calculator Keystrokes	Display
a. $2^{-3.1}$	2 ^ (−) 3.1 ENTER	0.1166291
b. $2^{-\pi}$	2 ^ (−) π ENTER	0.1133147
c. $12^{5/7}$	12 ^ (5 ÷ 7) ENTER	5.8998877
d. $(0.6)^{3/2}$	.6 ^ (3 ÷ 2) ENTER	0.4647580

What You Should Learn:

- How to recognize and evaluate exponential functions with base a
- How to graph exponential functions
- How to recognize, evaluate, and graph exponential functions with base e
- How to use exponential functions to model and solve real-life problems

Why You Should Learn It:

Exponential functions are useful in modeling data that increase or decrease quickly. For instance, Exercise 77 on page 227 shows how to use an exponential function to model the amount of defoliation caused by a gypsy moth.

Jenny Hager/The Image Works

Graphs of Exponential Functions

The graphs of all exponential functions have similar characteristics, as shown in Examples 2, 3, and 4.

EXAMPLE 2   Graphs of $y = a^x$

In the same coordinate plane, sketch the graph of each function.

a. $f(x) = 2^x$ **b.** $g(x) = 4^x$

Solution

The table below lists some values for each function, and Figure 3.1 shows the graphs of both functions. Note that both graphs are increasing. Moreover, the graph of $g(x) = 4^x$ is increasing more rapidly than the graph of $f(x) = 2^x$.

x	-2	-1	0	1	2	3
2^x	$\frac{1}{4}$	$\frac{1}{2}$	1	2	4	8
4^x	$\frac{1}{16}$	$\frac{1}{4}$	1	4	16	64

The table in Example 2 was evaluated by hand. You could, of course, use a graphing utility to construct tables with even more values.

EXAMPLE 3   Graphs of $y = a^{-x}$

In the same coordinate plane, sketch the graph of each function.

a. $F(x) = 2^{-x}$ **b.** $G(x) = 4^{-x}$

Solution

The table below lists some values for each function, and Figure 3.2 shows the graphs of both functions. Note that both graphs are decreasing. Moreover, the graph of $G(x) = 4^{-x}$ is decreasing more rapidly than the graph of $F(x) = 2^{-x}$.

x	-3	-2	-1	0	1	2
2^{-x}	8	4	2	1	$\frac{1}{2}$	$\frac{1}{4}$
4^{-x}	64	16	4	1	$\frac{1}{4}$	$\frac{1}{16}$

In Example 3, note that the functions $F(x) = 2^{-x}$ and $G(x) = 4^{-x}$ can be rewritten with positive exponents.

$$F(x) = 2^{-x} = \left(\frac{1}{2}\right)^x \quad \text{and} \quad G(x) = 4^{-x} = \left(\frac{1}{4}\right)^x$$

In general,

$$a^{-x} = \left(\frac{1}{a}\right)^x.$$

Library of Functions

The exponential function $f(x) = a^x$ is different from all the functions you have studied so far because the variable x is an *exponent*. The domain, like those of polynomial functions, is the set of all real numbers.

Consult the Library of Functions Summary inside the front cover for a description of the exponential function.

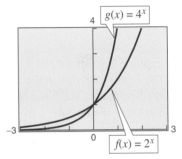

Figure 3.1

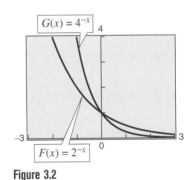

Figure 3.2

Comparing the functions in Examples 2 and 3, observe that

$$F(x) = 2^{-x} = f(-x) \qquad \text{and} \qquad G(x) = 4^{-x} = g(-x).$$

Consequently, the graph of F is a reflection (in the y-axis) of the graph of f, as shown in Figure 3.3(a). The graph of G and g have the same relationship, as shown in Figure 3.3(b).

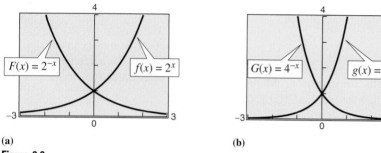

(a)   (b)

Figure 3.3

The graphs in Figures 3.1 and 3.2 are typical of the exponential functions a^x and a^{-x}. They have one y-intercept and one horizontal asymptote (the x-axis), and they are continuous. The basic characteristics of these exponential functions are summarized in Figure 3.4.

Graph of $y = a^x$, $a > 1$

- Domain: $(-\infty, \infty)$
- Range: $(0, \infty)$
- Intercept: $(0, 1)$
- Increasing
- x-axis is a horizontal asymptote $(a^x \to 0 \text{ as } x \to -\infty)$
- Continuous

Graph of $y = a^{-x}$, $a > 1$

- Domain: $(-\infty, \infty)$
- Range: $(0, \infty)$
- Intercept: $(0, 1)$
- Decreasing
- x-axis is a horizontal asymptote $(a^{-x} \to 0 \text{ as } x \to \infty)$
- Continuous

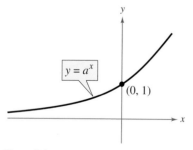

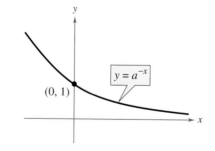

Figure 3.4

Exploration

Use a graphing utility to graph $y = a^x$ with $a = 3$, 5, and 7 in the same viewing window. (Use a viewing window in which $-2 \le x \le 1$ and $0 \le y \le 2$.) How do the graphs compare with each other? Which graph is on the top in the interval $(-\infty, 0)$? Which is on the bottom? Which graph is on the top in the interval $(0, \infty)$? Which is on the bottom? Repeat this experiment with the graphs of $y = a^x$ for $a = \frac{1}{3}, \frac{1}{5}$, and $\frac{1}{7}$. (Use a viewing window in which $-1 \le x \le 2$ and $0 \le y \le 2$.) What can you conclude about the shape of the graph of $y = a^x$ and the value of a?

In the following example, notice how the graph of $y = a^x$ can be used to sketch the graphs of functions of the form

$$f(x) = b \pm a^{x+c}.$$

A computer animation of this example appears in the *Interactive* CD-ROM and *Internet* versions of this text.

EXAMPLE 4   Transformations of Graphs of Exponential Functions

Each of the following graphs is a transformation of the graph of $f(x) = 3^x$, as shown in Figure 3.5.

a. Because $g(x) = 3^{x+1} = f(x + 1)$, the graph of g can be obtained by shifting the graph of f one unit to the left.

b. Because $h(x) = 3^x - 2 = f(x) - 2$, the graph of h can be obtained by shifting the graph of f down two units.

c. Because $k(x) = -3^x = -f(x)$, the graph of k can be obtained by reflecting the graph of f in the x-axis.

d. Because $j(x) = 3^{-x} = f(-x)$, the graph of j can be obtained by reflecting the graph of f in the y-axis.

STUDY T!P

The following table shows some points of the graphs in Figure 3.5(a). The functions $f(x)$ and $g(x)$ are represented by Y1 and Y2, respectively. Explain how you can use the table to describe the transformation.

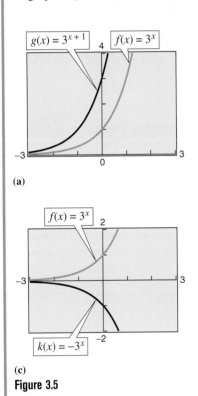

(a)

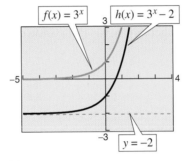

(b)

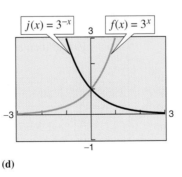

X	Y1	Y2
-3	0.03704	0.1111
-2	0.11111	0.3333
-1	0.33333	1
0	1	3
1	3	9
2	9	27
3	27	81
X=-3		

(c)

(d)

Figure 3.5

In Figure 3.5, notice that the transformations in parts (a), (c), and (d) keep the x-axis as a horizontal asymptote, but the transformation in part (b) yields a new horizontal asymptote of $y = -2$. Also, be sure to note how the y-intercept is affected by each transformation.

The Natural Base e

For many applications, the convenient choice for a base is the irrational number

$$e \approx 2.71828 \ldots .$$

This number is called the **natural base.** The function $f(x) = e^x$ is the **natural exponential function.** Its graph is shown in Figure 3.6. Be sure you see that for the exponential function $f(x) = e^x$, e is the constant 2.71828 . . . , whereas x is the variable.

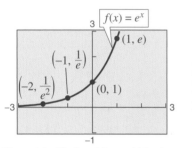

Domain: $(-\infty, \infty)$
Range: $(0, \infty)$
Intercept: $(0, 1)$

Figure 3.6   *The Natural Exponential Function*

Exploration

Use your graphing utility to graph the functions

$$y_1 = 2^x$$
$$y_2 = e^x$$
$$y_3 = 3^x$$

in the same viewing window. From the relative positions of these graphs, make a guess as to the value of the real number e. Then try to find a number a such that the graphs of $y_2 = e^x$ and $y_4 = a^x$ are as close as possible.

In Example 5, you will see that the number e can be approximated by the expression

$$\left(1 + \frac{1}{x}\right)^x \text{ for large values of } x.$$

EXAMPLE 5   Approximation of the Number e

Evaluate the expression $[1 + (1/x)]^x$ for several large values of x to see that the values approach $e \approx 2.71828$ as x increases without bound.

Graphical Solution

Use a graphing utility to graph

$$y_1 = [1 + (1/x)]^x \quad \text{and} \quad y_2 = e$$

in the same viewing window, as shown in Figure 3.7. Use the *trace* feature of the graphing utility to verify that as x increases, the graph of y_1 gets closer and closer to the line $y_2 = e$.

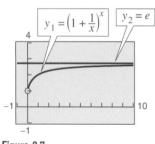

Figure 3.7

Numerical Solution

Create a table of values for the function $y = [1 + (1/x)]^x$, beginning at $x = 10$ and increasing the x-values as shown in the table below.

x	10	100	1000
$\left(1 + \dfrac{1}{x}\right)^x$	2.59374	2.70481	2.71692

x	10,000	100,000	1,000,000
$\left(1 + \dfrac{1}{x}\right)^x$	2.71815	2.71827	2.71828

From the table, it seems reasonable to conclude that

$$\left(1 + \frac{1}{x}\right)^x \to e \text{ as } x \to \infty.$$

EXAMPLE 6   Evaluating the Natural Exponential Function

Use a calculator to evaluate each expression.

a. e^{-2} **b.** e^{-1} **c.** e^{1} **d.** e^{2}

Solution

Number	Graphing Calculator Keystrokes	Display
a. e^{-2}	e^x $(-)$ 2 $\boxed{\text{ENTER}}$	0.1353353
b. e^{-1}	e^x $(-)$ 1 $\boxed{\text{ENTER}}$	0.3678794
c. e^{1}	e^x 1 $\boxed{\text{ENTER}}$	2.7182818
d. e^{2}	e^x 2 $\boxed{\text{ENTER}}$	7.3890561

Exploration

Use a graphing utility to graph $y = (1 + x)^{1/x}$. Describe the behavior of the graph near $x = 0$. Is there a y-intercept? How does the behavior of the graph near $x = 0$ relate to the result of Example 5? Create a table that shows values of y for values of x near $x = 0$ to help you describe the behavior of the graph near this point.

EXAMPLE 7   Graphing Natural Exponential Functions

Sketch the graph of each natural exponential function.

a. $f(x) = 2e^{0.24x}$ **b.** $g(x) = \frac{1}{2}e^{-0.58x}$

Solution

To sketch these two graphs, you can use a calculator to construct a table of values, as shown below.

x	-3	-2	-1	0	1	2	3
$f(x)$	0.974	1.238	1.573	2.000	2.542	3.232	4.109
$g(x)$	2.849	1.595	0.893	0.500	0.280	0.157	0.088

After constructing the table, plot the points and connect them with smooth curves, as shown in Figure 3.8. Note that the graph in part (a) is increasing, whereas the graph in part (b) is decreasing. Use a graphing calculator to verify these graphs.

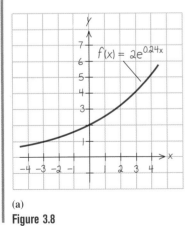

(a)

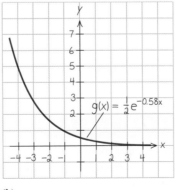

(b)

Figure 3.8

The *Interactive* CD-ROM and *Internet* versions of this text show every example with its solution; clicking on the *Try It!* button brings up similar problems. Guided Examples and Integrated Examples show step-by-step solutions to additional examples. Integrated Examples are related to several concepts in the section.

Compound Interest

One of the most familiar examples of exponential growth is that of an investment earning *continuously compounded interest.* Suppose a principal P is invested at an annual interest rate r, compounded once a year. If the interest is added to the principal at the end of the year, the balance is $P_1 = P + Pr = P(1 + r)$. This pattern of multiplying the previous principal by $1 + r$ is then repeated each successive year, as shown in the table.

Time in Years	Balance After Each Compounding
0	$P = P$
1	$P_1 = P(1 + r)$
2	$P_2 = P_1(1 + r) = P(1 + r)(1 + r) = P(1 + r)^2$
$\vdots$	$\vdots$
n	$P_n = P(1 + r)^n$

To accommodate more frequent (quarterly, monthly, or daily) compounding of interest, let n be the number of compoundings per year and let t be the number of years. (The product nt represents the total number of times the interest will be compounded.) Then the interest rate per compounding period is r/n, and the account balance after t years is

$$A = P\left(1 + \frac{r}{n}\right)^{nt}.$$ Amount with n compoundings per year

If you let the number of compoundings n increase without bound, you approach **continuous compounding.** In the formula for n compoundings per year, let $m = n/r$. This produces

$$A = P\left(1 + \frac{r}{n}\right)^{nt} = P\left(1 + \frac{1}{m}\right)^{mrt} = P\left[\left(1 + \frac{1}{m}\right)^{m}\right]^{rt}.$$

As m increases without bound, you know from Example 5 that $[1 + (1/m)]^m$ approaches e. So, for continuous compounding, it follows that

$$P\left[\left(1 + \frac{1}{m}\right)^{m}\right]^{rt} \rightarrow P[e]^{rt}$$

and you can write $A = Pe^{rt}$. This result is part of the reason that e is the "natural" choice for a base of an exponential function.

Formulas for Compound Interest

After t years, the balance A in an account with principal P and annual interest rate r (expressed as a decimal) is given by the following formulas.

1. For n compoundings per year: $A = P\left(1 + \dfrac{r}{n}\right)^{nt}$

2. For continuous compounding: $A = Pe^{rt}$

EXAMPLE 8   Finding the Balance for Compound Interest

A sum of $9000 is invested at an annual interest rate of 8.5%, compounded annually. Find the balance in the account after 3 years.

Algebraic Solution

In this case,

$$P = 9000, r = 8.5\% = 0.085, n = 1, t = 3.$$

Using the formula for compound interest with n compoundings per year, you have

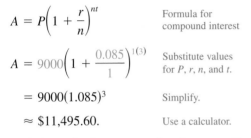

$A = P\left(1 + \dfrac{r}{n}\right)^{nt}$ Formula for compound interest

$A = 9000\left(1 + \dfrac{0.085}{1}\right)^{1(3)}$ Substitute values for P, r, n, and t.

$\quad = 9000(1.085)^3$ Simplify.

$\quad \approx \$11,495.60.$ Use a calculator.

So, the balance in the account after 3 years will be about $11,495.60.

Graphical Solution

Substitute values for P, r, and n into the formula for compound interest with n compoundings per year as follows.

$A = P\left(1 + \dfrac{r}{n}\right)^{nt}$ Formula for compound interest

$\quad = 9000\left(1 + \dfrac{0.085}{1}\right)^{(1)t}$ Substitute values for P, r, and n.

$\quad = 9000(1 + 0.085)^t$ Simplify.

Use a graphing utility to graph $y = 9000(1 + 0.085)^x$, as shown in Figure 3.9. Using the *value* feature or *zoom* and *trace* features, you can approximate the value of y when $x = 3$ to be about 11,495.60. So, the balance in the account after 3 years will be about $11,495.60.

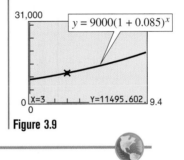

Figure 3.9

EXAMPLE 9   Finding Compound Interest

A total of $12,000 is invested at an annual interest rate of 9%. Find the balance after 5 years if it is compounded

a. quarterly.   **b.** monthly.   **c.** continuously.

Solution

a. For quarterly compoundings, $n = 4$. So, in 5 years at 9%, the balance is

$$A = P\left(1 + \frac{r}{n}\right)^{nt} = 12,000\left(1 + \frac{0.09}{4}\right)^{4(5)} = \$18,726.11.$$

b. For monthly compoundings, $n = 12$. So, in 5 years at 9%, the balance is

$$A = P\left(1 + \frac{r}{n}\right)^{nt} = 12,000\left(1 + \frac{0.09}{12}\right)^{12(5)} = \$18,788.17.$$

c. For continuous compounding, the balance is

$$A = Pe^{rt} = 12,000e^{0.09(5)} = \$18,819.75.$$

Note that continuous compounding yields more than quarterly or monthly compounding.

Other Applications

Exponential functions are used in various other applications.

EXAMPLE 10   Radioactive Decay

Let y represent the mass of a quantity of a radioactive element whose half-life is 25 years. After t years, the mass (in grams) is $y = 10\left(\frac{1}{2}\right)^{t/25}$.

a. What is the initial mass (when $t = 0$)?

b. How much of the initial mass is present after 80 years?

Algebraic Solution

a. $y = 10\left(\frac{1}{2}\right)^{t/25}$ Write original equation.

$= 10\left(\frac{1}{2}\right)^{0/25}$ Substitute 0 for t.

$= 10$ Simplify.

So, the initial mass is 10 grams.

b. $y = 10\left(\frac{1}{2}\right)^{t/25}$ Write original equation.

$= 10\left(\frac{1}{2}\right)^{80/25}$ Substitute 80 for t.

$= 10\left(\frac{1}{2}\right)^{3.2}$ Simplify.

≈ 1.088 Use a calculator.

So, about 1.088 grams is present after 80 years.

Graphical Solution

Use a graphing utility to graph $y = 10\left(\frac{1}{2}\right)^{x/25}$.

a. Use the *value* feature or *zoom* and *trace* features of the graphing utility to determine that the value of y when $x = 0$ is 10, as shown in Figure 3.10. So, the initial mass is 10 grams.

b. Use the *value* feature or *zoom* and *trace* features of the graphing utility to determine that the value of y when $x = 80$ is about 1.088, as shown in Figure 3.11. So, about 1.088 grams is present after 80 years.

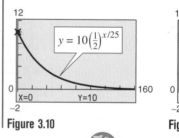

Figure 3.10

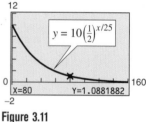

Figure 3.11

EXAMPLE 11 • Population Growth

The approximate number of fruit flies in an experimental population after t hours is

$$Q(t) = 20e^{0.03t}, \qquad t \geq 0.$$

a. Find the initial number of fruit flies in the population.

b. How large is the population of fruit flies after 72 hours?

c. Sketch the graph of Q.

Solution

a. To find the initial population, evaluate $Q(t)$ at $t = 0$.

$$Q(0) = 20e^{0.03(0)} = 20e^0 = 20(1) = 20 \text{ flies}$$

b. After 72 hours, the population size is

$$Q(72) = 20e^{0.03(72)} = 20e^{2.16} \approx 173 \text{ flies}.$$

c. The graph of Q is shown in Figure 3.12.

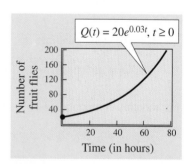

Figure 3.12

In Exercises 1–10, use a calculator to evaluate the expression. Round your result to three decimal places.

1. $(3.4)^{6.8}$ **2.** $5000(2^{-1.5})$

3. $6^{2\pi}$ **4.** $5^{-\pi}$

5. $\sqrt[3]{7493}$ **6.** $100^{\sqrt{2}}$

7. $e^{1/2}$ **8.** $e^{-3/4}$

9. $e^{9.2}$ **10.** $e^{3.78}$

Think About It In Exercises 11–14, use properties of exponents to determine which functions (if any) are the same.

11. $f(x) = 3^{x-2}$ **12.** $f(x) = 4^x + 12$
     $g(x) = 3^x - 9$ $g(x) = 2^{2x+6}$
     $h(x) = \frac{1}{9}(3^x)$ $h(x) = 64(4^x)$

13. $f(x) = 16(4^{-x})$ **14.** $f(x) = 5^{-x} + 3$
     $g(x) = \left(\frac{1}{4}\right)^{x-2}$ $g(x) = 5^{3-x}$
     $h(x) = 16(2^{-2x})$ $h(x) = -5^{x-3}$

In Exercises 15–22, match the exponential function with its graph. [The graphs are labeled (a), (b), (c), (d), (e), (f), (g), and (h).]

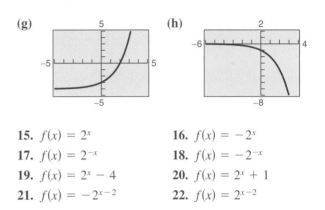

15. $f(x) = 2^x$ **16.** $f(x) = -2^x$

17. $f(x) = 2^{-x}$ **18.** $f(x) = -2^{-x}$

19. $f(x) = 2^x - 4$ **20.** $f(x) = 2^x + 1$

21. $f(x) = -2^{x-2}$ **22.** $f(x) = 2^{x-2}$

In Exercises 23–26, use the graph of f to describe the transformation that yields the graph of g.

23. $f(x) = 3^x$, $g(x) = 3^{x-5}$

24. $f(x) = -2^x$, $g(x) = 5 - 2^x$

25. $f(x) = \left(\frac{3}{5}\right)^x$, $g(x) = -\left(\frac{3}{5}\right)^{x+4}$

26. $f(x) = 0.3^x$, $g(x) = -0.3^x + 5$

In Exercises 27–34, graph the exponential function *by hand*. Identify the following features of the graph.

(a) Asymptotes

(b) Intercepts

(c) Increasing or decreasing

27. $g(x) = 5^x$ **28.** $f(x) = \left(\frac{3}{2}\right)^x$

29. $f(x) = \left(\frac{1}{5}\right)^x = 5^{-x}$ **30.** $h(x) = \left(\frac{3}{2}\right)^{-x}$

31. $h(x) = 5^{x-2}$ **32.** $g(x) = \left(\frac{3}{2}\right)^{x+2}$

33. $g(x) = 5^{-x} - 3$ **34.** $f(x) = \left(\frac{3}{2}\right)^{-x} + 2$

In Exercises 35–44, use a graphing utility to construct a table of values for the function. Then sketch the graph of the function.

35. $f(x) = \left(\frac{5}{2}\right)^x$ **36.** $f(x) = \left(\frac{5}{2}\right)^{-x}$

37. $f(x) = 6^x$ **38.** $f(x) = 2^{x-1}$

39. $f(x) = 3^{x+2}$ **40.** $f(x) = e^{-x}$

41. $f(x) = 3e^{x+4}$ **42.** $f(x) = 2e^{-0.5x}$

43. $f(x) = 2 + e^{x-5}$ **44.** $f(x) = 4^{x-3} + 3$

The *Interactive* CD-ROM and *Internet* versions of this text contain step-by-step solutions to all odd-numbered Section and Review Exercises. They also provide Tutorial Exercises, which link to Guided Examples for additional help.

In Exercises 45–52, use a graphing utility to graph the exponential function. Identify any asymptotes of the graph.

45. $y = 2^{-x^2}$

46. $y = 3^{-|x|}$

47. $y = 3^{x-2} + 1$

48. $y = 4^{x+1} - 2$

49. $y = 1.08^{-5x}$

50. $s(t) = 2e^{0.12t}$

51. $s(t) = 3e^{-0.2t}$

52. $g(x) = 1 + e^{-x}$

53. *Exploration* Consider the functions $f(x) = 3^x$ and $g(x) = 4^x$.

   (a) Use a graphing utility to complete the table, and use the table to estimate the solution of the inequality $4^x < 3^x$.

x	-1	-0.5	0	0.5	1
$f(x)$					
$g(x)$					

   (b) Use a graphing utility to graph $f(x)$ and $g(x)$ in the same viewing window. Use the graphs to solve the inequalities.

     (i) $4^x < 3^x$ (ii) $4^x > 3^x$

54. *Exploration* Consider the functions $f(x) = \left(\frac{1}{2}\right)^x$ and $g(x) = \left(\frac{1}{4}\right)^x$.

   (a) Use a graphing utility to complete the table, and use the table to estimate the solution of the inequality $\left(\frac{1}{4}\right)^x < \left(\frac{1}{2}\right)^x$.

x	-1	-0.5	0	0.5	1
$f(x)$					
$g(x)$					

   (b) Use a graphing utility to graph $f(x)$ and $g(x)$ in the same viewing window. Use the graphs to solve the inequalities.

     (i) $\left(\frac{1}{4}\right)^x < \left(\frac{1}{2}\right)^x$ (ii) $\left(\frac{1}{4}\right)^x > \left(\frac{1}{2}\right)^x$

In Exercises 55–58, use a graphing utility to (a) graph the function and (b) find any asymptotes numerically by creating a table of values for the function.

55. $f(x) = \dfrac{8}{1 + e^{-0.5x}}$

56. $g(x) = \dfrac{8}{1 + e^{-0.5/x}}$

57. $f(x) = -\dfrac{6}{2 - e^{0.2x}}$

58. $f(x) = \dfrac{6}{2 - e^{0.2/x}}$

In Exercises 59–62, (a) use a graphing utility to graph the function, (b) use the graph to find the open intervals on which the function is increasing and decreasing, and (c) approximate any relative maximum or minimum values.

59. $f(x) = x^2 e^{-x}$

60. $f(x) = 2x^2 e^{x+1}$

61. $f(x) = x(2^{3-x})$

62. $f(x) = -\left(\frac{1}{2}x\right)3^{x+4}$

Compound Interest **In Exercises 63–66, complete the table to determine the balance A for P dollars invested at rate r for t years and compounded n times per year.**

n	1	2	4	12	365	Continuous
A						

63. $P = \$2500$, $r = 8\%$, $t = 10$ years

64. $P = \$1000$, $r = 6\%$, $t = 10$ years

65. $P = \$2500$, $r = 8\%$, $t = 20$ years

66. $P = \$1000$, $r = 6\%$, $t = 40$ years

Compound Interest **In Exercises 67–70, complete the table to determine the balance A for \$12,000 invested at a rate r for t years.**

t	1	10	20	30	40	50
A						

67. $r = 8\%$, compounded continuously

68. $r = 6\%$, compounded continuously

69. $r = 6.5\%$, compounded monthly

70. $r = 7.5\%$, compounded daily

71. *Demand Function* The demand equation for a certain product is

$$p = 5000\left(1 - \frac{4}{4 + e^{-0.002x}}\right)$$

   where p is the price and x is the number of units.

   (a) Use a graphing utility to graph the demand function for $x > 0$ and $p > 0$.

   (b) Find the price p for a demand of $x = 500$ units.

   (c) Use the graph in part (a) to approximate the highest price that will still yield a demand of at least 600 units.

   (d) Verify your answers to parts (b) and (c) numerically by creating a table of values for the function.

72. *Graphical Reasoning* There are two options for investing $500. The first earns 7% compounded annually, and the second earns 7% simple interest. The figure shows the growth of each investment over a 30-year period.

(a) Identify the two types of investments in the figure. Explain your reasoning.

(b) Verify your answer in part (a) by finding the equations that model the investment growth and using a graphing utility to graph the models.

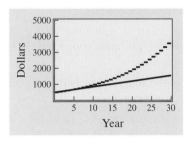

73. *Bacteria Growth* A certain type of bacteria increases according to the model

$$P(t) = 100e^{0.2197t}$$

where t is the time in hours.

(a) Use a graphing utility to graph the model.

(b) Use a graphing utility to approximate $P(0)$, $P(5)$, and $P(10)$.

(c) Verify your answers in part (b) algebraically.

74. *Population Growth* The population of a town increases according to the model

$$P(t) = 2500e^{0.0293t}$$

where t is the time in years, with $t = 0$ corresponding to 1990.

(a) Find the population in 1992, 1995, and 1998.

(b) Use a graphing utility to graph the function for the years 1990 through 2015.

(c) Use a graphing utility to approximate the population in 2005 and 2010.

(d) Verify your answers in part (c) algebraically.

75. *Radioactive Decay* Let Q (in grams) represent the mass of a quantity of radium 226, which has a half-life of 1620 years. The quantity of radium present after t years is

$$Q = 25\left(\tfrac{1}{2}\right)^{t/1620}.$$

(a) Determine the initial quantity (when $t = 0$).

(b) Determine the quantity present after 1000 years.

(c) Use a graphing utility to graph the function over the interval $t = 0$ to $t = 5000$.

(d) When will the quantity of radium be 0 grams? Explain.

76. *Radioactive Decay* Let Q (in grams) represent the mass of a quantity of carbon 14, which has a half-life of 5730 years. The quantity present after t years is $Q = 10\left(\tfrac{1}{2}\right)^{t/5730}$.

(a) Determine the initial quantity (when $t = 0$).

(b) Determine the quantity present after 2000 years.

(c) Sketch the graph of the function over the interval $t = 0$ to $t = 10,000$.

77. *Data Analysis* To estimate the amount of defoliation caused by gypsy moths, a forester counts the number x of egg masses on $\frac{1}{40}$ of an acre (circle of radius 18.6 feet) during the fall. The percent of defoliation y the next spring is shown in the table. (Source: USDA Forest Service)

x	0	25	50	75	100
y	12	44	81	96	99

(a) Use a graphing utility to plot the data points.

(b) A model for the data is $y = \dfrac{300}{3 + 17e^{-0.065x}}$.
Use a graphing utility to graph the model in the same viewing window used in part (a). How well does the model fit the data?

(c) Use a graphing utility to create a table comparing the model with the sample data.

(d) Estimate the percent of defoliation if 36 egg masses are counted on $\frac{1}{40}$ acre.

(e) Use the graph to estimate the number of egg masses per $\frac{1}{40}$ acre if approximately $\frac{2}{3}$ of the forest is defoliated the next spring.

78. *Data Analysis* A cup of water at an initial temperature of 78°C is placed in a room at a constant temperature of 21°C. The temperature of the water is measured every 5 minutes for a period of $\frac{1}{2}$ hour. The results are recorded in the table, where t is the time (in minutes) and T is the temperature (in degrees Celsius).

t	0	5	10	15	20	25	30
T	78.0°	66.0°	57.5°	51.2°	46.3°	42.5°	39.6°

(a) Use the regression capabilities of a graphing utility to fit a line to the data. Use the graphing utility to plot the data points and the regression line in the same viewing window. Does the data appear linear? Explain.

(b) Use the regression capabilities of a graphing utility to fit a parabola to the data. Use the graphing utility to plot the data points and the regression parabola in the same viewing window. Does the data appear quadratic? Even though the quadratic model appears to be a "good" fit, explain why it may not be a good model for predicting the temperature of the water when $t = 60$.

(c) The graph of the model should be asymptotic with the temperature of the room. Subtract the room temperature from each of the temperatures in the table. Use a graphing utility to fit an exponential model to the revised data. Add the room temperature to this regression model. Use a graphing utility to plot the original data points and the model in the same viewing window.

(d) Explain why the procedure in part (c) was necessary for finding the exponential model.

79. *Inflation* If the annual rate of inflation averages 4% over the next 10 years, the approximate cost C of goods or services during any year in that decade will be $C(t) = P(1.04)^t$, where t is the time (in years) and P is the present cost. Assume the price of an oil change for your car is presently $23.95.

(a) Use a graphing utility to graph the function.

(b) Use the graph in part (a) to approximate the price of an oil change 10 years from now.

(c) Verify your answer in part (b) algebraically.

80. *Depreciation* After t years, the value of a car that costs $20,000 is $V(t) = 20,000\left(\frac{3}{4}\right)^t$.

(a) Use a graphing utility to graph the function.

(b) Use a graphing utility to create a table of values that shows the value V for $t = 1$ to 10 years.

Synthesis

True or False? **In Exercises 81 and 82, determine whether the statement is true or false. Justify your answer.**

81. The x-axis is an asymptote for the graph of $f(x) = 10^x$.

82. $e = \dfrac{271,801}{99,990}$

83. *Exploration* Use a graphing utility to graph $y_1 = e^x$ and each of the functions $y_2 = x^2$, $y_3 = x^3$, $y_4 = \sqrt{x}$, and $y_5 = |x|$.

(a) Which function increases at the fastest rate for "large" values of x?

(b) Use the result of part (a) to make a conjecture about the rates of growth of $y_1 = e^x$ and $y = x^n$ where n is a natural number and x is "large."

(c) Use the results of parts (a) and (b) to describe what is implied when it is stated that a quantity is growing exponentially.

84. *Conjecture* Use a graphing utility to graph $f(x) = (1 + 0.5/x)^x$ and $g(x) = e^{0.5}$ in the same viewing window.

(a) What is the relationship between f and g as x increases without bound?

(b) Use the result of part (a) to make a conjecture about the value of $(1 + r/x)^x$ as x increases without bound.

85. *Think About It* Without a graphing utility, explain why you know $2^{\sqrt{2}}$ is greater than 2, but less than 4.

86. *Think About It* Which functions are exponential? Why?

  (a) $3x$ (b) $3x^2$ (c) 3^x (d) 2^{-x}

87. *Pattern Recognition* Use a graphing utility to compare the graph of the function $y = e^x$ with the graphs of the following functions.

(a) $y_1 = 1 + \dfrac{x}{1!}$ (b) $y_2 = 1 + \dfrac{x}{1!} + \dfrac{x^2}{2!}$

(c) $y_3 = 1 + \dfrac{x}{1!} + \dfrac{x^2}{2!} + \dfrac{x^3}{3!}$

In your opinion, which function is the best approximation of $y = e^x$?

Review

In Exercises 88–91, determine whether the function has an inverse. If it does, find f^{-1}.

88. $f(x) = 5x - 7$

89. $f(x) = -\frac{2}{3}x + \frac{5}{2}$

90. $f(x) = \sqrt[3]{x + 8}$

91. $f(x) = \sqrt{x^2 + 6}$

In Exercises 92–95, sketch the graph of the rational function.

92. $f(x) = \dfrac{2x}{x - 7}$

93. $f(x) = \dfrac{x^2 + 3}{x + 1}$

94. $f(x) = \dfrac{4x}{x^2 + 11x + 24}$

95. $f(x) = \dfrac{2x^3}{(x - 2)^2}$

3.2 Logarithmic Functions and Their Graphs

Logarithmic Functions

In Section 1.5, you studied the concept of the inverse of a function. There, you learned if a function has the property such that no horizontal line intersects its graph more than once, the function must have an inverse. By looking back at the graphs of the exponential functions introduced in Section 3.1, you will see that every function of the form

$$f(x) = a^x, \qquad a > 0, a \neq 1$$

passes the Horizontal Line Test and therefore must have an inverse. This inverse function is called the **logarithmic function with base a**.

Definition of Logarithmic Function

For $x > 0$ and $0 < a \neq 1$,

$$y = \log_a x \qquad \text{if and only if} \qquad x = a^y.$$

The function

$$f(x) = \log_a x$$

is called the **logarithmic function with base a**.

The equations

$$y = \log_a x \qquad \text{Logarithmic form}$$

and

$$x = a^y \qquad \text{Exponential form}$$

are equivalent. The first equation is in logarithmic form and the second is in exponential form.

When evaluating logarithms, remember that *a logarithm is an exponent*. This means that $\log_a x$ is the exponent to which a must be raised to obtain x. For instance, $\log_2 8 = 3$ because 2 must be raised to the third power to get 8.

What You Should Learn:

- How to recognize and evaluate logarithmic functions with base a
- How to graph logarithmic functions
- How to recognize, evaluate, and graph natural logarithmic functions
- How to use logarithmic functions to model and solve real-life problems

Why You Should Learn It:

Logarithmic functions are useful in modeling data that increase or decrease slowly. For instance, Exercise 81 on page 238 shows how to use a logarithmic function to model the minimum required ventilation rates in public school classrooms.

Mark Richards/PhotoEdit

EXAMPLE 1   Evaluating Logarithms

Evaluate each expression.

a. $\log_2 32$ **b.** $\log_3 27$ **c.** $\log_4 2$
d. $\log_{10} \frac{1}{100}$ **e.** $\log_3 1$ **f.** $\log_2 2$

Solution

a. $\log_2 32 = 5$ because $2^5 = 32.$ **e.** $\log_3 1 = 0$ because $3^0 = 1.$
b. $\log_3 27 = 3$ because $3^3 = 27.$ **f.** $\log_2 2 = 1$ because $2^1 = 2.$
c. $\log_4 2 = \frac{1}{2}$ because $4^{1/2} = \sqrt{4} = 2.$
d. $\log_{10} \frac{1}{100} = -2$ because $10^{-2} = \frac{1}{10^2} = \frac{1}{100}.$

The logarithmic function with base 10 is called the **common logarithmic function.** On most calculators, this function is denoted by $\boxed{\text{LOG}}$. Because $\log_a x$ is the inverse function of a^x, it follows that the domain of $\log_a x$ is the range of a^x, $(0, \infty)$. In other words, $\log_a x$ is defined only if x is positive.

EXAMPLE 2   Evaluating Logarithms on a Calculator

Use a calculator to evaluate each expression.

a. $\log_{10} 10$ **b.** $2 \log_{10} 2.5$ **c.** $\log_{10}(-2)$

Solution

Number	Graphing Calculator Keystrokes	Display
a. $\log_{10} 10$	$\boxed{\text{LOG}}$ 10 $\boxed{\text{ENTER}}$	1
b. $2 \log_{10} 2.5$	2 $\boxed{\text{LOG}}$ 2.5 $\boxed{\text{ENTER}}$	0.7958800
c. $\log_{10}(-2)$	$\boxed{\text{LOG}}$ $\boxed{(-)}$ 2 $\boxed{\text{ENTER}}$	ERROR

Note that the calculator displays an error message when you try to evaluate $\log_{10}(-2)$. The reason for this is that the domain of every logarithmic function is the set of *positive* real numbers. In this case, there is no *real* power to which 10 can be raised to get -2.

The following properties follow directly from the definition of the logarithmic function with base a.

Properties of Logarithms

1. $\log_a 1 = 0$ because $a^0 = 1$.

2. $\log_a a = 1$ because $a^1 = a$.

3. $\log_a a^x = x$ and $a^{\log_a x} = x$.   Inverse Properties

4. If $\log_a x = \log_a y$, then $x = y$.   One-to-One Property

---

EXAMPLE 3   Using Properties of Logarithms

Solve each equation for x.

a. $\log_2 x = \log_2 3$

b. $\log_4 4 = x$

Solution

a. Using the One-to-One Property (Property 4), you can conclude that $x = 3$.

b. Using Property 2, you can conclude that $x = 1$.

STUDY T!P

Some graphing utilities do not give an error message for $\log_{10}(-2)$. Instead, the graphing utility will display a complex number. For the purpose of this text, however, it will still be said that the domain of a logarithmic function is the set of positive *real* numbers.

Library of Functions

The logarithmic function is the inverse of the exponential function. Its domain is the set of positive real numbers and its range is the set of all real numbers. Because of the inverse properties of logarithms and exponents, the exponential equation $a^0 = 1$ implies that $\log_a 1 = 0$.

Consult the Library of Functions Summary inside the front cover for a description of the logarithmic function.

Graphs of Logarithmic Functions

To sketch the graph of $y = \log_a x$, you can use the fact that the graphs of inverse functions are reflections of each other in the line $y = x$.

EXAMPLE 4   Graphs of Exponential and Logarithmic Functions

In the same coordinate plane, sketch the graph of each function.

a. $f(x) = 2^x$ **b.** $g(x) = \log_2 x$

Solution

a. For $f(x) = 2^x$, construct a table of values.

x	-2	-1	0	1	2	3
2^x	$\frac{1}{4}$	$\frac{1}{2}$	1	2	4	8

By plotting these points and connecting them with a smooth curve, you obtain the graph of $f(x)$ shown in Figure 3.13.

b. Because $g(x) = \log_2 x$ is the inverse of $f(x) = 2^x$, the graph of g is obtained by plotting the points $(2^x, x)$ and connecting them with a smooth curve. The graph of g is a reflection of the graph of f in the line $y = x$, as shown in Figure 3.13.

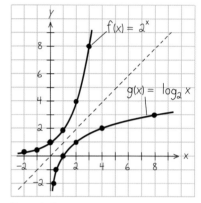

Figure 3.13

Before you can confirm the result of Example 4 using a graphing utility, you need to know how to enter $\log_2 x$. You will learn how to do this using the *change-of-base formula* discussed in Section 3.3.

EXAMPLE 5   Sketching the Graph of a Logarithmic Function

Sketch the graph of the common logarithmic function $f(x) = \log_{10} x$.

Solution

Begin by constructing a table of values. Note that some of the values can be obtained without a calculator using the Inverse Property of logarithms. Others require a calculator. Next, plot the points and connect them with a smooth curve, as shown in Figure 3.14.

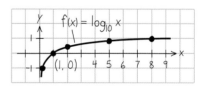

Figure 3.14

	Without Calculator				With Calculator		
x	$\frac{1}{100}$	$\frac{1}{10}$	1	10	2	5	8
$\log_{10} x$	-2	-1	0	1	0.301	0.699	0.903

STUDY TIP

Compare Figure 3.14 with that obtained using a graphing utility. Note that the domain is $(0, \infty)$ and the range is $(-\infty, \infty)$.

In Example 5, you can also sketch the graph of $f(x) = \log_{10} x$ as follows. Evaluate the inverse of f, $g(x) = 10^x$ for several values of x. Plot the points, sketch the graph of g, then reflect the graph in the line $y = x$ to obtain the graph of f.

The nature of the graph in Figure 3.14 is typical of functions of the form $f(x) = \log_a x, a > 1$. They have one x-intercept and one vertical asymptote. Notice how slowly the graph rises for $x > 1$. In Figure 3.14, you would need to move out to $x = 100$ before the graph rose to $y = 2$, and $x = 1000$ before the graph rose to $y = 3$. The basic characteristics of logarithmic graphs are summarized in Figure 3.15.

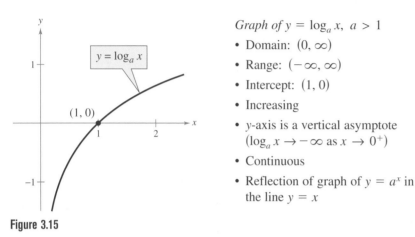

Figure 3.15

Graph of $y = \log_a x,\ a > 1$

- Domain: $(0, \infty)$
- Range: $(-\infty, \infty)$
- Intercept: $(1, 0)$
- Increasing
- y-axis is a vertical asymptote ($\log_a x \to -\infty$ as $x \to 0^+$)
- Continuous
- Reflection of graph of $y = a^x$ in the line $y = x$

In Figure 3.15, the vertical asymptote occurs at $x = 0$, where $\log_a x$ is *undefined*.

EXAMPLE 6 Transformations of Graphs of Logarithmic Functions

The graph of each of the following functions is similar to the graph of $f(x) = \log_{10} x$, as shown in Figure 3.16.

a. Because $g(x) = \log_{10}(x - 1) = f(x - 1)$, the graph of g can be obtained by shifting the graph of f one unit to the right.

b. Because $h(x) = 2 + \log_{10} x = 2 + f(x)$, the graph of h can be obtained by shifting the graph of f two units up.

A computer animation of this example appears in the *Interactive* CD-ROM and *Internet* versions of this text.

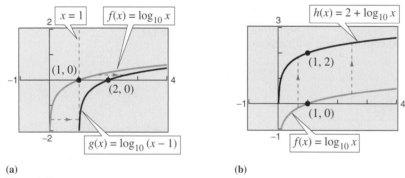

(a)   **(b)**

Figure 3.16

In Figure 3.16, notice that the transformation in part (b) keeps the y-axis as a vertical asymptote, but the transformation in part (a) yields the new vertical asymptote $x = 1$.

The Natural Logarithmic Function

The most widely used base for logarithmic functions is the number e, where

$$e \approx 2.718281828\ldots\ .$$

The logarithmic function with base e is the **natural logarithmic function** and is denoted by the special symbol $\ln x$, read as "the natural log of x" or "el en of x."

The Natural Logarithmic Function

The function defined by

$$f(x) = \log_e x = \ln x, \qquad x > 0$$

is called the **natural logarithmic function.**

The *Interactive* CD-ROM and *Internet* versions of this text offer a built-in graphing calculator, which can be used with the Examples, Explorations, and Exercises.

On most calculators, the natural logarithm is denoted by $\boxed{\text{LN}}$, as illustrated in Example 7.

EXAMPLE 7   Evaluating the Natural Logarithmic Function

Use a calculator to evaluate each expression.

a. $\ln 2$ **b.** $\ln 0.3$ **c.** $\ln e^2$ **d.** $\ln(-1)$

Solution

Number	Graphing Calculator Keystrokes	Display
a. $\ln 2$	$\boxed{\text{LN}}$ 2 $\boxed{\text{ENTER}}$	0.6931472
b. $\ln 0.3$	$\boxed{\text{LN}}$ .3 $\boxed{\text{ENTER}}$	-1.2039728
c. $\ln e^2$	$\boxed{\text{LN}}$ $\boxed{e^x}$ 2 $\boxed{\text{ENTER}}$	2
d. $\ln(-1)$	$\boxed{\text{LN}}$ $\boxed{(-)}$ 1 $\boxed{\text{ENTER}}$	ERROR

In Example 7, be sure you see that $\ln(-1)$ gives an error message on most calculators. This occurs because the domain of $\ln x$ is the set of *positive* real numbers. So, $\ln(-1)$ is undefined.

The four properties of logarithms listed on page 230 are also valid for natural logarithms.

Exploration

Because the natural exponential function $y = e^x$ passes the Horizontal Line Test, it has an inverse. Use the *draw inverse* feature of your graphing utility to graph $y = e^x$ and its inverse. Compare this graph with that of the natural logarithmic function. What can you conclude?

Properties of Natural Logarithms

1. $\ln 1 = 0$ because $e^0 = 1$.
2. $\ln e = 1$ because $e^1 = e$.
3. $\ln e^x = x$ and $e^{\ln x} = x$.   Inverse Properties
4. If $\ln x = \ln y$, then $x = y$.   One-to-One Property

EXAMPLE 8   Using Properties of Natural Logarithms

Use the properties of natural logarithms to rewrite each expression.

a. $\ln \dfrac{1}{e}$ **b.** $\ln e^2$ **c.** $\ln e^0$ **d.** $2 \ln e$

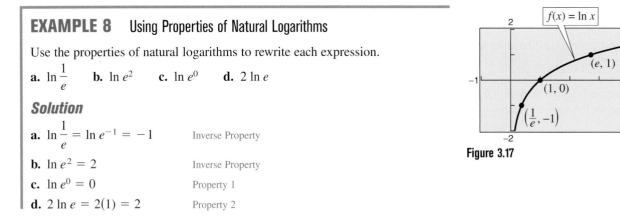

Figure 3.17

Solution

a. $\ln \dfrac{1}{e} = \ln e^{-1} = -1$ Inverse Property

b. $\ln e^2 = 2$ Inverse Property

c. $\ln e^0 = 0$ Property 1

d. $2 \ln e = 2(1) = 2$ Property 2

The graph of the natural logarithmic function is shown in Figure 3.17. Try using a graphing utility to confirm this graph and to verify that the domain of the natural logarithmic function is $(0, \infty)$.

EXAMPLE 9   Finding the Domains of Logarithmic Functions

Find the domain of each function.

a. $f(x) = \ln(x - 2)$ **b.** $g(x) = \ln(2 - x)$ **c.** $h(x) = \ln x^2$

Algebraic Solution

a. Because $\ln(x - 2)$ is defined only if

$$x - 2 > 0,$$

it follows that the domain of f is $(2, \infty)$.

b. Because $\ln(2 - x)$ is defined only if

$$2 - x > 0,$$

it follows that the domain of g is $(-\infty, 2)$.

c. Because $\ln x^2$ is defined only if

$$x^2 > 0,$$

it follows that the domain of h is all real numbers except $x = 0$.

Graphical Solution

a. Use a graphing utility to graph the equation $y = \ln(x - 2)$ using a decimal setting, as shown in Figure 3.18. Use the *trace* feature to see that the x-coordinates of the points on the graph appear to extend from the right of 2 to $+\infty$. So, you can estimate the domain to be $(2, \infty)$.

b. Use a graphing utility to graph the equation $y = \ln(2 - x)$ using a decimal setting, as shown in Figure 3.19. Use the *trace* feature to see that the x-coordinates of the points on the graph appear to extend from $-\infty$ to the left of 2. So, you can estimate the domain to be $(-\infty, 2)$.

c. Use a graphing utility to graph the equation $y = \ln x^2$ using a decimal setting, as shown in Figure 3.20. Use the *trace* feature to see that the x-coordinates of the points on the graph appear to include all real numbers except $x = 0$. So, you can estimate the domain to be all real numbers except $x = 0$.

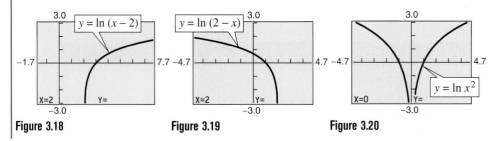

Figure 3.18   **Figure 3.19**   **Figure 3.20**

In Example 9, suppose you had been asked to analyze the function $h(x) = \ln|x - 2|$. How would the domain of this function compare with the domains of the functions given in parts (a) and (b) of the example?

Application

Exponential and logarithmic functions are used to model many situations in real life, as shown in the next example.

EXAMPLE 10   Human Memory Model

Students participating in a psychological experiment attended several lectures on a subject. At the end of the last lecture, and every month for the next year, the students were tested to see how much of the material they remembered. The average scores for the group were given by the *human memory model*

$$f(t) = 75 - 6\ln(t + 1), \qquad 0 \le t \le 12$$

where t is the time in months.

a. What was the average score on the original ($t = 0$) exam?

b. What was the average score at the end of $t = 2$ months?

c. What was the average score at the end of $t = 6$ months?

Algebraic Solution

a. The original average score was

$$f(0) = 75 - 6\ln 1$$

$$= 75 - 6(0)$$

$$= 75.$$

b. After 2 months, the average score was

$$f(2) = 75 - 6\ln 3$$

$$\approx 75 - 6(1.0986)$$

$$\approx 68.4.$$

c. After 6 months, the average score was

$$f(6) = 75 - 6\ln 7$$

$$\approx 75 - 6(1.9459)$$

$$\approx 63.3.$$

Graphical Solution

Use a graphing utility to graph the function $y = 75 - 6\ln(x + 1)$ as shown in Figure 3.21. Then use the *value* or *trace* feature to approximate the following.

a. When $x = 0$, $y = 75$. So, the original average score was 75.

b. When $x = 2$, $y \approx 68.4$. So, the average score after 2 months was about 68.4.

c. When $x = 6$, $y \approx 63.3$. So, the average score after 6 months was about 63.3.

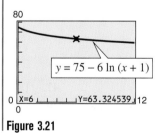

Figure 3.21

Writing About Math   *Analyzing a Human Memory Model*

Use a graphing utility to determine how many months it would take for the average score in Example 10 to decrease to 60. Explain your method of solving the problem. Describe another way that you can use a graphing utility to determine the answer. Also, make a statement about the general shape of the model. Would the students forget more quickly soon after the test or as time passes? Explain your reasoning.

3.2   E x e r c i s e s

In Exercises 1–8, write the logarithmic equation in exponential form. For example, the exponential form of $\log_5 25 = 2$ is $5^2 = 25$.

1. $\log_4 64 = 3$ **2.** $\log_3 81 = 4$

3. $\log_7 \frac{1}{49} = -2$ **4.** $\log_{10} \frac{1}{1000} = -3$

5. $\log_{32} 4 = \frac{2}{5}$ **6.** $\log_{16} 8 = \frac{3}{4}$

7. $\ln 1 = 0$ **8.** $\ln 4 = 1.386\ldots$

In Exercises 9–18, write the exponential equation in logarithmic form.

9. $5^3 = 125$ **10.** $8^2 = 64$

11. $81^{1/4} = 3$ **12.** $9^{3/2} = 27$

13. $6^{-2} = \frac{1}{36}$ **14.** $10^{-3} = 0.001$

15. $e^3 = 20.0855\ldots$ **16.** $e^x = 4$

17. $e^{2.6} = 13.463\ldots$ **18.** $e_\pi = 23.140\ldots$

In Exercises 19–24, evaluate the expression without using a calculator.

19. $\log_2 16$ **20.** $\log_{27} 9$

21. $\log_{16}\left(\frac{1}{4}\right)$ **22.** $\log_2\left(\frac{1}{8}\right)$

23. $\log_{10} 0.01$ **24.** $\log_{10} 0.1$

In Exercises 25–30, solve the equation for x.

25. $\log_7 x = \log_7 9$ **26.** $\log_5 5 = x$

27. $\ln e^8 = x$ **28.** $\ln 1 = \ln x$

29. $\log_6 6^2 = x$ **30.** $\log_2 2^{-1} = x$

In Exercises 31–42, use a calculator to evaluate the logarithm. Round to three decimal places.

31. $\log_{10} 345$ **32.** $\log_{10} 145$

33. $\log_{10}\left(\frac{4}{5}\right)$ **34.** $\log_{10} \frac{25}{2}$

35. $\ln\left(4 + \sqrt{3}\right)$ **36.** $\ln\left(\sqrt{5} - 2\right)$

37. $\ln \sqrt{42}$ **38.** $\ln \sqrt{752}$

39. $6 \log_{10} 14.8$ **40.** $-3 \log_{10} 0.09$

41. $12 \ln 6.4$ **42.** $-5.5 \ln 34$

In Exercises 43–46, describe the relationship between the graphs of f and g. What is the relationship between the functions f and g?

43. $f(x) = 3^x$ **44.** $f(x) = 5^x$
$g(x) = \log_3 x$ $g(x) = \log_5 x$

45. $f(x) = e^x$ **46.** $f(x) = 10^x$
$g(x) = \ln x$ $g(x) = \log_{10} x$

In Exercises 47–52, use the graph of $y = \log_3 x$ to match the given function with its graph. [The graphs are labeled (a), (b), (c), (d), (e), and (f).]

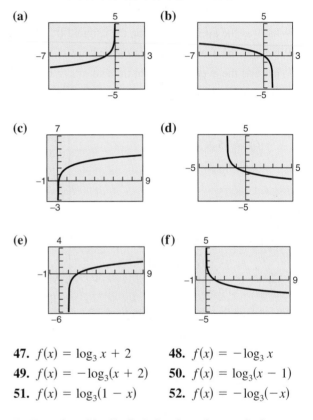

47. $f(x) = \log_3 x + 2$ **48.** $f(x) = -\log_3 x$

49. $f(x) = -\log_3(x + 2)$ **50.** $f(x) = \log_3(x - 1)$

51. $f(x) = \log_3(1 - x)$ **52.** $f(x) = -\log_3(-x)$

In Exercises 53–60, find the domain, vertical asymptote, and x-intercept of the logarithmic function, and sketch its graph by hand. Verify using a graphing utility.

53. $f(x) = \log_4 x$ **54.** $g(x) = \log_6 x$

55. $h(x) = \log_4(x - 3)$ **56.** $f(x) = -\log_6(x + 2)$

57. $y = -\log_3 x + 2$ **58.** $y = \log_5(x - 1) + 4$

59. $f(x) = 6 + \log_6(x - 3)$

60. $f(x) = -\log_3(x + 2) - 4$

In Exercises 61–66, use a graphing utility to graph the logarithmic function. Determine the domain and identify any vertical asymptote and *x*-intercept.

61. $y = \log_{10}\left(\dfrac{x}{5}\right)$ **62.** $y = \log_{10}(-x)$

63. $f(x) = \ln(x - 2)$ **64.** $h(x) = \ln(x + 1)$

65. $g(x) = \ln(-x)$ **66.** $f(x) = \ln(3 - x)$

In Exercises 67–70, (a) use a graphing utility to graph the function, (b) find the domain, (c) use the graph to determine the intervals in which the function is increasing and decreasing, and (d) approximate any relative maximum or minimum values of the function.

67. $f(x) = \dfrac{x}{2} - \ln\dfrac{x}{4}$ **68.** $g(x) = \dfrac{12 \ln x}{x}$

69. $h(x) = 4x \ln x$ **70.** $f(x) = \dfrac{x}{\ln x}$

71. *Population Growth* The population of a town will double in

$$t = \frac{10 \ln 2}{\ln 67 - \ln 50} \text{ years.}$$

Find t.

72. *Work* The work (in foot-pounds) done in compressing an initial volume of 9 cubic feet of air at a pressure of 15 pounds per square inch to a volume of 3 cubic feet is $W = 19{,}440(\ln 9 - \ln 3)$. Find W.

73. *Human Memory Model* Students in a mathematics class were given an exam and then tested monthly with an equivalent exam. The average score for the class was given by the human memory model

$$f(t) = 80 - 17 \log_{10}(t + 1), \quad 0 \le t \le 12$$

where t is the time in months.

(a) What was the average score on the original exam ($t = 0$)?

(b) What was the average score after 4 months?

(c) What was the average score after 10 months?

(d) Verify your answers in parts (a), (b), and (c) using a graphing utility.

74. *Investment Time* A principal P, invested at $5\frac{1}{2}\%$ and compounded continuously, increases to an amount K times the original principal after t years, where $t = (\ln K)/0.055$.

(a) Complete the table and interpret your results.

K	1	2	4	6	8	10	12
t							

(b) Use a graphing utility to graph the function.

75. *Data Analysis* The table shows the temperatures T (in °F) at which water boils at selected pressures p (in pounds per square inch). (Source: Standard Handbook of Mechanical Engineers)

p	5	10	14.696 (1 atm)
T	162.24°	193.21°	212.00°

p	20	30	40
T	227.96°	250.33°	267.25°

p	60	80	100
T	292.71°	312.03°	327.81°

A model that approximates this data is

$$T = 87.97 + 34.96 \ln p + 7.91 \sqrt{p}.$$

(a) Use a graphing utility to plot the data points and graph the model in the same viewing window. How well does the model fit the data?

(b) Use the graph to estimate the pressure required for the boiling point of water to exceed 300°F.

(c) Calculate T when the pressure is 74 pounds per square inch. Verify your answer graphically.

76. *Data Analysis* A meteorologist measures the atmospheric pressure P (in Pascals) at altitude h (in kilometers). The data is shown below.

h	0	5	10	15	20
P	101,293	54,735	23,294	12,157	5069

A model for the data is

$$P = 102{,}303 e^{-0.137h}.$$

(a) Use a graphing utility to plot the data points and graph the model in the same viewing window.

(b) Use a graphing utility to plot the points $(h, \ln P)$. Use the regression capabilities of the graphing utility to fit a regression line to the revised data points.

(c) The line in part (b) has the form $\ln P = ah + b$. Write the line in exponential form.

(d) Verify graphically and algebraically that the result of part (c) is equivalent to the given exponential model for the data.

77. **World Population Growth** The time t in years for the world population to double if it is increasing at a continuous rate of r is

$$t = \frac{\ln 2}{r}.$$

Complete the table. What do your results imply?

r	0.005	0.010	0.015	0.020	0.025	0.030
t						

78. **Tractrix** A person walking along a dock (the y-axis) drags a boat by a 10-foot rope. The boat travels along a path known as a tractrix. The equation of this path is

$$y = 10 \ln\left(\frac{10 + \sqrt{100 - x^2}}{x}\right) - \sqrt{100 - x^2}.$$

(a) Use a graphing utility to obtain a graph of the function. What is the domain of the function?

(b) Identify any asymptotes of the graph.

(c) Determine the position of the person when the x-coordinate of the position of the boat is $x = 2$.

(d) Let $(0, p)$ be the position of the person. Determine p as a function of x, the x-coordinate of the position of the boat.

(e) Use a graphing utility to graph the function p. When does the position of the person change most for a small change in the position of the boat? Explain.

79. **Sound Intensity** The relationship between the number of decibels β and the intensity of a sound I in watts per square meter is

$$\beta = 10 \log_{10}\left(\frac{I}{10^{-12}}\right).$$

(a) Determine the number of decibels of a sound with an intensity of 1 watt per square meter.

(b) Determine the number of decibels of a sound with an intensity of 10^{-2} watt per square meter.

(c) The intensity of the sound in part (a) is 100 times as great as that in part (b). Is the number of decibels 100 times as great? Explain.

Ventilation Rates **In Exercises 80 and 81, use the model**

$$y = 80.4 - 11 \ln x, \quad 100 \le x \le 1500$$

which approximates the minimum required ventilation rate in terms of the air space per child in a public school classroom. In the model, x is the air space per child (in cubic feet) and y is the ventilation rate (in cubic feet per minute).

80. Use a graphing utility to graph the function and approximate the required ventilation rate if there is 300 cubic feet of air space per child.

81. A classroom is designed for 30 students. The air-conditioning system in the room has the capacity to move 450 cubic feet of air per minute.

(a) Determine the ventilation rate per child, assuming that the room is filled to capacity.

(b) Use the graph of Exercise 80 to estimate the air space required per child.

(c) Determine the minimum number of square feet of floor space required for the room if the ceiling height is 30 feet.

Monthly Payment **In Exercises 82–85, use the model**

$$t = 16.625 \ln\left(\frac{x}{x - 750}\right), \quad x > 750$$

which approximates the length of a home mortgage of $150,000 at 6% in terms of the monthly payment. In the model, t is the length of the mortgage in years and x is the monthly payment in dollars.

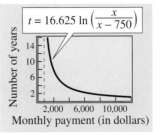

82. Use the model to approximate the length of a $150,000 mortgage (at 6%) if the monthly payment is $897.72.

83. Use the model to approximate the length of a $150,000 mortgage (at 6%) if the monthly payment is $1659.24.

84. Approximate the total amount paid over the term of the mortgage in Exercise 82 with a monthly payment of $897.72. What amount of the total is interest costs?

85. Approximate the total amount paid over the term of the mortgage in Exercise 83 with a monthly payment of $1659.24. What amount of the total is interest costs?

Synthesis

True or False? **In Exercises 86 and 87, determine whether the statement is true or false. Justify your answer.**

86. You can determine the graph of $f(x) = \log_6 x$ by graphing $g(x) = 6^x$ and reflecting it in the x-axis.

87. The graph of $f(x) = \log_3 x$ contains the point $(27, 3)$.

88. ***Exploration*** The table of values was obtained by evaluating a function. Determine which of the statements may be true and which must be false.

x	1	2	8
y	0	1	3

(a) y is an exponential function of x.

(b) y is a logarithmic function of x.

(c) x is an exponential function of y.

(d) y is a linear function of x.

89. ***Exploration*** Use a graphing utility to compare the graph of the function $y = \ln x$ with the graphs of the following functions.

(a) $y = x - 1$

(b) $y = (x - 1) - \frac{1}{2}(x - 1)^2$

(c) $y = (x - 1) - \frac{1}{2}(x - 1)^2 + \frac{1}{3}(x - 1)^3$

90. ***Finding a Pattern*** Identify the pattern of successive polynomials given in Exercise 89. Extend the pattern one more term and compare the graph of the resulting polynomial function with the graph of $y = \ln x$. What do you think the pattern implies?

91. ***Numerical and Graphical Analysis*** (a) Use a graphing utility to complete the table for the function

$$f(x) = \frac{\ln x}{x}.$$

x	1	5	10	10^2	10^4	10^6
$f(x)$						

(b) Use the table to determine what $f(x)$ approaches as x increases without bound.

(c) Use a graphing utility to check the result of part (b).

Review

In Exercises 92–95, find the vertical and horizontal asymptotes of the rational function.

92. $f(x) = \dfrac{4}{-8 - x}$

93. $f(x) = \dfrac{2x^3 - 3}{x^2}$

94. $f(x) = \dfrac{x + 5}{2x^2 + x - 15}$

95. $f(x) = \dfrac{2x^2(x - 5)}{x - 7}$

In Exercises 96–99, evaluate the expression. Round your answer to three decimal places.

96. e^{12}

97. $e^{7/2}$

98. e^{-5}

99. $6e^{-8}$

3.3 Properties of Logarithms

Change of Base

Most calculators have only two types of log keys, one for common logarithms (base 10) and one for natural logarithms (base e). Although common logs and natural logs are the most frequently used, you may occasionally need to evaluate logarithms to other bases. To do this, you can use the following **change-of-base formula.**

Change-of-Base Formula

Let a, b, and x be positive real numbers such that $a \neq 1$ and $b \neq 1$. Then $\log_a x$ can be converted to a different base using any of the following formulas.

Base b	Base 10	Base e
$\log_a x = \dfrac{\log_b x}{\log_b a}$	$\log_a x = \dfrac{\log_{10} x}{\log_{10} a}$	$\log_a x = \dfrac{\ln x}{\ln a}$

One way to look at the change-of-base formula is that logarithms to base a are simply *constant multiples* of logarithms to base b. The constant multiplier is $1/(\log_b a)$.

EXAMPLE 1   Changing Bases Using Common Logarithms

a. $\log_4 30 = \dfrac{\log_{10} 30}{\log_{10} 4}$ $\qquad \log_a x = \dfrac{\log_{10} x}{\log_{10} a}$

$\qquad \approx \dfrac{1.47712}{0.60206}$ $\qquad$ Use a calculator.

$\qquad \approx 2.4534$ $\qquad$ Use a calculator.

b. $\log_2 x = \dfrac{\log_{10} x}{\log_{10} 2} \approx \dfrac{\log_{10} x}{0.30103} \approx 3.3219 \log_{10} x$

EXAMPLE 2   Changing Bases Using Natural Logarithms

a. $\log_4 30 = \dfrac{\ln 30}{\ln 4}$ $\qquad \log_a x = \dfrac{\ln x}{\ln a}$

$\qquad \approx \dfrac{3.40120}{1.38629}$ $\qquad$ Use a calculator.

$\qquad \approx 2.4535$ $\qquad$ Use a calculator.

b. $\log_2 x = \dfrac{\ln x}{\ln 2} \approx \dfrac{\ln x}{0.693147} \approx 1.4427 \ln x$

In Examples 1 and 2, the result is the same whether common logarithms or natural logarithms are used in the change-of-base formula.

Gary Conner/PhotoEdit

Properties of Logarithms

You know from the previous section that the logarithmic function with base a is the *inverse* of the exponential function with base a. So, it makes sense that the properties of exponents should have corresponding properties involving logarithms. For instance, the exponential property $a^0 = 1$ has the corresponding logarithmic property $\log_a 1 = 0$.

Properties of Logarithms

Let a be a positive number such that $a \neq 1$, and let n be a real number. If u and v are positive real numbers, the following properties are true.

1. $\log_a(uv) = \log_a u + \log_a v$ **1.** $\ln(uv) = \ln u + \ln v$

2. $\log_a \dfrac{u}{v} = \log_a u - \log_a v$ **2.** $\ln \dfrac{u}{v} = \ln u - \ln v$

3. $\log_a u^n = n \log_a u$ **3.** $\ln u^n = n \ln u$

For a proof of Property 1, see Appendix A.

EXAMPLE 3   Using Properties of Logarithms

Write the logarithm in terms of $\ln 2$ and $\ln 3$.

a. $\ln 6$ **b.** $\ln \dfrac{2}{27}$

Solution

a. $\ln 6 = \ln(2 \cdot 3)$ Rewrite 6 as $2 \cdot 3$.

       $= \ln 2 + \ln 3$ Property 1

b. $\ln \dfrac{2}{27} = \ln 2 - \ln 27$ Property 2

       $= \ln 2 - \ln 3^3$ Rewrite 27 as 3^3.

       $= \ln 2 - 3 \ln 3$ Property 3

EXAMPLE 4   Using Properties of Logarithms

Use the properties of logarithms to verify that $-\ln \frac{1}{2} = \ln 2$.

Solution

     $-\ln \frac{1}{2} = -\ln(2^{-1})$ Rewrite $\frac{1}{2}$ as 2^{-1}.

        $= -(-1)\ln 2$ Property 3

        $= \ln 2$ Simplify.

Try checking this result on your calculator.

STUDY TIP

There is no general property that can be used to rewrite $\log_a(u \pm v)$. Specifically, $\log_a(x + y)$ is *not* equal to $\log_a x + \log_a y$.

John Napier (1550–1617), a Scottish mathematician, developed logarithms as a way to simplify some of the tedious calculations of his day. Beginning in 1594, Napier worked about 20 years on the invention of logarithms. Napier was only partially successful in his quest to simplify tedious calculations. Nonetheless, the development of logarithms was a step forward and received immediate recognition.

Rewriting Logarithmic Expressions

The properties of logarithms are useful for rewriting logarithmic expressions in forms that simplify the operations of algebra. This is true because they convert complicated products, quotients, and exponential forms into simpler sums, differences, and products, respectively.

EXAMPLE 5   Expanding the Logarithm of a Product

Use the properties of logarithms to expand $\log_{10} 5x^3y$.

Solution

$$\log_{10} 5x^3y = \log_{10} 5 + \log_{10} x^3y \qquad \text{Property 1}$$
$$= \log_{10} 5 + \log_{10} x^3 + \log_{10} y \qquad \text{Property 1}$$
$$= \log_{10} 5 + 3 \log_{10} x + \log_{10} y \qquad \text{Property 3}$$

EXAMPLE 6   Expanding the Logarithm of a Quotient

Use the properties of logarithms to expand $\ln \dfrac{\sqrt{3x - 5}}{7}$.

Solution

$$\ln \frac{\sqrt{3x - 5}}{7} = \ln\left[\frac{(3x - 5)^{1/2}}{7}\right] \qquad \text{Rewrite rational exponent.}$$
$$= \ln(3x - 5)^{1/2} - \ln 7 \qquad \text{Property 2}$$
$$= \frac{1}{2}\ln(3x - 5) - \ln 7 \qquad \text{Property 3}$$

In Examples 5 and 6, the properties of logarithms were used to *expand* logarithmic expressions. In Example 7, this procedure is reversed and the properties of logarithms are used to *condense* logarithmic expressions.

EXAMPLE 7   Condensing a Logarithmic Expression

Use the properties of logarithms to condense each logarithmic expression.

a. $\frac{1}{2}\log_{10} x + 3 \log_{10}(x + 1)$ **b.** $2 \ln(x + 2) - \ln x$

Solution

a. $\frac{1}{2}\log_{10} x + 3 \log_{10}(x + 1) = \log_{10} x^{1/2} + \log_{10}(x + 1)^3 \qquad$ Property 3

$$= \log_{10}\left[\sqrt{x} \cdot (x + 1)^3\right] \qquad \text{Property 1}$$

b. $2 \ln(x + 2) - \ln x = \ln(x + 2)^2 - \ln x \qquad$ Property 3

$$= \ln \frac{(x + 2)^2}{x} \qquad \text{Property 2}$$

Exploration

Use a graphing utility to graph the functions

$$y = \ln x - \ln(x - 3)$$

and

$$y = \ln \frac{x}{x - 3}$$

in the same viewing window. Does the graphing utility show the functions with the same domain? If so, should it? Explain your reasoning.

Application

EXAMPLE 8   Finding a Mathematical Model

The table shows the mean distance from the sun x and the orbital period y of the six planets that are closest to the sun. In the table, the mean distance is given in terms of astronomical units (where the earth's mean distance is defined as 1.0), and the period is given in terms of years. Find an equation that expresses y as a function of x.

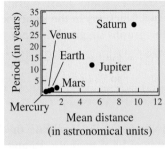

Figure 3.22

Planet	Mercury	Venus	Earth	Mars	Jupiter	Saturn
Mean Distance, x	0.387	0.723	1.0	1.524	5.203	9.539
Period, y	0.241	0.615	1.0	1.881	11.862	29.458

Algebraic Solution

The points in the table are plotted in Figure 3.22. From this figure it is not clear how to find an equation that relates y and x. To solve this problem, take the natural log of each of the x- and y-values shown in the table. This produces the following results.

Planet	Mercury	Venus	Earth
$\ln x = X$	-0.949	-0.324	0.0
$\ln y = Y$	-1.423	-0.486	0.0

Planet	Mars	Jupiter	Saturn
$\ln x = X$	0.421	1.649	2.255
$\ln y = Y$	0.632	2.473	3.383

Now, by plotting the points in the table, you can see that all six of the points appear to lie in a line.

Choose any two points to determine the slope of the line. Using the two points $(0.421, 0.632)$ and $(0, 0)$, you can determine the slope of the line.

$$m = \frac{0.632 - 0}{0.421 - 0} \approx 1.5 = \frac{3}{2}$$

By the point-slope form, the equation of the line is $Y = \frac{3}{2}X$, where $Y = \ln y$ and $X = \ln x$. You can therefore conclude that $\ln y = \frac{3}{2}\ln x$.

Graphical Solution

The points in the table are plotted in Figure 3.22. From this figure it is not clear how to find an equation that relates y and x. To solve this problem, take the natural log of each of the x- and y-values given in the table. This produces the following results.

Planet	Mercury	Venus	Earth	Mars	Jupiter	Saturn
$\ln x = X$	-0.949	-0.324	0.0	0.421	1.649	2.255
$\ln y = Y$	-1.423	-0.486	0.0	0.632	2.473	3.383

Now, by plotting the points in the table, you can see that all six of the points appear to lie in a line, as shown in Figure 3.23.

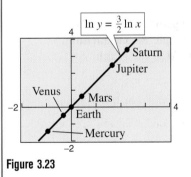

Figure 3.23

Using the regression features of a graphing utility, you can find a linear model for the data $Y = 1.5X = \frac{3}{2}X$, where $Y = \ln y$ and $X = \ln x$. From the model, you can see that the slope of the line is $\frac{3}{2}$. So, you can conclude that $\ln y = \frac{3}{2}\ln x$.

In Example 8, try to convert this to $y = f(x)$ form. You will get a function of the form $y = ax^b$, which is a power model.

3.3   E x e r c i s e s

In Exercises 1 and 2, use a graphing utility to graph the two functions in the same viewing window. What do the graphs suggest? Explain your reasoning.

1. $f(x) = \log_{10} x$

$g(x) = \dfrac{\ln x}{\ln 10}$

2. $f(x) = \ln x$

$g(x) = \dfrac{\log_{10} x}{\log_{10} e}$

In Exercises 3–10, evaluate the logarithm using the change-of-base formula. Round your result to three decimal places.

3. $\log_3 7$

4. $\log_7 4$

5. $\log_{1/2} 4$

6. $\log_{1/8} 64$

7. $\log_9(0.8)$

8. $\log_{1/3}(0.015)$

9. $\log_{15} 1460$

10. $\log_{20} 135$

In Exercises 11–18, rewrite the logarithm as a multiple of (a) a common logarithm and (b) a natural logarithm.

11. $\log_5 x$

12. $\log_3 x$

13. $\log_{1/5} x$

14. $\log_{1/3} x$

15. $\log_x \frac{3}{10}$

16. $\log_x \frac{3}{4}$

17. $\log_{2.6} x$

18. $\log_{7.1} x$

In Exercises 19–26, use the change-of-base formula to rewrite the logarithm as a multiple of a logarithm. Then use a graphing utility to sketch the graph.

19. $f(x) = \log_2 x$

20. $f(x) = \log_4 x$

21. $f(x) = \log_{1/2} x$

22. $f(x) = \log_{1/4} x$

23. $f(x) = \log_{11.8} x$

24. $f(x) = \log_{12.4} x$

25. $f(x) = \log_3 x^{1/2}$

26. $f(x) = \log_5 \dfrac{x}{3}$

In Exercises 27–46, use the properties of logarithms to write the expression as a sum, difference, and/or constant multiple of logarithms. (Assume all variables are positive.)

27. $\log_{10} 5x$

28. $\log_{10} 10z$

29. $\log_{10} \dfrac{5}{x}$

30. $\log_{10} \dfrac{y}{2}$

31. $\log_8 x^4$

32. $\log_6 z^{-3}$

33. $\ln \sqrt{z}$

34. $\ln \sqrt[3]{t}$

35. $\ln xyz$

36. $\ln \dfrac{xy}{z}$

37. $\ln \sqrt{a-1}, \quad a > 1$

38. $\ln\left(\dfrac{x^2-1}{x^3}\right), \quad x > 1$

39. $\ln z(z-1)^2, \quad z > 1$

40. $\ln \sqrt{\dfrac{x^2}{y^3}}$

41. $\ln \sqrt[3]{\dfrac{x}{y}}$

42. $\ln \dfrac{x}{\sqrt{x^2+1}}$

43. $\ln \dfrac{x^4\sqrt{y}}{z^5}$

44. $\ln \sqrt{x^2(x+2)}$

45. $\log_b \dfrac{x^2}{y^2 z^3}$

46. $\log_b \dfrac{\sqrt{x} y^4}{z^4}$

Graphical Analysis   In Exercises 47 and 48, use a graphing utility to graph the two equations in the same viewing window. What do the graphs suggest? Explain your reasoning.

47. $y_1 = \ln[x^3(x+4)], \qquad y_2 = 3\ln x + \ln(x+4)$

48. $y_1 = \ln\left(\dfrac{\sqrt{x}}{x-2}\right), \qquad y_2 = \frac{1}{2}\ln x - \ln(x-2)$

In Exercises 49–68, write the expression as the logarithm of a single quantity.

49. $\ln x + \ln 4$

50. $\ln y + \ln z$

51. $\log_4 z - \log_4 y$

52. $\log_5 8 - \log_5 t$

53. $2\log_2(x+3)$

54. $-6\log_6 2x$

55. $\frac{1}{3}\log_3 7x$

56. $\frac{5}{2}\log_7(z-4)$

57. $\ln x - 3\ln(x+1)$

58. $2\ln 8 + 5\ln z$

59. $\ln(x-2) - \ln(x+2)$

60. $3\ln x + 2\ln y - 4\ln z$

61. $\ln x - 2[\ln(x+2) + \ln(x-2)]$

62. $4[\ln z + \ln(z+5)] - 2\ln(z-5)$

63. $\frac{1}{3}[2\ln(x+3) + \ln x - \ln(x^2-1)]$

64. $2[\ln x - \ln(x+1) - \ln(x-1)]$

65. $\frac{1}{3}[\ln y + 2\ln(y+4)] - \ln(y-1)$

66. $\frac{1}{2}[\ln(x+1) + 2\ln(x-1)] + 3\ln x$

67. $2\ln 3 - \frac{1}{2}\ln(x^2+1)$

68. $\frac{3}{2}\ln 5t^6 - \frac{3}{4}\ln t^4$

Graphical Analysis   In Exercises 69 and 70, use a graphing utility to graph the two equations in the same viewing window. What do the graphs suggest? Verify your conclusion algebraically.

69. $y_1 = 2[\ln 8 - \ln(x^2 + 1)], \quad y_2 = \ln\left[\dfrac{64}{(x^2 + 1)^2}\right]$

70. $y_1 = \ln x + \frac{1}{3}\ln(x + 1), \quad y_2 = \ln\left(x\sqrt[3]{x + 1}\right)$

Think About It   In Exercises 71 and 72, use a graphing utility to graph the two equations in the same viewing window. Are the expressions equivalent? Explain.

71. $y_1 = \ln x^2, \qquad y_2 = 2\ln x$

72. $y_1 = \frac{1}{4}\ln[x^4(x^2 + 1)], \qquad y_2 = \ln x + \frac{1}{4}\ln(x^2 + 1)$

In Exercises 73–86, find the exact value of the logarithm without using a calculator. (If this is not possible, state the reason.)

73. $\log_3 9$

74. $\log_6 \sqrt[3]{6}$

75. $\log_4 16^{3.4}$

76. $\log_5\left(\frac{1}{125}\right)$

77. $\log_2(-4)$

78. $\log_4(-16)$

79. $\log_5 75 - \log_5 3$

80. $\log_4 2 + \log_4 32$

81. $\ln e^3 - \ln e^7$

82. $3\ln e^4$

83. $\log_{10} 0$

84. $\ln 1$

85. $\ln e^{8.5}$

86. $\ln \sqrt[5]{e^3}$

In Exercises 87–94, use the properties of logarithms to simplify the logarithmic expression.

87. $\log_4 8$

88. $\log_5\left(\frac{1}{15}\right)$

89. $\log_7 \sqrt{70}$

90. $\log_2(4^2 \cdot 3^4)$

91. $\log_5\left(\frac{1}{250}\right)$

92. $\log_{10}\left(\frac{9}{300}\right)$

93. $\ln(5e^6)$

94. $\ln \dfrac{6}{e^2}$

95. *Sound Intensity*   The relationship between the number of decibels β and the intensity of a sound I in watts per square meter is

$$\beta = 10 \log_{10}\left(\frac{I}{10^{-12}}\right).$$

(a) Use the properties of logarithms to write the formula in a simpler form.

(b) Use a graphing utility to complete the table.

I	10^{-4}	10^{-6}	10^{-8}	10^{-10}	10^{-12}	10^{-14}
β						

(c) Verify your answers in part (b) algebraically.

96. *Human Memory Model*   Students participating in a psychological experiment attended several lectures. After the last lecture, and every month for the next year, the students were tested to see how much of the material they remembered. The average scores for the group were given by the memory model

$$f(t) = 90 - 15 \log_{10}(t + 1), \quad 0 \le t \le 12$$

where t is the time (in months).

(a) Use a graphing utility to graph the function over the specified domain.

(b) What was the average score on the original exam ($t = 0$)?

(c) What was the average score after 6 months?

(d) What was the average score after 12 months?

(e) When would the average score have decreased to 75?

97. *Comparing Models*   A cup of water at an initial temperature of 78°C is placed in a room at a constant temperature of 21°C. The temperature of the water is measured every 5 minutes during a half-hour period. The results are recorded as ordered pairs of the form (t, T), where t is the time (in minutes) and T is the temperature (in degrees Celsius).

$(0, 78.0°), (5, 66.0°), (10, 57.5°), (15, 51.2°),$
$(20, 46.3°), (25, 42.5°), (30, 39.6°)$

(a) The graph of the model for the data should be asymptotic with the temperature of the room. Subtract the room temperature from each of the temperatures in the ordered pairs. Use a graphing utility to plot the data points (t, T) and $(t, T - 21)$.

(b) Use the regression capabilities of a graphing utility to fit an exponential model to the revised data. This model will be of the form

$$T - 21 = ab^x.$$

Solve for T and graph the model. Compare the result with the plot of the original data.

(c) Take the natural logarithms of the revised temperatures. Use a graphing utility to plot the points $(t, \ln(T - 21))$ and observe that the points appear linear. Use the regression capabilities of a graphing utility to fit a line to this data. The resulting line has the form

$$\ln(T - 21) = at + b.$$

Use the properties of logarithms to solve for T. Verify that the result is equivalent to the model in part (b).

(d) Fit a rational model to the data. Take the reciprocals of the y-coordinates of the revised data to generate the points

$$\left(t, \frac{1}{T - 21}\right).$$

Use a graphing utility to plot these points and observe that they appear linear. Use the regression capabilities of a graphing utility to fit a line to this data. The resulting line has the form

$$\frac{1}{T - 21} = at + b.$$

Solve for T, and use a graphing utility to graph the rational function and the original data points.

98. Writing Write a short paragraph explaining why the transformations of the data in Exercise 97 were necessary to obtain the models. Why did taking the logarithms of the temperatures lead to a linear scatter plot? Why did taking the reciprocals of the temperatures lead to a linear scatter plot?

Synthesis

True or False? In Exercises 99–104, determine whether the statement is true or false given that $f(x) = \ln x$. Justify your answer.

99. $f(0) = 0$

100. $f(ax) = f(a) + f(x), \quad a > 0, x > 0$

101. $f(x - 2) = f(x) - f(2), \quad x > 2$

102. $\sqrt{f(x)} = \frac{1}{2}f(x)$

103. If $f(u) = 2f(v)$, then $v = u^2$.

104. If $f(x) < 0$, then $0 < x < 1$.

105. Think About It Use a graphing utility to graph

$$f(x) = \ln \frac{x}{2}, \quad g(x) = \frac{\ln x}{\ln 2}, \quad h(x) = \ln x - \ln 2$$

in the same viewing window. Which two functions have identical graphs? Explain why.

106. Exploration Approximate the natural logarithms of as many integers as possible between 1 and 20, given that $\ln 2 \approx 0.6931$, $\ln 3 \approx 1.0986$, and $\ln 5 \approx 1.6094$. (Do not use a calculator.)

107. Prove that $\log_b \dfrac{u}{v} = \log_b u - \log_b v$.

108. Prove that $\log_b u^n = n \log_b u$.

Review

In Exercises 109–112, sketch the graph of the function. Use a graphing utility to verify the graph.

109. $f(x) = -\frac{1}{2}(x^2 + 4x)$ **110.** $f(x) = \frac{1}{4}x(x - 2)^2$

111. $f(x) = x^3 + 2x^2 - 9x - 18$

112. $f(x) = -(x + 2)^3 + 1$

In Exercises 113–118, find all solutions of the equation. Be sure to check all your solutions.

113. $x^2 - 6x + 2 = 0$ **114.** $2x^3 + 20x^2 + 50x = 0$

115. $x^4 - 19x^2 + 48 = 0$ **116.** $9x^4 - 37x^2 + 4 = 0$

117. $x^3 - 6x^2 - 4x + 24 = 0$

118. $9x^4 - 226x^2 + 25 = 0$

In Exercises 119–126, use a calculator to evaluate the expression. Round the result to three decimal places.

119. $1.6^{-2\pi}$ **120.** $\sqrt[5]{8251}$

121. $260^{\sqrt{3}}$ **122.** $170(4^{-1.1})$

123. $\log_{10}(220)$ **124.** $\log_{10}\left(\frac{7}{5}\right)$

125. $\ln 2.008$ **126.** $\ln\left(5 - \sqrt{7}\right)$

3.4 Solving Exponential and Logarithmic Equations

Introduction

So far in this chapter, you have studied the definitions, graphs, and properties of exponential and logarithmic functions. In this section, you will study procedures for *solving equations* involving exponential and logarithmic functions.

There are two basic strategies for solving exponential or logarithmic equations. The first is based on the One-to-One Properties and the second is based on the Inverse Properties. For $a > 0$ and $a \neq 1$, the following properties are true for all x and y for which $\log_a x$ and $\log_a y$ are defined.

One-to-One Properties

$a^x = a^y$ if and only if $x = y$.

$\log_a x = \log_a y$ if and only if $x = y$.

Inverse Properties

$a^{\log_a x} = x$

$\log_a a^x = x$

What You Should Learn:

- How to solve simple exponential and logarithmic equations
- How to solve more complicated exponential equations
- How to solve more complicated logarithmic equations
- How to use exponential and logarithmic equations to model and solve real-life problems

Why You Should Learn It:

Exponential and logarithmic equations can be used to model and solve real-life problems. For instance, Exercise 126 on page 257 shows how to use a logarithmic function to model crumple zones for automobile crash tests.

Index Stock

EXAMPLE 1   Solving Simple Exponential and Logarithmic Equations

Original Equation	Rewritten Equation	Solution	Property
a. $2^x = 32$	$2^x = 2^5$	$x = 5$	One-to-One
b. $\ln x - \ln 3 = 0$	$\ln x = \ln 3$	$x = 3$	One-to-One
c. $\left(\frac{1}{3}\right)^x = 9$	$3^{-x} = 3^2$	$x = -2$	One-to-One
d. $e^x = 7$	$\ln e^x = \ln 7$	$x = \ln 7$	Inverse
e. $\ln x = -3$	$e^{\ln x} = e^{-3}$	$x = e^{-3}$	Inverse
f. $\log x = -1$	$10^{\log x} = 10^{-1}$	$x = 10^{-1} = \frac{1}{10}$	Inverse

The strategies used in Example 1 are summarized as follows.

Strategies for Solving Exponential and Logarithmic Equations

1. Rewrite the given equation in a form to use the One-to-One Properties of exponential or logarithmic functions.
2. Rewrite an *exponential* equation in logarithmic form and apply the Inverse Property of logarithmic functions.
3. Rewrite a *logarithmic* equation in exponential form and apply the Inverse Property of exponential functions.

Solving Exponential Equations

EXAMPLE 2   Solving Exponential Equations

Solve each equation.   **a.** $e^x = 72$ **b.** $3(2^x) = 42$

Algebraic Solution

a. $e^x = 72$ Write original equation.

   $\ln e^x = \ln 72$ Take natural log of each side.

   $x = \ln 72$ Inverse Property

   $x \approx 4.277$ Use a calculator.

The solution is $\ln 72 \approx 4.277$. Check this solution in the original equation.

b. $3(2^x) = 42$ Write original equation.

   $2^x = 14$ Divide each side by 3.

   $\log_2 2^x = \log_2 14$ Take log (base 2) of each side.

   $x = \log_2 14$ Inverse Property

   $x = \dfrac{\ln 14}{\ln 2}$ Change-of-base formula

   $x \approx 3.807$ Use a calculator.

The solution is $\log_2 14 \approx 3.807$. Check this solution in the original equation.

Graphical Solution

a. Use a graphing utility to graph the left- and right-hand sides of the equation as $y_1 = e^x$ and $y_2 = 72$ in the same viewing window. Use the *intersect* feature or the *zoom* and *trace* features of the graphing utility to approximate the intersection point, as shown in Figure 3.24. So, the approximate solution is 4.277.

b. Use a graphing utility to graph $y_1 = 3(2^x)$ and $y_2 = 42$ in the same viewing window. Use the *intersect* feature or the *zoom* and *trace* features to approximate the intersection point, as shown in Figure 3.25. So, the approximate solution is 3.807.

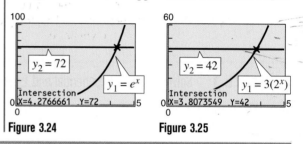

Figure 3.24   **Figure 3.25**

EXAMPLE 3   Solving an Exponential Equation

Solve $4e^{2x} = 5$.

Algebraic Solution

   $4e^{2x} = 5$ Write original equation.

   $e^{2x} = \dfrac{5}{4}$ Divide each side by 4.

   $\ln e^{2x} = \ln \dfrac{5}{4}$ Take logarithm of each side.

   $2x = \ln \dfrac{5}{4}$ Inverse Property

   $x = \dfrac{1}{2} \ln \dfrac{5}{4}$ Solve for x.

   $x \approx 0.112$ Use a calculator.

The solution is $\frac{1}{2} \ln \frac{5}{4} \approx 0.112$. Check this solution in the original equation.

Graphical Solution

Rather than graph both sides of the equation you are solving as separate graphs, as you did in Example 2, another way to graphically solve the equation is to first rewrite the equation as $4e^{2x} - 5 = 0$, then use a graphing utility to graph $y = 4e^{2x} - 5$. Use the *zero* or *root* feature or the *zoom* and *trace* features of the graphing utility to approximate the value of x for which $y = 0$. From Figure 3.26, you can see that the zero occurs when $x \approx 0.112$. So, the approximate solution is 0.112.

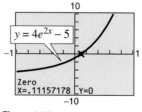

Figure 3.26

EXAMPLE 4   Solving an Exponential Equation

Solve $2(3^{2t-5}) - 4 = 11$.

Solution

$2(3^{2t-5}) - 4 = 11$	Write original equation.
$2(3^{2t-5}) = 15$	Add 4 to each side.
$3^{2t-5} = \frac{15}{2}$	Divide each side by 2.
$\log_3 3^{2t-5} = \log_3 \frac{15}{2}$	Take log (base 3) of each side.
$2t - 5 = \log_3 \frac{15}{2}$	Inverse Property
$2t = 5 + \log_3 7.5$	Add 5 to each side.
$t = \frac{5}{2} + \frac{1}{2}\log_3 7.5$	Divide each side by 2.
$t \approx 3.417$	Use a calculator.

The solution is $\frac{5}{2} + \frac{1}{2}\log_3 7.5 \approx 3.417$. Check this in the original equation.

When an equation involves two or more exponential expressions, you can still use a procedure similar to that demonstrated in the previous three examples. However, the algebra is a bit more complicated.

EXAMPLE 5   Solving an Exponential Equation in Quadratic Form

Solve $e^{2x} - 3e^x + 2 = 0$.

Algebraic Solution

$e^{2x} - 3e^x + 2 = 0$	Write original equation.
$(e^x)^2 - 3e^x + 2 = 0$	Write in quadratic form.
$(e^x - 2)(e^x - 1) = 0$	Factor.
$e^x - 2 = 0$	Set 1st factor equal to 0.
$e^x = 2$	Add 2 to each side.
$x = \ln 2$	Inverse Property
$e^x - 1 = 0$	Set 2nd factor equal to 0.
$e^x = 1$	Add 1 to each side.
$x = \ln 1$	Inverse Property
$x = 0$	Simplify.

The solutions are $\ln 2 \approx 0.693$ and 0. Check these in the original equation.

Graphical Solution

Use a graphing utility to graph $y = e^{2x} - 3e^x + 2$. Use the *zero* or *root* feature or the *zoom* and *trace* features of the graphing utility to approximate the values of x for which $y = 0$. In Figure 3.27, you can see that the zeros occur when $x = 0$ and when $x \approx 0.693$. So, the approximate solutions are 0 and 0.693.

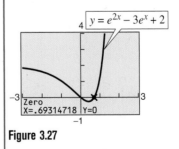

Figure 3.27

Solving Logarithmic Equations

To solve a logarithmic equation, you can write it in exponential form.

$\ln x = 3$	Logarithmic form
$e^{\ln x} = e^3$	Exponentiate each side.
$x = e^3$	Exponential form

This procedure is called *exponentiating* both sides of an equation. It is applied after the logarithmic expression has been isolated.

EXAMPLE 6   Solving Logarithmic Equations

Solve each logarithmic equation.

a. $\ln x = 2$ **b.** $\log_3(5x - 1) = \log_3(x + 7)$

Solution

a.

$\ln x = 2$	Write original equation.
$e^{\ln x} = e^2$	Exponentiate each side.
$x = e^2$	Inverse Property
$x \approx 7.389$	Use a calculator.

The solution is $e^2 \approx 7.389$. Check this in the original equation.

b.

$\log_3(5x - 1) = \log_3(x + 7)$	Write original equation.
$5x - 1 = x + 7$	One-to-One Property
$4x = 8$	Add $-x$ and 1 to each side.
$x = 2$	Divide each side by 4.

The solution is 2. Check this in the original equation.

EXAMPLE 7   Solving a Logarithmic Equation

Solve $5 + 2 \ln x = 4$.

Algebraic Solution

$5 + 2 \ln x = 4$	Write original equation.
$2 \ln x = -1$	Subtract 5 from each side.
$\ln x = -\frac{1}{2}$	Divide each side by 2.
$e^{\ln x} = e^{-1/2}$	Exponentiate each side.
$x = e^{-1/2}$	Inverse property
$x \approx 0.607$	Use a calculator.

The solution is $e^{-1/2} \approx 0.607$. Check this in the original equation.

Graphical Solution

Use a graphing utility to graph $y_1 = 5 + 2 \ln x$ and $y_2 = 4$ in the same viewing window. Use the *intersect* feature or the *zoom* and *trace* features to approximate the intersection point, as shown in Figure 3.28. So, the approximate solution is 0.607.

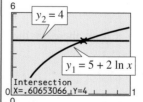

Figure 3.28

EXAMPLE 8   Solving a Logarithmic Equation

Solve $2 \log_5 3x = 4$.

Solution

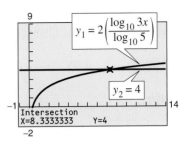

$2 \log_5 3x = 4$	Write original equation.
$\log_5 3x = 2$	Divide each side by 2.
$5^{\log_5 3x} = 5^2$	Exponentiate each side (base 5).
$3x = 25$	Inverse Property.
$x = \frac{25}{3}$	Divide each side by 3.

The solution is $\frac{25}{3}$. Check this in the original equation. Or, perform a graphical check by graphing

$$y_1 = 2 \log_5 3x = 2\left(\frac{\log_{10} 3x}{\log_{10} 5}\right) \qquad \text{and} \qquad y_2 = 4$$

Figure 3.29

in the same viewing window. The two graphs should intersect when $x = \frac{25}{3}$ and $y = 4$, as shown in Figure 3.29.

Because the domain of a logarithmic function generally does not include all real numbers, you should be sure to check for extraneous solutions of logarithmic equations, as shown in the next example.

EXAMPLE 9   Checking for Extraneous Solutions

Solve for x in the equation $\ln(x - 2) + \ln(2x - 3) = 2 \ln x$.

Algebraic Solution

$\ln(x - 2) + \ln(2x - 3) = 2 \ln x$	Write original equation.
$\ln[(x - 2)(2x - 3)] = \ln x^2$	Use properties of logarithms.
$\ln(2x^2 - 7x + 6) = \ln x^2$	
$2x^2 - 7x + 6 = x^2$	One-to-One Property
$x^2 - 7x + 6 = 0$	Write in general form.
$(x - 6)(x - 1) = 0$	Factor.
$x - 6 = 0 \implies x = 6$	Set 1st factor equal to 0.
$x - 1 = 0 \implies x = 1$	Set 2nd factor equal to 0.

Finally, by checking these two "solutions" in the original equation, you can conclude that $x = 1$ is not valid. This is because when $x = 1$, $\ln(x - 2) + \ln(2x - 3) = \ln(-1) + \ln(-1)$, which is invalid because -1 is not in the domain of the natural log function. So, the only solution is 6.

Graphical Solution

First rewrite the original equation as $\ln(x - 2) + \ln(2x - 3) - 2 \ln x = 0$. Then use a graphing utility to graph $y = \ln(x - 2) + \ln(2x - 3) - 2 \ln x$, as shown in Figure 3.30. Use the *zero* or *root* feature or the *zoom* and *trace* features of the graphing utility to determine that 6 is an approximate solution. You can verify that 6 is an exact solution by substituting $x = 6$ in the original equation.

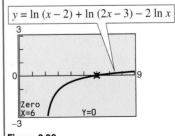

Figure 3.30

EXAMPLE 10   The Change-of-Base Formula

Prove the change-of-base formula: $\log_a x = \dfrac{\log_b x}{\log_b a}$.

Solution

Begin by letting $y = \log_a x$ and writing the equivalent exponential form $a^y = x$.
Now, taking the logarithms *with base b* of both sides produces the following.

$$\log_b a^y = \log_b x$$

$$y \log_b a = \log_b x$$

$$y = \frac{\log_b x}{\log_b a}$$

$$\log_a x = \frac{\log_b x}{\log_a a}$$

STUDY T!P

To solve exponential equations,
it is useful to first isolate the
exponential expression, then
take the logarithm of both sides
and solve for the variable.

To solve logarithmic equations,
condense the logarithmic part
into a single logarithm, then
rewrite in exponential form and
solve for the variable.

Approximating Solutions

Equations that involve combinations of algebraic functions, exponential functions, and/or logarithmic functions can be very difficult to solve by algebraic
procedures. Here again, you can take advantage of a graphing utility.

EXAMPLE 11   Approximating the Solution of an Equation

Approximate the solution of $\ln x = x^2 - 2$.

Solution

To begin, write the equation so that all terms on one side are equal to 0.

$$\ln x - x^2 + 2 = 0$$

Then use a graphing utility to graph

$$y = -x^2 + 2 + \ln x$$

as shown in Figure 3.31. From this graph, you can see that the equation has two
solutions. Next, using the *zero* or *root* feature or the *zoom* and *trace* features, you
can approximate the two solutions to be 0.138 and 1.564.

Check

$$\ln x = x^2 - 2 \qquad \text{Write original equation.}$$

$$\ln(0.138) \stackrel{?}{\approx} (0.138)^2 - 2 \qquad \text{Substitute 0.138 for } x.$$

$$-1.9805 \approx -1.9810 \qquad \text{Solution checks. } \checkmark$$

$$\ln(1.564) \stackrel{?}{\approx} (1.564)^2 - 2 \qquad \text{Substitute 1.564 for } x.$$

$$0.4472 \approx 0.4461 \qquad \text{Solution checks. } \checkmark$$

So, the two solutions 0.138 and 1.564 seem reasonable.

Figure 3.31

Applications

EXAMPLE 12   Doubling an Investment

You have deposited $500 in an account that pays 6.75% interest, compounded continuously. How long will it take your money to double?

Solution

Using the formula for continuous compounding, you can find that the balance in the account is

$$A = Pe^{rt}$$

$$= 500e^{0.0675t}.$$

To find the time required for the balance to double, let $A = 1000$, and solve the resulting equation for t.

$500e^{0.0675t} = 1000$	Substitute 1000 for A.
$e^{0.0675t} = 2$	Divide each side by 500.
$\ln e^{0.0675t} = \ln 2$	Take natural log of each side.
$0.0675t = \ln 2$	Inverse Property
$t = \dfrac{\ln 2}{0.0675}$	Divide each side by 0.0675.
$t \approx 10.27$	Use a calculator.

The balance in the account will double after approximately 10.27 years. This result is demonstrated graphically in Figure 3.32.

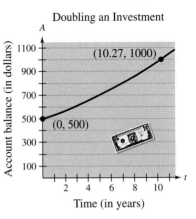

Doubling an Investment

Figure 3.32

EXAMPLE 13   Consumer Price Index for Sugar

From 1970 to 1997, the Consumer Price Index (CPI) value y for a fixed amount of sugar for the year t can be modeled by the equation

$$y = -171.8 + 87.1 \ln t$$

where $t = 10$ represents 1970 (see Figure 3.33). During which year did the price of sugar reach 4.5 times its 1970 price of 28.8 on the CPI?   (Source: U.S. Bureau of Labor Statistics)

Solution

$-171.8 + 87.1 \ln t = y$	Write original equation.
$-171.8 + 87.1 \ln t = 129.6$	Substitute $(4.5)(28.8) = 129.6$ for y.
$87.1 \ln t = 301.4$	Add 171.8 to each side.
$\ln t = 3.460$	Divide each side by 87.1.
$e^{\ln t} = e^{3.460}$	Exponentiate each side.
$t \approx 31.8$	Inverse Property

The solution is 31.8 years. Because $t = 10$ represents 1970, it follows that the price of sugar reached 4.5 times its 1970 price in late 1991.

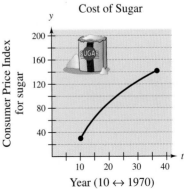

Cost of Sugar

Figure 3.33

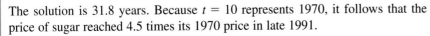

3.4 Exercises

In Exercises 1–8, determine whether each *x*-value is a solution of the equation.

1. $4^{2x-7} = 64$

  (a) $x = 5$

  (b) $x = 2$

2. $2^{3x+1} = 32$

  (a) $x = -1$

  (b) $x = 2$

3. $3e^{x+2} = 75$

  (a) $x = -2 + e^{25}$

  (b) $x = -2 + \ln 25$

  (c) $x \approx 1.2189$

4. $-4e^{x-1} = -60$

  (a) $x = 1 + \ln 15$

  (b) $x \approx 3.7081$

  (c) $x = \ln 16$

5. $\log_4(3x) = 3$

  (a) $x \approx 20.3560$

  (b) $x = -4$

  (c) $x = \frac{64}{3}$

6. $\log_6\left(\frac{5}{3}x\right) = 2$

  (a) $x \approx 20.2882$

  (b) $x = \frac{108}{5}$

  (c) $x = 7.2$

7. $\ln(x - 1) = 3.8$

  (a) $x = 1 + e^{3.8}$

  (b) $x \approx 45.7012$

  (c) $x = 1 + \ln 3.8$

8. $\frac{1}{5}\ln(2 + x) = \frac{1}{2}$

  (a) $x = e^{2.5} - 2$

  (b) $x \approx \frac{4073}{400}$

  (c) $x = \frac{1}{2}$

In Exercises 9–14, use a graphing utility to graph *f* and *g* in the same viewing window. Approximate the point of intersection of the graphs of *f* and *g*. Then solve the equation $f(x) = g(x)$ algebraically.

9. $f(x) = 2^x$

  $g(x) = 8$

10. $f(x) = 27^x$

  $g(x) = 9$

11. $f(x) = \log_3 x$

  $g(x) = 2$

12. $f(x) = 3\log_5 x$

  $g(x) = 6$

13. $f(x) = \ln(x - 4)$

  $g(x) = 0$

14. $f(x) = \ln e^{x-2}$

  $g(x) = 3x + 2$

In Exercises 15–36, solve for *x*.

15. $4^x = 16$

16. $3^x = 243$

17. $5^x = 625$

18. $7^x = \frac{1}{49}$

19. $8^x = 4$

20. $\left(\frac{1}{2}\right)^x = 32$

21. $\left(\frac{1}{4}\right)^x = 64$

22. $\left(\frac{3}{4}\right)^x = \frac{27}{64}$

23. $3^{x-1} = 27$

24. $\ln x - \ln 2 = 0$

25. $\ln x - \ln 5 = 0$

26. $e^x = 2$

27. $e^x = 4$

28. $\ln x = -1$

29. $\ln x = -7$

30. $\log_4 x = 3$

31. $\log_x 625 = 4$

32. $\log_{10} x - 2 = 0$

33. $\log_{10} x = -1$

34. $\log_{10} x = -\frac{1}{2}$

35. $\ln(2x - 1) = 0$

36. $\ln(3x + 5) = 8$

In Exercises 37–42, simplify the expression.

37. $\ln e^{x^2}$

38. $\ln e^{2x-1}$

39. $e^{\ln(5x+2)}$

40. $-1 + \ln e^{2x}$

41. $e^{\ln x^2}$

42. $-8 + e^{\ln x^3}$

In Exercises 43–60, solve the exponential equation algebraically. Round your result to three decimal places. Use a graphing utility to verify your answer.

43. $10^x = 570$

44. $6^{5x} = 3000$

45. $e^x = 10$

46. $4e^{2x} = 40$

47. $5^{-t/2} = 0.20$

48. $4^{-3t} = 0.10$

49. $2^{3-x} = 565$

50. $8^{-2-x} = 431$

51. $500e^{-x} = 300$

52. $1000e^{-4x} = 75$

53. $7 - 2e^x = 5$

54. $-14 + 3e^x = 11$

55. $e^{2x} - 4e^x - 5 = 0$

56. $e^{2x} - 5e^x + 6 = 0$

57. $50(120 - e^{x/2}) = 600$

58. $\dfrac{525}{1 + e^{-x}} = 275$

59. $\left(1 + \dfrac{0.10}{12}\right)^{12t} = 2$

60. $\left(16 + \dfrac{0.878}{26}\right)^{3t} = 30$

In Exercises 61–64, let $f(x)$ represent the left side of the equation. Complete the table to obtain a rough estimate of the solution, and then use a graphing utility to graph both sides of the equation to obtain a better estimate of the solution. Round your result to three decimal places.

61. $e^{3x} = 12$

x	0.6	0.7	0.8	0.9	1.0
$f(x)$					

62. $e^{2x} = 50$

x	1.6	1.7	1.8	1.9	2.0
$f(x)$					

63. $20(100 - e^{x/2}) = 500$

x	5	6	7	8	9
$f(x)$					

64. $\dfrac{400}{1 + e^{-x}} = 350$

x	0	1	2	3	4
$f(x)$					

In Exercises 65–74, use the *zero* or *root* feature or the *zoom* and *trace* features of a graphing utility to find an approximate solution of the exponential equation accurate to three decimal places.

65. $2^{3x} = 50$ **66.** $4^{-x/2} = 0.10$

67. $2^{-3x} = 0.90$ **68.** $8(10^{3x}) = 12$

69. $5(10^{x-6}) = 7$ **70.** $8(3^{6-x}) = 40$

71. $\left(1 + \dfrac{0.065}{365}\right)^{365t} = 4$ **72.** $\left(4 - \dfrac{2.471}{40}\right)^{9t} = 21$

73. $\dfrac{3000}{2 + e^{2x}} = 2$ **74.** $\dfrac{119}{e^{6x} - 14} = 7$

In Exercises 75–78, use a graphing utility to graph the function and approximate its zero accurate to three decimal places.

75. $g(x) = 6e^{1-x} - 25$ **76.** $f(x) = 3e^{3x/2} - 962$

77. $g(t) = e^{0.09t} - 3$ **78.** $h(t) = e^{0.125t} - 8$

In Exercises 79–98, solve the logarithmic equation algebraically. Round the result to three decimal places. Verify your answer using a graphing utility.

79. $\ln x = -3$ **80.** $\ln x = 2$

81. $\ln 4x = 2.1$ **82.** $\ln 4x = 1$

83. $2 \ln 3x = 19$ **84.** $2 \ln x = 7$

85. $\log_{10}(z - 3) = 2$ **86.** $\log_{10} x^2 = 6$

87. $7 \log_4(0.6x) = 12$ **88.** $4 \log_{10}(x - 6) = 11$

89. $\ln \sqrt{x + 2} = 1$ **90.** $\ln \sqrt{x - 8} = 5$

91. $\ln(x + 1)^2 = 2$

92. $\ln(x^2 + 1) = 8$

93. $\log_4 x - \log_4(x - 1) = \frac{1}{2}$

94. $\log_3 x + \log_3(x^2 - 8) = \log_3 8x$

95. $\ln(x + 5) = \ln(x - 1) - \ln(x + 1)$

96. $\ln(x + 1) - \ln(x - 2) = \ln x^2$

97. $\log_{10} 8x - \log_{10}(1 + \sqrt{x}) = 2$
98. $\log_{10} 4x - \log_{10}(12 + \sqrt{x}) = 2$

In Exercises 99–102, let $f(x)$ represent the left side of the equation. Complete the table to obtain a rough estimate of the solution, and then use a graphing utility to graph both sides of the equation to obtain a better estimate of the solution. Round your result to three decimal places.

99. $\ln 2x = 2.4$

x	2	3	4	5	6
$f(x)$					

100. $3 \ln 5x = 10$

x	4	5	6	7	8
$f(x)$					

101. $6 \log_3(0.5x) = 11$

x	12	13	14	15	16
$f(x)$					

102. $5 \log_{10}(x - 2) = 11$

x	150	155	160	165	170
$f(x)$					

In Exercises 103–110, use the *zero* or *root* feature or the *zoom* and *trace* features of a graphing utility to find an approximate solution of the logarithmic equation accurate to three decimal places.

103. $\log_{10}(z - 4) = 1$ **104.** $\log_{10} x^2 = 4$

105. $3 \ln x = 5$ **106.** $\ln 4x = 3$

107. $\ln x + \ln(x - 3) = 1$

108. $\log_2 x + \log_2(x + 5) = \log_2(x + 4)$

109. $\ln(x - 5) = \ln(x - 3) - \ln(x + 3)$

110. $\ln x + \ln(x^2 + 4) = 10$

In Exercises 111–116, use a graphing utility to approximate the point of intersection of the graphs. Round your result to three decimal places.

111. $y_1 = 7$ **112.** $y_1 = -4$
$\quad\ y_2 = 2^x$ $\qquad\ y_2 = -3^{x+1} - 2$

113. $y_1 = 8$

$y_2 = 4e^{-0.2x}$

115. $y_1 = 3$

$y_2 = \ln x$

114. $y_1 = 500$

$y_2 = 1500e^{-x/2}$

116. $y_1 = 10$

$y_2 = 4 \ln(x - 2)$

Compound Interest **In Exercises 117 and 118, find the time required for a $1000 investment to (a) double at interest rate r, compounded continuously, and (b) triple at interest rate r, compounded continuously.**

117. $r = 0.085$

118. $r = 0.12$

119. *Average Heights* The percentages of American males and females between the ages of 18 and 24 who are at least x inches tall are

$$m(x) = \frac{100}{1 + e^{-0.6114(x - 69.71)}} \quad \text{Males}$$

$$f(x) = \frac{100}{1 + e^{-0.66607(x - 64.51)}} \quad \text{Females}$$

where m and f are the percentages and x is the height in inches. (Source: U.S. National Center for Health Statistics)

(a) Use a graphing utility to graph the two functions in the same viewing window.

(b) Use the graph to determine the horizontal asymptotes of the functions.

(c) What is the median height for each sex?

120. *Human Memory Model* In a group project in learning theory, a mathematical model for the proportion P of correct responses after n trials was found to be

$$P = \frac{0.83}{1 + e^{-0.2n}}.$$

(a) Use a graphing utility to graph the function.

(b) Use the graph to determine the horizontal asymptotes of the function. Interpret the meaning of the upper asymptote in the context of the problem.

(c) After how many trials will 60% of the responses be correct?

121. *Demand Function* The demand equation for a camera is

$$p = 500 - 0.5(e^{0.004x}).$$

Find the demands x for prices of (a) $p = \$350$ and (b) $p = \$300$.

122. *Demand Function* The demand equation for a camcorder is

$$p = 5000\left(1 - \frac{4}{4 + e^{-0.002x}}\right).$$

Find the demands x for prices of (a) $p = \$600$ and (b) $p = \$400$.

123. *Forest Yield* The yield V (in millions of cubic feet per acre) for a forest at age t years is

$$V = 6.7e^{-48.1/t}.$$

(a) Use a graphing utility to graph the function.

(b) Determine the horizontal asymptote of the function. Interpret its meaning in the context of the problem.

(c) Find the time necessary to obtain a yield of 1.3 million cubic feet.

124. *Trees per Acre* The number of trees per acre N of a certain species is approximated by the model

$$N = 68(10^{-0.04x}), \quad 5 \le x \le 40$$

where x is the average diameter of the trees (in inches) three feet above the ground. Use the model to approximate the average diameter of the trees in a test plot when $N = 21$.

125. *Data Analysis* An object at a temperature of 160°C is removed from a furnace and placed in a room at 20°C. The temperature T of the object is measured each hour h and recorded in the table.

h	0	1	2	3	4	5
T	160°	90°	56°	38°	29°	24°

A model for this data is

$$T = 20[1 + 7(2^{-h})].$$

(a) Use a graphing utility to plot the data points and graph the model in the same viewing window.

(b) Use the graph to identify the horizontal asymptote of the model and interpret the asymptote in the context of the problem.

(c) Approximate the time when the temperature of the object is 100°C.

126. *Automobiles* Automobiles are designed with crumple zones that help protect their occupants in crashes. The crumple zones allow the occupants to move short distances when the automobiles come to abrupt stops. The greater the distance moved, the less g's the crash victims experience. (One g is equal to the acceleration due to gravity. For very short periods of time, humans have withstood as much as 40 g's.) In crash tests with vehicles moving at 90 kilometers per hour, analysts measured the number of g's experienced during deceleration y by crash dummies that were permitted to move x meters during impact. The data is shown in the table.

x	0.2	0.4	0.6	0.8	1
y	158	80	53	40	32

(a) Use the regression capabilities of a graphing utility to find a natural logarithmic model for the data. A second model for the data is

$$y = -3.00 + 11.88 \ln x + (36.94/x).$$

Graph both models with the data points in the same viewing window. Which model do you think is a better choice? Comment on accuracy and simplicity.

(b) Use both models to estimate the distance traveled during impact if the passenger deceleration must not exceed 30 g's. Comment on the difference in distances traveled given by the two models and the practical implications of choosing a model when designing the crumple zones.

127. *Comparing Mathematical Models* The table shows the number y (in millions) of single compact discs (CDs) shipped annually by manufacturers from 1993 through 1997, where $x = 3$ represents 1993. (Source: Recording Industry Association of America)

x	3	4	5	6	7
y	7.8	9.3	21.5	43.2	66.7

(a) Create a scatter plot of the data. Find a linear model for the data, and add its graph to your scatter plot. Using this model, approximate the year that corresponds to an annual shipment of single CDs of 100 million.

(b) Create a new table giving values for $\ln x$ and $\ln y$ and create a scatter plot of this transformed data. Use the method illustrated in Example 8 in Section 3.3 to find a model for the transformed data, and add its graph to your scatter plot. According to this model, approximate the year that corresponds to an annual shipment of single CDs of 100 million.

(c) Solve the model in part (b) for y, and add its graph to your scatter plot in part (a). Which model better fits the original data? Which model will better predict future shipping levels? Explain.

Synthesis

True or False? In Exercises 128 and 129, determine whether the statement is true or false. Justify your answer.

128. You can approximate the solution to the equation $\frac{2}{3}e^x = 42$ by graphing $y = \frac{2}{3}e^x - 42$ and finding its x-intercept.

129. A logarithmic equation can have, at most, one extraneous solution.

130. *Think About It* Is the time required for an investment to quadruple twice as long as the time for it to double? Give a reason for your answer and give an example to verify your answer algebraically.

131. *Writing* Write a paragraph explaining whether or not the time required for an investment to double is dependent on the size of the investment.

Review

In Exercises 132–135, sketch the graph of the function.

132. $f(x) = 3^{x+3} - 5$ **133.** $f(x) = -3^{-x-3} + 5$

134. $f(x) = -2^{x+2} - 6$ **135.** $f(x) = \left(\frac{1}{2}\right)^{-x} + 3$

In Exercises 136–139, evaluate the logarithm using the change-of-base formula. Round your result to three decimal places.

136. $\log_7 11$ **137.** $\log_3 22$

138. $\log_9 6$ **139.** $\log_{21} 140$

3.5 Exponential and Logarithmic Models

Introduction

The five most common types of mathematical models involving exponential functions and logarithmic functions are as follows.

1. **Exponential growth model:** $\quad y = ae^{bx}, \qquad b > 0$

2. **Exponential decay model:** $\quad y = ae^{-bx}, \qquad b > 0$

3. **Gaussian model:** $\qquad\qquad y = ae^{-(x-b)^2/c}$

4. **Logistic growth model:** $\qquad y = \dfrac{a}{1 + be^{-rx}}$

5. **Logarithmic models:** $\qquad y = a + b \ln x, \qquad y = a + b \log_{10} x$

The basic shapes of these graphs are shown in Figure 3.34.

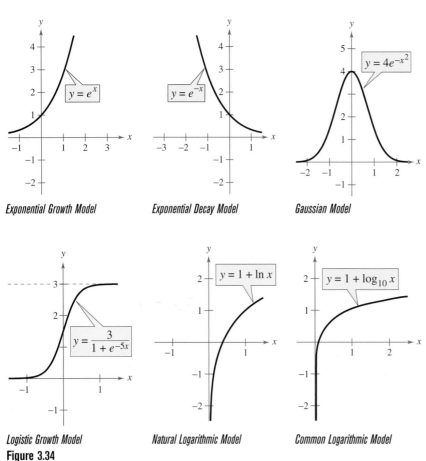

$y = e^x$

Exponential Growth Model

$y = e^{-x}$

Exponential Decay Model

$y = 4e^{-x^2}$

Gaussian Model

$y = \dfrac{3}{1 + e^{-5x}}$

Logistic Growth Model

$y = 1 + \ln x$

Natural Logarithmic Model

$y = 1 + \log_{10} x$

Common Logarithmic Model

Figure 3.34

What You Should Learn:

- How to recognize the five most common types of models involving exponential or logarithmic functions
- How to use exponential growth and decay functions to model and solve real-life problems
- How to use Gaussian functions to model and solve real-life problems
- How to use logistic growth functions to model and solve real-life problems
- How to use logarithmic functions to model and solve real-life problems
- How to fit exponential and logarithmic models to sets of data

Why You Should Learn It:

Exponential and logarithmic functions can be used to model and solve a variety of business applications. In Exercise 44 on page 268, you will compare an exponential decay model and a linear model for the depreciation of a computer over three years.

You can often gain quite a bit of insight into a situation modeled by an exponential or logarithmic function by identifying and interpreting the function's asymptotes. Use the graphs in Figure 3.34 to identify the asymptotes of each function.

Exponential Growth and Decay

EXAMPLE 1   Population Growth

Estimates of the world population (in millions) from 1992 through 2000 are shown in the table. The scatter plot of the data is shown in Figure 3.35.   (Source: U.S. Bureau of the Census)

Year	1992	1993	1994	1995	1996	1997	1998	1999	2000
Population	5445	5527	5607	5688	5767	5847	5926	6005	6083

An exponential growth model that approximates this data is

$$P = 5303e^{0.013838t}, \qquad 2 \le t \le 10$$

where P is the population (in millions) and $t = 2$ represents 1992. Compare the values given by the model with the estimates given by the U.S. Bureau of the Census. According to this model, what year corresponds to a world population of 6.5 billion?

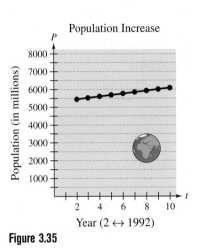

Figure 3.35

Algebraic Solution

The following table compares the two sets of population figures.

Year	1992	1993	1994	1995	1996	1997	1998	1999	2000
Population	5445	5527	5607	5688	5767	5847	5926	6005	6083
Model	5452	5528	5605	5683	5762	5842	5924	6006	6090

To find the year for which the world population is 6.5 billion, let $P = 6500$ in the model and solve for t.

$5303e^{0.013838t} = P$	Write original model.
$5303e^{0.013838t} = 6500$	Substitute 6500 for P.
$e^{0.013838t} \approx 1.2257$	Divide each side by 5303.
$\ln e^{0.013838t} \approx \ln 1.2257$	Take natural log of each side.
$0.013838t \approx 0.2035$	Inverse Property
$t \approx 14.71$	Divide each side by 0.013838.

According to the model, the year 2004 corresponds to a population of 6.5 billion.

Graphical Solution

Use a graphing utility to graph the model $y = 5303e^{0.013838x}$ and the data in the same viewing window. You can see in Figure 3.36 that the model appears to closely fit the data.

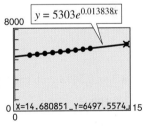

Figure 3.36

Use the *zoom* and *trace* features of the graphing utility to find that the approximate value of x for which $y = 6500$ is $x \approx 14.71$. So, according to the model, the year 2004 corresponds to a world population of 6.5 billion.

An exponential model increases (or decreases) by the same percent each year. What is the annual percent increase for the model above?

In Example 1, you were given the exponential growth model. But suppose this model had not been given; how could you have found such a model? One technique for doing this is demonstrated in Example 2.

EXAMPLE 2   Modeling Population Growth

In a research experiment, a population of fruit flies is increasing in accordance with the law of exponential growth. After 2 days there are 100 flies, and after 4 days there are 300 flies. How many flies will there be after 5 days?

Solution

Let y be the number of flies at time t. From the given information, you know that $y = 100$ when $t = 2$ and $y = 300$ when $t = 4$. Substituting this information into the model $y = ae^{bt}$ produces

$$100 = ae^{2b} \quad \text{and} \quad 300 = ae^{4b}.$$

To solve for b, solve for a in the first equation.

$$100 = ae^{2b} \quad \Longrightarrow \quad a = \frac{100}{e^{2b}} \qquad \text{Solve for } a \text{ in the first equation.}$$

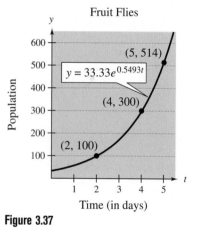

Figure 3.37

Then substitute the result into the second equation.

$300 = ae^{4b}$	Write second equation.
$300 = \left(\dfrac{100}{e^{2b}}\right)e^{4b}$	Substitute $100/e^{2b}$ for a.
$\dfrac{300}{100} = e^{2b}$	Divide each side by 100.
$\ln \dfrac{300}{100} = \ln e^{2b}$	Take natural log of each side.
$\ln 3 = 2b$	Inverse Property
$\dfrac{1}{2} \ln 3 = b$	Solve for b.

Using $b = \frac{1}{2} \ln 3 \approx 0.5493$, you can determine that

$a = \dfrac{100}{e^{2[(1/2)\ln 3]}}$	Substitute $\frac{1}{2} \ln 3$ for b.
$= \dfrac{100}{e^{\ln 3}}$	Simplify.
$= \dfrac{100}{3}$	Inverse Property
≈ 33.33	Simplify.

So, the exponential growth model is

$$y = 33.33e^{0.5493t},$$

as shown in Figure 3.37. This implies that after 5 days, the population is

$$y = 33.33e^{0.5493(5)} \approx 520 \text{ flies.}$$

In living organic material, the ratio of the content of radioactive carbon (carbon 14) to the content of nonradioactive carbon (carbon 12) is about 1 to 10^{12}. When organic material dies, its carbon 12 content remains fixed, whereas its radioactive carbon 14 begins to decay with a half-life of about 5700 years. To estimate the age of dead organic material, scientists use the following formula, which denotes the ratio of carbon 14 to carbon 12 present at any time t (in years).

$$R = \frac{1}{10^{12}}e^{-t/8223}$$

The graph of R is shown in Figure 3.38. Note that R decreases as t increases.

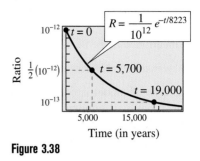

Figure 3.38

EXAMPLE 3   Carbon Dating

The ratio of carbon 14 to carbon 12 in a newly discovered fossil is

$$R = \frac{1}{10^{13}}.$$

Estimate the age of the fossil.

Algebraic Solution

In the carbon dating model, substitute the given value of R to obtain the following.

$\dfrac{1}{10^{12}}e^{-t/8223} = R$ Write original formula.

$\dfrac{e^{-t/8223}}{10^{12}} = \dfrac{1}{10^{13}}$ Substitute $\frac{1}{10^{13}}$ for R.

$e^{-t/8223} = \dfrac{1}{10}$ Multiply each side by 10^{12}.

$\ln e^{-t/8223} = \ln \dfrac{1}{10}$ Take natural log of each side.

$-\dfrac{t}{8223} \approx -2.3026$ Inverse Property

$t \approx 18{,}934$ Multiply each side by -8223.

So, to the nearest thousand years, you can estimate the age of the fossil to be 19,000 years.

Graphical Solution

Use a graphing utility to graph the formula for the ratio of carbon 14 to carbon 12 at any time t as

$$y_1 = \frac{1}{10^{12}}e^{-x/8223}.$$

In the same viewing window, graph $y_2 = 1/(10^{13})$. Use the *intersect* feature or the *zoom* and *trace* features of the graphing utility to estimate that $x \approx 18{,}934$ when $y = 1/(10^{13})$, as shown in Figure 3.39.

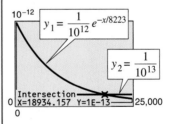

Figure 3.39

So, to the nearest thousand years, you can estimate the age of the fossil to be 19,000 years.

The carbon dating model in Example 3 assumed that the carbon 14 to carbon 12 ratio was 1 part in 10,000,000,000,000. Suppose an error in measurement occurred and the actual ratio was only 1 part in 8,000,000,000,000. The fossil age corresponding to the actual ratio would then be approximately 17,000 years. Try checking this result.

Gaussian Models

As mentioned at the beginning of this section, Gaussian models are of the form

$$y = ae^{-(x-b)^2/c}.$$

This type of model is commonly used in probability and statistics to represent populations that are **normally distributed.** For *standard* normal distributions, the model takes the form

$$y = \frac{1}{\sigma\sqrt{2\pi}}e^{-x^2/2\sigma^2}$$

where σ is the standard deviation (σ is the lowercase Greek letter sigma). The graph of a Gaussian model is called a **bell-shaped curve.** Try assigning a value to σ and sketching a normal distribution curve with a graphing utility. Can you see why it is called a bell-shaped curve?

EXAMPLE 4   SAT Scores

In 1997, the Scholastic Aptitude Test (SAT) math scores for college-bound seniors roughly followed the normal distribution

$$y = 0.0036e^{-(x-511)^2/25,088}, \qquad 200 \le x \le 800$$

where x is the SAT score for mathematics. Use a graphing utility to graph this function. From the graph, estimate the average SAT score.   (Source: College Board)

Solution

The graph of the function is given in Figure 3.40. On this bell-shaped curve, the maximum value of the curve represents the average score. Using the *maximum* feature or the *zoom* and *trace* features, you can see that the average mathematics score for college-bound seniors in 1997 was 511.

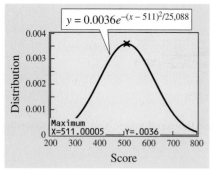

Figure 3.40

In Example 4, note that 50% of the seniors who took the test received a score lower than 511.

Logistic Growth Models

Some populations initially have rapid growth, followed by a declining rate of growth, as indicated by the graph in Figure 3.41. One model for describing this type of growth pattern is the **logistic curve** given by the function

$$y = \frac{a}{1 + be^{-rx}}$$

where y is the population size and x is the time. An example is a bacteria culture allowed to grow initially under ideal conditions and then under less favorable conditions that inhibit growth. A logistic growth curve is also called a **sigmoidal curve.**

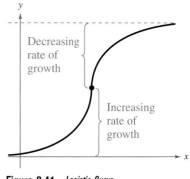

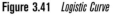

Figure 3.41 *Logistic Curve*

EXAMPLE 5   Spread of a Virus

On a college campus of 5000 students, one student returns from vacation with a contagious flu virus. The spread of the virus is modeled by

$$y = \frac{5000}{1 + 4999e^{-0.8t}}, \qquad t \geq 0$$

where y is the total number infected after t days. The college will cancel classes when 40% or more of the students are ill.

a. How many students are infected after 5 days?

b. After how many days will the college cancel classes?

Algebraic Solution

a. After 5 days, the number infected is

$$y = \frac{5000}{1 + 4999e^{-0.8(5)}} = \frac{5000}{1 + 4999e^{-4}} \approx 54.$$

b. In this case, the number of infected students is $(0.40)(5000) = 2000$. Therefore, you solve for t in the following equation.

$$2000 = \frac{5000}{1 + 4999e^{-0.8t}}$$

$$1 + 4999e^{-0.8t} = 2.5$$

$$e^{-0.8t} \approx \frac{1.5}{4999}$$

$$\ln e^{-0.8t} \approx \ln \frac{1.5}{4999}$$

$$-0.8t \approx \ln \frac{1.5}{4999}$$

$$t = -\frac{1}{0.8} \ln \frac{1.5}{4999}$$

$$t \approx 10.14$$

So, after about 10 days, at least 40% of the students will be infected, and classes will be canceled.

Graphical Solution

a. Use a graphing utility to graph $y = \dfrac{5000}{1 + 4999e^{-0.8x}}$.
Use the *value* feature or *zoom* and *trace* features of the graphing utility to estimate that $y = 54$ when $x = 5$. So, after 5 days, about 54 people will be infected.

b. In this case, the number infected must be $(0.40)(5000) = 2000$ to cancel classes. Use a graphing utility to graph

$$y_1 = \frac{5000}{1 + 4999e^{-0.8x}}. \qquad \text{and} \qquad y_2 = 2000$$

in the same viewing window. Use the *intersect* feature or the *zoom* and *trace* features of the graphing utility to find the point of intersection of the graphs. In Figure 3.42, you can see that the point of intersection occurs near $x \approx 10.14$. So, a short time after 10 days, at least 40% of the students will be infected, and the college will cancel classes.

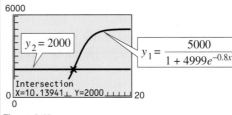

Figure 3.42

Logarithmic Models

EXAMPLE 6   Magnitudes of Earthquakes

On the Richter scale, the magnitude R of an earthquake of intensity I is

$$R = \log_{10} \frac{I}{I_0}$$

where $I_0 = 1$ is the minimum intensity used for comparison. Find the intensities per unit of area for the following earthquakes. (Intensity is a measure of the wave energy of the earthquake.)

a. Tokyo and Yokohama, Japan, in 1923; $R = 8.3$

b. Izmit, Turkey, in 1999; $R = 7.4$

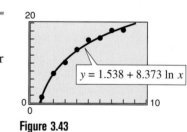

More than 15,000 people were killed in the 7.4 magnitude earthquake in Izmit, Turkey, on August 17, 1999.

Solution

a. Because $I_0 = 1$ and $R = 8.3$, you have

$$8.3 = \log_{10} I$$

$$I = 10^{8.3} \approx 199{,}526{,}000. \qquad \text{Intensity}$$

b. For $R = 7.4$, you have $7.4 = \log_{10} I$, and $I = 10^{7.4} \approx 25{,}119{,}000$.

Note that an increase of 0.9 units on the Richter scale (from 7.4 to 8.3) represents an intensity change by a factor of

$$\frac{199{,}526{,}000}{25{,}119{,}000} \approx 8.$$

In other words, the earthquake in 1923 had a magnitude about 8 times greater than that of the 1999 quake.

Fitting Models to Data

EXAMPLE 7   Fitting a Logarithmic Model

The data in the table gives the yield y (in milligrams) of a chemical reaction after x minutes.

x	1	2	3	4	5	6	7	8
y	1.5	7.4	10.2	13.4	15.8	16.3	18.2	18.3

Use a graphing utility to fit a logarithmic model to the data.

Solution

Enter the data in the table into a graphing utility. Use the regression capabilities of the graphing utility to find the following logarithmic model.

$$y = 1.538 + 8.373 \ln x$$

When you graph the data and the model in the same viewing window, as shown in Figure 3.43, you can see that the model is a good fit to the data.

Figure 3.43

$y = 1.538 + 8.373 \ln x$

EXAMPLE 8   Fitting an Exponential Model

The total amount A (in billions of dollars) spent on health care in the United States in the years 1970 through 1996 are shown below. Find a model for the data, and use the model to estimate the amount spent in 2004. In the list of data points (t, A), t represents the year, with $t = 0$ corresponding to 1970.   (Source: U.S. Health Care Financing Administration)

(0, 74.3), (1, 82.2), (2, 92.3), (3, 102.4), (4, 115.9), (5, 132.6),
(6, 151.9), (7, 172.6), (8, 193.2), (9, 218.3), (10, 247.3), (11, 291.4),
(12, 328.2), (13, 360.8), (14, 396.0), (15, 428.7), (16, 466.0), (17, 506.2),
(18, 562.3), (19, 623.9), (20, 699.5), (21, 766.8), (22, 836.6), (23, 895.1),
(24, 945.7), (25, 991.4), (26, 1035.1)

Solution

Enter the data into a graphing utility. Then graph the data, as shown in Figure 3.44. From the scatter plot, it appears that an exponential model is a good fit. Use the regression capabilities of the graphing utility to find the following exponential model.

$$A = 81.10(1.11)^t \quad \text{or} \quad A = 81.10e^{0.1044t}$$

From the model, you can see that the amount spent on health care from 1970 through 1996 had an average annual increase of 10%. From this model, you can estimate the 2004 amount to be

$$A = 81.10(1.11)^{34} \approx 2818.4 \text{ billion dollars}$$

which is more than twice the amount spent in 1996.

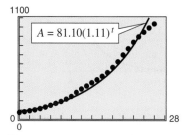

1100

$A = 81.10(1.11)^t$

0   28
0

Figure 3.44

Writing About Math   *Fitting a Model to Data*

The numbers y (in millions) of vinyl single records sold in the United States in the years 1984 through 1997 are listed below. The data is given as ordered pairs of the form (t, y), where t is the year, with $t = 4$ representing 1984. Create a scatter plot of the data. Decide which type of model best fits this data. Then find the model. Write a paragraph explaining why you think the model you chose is a good fit to the data.

(4, 131.5), (5, 120.7), (6, 93.9), (7, 82.0),
(8, 65.6), (9, 36.6), (10, 27.6), (11, 22.0),
(12, 19.8), (13, 15.1), (14, 11.7), (15, 10.2),
(16, 10.1), (17, 7.5)

3.5   E x e r c i s e s

In Exercises 1–6, match the function with its graph. [The graphs are labeled (a), (b), (c), (d), (e), and (f).]

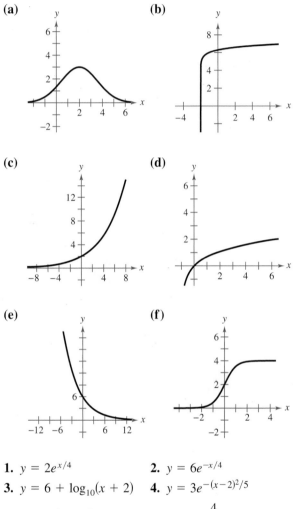

(a)

(b)

(c)

(d)

(e)

(f)

1. $y = 2e^{x/4}$

2. $y = 6e^{-x/4}$

3. $y = 6 + \log_{10}(x + 2)$

4. $y = 3e^{-(x-2)^2/5}$

5. $y = \ln(x + 1)$

6. $y = \dfrac{4}{1 + e^{-2x}}$

In Exercises 7–14, determine whether the scatter plot could best be modeled by a linear model, a quadratic model, an exponential model, a logarithmic model, a Gaussian model, or a logistic model.

7.

8.

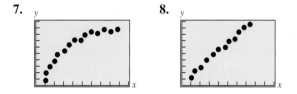

9.

10.

11.

12.

13.

14.

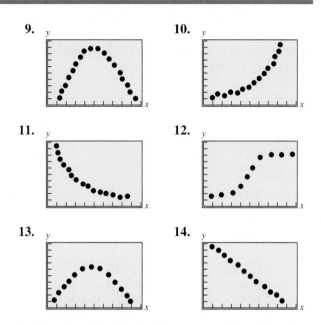

In Exercises 15–18, sketch a scatter plot of the data. Decide whether the data could best be modeled by a linear, an exponential, or a logarithmic model.

15. $(1, 2.0), (1.5, 3.5), (2, 4.0), (4, 5.8), (6, 7.0), (8, 7.8)$

16. $(1, 11.0), (1.5, 9.6), (2, 8.2), (4, 4.5), (6, 2.5), (8, 1.4)$

17. $(1, 7.5), (1.5, 7.0), (2, 6.8), (4, 5.0), (6, 3.5), (8, 2.0)$

18. $(1, 5.0), (1.5, 6.0), (2, 6.4), (4, 7.8), (6, 8.6), (8, 9.0)$

Compound Interest   **In Exercises 19–26, complete the table for a savings account in which interest is compounded continuously.**

Initial Investment	Annual % Rate	Time to Double	Amount After 10 Years
19. $1000	12%		
20. $20,000	$10\frac{1}{2}\%$		
21. $750		$7\frac{3}{4}$ yr	
22. $10,000		12 yr	
23. $500			$1292.85
24. $600			$1505.00
25.	4.5%		$10,000.00
26.	8%		$20,000.00

27. *Compound Interest*   (a) Complete the table for the time t necessary for P dollars to triple if interest is compounded continuously at rate r.

r	2%	4%	6%	8%	10%	12%
t						

(b) Draw a scatter plot of the data in part (a). Use the regression capabilities of a graphing utility to find a model for the data.

28. *Compound Interest*   (a) Complete the table for the time t necessary for P dollars to triple if interest is compounded annually at rate r.

r	2%	4%	6%	8%	10%	12%
t						

(b) Draw a scatter plot of the data in part (a). Use the regression capabilities of a graphing utility to find a model for the data.

29. *Comparing Investments*   If $1 is invested in an account over a 10-year period, the amount in the account, where t represents the time in years, is

$$A = 1 + 0.075[\![t]\!] \quad \text{or} \quad A = e^{0.07t}$$

depending on whether the account pays simple interest at $7\frac{1}{2}\%$ or continuous compound interest at 7%. Use a graphing utility to graph each function in the same viewing window. Which grows at a faster rate?

30. *Comparing Investments*   If $1 is invested in an account over a 10-year period, the amount in the account, where t represents the time in years, is

$$A = 1 + 0.06[\![t]\!] \quad \text{or} \quad A = \left(1 + \frac{0.055}{365}\right)^{[\![365t]\!]}$$

depending on whether the account pays simple interest at 6% or compound interest at $5\frac{1}{2}\%$ compounded daily. Use a graphing utility to graph each function in the same viewing window. Which grows at a faster rate?

In Exercises 31–34, complete the table for the given radioactive isotope.

Isotope	Half-Life (years)	Initial Quantity	Amount After 1000 Years
31. ^{226}Ra	1620	10 g	
32. ^{226}Ra	1620		1.5 g
33. ^{14}C	5730	3 g	
34. ^{230}Pu	24,360		0.4 g

35. *Population*   The population P of a city is

$$P = 105,300e^{0.015t}$$

where $t = 0$ represents 2000. According to this model, when will the population reach 150,000?

36. *Population*   The population P of a city is

$$P = 240,360e^{0.012t}$$

where $t = 0$ represents 2000. According to this model, when will the population reach 275,000?

37. *Population*   The population P of a city is

$$P = 2500e^{kt}$$

where $t = 0$ represents 2000. In 1945, the population was 1350. Find the value of k, and use this value to estimate the population in the year 2010.

38. *Population*   The table shows the population (in millions) of a country in 1997 and the projected population (in millions) for the year 2020.   (Source: U.S. Bureau of the Census)

Country	1997	2020
Croatia	5.0	4.8
Mali	9.9	20.4
Singapore	3.5	4.3
Sweden	8.9	9.5

(a) Find the exponential growth model $y = ae^{bt}$ for the population in each country by letting $t = 0$ correspond to 1997. Use the model to estimate the population of each country in 2030.

(b) You can see that the populations of Mali and Sweden are growing at different rates. What constant in the equation $y = ae^{bt}$ is determined by these different growth rates? Discuss the relationship between the different growth rates and the magnitude of the constant.

(c) You can see that the population of Singapore is increasing while the population of Croatia is decreasing. What constant in the equation $y = ae^{bt}$ reflects this difference? Explain.

39. Bacteria Growth The number of bacteria N in a culture is given by the model

$$N = 100e^{kt}$$

where t is the time (in hours). If $N = 300$ when $t = 5$, estimate the time required for the population to double in size. Verify your estimate graphically.

40. Bacteria Growth The number of bacteria N in a culture is given by the model

$$N = 250e^{kt}$$

where t is the time (in hours). If $N = 280$ when $t = 10$, estimate the time required for the population to double in size. Verify your estimate graphically.

41. Radioactive Decay The half-life of radioactive radium (radium 226) is 1620 years. What percent of a present amount of radioactive radium will remain after 100 years?

42. Radioactive Decay Carbon 14 dating assumes that the carbon dioxide on earth today has the same radioactive content as it did centuries ago. If this is true, the amount of carbon 14 absorbed by a tree that grew several centuries ago should be the same as the amount of carbon 14 absorbed by a tree growing today. A piece of ancient charcoal contains only 15% as much radioactive carbon as a piece of modern charcoal. How long ago was the tree burned to make the ancient charcoal if the half-life of carbon 14 is 5730 years?

43. Depreciation A car that cost $22,000 new has a book value of $13,000 after 2 years.

(a) Find the linear model $V = mt + b$.

(b) Find the exponential model $V = ae^{kt}$.

(c) Use a graphing utility to graph the two models in the same viewing window. Which model depreciates faster in the first 2 years?

(d) Use each model to find the book values of the car after 1 year and after 3 years.

(e) Interpret the slope of the linear model.

44. Depreciation A computer that cost $4600 new has a book value of $3000 after 2 years.

(a) Find the linear model $V = mt + b$.

(b) Find the exponential model $V = ae^{kt}$.

(c) Use a graphing utility to graph the two models in the same viewing window. Which model depreciates faster in the first year?

(d) Use each model to find the book values of the computer after 1 year and after 3 years.

(e) Interpret the slope of the linear model.

45. Sales The sales S (in thousands of units) of a new product after it has been on the market t years are

$$S(t) = 100(1 - e^{kt}).$$

Fifteen thousand units of the new product were sold the first year.

(a) Complete the model by solving for k.

(b) Use a graphing utility to graph the model.

(c) Use the graph to estimate the number of units sold after 5 years.

46. Sales and Advertising The sales S (in thousands of units) of a product after x hundred dollars is spent on advertising is

$$S = 10(1 - e^{kx}).$$

When $500 is spent on advertising, 2500 units are sold.

(a) Complete the model by solving for k.

(b) Estimate the number of units that will be sold if advertising expenditures are raised to $700.

47. Learning Curve The management at a factory has found that the maximum number of units a worker can produce in a day is 30. The learning curve for the number of units N produced per day after a new employee has worked t days is

$$N = 30(1 - e^{kt}).$$

After 20 days on the job, a new employee produces 19 units.

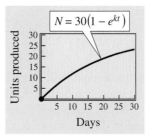

(a) Find the learning curve for this employee (first find the value of k).

(b) How many days should pass before this employee is producing 25 units per day?

(c) Is the employee's production increasing at a linear rate? Explain your reasoning.

48. *Endangered Species* A conservation organization releases 100 animals of an endangered species into a game preserve. The organization believes that the preserve has a carrying capacity of 1000 animals and that the growth of the herd will follow the logistic curve

$$p(t) = \frac{1000}{1 + 9e^{-0.1656t}}$$

where t is measured in months.

(a) Use a graphing utility to graph the function. Use the graph to determine the horizontal asymptotes and interpret the meaning of the larger asymptote in the context of the problem.

(b) Estimate the population after 5 months.

(c) When will the population reach 500?

Earthquake Magnitudes **In Exercises 49 and 50, use the Richter scale (see page 264) for measuring the magnitudes of earthquakes.**

49. Find the magnitude R of an earthquake of intensity I (let $I_0 = 1$).

(a) Taiwan in 1999, $I = 39{,}811{,}000$

(b) Hector Mine, California, in 1999,

$I = 12{,}589{,}000$

50. Find the intensity I of an earthquake measuring R on the Richter scale (let $I_0 = 1$).

(a) Chile in 1906, $R = 8.6$

(b) Los Angeles in 1971, $R = 6.7$

Intensity of Sound **In Exercises 51–54, use the following information. For Exercises 51 and 52, determine the level of sound (in decibels) for the given sound intensity.**

The level of sound β (in decibels) with an intensity I is $\beta(I) = 10 \log_{10}(I/I_0)$, where I_0 is an intensity of 10^{-12} watt per square meter, corresponding roughly to the faintest sound that can be heard by the human ear.

51. (a) $I = 10^{-10}$ watt per m^2 (faint whisper)

(b) $I = 10^{-5}$ watt per m^2 (busy street corner)

(c) $I = 10^0$ watt per m^2 (threshold of pain)

52. (a) $I = 10^{-3.5}$ watt per m^2 (jet 4 miles from takeoff)

(b) $I = 10^{-3}$ watt per m^2 (diesel truck at 25 feet)

(c) $I = 10^{-1.5}$ watt per m^2 (auto horn at 3 feet)

53. *Noise Level* As a result of the installation of noise suppression materials, the noise level in an auditorium was reduced from 93 to 80 decibels. Find the percent decrease in the intensity level of the noise due to the installation of these materials.

54. *Noise Level* As a result of the installation of a muffler, the level of noise emitted by an engine was reduced from 88 to 72 decibels. Find the percent decrease in the intensity level of the noise due to the installation of the muffler.

Acidity **In Exercises 55–58, use the acidity model given by pH $= -\log_{10}[H^+]$, where acidity (pH) is a measure of the hydrogen ion concentration $[H^+]$ (measured in moles of hydrogen per liter) of a solution.**

55. Find the pH if $[H^+] = 2.3 \times 10^{-5}$.

56. Compute $[H^+]$ for a solution in which pH $= 5.8$.

57. A certain fruit has a pH of 2.5, and an antacid tablet has a pH of 9.5. The hydrogen ion concentration of the fruit is how many times that of the tablet?

58. If the pH of a solution is decreased by one unit, the hydrogen ion concentration is increased by what factor?

59. *Finance* A \$120,000 home mortgage for 30 years at $7\frac{1}{2}\%$ has a monthly payment of \$839.06. Part of the monthly payment goes for the interest charge on the unpaid balance, and the remainder of the payment is used to reduce the principal. The amount that goes for interest is

$$u = M - \left(M - \frac{Pr}{12}\right)\left(1 + \frac{r}{12}\right)^{12t}$$

and the amount that goes toward reduction of the principal is

$$v = \left(M - \frac{Pr}{12}\right)\left(1 + \frac{r}{12}\right)^{12t}.$$

In these formulas, P is the size of the mortgage, r is the interest rate, M is the monthly payment, and t is the time (in years).

(a) Use a graphing utility to graph each function in the same viewing window. (The viewing window should show all 30 years of mortgage payments.)

(b) In the early years of the mortgage, the larger part of the monthly payment goes for what purpose? Approximate the time when the monthly payment is evenly divided between interest and principal reduction.

(c) Repeat (a) and (b) for a repayment period of 20 years ($M = \$966.71$). What can you conclude?

60. *Finance*   The total interest u paid on a home mortgage of P dollars at interest rate r for t years is

$$u = P\left[\dfrac{rt}{1 - \left(\dfrac{1}{1 + r/12}\right)^{12t}} - 1\right].$$

Consider a \$120,000 home mortgage at $7\frac{1}{2}\%$.

(a) Use a graphing utility to graph the total interest function.

(b) Approximate the length of the mortgage when the total interest paid is the same as the size of the mortgage. Is it possible that a person could pay twice as much in interest charges as the size of one's mortgage?

In Exercises 61–64, find the exponential model $y = ae^{bx}$ that fits the points given in the graph or table.

61.

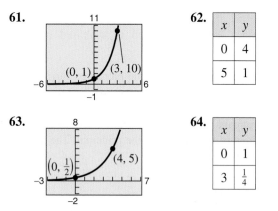

62.

x	y
0	4
5	1

63.

64.

x	y
0	1
3	$\frac{1}{4}$

65. *Data Analysis*   The time t (in seconds) required to attain a speed of s miles per hour from a standing start for a particular car is given in the table. (Source: *Road & Track*, March 1995)

s	30	40	50	60	70	80	90
t	3.4	5.0	7.0	9.3	12.0	15.8	20.0

Two models for this data are as follows.

$$t_1 = 40.757 + 0.556s - 15.817 \ln s$$

$$t_2 = 1.2259 + 0.0023s^2$$

(a) Use a graphing utility to fit a linear model t_3 and an exponential model t_4 to the data.

(b) Use a graphing utility to graph the data points and each model.

(c) Create a table to compare the data with estimates obtained from each model.

(d) Use the results of part (c) to find the sum of the absolute values of the differences between the data and the estimated values given by each model. Based on the four sums, which model do you think best fits the data? Explain.

66. *Comparing Models*   Suppose that in 1995 a car had a retail price of \$21,800. A local dealership used the following guide for the approximate value of the car for the years 1995 through 2000.

Year	1995	1996	1997
Value	\$21,800	\$19,530	\$19,210

Year	1998	1999	2000
Value	\$17,945	\$17,120	\$16,080

Let V represent the value of the automobile in the year t, with $t = 0$ corresponding to 1995.

(a) Use the regression capabilities of a graphing utility to find linear and quadratic models for the data. Use the graphing utility to plot the data and graph the models in the same viewing window.

(b) What does the slope represent in the linear model in part (a)?

(c) Do you think the quadratic model is realistic? Explain.

(d) Use the regression capabilities of a graphing utility to fit an exponential model to the data. Use the graphing utility to plot the data and graph the model in the same viewing window. How well does the model fit the data?

(e) Fit a rational model to the data. Take the reciprocal of the depreciated values of the automobile to generate the points $(t, 1/V)$. Use the regression capabilities of a graphing utility to fit a line to this data. The resulting line has the form

$$\frac{1}{V} = at + b.$$

Solve for V and use the graphing utility to graph the rational function and the original data points in the same viewing window. How well does the model fit the data?

(f) Determine the horizontal asymptotes of the exponential and rational models. Interpret their meanings in the context of the problem.

67. *Forensics*   At 8:30 A.M., a coroner was called to the home of a person who had died during the night. In order to estimate the time of death, the coroner took the person's temperature twice. At 9:00 A.M. the temperature was 85.7°F, and at 9:30 A.M. the temperature was 82.8°F. From these two temperatures the coroner was able to determine that the time elapsed since death and the body temperature were related by the formula

$$t = -2.5 \ln \frac{T - 70}{98.6 - 70}$$

where t is the time (in hours elapsed since the person died) and T is the temperature (in degrees Fahrenheit) of the person's body. Assume the person had a normal body temperature of 98.6°F at death, and the room temperature was a constant 70°F. (This formula is derived from a general cooling principle called Newton's Law of Cooling.) Use the formula to estimate the time of death of the person.

68. *Foreign Travel*   The numbers of United States citizens y (in millions) who traveled to foreign countries in the years 1988 through 1996 are shown in the table, where $t = 8$ represents the year 1988. (Source: U.S. Travel and Tourism Administration, U.S. Department of Commerce)

t	8	9	10	11	12
y	40.7	41.1	44.6	41.6	43.9

t	13	14	15	16
y	44.4	46.5	50.8	52.3

(a) Use the regression capabilities of a graphing utility to find an exponential model that fits the data.

(b) Use the graphing utility to plot the data and graph the model in the same viewing window.

(c) Use the model to estimate the number of foreign travelers in the year 2005.

69. *World Population*   The world population y (in billions) for the years 1987 through 1998 is shown in the table, where $x = 7$ corresponds to 1987. (Source: U.S. Bureau of the Census)

x	7	8	9	10	11	12
y	5.02	5.11	5.19	5.28	5.37	5.45

x	13	14	15	16	17	18
y	5.53	5.61	5.69	5.77	5.85	5.93

(a) Use the regression capabilities of a graphing utility to fit a linear model to the data.

(b) Use the regression capabilities of a graphing utility to fit an exponential model to the data.

(c) Population growth is often exponential. For the 12 years of data in the table, is the exponential model better than the linear model? Explain.

(d) Use each model to estimate the population in the year 2005.

70. *Atmospheric Pressure*   The atmospheric pressure decreases with increasing altitude. At sea level, the average air pressure is 1.033227 kilograms per square centimeter, and this pressure is called one atmosphere. Variations in weather conditions cause changes in the atmospheric pressure of up to ±5 percent. The table shows the pressures p (in atmospheres) at given altitudes h (in kilometers).

h	0	5	10	15	20	25
p	1	0.55	0.25	0.12	0.06	0.02

(a) Use a graphing utility to attempt to find the logarithmic model $p = a + b \ln h$ for the data. Explain why the result is an error message.

(b) Use a graphing utility to find the logarithmic model $h = a + b \ln p$ for the data.

(c) Use a graphing utility to plot the data and graph the logarithmic model.

(d) Use the model to estimate the altitude at which the pressure is 0.75 atmosphere.

(e) Use the graph to estimate the pressure at an altitude of 13 kilometers.

71. *National Health Expenditures*   The table shows the national health expenditures y (in billions of dollars) for the years 1987 through 1996. The years are represented by x, where $x = 7$ corresponds to 1987. (Source: U.S. Health Care Financing Administration)

x	7	8	9	10	11
y	500.1	559.6	622.0	699.5	766.8

x	12	13	14	15	16
y	836.6	895.1	945.7	991.4	1035.1

(a) Use a graphing utility to find an exponential model for the data. Use the graphing utility to plot the data and graph the exponential model in the same viewing window.

(b) Use a graphing utility to find a logarithmic model for the data. Use the graphing utility to plot the data and graph the logarithmic model in the same viewing window.

(c) Use the graphs plotted in parts (a) and (b) to determine which model is better. If the rate of growth of health care costs could be slowed, which model may be better for the future? Explain.

72. *Comparing Models* The amounts y (in billions of dollars) donated to charity (by individuals, foundations, corporations, and charitable bequests) in the years 1987 through 1996 in the United States are shown in the table, where $x = 7$ corresponds to 1987. (Source: AAFRC Trust for Philanthropy)

x	7	8	9	10	11
y	90.0	98.1	106.7	111.5	117.2

x	12	13	14	15	16
y	121.1	126.5	129.3	140.5	150.7

(a) Use the regression capabilities of a graphing utility to find the following models for the data.

$y_1 = ax + b$

$y_2 = a + b \ln x$

$y_3 = ab^x$

(b) Use the graphing utility to graph the data and each of the models. Use the graphs to select the model that you think best fits the data.

(c) For each of the models y_i ($i = 1, 2,$ and 3), complete the table.

x	y	$y - y_i$	$(y - y_i)^2$
7	90.0		
8	98.1		
9	106.7		
10	111.5		
11	117.2		
12	121.1		
13	126.5		
14	129.3		
15	140.5		
16	150.7		

(d) For each model, find the sum of the entries in the last column of the table in part (c). Use the results to select the best model for the data.

(e) Explain what the sums in part (d) represent.

73. *Comparing Models* A V-8 car engine is coupled to a dynamometer, and the horsepower y is measured at different engine speeds x (in thousands of revolutions per minute). The results are shown in the table.

x	1	2	3	4	5	6
y	40	85	140	200	225	245

(a) Use the regression capabilities of a graphing utility to find the following models for the data.

$y_1 = ax^3 + bx^2 + cx + d$

$y_2 = a + b \ln x$

$y_3 = ab^x$

(b) Use the graphing utility to graph the data and each of the models. Use the graphs to select the model that you think best fits the data.

3 Review Exercises

3.1 In Exercises 1–4, use a calculator to evaluate the expression. Round your answer to three decimal places.

1. $(1.45)^{2\pi}$

2. $\sqrt[5]{6240}$

3. $1.59^{-2\sqrt{3}}$

4. $136(5^{-1.1})$

In Exercises 5–10, match the function with its graph. [The graphs are labeled (a), (b), (c), (d), (e), and (f).]

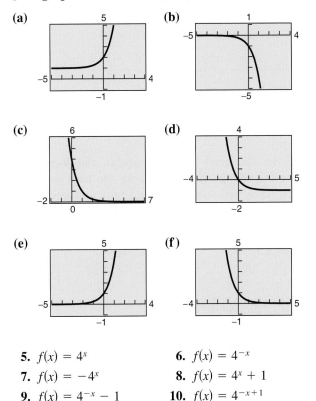

(a)

(b)

(c)

(d)

(e)

(f)

5. $f(x) = 4^x$

6. $f(x) = 4^{-x}$

7. $f(x) = -4^x$

8. $f(x) = 4^x + 1$

9. $f(x) = 4^{-x} - 1$

10. $f(x) = 4^{-x+1}$

In Exercises 11–20, sketch the graph of the function. Use a graphing utility to verify your graph.

11. $f(x) = 6^x$

12. $f(x) = 0.3^x$

13. $g(x) = 6^{-x}$

14. $g(x) = 0.3^{-x}$

15. $h(x) = e^{-x/2}$

16. $f(x) = e^{x+2}$

17. $h(x) = 2 - e^{-x/2}$

18. $f(x) = 3 - e^{-x-2}$

19. $s(t) = 4e^{-2/t}, \ t > 0$

20. $f(x) = 6e^{-3/x}, \ x > 0$

In Exercises 21–26, use a graphing utility to graph the function. Identify any asymptotes.

21. $g(t) = 8 - 0.5e^{-t/4}$

22. $h(x) = 12(1 + e^{-x/2})$

23. $g(x) = 200e^{4/x}$

24. $f(x) = -8e^{-4/x}$

25. $f(x) = \dfrac{10}{1 + 2^{-0.05x}}$

26. $f(x) = -\dfrac{12}{1 + 4^{-x}}$

In Exercises 27–30, use a calculator to evaluate the expression. Round your result to three decimal places.

27. $e^{-8/3}$

28. $e^{\sqrt{5}}$

29. $6e^{4.12}$

30. $-\frac{5}{3}e^{-3/5}$

In Exercises 31 and 32, complete the table to determine the amount of money P that should be invested at rate r to produce a final balance of \$200,000 in t years.

t	1	10	20	30	40	50
P						

31. $r = 8\%$, compounded continuously

32. $r = 10\%$, compounded monthly

33. *Depreciation* After t years, the value of a car that originally cost \$26,000 is

$$V(t) = 26{,}000\left(\frac{3}{4}\right)^t.$$

(a) Use a graphing utility to graph the function.

(b) Find the value of the car 2 years after it was purchased.

(c) According to the model, when does the car depreciate most rapidly? Is this realistic? Explain.

34. *Fuel Efficiency* A certain automobile gets 28 miles per gallon of gasoline for speeds up to 50 miles per hour. Over 50 miles per hour, the number of miles per gallon drops at the rate of 12% for each additional 10 miles per hour. If s is the speed and y is the number of miles per gallon, then

$$y = 28e^{0.6 - 0.012s}, \quad s \ge 50.$$

Use the model to complete the table.

s	50	55	60	65	70
y	28				

3.2 In Exercises 35–38, write the exponential equation in logarithmic form.

35. $4^3 = 64$

36. $3^5 = 243$

37. $25^{3/2} = 125$

38. $12^{-1} = \frac{1}{12}$

In Exercises 39–44, evaluate the expression without using a graphing utility.

39. $\log_8 8^{-0.34}$

40. $\log_7 1$

41. $\log_6 216$

42. $\log_4 1024$

43. $\log_{36} \frac{1}{6}$

44. $\log_{10} 0.001$

In Exercises 45–52, sketch the graph of the function. Use a graphing utility to verify your graph.

45. $g(x) = -\log_2 x + 5$

46. $g(x) = \log_5(x - 3)$

47. $f(x) = \log_2(x - 1) + 6$

48. $f(x) = \log_5(x + 2) - 3$

49. $f(x) = \ln x + 3$

50. $f(x) = \ln(x - 3)$

51. $h(x) = \frac{1}{2} \ln x$

52. $f(x) = \frac{1}{4} \ln x$

In Exercises 53–56, use a graphing utility to graph the function and determine its domain.

53. $y = \log_{10}(x^2 + 1)$

54. $y = -2 \log_{10}(4 - x^2)$

55. $y = \sqrt{x} \ln(x + 1)$

56. $y = \dfrac{14 \ln x}{x}$

In Exercises 57–60, evaluate the expression without using a graphing utility.

57. $\ln e^7$

58. $\ln 1$

59. $6 \ln e^{-3}$

60. $-\frac{3}{2} \ln e^{-10/11}$

61. *Climb Rate* The time t (in minutes) for a small plane to climb to an altitude of h feet is

$$t = 50 \log_{10} \frac{18{,}000}{18{,}000 - h}$$

where 18,000 feet is the plane's absolute ceiling.

(a) Determine the domain of the function appropriate for the context of the problem.

(b) Use a graphing utility to graph the time function and identify any asymptotes.

(c) As the plane approaches its absolute ceiling, what can be said about the time required to further increase its altitude?

(d) Find the time for the plane to climb to an altitude of 4000 feet.

62. *Earthquake Magnitudes* The magnitude M of an earthquake measured on the Richter scale can be found using the model

$$M = \tfrac{2}{3}(\log_{10} E - 11.4)$$

where E is the energy of the earthquake (in ergs). Use a graphing utility and the table shown below to find the ratio of the magnitudes of the earthquakes.

Earthquake	E (ergs)
San Francisco, 1906	7.079×10^{23}
Japan, 1933	5.623×10^{24}
Armenia, 1988	3.981×10^{21}
Russia, 1995	6.310×10^{22}
Indonesia, 1996	4.467×10^{22}

(a) Japan and San Francisco

(b) Russia and Indonesia

(c) Indonesia and Armenia

3.3 In Exercises 63–66, evaluate the logarithm using the change-of-base formula. Do each problem twice, once with common logarithms and once with natural logarithms. Round the result to three decimal places.

63. $\log_4 9$

64. $\log_{1/2} 5$

65. $\log_{12} 200$

66. $\log_3 0.28$

In Exercises 67–70, rewrite the expression in terms of $\ln 4$ and $\ln 5$.

67. $\ln 20$

68. $\ln 45 - 2 \ln 3$

69. $\ln \frac{5}{64}$

70. $\ln \frac{2}{5}$

In Exercises 71–74, approximate the logarithm using the properties of logarithms, given that $\log_b 2 \approx 0.3562$, $\log_b 3 \approx 0.5646$, and $\log_b 5 \approx 0.8271$.

71. $\log_b 25$

72. $\log_b\left(\frac{25}{9}\right)$

73. $\log_b \sqrt{3}$

74. $\log_b 30$

In Exercises 75–80, rewrite the expression as a sum, difference, and/or multiple of logarithms.

75. $\log_5 5x^2$

76. $\log_7 \dfrac{\sqrt{x}}{4}$

77. $\log_{10} \dfrac{5\sqrt{y}}{x^2}$

78. $\ln \left| \dfrac{x - 1}{x + 1} \right|$

79. $\ln[(x^2 + 1)(x - 1)]$ **80.** $\ln \sqrt[5]{\dfrac{4x^2 - 1}{4x^2 + 1}}$

In Exercises 81–86, write the expression as the logarithm of a single quantity.

81. $\log_2 5 + \log_2 x$ **82.** $\log_6 y - 2 \log_6 z$

83. $\frac{1}{2} \ln|2x - 1| - 2 \ln|x + 1|$

84. $5 \ln|x - 2| - \ln|x + 2| - 3 \ln|x|$

85. $\ln 3 + \frac{1}{3} \ln(4 - x^2) - \ln x$

86. $3[\ln x - 2 \ln(x^2 + 1)] + 2 \ln 5$

87. *Snow Removal* The number of miles s of roads cleared of snow is approximated by the model

$$s = 25 - \frac{13 \ln(h/12)}{\ln 3}, \quad 2 \le h \le 15$$

where h is the depth of the snow (in inches).

(a) Use a graphing utility to graph the function.

(b) Complete the table.

h	4	6	8	10	12	14
s						

(c) Using the graph of the function and the table, what conclusion can you make as the depth of the snow increases?

88. *Human Memory Model* Students in a sociology class were given an exam and then were retested monthly with an equivalent exam. The average score for the class was given by the human memory model, $f(t) = 85 - 14 \log_{10}(t + 1)$, $0 \le t \le 4$, where t is the time in months. How did the average score change over the 4-month period?

3.4 **In Exercises 89–94, solve for x.**

89. $8^x = 512$ **90.** $3^x = 729$

91. $6^x = \frac{1}{216}$ **92.** $6^{x-2} = 1296$

93. $\log_7 x = 4$ **94.** $\log_x 243 = 5$

In Exercises 95–104, solve the exponential equation. Round the result to three decimal places.

95. $e^x = 12$ **96.** $e^{3x} = 25$

97. $3e^{-5x} = 132$ **98.** $14e^{3x+2} = 560$

99. $e^x + 13 = 35$ **100.** $e^x - 28 = -8$

101. $-4(5^x) = -68$ **102.** $2(12^x) = 190$

103. $e^{2x} - 7e^x + 10 = 0$ **104.** $e^{2x} - 6e^x + 8 = 0$

In Exercises 105–116, solve the logarithmic equation. Round the result to three decimal places.

105. $\ln 3x = 8.2$ **106.** $\ln 5x = 7.2$

107. $2 \ln 4x = 15$ **108.** $4 \ln 3x = 15$

109. $\ln x - \ln 3 = 2$ **110.** $\ln \sqrt{x + 8} = 3$

111. $\ln \sqrt{x + 1} = 2$ **112.** $\ln x - \ln 5 = 4$

113. $\log_{10}(x - 1) = \log_{10}(x - 2) - \log_{10}(x + 2)$

114. $\log_{10}(x + 2) - \log_{10} x = \log_{10}(x + 5)$

115. $\log_{10}(1 - x) = -1$ **116.** $\log_{10}(-x - 4) = 2$

117. *Finance* You deposit $7550 into an account that pays 7.25% interest, compounded continuously. How long will it take the money to triple?

118. *Economics* The demand equation for a certain product is modeled by $p = 500 - 0.5e^{0.004x}$. Find the demand x that corresponds to a price of (a) $p = \$450$ and (b) $p = \$400$.

3.5 **In Exercises 119–124, match the function on page 360 with its graph. [The graphs are labeled (a), (b), (c), (d), (e), and (f).]**

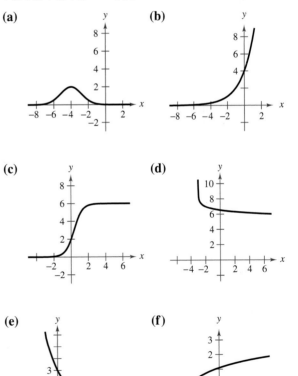

(a)

(b)

(c)

(d)

(e)

(f)

119. $y = 3e^{-2x/3}$ **120.** $y = 4e^{2x/3}$

121. $y = \ln(x + 3)$ **122.** $y = 7 - \log_{10}(x + 3)$

123. $y = 2e^{-(x+4)^2/3}$ **124.** $y = \dfrac{6}{1 + 2e^{-2x}}$

125. *Population* The population of a town is modeled by

$$P = 12,620e^{0.0118t}$$

where $t = 0$ represents 2000. According to this model, when will the population reach 17,000?

126. *Radioactive Decay* The half-life of radioactive uranium II (^{234}U) is 250,000 years. What percent of the present amount of radioactive uranium II will remain after 5000 years?

127. *Compound Interest* A deposit of $10,000 is made in a savings account for which the interest is compounded continuously. The balance will double in 12 years.

(a) What is the annual interest rate for this account?

(b) Find the balance after 1 year.

(c) The *effective yield* of a savings plan is the percent increase in the balance after 1 year. Find the effective yield.

128. *Test Scores* The test scores for a biology test follow a normal distribution modeled by

$$y = 0.0499e^{-(x-71)^2/128}.$$

(a) Use a graphing utility to graph the equation.

(b) From the graph, estimate the average test score.

129. *Typing Speed* In a typing class, the average number of words per minute N typed after t weeks of lessons was found to be

$$N = \frac{157}{1 + 5.4e^{-0.12t}}.$$

Find the time necessary to type (a) 50 words per minute and (b) 75 words per minute.

130. *Earthquake Magnitude* On the Richter scale, the magnitude R of an earthquake I is

$$R = \log_{10} \frac{I}{I_0}$$

where $I_0 = 1$ is the minimum intensity used for comparison. Find the intensity per unit of area for the following values of R.

(a) $R = 8.4$ (b) $R = 6.85$ (c) $R = 9.1$

In Exercises 131–134, find the exponential function $y = Ae^{bx}$ that passes through the two points.

131. $(0, 2), (4, 3)$ **132.** $(0, 2), (5, 1)$

133. $\left(0, \frac{1}{2}\right), (5, 5)$ **134.** $(0, 4), \left(5, \frac{1}{2}\right)$

135. *Exponential Model* Use a graphing utility to find an exponential model $y = ab^x$ through the points $(0, 250)$, $(4, 135)$, $(6, 92)$, and $(10, 67)$. Graph the data and exponential model in the same viewing window.

136. *Net Profit* The table shows the net profit P (in millions of dollars) of the Bristol-Myers Squibb Company for 1990 through 1998. (Source: Bristol-Myers Squibb Company)

Year	1990	1991	1992	1993	1994
P	1748.0	2056.0	2108.0	2269.0	2330.6

Year	1995	1996	1997	1998
P	2600.0	2850.0	3205.0	3141.0

(a) Make a scatter plot of the data. Let t represent the year, with $t = 0$ corresponding to 1990.

(b) Use a graphing utility to find an exponential model $P = ab^t$ for the data. Let $t = 0$ represent 1990.

Synthesis

True or False? **In Exercises 137–142, determine whether the equation or statement is true or false. Justify your answer.**

137. $\log_b b^{2x} = 2x$ **138.** $e^{x-1} = \dfrac{e^x}{e}$

139. $\ln(x + y) = \ln x + \ln y$

140. $\ln(x + y) = \ln(xy)$

141. $\log_{10}\left(\dfrac{10}{x}\right) = 1 - \log_{10} x$

142. The domain of the function $f(x) = \ln x$ is the set of all real numbers.

Chapter Project *A Graphical Approach to Compound Interest*

In this project, you will use a graphing utility to compare savings plans. For instance, suppose you are depositing $1000 in a savings account and are given the following options.

- 6.2% annual interest rate, compounded annually
- 6.1% annual interest rate, compounded quarterly
- 6.0% annual interest rate, compounded continuously

a. For each option, write a function that gives the balance as a function of the time t (in years).

b. Graph all three functions in the same viewing window. Can you find a viewing window that distinguishes among the graphs of the three functions? If so, describe the viewing window.

c. Find the balances for the three options after 25, 50, 75, and 100 years. Is the option that yields the greatest balance after 25 years the same option that yields the greatest balances after 50, 75, and 100 years? Explain.

d. The *effective yield* of a savings plan is the percent increase in the balance after 1 year. Find the effective yields for the three options listed above. How can the effective yield be used to decide which option is best?

Questions for Further Exploration

1. You deposit $25,000 in an account to accrue interest for 40 years. The account pays 4% compounded annually. Assume that the income tax on the earned interest is 30%. Which of the following plans produces a larger balance after all income tax is paid?

   (a) *Deferred*   The income tax on the interest that is earned is paid in one lump sum at the end of 40 years.

   (b) *Not Deferred*   The income tax on the interest that is earned each year is paid at the end of each year.

2. Which of the following would produce a larger balance? Explain.

   (a) 4.02% annual interest rate, compounded monthly

   (b) 4% annual interest rate, compounded continuously

3. You deposit $1000 in each of two savings accounts. The interest for the accounts is paid according to the two options described in Question 2. How long would it take for the balance in one of the accounts to exceed the balance in the other account by $100? By $100,000?

4. No income tax is due on the interest earned in some types of investments. You deposit $25,000 into an account. Which of the following plans is better? Explain.

   (a) *Tax-Free*   The account pays 5%, compounded annually. There is no income tax on the earned interest.

   (b) *Tax-Deferred*   The account pays 7%, compounded annually. At maturity, the earned interest is taxable at a rate of 40%.

3 Chapter Test

Take this test as you would take a test in class. After you are done, check your work against the answers in the back of the book.

The *Interactive* CD-ROM and *Internet* versions of this text provide answers to the Chapter Tests and Cumulative Tests. They also offer Chapter Pre-Tests (that test key skills and concepts covered in previous chapters) and Chapter Post-Tests, both of which have randomly generated exercises with diagnostic capabilities.

1. Sketch the graph of the function $f(x) = 2^{-x+1} - 3$ by hand.

2. Use a graphing utility to graph f and determine its horizontal asymptotes.
$$f(x) = \frac{1000}{1 + 4e^{-0.2x}}$$

3. Determine the principal that will yield $200,000 when invested at 8% compounded daily for 20 years.

4. Write the logarithmic equation $\log_4 64 = 3$ in exponential form.

5. Sketch the graph of the function $g(x) = \log_3(x - 2)$ by hand.

6. Use the properties of logarithms to expand $\ln\left(\dfrac{6x^2}{\sqrt{x^2 + 1}}\right)$.

In Exercises 7 and 8, evaluate the expression without using a calculator.

7. $\log_5 25$

8. $-2 \ln e^2 + 1$

In Exercises 9–11, solve the equation. Round to three decimal places.

9. $8 + \frac{1}{4}e^{x/2} = 450$

10. $\left(1 + \dfrac{0.06}{4}\right)^{4t} = 3$

11. $3.6 - 5 \ln(x + 4) = 22$

12. A truck that costs $28,000 new has a depreciated value of $20,000 after 1 year. Find the value of the truck when it is 3 years old by using the exponential model $y = ae^{bx}$.

13. The average time between incoming calls at a switchboard is 3 minutes. The probability of waiting less than t minutes for the next incoming call is approximated by the model $F(t) = 1 - e^{-t/3}$. If a call has just come in, find the probability that the next call will come within (a) $\frac{1}{2}$ minute, (b) 2 minutes, and (c) 5 minutes.

14. The population of a certain species t years after it is introduced into a new habitat is $p(t) = \dfrac{1200}{1 + 3e^{-t/5}}$. (a) Determine the initial size of the population. (b) Determine the population after 5 years. (c) After how many years will the population be 800?

15. By observation, identify the equation that corresponds to the graph shown at the right. Explain your reasoning.

   (a) $y = 6e^{-x^2/2}$ (b) $y = \dfrac{6}{1 + e^{-x/2}}$ (c) $y = 6(1 - e^{-x^2/2})$

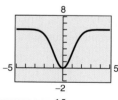

FIGURE FOR 15

16. The numbers of cellular-phone subscribers y (in millions) for the years 1993 through 1997 are $(3, 16.0)$, $(4, 24.1)$, $(5, 33.8)$, $(6, 44.0)$, and $(7, 55.3)$, where x is the time (in years) and $x = 3$ corresponds to 1993. Use a graphing utility to fit an exponential model to the data. Sketch a scatter plot of the data and graph the model in the same viewing window. (Source: Cellular Telecommunications Industry Association)

1–3 Cumulative Test

Take this test to review the material from earlier chapters. After you are done, check your work against the answers in the back of the book.

1. Does the graph at the right represent y as a function of x? Explain.

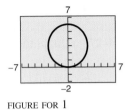

FIGURE FOR 1

In Exercises 2 and 3, evaluate the function at each specified value of the independent variable and simplify.

2. $f(x) = \dfrac{x}{x - 2}$

    (a) $f(6)$ (b) $f(2)$ (c) $f(s + 2)$

3. $f(x) = \begin{cases} 3x - 8, & x \le -\frac{5}{3} \\ 3x^2 + 9x - 8, & x > -\frac{5}{3} \end{cases}$

    (a) $f\left(-\frac{5}{3}\right)$ (b) $f(-1)$ (c) $f(0)$

4. Use a graphing utility to graph the function $f(x) = 2|x - 5| - |x + 5|$. Then determine the open intervals over which the function is increasing, decreasing, or constant.

5. Use a graphing utility to graph the function $f(x) = 2x^3 - x + 5$. Approximate any relative minimum or maximum values of the function.

6. Compare the graphs of each function with the graph of $y = \sqrt[3]{x}$.

    (a) $r(x) = \frac{1}{2}\sqrt[3]{x}$ (b) $h(x) = \sqrt[3]{x} + 2$ (c) $g(x) = \sqrt[3]{x + 2}$

In Exercises 7–10, evaluate the indicated function for

$$f(x) = -x^2 + 3x - 10 \quad \text{and} \quad g(x) = 4x + 1.$$

7. $(f + g)(-4)$ **8.** $(g - f)\left(\frac{3}{4}\right)$ **9.** $(g \circ f)(-2)$ **10.** $(fg)(-1)$

11. Find the domain of the composition $(f \circ g)(x)$ for the functions $f(x) = \sqrt{x}$ and $g(x) = \dfrac{1}{x^2 - 1}$.

12. Find the inverse of $f(x) = \sqrt{x - 2}$, if it exists.

13. Identify the vertex and intercepts of the graph of $y = 2x^2 - 3x - 5$.

14. Find the quadratic function whose graph has a vertex at $(-2, 3)$ and passes through the point $(-4, 9)$.

15. Show that $(x + 3)$ is a factor of $f(x) = 2x^4 + 5x^3 + 7x - 6$.

16. Plot the complex number $-5 + 4i$ in the complex plane.

In Exercises 17–20, perform the operation.

17. $(1 - 3i) - (2i - 3)$ **18.** $\sqrt{-20}\left(\sqrt{18} - \sqrt{-12}\right)$

19. $\dfrac{3i + 7}{1 - 3i}$ **20.** $(6 - i)(5i + 4)$

21. Find a polynomial function with integer coefficients that has $-1, 3, 3,$ and $4i$ as its zeros.

22. Find all the zeros of the function $f(x) = 2x^6 + x^5 - 13x^4 + 18x^3 + 12x^2 - 63x + 27$. Write the polynomial as a product of linear factors.

23. Sketch the graph of $f(x) = \dfrac{2x - 1}{x^2 + 3x - 10}$. Identify the domain, x-intercepts, and vertical and horizontal asymptotes.

In Exercises 24–27, evaluate the expression without using a calculator.

24. $\log_2 64$ **25.** $\log_2\left(\frac{1}{16}\right)$ **26.** $\ln e^{10}$ **27.** $\ln \dfrac{1}{e^3}$

In Exercises 28–31, use a calculator to evaluate the expression. Round your answer to three decimal places.

28. $(1.85)^{3.1}$ **29.** $58^{\sqrt{5}}$ **30.** $e^{-20/11}$ **31.** $4e^{2.56}$

In Exercises 32–35, sketch the graph of the function by hand. Use a graphing utility to verify the graph.

32. $f(x) = -3^{x+4} - 5$ **33.** $f(x) = -\left(\frac{1}{2}\right)^{-x} - 3$
34. $f(x) = 4 + \log_{10}(x - 3)$ **35.** $f(x) = \ln(4 - x)$

In Exercises 36–39, evaluate the logarithm using the change-of-base formula. Round your answer to three decimal places.

36. $\log_5 21$ **37.** $\log_9 6.8$ **38.** $\log_{3/4}(8.61)$ **39.** $\log_{7/8}\left(\frac{3}{2}\right)$

40. Write $2 \ln x - \frac{1}{2}\ln(x + 5)$ as a logarithm of a single quantity.

41. Expand the logarithmic expression: $\ln\left(x\sqrt[3]{x - 5}\right)$.

In Exercises 42–47, solve the equation algebraically. Round your result to three decimal places. Use a graphing utility to verify your answer graphically.

42. $6e^{2x} = 72$ **43.** $4e^{x-3} + 21 = 30$ **44.** $\log_2 x + \log_2 5 = 6$
45. $-3 + \ln 4x = 0$ **46.** $\ln\sqrt{x + 2} = 3$ **47.** $\ln 4x^2 = 7$

48. A rectangular plot of land with a perimeter of 546 feet has a width of x.

(a) Express the area of the plot as a function of x.

(b) Use a graphing utility to graph the area function. What is the domain of the function?

(c) Approximate the dimensions of the plot if its area is 15,000 square feet.

49. The numbers of subscribers of cellular phones y (in millions) for the years 1990 through 1997 are

$(0, 5.3), (1, 7.6), (2, 11.0), (3, 16.0), (4, 24.1), (5, 33.8), (6, 44.0), (7, 55.3)$

where x is the time (in years) with $x = 0$ corresponding to 1990. Use the regression capabilities of a graphing utility to fit an exponential model to the data. Sketch a scatter plot and graph the model in the same viewing window. (Source: Cellular Telecommunications Industry Association)

Trigonometric Functions

The Big Picture

In this chapter you will learn how to

❏ describe an angle and convert between degree and radian measures.
❏ identify a unit circle and its relationship to real numbers.
❏ evaluate trigonometric functions of any angle.
❏ use fundamental trigonometric identities.
❏ sketch graphs of trigonometric functions.
❏ evaluate inverse trigonometric functions.
❏ evaluate the composition of trigonometric functions.
❏ use trigonometric functions to model and solve real-life problems.

The Mauna Loa Observatory in Hawaii conducts research to understand the global carbon cycle. It is located far from pollution sources that would affect the gases being measured. (Source: NOAA/Climate Monitoring and Diagnostic Laboratory)

Ron Watts/CORBIS

Important Vocabulary

As you encounter each new vocabulary term in this chapter, add the term and its definition to your notebook glossary.

Additional Resources Text-specific additional resources are available to help you do well in this course. See page xvi for details.

4.1 Radian and Degree Measure

Angles

As derived from the Greek language, the word **trigonometry** means *"measurement of triangles."* Initially, trigonometry dealt with relationships among the sides and angles of triangles and was used in the development of astronomy, navigation, and surveying. With the development of calculus and the physical sciences in the 17th century, a different perspective arose—one that viewed the classic trigonometric relationships as *functions* with the set of real numbers as their domains. Consequently, the applications of trigonometry expanded to include a vast number of physical phenomena involving rotations and vibrations, including sound waves, light rays, planetary orbits, vibrating strings, pendulums, and orbits of atomic particles. The approach in this text incorporates *both* perspectives, starting with angles and their measure.

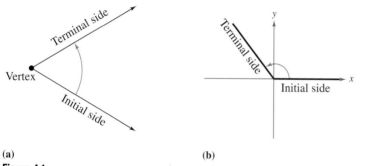

(a) **(b)**

Figure 4.1

Bob Martin/Allsport

An **angle** is determined by rotating a ray (half-line) about its endpoint. The starting position of the ray is the **initial side** of the angle, and the position after rotation is the **terminal side,** as shown in Figure 4.1(a). The endpoint of the ray is the **vertex** of the angle. This perception of an angle fits a coordinate system in which the origin is the vertex and the initial side coincides with the positive *x*-axis. Such an angle is in **standard position,** as shown in Figure 4.1(b). **Positive angles** are generated by counterclockwise rotation, and **negative angles** by clockwise rotation, as shown in Figure 4.2. Angles are labeled with Greek letters α (alpha), β (beta), and θ (theta), as well as uppercase letters *A*, *B*, and *C*. In Figure 4.3, note that angles α and β have the same initial and terminal sides. Such angles are **coterminal.**

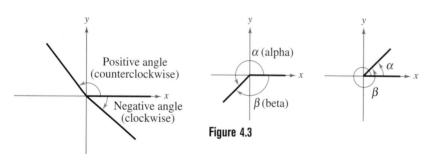

Figure 4.3

Figure 4.2

Radian Measure

One way to measure angles is in radians. This type of measure is especially useful in calculus. To define a radian, you can use a **central angle** of a circle, one whose vertex is the center of the circle, as shown in Figure 4.4.

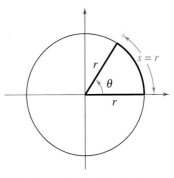

Arc Length = radius when θ = 1 radian.
Figure 4.4

> ### Definition of a Radian
>
> One **radian** is the measure of a central angle θ that intercepts an arc s equal in length to the radius r of the circle. See Figure 4.4.

Because the circumference of a circle is $2\pi r$, it follows that a central angle of one full revolution (counterclockwise) corresponds to an arc length of $s = 2\pi r$. Therefore, 2π radians corresponds to $360°$, π radians corresponds to $180°$, and $\pi/2$ radians corresponds to $90°$. Moreover, because $2\pi \approx 6.28$, there are just over six radius lengths in a full circle, as shown in Figure 4.5. In general, the radian measure of a central angle θ is obtained by dividing the arc length s by r. That is, $s/r = \theta$, where θ is *measured in radians*. Because the units of measure for s and r are the same, this ratio is unitless—it is simply a real number.

Because the radian measure of an angle of one full revolution is 2π, you can obtain the following.

$$\frac{1}{2}\text{ revolution} = \frac{2\pi}{2} = \pi \text{ radians}$$

$$\frac{1}{4}\text{ revolution} = \frac{2\pi}{4} = \frac{\pi}{2} \text{ radians}$$

$$\frac{1}{6}\text{ revolution} = \frac{2\pi}{6} = \frac{\pi}{3} \text{ radians}$$

These and other common angles are shown in Figure 4.6.

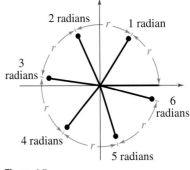

Figure 4.5

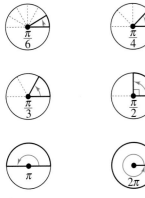

Figure 4.6

> ## STUDY T!P
>
> One revolution around a circle of radius r corresponds to an angle of 2π radians because
>
> $$\frac{s}{r} = \frac{2\pi r}{r} = 2\pi \text{ radians.}$$

Recall that the four quadrants in a coordinate system are numbered I, II, III, and IV. Figure 4.7 shows which angles between 0 and 2π lie in each of the four quadrants. Note that angles between 0 and $\pi/2$ are acute and that angles between $\pi/2$ and π are obtuse.

$$\theta = \frac{\pi}{2}$$

Quadrant II	Quadrant I
$\frac{\pi}{2} < \theta < \pi$	$0 < \theta < \frac{\pi}{2}$

$\theta = \pi$ ——————————————→ $\theta = 0$

Quadrant III	Quadrant IV
$\pi < \theta < \frac{3\pi}{2}$	$\frac{3\pi}{2} < \theta < 2\pi$

$$\theta = \frac{3\pi}{2}$$

Figure 4.7

> ## STUDY T!P
>
> The phrase "the terminal side of θ lies in a quadrant" is often abbreviated by simply saying "θ lies in a quadrant." The terminal sides of the "quadrant angles" 0, $\pi/2$, π, and $3\pi/2$ do not lie within quadrants.

Two angles are coterminal if they have the same initial and terminal sides. For instance, the angles 0 and 2π are coterminal, as are the angles $\pi/6$ and $13\pi/6$. You can find an angle that is coterminal to a given angle θ by adding or subtracting 2π (one revolution), as demonstrated in Example 1. A given angle θ has infinitely many coterminal angles. For instance, $\theta = \pi/6$ is coterminal with

$$\frac{\pi}{6} + 2n\pi, \text{ where } n \text{ is an integer.}$$

EXAMPLE 1 Sketching and Finding Coterminal Angles

a. For the positive angle $13\pi/6$, subtract 2π to obtain a coterminal angle

$$\frac{13\pi}{6} - 2\pi = \frac{\pi}{6}. \qquad \text{See Figure 4.8(a).}$$

b. For the positive angle $3\pi/4$, subtract 2π to obtain a coterminal angle

$$\frac{3\pi}{4} - 2\pi = -\frac{5\pi}{4}. \qquad \text{See Figure 4.8(b).}$$

c. For the negative angle $-2\pi/3$, add 2π to obtain a coterminal angle

$$-\frac{2\pi}{3} + 2\pi = \frac{4\pi}{3}. \qquad \text{See Figure 4.8(c).}$$

The *Interactive* CD-ROM and *Internet* versions of this text show every example with its solution; clicking on the *Try It!* button brings up similar problems. Guided Examples and Integrated Examples show step-by-step solutions to additional examples. Integrated Examples are related to several concepts in the section.

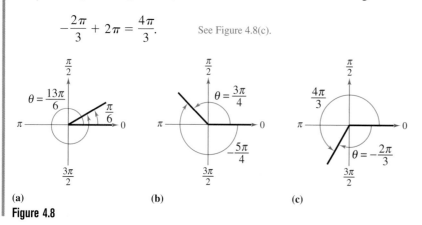

(a)   **(b)**   **(c)**

Figure 4.8

Two positive angles α and β are **complementary** (complements of each other) if their sum is $\pi/2$. Two positive angles are **supplementary** (supplements of each other) if their sum is π. See Figure 4.9.

Complimentary angles
Figure 4.9   *Supplementary angles*

EXAMPLE 2   Complementary and Supplementary Angles

If possible, find the complement and the supplement of (a) $2\pi/5$ and (b) $4\pi/5$.

Solution

a. The complement of $2\pi/5$ is

$$\frac{\pi}{2} - \frac{2\pi}{5} = \frac{5\pi}{10} - \frac{4\pi}{10} = \frac{\pi}{10}.$$

The supplement of $2\pi/5$ is

$$\pi - \frac{2\pi}{5} = \frac{3\pi}{5}.$$

b. Because $4\pi/5$ is greater than $\pi/2$, it has no complement. (Remember to use only *positive* angles for complements.) The supplement is

$$\pi - \frac{4\pi}{5} = \frac{\pi}{5}.$$

Degree Measure

A second way to measure angles is in terms of degrees. A measure of **one degree (1°)** is equivalent to a rotation of $1/360$ of a complete revolution about the vertex. To measure angles, it is convenient to mark degrees on the circumference of a circle, as shown in Figure 4.10. So, a full revolution (counterclockwise) corresponds to 360°, a half revolution to 180°, a quarter revolution to 90°, and so on.

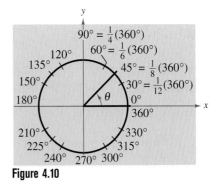

Figure 4.10

Because 2π radians corresponds to one complete revolution, degrees and radians are related by the equations

$$360° = 2\pi \text{ rad} \quad \text{and} \quad 180° = \pi \text{ rad}.$$

From the latter equation, you obtain

$$1° = \frac{\pi}{180} \text{ rad} \quad \text{and} \quad 1 \text{ rad} = \frac{180°}{\pi}$$

which lead to the conversion rules at the top of the next page.

Conversions Between Degrees and Radians

1. To convert degrees to radians, multiply degrees by $\dfrac{\pi \, \text{rad}}{180°}$.

2. To convert radians to degrees, multiply radians by $\dfrac{180°}{\pi \, \text{rad}}$.

To apply these two conversion rules, use the basic relationship $\pi \, \text{rad} = 180°$. (See Figure 4.11.)

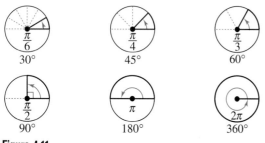

$\dfrac{\pi}{6}$
30°

$\dfrac{\pi}{4}$
45°

$\dfrac{\pi}{3}$
60°

$\dfrac{\pi}{2}$
90°

π
180°

2π
360°

Figure 4.11

When no units of angle measure are specified, *radian measure is implied*. For instance, if you write $\theta = \pi$ or $\theta = 2$, you should mean $\theta = \pi$ radians or $\theta = 2$ radians.

EXAMPLE 3   Converting from Degrees to Radians

Express each angle in radian measure.

a. 135°   **b.** −270°

Solution

a. $135° = (135 \deg)\left(\dfrac{\pi \, \text{rad}}{180 \deg}\right) = \dfrac{3\pi}{4} \, \text{rad}$ Multiply by $\frac{\pi}{180}$.

b. $-270° = (-270 \deg)\left(\dfrac{\pi \, \text{rad}}{180 \deg}\right) = -\dfrac{3\pi}{2} \, \text{rad}$ Multiply by $\frac{\pi}{180}$.

EXAMPLE 4   Converting from Radians to Degrees

Express each angle in degrees.

a. $-\dfrac{\pi}{2} \, \text{rad}$ **b.** 2 rad

Solution

a. $-\dfrac{\pi}{2} \, \text{rad} = \left(-\dfrac{\pi}{2} \, \text{rad}\right)\left(\dfrac{180 \deg}{\pi \, \text{rad}}\right) = -90°$ Multiply by $\frac{180}{\pi}$.

b. $2 \, \text{rad} = (2 \, \text{rad})\left(\dfrac{180 \deg}{\pi \, \text{rad}}\right) = \dfrac{360}{\pi} \approx 114.59°$ Multiply by $\frac{180}{\pi}$.

Applications

The *radian measure* formula $\theta = s/r$ can be used to measure arc length along a circle. Specifically, for a circle of radius r, a central angle θ intercepts an arc of length s given by

$$s = r\theta \qquad \text{Length of circular arc}$$

where θ is measured in radians.

A computer simulation to accompany this concept appears in the *Interactive* CD-ROM and *Internet* versions of this text.

EXAMPLE 5   Finding Arc Length

A circle has a radius of 4 inches. Find the length of the arc intercepted by a central angle of 240°, as shown in Figure 4.12.

Solution

To use the formula $s = r\theta$, first convert 240° to radian measure.

$$240° = (240 \text{ deg})\left(\frac{\pi \text{ rad}}{180 \text{ deg}}\right) \qquad \text{Convert from degrees to radians.}$$

$$= \frac{4\pi}{3} \text{ radians} \qquad \text{Simplify.}$$

Then, using a radius of $r = 4$ inches, you can find the arc length to be

$$s = r\theta \qquad \text{Length of circular arc}$$

$$= 4\left(\frac{4\pi}{3}\right) \qquad \text{Substitute for } r \text{ and } \theta.$$

$$= \frac{16\pi}{3} \qquad \text{Simplify.}$$

$$\approx 16.76 \text{ inches.} \qquad \text{Use a calculator.}$$

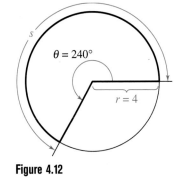

Figure 4.12

Note that the units for $r\theta$ are determined by the units for r because θ is given in radian measure and therefore has no units.

The formula for the length of a circular arc can be used to analyze the motion of a particle moving at a *constant speed* along a circular path.

Linear and Angular Speed

Consider a particle moving at a constant speed along a circular arc of radius r. If s is the length of the arc traveled in time t, then the **linear speed** of the particle is

$$\text{Linear speed} = \frac{\text{arc length}}{\text{time}} = \frac{s}{t}.$$

Moreover, if θ is the angle (in radian measure) corresponding to the arc length s, the **angular speed** of the particle is

$$\text{Angular speed} = \frac{\text{central angle}}{\text{time}} = \frac{\theta}{t}.$$

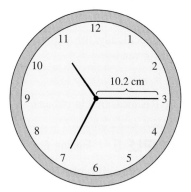

EXAMPLE 6   Finding Linear Speed

The second hand of a clock is 10.2 centimeters long, as shown in Figure 4.13. Find the linear speed of the tip of this second hand.

Solution

In one revolution, the arc length traveled is

$$s = 2\pi r$$

$$= 2\pi(10.2)$$

$$= 20.4\pi \text{ centimeters.}$$

The time required for the second hand to travel this distance is

$$t = 60 \text{ seconds} = 1 \text{ minute.}$$

So, the linear speed of the tip of the second hand is

$$\text{Linear speed} = \frac{s}{t}$$

$$= \frac{20.4\pi \text{ centimeters}}{60 \text{ seconds}}$$

$$\approx 1.068 \text{ cm/sec.}$$

Figure 4.13

EXAMPLE 7   Finding Angular and Linear Speed

A 10 inch radius lawn roller makes 1.2 revolutions per second, as shown in Figure 4.14.

a. Find the angular speed of the roller in radians per second.

b. Find the speed of the tractor that is pulling the roller.

Solution

a. Because each revolution generates 2π radians, it follows that the roller turns $(1.2)(2\pi) = 2.4\pi$ radians per second. In other words, the angular speed is

$$\text{Angular speed} = \frac{\theta}{t}$$

$$= \frac{2.4\pi \text{ radians}}{1 \text{ second}}$$

$$= 2.4\pi \text{ rad/sec.}$$

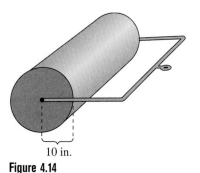

Figure 4.14

b. The linear speed is

$$\text{Linear speed} = \frac{s}{t}$$

$$= \frac{r\theta}{t}$$

$$= \frac{10(2.4\pi) \text{ inches}}{1 \text{ second}} \approx 75.4 \text{ in./sec.}$$

4.1 Exercises

In Exercises 1–4, estimate the angle to the nearest one-half radian.

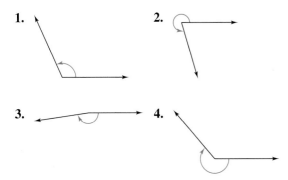

1.

2.

3.

4.

In Exercises 5–10, determine the quadrant in which the angle lies. (The angle is given in radian measure.)

5. (a) $\dfrac{\pi}{5}$ (b) $\dfrac{7\pi}{5}$

6. (a) $\dfrac{7\pi}{4}$ (b) $\dfrac{11\pi}{4}$

7. (a) $-\dfrac{\pi}{12}$ (b) $-\dfrac{11\pi}{9}$

8. (a) -1 (b) -2

9. (a) 3.5 (b) 2.25

10. (a) 5.63 (b) -2.25

In Exercises 11–14, sketch the angle in standard position.

11. (a) $\dfrac{3\pi}{4}$ (b) $\dfrac{5\pi}{3}$

12. (a) $-\dfrac{7\pi}{4}$ (b) $-\dfrac{7\pi}{2}$

13. (a) $\dfrac{11\pi}{6}$ (b) 9π

14. (a) 4 (b) -3

In Exercises 15–18, determine two coterminal angles in radian measure (one positive and one negative) for the given angle.

15. (a)   (b)

16. (a)   (b)

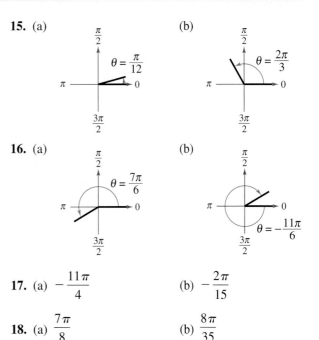

17. (a) $-\dfrac{11\pi}{4}$ (b) $-\dfrac{2\pi}{15}$

18. (a) $\dfrac{7\pi}{8}$ (b) $\dfrac{8\pi}{35}$

In Exercises 19–22, find (if possible) the complement and supplement of the angle.

19. (a) $\dfrac{\pi}{3}$ (b) $\dfrac{3\pi}{4}$

20. (a) $\dfrac{\pi}{12}$ (b) $\dfrac{11\pi}{12}$

21. (a) 1 (b) 2

22. (a) 3 (b) 1.5

In Exercises 23–26, estimate the number of degrees in the angle.

23.   **24.**

25.   **26.**

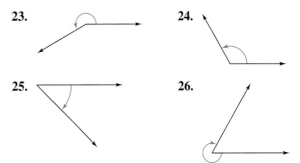

In Exercises 27–30, determine the quadrant in which the angle lies.

27. (a) 150°   (b) 282°

28. (a) 7.9°   (b) 257.5°

29. (a) −132° 50′   (b) −336° 30′

30. (a) −260.25°   (b) −2.4°

In Exercises 31–34, sketch the angle in standard position.

31. (a) 60°   (b) 150°

32. (a) −270°   (b) −120°

33. (a) 405°   (b) 780°

34. (a) −450°   (b) −570°

In Exercises 35–38, determine two coterminal angles in degree measure (one positive and one negative) for the given angle.

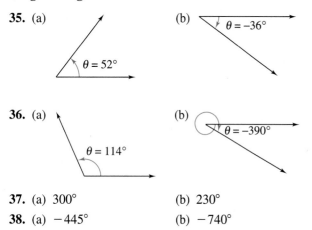

35. (a)   (b) $\theta = -36°$

       $\theta = 52°$

36. (a)   (b) $\theta = -390°$

       $\theta = 114°$

37. (a) 300°   (b) 230°

38. (a) −445°   (b) −740°

In Exercises 39–42, find (if possible) the complement and supplement of the angle.

39. (a) 24°   (b) 126°

40. (a) 87°   (b) 167°

41. (a) 79°   (b) 150°

42. (a) 130°   (b) 170°

In Exercises 43–46, express the angle in radian measure as a multiple of π. (Do not use a calculator.)

43. (a) 30°   (b) 150°

44. (a) 315°   (b) 120°

45. (a) −20°   (b) −240°

46. (a) −270°   (b) 144°

In Exercises 47–54, convert the angle measure from degrees to radians. Round your answer to three decimal places.

47. 115°   **48.** 83.7°

49. −216.35°   **50.** −46.52°

51. 642°   **52.** 0.54°

53. −0.78°   **54.** 395°

In Exercises 55–58, express the angle in degree measure. (Do not use a calculator.)

55. (a) $\dfrac{3\pi}{2}$ (b) $-\dfrac{7\pi}{6}$

56. (a) $-\dfrac{7\pi}{12}$ (b) $\dfrac{\pi}{9}$

57. (a) $\dfrac{7\pi}{3}$ (b) $-\dfrac{13\pi}{60}$

58. (a) $\dfrac{15\pi}{6}$ (b) $\dfrac{28\pi}{15}$

In Exercises 59–66, convert the angle measure from radians to degrees. Round your answer to three decimal places.

59. $\dfrac{\pi}{7}$ **60.** $\dfrac{8\pi}{13}$ **61.** $\dfrac{25\pi}{8}$

62. 6.5π **63.** -4.2π **64.** 4.8

65. −2   **66.** −0.48

In Exercises 67–70, use the angle-conversion capabilities of a graphing utility to convert the angle measure to decimal degree form. Round your answer to three decimal places if necessary.

67. (a) 64° 45′   (b) −124° 30′

68. (a) 275° 10′   (b) 9° 12′

69. (a) 85° 18′ 30″   (b) −408° 16′ 25″

70. (a) −125° 36″   (b) 330° 25″

In Exercises 71–74, use the angle-conversion capabilities of a graphing utility to convert the angle measure to D° M′ S″ form.

71. (a) 280.6°   (b) −115.8°

72. (a) −345.12°   (b) 310.75°

73. (a) 4.5   (b) −3.58

74. (a) −0.355   (b) 0.7865

In Exercises 75–78, find the angle in radians.

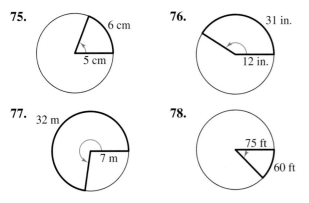

75. 6 cm
    5 cm

76. 31 in.
    12 in.

77. 32 m
    7 m

78. 75 ft
    60 ft

In Exercises 79–82, find the radian measure of the central angle of a circle of radius *r* that intercepts an arc of length *s*.

Radius *r*	Arc Length *s*
79. 15 inches	8 inches
80. 22 feet	10 feet
81. 14.5 centimeters	35 centimeters
82. 80 kilometers	160 kilometers

In Exercises 83–86, find the length of the arc on a circle of radius *r* intercepted by a central angle *θ*.

Radius *r*	Central Angle *θ*
83. 14 inches	180°
84. 9 feet	60°
85. 6 meters	$\dfrac{2\pi}{3}$ radians
86. 40 centimeters	$\dfrac{3\pi}{4}$ radians

Distance Between Cities **In Exercises 87 and 88, find the distance between the cities. Assume that earth is a sphere of radius 4000 miles and the cities are on the same meridian (one city is due north of the other).**

City	Latitude
87. Miami	25° 46′ 37″ N
Erie	42° 7′ 15″ N
88. Johannesburg, South Africa	26° 10′ S
Jerusalem, Israel	31° 47′ N

89. *Difference in Latitudes* Assuming that earth is a sphere of radius 6378 kilometers, what is the differ- ence in latitude of two cities, one of which is 600 kilometers due north of the other?

90. *Difference in Latitudes* Assuming that earth is a sphere of radius 6378 kilometers, what is the differ- ence in latitude of two cities, one of which is 800 kilometers due north of the other?

91. *Instrumentation* The pointer on a voltmeter is 6 centimeters in length. Find the angle through which the pointer rotates when it moves 2.5 centimeters on the scale.

6 cm

92. *Electric Hoist* An electric hoist is used to lift a piece of equipment 1 foot. The diameter of the drum on the hoist is 10 inches. Find the number of degrees through which the drum must rotate.

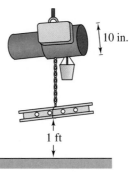

10 in.

1 ft

93. *Figure Skating* The number of revolutions made by a figure skater for each type of axel jump is given. Determine the measure of the angle generated as the skater performs each jump. Give the answer in both degrees and radians.

(a) single axel: $1\frac{1}{2}$

(b) double axel: $2\frac{1}{2}$

(c) triple axel: $3\frac{1}{2}$

94. *Linear Speed of an Earth Satellite* An earth satellite in circular orbit 1250 kilometers high makes one complete revolution every 90 minutes. What is its linear speed? Use 6400 kilometers for the radius of the earth.

95. *Angular Speed* A car is moving at a rate of 40 miles per hour, and the diameter of its wheels is 2.5 feet.

(a) Find the rotational speed of the wheels in revolutions per minute.

(b) Find the angular speed of the wheels in radians per minute.

96. *Circular Saw Speed* The circular blade on a saw has a diameter of 7.5 inches and rotates at 2400 revolutions per minute.

(a) Find the angular speed in radians per second.

(b) Find the linear speed of the saw teeth (in feet per second) as they contact the wood being cut.

———— 7.5 in. ————

97. *Floppy Disk* The radius of the magnetic disk in a 3.5-inch diskette is 1.68 inches. Find the linear speed of a point on the circumference of the disk if it is rotating at a speed of 360 revolutions per minute.

98. *Speed of a Bicycle* The radii of the sprocket assemblies and the wheel of a bicycle are 4 inches, 2 inches, and 14 inches, respectively. If the cyclist is pedaling at a rate of 1 revolution per second, find the speed of the bicycle in (a) feet per second and (b) miles per hour.

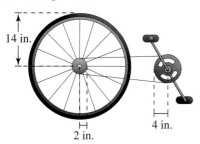

Synthesis

True or False? **In Exercises 99–101, determine whether the statement is true or false. Justify your answer.**

99. A degree is a larger unit of measure than a radian.

100. An angle that measures $-1260°$ lies in Quadrant III.

101. The angles of a triangle can have radian measures $\frac{2\pi}{3}, \frac{\pi}{4}$, and $\frac{\pi}{12}$.

102. *Writing* If the radius of a circle is increasing and the magnitude of a central angle is held constant, how is the length of the intercepted arc changing? Explain your reasoning.

103. *Writing* Write a short paragraph defining and giving examples of coterminal angles.

104. *Geometry* Show that the area of a circular sector of radius r with central angle θ is $A = \frac{1}{2}r^2\theta$, where θ is measured in radians.

Area of a Circular Sector **In Exercises 105 and 106, use the result of Exercise 104 to find the area of the sector.**

105.

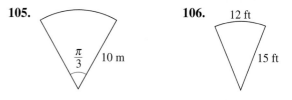

$\frac{\pi}{3}$ / 10 m

106. 12 ft / 15 ft

107. *Graphical Reasoning* The formulas for the area of a circular sector and arc length are $A = \frac{1}{2}r^2\theta$ and $s = r\theta$, respectively. (r is the radius and θ is the angle measured in radians.)

(a) If $\theta = 0.8$, express the area and arc length as functions of r. What is the domain of each function? Graph the functions using a graphing utility. Use the graphs to determine which function changes more rapidly for changes in r when $r > 1$. Explain.

(b) If $r = 10$ centimeters, express the area and arc length as functions of θ. What is the domain of each function? Use a graphing utility to graph and identify the functions.

Review

In Exercises 108–111, sketch the graph of $y = x^5$ and the specified transformations.

108. $f(x) = (x - 2)^5$ **109.** $f(x) = x^5 - 4$

110. $f(x) = 2 - x^5$ **111.** $f(x) = -(x + 3)^5$

In Exercises 112–115, graph the exponential function.

112. $f(x) = 6^x$ **113.** $f(x) = 6^x - 2$

114. $f(x) = 6^{-x}$ **115.** $f(x) = 6^{x+1}$

Trigonometric Functions: The Unit Circle

The Unit Circle

The two historical perspectives of trigonometry incorporate different methods for introducing the trigonometric functions. Our first introduction to these functions is based on the unit circle.

Consider the **unit circle** given by

$$x^2 + y^2 = 1 \qquad \text{Unit circle}$$

as shown in Figure 4.15.

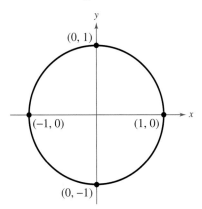

Figure 4.15

Imagine that the real number line is wrapped around this circle, with positive numbers corresponding to a counterclockwise wrapping and negative numbers corresponding to a clockwise wrapping, as shown in Figure 4.16.

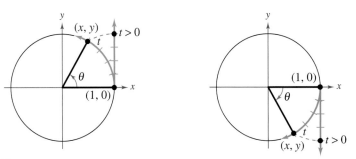

Figure 4.16

As the real number line is wrapped around the unit circle, each real number t corresponds to a point (x, y) on the circle. For example, the real number 0 corresponds to the point $(1, 0)$. Moreover, because the unit circle has a circumference of 2π, the real number 2π also corresponds to the point $(1, 0)$.

In general, each real number t also corresponds to a central angle θ (in standard position) whose radian measure is t. With this interpretation of t, the arc length formula $s = r\theta$ (with $r = 1$) indicates that the real number t is the length of the arc intercepted by the angle θ, given in radians.

What You Should Learn:

- How to identify a unit circle and its relationship to real numbers
- How to evaluate trigonometric functions using the unit circle
- How to use the domain and period to evaluate sine and cosine functions
- How to use a calculator to evaluate trigonometric functions

Why You Should Learn It:

Trigonometric functions are used to model the movement of an oscillating weight. For instance, in Exercise 59 on page 301, the displacement from equilibrium of an oscillating weight suspended by a spring is modeled as a function of time.

Richard Megna/Fundamental Photographs

A computer animation of this concept appears in the *Interactive* CD-ROM and *Internet* versions of this text.

The Trigonometric Functions

From the preceding discussion, it follows that the coordinates x and y are two functions of the real variable t. You can use these coordinates to define the six trigonometric functions of t.

sine	**cosecant**
cosine	**secant**
tangent	**cotangent**

These six functions are normally abbreviated sin, csc, cos, sec, tan, and cot, respectively.

Definitions of Trigonometric Functions

Let t be a real number and let (x, y) be the point on the unit circle corresponding to t.

$$\sin t = y \qquad\qquad \csc t = \frac{1}{y}, \qquad y \neq 0$$

$$\cos t = x \qquad\qquad \sec t = \frac{1}{x}, \qquad x \neq 0$$

$$\tan t = \frac{y}{x}, \qquad x \neq 0 \qquad\qquad \cot t = \frac{x}{y}, \qquad y \neq 0$$

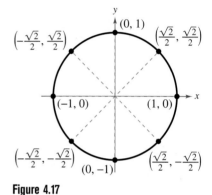

Figure 4.17

Note that the functions in the second column are the *reciprocals* of the corresponding functions in the first column.

In the definitions of the trigonometric functions, note that the tangent and secant are not defined when $x = 0$. For instance, because $t = \pi/2$ corresponds to $(x, y) = (0, 1)$, it follows that $\tan(\pi/2)$ and $\sec(\pi/2)$ are *undefined*. Similarly, the cotangent and cosecant are not defined when $y = 0$. For instance, because $t = 0$ corresponds to $(x, y) = (1, 0)$, cot 0 and csc 0 are *undefined*.

In Figure 4.17, the unit circle has been divided into eight equal arcs, corresponding to t-values of

$$0, \frac{\pi}{4}, \frac{\pi}{2}, \frac{3\pi}{4}, \pi, \frac{5\pi}{4}, \frac{3\pi}{2}, \frac{7\pi}{4}, \text{ and } 2\pi.$$

Similarly, in Figure 4.18, the unit circle has been divided into twelve equal arcs, corresponding to t-values of

$$0, \frac{\pi}{6}, \frac{\pi}{3}, \frac{\pi}{2}, \frac{2\pi}{3}, \frac{5\pi}{6}, \pi, \frac{7\pi}{6}, \frac{4\pi}{3}, \frac{3\pi}{2}, \frac{5\pi}{3}, \frac{11\pi}{6}, \text{ and } 2\pi.$$

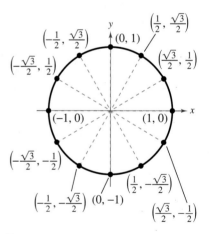

Figure 4.18

Using the (x, y) coordinates in Figures 4.17 and 4.18, you can easily evaluate the trigonometric functions for common t-values. This procedure is demonstrated in Examples 1 and 2.

A computer animation of this concept appears in the *Interactive* CD-ROM and *Internet* versions of this text.

EXAMPLE 1   Evaluating Trigonometric Functions

Evaluate the six trigonometric functions at each real number.

a. $t = \dfrac{\pi}{6}$ **b.** $t = \dfrac{5\pi}{4}$ **c.** $t = 0$ **d.** $t = \pi$

Solution

For each t-value, begin by finding the corresponding point (x, y) on the unit circle. Then use the definitions of trigonometric functions listed on page 296.

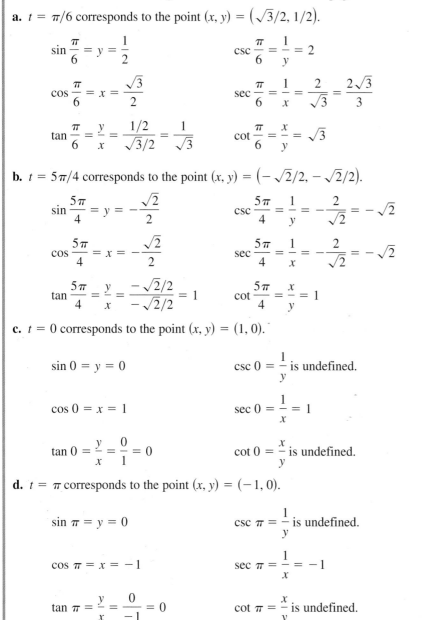

a. $t = \pi/6$ corresponds to the point $(x, y) = \left(\sqrt{3}/2,\, 1/2\right)$.

$$\sin \frac{\pi}{6} = y = \frac{1}{2} \qquad\qquad \csc \frac{\pi}{6} = \frac{1}{y} = 2$$

$$\cos \frac{\pi}{6} = x = \frac{\sqrt{3}}{2} \qquad\qquad \sec \frac{\pi}{6} = \frac{1}{x} = \frac{2}{\sqrt{3}} = \frac{2\sqrt{3}}{3}$$

$$\tan \frac{\pi}{6} = \frac{y}{x} = \frac{1/2}{\sqrt{3}/2} = \frac{1}{\sqrt{3}} \qquad \cot \frac{\pi}{6} = \frac{x}{y} = \sqrt{3}$$

b. $t = 5\pi/4$ corresponds to the point $(x, y) = \left(-\sqrt{2}/2,\, -\sqrt{2}/2\right)$.

$$\sin \frac{5\pi}{4} = y = -\frac{\sqrt{2}}{2} \qquad\qquad \csc \frac{5\pi}{4} = \frac{1}{y} = -\frac{2}{\sqrt{2}} = -\sqrt{2}$$

$$\cos \frac{5\pi}{4} = x = -\frac{\sqrt{2}}{2} \qquad\qquad \sec \frac{5\pi}{4} = \frac{1}{x} = -\frac{2}{\sqrt{2}} = -\sqrt{2}$$

$$\tan \frac{5\pi}{4} = \frac{y}{x} = \frac{-\sqrt{2}/2}{-\sqrt{2}/2} = 1 \qquad \cot \frac{5\pi}{4} = \frac{x}{y} = 1$$

c. $t = 0$ corresponds to the point $(x, y) = (1, 0)$.

$$\sin 0 = y = 0 \qquad\qquad \csc 0 = \frac{1}{y} \text{ is undefined.}$$

$$\cos 0 = x = 1 \qquad\qquad \sec 0 = \frac{1}{x} = 1$$

$$\tan 0 = \frac{y}{x} = \frac{0}{1} = 0 \qquad\qquad \cot 0 = \frac{x}{y} \text{ is undefined.}$$

d. $t = \pi$ corresponds to the point $(x, y) = (-1, 0)$.

$$\sin \pi = y = 0 \qquad\qquad \csc \pi = \frac{1}{y} \text{ is undefined.}$$

$$\cos \pi = x = -1 \qquad\qquad \sec \pi = \frac{1}{x} = -1$$

$$\tan \pi = \frac{y}{x} = \frac{0}{-1} = 0 \qquad\qquad \cot \pi = \frac{x}{y} \text{ is undefined.}$$

EXAMPLE 2   Evaluating Trigonometric Functions

Evaluate the six trigonometric functions at $t = -\dfrac{\pi}{3}$.

Solution

Moving *clockwise* around the unit circle, it follows that $t = -\pi/3$ corresponds to the point $(x, y) = \left(1/2, -\sqrt{3}/2\right)$.

$$\sin\left(-\frac{\pi}{3}\right) = -\frac{\sqrt{3}}{2} \qquad \csc\left(-\frac{\pi}{3}\right) = -\frac{2}{\sqrt{3}}$$

$$\cos\left(-\frac{\pi}{3}\right) = \frac{1}{2} \qquad \sec\left(-\frac{\pi}{3}\right) = 2$$

$$\tan\left(-\frac{\pi}{3}\right) = -\sqrt{3} \qquad \cot\left(-\frac{\pi}{3}\right) = -\frac{1}{\sqrt{3}}$$

Domain and Period of Sine and Cosine

The *domain* of the sine and cosine functions is the set of all real numbers. To determine the *range* of these two functions, consider the unit circle shown in Figure 4.19. Because $r = 1$, it follows that $\sin t = y$ and $\cos t = x$. Moreover, because (x, y) is on the unit circle, you know that $-1 \le y \le 1$ and $-1 \le x \le 1$. So, the values of sine and cosine also range between -1 and 1.

$$\begin{matrix} -1 \le & y & \le 1 \\ -1 \le \sin t \le 1 \end{matrix} \quad \text{and} \quad \begin{matrix} -1 \le & x & \le 1 \\ -1 \le \cos t \le 1 \end{matrix}$$

Adding 2π to each value of t in the interval $[0, 2\pi]$ completes a second revolution around the unit circle, as shown in Figure 4.20. The values of $\sin(t + 2\pi)$ and $\cos(t + 2\pi)$ correspond to those of $\sin t$ and $\cos t$. Similar results can be obtained for repeated revolutions (positive or negative) on the unit circle. This leads to the general result

$$\sin(t + 2\pi n) = \sin t$$

and

$$\cos(t + 2\pi n) = \cos t$$

for any integer n and real number t. Functions that behave in such a repetitive (or cyclic) manner are called **periodic.**

Definition of a Periodic Function

A function f is **periodic** if there exists a positive real number c such that

$$f(t + c) = f(t)$$

for all t in the domain of f. The smallest number c for which f is periodic is called the **period** of f.

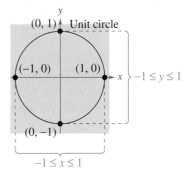

Figure 4.19

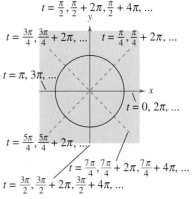

Figure 4.20

EXAMPLE 3   Using the Period to Evaluate the Sine and Cosine

a. Because $\dfrac{13\pi}{6} = 2\pi + \dfrac{\pi}{6}$, you have

$$\sin\frac{13\pi}{6} = \sin\left(2\pi + \frac{\pi}{6}\right) = \sin\frac{\pi}{6} = \frac{1}{2}.$$

b. Because $-\dfrac{7\pi}{2} = -4\pi + \dfrac{\pi}{2}$, you have

$$\cos\left(-\frac{7\pi}{2}\right) = \cos\left(-4\pi + \frac{\pi}{2}\right) = \cos\frac{\pi}{2} = 0.$$

Recall from Section 1.2 that a function f is *even* if $f(-t) = f(t)$, and it is *odd* if $f(-t) = -f(t)$.

> ### Even and Odd Trigonometric Functions
>
> The cosine and secant functions are *even*.
>
> $$\cos(-t) = \cos t \qquad \sec(-t) = \sec t$$
>
> The sine, cosecant, tangent, and cotangent functions are *odd*.
>
> $$\sin(-t) = -\sin t \qquad \csc(-t) = -\csc t$$
> $$\tan(-t) = -\tan t \qquad \cot(-t) = -\cot t$$

Evaluating Trigonometric Functions with a Calculator

When evaluating a trigonometric function with a calculator, you need to set the calculator to the desired *mode* of measurement (degrees or radians).

Most calculators do not have keys for the cosecant, secant, and cotangent functions. To evaluate these functions, you can use the $\boxed{x^{-1}}$ key with their respective reciprocal functions sine, cosine, and tangent. For example, to evaluate $\csc(\pi/8)$, use the fact that

$$\csc\frac{\pi}{8} = \frac{1}{\sin(\pi/8)}$$

and enter the following keystroke sequence in radian mode.

$\boxed{(}\ \boxed{\text{SIN}}\ \boxed{(}\ \boxed{\pi}\ \boxed{\div}\ 8\ \boxed{)}\ \boxed{)}\ \boxed{x^{-1}}\ \boxed{\text{ENTER}}$ Display 2.6131259

EXAMPLE 4   Using a Calculator

Function	Mode	Calculator Keystrokes	Display
a. $\sin 2\pi/3$	Radian	$\boxed{\text{SIN}}\ \boxed{(}\ 2\ \boxed{\pi}\ \boxed{\div}\ 3\ \boxed{)}\ \boxed{\text{ENTER}}$	0.8660254
b. $\cot 1.5$	Radian	$\boxed{(}\ \boxed{\text{TAN}}\ 1.5\ \boxed{(}\ \boxed{x^{-1}}\ \boxed{\text{ENTER}}$	0.0709148

STUDY TIP

From the definition of periodic function it follows that the sine and cosine functions are periodic and have a period of 2π. The other four trigonometric functions are also periodic, and more will be said about that in Section 4.6.

STUDY TIP

When evaluating trigonometric functions with a calculator, remember to enclose all fractional angle measures in parentheses. For instance, if you want to evaluate $\sin\theta$ for $\theta = \pi/6$, you should enter

$\boxed{\text{SIN}}\ \boxed{(}\ \boxed{\pi}\ \boxed{\div}\ 6\ \boxed{)}\ \boxed{\text{ENTER}}$

rather than

$\boxed{\text{SIN}}\ \boxed{\pi}\ \boxed{\div}\ 6\ \boxed{\text{ENTER}}$.

The first set of keystrokes yields the correct value of 0.5. The second set yields the incorrect value of 0.

4.2 Exercises

In Exercises 1–4, determine the exact values of the six trigonometric functions of the angle θ.

1.   **2.**

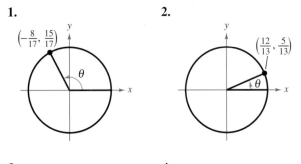

3.   **4.**

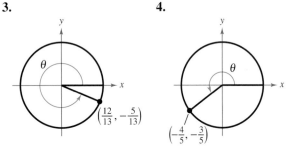

In Exercises 5–12, find the point (x, y) on the unit circle that corresponds to the real number t.

5. $t = \dfrac{\pi}{4}$ **6.** $t = \dfrac{\pi}{3}$

7. $t = \dfrac{7\pi}{6}$ **8.** $t = \dfrac{5\pi}{4}$

9. $t = \dfrac{4\pi}{3}$ **10.** $t = \dfrac{5\pi}{3}$

11. $t = \dfrac{3\pi}{2}$ **12.** $t = \pi$

In Exercises 13–22, evaluate (if possible) the sine, cosine, and tangent of the real number.

13. $t = \dfrac{\pi}{4}$ **14.** $t = \dfrac{\pi}{3}$

15. $t = -\dfrac{\pi}{6}$ **16.** $t = -\dfrac{\pi}{4}$

17. $t = -\dfrac{7\pi}{4}$ **18.** $t = -\dfrac{4\pi}{3}$

19. $t = \dfrac{11\pi}{6}$ **20.** $t = \dfrac{5\pi}{3}$

21. $t = -\dfrac{3\pi}{2}$ **22.** $t = -2\pi$

In Exercises 23–28, evaluate (if possible) the six trigonometric functions of the real number.

23. $t = \dfrac{3\pi}{4}$ **24.** $t = \dfrac{5\pi}{6}$

25. $t = \dfrac{\pi}{2}$ **26.** $t = \dfrac{3\pi}{2}$

27. $t = -\dfrac{\pi}{3}$ **28.** $t = -\dfrac{3\pi}{2}$

In Exercises 29–36, evaluate the trigonometric function using its period as an aid.

29. $\sin 5\pi$ **30.** $\cos 5\pi$

31. $\cos \dfrac{8\pi}{3}$ **32.** $\sin \dfrac{9\pi}{4}$

33. $\cos(-3\pi)$ **34.** $\sin(-3\pi)$

35. $\sin\left(-\dfrac{9\pi}{4}\right)$ **36.** $\cos\left(-\dfrac{8\pi}{3}\right)$

In Exercises 37–42, use the value of the trigonometric function to evaluate the indicated functions.

37. $\sin t = \frac{1}{3}$ **38.** $\sin(-t) = \frac{3}{8}$
    (a) $\sin(-t)$ (a) $\sin t$
    (b) $\csc(-t)$ (b) $\csc t$

39. $\cos(-t) = -\frac{1}{5}$ **40.** $\cos t = -\frac{3}{4}$
    (a) $\cos t$ (a) $\cos(-t)$
    (b) $\sec(-t)$ (b) $\sec(-t)$

41. $\sin t = \frac{4}{5}$ **42.** $\cos t = \frac{4}{5}$
    (a) $\sin(\pi - t)$ (a) $\cos(\pi - t)$
    (b) $\sin(t + \pi)$ (b) $\cos(t + \pi)$

In Exercises 43–52, use a calculator to evaluate the expression. Round to four decimal places.

43. $\sin \dfrac{\pi}{4}$ **44.** $\tan \dfrac{\pi}{3}$

45. $\csc 1.3$ **46.** $\cot 1$

47. $\cos(-1.7)$ **48.** $\cos(-2.5)$

49. $\csc 0.8$ **50.** $\sec 1.8$

51. $\sec 22.8$ **52.** $\sin(-0.9)$

Estimation **In Exercises 53 and 54, use the figure and a straightedge to approximate the value of the trigonometric function.**

53. (a) sin 5   (b) cos 2

54. (a) sin 0.75   (b) cos 2.5

Estimation **In Exercises 55 and 56, use the figure and a straightedge to approximate the solution of the equation, where $0 \leq t < 2\pi$.**

55. (a) sin t = 0.25   (b) cos t = −0.25

56. (a) sin t = −0.75   (b) cos t = 0.75

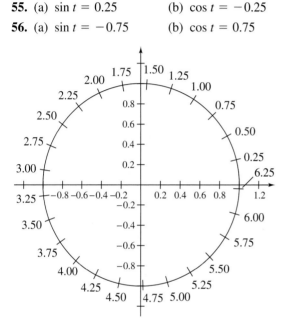

FIGURE FOR 53–56

57. *Electric Circuits* The initial current and charge in the electrical circuit shown in the figure are zero. When 100 volts is applied to the circuit, the current is

$$I = 5e^{-2t} \sin t$$

if the resistance, inductance, and capacitance are 80 ohms, 20 henrys, and 0.01 farads, respectively. Approximate the current $t = 0.7$ seconds after the voltage is applied.

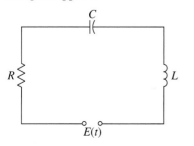

58. *Electrical Circuits* Approximate the current in the electrical circuit in Exercise 57 at $t = 1.4$ seconds after the voltage is applied.

59. *Harmonic Motion* The displacement from equilibrium of an oscillating weight suspended by a spring is

$$y(t) = \frac{1}{4} \cos 6t$$

where y is the displacement in feet and t is the time in seconds (see figure). Find the displacement when (a) $t = 0$, (b) $t = \frac{1}{4}$, and (c) $t = \frac{1}{2}$.

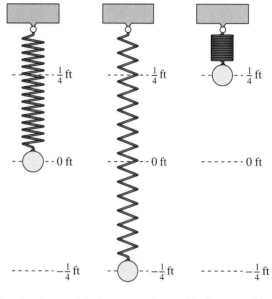

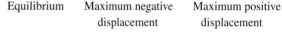

Equilibrium   Maximum negative   Maximum positive
displacement   displacement

Simple Harmonic Motion

60. *Harmonic Motion* The displacement from equilibrium of an oscillating weight suspended by a spring and subject to the damping effect of friction is

$$y(t) = \frac{1}{4}e^{-t} \cos 6t$$

where y is the displacement in feet and t is the time in seconds. Find the displacement when (a) $t = 0$, (b) $t = \frac{1}{4}$, and (c) $t = \frac{1}{2}$.

61. Verify that $\cos 2t \neq 2 \cos t$ by approximating $\cos 1.5$ and $2 \cos 0.75$.

62. Verify that $\sin(t_1 + t_2) \neq \sin t_1 + \sin t_2$ by approximating $\sin 0.25$, $\sin 0.75$, and $\sin 1$.

Synthesis

True or False? **In Exercises 63 and 64, determine whether the statement is true or false. Justify your answer.**

63. Because $\sin(-t) = -\sin t$, it can be said that the sine of a negative angle is a negative number.

64. $\tan a = \tan(a - 6\pi)$

65. *Exploration* Let (x_1, y_1) and (x_2, y_2) be points on the unit circle corresponding to $t = t_1$ and $t = \pi - t_1$, respectively.

(a) Identify the symmetry of the points (x_1, y_1) and (x_2, y_2).

(b) Make a conjecture about any relationship between $\sin t_1$ and $\sin(\pi - t_1)$.

(c) Make a conjecture about any relationship between $\cos t_1$ and $\cos(\pi - t_1)$.

66. *Exploration* Let (x_1, y_1) and (x_2, y_2) be points on the unit circle corresponding to $t = t_1$ and $t = t_1 + \pi$, respectively.

(a) Identify the symmetry of the points (x_1, y_1) and (x_2, y_2).

(b) Make a conjecture about any relationship between $\sin t_1$ and $\sin(t_1 + \pi)$.

(c) Make a conjecture about any relationship between $\cos t_1$ and $\cos(t_1 + \pi)$.

67. Use the unit circle to verify that the cosine and secant functions are even and that the sine, cosecant, tangent, and cotangent functions are odd.

68. *Think About It* Because $f(t) = \sin t$ is an odd function and $g(t) = \cos t$ is an even function, what can be said about the function $h(t) = f(t)g(t)$?

69. *Think About It* Because $f(t) = \sin t$ and $g(t) = \tan t$ are odd functions, what can be said about the function $h(t) = f(t)g(t)$?

Review

In Exercises 70–73, find the inverse of the one-to-one function f. Use a graphing utility to graph both f and f^{-1} in the same viewing window.

70. $f(x) = \frac{1}{2}(3x - 2)$

71. $f(x) = \frac{1}{4}x^3 + 1$

72. $f(x) = \sqrt{x^2 - 4}, \quad x \geq 2$

73. $f(x) = \frac{2x}{x + 1}, \quad x > -1$

In Exercises 74–77, sketch the graph of the rational function. Be sure to label all asymptotes. Use a graphing utility to verify the graph.

74. $f(x) = \frac{2x}{x - 3}$

75. $f(x) = \frac{5x}{x^2 + x - 6}$

76. $f(x) = \frac{x^2 - 3x + 8}{x - 2}$

77. $f(x) = \frac{x^3 - 6x^2 + x - 1}{2x^2 - 5x - 8}$

78. *Average Cost* The average cost (in dollars per pound) of recycling a waste product x (in pounds) is

$$\overline{C}(x) = \frac{450,000 + 5x}{x}, \quad x > 0.$$

Find the average cost of recycling $x = 10,000$ pounds, $x = 100,000$ pounds, and $x = 1,000,000$ pounds. According to this model, what is the limiting average cost as the number of pounds increases?

79. *Inflation* If the inflation rate averages 4.5% over the next 10 years, the approximate cost C of goods or services t years from now is

$$C(t) = P(1.045)^t$$

where P is the present cost. If the price of a tire is presently $69.95, estimate the price 10 years from now.

80. *Population* The population P of a city is $P = 135,200e^{0.018t}$, where $t = 0$ represents 1990. Use a graphing utility to graph the model and determine when the population reaches 190,000. Verify your answer algebraically.

81. *Population* The population p of a certain species t years after it is introduced into a new habitat is given by $p(t) = 1200/(1 + 3e^{-t/5})$.

(a) Determine the population size that was introduced into the habitat.

(b) Determine the population after 5 years.

(c) After how many years will the population be 800?

4.3 Right Triangle Trigonometry

The Six Trigonometric Functions

Our second look at the trigonometric functions is from a *right triangle* perspective. Consider a right triangle, one of whose acute angles is labeled θ, as shown in Figure 4.21. Relative to the angle θ, the three sides of the triangle are the **hypotenuse**, the **opposite side** (the side opposite the angle θ), and the **adjacent side** (the side adjacent to the angle θ).

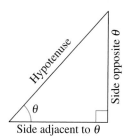

Figure 4.21

Using the lengths of these three sides, you can form six ratios that define the six trigonometric functions of the acute angle θ.

sine   cosecant   cosine   secant   tangent   cotangent

In the following definitions it is important to see that $0° < \theta < 90°$ and that for such angles the value of each trigonometric function is *positive*.

Right Triangle Definitions of Trigonometric Functions

Let θ be an *acute* angle of a right triangle. Then the six trigonometric functions *of the angle* θ are defined as follows.

$$\sin \theta = \frac{\text{opp}}{\text{hyp}} \qquad \cos \theta = \frac{\text{adj}}{\text{hyp}} \qquad \tan \theta = \frac{\text{opp}}{\text{adj}}$$

$$\csc \theta = \frac{\text{hyp}}{\text{opp}} \qquad \sec \theta = \frac{\text{hyp}}{\text{adj}} \qquad \cot \theta = \frac{\text{adj}}{\text{opp}}$$

The abbreviations "opp", "adj", and "hyp" represent the lengths of the three sides of a right triangle.

opp = the length of the side *opposite* θ
adj = the length of the side *adjacent* to θ
hyp = the length of the *hypotenuse*

Note that the functions in the second row above are the *reciprocals* of the corresponding functions in the first row.

What You Should Learn:

- How to evaluate trigonometric functions of acute angles
- How to use the fundamental trigonometric identities
- How to use a calculator to evaluate trigonometric functions
- How to use trigonometric functions to model and solve real-life problems

Why You Should Learn It:

You can use trigonometric functions to model and solve real-life problems. For instance, Exercise 69 on page 312 shows you how trigonometric functions can be used to approximate the height of a buidling.

Chen Yixin/China Stock

A computer animation of this concept appears in the *Interactive* CD-ROM and *Internet* versions of this text.

EXAMPLE 1 Evaluating Trigonometric Functions

Find the exact values of the six trigonometric functions of θ, as shown in Figure 4.22.

Figure 4.22

> **Georg Joachim Rhaeticus (1514–1576),** the leading Teutonic mathematical astronomer of the 16th century, was the first to define the trigonometric functions as ratios of the sides of a right triangle.

Solution

By the Pythagorean Theorem, $(\text{hyp})^2 = (\text{opp})^2 + (\text{adj})^2$, it follows that

$$\text{hyp} = \sqrt{4^2 + 3^2} = \sqrt{25} = 5.$$

So, the six trigonometric functions of θ are

$$\sin \theta = \frac{\text{opp}}{\text{hyp}} = \frac{4}{5} \qquad \cos \theta = \frac{\text{adj}}{\text{hyp}} = \frac{3}{5} \qquad \tan \theta = \frac{\text{opp}}{\text{adj}} = \frac{4}{3}$$

$$\csc \theta = \frac{\text{hyp}}{\text{opp}} = \frac{5}{4} \qquad \sec \theta = \frac{\text{hyp}}{\text{adj}} = \frac{5}{3} \qquad \cot \theta = \frac{\text{adj}}{\text{opp}} = \frac{3}{4}.$$

In Example 1, you were given the lengths of the sides of the right triangle, but not the angle θ. A more common problem in trigonometry is finding the trigonometric functions for a *given* acute angle θ. To do this, you can construct a right triangle having θ as one of its angles.

EXAMPLE 2 Evaluating Trigonometric Functions of 45°

Find the exact values of $\sin 45°$, $\cos 45°$, and $\tan 45°$.

Solution

Construct a right triangle having 45° as one of its acute angles, as shown in Figure 4.23. Choose 1 as the length of the adjacent side. From geometry, you know that the other acute angle is also 45°. So, the triangle is isosceles and the length of the opposite side is also 1. Using the Pythagorean Theorem, you find the length of the hypotenuse to be $\sqrt{2}$.

$$\sin 45° = \frac{\text{opp}}{\text{hyp}} = \frac{1}{\sqrt{2}} = \frac{\sqrt{2}}{2} \qquad \cos 45° = \frac{\text{adj}}{\text{hyp}} = \frac{1}{\sqrt{2}} = \frac{\sqrt{2}}{2}$$

$$\tan 45° = \frac{\text{opp}}{\text{adj}} = \frac{1}{1} = 1$$

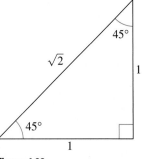

Figure 4.23

EXAMPLE 3   Evaluating Trigonometric Functions of 30° and 60°

Use the equilateral triangle shown in Figure 4.24 to find the exact values of sin 60°, cos 60°, sin 30°, and cos 30°.

Solution

Try using the Pythagorean Theorem and the equilateral triangle to verify the lengths of the sides given in Figure 4.24. For $\theta = 60°$, you have adj = 1, opp = $\sqrt{3}$, and hyp = 2. Therefore,

$$\sin 60° = \frac{\text{opp}}{\text{hyp}} = \frac{\sqrt{3}}{2}$$

and

$$\cos 60° = \frac{\text{adj}}{\text{hyp}} = \frac{1}{2}.$$

For $\theta = 30°$, adj = $\sqrt{3}$, opp = 1, and hyp = 2. So,

$$\sin 30° = \frac{\text{opp}}{\text{hyp}} = \frac{1}{2}$$

and

$$\cos 30° = \frac{\text{adj}}{\text{hyp}} = \frac{\sqrt{3}}{2}.$$

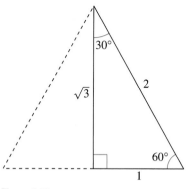

Figure 4.24

Because the angles 30°, 45°, and 60° ($\pi/6$, $\pi/4$, and $\pi/3$) occur frequently in trigonometry, you should learn to construct the triangles shown in Figures 4.23 and 4.24.

Sines, Cosines, and Tangents of Special Angles

$$\sin 30° = \sin \frac{\pi}{6} = \frac{1}{2} \qquad \cos 30° = \cos \frac{\pi}{6} = \frac{\sqrt{3}}{2} \qquad \tan 30° = \tan \frac{\pi}{6} = \frac{\sqrt{3}}{3}$$

$$\sin 45° = \sin \frac{\pi}{4} = \frac{\sqrt{2}}{2} \qquad \cos 45° = \cos \frac{\pi}{4} = \frac{\sqrt{2}}{2} \qquad \tan 45° = \tan \frac{\pi}{4} = 1$$

$$\sin 60° = \sin \frac{\pi}{3} = \frac{\sqrt{3}}{2} \qquad \cos 60° = \cos \frac{\pi}{3} = \frac{1}{2} \qquad \tan 60° = \tan \frac{\pi}{3} = \sqrt{3}$$

A computer animation of this concept appears in the *Interactive* CD-ROM and *Internet* versions of this text.

In the box, note that $\sin 30° = \frac{1}{2} = \cos 60°$. This occurs because 30° and 60° are complementary angles, and, in general, it can be shown from the right triangle definitions that *cofunctions of complementary angles are equal.* That is, if θ is an acute angle, the following relationships are true.

$$\sin(90° - \theta) = \cos \theta \qquad \cos(90° - \theta) = \sin \theta$$

$$\tan(90° - \theta) = \cot \theta \qquad \cot(90° - \theta) = \tan \theta$$

$$\sec(90° - \theta) = \csc \theta \qquad \csc(90° - \theta) = \sec \theta$$

Trigonometric Identities

In trigonometry, a great deal of time is spent studying relationships between trigonometric functions (identities).

Fundamental Trigonometric Identities

Reciprocal Identities

$$\sin \theta = \frac{1}{\csc \theta} \qquad \cos \theta = \frac{1}{\sec \theta} \qquad \tan \theta = \frac{1}{\cot \theta}$$

$$\csc \theta = \frac{1}{\sin \theta} \qquad \sec \theta = \frac{1}{\cos \theta} \qquad \cot \theta = \frac{1}{\tan \theta}$$

Quotient Identities

$$\tan \theta = \frac{\sin \theta}{\cos \theta} \qquad \cot \theta = \frac{\cos \theta}{\sin \theta}$$

Pythagorean Identities

$$\sin^2 \theta + \cos^2 \theta = 1$$

$$1 + \tan^2 \theta = \sec^2 \theta$$

$$1 + \cot^2 \theta = \csc^2 \theta$$

Note that $\sin^2 \theta$ represents $(\sin \theta)^2$ not $\sin(\theta^2)$; $\cos^2 \theta$ represents $(\cos \theta)^2$, not $\cos (\theta^2)$; and so on.

EXAMPLE 4   Applying Trigonometric Identities

Let θ be an acute angle such that $\sin \theta = 0.6$. Find the values of (a) $\cos \theta$ and (b) $\tan \theta$ using trigonometric identities.

Solution

a. To find the value of $\cos \theta$, use the Pythagorean identity

$$\sin^2 \theta + \cos^2 \theta = 1.$$

So, you have

$$(0.6)^2 + \cos^2 \theta = 1$$

$$\cos^2 \theta = 1 - (0.6)^2 = 0.64$$

$$\cos \theta = \sqrt{0.64} = 0.8.$$

b. Now, knowing the sine and cosine of θ, you can find the tangent of θ to be

$$\tan \theta = \frac{\sin \theta}{\cos \theta} = \frac{0.6}{0.8} = 0.75.$$

Try using the definitions of $\cos \theta$ and $\tan \theta$, and the triangle shown in Figure 4.25, to check these results.

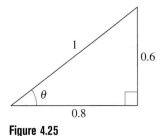

Figure 4.25

EXAMPLE 5 Using Trigonometric Identities

Use trigonometric identities to transform one side of the equation into the other.

a. $\cos \theta \sec \theta = 1$ **b.** $(\sec \theta + \tan \theta)(\sec \theta - \tan \theta) = 1$

Solution

a. $\cos \theta \sec \theta = 1$ Write original equation.

$\left(\dfrac{1}{\sec \theta}\right)\sec \theta = 1$ Reciprocal identity

$1 = 1$ Divide out common factor.

b. $(\sec \theta + \tan \theta)(\sec \theta - \tan \theta) = 1$ Write original equation.

$\sec^2 \theta - \sec\theta \tan \theta + \sec \theta \tan \theta - \tan^2 \theta = 1$ Distributive Property

$\sec^2\theta - \tan^2\theta = 1$ Simplify.

$1 = 1$ Pythagorean identity

Evaluating Trigonometric Functions with a Calculator

To use a calculator to evaluate trigonometric functions of angles measured in degrees, first set the calculator to degree mode and then proceed as demonstrated in Section 4.2.

EXAMPLE 6 Using a Calculator

Use a calculator to evaluate $\sec(5° \, 40' \, 12'')$.

Solution

Begin by converting to decimal form.

$$5° \, 40' \, 12'' = 5° + \left(\frac{40}{60}\right)° + \left(\frac{12}{3600}\right)° = 5.67°$$

Then use a calculator to evaluate $\sec 5.67°$.

$$\sec(5° \, 40' \, 12'') = \sec 5.67° = \frac{1}{\cos 5.67°} \approx 1.00492$$

STUDY TIP

Remember that throughout this text, it is assumed that angles are measured in radians unless noted otherwise. For example, sin 1 means the sine of 1 radian and sin 1° means the sine of one degree.

Applications Involving Right Triangles

Many applications of trigonometry involve a process called **solving right triangles.** In this type of application, you are usually given one side of a right triangle and one of the acute angles and asked to find one of the other sides, *or* you are given two sides and asked to find one of the acute angles. In Example 7, the angle you are given is the **angle of elevation**, which denotes the angle from the horizontal upward to the object. In other applications you may be given the **angle of depression**, which denotes the angle from the horizontal downward to the object.

EXAMPLE 7 Solving a Right Triangle

A surveyor is standing 50 feet from the base of a large tree, as shown in Figure 4.26. The surveyor measures the angle of elevation to the top of the tree as 71.5°. How tall is the tree?

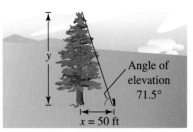

Figure 4.26

Algebraic Solution

From Figure 4.26, you can see that

$$\tan 71.5° = \frac{\text{opp}}{\text{adj}}$$

$$= \frac{y}{x}$$

where $x = 50$ and y is the height of the tree. So, the height of the tree is

$$y = x \tan 71.5°$$

$$\approx 50(2.98868)$$

$$\approx 149.4 \text{ feet.}$$

Numerical Solution

Use a graphing utility to create a table, as shown in Figure 4.27(a). The first column is the base of the triangle, $x = 50$. The second column is the height y, which varies from 10 to 200. The quotient y/x is shown in the third column. Because $\tan 71.5 = y/x \approx 2.9887$, you can compare 2.9887 to the value of y/x in the table. Because 2.9887 is between 2.8 and 3, the height y is between 140 and 150. Create a second table that shows the value of y/x when the values of y increase from 140 to 150, as shown in Figure 4.27(b). Because 2.9887 is between 2.98 and 3, the height is between 149 and 150 feet. Repeating the process to create a table that shows values of y/x when the values of y increase from 149.1 to 150 results in an estimate of the tree height of 149.4 feet.

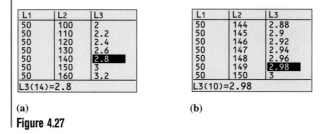

(a)   (b)

Figure 4.27

EXAMPLE 8 Solving a Right Triangle

A person is 200 yards from a river. Rather than walk directly to the river, the person walks 400 yards along a straight path to the river's edge. Find the acute angle θ between this path and the river's edge, as indicated in Figure 4.28.

Solution

From Figure 4.28, you can see that the sine of the angle θ is

$$\sin \theta = \frac{\text{opp}}{\text{hyp}} = \frac{200}{400} = \frac{1}{2}.$$

Now, you should recognize that $\theta = 30°$.

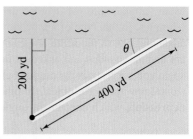

Figure 4.28

In Example 8, you were able to recognize that $\theta = 30°$ is the acute angle that satisfies the equation $\sin \theta = \frac{1}{2}$. Suppose, however, that you were given the equation $\sin \theta = 0.6$ and asked to find the acute angle θ. Because

$$\sin 30° = \frac{1}{2} = 0.5000 \quad \text{and} \quad \sin 45° = \frac{1}{\sqrt{2}} \approx 0.7071,$$

you might guess that θ lies somewhere between $30°$ and $45°$. A more precise value of θ can be found by using the *inverse* key on a calculator. To do this, you can use the following keystroke sequence in *degree* mode.

SIN⁻¹ .6 ENTER   Display 36.8699

So, you can conclude that if $\sin \theta = 0.6$, then $\theta \approx 36.87°$.

> ### STUDY T!P
>
> Calculators and graphing utilities have both *degree* and *radian* modes. As you progress through this chapter, be sure you use the correct mode.

EXAMPLE 9 Solving a Right Triangle

A skateboard ramp requires a rise of one foot for each three feet of horizontal length. In Figure 4.29, find the lengths of sides b and c and find the measure of θ.

Figure 4.29

Solution

From the given specifications, you can write

$$\frac{\text{rise}}{\text{run}} = \frac{1}{3} = \frac{4}{b} \text{ ft}$$

which implies that $b = 12$ feet. Using the Pythagorean Theorem, you can write

$$c^2 = 4^2 + 12^2 \qquad \text{Pythagorean Theorem}$$

$$c^2 = 160 \qquad \text{Simplify.}$$

$$c = 4\sqrt{10}. \qquad \text{Extract positive square root.}$$

So, $c = 4\sqrt{10} \approx 12.65$ feet. To solve for θ, you can write

$$\tan \theta = \frac{4}{12} = \frac{1}{3}.$$

Then, using the calculator keystrokes

tan⁻¹ (1 ÷ 3) ENTER

you obtain $\theta \approx 18.43°$.

4.3 Exercises

In Exercises 1–6, find the exact values of the six trigonometric functions of the angle θ given in the figure. (Use the Pythagorean Theorem to find the third side of the triangle.)

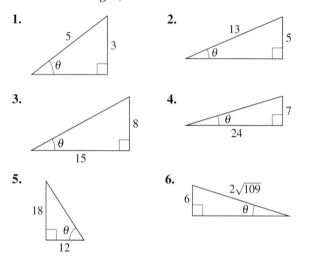

In Exercises 7–10, find the exact values of the six trigonometric functions of the angle θ for each of the triangles. Explain why the function values are the same.

In Exercises 11–18, sketch a right triangle corresponding to the trigonometric function of the acute angle θ. Use the Pythagorean Theorem to determine the third side and then find the other five trigonometric functions of θ.

11. $\sin \theta = \frac{5}{6}$ **12.** $\cot \theta = 5$

13. $\sec \theta = 4$ **14.** $\cos \theta = \frac{3}{7}$

15. $\tan \theta = 3$ **16.** $\csc \theta = \frac{17}{4}$

17. $\cot \theta = \frac{9}{4}$ **18.** $\sin \theta = \frac{3}{8}$

In Exercises 19–24, use the given function value(s) and the trigonometric identities to find the indicated trigonometric functions.

19. $\sin 60° = \dfrac{\sqrt{3}}{2}, \quad \cos 60° = \dfrac{1}{2}$

    (a) $\tan 60°$ (b) $\sin 30°$

    (c) $\cos 30°$ (d) $\cot 60°$

20. $\sin 30° = \dfrac{1}{2}, \quad \tan 30° = \dfrac{\sqrt{3}}{3}$

    (a) $\csc 30°$ (b) $\cot 60°$

    (c) $\cos 30°$ (d) $\cot 30°$

21. $\csc \theta = 3, \quad \sec \theta = \dfrac{3\sqrt{2}}{4}$

    (a) $\sin \theta$ (b) $\cos \theta$

    (c) $\tan \theta$ (d) $\sec(90° - \theta)$

22. $\sec \theta = 5, \quad \tan \theta = 2\sqrt{6}$

    (a) $\cos \theta$ (b) $\cot \theta$

    (c) $\cot(90° - \theta)$ (d) $\sin \theta$

23. $\cos \alpha = \dfrac{1}{4}$

    (a) $\sec \alpha$ (b) $\sin \alpha$

    (c) $\cot \alpha$ (d) $\sin(90° - \alpha)$

24. $\tan \beta = 5$

    (a) $\cot \beta$ (b) $\cos \beta$

    (c) $\tan(90° - \beta)$ (d) $\csc \beta$

In Exercises 25–32, use trigonometric identities to transform one side of the equation into the other.

25. $\tan \theta \cot \theta = 1$ **26.** $\csc \theta \tan \theta = \sec \theta$

27. $\tan \alpha \cos \alpha = \sin \alpha$ **28.** $\cot \alpha \sin \alpha = \cos \alpha$

29. $(1 + \cos \theta)(1 - \cos \theta) = \sin^2 \theta$

30. $(\csc \theta + \cot \theta)(\csc \theta - \cot \theta) = 1$

31. $\dfrac{\sin \theta}{\cos \theta} + \dfrac{\cos \theta}{\sin \theta} = \csc \theta \sec \theta$

32. $\dfrac{\tan \beta + \cot \beta}{\tan \beta} = \csc^2 \beta$

In Exercises 33–38, evaluate the trigonometric function by hand.

33. (a) $\cos 60°$

(b) $\tan \dfrac{\pi}{6}$

34. (a) $\csc 30°$

(b) $\sin \dfrac{\pi}{4}$

35. (a) $\cot \dfrac{\pi}{4}$

(b) $\cos 45°$

36. (a) $\sin \dfrac{\pi}{3}$

(b) $\csc 45°$

37. (a) $\cos \dfrac{\pi}{6}$

(b) $\sec 60°$

38. (a) $\tan \dfrac{\pi}{3}$

(b) $\cot 30°$

In Exercises 39–46, use a calculator to evaluate each function. Round your answers to four decimal places.

39. (a) $\sin 25°$ (b) $\cos 65°$

40. (a) $\tan 18.5°$ (b) $\cot 71.5°$

41. (a) $\sec 42° \, 12'$ (b) $\csc 48° \, 7'$

42. (a) $\cos 8° \, 50' \, 25''$ (b) $\sec 8° \, 50' \, 25''$

43. (a) $\cot \dfrac{\pi}{16}$ (b) $\tan \dfrac{\pi}{16}$

44. (a) $\sec 1.25$ (b) $\cos 1.25$

45. (a) $\csc 1$ (b) $\tan \frac{1}{2}$

46. (a) $\sec\left(\dfrac{\pi}{2} - 1\right)$ (b) $\cot\left(\dfrac{\pi}{2} - \dfrac{1}{2}\right)$

In Exercises 47–52, find the value of θ in degrees ($0° < \theta < 90°$) and radians ($0 < \theta < \pi/2$) without using a calculator.

47. (a) $\sin \theta = \frac{1}{2}$ (b) $\csc \theta = 2$

48. (a) $\cos \theta = \dfrac{\sqrt{2}}{2}$ (b) $\tan \theta = 1$

49. (a) $\sec \theta = 2$ (b) $\cot \theta = 1$

50. (a) $\tan \theta = \sqrt{3}$ (b) $\cos \theta = \frac{1}{2}$

51. (a) $\csc \theta = \dfrac{2\sqrt{3}}{3}$ (b) $\sin \theta = \dfrac{\sqrt{2}}{2}$

52. (a) $\cot \theta = \dfrac{\sqrt{3}}{3}$ (b) $\sec \theta = \sqrt{2}$

In Exercises 53–56, find the value of θ in degrees ($0° < \theta < 90°$) and radians ($0 < \theta < \pi/2$) by using a calculator.

53. (a) $\sin \theta = 0.8191$ (b) $\cos \theta = 0.0175$

54. (a) $\cos \theta = 0.9848$ (b) $\cos \theta = 0.8746$

55. (a) $\tan \theta = 1.1920$ (b) $\tan \theta = 0.4663$

56. (a) $\sin \theta = 0.3746$ (b) $\cos \theta = 0.3746$

In Exercises 57–62, solve for $x, y,$ or $r,$ as indicated.

57. Solve for y.

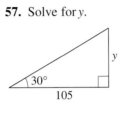

58. Solve for x.

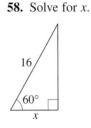

59. Solve for x.

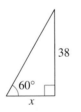

60. Solve for r.

61. Solve for y.

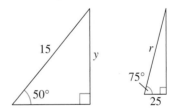

62. Solve for r.

63. *Height*   A 6-foot person walks from the base of a broadcasting tower directly toward the tip of the shadow cast by the tower. When the person is 132 feet from the tower and 3 feet from the tip of the tower's shadow, the person's shadow starts to appear beyond the tower's shadow.

(a) Draw a right triangle to represent the problem. Show the known quantities on the triangle and use a variable to indicate the unknown height of the tower.

(b) Write an equation involving the unknown.

(c) What is the height of the tower?

64. Length A 30-meter line is used to tether a helium-filled balloon. Because of a breeze, the line makes an angle of approximately 75° with the ground.

(a) Draw a right triangle to represent the problem. Show the known quantities on the triangle and use a variable to indicate the unknown height of the balloon.

(b) Use a trigonometric function to write an equation involving the unknown.

(c) What is the height of the balloon?

65. Width of a River A biologist wants to know the width *w* of a river in order to properly set instruments for studying the pollutants in the water. From point *A*, the biologist walks downstream 100 feet and sights to point *C*. From this sighting, it is determined that $\theta = 58°$. How wide is the river? Verify your result numerically using a graphing utility.

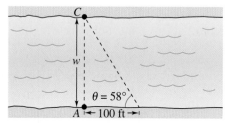

66. Distance From a 60-foot observation tower on the coast, a Coast Guard officer sights a boat in difficulty. The angle of depression of the boat is 4.5°. How far is the boat from the shoreline? Verify your result numerically using a graphing utility.

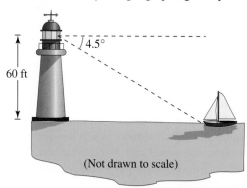

(Not drawn to scale)

67. Angle of Elevation A ramp 20 feet in length rises to a loading platform that is $3\frac{1}{3}$ feet off the ground.

(a) Draw a right triangle to represent the problem. Show the known quantities on the triangle and use a variable to indicate the unknown angle of elevation of the ramp.

(b) Use a trigonometric function to write an equation involving the unknown.

(c) Find the angle of elevation of the ramp algebraically.

(d) Use the *table* feature of a graphing utility to approximate the angle of elevation numerically.

68. Duquesne Incline The track of the Duquesne Incline in Pittsburgh, Pennsylvania, is about 800 feet long, and the angle of elevation is 30°. The average speed of the cable cars is about 320 feet per minute.

(a) How high does the Duquesne Incline rise?

(b) What is the vertical speed of the cable cars (in feet per minute)?

69. Jin Mao Building You are standing 75 meters from the base of the Jin Mao Building in Shanghai, China. You estimate that the angle of elevation to the top of the building is 80°. What is the approximate height of the building? Suppose one of your friends is at the top of the building. What is the distance between you and your friend?

70. Machine Shop Calculations A steel plate has the form of one fourth of a circle with a radius of 60 centimeters. Two 2-centimeter holes are to be drilled in the plate positioned as shown in the figure. Find the coordinates of the center of each hole.

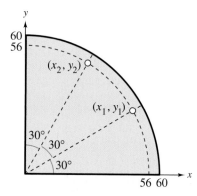

71. Machine Shop Calculations A tapered shaft has a diameter of 5 centimeters at the small end and is 15 centimeters long. If the taper is 3°, find the diameter *d* of the large end of the shaft.

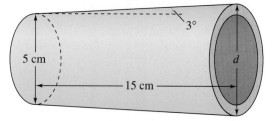

72. Geometry Use a compass to sketch a quarter of a circle of radius 10 centimeters. Using a protractor, construct an angle of 20° in standard position. Drop a perpendicular from the point of intersection of the terminal side of the angle and the arc of the circle. By actual measurement, calculate the coordinates (x, y) of the point of intersection and use these measurements to approximate the six trigonometric functions of a 20° angle.

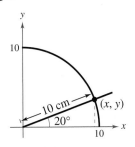

73. Geometry Repeat Exercise 72 using a 75° angle.

Synthesis

True or False? In Exercises 74–77, determine whether the statement is true or false. Justify your answer.

74. $\sin 60° \csc 60° = 1$ **75.** $\sec 30° = \csc 60°$

76. $\sin 45° + \cos 45° = 1$

77. $\cot^2 10° - \csc^2 10° = -1$

78. Exploration

(a) Use a graphing utility to complete the table. Round your results to four decimal places.

θ	0°	20°	40°	60°	80°
$\sin \theta$					
$\cos \theta$					
$\tan \theta$					

(b) Classify each of the three trigonometric functions as increasing or decreasing for the table values.

(c) For the values in the table, verify that the tangent function is the quotient of the sine and cosine functions.

79. Exploration Use a graphing utility to complete the table and make a conjecture about the relationship between $\cos \theta$ and $\sin(90° - \theta)$. What are the angles θ and $90° - \theta$ called?

θ	0°	20°	40°	60°	80°
$\cos \theta$					
$\sin (90° - \theta)$					

80. Numerical Analysis A 3000-pound automobile is negotiating a circular interchange of radius 300 feet at a speed of s miles per hour. The relationship between the speed and the angle θ (in degrees) at which the roadway should be banked so that no lateral frictional force is exerted on the tires is $\tan \theta = 0.672s^2/3000$.

(a) Use a graphing utility to complete the table.

s	10	20	30	40	50	60
θ						

(b) The speeds are incremented by 10 miles per hour in the table, but θ does not increase by equal increments. Explain.

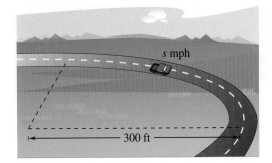

Review

In Exercises 81–84, sketch the graph of the equation and identify all x- and y-intercepts.

81. $y = -x - 9$ **82.** $2x + y = 10$

83. $-3x + 8y = 16$ **84.** $12x - 7y = 22$

In Exercises 85–88, determine the quadrant in which the angle lies.

85. $146°$ **86.** $-290.8°$

87. $-310° 30'$ **88.** $-122° 50'$

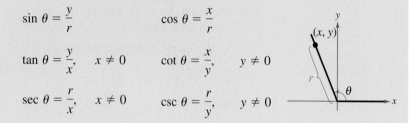

4.4 Trigonometric Functions of Any Angle

Introduction

In Section 4.3, the definitions of trigonometric functions were restricted to acute angles. In this section, the definitions are extended to cover *any* angle. If θ is an *acute* angle, the definitions here coincide with those given in the previous section.

What You Should Learn:

- How to evaluate trigonometric functions of any angle
- How to use reference angles to evaluate trigonometric functions
- How to evaluate trigonometric functions of real numbers

Why You Should Learn It:

You can use trigonometric functions to model and solve real-life problems. For instance, Exercise 103 on page 321 shows you how trigonometric functions can be used to model the average daily temperature in a city.

Henryk Kaiser/Leo de Wys

Definitions of Trigonometric Functions of Any Angle

Let θ be an angle in standard position with (x, y) a point on the terminal side of θ and $r = \sqrt{x^2 + y^2} \neq 0$.

$$\sin\theta = \frac{y}{r} \qquad \cos\theta = \frac{x}{r}$$

$$\tan\theta = \frac{y}{x}, \quad x \neq 0 \qquad \cot\theta = \frac{x}{y}, \quad y \neq 0$$

$$\sec\theta = \frac{r}{x}, \quad x \neq 0 \qquad \csc\theta = \frac{r}{y}, \quad y \neq 0$$

Because $r = \sqrt{x^2 + y^2}$ *cannot* be zero, it follows that the sine and cosine functions are defined for any real value of θ. However, if $x = 0$, the tangent and secant of θ are undefined. For example, the tangent of $90°$ is undefined. Similarly, if $y = 0$, the cotangent and cosecant of θ are undefined.

EXAMPLE 1 Evaluating Trigonometric Functions

Let $(-3, 4)$ be a point on the terminal side of θ. Find the sine, cosine, and tangent of θ.

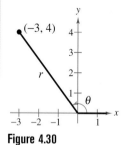

Figure 4.30

Solution

Referring to Figure 4.30, you see that $x = -3$, $y = 4$, and

$$r = \sqrt{x^2 + y^2} = \sqrt{(-3)^2 + 4^2} = \sqrt{25} = 5.$$

So, you have $\sin\theta = \dfrac{y}{r} = \dfrac{4}{5}$, $\cos\theta = \dfrac{x}{r} = -\dfrac{3}{5}$, and $\tan\theta = \dfrac{y}{x} = -\dfrac{4}{3}$.

The *signs* of the trigonometric functions in the four quadrants can be determined easily from the definitions of the functions. For instance, because $\cos\theta = x/r$, it follows that $\cos\theta$ is positive wherever $x > 0$, which is in Quadrants I and IV. (Remember, r is always positive.) In a similar manner you can verify the results shown in Figure 4.31.

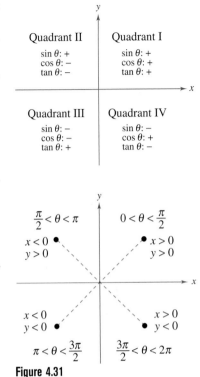

Quadrant II	Quadrant I
$\sin\theta$: +	$\sin\theta$: +
$\cos\theta$: −	$\cos\theta$: +
$\tan\theta$: −	$\tan\theta$: +

Quadrant III	Quadrant IV
$\sin\theta$: −	$\sin\theta$: −
$\cos\theta$: −	$\cos\theta$: +
$\tan\theta$: +	$\tan\theta$: −

EXAMPLE 2   Evaluating Trigonometric Functions

Given $\tan\theta = -\dfrac{5}{4}$ and $\cos\theta > 0$, find $\sin\theta$ and $\sec\theta$.

Solution
Note that θ lies in Quadrant IV because that is the only quadrant in which the tangent is negative and the cosine is positive. Moreover, using

$$\tan\theta = \frac{y}{x} = -\frac{5}{4}$$

and the fact that y is negative in Quadrant IV, you can let $y = -5$ and $x = 4$. So, $r = \sqrt{16 + 25} = \sqrt{41}$, and you have

$$\sin\theta = \frac{y}{r} = \frac{-5}{\sqrt{41}} \qquad \text{Exact value}$$

$$\approx -0.7809 \qquad \text{Approximate value}$$

$$\sec\theta = \frac{r}{x} = \frac{\sqrt{41}}{4} \qquad \text{Exact value}$$

$$\approx 1.6008. \qquad \text{Approximate value}$$

Figure 4.31

$\dfrac{\pi}{2} < \theta < \pi$ $0 < \theta < \dfrac{\pi}{2}$
$x < 0$ $x > 0$
$y > 0$ $y > 0$

$x < 0$ $x > 0$
$y < 0$ $y < 0$
$\pi < \theta < \dfrac{3\pi}{2}$ $\dfrac{3\pi}{2} < \theta < 2\pi$

EXAMPLE 3   Trigonometric Functions of Quadrant Angles

Evaluate the sine and cosine functions at the four quadrant angles 0, $\pi/2$, π, and $3\pi/2$.

Solution
To begin, choose a point on the terminal side of each angle, as shown in Figure 4.32. For each of the four given points, $r = 1$, and you have the following.

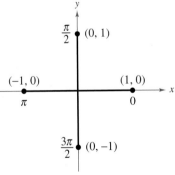

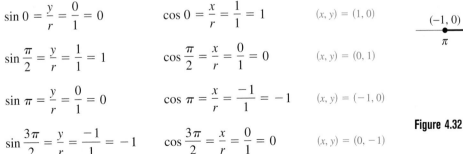

$$\sin 0 = \frac{y}{r} = \frac{0}{1} = 0 \qquad \cos 0 = \frac{x}{r} = \frac{1}{1} = 1 \qquad (x, y) = (1, 0)$$

$$\sin\frac{\pi}{2} = \frac{y}{r} = \frac{1}{1} = 1 \qquad \cos\frac{\pi}{2} = \frac{x}{r} = \frac{0}{1} = 0 \qquad (x, y) = (0, 1)$$

$$\sin\pi = \frac{y}{r} = \frac{0}{1} = 0 \qquad \cos\pi = \frac{x}{r} = \frac{-1}{1} = -1 \qquad (x, y) = (-1, 0)$$

$$\sin\frac{3\pi}{2} = \frac{y}{r} = \frac{-1}{1} = -1 \qquad \cos\frac{3\pi}{2} = \frac{x}{r} = \frac{0}{1} = 0 \qquad (x, y) = (0, -1)$$

Figure 4.32

Reference Angles

The values of the trigonometric functions of angles greater than 90° (or less than 0°) can be determined from their values at corresponding acute angles called **reference angles.**

Definition of Reference Angle

Let θ be an angle in standard position. Its **reference angle** is the acute angle θ' formed by the terminal side of θ and the horizontal axis.

A computer simulation to accompany this concept appears in the *Interactive* CD-ROM and *Internet* versions of this text.

Figure 4.33 shows the reference angles for θ in Quadrants II, III, and IV.

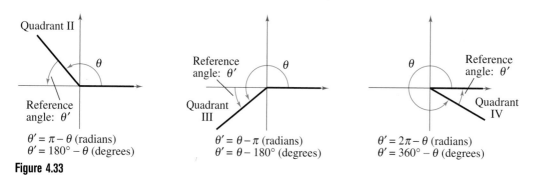

$\theta' = \pi - \theta$ (radians)
$\theta' = 180° - \theta$ (degrees)

$\theta' = \theta - \pi$ (radians)
$\theta' = \theta - 180°$ (degrees)

$\theta' = 2\pi - \theta$ (radians)
$\theta' = 360° - \theta$ (degrees)

Figure 4.33

EXAMPLE 4   Finding Reference Angles

Find the reference angle θ'.

a. $\theta = 300°$ **b.** $\theta = 2.3$ **c.** $\theta = -135°$

Solution

a. Because 300° lies in Quadrant IV, the angle it makes with the *x*-axis is

$$\theta' = 360° - 300° = 60°. \qquad \text{Degrees}$$

b. Because 2.3 lies between $\pi/2 \approx 1.5708$ and $\pi \approx 3.1416$, it follows that it is in Quadrant II and its reference angle is

$$\theta' = \pi - 2.3 \approx 0.8416. \qquad \text{Radians}$$

c. First, determine that $-135°$ is coterminal with 225°, which lies in Quadrant III. So, the reference angle is

$$\theta' = 225° - 180° = 45°. \qquad \text{Degrees}$$

Figure 4.34 shows each angle θ and its reference angle θ'.

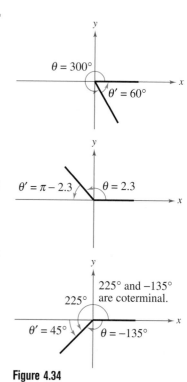

Figure 4.34

Trigonometric Functions of Real Numbers

To see how a reference angle is used to evaluate a trigonometric function, consider the point (x, y) on the terminal side of θ as shown in Figure 4.35. By definition, you know that

$$\sin \theta = \frac{y}{r} \quad \text{and} \quad \tan \theta = \frac{y}{x}.$$

For the right triangle with acute angle θ' and sides of lengths $|x|$ and $|y|$, you have

$$\sin \theta' = \frac{\text{opp}}{\text{hyp}} = \frac{|y|}{r}$$

and

$$\tan \theta' = \frac{\text{opp}}{\text{adj}} = \frac{|y|}{|x|}.$$

So, it follows that $\sin \theta$ and $\sin \theta'$ are equal, *except possibly in sign*. The same is true for $\tan \theta$ and $\tan \theta'$ *and* for the other four trigonometric functions. In all cases, the sign of the function value can be determined by the quadrant in which θ lies.

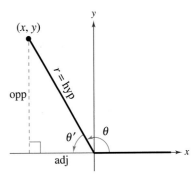

Figure 4.35

Evaluating Trigonometric Functions of Any Angle

To find the value of a trigonometric function of any angle θ:

1. Determine the function value for the associated reference angle θ'.
2. Depending on the quadrant in which θ lies, prefix the appropriate sign to the function value.

By using reference angles and the special angles discussed in the previous section, you can greatly extend the scope of *exact* trigonometric values. For instance, knowing the function values of 30° means that you know the function values of all angles for which 30° is a reference angle. For convenience, the following table gives the exact values of the trigonometric functions of special angles and quadrant angles.

Trigonometric Values of Common Angles

θ (degrees)	0°	30°	45°	60°	90°	180°	270°
θ (radians)	0	$\dfrac{\pi}{6}$	$\dfrac{\pi}{4}$	$\dfrac{\pi}{3}$	$\dfrac{\pi}{2}$	π	$\dfrac{3\pi}{2}$
$\sin \theta$	0	$\dfrac{1}{2}$	$\dfrac{\sqrt{2}}{2}$	$\dfrac{\sqrt{3}}{2}$	1	0	-1
$\cos \theta$	1	$\dfrac{\sqrt{3}}{2}$	$\dfrac{\sqrt{2}}{2}$	$\dfrac{1}{2}$	0	-1	0
$\tan \theta$	0	$\dfrac{\sqrt{3}}{3}$	1	$\sqrt{3}$	Undef.	0	Undef.

Library of Functions

The trigonometric functions are important for modeling periodic behavior, such as business cycles, planetary orbits, pendulums, and light rays. These functions play a prominent role in calculus.

Consult the Library of Functions Summary inside the front cover for a description of the trigonometric functions.

STUDY T!P

Learning the table of values at the left is worth the effort. Doing so will increase both your efficiency and your confidence. Here is a pattern for the sine function that may help you remember the values.

θ	0°	30°	45°	60°	90°
$\sin \theta$	$\dfrac{\sqrt{0}}{2}$	$\dfrac{\sqrt{1}}{2}$	$\dfrac{\sqrt{2}}{2}$	$\dfrac{\sqrt{3}}{2}$	$\dfrac{\sqrt{4}}{2}$

Reverse the order to get cosine values of the same angles.

EXAMPLE 5   Trigonometric Functions of Nonacute Angles

Evaluate the trigonometric functions.

a. $\cos \dfrac{4\pi}{3}$ **b.** $\tan(-210°)$ **c.** $\csc \dfrac{11\pi}{4}$

Solution

a. Because $\theta = 4\pi/3$ lies in Quadrant III, the reference angle is $\theta' = (4\pi/3) - \pi = \pi/3$, as shown in Figure 4.36(a). Moreover, the cosine is negative in Quadrant III, so

$$\cos \frac{4\pi}{3} = (-)\cos \frac{\pi}{3}$$

$$= -\frac{1}{2}.$$

b. Because $-210° + 360° = 150°$, it follows that $-210°$ is coterminal with the second-quadrant angle $150°$. Therefore, the reference angle is $\theta' = 180° - 150° = 30°$, as shown in Figure 4.36(b). Finally, because the tangent is negative in Quadrant II, you have

$$\tan(-210°) = (-)\tan 30°$$

$$= -\frac{\sqrt{3}}{3}.$$

c. Because $(11\pi/4) - 2\pi = 3\pi/4$, it follows that $11\pi/4$ is coterminal with the second-quadrant angle $3\pi/4$. Therefore, the reference angle is $\theta' = \pi - (3\pi/4) = \pi/4$, as shown in Figure 4.36(c). Because the cosecant is positive in Quadrant II, you have

$$\csc \frac{11\pi}{4} = (+)\csc \frac{\pi}{4}$$

$$= \frac{1}{\sin(\pi/4)}$$

$$= \sqrt{2}.$$

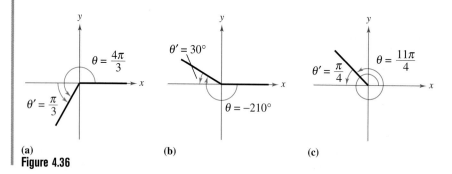

(a)   **(b)**   **(c)**

Figure 4.36

The fundamental trigonometric identities listed in the previous section (for an acute angle θ) are also valid when θ is any angle in the domain of the function.

EXAMPLE 6   Using Trigonometric Identities

Let θ be an angle in Quadrant II such that $\sin \theta = \frac{1}{3}$. Find (a) $\cos \theta$ and (b) $\tan \theta$ by using trigonometric identities.

Solution

a. Using the Pythagorean identity $\sin^2 \theta + \cos^2 \theta = 1$, you obtain

$$\left(\frac{1}{3}\right)^2 + \cos^2 \theta = 1$$

$$\cos^2 \theta = 1 - \frac{1}{9} = \frac{8}{9}.$$

Because $\cos \theta < 0$ in Quadrant II, you can use the negative root to obtain

$$\cos \theta = -\frac{\sqrt{8}}{\sqrt{9}} = -\frac{2\sqrt{2}}{3}.$$

b. Using the trigonometric identity $\tan \theta = \sin \theta / \cos \theta$, you obtain

$$\tan \theta = \frac{1/3}{-2\sqrt{2}/3}$$

$$= -\frac{1}{2\sqrt{2}}$$

$$= -\frac{\sqrt{2}}{4}.$$

Exploration

Set your graphing utility to *degree* mode and enter tan 90. What happens? Why? Now set your graphing utility to *radian* mode and enter $\tan(\pi/2)$. Explain the graphing utility's answer.

The *Interactive* CD-ROM and *Internet* versions of this text offer a built-in graphing calculator, which can be used with the Examples, Explorations, and Exercises.

You can use a calculator to evaluate trigonometric functions, as shown in the next example.

EXAMPLE 7   Using a Calculator

a. Use a calculator to evaluate $\cot 410°$ and $\sin(-7)$.

b. Use a calculator to solve $\tan \theta = 4.812, \ 0 \le \theta \le 2\pi$.

Solution

Function	Mode	Graphing Calculator Keystrokes	Display
a. $\cot 410°$	Degree	`(` `TAN` `410` `)` `x⁻¹` `ENTER`	0.8390996
$\sin(-7)$	Radian	`SIN` `(` `(-)` `7` `)` `ENTER`	-0.6569866

b. To solve the equation $\tan \theta = 4.812$, you can use the inverse tangent key, as follows.

Equation	Mode	Graphing Calculator Keystrokes	Display
$\tan \theta = 4.812$	Radian	`TAN⁻¹` `4.812` `ENTER`	1.365898912

The angle $\theta \approx 1.366$ lies in Quadrant I. A second value of θ lies in Quadrant III (tangent is positive) and is

$$\theta = \pi + 1.366 \approx 4.507.$$

4.4   Exercises

In Exercises 1–4, determine the exact values of the six trigonometric functions of the angle θ.

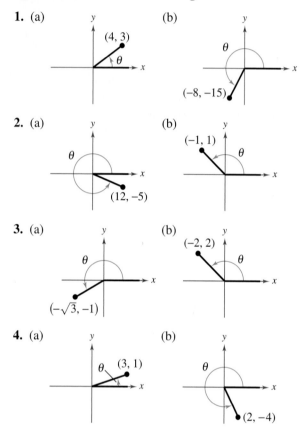

1. (a)   (b)

2. (a)   (b)

3. (a)   (b)

4. (a)   (b)

In Exercises 5–12, the given point is on the terminal side of an angle in standard position. Determine the exact values of the six trigonometric functions of the angle.

5. $(7, 24)$
6. $(8, 15)$
7. $(5, -12)$
8. $(-24, 10)$
9. $(-4, 10)$
10. $(-5, -6)$
11. $(-2, 9)$
12. $(6, -14)$

In Exercises 13–18, state the quadrant in which θ lies.

13. $\sin \theta < 0$ and $\cos \theta < 0$
14. $\sin \theta > 0$ and $\cos \theta > 0$
15. $\sin \theta > 0$ and $\tan \theta < 0$
16. $\sec \theta > 0$ and $\cot \theta < 0$
17. $\cot \theta > 0$ and $\cos \theta > 0$
18. $\tan \theta > 0$ and $\csc \theta < 0$

In Exercises 19–26, find the values of the six trigonometric functions of θ.

Function Value	Constraint
19. $\sin \theta = \frac{3}{5}$	θ lies in Quadrant II.
20. $\cos \theta = -\frac{4}{5}$	θ lies in Quadrant III.
21. $\tan \theta = -\frac{15}{8}$	$\sin \theta < 0$
22. $\csc \theta = 4$	$\cot \theta < 0$
23. $\sec \theta = -2$	$0 \le \theta \le \pi$
24. $\cot \theta$ is undefined.	$\frac{\pi}{2} \le \theta \le \frac{3\pi}{2}$
25. $\sin \theta = 0$	$\sec \theta = -1$
26. $\tan \theta$ is undefined.	$\csc \theta = -1$

In Exercises 27–30, the terminal side of θ lies on the given line in the specified quadrant. Find the values of the six trigonometric functions of θ.

Line	Quadrant
27. $y = -x$	II
28. $y = \frac{1}{3}x$	III
29. $y = 2x$	III
30. $4x + 3y = 0$	IV

In Exercises 31–38, evaluate the trigonometric function of the quadrant angle.

31. $\sec \pi$ 32. $\tan \dfrac{\pi}{2}$ 33. $\cot \dfrac{\pi}{2}$

34. $\csc \pi$ 35. $\sec 0$ 36. $\csc \dfrac{3\pi}{2}$

37. $\cot \pi$ 38. $\csc \dfrac{\pi}{2}$

In Exercises 39–50, find the reference angle θ' and sketch θ and θ' in standard position.

39. $\theta = 208°$ 40. $\theta = 322°$
41. $\theta = -245°$ 42. $\theta = -145°$
43. $\theta = -292°$ 44. $\theta = -95°$
45. $\theta = \dfrac{11\pi}{3}$ 46. $\theta = \dfrac{17\pi}{6}$
47. $\theta = 3.5$ 48. $\theta = 4.8$
49. $\theta = -3.68$ 50. $\theta = -1.72$

In Exercises 51–64, evaluate the sine, cosine, and tangent of the angle without using a calculator.

51. 225° **52.** 300°

53. −750° **54.** −495°

55. −240° **56.** −330°

57. $\dfrac{5\pi}{3}$ **58.** $\dfrac{\pi}{4}$

59. $-\dfrac{\pi}{6}$ **60.** $-\dfrac{\pi}{2}$

61. $\dfrac{11\pi}{4}$ **62.** $\dfrac{10\pi}{3}$

63. $-\dfrac{7\pi}{6}$ **64.** $-\dfrac{20\pi}{3}$

In Exercises 65–78, use a calculator to evaluate the trigonometric function. Round you answer to four decimal places.

65. sin 10° **66.** sec 225°

67. tan 240° **68.** csc 330°

69. cos(−110°) **70.** cot(−220°)

71. sec(−280°) **72.** sin(−195°)

73. sin 0.65 **74.** sin(−0.65)

75. $\tan\dfrac{\pi}{9}$ **76.** $\tan\left(-\dfrac{\pi}{9}\right)$

77. $\csc\left(-\dfrac{8\pi}{9}\right)$ **78.** $\cos\left(-\dfrac{15\pi}{14}\right)$

In Exercises 79–84 find two solutions of the equation. Give your answers in both degrees (0° ≤ θ ≤ 360°) and radians (0 ≤ θ ≤ 2π). Do not use a calculator.

79. (a) $\sin\theta = \dfrac{1}{2}$ (b) $\sin\theta = -\dfrac{1}{2}$

80. (a) $\cos\theta = \dfrac{\sqrt{2}}{2}$ (b) $\cos\theta = -\dfrac{\sqrt{2}}{2}$

81. (a) $\csc\theta = \dfrac{2\sqrt{3}}{3}$ (b) $\cot\theta = -1$

82. (a) $\csc\theta = -\sqrt{2}$ (b) $\csc\theta = 2$

83. (a) $\sec\theta = -\dfrac{2\sqrt{3}}{3}$ (b) $\cos\theta = -\dfrac{1}{2}$

84. (a) $\cot\theta = -\sqrt{3}$ (b) $\sec\theta = \sqrt{2}$

In Exercises 85–96, use a calculator to approximate two values of θ (0° ≤ θ < 360°) that satisfy the equation. Round your answers to two decimal places.

85. sin θ = 0.8191 **86.** cos θ = 0.8746

87. tan θ = 0.6524 **88.** cot θ = 0.7521

89. sec θ = −1.2241 **90.** csc θ = −1.0038

91. sin θ = −0.4793 **92.** tan θ = −2.1832

93. cos θ = 0.9848 **94.** sin θ = 0.0175

95. tan θ = 1.192 **96.** cot θ = 5.671

In Exercises 97–102, find the indicated trigonometric value in the specified quadrant.

	Function	*Quadrant*	*Trigonometric Value*
97.	$\sin\theta = -\dfrac{3}{5}$	IV	cos θ
98.	$\cot\theta = -3$	II	sin θ
99.	$\tan\theta = \dfrac{3}{2}$	III	sec θ
100.	$\csc\theta = -2$	IV	cot θ
101.	$\cos\theta = \dfrac{5}{8}$	I	sec θ
102.	$\sec\theta = -\dfrac{9}{4}$	III	tan θ

103. *Average Temperature* The average daily temperature T (in degrees Fahrenheit) for a certain city is

$$T = 45 - 23\cos\left[\dfrac{2\pi}{365}(t - 32)\right]$$

where t is the time in days, with $t = 1$ corresponding to January 1. Find the average daily temperatures on the following days.

(a) January 1

(b) July 4 ($t = 185$)

(c) October 18 ($t = 291$)

104. *Sales* A company that sells seasonal products forecasts monthly sales over a 2-year period to be

$$S = 23.1 + 0.442t + 4.3\sin\dfrac{\pi t}{6}$$

where S is measured in thousands of units and t is the time (in months), with $t = 1$ representing January 2000. Estimate sales for the following months.

(a) February 2000 (b) February 2001

(c) September 2000 (d) September 2001

105. *Distance*   An airplane, flying at an altitude of 6 miles, is on a flight path that passes directly over an observer. If θ is the angle of elevation from the observer to the plane, find the distance from the observer to the plane when (a) $\theta = 30°$, (b) $\theta = 90°$, and (c) $\theta = 120°$.

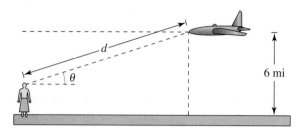

Synthesis

True or False? **In Exercises 106–108, determine whether the statement is true or false. Justify your answer.**

106. $\sin 151° = \sin 29°$

107. $\csc\left(-\dfrac{7\pi}{6}\right) = \csc\left(-\dfrac{11\pi}{6}\right)$

108. $-\cot\left(\dfrac{3\pi}{4}\right) = \cot\left(-\dfrac{\pi}{4}\right)$

109. *Conjecture*

(a) Use a graphing utility to complete the table.

θ	0°	20°	40°	60°	80°
$\sin\theta$					
$\sin(180° - \theta)$					

(b) Make a conjecture about the relationship between $\sin\theta$ and $\sin(180° - \theta)$.

110. *Conjecture*

(a) Use a graphing utility to complete the table.

θ	0	0.3	0.6	0.9	1.2	1.5
$\cos\left(\dfrac{3\pi}{2} - \theta\right)$						
$-\sin\theta$						

(b) Make a conjecture about the relationship between $\cos\left(\dfrac{3\pi}{2} - \theta\right)$ and $-\sin\theta$.

111. Explain how reference angles are used to find the value of trigonometric functions of any angle θ.

112. *Writing*   Consider an angle in standard position with $r = 12$ centimeters, as shown in the figure. Write a short paragraph describing the change in the magnitudes of x, y, $\sin\theta$, $\cos\theta$, and $\tan\theta$ as θ increases continually from 0° to 90°.

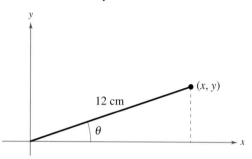

Review

In Exercises 113–116, sketch the graph the function. Identify any intercepts and asymptotes.

113. $y = 2^{x-1}$

114. $y = 3^{x+2}$

115. $y = \ln(x - 1)$

116. $y = \ln(x + 1)$

In Exercises 117–120, solve the equation. Round your answer to three decimal places.

117. $4^{3-x} = 726$

118. $\dfrac{4500}{4 + e^{2x}} = 50$

119. $\ln x = -6$

120. $\ln\sqrt{x + 10} = 1$

In Exercises 121–124, sketch a right triangle corresponding to the trigonometric function of the acute angle θ. Use the Pythagorean Theorem to determine the third side and then find the other five trigonometric functions of θ.

121. $\cot\theta = \dfrac{8}{3}$

122. $\sec\theta = 8$

123. $\tan\theta = \dfrac{7}{24}$

124. $\cos\theta = \dfrac{2}{9}$

4.5 Graphs of Sine and Cosine Functions

Basic Sine and Cosine Curves

In this section you will study techniques for sketching the graphs of the sine and cosine functions. The graph of the sine function is a **sine curve**. In Figure 4.37, the black portion of the graph represents one period of the function and is called **one cycle** of the sine curve. The gray portion of the graph indicates that the basic sine wave repeats indefinitely to the right and left. The graph of the cosine function is shown in Figure 4.38. To produce these graphs with a graphing utility, make sure you have set the graphing utility to *radian* mode.

Recall from Section 4.2 that the domain of the sine and cosine functions is the set of all real numbers. Moreover, the range of each function is the interval $[-1, 1]$, and each function has a period of 2π. Do you see how this information is consistent with the basic graphs given in Figures 4.37 and 4.38?

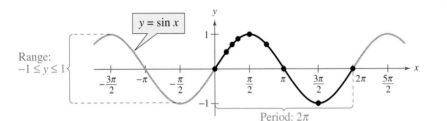

Figure 4.37

Jim Corwin/Tony Stone Images

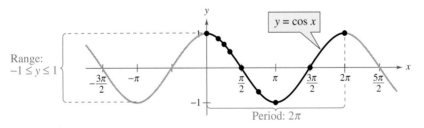

Figure 4.38

The table below lists key points on the graphs of $y = \sin x$ and $y = \cos x$.

x	0	$\dfrac{\pi}{6}$	$\dfrac{\pi}{4}$	$\dfrac{\pi}{3}$	$\dfrac{\pi}{2}$	$\dfrac{3\pi}{4}$	π	$\dfrac{3\pi}{2}$	2π
$\sin x$	0	$\dfrac{1}{2}$	$\dfrac{\sqrt{2}}{2}$	$\dfrac{\sqrt{3}}{2}$	1	$\dfrac{\sqrt{2}}{2}$	0	-1	0
$\cos x$	1	$\dfrac{\sqrt{3}}{2}$	$\dfrac{\sqrt{2}}{2}$	$\dfrac{1}{2}$	0	$-\dfrac{\sqrt{2}}{2}$	-1	0	1

Note from Figures 4.37 and 4.38 that the sine graph is symmetric with respect to the *origin,* whereas the cosine graph is symmetric with respect to the *y-axis.* These properties of symmetry follow from the fact that the sine function is odd whereas the cosine function is even.

To sketch the graphs of the basic sine and cosine functions by hand, it helps to note five *key points* in one period of each graph: the *intercepts*, *maximum points*, and *minimum points*. See Figure 4.39

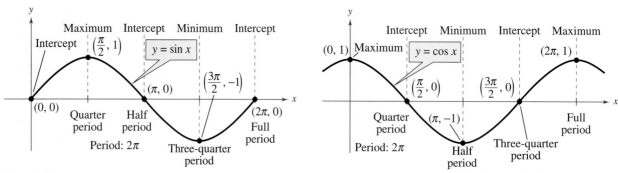

Figure 4.39

EXAMPLE 1   Using Key Points to Sketch a Sine Curve

Sketch the graph of $y = 2 \sin x$ on the interval $[-\pi, 4\pi]$.

Solution
Note that

$$y = 2 \sin x = 2(\sin x)$$

indicates that the y-values for the key points will have twice the magnitude of the graph of $y = \sin x$. Divide the period 2π into four equal parts to get the key points for $y = 2 \sin x$.

$$(0, 0), \qquad \left(\frac{\pi}{2}, 2\right), \qquad (\pi, 0), \qquad \left(\frac{3\pi}{2}, -2\right), \qquad \text{and} \qquad (2\pi, 0)$$

By connecting these key points with a smooth curve and extending the curve in both directions over the interval $[-\pi, 4\pi]$, you obtain the graph shown in Figure 4.40. Use a graphing utility to confirm this graph. Be sure to set the graphing utility to *radian* mode.

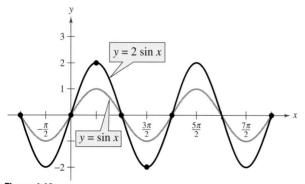

Figure 4.40

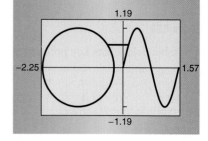

Amplitude and Period of Sine and Cosine Curves

In the rest of this section you will study the graphic effect of each of the constants a, b, c, and d in equations of the forms

$$y = d + a \sin(bx - c)$$

and

$$y = d + a \cos(bx - c).$$

A quick review of the transformations studied in Section 1.3 should help in this investigation.

The constant factor a in $y = a \sin x$ acts as a *scaling factor*—a *vertical stretch* or *vertical shrink* of the basic sine curve. If $|a| > 1$, the basic sine curve is stretched, and if $|a| < 1$, the basic sine curve is shrunk. The result is that the graph of $y = a \sin x$ ranges between $-a$ and a instead of between -1 and 1. The absolute value of a is the **amplitude** of the function $y = a \sin x$. The range of the function $y = a \sin x$ is $-a \le y \le a$.

Definition of Amplitude of Sine and Cosine Curves

The **amplitude** of $y = a \sin x$ and $y = a \cos x$ represents half the distance between the maximum and minimum values of the function and is given by

$$\text{Amplitude} = |a|.$$

EXAMPLE 2   Scaling: Vertical Shrinking and Stretching

On the same coordinate axes, sketch the graph of each function.

a. $y = \dfrac{1}{2} \cos x$ **b.** $y = 3 \cos x$

Solution

a. Because the amplitude of $y = \frac{1}{2} \cos x$ is $\frac{1}{2}$, the maximum value is $\frac{1}{2}$ and the minimum value is $-\frac{1}{2}$. Divide one cycle, $0 \le x \le 2\pi$, into four equal parts to get the key points

$$\left(0, \frac{1}{2}\right), \qquad \left(\frac{\pi}{2}, 0\right), \qquad \left(\pi, -\frac{1}{2}\right), \qquad \left(\frac{3\pi}{2}, 0\right), \qquad \text{and} \qquad \left(2\pi, \frac{1}{2}\right).$$

b. A similar analysis shows that the amplitude of $y = 3 \cos x$ is 3, and the key points are

$$(0, 3), \qquad \left(\frac{\pi}{2}, 0\right), \qquad (\pi, -3), \qquad \left(\frac{3\pi}{2}, 0\right), \qquad \text{and} \qquad (2\pi, 3).$$

The graphs of these two functions are shown in Figure 4.41. Use a graphing utility to confirm these graphs.

Exploration

Use a graphing utility to graph $y = a \sin x$, where $a = 0.5, 1, 2,$ and 3. How does the value of a affect the shape of the graph? What happens to the graph if a is negative?

STUDY T!P

When using a graphing utility to graph trigonometric functions, pay special attention to the viewing window you use. For instance, try graphing $y = [\sin(10x)]/10$ in the standard viewing window in *radian* mode. What do you observe? Use the *zoom* feature to find a viewing window that displays a good view of the graph.

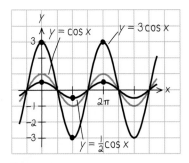

Figure 4.41

You know from Section 1.3 that the graph of $y = -f(x)$ is a *reflection* in the *x*-axis of the graph of $y = f(x)$. For instance, the graph of

$$y = -3 \cos x$$

is a reflection of the graph of

$$y = 3 \cos x,$$

as shown in Figure 4.42.

Because $y = a \sin x$ completes one cycle from $x = 0$ to $x = 2\pi$, it follows that $y = a \sin bx$ completes one cycle from $x = 0$ to $x = 2\pi/b$.

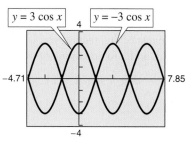

Figure 4.42

Period of Sine and Cosine Functions

Let b be a positive real number. The **period** of $y = a \sin bx$ and $y = a \cos bx$ is given by

$$\text{Period} = \frac{2\pi}{b}.$$

Note that if $0 < b < 1$, the period of $y = a \sin bx$ is greater than 2π and represents a *horizontal stretching* of the graph of $y = a \sin x$. Similarly, if $b > 1$, the period of $y = a \sin bx$ is less than 2π and represents a *horizontal shrinking* of the graph of $y = a \sin x$. If b is negative, the identities $\sin(-x) = -\sin x$ and $\cos(-x) = \cos x$ are used to rewrite the function.

EXAMPLE 3 Scaling: Horizontal Stretching

Sketch the graph of $y = \sin \dfrac{x}{2}$.

Solution

The amplitude is 1. Moreover, because $b = \frac{1}{2}$, the period is

$$\frac{2\pi}{b} = \frac{2\pi}{\frac{1}{2}} = 4\pi.$$

Now, divide the period-interval $[0, 4\pi]$ into four equal parts with the values π, 2π, and 3π, to obtain the key points on the graph

$$(0, 0), \quad (\pi, 1), \quad (2\pi, 0), \quad (3\pi, -1), \quad \text{and} \quad (4\pi, 0).$$

The graph is shown in Figure 4.43. Use a graphing utility to confirm this graph.

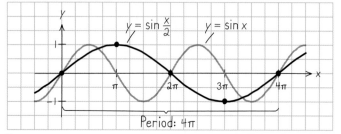

Figure 4.43

Exploration

Use a graphing utility to graph $y = \sin bx$, where $b = 0.5$, 1, and 2. How does the value of b affect the graph?

STUDY T!P

In general, to divide a period-interval into four equal parts, successively add "period/4," starting with the left endpoint of the interval. For instance, for the period-interval $[-\pi/6, \pi/2]$ of length $2\pi/3$, you would successively add

$$\frac{2\pi/3}{4} = \frac{\pi}{6}$$

to get $-\pi/6$, 0, $\pi/6$, $\pi/3$, and $\pi/2$.

Translations of Sine and Cosine Curves

The constant c in the general equations

$$y = a \sin(bx - c) \qquad \text{and} \qquad y = a \cos(bx - c)$$

creates *horizontal translations* (shifts) of the basic sine and cosine curves. Comparing $y = a \sin bx$ with $y = a \sin(bx - c)$, you find that the graph of $y = a \sin(bx - c)$ completes one cycle from $bx - c = 0$ to $bx - c = 2\pi$. By solving for x, you can find the interval for one cycle to be

$$\overbrace{\frac{c}{b}}^{\text{Left endpoint}} \leq x \leq \overbrace{\frac{c}{b}}^{\text{Right endpoint}} + \underbrace{\frac{2\pi}{b}}_{\text{Period}}.$$

This implies that the period of $y = a \sin(bx - c)$ is $2\pi/b$, and the graph of $y = a \sin bx$ is shifted by an amount c/b. The number c/b is the **phase shift.**

Graphs of Sine and Cosine Functions

The graphs of $y = a \sin(bx - c)$ and $y = a \cos(bx - c)$ have the following characteristics. (Assume $b > 0$.)

  Amplitude $= |a|$ Period $= 2\pi/b$

The left and right endpoints of a one-cycle interval can be determined by solving the equations $bx - c = 0$ and $bx - c = 2\pi$.

EXAMPLE 4   Horizontal Translation

Analyze the graph of $y = \dfrac{1}{2} \sin\left(x - \dfrac{\pi}{3}\right)$.

Algebraic Solution

The amplitude is $\frac{1}{2}$ and the period is 2π. By solving the equations

$$x - \frac{\pi}{3} = 0 \qquad \text{and} \qquad x - \frac{\pi}{3} = 2\pi$$

$$x = \frac{\pi}{3} \qquad\qquad\qquad x = \frac{7\pi}{3}$$

you see that the interval $[\pi/3, 7\pi/3]$ corresponds to one cycle of the graph. Dividing this interval into four equal parts produces the following key points.

$$\left(\frac{\pi}{3}, 0\right), \quad \left(\frac{5\pi}{6}, \frac{1}{2}\right), \quad \left(\frac{4\pi}{3}, 0\right), \quad \left(\frac{11\pi}{6}, -\frac{1}{2}\right), \quad \left(\frac{7\pi}{3}, 0\right)$$

Graphical Solution

Use a graphing utility set in *radian* mode to graph $y = (1/2) \sin(x - \pi/3)$, as shown in Figure 4.44. Use the *minimum*, *maximum*, and *zero* or *root* features of the graphing utility to approximate the key points $(1.047, 0)$, $(2.618, 0.5)$, $(4.189, 0)$, $(5.760, -0.5)$, and $(7.330, 0)$.

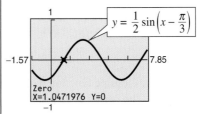

Figure 4.44

EXAMPLE 5 Horizontal Translation

Use a graphing utility to analyze the graph of $y = -3 \cos(2\pi x + 4\pi)$.

Solution

The amplitude is 3 and the period is $2\pi/2\pi = 1$. By solving the equations

$$2\pi x + 4\pi = 0 \qquad \text{and} \qquad 2\pi x + 4\pi = 2\pi$$

$$2\pi x = -4\pi \qquad\qquad\qquad 2\pi x = -2\pi$$

$$x = -2 \qquad\qquad\qquad\qquad x = -1$$

you see that the interval $[-2, -1]$ corresponds to one cycle of the graph. Dividing this interval into four equal parts produces the key points

$$(-2, -3), \quad (-7/4, 0), \quad (-3/2, 3), \quad (-5/4, 0), \quad \text{and} \quad (-1, -3).$$

The graph is shown in Figure 4.45.

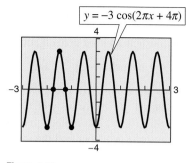

Figure 4.45

A computer animation of this example appears in the *Interactive* CD-ROM and *Internet* versions of this text.

The final type of transformation is the *vertical translation* caused by the constant d in the equations

$$y = d + a \sin(bx - c) \qquad \text{and} \qquad y = d + a \cos(bx - c).$$

The shift is d units upward for $d > 0$ and d units downward for $d < 0$. In other words, the graph oscillates about the horizontal line $y = d$ instead of about the x-axis.

EXAMPLE 6 Vertical Translation

Use a graphing utility to analyze the graph of $y = 2 + 3 \cos 2x$.

Solution

The amplitude is 3 and the period is π. The key points over the interval $[0, \pi]$ are

$$(0, 5), \qquad (\pi/4, 2), \qquad (\pi/2, -1), \qquad (3\pi/4, 2), \qquad \text{and} \qquad (\pi, 5).$$

The graph is shown in Figure 4.46. Compared with the graph of $f(x) = 3 \cos 2x$, the graph of $y = 2 + 3 \cos 2x$ is shifted upward two units.

Exploration

Use a graphing utility to graph $y = d + \sin x$, where $d = -2, 1,$ and 3. How does the value of d affect the graph?

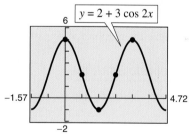

Figure 4.46

EXAMPLE 7 Finding an Equation for a Graph

Find the amplitude, period, and phase shift for the sine function whose graph is shown in Figure 4.47. Write an equation for this graph.

Solution

The amplitude for this sine curve is 2. The period is 2π, and there is a right phase shift of $\pi/2$. So, you can write

$$y = 2 \sin\left(x - \frac{\pi}{2}\right).$$

In Example 7, you can find a cosine function with the same graph. The amplitude for this cosine curve is 2. The period is 2π, and there is a right phase shift of π. So, you can write

$$y = 2 \cos(x - \pi).$$

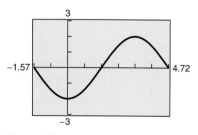

Figure 4.47

Mathematical Modeling

Sine and cosine functions can be used to model many real-life situations, including electric currents, musical tones, radio waves, tides, and weather patterns.

EXAMPLE 8   Finding a Trigonometric Model

Throughout the day, the depth of water at the end of a dock varies with the tides. The table shows the depths (in meters) at various times during the morning.

t (time)	Midnight	2 A.M.	4 A.M.	6 A.M.	8 A.M.	10 A.M.	Noon
y (depth)	2.55	3.80	4.40	3.80	2.55	1.80	2.27

a. Use a trigonometric function to model this data.

b. Find the depths at 9 A.M. and 3 P.M.

c. A boat needs at least 3 meters of water to moor at the dock. During what times in the afternoon can it safely dock?

Solution

a. Begin by graphing the data, as shown in Figure 4.48. You can use either a sine or cosine model. Suppose you use a cosine model of the form

$$y = a\cos(bt - c) + d.$$

The difference between the maximum height and minimum height of the graph is twice the amplitude of the function. So, the amplitude is

$$a = \tfrac{1}{2}[(\text{maximum depth}) - (\text{minimum depth})] = \tfrac{1}{2}(4.4 - 1.8) = 1.3.$$

The cosine function completes one half of a cycle between when the maximum and minimum depths occur. So, the period is

$$p = 2[(\text{time of min. depth}) - (\text{time of max. depth})] = 2(10 - 4) = 12$$

which implies that $b = 2\pi/p \approx 0.524$. Because high tide occurs 4 hours after midnight, consider the left endpoint to be $c/b = 4$, so $c \approx 2.094$. Moreover, because the average depth is $\tfrac{1}{2}(4.4 + 1.8) = 3.1$, it follows that $d = 3.1$. So, you can model the depth with the function

$$y = 1.3\cos(0.524t - 2.094) + 3.1.$$

b. The depths at 9 A.M. and 3 P.M. are as follows:

$$y = 1.3\cos(0.524 \cdot 9 - 2.094) + 3.1 \approx 1.97 \text{ meters} \qquad \text{9 A.M.}$$

$$y = 1.3\cos(0.524 \cdot 15 - 2.094) + 3.1 \approx 4.23 \text{ meters.} \qquad \text{3 P.M.}$$

c. Using a graphing utility, graph the model with the line $y = 3$, as shown in Figure 4.49. From the graph, it follows that the depth is at least 3 meters between 12:54 P.M. ($t \approx 12.9$) and 7:06 P.M. ($t \approx 19.1$).

Note that a sine model for the data in Example 8 is

$y = 1.3\sin(0.524t - 0.524) + 3.1$. Can you see how to find this?

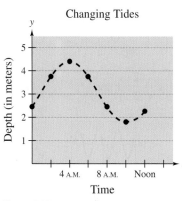

Figure 4.48

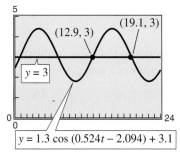

Figure 4.49

4.5   E x e r c i s e s

In Exercises 1–14, find the period and amplitude.

1. $y = 3 \sin 2x$

2. $y = 2 \cos 3x$

3. $y = \dfrac{5}{2} \cos \dfrac{x}{2}$

4. $y = -3 \sin \dfrac{x}{3}$

5. $y = \dfrac{2}{3} \sin \pi x$

6. $y = \dfrac{3}{2} \cos \dfrac{\pi x}{2}$

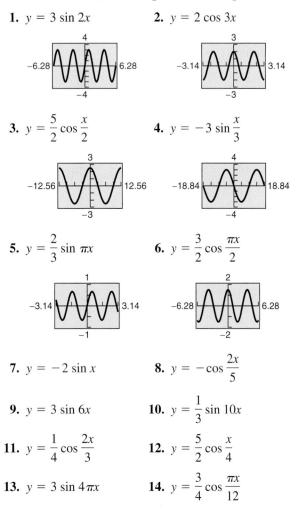

7. $y = -2 \sin x$

8. $y = -\cos \dfrac{2x}{5}$

9. $y = 3 \sin 6x$

10. $y = \dfrac{1}{3} \sin 10x$

11. $y = \dfrac{1}{4} \cos \dfrac{2x}{3}$

12. $y = \dfrac{5}{2} \cos \dfrac{x}{4}$

13. $y = 3 \sin 4\pi x$

14. $y = \dfrac{3}{4} \cos \dfrac{\pi x}{12}$

In Exercises 15–22, describe the relationship between the graphs of f and g.

15. $f(x) = \sin x$
     $g(x) = \sin(x - \pi)$

16. $f(x) = \cos x$
     $g(x) = \cos(x + \pi)$

17. $f(x) = \cos 2x$
     $g(x) = -\cos 2x$

18. $f(x) = \sin 3x$
     $g(x) = \sin(-3x)$

19. $f(x) = \cos x$
     $g(x) = -5 \cos x$

20. $f(x) = \sin x$
     $g(x) = \sin 3x$

21. $f(x) = \sin x$
     $g(x) = 4 + \sin x$

22. $f(x) = \cos 4x$
     $g(x) = -6 + \cos 4x$

In Exercises 23–26, describe the relationship between the graphs of f and g.

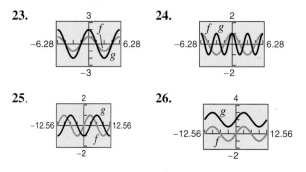

23.   **24.**

25.   **26.**

In Exercises 27–34, sketch the graphs of the two functions in the same coordinate plane. (Include two full periods.)

27. $f(x) = -2 \sin x$
     $g(x) = 4 \sin x$

28. $f(x) = \sin x$
     $g(x) = \sin \dfrac{x}{3}$

29. $f(x) = \cos x$
     $g(x) = 4 + \cos x$

30. $f(x) = 2 \cos 2x$
     $g(x) = -\cos 4x$

31. $f(x) = -\dfrac{1}{2} \sin \dfrac{x}{2}$
     $g(x) = 5 - \dfrac{1}{2} \sin \dfrac{x}{2}$

32. $f(x) = 4 \sin \pi x$
     $g(x) = 4 \sin \pi x - 1$

33. $f(x) = 2 \cos x$
     $g(x) = 2 \cos(x + \pi)$

34. $f(x) = -\cos x$
     $g(x) = -\cos\left(x - \dfrac{\pi}{2}\right)$

Conjecture   **In Exercises 35–38, use a graphing utility to graph f and g in the same viewing window. (Include two full periods.) Make a conjecture about the functions.**

35. $f(x) = \sin x$
     $g(x) = \cos\left(x - \dfrac{\pi}{2}\right)$

36. $f(x) = \sin x$
     $g(x) = -\cos\left(x + \dfrac{\pi}{2}\right)$

37. $f(x) = \cos x$
     $g(x) = -\sin\left(x - \dfrac{\pi}{2}\right)$

38. $f(x) = \cos x$
     $g(x) = -\cos(x - \pi)$

In Exercises 39–54, sketch the graph of the function by hand. Use a graphing utility to verify your sketch. (Include two full periods.)

39. $y = -2 \sin 4x$

40. $y = -3 \cos 2x$

41. $y = \cos 2\pi x$

42. $y = \dfrac{5}{2} \sin \dfrac{\pi x}{4}$

43. $y = -2 \sin \dfrac{2\pi x}{3}$

44. $y = -10 \cos \dfrac{\pi x}{6}$

45. $y = \sin\left(x - \dfrac{\pi}{4}\right)$

46. $y = \dfrac{1}{2} \sin(x - \pi)$

47. $y = 8 \cos(x + \pi)$

48. $y = 6 \cos\left(x + \dfrac{\pi}{6}\right)$

49. $y = \dfrac{1}{10} \cos 60\pi x$

50. $y = -4 + 5 \cos \dfrac{\pi t}{12}$

51. $y = 2 - 2 \sin \dfrac{2\pi x}{3}$

52. $y = 2 \cos x - 3$

53. $y = \dfrac{2}{3} \cos\left(\dfrac{x}{2} - \dfrac{\pi}{4}\right)$

54. $y = -3 \cos(6x + \pi)$

In Exercises 55–62, use a graphing utility to graph the function. (Include two full periods.) Identify the amplitude and period of the graph.

55. $y = -2 \sin(4x + \pi)$

56. $y = -4 \sin\left(\dfrac{2}{3}x - \dfrac{\pi}{3}\right)$

57. $y = \cos\left(2\pi x - \dfrac{\pi}{2}\right) + 1$

58. $y = 3 \cos\left(\dfrac{\pi x}{2} + \dfrac{\pi}{2}\right) - 3$

59. $y = 5 \sin(\pi - 2x) + 10$

60. $y = 5 \cos(\pi - 2x) + 6$

61. $y = \frac{1}{100} \sin 120\pi t$

62. $y = -\frac{97}{100} \cos 50\pi t$

Graphical Reasoning In Exercises 63–66, find a and d for the function $f(x) = a \cos x + d$ so that the graph of f matches the figure.

63.

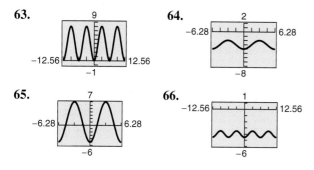

64.

65.

66.

Graphical Reasoning In Exercises 67–70, find a, b, and c for the function $f(x) = a \sin(bx - c)$ so that the graph of f matches the figure.

67.

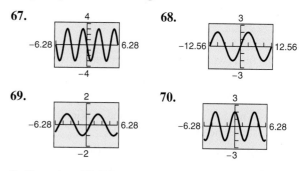

68.

69.

70.

In Exercises 71–74, use a graphing utility to graph y_1 and y_2 for all real numbers x in the interval $[-2\pi, 2\pi]$. Use the graphs to find the real numbers x such that $y_1 = y_2$.

71. $y_1 = \sin x$
$y_2 = -\dfrac{1}{2}$

72. $y_1 = \cos x$
$y_2 = -1$

73. $y_1 = \cos x$
$y_2 = \dfrac{\sqrt{2}}{2}$

74. $y_1 = \sin x$
$y_2 = \dfrac{\sqrt{3}}{2}$

75. *Respiratory Cycle* For a person at rest, the velocity v (in liters per second) of air flow during a respiratory cycle is

$$v = 0.85 \sin \dfrac{\pi t}{3}$$

where t is the time (in seconds). (Inhalation occurs when $v > 0$, and exhalation occurs when $v < 0$.)

(a) Use a graphing utility to graph v.

(b) Find the time for one full respiratory cycle.

(c) Find the number of cycles per minute.

76. *Respiratory Cycle* The model in Exercise 75 is for a person at rest. How might the model change for a person who is exercising? Explain.

Sales In Exercises 77 and 78, use a graphing utility to graph the sales function over 1 year, where S is the sales in thousands of units and t is the time in months, with $t = 1$ corresponding to January. Determine the months of maximum and minimum sales.

77. $S = 22.3 - 3.4 \cos \dfrac{\pi t}{6}$

78. $S = 74.50 + 43.75 \sin \dfrac{\pi t}{6}$

79. Ferris wheel You are riding a Ferris wheel. Your height h (in feet) above the ground at any time t (in seconds) can be modeled by

$$h = 25 \sin \frac{\pi}{15}(t - 75) + 30.$$

The ferris wheel turns for 135 seconds before it stops to let the first passengers off.

(a) Graph the model.

(b) What are your minimum and maximum heights above the ground?

80. Blood Pressure The pressure P (in millimeters of mercury) against the walls of the blood vessels of a certain person is modeled by

$$P = 100 - 20 \cos \frac{8\pi}{3}t$$

where t is the time (in seconds). Graph the model. If one cycle is equivalent to one heartbeat, what is the person's pulse rate in heartbeats per minute?

81. Fuel Consumption The daily consumption C (in gallons) of diesel fuel on a farm is modeled by

$$C = 30.3 + 21.6 \sin\left(\frac{2\pi t}{365} + 10.9\right)$$

where t is the time in days, with $t = 1$ corresponding to January 1.

(a) What is the period of the model? Is it what you expected? Explain.

(b) What is the average daily fuel consumption? Which term of the model did you use? Explain.

(c) Use a graphing utility to graph the model. Use the graph to approximate the time of the year when consumption exceeds 40 gallons per day.

82. Data Analysis The motion of an oscillating weight suspended by a spring was measured by a motion detector. The data was collected, and the approximate maximum displacements from equilibrium ($y = 2$) are labeled in the figure. The distance y from the motion detector is measured in centimeters and the time t is measured in seconds.

(a) Is y a function of t? Explain.

(b) Approximate the amplitude and period.

(c) Find a model for the data.

(d) Use a graphing utility to graph the model in part (c). Compare the result with the data in the figure.

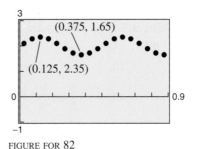

FIGURE FOR 82

83. Data Analysis The normal monthly precipitation p (in inches) for Seattle, Washington, for month t is shown in the table, with $t = 1$ corresponding to January. (Source: National Climatic Data Center)

t	1	2	3	4	5	6
p	5.4	4.0	3.8	2.5	1.8	1.6

t	7	8	9	10	11	12
p	0.9	1.2	1.9	3.3	5.7	6.0

(a) Find a trigonometric model for the normal monthly precipitation in Seattle.

(b) Use a graphing utility to graph the data and the model found in part (a). How well does the model fit the data?

84. Data Analysis The table shows the normal daily high temperatures for Honolulu H and Chicago C (in degrees Fahrenheit) for month t, with $t = 1$ corresponding to January. (Source: NOAA)

t	1	2	3	4	5	6
H	80.1	80.5	81.6	82.8	84.7	86.5
C	29.0	33.5	45.8	58.6	70.1	79.6

t	7	8	9	10	11	12
H	87.5	88.7	88.5	86.9	84.1	81.2
C	83.7	81.8	74.8	63.3	48.4	34.0

(a) A model for Honolulu is

$$H(t) = 84.40 + 4.28 \sin\left(\frac{\pi t}{6} + 3.86\right).$$

Find a trigonometric model for Chicago.

(b) Use a graphing utility to graph the data points and the model for the temperatures in Honolulu. How well does the model fit the data?

(c) Use a graphing utility to graph the data points and the model for the temperatures in Chicago. How well does the model fit the data?

(d) Use the models to estimate the average annual temperature in each city. Which term of the models did you use? Explain.

(e) What are the periods of the two models? Are they what you expected? Explain.

(f) Which city has the greater variability in temperature throughout the year? Which factor of the models determines this variability? Explain.

Synthesis

True or False? **In Exercises 85 and 86, determine whether the statement is true or false. Justify your answer.**

85. The graph of $y = 6 - \dfrac{3}{4} \sin \dfrac{3x}{10}$ has a period of $\dfrac{20\pi}{3}$.

86. The graph of $y = -\cos x$ is a reflection of the graph of $y = \sin(x + \pi/2)$ in the x-axis.

87. *Writing* Use a graphing utility to graph the function $y = a \sin x$ for $a = \frac{1}{2}$, $a = \frac{3}{2}$, and $a = -3$. Write a paragraph describing the changes in the graph for the specified changes in a.

88. *Writing* Use a graphing utility to graph the function $y = d + \sin x$ for $d = 2$, $d = 3.5$, and $d = -2$. Write a paragraph describing the changes in the graph for the specified changes in d.

89. *Writing* Use a graphing utility to graph the function $y = \sin bx$ for $b = \frac{1}{2}$, $b = \frac{3}{2}$, and $b = 4$. Write a paragraph describing the changes in the graph for the specified changes in b.

90. *Writing* Use a graphing utility to graph the function $y = \sin(x - c)$ for $c = 1$, $c = 3$, and $c = -2$. Write a paragraph describing the changes in the graph for the specified changes in c.

91. *Exploration* In Section 4.2 it was shown that $f(x) = \cos x$ is an even function and $g(x) = \sin x$ is an odd function. Use a graphing utility to graph h and use the graph to determine whether h is even, odd, or neither.

(a) $h(x) = \cos^2 x$

(b) $h(x) = \sin^2 x$

(c) $h(x) = \sin x \cos x$

92. *Conjecture* If f is an even function and g is an odd function, use the results of Exercise 91 to make a conjecture about the following.

(a) $h(x) = [f(x)]^2$

(b) $h(x) = [g(x)]^2$

(c) $h(x) = f(x)g(x)$

93. *Graphical Reasoning* The figure shows the graphs of the functions $f(x) = 1 - \frac{1}{2}x^2$ and $g(x) = \cos x$. Identify the graphs and explain your reasoning.

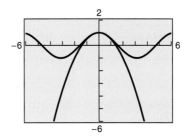

Review

In Exercises 94–97, sketch the graph of the rational function and identify its asymptotes. Use a graphing utility to verify your results.

94. $f(x) = \dfrac{2}{11 + x}$

95. $f(x) = \dfrac{5x - 3}{x}$

96. $f(x) = \dfrac{10}{x^2 + x - 12}$

97. $f(x) = \dfrac{x^2 - 7x + 12}{x + 2}$

In Exercises 98–101, convert the angle measure from radians to degrees. Round your answer to three decimal places.

98. $\dfrac{13\pi}{2}$

99. $-\dfrac{\pi}{9}$

100. 8.5π

101. -0.48

4.6   Graphs of Other Trigonometric Functions

Graph of the Tangent Function

Recall that the tangent function is odd. That is, $\tan(-x) = -\tan x$. Consequently, the graph of $y = \tan x$ is symmetric with respect to the origin. You also know from the identity $\tan x = \sin x / \cos x$ that the tangent is undefined when $\cos x = 0$. Two such values are $x = \pm \pi/2 \approx \pm 1.5708$.

x	$-\dfrac{\pi}{2}$	-1.57	-1.5	-1	0	1	1.5	1.57	$\dfrac{\pi}{2}$
$\tan x$	Undef.	-1255.8	-14.1	-1.56	0	1.56	14.1	1255.8	Undef.

$\tan x$ approaches $-\infty$ as x approaches $-\pi/2$ from the right.

$\tan x$ approaches ∞ as x approaches $\pi/2$ from the left.

As indicated in the table, $\tan x$ increases without bound as x approaches $\pi/2$ from the left, and it decreases without bound as x approaches $-\pi/2$ from the right. So, the graph of $y = \tan x$ has *vertical asymptotes* at $x = \pi/2$ and $-\pi/2$, as shown in Figure 4.50. Moreover, because the period of the tangent function is π, vertical asymptotes also occur when $x = \pi/2 + n\pi$, where n is an integer. The domain of the tangent function is the set of all real numbers other than $x = \pi/2 + n\pi$, and the range is the set of all real numbers.

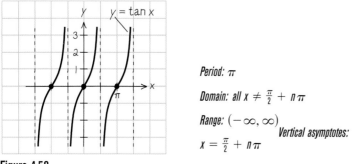

Period: π

Domain: all $x \neq \frac{\pi}{2} + n\pi$

Range: $(-\infty, \infty)$

Vertical asymptotes:
$x = \frac{\pi}{2} + n\pi$

Figure 4.50

Sketching the graph of a function of the form $y = a\tan(bx - c)$ is similar to sketching the graph of $y = a\sin(bx - c)$ in that you locate key points that identify the intercepts and asymptotes. Two consecutive asymptotes can be found by solving the equations $bx - c = -\pi/2$ and $bx - c = \pi/2$. The midpoint between two consecutive asymptotes is an x-intercept of the graph. The period of the function $y = a\tan(bx - c)$ is the distance between two consecutive asymptotes. The amplitude of a tangent function is not defined. After plotting the asymptotes and the x-intercept, plot a few additional points between the two asymptotes and sketch one cycle. Finally, sketch one or two additional cycles to the left and right.

What You Should Learn:

- How to sketch the graphs of tangent functions
- How to sketch the graphs of cotangent functions
- How to sketch the graphs of secant and cosecant functions
- How to sketch the graphs of damped trigonometric functions

Why You Should Learn It:

You can use tangent, cotangent, secant, and cosecant functions to model real-life data. For instance, Exercise 70 on page 342 shows you how a tangent function can be used to model and analyze the distance between a television camera and a parade unit.

A. Ramey/PhotoEdit

EXAMPLE 1   Sketching the Graph of a Tangent Function

Sketch the graph of $y = \tan \dfrac{x}{2}$.

Solution

By solving the equations

$$\frac{x}{2} = -\frac{\pi}{2} \quad \text{and} \quad \frac{x}{2} = \frac{\pi}{2}$$

$$x = -\pi \qquad\qquad x = \pi$$

you can see that two consecutive asymptotes occur at $x = -\pi$ and $x = \pi$. Between these two asymptotes, plot a few points, including the x-intercept, as shown in the table. Three cycles of the graph are shown in Figure 4.51. Use a graphing utility to confirm this graph.

x	$-\pi$	$-\dfrac{\pi}{2}$	0	$\dfrac{\pi}{2}$	π
$\tan \dfrac{x}{2}$	Undef.	-1	0	1	Undef.

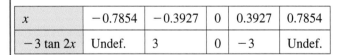

Figure 4.51

EXAMPLE 2   Sketching the Graph of a Tangent Function

Sketch the graph of $y = -3 \tan 2x$.

A computer animation of this example appears in the *Interactive* CD-ROM and Internet versions of this text.

Algebraic Solution

By solving the equations $2x = -\pi/2$ and $2x = \pi/2$, you can see that two consecutive asymptotes occur at $x = -\pi/4$ and $x = \pi/4$. Between these two asymptotes, plot a few points, including the x-intercept, as shown in the table. Three complete cycles of the graph are shown in Figure 4.52.

x	$-\dfrac{\pi}{4}$	$-\dfrac{\pi}{8}$	0	$\dfrac{\pi}{8}$	$\dfrac{\pi}{4}$
$-3 \tan 2x$	Undef.	3	0	-3	Undef.

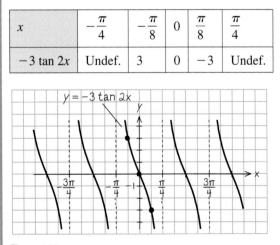

Figure 4.52

Graphical Solution

Use a graphing utility set in *radian* mode and *dot* mode to graph $y = -3 \tan 2x$, as shown in Figure 4.53. Use the *zero* or *root* feature or the *zoom* and *trace* features of the graphing utility to approximate the key points as shown in the table.

x	-0.7854	-0.3927	0	0.3927	0.7854
$-3 \tan 2x$	Undef.	3	0	-3	Undef.

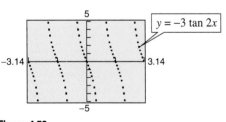

Figure 4.53

By comparing the graphs in Examples 1 and 2, you can see that the graph of $y = a \tan(bx - c)$ is increasing between consecutive vertical asymptotes if $a > 0$ and decreasing between consecutive vertical asymptotes if $a < 0$. In other words, the graph for $a < 0$ is a reflection in the x-axis of the graph for $a > 0$.

Graph of the Cotangent Function

The graph of the cotangent function is similar to the graph of the tangent function. It also has a period of π. However, from the identity

$$y = \cot x = \frac{\cos x}{\sin x}$$

you can see that the cotangent function has vertical asymptotes $x = n\pi$, where n is an integer, because $\sin x$ is zero at these x-values. The graph of the cotangent function is shown in Figure 4.54.

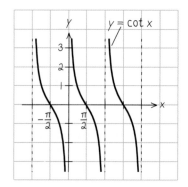

Period: π
Domain: all $x \neq n\pi$
Range: $(-\infty, \infty)$
Vertical asymptotes: $x = n\pi$
Figure 4.54

EXAMPLE 3   Sketching the Graph of a Cotangent Function

Sketch the graph of $y = 2 \cot \dfrac{x}{3}$.

Solution

To locate two consecutive vertical asymptotes of the graph, solve the equations $x/3 = 0$ and $x/3 = \pi$ to see that two consecutive asymptotes occur at $x = 0$ and $x = 3\pi$.

Then, between these two asymptotes, plot a few points, including the x-intercept, as shown in the table. Three cycles of the graph are shown in Figure 4.55. Use a graphing utility to confirm this graph. [Enter the function as $y = 2/\tan(x/3)$.] Note that the period is 3π, the distance between consecutive asymptotes.

x	0	$\dfrac{3\pi}{4}$	$\dfrac{3\pi}{2}$	$\dfrac{9\pi}{4}$	3π
$2 \cot \dfrac{x}{3}$	Undef.	2	0	-2	Undef.

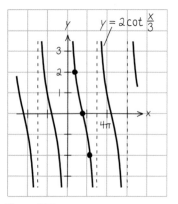

Figure 4.55

Graphs of the Reciprocal Functions

The graphs of the two remaining trigonometric functions can be obtained from the graphs of the sine and cosine functions using the reciprocal identities

$$\csc x = \frac{1}{\sin x} \qquad \text{and} \qquad \sec x = \frac{1}{\cos x}.$$

For instance, at a given value of x, the y-coordinate for sec x is the reciprocal of the y-coordinate for cos x. Of course, when $\cos x = 0$, the reciprocal does not exist. Near such values of x, the behavior of the secant function is similar to that of the tangent function. In other words, the graphs of

$$\tan x = \frac{\sin x}{\cos x} \qquad \text{and} \qquad \sec x = \frac{1}{\cos x}$$

have vertical asymptotes at $x = \pi/2 + n\pi$, where n is an integer and the cosine is zero at these x-values. Similarly,

$$\cot x = \frac{\cos x}{\sin x} \qquad \text{and} \qquad \csc x = \frac{1}{\sin x}$$

have vertical asymptotes where $\sin x = 0$, that is, at $x = n\pi$.

To sketch the graph of a secant or cosecant function, you should first make a sketch of its reciprocal function. For instance, to sketch the graph of $y = \csc x$, first sketch the graph of $y = \sin x$. Then take reciprocals of the y-coordinates to obtain points on the graph of $y = \csc x$. You can use this procedure to obtain the graphs shown in Figure 4.56.

Exploration

Use a graphing utility to graph the functions $y_1 = \sin x$ and $y_2 = \csc x = 1/\sin x$ in the same viewing window. How are the graphs related? What happens to the graph of the cosecant function as x approaches the zeros of the sine function?

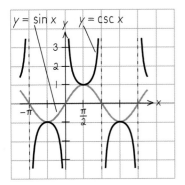

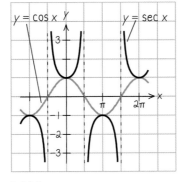

Period: 2π

Domain: all $x \neq n\pi$

Range: all y not in $(-1, 1)$

Vertical asymptotes: $x = n\pi$

Symmetry: origin

Figure 4.56

Period: 2π

Domain: all $x \neq \dfrac{\pi}{2} + n\pi$

Range: all y not in $(-1, 1)$

Vertical asymptotes: $x = \dfrac{\pi}{2} + n\pi$

Symmetry: y-axis

In comparing the graphs of the secant and cosecant functions with those of the sine and cosine functions, note that the "hills" and "valleys" are interchanged. For example, a hill (or maximum point) on the sine curve corresponds to a valley (a local minimum) on the cosecant curve. Similarly, a valley (or minimum point) on the sine curve corresponds to a hill (a local maximum) on the cosecant curve, as shown in Figure 4.57.

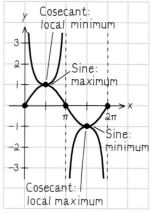

Figure 4.57

EXAMPLE 4   Comparing Trigonometric Graphs

Use a graphing utility to compare the graphs of

$$y = 2 \sin\left(x + \frac{\pi}{4}\right) \quad \text{and} \quad y = 2 \csc\left(x + \frac{\pi}{4}\right).$$

Solution

The two graphs are shown in Figure 4.58. Note how the hills and valleys of the graphs are related. For the function $y = 2 \sin[x + (\pi/4)]$, the amplitude is 2 and the period is 2π. By solving the double inequality

$$0 < x + \frac{\pi}{4} < 2\pi \quad \Longrightarrow \quad -\frac{\pi}{4} < x < \frac{7\pi}{4}$$

you can see that one cycle of the sine function corresponds to the interval from $x = -\pi/4$ to $x = 7\pi/4$. The graph of this sine function is represented by the gray curve in Figure 4.58. Because the sine function is zero at the endpoints of this interval, the corresponding cosecant function

$$y = 2 \csc\left(x + \frac{\pi}{4}\right) = 2\left(\frac{1}{\sin[x + (\pi/4)]}\right)$$

has vertical asymptotes at $x = -\pi/4, 3\pi/4, 7\pi/4$, etc. The graph of the cosecant function is represented by the black curve.

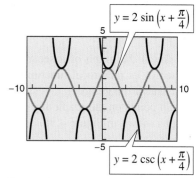

Figure 4.58

EXAMPLE 5   Comparing Trigonometric Graphs

Use a graphing utility to compare the graphs of

$$y = \cos 2x \quad \text{and} \quad y = \sec 2x.$$

Solution

Begin by graphing $y_1 = \cos 2x$ and $y_2 = \sec 2x = 1/\cos 2x$ in the same viewing window, as shown in Figure 4.59. Note that the x-intercepts of $y = \cos 2x$

$$\left(-\frac{\pi}{4}, 0\right), \quad \left(\frac{\pi}{4}, 0\right), \quad \left(\frac{3\pi}{4}, 0\right), \ldots$$

correspond to the vertical asymptotes

$$x = -\frac{\pi}{4}, \quad x = \frac{\pi}{4}, \quad x = \frac{3\pi}{4}, \ldots$$

of the graph of $y = \sec 2x$.

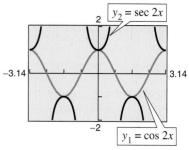

Figure 4.59

Damped Trigonometric Graphs

A *product* of two functions can be graphed using properties of the individual functions. For instance, consider the function

$$f(x) = x \sin x$$

as the product of the functions $y = x$ and $y = \sin x$. Using properties of absolute value and the fact that $|\sin x| \leq 1$, you have $0 \leq |x| \, |\sin x| \leq |x|$. Consequently,

$$-|x| \leq x \sin x \leq |x|$$

which means that the graph of $f(x) = x \sin x$ lies between the lines $y = -x$ and $y = x$. Furthermore, because

$$f(x) = x \sin x = \pm x \quad \text{at} \quad x = \frac{\pi}{2} + n\pi$$

$$f(x) = x \sin x = 0 \quad \text{at} \quad x = n\pi$$

the graph of f touches the line $y = -x$ or the line $y = x$ at $x = \pi/2 + n\pi$ and has x-intercepts at $x = n\pi$. A sketch of f is shown in Figure 4.60. In the function $f(x) = x \sin x$, the factor x is called the **damping factor.**

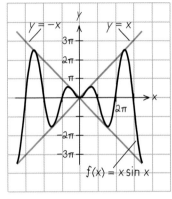

Figure 4.60

EXAMPLE 6   Analyzing a Damped Sine Curve

Analyze the graph of

$$f(x) = e^{-x} \sin 3x.$$

Solution

Consider $f(x)$ as the product of the two functions

$$y = e^{-x} \quad \text{and} \quad y = \sin 3x$$

each of which has the set of real numbers as its domain. For any real number x, you know that $e^{-x} \geq 0$ and $|\sin 3x| \leq 1$. Therefore, $|e^{-x}| \, |\sin 3x| \leq e^{-x}$, which means that

$$-e^{-x} \leq e^{-x} \sin 3x \leq e^{-x}.$$

Furthermore, because

$$f(x) = e^{-x} \sin 3x = \pm e^{-x} \quad \text{at} \quad x = \frac{\pi}{6} + \frac{n\pi}{3}$$

and

$$f(x) = e^{-x} \sin 3x = 0 \quad \text{at} \quad x = \frac{n\pi}{3}$$

the graph of f touches the curves $y = -e^{-x}$ and $y = e^{-x}$ at $x = \frac{\pi}{6} + \frac{n\pi}{3}$ and has intercepts at $x = \frac{n\pi}{3}$. The graph is shown in Figure 4.61.

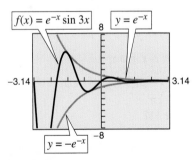

Figure 4.61

Figure 4.62 summarizes the graphs, domains, ranges, and periods of the six basic trigonometric functions.

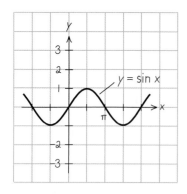

Domain: all reals

Range: $[-1, 1]$

Period: 2π

Domain: all reals

Range: $[-1, 1]$

Period: 2π

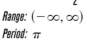

Domain: all $x \neq \dfrac{\pi}{2} + n\pi$

Range: $(-\infty, \infty)$

Period: π

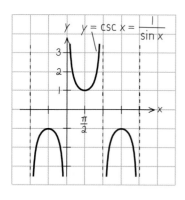

Domain: all $x \neq n\pi$

Range: $(-\infty, -1]$ and $[1, \infty)$

Period: 2π

Figure 4.62

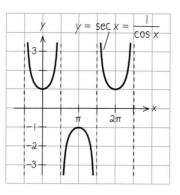

Domain: all $x \neq \dfrac{\pi}{2} + n\pi$

Range: $(-\infty, -1]$ and $[1, \infty)$

Period: 2π

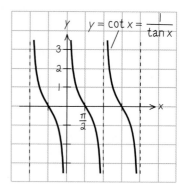

Domain: all $x \neq n\pi$

Range: $(-\infty, \infty)$

Period: π

Writing About Math   *Combining Trigonometric Functions*

Recall from Section 1.4 that functions can be combined arithmetically. This also applies to trigonometric functions. For each of the functions $h(x) = x + \sin x$ and $h(x) = \cos x - \sin 3x$, (a) identify two simpler functions f and g that comprise the combination, (b) use a table to show how to obtain the numerical values of $h(x)$ from the numerical values of $f(x)$ and $g(x)$, and (c) use graphing utility graphs of f and g to show how h may be formed.

Can you find functions

$$f(x) = d + a\sin(bx - c) \qquad \text{and} \qquad g(x) = d + a\cos(bx - c)$$

such that $f(x) + g(x) = 0$ for all x? If so, write a short paragraph explaining how you found the functions.

4.6 Exercises

In Exercises 1–8, match the function with its graph. State the period of the function. [The graphs are labeled (a), (b), (c), (d), (e), (f), (g), and (h).]

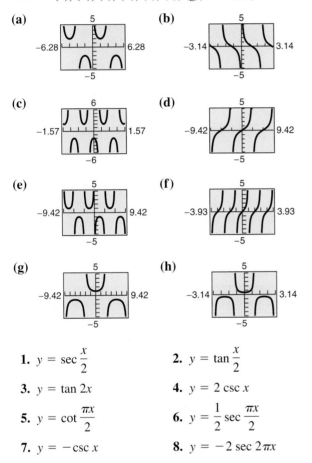

(a)

(b)

(c)

(d)

(e)

(f)

(g)

(h)

1. $y = \sec \dfrac{x}{2}$

2. $y = \tan \dfrac{x}{2}$

3. $y = \tan 2x$

4. $y = 2 \csc x$

5. $y = \cot \dfrac{\pi x}{2}$

6. $y = \dfrac{1}{2} \sec \dfrac{\pi x}{2}$

7. $y = -\csc x$

8. $y = -2 \sec 2\pi x$

In Exercises 9–30, sketch the graph of the function. (Include two full periods.) Use a graphing utility to verify your result.

9. $y = \frac{1}{3} \tan x$

10. $y = \frac{1}{4} \tan 2x$

11. $y = -2 \tan 2x$

12. $y = -3 \tan \pi x$

13. $y = -\frac{1}{2} \sec x$

14. $y = \frac{1}{4} \sec 2x$

15. $y = -\sec \pi x$

16. $y = 2 \sec 4x$

17. $y = \sec \pi x - 3$

18. $y = -2 \sec 4x + 2$

19. $y = \csc \dfrac{x}{2}$

20. $y = -\csc \dfrac{x}{3}$

21. $y = \dfrac{1}{2} \cot \dfrac{x}{2}$

22. $y = 3 \cot \dfrac{\pi x}{2}$

23. $y = \frac{1}{2} \sec 2x$

24. $y = -\frac{1}{2} \tan \pi x$

25. $y = 2 \tan \dfrac{\pi x}{4}$

26. $y = \sec(x + \pi)$

27. $y = \csc(\pi - x)$

28. $y = \sec(\pi - x)$

29. $y = 2 \cot\left(x - \dfrac{\pi}{2}\right)$

30. $y = \dfrac{1}{4} \csc\left(x + \dfrac{\pi}{4}\right)$

In Exercises 31–40, use a graphing utility to graph the function. (Include two full periods.) Describe your viewing window.

31. $y = \dfrac{1}{3} \tan \dfrac{x}{3}$

32. $y = -2 \tan 2\pi x$

33. $y = -2 \sec 4x$

34. $y = \dfrac{1}{4} \sec \pi x$

35. $y = \tan\left(x - \dfrac{\pi}{4}\right)$

36. $y = -\csc(4x - \pi)$

37. $y = \dfrac{1}{4} \cot\left(x + \dfrac{\pi}{2}\right)$

38. $y = 0.1 \tan\left(\dfrac{\pi x}{4} + \dfrac{\pi}{4}\right)$

39. $y = \dfrac{1}{2} \sec(2x - \pi)$

40. $y = \dfrac{1}{3} \sec\left(\dfrac{\pi x}{2} + \dfrac{\pi}{2}\right)$

In Exercises 41–44, use a graph to solve the equation in the interval $[-2\pi, 2\pi]$.

41. $\tan x = 1$

42. $\cot x = -\sqrt{3}$

43. $\sec x = -2$

44. $\csc x = \sqrt{2}$

In Exercises 45 and 46, use the graph of the function to determine whether the function is even, odd, or neither.

45. $f(x) = \sec x$

46. $f(x) = \tan x$

In Exercises 47–50, use a graphing utility to graph the two equations in the same viewing window. Use the graphs to lend evidence that the expressions are equivalent. Verify the results algebraically.

47. $y_1 = \sin x \csc x, \quad y_2 = 1$

48. $y_1 = \sin x \sec x, \quad y_2 = \tan x$

49. $y_1 = \dfrac{\cos x}{\sin x}, \quad y_2 = \cot x$

50. $y_1 = \sec^2 x - 1, \quad y_2 = \tan^2 x$

In Exercises 51–54, match the function with its graph. Describe the behavior of the function as x approaches zero. [The graphs are labeled (a), (b), (c), and (d).]

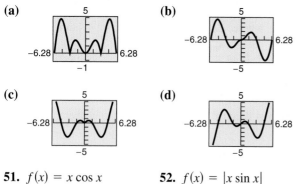

(a)

(b)

(c)

(d)

51. $f(x) = x \cos x$

52. $f(x) = |x \sin x|$

53. $g(x) = |x| \sin x$

54. $g(x) = |x| \cos x$

Conjecture In Exercises 55–58, use a graphing utility to graph the functions f and g. Use the graphs to make a conjecture about the relationship between the functions.

55. $f(x) = \sin x + \cos\left(x + \dfrac{\pi}{2}\right), \quad g(x) = 0$

56. $f(x) = \sin x - \cos\left(x + \dfrac{\pi}{2}\right), \quad g(x) = 2 \sin x$

57. $f(x) = \sin^2 x, \quad g(x) = \frac{1}{2}(1 - \cos 2x)$

58. $f(x) = \cos^2 \dfrac{\pi x}{2}, \quad g(x) = \dfrac{1}{2}(1 + \cos \pi x)$

In Exercises 59–62, use a graphing utility to graph the function and the damping factor of the function in the same viewing window. Describe the behavior of the function as x increases without bound.

59. $f(x) = e^{-x} \cos x$

60. $g(x) = e^{-x^2/2} \sin x$

61. $f(x) = 2^{-x/4} \cos \pi x$

62. $h(x) = 2^{-x^2/4} \sin x$

Exploration In Exercises 63–68, use a graphing utility to graph the function. Describe the behavior of the function as x approaches zero.

63. $f(x) = \dfrac{6}{x} + \cos x$

64. $f(x) = \sin x - \dfrac{4}{x}$

65. $f(x) = \dfrac{\sin x}{x}$

66. $f(x) = \dfrac{1 - \cos x}{x}$

67. $f(x) = \dfrac{\tan x}{x}$

68. $f(x) = \dfrac{x}{\cot x}$

69. *Distance* A plane flying at an altitude of 5 miles over level ground will pass directly over a radar antenna. Let d be the ground distance from the antenna to the point directly under the plane and let x be the angle of elevation to the plane from the antenna. Write d as a function of x and graph the function over the interval $0 < x < \pi$.

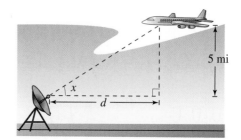

70. *Television Coverage* A television camera is on a reviewing platform 36 meters from the street on which a parade will be passing from left to right. Express the distance d from the camera to a particular unit in the parade as a function of the angle x, and graph the function over the interval $-\pi/2 < x < \pi/2$. (Consider x as negative when a unit in the parade approaches from the left.)

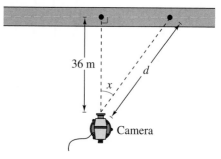

71. *Predatory-Prey Model* Suppose the population of a certain predator at time t (in months) in a region is estimated to be $P = 10{,}000 + 3000 \sin (2\pi t/24)$ and the population of its primary food source (its prey) is estimated to be $p = 15{,}000 + 5000 \cos (2\pi t/24)$. Use the graph of the models to explain the oscillations in the size of each population.

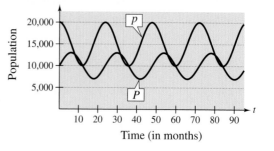

Time (in months)

72. Normal Temperatures   The normal monthly high temperatures in degrees Fahrenheit for Erie, Pennsylvania, are approximated by

$$H(t) = 54.33 - 20.38 \cos \frac{\pi t}{6} - 15.69 \sin \frac{\pi t}{6}$$

and the normal monthly low temperatures are approximated by

$$L(t) = 39.36 - 15.70 \cos \frac{\pi t}{6} - 14.16 \sin \frac{\pi t}{6}$$

where t is the time (in months), with $t = 1$ corresponding to January. (Source: National Oceanic and Atmospheric Association)

(a) Use a graphing utility to graph each function. What is the period of each function?

(b) During what part of the year is the difference between the normal high and low temperatures greatest? When is it smallest?

(c) The sun is the farthest north in the sky around June 21, but the graph shows the warmest temperatures at a later date. Approximate the lag time of the temperatures relative to the position of the sun.

73. Harmonic Motion   An object weighing W pounds is suspended from a ceiling by a steel spring. The weight is pulled downward (positive direction) from its equilibrium position and released. The resulting motion of the weight is described by the function

$$y = \frac{1}{2} e^{-t/4} \cos 4t, \quad t > 0$$

where y is the distance (in feet) and t is the time (in seconds).

(a) Use a graphing utility to graph the function.

(b) Describe the behavior of the displacement function for increasing values of time t.

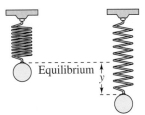

Equilibrium

74. Sales   The projected monthly sales S (in thousands of units) of a seasonal product is modeled by

$$S = 74 + 3t + 40 \sin \frac{\pi t}{6}$$

where t is the time (in months), with $t = 1$ corresponding to January.

(a) Graph the sales function over 1 year. Use a graphing utility to verify your graph.

(b) Which month should have the greatest number of sales? Which month should have the least number of sales? Justify your answers.

75. Sales   The projected monthly sales S (in thousands of units) of a seasonal product is modeled by

$$S = 52 + 5t - 28 \cos \frac{\pi t}{6}$$

where t is the time (in months), with $t = 1$ corresponding to January.

(a) Graph the sales function over 1 year. Use a graphing utility to verify your graph.

(b) Which month should have the greatest number of sales? Which month should have the least number of sales? Justify your answers.

76. Data Analysis   The motion of an oscillating weight suspended by a spring was measured by a motion detector. The data was collected, and the approximate maximum (positive and negative) displacements from equilibrium are shown in the graph. The displacement y is measured in centimeters and the time t is measured in seconds.

(a) Is y a function of t? Explain.

(b) Approximate the frequency of the oscillations.

(c) Fit a model of the form $y = ab^t \cos ct$ to the data. Use the result of part (b) to approximate c. Use the regression capabilities of a graphing utility to fit an exponential model to the positive maximum displacements of the weight.

(d) Rewrite the model in the form $y = ae^{kt} \cos ct$.

(e) Use a graphing utility to graph the model. Compare the result with the data in the graph.

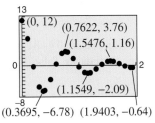

77. Writing   Write a short paragraph describing the specified change in the physical system of Exercise 76.

(a) A spring of less stiffness is used, and so the length of time in each oscillation is greater.

(b) The effect of friction is decreased.

78. Numerical and Graphical Reasoning A crossed belt connects a 10-centimeter pulley on an electric motor with a 20-centimeter pulley on a saw arbor. The electric motor runs at 1700 revolutions per minute.

(a) Determine the number of revolutions per minute of the saw.

(b) How does crossing the belt affect the saw in relation to the motor?

(c) Let L be the total length of the belt. Write L as a function of ϕ, where ϕ is measured in radians. What is the domain of the function? (*Hint:* Add the lengths of the straight sections of the belt and the length of belt around each pulley.)

(d) Use a graphing utility to complete the table.

ϕ	0.3	0.6	0.9	1.2	1.5
L					

(e) As ϕ increases, do the lengths of the straight sections of the belt change faster or slower than the lengths of belt around each pulley?

(f) Use a graphing utility to graph the function over the appropriate domain.

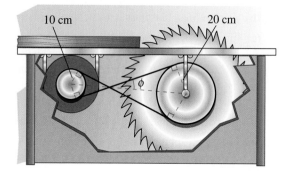

Synthesis

True or False? **In Exercises 79–81, determine whether the statement is true or false. Justify your answer.**

79. The graph of $y = -\dfrac{1}{8}\tan\left(\dfrac{x}{2} + \pi\right)$ has an asymptote at $x = -3\pi$.

80. The graph of $y = -2\csc\left(x + \dfrac{\pi}{3}\right)$ has an asymptote at $x = -\dfrac{\pi}{3}$.

81. In the graph of $y = 2^x \sin x$, as x approaches $-\infty$, y approaches 0.

82. Writing Describe the behavior of $f(x) = \tan x$ as x approaches $\pi/2$ from the left and from the right.

83. Writing Describe the behavior of $f(x) = \csc x$ as x approaches π from the left and from the right.

84. Graphical Reasoning Consider the two functions $f(x) = 2\sin x$ and $g(x) = \frac{1}{2}\csc x$ on the interval $(0, \pi)$.

(a) Use a graphing utility to graph f and g in the same viewing window.

(b) Approximate the interval where $f > g$.

(c) Describe the behavior of each of the functions as x approaches π. How is the behavior of g related to the behavior of f as x approaches π?

85. Pattern Recognition

(a) Use a graphing utility to graph each function.

$$y_1 = \frac{4}{\pi}\left(\sin \pi x + \frac{1}{3}\sin 3\pi x\right)$$

$$y_2 = \frac{4}{\pi}\left(\sin \pi x + \frac{1}{3}\sin 3\pi x + \frac{1}{5}\sin 5\pi x\right)$$

(b) Identify the pattern in part (a) and find a function y_3 that continues the pattern one more term. Use a graphing utility to graph y_3.

(c) The graphs of parts (a) and (b) approximate the periodic function in the figure. Find a function y_4 that is a better approximation.

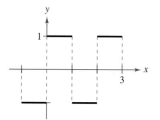

Review

In Exercises 86–89, determine whether the function is one-to-one. If it is, find its inverse.

86. $f(x) = -10$ **87.** $f(x) = (x - 7)^2 + 3$

88. $f(x) = \sqrt{3x - 14}$ **89.** $f(x) = \sqrt[3]{x - 5}$

In Exercises 90 and 91, find the exact values of the six trigonometric functions of the angle θ.

90. **91.**

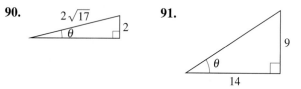

4.7 Inverse Trigonometric Functions

Inverse Sine Function

Recall from Section 1.5 that for a function to have an inverse, it must pass the Horizontal Line Test. From Figure 4.63 it is obvious that $y = \sin x$ does not pass the test because different values of x yield the same y-value.

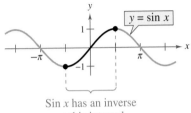

Sin x has an inverse
on this interval.

Figure 4.63

However, if you restrict the domain to the interval $-\pi/2 \le x \le \pi/2$ (corresponding to the black portion of the graph in Figure 4.63), the following properties hold.

1. On the interval $[-\pi/2, \pi/2]$, the function $y = \sin x$ is increasing.
2. On the interval $[-\pi/2, \pi/2]$, $y = \sin x$ takes on its full range of values, $-1 \le \sin x \le 1$.
3. On the interval $[-\pi/2, \pi/2]$, $y = \sin x$ passes the Horizontal Line Test.

So, on the restricted domain $-\pi/2 \le x \le \pi/2$, $y = \sin x$ has a unique inverse called the **inverse sine function.** It is denoted by

$$y = \arcsin x \qquad \text{or} \qquad y = \sin^{-1} x.$$

The notation $\sin^{-1} x$ is consistent with the inverse function notation $f^{-1}(x)$. The arcsin x notation (read as "the arcsine of x") comes from the association of a central angle with its intercepted *arc length* on a unit circle. So, arcsin x means the angle (or arc) whose sine is x. Both notations, arcsin x and $\sin^{-1} x$, are commonly used in mathematics, so remember that $\sin^{-1} x$ denotes the *inverse* sine function rather than $1/\sin x$. The values of arcsin x lie in the interval $-\pi/2 \le \arcsin x \le \pi/2$. The graph of $y = \arcsin x$ is shown in Example 2.

What You Should Learn:

- How to evaluate inverse sine functions
- How to evaluate other inverse trigonometric functions
- How to evaluate compositions of trigonometric functions

Why You Should Learn It:

Inverse trigonometric functions can be useful in exploring how two aspects of a real-life problem relate to each other. Exercise 70 on page 353 investigates the relationship between the length of rope from a winch to a boat and the angle of elevation between them.

Arlene Collins

Definition of Inverse Sine Function

The **inverse sine function** is defined by

$$y = \arcsin x \qquad \text{if and only if} \qquad \sin y = x$$

where $-1 \le x \le 1$ and $-\pi/2 \le y \le \pi/2$. The domain of $y = \arcsin x$ is $[-1, 1]$ and the range is $[-\pi/2, \pi/2]$.

When evaluating the inverse sine function, it helps to remember the phrase "the arcsine of x is the angle (or number) whose sine is x."

EXAMPLE 1   Evaluating the Inverse Sine Function

If possible, find the exact value.

a. $\arcsin\left(-\dfrac{1}{2}\right)$ **b.** $\sin^{-1}\dfrac{\sqrt{3}}{2}$ **c.** $\sin^{-1} 2$

Solution

a. Because $\sin(-\pi/6) = -\dfrac{1}{2}$ for $-\pi/2 \leq y \leq \pi/2$, it follows that

$$\arcsin\left(-\frac{1}{2}\right) = -\frac{\pi}{6}.$$ *Angle whose sine is $-\frac{1}{2}$*

b. Because $\sin(\pi/3) = \sqrt{3}/2$ for $-\pi/2 \leq y \leq \pi/2$, it follows that

$$\sin^{-1}\frac{\sqrt{3}}{2} = \frac{\pi}{3}.$$ *Angle whose sine is $\frac{\sqrt{3}}{2}$*

c. It is not possible to evaluate $y = \sin^{-1} x$ when $x = 2$ because there is no angle whose sine is 2. Remember that the domain of the inverse sine function is $[-1, 1]$.

EXAMPLE 2   Graphing the Arcsine Function

Sketch a graph of $y = \arcsin x$.

Solution

By definition, the equations

$$y = \arcsin x \qquad \text{and} \qquad \sin y = x$$

are equivalent for $-\pi/2 \leq y \leq \pi/2$. So, their graphs are the same. From the interval $[-\pi/2, \pi/2]$, you can assign values to y in the second equation to make a table of values.

y	$-\dfrac{\pi}{2}$	$-\dfrac{\pi}{4}$	$-\dfrac{\pi}{6}$	0	$\dfrac{\pi}{6}$	$\dfrac{\pi}{4}$	$\dfrac{\pi}{2}$
$x = \sin y$	-1	$-\dfrac{\sqrt{2}}{2}$	$-\dfrac{1}{2}$	0	$\dfrac{1}{2}$	$\dfrac{\sqrt{2}}{2}$	1

The resulting graph for

$$y = \arcsin x$$

is shown in Figure 4.64. Note that it is the reflection (in the line $y = x$) of the black portion of the graph in Figure 4.63. Use a graphing utility to confirm this graph. Be sure you see that Figure 4.64 shows the *entire* graph of the inverse sine function. Remember that the range of $y = \arcsin x$ is the closed interval $[-\pi/2, \pi/2]$.

STUDY TIP

As with the trigonometric functions, much of the work with the inverse trigonometric functions can be done by *exact* calculations rather than by calculator approximations. Exact calculations help to increase your understanding of the inverse functions by relating them to the triangle definitions of the trigonometric functions.

Library of Functions

The inverse trigonometric functions are obtained from the trigonometric functions in much the same way that the logarithmic function was developed from the exponential function. However, unlike the exponential function, the trigonometric functions are not one-to-one, and so it is necessary to restrict their domains to regions that pass the Horizontal Line Test. Consult the Library of Functions Summary inside the front cover for a description of the inverse trigonometric functions.

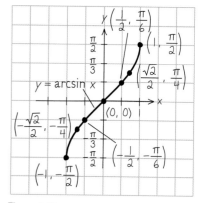

Figure 4.64

Other Inverse Trigonometric Functions

The cosine function is decreasing on the interval $0 \le x \le \pi$, as shown in Figure 4.65.

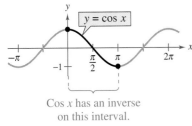

Cos x has an inverse on this interval.

Figure 4.65

Consequently, on this interval the cosine function has an inverse function—the **inverse cosine function**—denoted by

$$y = \arccos x \qquad \text{or} \qquad y = \cos^{-1} x.$$

Because these equations are equivalent for $0 \le y \le \pi$, their values are the same, as shown in the table below.

y	0	$\dfrac{\pi}{6}$	$\dfrac{\pi}{3}$	$\dfrac{\pi}{2}$	$\dfrac{2\pi}{3}$	$\dfrac{5\pi}{6}$	π
$x = \cos y$	1	$\dfrac{\sqrt{3}}{2}$	$\dfrac{1}{2}$	0	$-\dfrac{1}{2}$	$-\dfrac{\sqrt{3}}{2}$	-1

Similarly, you can define an **inverse tangent function** by restricting the domain of $y = \tan x$ to the interval $(-\pi/2, \pi/2)$. The following list summarizes the definitions of the three most common inverse trigonometric functions. The remaining three are discussed in Exercises 80–82 in this section.

Definition of Inverse Trigonometric Functions

Function	Domain	Range
$y = \arcsin x$ if and only if $\sin y = x$	$-1 \le x \le 1$	$-\dfrac{\pi}{2} \le y \le \dfrac{\pi}{2}$
$y = \arccos x$ if and only if $\cos y = x$	$-1 \le x \le 1$	$0 \le y \le \pi$
$y = \arctan x$ if and only if $\tan y = x$	$-\infty < x < \infty$	$-\dfrac{\pi}{2} < y < \dfrac{\pi}{2}$

The graphs of these three inverse trigonometric functions are shown in Figure 4.66.

Domain: $[-1, 1]$; *Range:* $\left[-\dfrac{\pi}{2}, \dfrac{\pi}{2}\right]$

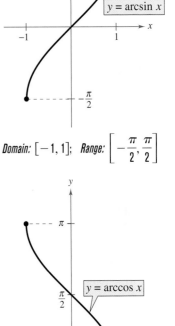

Domain: $[-1, 1]$; *Range:* $[0, \pi]$

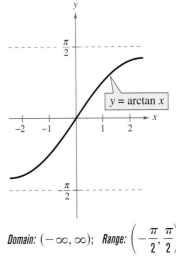

Domain: $(-\infty, \infty)$; *Range:* $\left(-\dfrac{\pi}{2}, \dfrac{\pi}{2}\right)$

Figure 4.66

 A computer animation of this concept appears in the *Interactive* CD-ROM and *Internet* versions of this text.

EXAMPLE 3   Evaluating Inverse Trigonometric Functions

Find the exact value.

a. $\arccos \dfrac{\sqrt{2}}{2}$ **b.** $\arccos(-1)$ **c.** $\arctan 0$ **d.** $\arctan(-1)$

Solution

a. Because $\cos(\pi/4) = \sqrt{2}/2$, and $\pi/4$ lies in $[0, \pi]$, it follows that

$$\arccos \dfrac{\sqrt{2}}{2} = \dfrac{\pi}{4}. \qquad \text{Angle whose cosine is } \dfrac{\sqrt{2}}{2}$$

b. Because $\cos \pi = -1$, and π lies in $[0, \pi]$, it follows that

$$\arccos(-1) = \pi. \qquad \text{Angle whose cosine is } -1$$

c. Because $\tan 0 = 0$, and 0 lies in $(-\pi/2, \pi/2)$, it follows that

$$\arctan 0 = 0. \qquad \text{Angle whose tangent is } 0$$

d. Because $\tan(-\pi/4) = -1$ and $-\pi/4$ lies in $(-\pi/2, \pi/2)$, it follows that

$$\arctan(-1) = -\dfrac{\pi}{4}. \qquad \text{Angle whose tangent is } -1$$

You can use the $\boxed{\text{SIN}^{-1}}$, $\boxed{\text{COS}^{-1}}$, and $\boxed{\text{TAN}^{-1}}$ keys on your calculator to approximate values of trigonometric functions.

STUDY T!P

Most graphing utilities do not have function keys for the inverse cosecant function, the inverse secant function, or the inverse cotangent function. To evaluate these functions, you can use the inverse sine, inverse cosine, and inverse tangent functions, respectively. For instance, to evaluate arcsec 3.4, enter the expression as arccos(1/3.4).

EXAMPLE 4   Calculators and Inverse Trigonometric Functions

Use a calculator to approximate the value (if possible).

a. $\arctan(-8.45)$ **b.** $\arcsin 0.2447$ **c.** $\arccos 2$

Solution

Function	Mode	Graphing Calculator Keystrokes
a. $\arctan(-8.45)$	Radian	$\boxed{\text{TAN}^{-1}}$ $\boxed{(}$ $\boxed{(-)}$ 8.45 $\boxed{)}$ $\boxed{\text{ENTER}}$

From the display, it follows that $\arctan(-8.45) \approx -1.453001$.

b. $\arcsin 0.2447$	Radian	$\boxed{\text{SIN}^{-1}}$ 0.2447 $\boxed{\text{ENTER}}$

From the display, it follows that $\arcsin 0.2447 \approx 0.2472103$.

c. $\arccos 2$	Radian	$\boxed{\text{COS}^{-1}}$ 2 $\boxed{\text{ENTER}}$

In real number mode, the calculator should display an error message because the domain of the inverse cosine function is $[-1, 1]$.

In Example 4, if you had set the calculator to *degree* mode, the display would have been in degrees rather than in radians. This convention is peculiar to calculators. By definition, the values of inverse trigonometric functions are always in *radians*.

Compositions of Functions

Recall from Section 1.5 that inverse functions possess the properties

$$f(f^{-1}(x)) = x \quad \text{and} \quad f^{-1}(f(x)) = x.$$

The inverse trigonometric versions of these properties are given below.

Inverse Properties

If $-1 \leq x \leq 1$ and $-\pi/2 \leq y \leq \pi/2$, then

$$\sin(\arcsin x) = x \quad \text{and} \quad \arcsin(\sin y) = y.$$

If $-1 \leq x \leq 1$ and $0 \leq y \leq \pi$, then

$$\cos(\arccos x) = x \quad \text{and} \quad \arccos(\cos y) = y.$$

If x is a real number and $-\pi/2 < y < \pi/2$, then

$$\tan(\arctan x) = x \quad \text{and} \quad \arctan(\tan y) = y.$$

Keep in mind that these inverse properties do not apply for arbitrary values of x and y. For instance,

$$\arcsin\left(\sin \frac{3\pi}{2}\right) = \arcsin(-1) = -\frac{\pi}{2} \neq \frac{3\pi}{2}.$$

In other words, the property

$$\arcsin(\sin y) = y$$

is not valid for values of y outside the interval $[-\pi/2, \pi/2]$.

Exploration

Use a graphing utility to graph $y = \arcsin(\sin x)$. What are the domain and range of this function? Explain why $\arcsin(\sin 4)$ does not equal 4.

Now graph $y = \sin(\arcsin x)$ and determine the domain and range. Explain why $\sin(\arcsin 4)$ is not defined.

EXAMPLE 5   Using Inverse Properties

If possible, find the exact value.

a. $\tan[\arctan(-5)]$ **b.** $\arcsin\left(\sin \dfrac{5\pi}{3}\right)$ **c.** $\cos(\cos^{-1} \pi)$

Solution

a. Because -5 lies in the domain of $\arctan x$, the inverse property applies, and you have

$$\tan[\arctan(-5)] = -5.$$

b. In this case, $5\pi/3$ does not lie within the range of the arcsine function, $-\pi/2 \leq y \leq \pi/2$. However, $5\pi/3$ is coterminal with

$$\frac{5\pi}{3} - 2\pi = -\frac{\pi}{3}$$

which does lie in the range of the arcsine function, and you have

$$\arcsin\left(\sin \frac{5\pi}{3}\right) = \arcsin\left[\sin\left(-\frac{\pi}{3}\right)\right] = -\frac{\pi}{3}.$$

c. The expression $\cos(\cos^{-1} \pi)$ is not defined because $\cos^{-1} \pi$ is not defined. Remember that the domain of the inverse cosine function is $[-1, 1]$.

Example 6 shows how to use right triangles to find exact values of compositions of inverse functions. Then, Example 7 shows how to use triangles to convert a trigonometric expression into an algebraic expression. This conversion technique is used frequently in calculus.

EXAMPLE 6   Evaluating Compositions of Functions

Find the exact value of (a) $\tan\left(\arccos\dfrac{2}{3}\right)$ and (b) $\cos\left[\arcsin\left(-\dfrac{3}{5}\right)\right]$.

Algebraic Solution

a. If you let $u = \arccos\frac{2}{3}$, then $\cos u = \frac{2}{3}$. Because $\cos u$ is positive, u is a first-quadrant angle. You can sketch and label angle u as shown in Figure 4.67(a). Consequently,

$$\tan\left(\arccos\frac{2}{3}\right) = \tan u = \frac{\text{opp}}{\text{adj}} = \frac{\sqrt{5}}{2}.$$

b. If you let $u = \arcsin\left(-\frac{3}{5}\right)$, then $\sin u = -\frac{3}{5}$. Because $\sin u$ is negative, u is a fourth-quadrant angle. You can sketch and label angle u as shown in Figure 4.67(b). Consequently,

$$\cos\left[\arcsin\left(-\frac{3}{5}\right)\right] = \cos u = \frac{\text{adj}}{\text{hyp}} = \frac{4}{5}.$$

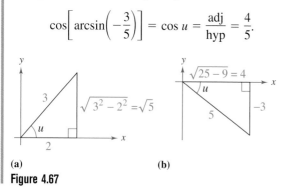

(a)   **(b)**

Figure 4.67

Graphical Solution

a. Use a graphing utility set in *radian* mode to graph $y = \tan(\arccos x)$, as shown in Figure 4.68(a). Use the *value* feature or *zoom* and *trace* features of the graphing utility to find that the value of the composition of functions when $x = \frac{2}{3} \approx 0.66667$ is $y = 1.118 \approx \sqrt{5}/2$.

b. Use a graphing utility set in *radian* mode to graph $y = \cos(\arcsin x)$, as shown in Figure 4.68(b). Use the *value* feature or *zoom* and *trace* features of the graphing utility to find that the value of the composition of functions when $x = -\frac{3}{5} = -0.6$ is $y = 0.8 = \frac{4}{5}$.

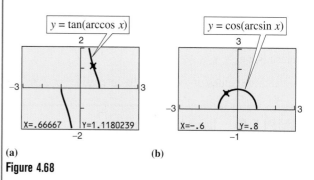

(a)   **(b)**

Figure 4.68

EXAMPLE 7   Some Problems from Calculus

Write each of the following as an algebraic expression in x.

a. $\sin(\arccos 3x), \quad 0 \le x \le \dfrac{1}{3}$ **b.** $\cot(\arccos 3x), \quad 0 \le x \le \dfrac{1}{3}$

Solution

If you let $u = \arccos 3x$, then $\cos u = 3x$. Because $\cos u = 3x/1 = \text{adj}/\text{hyp}$ you can sketch a right triangle with acute angle u, as shown in Figure 4.69. From this triangle, you can easily convert each expression to algebraic form.

a. $\sin(\arccos 3x) = \sin u = \dfrac{\text{opp}}{\text{hyp}} = \sqrt{1 - 9x^2}, \qquad 0 \le x < \dfrac{1}{3}$

b. $\cot(\arccos 3x) = \cot u = \dfrac{\text{adj}}{\text{opp}} = \dfrac{3x}{\sqrt{1 - 9x^2}}, \qquad 0 \le x < \dfrac{1}{3}$

Figure 4.69

A similar argument can be made here for x-values lying in the interval $\left[-\frac{1}{3}, 0\right]$.

4.7 Exercises

1. *Numerical and Graphical Analysis* Consider the function $y = \arcsin x$.

   (a) Use a graphing utility to complete the table.

x	-1	-0.8	-0.6	-0.4	-0.2
y					

x	0	0.2	0.4	0.6	0.8	1
y						

   (b) Plot the points from the table in part (a) and graph the function. (Do not use a graphing utility.)

   (c) Use a graphing utility to graph the inverse sine function and compare the result with your hand-drawn graph in part (b).

   (d) Determine any intercepts and symmetry of the graph.

2. *Numerical and Graphical Analysis* Consider the function $y = \arccos x$.

   (a) Use a graphing utility to complete the table.

x	-1	-0.8	-0.6	-0.4	-0.2
y					

x	0	0.2	0.4	0.6	0.8	1
y						

   (b) Plot the points from the table in part (a) and graph the function. (Do not use a graphing utility.)

   (c) Use a graphing utility to graph the inverse cosine function and compare the result with your hand-drawn graph in part (b).

   (d) Determine any intercepts and symmetry of the graph.

3. *Numerical and Graphical Analysis* Consider the function $y = \arctan x$.

   (a) Use a graphing utility to complete the table.

x	-10	-8	-6	-4	-2
y					

x	0	2	4	6	8	10
y						

   (b) Plot the points from the table in part (a) and graph the function. (Do not use a graphing utility.)

   (c) Use a graphing utility to graph the inverse tangent function and compare the result with your hand-drawn graph in part (b).

   (d) Determine the horizontal asymptotes of the graph.

In Exercises 4–7, write each trigonometric expression in inverse function form, or vice versa. For example,

$$\sin \frac{\pi}{2} = 1 \quad \Longrightarrow \quad \arcsin 1 = \frac{\pi}{2}.$$

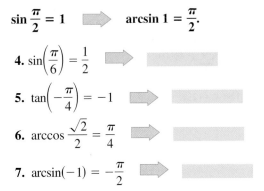

4. $\sin\left(\dfrac{\pi}{6}\right) = \dfrac{1}{2}$ $\Longrightarrow$

5. $\tan\left(-\dfrac{\pi}{4}\right) = -1$ $\Longrightarrow$

6. $\arccos \dfrac{\sqrt{2}}{2} = \dfrac{\pi}{4}$ $\Longrightarrow$

7. $\arcsin(-1) = -\dfrac{\pi}{2}$ $\Longrightarrow$

In Exercises 8–15, evaluate the expression using a calculator.

8. (a) $\arcsin \dfrac{1}{2}$ (b) $\arcsin 0$

9. (a) $\arccos \dfrac{1}{2}$ (b) $\arccos 0$

10. (a) $\arctan \dfrac{\sqrt{3}}{3}$ (b) $\arctan 1$

11. (a) $\arccos\left(-\dfrac{\sqrt{2}}{2}\right)$ (b) $\arcsin\left(-\dfrac{\sqrt{2}}{2}\right)$

12. (a) $\arctan\left(-\sqrt{3}\right)$ (b) $\arctan \sqrt{3}$

13. (a) $\arccos\left(-\dfrac{1}{2}\right)$ (b) $\arcsin \dfrac{\sqrt{2}}{2}$

14. (a) $\arcsin\left(-\dfrac{\sqrt{3}}{2}\right)$ (b) $\arctan\left(-\dfrac{\sqrt{3}}{3}\right)$

15. (a) $\arcsin(-1)$ (b) $\arccos 1$

In Exercises 16 and 17, determine the missing coordinates of the points on the graph of the function.

16. **17.**

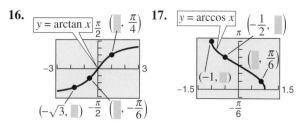

In Exercises 18–23, use a calculator to approximate the value of the expression. (Round your answer to two decimal places.)

18. (a) $\arccos 0.22$ (b) $\arcsin 0.45$

19. (a) $\arcsin(-0.75)$ (b) $\arccos(-0.7)$

20. (a) $\arctan(-6)$ (b) $\arctan 18$

21. (a) $\arcsin 0.41$ (b) $\arccos 0.36$

22. (a) $\arccos(-0.51)$ (b) $\arcsin(-0.125)$

23. (a) $\arctan 0.98$ (b) $\arctan 4.7$

In Exercises 24 and 25, use a graphing utility to graph f, g, and $y = x$ in the same viewing window to verify geometrically that g is the inverse of f. (Be sure to properly restrict the domain of f.)

24. $f(x) = \tan x, \quad g(x) = \arctan x$

25. $f(x) = \sin x, \quad g(x) = \arcsin x$

In Exercises 26–29, use an inverse trigonometric function to write θ as a function of x.

26. **27.**

28. **29.**

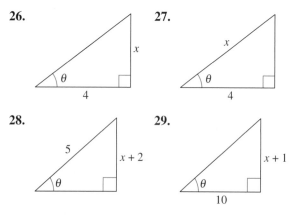

In Exercises 30–37, use the properties of inverse functions to evaluate the expression.

30. $\sin(\arcsin 0.7)$ **31.** $\tan(\arctan 35)$

32. $\cos[\arccos(-0.3)]$ **33.** $\sin[\arcsin(-0.1)]$

34. $\arcsin(\sin 3\pi)$ **35.** $\arccos\left(\cos\dfrac{7\pi}{2}\right)$

36. $\arctan\left(\tan\dfrac{11\pi}{6}\right)$ **37.** $\arcsin\left(\sin\dfrac{7\pi}{4}\right)$

In Exercises 38–45, find the exact value of the expression. Use a graphing utility to verify your result. (*Hint:* Make a sketch of a right triangle.)

38. $\sin\left(\arctan\dfrac{4}{3}\right)$ **39.** $\sec\left(\arcsin\dfrac{3}{5}\right)$

40. $\cos\left(\arcsin\dfrac{24}{25}\right)$ **41.** $\csc\left[\arctan\left(-\dfrac{12}{5}\right)\right]$

42. $\sec\left[\arctan\left(-\dfrac{3}{5}\right)\right]$ **43.** $\tan\left[\arcsin\left(-\dfrac{3}{8}\right)\right]$

44. $\sin\left[\arccos\left(-\dfrac{2}{3}\right)\right]$ **45.** $\cot\left(\arctan\dfrac{6}{11}\right)$

In Exercises 46–53, write an algebraic expression that is equivalent to the expression. (*Hint:* Sketch a right triangle, as demonstrated in Example 7.)

46. $\cot(\arctan x)$ **47.** $\sin(\arctan x)$

48. $\sin(\arccos x)$ **49.** $\sec[\arcsin(x - 1)]$

50. $\tan\left(\arccos\dfrac{x}{5}\right)$ **51.** $\cot\left(\arctan\dfrac{1}{x}\right)$

52. $\csc\left(\arctan\dfrac{x}{\sqrt{7}}\right)$ **53.** $\cos\left(\arcsin\dfrac{x - h}{r}\right)$

In Exercises 54 and 55, use a graphing utility to graph f and g in the same viewing window to verify that the two are equal. Explain why they are equal. Identify any asymptotes of the graphs.

54. $f(x) = \sin(\arctan 2x), \quad g(x) = \dfrac{2x}{\sqrt{1 + 4x^2}}$

55. $f(x) = \tan\left(\arccos\dfrac{x}{2}\right), \quad g(x) = \dfrac{\sqrt{4 - x^2}}{x}$

In Exercises 56–59, complete the equation.

56. $\arctan\dfrac{14}{x} = \arcsin(\quad), \quad x \neq 0$

57. $\arcsin\dfrac{\sqrt{36 - x^2}}{6} = \arccos(\quad), \quad 0 \leq x \leq 6$

58. $\arccos \dfrac{3}{\sqrt{x^2 - 2x + 10}} = \arcsin\left(\rule{1.5em}{0.9em}\right)$

59. $\arccos \dfrac{x - 2}{2} = \arctan\left(\rule{1.5em}{0.9em}\right), \quad |x - 2| < 2$

In Exercises 60–67, use a graphing utility to graph the function.

60. $y = 2\arccos x$

61. $y = \arcsin \dfrac{x}{2}$

62. $f(x) = \arcsin(x - 2)$

63. $g(t) = \arccos(t + 2)$

64. $f(x) = \arctan 2x$

65. $f(x) = \pi + \arctan x$

66. $h(v) = \tan(\arccos v)$

67. $f(x) = \arccos \dfrac{x}{4}$

In Exercises 68 and 69, write the given function in terms of the sine function by using the identity

$$A\cos \omega t + B\sin \omega t = \sqrt{A^2 + B^2} \sin\left(\omega t + \arctan \dfrac{A}{B}\right).$$

Use a graphing utility to graph both forms of the function. What does the graph imply?

68. $f(t) = 3\cos 2t + 3\sin 2t$

69. $f(t) = 4\cos \pi t + 3\sin \pi t$

70. *Docking a Boat*   A boat is pulled in by means of a winch located on a dock 10 feet above the deck of the boat. Let θ be the angle of elevation from the boat to the winch and let s be the length of the rope from the winch to the boat.

(a) Write θ as a function of s.

(b) Find θ when $s = 52$ feet and when $s = 26$ feet.

71. *Photography*   A television camera at ground level is filming the lift-off of a space shuttle at a point 750 meters from the launch pad. Let θ be the angle of elevation to the shuttle and let s be the height of the shuttle.

(a) Write θ as a function of s.

(b) Find θ when $s = 400$ meters and when $s = 1600$ meters.

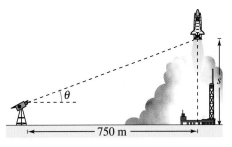

FIGURE FOR 71

72. *Rock Salt*   Different types of granular substances naturally settle at different angles when stored in cone-shaped piles. This angle θ is called the *angle of repose*. When rock salt is stored in a cone-shaped pile 11 feet high, the diameter of the pile's base is about 34 feet.   (Source: Bulk-Store Structures, Inc.)

(a) Find the angle of repose for rock salt.

(b) How tall is a pile of rock salt that has a base diameter of 40 feet?

73. *Granular Angle of Repose*   When whole corn is stored in a cone-shaped pile 20 feet high, the diameter of the pile's base is about 82 feet.

(a) Find the angle of repose for whole corn.

(b) How tall is a pile of corn that has a base diameter of 100 feet?

74. *Photography*   A photographer is taking a picture of a 3-foot painting hung in an art gallery. The camera lens is 1 foot below the lower edge of the painting. The angle β subtended by the camera lens x feet from the painting is

$$\beta = \arctan \dfrac{3x}{x^2 + 4}, \quad x > 0.$$

(a) Use a graphing utility to graph β as a function of x.

(b) Move the cursor to approximate the distance from the picture when β is maximum.

(c) Identify the asymptote of the graph and discuss its meaning in the context of the problem.

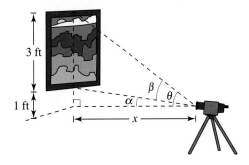

75. _Area_ In calculus, it is shown that the area of the region bounded by the graphs of $y = 0$, $y = 1/(x^2 + 1)$, $x = a$, and $x = b$ is

Area $= \arctan b - \arctan a$.

Find the areas for the following values of a and b.

(a) $a = 0, b = 1$ (b) $a = -1, b = 1$

(c) $a = 0, b = 3$ (d) $a = -1, b = 3$

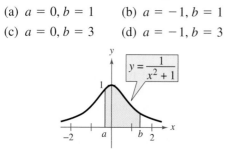

76. _Angle of Elevation_ An airplane flies at an altitude of 5 miles toward a point directly over an observer. Consider θ and x as shown in the figure.

(a) Write θ as a function of x.

(b) Find θ when $x = 10$ miles and $x = 3$ miles.

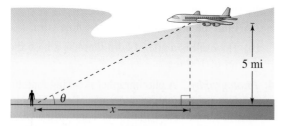

77. _Security Patrol_ A security car with its spotlight on is parked 20 meters from a long warehouse. Consider θ and x as shown in the figure.

(a) Write θ as a function of x.

(b) Find θ when $x = 5$ meters and when $x = 12$ meters.

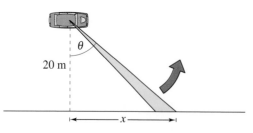

Synthesis

True or False? In Exercises 78 and 79, determine whether the statement is true or false. Justify your answer.

78. $\sin \dfrac{5\pi}{6} = \dfrac{1}{2}$ ➡ $\arcsin \dfrac{1}{2} = \dfrac{5\pi}{6}$

79. $\arctan x = \dfrac{\arcsin x}{\arccos x}$

80. Define the inverse cotangent function by restricting the domain of the cotangent function to the interval $(0, \pi)$, and sketch the inverse function's graph.

81. Define the inverse secant function by restricting the domain of the secant function to the intervals $[0, \pi/2)$ and $(\pi/2, \pi]$, and sketch the inverse function's graph.

82. Define the inverse cosecant function by restricting the domain of the cosecant function to the intervals $[-\pi/2, 0)$ and $(0, \pi/2]$, and sketch the inverse function's graph.

83. Use the results of Exercises 80–82 to evaluate the following without using a calculator.

(a) $\operatorname{arcsec} \sqrt{2}$ (b) $\operatorname{arcsec} 1$

(c) $\operatorname{arccot}\left(-\sqrt{3}\right)$ (d) $\operatorname{arccsc} 2$

In Exercises 84–89, prove the identity.

84. $\arcsin(-x) = -\arcsin x$

85. $\arctan(-x) = -\arctan x$

86. $\arccos(-x) = \pi - \arccos x$

87. $\arctan x + \arctan \dfrac{1}{x} = \dfrac{\pi}{2}, \quad x > 0$

88. $\arcsin x + \arccos x = \dfrac{\pi}{2}$

89. $\arcsin x = \arctan \dfrac{x}{\sqrt{1 - x^2}}$

Review

In Exercises 90–93, evaluate the sine, cosine, and tangent of the angle without using a calculator.

90. $840°$ **91.** $-585°$

92. $\dfrac{17\pi}{3}$ **93.** $-\dfrac{19\pi}{4}$

In Exercises 94–97, sketch the graph of the function. (Include two full periods.) Use a graphing utility to verify your result.

94. $y = \dfrac{1}{2}\sin(x + \pi)$ **95.** $y = 2 - \cos(x + \pi)$

96. $y = \dfrac{1}{2}\tan\left(x + \dfrac{\pi}{2}\right)$ **97.** $y = 4\cot \dfrac{1}{2}x$

4.8 Applications and Models

Applications Involving Right Triangles

In this section, the three angles of a right triangle are denoted by the letters A, B, and C (where C is the right angle), and the lengths of the sides opposite these angles by the letters a, b, and c (where c is the hypotenuse).

EXAMPLE 1   Solving a Right Triangle

Solve the right triangle shown in Figure 4.70.

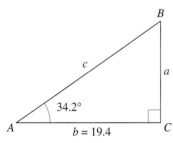

Figure 4.70

Solution

Because $C = 90°$, it follows that $A + B = 90°$ and $B = 90° - 34.2° = 55.8°$. To solve for a, use the fact that

$$\tan A = \frac{\text{opp}}{\text{adj}} = \frac{a}{b} \qquad \Longrightarrow \qquad a = b \tan A.$$

So, $a = 19.4 \tan 34.2° \approx 13.18$. Similarly, to solve for c, use the fact that

$$\cos A = \frac{\text{adj}}{\text{hyp}} = \frac{b}{c} \qquad \Longrightarrow \qquad c = \frac{b}{\cos A}.$$

So, $c = \dfrac{19.4}{\cos 34.2°} \approx 23.46$.

Recall from Section 4.3 that the term *angle of elevation* denotes the angle from the horizontal upward to an object and that the term *angle of depression* denotes the angle from the horizontal downward to an object. An angle of elevation and an angle of depression are shown in Figure 4.71.

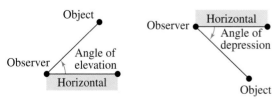

Figure 4.71

What You Should Learn:

- How to solve real-life problems involving right triangles
- How to solve real-life problems involving directional bearings
- How to solve real-life problems involving harmonic motion

Why You Should Learn It:

You can use trigonometric functions to model and solve real-life problems. For instance, Exercise 60 on page 365 shows you how a trigonometric function can be used to model the harmonic motion of a buoy.

Mary Kate Denny/PhotoEdit

EXAMPLE 2   Finding a Side of a Right Triangle

A safety regulation states that the maximum angle of elevation for a rescue ladder is 72°. If a fire department's longest ladder is 110 feet, what is the maximum safe rescue height?

Solution

A sketch is shown in Figure 4.72. From the equation $\sin A = a/c$, it follows that

$$a = c \sin A = 110 \sin 72° \approx 104.6.$$

So, the maximum safe rescue height is about 104.6 feet above the height of the fire truck.

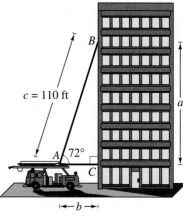

Figure 4.72

EXAMPLE 3   Finding a Side of a Right Triangle

At a point 200 feet from the base of a building, the angle of elevation to the *bottom* of a smokestack is 35°, and the angle of elevation to the *top* is 53°, as shown in Figure 4.73. Find the height s of the smokestack alone.

Solution

Note from Figure 4.73 that this problem involves two right triangles. In the smaller right triangle, use the fact that $\tan 35° = a/200$ to conclude that the height of the building is

$$a = 200 \tan 35°.$$

Now, in the larger right triangle, use the equation

$$\tan 53° = \frac{a + s}{200}$$

to conclude that $s = 200 \tan 53° - a$. So, the height of the smokestack is

$$s = 200 \tan 53° - a = 200 \tan 53° - 200 \tan 35° \approx 125.4 \text{ feet.}$$

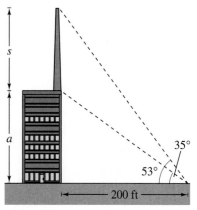

Figure 4.73

EXAMPLE 4   Finding an Angle of Depression

A swimming pool is 20 meters long and 12 meters wide. The bottom of the pool has a constant slant so that the water depth is 1.3 meters at the shallow end and 4 meters at the deep end, as shown in Figure 4.74. Find the angle of depression of the bottom of the pool.

Solution

Using the tangent function, you see that

$$\tan A = \frac{\text{opp}}{\text{adj}} = \frac{2.7}{20} = 0.135.$$

So, the angle of depression is

$$A = \arctan 0.135 \approx 0.13419 \text{ radian} \approx 7.69°.$$

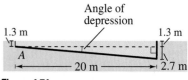

Figure 4.74

Trigonometry and Bearings

In surveying and navigation, directions are generally given in terms of **bearings.** A bearing measures the acute angle a path or line of sight makes with a fixed north–south line, as shown in Figure 4.75. For instance, the bearing of S 35° E in Figure 4.75(a) means 35 degrees east of south.

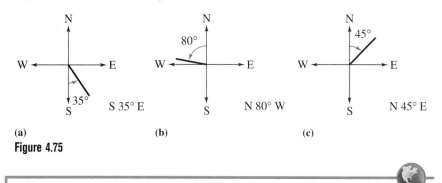

(a)   **(b)**   **(c)**

Figure 4.75

EXAMPLE 5   Finding Directions in Terms of Bearings

A ship leaves port at noon and heads due west at 20 knots (nautical miles per hour). At 2 P.M. the ship changes course to N 54° W, as shown in Figure 4.76. Find the ship's bearing and distance from the port of departure at 3 P.M.

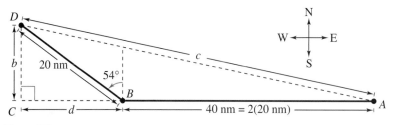

Figure 4.76

Solution

In triangle *BCD*, you have $B = 90° - 54° = 36°$. The two sides of this triangle can be determined to be

$$b = 20 \sin 36° \qquad \text{and} \qquad d = 20 \cos 36°.$$

In triangle *ACD*, you can find angle *A* as follows.

$$\tan A = \frac{b}{d + 40} = \frac{20 \sin 36°}{20 \cos 36° + 40} \approx 0.2092494$$

$$A \approx \arctan 0.2092494 \approx 0.2062732 \text{ radian} \approx 11.82°$$

The angle with the north–south line is $90° - 11.82° = 78.18°$. Therefore, the bearing of the ship is

$$\text{N } 78.18° \text{ W.} \qquad\qquad \text{\small Bearing}$$

Finally, from triangle *ACD*, you have $\sin A = b/c$, which yields

$$c = \frac{b}{\sin A} = \frac{20 \sin 36°}{\sin 11.82°} \approx 57.4 \text{ nautical miles.} \qquad \text{\small Distance from port}$$

Harmonic Motion

The periodic nature of the trigonometric functions is useful for describing the motion of a point on an object that vibrates, oscillates, rotates, or is moved by wave motion.

For example, consider a ball that is bobbing up and down on the end of a spring, as shown in Figure 4.77. Suppose that 10 centimeters is the maximum distance the ball moves vertically upward or downward from its equilibrium (at-rest) position. Suppose further that the time it takes for the ball to move from its maximum displacement above zero to its maximum displacement below zero and back again is $t = 4$ seconds. Assuming the ideal conditions of perfect elasticity and no friction or air resistance, the ball would continue to move up and down in a uniform and regular manner.

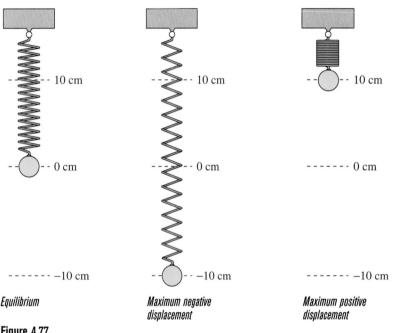

Equilibrium

Maximum negative displacement

Maximum positive displacement

Figure 4.77

From this spring you can conclude that the period (time for one complete cycle) of the motion is

Period = 4 seconds

and that its amplitude (maximum displacement from equilibrium) is

Amplitude = 10 centimeters.

Motion of this nature can be described by a sine or cosine function, and is called **simple harmonic motion.**

Definition of Simple Harmonic Motion

A point that moves on a coordinate line is said to be in **simple harmonic motion** if its distance d from the origin at time t is given by either

$$d = a \sin \omega t \qquad \text{or} \qquad d = a \cos \omega t$$

where a and ω are real numbers such that $\omega > 0$. The motion has amplitude $|a|$, period $2\pi/\omega$, and frequency $\omega/2\pi$.

EXAMPLE 6   Simple Harmonic Motion

Write the equation for the simple harmonic motion of the ball illustrated in Figure 4.77, where the period is four seconds. What is the frequency of this motion?

Solution
Because the spring is at equilibrium ($d = 0$) when $t = 0$, you use the equation

$$d = a \sin \omega t.$$

Moreover, because the maximum displacement from zero is 10 and the period is 4, you have

$$\text{Amplitude} = |a| = 10$$

$$\text{Period} = \frac{2\pi}{\omega} = 4 \quad \Longrightarrow \quad \omega = \frac{\pi}{2}.$$

Consequently, the equation of motion is

$$d = 10 \sin \frac{\pi}{2} t.$$

Note that the choice of $a = 10$ or $a = -10$ depends on whether the ball initially moves up or down. The frequency is

$$\begin{aligned}
\text{Frequency} &= \frac{\omega}{2\pi} \\
&= \frac{\pi/2}{2\pi} \\
&= \frac{1}{4} \quad \text{cycle per second.}
\end{aligned}$$

One illustration of the relationship between sine waves and harmonic motion is the wave motion that results when a stone is dropped into a calm pool of water. The waves move outward in roughly the shape of sine (or cosine) waves, as shown in Figure 4.78. As an example, suppose you are fishing and your fishing bob is attached so that it does not move horizontally. As the waves move outward from the dropped stone, your fishing bob will move up and down in simple harmonic motion, as shown in Figure 4.79.

Figure 4.78

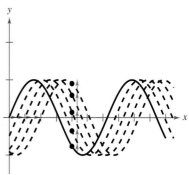

Figure 4.79

EXAMPLE 7   Simple Harmonic Motion

Given the equation for simple harmonic motion

$$d = 6 \cos \frac{3\pi}{4} t$$

find (a) the maximum displacement, (b) the frequency, (c) the value of d when $t = 4$, and (d) the least positive value of t for which $d = 0$.

Algebraic Solution

The given equation has the form $d = a \cos \omega t$, with $a = 6$ and $\omega = 3\pi/4$.

a. The maximum displacement (from the point of equilibrium) is given by the amplitude. So, the maximum displacement is 6.

b. Frequency $= \dfrac{\omega}{2\pi} = \dfrac{3\pi/4}{2\pi} = \dfrac{3}{8}$ cycle per unit of time

c. $d = 6 \cos\left[\dfrac{3\pi}{4}(4)\right] = 6 \cos 3\pi = 6(-1) = -6$

d. To find the least positive value of t for which $d = 0$, solve the equation

$$d = 6 \cos \frac{3\pi}{4} t = 0$$

to obtain

$$\frac{3\pi}{4} t = \frac{\pi}{2}, \frac{3\pi}{2}, \frac{5\pi}{2}, \dots \quad \Longrightarrow \quad t = \frac{2}{3}, 2, \frac{10}{3}, \dots \dots$$

So, the least positive value of t is $t = \frac{2}{3}$.

Graphical Solution

Use a graphing utility set in *radian* mode to graph

$$y = 6 \cos \frac{3\pi}{4} x.$$

a. Use the *maximum* feature of the graphing utility to estimate that the maximum displacement from the point of equilibrium $y = 0$ is 6, as shown in Figure 4.80(a).

b. The period is the time for the graph to complete one cycle, which is $x \approx 2.667$, as shown in Figure 4.80(a). You can estimate the frequency as follows.

$$\text{Frequency} \approx \frac{1}{2.667} \approx 0.375 \text{ cycles per unit of time}$$

c. Use the *value* or *trace* feature to estimate that the value of y when $x = 4$ is $y = -6$.

d. Use the *zero* or *root* feature to estimate that the least positive value of x for which $y = 0$ is $x \approx 0.6667$, as shown in Figure 4.80(b).

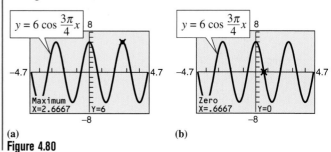

(a)   (b)
Figure 4.80

Writing About Math   *Radio Waves*

Many different physical phenomena can be characterized by wave motion. These include electromagnetic waves such as radio waves, television waves, and microwaves. Radio waves transmit sound in two different ways. For an AM station, the *amplitude* of the wave is modified to carry sound. The letters AM stand for *amplitude modulation*. An FM radio signal has its *frequency* modified in order to carry sound, hence the term *frequency modulation*. Of the two graphs at the left, one shows an AM wave and the other shows an FM wave. Which is which? Write a short paragraph explaining your reasoning.

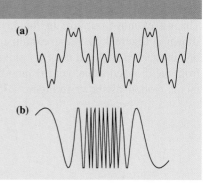

4.8 Exercises

In Exercises 1–10, solve the right triangle shown in the figure below. (Round your answers to two decimal places.)

1. $A = 25°$, $b = 10$
2. $B = 56°$, $c = 15$
3. $B = 71°$, $b = 24$
4. $A = 7.4°$, $a = 40.5$
5. $a = 6$, $b = 16$
6. $a = 25$, $c = 35$
7. $b = 18$, $c = 42$
8. $b = 1.72$, $c = 8.35$
9. $A = 12° 15'$, $c = 430.5$
10. $B = 65° 12'$, $a = 14.2$

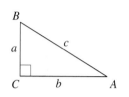

In Exercises 11–14, find the altitude of the isosceles triangle shown in the figure below. (Round your answers to two decimal places.)

11. $\theta = 52°$, $b = 4$ inches
12. $\theta = 18°$, $b = 10$ meters
13. $\theta = 41.6°$, $b = 14.2$ feet
14. $\theta = 72.94°$, $b = 3.36$ centimeters

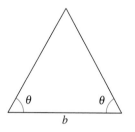

15. **Length of a Shadow** A shadow of length L is created by a 60-foot silo when the sun is $\theta°$ above the horizon.

    (a) Write L as a function of θ.

    (b) Use a graphing utility to complete the table.

θ	10°	20°	30°	40°	50°
L					

(c) The angle measure increases in equal increments in the table. Does the length of the shadow change in equal increments? Explain.

16. **Length of a Shadow** A shadow of length L is created by an 850-foot building when the sun is $\theta°$ above the horizon.

    (a) Write L as a function of θ.

    (b) Use a graphing utility to complete the table.

θ	10°	20°	30°	40°	50°
L					

(c) The angle measure increases in equal increments in the table. Does the length of the shadow change in equal increments? Explain.

17. **Height** A ladder of length 20 feet leans against the side of a house. The angle of elevation of the ladder is $\theta°$.

    (a) Write the height h of the ladder as a function of θ.

    (b) Use a graphing utility to complete the table.

θ	60°	65°	70°	75°	80°
h					

18. **Height** The length of a shadow of a tree is 130 feet when the angle of elevation of the sun is $\theta°$.

    (a) Write the height h of the tree as a function of θ.

    (b) Use a graphing utility to complete the table.

θ	10°	15°	20°	25°	30°
h					

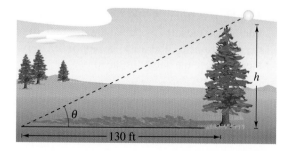

19. *Height*   From a point 50 feet in front of a church, the angles of elevation to the base of the steeple and the top of the steeple are 35° and 47° 40′, respectively.

   (a) Draw right triangles that give a visual representation of the problem. Label the known quantities and the unknown height of the steeple.

   (b) Use a trigonometric function to write an equation involving the unknown quantity.

   (c) Find the height of the steeple.

20. *Height*   From a point 100 feet in front of a public library, the angles of elevation to the base of the flagpole and the top of the flagpole are 28° and 39° 45′, respectively. The flagpole is mounted on the front of the library's roof. Find the height of the pole.

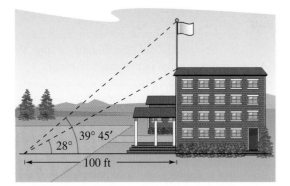

21. *Depth of a Submarine*   The sonar of a navy cruiser detects a nuclear submarine that is 4000 feet from the cruiser. The angle between the water level and the submarine is 31.5°. How deep is the submarine?

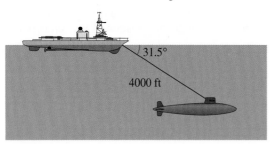

22. *Height of a Kite*   A 100-foot line is attached to a kite. When the kite has pulled the line taut, the angle of elevation to the kite is approximately 50°. Approximate the height of the kite.

23. *Angle of Elevation*   An engineer erects a 75-foot vertical cellular-phone tower. Find the angle of elevation to the top of the tower from a point on level ground 95 feet from its base.

24. *Angle of Elevation*   The height of an outdoor basketball backboard is $12\frac{1}{2}$ feet, and the backboard casts a shadow $17\frac{1}{3}$ feet long.

   (a) Draw a right triangle that represents the problem. Label the known and unknown quantities.

   (b) Use a trigonometric function to write an equation involving the unknown quantity.

   (c) Find the angle of elevation of the sun.

25. *Angle of Depression*   A television satellite is traveling in a circular orbit 150 miles above the surface of the earth. Find the angle of depression from the television satellite to the horizon. Assume that the radius of the earth is 4000 miles.

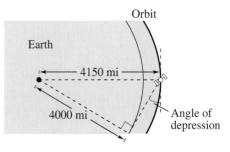

26. *Angle of Depression*   Find the angle of depression from the top of a lighthouse 250 feet above water level to the water line of a ship $2\frac{1}{2}$ miles offshore.

27. *Airplane Ascent*   When an airplane leaves the runway, its angle of climb is 18° and its speed is 275 feet per second. Find the plane's altitude after 1 minute.

28. *Airplane Ascent*   How long will it take the plane in Exercise 27 to climb to an altitude of 10,000 feet? 16,000 feet?

29. *Mountain Descent*   A sign on the roadway at the top of a mountain indicates that for the next 4 miles the grade is 9.5°. Find the change in elevation for a car descending the mountain.

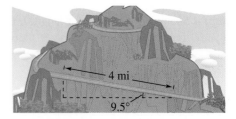

30. *Mountain Descent*   A road sign at the top of a mountain indicates that for the next 4 miles the grade is 12%. Find the angle of the grade and the change in elevation for a car descending the mountain.

31. *Navigation*   An airplane flying at 550 miles per hour has a bearing of N 58° E. After flying 1.5 hours, how far north and how far east has the plane traveled from its point of departure?

32. *Navigation*   A ship leaves port at noon and has a bearing of S 29° W. If the ship sails at 20 knots, how many nautical miles south and how many nautical miles west will the ship have traveled by 6:00 P.M.?

33. *Surveying*   A surveyor wishes to find the distance across a swamp. The bearing from *A* to *B* is N 32° W. The surveyor walks 50 meters from *A*, and at the point *C* the bearing to *B* is N 68° W. Find (a) the bearing from *A* to *C* and (b) the distance from *A* to *B*.

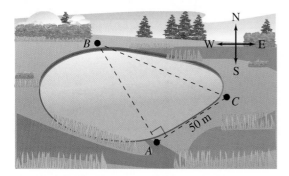

34. *Location of a Fire*   Two fire towers are 30 kilometers apart, tower *A* being due west of tower *B*. A fire is spotted from the towers, and the bearings from *A* and *B* are E 14° N and W 34° N, respectively. Find the distance *d* of the fire from the line segment *AB*.

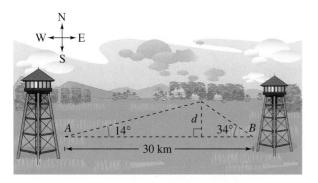

35. *Navigation*   A ship is 50 miles east and 35 miles south of port. If the captain wants to sail directly to port, what bearing should be taken?

36. *Navigation*   A plane is 160 miles north and 85 miles east of an airport. If the pilot wants to fly directly to the airport, what bearing should be taken?

37. *Distance Between Ships*   An observer in a lighthouse 350 feet above sea level observes two ships directly offshore. The angles of depression to the ships are 4° and 6.5°. How far apart are the ships?

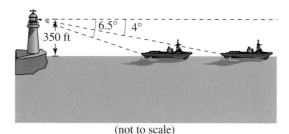

(not to scale)

38. *Distance Between Towns*   A passenger in an airplane flying at an altitude of 10 kilometers sees two towns directly to the left of the plane. The angles of depression to the towns are 28° and 55°. How far apart are the towns?

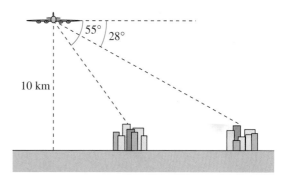

39. *Altitude of a Plane*   A plane is observed approaching your home and you assume its speed is 550 miles per hour. If the angle of elevation of the plane is 16° at one time and 57° 1 minute later, approximate the altitude of the plane.

40. *Height of a Mountain*   While traveling across flat land, you notice a mountain directly in front of you. The angle of elevation to the peak is 2.5°. After you drive 18 miles closer to the mountain, the angle of elevation is 10°. Approximate the height of the mountain.

Geometry In Exercises 41 and 42, find the angle α between two nonvertical lines L_1 and L_2. The angle α satisfies the equation

$$\tan \alpha = \left| \frac{m_2 - m_1}{1 + m_2 m_1} \right|$$

where m_1 and m_2 are the slopes of L_1 and L_2, respectively. (Assume $m_1 m_2 \neq -1$.)

41. $3x - 2y = 5$
    $x + y = 1$

42. $2x + y = 8$
    $x - 5y = -4$

43. *Geometry* Determine the angle between the diagonal of a cube and the diagonal of its base, as shown in the figure.

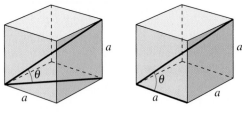

FIGURE FOR 43   FIGURE FOR 44

44. *Geometry* Determine the angle between the diagonal of a cube and its edge, as shown in the figure.

45. *Wrench Size* Express the distance y across the flat sides of a hexagonal nut as a function of r, as shown in the figure.

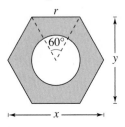

46. *Bolt Circle* The figure shows a circular sheet of diameter 40 centimeters. The sheet contains 12 equally spaced bolt holes. Determine the straight-line distance between the centers of the bolt holes.

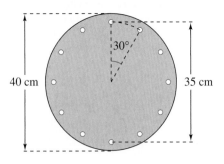

47. *Geometry* A regular pentagon (a pentagon with congruent sides and angles) is inscribed in a circle of radius 24 inches. Find the length of the sides of the pentagon.

48. *Geometry* A regular hexagon (a hexagon with congruent sides and angles) is inscribed in a circle of radius 24 inches. Find the length of the sides of the hexagon.

Trusses In Exercises 49 and 50, find the lengths of all the unknown members of the truss.

49.

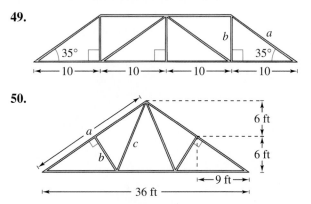

50.

Harmonic Motion In Exercises 51–54, for the simple harmonic motion described by the trigonometric function, find (a) the maximum displacement, (b) the frequency, and (c) the least positive value of t for which $d = 0$. Use a graphing utility to verify your results.

51. $d = 4 \cos 8\pi t$ **52.** $d = \frac{1}{2} \cos 20\pi t$

53. $d = \frac{1}{16} \sin 140\pi t$ **54.** $d = \frac{1}{64} \sin 792\pi t$

Harmonic Motion In Exercises 55–58, find a model for simple harmonic motion satisfying the specified conditions.

	Displacement ($t = 0$)	Amplitude	Period
55.	0	8 cm	2 sec
56.	0	3 m	6 sec
57.	3 in.	3 in.	1.8 sec
58.	2 ft	2 ft	10 sec

59. *Tuning Fork* A point on the end of a tuning fork moves in the simple harmonic motion described by $d = a \sin \omega t$. Find ω given that the tuning fork for middle C has a frequency of 264 vibrations per second.

60. Wave Motion A buoy oscillates in simple harmonic motion as waves go past. At a given time it is noted that the buoy moves a total of 3.5 feet from its low point to its high point, and that it returns to its high point every 10 seconds. Write an equation that describes the motion of the buoy if it is at its high point at $t = 0$.

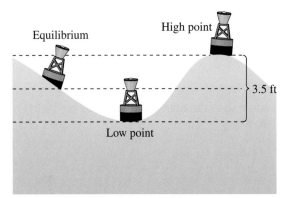

61. Springs A weight stretches a spring 1.5 inches. The weight is pushed 3 inches above the point of equilibrium and released. Its motion is modeled by

$$y = \tfrac{1}{4} \cos 16t, \quad t > 0$$

where y is in feet and t is in seconds.

(a) Use a graphing utility to graph the function.

(b) What is the period of the oscillations?

(c) Determine the first time the weight passes the point of equilibrium $(y = 0)$.

62. Numerical and Graphical Analysis A 2-meter-high fence is 3 meters from the side of a grain storage bin. A grain elevator must reach from ground level outside the fence to the storage bin. The objective is to determine the shortest elevator meeting the constraints.

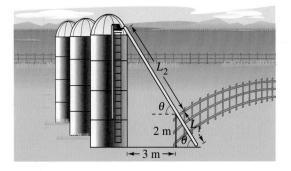

(a) Complete four rows of the table.

θ	L_1	L_2	$L_1 + L_2$
0.1	$\dfrac{2}{\sin 0.1}$	$\dfrac{3}{\cos 0.1}$	23.0
0.2	$\dfrac{2}{\sin 0.2}$	$\dfrac{3}{\cos 0.2}$	13.1

(b) Use a graphing utility to generate additional rows of the table. Use the table to estimate the minimum length.

(c) Write the length L as a function of θ.

(d) Use a graphing utility to graph the function. Use the graph to estimate the minimum length. How does your estimate compare with that of part (b)?

63. Data Analysis The times S of sunset (Greenwich Mean Time) at 40° north latitude on the 15th of each month are: 1(16:59), 2(17:35), 3(18:06), 4(18:38), 5(19:08), 6(19:30), 7(19:28), 8(18:57), 9(18:09), 10(17:21), 11(16:44), 12(16:36). The month is represented by t, with $t = 1$ corresponding to January. A model (where minutes have been converted to the decimal part of an hour) for this data is

$$S(t) = 18.09 + 1.41 \sin\left(\frac{\pi t}{6} + 4.60\right).$$

(a) Use a graphing utility to graph the data points and the model in the same viewing window.

(b) What is the period of the model? Is it what you expected? Explain.

(c) What is the amplitude of the function? What does it represent in the model? Explain.

64. Numerical and Graphical Analysis The cross sections of an irrigation canal are isosceles trapezoids where the length of three of the sides is 8 feet. The objective is to find the angle θ that maximizes the area of the cross sections.

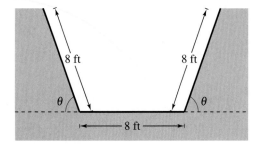

Base 1	Base 2	Altitude	Area
8	8 + 16 cos 10°	8 sin 10°	22.1
8	8 + 16 cos 20°	8 sin 20°	42.5

(a) Complete six rows of the table.

(b) Use a graphing utility to generate additional rows of the table. Use the table to estimate the maximum cross-sectional area.

(c) Write the area A as a function of θ.

(d) Use a graphing utility to graph the function. Use the graph to estimate the maximum cross-sectional area. How does your estimate compare with that of part (b)?

65. *Data Analysis* The table shows the average sales S (in millions of dollars) of an outerwear manufacturer for each month t, where $t = 1$ represents January.

t	1	2	3	4	5	6
S	13.46	11.15	8.00	4.85	2.54	1.70

t	7	8	9	10	11	12
S	2.54	4.85	8.00	11.15	13.46	14.30

(a) Create a scatter plot of the data.

(b) Find a trigonometric model that fits the data. Graph the model on your scatter plot. How well does the model fit?

(c) What is the period of the model? Do you think it is reasonable given the context? Explain your reasoning.

(d) Interpret the meaning of the model's amplitude in the context of the problem.

66. *Numerical and Graphical Analysis* Consider the shaded region outside the sector of the circle of radius 10 meters and inside the right triangle.

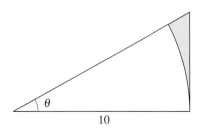

(a) Write the area A of the region as a function of θ. Determine the domain of the function.

(b) Use a graphing utility to complete the table.

θ	0	0.3	0.6	0.9	1.2	1.5
A						

(c) Use a graphing utility to graph the function over the appropriate domain.

(d) What does the area approach as θ approaches $\pi/2$?

Synthesis

True or False? **In Exercises 67 and 68, determine whether the statement is true or false. Justify your answer.**

67. In the right triangle shown below, $a = \dfrac{22.56}{\tan 41.9°}$.

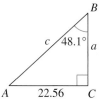

68. N 24° E means 24 degrees north of east.

Review

In Exercises 69–72, solve the equation. (Round your answer to three decimal places.)

69. $e^{2x} = 54$

70. $\dfrac{300}{1 + e^{-x}} = 100$

71. $\ln(x^2 + 1) = 3.2$

72. $\log_8 x + \log_8(x - 1) = \frac{1}{3}$

In Exercises 73–76, use a calculator to approximate the value of the expression. Round your answer to two decimal places.

73. arccos 0.13

74. arcsin 0.96

75. arcsin(−0.11)

76. arctan(−12)

4   Chapter Summary

What did you learn?

Section 4.1	Review Exercises
☐ How to describe angles	1–4
☐ How to use radian and degree measure	5–36
☐ How to use angles to model and solve real-life problems	37–42

Section 4.2

☐ How to identify a unit circle and its relationship to real numbers	43–46
☐ How to evaluate trigonometric functions using the unit circle	47–50
☐ How to use the domain and period to evaluate sine and cosine functions	51–54
☐ How to use a calculator to evaluate trigonometric functions	55–58

Section 4.3

☐ How to evaluate trigonometric functions of acute angles	59–62
☐ How to use the fundamental trigonometric identities	63, 64
☐ How to use a calculator to evaluate trigonometric functions	65–68
☐ How to use trigonometric functions to model and solve real-life problems	69, 70

Section 4.4

☐ How to evaluate trigonometric functions of any angle	71–90
☐ How to use reference angles to evaluate trigonometric functions	91–102
☐ How to evaluate trigonometric functions of real numbers	103–106

Section 4.5

☐ How to use amplitude and period to sketch the graphs of sine and cosine functions	107–120
☐ How to sketch translations of graphs of sine and cosine functions	121–130
☐ How to use sine and cosine functions to model real-life data	131, 132

Section 4.6

☐ How to sketch the graphs of tangent and cotangent functions	133–144
☐ How to sketch the graphs of secant and cosecant functions	145–154
☐ How to sketch the graphs of damped trigonometric functions	155–158

Section 4.7

☐ How to evaluate inverse trigonometric functions	159–168
☐ How to evaluate compositions of trigonometric functions	169–172

Section 4.8

☐ How to solve real-life problems involving right triangles	173, 174
☐ How to solve real-life problems involving directional bearings	175, 176
☐ How to solve real-life problems involving harmonic motion	177

4.1 In Exercises 1–4, estimate the angle to the nearest one-half radian.

1.   **2.**

3.   **4.**

In Exercises 5–8, (a) sketch the angle in standard position, (b) determine the quadrant in which the angle lies, and (c) list one positive and one negative coterminal angle.

5. $\dfrac{\pi}{16}$

6. $\dfrac{40\pi}{47}$

7. $-\dfrac{9\pi}{15}$

8. $-\dfrac{11\pi}{3}$

In Exercises 9–12, find the complement (if possible) and supplement of each angle.

9. $\dfrac{\pi}{8}$

10. $\dfrac{13\pi}{18}$

11. $\dfrac{3\pi}{10}$

12. $\dfrac{2\pi}{21}$

In Exercises 13–16, convert the measure from radians to degrees. Round your answer to two decimal places.

13. $\dfrac{5\pi}{7}$

14. $-\dfrac{3\pi}{5}$

15. -3.5

16. 1.55

In Exercises 17–20, (a) sketch the angle in standard position, (b) determine the quadrant in which the angle lies, and (c) list one positive and one negative coterminal angle.

17. $40°$

18. $190°$

19. $-110°$

20. $-405°$

In Exercises 21–24, find the complement (if possible) and supplement of each angle.

21. $8°$ **22.** $94°$ **23.** $171°$ **24.** $49°$

In Exercises 25–28, convert the measure to decimal degree form. Round your answer to two decimal places.

25. $135°\,16'\,65''$

26. $-234°\,40''$

27. $5°\,22'\,53''$

28. $280°\,8'\,50''$

In Exercises 29–32, convert the measure to D°M′S″ form.

29. $135.29°$

30. $25.8°$

31. $-85.36°$

32. $-327.93°$

In Exercises 33–36, convert the measure from degrees to radians. Round your answer to four decimal places.

33. $480°$

34. $-16.5°$

35. $-33°\,45'$

36. $84°\,15'$

37. Find the radian measure of the central angle of a circle with a radius of 12 feet that intercepts an arc of length 25 feet.

38. Find the radian measure of the central angle of a circle with a radius of 60 inches that intercepts an arc of length 245 inches.

39. Find the length of the arc on a circle with a radius of 20 meters intercepted by a central angle of $138°$.

40. Find the radius of a circle with a central angle of $\pi/3$ that intercepts an arc of length 8.5 kilometers.

41. *Difference in Latitudes* Assuming that earth is a sphere of radius 6378 kilometers, what is the difference in latitude of two cities, one of which is 600 kilometers due north of the other?

42. *Angular Speed* A car is moving at a rate of 28 miles per hour, and the diameter of its wheels is about $2\frac{1}{3}$ feet.

  (a) Find the rotational speed of the wheels in revolutions per minute.

  (b) Find the angular speed of the wheels in radians per minute.

4.2 In Exercises 43–46, find the point (x, y) on the unit circle that corresponds to the real number t.

43. $t = \dfrac{2\pi}{3}$ **44.** $t = \dfrac{3\pi}{4}$

45. $t = \dfrac{5\pi}{6}$ **46.** $t = -\dfrac{4\pi}{3}$

In Exercises 47–50, evaluate (if possible) the six trigonometric functions of the real number.

47. $t = \dfrac{7\pi}{6}$ **48.** $t = \dfrac{\pi}{4}$

49. $t = -\dfrac{2\pi}{3}$ **50.** $t = 2\pi$

In Exercises 51–54, evaluate the trigonometric function using its period as an aid.

51. $\sin \dfrac{11\pi}{4}$ **52.** $\cos 4\pi$

53. $\sin\left(-\dfrac{17\pi}{6}\right)$ **54.** $\cos\left(-\dfrac{13\pi}{3}\right)$

In Exercises 55–58, use a calculator to evaluate the trigonometric function. Round to two decimal places.

55. $\cot 2.3$ **56.** $\sec 4$

57. $\cos \dfrac{5\pi}{3}$ **58.** $\tan\left(-\dfrac{11\pi}{6}\right)$

4.3 In Exercises 59–62, find the exact values of the six trigonometric functions of the angle θ given in the figure.

59. **60.**

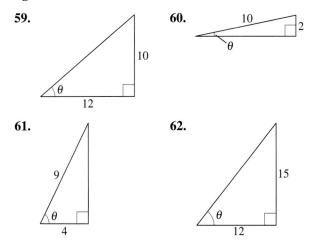

61. **62.**

In Exercises 63 and 64, use trigonometric identities to transform one side of the equation into the other.

63. $\csc \theta \tan \theta = \sec \theta$

64. $\dfrac{\cot \theta + \tan \theta}{\cot \theta} = \sec^2 \theta$

In Exercises 65–68, use a calculator to evaluate each function. Round your answer to four decimal places.

65. (a) $\cos 84°$ (b) $\sin 6°$

66. (a) $\csc 52° \, 12'$ (b) $\sec 54° \, 7'$

67. (a) $\cos \dfrac{\pi}{4}$ (b) $\sec \dfrac{\pi}{4}$

68. (a) $\tan\left(\dfrac{3\pi}{20}\right)$ (b) $\cot\left(\dfrac{3\pi}{20}\right)$

69. *Altitude of Elevation* Find the altitude of the triangle shown.

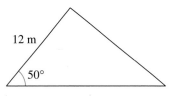

70. *Width of a River* An engineer is trying to determine the width of a river. From point P, the engineer walks downstream 125 feet and sights to point Q. From this sighting, it is determined that $\theta = 62°$. How wide is the river?

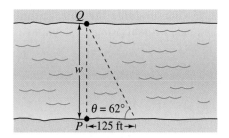

4.4 In Exercises 71–76, find the six trigonometric functions of the angle θ (in standard position) whose terminal side passes through the given point.

71. $(12, 16)$ **72.** $(-4, -6)$

73. $(-7, 2)$ **74.** $(4, -8)$

75. $\left(\dfrac{2}{3}, \dfrac{5}{2}\right)$ **76.** $(x, 4x)$, $x > 0$

In Exercises 77–80, find the remaining five trigono-metric functions of θ satisfying the given conditions.

77. $\sec \theta = \frac{6}{5}$, $\tan \theta < 0$

78. $\tan \theta = -\frac{12}{5}$, $\sin \theta > 0$

79. $\sin \theta = \frac{3}{8}$, $\cos \theta < 0$

80. $\cos \theta = -\frac{2}{5}$, $\sin \theta > 0$

In Exercises 81–86, evaluate the trigonometric func-tion without using a calculator.

81. $\tan \dfrac{\pi}{3}$

82. $\sec \dfrac{\pi}{4}$

83. $\sin\left(-\dfrac{5\pi}{3}\right)$

84. $\cot\left(-\dfrac{5\pi}{6}\right)$

85. $\cos 495°$

86. $\csc 270°$

In Exercises 87–90, use a calculator to evaluate the trigonometric function. Round to two decimal places.

87. $\tan 33°$

88. $\csc 105°$

89. $\sec \dfrac{12\pi}{5}$

90. $\sin\left(-\dfrac{\pi}{9}\right)$

In Exercises 91–94, find the reference angle for the given angle.

91. $264°$

92. $635°$

93. $-\dfrac{6\pi}{5}$

94. $\dfrac{17\pi}{3}$

In Exercises 95–102, evaluate the sine, cosine, and tangent of the angle without using a calculator.

95. $240°$

96. $315°$

97. $-210°$

98. $-315°$

99. $-\dfrac{9\pi}{4}$

100. $-\dfrac{11\pi}{6}$

101. $-\dfrac{\pi}{2}$

102. $-\dfrac{\pi}{3}$

In Exercises 103–106, find the point (x, y) on the unit circle that corresponds to the given real number t. Use the result to evaluate $\sin t$, $\cos t$, and $\tan t$.

103. $\dfrac{2\pi}{3}$

104. $\dfrac{7\pi}{4}$

105. $\dfrac{7\pi}{6}$

106. $\dfrac{3\pi}{4}$

4.5 In Exercises 107–110, sketch the graph of the function.

107. $f(x) = 3 \sin x$

108. $f(x) = 2 \cos x$

109. $f(x) = \frac{7}{2} \sin x$

110. $f(x) = \frac{1}{4} \cos x$

In Exercises 111–114, find the period and amplitude.

111.

112.

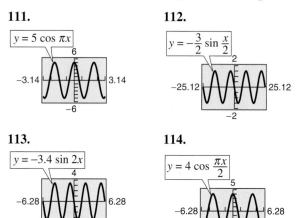

113.

114.

In Exercises 115–126, sketch the graph of the func-tion.

115. $f(x) = 3 \cos 2\pi x$

116. $f(x) = -2 \sin \pi x$

117. $f(x) = 5 \sin \dfrac{2x}{5}$

118. $f(x) = 8 \cos\left(-\dfrac{x}{4}\right)$

119. $f(x) = -\dfrac{5}{2} \cos \dfrac{x}{4}$

120. $f(x) = -\dfrac{1}{2} \sin \dfrac{\pi x}{4}$

121. $f(x) = \frac{5}{2} \sin(x - \pi)$

122. $f(x) = 3 \cos(x + \pi)$

123. $f(x) = 4 \sin\left(x - \dfrac{\pi}{2}\right)$

124. $f(x) = \frac{1}{2} \sin \pi x - 3$

125. $f(x) = -3 \cos\left(\dfrac{x}{2} - \dfrac{\pi}{4}\right)$

126. $f(x) = 4 - 2 \cos(4x + \pi)$

In Exercises 127–130, find a, b, and c for $f(x) = a \cos(bx - c)$ so that the graph of f matches the graph shown.

127.

128.

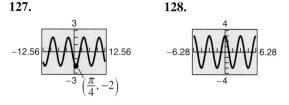

129. **130.**

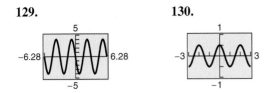

Sales In Exercises 131 and 132, use a graphing utility to graph the sales function over one year, where S is the sales (in thousands of units) and t is the time (in months), with $t = 1$ corresponding to January. Determine the months of maximum and minimum sales.

131. $S = 48.4 - 6.1 \cos \dfrac{\pi t}{6}$

132. $S = 56.25 + 9.50 \sin \dfrac{\pi t}{6}$

4.6 In Exercises 133 and 134, find a and b for $f(x) = a \tan bx$ so that the graph of f matches the graph shown.

133. **134.**

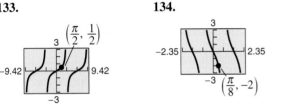

In Exercises 135–154, sketch the graph of the function.

135. $f(x) = -\tan \dfrac{\pi x}{4}$ **136.** $f(x) = 4 \tan \pi x$

137. $f(x) = -\dfrac{1}{4} \tan \dfrac{\pi x}{2}$ **138.** $f(x) = \tan\left(x + \dfrac{\pi}{4}\right)$

139. $f(x) = \dfrac{1}{4} \tan\left(x - \dfrac{\pi}{2}\right)$ **140.** $f(x) = 2 + 2 \tan \dfrac{x}{3}$

141. $f(x) = 3 \cot \dfrac{x}{2}$ **142.** $f(x) = \dfrac{1}{2} \cot \dfrac{\pi x}{2}$

143. $f(x) = \dfrac{1}{2} \cot\left(x - \dfrac{\pi}{2}\right)$

144. $f(x) = 4 \cot\left(x + \dfrac{\pi}{4}\right)$

145. $f(x) = \dfrac{1}{4} \sec x$ **146.** $f(x) = \dfrac{1}{2} \csc x$

147. $f(x) = \dfrac{1}{4} \csc 2x$ **148.** $f(x) = \dfrac{1}{2} \sec 2\pi x$

149. $f(x) = \sec\left(x - \dfrac{\pi}{4}\right)$

150. $f(x) = \dfrac{1}{2} \csc(2x + \pi)$

151. $f(x) = 2 \sec(x - \pi)$

152. $f(x) = -2\csc(x - \pi)$

153. $f(x) = \csc\left(3x - \dfrac{\pi}{2}\right)$

154. $f(x) = 3 \csc\left(2x + \dfrac{\pi}{4}\right)$

In Exercises 155–158, sketch the graph of the function and the damping factor of the function by hand. Use a graphing utility to verify your graph.

155. $f(x) = e^x \sin 2x$ **156.** $f(x) = 2x \cos x$

157. $f(x) = x \sin \pi x$

158. $f(x) = 2.5e^{-x/4} \sin 2\pi x$

4.7 In Exercises 159–162, evaluate the expression without using a calculator.

159. (a) $\arcsin 1$ (b) $\arcsin 4$

160. (a) $\arcsin\left(-\dfrac{\sqrt{2}}{2}\right)$ (b) $\arcsin \dfrac{\sqrt{3}}{2}$

161. (a) $\arccos \dfrac{\sqrt{2}}{2}$ (b) $\arccos\left(-\dfrac{\sqrt{3}}{2}\right)$

162. (a) $\arctan\left(-\sqrt{3}\right)$ (b) $\arctan(-1)$

In Exercises 163–166, use a calculator to approximate the value of the expression. (Round your answer to three decimal places.)

163. (a) $\arccos 0.42$ (b) $\arcsin 0.63$

164. (a) $\arcsin(-0.94)$ (b) $\arccos(-0.12)$

165. (a) $\arctan(-12)$ (b) $\arctan 21$

166. (a) $\arctan 0.81$ (b) $\arctan 6.4$

In Exercises 167 and 168, use an inverse trigonometric function to write θ as a function of x.

167. **168.**

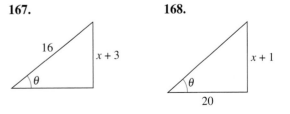

In Exercises 169–172, write an algebraic expression for the given expression.

169. $\sec[\arcsin(x-1)]$

170. $\tan\left(\arccos\dfrac{x}{2}\right)$

171. $\sin\left(\arccos\dfrac{x^2}{4-x^2}\right)$

172. $\csc(\arcsin 10x)$

4.8 **173.** *Railroad Grade* A train travels 3.5 kilometers on a straight track with a grade of $1°\,10'$. What is the vertical rise of the train in that distance?

174. *Distance Between Towns* A passenger in an airplane flying at an altitude of 37,000 feet sees two towns directly to the left of the airplane. The angles of depression to the towns are $32°$ and $76°$. How far apart are the towns?

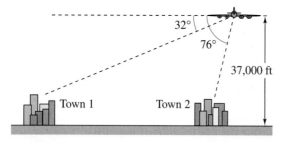

175. *Navigation* An airplane flying at 490 miles per hour has a bearing of N 46° W. After flying 1.8 hours, how far north and how far west has the plane traveled from its point of departure?

176. *Distance* From city A to city B, a plane flies 650 miles at a bearing of N 48° E. From city B to city C, the plane flies 810 miles at a bearing of S 65° E. Find the distance from A to C and the bearing from A to C.

177. *Wave Motion* A buoy oscillates in simple harmonic motion as waves go past. At a given time it is noted that the buoy moves a total of 6 feet from its low point to its high point, returning to its high point every 15 seconds (see figure). Write an equation that describes the motion of the buoy if at $t = 0$ it is at its high point.

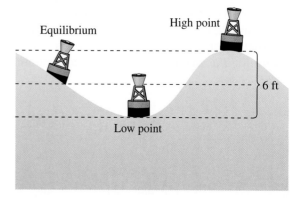

Synthesis

True or False? **In Exercises 178–181, determine whether the statement is true or false. Justify your answer.**

178. $\dfrac{\sin 60°}{\sin 30°} = \sin 2°$

179. $\tan[(0.8)^2] = \tan^2(0.8)$

180. $y = \sin\theta$ is not a function because $\sin 30° = \sin 150°$.

181. $\tan\dfrac{3\pi}{4} = -1 \implies \arctan(-1) = \dfrac{3\pi}{4}$

182. In your own words, explain the meaning of (a) an angle in standard position, (b) a negative angle, (c) coterminal angles, and (d) an obtuse angle.

183. A fan motor turns at a given angular speed. How does the speed of the tips of the blades change if a fan of greater diameter is installed on the motor? Explain.

Chapter Project *Fitting a Model to Data*

In this project, you will find and use models relating to the carbon dioxide level of the earth's atmosphere.

Since 1958, the Mauna Loa Climate Observatory in Hawaii has been collecting data on the carbon dioxide level of the earth's atmosphere. The table shows the average monthly readings for January of each year from 1974 through 1999. The readings measure the carbon dioxide concentration in parts per million. (Source: National Oceanic Atmospheric Administration, Climate Monitoring and Diagnostic Laboratory, CarbonCycle-Greenhouse Gases)

1974	1975	1976	1977	1978	1979	1980	1981	1982	1983	1984	1985	1986
329.7	331.1	332.0	333.8	335.4	336.7	338.7	340.1	341.2	342.8	344.5	345.9	347.2

1987	1988	1989	1990	1991	1992	1993	1994	1995	1996	1997	1998	1999
349.0	351.5	352.9	354.2	355.5	356.3	357.0	358.6	360.6	362.4	363.5	366.5	368.6

a. Enter the data in the table into a graphing utility. (Let $t = 0$ represent 1970.) Draw a scatter plot for the data. Does the data appear to be modeled best with a linear, quadratic, or exponential model?

b. Find the linear, quadratic, or exponential model that you think best fits the data.

Questions for Further Exploration

1. The data in the table represents the carbon dioxide levels for January of each year. Throughout each year, the level oscillated as follows.

   • In April, the average reading was about 2.5 parts per million higher than the average reading given by the model in part (b) above.

   • In July, the average reading was the same as the average reading given by the model in part (b) above.

   • In October, the average reading was about 2.5 parts per million lower than the average reading given by the model in part (b) above.

   Use a sine function to rewrite the model found in part (b) above so that the model incorporates the described oscillations.

2. Use a graphing utility to graph the revised model.

3. Make a careful sketch of the model for 1 year. What physical factors on the earth would contribute to the oscillation in the carbon dioxide level during the year?

4. Is the model you found periodic? Explain your reasoning.

5. Use the model to estimate the level of carbon dioxide in the earth's atmosphere in the following years.

   (a) 2000

   (b) 2010

   (c) 2020

6. What significance does the trend in the carbon dioxide level in the earth's atmosphere have?

4 Chapter Test

Take this test as you would take a test in class. After you are done, check your work against the answers in the back of the book.

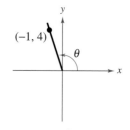

The *Interactive* CD-ROM and *Internet* versions of this text provide answers to the Chapter Tests and Cumulative Tests. They also offer Chapter Pre-Tests (that test key skills and concepts covered in previous chapters) and Chapter Post-Tests, both of which have randomly generated exercises with diagnostic capabilities.

1. Consider the angle of magnitude $5\pi/4$ radians.

   (a) Sketch the angle in standard position.

   (b) Determine two coterminal angles (one positive and one negative).

   (c) Convert the angle to degree measure.

2. A truck is moving at a rate of 90 kilometers per hour, and the diameter of its wheels is 1 meter. Find the angular speed of the wheels in radians per minute.

3. Find the exact values of the six trigonometric functions of the angle θ shown in the figure.

4. Given that $\tan \theta = \frac{11}{6}$, find the other five trigonometric functions of θ.

5. Determine the reference angle θ' of the angle $\theta = 290°$ and sketch θ and θ' in standard position.

6. Determine the quadrant in which θ lies if $\sec \theta < 0$ and $\tan \theta > 0$.

7. Find two values of θ in degrees ($0 \le \theta < 360°$) if $\cos \theta = -\sqrt{3}/2$. (Do not use a calculator.)

8. Use a calculator to approximate two values of θ in radians ($0 \le \theta < 2\pi$) if $\csc \theta = 1.030$. Round the result to two decimal places.

FIGURE FOR 3

9. Find the five remaining trigonometric functions of θ, given that $\sec \theta = \frac{12}{10}$ and $\tan \theta < 0$.

In Exercises 10–15, graph the function through two full periods without using a graphing utility.

10. $g(x) = -2 \sin\left(x - \frac{\pi}{4}\right)$ 11. $f(\alpha) = \frac{1}{2} \tan 2\alpha$

12. $f(x) = \frac{1}{2} \sec(x - \pi)$ 13. $f(x) = 2 \cos(\pi - 2x) + 3$

14. $f(x) = 2 \csc\left(x + \frac{\pi}{2}\right)$ 15. $f(x) = \frac{1}{4} \cot\left(x - \frac{\pi}{2}\right)$

In Exercises 16 and 17, use a graphing utility to graph the function. If the function is periodic, find its period.

16. $y = \sin 2\pi x + 2 \cos \pi x$ 17. $y = 6e^{-0.12t} \cos(0.25t), \quad 0 \le t \le 32$

18. Find a, b, and c for the function $f(x) = a \sin(bx + c)$ so that the graph of f matches the graph at the right.

19. Find the exact value of $\tan\left(\arccos \frac{2}{3}\right)$ without using a calculator.

In Exercises 20–22, sketch the graph of the function

20. $f(x) = 2 \arcsin\left(\frac{1}{2}x\right)$ 21. $f(x) = 2 \arccos x$ 22. $f(x) = \arctan \frac{x}{2}$

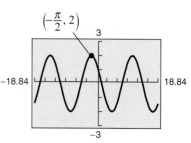

FIGURE FOR 18

23. A plane is 160 miles north and 110 miles east of an airport. If the pilot wants to fly directly to the airport, what bearing should he take?

24. From a 100-foot roof of a condominium on the coast, a tourist sights a cruise ship. The angle of depression is 2.5°. How far is the ship from the shore line?

Analytic Trigonometry

5.1 Using Fundamental Identities
5.2 Verifying Trigonometric Identities
5.3 Solving Trigonometric Equations
5.4 Sum and Difference Formulas
5.5 Multiple-Angle and Product-Sum Formulas

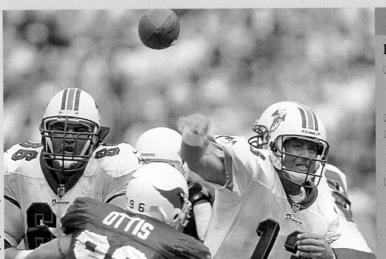

Andy Lyons/Allsport

The Big Picture

In this chapter you will learn how to

❏ use fundamental trigonometric identities to evaluate trigonometric functions and simplify trigonometric expressions.
❏ verify trigonometric identities.
❏ use standard algebraic techniques and inverse trigonometric functions to solve trigonometric equations.
❏ use sum and difference formulas, multiple-angle formulas, power-reducing formulas, half-angle formulas, and product-sum formulas to rewrite and evaluate trigonometric functions.

A football thrown by a quarterback follows a parabolic path. The horizontal distance the football travels depends not only on the speed of the throw, but on the angle at which the ball is thrown.

Important Vocabulary

As you encounter each new vocabulary term in this chapter, add the term and its definition to your notebook glossary.

- reduction formulas (p. 406)
- double-angle formulas (p. 411)
- power-reducing formulas (p. 413)
- half-angle formulas (p. 414)
- product-to-sum formulas (p. 415)
- sum-to-product formulas (p. 416)

Additional Resources   Text-specific additional resources are available to help you do well in this course. See page xvi for details.

5.1 Using Fundamental Identities

Introduction

In Chapter 4, you studied the basic definitions, properties, graphs, and applications of the individual trigonometric functions. In this chapter, you will learn how to use the fundamental identities to

1. evaluate trigonometric functions,
2. simplify trigonometric expressions,
3. develop additional trigonometric identities, and
4. solve trigonometric equations.

What You Should Learn:

- How to recognize and write fundamental trigonometric identities
- How to use fundamental trigonometric identities to evaluate trigonometric functions, simplify trigonometric expressions, and rewrite trigonometric expressions

Why You Should Learn It:

Fundamental trigonometric identities can be used to simplify trigonometric expressions. For instance, Exercise 109 on page 383 shows you how trigonometric identities can be used to simplify an expression for the rate of change of a function, a concept used in calculus.

Fundamental Trigonometric Identities

Reciprocal Identities

$$\sin u = \frac{1}{\csc u} \qquad \cos u = \frac{1}{\sec u} \qquad \tan u = \frac{1}{\cot u}$$

$$\csc u = \frac{1}{\sin u} \qquad \sec u = \frac{1}{\cos u} \qquad \cot u = \frac{1}{\tan u}$$

Quotient Identities

$$\tan u = \frac{\sin u}{\cos u} \qquad \cot u = \frac{\cos u}{\sin u}$$

Pythagorean Identities

$$\sin^2 u + \cos^2 u = 1 \qquad 1 + \tan^2 u = \sec^2 u \qquad 1 + \cot^2 u = \csc^2 u$$

Cofunction Identities

$$\sin\left(\frac{\pi}{2} - u\right) = \cos u \qquad \cos\left(\frac{\pi}{2} - u\right) = \sin u$$

$$\tan\left(\frac{\pi}{2} - u\right) = \cot u \qquad \cot\left(\frac{\pi}{2} - u\right) = \tan u$$

$$\sec\left(\frac{\pi}{2} - u\right) = \csc u \qquad \csc\left(\frac{\pi}{2} - u\right) = \sec u$$

Even/Odd Identities

$$\sin(-u) = -\sin u \qquad \sec(-u) = \sec u \qquad \tan(-u) = -\tan u$$

$$\csc(-u) = -\csc u \qquad \cos(-u) = \cos u \qquad \cot(-u) = -\cot u$$

Pythagorean identities are sometimes used in radical form such as

$$\sin u = \pm\sqrt{1 - \cos^2 u} \quad \text{or} \quad \tan u = \pm\sqrt{\sec^2 u - 1}$$

where the sign depends on the choice of u.

Using the Fundamental Identities

One common use of trigonometric identities is to use given values of trigonometric functions to evaluate other trigonometric functions.

The *Interactive* CD-ROM and *Internet* versions of this text show every example with its solution; clicking on the *Try It!* button brings up similar problems. Guided Examples and Integrated Examples show step-by-step solutions to additional examples. Integrated Examples are related to several concepts in the section.

EXAMPLE 1   Using Identities to Evaluate a Function

Use the values $\sec u = -\frac{3}{2}$ and $\tan u > 0$ to find the values of all six trigonometric functions.

Solution

Using a reciprocal identity, you have

$$\cos u = \frac{1}{\sec u} = \frac{1}{-3/2} = -\frac{2}{3}.$$

Using a Pythagorean identity, you have

$$\sin^2 u = 1 - \cos^2 u \qquad \text{Pythagorean identity}$$

$$= 1 - \left(-\frac{2}{3}\right)^2 \qquad \text{Substitute } -\frac{2}{3} \text{ for } \cos u.$$

$$= 1 - \frac{4}{9} \qquad \text{Simplify.}$$

$$= \frac{5}{9}. \qquad \text{Simplify.}$$

Because $\sec u < 0$ and $\tan u > 0$, it follows that u lies in Quadrant III. Moreover, because $\sin u$ is negative when u is in Quadrant III, you can choose the negative root and obtain $\sin u = -\sqrt{5}/3$. Now, knowing the values of the sine and cosine, you can find the values of all six trigonometric functions.

$$\sin u = -\frac{\sqrt{5}}{3} \qquad\qquad \csc u = \frac{1}{\sin u} = -\frac{3}{\sqrt{5}}$$

$$\cos u = -\frac{2}{3} \qquad\qquad \sec u = \frac{1}{\cos u} = -\frac{3}{2}$$

$$\tan u = \frac{\sin u}{\cos u} = \frac{-\sqrt{5}/3}{-2/3} = \frac{\sqrt{5}}{2} \qquad \cot u = \frac{1}{\tan u} = \frac{2}{\sqrt{5}}$$

EXAMPLE 2   Simplifying a Trigonometric Expression

Simplify $\sin x \cos^2 x - \sin x$.

Solution

First factor out a common monomial factor and then use a fundamental identity.

$$\sin x \cos^2 x - \sin x = \sin x(\cos^2 x - 1) \qquad \text{Factor out monomial factor.}$$

$$= -\sin x(1 - \cos^2 x) \qquad \text{Distributive Property}$$

$$= -\sin x(\sin^2 x) \qquad \text{Pythagorean identity}$$

$$= -\sin^3 x \qquad \text{Multiply.}$$

STUDY T!P

You can use a graphing utility to check the result of Example 2. To do this, graph

$$y_1 = \sin x \cos^2 x - \sin x$$

and

$$y_2 = -\sin^3 x$$

in the same viewing window, as shown below. The two graphs *appear* to coincide, so the expressions *appear* to be equivalent. Remember that in order to be certain that two expressions are equivalent, you need to show their equivalence algebraically, as in Example 2.

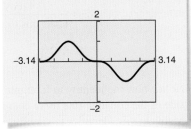

EXAMPLE 3 Verifying Trigonometric Identities

Determine which of the following appears to be an identity.

a. $\cos 3x \overset{?}{=} 4\cos^3 x - 3\cos x$ **b.** $\cos 3x \overset{?}{=} \sin\left(3x - \dfrac{\pi}{2}\right)$

Numerical Solution

a. Use the *table* feature of a graphing utility set in *radian* mode to create a table that shows the values of $y_1 = \cos 3x$ and $y_2 = 4\cos^3 x - 3\cos x$ for different values of x, as shown in Figure 5.1(a). The values of y_1 and y_2 appear to be identical, so $\cos 3x = 4\cos^3 x - 3\cos x$ appears to be an identity.

b. Create a table that shows the values of $y_1 = \cos 3x$ and $y_2 = \sin(3x - \pi/2)$ for different values of x, as shown in Figure 5.1(b). The values of y_1 and y_2 are not identical, so $\cos 3x = \sin(3x - \pi/2)$ is not an identity.

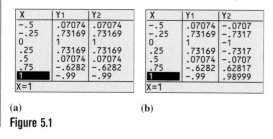

X	Y1	Y2
-.5	.07074	.07074
-.25	.73169	.73169
0	1	1
.25	.73169	.73169
.5	.07074	.07074
.75	-.6282	-.6282
1	-.99	-.99

X=1

X	Y1	Y2
-.5	.07074	-.0707
-.25	.73169	-.7317
0	1	-1
.25	.73169	-.7317
.5	.07074	-.0707
.75	-.6282	.62817
1	-.99	.98999

X=1

(a)   (b)

Figure 5.1

Graphical Solution

a. Use a graphing utility set in *radian* mode to graph $y_1 = \cos 3x$ and $y_2 = 4\cos^3 x - 3\cos x$ in the same viewing window, as shown in Figure 5.2(a). Because the graphs appear to coincide, $\cos 3x = 4\cos^3 x - 3\cos x$ appears to be an identity.

b. Graph $y_1 = \cos 3x$ and $y_2 = \sin(3x - \pi/2)$ in the same viewing window, as shown in Figure 5.2(b). Because the graphs do not coincide, $\cos 3x = \sin(3x - \pi/2)$ is not an identity.

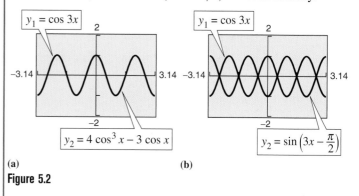

(a)   (b)

Figure 5.2

EXAMPLE 4 Verifying a Trigonometric Identity

Verify the identity $\dfrac{\sin\theta}{1+\cos\theta} + \dfrac{\cos\theta}{\sin\theta} = \csc\theta.$

Algebraic Solution

$$\frac{\sin\theta}{1+\cos\theta} + \frac{\cos\theta}{\sin\theta} = \frac{(\sin\theta)(\sin\theta) + (\cos\theta)(1+\cos\theta)}{(1+\cos\theta)(\sin\theta)}$$

$$= \frac{\sin^2\theta + \cos^2\theta + \cos\theta}{(1+\cos\theta)(\sin\theta)} \quad \text{Multiply.}$$

$$= \frac{1+\cos\theta}{(1+\cos\theta)(\sin\theta)} \quad \text{Pythagorean identity}$$

$$= \csc\theta \quad \begin{array}{l}\text{Divide out common factor}\\ \text{and use reciprocal identity.}\end{array}$$

Notice how the identity in Example 4 is verified. You start with the left side of the equation (the more complicated side) and use the fundamental trigonometric identities to simplify it until you obtain the right side.

Graphical Solution

Use a graphing utility set in *radian* and *dot* modes to graph y_1 and y_2 in the same viewing window, as shown in Figure 5.3. Because the graphs appear to coincide, this equation appears to be an identity.

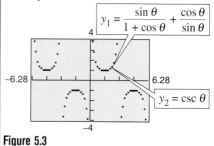

Figure 5.3

When factoring trigonometric expressions, it is helpful to find a polynomial form that fits the expression, as shown in Example 5.

EXAMPLE 5   Factoring Trigonometric Expressions

Factor each expression.

a. $\sec^2 \theta - 1$

b. $4 \tan^2 \theta + \tan \theta - 3$

Solution

a. Here the expression has the polynomial form $u^2 - v^2$ (the difference of two squares), which factors as

$$\sec^2 \theta - 1 = (\sec \theta - 1)(\sec \theta + 1).$$

b. This expression has the polynomial form, $ax^2 + bx + c$, and it factors as

$$4 \tan^2 \theta + \tan \theta - 3 = (4 \tan \theta - 3)(\tan \theta + 1).$$

On occasion, factoring or simplifying can best be done by first rewriting the expression in terms of just *one* trigonometric function or in terms of *sine* or *cosine alone*. These strategies are illustrated in Examples 6 and 7.

EXAMPLE 6   Factoring a Trigonometric Expression

Factor $\csc^2 x - \cot x - 3$.

Solution

You can use the identity $\csc^2 x = 1 + \cot^2 x$ to rewrite the expression in terms of the cotangent alone.

$$\csc^2 x - \cot x - 3 = (1 + \cot^2 x) - \cot x - 3 \qquad \text{Pythagorean identity}$$

$$= \cot^2 x - \cot x - 2 \qquad \text{Combine like terms.}$$

$$= (\cot x - 2)(\cot x + 1) \qquad \text{Factor.}$$

EXAMPLE 7   Simplifying a Trigonometric Expression

Simplify $\sin t + \cot t \cos t$.

Solution

Begin by rewriting the expression in terms of sine and cosine.

$$\sin t + \cot t \cos t = \sin t + \left(\frac{\cos t}{\sin t} \right) \cos t \qquad \text{Quotient identity}$$

$$= \frac{\sin^2 t + \cos^2 t}{\sin t} \qquad \text{Add fractions.}$$

$$= \frac{1}{\sin t} \qquad \text{Pythagorean identity}$$

$$= \csc t \qquad \text{Reciprocal identity}$$

The last two examples in this section involve techniques for rewriting expressions into forms that are used in calculus.

EXAMPLE 8 Rewriting a Trigonometric Expression

Rewrite $\dfrac{1}{1 + \sin x}$ so that it is *not* in fractional form.

Solution

From the Pythagorean identity $\cos^2 x = 1 - \sin^2 x = (1 - \sin x)(1 + \sin x)$, you can see that by multiplying both the numerator and the denominator by $(1 - \sin x)$ you will produce a monomial denominator.

$$\frac{1}{1 + \sin x} = \frac{1}{1 + \sin x} \cdot \frac{1 - \sin x}{1 - \sin x} \qquad \text{Multiply numerator and denominator by } (1 - \sin x).$$

$$= \frac{1 - \sin x}{1 - \sin^2 x} \qquad \text{Multiply.}$$

$$= \frac{1 - \sin x}{\cos^2 x} \qquad \text{Pythagorean identity}$$

$$= \frac{1}{\cos^2 x} - \frac{\sin x}{\cos^2 x} \qquad \text{Write as separate fractions.}$$

$$= \frac{1}{\cos^2 x} - \frac{\sin x}{\cos x} \cdot \frac{1}{\cos x} \qquad \text{Write as separate fractions.}$$

$$= \sec^2 x - \tan x \sec x \qquad \text{Reciprocal and quotient identities}$$

EXAMPLE 9 Trigonometric Substitution

Use the substitution $x = 2 \tan \theta$, $0 < \theta < \pi/2$, to express $\sqrt{4 + x^2}$ as a trigonometric function of θ.

Solution

Begin by letting $x = 2 \tan \theta$. Then you can obtain

$$\sqrt{4 + x^2} = \sqrt{4 + (2 \tan \theta)^2} \qquad \text{Substitute } 2 \tan \theta \text{ for } x.$$

$$= \sqrt{4(1 + \tan^2 \theta)} \qquad \text{Distributive Property}$$

$$= \sqrt{4 \sec^2 \theta} \qquad \text{Pythagorean identity}$$

$$= 2 \sec \theta. \qquad \sec \theta > 0 \text{ for } 0 < \theta < \tfrac{\pi}{2}$$

Figure 5.4 shows the right triangle illustration of this substitution. For $0 < \theta < \pi/2$, you have

opposite $= x$, adjacent $= 2$, and hypotenuse $= \sqrt{4 + x^2}$.

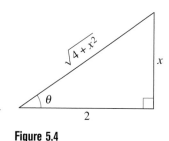

Figure 5.4

Try using these expressions to obtain the result shown above.

5.1 Exercises

In Exercises 1–14, use the given values to evaluate (if possible) the remaining trigonometric functions.

1. $\sin x = \dfrac{\sqrt{3}}{2}, \quad \cos x = \dfrac{1}{2}$

2. $\csc \theta = \dfrac{5}{3}, \quad \tan \theta = \dfrac{3}{4}$

3. $\sec \theta = \sqrt{2}, \quad \sin \theta = -\dfrac{\sqrt{2}}{2}$

4. $\tan x = \dfrac{\sqrt{3}}{3}, \quad \cos x = -\dfrac{\sqrt{3}}{2}$

5. $\tan x = \dfrac{7}{24}, \quad \sec x = -\dfrac{25}{24}$

6. $\cot \phi = -5, \quad \sin \phi = \dfrac{\sqrt{26}}{26}$

7. $\sec \phi = -1, \quad \sin \phi = 0$

8. $\cos\left(\dfrac{\pi}{2} - x\right) = \dfrac{3}{5}, \quad \cos x = \dfrac{4}{5}$

9. $\sin(-x) = -\dfrac{2}{3}, \quad \tan x = -\dfrac{2\sqrt{5}}{5}$

10. $\csc x = 5, \quad \cos x > 0$

11. $\tan \theta = 4, \quad \sin \theta < 0$

12. $\sec \theta = -3, \quad \tan \theta < 0$

13. $\sin \theta = -1, \quad \cot \theta = 0$

14. $\tan \theta$ is undefined, $\quad \sin \theta > 0$

In Exercises 15–18, fill in the blanks. (*Note:* $x \to c^+$ indicates that x approaches c from the right, and $x \to c^-$ indicates that x approaches c from the left.)

15. As $x \to \dfrac{\pi^-}{2}$, $\sin x \to$ ▢ and $\csc x \to$ ▢.

16. As $x \to 0^+$, $\cos x \to$ ▢ and $\sec x \to$ ▢.

17. As $x \to \dfrac{\pi^-}{2}$, $\tan x \to$ ▢ and $\cot x \to$ ▢.

18. As $x \to \pi^+$, $\sin x \to$ ▢ and $\csc x \to$ ▢.

In Exercises 19–24, match the trigonometric expression with one of the following.

(a) -1 (b) $\cos x$ (c) $\cot x$

(d) 1 (e) $-\tan x$ (f) $\sin x$

19. $\csc x \sin x$

20. $\tan x \cos x$

21. $\tan^2 x - \sec^2 x$

22. $(1 - \sin^2 x)(\sec x)$

23. $\dfrac{\sin(-x)}{\cos(-x)}$

24. $\dfrac{\sin[(\pi/2) - x]}{\cos[(\pi/2) - x]}$

In Exercises 25–30, match the trigonometric expression with one of the following.

(a) $\csc x$ (b) $\cot x$ (c) $\cos^2 x$

(d) $\sin x \tan x$ (e) $\sec^2 x$ (f) $\sec^2 x + \tan^2 x$

25. $\cos x \csc x$

26. $\sin^2 x(\csc^2 x - 1)$

27. $\sec^4 x - \tan^4 x$

28. $\cot x \sec x$

29. $\dfrac{\sec^2 x - 1}{\sin^2 x}$

30. $\dfrac{\cos^2[(\pi/2) - x]}{\cos x}$

In Exercises 31–44, use the fundamental identities to simplify the expression. Use the *table* feature of a graphing utility to check your result numerically.

31. $\cot x \sin x$

32. $\cos \beta \tan \beta$

33. $\sin \phi(\csc \phi - \sin \phi)$

34. $\sec^2 x(1 - \sin^2 x)$

35. $\dfrac{\cot x}{\csc x}$

36. $\dfrac{\sec \theta}{\csc \theta}$

37. $\sec \alpha \cdot \dfrac{\sin \alpha}{\tan \alpha}$

38. $\dfrac{1}{\tan^2 x + 1}$

39. $\dfrac{\sin(-x)}{\cos x}$

40. $\dfrac{\tan^2 \theta}{\sec^2 \theta}$

41. $\sin\left(\dfrac{\pi}{2} - x\right)\csc x$

42. $\cot\left(\dfrac{\pi}{2} - x\right)\cos x$

43. $\dfrac{\cos^2 y}{1 - \sin y}$

44. $\cos t(1 + \tan^2 t)$

In Exercises 45–52, verify the identity algebraically. Then use a graphing utility to check your result graphically.

45. $\csc \theta \tan \theta = \sec \theta$

46. $\cos \theta \sec \theta - \cos^2 \theta = \sin^2 \theta$

47. $\dfrac{\csc \theta}{\sec \theta} + \dfrac{\cos \theta}{\sin \theta} = 2 \cot \theta$

48. $\dfrac{\sec^2 \theta - \tan^2 \theta + \tan \theta}{\sec \theta} = \cos \theta + \sin \theta$

49. $1 - \dfrac{\sin^2 \theta}{1 - \cos \theta} = -\cos \theta$

50. $\dfrac{\tan \theta}{1 + \sec \theta} + \dfrac{1 + \sec \theta}{\tan \theta} = 2 \csc \theta$

51. $\dfrac{\cot(-\theta)}{\csc \theta} = -\cos \theta$ **52.** $\dfrac{\csc\left(\dfrac{\pi}{2} - \theta\right)}{\tan(-\theta)} = -\csc \theta$

In Exercises 53–62, verify the identity algebraically. Then use the *table* feature of a graphing utility to check your result numerically.

53. $\sin \theta + \cos \theta \cot \theta = \csc \theta$

54. $(\sec \theta - \tan \theta)(\csc \theta + 1) = \cot \theta$

55. $\dfrac{\cos \theta}{1 - \sin \theta} = \sec \theta + \tan \theta$

56. $\dfrac{1 + \csc \theta}{\cot \theta + \cos \theta} = \sec \theta$

57. $\dfrac{\sin \theta}{\csc \theta} + \dfrac{\cos \theta}{\sec \theta} = 1$

58. $\dfrac{1 + \csc \theta}{\sec \theta} - \cot \theta = \cos \theta$

59. $\dfrac{1 + \cos \theta}{\sin \theta} + \dfrac{\sin \theta}{1 + \cos \theta} = 2 \csc \theta$

60. $\dfrac{\sin \theta + \cos \theta}{\sin \theta} - \dfrac{\cos \theta - \sin \theta}{\cos \theta} = \sec \theta \csc \theta$

61. $\ln|\csc \theta| = -\ln|\sin \theta|$

62. $\ln|1 + \cos \theta| + \ln|1 - \cos \theta| = 2 \ln|\sin \theta|$

In Exercises 63–70, factor the expression and use the fundamental identities to simplify. Use a graphing utility to check your result graphically.

63. $\cot^2 x - \cot^2 x \cos^2 x$

64. $\sec^2 x \tan^2 x + \sec^2 x$

65. $\sin^2 x \sec^2 x - \sin^2 x$

66. $\dfrac{\csc^2 x - 1}{\csc x - 1}$

67. $\tan^4 x + 2 \tan^2 x + 1$

68. $1 - 2 \sin^2 x + \sin^4 x$

69. $\sin^4 x - \cos^4 x$

70. $\sec^3 x - \sec^2 x - \sec x + 1$

In Exercises 71–74, perform the multiplication and use the fundamental identities to simplify.

71. $(\sin x + \cos x)^2$

72. $(\cot x + \csc x)(\cot x - \csc x)$

73. $(\sec x + 1)(\sec x - 1)$

74. $(3 - 3 \sin x)(3 + 3 \sin x)$

In Exercises 75–78, perform the addition or subtraction and use the fundamental identities to simplify.

75. $\dfrac{1}{1 + \cos x} + \dfrac{1}{1 - \cos x}$

76. $\dfrac{1}{\sec x + 1} - \dfrac{1}{\sec x - 1}$

77. $\dfrac{\cos x}{1 + \sin x} + \dfrac{1 + \sin x}{\cos x}$

78. $\tan x - \dfrac{\sec^2 x}{\tan x}$

In Exercises 79–82, rewrite the expression so that it is *not* in fractional form.

79. $\dfrac{\sin^2 y}{1 - \cos y}$ **80.** $\dfrac{5}{\tan x + \sec x}$

81. $\dfrac{3}{\sec x - \tan x}$ **82.** $\dfrac{\tan^2 x}{\csc x + 1}$

Numerical and Graphical Analysis **In Exercises 83–86, use a graphing utility to complete the table and graph the functions. Make a conjecture about y_1 and y_2.**

x	0.2	0.4	0.6	0.8	1.0	1.2	1.4
y_1							
y_2							

83. $y_1 = \cos\left(\dfrac{\pi}{2} - x\right), \quad y_2 = \sin x$

84. $y_1 = \cos x + \sin x \tan x, \quad y_2 = \sec x$

85. $y_1 = \dfrac{\cos x}{1 - \sin x}, \quad y_2 = \dfrac{1 + \sin x}{\cos x}$

86. $y_1 = \sec^4 x - \sec^2 x, \quad y_2 = \tan^2 x + \tan^4 x$

In Exercises 87–90, use a graphing utility to determine which of the six trigonometric functions is equal to the expression.

87. $\cos x \cot x + \sin x$ **88.** $\sin x(\cot x + \tan x)$

89. $\sec x - \dfrac{\cos x}{1 + \sin x}$

90. $\dfrac{1}{2}\left(\dfrac{1 + \sin \theta}{\cos \theta} + \dfrac{\cos \theta}{1 + \sin \theta}\right)$

In Exercises 91–96, use the trigonometric substitution to write the algebraic expression as a trigonometric function of θ, where $0 < \theta < \pi/2$.

91. $\sqrt{25 - x^2}, \quad x = 5 \sin \theta$

92. $\sqrt{16 - 4x^2}, \quad x = 2 \sin \theta$

93. $\sqrt{x^2 - 9}, \quad x = 3 \sec \theta$

94. $\sqrt{x^2 - 4}, \quad x = 2 \sec \theta$

95. $\sqrt{x^2 + 25}, \quad x = 5 \tan \theta$

96. $\sqrt{x^2 + 100}, \quad x = 10 \tan \theta$

In Exercises 97–100, use a graphing utility to solve the equation for θ, where $0 \le \theta < 2\pi$.

97. $\sin \theta = \sqrt{1 - \cos^2 \theta}$ **98.** $\cos \theta = -\sqrt{1 - \sin^2 \theta}$

99. $\sec \theta = \sqrt{1 + \tan^2 \theta}$ **100.** $\tan \theta = \sqrt{\sec^2 \theta - 1}$

In Exercises 101–104, rewrite the expression as a single logarithm and simplify the result.

101. $\ln|\cos \theta| - \ln|\sin \theta|$ **102.** $\ln|\csc \theta| + \ln|\tan \theta|$

103. $\ln(1 + \sin x) - \ln|\sec x|$

104. $\ln|\cot t| + \ln(1 + \tan^2 t)$

In Exercises 105–108, use the *table* feature of a graphing utility to demonstrate the identity for the given values of θ.

105. $\csc^2 \theta - \cot^2 \theta = 1$, (a) $\theta = 132°$, (b) $\theta = \dfrac{2\pi}{7}$

106. $\tan^2 \theta + 1 = \sec^2 \theta$, (a) $\theta = 346°$, (b) $\theta = 3.1$

107. $\cos\left(\dfrac{\pi}{2} - \theta\right) = \sin \theta$, (a) $\theta = 80°$, (b) $\theta = 0.8$

108. $\sin(-\theta) = -\sin \theta$, (a) $\theta = 250°$, (b) $\theta = \frac{1}{2}$

109. *Rate of Change* The rate of change of the function $f(x) = -\csc x - \sin x$ is given by the expression $\csc x \cot x - \cos x$. Show that this expression can also be written as $\cos x \cot^2 x$.

Synthesis

True or False? In Exercises 110–113, determine whether the statement is true or false. Justify your answer.

110. $\dfrac{\sin k\theta}{\cos k\theta} = \tan \theta$, k is a constant.

111. $\dfrac{1}{5 \cos \theta} = 5 \sec \theta$

112. $\sin \theta \csc \theta = 1$ **113.** $\sin \theta \csc \phi = 1$

114. Express each of the other trigonometric functions of θ in terms of $\sin \theta$.

115. Express each of the other trigonometric functions of θ in terms of $\cos \theta$.

Review

In Exercises 116–121, find the reference angle θ' and sketch θ and θ' in standard position.

116. $\theta = 254°$ **117.** $\theta = 341°$

118. $\theta = -178°$ **119.** $\theta = -212°$

120. $\theta = \dfrac{13\pi}{15}$ **121.** $\theta = \dfrac{35\pi}{6}$

In Exercises 122–125, sketch the graph of the function.

122. $f(x) = \dfrac{1}{2} \sin \pi x$ **123.** $f(x) = -2 \tan \dfrac{\pi x}{2}$

124. $f(x) = \dfrac{1}{2} \cot\left(x + \dfrac{\pi}{4}\right)$

125. $f(x) = \dfrac{3}{2} \cos(x - \pi) + 3$

In Exercises 126–129, solve the right triangle shown in the figure. (Round your answers to two decimal places.)

126. $B = 80°$, $a = 16$ **127.** $A = 28°$, $c = 20$

128. $a = 14$, $b = 8$ **129.** $b = 6.2$, $c = 12.54$

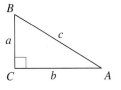

5.2 Verifying Trigonometric Identities

Introduction

In this section, you will study techniques for verifying trigonometric identities. In the next section, you will study techniques for solving trigonometric equations. The key to both verifying identities *and* solving equations is the ability to use the fundamental identities and the rules of algebra to rewrite trigonometric expressions.

Remember that a *conditional equation* is an equation that is true for only some of the values in its domain. For example, the conditional equation

$$\sin x = 0 \qquad \text{Conditional equation}$$

is true only for $x = n\pi$, where n is an integer. When you find these values, you are *solving* the equation.

On the other hand, an equation that is true for all real values in the domain of the variable is an *identity*. For example, the familiar equation

$$\sin^2 x = 1 - \cos^2 x \qquad \text{Identity}$$

is true for all real numbers x. So, it is an identity.

Verifying Trigonometric Identities

Although there are similarities, proving that a trigonometric equation is an identity is quite different from solving an equation. There is no well-defined set of rules to follow in verifying trigonometric identities, and the process is best learned by practice.

Guidelines for Verifying Trigonometric Identities

1. Work with one side of the equation at a time. It is often better to work with the more complicated side first.

2. Look for opportunities to factor an expression, add fractions, square a binomial, or create a monomial denominator.

3. Look for opportunities to use the fundamental identities. Note which functions are in the final expression you want. Sines and cosines pair up well, as do secants and tangents, and cosecants and cotangents.

4. If the preceding guidelines do not help, try converting all terms to sines and cosines.

5. Try something! Even making an attempt that leads to a dead end gives insight.

What You Should Learn:

• How to verify trigonometric identities

Why You Should Learn It:

You can use trigonometric identities to rewrite trigonometric expressions. For instance, Exercise 69 on page 391, shows you how trigonometric identities can be used to solve a problem about the coefficient of friction for an object on an inclined plane.

Nancy Dudley/Stock Boston

A computer animation of this concept appears in the *Interactive* CD-ROM and *Internet* versions of this text.

EXAMPLE 1   Verifying a Trigonometric Identity

Verify the identity $\dfrac{\sec^2 \theta - 1}{\sec^2 \theta} = \sin^2 \theta$.

Solution
Because the left side is more complicated, start with it.

$$\frac{\sec^2 \theta - 1}{\sec^2 \theta} = \frac{(\tan^2 \theta + 1) - 1}{\sec^2 \theta} \qquad \text{Pythagorean identity}$$

$$= \frac{\tan^2 \theta}{\sec^2 \theta} \qquad \text{Simplify.}$$

$$= \tan^2 \theta (\cos^2 \theta) \qquad \text{Reciprocal identity}$$

$$= \frac{\sin^2 \theta}{\cos^2 \theta} (\cos^2 \theta) \qquad \text{Quotient identity}$$

$$= \sin^2 \theta \qquad \text{Simplify.}$$

The *Interactive* CD-ROM and *Internet* versions of this text offer a built-in graphing calculator, which can be used with the Examples, Explorations, and Exercises.

Here is another way to verify the identity in Example 1.

$$\frac{\sec^2 \theta - 1}{\sec^2 \theta} = \frac{\sec^2 \theta}{\sec^2 \theta} - \frac{1}{\sec^2 \theta}$$

$$= 1 - \cos^2 \theta$$

$$= \sin^2 \theta$$

There can be more than one way to verify an identity. Your method may differ from that used by your instructor or fellow students. This is a good chance to be creative and establish your own style, but try to be as efficient as possible.

EXAMPLE 2   Combining Fractions Before Using Identities

Verify the identity $\dfrac{1}{1 - \sin \alpha} + \dfrac{1}{1 + \sin \alpha} = 2 \sec^2 \alpha$.

Algebraic Solution

$$\frac{1}{1 - \sin \alpha} + \frac{1}{1 + \sin \alpha} = \frac{1 + \sin \alpha + 1 - \sin \alpha}{(1 - \sin \alpha)(1 + \sin \alpha)} \qquad \text{Add fractions.}$$

$$= \frac{2}{1 - \sin^2 \alpha} \qquad \text{Simplify.}$$

$$= \frac{2}{\cos^2 \alpha} \qquad \text{Pythagorean identity}$$

$$= 2 \sec^2 \alpha \qquad \text{Reciprocal identity}$$

Numerical Solution

Use the *table* feature of a graphing utility set in *radian* mode to create a table that shows the values of $y_1 = 1/(1 - \sin x) + 1/(1 + \sin x)$ and $y_2 = 2/\cos^2 x$ for different values of x, as shown in Figure 5.5. From the table, you can see that the values appear to be identical, so $1/(1 - \sin x) + 1/(1 + \sin x) = 2 \sec^2$ appears to be an identity.

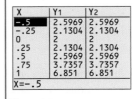

X	Y1	Y2
-.5	2.5969	2.5969
-.25	2.1304	2.1304
0	2	2
.25	2.1304	2.1304
.5	2.5969	2.5969
.75	3.7357	3.7357
1	6.851	6.851
X=-.5		

Figure 5.5

EXAMPLE 3   Verifying a Trigonometric Identity

Verify the identity $(\tan^2 x + 1)(\cos^2 x - 1) = -\tan^2 x$.

Solution

By applying identities before multiplying, you obtain the following.

$$(\tan^2 x + 1)(\cos^2 x - 1) = (\sec^2 x)(-\sin^2 x) \qquad \text{Pythagorean identities}$$

$$= -\frac{\sin^2 x}{\cos^2 x} \qquad \text{Reciprocal identity}$$

$$= -\left(\frac{\sin x}{\cos x}\right)^2 \qquad \text{Rule of exponents}$$

$$= -\tan^2 x \qquad \text{Quotient identity}$$

You can use a graphing utility to verify your results by graphing both sides of the original identity in the same viewing window, as shown in Figure 5.6. The graphs appear to coincide.

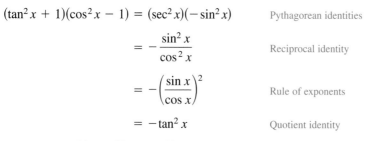

Figure 5.6

> ## STUDY TIP
>
> Although a graphing utility can be useful in helping to verify an identity, you must use algebraic techniques to produce a valid proof. For example, graph the two functions
>
> $$y_1 = \sin 50x$$
>
> $$y_2 = \sin 2x$$
>
> in a trigonometric viewing window. Although their graphs seem identical, $\sin 50x \neq \sin 2x$.

EXAMPLE 4   Converting to Sines and Cosines

Verify the identity $\tan x + \cot x = \sec x \csc x$.

Solution

In this case there appear to be no fractions to add, no products to find, and no opportunity to use one of the Pythagorean identities. So, try converting the left side into sines and cosines to see what happens.

$$\tan x + \cot x = \frac{\sin x}{\cos x} + \frac{\cos x}{\sin x} \qquad \text{Quotient identities}$$

$$= \frac{\sin^2 x + \cos^2 x}{\cos x \sin x} \qquad \text{Add fractions.}$$

$$= \frac{1}{\cos x \sin x} \qquad \text{Pythagorean identity}$$

$$= \frac{1}{\cos x} \cdot \frac{1}{\sin x} \qquad \text{Product of fractions}$$

$$= \sec x \csc x \qquad \text{Reciprocal identities}$$

Recall from algebra that *rationalizing the denominator* is, on occasion, a powerful simplification technique. A related form of this technique works for simplifying trigonometric expressions as well.

EXAMPLE 5   Verifying a Trigonometric Identity

Verify the identity $\sec x + \tan x = \dfrac{\cos x}{1 - \sin x}$.

Algebraic Solution

Work with the *right* side because you can create a monomial denominator by multiplying the numerator and denominator by $(1 + \sin x)$.

$$\frac{\cos x}{1 - \sin x} = \frac{\cos x}{1 - \sin x}\left(\frac{1 + \sin x}{1 + \sin x}\right) \qquad \text{Multiply numerator and denominator by } (1 + \sin x).$$

$$= \frac{\cos x + \cos x \sin x}{1 - \sin^2 x} \qquad \text{Multiply.}$$

$$= \frac{\cos x + \cos x \sin x}{\cos^2 x} \qquad \text{Pythagorean identity}$$

$$= \frac{\cos x}{\cos^2 x} + \frac{\cos x \sin x}{\cos^2 x} \qquad \text{Separate fractions.}$$

$$= \frac{1}{\cos x} + \frac{\sin x}{\cos x} \qquad \text{Simplify.}$$

$$= \sec x + \tan x \qquad \text{Identities}$$

Graphical Solution

Use a graphing utility set in *radian* and *dot* modes to graph $y_1 = \sec x + \tan x = 1/\cos x + \tan x$ and $y_2 = \cos x/(1 - \sin x)$ in the same viewing window, as shown in Figure 5.7. Because the graphs appear to coincide, $\sec x + \tan x = \cos x/(1 - \sin x)$ appears to be an identity.

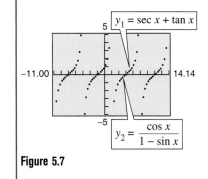

Figure 5.7

In Examples 1 through 5, you have been verifying trigonometric identities by working with one side of the equation and converting to the form given on the other side. On occasion it is practical to work with each side *separately* to obtain one common form equivalent to both sides. This is illustrated in Example 6.

EXAMPLE 6   Working with Each Side Separately

Verify the identity $\dfrac{\cot^2 \theta}{1 + \csc \theta} = \dfrac{1 - \sin \theta}{\sin \theta}$.

Solution
Working with the left side, you have

$$\frac{\cot^2 \theta}{1 + \csc \theta} = \frac{\csc^2 \theta - 1}{1 + \csc \theta} \qquad \text{Pythagorean identity}$$

$$= \frac{(\csc \theta - 1)(\csc \theta + 1)}{1 + \csc \theta} \qquad \text{Factor.}$$

$$= \csc \theta - 1. \qquad \text{Simplify.}$$

Now, simplifying the right side, you have

$$\frac{1 - \sin \theta}{\sin \theta} = \frac{1}{\sin \theta} - \frac{\sin \theta}{\sin \theta} \qquad \text{Separate fractions.}$$

$$= \csc \theta - 1. \qquad \text{Reciprocal identity}$$

The identity is verified because both sides are equal to $\csc \theta - 1$.

In Example 7, powers of trigonometric functions are rewritten as more complicated sums of products of trigonometric functions. This is a common procedure used in calculus.

EXAMPLE 7   Three Examples from Calculus

Verify each identity.

a. $\tan^4 x = \tan^2 x \sec^2 x - \tan^2 x$

b. $\sin^3 x \cos^4 x = (\cos^4 x - \cos^6 x)\sin x$

c. $\sec^4 x \tan^2 x = (\tan^2 x + \tan^4 x)\sec^2 x$

Solution

a. $\tan^4 x = (\tan^2 x)(\tan^2 x)$ Write as separate factors.

$\qquad = \tan^2 x(\sec^2 x - 1)$ Pythagorean identity

$\qquad = \tan^2 x \sec^2 x - \tan^2 x$ Multiply.

b. $\sin^3 x \cos^4 x = \sin^2 x \cos^4 x \sin x$ Write as separate factors.

$\qquad = (1 - \cos^2 x)\cos^4 x \sin x$ Pythagorean identity

$\qquad = (\cos^4 x - \cos^6 x)\sin x$ Multiply.

c. $\sec^4 x \tan^2 x = \sec^2 x \tan^2 x \sec^2 x$ Write as separate facors.

$\qquad = (1 + \tan^2 x)\tan^2 x \sec^2 x$ Pythagorean identity

$\qquad = (\tan^2 x + \tan^4 x)\sec^2 x$ Multiply.

Writing About Math   *Error Analysis*

Suppose you are tutoring a student in trigonometry. One of the homework problems your student encounters asks whether the following statement is an identity.

$$\tan^2 x \sin^2 x \overset{?}{=} \frac{5}{6} \tan^2 x$$

Your student does not attempt to verify the equivalence algebraically, but mistakenly uses only a graphical approach. Using window settings of $-3\pi \le x \le 3\pi$ with an x-scale of $\pi/2$, and $-20 \le y \le 20$ with a y-scale of 1, your student graphs both sides of the expression on a graphing utility and concludes that the statement is an identity.

Write a short paragraph explaining what is wrong with your student's reasoning.

5.2 Exercises

In Exercises 1–10, verify the identity.

1. $\sin t \csc t = 1$
2. $\tan y \cot y = 1$
3. $\dfrac{\csc^2 x}{\cot x} = \csc x \sec x$
4. $\cot^2 y (\sec^2 y - 1) = 1$
5. $\cos^2 \beta - \sin^2 \beta = 1 - 2 \sin^2 \beta$
6. $\cos^2 \beta - \sin^2 \beta = 2 \cos^2 \beta - 1$
7. $\tan^2 \theta + 6 = \sec^2 \theta + 5$
8. $2 - \csc^2 z = 1 - \cot^2 z$
9. $\cos x + \sin x \tan x = \sec x$
10. $\dfrac{\cot^3 t}{\csc t} = \cos t (\csc^2 t - 1)$

Numerical, Graphical, and Algebraic Analysis **In Exercises 11–18, use a graphing utility to complete the table and graph the functions. Use both as evidence that $y_1 = y_2$. Then verify the identity algebraically.**

x	0.2	0.4	0.6	0.8	1.0	1.2	1.4
y_1							
y_2							

11. $y_1 = \dfrac{1}{\sec x \tan x}$, $\quad y_2 = \csc x - \sin x$
12. $y_1 = \dfrac{\csc x - 1}{1 - \sin x}$, $\quad y_2 = \csc x$
13. $y_1 = \csc x - \sin x$, $\quad y_2 = \cos x \cot x$
14. $y_1 = \sec x - \cos x$, $\quad y_2 = \sin x \tan x$
15. $y_1 = \sin x + \cos x \cot x$, $\quad y_2 = \csc x$
16. $y_1 = \dfrac{\sec x + \tan x}{\sec x - \tan x}$, $\quad y_2 = (\sec x + \tan x)^2$
17. $y_1 = \dfrac{1}{\tan x} + \dfrac{1}{\cot x}$, $\quad y_2 = \tan x + \cot x$
18. $y_1 = \dfrac{1}{\cos x} - \dfrac{1}{\sec x}$, $\quad y_2 = \sec x - \cos x$

In Exercises 19 and 20, find the error.

19.
$$(1 + \tan x)[1 + \cot(-x)]$$
$$= (1 + \tan x)(1 + \cot x)$$
$$= 1 + \cot x + \tan x + \tan x \cot x$$
$$= 1 + \cot x + \tan x + 1$$
$$= 2 + \cot x + \tan x$$

20.
$$\dfrac{1 + \sec(-\theta)}{\sin(-\theta) + \tan(-\theta)} = \dfrac{1 - \sec \theta}{\sin \theta - \tan \theta}$$
$$= \dfrac{1 - \sec \theta}{(\sin \theta)\left[1 - \left(\dfrac{1}{\cos \theta}\right)\right]}$$
$$= \dfrac{1 - \sec \theta}{\sin \theta (1 - \sec \theta)}$$
$$= \dfrac{1}{\sin \theta}$$
$$= \csc \theta$$

In Exercises 21 and 22, fill in the missing step(s).

21. $\sec^4 x - 2 \sec^2 x + 1 = (\sec^2 x - 1)^2$
$$= \rule{2cm}{0.4pt}$$
$$= \tan^4 x$$

22. $\dfrac{\tan x - \cot x}{\tan x + \cot x} = \dfrac{\dfrac{\sin x}{\cos x} - \dfrac{\cos x}{\sin x}}{\dfrac{\sin x}{\cos x} + \dfrac{\cos x}{\sin x}}$
$$= \rule{2cm}{0.4pt}$$
$$= \dfrac{\sin^2 x - \cos^2 x}{1}$$
$$= \sin^2 x - \cos^2 x$$
$$= \rule{2cm}{0.4pt}$$
$$= 1 - 2 \cos^2 x$$

In Exercises 23–32, verify the identity.

23. $\sin^{1/2} x \cos x - \sin^{5/2} x \cos x = \cos^3 x \sqrt{\sin x}$
24. $\sec^6 x (\sec x \tan x) - \sec^4 x (\sec x \tan x) = \sec^5 x \tan^3 x$

25. $\tan\left(\dfrac{\pi}{2} - x\right)\sec x = \csc x$

26. $\dfrac{\sin[(\pi/2) - x]}{\cos[(\pi/2) - x]} = \cot x$

27. $\dfrac{\sec(-x)}{\csc(-x)} = -\tan x$

28. $(1 + \sin y)[1 + \sin(-y)] = \cos^2 y$

29. $\dfrac{\cos(-\theta)}{1 + \sin(-\theta)} = \sec\theta + \tan\theta$

30. $\dfrac{1 + \csc(-\theta)}{\cos(-\theta) + \cot(-\theta)} = \sec\theta$

31. $\dfrac{\sin x \cos y + \cos x \sin y}{\cos x \cos y - \sin x \sin y} = \dfrac{\tan x + \tan y}{1 - \tan x \tan y}$

32. $\dfrac{\tan x + \tan y}{1 - \tan x \tan y} = \dfrac{\cot x + \cot y}{\cot x \cot y - 1}$

In Exercises 33–40, verify the identity algebraically, and use the *table* feature of a graphing utility to confirm it numerically.

33. $\dfrac{\tan x + \cot y}{\tan x \cot y} = \tan y + \cot x$

34. $\dfrac{\cos x - \cos y}{\sin x + \sin y} + \dfrac{\sin x - \sin y}{\cos x + \cos y} = 0$

35. $\sqrt{\dfrac{1 + \sin\theta}{1 - \sin\theta}} = \dfrac{1 + \sin\theta}{|\cos\theta|}$

36. $\sqrt{\dfrac{1 - \cos\theta}{1 + \cos\theta}} = \dfrac{1 - \cos\theta}{|\sin\theta|}$

37. $\cos^2 x + \cos^2\left(\dfrac{\pi}{2} - x\right) = 1$

38. $\sec^2 y - \cot^2\left(\dfrac{\pi}{2} - y\right) = 1$

39. $\sec x \sin\left(\dfrac{\pi}{2} - x\right) = 1$

40. $\csc^2\left(\dfrac{\pi}{2} - x\right) - 1 = \tan^2 x$

In Exercises 41–52, verify the identity algebraically, and use a graphing utility to confirm it graphically.

41. $2\sec^2 x - 2\sec^2 x \sin^2 x - \sin^2 x - \cos^2 x = 1$

42. $\csc x(\csc x - \sin x) + \dfrac{\sin x - \cos x}{\sin x} + \cot x = \csc^2 x$

43. $2 + \cos^2 x - 3\cos^4 x = \sin^2 x(2 + 3\cos^2 x)$

44. $4\tan^4 x + \tan^2 x - 3 = \sec^2 x(4\tan^2 x - 3)$

45. $\csc^4 x - 2\csc^2 x + 1 = \cot^4 x$

46. $\sin x(1 - 2\cos^2 x + \cos^4 x) = \sin^5 x$

47. $\sec^4\theta - \tan^4\theta = 1 + 2\tan^2\theta$

48. $\csc^4\theta - \cot^4\theta = 2\csc^2\theta - 1$

49. $\dfrac{\sin\beta}{1 - \cos\beta} = \dfrac{1 + \cos\beta}{\sin\beta}$

50. $\dfrac{\cot\alpha}{\csc\alpha - 1} = \dfrac{\csc\alpha + 1}{\cot\alpha}$

51. $\dfrac{\tan^3\alpha - 1}{\tan\alpha - 1} = \tan^2\alpha + \tan\alpha + 1$

52. $\dfrac{\sin^3\beta + \cos^3\beta}{\sin\beta + \cos\beta} = 1 - \sin\beta\cos\beta$

Conjecture **In Exercises 53–56, use a graphing utility to graph the trigonometric function. Use the graph to make a conjecture about a simplification of the expression. Verify the resulting identity algebraically.**

53. $y = \dfrac{1}{\cot x + 1} + \dfrac{1}{\tan x + 1}$

54. $y = \dfrac{\cos x}{1 - \tan x} + \dfrac{\sin x \cos x}{\sin x - \cos x}$

55. $y = \dfrac{1}{\sin x} - \dfrac{\cos^2 x}{\sin x}$

56. $y = \sin t + \dfrac{\cot^2 t}{\csc t}$

In Exercises 57–60, use the properties of logarithms and trigonometric identities to verify the identity.

57. $\ln|\cot\theta| = \ln|\cos\theta| - \ln|\sin\theta|$

58. $\ln|\sec\theta| = -\ln|\cos\theta|$

59. $-\ln(1 + \cos\theta) = \ln(1 - \cos\theta) - 2\ln|\sin\theta|$

60. $-\ln|\csc\theta + \cot\theta| = \ln|\csc\theta - \cot\theta|$

In Exercises 61–64, use the cofunction identities to evaluate the expression without using a calculator.

61. $\sin^2 25° + \sin^2 65°$

62. $\cos^2 14° + \cos^2 76°$

63. $\cos^2 20° + \cos^2 52° + \cos^2 38° + \cos^2 70°$

64. $\sin^2 12° + \sin^2 40° + \sin^2 50° + \sin^2 78°$

In Exercises 65–68, the powers of trigonometric functions are rewritten to be useful in calculus. Verify the identity.

65. $\tan^5 x = \tan^3 x \sec^2 x - \tan^3 x$

66. $\tan^3 x \sec^3 x = \sec^4 x(\sec x \tan x) - \sec^2 x(\sec x \tan x)$

67. $\cos^3 x \sin^2 x = (\sin^2 x - \sin^4 x)\cos x$

68. $\sin^4 x + \cos^4 x = 1 - 2\cos^2 x + 2\cos^4 x$

69. *Friction* The forces acting on an object weighing W units on an inclined plane positioned at an angle of θ with the horizontal are modeled by

$$\mu W \cos \theta = W \sin \theta$$

where μ is the coefficient of friction. Solve the equation for μ and simplify the result.

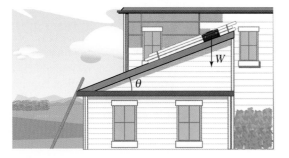

70. *Shadow of a Sundial* The length s of a shadow cast by a vertical *gnomon* (column or shaft on a sundial) of height h when the angle of the sun above the horizon is θ can be modeled by the equation

$$s = \frac{h \sin(90° - \theta)}{\sin \theta}.$$

Show that the equation is equivalent to $s = h \cot \theta$.

71. *Rate of Change* The rate of change of the function $f(x) = \sin x + \csc x$ is given by $\cos x - \csc x \cot x$. Show that the expression for the rate of change can also be given by $-\cos x \cot^2 x$.

Synthesis

True or False? In Exercises 72–75, determine whether the statement is true or false. Justify your answer.

72. There can be more than one way to verify a trigonometric identity.

73. Of the six trigonometric functions, two are even.

74. $\ln(\cos x) = \cos(\ln x)$

75. $\sin(x^2) = \sin^2(x)$

Think About It In Exercises 76–79, explain why the equation is *not* an identity and find one value of the variable for which the equation is not true.

76. $\sqrt{\tan^2 x} = \tan x$ **77.** $\sin \theta = \sqrt{1 - \cos^2 \theta}$

78. $\tan \theta = \sqrt{\sec^2 \theta - 1}$

79. $\sqrt{\sin^2 x + \cos^2 x} = \sin x + \cos x$

80. Verify that for all integers n

$$\cos\left[\frac{(2n + 1)\pi}{2}\right] = 0.$$

81. Verify that for all integers n

$$\sin\left[\frac{(12n + 1)\pi}{6}\right] = \frac{1}{2}.$$

Review

In Exercises 82–85, find a polynomial function with integer coefficients that has the given zeros. (There are many correct answers.)

82. $1, 8i, -8i$ **83.** $i, -i, 4i, -4i$

84. $4, 6 + i, 6 - i$ **85.** $0, 0, 2, 1 - i$

In Exercises 86–89, sketch the graph of the function by hand.

86. $f(x) = 2^x + 3$ **87.** $f(x) = -2^{x-3}$

88. $f(x) = \left(\frac{3}{2}\right)^{x+1}$ **89.** $f(x) = 5^{-x} - 2$

In Exercises 90 and 91, find the measure of the angle in radians.

90.   **91.**

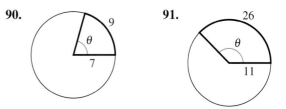

In Exercises 92–95, state the quadrant in which θ lies.

92. $\csc \theta > 0$ and $\tan \theta < 0$

93. $\cot \theta > 0$ and $\cos \theta < 0$

94. $\sec \theta > 0$ and $\sin \theta < 0$

95. $\cot \theta > 0$ and $\sec \theta < 0$

5.3   Solving Trigonometric Equations

Introduction

To solve a trigonometric equation, use standard algebraic techniques such as collecting like terms and factoring. Your preliminary goal is to isolate the trigonometric function involved in the equation.

EXAMPLE 1   Solving a Trigonometric Equation

$$2 \sin x - 1 = 0 \qquad \text{Original equation}$$

$$2 \sin x = 1 \qquad \text{Add 1 to each side.}$$

$$\sin x = \tfrac{1}{2} \qquad \text{Divide each side by 2.}$$

To solve for x, note that the equation $\sin x = \frac{1}{2}$ has solutions $x = \pi/6$ and $x = 5\pi/6$ in the interval $[0, 2\pi)$. Moreover, because $\sin x$ has a period of 2π, there are infinitely many other solutions, which can be written as

$$x = \frac{\pi}{6} + 2n\pi \qquad \text{and} \qquad x = \frac{5\pi}{6} + 2n\pi \qquad \text{General solution}$$

where n is an integer, as shown in Figure 5.8.

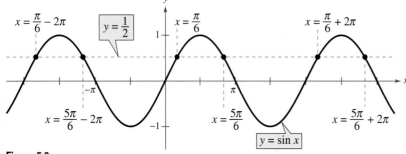

Figure 5.8

Another way to see that the equation $\sin x = \frac{1}{2}$ has infinitely many solutions is indicated in Figure 5.9. Any angles that are coterminal with $\pi/6$ or $5\pi/6$ are also solutions of the equation.

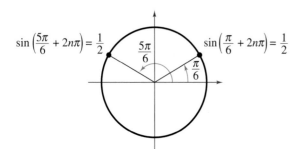

Figure 5.9

What You Should Learn:

- How to use standard algebraic techniques to solve trigonometric equations
- How to solve trigonometric equations of quadratic type
- How to solve trigonometric equations involving multiple angles
- How to use inverse trigonometric functions to solve trigonometric equations

Why You Should Learn It:

You can use trigonometric equations to solve a variety of real-life problems. For instance, Exercise 75 on page 402 shows you how solving a trigonometric equation can help answer questions about the position of the sun in Cheyenne, Wyoming.

Superstock

EXAMPLE 2 Collecting Like Terms

Find all solutions of $\sin x + \sqrt{2} = -\sin x$ in the interval $[0, 2\pi)$.

Algebraic Solution

Rewrite the equation so that $\sin x$ is isolated on one side of the equation.

$$\sin x + \sqrt{2} = -\sin x \qquad \text{Write original equation.}$$

$$\sin x + \sin x = -\sqrt{2} \qquad \begin{array}{l}\text{Add } \sin x \text{ and subtract} \\ \sqrt{2} \text{ from each side.}\end{array}$$

$$2 \sin x = -\sqrt{2} \qquad \text{Combine like terms.}$$

$$\sin x = -\frac{\sqrt{2}}{2} \qquad \text{Divide each side by 2.}$$

The solutions in the interval $[0, 2\pi)$ are

$$x = \frac{5\pi}{4} \quad \text{and} \quad x = \frac{7\pi}{4}.$$

Numerical Solution

Use the *table* feature of a graphing utility set in *radian* mode to create a table that shows the values of $y_1 = \sin x + \sqrt{2}$ and $y_2 = -\sin x$ for different values of x. Your table should go from $x = 0$ to $x = 2\pi$ using increments in the table of $\pi/8$, as shown in Figure 5.10. From the table, you can see that the values of y_1 and y_2 appear to be identical when $x \approx 3.927 \approx 5\pi/4$ and $x \approx 5.4978 \approx 7\pi/4$. These values are the approximate solutions of $\sin x + \sqrt{2} = -\sin x$.

X	Y1	Y2
3.1416	1.4142	1E-13
3.5343	1.0315	.38268
3.927	.70711	.70711
4.3197	.49033	.92388
4.7124	.41421	1
5.1051	.49033	.92388
5.4978	.70711	.70711

X=5.497787143782

Figure 5.10

EXAMPLE 3 Extracting Square Roots

Solve $3 \tan^2 x - 1 = 0$.

Solution

Rewrite the equation so that $\tan x$ is isolated on one side of the equation.

$$3 \tan^2 x - 1 = 0 \qquad \text{Write original equation.}$$

$$3 \tan^2 x = 1 \qquad \text{Add 1 to each side.}$$

$$\tan^2 x = \frac{1}{3} \qquad \text{Divide each side by 3.}$$

$$\tan x = \pm\frac{1}{\sqrt{3}} \qquad \text{Extract square roots.}$$

Because $\tan x$ has a period of π, first find all solutions in the interval $[0, \pi)$. These are $x = \pi/6$ and $x = 5\pi/6$. Finally, add $n\pi$ to each of these solutions to get the general form

$$x = \frac{\pi}{6} + n\pi \quad \text{and} \quad x = \frac{5\pi}{6} + n\pi \qquad \text{General solution}$$

where n is an integer. The graph of $y = 3 \tan^2 x - 1$, shown in Figure 5.11, confirms this result.

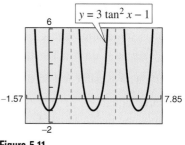

$y = 3 \tan^2 x - 1$

Figure 5.11

The equations in Examples 1, 2, and 3 involved only one trigonometric function. When two or more functions occur in the same equation, collect all terms on one side and try to separate the functions by factoring or by using appropriate identities. This may produce factors that yield no solutions, as illustrated in Example 4.

EXAMPLE 4 Factoring

Solve $\cot x \cos^2 x = 2 \cot x$.

Solution

$$\cot x \cos^2 x = 2 \cot x \qquad \text{Write original equation.}$$

$$\cot x \cos^2 x - 2 \cot x = 0 \qquad \text{Subtract 2 cot } x \text{ from each side.}$$

$$\cot x(\cos^2 x - 2) = 0 \qquad \text{Factor.}$$

By setting each of these factors equal to zero, you obtain the following.

$$\cot x = 0 \qquad \text{and} \qquad \cos^2 x - 2 = 0$$

$$x = \frac{\pi}{2} \qquad\qquad\qquad \cos^2 x = 2$$

$$\cos x = \pm\sqrt{2}$$

The equation $\cot x = 0$ has the solution $x = \pi/2$. No solution is obtained from $\cos x = \pm\sqrt{2}$ because $\pm\sqrt{2}$ are outside the range of the cosine function. Therefore, the general form of the solution is obtained by adding multiples of π to $x = \pi/2$, to get

$$x = \frac{\pi}{2} + n\pi \qquad \text{General solution}$$

where n is an integer. The graph of $y = \cot x \cos^2 x - 2 \cot x$, shown in Figure 5.12, confirms this result.

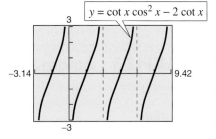

Figure 5.12

In Example 4, don't make the mistake of dividing each side of the equation by $\cot x$. Doing this would lose the solutions. Can you see why?

Equations of Quadratic Type

Many trigonometric equations are of quadratic type. Here are a few examples.

Quadratic in sin x	*Quadratic in sec x*
$2 \sin^2 x - \sin x - 1 = 0$	$\sec^2 x - 3 \sec x - 2 = 0$
$2(\sin x)^2 - \sin x - 1 = 0$	$(\sec x)^2 - 3 \sec x - 2 = 0$

To solve equations of this type, factor the quadratic or, if factoring is not possible, use the Quadratic Formula.

EXAMPLE 5 Factoring an Equation of Quadratic Type

Find all solutions of $2 \sin^2 x - \sin x - 1 = 0$ in the interval $[0, 2\pi)$.

Algebraic Solution

Treating the equation as a quadratic in $\sin x$ and factoring produces the following.

$$2 \sin^2 x - \sin x - 1 = 0 \qquad \text{Write original equation.}$$

$$(2 \sin x + 1)(\sin x - 1) = 0 \qquad \text{Factor.}$$

Setting each factor equal to zero, you obtain the following solutions in the interval $[0, 2\pi)$.

$$2 \sin x + 1 = 0 \qquad \text{and} \qquad \sin x - 1 = 0$$

$$\sin x = -\frac{1}{2} \qquad\qquad \sin x = 1$$

$$x = \frac{7\pi}{6}, \frac{11\pi}{6} \qquad\qquad x = \frac{\pi}{2}$$

So, the solutions in the interval $[0, 2\pi)$ are $\pi/2$, $7\pi/6$, and $11\pi/6$.

Graphical Solution

Use a graphing utility set in *radian* mode to graph $y = 2 \sin^2 x - \sin x - 1$ for $0 \le x < 2\pi$, as shown in Figure 5.13. Use the *zero* or *root* feature or the *zoom* and *trace* features to approximate the x-intercepts to be

$$x \approx 1.571 \approx \frac{\pi}{2}, \quad x \approx 3.665 \approx \frac{7\pi}{6}, \quad x \approx 5.760 \approx \frac{11\pi}{6}.$$

These values are the approximate solutions of $2 \sin^2 x - \sin x - 1 = 0$.

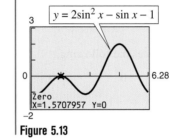

Figure 5.13

In Example 5, the general solution would be

$$x = \frac{\pi}{2} + 2n\pi, \quad x = \frac{7\pi}{6} + 2n\pi, \quad x = \frac{11\pi}{6} + 2n\pi \qquad \text{General solution}$$

where n is an integer.

When working with an equation of quadratic type, be sure that the equation involves a *single* trigonometric function, as shown in the next example.

EXAMPLE 6 Rewriting with a Single Trigonometric Function

Solve $2 \sin^2 x + 3 \cos x - 3 = 0$.

Solution

Begin by rewriting the equation so that it has only cosine functions.

$$2 \sin^2 x + 3 \cos x - 3 = 0 \qquad \text{Write original equation.}$$

$$2(1 - \cos^2 x) + 3 \cos x - 3 = 0 \qquad \text{Pythagorean identity}$$

$$2 \cos^2 x - 3 \cos x + 1 = 0 \qquad \text{Combine like terms and multiply each side by } -1.$$

$$(2 \cos x - 1)(\cos x - 1) = 0 \qquad \text{Factor.}$$

By setting each factor equal to zero, you can find the solutions in the interval $[0, 2\pi)$ to be $x = 0$, $x = \pi/3$, and $x = 5\pi/3$. The general solution is therefore

$$x = 2n\pi, \quad x = \frac{\pi}{3} + 2n\pi, \quad x = \frac{5\pi}{3} + 2n\pi \qquad \text{General solution}$$

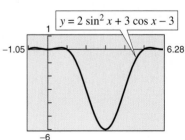

Figure 5.14

where n is an integer. The graph of $y = 2 \sin^2 x + 3 \cos x - 3$, shown in Figure 5.14, confirms this result.

Sometimes you must square both sides of an equation to obtain a quadratic. This can introduce extraneous solutions.

EXAMPLE 7 Squaring and Converting to Quadratic Type

Find all solutions of $\cos x + 1 = \sin x$ in the interval $[0, 2\pi)$.

Solution

It is not clear how to rewrite this equation in terms of a single trigonometric function. See what happens when you square both sides of the equation.

$\cos x + 1 = \sin x$	Write original equation.
$\cos^2 x + 2\cos x + 1 = \sin^2 x$	Square each side.
$\cos^2 x + 2\cos x + 1 = 1 - \cos^2 x$	Pythagorean identity
$2\cos^2 x + 2\cos x = 0$	Combine like terms.
$2\cos x(\cos x + 1) = 0$	Factor.

Setting each factor equal to zero produces the following.

$$2\cos x = 0 \quad \text{and} \quad \cos x + 1 = 0$$

$$\cos x = 0 \qquad\qquad\qquad \cos x = -1$$

$$x = \frac{\pi}{2}, \frac{3\pi}{2} \qquad\qquad\qquad x = \pi$$

Because you squared the original equation, check for extraneous solutions. Of the three possible solutions, $x = 3\pi/2$ is extraneous. (Try checking this.) So, in the interval $[0, 2\pi)$, the only solutions are $x = \pi/2$ and $x = \pi$. The graph of $y = \cos x + 1 - \sin x$, shown in Figure 5.15, confirms this result.

In Example 7, the general solution is

$$x = \frac{\pi}{2} + 2n\pi, \, x = \pi + 2n\pi \qquad \text{General solution}$$

where n is an integer.

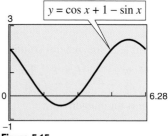

Figure 5.15

Exploration

Use a graphing utility to confirm the solutions found in Example 7 in two different ways.

1. Graph both sides of the equation and find the x-coordinates of the points at which the graphs intersect.

  Left side: $y = \cos x + 1$ *Right side:* $y = \sin x$

2. Graph the equation $y = \cos x + 1 - \sin x$ and find the x-intercepts of the graph.

Do both methods produce the same x-values? Which method do you prefer? Why?

Functions Involving Multiple Angles

EXAMPLE 8 Functions of Multiple Angles

Find all solutions of $2 \cos 3t - 1 = 0$.

Solution

$$2 \cos 3t - 1 = 0 \qquad \text{Write original equation.}$$

$$2 \cos 3t = 1 \qquad \text{Add 1 to each side.}$$

$$\cos 3t = \frac{1}{2} \qquad \text{Divide each side by 2.}$$

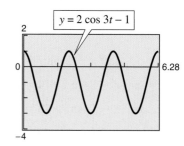

Figure 5.16

In the interval $[0, 2\pi)$, you know that $3t = \pi/3$ and $3t = 5\pi/3$ are the only solutions. So in general, you have

$$3t = \frac{\pi}{3} + 2n\pi \qquad \text{and} \qquad 3t = \frac{5\pi}{3} + 2n\pi.$$

Dividing this result by 3, you obtain the general solution

$$t = \frac{\pi}{9} + \frac{2n\pi}{3} \qquad \text{and} \qquad t = \frac{5\pi}{9} + \frac{2n\pi}{3} \qquad \text{General solution}$$

where n is an integer. This is confirmed graphically in Figure 5.16.

EXAMPLE 9 Functions of Multiple Angles

Find all solutions of $3 \tan\left(\dfrac{x}{2}\right) + 3 = 0$.

Solution

$$3 \tan \frac{x}{2} + 3 = 0 \qquad \text{Write original equation.}$$

$$3 \tan \frac{x}{2} = -3 \qquad \text{Subtract 3 from each side.}$$

$$\tan \frac{x}{2} = -1 \qquad \text{Divide each side by 3.}$$

In the interval $[0, \pi)$, you know that $x/2 = 3\pi/4$ is the only solution. So in general, you have

$$\frac{x}{2} = \frac{3\pi}{4} + n\pi.$$

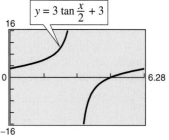

Figure 5.17

Multiplying this result by 2, you obtain the general solution

$$x = \frac{3\pi}{2} + 2n\pi \qquad \text{General solution}$$

where n is an integer. This is confirmed graphically in Figure 5.17.

Using Inverse Functions

EXAMPLE 10   Using Inverse Functions

Find all solutions of $\sec^2 x - 2 \tan x = 4$.

Solution

$$\sec^2 x - 2 \tan x = 4 \qquad \text{Write original equation.}$$
$$1 + \tan^2 x - 2 \tan x - 4 = 0 \qquad \text{Pythagorean identity}$$
$$\tan^2 x - 2 \tan x - 3 = 0 \qquad \text{Combine like terms.}$$
$$(\tan x - 3)(\tan x + 1) = 0 \qquad \text{Factor.}$$

Setting each factor equal to zero, you obtain two solutions in the interval $(-\pi/2, \pi/2)$. [Recall that the range of the inverse tangent function is $(-\pi/2, \pi/2)$.]

$$\tan x = 3, \qquad\qquad \tan x = -1$$
$$x = \arctan 3 \qquad\quad x = \arctan(-1)$$
$$= -\frac{\pi}{4}$$

Finally, by adding multiples of π, you obtain the general solution

$$x = \arctan 3 + n\pi \quad \text{and} \quad x = -\frac{\pi}{4} + n\pi \quad \text{General solution}$$

where n is an integer. This is confirmed graphically in Figure 5.18.

With some trigonometric equations, there is no reasonable way to find the solutions algebraically. In such cases, you can still use a graphing utility to approximate the solutions.

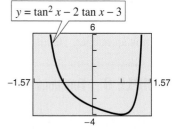

Figure 5.18

EXAMPLE 11   Approximating Solutions

Approximate the solutions of $x = 2 \sin x$.

Solution

The graph of $y = x - 2 \sin x$ is shown in Figure 5.19. From the graph, there appear to be three solutions. Using the *zero* or *root* feature or the *zoom* and *trace* features of a graphing utility, you can see that one solution is $x = 0$, and you can approximate the positive solution to be $x \approx 1.8955$. Finally, by symmetry, you can approximate the negative solution to be $x \approx -1.8955$.

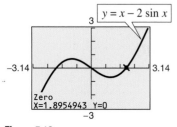

Figure 5.19

EXAMPLE 12   Surface Area of a Honeycomb

The surface area of a honeycomb is given by the equation

$$S = 6hs + \frac{3}{2}s^2\left(\frac{\sqrt{3} - \cos \theta}{\sin \theta}\right), \qquad 0 < \theta \le 90°$$

where $h = 2.4$ inches, $s = 0.75$ inch, and θ is the angle indicated in Figure 5.20.

a. What value of θ gives a surface area of 12 square inches?

b. What value of θ gives the minimum surface area?

Solution

a. Let $h = 2.4$, $s = 0.75$, and $S = 12$.

$$S = 6hs + \frac{3}{2}s^2\left(\frac{\sqrt{3} - \cos \theta}{\sin \theta}\right)$$

$$12 = 6(2.4)(0.75) + \frac{3}{2}(0.75)^2\left(\frac{\sqrt{3} - \cos \theta}{\sin \theta}\right)$$

$$12 = 10.8 + 0.84375\left(\frac{\sqrt{3} - \cos \theta}{\sin \theta}\right)$$

$$0 = 0.84375\left(\frac{\sqrt{3} - \cos \theta}{\sin \theta}\right) - 1.2$$

Using a graphing utility set to *degree* mode, you can graph the function

$$y = 0.84375\left(\frac{\sqrt{3} - \cos \theta}{\sin \theta}\right) - 1.2$$

as shown in Figure 5.21. Using the *zero* or *root* feature or the *zoom* and *trace* features you can determine that $\theta \approx 59.9°$ and $\theta \approx 49.9°$.

b. From part (a), let $h = 2.4$, $s = 0.75$, and $S = 12$ to obtain

$$S = 10.8 + 0.84375\left(\frac{\sqrt{3} - \cos \theta}{\sin \theta}\right).$$

The graph of this function is shown in Figure 5.22. Use the *minimum* feature or the *zoom* and *trace* features of a graphing utility to approximate the minimum point on the graph, which occurs at $\theta \approx 54.7°$. By using calculus, it can be shown that

$$\theta = \arccos\left(\frac{1}{\sqrt{3}}\right) \approx 54.7356°$$

is the exact minimum value.

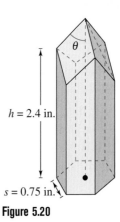

$h = 2.4$ in.

$s = 0.75$ in.

Figure 5.20

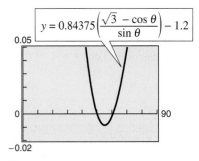

$y = 0.84375\left(\dfrac{\sqrt{3} - \cos \theta}{\sin \theta}\right) - 1.2$

Figure 5.21

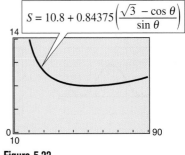

$S = 10.8 + 0.84375\left(\dfrac{\sqrt{3} - \cos \theta}{\sin \theta}\right)$

Figure 5.22

5.3 Exercises

In Exercises 1–6, verify that the *x*-values are solutions of the equation.

1. $2 \cos x - 1 = 0$

  (a) $x = \dfrac{\pi}{3}$ (b) $x = \dfrac{5\pi}{3}$

2. $\csc x - 2 = 0$

  (a) $x = \dfrac{\pi}{6}$ (b) $x = \dfrac{5\pi}{6}$

3. $3 \tan^2 2x - 1 = 0$

  (a) $x = \dfrac{\pi}{12}$ (b) $x = \dfrac{5\pi}{12}$

4. $4 \cos^2 2x - 2 = 0$

  (a) $x = \dfrac{\pi}{8}$ (b) $x = \dfrac{7\pi}{8}$

5. $2 \cos^2 x + 3 \cos x + 1 = 0$

  (a) $x = \dfrac{4\pi}{3}$ (b) $x = \pi$

6. $\sec^4 x - 3 \sec^2 x - 4 = 0$

  (a) $x = \dfrac{2\pi}{3}$ (b) $x = \dfrac{5\pi}{3}$

In Exercises 7–10, find the *x*-intercepts of the graph. Use a graphing utility to check your solutions.

7. $y = \sin \dfrac{\pi x}{2} + 1$ **8.** $y = \sin \pi x + \cos \pi x$

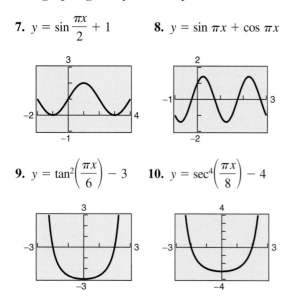

9. $y = \tan^2\left(\dfrac{\pi x}{6}\right) - 3$ **10.** $y = \sec^4\left(\dfrac{\pi x}{8}\right) - 4$

In Exercises 11–24, solve the equation for $0 \le x < 2\pi$.

11. $2 \cos x + 1 = 0$ **12.** $\sqrt{2} \sin x + 1 = 0$

13. $\sqrt{3} \sec x - 2 = 0$ **14.** $\cot x + 1 = 0$

15. $3 \csc^2 x - 4 = 0$ **16.** $\csc^2 x - 2 = 0$

17. $2 \sin^2 2x = 1$ **18.** $\tan^2 3x = 3$

19. $4 \cos^2 x - 3 = 0$ **20.** $\cos x(\cos x - 1) = 0$

21. $\sin^2 x = 3 \cos^2 x$ **22.** $\tan 3x(\tan x - 1) = 0$

23. $(3 \tan^2 x - 1)(\tan^2 x - 3) = 0$

24. $\cos 2x(2 \cos x + 1) = 0$

In Exercises 25–42, find all solutions of the equation in the interval $[0, 2\pi)$ algebraically. Use the *table* feature of a graphing utility to confirm your answers numerically.

25. $\cos^3 x = \cos x$ **26.** $\tan^2 x - 1 = 0$

27. $3 \tan^3 x = \tan x$ **28.** $2 \sin^2 x = 2 + \cos x$

29. $\sec^2 x - \sec x = 2$ **30.** $\sec x \csc x = 2 \csc x$

31. $2 \sin x + \csc x = 0$ **32.** $\sin 2x = -\dfrac{\sqrt{3}}{2}$

33. $\csc x + \cot x = 1$ **34.** $\tan 3x = 1$

35. $\cos \dfrac{x}{2} = \dfrac{\sqrt{2}}{2}$ **36.** $\sec 4x = 2$

37. $\dfrac{1 + \cos x}{1 - \cos x} = 0$

38. $2 \sin^2 x + 3 \sin x + 1 = 0$

39. $2 \sec^2 x + \tan^2 x - 3 = 0$

40. $\cos x + \sin x \tan x = 2$

41. $\sec^2 x + \tan x = 3$

42. $\csc^2 x - 4 \cot x = -2$

In Exercises 43–48, use a graphing utility to approximate the solutions of the equation in the interval $[0, 2\pi)$ by setting the equation equal to 0, graphing the new equation, and using the *zero* or *root* feature to approximate the *x*-intercepts of the graph.

43. $9 \cos x + 2 = 3$ **44.** $4 \sin x = \cos x - 2$

45. $4 \sin^2 x = 2 \cos x + 1$ **46.** $\csc^2 x = 3 \csc x + 4$

47. $4 \sin^3 x + 2 \sin^2 x = 2 \sin x + 1$

48. $\dfrac{1 + \sin x}{\cos x} + \dfrac{\cos x}{1 + \sin x} = 4$

In Exercises 49–56, use a graphing utility to approximate the solutions of the equation in the interval $[0, 2\pi)$.

49. $\dfrac{\cos x \cot x}{1 - \sin x} = 3$ **50.** $2 \cos x - \sin x = 0$

51. $x \cos x - 1 = 0$ **52.** $2x \sin x - 2 = 0$

53. $\sec^2 x + 0.5 \tan x = 1$

54. $\csc^2 x + 0.5 \cot x = 5$

55. $12 \sin^2 x - 13 \sin x + 3 = 0$

56. $3 \tan^2 x + 4 \tan x - 4 = 0$

In Exercises 57–60, approximate the solutions (to three decimal places) of the equation in the given interval. Use a graphing utility to check your result.

57. $3 \tan^2 x + 5 \tan x - 4 = 0$, $\left[-\dfrac{\pi}{2}, \dfrac{\pi}{2}\right]$

58. $\cos^2 x - 2 \cos x - 1 = 0$, $[0, \pi]$

59. $4 \cos^2 x - 2 \sin x + 1 = 0$, $\left[-\dfrac{\pi}{2}, \dfrac{\pi}{2}\right]$

60. $2 \sec^2 x + \tan x - 6 = 0$, $\left[-\dfrac{\pi}{2}, \dfrac{\pi}{2}\right]$

In Exercises 61–64, consider the function in the interval $[0, 2\pi)$. **(a) Use a graphing utility to complete the table. Then make a conjecture about any unit intervals containing a zero of the function. Give a reason for your answer. (b) Use a graphing utility to graph the function and determine any unit intervals containing a zero of the function. Do the results agree with those of part (a)? Explain. (c) Use a graphing utility to approximate the zero of the function.**

x	0	1	2	3	4	5	6
$f(x)$							

61. $f(x) = \cot \dfrac{x}{2} - x$ **62.** $f(x) = x \sec x - 1$

63. $f(x) = \sin^2 x + 2 \sin x - 1$

64. $f(x) = 4 \cos^2 x - 4 \cos x - 1$

⚫ In Exercises 65–68, **(a) use a graphing utility to graph the function and approximate the maximum and minimum points on the graph in the interval $[0, 2\pi]$, and (b) solve the trigonometric equation and verify that the x-coordinates of the maximum and minimum points of f are among its solutions (the trigonometric equation is found using calculus).**

	Function	*Trigonometric Equation*
65.	$f(x) = \sin x + \cos x$	$\cos x - \sin x = 0$
66.	$f(x) = 3 \sin x - 5 \cos x$	$3 \cos x + 5 \sin x = 0$
67.	$f(x) = x \sin 2x$	$2x \cos 2x + \sin 2x = 0$
68.	$f(x) = 2 \sin x + \cos 2x$	$2 \cos x - 4 \sin x \cos x = 0$

Fixed Point In Exercises 69 and 70, find the smallest positive fixed point of the function f. [A fixed point of a function f is a real number c such that $f(c) = c$.]

69. $f(x) = \tan \dfrac{\pi x}{4}$ **70.** $f(x) = \cos x$

71. *Graphical Reasoning* Consider the function

$$f(x) = \cos \frac{1}{x}$$

and its graph shown below.

(a) What is the domain of the function?

(b) Identify any symmetry or asymptotes of the graph.

(c) Describe the behavior of the function as $x \to 0$.

(d) How many solutions does the equation

$$\cos \frac{1}{x} = 0$$

have in the interval $[-1, 1]$? Find the solutions.

(e) Does the equation $\cos(1/x) = 0$ have a greatest solution? If so, approximate the solution. If not, explain.

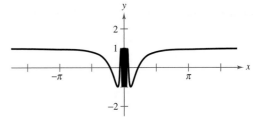

72. *Graphical Reasoning* Consider the function

$$f(x) = \frac{\sin x}{x}$$ and its graph on the following page.

(a) What is the domain of the function?

(b) Identify any symmetry or asymptotes of the graph.

(c) Describe the behavior of the function as $x \to 0$.

(d) How many solutions does the equation

$$\frac{\sin x}{x} = 0$$ have in the interval $[-8, 8]$?

Find the solutions.

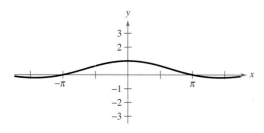

73. Harmonic Motion A weight is oscillating on the end of a spring. The position of the weight relative to the point of equilibrium is

$$y = \tfrac{1}{12}(\cos 8t - 3 \sin 8t)$$

where y is the displacement (in meters) and t is the time (in seconds). Find the times when the weight is at the point of equilibrium ($y = 0$) for $0 \le t \le 1$.

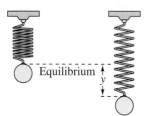

74. Sales The monthly sales (in thousands of units) of a seasonal product are approximated by

$$S = 74.50 + 43.75 \sin \frac{\pi t}{6}$$

where t is the time (in months), with $t = 1$ corresponding to January. Determine the months when sales exceed 100,000 units.

75. Position of the Sun Cheyenne, Wyoming, has a latitude of $41°$ N. At this latitude, the position of the sun at sunrise can be modeled by

$$D = 31 \sin\left(\frac{2\pi}{365} t - 1.4\right)$$

where t is the time (in days) and $t = 1$ represents January 1. In this model, D represents the number of degrees north or south of due east that the sun rises. Use a graphing calculator to determine the days that the sun is more than $20°$ north of due east at sunrise.

76. Projectile Motion A batted baseball leaves the bat at an angle of θ with the horizontal and an initial velocity of $v_0 = 100$ feet per second. The ball is caught by an outfielder 300 feet from home plate. Find θ if the range r of a projectile is

$$r = \tfrac{1}{32}v_0{}^2 \sin 2\theta.$$

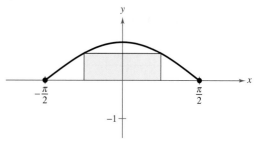

77. Projectile Motion A sharpshooter intends to hit a target 1000 yards away with a gun that has a muzzle velocity of 1200 feet per second. Neglecting air resistance, determine the minimum angle of elevation of the gun if the range is $r = \tfrac{1}{32}v_0{}^2 \sin 2\theta.$

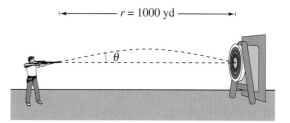

78. Area The area of a rectangle inscribed in one arch of the graph of $y = \cos x$ is

$$A = 2x \cos x, \quad -\frac{\pi}{2} < x < \frac{\pi}{2}.$$

(a) Use a graphing utility to graph the area function, and approximate the area of the largest inscribed rectangle.

(b) Determine the values of x for which $A \ge 1$.

79. Damped Harmonic Motion The displacement from equilibrium of a weight oscillating on the end of a spring is $y = 1.56e^{-0.22t} \cos 4.9t$, where y is the displacement (in feet) and t is the time (in seconds). Use a graphing utility to graph the displacement function for $0 \le t \le 10$. Find the time beyond which the displacement does not exceed 1 foot from equilibrium.

80. **Data Analysis** The table gives the unemployment rate r for the years 1989 through 1996 in the United States. The time t is measured in years, with $t = 0$ corresponding to 1990. (Source: U.S. Bureau of Labor Statistics)

t	-1	0	1	2
r	5.3	5.6	6.8	7.5

t	3	4	5	6
r	6.9	6.1	5.6	5.4

(a) Create a scatter plot of the data.

(b) Which of the following models best represents the data? Explain your reasoning.

   (i) $r = 1.10 \cos(t + 0.93) + 6.35$

   (ii) $r = 1.02 \sin(0.98t + 5.12) + 6.26$

   (iii) $r = 1.04 \sin[0.98(t + 5.81)] + 6.34$

   (iv) $r = 1.20 \sin[0.95(t + 5.48)] + 6.20$

(c) What term in the model gives the average unemployment rate? What is the rate?

(d) Economists study the lengths of business cycles, such as cycles in the unemployment rate. Based on this short span of time, use the model to give the length of this cycle.

(e) Use the model to estimate the next time the unemployment rate will be 5.5% or less.

81. **Quadratic Approximation** Consider the function

$$f(x) = 3 \sin(0.6x - 2).$$

(a) Approximate the zero of the function in the interval $[0, 6]$.

(b) A quadratic approximation agreeing with f at $x = 5$ is

$$g(x) = -0.45x^2 + 5.52x - 13.70.$$

Use a graphing utility to graph f and g in the same viewing window. Describe the result.

(c) Use the Quadratic Formula to find the zeros of g. Compare the zero in the interval $[0, 6]$ with the result of part (a).

Synthesis

True or False? **In Exercises 82 and 83, determine whether the statement is true or false. Justify your answer.**

82. All trigonometric equations have either an infinite number of solutions or no solution.

83. The solutions of any trigonometric equation can always be found from its solutions in the interval $[0, 2\pi)$.

84. *Writing* Describe the difference between verifying an identity and solving an equation.

Review

In Exercises 85–88, convert the angle measure from degrees to radians. Round your result to three decimal places.

85. $124°$

86. $486°$

87. $-0.41°$

88. $-210.55°$

In Exercises 89–92, solve for x.

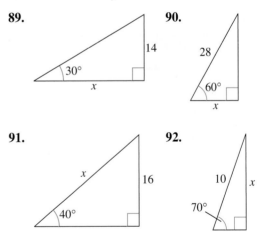

89.

90.

91.

92.

In Exercises 93–96, sketch the graph of the function.

93. $f(x) = \dfrac{1}{4} \sin\left(x - \dfrac{\pi}{2}\right)$

94. $f(x) = -6 \cos \dfrac{\pi x}{4}$

95. $f(x) = \dfrac{1}{2} \cot\left(x - \dfrac{\pi}{4}\right)$

96. $f(x) = \dfrac{1}{2} \sec(2x + \pi)$

5.4 Sum and Difference Formulas

Using Sum and Difference Formulas

In this and the following section, you will study the uses of several trigonometric identities and formulas. For proofs of these formulas see Appendix A.

Sum and Difference Formulas

$$\sin(u + v) = \sin u \cos v + \cos u \sin v$$

$$\sin(u - v) = \sin u \cos v - \cos u \sin v$$

$$\cos(u + v) = \cos u \cos v - \sin u \sin v$$

$$\cos(u - v) = \cos u \cos v + \sin u \sin v$$

$$\tan(u + v) = \frac{\tan u + \tan v}{1 - \tan u \tan v}$$

$$\tan(u - v) = \frac{\tan u - \tan v}{1 + \tan u \tan v}$$

Exploration

Use a graphing utility to graph $y = \cos(x + 2)$ and $y = \cos x + \cos 2$ in the same viewing window. What can you conclude about the graphs? Is it true that $\cos(x + 2) = \cos x + \cos 2$?

Use a graphing utility to graph $y = \sin(x + 4)$ and $y = \sin x + \sin 4$ in the same viewing window. What can you conclude about the graphs? Is it true that $\sin(x + 4) = \sin x + \sin 4$?

Examples 1 and 2 show how sum and difference formulas can be used to find exact values of trigonometric functions involving sums or differences of special angles.

EXAMPLE 1   Evaluating a Trigonometric Function

Find the exact value of $\cos 75°$.

Solution

To find the exact value of $\cos 75°$, use the fact that $75° = 30° + 45°$. Consequently, the formula for $\cos(u + v)$ yields

$$\cos 75° = \cos(30° + 45°)$$

$$= \cos 30° \cos 45° - \sin 30° \sin 45°$$

$$= \frac{\sqrt{3}}{2}\left(\frac{\sqrt{2}}{2}\right) - \frac{1}{2}\left(\frac{\sqrt{2}}{2}\right)$$

$$= \frac{\sqrt{6} - \sqrt{2}}{4}.$$

Try checking this result on your calculator. You will find that $\cos 75° \approx 0.259$.

What You Should Learn:

- How to use sum and difference formulas to evaluate trigonometric functions
- How to use sum and difference formulas to verify identities and solve trigonometric equations

Why You Should Learn It:

You can use sum and difference formulas to rewrite trigonometric expressions. For instance, Exercise 65 on page 409 shows how to use sum and difference formulas to rewrite a trigonometric expression in a form that helps you find the equation of a standing wave.

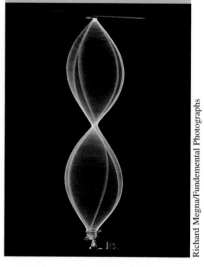

Richard Megna/Fundemental Photographs

EXAMPLE 2 Evaluating a Trigonometric Function

Find the exact value of $\sin \dfrac{\pi}{12}$.

Solution

Using the fact that $\pi/12 = \pi/3 - \pi/4$ together with the formula for $\sin(u - v)$, you obtain

$$\sin \frac{\pi}{12} = \sin\left(\frac{\pi}{3} - \frac{\pi}{4}\right) = \sin \frac{\pi}{3} \cos \frac{\pi}{4} - \cos \frac{\pi}{3} \sin \frac{\pi}{4}$$

$$= \frac{\sqrt{3}}{2}\left(\frac{\sqrt{2}}{2}\right) - \frac{1}{2}\left(\frac{\sqrt{2}}{2}\right) = \frac{\sqrt{6} - \sqrt{2}}{4}.$$

EXAMPLE 3 Evaluating a Trigonometric Expression

Find the exact value of $\sin(u + v)$ given that

$$\sin u = \frac{4}{5}, \text{ where } 0 < u < \frac{\pi}{2} \quad \text{and} \quad \cos v = -\frac{12}{13}, \text{ where } \frac{\pi}{2} < v < \pi.$$

Solution

Because $\sin u = 4/5$ and u is in Quadrant I, $\cos u = 3/5$, as shown in Figure 5.23(a). Because $\cos v = -12/13$ and v is in Quadrant II, $\sin v = 5/13$, as shown in Figure 5.23(b). You can find $\sin(u + v)$ as follows.

$$\sin(u + v) = \sin u \cos v + \cos u \sin v$$

$$= \left(\frac{4}{5}\right)\left(-\frac{12}{13}\right) + \left(\frac{3}{5}\right)\left(\frac{5}{13}\right) = -\frac{48}{65} + \frac{15}{65} = -\frac{33}{65}$$

EXAMPLE 4 An Application of a Sum Formula

Evaluate $\cos(\arctan 1 + \arccos x)$.

Solution

This expression fits the formula for $\cos(u + v)$. Angles $u = \arctan 1$ and $v = \arccos x$ are shown in Figure 5.24.

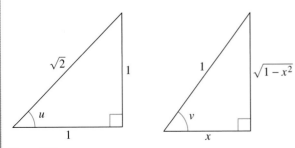

Figure 5.24

$$\cos(u + v) = \cos(\arctan 1)\cos(\arccos x) - \sin(\arctan 1)\sin(\arccos x)$$

$$= \frac{1}{\sqrt{2}} \cdot x - \frac{1}{\sqrt{2}} \cdot \sqrt{1 - x^2} = \frac{x - \sqrt{1 - x^2}}{\sqrt{2}}.$$

Hipparchus, considered the most eminent of Greek astronomers, was born about 160 B.C. in Nicea. He is credited with the invention of trigonometry. He also derived the sum and difference formulas for $\sin(A \pm B)$ and $\cos(A \pm B)$.

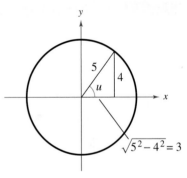

(a)

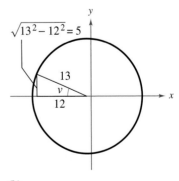

(b)
Figure 5.23

EXAMPLE 5   Proving a Cofunction Identity

Prove the cofunction identity $\cos\left(\dfrac{\pi}{2} - x\right) = \sin x$.

Solution

Using the formula for $\cos(u - v)$, you have

$$\cos\left(\frac{\pi}{2} - x\right) = \cos\frac{\pi}{2}\cos x + \sin\frac{\pi}{2}\sin x$$

$$= (0)(\cos x) + (1)(\sin x)$$

$$= \sin x.$$

Sum and difference formulas can be used to derive **reduction formulas** involving expressions such as

$$\sin\left(\theta + \frac{n\pi}{2}\right) \quad \text{and} \quad \cos\left(\theta + \frac{n\pi}{2}\right)$$

where n is an integer, as shown in the following example.

EXAMPLE 6   Deriving Reduction Formulas

Simplify each expression.

a. $\cos\left(\theta - \dfrac{3\pi}{2}\right)$ **b.** $\tan(\theta + 3\pi)$

Solution

a. Using the formula for $\cos(u - v)$, you have

$$\cos\left(\theta - \frac{3\pi}{2}\right) = \cos\theta\cos\frac{3\pi}{2} + \sin\theta\sin\frac{3\pi}{2}$$

$$= (\cos\theta)(0) + (\sin\theta)(-1)$$

$$= -\sin\theta.$$

b. Using the formula for $\tan(u + v)$, you have

$$\tan(\theta + 3\pi) = \frac{\tan\theta + \tan 3\pi}{1 - \tan\theta\tan 3\pi}$$

$$= \frac{\tan\theta + 0}{1 - (\tan\theta)(0)}$$

$$= \tan\theta.$$

Note that the period of $\tan\theta$ is π, so the period of $\tan(\theta + 3\pi)$ is the same as the period of $\tan\theta$.

The next example was taken from calculus. It is used to derive the formula for the derivative of the sine function.

EXAMPLE 7   An Application from Calculus

Verify that

$$\frac{\sin(x + h) - \sin x}{h} = (\cos x)\left(\frac{\sin h}{h}\right) - (\sin x)\left(\frac{1 - \cos h}{h}\right)$$

where $h \neq 0$.

Solution

Using the formula for $\sin(u + v)$, you have

$$\frac{\sin(x + h) - \sin x}{h} = \frac{\sin x \cos h + \cos x \sin h - \sin x}{h}$$

$$= \frac{\cos x \sin h - \sin x(1 - \cos h)}{h}$$

$$= (\cos x)\left(\frac{\sin h}{h}\right) - (\sin x)\left(\frac{1 - \cos h}{h}\right).$$

EXAMPLE 8   Solving a Trigonometric Equation

Find all solutions of $\sin\left(x + \frac{\pi}{4}\right) + \sin\left(x - \frac{\pi}{4}\right) = -1$

in the interval $[0, 2\pi)$.

Algebraic Solution

Using sum and difference formulas, rewrite the equation as

$$\sin x \cos \frac{\pi}{4} + \cos x \sin \frac{\pi}{4} + \sin x \cos \frac{\pi}{4} - \cos x \sin \frac{\pi}{4} = -1$$

$$2 \sin x \cos \frac{\pi}{4} = -1$$

$$2(\sin x)\left(\frac{\sqrt{2}}{2}\right) = -1$$

$$\sin x = -\frac{1}{\sqrt{2}}$$

$$\sin x = -\frac{\sqrt{2}}{2}.$$

Therefore, the only solutions in the interval $[0, 2\pi)$ are

$$x = \frac{5\pi}{4} \quad \text{and} \quad x = \frac{7\pi}{4}.$$

Graphical Solution

Use a graphing utility set in *radian* mode to graph $y = \sin\left(x + \frac{\pi}{4}\right) + \sin\left(x - \frac{\pi}{4}\right) + 1$, as shown in Figure 5.25. Use the *zero* or *root* feature or the *zoom* and *trace* features to approximate the x-intercepts in the interval $[0, 2\pi)$ to be

$$x \approx 3.927 \approx \frac{5\pi}{4} \text{ and } x \approx 5.498 \approx \frac{7\pi}{4}.$$

These values are the approximate solutions in the interval $[0, 2\pi)$ of $\sin\left(x + \frac{\pi}{4}\right) + \sin\left(x - \frac{\pi}{4}\right) = -1$.

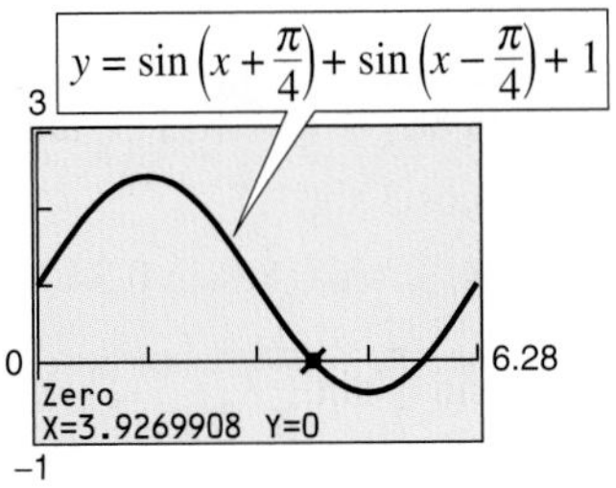

Figure 5.25

5.4   E x e r c i s e s

In Exercises 1–8, find the exact value of each expression.

1. (a) $\cos\left(\dfrac{\pi}{6} + \dfrac{\pi}{3}\right)$ (b) $\cos\dfrac{\pi}{6} + \cos\dfrac{\pi}{3}$

2. (a) $\sin\left(\dfrac{2\pi}{3} + \dfrac{3\pi}{4}\right)$ (b) $\sin\dfrac{2\pi}{3} + \sin\dfrac{3\pi}{4}$

3. (a) $\sin\left(\dfrac{7\pi}{6} - \dfrac{\pi}{3}\right)$ (b) $\sin\dfrac{7\pi}{6} - \sin\dfrac{\pi}{3}$

4. (a) $\cos\left(\dfrac{5\pi}{4} - \dfrac{\pi}{6}\right)$ (b) $\cos\dfrac{5\pi}{4} - \cos\dfrac{\pi}{6}$

5. (a) $\cos(0° + 135°)$ (b) $\cos 0° + \cos 135°$
6. (a) $\cos(240° - 0°)$ (b) $\cos 240° - \cos 0°$
7. (a) $\sin(315° - 60°)$ (b) $\sin 315° - \sin 60°$
8. (a) $\sin(390° + 120°)$ (b) $\sin 390° + \sin 120°$

In Exercises 9–18, use the sum and difference formulas to find the exact values of the sine, cosine, and tangent of the angle.

9. $75° = 30° + 45°$ 10. $15° = 45° - 30°$
11. $105° = 60° + 45°$ 12. $165° = 135° + 30°$
13. $195° = 225° - 30°$ 14. $285° = 330° - 45°$
15. $\dfrac{11\pi}{12} = \dfrac{3\pi}{4} + \dfrac{\pi}{6}$ 16. $\dfrac{17\pi}{12} = \dfrac{7\pi}{6} + \dfrac{\pi}{4}$
17. $-\dfrac{\pi}{12} = \dfrac{\pi}{6} - \dfrac{\pi}{4}$ 18. $-\dfrac{19\pi}{12} = \dfrac{2\pi}{3} - \dfrac{9\pi}{4}$

In Exercises 19–28, use the sum and difference formulas to write the expression as the sine, cosine, or tangent of an angle.

19. $\cos 40° \cos 15° - \sin 40° \sin 15°$
20. $\sin 110° \cos 80° + \cos 110° \sin 80°$
21. $\sin 340° \cos 50° - \cos 340° \sin 50°$
22. $\cos 20° \cos 30° + \sin 20° \sin 30°$
23. $\dfrac{\tan 325° - \tan 86°}{1 + \tan 325° \tan 86°}$ 24. $\dfrac{\tan 140° - \tan 60°}{1 + \tan 140° \tan 60°}$
25. $\sin 3 \cos 1.2 - \cos 3 \sin 1.2$
26. $\cos 0.88 \cos 0.34 + \sin 0.88 \sin 0.34$
27. $\cos\dfrac{\pi}{7} \cos\dfrac{\pi}{5} - \sin\dfrac{\pi}{7} \sin\dfrac{\pi}{5}$
28. $\sin\dfrac{2\pi}{9} \cos\dfrac{\pi}{10} + \cos\dfrac{2\pi}{9} \sin\dfrac{\pi}{10}$

Numerical, Graphical, and Algebraic Analysis   In Exercises 29–34, use a graphing utility to complete the table and graph the two functions. Use both table and graph as evidence that $y_1 = y_2$. Then verify the identity algebraically.

x	0.2	0.4	0.6	0.8	1.0	1.2	1.4
y_1							
y_2							

29. $y_1 = \sin\left(\dfrac{\pi}{2} + x\right),\quad y_2 = \cos x$

30. $y_1 = \sin(3\pi - x),\quad y_2 = \sin x$

31. $y_1 = \sin\left(\dfrac{\pi}{6} + x\right),\quad y_2 = \dfrac{1}{2}(\cos x + \sqrt{3}\sin x)$

32. $y_1 = \cos\left(\dfrac{5\pi}{4} - x\right),\quad y_2 = -\dfrac{\sqrt{2}}{2}(\cos x + \sin x)$

33. $y_1 = \cos(x + \pi)\cos(x - \pi),\quad y_2 = \cos^2 x$
34. $y_1 = \sin(x + \pi)\sin(x - \pi),\quad y_2 = \sin^2 x$

In Exercises 35–38, find the exact value of the trigonometric function given that

$\sin u = 5/13,\quad$ where $0 < u < \pi/2$
$\cos v = -3/5,\quad$ where $\pi/2 < v < \pi.$

35. $\sin(u + v)$ 36. $\cos(v - u)$
37. $\cos(u + v)$ 38. $\sin(u - v)$

In Exercises 39–42, find the exact value of the trigonometric function given that

$\sin u = 7/25,\quad$ where $\pi/2 < u < \pi$
$\cos v = 4/5,\quad$ where $3\pi/2 < v < 2\pi.$

39. $\cos(u + v)$ 40. $\sin(u + v)$
41. $\sin(v - u)$ 42. $\cos(u - v)$

In Exercises 43–50, verify the identity.

43. $\cos(\pi - \theta) + \sin\left(\dfrac{\pi}{2} + \theta\right) = 0$

44. $\sin(\theta + \pi) + \cos\left(\theta - \dfrac{\pi}{2}\right) = 0$

45. $\tan(x + \pi) - \tan(\pi - x) = 2 \tan x$

46. $\tan\left(\dfrac{\pi}{4} - \theta\right) = \dfrac{1 - \tan \theta}{1 + \tan \theta}$

47. $\sin(x + y) + \sin(x - y) = 2 \sin x \cos y$

48. $\cos(x + y) + \cos(x - y) = 2 \cos x \cos y$

49. $\cos(x + y) \cos(x - y) = \cos^2 x - \sin^2 y$

50. $\sin(x + y) \sin(x - y) = \sin^2 x - \sin^2 y$

In Exercises 51–54, write the trigonometric expression as an algebraic expression.

51. $\sin(\arcsin x + \arccos x)$

52. $\cos(\arccos x - \arcsin x)$

53. $\sin(\arctan 2x - \arccos x)$

54. $\cos(\arcsin x - \arctan 2x)$

In Exercises 55–60, find all solutions of the equation in the interval $[0, 2\pi)$. Use a graphing utility to verify your results.

55. $\sin\left(x + \dfrac{\pi}{3}\right) + \sin\left(x - \dfrac{\pi}{3}\right) = 1$

56. $\sin\left(x + \dfrac{\pi}{6}\right) - \sin\left(x - \dfrac{\pi}{6}\right) = \dfrac{1}{2}$

57. $\cos\left(x + \dfrac{\pi}{4}\right) - \cos\left(x - \dfrac{\pi}{4}\right) = 1$

58. $\cos\left(x + \dfrac{\pi}{6}\right) - \cos\left(x - \dfrac{\pi}{6}\right) = 1$

59. $\tan(x + \pi) + 2 \sin(x + \pi) = 0$

60. $2 \sin\left(x + \dfrac{\pi}{2}\right) + 3 \tan(\pi - x) = 0$

In Exercises 61–64, use a graphing utility to approximate all solutions of the equation in the interval $[0, 2\pi)$.

61. $\cos\left(x + \dfrac{\pi}{4}\right) + \cos\left(x - \dfrac{\pi}{4}\right) = 1$

62. $\sin\left(x + \dfrac{\pi}{2}\right) - \cos\left(x + \dfrac{3\pi}{2}\right) = 0$

63. $\tan(x + \pi) - \cos\left(x + \dfrac{\pi}{2}\right) = 0$

64. $\tan(\pi - x) + 2 \cos\left(x + \dfrac{3\pi}{2}\right) = 0$

65. *Standing Waves* The equation of a standing wave is obtained by adding the displacements of two waves traveling in opposite directions. Assume that each of the waves has amplitude A, period T, and wavelength λ. If the models for these waves are y_1 and y_2 given below, show that

$$y_1 + y_2 = 2A \cos \dfrac{2\pi t}{T} \cos \dfrac{2\pi x}{\lambda}.$$

$$y_1 = A \cos 2\pi\left(\dfrac{t}{T} - \dfrac{x}{\lambda}\right), \quad y_2 = A \cos 2\pi\left(\dfrac{t}{T} + \dfrac{x}{\lambda}\right)$$

66. *Harmonic Motion* A weight is attached to a spring suspended vertically from a ceiling. When a driving force is applied to the system, the weight moves vertically from its equilibrium position, and this motion is modeled by

$$y = \tfrac{1}{3} \sin 2t + \tfrac{1}{4} \cos 2t$$

where y is the distance from equilibrium (in feet) and t is the time (in seconds).

(a) Use a graphing utility to graph the model.

(b) Write the model in the form

$$y = \sqrt{a^2 + b^2} \sin(Bt + C).$$

Use a graphing utility to verify your result.

(c) Find the amplitude of the oscillations of the weight.

(d) Find the frequency of the oscillations of the weight.

Synthesis

True or False? **In Exercises 67–70, determine whether the statement is true or false. Justify your answer.**

67. $\sin(u \pm v) = \sin u \pm \sin v$

68. $\cos(u \pm v) = \cos u \pm \cos v$

69. It is not possible to find the exact value of $\sin 75°$.

70. $\sin\left(x - \dfrac{11\pi}{2}\right) = \cos x$

In Exercises 71–74, verify the identity.

71. $\cos(n\pi + \theta) = (-1)^n \cos \theta,$ n is an integer.

72. $\sin(n\pi + \theta) = (-1)^n \sin \theta,$ n is an integer.

73. $a \sin B\theta + b \cos B\theta = \sqrt{a^2 + b^2}\, \sin(B\theta + C)$
where $C = \arctan(b/a)$ and $a > 0$.

74. $a \sin B\theta + b \cos B\theta = \sqrt{a^2 + b^2}\, \cos(B\theta - C)$
where $C = \arctan(a/b)$ and $b > 0$.

In Exercises 75–78, use the formulas given in Exercises 73 and 74 to write the expression in the following forms. Use a graphing utility to verify your results.

(a) $\sqrt{a^2 + b^2}\, \sin(B\theta + C)$

(b) $\sqrt{a^2 + b^2}\, \cos(B\theta - C)$

75. $\sin \theta + \cos \theta$

76. $3 \sin 2\theta + 4 \cos 2\theta$

77. $12 \sin 3\theta + 5 \cos 3\theta$

78. $\sin 2\theta - \cos 2\theta$

In Exercises 79 and 80, use the formulas given in Exercises 73 and 74 to write the trigonometric expression in the form $a \sin B\theta + b \cos B\theta$.

79. $2 \sin\left(\theta + \dfrac{\pi}{2}\right)$

80. $5 \cos\left(\theta + \dfrac{\pi}{4}\right)$

81. *Conjecture*   Three squares of side s are placed side by side. Make a conjecture about the relationship between the sum $u + v$ and w. Prove your conjecture by using the identity for the tangent of the sum of two angles.

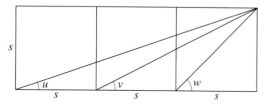

Conjecture   **In Exercises 82 and 83, use a graphing utility to graph the trigonometric function. Use the graph to make a conjecture about a simplification of the expression. Verify the resulting identity algebraically.**

82. $g(x) = \cos(\pi + x)$

83. $h(\theta) = \tan(\pi + \theta)$

84. Verify the identity used in calculus.
$$\frac{\cos(x + h) - \cos x}{h} = \frac{\cos x(\cos h - 1)}{h} - \frac{\sin x \sin h}{h}$$

85. *Exploration*   Let $x = \pi/6$ in the identity in Exercise 84 and define the functions f and g as follows.
$$f(h) = \frac{\cos(\pi/6 + h) - \cos(\pi/6)}{h}$$
$$g(h) = \cos\frac{\pi}{6}\left(\frac{\cos h - 1}{h}\right) - \sin\frac{\pi}{6}\left(\frac{\sin h}{h}\right)$$

(a) What are the domains of these functions?

(b) Use a graphing utility to complete the table.

h	0.01	0.02	0.05	0.1	0.2	0.5
$f(h)$						
$g(h)$						

(c) Use a graphing utility to graph the functions.

(d) Use the table and graph to make a conjecture about the values of the functions as $h \to 0$.

86. Use the sum formulas for sine and cosine to derive the formula
$$\tan(u + v) = \frac{\tan u + \tan v}{1 - \tan u \tan v}.$$

Review

In Exercises 87–90, find the x- and y-intercepts of the graph of the equation. Use a graphing utility to verify your results.

87. $y = -\frac{1}{2}(x - 10) + 14$

88. $y = x^2 - 3x - 40$

89. $y = |2x - 9| - 5$

90. $y = 2x\sqrt{x + 7}$

In Exercises 91–94, evaluate the expression without using a calculator.

91. $\arccos\left(\dfrac{\sqrt{3}}{2}\right)$

92. $\arctan\left(-\sqrt{3}\right)$

93. $\arcsin 1$

94. $\arctan 0$

5.5 Multiple-Angle and Product-Sum Formulas

Multiple-Angle Formulas

In this section you will study four other categories of trigonometric identities.

1. The first category involves functions of multiple angles such as $\sin ku$ and $\cos ku$.
2. The second category involves squares of trigonometric functions such as $\sin^2 u$.
3. The third category involves functions of half-angles such as $\sin(u/2)$.
4. The fourth category involves products of trigonometric functions such as $\sin u \cos v$.

The most commonly used multiple-angle formulas are the **double-angle formulas.** They are used often, so you should learn them. (Proofs of the double-angle formulas are given in Appendix A.)

Double-Angle Formulas

$$\sin 2u = 2 \sin u \cos u \qquad\qquad \tan 2u = \frac{2 \tan u}{1 - \tan^2 u}$$

$$\cos 2u = \cos^2 u - \sin^2 u$$

$$= 2 \cos^2 u - 1$$

$$= 1 - 2 \sin^2 u$$

Note that $\sin 2u \neq 2 \sin u$, $\cos 2u \neq 2 \cos u$, and $\tan 2u \neq 2 \tan u$.

EXAMPLE 1   Solving a Multiple-Angle Equation

Find all solutions of $2 \cos x + \sin 2x = 0$.

Solution

Begin by rewriting the equation so that it involves functions of x (rather than $2x$). Then factor and solve as usual.

$2 \cos x + \sin 2x = 0$	Write original equation.
$2 \cos x + 2 \sin x \cos x = 0$	Double-angle formula
$2 \cos x(1 + \sin x) = 0$	Factor.
$\cos x = 0, \quad 1 + \sin x = 0$	Set factors equal to zero.
$x = \dfrac{\pi}{2}, \dfrac{3\pi}{2} \qquad x = \dfrac{3\pi}{2}$	Solutions in $[0, 2\pi)$

Therefore, the general solution is

$$x = \frac{\pi}{2} + 2n\pi \quad \text{and} \quad x = \frac{3\pi}{2} + 2n\pi \qquad \text{General solution}$$

where n is an integer. Try verifying these solutions graphically.

NASA-Liaison Agency

A computer animation of this concept appears in the *Interactive* CD-ROM and *Internet* versions of this text.

EXAMPLE 2   Using Double-Angle Formulas in Sketching Graphs

Analyze the graph of $y = 4\cos^2 x - 2$ over the interval $[0, 2\pi]$.

Solution

Using a double-angle formula, you can rewrite the given function as

$$y = 4\cos^2 x - 2$$
$$= 2(2\cos^2 x - 1)$$
$$= 2\cos 2x.$$

Using the techniques discussed in Section 4.5, you can recognize that the graph of this function has an amplitude of 2 and a period of π. The key points in the interval $[0, \pi]$ are as follows.

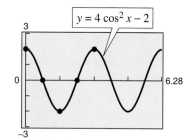

Figure 5.26

Maximum	*Intercept*	*Minimum*	*Intercept*	*Maximum*
$(0, 2)$	$\left(\dfrac{\pi}{4}, 0\right)$	$\left(\dfrac{\pi}{2}, -2\right)$	$\left(\dfrac{3\pi}{4}, 0\right)$	$(\pi, 2)$

Two cycles of the graph are shown in Figure 5.26.

EXAMPLE 3   Evaluating Functions Involving Double Angles

Use the following to find $\sin 2\theta$, $\cos 2\theta$, and $\tan 2\theta$.

$$\cos\theta = \frac{5}{13}, \qquad \frac{3\pi}{2} < \theta < 2\pi$$

Solution

From Figure 5.27, you can see that

$$\sin\theta = \frac{y}{r} = -\frac{12}{13}.$$

Consequently, you can write

$$\sin 2\theta = 2\sin\theta\cos\theta = 2\left(-\frac{12}{13}\right)\left(\frac{5}{13}\right) = -\frac{120}{169}$$

$$\cos 2\theta = 2\cos^2\theta - 1 = 2\left(\frac{25}{169}\right) - 1 = -\frac{119}{169}$$

$$\tan 2\theta = \frac{\sin 2\theta}{\cos 2\theta} = \frac{120}{119}.$$

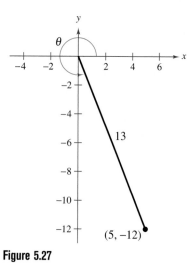

Figure 5.27

The double-angle formulas are not restricted to angles 2θ and θ. Other *double* combinations, such as 4θ and 2θ or 6θ and 3θ, are also valid. Here are two examples.

$$\sin 4\theta = 2\sin 2\theta\cos 2\theta \qquad \text{and} \qquad \cos 6\theta = \cos^2 3\theta - \sin^2 3\theta$$

By using double-angle formulas together with the sum formulas derived in the previous section, you can form other multiple-angle formulas.

STUDY T!P

Notice that you cannot solve Example 3 by simply evaluating $\theta = \cos^{-1}(5/13)$ and then calculating $\sin 2\theta$, $\cos 2\theta$, and $\tan 2\theta$. Do you see why?

EXAMPLE 4 Deriving a Triple-Angle Formula

Express $\sin 3x$ in terms of $\sin x$.

Solution

$$\sin 3x = \sin(2x + x)$$

$= \sin 2x \cos x + \cos 2x \sin x$	Sum formula
$= 2 \sin x \cos x \cos x + (1 - 2 \sin^2 x)\sin x$	Double-angle formula
$= 2 \sin x \cos^2 x + \sin x - 2 \sin^3 x$	Multiply.
$= 2 \sin x(1 - \sin^2 x) + \sin x - 2 \sin^3 x$	Pythagorean identity
$= 2 \sin x - 2 \sin^3 x + \sin x - 2 \sin^3 x$	Multiply.
$= 3 \sin x - 4 \sin^3 x$	Simplify.

Power-Reducing Formulas

The double-angle formulas can be used to obtain the following **power-reducing formulas.** (Proofs of the power-reducing formulas are given in Appendix A.)

A computer animation of this concept appears in the *Interactive* CD-ROM and *Internet* versions of this text.

Power-Reducing Formulas

$$\sin^2 u = \frac{1 - \cos 2u}{2} \qquad \cos^2 u = \frac{1 + \cos 2u}{2} \qquad \tan^2 u = \frac{1 - \cos 2u}{1 + \cos 2u}$$

Example 5 shows a typical power reduction that is used in calculus.

EXAMPLE 5 Reducing a Power

Rewrite $\sin^4 x$ as a sum of first powers of the cosines of multiple angles.

Solution

Note the repeated use of power-reducing formulas.

$\sin^4 x = (\sin^2 x)^2$	Property of exponents
$= \left(\dfrac{1 - \cos 2x}{2}\right)^2$	Power-reducing formula
$= \dfrac{1}{4}(1 - 2 \cos 2x + \cos^2 2x)$	Expand binomial.
$= \dfrac{1}{4}\left(1 - 2 \cos 2x + \dfrac{1 + \cos 4x}{2}\right)$	Power-reducing formula
$= \dfrac{1}{4} - \dfrac{1}{2}\cos 2x + \dfrac{1}{8} + \dfrac{1}{8}\cos 4x$	Distributive Property
$= \dfrac{3}{8} - \dfrac{1}{2}\cos 2x + \dfrac{1}{8}\cos 4x$	Simplify.
$= \dfrac{1}{8}(3 - 4 \cos 2x + \cos 4x)$	Factor.

Half-Angle Formulas

You can derive some useful alternative forms of the power-reducing formulas by replacing u with $u/2$. The results are called **half-angle formulas.**

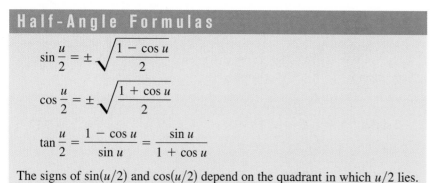

Half-Angle Formulas

$$\sin\frac{u}{2} = \pm\sqrt{\frac{1 - \cos u}{2}}$$

$$\cos\frac{u}{2} = \pm\sqrt{\frac{1 + \cos u}{2}}$$

$$\tan\frac{u}{2} = \frac{1 - \cos u}{\sin u} = \frac{\sin u}{1 + \cos u}$$

The signs of $\sin(u/2)$ and $\cos(u/2)$ depend on the quadrant in which $u/2$ lies.

A computer animation of this concept appears in the *Interactive* CD-ROM and *Internet* versions of this text.

EXAMPLE 6   Using a Half-Angle Formula

Find the exact value of $\sin 105°$.

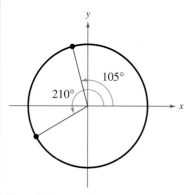

Figure 5.28

Solution

Begin by noting that $105°$ is half of $210°$. Then, using the half-angle formula for $\sin(u/2)$ and the fact that $105°$ lies in Quadrant II (see Figure 5.28), you have

$$\sin 105° = \sqrt{\frac{1 - \cos 210°}{2}}$$

$$= \sqrt{\frac{1 - (-\cos 30°)}{2}}$$

$$= \sqrt{\frac{1 + \left(\sqrt{3}/2\right)}{2}} = \frac{\sqrt{2 + \sqrt{3}}}{2}.$$

The positive square root is chosen because $\sin\theta$ is positive in Quadrant II.

Use your calculator to verify the result obtained in Example 6. That is, evaluate $\sin 105°$ and $\left(\sqrt{2 + \sqrt{3}}\right)/2$.

EXAMPLE 7 Solving a Trigonometric Equation

Find all solutions of

$$2 - \sin^2 x = 2\cos^2 \frac{x}{2}$$

in the interval $[0, 2\pi)$.

Algebraic Solution

$$2 - \sin^2 x = 2\cos^2 \frac{x}{2} \qquad \text{Write original equation.}$$

$$2 - \sin^2 x = 2\left(\pm\sqrt{\frac{1 + \cos x}{2}}\right)^2 \qquad \text{Half-angle formula}$$

$$2 - \sin^2 x = 2\left(\frac{1 + \cos x}{2}\right) \qquad \text{Simplify.}$$

$$2 - \sin^2 x = 1 + \cos x \qquad \text{Simplify.}$$

$$2 - (1 - \cos^2 x) = 1 + \cos x \qquad \text{Pythagorean identity}$$

$$\cos^2 x - \cos x = 0 \qquad \text{Simplify.}$$

$$\cos x(\cos x - 1) = 0 \qquad \text{Factor.}$$

By setting the factors $\cos x$ and $(\cos x - 1)$ equal to zero, you find that the solutions in the interval $[0, 2\pi)$ are

$$x = \frac{\pi}{2}, \quad x = \frac{3\pi}{2}, \quad \text{and} \quad x = 0.$$

Graphical Solution

Use a graphing utility set in *radian* mode to graph $y = 2 - \sin^2 x - 2\cos^2(x/2)$, as shown in Figure 5.29. Use the *zero* or *root* feature or the *zoom* and *trace* features to approximate the x-intercepts in the interval $[0, 2\pi)$ to be

$$x = 0, \quad x \approx 1.5708 \approx \frac{\pi}{2}, \quad \text{and} \quad x \approx 4.7124 \approx \frac{3\pi}{2}.$$

These values are the approximate solutions in the interval $[0, 2\pi)$ of $2 - \sin^2 x = \cos^2 \frac{x}{2}$.

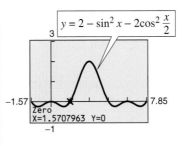

Figure 5.29

Product-to-Sum Formulas

Each of the following **product-to-sum formulas** is easily verified using the sum and difference formulas discussed in the preceding section.

Product-to-Sum Formulas
$\sin u \sin v = \dfrac{1}{2}[\cos(u - v) - \cos(u + v)]$
$\cos u \cos v = \dfrac{1}{2}[\cos(u - v) + \cos(u + v)]$
$\sin u \cos v = \dfrac{1}{2}[\sin(u + v) + \sin(u - v)]$
$\cos u \sin v = \dfrac{1}{2}[\sin(u + v) - \sin(u - v)]$

EXAMPLE 8   Writing Products as Sums

Rewrite the product as a sum or difference.

$$\cos 5x \sin 4x$$

Solution

$$\cos 5x \sin 4x = \frac{1}{2}[\sin(5x + 4x) - \sin(5x - 4x)]$$

$$= \frac{1}{2} \sin 9x - \frac{1}{2} \sin x$$

Occasionally, it is useful to reverse the procedure and write a sum of trigonometric functions as a product. This can be accomplished with the following **sum-to-product formulas.** (A proof of the first formula is given in Appendix A.)

Sum-to-Product Formulas

$$\sin x + \sin y = 2 \sin\left(\frac{x + y}{2}\right) \cos\left(\frac{x - y}{2}\right)$$

$$\sin x - \sin y = 2 \cos\left(\frac{x + y}{2}\right) \sin\left(\frac{x - y}{2}\right)$$

$$\cos x + \cos y = 2 \cos\left(\frac{x + y}{2}\right) \cos\left(\frac{x - y}{2}\right)$$

$$\cos x - \cos y = -2 \sin\left(\frac{x + y}{2}\right) \sin\left(\frac{x - y}{2}\right)$$

EXAMPLE 9   Using a Sum-to-Product Formula

Find the exact value of

$$\cos 195° + \cos 105°.$$

Solution
Using the appropriate sum-to-product formula, you obtain

$$\cos 195° + \cos 105° = 2 \cos\left(\frac{195° + 105°}{2}\right) \cos\left(\frac{195° - 105°}{2}\right)$$

$$= 2 \cos 150° \cos 45°$$

$$= 2\left(-\frac{\sqrt{3}}{2}\right)\left(\frac{\sqrt{2}}{2}\right)$$

$$= -\frac{\sqrt{6}}{2}.$$

EXAMPLE 10   Solving a Trigonometric Equation

Find all solutions of $\sin 5x + \sin 3x = 0$ in the interval $[0, 2\pi)$.

Algebraic Solution

$$\sin 5x + \sin 3x = 0 \qquad \text{Write original equation.}$$

$$2 \sin\left(\frac{5x + 3x}{2}\right) \cos\left(\frac{5x - 3x}{2}\right) = 0 \qquad \text{Sum-to-product formula}$$

$$2 \sin 4x \cos x = 0 \qquad \text{Simplify.}$$

By setting the factor $\sin 4x$ equal to zero, you can find that the solutions in the interval $[0, 2\pi)$ are

$$x = 0, \frac{\pi}{4}, \frac{\pi}{2}, \frac{3\pi}{4}, \pi, \frac{5\pi}{4}, \frac{3\pi}{2}, \frac{7\pi}{4}.$$

Moreover, the equation $\cos x = 0$ yields no additional solutions.

Graphical Solution

Use a graphing utility set in *radian* mode to graph $y = \sin 5x + \sin 3x$, as shown in Figure 5.30. Use the *zero* or *root* feature or the *zoom* and *trace* features to approximate the x-intercepts in the interval $[0, 2\pi)$ to be

$$x \approx 0, \ x \approx 0.7854 \approx \frac{\pi}{4}, x \approx 1.5708 \approx \frac{\pi}{2},$$

$$x \approx 2.3562 \approx \frac{3\pi}{4}, \ x \approx 3.1416 \approx \pi, x \approx 3.9270 \approx \frac{5\pi}{4},$$

$$x \approx 4.7124 \approx \frac{3\pi}{2}, \ x \approx 5.4978 \approx \frac{7\pi}{4}.$$

These values are the approximate solutions in the interval $[0, 2\pi)$ of $\sin 5x + \sin 3x = 0$.

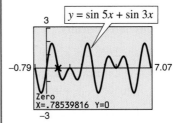

Figure 5.30

Note in Example 10 that the general solution would be $x = \frac{n\pi}{4}$, where n is an integer.

EXAMPLE 11   Verifying a Trigonometric Identity

Verify the identity $\dfrac{\sin t + \sin 3t}{\cos t + \cos 3t} = \tan 2t.$

Solution

Using appropriate sum-to-product formulas, you have

$$\frac{\sin t + \sin 3t}{\cos t + \cos 3t} = \frac{2 \sin 2t \cos(-t)}{2 \cos 2t \cos(-t)} = \frac{\sin 2t}{\cos 2t} = \tan 2t.$$

Writing About Math   *Deriving an Area Formula*

Describe how you can use a double-angle formula or a half-angle formula to derive a formula for the area of an isosceles triangle. Use a labeled sketch to illustrate your derivation. Then write two examples that show how your formula can be used.

5.5 Exercises

In Exercises 1–8, use the figure to find the exact value of the trigonometric function.

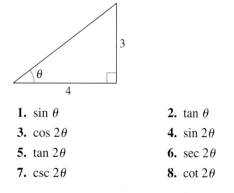

1. $\sin \theta$
2. $\tan \theta$

3. $\cos 2\theta$
4. $\sin 2\theta$

5. $\tan 2\theta$
6. $\sec 2\theta$

7. $\csc 2\theta$
8. $\cot 2\theta$

In Exercises 9–16, use a graphing utility to approximate the equation's solutions in the interval $[0, 2\pi)$. If possible, find the exact solutions algebraically.

9. $\sin 2x - \sin x = 0$
10. $\sin 2x + \cos x = 0$

11. $8 \sin x \cos x = 1$
12. $\sin 2x \sin x = \cos x$

13. $\cos 2x + \cos x = 0$
14. $\tan 2x - \cot x = 0$

15. $\sin 4x = -2 \sin 2x$
16. $(\sin 2x + \cos 2x)^2 = 1$

In Exercises 17–20, use a double-angle formula to rewrite the expression. Use a graphing utility to graph both expressions to verify that both forms are the same.

17. $8 \sin x \cos x$
18. $6 \sin x \cos x + 4$

19. $5 - 10 \sin^2 x$

20. $(\cos x + \sin x)(\cos x - \sin x)$

In Exercises 21–24, find the exact values of $\sin 2u$, $\cos 2u$, and $\tan 2u$ using the double-angle formulas.

21. $\sin u = \frac{3}{5}, \quad 0 < u < \pi/2$

22. $\cos u = -\frac{2}{7}, \quad \pi/2 < u < \pi$

23. $\tan u = \frac{1}{2}, \quad \pi < u < 3\pi/2$

24. $\cot u = -6, \quad 3\pi/2 < u < 2\pi$

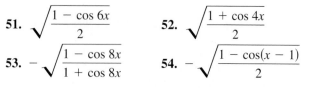 In Exercises 25–30, rewrite the expression in terms of the first power of the cosine. Use a graphing utility to graph both expressions and verify that both forms are the same.

25. $\cos^4 x$
26. $\sin^4 x$

27. $\sin^2 x \cos^2 x$
28. $\cos^6 x$

29. $\sin^2 x \cos^4 x$
30. $\sin^4 x \cos^2 x$

In Exercises 31–38, use the figure to find the exact value of the trigonometric function.

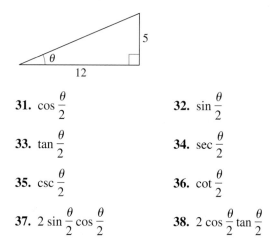

31. $\cos \dfrac{\theta}{2}$
32. $\sin \dfrac{\theta}{2}$

33. $\tan \dfrac{\theta}{2}$
34. $\sec \dfrac{\theta}{2}$

35. $\csc \dfrac{\theta}{2}$
36. $\cot \dfrac{\theta}{2}$

37. $2 \sin \dfrac{\theta}{2} \cos \dfrac{\theta}{2}$
38. $2 \cos \dfrac{\theta}{2} \tan \dfrac{\theta}{2}$

In Exercises 39–46, use the half-angle formulas to determine the exact values of the sine, cosine, and tangent of the angle.

39. $15°$
40. $165°$

41. $112° \, 30'$
42. $157° \, 30'$

43. $\dfrac{\pi}{8}$
44. $\dfrac{\pi}{12}$

45. $\dfrac{3\pi}{8}$
46. $\dfrac{7\pi}{12}$

In Exercises 47–50, find the exact values of $\sin(u/2)$, $\cos(u/2)$ and $\tan(u/2)$ using the half-angle formulas.

47. $\sin u = \frac{5}{13}, \quad \pi/2 < u < \pi$

48. $\cos u = \frac{7}{25}, \quad 0 < u < \pi/2$

49. $\tan u = -\frac{8}{5}, \quad 3\pi/2 < u < 2\pi$

50. $\cot u = 7, \quad \pi < u < 3\pi/2$

In Exercises 51–54, use the half-angle formulas to simplify the expression.

51. $\sqrt{\dfrac{1 - \cos 6x}{2}}$
52. $\sqrt{\dfrac{1 + \cos 4x}{2}}$

53. $-\sqrt{\dfrac{1 - \cos 8x}{1 + \cos 8x}}$
54. $-\sqrt{\dfrac{1 - \cos(x - 1)}{2}}$

In Exercises 55–58, find the exact zeros of the function in the interval $[0, 2\pi)$. Use a graphing utility to graph the function and verify your answer.

55. $f(x) = \sin\dfrac{x}{2} - \cos x$

56. $h(x) = \sin\dfrac{x}{2} + \cos x - 1$

57. $h(x) = \cos\dfrac{x}{2} - \sin x$ **58.** $g(x) = \tan\dfrac{x}{2} - \sin x$

In Exercises 59–64, use the product-to-sum formulas to write the product as a sum or difference.

59. $6 \sin\dfrac{\pi}{3} \cos\dfrac{\pi}{3}$ **60.** $4 \sin\dfrac{\pi}{3} \cos\dfrac{5\pi}{6}$

61. $\sin 5\theta \cos 3\theta$ **62.** $5 \sin 3\alpha \sin 4\alpha$

63. $5 \cos(-5\beta) \cos 3\beta$ **64.** $\cos 2\theta \cos 4\theta$

In Exercises 65–72, use the sum-to-product formulas to write the sum or difference as a product.

65. $\sin 60° + \sin 30°$ **66.** $\cos 120° + \cos 30°$

67. $\sin 5\theta - \sin \theta$ **68.** $\sin x + \sin 7x$

69. $\sin(\alpha + \beta) - \sin(\alpha - \beta)$

70. $\cos(\phi + 2\pi) + \cos \phi$

71. $\cos\left(\theta + \dfrac{\pi}{2}\right) - \cos\left(\theta - \dfrac{\pi}{2}\right)$

72. $\sin\left(x + \dfrac{\pi}{2}\right) + \sin\left(x - \dfrac{\pi}{2}\right)$

In Exercises 73–76, find the exact zeros of the function in the interval $[0, 2\pi)$. Use a graphing utility to graph the function and verify your answer.

73. $g(x) = \sin 6x + \sin 2x$

74. $h(x) = \cos 2x - \cos 6x$

75. $f(x) = \dfrac{\cos 2x}{\sin 3x - \sin x} - 1$

76. $f(x) = \sin^2 3x - \sin^2 x$

In Exercises 77–80, use the figure to find the exact value of the trigonometric function in two ways.

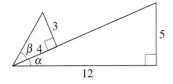

77. $\sin^2 \alpha$ **78.** $\cos^2 \alpha$

79. $\sin \alpha \cos \beta$ **80.** $\cos \alpha \sin \beta$

In Exercises 81–92, verify the identity algebraically. Use a graphing utility to confirm the identity graphically.

81. $\csc 2\theta = \dfrac{\csc \theta}{2 \cos \theta}$ **82.** $\sec 2\theta = \dfrac{\sec^2 \theta}{2 - \sec^2 \theta}$

83. $\cos^2 2\alpha - \sin^2 2\alpha = \cos 4\alpha$

84. $\cos^4 x - \sin^4 x = \cos 2x$

85. $(\sin x + \cos x)^2 = 1 + \sin 2x$

86. $1 + \cos 10y = 2 \cos^2 5y$

87. $\sec\dfrac{u}{2} = \pm\sqrt{\dfrac{2 \tan u}{\tan u + \sin u}}$

88. $\tan\dfrac{u}{2} = \csc u - \cot u$

89. $\cos 3\beta = \cos^3 \beta - 3 \sin^2 \beta \cos \beta$

90. $\sin 4\beta = 4 \sin \beta \cos \beta(1 - 2 \sin^2 \beta)$

91. $\dfrac{\cos 4x - \cos 2x}{2 \sin 3x} = -\sin x$

92. $\dfrac{\cos 3x - \cos x}{\sin 3x - \sin x} = -\tan 2x$

In Exercises 93 and 94, graph the function by using the power-reducing formulas.

93. $f(x) = \sin^2 x$ **94.** $f(x) = \cos^2 x$

In Exercises 95–98, (a) use a graphing utility to graph the function and approximate the maximum and minimum points on the graph in the interval $[0, 2\pi]$, and (b) solve the trigonometric equation and verify that the x-coordinates of the maximum and minimum points of f are among its solutions (calculus is required to find the trigonometric equation).

Function	*Trigonometric Equation*
95. $f(x) = 4 \sin\dfrac{x}{2} + \cos x$	$2 \cos\dfrac{x}{2} - \sin x = 0$
96. $f(x) = \cos 2x - 2 \sin x$	$-2 \cos x(2 \sin x + 1) = 0$
97. $f(x) = 2 \cos\dfrac{x}{2} + \sin 2x$	$2 \cos 2x - \sin\dfrac{x}{2} = 0$
98. $f(x) = 2 \sin\dfrac{x}{2}$ $- 5 \cos\left(2x - \dfrac{\pi}{4}\right)$	$10 \sin\left(2x - \dfrac{\pi}{4}\right)$ $+ \cos\dfrac{x}{2} = 0$

In Exercises 99–102, write the trigonometric expression as an algebraic expression.

99. $\sin(2 \arcsin x)$ **100.** $\cos(2 \arccos x)$

101. $\cos(2 \arcsin x)$ **102.** $\sin(2 \arctan x)$

103. *Projectile Motion*   The range of a projectile fired at an angle θ with the horizontal and with an initial velocity of v_0 feet per second is

$$r = \tfrac{1}{32} v_0 \sin 2\theta$$

where r is measured in feet. Determine the expression for the range in terms of θ.

104. *Area*   The length of each of the two equal sides of an isosceles triangle is 10 meters. The angle between the equal sides is θ.

(a) Express the area of the triangle as a function of $\theta/2$.

(b) Express the area of the triangle as a function of θ and determine the value of θ such that the area is a maximum.

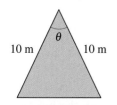

105. *Mach Number*   The mach number M of an airplane is the ratio of its speed to the speed of sound. When an airplane travels faster than the speed of sound, the sound waves form a cone behind the airplane. The mach number is related to the apex angle θ of the cone by

$$\sin \frac{\theta}{2} = \frac{1}{M}.$$

Find the angle θ that corresponds to a mach number of 4.5.

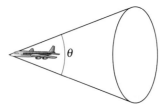

106. *Railroad Track*   When two railroad tracks merge, the overlapping portions of the tracks are in the shape of a circular arcs. The radius of the arcs r (in

feet) and the angle θ are related by

$$\frac{x}{2} = 2r \sin^2 \frac{\theta}{2}.$$

Write a formula for x in terms of $\cos \theta$.

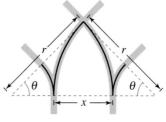

FIGURE FOR 106

Synthesis

True or False?   **In Exercises 107 and 108, determine whether the statement is true or false. Justify your answer.**

107. $\sin \dfrac{x}{2} = -\sqrt{\dfrac{1 - \cos x}{2}},\qquad \pi \le x \le 2\pi.$

108. The graph of $y = 4 - 8\sin^2 x$ has a maximum at $(\pi, 4)$.

109. *Conjecture*   Consider the function

$$f(x) = 2 \sin x \left[2 \cos^2 \left(\frac{x}{2} \right) - 1 \right].$$

(a) Use a graphing utility to graph the function.

(b) Make a conjecture about the function that is an identity with f.

(c) Verify your conjecture algebraically.

Review

In Exercises 110 and 111, find (if possible) the complement and supplement of each angle.

110. (a) $55°$ (b) $162°$

111. (a) $\dfrac{\pi}{18}$ (b) $\dfrac{9\pi}{20}$

In Exercises 112–115, sketch the graph of the function. Use a graphing utility to verify your graph.

112. $f(x) = \dfrac{3}{2} \cos 2x$ **113.** $f(x) = \dfrac{5}{2} \sin \dfrac{1}{2}x$

114. $f(x) = \dfrac{1}{2} \tan 2\pi x$ **115.** $f(x) = \dfrac{1}{4} \sec \dfrac{\pi x}{2}$

5   Chapter Summary

What did you learn?

Section 5.1	**Review Exercises**
☐ How to recognize and write the fundamental trigonometric identities	1–10
☐ How to use fundamental trigonometric identities to evaluate trigonometric functions, simplify trigonometric expressions, and rewrite trigonometric expressions	11–24

Section 5.2

☐ How to verify trigonometric identities	25–36

Section 5.3

☐ How to use standard algebraic techniques to solve trigonometric equations	37–50
☐ How to solve trigonometric equations of quadratic type	51–54
☐ How to solve trigonometric equations involving multiple angles	55–60
☐ How to use inverse trigonometric functions to solve trigonometric equations	61–64

Section 5.4

☐ How to use sum and difference formulas to evaluate trigonometric functions	65–78
☐ How to use sum and difference formulas to verify identities and solve trigonometric equations	79–86

Section 5.5

☐ How to use multiple-angle formulas to rewrite and evaluate trigonometric formulas	87–95
☐ How to use power-reducing formulas to rewrite and evaluate trigonometric formulas	96–99
☐ How to use half-angle formulas to rewrite and evaluate trigonometric functions	100–108
☐ How to use product-sum formulas to rewrite and evaluate trigonometric functions	109–117

5 Review Exercises

5.1 In Exercises 1–10, name the trigonometric function equivalent to the expression.

1. $\dfrac{1}{\cos x}$

2. $\dfrac{1}{\sin x}$

3. $\dfrac{1}{\sec x}$

4. $\dfrac{1}{\tan x}$

5. $\dfrac{\cos x}{\sin x}$

6. $\sqrt{1 + \tan^2 x}$

7. $\csc\left(\dfrac{\pi}{2} - x\right)$

8. $\cot\left(\dfrac{\pi}{2} - x\right)$

9. $\sec(-x)$

10. $\tan(-x)$

In Exercises 11–14, use the given values and trigonometric identities to evaluate (if possible) the other trigonometric functions of the angle.

11. $\sin x = \dfrac{3}{5}, \quad \cos x = \dfrac{4}{5}$

12. $\tan \theta = \dfrac{3}{2}, \quad \sec \theta = \dfrac{\sqrt{13}}{2}$

13. $\sin\left(\dfrac{\pi}{2} - x\right) = \dfrac{1}{\sqrt{2}}, \quad \sin x = -\dfrac{1}{\sqrt{2}}$

14. $\csc\left(\dfrac{\pi}{2} - \theta\right) = 3, \quad \sin \theta = \dfrac{2\sqrt{2}}{3}$

In Exercises 15–22, use the fundamental identities to simplify the trigonometric expression. Use the *table* feature of a graphing utility to check your result numerically.

15. $\dfrac{1}{\cot^2 x + 1}$

16. $\dfrac{\sec^2 x - 1}{\sec x - 1}$

17. $\dfrac{\sin^2 \alpha - \cos^2 \alpha}{\sin^2 \alpha - \sin \alpha \cos \alpha}$

18. $\dfrac{\sin^3 \beta + \cos^3 \beta}{\sin \beta + \cos \beta}$

19. $\tan^2 \theta(\csc^2 \theta - 1)$

20. $\csc^2 x(1 - \cos^2 x)$

21. $\tan\left(\dfrac{\pi}{2} - x\right) \sec x$

22. $\dfrac{\sin(-x) \cot x}{\sin\left(\dfrac{\pi}{2} - x\right)}$

23. *Rate of Change* The rate of change of the function $f(x) = 2\sqrt{\sin x}$ is given by the expression $\sin^{-1/2} x \cos x$. Show that this expression can also be written as $\cot x \sqrt{\sin x}$.

24. *Rate of Change* The rate of change of the function $f(x) = \csc x - \cot x$ is the expression $\csc^2 x - \csc x \cot x$. Show that this expression can also be written as

$$\dfrac{(1 - \cos x)}{\sin^2 x}.$$

5.2 In Exercises 25–36, verify the identity.

25. $\cos x(\tan^2 x + 1) = \sec x$

26. $\sec^2 x \cot x - \cot x = \tan x$

27. $\sin^3 \theta + \sin \theta \cos^2 \theta = \sin \theta$

28. $\cot^2 x - \cos^2 x = \cot^2 x \cos^2 x$

29. $\sin^5 x \cos^2 x = (\cos^2 x - 2\cos^4 x + \cos^6 x)\sin x$

30. $\cos^3 x \sin^2 x = (\sin^2 x - \sin^4 x)\cos x$

31. $\sqrt{\dfrac{1 - \sin \theta}{1 + \sin \theta}} = \dfrac{1 - \sin \theta}{|\cos \theta|}$

32. $\sqrt{1 - \cos x} = \dfrac{|\sin x|}{\sqrt{1 + \cos x}}$

33. $\dfrac{\csc(-x)}{\sec(-x)} = -\cot x$

34. $\dfrac{1 + \sec(-x)}{\sin(-x) + \tan(-x)} = -\csc x$

35. $\sin^2 x + \sin^2\left(\dfrac{\pi}{2} - x\right) = 1$

36. $\csc x \sin\left(\dfrac{\pi}{2} - x\right) = \cot x$

5.3 In Exercises 37 and 38, verify that the *x*-values are solutions of the equation.

37. $2\cos^2 4x - 1 = 0$

    (a) $x = \dfrac{\pi}{16}$ (b) $x = \dfrac{3\pi}{16}$

38. $2\sin^2 x - \sin x - 1 = 0$

    (a) $x = \dfrac{\pi}{2}$ (b) $x = \dfrac{7\pi}{6}$

In Exercises 39–50, solve the equation for $0 \le x < 2\pi$.

39. $2\sin x - 1 = 0$

40. $\tan x + 1 = 0$

41. $\sin x = \sqrt{3} - \sin x$

42. $4\cos x = 1 + 2\cos x$

43. $3\sqrt{3}\tan x = 3$

44. $\frac{1}{2}\sec x - 1 = 0$

45. $3 \csc^2 x = 4$

46. $4 \tan^2 x - 1 = \tan^2 x$

47. $4 \cos^2 x - 3 = 0$

48. $\sin x(\sin x + 1) = 0$

49. $\sin x - \tan x = 0$

50. $\csc x - 2 \cot x = 0$

In Exercises 51–60, find all solutions of the equation in the interval $[0, 2\pi)$. Use a graphing utility to verify your answers.

51. $2 \cos^2 x - \cos x = 1$

52. $2 \sin^2 x - 3 \sin x = -1$

53. $\cos^2 x + \sin x = 1$

54. $\sin^2 x + 2 \cos x = 2$

55. $2 \sin 2x - \sqrt{2} = 0$

56. $\sqrt{3} \tan 3x = 0$

57. $\cos 4x(\cos x - 1) = 0$

58. $3 \csc^2 5x = -4$

59. $\cos 4x - 7 \cos 2x = 8$

60. $\sin 4x - \sin 2x = 0$

In Exercises 61–64, use the inverse functions where necessary to find all solutions of the equation in the interval $[0, 2\pi)$. Use a graphing utility to verify your answers.

61. $\sin^2 x - 2 \sin x = 0$

62. $2 \cos^2 x + 3 \cos x = 0$

63. $\tan^2 \theta + \tan \theta - 12 = 0$

64. $\sec^2 x + 6 \tan x + 4 = 0$

5.4 **In Exercises 65–68, find the exact value of the sine, cosine, and tangent of the angle by using the sum or difference formula.**

65. $285° = 225° + 60°$

66. $345° = 300° + 45°$

67. $\dfrac{5\pi}{12} = \dfrac{2\pi}{3} - \dfrac{\pi}{4}$

68. $\dfrac{19\pi}{12} = \dfrac{11\pi}{6} - \dfrac{\pi}{4}$

In Exercises 69–72, write the expression as the sine, cosine, or tangent of an angle.

69. $\sin 140° \cos 50° + \cos 140° \sin 50°$

70. $\cos 25° \cos 15° - \sin 25° \sin 15°$

71. $\dfrac{\tan 25° + \tan 10°}{1 - \tan 25° \tan 10°}$

72. $\dfrac{\tan 68° - \tan 115°}{1 + \tan 68° \tan 115°}$

In Exercises 73–78, find the exact value of the trigonometric function given that $\sin u = \frac{3}{4}$, $\cos v = -\frac{5}{13}$, and u and v are in Quadrant II.

73. $\sin(u + v)$

74. $\tan(u + v)$

75. $\cos(u - v)$

76. $\sin(u - v)$

77. $\cos(u + v)$

78. $\tan(u - v)$

In Exercises 79–84, verify the identity.

79. $\cos\left(x + \dfrac{\pi}{2}\right) = -\sin x$

80. $\sin\left(x - \dfrac{3\pi}{2}\right) = \cos x$

81. $\cot\left(\dfrac{\pi}{2} - x\right) = \tan x$

82. $\sin(\pi - x) = \sin x$

83. $\cos 3x = 4 \cos^3 x - 3 \cos x$

84. $\dfrac{\sin(\alpha + \beta)}{\cos \alpha \cos \beta} = \tan \alpha + \tan \beta$

In Exercises 85 and 86, find all solutions of the equation in the interval $[0, 2\pi)$.

85. $\sin\left(x + \dfrac{\pi}{2}\right) - \sin\left(x - \dfrac{\pi}{2}\right) = \sqrt{3}$

86. $\cos\left(x + \dfrac{3\pi}{4}\right) - \cos\left(x - \dfrac{3\pi}{4}\right) = 0$

5.5 **In Exercises 87–92, use double-angle formulas to verify the identity algebraically and use a graphing utility to confirm it graphically.**

87. $6 \sin x \cos x = 3 \sin 2x$

88. $4 \sin x \cos x + 2 = 2 \sin 2x + 2$

89. $1 - 4 \sin^2 x \cos^2 x = \cos^2 2x$

90. $\sin 4x = 8 \cos^3 x \sin x - 4 \cos x \sin x$

91. $\dfrac{\sin 2\alpha}{\cos^2 \alpha - \sin^2 \alpha} = \tan 2\alpha$

92. $\tan^2 x = \dfrac{1 - \cos 2x}{1 + \cos 2x}$

In Exercises 93 and 94, find the exact values of $\sin 2u$, $\cos 2u$, and $\tan 2u$ using the double-angle formulas.

93. $\sin u = -\dfrac{5}{7}, \quad \pi < u < \dfrac{3\pi}{2}$

94. $\cos u = -\dfrac{2}{\sqrt{5}}, \quad \dfrac{\pi}{2} < u < \pi$

95. *Projectile Motion* A baseball leaves the hand of the first baseman at an angle of θ with the horizontal and an initial velocity of $v_0 = 80$ feet per second. The ball is caught by the second baseman 100 feet away. Find θ if the range r of a projectile is

$$r = \dfrac{1}{32} v_0{}^2 \sin 2\theta.$$

In Exercises 96–99, use the power-reducing formulas to rewrite the expression in terms of the first power of the cosine of the angle.

96. $\sin^6 x$

97. $\cos^4 x \sin^4 x$

98. $\cos^4 2x$

99. $\sin^4 2x$

In Exercises 100–103, use the half-angle formulas to determine the exact value of the sine, cosine, and tangent of the angle.

100. $105°$

101. $67° \, 30'$

102. $\dfrac{7\pi}{8}$

103. $\dfrac{11\pi}{12}$

In Exercises 104 and 105, use the half-angle formulas to simplify the expression.

104. $-\sqrt{\dfrac{1 + \cos 10x}{2}}$

105. $\dfrac{\sin 6x}{1 + \cos 6x}$

In Exercises 106 and 107, use the half-angle formulas to verify the identity.

106. $\sqrt{\dfrac{1 - \cos^2 x}{1 + \cos x}} = \sqrt{2}\left|\sin\dfrac{x}{2}\right|$

107. $\dfrac{\sec x - 1}{\tan x} = \tan\dfrac{x}{2}$

108. *Volume*   A trough for feeding cattle is 4 meters long and its cross sections are isosceles triangles with the two equal sides being $\frac{1}{2}$ meter. The angle between the equal sides is θ.

   (a) Express the trough's volume as a function of $\theta/2$.

   (b) Express the volume of the trough as a function of θ and determine the value of θ such that the volume is maximum.

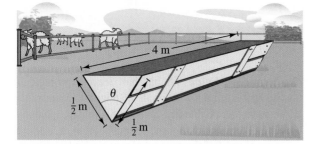

In Exercises 109–112, write the trigonometric expression as a product.

109. $\cos 3\theta + \cos 2\theta$ **110.** $\sin 5\theta - \sin 3\theta$

111. $\cos\dfrac{3\pi}{4} - \cos\dfrac{\pi}{4}$ **112.** $\sin\left(x + \dfrac{\pi}{4}\right) - \sin\left(x - \dfrac{\pi}{4}\right)$

In Exercises 113–116, write the trigonometric expression as a sum or difference.

113. $\sin 3\alpha \sin 2\alpha$

114. $3 \sin 2x \sin 3x$

115. $6 \sin\dfrac{\pi}{4} \cos\dfrac{\pi}{4}$

116. $\cos\dfrac{x}{2} \cos\dfrac{x}{4}$

117. *Harmonic Motion*   A weight is attached to a spring suspended vertically from a ceiling. When a driving force is applied to the system, the weight moves vertically from its equilibrium position, and this motion is described by the model

$$y = 1.5 \sin 8t - 0.5 \cos 8t$$

where y is the distance from equilibrium measured in feet and t is the time in seconds.

   (a) Write the model in the form
$$y = \sqrt{a^2 + b^2}\,\sin(Bt + C).$$

   (b) Use a graphing utility to graph the model.

   (c) Find the amplitude of the weight's oscillations.

   (d) Find the frequency of the weight's oscillations.

Synthesis

True or False?   **In Exercises 118–121, determine whether the statement is true or false. Justify your answer.**

118. If $\dfrac{\pi}{2} < \theta < \pi$, then $\cos\dfrac{\theta}{2} < 0$.

119. $\sin(x + y) = \sin x + \sin y$

120. $4 \sin(-x) \cos(-x) = -2 \sin 2x$

121. $4 \sin 45° \cos 15° = 1 + \sqrt{3}$

122. List the reciprocal identities, quotient identities, and Pythagorean identities from memory.

123. Is $\cos \theta = \sqrt{1 - \sin^2 \theta}$ an identity? Explain.

In Exercises 124 and 125, use the graphs of y_1 and y_2 to determine how to change y_2 to a new function y_3 so that $y_1 = y_3$.

124. $y_1 = \sec^2\left(\dfrac{\pi}{2} - x\right)$
$y_2 = \cot^2 x$

125. $y_1 = \dfrac{\cos 3x}{\cos x}$
$y_2 = (2 \sin x)^2$

Chapter Project *Projectile Motion*

Parametric equations are equations that express two variables in terms of a third variable, called the parameter. In this project, you will use parametric equations to model the path of a projectile. These parametric equations express position x and position y in terms of time t. For any time t, the horizontal position $x(t)$ and vertical position $y(t)$ of a projectile (ignoring air resistance) launched at ground level is given by the parameter equations

$$x(t) = (v_0 \cos \theta)t$$

$$y(t) = (v_0 \sin \theta)t - 16t^2.$$

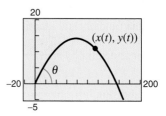

In these equations, θ is the angle with the horizontal and v_0 is the initial velocity in feet per second, as indicated in the figure at the right.

Set your graphing utility to parametric and degree modes, and use the viewing window

$$0 \leq t \leq 5$$

$$-20 \leq x \leq 200$$

$$-5 \leq y \leq 20.$$

Letting $v_0 = 88$ feet per second and $\theta = 20°$, graph the parametric equations

$$x(t) = (88 \cos 20°)t$$

$$y(t) = (88 \sin 20°)t - 16t^2.$$

Use the *zoom* and *trace* features to (a) find the maximum height attained by the projectile, (b) find the time at which the maximum height occurs, (c) determine the length of time that the projectile is in the air, and (d) determine the range of the projectile.

Questions for Further Exploration

1. Verify algebraically the range of the projectile by solving the equation $(88 \sin 20°)t - 16t^2 = 0$ for t and then evaluating $x(t)$ at this t-value.

2. Use a graphing utility to find the maximum height and range of the projectile when $\theta = 30°$ and $v_0 = 132$ feet per second.

3. Let $v_0 = 60$ feet per second. Find the maximum range for the angles $\theta = 20°, 30°, 40°, 50°,$ and $60°$. In general, what angle should you use to produce the maximum range?

4. What is the relationship between the time the projectile reaches its maximum height and the time it takes for the projectile to return to the ground? Explain.

5. Eliminate t from the parametric equations

$$x(t) = (v_0 \cos \theta)t \quad \text{and} \quad y(t) = (v_0 \sin \theta)t - 16t^2$$

by solving for t in the first equation and substituting this value into the equation for y. Show in this case that the height of the projectile is given by the equation

$$y = (\tan \theta)x - \frac{16 \sec^2 \theta}{v_0^2}x^2.$$

Use this equation to find the angle θ corresponding to a maximum range of 200 feet and initial velocity of $v_0 = 80$ feet per second.

5 Chapter Test

Take this test as you would take a test in class. After you are done, check your work against the answers given in the back of the book.

1. If $\tan \theta = \frac{6}{5}$ and $\cos \theta < 0$, use the fundamental identities to evaluate the other five trigonometric functions of θ.

2. Use the fundamental identities to simplify $\csc^2 \beta (1 - \cos^2 \beta)$.

3. Factor and simplify $\dfrac{\sec^4 x - \tan^4 x}{\sec^2 x + \tan^2 x}$.

4. Add and simplify $\dfrac{\cos \theta}{\sin \theta} + \dfrac{\sin \theta}{\cos \theta}$.

5. Determine the values of θ, $0 \le \theta < 2\pi$, for which $\tan \theta = -\sqrt{\sec^2 \theta - 1}$ is true.

6. Use a graphing utility to graph the functions $y_1 = \cos x + \sin x \tan x$ and $y_2 = \sec x$. Make a conjecture about y_1 and y_2. Verify the result algebraically.

In Exercises 7–12, verify the identity.

7. $\sin \theta \sec \theta = \tan \theta$

8. $\sec^2 x \tan^2 x + \sec^2 x = \sec^4 x$

9. $\dfrac{\csc \alpha + \sec \alpha}{\sin \alpha + \cos \alpha} = \cot \alpha + \tan \alpha$

10. $\cos\left(x + \dfrac{\pi}{2}\right) = -\sin x$

11. $\sin(n\pi + \theta) = (-1)^n \sin \theta$, where n is an integer.

12. $(\sin x + \cos x)^2 = 1 + \sin 2x$

13. Find the exact values of the sine, cosine, and tangent of the angle $-\dfrac{7\pi}{12}$.

14. Use the fundamental identities to simplify the expression $\dfrac{\sin^4 x}{\tan^2 x}$.

15. Write $3 \sin 2\theta \sin 6\theta$ as a sum or difference.

16. Write $\cos 5\theta + \cos 3\theta$ as a product.

In Exercises 17–20, find all solutions of the equation in the interval $[0, 2\pi)$.

17. $\tan^2 x + \tan x = 0$

18. $\sin 2\alpha - \cos \alpha = 0$

19. $4 \cos^2 x - 3 = 0$

20. $\csc^2 x - \csc x - 2 = 0$

21. Use a graphing utility to approximate the solutions of the equation $5 \cos x - x = 0$ in the interval $[0, 2\pi)$ accurate to three decimal places.

22. Use the figure at the right to find the exact values of $\sin 2u$ and $\tan 2u$.

23. Explain why the equation $\cos^2 x + \cos x - 6 = 0$ has no solution.

24. Find the exact value of $\tan 105°$.

25. The index of refraction n of a transparent material is the ratio of the speed of light in a vacuum to the speed of light in the material. For the glass triangular prism in the figure at the right, $n = 1.5$ and $\alpha = 60°$. Find the angle θ for the glass prism if

$$n = \frac{\sin(\theta/2 + \alpha/2)}{\sin(\theta/2)}.$$

The *Interactive* CD-ROM and *Internet* versions of this text provide answers to the Chapter Tests and Cumulative Tests. They also offer Chapter Pre-Tests (that test key skills and concepts covered in previous chapters) and Chapter Post-Tests, both of which have randomly generated exercises with diagnostic capabilities.

FIGURE FOR 22

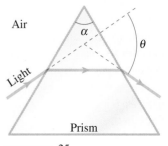

FIGURE FOR 25

Additional Topics in Trigonometry

6

Supertock

The Big Picture

In this chapter you will learn how to

❑ use the Law of Sines and the Law of Cosines to solve oblique triangles.

❑ find areas of oblique triangles.

❑ represent vectors as directed line segments and perform mathematical operations on vectors.

❑ find direction angles of vectors.

❑ find the dot product of two vectors and use properties of the dot product.

❑ multiply and divide complex numbers written in trigonometric form.

❑ find powers and nth roots of complex numbers.

In November of 1999, 81.4% of U.S. airline flights arrived on time. Factors such as severe weather, aircraft maintenance, and air traffic control decisions can cause flight delays. (Source: U.S. Department of Transportation)

Important Vocabulary

As you encounter each new vocabulary term in this chapter, add the term and its definition to your notebook glossary.

- oblique triangle (p. 428)
- Law of Sines (p. 428)
- Law of Cosines (p. 437)
- Heron's Area Formula (p. 440)
- directed line segment (p. 444)
- initial point (p. 444)
- terminal point (p. 444)
- vector **v** in the plane (p. 444)
- standard position (p. 445)
- component form of vector **v** (p. 445)
- zero vector (p. 445)

- magnitude of **v** (p. 445)
- unit vector (p. 445)
- scalar multiplication (p. 446)
- vector addition (p. 446)
- standard unit vectors (p. 449)
- horizontal and vertical components of **v** (p. 449)
- linear combination of vectors (p. 449)
- direction angle (p. 450)
- dot product (p. 458)

- angle between two nonzero vectors (p. 459)
- orthogonal vectors (p. 460)
- vector components (p. 461)
- projection (p. 461)
- trigonometric form of a complex number (p. 468)
- DeMoivre's Theorem (p. 471)
- nth root of a complex number (p. 472)
- nth roots of unity (p. 474)

Additional Resources   Text-specific additional resources are available to help you do well in this course. See page xvi for details.

6.1 Law of Sines

Introduction

In Chapter 4 you looked at techniques for solving right triangles. In this section and the next, you will solve **oblique triangles**—triangles that have no right angles. As standard notation, the angles of a triangle are labeled A, B, and C, and their opposite sides are labeled a, b, and c, as shown in Figure 6.1.

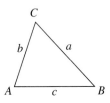

Figure 6.1

To solve an oblique triangle, you need to know the measure of at least one side and any two other parts of the triangle—two sides, two angles, or one angle and one side. This breaks down into the following four cases.

1. Two angles and any side (AAS or ASA)
2. Two sides and an angle opposite one of them (SSA)
3. Three sides (SSS)
4. Two sides and their included angle (SAS)

The first two cases can be solved using the **Law of Sines,** whereas the last two cases require the Law of Cosines (Section 6.2). A proof of the Law of Sines is given in Appendix A.

What You Should Learn:
- How to use the Law of Sines to solve oblique triangles (AAS, ASA, or SSA)
- How to find areas of oblique triangles
- How to use the Law of Sines to model and solve real-life problems

Why You Should Learn It:
You can use the Law of Sines to solve real-life problems involving oblique triangles. For instance, Exercise 35 on page 435 shows how the Law of Sines can be used to help determine the distance from a ranger station to a forest fire.

Law of Sines

If ABC is a triangle with sides a, b, and c, then

$$\frac{a}{\sin A} = \frac{b}{\sin B} = \frac{c}{\sin C}.$$

Oblique Triangles

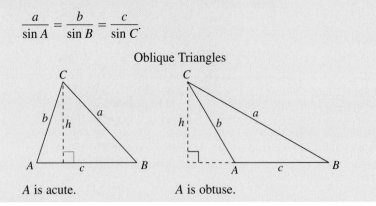

A is acute. A is obtuse.

The Law of Sines can also be written in the reciprocal form

$$\frac{\sin A}{a} = \frac{\sin B}{b} = \frac{\sin C}{c}.$$

EXAMPLE 1 Given Two Angles and One Side—AAS

For the triangle in Figure 6.2, $C = 102.3°$, $B = 28.7°$, and $b = 27.4$ feet. Find the remaining angle and sides.

Solution

The third angle of the triangle is

$$A = 180° - B - C$$

$$= 180° - 28.7° - 102.3°$$

$$= 49.0°.$$

By the Law of Sines, you have

$$\frac{a}{\sin 49°} = \frac{b}{\sin 28.7°} = \frac{c}{\sin 102.3°}. \qquad \frac{a}{\sin A} = \frac{b}{\sin B} = \frac{c}{\sin C}$$

Using $b = 27.4$ produces

$$a = \frac{27.4}{\sin 28.7°}(\sin 49°) \approx 43.06 \text{ feet}$$

and

$$c = \frac{27.4}{\sin 28.7°}(\sin 102.3°) \approx 55.75 \text{ feet}.$$

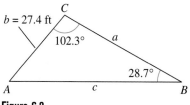

Figure 6.2

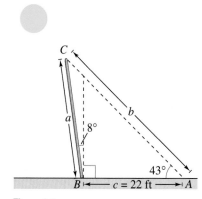

EXAMPLE 2 Given Two Angles and One Side—ASA

A pole tilts *toward* the sun at an 8° angle from the vertical, and it casts a 22-foot shadow. The angle of elevation from the tip of the shadow to the top of the pole is 43°. How tall is the pole?

Solution

From Figure 6.3, note that $A = 43°$ and $B = 90° + 8° = 98°$. So, the third angle is

$$C = 180° - A - B$$

$$= 180° - 43° - 98°$$

$$= 39°.$$

By the Law of Sines, you have

$$\frac{a}{\sin 43°} = \frac{c}{\sin 39°}. \qquad \frac{a}{\sin A} = \frac{c}{\sin C}$$

Because $c = 22$ feet, the length of the pole is

$$a = \frac{22}{\sin 39°}(\sin 43°) \approx 23.84 \text{ feet}.$$

For practice, try reworking Example 2 for a pole that tilts *away from* the sun under the same conditions.

> **STUDY T!P**
>
> When you are solving triangles, a careful sketch is useful as a quick test for the feasibility of an answer. Remember that the longest side lies opposite the largest angle, and the shortest side lies opposite the smallest angle.

Figure 6.3

The Ambiguous Case (SSA)

In Examples 1 and 2 you saw that two angles and one side determine a unique triangle. However, if two sides and one opposite angle are given, three possible situations can occur: (1) no such triangle exists, (2) one such triangle exists, or (3) two distinct triangles may satisfy the conditions.

A computer simulation to accompany this concept appears in the *Interactive* CD-ROM and *Internet* versions of this text.

The Ambiguous Case (SSA)

Consider a triangle in which you are given a, b, and A ($h = b \sin A$).

	A is acute.	A is acute.	A is acute.	A is acute.	A is obtuse.	A is obtuse.
Sketch						
Necessary condition	$a < h$	$a = h$	$a > b$	$h < a < b$	$a \le b$	$a > b$
Possible triangles	None	One	One	Two	None	One

EXAMPLE 3   Single-Solution Case—SSA

For the triangle in Figure 6.4, $a = 22$ inches, $b = 12$ inches, and $A = 42°$. Find the remaining side and angles.

Solution

By the Law of Sines, you have

$$\frac{22}{\sin 42°} = \frac{12}{\sin B} \qquad \frac{a}{\sin A} = \frac{b}{\sin B}$$

$$\sin B = 12\left(\frac{\sin 42°}{22}\right) \approx 0.3649803$$

$$B \approx 21.41°. \qquad \text{\textit{B} is acute.}$$

Now you can determine that

$$C \approx 180° - 42° - 21.41°$$

$$= 116.59°.$$

Then the remaining side is given by

$$\frac{c}{\sin 116.59°} = \frac{22}{\sin 42°} \qquad \frac{c}{\sin C} = \frac{a}{\sin A}$$

$$c = \sin 116.59°\left(\frac{22}{\sin 42°}\right) \approx 29.40 \text{ inches.}$$

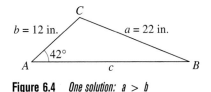

Figure 6.4 *One solution: $a > b$*

Notice in Example 3 that A is acute and $a > b$, which results in one possible triangle.

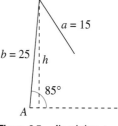

Figure 6.5 *No solution: a < h*

EXAMPLE 4   No-Solution Case—SSA

Show that there is no triangle for which $a = 15$, $b = 25$, and $A = 85°$.

Solution

Begin by making the sketch shown in Figure 6.5. From this figure it appears that no triangle is formed. You can verify this by using the Law of Sines.

$$\frac{a}{\sin A} = \frac{b}{\sin B}$$

$$\frac{15}{\sin 85°} = \frac{25}{\sin B}$$

$$\sin B = 25\left(\frac{\sin 85°}{15}\right) \approx 1.660 > 1$$

This contradicts the fact that $|\sin B| \leq 1$. So, no triangle can be formed having sides $a = 15$ and $b = 25$ and an angle of $A = 85°$.

The *Interactive* CD-ROM and *Internet* versions of this text show every example with its solution; clicking on the *Try It!* button brings up similar problems. Guided Examples and Integrated Examples show step-by-step solutions to additional examples. Integrated Examples are related to several concepts in the section.

EXAMPLE 5   Two-Solution Case—SSA

Find two triangles for which $a = 12$ meters, $b = 31$ meters, and $A = 20.5°$.

Solution

Because $h = b \sin A = 31(\sin 20.5°) \approx 10.86$ meters, you can conclude that there are two possible triangles (because $h < a < b$). By the Law of Sines, you have

$$\frac{a}{\sin A} = \frac{b}{\sin B}$$

$$\sin B = b\left(\frac{\sin A}{a}\right) = 31\left(\frac{\sin 20.5°}{12}\right) \approx 0.9047.$$

There are two angles between $0°$ and $180°$ whose sine is 0.9047: $B_1 \approx 64.8°$ and $B_2 \approx 180° - 64.8° = 115.2°$. For $B_1 \approx 64.8°$, you obtain

$$C \approx 180° - 20.5° - 64.8° = 94.7°$$

$$c = \frac{a}{\sin A}(\sin C) = \frac{12}{\sin 20.5°}(\sin 94.7°) \approx 34.15 \text{ meters.}$$

For $B_2 \approx 115.2°$, you obtain

$$C \approx 180° - 20.5° - 115.2° = 44.3°$$

$$c = \frac{a}{\sin A}(\sin C) = \frac{12}{\sin 20.5°}(\sin 44.3°) \approx 23.93 \text{ meters.}$$

The resulting triangles are shown in Figure 6.6.

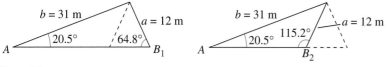

Figure 6.6

Area of an Oblique Triangle

The procedure used to prove the Law of Sines leads to a simple formula for the area of an oblique triangle. Referring to Figure 6.7, note that each triangle has a height of

$$h = b \sin A.$$

Consequently, the area of each triangle is

$$\text{Area} = \frac{1}{2}(\text{base})(\text{height}) = \frac{1}{2}(c)(b \sin A) = \frac{1}{2}bc \sin A.$$

By similar arguments, you can develop the formulas

$$\text{Area} = \frac{1}{2}ab \sin C = \frac{1}{2}ac \sin B.$$

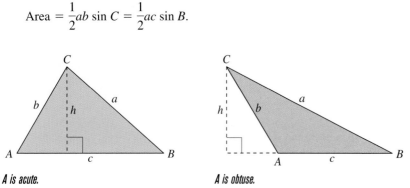

A is acute.   *A is obtuse.*

Figure 6.7

Area of an Oblique Triangle

The area of any triangle is one-half the product of the lengths of two sides times the sine of their included angle. That is,

$$\text{Area} = \frac{1}{2}bc \sin A = \frac{1}{2}ab \sin C = \frac{1}{2}ac \sin B.$$

Note that if angle A is $90°$, the formula gives the area for a right triangle as

$$\text{Area} = \frac{1}{2}bc = \frac{1}{2}(\text{base})(\text{height}).$$

Similar results are obtained for angles C and B equal to $90°$.

EXAMPLE 6 Finding the Area of an Oblique Triangle

Find the area of a triangular lot having two sides of lengths 90 meters and 52 meters and an included angle of $102°$.

Solution

Consider $a = 90$ m, $b = 52$ m, and angle $C = 102°$, as shown in Figure 6.8. Then the area of the triangle is

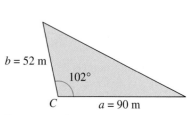

Figure 6.8

$$\text{Area} = \frac{1}{2}ab \sin C = \frac{1}{2}(90)(52)(\sin 102°) \approx 2289 \text{ square meters.}$$

Application

EXAMPLE 7   An Application of the Law of Sines

The course for a boat race starts at point A and proceeds in the direction S 52° W to point B, then in the direction S 40° E to point C, and finally back to A, as shown in Figure 6.9. The point C lies 8 kilometers directly south of point A. Approximate the total distance of the race course.

Solution

Because lines BD and AC are parallel, it follows that $\angle BCA \cong \angle DBC$. Consequently, triangle ABC has the measures shown in Figure 6.10. For angle B, you have $B = 180° - 52° - 40° = 88°$. Using the Law of Sines

$$\frac{a}{\sin 52°} = \frac{b}{\sin 88°} = \frac{c}{\sin 40°}$$

you can let $b = 8$ and obtain

$$a = \frac{8}{\sin 88°}(\sin 52°) \approx 6.308$$

and

$$c = \frac{8}{\sin 88°}(\sin 40°) \approx 5.145.$$

The total length of the course is approximately

$$\text{Length} \approx 8 + 6.308 + 5.145 = 19.453 \text{ kilometers.}$$

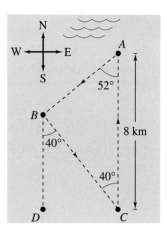

Figure 6.9

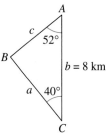

Figure 6.10

W r i t i n g   A b o u t   M a t h   *Using the Law of Sines*

In this section, you have been using the Law of Sines to solve *oblique* triangles. Can the Law of Sines also be used to solve a right triangle? If so, write a short paragraph explaining how to use the Law of Sines to solve the following two triangles. Is there an easier way to solve these triangles?

a. (ASA)   **b.** (SSA)

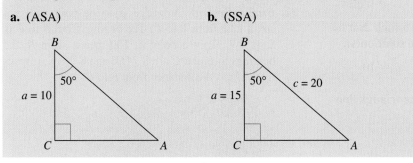

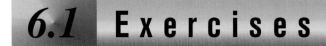

6.1 Exercises

In Exercises 1–14, use the given information to solve the triangle.

1. $A = 30°$, $a = 12$, $B = 45°$
2. $C = 120°$, $c = 15$, $B = 45°$
3. $A = 10°$, $a = 4.5$, $B = 60°$
4. $C = 135°$, $c = 45$, $B = 10°$
5. $A = 36°$, $a = 10$, $b = 4$
6. $A = 60°$, $a = 9$, $c = 10$
7. $A = 150°$, $C = 20°$, $a = 200$
8. $A = 24.3°$, $C = 54.6°$, $c = 2.68$
9. $A = 83° 20'$, $C = 54.6°$, $c = 18.1$
10. $A = 5° 40'$, $B = 8° 15'$, $b = 4.8$
11. $B = 15° 30'$, $a = 4.5$, $b = 6.8$
12. $C = 85° 20'$, $a = 35$, $c = 50$
13. $A = 110° 15'$, $a = 48$, $b = 16$
14. $B = 2° 45'$, $b = 6.2$, $c = 5.8$

In Exercises 15–20, use the given information to solve the triangle. If two solutions exist, find both.

15. $A = 58°$, $a = 4.5$, $b = 12.8$
16. $A = 94°$, $a = 14.6$, $b = 14.6$
17. $A = 58°$, $a = 11.4$, $b = 12.8$
18. $A = 58°$, $a = 4.5$, $b = 5$
19. $A = 110°$, $a = 125$, $b = 100$
20. $A = 140°$, $a = 48$, $b = 46$

In Exercises 21 and 22, find a value for b such that the triangle has (a) one, (b) two, and (c) no solution(s).

21. $A = 36°$, $a = 5$ 22. $A = 60°$, $a = 10$

In Exercises 23–28, find the area of the triangle having the indicated sides and angle.

23. $C = 110°$, $a = 6$, $b = 10$
24. $B = 74° 30'$, $a = 103$, $c = 58$
25. $A = 38° 45'$, $b = 67$, $c = 85$
26. $A = 5° 15'$, $b = 4.5$, $c = 22$
27. $B = 130°$, $a = 92$, $c = 30$
28. $C = 84° 30'$, $a = 16$, $b = 20$

29. **Height**   A flagpole is located on a slope that makes an angle of 14° with the horizontal. The pole casts a 16-meter shadow up the slope when the angle of elevation of the sun is 20°.

    (a) Draw a triangle that represents the problem.

    (b) Write an equation involving the unknown quantity.

    (c) Find the height of the flagpole.

30. **Height**   You are standing 40 meters from the base of a tree that is leaning 8° from vertical away from you. The angle of elevation from your feet to the top of the tree is 20° 50′.

    (a) Draw a triangle that represents the problem.

    (b) Write an equation involving the unknown slant height of the tree.

    (c) Find the slant height of the tree.

31. **Flight Path**   A plane flies 500 kilometers with a bearing of N 44° W from B to C. The plane then flies southwest 840 kilometers from C to A. Find the bearing of the flight from C to A.

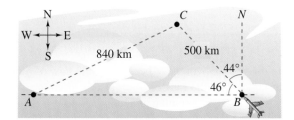

32. **Bridge Design**   A bridge is to be built across a small lake from B to C. The bearing from B to C is S 41° W. From a point A, 100 meters from B, the bearings to B and C are S 74° E and S 28° E, respectively. Find the distance from B to C.

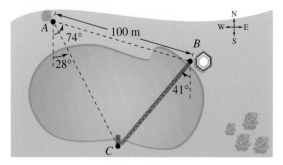

33. *Railroad Track Design* The circular arc of a railroad curve has a chord of length 3000 feet and a central angle of 40°.

(a) Draw a diagram to represent the problem. Show the known quantities on the diagram and use variables r and s to represent the radius of the arc and the length of the arc, respectively.

(b) Find the radius r of the circular arc.

(c) Find the length s of the circular arc.

34. *Glide Path* A pilot has just started on the glide path for landing at an airport where the length of the runway is 9000 feet. The angles of depression from the plane to the ends of the runway are 17.5° and 18.8°.

(a) Draw a diagram to represent the problem.

(b) Find the air distance the plane must travel until touching down on the near end of the runway.

(c) Find the ground distance the plane must travel until touching down.

(d) Find the altitude of the plane when the pilot begins the descent.

35. *Locating a Fire* Two fire towers A and B are 30 kilometers apart. The bearing from A to B is N 65° E. A fire is spotted by a ranger in each tower, and its bearings from A and B are N 28° E and N 16.5° W, respectively. Find the distance of the fire from each tower.

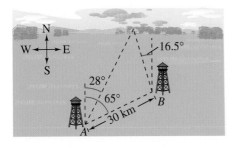

36. *Distance* A boat is sailing due east parallel to the shoreline at a speed of 10 miles per hour. At a given time the bearing to the lighthouse is S 70° E, and 15 minutes later the bearing is S 63° E. Find the distance from the boat to the shoreline if the lighthouse is at the shoreline.

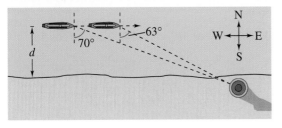

37. *Distance* A family is traveling due west on a road that passes a famous landmark. At a given time the bearing to the landmark is N 62° W, and after the family travels 5 miles farther the bearing is N 38° W. What is the closest the family will come to the landmark while on the road?

38. *Altitude* The angles of elevation to an airplane from two points A and B on level ground are 55° and 72°, respectively. The points A and B are 2.2 miles apart, and the airplane is east of both points in the same vertical plane. Find the altitude of the plane.

39. *Engine Design* The connecting rod in an engine is 6 inches in length and the radius of the crankshaft is $1\frac{1}{2}$ inches. Let d be the distance the piston is from the top of its stroke for the angle θ.

(a) Use a graphing utility to complete the table.

θ	0°	45°	90°	135°	180°
d					

(b) The spark plug fires at $\theta = 5°$ before top dead center. How far is the piston from the top of its stroke at this time?

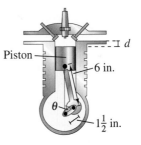

40. *Angle of Elevation* A 10-meter telephone pole casts a 17-meter shadow directly down a slope when the angle of elevation of the sun is 42°. Find θ, the angle of elevation of the ground.

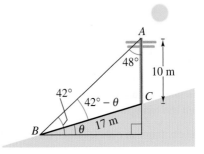

41. *Distance* The angles of elevation θ and ϕ to an airplane are being continuously monitored at two observation points A and B, respectively, which are 2 miles apart. Write an equation giving the distance d between the plane and point B in terms of θ and ϕ.

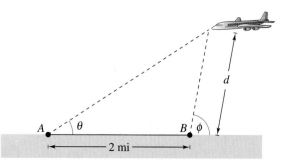

42. *Graphical and Numerical Analysis* In the figure, α and β are positive angles.

(a) Write α as a function of β.

(b) Use a graphing utility to graph the function. Determine its domain and range.

(c) Write c as a function of β.

(d) Use a graphing utility to graph the function in part (c). Determine its domain and range.

(e) Use a graphing utility to complete the table. What can you conclude?

β	0.4	0.8	1.2	1.6	2.0	2.4	2.8
α							
c							

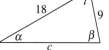

Synthesis

True or False? **In Exercises 43–45, determine whether the statement is true or false. Justify your answer.**

43. If any three sides or angles of an oblique triangle are known, then the triangle can be solved.

44. If a triangle contains an obtuse angle, then it must be oblique.

45. Two sides and an opposite angle determine a unique triangle.

46. *Graphical Analysis*

(a) Write the area A of the shaded region in the figure as a function of θ.

(b) Use a graphing utility to graph the area function.

(c) Determine the domain of the area function. Explain how the area of the region and the domain of the function would change if the 8-centimeter line segment were decreased in length.

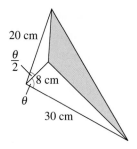

Review

In Exercises 47–50, use the given values to evaluate (if possible) the other four trigonometric functions.

47. $\cos \theta = \frac{5}{13}$, $\sin \theta = -\frac{12}{13}$

48. $\tan \theta = \frac{2}{9}$, $\csc \theta = -\frac{\sqrt{85}}{2}$

49. $\sec \theta = -\frac{\sqrt{122}}{11}$, $\cot \theta = -11$

50. $\cot \theta$ is undefined, $\cos \theta < 0$

In Exercises 51 and 52, verify the identity.

51. $\sec^2 x(\csc^2 x - 1) = \csc^2 x$

52. $\csc \theta \tan \theta - \dfrac{1 - \sin^2 \theta}{\cos \theta} = \dfrac{1 - \cos^2 \theta}{\cos \theta}$

In Exercises 53–56, write the product as a sum or difference.

53. $6 \sin 8\theta \cos 3\theta$

54. $2 \cos 2\theta \cos 5\theta$

55. $3 \cos \dfrac{\pi}{6} \sin \dfrac{5\pi}{3}$

56. $\dfrac{5}{2} \sin \dfrac{3\pi}{4} \sin \dfrac{5\pi}{6}$

6.2 Law of Cosines

Introduction

Two cases remain in the list of conditions needed to solve an oblique triangle—SSS and SAS. To use the Law of Sines, you must know at least one side and its opposite angle. If you are given three sides (SSS), or two sides and their included angle (SAS), none of the ratios in the Law of Sines would be complete. In such cases you can use the **Law of Cosines.** A proof of the Law of Cosines is given in Appendix A.

Law of Cosines

Standard Form	Alternative Form
$a^2 = b^2 + c^2 - 2bc \cos A$	$\cos A = \dfrac{b^2 + c^2 - a^2}{2bc}$
$b^2 = a^2 + c^2 - 2ac \cos B$	$\cos B = \dfrac{a^2 + c^2 - b^2}{2ac}$
$c^2 = a^2 + b^2 - 2ab \cos C$	$\cos C = \dfrac{a^2 + b^2 - c^2}{2ab}$

What You Should Learn:

- How to use the Law of Cosines to solve oblique triangles (SSS or SAS)
- How to use the Law of Cosines to model and solve real-life problems
- How to use Heron's Area Formula to find areas of triangles

Why You Should Learn It:

You can use the Law of Cosines to solve real-life problems involving oblique triangles. For instance, Exercise 32 on page 442 shows you how the Law of Cosines can be used to approximate the number of feet a baseball player must run to catch a ball.

EXAMPLE 1   Three Sides of a Triangle—SSS

Find the three angles of the triangle in Figure 6.11.

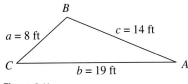

Figure 6.11

Solution

It is a good idea first to find the angle opposite the longest side—side b in this case. Using the Law of Cosines, you find that

$$\cos B = \frac{a^2 + c^2 - b^2}{2ac} = \frac{8^2 + 14^2 - 19^2}{2(8)(14)} \approx -0.45089.$$

Because $\cos B$ is negative, you know that B is an *obtuse* angle given by $B \approx 116.80°$. At this point knowing that $B \approx 116.80°$, it is simpler to use the Law of Sines to determine A.

$$\sin A = a\left(\frac{\sin B}{b}\right) \approx 8\left(\frac{\sin 116.80°}{19}\right) \approx 0.37582$$

Because B is obtuse, you know that A must be acute, because a triangle can have at most one obtuse angle. So, $A \approx 22.08°$ and

$$C \approx 180° - 22.08° - 116.80° = 41.12°.$$

Do you see why it was wise to find the largest angle *first* in Example 1? Knowing the cosine of an angle, you can determine whether the angle is acute or obtuse. That is,

$\cos\theta > 0$ for $0° < \theta < 90°$ Acute

$\cos\theta < 0$ for $90° < \theta < 180°$.   Obtuse

So, in Example 1, once you found that angle *B* was obtuse, you knew that angles *A* and *C* were both acute. If the largest angle is acute, the remaining two angles will be acute also.

Exploration

What familiar formula do you obtain when you use the third form of the Law of Cosines

$$c^2 = a^2 + b^2 - 2ab \cos C$$

and you let $C = 90°$? What is the relationship between the Law of Cosines and this formula?

EXAMPLE 2   Two Sides and the Included Angle—SAS

Find the remaining two angles and the side of the triangle in Figure 6.12.

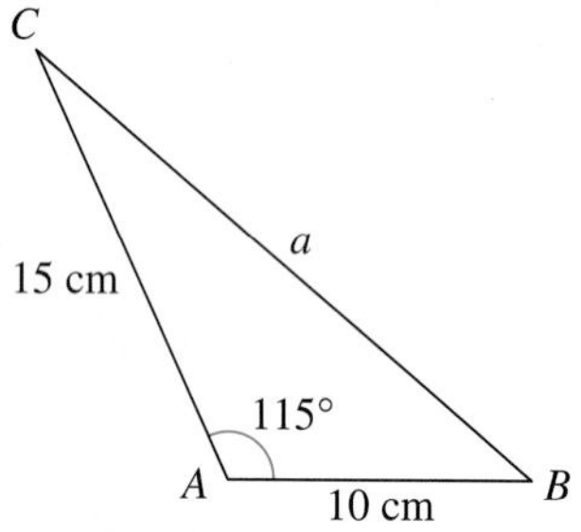

Figure 6.12

Solution

Use the Law of Cosines to find the unknown side *a* in the figure.

$a^2 = b^2 + c^2 - 2bc \cos A$

$a^2 = 15^2 + 10^2 - 2(15)(10) \cos 115°$

$a^2 \approx 451.79$

$a \approx 21.26$

Because $a \approx 21.26$ cm, you now know the ratio $a/\sin A$ and you can use the Law of Sines

$$\frac{a}{\sin A} = \frac{b}{\sin B}$$

to solve for *B*.

$$\sin B = b\left(\frac{\sin A}{a}\right)$$

$$= 15\left(\frac{\sin 115°}{21.26}\right)$$

$$\approx 0.63945$$

So, $B = \arcsin 0.63945 \approx 39.75°$, and $C \approx 180 - 115° - 39.75° = 25.25°$.

EXAMPLE 3   Two Sides and the Included Angle—SAS

The pitcher's mound on a softball field is 46 feet from home plate and the distance between the bases is 60 feet, as shown in Figure 6.13. (The pitcher's mound is *not* halfway between home plate and second base.) How far is the pitcher's mound from first base?

Solution

In triangle *HPF*, $H = 45°$ (line *HP* bisects the right angle at *H*), $f = 46$, and $p = 60$. Using the Law of Cosines for this SAS case, you have

$$h^2 = f^2 + p^2 - 2fp \cos H$$

$$= 46^2 + 60^2 - 2(46)(60) \cos 45° \approx 1812.8.$$

Therefore, the approximate distance from the pitcher's mound to first base is

$$h \approx \sqrt{1812.8} \approx 42.58 \text{ feet.}$$

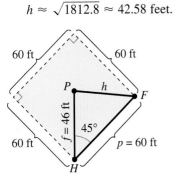

Figure 6.13

EXAMPLE 4   Two Sides and the Included Angle—SAS

A ship travels 60 miles due east, then adjusts its course 15° northward, as shown in Figure 6.14. After traveling 80 miles in the new direction, how far is the ship from its point of departure?

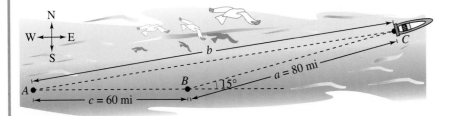

Figure 6.14

Solution

You have $c = 60$, $B = 180° - 15° = 165°$, and $a = 80$. Consequently, by the Law of Cosines, you have

$$b^2 = a^2 + c^2 - 2ac \cos B$$

$$= 80^2 + 60^2 - 2(80)(60) \cos 165° \approx 19,273.$$

Therefore, the distance *b* is $b \approx \sqrt{19,273} \approx 138.8$ miles.

Heron's Area Formula

The Law of Cosines can be used to establish the following formula for the area of a triangle. This formula is called **Heron's Area Formula** after the Greek mathematician Heron (ca. 100 B.C.). For a proof of Heron's Area Formula, see Appendix A.

Heron of Alexandria (ca. 100 B.C.) was a Greek geometer and inventor. His works include the means of finding the areas of triangles, quadrilaterals, regular polygons having 3 to 12 sides, and circles, as well as finding the surface area and volume of three-dimensional objects.

Heron's Area Formula

Given any triangle with sides of lengths a, b, and c, the area of the triangle is

$$\text{Area} = \sqrt{s(s-a)(s-b)(s-c)}$$

where $s = (a + b + c)/2$.

EXAMPLE 5   Using Heron's Area Formula

Find the area of the triangular region having sides of lengths $a = 43$ meters, $b = 53$ meters, and $c = 72$ meters.

Solution

Because $s = (a + b + c)/2 = 168/2 = 84$, Heron's Area Formula yields

$$\text{Area} = \sqrt{s(s-a)(s-b)(s-c)}$$
$$= \sqrt{84(41)(31)(12)} \approx 1131.89 \text{ square meters.}$$

Writing About Math   *The Area of a Triangle*

You have now studied three different formulas for the area of a triangle.

Standard Formula: $\text{Area} = \frac{1}{2}bh$

Oblique Triangle: $\text{Area} = \frac{1}{2}bc \sin A = \frac{1}{2}ab \sin C = \frac{1}{2}ac \sin B$

Heron's Area Formula: $\text{Area} = \sqrt{s(s-a)(s-b)(s-c)}$

Use the most appropriate formula to find the area of each triangle. Show your work and write a short paragraph stating your reasons for choosing each formula.

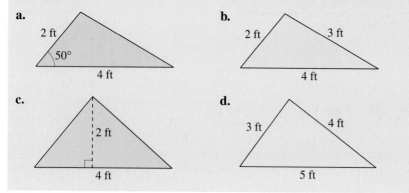

a.   2 ft   50°   4 ft

b.   2 ft   3 ft   4 ft

c.   2 ft   4 ft

d.   3 ft   4 ft   5 ft

6.2 E x e r c i s e s

In Exercises 1–10, use the Law of Cosines to solve the triangle.

1. $a = 6$, $b = 8$, $c = 12$
2. $a = 9$, $b = 3$, $c = 11$
3. $A = 50°$, $b = 15$, $c = 30$
4. $C = 108°$, $a = 10$, $b = 6.5$
5. $a = 9$, $b = 12$, $c = 15$
6. $a = 45$, $b = 30$, $c = 72$
7. $a = 75.4$, $b = 48$, $c = 48$
8. $a = 1.42$, $b = 0.75$, $c = 1.25$
9. $B = 8° \, 15'$, $a = 26$, $c = 18$
10. $B = 75° \, 20'$, $a = 6.2$, $c = 9.5$

In Exercises 11–16, solve the parallelogram. (The lengths of the diagonals are given by c and d.)

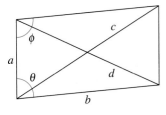

	a	b	c	d	θ	ϕ
11.	4	8			30°	
12.	25	36				110°
13.	10	14	20			
14.	40	60		80		
15.	10		18	12		
16.		25	50	35		

In Exercises 17–22, use Heron's Area Formula to find the area of the triangle.

17. $a = 5$, $b = 9$, $c = 10$
18. $a = 14$, $b = 17$, $c = 7$
19. $a = 3.5$, $b = 10.2$, $c = 9$
20. $a = 75.4$, $b = 52$, $c = 52$
21. $a = 20$, $b = 20$, $c = 10$
22. $a = 4.45$, $b = 1.85$, $c = 3.00$

23. **Navigation**   A plane flies 810 miles from A to B with a bearing of N 75° E. Then it flies 648 miles from B to C with a bearing of N 32° E. Draw a diagram to represent the problem and find the straight-line distance and bearing from C to A.

24. **Navigation**   A boat race runs along a triangular course marked by buoys A, B, and C. The race starts with the boats headed west for 2500 meters. The other two sides of the course lie to the north of the first side, and their lengths are 1100 meters and 2000 meters. Draw a diagram to represent the problem and find the bearings for the last two legs of the race.

25. **Surveying**   To approximate the length of a marsh, a surveyor walks 380 meters from point A to point B, then turns 80° and walks 240 meters to point C. Approximate the length AC of the marsh.

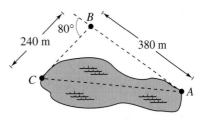

26. **Streetlight Design**   Determine the angle θ in the design of the streetlight shown in the figure.

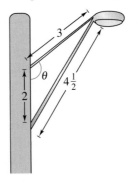

27. **Distance**   Two ships leave a port at 9 A.M. One travels at a bearing of N 53° W at 12 miles per hour and the other travels at a bearing of S 67° W at 16 miles per hour. Approximate how far apart they are at noon that day.

28. Distance A 100-foot vertical tower is to be erected on the side of a hill that makes a 6° angle with the horizontal. Find the length of each of the two guy wires that will be anchored 75 feet uphill and downhill from the base of the tower.

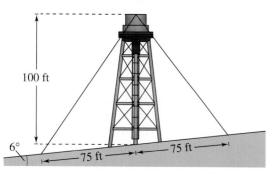

100 ft

6°

75 ft   75 ft

29. Navigation On a map, Orlando is 178 millimeters due south of Niagara Falls, Denver is 273 millimeters from Orlando, and Denver is 235 millimeters from Niagara Falls.

(a) Find the bearing of Denver from Orlando.

(b) Find the bearing of Denver from Niagara Falls.

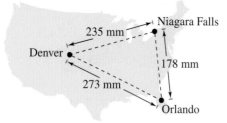

Niagara Falls

235 mm

Denver

178 mm

273 mm

Orlando

30. Navigation On a map, Minneapolis is 165 millimeters due west of Albany, Phoenix is 216 millimeters from Minneapolis, and Phoenix is 368 millimeters from Albany.

(a) Find the bearing of Minneapolis from Phoenix.

(b) Find the bearing of Albany from Phoenix.

31. Baseball On a baseball diamond with 90-foot sides, the pitcher's mound is 60.5 feet from home plate. How far is it from the pitcher's mound to third base?

32. Baseball The baseball player in center field is playing approximately 330 feet from the television camera that is behind home plate. A batter hits a fly ball that goes to the wall 420 feet from the camera. Approximate the number of feet the center fielder has to run to make the catch if the camera turns 6° to follow the play.

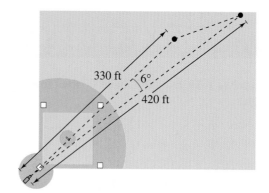

330 ft   6°

420 ft

FIGURE FOR 32

33. Engineering If Q is the midpoint of the line segment $\overline{PR}$ in the truss rafter shown in the figure, find the lengths of the line segments $\overline{PQ}$, $\overline{QS}$, and $\overline{RS}$.

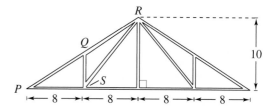

R

Q

10

P   S

8   8   8   8

34. Landau Building The Landau Building in Cambridge, Massachusetts, is a building with a triangular-shaped base. The lengths of the sides of the triangular base are 145 feet, 257 feet, and 290 feet. Find the area of the base of the building.

35. Engine Design An engine has a 7-inch connecting rod fastened to a crank.

(a) Use the Law of Cosines to write an equation giving the relationship between x and θ.

(b) Write x as a function of θ. (Select the sign that yields positive values of x.)

(c) Use a graphing utility to graph the function in part (b).

(d) Use the graph in part (c) to determine the total distance the piston moves in one cycle.

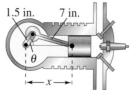

1.5 in.   7 in.

θ

x

36. Paper Manufacturing In a certain process with continuous paper, the paper passes across three rollers of radii 3 inches, 4 inches, and 6 inches. The centers of the 3-inch and 6-inch rollers are d inches apart, and the length of the arc in contact with the paper on the 4-inch roller is s inches.

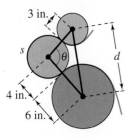

(a) Use the Law of Cosines to write an equation giving the relationship between d and θ.

(b) Write θ as a function of d.

(c) Write s as a function of θ.

(d) Complete the following table.

d (inches)	9	10	12	13	14	15	16
θ (degrees)							
s (inches)							

37. Awning Design A retractable awning lowers at an angle of 50° from the top of a patio door that is 7 feet high. Find the length x of the awning if no direct sunlight is to enter the door when the angle of elevation of the sun is greater than 70°.

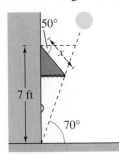

38. Awning Design Suppose the retractable awning in Exercise 37 lowers at an angle of 40° from the top of the patio door and no direct sunlight is to enter the door when the angle of elevation of the sun is greater than 65°. What is the length of the awning?

Synthesis

True or False? **In Exercises 39 and 40, determine whether the statement is true or false. Justify your answer.**

39. A triangle with side lengths of 10 cm, 16 cm, and 5 cm can be solved using the Law of Cosines.

40. Two sides and their included angle determine a unique triangle.

41. Circumscribed and Inscribed Circles Let R and r be the radii of the circumscribed and inscribed circles of a triangle ABC, respectively, and let $s = (a + b + c)/2$. Prove the following.

(a) $2R = \dfrac{a}{\sin A} = \dfrac{b}{\sin B} = \dfrac{c}{\sin C}$

(b) $r = \sqrt{\dfrac{(s - a)(s - b)(s - c)}{s}}$

Circumscribed and Inscribed Circles **In Exercises 42 and 43, use the results of Exercise 41.**

42. Given a triangle with $a = 25$, $b = 55$, and $c = 72$, find the area of (a) the triangle, (b) the circumscribed circle, and (c) the inscribed circle.

43. Find the length of the largest circular track that can be built on a triangular piece of property whose sides are 200 feet, 250 feet, and 325 feet.

44. Use the Law of Cosines to prove that

$$\frac{1}{2} bc (1 + \cos A) = \left(\frac{a + b + c}{2}\right)\left(\frac{-a + b + c}{2}\right).$$

Review

In Exercises 45–48, solve the equation for $0 \le x < 2\pi$.

45. $3 \sec x + 4 = 10$

46. $2 \csc^2 x - 3 = \csc^2 x - 1$

47. $\dfrac{3}{\cos x \sin x} = 2\sqrt{3} \csc x$

48. $\cos x \cot x - \cos x = 0$

In Exercises 49 and 50, write the sum or difference as a product.

49. $\cos \dfrac{5\pi}{6} - \cos \dfrac{\pi}{3}$

50. $\sin\left(x - \dfrac{\pi}{2}\right) - \sin\left(x + \dfrac{\pi}{2}\right)$

6.3 Vectors in the Plane

Introduction

Many quantities in geometry and physics, such as area, time, and temperature, can be represented by a single real number. Other quantities, such as force and velocity, involve both *magnitude* and *direction* and cannot be completely characterized by a single real number. To represent such a quantity, you can use a **directed line segment,** as shown in Figure 6.15. The directed line segment $\overrightarrow{PQ}$ has **initial point** P and **terminal point** Q. Its **magnitude, or length,** is denoted by $\|PQ\|$.

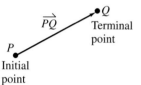

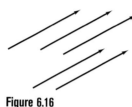

Figure 6.15

Figure 6.16

Two directed line segments that have the same magnitude and direction are equivalent. For example, the directed line segments in Figure 6.16 are all equivalent. The set of all directed line segments that are equivalent to a given directed line segment $\overrightarrow{PQ}$ is a **vector v in the plane,** written $\mathbf{v} = \overrightarrow{PQ}$. Vectors are denoted by lowercase, boldface letters such as $\mathbf{u}$, $\mathbf{v}$, and $\mathbf{w}$.

EXAMPLE 1 Vector Representation by Directed Line Segments

Let $\mathbf{u}$ be represented by the directed line segment from $P = (0, 0)$ to $Q = (3, 2)$, and let $\mathbf{v}$ be represented by the directed line segment from $R = (1, 2)$ to $S = (4, 4)$, as shown in Figure 6.17. Show that $\mathbf{u} = \mathbf{v}$.

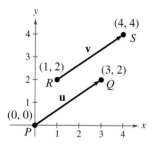

Figure 6.17

Solution

From the distance formula, it follows that $\overrightarrow{PQ}$ and $\overrightarrow{RS}$ have the *same magnitude*.

$$\|\overrightarrow{PQ}\| = \sqrt{(3-0)^2 + (2-0)^2} = \sqrt{13}$$

$$\|\overrightarrow{RS}\| = \sqrt{(4-1)^2 + (4-2)^2} = \sqrt{13}$$

Moreover, both line segments have the *same direction*, because they are both directed toward the upper right on lines having a slope of $\frac{2}{3}$. So, $\overrightarrow{PQ}$ and $\overrightarrow{RS}$ have the same magnitude and direction, and it follows that $\mathbf{u} = \mathbf{v}$.

Sandra Baker/Tony Stone Images

Component Form of a Vector

The directed line segment whose initial point is the origin is often the most convenient representative of a set of equivalent directed line segments. This representative of the vector **v** is in **standard position.**

A vector whose initial point is at the origin $(0, 0)$ can be uniquely represented by the coordinates of its terminal point (v_1, v_2). This is the **component form of a vector v,** written

$$\mathbf{v} = \langle v_1, v_2 \rangle.$$

The coordinates v_1 and v_2 are the *components* of **v**. If both the initial point and the terminal point lie at the origin, **v** is the **zero vector** and is denoted by $\mathbf{0} = \langle 0, 0 \rangle$.

STUDY T!P

You can use the *draw* feature of a graphing utility to graph a vector. Consult your user's manual for instructions.

Component Form of a Vector

The component form of the vector with initial point $P = (p_1, p_2)$ and terminal point $Q = (q_1, q_2)$ is

$$\overrightarrow{PQ} = \langle q_1 - p_1, q_2 - p_2 \rangle = \langle v_1, v_2 \rangle = \mathbf{v}.$$

The **magnitude** (or length) of **v** is

$$\|\mathbf{v}\| = \sqrt{(q_1 - p_1)^2 + (q_2 - p_2)^2} = \sqrt{v_1^2 + v_2^2}.$$

If $\|\mathbf{v}\| = 1$, **v** is a **unit vector.** Moreover, $\|\mathbf{v}\| = 0$ if and only if **v** is the zero vector **0**.

Two vectors $\mathbf{u} = \langle u_1, u_2 \rangle$ and $\mathbf{v} = \langle v_1, v_2 \rangle$ are *equal* if and only if $u_1 = v_1$ and $u_2 = v_2$. For instance, in Example 1, the vector **u** from $P = (0, 0)$ to $Q = (3, 2)$ is

$$\mathbf{u} = \overrightarrow{PQ} = \langle 3 - 0, 2 - 0 \rangle = \langle 3, 2 \rangle$$

and the vector **v** from $R = (1, 2)$ to $S = (4, 4)$ is

$$\mathbf{v} = \overrightarrow{RS} = \langle 4 - 1, 4 - 2 \rangle = \langle 3, 2 \rangle.$$

EXAMPLE 2   Finding the Component Form of a Vector

Find the component form and length of the vector **v** that has initial point $(4, -7)$ and terminal point $(-1, 5)$.

Solution

Let $P = (4, -7) = (p_1, p_2)$ and $Q = (-1, 5) = (q_1, q_2)$ as shown in Figure 6.18. Then, the components of $\mathbf{v} = \langle v_1, v_2 \rangle$ are

$$v_1 = q_1 - p_1 = -1 - 4 = -5$$

$$v_2 = q_2 - p_2 = 5 - (-7) = 12.$$

So, $\mathbf{v} = \langle -5, 12 \rangle$ and the magnitude of **v** is

$$\|\mathbf{v}\| = \sqrt{(-5)^2 + 12^2} = \sqrt{169} = 13.$$

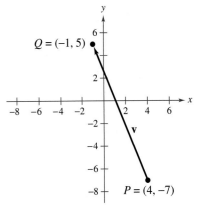

Figure 6.18

Vector Operations

The two basic vector operations are **scalar multiplication** and **vector addition.** Geometrically, the product of a vector **v** and a scalar k is the vector that is $|k|$ times as long as **v**. If k is positive, $k\mathbf{v}$ has the same direction as **v**, and if k is negative, $k\mathbf{v}$ has the opposite direction of **v**, as shown in Figure 6.19.

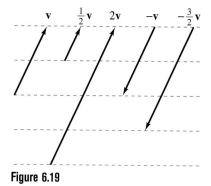

Figure 6.19

To add two vectors geometrically, position them (without changing length or direction) so that the initial point of one coincides with the terminal point of the other. The sum $\mathbf{u} + \mathbf{v}$ is formed by joining the initial point of the second vector **v** with the terminal point of the first vector **u**, as shown in Figure 6.20. This technique is called the **parallelogram law** for vector addition because the vector $\mathbf{u} + \mathbf{v}$, often called the **resultant** of vector addition, is the diagonal of a parallelogram having **u** and **v** as its adjacent sides.

A computer animation of this concept appears in the *Interactive* CD-ROM and *Internet* versions of this text.

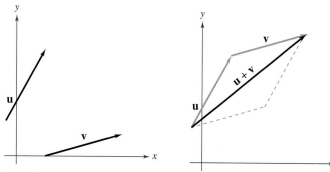

Figure 6.20

Definition of Vector Addition and Scalar Multiplication

Let $\mathbf{u} = \langle u_1, u_2 \rangle$ and $\mathbf{v} = \langle v_1, v_2 \rangle$ be vectors and let k be a scalar (a real number). Then the **sum** of **u** and **v** is the vector

$$\mathbf{u} + \mathbf{v} = \langle u_1 + v_1, u_2 + v_2 \rangle \qquad \text{Sum}$$

and the **scalar multiple** of k times **u** is the vector

$$k\mathbf{u} = k\langle u_1, u_2 \rangle = \langle ku_1, ku_2 \rangle. \qquad \text{Scalar multiple}$$

The **negative** of $\mathbf{v} = \langle v_1, v_2 \rangle$ is

$$-\mathbf{v} = (-1)\mathbf{v}$$

$$= \langle -v_1, -v_2 \rangle \qquad \text{Negative}$$

and the **difference** of **u** and **v** is

$$\mathbf{u} - \mathbf{v} = \mathbf{u} + (-\mathbf{v})$$

$$= \langle u_1 - v_1, u_2 - v_2 \rangle. \qquad \text{Difference}$$

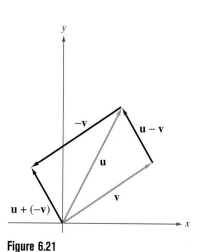

Figure 6.21

To represent $\mathbf{u} - \mathbf{v}$ geometrically, you can use directed line segments with the *same* initial points. The difference $\mathbf{u} - \mathbf{v}$ is the vector from the terminal point of **v** to the terminal point of **u**, as shown in Figure 6.21.

The component definitions of vector addition and scalar multiplication are illustrated in Example 3. In this example, notice that each of the vector operations can be interpreted geometrically.

EXAMPLE 3   Vector Operations

Let $\mathbf{v} = \langle -2, 5 \rangle$ and $\mathbf{w} = \langle 3, 4 \rangle$, and find each of the following vectors.

a. $2\mathbf{v}$ **b.** $\mathbf{w} - \mathbf{v}$ **c.** $\mathbf{v} + 2\mathbf{w}$

A computer animation of this example appears in the *Interactive* CD-ROM and *Internet* versions of this text.

Solution

a. Because $\mathbf{v} = \langle -2, 5 \rangle$, you have

$$2\mathbf{v} = 2\langle -2, 5 \rangle$$
$$= \langle 2(-2), 2(5) \rangle$$
$$= \langle -4, 10 \rangle.$$

A sketch of $2\mathbf{v}$ is shown in Figure 6.22(a).

b. The difference of $\mathbf{w}$ and $\mathbf{v}$ is

$$\mathbf{w} - \mathbf{v} = \langle 3 - (-2), 4 - 5 \rangle$$
$$= \langle 5, -1 \rangle.$$

A sketch of $\mathbf{w} - \mathbf{v}$ is shown in Figure 6.22(b).

c. The sum of $\mathbf{v} + 2\mathbf{w}$ is

$$\mathbf{v} + 2\mathbf{w} = \langle -2, 5 \rangle + 2\langle 3, 4 \rangle$$
$$= \langle -2, 5 \rangle + \langle 2(3), 2(4) \rangle$$
$$= \langle -2, 5 \rangle + \langle 6, 8 \rangle$$
$$= \langle -2 + 6, 5 + 8 \rangle$$
$$= \langle 4, 13 \rangle.$$

A sketch of $\mathbf{v} + 2\mathbf{w}$ is shown in Figure 6.22(c).

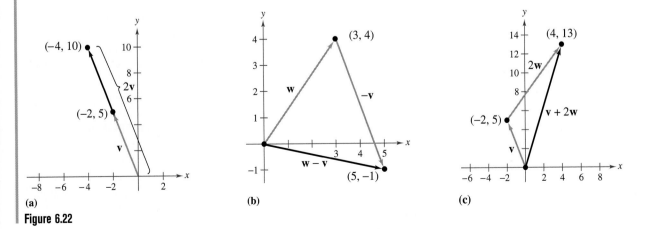

(a)   **(b)**   **(c)**

Figure 6.22

Note that Figure 6.22(b) shows the vector difference $\mathbf{w} - \mathbf{v}$ as the sum $\mathbf{w} + (-\mathbf{v})$.

Vector addition and scalar multiplication share many of the properties of ordinary arithmetic.

Properties of Vector Addition and Scalar Multiplication

Let **u, v,** and **w** be vectors and let c and d be scalars. Then the following properties are true.

1. $\mathbf{u} + \mathbf{v} = \mathbf{v} + \mathbf{u}$
2. $(\mathbf{u} + \mathbf{v}) + \mathbf{w} = \mathbf{u} + (\mathbf{v} + \mathbf{w})$
3. $\mathbf{u} + \mathbf{0} = \mathbf{u}$
4. $\mathbf{u} + (-\mathbf{u}) = \mathbf{0}$
5. $c(d\mathbf{u}) = (cd)\mathbf{u}$
6. $(c + d)\mathbf{u} = c\mathbf{u} + d\mathbf{u}$
7. $c(\mathbf{u} + \mathbf{v}) = c\mathbf{u} + c\mathbf{v}$
8. $1(\mathbf{u}) = \mathbf{u}, 0(\mathbf{u}) = \mathbf{0}$
9. $\|c\mathbf{v}\| = |c|\,\|\mathbf{v}\|$

Property 9 can be stated as follows: The length of the vector $c\mathbf{v}$ is the absolute value of c times the length of **v**.

Unit Vectors

In many applications of vectors it is useful to find a unit vector that has the same direction as a given nonzero vector **v**. To do this, you can divide **v** by its length to obtain

$$\mathbf{u} = \text{unit vector} = \frac{\mathbf{v}}{\|\mathbf{v}\|} = \left(\frac{1}{\|\mathbf{v}\|}\right)\mathbf{v}.$$

Note that **u** is a scalar multiple of **v**. The vector **u** has length 1 and the same direction as **v**. The vector **u** is called a **unit vector in the direction of v.**

EXAMPLE 4   Finding a Unit Vector

Find a unit vector in the direction of $\mathbf{v} = \langle -2, 5 \rangle$ and verify that the result has length 1.

Solution
The unit vector in the direction of **v** is

$$\frac{\mathbf{v}}{\|\mathbf{v}\|} = \frac{\langle -2, 5 \rangle}{\sqrt{(-2)^2 + (5)^2}}$$

$$= \frac{1}{\sqrt{29}}\langle -2, 5 \rangle$$

$$= \left\langle \frac{-2}{\sqrt{29}}, \frac{5}{\sqrt{29}} \right\rangle.$$

This vector has length 1 because

$$\sqrt{\left(\frac{-2}{\sqrt{29}}\right)^2 + \left(\frac{5}{\sqrt{29}}\right)^2} = \sqrt{\frac{4}{29} + \frac{25}{29}} = \sqrt{\frac{29}{29}} = 1.$$

William Rowan Hamilton (1805–1865) performed some of the earliest work with vectors. The Irish mathematician spent many years developing a system of vectorlike quantities called quaternions. Although Hamilton was convinced of the benefits of quaternions, the operations he defined did not produce good models for physical phenomena. It wasn't until the latter half of the nineteenth century that the Scottish physicist **James Maxwell (1831–1879)** restructured Hamilton's quaternions in a form useful for representing physical quantities such as force, velocity, and acceleration.

The unit vectors $\langle 1, 0 \rangle$ and $\langle 0, 1 \rangle$ are called the **standard unit vectors** and are denoted by

$$\mathbf{i} = \langle 1, 0 \rangle \qquad \text{and} \qquad \mathbf{j} = \langle 0, 1 \rangle$$

as shown in Figure 6.23. (Note that the lowercase letter $\mathbf{i}$ is written in boldface to distinguish it from the imaginary number $i = \sqrt{-1}$.) These vectors can be used to represent any vector $\mathbf{v} = \langle v_1, v_2 \rangle$ as follows.

$$\mathbf{v} = \langle v_1, v_2 \rangle$$
$$= v_1 \langle 1, 0 \rangle + v_2 \langle 0, 1 \rangle$$
$$= v_1 \mathbf{i} + v_2 \mathbf{j}$$

The scalars v_1 and v_2 are called the **horizontal and vertical components of v,** respectively. The vector sum

$$v_1 \mathbf{i} + v_2 \mathbf{j}$$

is called a **linear combination** of the vectors $\mathbf{i}$ and $\mathbf{j}$. Any vector in the plane can be expressed as a linear combination of the standard unit vectors $\mathbf{i}$ and $\mathbf{j}$.

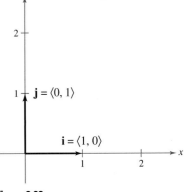

Figure 6.23

EXAMPLE 5   Writing a Linear Combination of Unit Vectors

Let $\mathbf{u}$ be the vector with initial point $(2, -5)$ and terminal point $(-1, 3)$. Write $\mathbf{u}$ as a linear combination of the standard unit vectors $\mathbf{i}$ and $\mathbf{j}$.

Solution
Begin by writing the component form of the vector $\mathbf{u}$.

$$\mathbf{u} = \langle -1 - 2, 3 + 5 \rangle$$
$$= \langle -3, 8 \rangle$$
$$= -3\mathbf{i} + 8\mathbf{j}$$

This result is shown graphically in Figure 6.24.

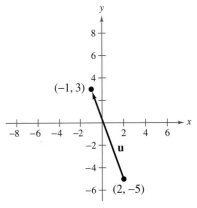

Figure 6.24

EXAMPLE 6   Vector Operations

Let $\mathbf{u} = -3\mathbf{i} + 8\mathbf{j}$ and $\mathbf{v} = 2\mathbf{i} - \mathbf{j}$. Find $2\mathbf{u} - 3\mathbf{v}$.

Solution
You could solve this problem by converting $\mathbf{u}$ and $\mathbf{v}$ to component form. This, however, is not necessary. It is just as easy to perform the operations in unit vector form.

$$2\mathbf{u} - 3\mathbf{v} = 2(-3\mathbf{i} + 8\mathbf{j}) - 3(2\mathbf{i} - \mathbf{j})$$
$$= -6\mathbf{i} + 16\mathbf{j} - 6\mathbf{i} + 3\mathbf{j}$$
$$= -12\mathbf{i} + 19\mathbf{j}$$

Direction Angles

If **u** is a *unit vector* such that θ is the angle (measured counterclockwise) from the positive x-axis to **u,** the terminal point of **u** lies on the unit circle and you have

$$\mathbf{u} = \langle x, y \rangle = \langle \cos \theta, \sin \theta \rangle = (\cos \theta)\mathbf{i} + (\sin \theta)\mathbf{j}$$

as shown in Figure 6.25. The angle θ is the **direction angle** of the vector **u**.

Suppose that **u** is a unit vector with direction angle θ. If **v** is any vector that makes an angle θ with the positive x-axis, then it has the same direction as **u** and you can write

$$\mathbf{v} = \|\mathbf{v}\| \langle \cos \theta, \sin \theta \rangle$$

$$= \|\mathbf{v}\| (\cos \theta)\mathbf{i} + \|\mathbf{v}\| (\sin \theta)\mathbf{j}.$$

Because $\mathbf{v} = a\mathbf{i} + b\mathbf{j} = \|\mathbf{v}\| (\cos \theta)\mathbf{i} + \|\mathbf{v}\| (\sin \theta)\mathbf{j}$, it follows that the direction angle θ for **v** is determined from

$$\tan \theta = \frac{\sin \theta}{\cos \theta}$$

$$= \frac{\|\mathbf{v}\| \sin \theta}{\|\mathbf{v}\| \cos \theta}$$

$$= \frac{b}{a}.$$

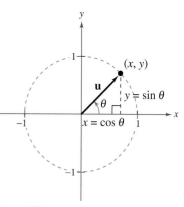

Figure 6.25

EXAMPLE 7   Finding Direction Angles of Vectors

Find the direction angle of each vector.

a. u $= 3\mathbf{i} + 3\mathbf{j}$ **b. v** $= 3\mathbf{i} - 4\mathbf{j}$

Solution

a. The direction angle is

$$\tan \theta = \frac{b}{a} = \frac{3}{3} = 1.$$

Therefore, $\theta = 45°$, as shown in Figure 6.26(a).

b. The direction angle is

$$\tan \theta = \frac{b}{a} = \frac{-4}{3}.$$

Moreover, because $\mathbf{v} = 3\mathbf{i} - 4\mathbf{j}$ lies in Quadrant IV, θ lies in Quadrant IV and its reference angle is

$$\theta' = \left| \arctan\left(-\frac{4}{3} \right) \right| \approx |-53.13°| = 53.13°.$$

Therefore, it follows that $\theta \approx 360° - 53.13° = 306.87°$, as shown in Figure 6.26(b).

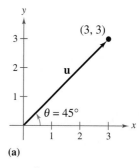

(a)

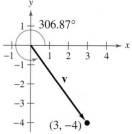

(b)

Figure 6.26

Applications of Vectors

EXAMPLE 8 Finding the Component Form of a Vector

Find the component form of the vector that represents the velocity of an airplane descending at a speed of 100 miles per hour at an angle 30° below the horizontal, as shown in Figure 6.27.

Solution

The velocity vector **v** has a magnitude of 100 and a direction angle of $\theta = 210°$.

$$\mathbf{v} = \|\mathbf{v}\| (\cos \theta)\mathbf{i} + \|\mathbf{v}\| (\sin \theta)\mathbf{j}$$

$$= 100(\cos 210°)\mathbf{i} + 100(\sin 210°)\mathbf{j}$$

$$= 100\left(\frac{-\sqrt{3}}{2}\right)\mathbf{i} + 100\left(\frac{-1}{2}\right)\mathbf{j}$$

$$= -50\sqrt{3}\,\mathbf{i} - 50\mathbf{j}$$

$$= \langle -50\sqrt{3}, -50 \rangle$$

You can check that **v** has a magnitude of 100.

$$\|\mathbf{v}\| = \sqrt{\left(-50\sqrt{3}\right)^2 + (-50)^2}$$

$$= \sqrt{7500 + 2500}$$

$$= \sqrt{10,000}$$

$$= 100 \qquad \text{Solution checks.} \checkmark$$

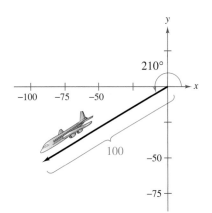

Figure 6.27

EXAMPLE 9 Using Vectors to Determine Weight

A force of 600 pounds is required to pull a boat and trailer up a ramp inclined at 15° from the horizontal. Find the combined weight of the boat and trailer.

Solution

Based on Figure 6.28, you can make the following observations.

$\|\overrightarrow{BA}\|$ = force of gravity = combined weight of boat and trailer

$\|\overrightarrow{BC}\|$ = force against ramp

$\|\overrightarrow{AC}\|$ = force required to move boat up ramp = 600 pounds

By construction, triangles *BWD* and *ABC* are similar. So, angle *ABC* is 15°. Therefore, in triangle *ABC* you have

$$\sin 15° = \frac{\|\overrightarrow{AC}\|}{\|\overrightarrow{BA}\|} = \frac{600}{\|\overrightarrow{BA}\|}$$

$$\|\overrightarrow{BA}\| = \frac{600}{\sin 15°} \approx 2318.$$

So, the combined weight is approximately 2318 pounds. (In Figure 6.28, note that $\overrightarrow{AC}$ is parallel to the ramp.)

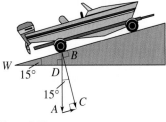

Figure 6.28

EXAMPLE 10   Using Vectors to Find Speed and Direction

An airplane is traveling at a fixed altitude with a negligible wind velocity. The airplane is headed N 30° W at a speed of 500 miles per hour, as shown in Figure 6.29. As the airplane reaches a certain point, it encounters a wind with a velocity of 70 miles per hour in the direction N 45° E. What are the resultant speed and direction of the airplane?

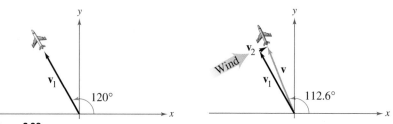

Figure 6.29

Solution

Using Figure 6.29, the velocity of the airplane (alone) is

$$\mathbf{v}_1 = 500\langle \cos 120°, \sin 120° \rangle$$
$$= \langle -250, 250\sqrt{3} \rangle$$

and the velocity of the wind is

$$\mathbf{v}_2 = 70\langle \cos 45°, \sin 45° \rangle$$
$$= \langle 35\sqrt{2}, 35\sqrt{2} \rangle.$$

So, the velocity of the airplane (in the wind) is

$$\mathbf{v} = \mathbf{v}_1 + \mathbf{v}_2$$
$$= \langle -250 + 35\sqrt{2}, 250\sqrt{3} + 35\sqrt{2} \rangle$$
$$\approx \langle -200.5, 482.5 \rangle$$

and the speed of the airplane is

$$\|\mathbf{v}\| = \sqrt{(-200.5)^2 + (482.5)^2}$$
$$\approx 522.5 \text{ miles per hour.}$$

Finally, if θ is the direction angle of the flight path, you have

$$\tan \theta = \frac{482.5}{-200.5}$$
$$\approx -2.4065$$

which implies that

$$\theta \approx 180° + \arctan(-2.4065)$$
$$\approx 180° - 67.4°$$
$$= 112.6°.$$

6.3 Exercises

In Exercises 1 and 2, show that u = v.

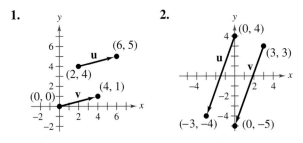

1.

2.

In Exercises 3–14, find the component form and the magnitude of the vector v.

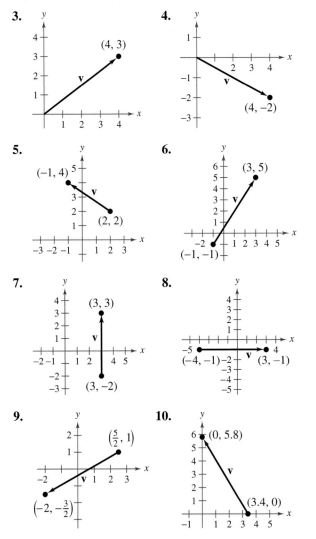

3.

4.

5.

6.

7.

8.

9.

10.

11. Initial point: $(-3, -5)$
Terminal point: $(5, 1)$

12. Initial point: $(-3, 11)$
Terminal point: $(9, 40)$

13. Initial point: $(-4.2, 5)$
Terminal point: $(3.7, -12.9)$

14. Initial point: $(1.64, 7.21)$
Terminal point: $(-2.33, 3.86)$

In Exercises 15–20, use the graph to sketch a graph of the specified vector.

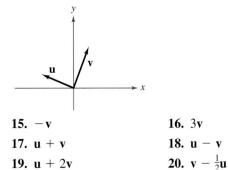

15. $-\mathbf{v}$ **16.** $3\mathbf{v}$

17. $\mathbf{u} + \mathbf{v}$ **18.** $\mathbf{u} - \mathbf{v}$

19. $\mathbf{u} + 2\mathbf{v}$ **20.** $\mathbf{v} - \frac{1}{2}\mathbf{u}$

In Exercises 21–26, find (a) u + v, (b) u − v, (c) 2u − 3v, and (d) v + 4u.

21. $\mathbf{u} = \langle 4, 2 \rangle, \quad \mathbf{v} = \langle 7, 1 \rangle$

22. $\mathbf{u} = \langle 5, 3 \rangle, \quad \mathbf{v} = \langle -4, 0 \rangle$

23. $\mathbf{u} = \langle -5, -2 \rangle, \quad \mathbf{v} = \langle 1, -3 \rangle$

24. $\mathbf{u} = \langle 0, -9 \rangle, \quad \mathbf{v} = \langle -6, 10 \rangle$

25. $\mathbf{u} = \mathbf{i} + \mathbf{j}, \quad \mathbf{v} = 2\mathbf{i} - 3\mathbf{j}$

26. $\mathbf{u} = 2\mathbf{i} - \mathbf{j}, \quad \mathbf{v} = -\mathbf{i} + \mathbf{j}$

In Exercises 27–36, find a unit vector in the direction of the given vector.

27. $\mathbf{u} = \langle 6, 0 \rangle$ **28.** $\mathbf{u} = \langle 0, -3 \rangle$

29. $\mathbf{v} = \langle -4, 4 \rangle$ **30.** $\mathbf{v} = \langle 5, -12 \rangle$

31. $\mathbf{v} = \langle -24, -7 \rangle$ **32.** $\mathbf{v} = \langle 8, -20 \rangle$

33. $\mathbf{v} = 4\mathbf{i} - 3\mathbf{j}$ **34.** $\mathbf{w} = \mathbf{i} - 2\mathbf{j}$

35. $\mathbf{w} = 2\mathbf{j}$ **36.** $\mathbf{w} = -3\mathbf{i}$

In Exercises 37–42, find the vector v with the given magnitude and the same direction as u.

	Magnitude	*Direction*
37.	$\|\mathbf{v}\| = 5$	$\mathbf{u} = \langle 3, 3 \rangle$
38.	$\|\mathbf{v}\| = 3$	$\mathbf{u} = \langle 4, -4 \rangle$
39.	$\|\mathbf{v}\| = 7$	$\mathbf{u} = 3\mathbf{i} + 4\mathbf{j}$
40.	$\|\mathbf{v}\| = 10$	$\mathbf{u} = 2\mathbf{i} - 3\mathbf{j}$
41.	$\|\mathbf{v}\| = 8$	$\mathbf{u} = -2\mathbf{i}$
42.	$\|\mathbf{v}\| = 4$	$\mathbf{u} = 5\mathbf{j}$

In Exercises 43–48, find the component form of v and sketch the specified vector operations geometrically, where u = 2i − j and w = i + 2j.

43. $\mathbf{v} = \frac{3}{2}\mathbf{u}$ **44.** $\mathbf{v} = 2\mathbf{w}$

45. $\mathbf{v} = \mathbf{u} + 2\mathbf{w}$ **46.** $\mathbf{v} = -\mathbf{u} + \mathbf{w}$

47. $\mathbf{v} = (3\mathbf{u} + \mathbf{w})$ **48.** $\mathbf{v} = \mathbf{u} - 2\mathbf{w}$

In Exercises 49–54, find the magnitude and direction angle of the vector v.

49. $\mathbf{v} = 5(\cos 30°\mathbf{i} + \sin 30°\mathbf{j})$

50. $\mathbf{v} = 8(\cos 135°\mathbf{i} + \sin 135°\mathbf{j})$

51. $\mathbf{v} = 6\mathbf{i} - 6\mathbf{j}$ **52.** $\mathbf{v} = -4\mathbf{i} - 7\mathbf{j}$

53. $\mathbf{v} = -2\mathbf{i} + 5\mathbf{j}$ **54.** $\mathbf{v} = 12\mathbf{i} + 15\mathbf{j}$

In Exercises 55–60, find the component form of v given its magnitude and the angle it makes with the positive x-axis. Sketch v.

	Magnitude	*Angle*
55.	$\|\mathbf{v}\| = 3$	$\theta = 0°$
56.	$\|\mathbf{v}\| = 1$	$\theta = 45°$
57.	$\|\mathbf{v}\| = 3\sqrt{2}$	$\theta = 150°$
58.	$\|\mathbf{v}\| = 9$	$\theta = 90°$
59.	$\|\mathbf{v}\| = 2$	$\mathbf{v}$ in the direction $\mathbf{i} + 3\mathbf{j}$
60.	$\|\mathbf{v}\| = 3$	$\mathbf{v}$ in the direction $3\mathbf{i} + 4\mathbf{j}$

In Exercises 61–64, find the component form of the sum of u and v with direction angles $\theta_{\mathbf{u}}$ and $\theta_{\mathbf{v}}$.

	Magnitude	*Angle*
61.	$\|\mathbf{u}\| = 5$	$\theta_{\mathbf{u}} = 60°$
	$\|\mathbf{v}\| = 5$	$\theta_{\mathbf{v}} = 90°$
62.	$\|\mathbf{u}\| = 2$	$\theta_{\mathbf{u}} = 30°$
	$\|\mathbf{v}\| = 2$	$\theta_{\mathbf{v}} = 90°$

	Magnitude	*Angle*
63.	$\|\mathbf{u}\| = 20$	$\theta_{\mathbf{u}} = 45°$
	$\|\mathbf{v}\| = 50$	$\theta_{\mathbf{v}} = 150°$
64.	$\|\mathbf{u}\| = 35$	$\theta_{\mathbf{u}} = 25°$
	$\|\mathbf{v}\| = 50$	$\theta_{\mathbf{v}} = 120°$

In Exercises 65–68, use the Law of Cosines to find the angle α between the given vectors. (Assume that $0° \le \alpha \le 180°$.)

65. $\mathbf{v} = \mathbf{i} + \mathbf{j}, \quad \mathbf{w} = 2(\mathbf{i} - \mathbf{j})$

66. $\mathbf{v} = 3\mathbf{i} + \mathbf{j}, \quad \mathbf{w} = 2\mathbf{i} - \mathbf{j}$

67. $\mathbf{v} = \mathbf{i} + \mathbf{j}, \quad \mathbf{w} = 3\mathbf{i} - \mathbf{j}$

68. $\mathbf{v} = \mathbf{i} + 2\mathbf{j}, \quad \mathbf{w} = 2\mathbf{i} - \mathbf{j}$

In Exercises 69 and 70, use a graphing utility to graph the vectors and the resultant of the vectors. Find the magnitude and direction of the resultant.

69.   **70.**

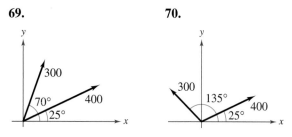

In Exercises 71 and 72, find the angle between the forces given the magnitude of their resultant. (*Hint:* Write Force 1 as a vector in the direction of the positive x-axis and Force 2 as a vector at an angle θ with the positive x-axis.)

	Force 1	*Force 2*	*Resultant Force*
71.	45 pounds	60 pounds	90 pounds
72.	3000 pounds	1000 pounds	3750 pounds

73. *Resultant Force* Forces with magnitudes of 2000 newtons and 900 newtons act on a machine part at angles of 30° and −45°, respectively, with the x-axis. Find the direction and magnitude of the resultant of these forces.

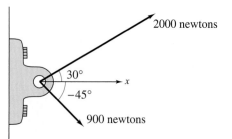

74. *Resultant Force*

(a) Three forces with magnitudes of 70 pounds, 40 pounds, and 60 pounds act on an object at angles of $-30°$, $45°$, and $135°$, respectively, with the positive x-axis. Find the direction and magnitude of the resultant of these forces.

(b) Three forces with magnitudes of 75 pounds, 100 pounds, and 125 pounds act on an object at angles of $30°$, $45°$, $120°$, respectively, with the positive x-axis. Find the direction and magnitude of the resultant of these forces.

75. *Horizontal and Vertical Components of Velocity* A ball is thrown with an initial velocity of 70 feet per second, at an angle of $40°$ with the horizontal. Find the vertical and horizontal components of the velocity.

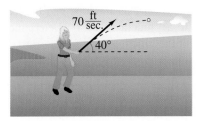

76. *Horizontal and Vertical Components of Velocity* A gun with a muzzle velocity of 1200 feet per second is fired at an angle of $4°$ with the horizontal. Find the vertical and horizontal components of the velocity.

77. *Cable Tension*   Use the figure to determine the tension in each cable supporting the given load.

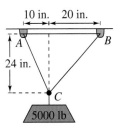

78. *Tension on Crane Cables*   The cranes in the figure below are lifting an object that weighs 20,240 pounds. Find the tension in the cable of each crane.

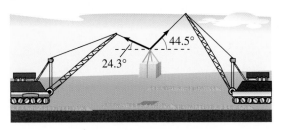

79. *Numerical and Graphical Analysis*   A loaded barge is being towed by two tugboats, and the magnitude of the resultant is 6000 pounds directed along the axis of the barge. Each tow line makes an angle of θ degrees with the axis of the barge.

(a) Find the tension in the tow lines if $\theta = 20°$.

(b) Write the tension T of each line as a function of θ. Determine the domain of the function.

(c) Use a graphing utility to complete the table.

θ	10°	20°	30°	40°	50°	60°
T						

(d) Use a graphing utility to graph the tension function.

(e) Explain why the tension increases as θ increases.

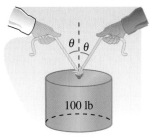

80. *Numerical and Graphical Analysis*   To carry a 100-pound cylindrical weight, two people lift on the ends of short ropes that are tied to an eyelet on the top center of the cylinder. Each rope makes an angle of θ degrees with the vertical.

(a) Find the tension in the ropes if $\theta = 30°$.

(b) Write the tension T of each rope as a function of θ. Determine the domain of the function.

(c) Use a graphing utility to complete the table.

θ	10°	20°	30°	40°	50°	60°
T						

(d) Use a graphing utility to graph the tension function.

(e) Explain why the tension increases as θ increases.

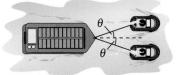

81. Navigation An airplane is flying in the direction S 32° E with an airspeed of 860 kilometers per hour. Because of the wind, its groundspeed and direction are respectively 800 kilometers per hour and S 40° E. Find the direction and speed of the wind.

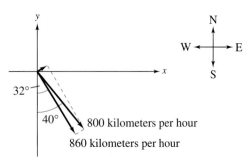

82. Navigation An airplane's velocity with respect to the air is 580 miles per hour, and it is heading N 60° W. The wind, at the altitude of the plane, is from the southwest and has a velocity of 60 miles per hour. Draw a diagram to represent the problem. What is the true direction of the plane, and what is its speed with respect to the ground?

83. Numerical and Graphical Analysis Forces with magnitudes of 150 newtons and 220 newtons act on a hook.

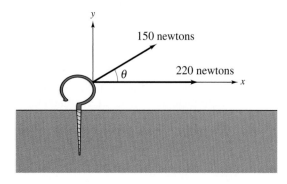

(a) If $\theta = 30°$, find the direction and magnitude of the resultant of these forces.

(b) Express the magnitude M of the resultant and the direction α of the resultant as functions of θ, where $0° \le \theta \le 180°$.

(c) Use a graphing utility to complete the table.

θ	0°	30°	60°	90°	120°	150°	180°
M							
α							

(d) Use a graphing utility to graph the two functions.

(e) Explain why one function decreases for increasing θ, whereas the other doesn't.

84. Numerical and Graphical Analysis A tetherball weighing 1 pound is pulled outward from the pole by a horizontal force **u** until the rope makes an angle of θ degrees with the pole.

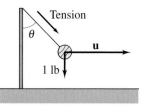

(a) Determine the resulting tension in the rope and the magnitude of **u** when $\theta = 30°$.

(b) Write the tension T in the rope and the magnitude of **u** as functions of θ. Determine the domains of the functions.

(c) Use a graphing utility to complete the table.

θ	0°	10°	20°	30°	40°	50°	60°
T							
$\|\mathbf{u}\|$							

(d) Use a graphing utility to graph the two functions for $0° \le \theta \le 60°$.

(e) Compare T and $\|\mathbf{u}\|$ as θ increases.

Synthesis

True or False? **In Exercises 85–88, determine whether the statement is true or false. Justify your answer.**

85. If **u** and **v** have the same magnitude and direction, then **u** = **v**.

86. If **u** is a unit vector in the direction of **v**, then **v** = $\|\mathbf{v}\|$ **u**.

87. If **v** = $a\mathbf{i} + b\mathbf{j} = \mathbf{0}$, then $a = -b$.

88. If **u** = $a\mathbf{i} + b\mathbf{j}$ is a unit vector, then $a^2 + b^2 = 1$.

89. *Think About It* Consider two forces of equal magnitude acting on a point.

(a) If the magnitude of the resultant is the sum of the magnitudes of the two forces, make a conjecture about the angle between the forces.

(b) If the resultant of the forces is **0**, make a conjecture about the angle between the forces.

(c) Can the magnitude of the resultant be greater than the sum of the magnitudes of the two forces? Explain.

90. *Graphical Reasoning* Consider two forces $\mathbf{F}_1 = \langle 10, 0 \rangle$ and $\mathbf{F}_2 = 5 \langle \cos \theta, \sin \theta \rangle$.

(a) Find $\|\mathbf{F}_1 + \mathbf{F}_2\|$ as a function of θ.

(b) Use a graphing utility to graph the function for $0 \le \theta < 2\pi$.

(c) Use the graph in part (b) to determine the range of the function. What is its maximum and for what value of θ does it occur? What is its minimum and for what value of θ does it occur?

(d) Explain why the magnitude of the resultant is never 0.

91. Prove that $(\cos \theta)\mathbf{i} + (\sin \theta)\mathbf{j}$ is a unit vector for any value of θ.

92. *Programming* Write a program for your graphing utility that graphs two vectors and their difference given the vectors in component form.

In Exercises 93 and 94, use the program in Exercise 92 to find the difference of the vectors shown in the graph.

93.

94.

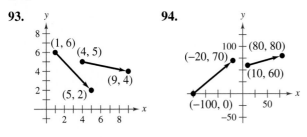

In Exercises 95–98, use the trigonometric substitution to write the algebraic expression as a trigonometric function of θ, where $0 < \theta < \pi/2$.

95. $\sqrt{49 - x^2}, \quad x = 7 \sin \theta$

96. $\sqrt{x^2 - 49}, \quad x = 7 \sec \theta$

97. $\sqrt{x^2 + 100}, \quad x = 10 \cot \theta$

98. $\sqrt{x^2 - 4}, \quad x = 2 \csc \theta$

In Exercises 99–102, use the given information to solve the triangle.

99. $A = 32°, \ a = 11, \ B = 61°$

100. $C = 106°, \ c = 34, \ B = 22°$

101. $a = 12, \ b = 15, \ c = 24$

102. $C = 96°, \ a = 14, \ b = 6.5$

6.4   Vectors and Dot Products

The Dot Product of Two Vectors

So far you have studied two vector operations—vector addition and multiplication by a scalar—each of which yields another vector. In this section you will study a third vector operation, the **dot product.** This product yields a scalar, rather than a vector.

Definition of a Dot Product

The **dot product** of $\mathbf{u} = \langle u_1, u_2 \rangle$ and $\mathbf{v} = \langle v_1, v_2 \rangle$ is

$$\mathbf{u} \cdot \mathbf{v} = u_1 v_1 + u_2 v_2.$$

Properties of the Dot Product

Let $\mathbf{u}$, $\mathbf{v}$, and $\mathbf{w}$ be vectors in the plane or in space and let c be a scalar.

1. $\mathbf{u} \cdot \mathbf{v} = \mathbf{v} \cdot \mathbf{u}$
2. $\mathbf{0} \cdot \mathbf{v} = 0$
3. $\mathbf{u} \cdot (\mathbf{v} + \mathbf{w}) = \mathbf{u} \cdot \mathbf{v} + \mathbf{u} \cdot \mathbf{w}$
4. $\mathbf{v} \cdot \mathbf{v} = \|\mathbf{v}\|^2$
5. $c(\mathbf{u} \cdot \mathbf{v}) = c\mathbf{u} \cdot \mathbf{v} = \mathbf{u} \cdot c\mathbf{v}$

Proofs of properties 1, 4, and 5 are given in Appendix A.

EXAMPLE 1   Finding Dot Products

Find each dot product.

a. $\langle 4, 5 \rangle \cdot \langle 2, 3 \rangle$
b. $\langle 2, -1 \rangle \cdot \langle 1, 2 \rangle$
c. $\langle 0, 3 \rangle \cdot \langle 4, -2 \rangle$

Solution

a. $\langle 4, 5 \rangle \cdot \langle 2, 3 \rangle = 4(2) + 5(3) = 8 + 15 = 23$
b. $\langle 2, -1 \rangle \cdot \langle 1, 2 \rangle = 2(1) + (-1)(2) = 2 - 2 = 0$
c. $\langle 0, 3 \rangle \cdot \langle 4, -2 \rangle = 0(4) + 3(-2) = 0 - 6 = -6$

In Example 1, be sure you see that the dot product of two vectors is a scalar (a real number), not a vector. Moreover, notice that the dot product can be positive, zero, or negative.

What You Should Learn:

- How to find the dot product of two vectors and use properties of the dot product
- How to find angles between vectors
- How to determine whether two vectors are orthogonal
- How to write vectors as sums of two vector components
- How to use vectors to find the work done by a force

Why You Should Learn It:

You can use the dot product of two vectors to solve real-life problems involving two vector quantities. For instance, Exercise 55 on page 465 shows you how the dot product can be used to find the force necessary to keep a truck from rolling down a hill.

Alan Thornton/Tony Stone Images

EXAMPLE 2   Using Properties of Dot Products

Let $\mathbf{u} = \langle -1, 3 \rangle$, $\mathbf{v} = \langle 2, -4 \rangle$, and $\mathbf{w} = \langle 1, -2 \rangle$. Find each dot product.

a. $(\mathbf{u} \cdot \mathbf{v})\mathbf{w}$ **b.** $\mathbf{u} \cdot 2\mathbf{v}$

Solution
Begin by finding the dot product of $\mathbf{u}$ and $\mathbf{v}$.

$$\mathbf{u} \cdot \mathbf{v} = \langle -1, 3 \rangle \cdot \langle 2, -4 \rangle$$
$$= (-1)(2) + 3(-4)$$
$$= -14$$

a. $(\mathbf{u} \cdot \mathbf{v})\mathbf{w} = -14\langle 1, -2 \rangle$
$$= \langle -14, 28 \rangle$$

b. $\mathbf{u} \cdot 2\mathbf{v} = 2(\mathbf{u} \cdot \mathbf{v})$
$$= 2(-14)$$
$$= -28$$

Notice that the first product is a vector, whereas the second is a scalar. Can you see why?

EXAMPLE 3   Dot Product and Length

The dot product of $\mathbf{u}$ with itself is 5. What is the length of $\mathbf{u}$?

Solution
Because $\|\mathbf{u}\|^2 = \mathbf{u} \cdot \mathbf{u} = 5$, it follows that

$$\|\mathbf{u}\| = \sqrt{\mathbf{u} \cdot \mathbf{u}}$$
$$= \sqrt{5}.$$

The Angle Between Two Vectors

The **angle between two nonzero vectors** is the angle θ, $0 \le \theta \le \pi$, between its respective standard position vectors, as shown in Figure 6.30. This angle can be found using the dot product. (Note that the angle between the zero vector and another vector is not defined.) See Appendix A for a proof.

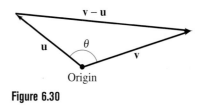

Figure 6.30

Angle Between Two Vectors

If θ is the angle between two nonzero vectors $\mathbf{u}$ and $\mathbf{v}$, then

$$\cos \theta = \frac{\mathbf{u} \cdot \mathbf{v}}{\|\mathbf{u}\| \, \|\mathbf{v}\|}.$$

EXAMPLE 4 Finding the Angle Between Two Vectors

Find the angle between $\mathbf{u} = \langle 4, 3 \rangle$ and $\mathbf{v} = \langle 3, 5 \rangle$.

Solution

$$\cos \theta = \frac{\mathbf{u} \cdot \mathbf{v}}{\|\mathbf{u}\| \|\mathbf{v}\|}$$

$$= \frac{\langle 4, 3 \rangle \cdot \langle 3, 5 \rangle}{\|\langle 4, 3 \rangle\| \|\langle 3, 5 \rangle\|}$$

$$= \frac{27}{5\sqrt{34}}$$

This implies that the angle between the two vectors is

$$\theta = \arccos \frac{27}{5\sqrt{34}} \approx 22.2°,$$

as shown in Figure 6.31.

Rewriting the expression for the angle between two vectors in the form

$$\mathbf{u} \cdot \mathbf{v} = \|\mathbf{u}\| \|\mathbf{v}\| \cos \theta \qquad \text{Alternative form of dot product}$$

produces an alternative way to calculate the dot product. From this form, you can see that because $\|\mathbf{u}\|$ and $\|\mathbf{v}\|$ are always positive, $\mathbf{u} \cdot \mathbf{v}$ and $\cos \theta$ will always have the same sign. Figure 6.32 shows the five possible orientations of two vectors.

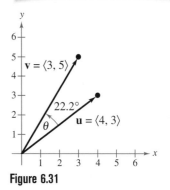

Figure 6.31

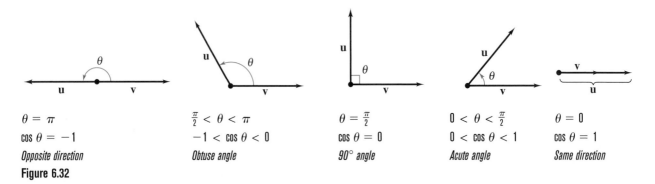

$\theta = \pi$
$\cos \theta = -1$
Opposite direction

$\frac{\pi}{2} < \theta < \pi$
$-1 < \cos \theta < 0$
Obtuse angle

$\theta = \frac{\pi}{2}$
$\cos \theta = 0$
90° angle

$0 < \theta < \frac{\pi}{2}$
$0 < \cos \theta < 1$
Acute angle

$\theta = 0$
$\cos \theta = 1$
Same direction

Figure 6.32

Definition of Orthogonal Vectors

The vectors $\mathbf{u}$ and $\mathbf{v}$ are **orthogonal** if $\mathbf{u} \cdot \mathbf{v} = 0$.

The terms *orthogonal* and *perpendicular* mean essentially the same thing—meeting at right angles. Even though the angle between the zero vector and another vector is not defined, it is convenient to extend the definition of orthogonality to include the zero vector. In other words, the zero vector is orthogonal to every vector $\mathbf{u}$ because $\mathbf{0} \cdot \mathbf{u} = 0$.

EXAMPLE 5   Determining Orthogonal Vectors

Are the vectors $\mathbf{u} = \langle 2, -3 \rangle$ and $\mathbf{v} = \langle 6, 4 \rangle$ orthogonal?

Solution

Begin by finding the dot product of the two vectors.

$$\mathbf{u} \cdot \mathbf{v} = \langle 2, -3 \rangle \cdot \langle 6, 4 \rangle$$
$$= 2(6) + (-3)(4)$$
$$= 0$$

The dot product is 0, so the two vectors are orthogonal, as shown in Figure 6.33.

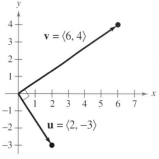

Figure 6.33

A computer simulation of this concept appears in the _Interactive_ CD-ROM and _Internet_ versions of this text.

Finding Vector Components

You have already seen applications in which two vectors are added to produce a resultant vector. Many applications in physics and engineering pose the reverse problem—decomposing a given vector into the sum of two **vector components.**

Consider a boat on an inclined ramp, as shown in Figure 6.34. The force $\mathbf{F}$ due to gravity pulls the boat _down_ the ramp and _against_ the ramp. These two orthogonal forces, $\mathbf{w}_1$ and $\mathbf{w}_2$, are vector components of $\mathbf{F}$. That is,

$$\mathbf{F} = \mathbf{w}_1 + \mathbf{w}_2. \qquad \text{Vector components of } \mathbf{F}$$

The negative of component $\mathbf{w}_1$ represents the force needed to keep the boat from rolling down the ramp, whereas $\mathbf{w}_2$ represents the force that the tires must withstand against the ramp. A procedure for finding $\mathbf{w}_1$ and $\mathbf{w}_2$ is shown below.

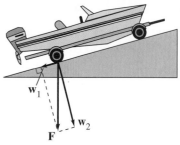

Figure 6.34

Definition of Vector Components

Let $\mathbf{u}$ and $\mathbf{v}$ be nonzero vectors such that

$$\mathbf{u} = \mathbf{w}_1 + \mathbf{w}_2$$

where $\mathbf{w}_1$ and $\mathbf{w}_2$ are orthogonal and $\mathbf{w}_1$ is parallel to (or a scalar multiple of) $\mathbf{v}$, as shown in Figure 6.35. The vectors $\mathbf{w}_1$ and $\mathbf{w}_2$ are called **vector components** of $\mathbf{u}$. The vector $\mathbf{w}_1$ is the **projection** of $\mathbf{u}$ onto $\mathbf{v}$ and is denoted by

$$\mathbf{w}_1 = \text{proj}_{\mathbf{v}}\mathbf{u}.$$

The vector $\mathbf{w}_2$ is given by $\mathbf{w}_2 = \mathbf{u} - \mathbf{w}_1$.

From the definition of vector components, you can see that it is easy to find the component $\mathbf{w}_2$ once you have found the projection of $\mathbf{u}$ onto $\mathbf{v}$. To find the projection, you can use the dot product.

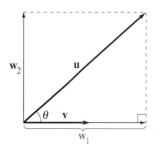

θ is acute.

Projection of u onto v

Let $\mathbf{u}$ and $\mathbf{v}$ be nonzero vectors. The projection of $\mathbf{u}$ onto $\mathbf{v}$ is

$$\text{proj}_{\mathbf{v}}\mathbf{u} = \left(\frac{\mathbf{u} \cdot \mathbf{v}}{\|\mathbf{v}\|^2} \right)\mathbf{v}.$$

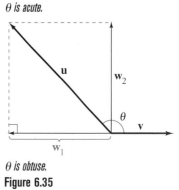

θ is obtuse.

Figure 6.35

EXAMPLE 6   Decomposing a Vector into Components

Find the projection of $\mathbf{u} = \langle 3, -5 \rangle$ onto $\mathbf{v} = \langle 6, 2 \rangle$. Then write $\mathbf{u}$ as the sum of two orthogonal vectors, one of which is $\text{proj}_\mathbf{v}\mathbf{u}$.

Solution

The projection of $\mathbf{u}$ onto $\mathbf{v}$ is

$$\mathbf{w}_1 = \text{proj}_\mathbf{v}\mathbf{u} = \left(\frac{\mathbf{u} \cdot \mathbf{v}}{\|\mathbf{v}\|^2}\right)\mathbf{v} = \left(\frac{8}{40}\right)\langle 6, 2 \rangle = \left\langle \frac{6}{5}, \frac{2}{5} \right\rangle$$

as shown in Figure 6.36. The other component, $\mathbf{w}_2$, is

$$\mathbf{w}_2 = \mathbf{u} - \mathbf{w}_1 = \langle 3, -5 \rangle - \left\langle \frac{6}{5}, \frac{2}{5} \right\rangle = \left\langle \frac{9}{5}, -\frac{27}{5} \right\rangle.$$

So, $\mathbf{u} = \mathbf{w}_1 + \mathbf{w}_2 = \left\langle \frac{6}{5}, \frac{2}{5} \right\rangle + \left\langle \frac{9}{5}, -\frac{27}{5} \right\rangle = \langle 3, -5 \rangle.$

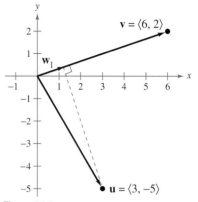

Figure 6.36

EXAMPLE 7   Finding a Force

A 600-pound boat sits on a ramp inclined at $30°$, as shown in Figure 6.37. What force is required to keep the boat from rolling down the ramp?

Solution

Because the force due to gravity is vertical and downward, you can represent the gravitational force by the vector

$$\mathbf{F} = -600\mathbf{j}.$$ Force due to gravity

To find the force required to keep the boat from rolling down the ramp, project $\mathbf{F}$ onto a unit vector $\mathbf{v}$ in the direction of the ramp, as follows.

$$\mathbf{v} = (\cos 30°)\mathbf{i} + (\sin 30°)\mathbf{j} = \frac{\sqrt{3}}{2}\mathbf{i} + \frac{1}{2}\mathbf{j}$$ Unit vector along ramp

Therefore, the projection of $\mathbf{F}$ onto $\mathbf{v}$ is

$$\mathbf{w}_1 = \text{proj}_\mathbf{v}\mathbf{F} = \left(\frac{\mathbf{F} \cdot \mathbf{v}}{\|\mathbf{v}\|^2}\right)\mathbf{v} = (\mathbf{F} \cdot \mathbf{v})\mathbf{v} = (-600)\left(\frac{1}{2}\right)\mathbf{v} = -300\left(\frac{\sqrt{3}}{2}\mathbf{i} + \frac{1}{2}\mathbf{j}\right).$$

The magnitude of this force is 300, and therefore a force of 300 pounds is required to keep the boat from rolling down the ramp.

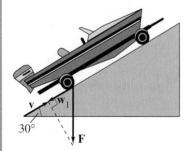

Figure 6.37

Work

The work W done by a constant force $\mathbf{F}$ acting along the line of motion of an object is

$$W = (\text{magnitude of force})(\text{distance})$$

$$= \|\mathbf{F}\| \|\overrightarrow{PQ}\|$$

as shown in Figure 6.38(a). If the constant force $\mathbf{F}$ is not directed along the line of motion, you can see from Figure 6.38(b) that the work W done by the force is

$$W = \|\text{proj}_{\overrightarrow{PQ}}\,\mathbf{F}\| \|\overrightarrow{PQ}\| = (\cos\theta)\|\mathbf{F}\| \|\overrightarrow{PQ}\| = \mathbf{F} \cdot PQ.$$

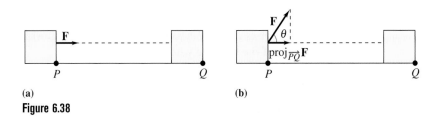

(a)   (b)

Figure 6.38

This notion of work is summarized in the following definition.

Definition of Work

The **work** W done by a constant force $\mathbf{F}$ as its point of application moves along the vector $\overrightarrow{PQ}$ is given by either of the following.

1. $W = \|\text{proj}_{\overrightarrow{PQ}}\,\mathbf{F}\| \|\overrightarrow{PQ}\|$ Projection form

2. $W = \mathbf{F} \cdot \overrightarrow{PQ}$ Dot product form

EXAMPLE 8   Finding Work

To close a barn's sliding door, a person pulls on a rope with a constant force of 50 pounds at a constant angle of $60°$, as shown in Figure 6.39. Find the work done in moving the door 12 feet to its closed position.

Solution

Using a projection, you can calculate the work as follows.

$$W = \|\text{proj}_{\overrightarrow{PQ}}\,\mathbf{F}\| \|\overrightarrow{PQ}\|$$

$$= (\cos 60°)\|\mathbf{F}\| \|\overrightarrow{PQ}\|$$

$$= \frac{1}{2}(50)(12)$$

$$= 300 \text{ ft-lb}$$

So, the work done is 300 foot-pounds. You can verify this result by finding the vectors $\mathbf{F}$ and $\overrightarrow{PQ}$ and calculating their dot product.

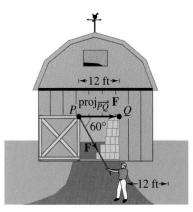

Figure 6.39

6.4 Exercises

In Exercises 1–4, find the dot product of u and v.

1. $\mathbf{u} = \langle 3, 6 \rangle$
   $\mathbf{v} = \langle 2, -4 \rangle$

2. $\mathbf{u} = \langle 5, 12 \rangle$
   $\mathbf{v} = \langle -3, 2 \rangle$

3. $\mathbf{u} = 4\mathbf{i} - 7\mathbf{j}$
   $\mathbf{v} = \mathbf{i} - \mathbf{j}$

4. $\mathbf{u} = 3\mathbf{i} + 9\mathbf{j}$
   $\mathbf{v} = 10\mathbf{i} - 3\mathbf{j}$

In Exercises 5–10, use the vectors $\mathbf{u} = \langle 2, 2 \rangle$, $\mathbf{v} = \langle -3, 4 \rangle$, and $\mathbf{w} = \langle 1, -4 \rangle$ to find the indicated quantity. State whether the result is a vector or a scalar.

5. $\mathbf{u} \cdot \mathbf{u}$

6. $\|\mathbf{u}\| - 2$

7. $(\mathbf{u} \cdot \mathbf{v})\mathbf{w}$

8. $(\mathbf{w} \cdot \mathbf{u})\mathbf{v}$

9. $\mathbf{u} \cdot 2\mathbf{v}$

10. $4\mathbf{u} \cdot \mathbf{v}$

In Exercises 11–16, use the dot product to find $\|\mathbf{u}\|$.

11. $\mathbf{u} = \langle -5, 12 \rangle$

12. $\mathbf{u} = \langle 2, -4 \rangle$

13. $\mathbf{u} = 20\mathbf{i} + 25\mathbf{j}$

14. $\mathbf{u} = 16\mathbf{i} - 10\mathbf{j}$

15. $\mathbf{u} = 6\mathbf{j}$

16. $\mathbf{u} = 9\mathbf{i}$

In Exercises 17–24, find the angle θ between the vectors.

17. $\mathbf{u} = \langle -1, 0 \rangle$
    $\mathbf{v} = \langle 0, 2 \rangle$

18. $\mathbf{u} = \langle 4, 4 \rangle$
    $\mathbf{v} = \langle -2, 0 \rangle$

19. $\mathbf{u} = 3\mathbf{i} + 4\mathbf{j}$
    $\mathbf{v} = -2\mathbf{i} + 3\mathbf{j}$

20. $\mathbf{u} = 2\mathbf{i} - 3\mathbf{j}$
    $\mathbf{v} = \mathbf{i} - 2\mathbf{j}$

21. $\mathbf{u} = 2\mathbf{i}$
    $\mathbf{v} = -3\mathbf{j}$

22. $\mathbf{u} = 4\mathbf{j}$
    $\mathbf{v} = -3\mathbf{i}$

23. $\mathbf{u} = \cos\left(\dfrac{\pi}{3}\right)\mathbf{i} + \sin\left(\dfrac{\pi}{3}\right)\mathbf{j}$
    $\mathbf{v} = \cos\left(\dfrac{3\pi}{4}\right)\mathbf{i} + \sin\left(\dfrac{3\pi}{4}\right)\mathbf{j}$

24. $\mathbf{u} = \cos\left(\dfrac{\pi}{4}\right)\mathbf{i} + \sin\left(\dfrac{\pi}{4}\right)\mathbf{j}$
    $\mathbf{v} = \cos\left(\dfrac{2\pi}{3}\right)\mathbf{i} + \sin\left(\dfrac{2\pi}{3}\right)\mathbf{j}$

In Exercises 25–28, use a graphing utility to sketch the vectors and find the degree measure of the angle between the vectors.

25. $\mathbf{u} = 3\mathbf{i} + 4\mathbf{j}$
    $\mathbf{v} = -7\mathbf{i} + 5\mathbf{j}$

26. $\mathbf{u} = -6\mathbf{i} - 3\mathbf{j}$
    $\mathbf{v} = -8\mathbf{i} + 4\mathbf{j}$

27. $\mathbf{u} = 5\mathbf{i} + 5\mathbf{j}$
    $\mathbf{v} = -8\mathbf{i} + 8\mathbf{j}$

28. $\mathbf{u} = 2\mathbf{i} - 3\mathbf{j}$
    $\mathbf{v} = 4\mathbf{i} + 3\mathbf{j}$

In Exercises 29 and 30, use vectors to find the interior angles of the triangle with the given vertices.

29. $(1, 2), (3, 4), (2, 5)$

30. $(-3, 0), (2, 2), (0, 6)$

In Exercises 31 and 32, find $\mathbf{u} \cdot \mathbf{v}$, where θ is the angle between u and v.

31. $\|\mathbf{u}\| = 4, \|\mathbf{v}\| = 10, \theta = \dfrac{2\pi}{3}$

32. $\|\mathbf{u}\| = 100, \|\mathbf{v}\| = 250, \theta = \dfrac{\pi}{6}$

In Exercises 33–38, determine whether u and v are orthogonal, parallel, or neither.

33. $\mathbf{u} = \langle -12, 30 \rangle$
    $\mathbf{v} = \langle \frac{1}{2}, -\frac{5}{4} \rangle$

34. $\mathbf{u} = \langle 15, 45 \rangle$
    $\mathbf{v} = \langle -5, 12 \rangle$

35. $\mathbf{u} = \frac{1}{4}(3\mathbf{i} - \mathbf{j})$
    $\mathbf{v} = 5\mathbf{i} + 6\mathbf{j}$

36. $\mathbf{u} = \mathbf{j}$
    $\mathbf{v} = \mathbf{i} - 2\mathbf{j}$

37. $\mathbf{u} = 2\mathbf{i} - 2\mathbf{j}$
    $\mathbf{v} = -\mathbf{i} - \mathbf{j}$

38. $\mathbf{u} = 8\mathbf{i} + 4\mathbf{j}$
    $\mathbf{v} = -2\mathbf{i} - \mathbf{j}$

In Exercises 39–42, find the projection of u onto v and the vector component of u orthogonal to v.

39. $\mathbf{u} = \langle 3, 4 \rangle$
    $\mathbf{v} = \langle 8, 2 \rangle$

40. $\mathbf{u} = \langle 4, 2 \rangle$
    $\mathbf{v} = \langle 1, -2 \rangle$

41. $\mathbf{u} = \langle 0, 3 \rangle$
    $\mathbf{v} = \langle 2, 15 \rangle$

42. $\mathbf{u} = \langle -5, -1 \rangle$
    $\mathbf{v} = \langle -1, 1 \rangle$

The *Interactive* CD-ROM and *Internet* versions of this text contain step-by-step solutions to all odd-numbered Section and Review Exercises. They also provide Tutorial Exercises which link to Guided Examples for additional help.

In Exercises 43–46, use the graph to mentally determine the projection of **u** onto **v**. (The coordinates of the terminal points of the vectors in standard position are given.) Use the formula for the projection of **u** onto **v** to verify your result.

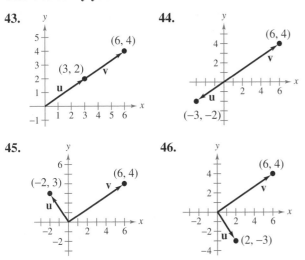

43.

44.

45.

46.

In Exercises 47–50, find two vectors in opposite directions that are orthogonal to the vector **u**. (There are many correct answers.)

47. $\mathbf{u} = \langle 4, 7 \rangle$

48. $\mathbf{u} = \langle -8, 1 \rangle$

49. $\mathbf{u} = \frac{1}{2}\mathbf{i} - \frac{3}{4}\mathbf{j}$

50. $\mathbf{u} = -\frac{5}{2}\mathbf{i} - 3\mathbf{j}$

Work **In Exercises 51 and 52, find the work done in moving a particle from *P* to *Q* if the magnitude and direction of the force are given by v.**

51. $P = (0, 0), \quad Q = (4, 7), \quad \mathbf{v} = \langle 1, 4 \rangle$

52. $P = (1, 3), \quad Q = (-3, 5), \quad \mathbf{v} = -2\mathbf{i} + 3\mathbf{j}$

53. *Revenue* The vector $\mathbf{u} = \langle 1245, 2600 \rangle$ gives the number of units of two products produced by a company. The vector $\mathbf{v} = \langle 12.20, 8.50 \rangle$ gives the price (in dollars) of each unit, respectively. Find the dot product $\mathbf{u} \cdot \mathbf{v}$, and explain what information it gives.

54. *Revenue* Repeat Exercise 53 after increasing the prices by 5%. Identify the vector operation used to increase the prices by 5%.

55. *Braking Load* A truck with a gross weight of 36,000 pounds is parked on a 10° slope. Assume the only force to overcome is that due to gravity.

    (a) Find the force required to keep the truck from rolling down the hill.

    (b) Find the force perpendicular to the hill.

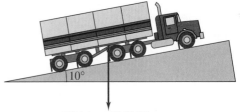

Weight = 36,000 lb

56. *Braking Load* Rework Exercise 55 for a truck that is parked on a 12° slope.

57. *Work* A 25-kilogram (245-newton) bag of sugar is lifted 3 meters. Determine the work done.

58. *Work* Determine the work done by a crane lifting a 2400-pound car 5 feet.

59. *Work* A force of 45 pounds in the direction of 30° above the horizontal is required to slide an implement across a floor. Find the work done if the implement is dragged 20 feet.

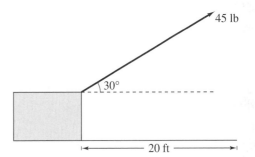

45 lb

30°

20 ft

60. *Work* A tractor pulls a log 800 meters and the tension in the cable connecting the tractor and log is approximately 1600 kilograms (15,691 newtons). Approximate the work done if the direction of the force is 35° above the horizontal.

61. *Work* A toy wagon is pulled by exerting a force of 20 pounds on a handle that makes a 25° angle with the horizontal. Find the work done in pulling the wagon 40 feet.

62. *Work* A mover exerts a horizontal force of 25 pounds on a crate as it is pushed up a ramp that is 12 feet long and inclined at an angle of 20° above the horizontal. Find the work done on the crate.

Synthesis

True or False? **In Exercises 63 and 64, determine whether the statement is true or false. Justify your answer.**

63. The vectors $\mathbf{u} = \langle 0, 0 \rangle$ and $\mathbf{v} = \langle -12, 6 \rangle$ are orthogonal.

64. The angle between $\mathbf{u} = \langle 5, 1 \rangle$ and $\mathbf{v} = \langle -4, 4 \rangle$ is obtuse.

65. If $\mathbf{u} = \langle \cos\theta, \sin\theta \rangle$ and $\mathbf{v} = \langle \sin\theta, -\cos\theta \rangle$, are $\mathbf{u}$ and $\mathbf{v}$ orthogonal, parallel, or neither? Explain.

66. *Think About It* What is known about θ, the angle between two nonzero vectors $\mathbf{u}$ and $\mathbf{v}$, if the following are true?

(a) $\mathbf{u} \cdot \mathbf{v} = 0$ (b) $\mathbf{u} \cdot \mathbf{v} > 0$ (c) $\mathbf{u} \cdot \mathbf{v} < 0$

67. *Think About It* What can be said about the vectors $\mathbf{u}$ and $\mathbf{v}$ if the following are true?

(a) The projection of $\mathbf{u}$ onto $\mathbf{v}$ equals $\mathbf{u}$.

(b) The projection of $\mathbf{u}$ onto $\mathbf{v}$ equals $\mathbf{0}$.

68. For nonzero vectors $\mathbf{u}$ and $\mathbf{v}$, prove that

$$\text{proj}_\mathbf{v}\mathbf{u} = \left(\frac{\mathbf{u} \cdot \mathbf{v}}{\|\mathbf{v}\|^2} \right)\mathbf{v}.$$

69. Use vectors to prove that the diagonals of a rhombus are perpendicular.

70. Prove the following.

$$\|\mathbf{u} - \mathbf{v}\|^2 = \|\mathbf{u}\|^2 + \|\mathbf{v}\|^2 - 2\mathbf{u} \cdot \mathbf{v}$$

71. Prove the following properties of the dot product.

(a) $\mathbf{0} \cdot \mathbf{v} = 0$

(b) $\mathbf{u} \cdot (\mathbf{v} + \mathbf{w}) = \mathbf{u} \cdot \mathbf{v} + \mathbf{u} \cdot \mathbf{w}$

72. Prove that if $\mathbf{u}$ is orthogonal to $\mathbf{v}$ and $\mathbf{w}$, then $\mathbf{u}$ is orthogonal to $c\mathbf{v} + d\mathbf{w}$ for any scalars c and d.

Review

In Exercises 73–76, find all solutions of the equation in the interval $[0, 2\pi)$. Use a graphing utility to verify your answers.

73. $2\cos(x + \pi) + 2\cos(x - \pi) = 0$

74. $\sin\left(x + \frac{3\pi}{2}\right) - \sin\left(x - \frac{3\pi}{2}\right) = 0$

75. $\sin\left(x - \frac{\pi}{3}\right) - \sin\left(x + \frac{\pi}{3}\right) = \frac{3}{2}$

76. $\tan(x + \pi) - \cos\left(x + \frac{5\pi}{2}\right) = 0$

In Exercises 77 and 78, use Heron's Area Formula (if possible) to find the area of the triangle.

77. $a = 8$, $b = 15$, $c = 16$

78. $a = 10.42$, $b = 6.83$, $c = 13.97$

In Exercises 79–82, find (a) $\mathbf{u} + \mathbf{v}$, (b) $2\mathbf{v} - \mathbf{u}$, and (c) $3\mathbf{u} - 5\mathbf{v}$.

79. $\mathbf{u} = \langle 3, 0 \rangle$, $\mathbf{v} = \langle 4, -1 \rangle$

80. $\mathbf{u} = \langle -5, 2 \rangle$, $\mathbf{v} = \langle 3, 8 \rangle$

81. $\mathbf{u} = \langle -2, -2 \rangle$, $\mathbf{v} = \langle -4, 5 \rangle$

82. $\mathbf{u} = \langle 8, 10 \rangle$, $\mathbf{v} = \langle 10, 2 \rangle$

83. *Buy Now or Wait?* A sales representative indicates that if a customer waits another month for a new car that currently costs $23,500, the price will increase by 4%. However, the customer will pay an interest penalty of $725 for the early withdrawal of a certificate of deposit if the car is purchased now. Determine whether the customer should buy now or wait another month.

84. *Insurance Premium* The annual insurance premium for a policyholder is normally $739. However, after having an automobile accident, the policyholder was charged an additional 30%. What is the new annual premium?

85. *Partnership Costs* A group of people agree to share equally in the cost of a $250,000 endowment to a college. If they could find two more people to join the group, each person's share of the cost would decrease by $6250. How many people are presently in the group?

86. *Speed* A boat travels at a speed of 18 miles per hour in still water. It travels 35 miles upstream and then returns to the starting point in a total of 4 hours. Find the speed of the current.

6.5 Trigonometric Form of a Complex Number

The Complex Plane

Recall from Section 2.4 that you can represent a complex number $z = a + bi$ as the point (a, b) in a coordinate plane (the complex plane). The horizontal axis is called the real axis and the vertical axis is called the imaginary axis, as shown in Figure 6.40.

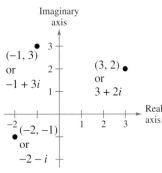

Figure 6.40

The **absolute value of a complex number** $a + bi$ is defined as the distance between the origin $(0, 0)$ and the point (a, b).

> **Definition of the Absolute Value of a Complex Number**
>
> The **absolute value** of the complex number $z = a + bi$ is
> $$|a + bi| = \sqrt{a^2 + b^2}.$$

If the complex number $a + bi$ is a real number (that is, if $b = 0$), this definition agrees with that given for the absolute value of a real number

$$|a + 0i| = \sqrt{a^2 + 0^2} = |a|.$$

EXAMPLE 1   Finding the Absolute Value of a Complex Number

Plot $z = -2 + 5i$ and find its absolute value.

Solution

The number is plotted in Figure 6.41. It has an absolute value of

$$|z| = \sqrt{(-2)^2 + 5^2} = \sqrt{29}.$$

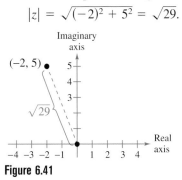

Figure 6.41

What You Should Learn:

- How to find absolute values of complex numbers
- How to write trigonometric forms of complex numbers
- How to multiply and divide complex numbers written in trigonometric form
- How to use DeMoivre's Theorem to find powers of complex numbers
- How to find *n*th roots of complex numbers

Why You Should Learn It:

You can use the trigonometric form of a complex number to perform operations with complex numbers. For instance, Exercises 107–114 on page 477 show you how the trigonometric form of a complex number can help you solve a polynomial equation.

Trigonometric Form of a Complex Number

In Section 2.4 you learned how to add, subtract, multiply, and divide complex numbers. To work effectively with *powers* and *roots* of complex numbers, it is helpful to write complex numbers in **trigonometric form.** In Figure 6.42, consider the nonzero complex number $a + bi$. By letting θ be the angle from the positive *x*-axis (measured counterclockwise) to the line segment connecting the origin and the point (a, b), you can write

$$a = r \cos \theta \qquad \text{and} \qquad b = r \sin \theta$$

where $r = \sqrt{a^2 + b^2}$. Consequently, you have

$$a + bi = (r \cos \theta) + (r \sin \theta)i$$

from which you can obtain the **trigonometric form of a complex number.**

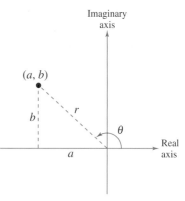

Figure 6.42

Trigonometric Form of a Complex Number

The **trigonometric form** of the complex number $z = a + bi$ is

$$z = r(\cos \theta + i \sin \theta)$$

where $a = r \cos \theta$, $b = r \sin \theta$, $r = \sqrt{a^2 + b^2}$, and $\tan \theta = b/a$. The number r is the **modulus** of z, and θ is called an **argument** of z.

The trigonometric form of a complex number is also called the *polar form*. Because there are infinitely many choices for θ, the trigonometric form of a complex number is not unique. Normally, θ is restricted to the interval $0 \le \theta < 2\pi$, although on occasion it is convenient to use $\theta < 0$.

EXAMPLE 2 Writing a Complex Number in Trigonometric Form

Write the complex number $z = -2 - 2\sqrt{3}i$ in trigonometric form.

Solution
The absolute value of z is

$$r = \left| -2 - 2\sqrt{3}i \right| = \sqrt{(-2)^2 + \left(-2\sqrt{3}\right)^2} = \sqrt{16} = 4$$

and the angle θ is

$$\tan \theta = \frac{b}{a} = \frac{-2\sqrt{3}}{-2} = \sqrt{3}.$$

Because $\tan(\pi/3) = \sqrt{3}$ and $z = -2 - 2\sqrt{3}i$ lies in Quadrant III, choose θ to be $\theta = \pi + \pi/3 = 4\pi/3$. So, the trigonometric form is

$$z = r(\cos \theta + i \sin \theta) = 4\left(\cos \frac{4\pi}{3} + i \sin \frac{4\pi}{3} \right).$$

See Figure 6.43.

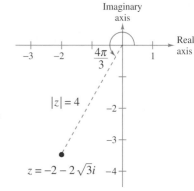

Figure 6.43

EXAMPLE 3   Writing a Complex Number in Standard Form

Write the complex number in standard form $a + bi$.

$$z = \sqrt{8}\left[\cos\left(-\frac{\pi}{3}\right) + i\sin\left(-\frac{\pi}{3}\right)\right]$$

Solution

Because $\cos(-\pi/3) = 1/2$ and $\sin(-\pi/3) = -\sqrt{3}/2$, you can write

$$z = \sqrt{8}\left[\cos\left(-\frac{\pi}{3}\right) + i\sin\left(-\frac{\pi}{3}\right)\right]$$

$$= \sqrt{8}\left[\frac{1}{2} - \frac{\sqrt{3}}{2}i\right]$$

$$= 2\sqrt{2}\left[\frac{1}{2} - \frac{\sqrt{3}}{2}i\right]$$

$$= \sqrt{2} - \sqrt{6}i.$$

STUDY T!P

A graphing utility can be used to convert a complex number in trigonometric (polar) form to standard (rectangular) form, and vice versa. Consult your user's manual for instructions.

Multiplication and Division of Complex Numbers

The trigonometric form adapts nicely to multiplication and division of complex numbers. Suppose you are given two complex numbers

$$z_1 = r_1(\cos\theta_1 + i\sin\theta_1) \qquad \text{and} \qquad z_2 = r_2(\cos\theta_2 + i\sin\theta_2).$$

The product of z_1 and z_2 is

$$z_1z_2 = r_1r_2(\cos\theta_1 + i\sin\theta_1)(\cos\theta_2 + i\sin\theta_2)$$

$$= r_1r_2[(\cos\theta_1\cos\theta_2 - \sin\theta_1\sin\theta_2) + i(\sin\theta_1\cos\theta_2 + \cos\theta_1\sin\theta_2)].$$

Using the sum and difference formulas for cosine and sine, you can rewrite this equation as

$$z_1z_2 = r_1r_2[\cos(\theta_1 + \theta_2) + i\sin(\theta_1 + \theta_2)].$$

This establishes the first part of the following rule. The second part is left to you (see Exercise 117).

The *Interactive* CD-ROM and *Internet* versions of this text offer a built-in graphing calculator, which can be used with the Examples, Explorations, and Exercises.

Product and Quotient of Two Complex Numbers

Let $z_1 = r_1(\cos\theta_1 + i\sin\theta_1)$ and $z_2 = r_2(\cos\theta_2 + i\sin\theta_2)$ be complex numbers.

$$z_1z_2 = r_1r_2[\cos(\theta_1 + \theta_2) + i\sin(\theta_1 + \theta_2)] \qquad \text{Product}$$

$$\frac{z_1}{z_2} = \frac{r_1}{r_2}[\cos(\theta_1 - \theta_2) + i\sin(\theta_1 - \theta_2)], \qquad z_2 \neq 0 \qquad \text{Quotient}$$

Note that this rule says that to *multiply* two complex numbers you multiply moduli and add arguments, whereas to *divide* two complex numbers you divide moduli and subtract arguments.

EXAMPLE 4   Multiplying Complex Numbers in Trigonometric Form

Find the product of the following complex numbers.

$$z_1 = 2\left(\cos\frac{2\pi}{3} + i\sin\frac{2\pi}{3}\right) \qquad z_2 = 8\left(\cos\frac{11\pi}{6} + i\sin\frac{11\pi}{6}\right)$$

Solution

$$z_1 z_2 = 2\left(\cos\frac{2\pi}{3} + i\sin\frac{2\pi}{3}\right) \cdot 8\left(\cos\frac{11\pi}{6} + i\sin\frac{11\pi}{6}\right)$$

$$= 16\left[\cos\left(\frac{2\pi}{3} + \frac{11\pi}{6}\right) + i\sin\left(\frac{2\pi}{3} + \frac{11\pi}{6}\right)\right]$$

$$= 16\left(\cos\frac{5\pi}{2} + i\sin\frac{5\pi}{2}\right)$$

$$= 16\left(\cos\frac{\pi}{2} + i\sin\frac{\pi}{2}\right)$$

$$= 16[0 + i(1)] = 16i$$

You can check this result by first converting to the standard forms $z_1 = -1 + \sqrt{3}i$ and $z_2 = 4\sqrt{3} - 4i$ and then multiplying algebraically, as in Section 2.4.

$$z_1 z_2 = \left(-1 + \sqrt{3}i\right)\left(4\sqrt{3} - 4i\right)$$

$$= -4\sqrt{3} + 4i + 12i + 4\sqrt{3}$$

$$= 16i$$

> **STUDY T!P**
>
> Some graphing utilities can multiply and divide complex numbers in trigonometric form. If you have access to such a graphing utility, use it to find $z_1 z_2$ and z_1/z_2 in Examples 4 and 5.

EXAMPLE 5   Dividing Complex Numbers in Trigonometric Form

Find z_1/z_2 for the following complex numbers.

$$z_1 = 24(\cos 300° + i\sin 300°) \qquad z_2 = 8(\cos 75° + i\sin 75°)$$

Solution

$$\frac{z_1}{z_2} = \frac{24(\cos 300° + i\sin 300°)}{8(\cos 75° + i\sin 75°)}$$

$$= \frac{24}{8}[\cos(300° - 75°) + i\sin(300° - 75°)]$$

$$= 3(\cos 225° + i\sin 225°)$$

$$= 3\left[\left(-\frac{\sqrt{2}}{2}\right) + i\left(-\frac{\sqrt{2}}{2}\right)\right]$$

$$= -\frac{3\sqrt{2}}{2} - \frac{3\sqrt{2}}{2}i$$

Powers of Complex Numbers

To raise a complex number to a power, consider repeated use of the multiplication rule.

$z = r(\cos \theta + i \sin \theta)$

$z^2 = r(\cos \theta + i \sin \theta)r(\cos \theta + i \sin \theta) = r^2(\cos 2\theta + i \sin 2\theta)$

$z^3 = r^2(\cos 2\theta + i \sin 2\theta)r(\cos \theta + i \sin \theta) = r^3(\cos 3\theta + i \sin 3\theta)$

$z^4 = r^4(\cos 4\theta + i \sin 4\theta)$

$z^5 = r^5(\cos 5\theta + i \sin 5\theta)$

$\vdots$

This pattern leads to **DeMoivre's Theorem,** which is named after the French mathematician Abraham DeMoivre (1667–1754).

DeMoivre's Theorem

If $z = r(\cos \theta + i \sin \theta)$ is a complex number and n is a positive integer, then

$$z^n = [r(\cos \theta + i \sin \theta)]^n$$

$$= r^n(\cos n\theta + i \sin n\theta).$$

Exploration

Plot the numbers $i, i^2, i^3, i^4,$ and i^5 in the complex plane. Write each number in trigonometric form and describe what happens to the angle θ as you form higher powers of i^n.

EXAMPLE 6   Finding Powers of a Complex Number

Use DeMoivre's Theorem to find $\left(-1 + \sqrt{3}i\right)^{12}$.

Solution

First convert to trigonometric form using

$$r = \sqrt{(-1)^2 + \left(\sqrt{3}\right)^2} = 2 \qquad r = \sqrt{a^2 + b^2}$$

and

$$\theta = \arctan \frac{\sqrt{3}}{-1} = \frac{2\pi}{3}. \qquad \theta = \arctan \frac{b}{a}$$

$$-1 + \sqrt{3}i = 2\left(\cos \frac{2\pi}{3} + i \sin \frac{2\pi}{3}\right) \qquad a + bi = r(\cos \theta + i \sin \theta)$$

Then, by DeMoivre's Theorem, you have

$$\left(-1 + \sqrt{3}i\right)^{12} = \left[2\left(\cos \frac{2\pi}{3} + i \sin \frac{2\pi}{3}\right)\right]^{12} \qquad z^n = [r(\cos \theta + i \sin \theta)]^n$$

$$= 2^{12}\left[\cos\left(12 \cdot \frac{2\pi}{3}\right) + i \sin\left(12 \cdot \frac{2\pi}{3}\right)\right] \qquad z^n = r^n(\cos n\theta + i \sin n\theta)$$

$$= 4096(\cos 8\pi + i \sin 8\pi)$$

$$= 4096(1 + 0)$$

$$= 4096.$$

Are you surprised to see a real number as the answer?

Roots of Complex Numbers

Recall that a consequence of the Fundamental Theorem of Algebra is that a polynomial equation of degree n has n solutions in the complex number system. So, an equation such as $x^6 = 1$ has six solutions, and in this particular case you can find the six solutions by factoring and using the Quadratic Formula.

$$x^6 - 1 = (x^3 - 1)(x^3 + 1)$$
$$= (x - 1)(x^2 + x + 1)(x + 1)(x^2 - x + 1) = 0$$

Consequently, the solutions are

$$x = \pm 1, \qquad x = \frac{-1 \pm \sqrt{3}i}{2}, \qquad \text{and} \qquad x = \frac{1 \pm \sqrt{3}i}{2}.$$

Each of these numbers is a sixth root of 1. In general, the **nth root of a complex number** is defined as follows.

Definition of nth Root of a Complex Number

The complex number $u = a + bi$ is an **nth root** of the complex number z if

$$z = u^n = (a + bi)^n.$$

To find a formula for an nth root of a complex number, let u be an nth root of z, where

$$u = s(\cos \beta + i \sin \beta)$$

and

$$z = r(\cos \theta + i \sin \theta).$$

By DeMoivre's Theorem and the fact that $u^n = z$, you have

$$s^n (\cos n\beta + i \sin n\beta) = r(\cos \theta + i \sin \theta).$$

Taking the absolute value of both sides of this equation, it follows that $s^n = r$. Substituting back into the previous equation and dividing by r, you get

$$\cos n\beta + i \sin n\beta = \cos \theta + i \sin \theta.$$

So, it follows that

$$\cos n\beta = \cos \theta$$

and

$$\sin n\beta = \sin \theta.$$

Because both sine and cosine have a period of 2π, these last two equations have solutions if and only if the angles differ by a multiple of 2π. Consequently, there must exist an integer k such that

$$n\beta = \theta + 2\pi k$$
$$\beta = \frac{\theta + 2\pi k}{n}.$$

By substituting this value of β into the trigonometric form of u, you get the result

Exploration

The nth roots of a complex number are useful for solving some polynomial equations. For instance, how can you use DeMoivre's Theorem to solve the polynomial equation

$$x^4 + 16 = 0?$$

[*Hint:* Write -16 as

$$16(\cos \pi + i \sin \pi).]$$

*n*th Roots of a Complex Number

For a positive integer n, the complex number $z = r(\cos \theta + i \sin \theta)$ has exactly n distinct nth roots given by

$$\sqrt[n]{r}\left(\cos \frac{\theta + 2\pi k}{n} + i \sin \frac{\theta + 2\pi k}{n}\right)$$

where $k = 0, 1, 2, \ldots, n - 1$.

When k exceeds $n - 1$, the roots begin to repeat. For instance, if $k = n$, the angle

$$\frac{\theta + 2\pi n}{n} = \frac{\theta}{n} + 2\pi$$

is coterminal with θ/n, which is also obtained when $k = 0$.

The formula for the nth roots of a complex number z has a nice geometrical interpretation, as shown in Figure 6.44. Note that because the nth roots of z all have the same magnitude $\sqrt[n]{r}$, they all lie on a circle of radius $\sqrt[n]{r}$ with center at the origin. Furthermore, because successive nth roots have arguments that differ by $2\pi/n$, the n roots are equally spaced along the circle.

You have already found the sixth roots of 1 by factoring and by using the Quadratic Formula. Example 7 shows how you can solve the same problem with the formula for nth roots.

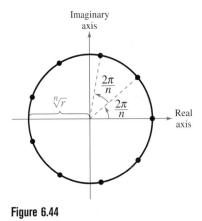

Figure 6.44

EXAMPLE 7   Finding the *n*th Roots of a Real Number

Find all the sixth roots of 1.

Solution

First write 1 in the trigonometric form $1 = 1(\cos 0 + i \sin 0)$. Then, by the nth root formula, with $n = 6$ and $r = 1$, the roots have the form

$$\sqrt[6]{1}\left(\cos \frac{0 + 2\pi k}{6} + i \sin \frac{0 + 2\pi k}{6}\right)$$

or simply $\cos(\pi k/3) + i \sin(\pi k/3)$. So, for $k = 0, 1, 2, 3, 4,$ and 5, the sixth roots are as follows. (See Figure 6.45.)

$$\cos 0 + i \sin 0 = 1$$

$$\cos \frac{\pi}{3} + i \sin \frac{\pi}{3} = \frac{1}{2} + \frac{\sqrt{3}}{2}i$$ Incremented by $\dfrac{2\pi}{n} = \dfrac{2\pi}{6} = \dfrac{\pi}{3}$

$$\cos \frac{2\pi}{3} + i \sin \frac{2\pi}{3} = -\frac{1}{2} + \frac{\sqrt{3}}{2}i$$

$$\cos \pi + i \sin \pi = -1$$

$$\cos \frac{4\pi}{3} + i \sin \frac{4\pi}{3} = -\frac{1}{2} - \frac{\sqrt{3}}{2}i$$

$$\cos \frac{5\pi}{3} + i \sin \frac{5\pi}{3} = \frac{1}{2} - \frac{\sqrt{3}}{2}i$$

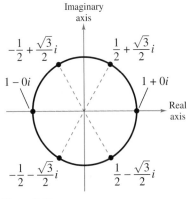

Figure 6.45

In Figure 6.45, notice that the roots obtained in Example 7 all have a magnitude of 1 and are equally spaced around this unit circle. Also notice that the complex roots occur in conjugate pairs, as discussed in Section 2.5. The n distinct nth roots of 1 are called the **nth roots of unity.**

EXAMPLE 8 Finding the nth Roots of a Complex Number

Find the three cube roots of $z = -2 + 2i$.

Solution
Because z lies in Quadrant II, the trigonometric form for z is

$$z = -2 + 2i$$

$$= \sqrt{8}(\cos 135° + i \sin 135°).$$

By the formula for nth roots, the cube roots have the form

$$\sqrt[6]{8}\left(\cos \frac{135° + 360°k}{3} + i \sin \frac{135° + 360°k}{3}\right).$$

Finally, for $k = 0, 1,$ and 2, you obtain the roots

$$\sqrt{2}(\cos 45° + i \sin 45°) = 1 + i$$

$$\sqrt{2}(\cos 165° + i \sin 165°) \approx -1.3660 + 0.3660i \qquad \text{Incremented by } \frac{360°}{3}$$

$$\sqrt{2}(\cos 285° + i \sin 285°) \approx 0.3660 - 1.3660i.$$

Writing About Math   *A Famous Mathematical Formula*

The famous formula

$$e^{a + bi} = e^a(\cos b + i \sin b)$$

is called Euler's Formula, after the German mathematician Leonhard Euler (1707–1783). Although the interpretation of this formula is beyond the scope of this text, it is included because it gives rise to one of the most wonderful equations in mathematics.

$$e^{\pi i} + 1 = 0$$

This elegant equation relates the five most famous numbers in mathematics—0, 1, π, e, and i—in a single equation. Show how Euler's Formula can be used to derive this equation. Write a short paragraph summarizing your work.

Exploration

Use a graphing utility, set in *parametric* and *radian* modes, to display the graph of

$$X_{1T} = \cos T$$

and

$$Y_{1T} = \sin T.$$

Set the viewing window so that $-1.5 \le X \le 1.5$ and $-1 \le Y \le 1$. Then, using $0 \le T \le 2\pi$, set the "Tstep" to $2\pi/n$ for various values of n. Explain how the graphing utility can be used to obtain the nth roots of unity.

6.5 Exercises

In Exercises 1–8, plot the complex number and find its absolute value.

1. $-6i$

2. $2i$

3. -4

4. 7

5. $-4 + 4i$

6. $-5 - 12i$

7. $9 - 7i$

8. $-10 + 3i$

In Exercises 9–12, write the complex number in trigonometric form.

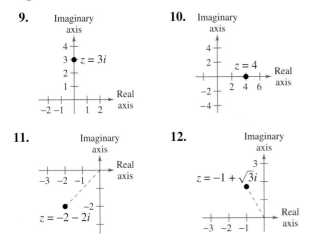

In Exercises 13–28, represent the complex number graphically, and find the trigonometric form of the number.

13. $5 - 5i$

14. $4 + 4i$

15. $\sqrt{3} + i$

16. $-1 + \sqrt{3}i$

17. $-2\left(1 + \sqrt{3}i\right)$

18. $\frac{5}{2}\left(\sqrt{3} - i\right)$

19. $8i$

20. $-6i$

21. $-7 + 4i$

22. $5 - i$

23. 7

24. 4

25. $1 + 10i$

26. $2\sqrt{2} - i$

27. $-7 - i$

28. $1 + 3i$

In Exercises 29–32, use a graphing utility to represent the complex number in trigonometric form.

29. $5 + 2i$

30. $-3 + i$

31. $3\sqrt{2} - 7i$

32. $-8 - 5\sqrt{3}i$

In Exercises 33–42, represent the complex number graphically, and find the standard form of the number.

33. $2(\cos 120° + i \sin 120°)$

34. $5(\cos 135° + i \sin 135°)$

35. $\frac{3}{2}(\cos 330° + i \sin 330°)$

36. $\frac{3}{4}(\cos 315° + i \sin 315°)$

37. $3.75\left(\cos \dfrac{3\pi}{4} + i \sin \dfrac{3\pi}{4}\right)$

38. $8\left(\cos \dfrac{5\pi}{6} + i \sin \dfrac{5\pi}{6}\right)$

39. $4\left(\cos \dfrac{3\pi}{2} + i \sin \dfrac{3\pi}{2}\right)$

40. $9(\cos 0 + i \sin 0)$

41. $3[\cos(18° \, 45') + i \sin(18° \, 45')]$

42. $6[\cos(230° \, 30') + i \sin(230° \, 30')]$

In Exercises 43–46, use a graphing utility to represent the complex number in standard form.

43. $5\left(\cos \dfrac{\pi}{9} + i \sin \dfrac{\pi}{9}\right)$

44. $12\left(\cos \dfrac{3\pi}{5} + i \sin \dfrac{3\pi}{5}\right)$

45. $9(\cos 58° + i \sin 58°)$

46. $4(\cos 216.5° + i \sin 216.5°)$

In Exercises 47 and 48, represent the powers z, z^2, z^3, and z^4 graphically. Describe the pattern.

47. $z = \dfrac{\sqrt{2}}{2}(1 + i)$

48. $z = \dfrac{1}{2}\left(1 + \sqrt{3}i\right)$

In Exercises 49–60, perform the operation and leave the result in trigonometric form.

49. $\left[3\left(\cos \dfrac{\pi}{3} + i \sin \dfrac{\pi}{3}\right)\right]\left[4\left(\cos \dfrac{\pi}{6} + i \sin \dfrac{\pi}{6}\right)\right]$

50. $\left[\dfrac{3}{2}\left(\cos \dfrac{\pi}{6} + i \sin \dfrac{\pi}{6}\right)\right]\left[6\left(\cos \dfrac{\pi}{4} + i \sin \dfrac{\pi}{4}\right)\right]$

51. $\left[\dfrac{5}{3}(\cos 140° + i \sin 140°)\right]\left[\dfrac{2}{3}(\cos 60° + i \sin 60°)\right]$

52. $\left[\frac{1}{2}(\cos 115° + i \sin 115°)\right]\left[\frac{4}{5}(\cos 300° + i \sin 300°)\right]$

53. $\left[\frac{11}{20}(\cos 290° + i \sin 290°)\right]\left[\frac{2}{5}(\cos 200° + i \sin 200°)\right]$

54. $(\cos 5° + i \sin 5°)(\cos 20° + i \sin 20°)$

55. $\dfrac{\cos 50° + i \sin 50°}{\cos 20° + i \sin 20°}$ **56.** $\dfrac{5(\cos 4.3 + i \sin 4.3)}{4(\cos 2.1 + i \sin 2.1)}$

57. $\dfrac{2(\cos 120° + i \sin 120°)}{4(\cos 40° + i \sin 40°)}$

58. $\dfrac{\cos\left(\frac{7\pi}{4}\right) + i \sin\left(\frac{7\pi}{4}\right)}{\cos \pi + i \sin \pi}$

59. $\dfrac{18(\cos 54° + i \sin 54°)}{3(\cos 102° + i \sin 102°)}$

60. $\dfrac{9(\cos 20° + i \sin 20°)}{5(\cos 75° + i \sin 75°)}$

In Exercises 61–66, (a) give the trigonometric form of the complex numbers, (b) perform the indicated operation using the trigonometric form, and (c) perform the indicated operation using the standard form and check your result with that of part (b).

61. $(2 + 2i)(1 - i)$ **62.** $\left(\sqrt{3} + i\right)(1 + i)$

63. $-2i(1 + i)$ **64.** $\dfrac{3 + 4i}{1 - \sqrt{3}i}$

65. $\dfrac{5}{2 + 3i}$ **66.** $\dfrac{4i}{-4 + 2i}$

In Exercises 67–70, sketch the graph of all complex numbers z satisfying the given condition.

67. $|z| = 2$ **68.** $|z| = 5$

69. $\theta = \dfrac{\pi}{6}$ **70.** $\theta = \dfrac{\pi}{4}$

In Exercises 71–86, use DeMoivre's Theorem to find the indicated power of the complex number. Express the result in standard form.

71. $(1 + i)^3$ **72.** $(2 + 2i)^6$

73. $(-1 + i)^{10}$ **74.** $(1 - i)^{12}$

75. $2(\sqrt{3} + i)^5$ **76.** $4(1 - \sqrt{3}i)^3$

77. $[5(\cos 20° + i \sin 20°)]^3$

78. $[3(\cos 150° + i \sin 150°)]^4$

79. $\left(\cos \dfrac{5\pi}{4} + i \sin \dfrac{5\pi}{4}\right)^{10}$

80. $\left[2\left(\cos \dfrac{\pi}{2} + i \sin \dfrac{\pi}{2}\right)\right]^{12}$

81. $[4(\cos 2.8 + i \sin 2.8)]^5$

82. $(\cos 0 + i \sin 0)^{20}$

83. $(3 - 2i)^5$ **84.** $\left(\sqrt{5} - 4i\right)^3$

85. $[3(\cos 15° + i \sin 15°)]^4$

86. $\left[2\left(\cos \dfrac{\pi}{10} + i \sin \dfrac{\pi}{10}\right)\right]^5$

87. Show that $-\frac{1}{2}\left(1 + \sqrt{3}i\right)$ is a sixth root of 1.

88. Show that $2^{-1/4}(1 - i)$ is a fourth root of -2.

Graphical Reasoning **In Exercises 89 and 90, use the graph of the roots of a complex number. (a) Write each of the roots in trigonometric form. (b) Identify the complex number whose roots are given. (c) Use a graphing utility to verify the results of part (b).**

89. **90.**

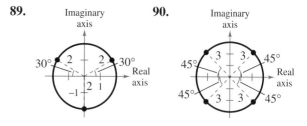

In Exercises 91–106, (a) use the theorem on page 473 to find the indicated roots of the complex number, (b) represent each of the roots graphically, and (c) express each of the roots in standard form.

91. Square roots of $5(\cos 120° + i \sin 120°)$

92. Square roots of $16(\cos 60° + i \sin 60°)$

93. Fourth roots of $16\left(\cos \dfrac{4\pi}{3} + i \sin \dfrac{4\pi}{3}\right)$

94. Fifth roots of $32\left(\cos \dfrac{5\pi}{6} + i \sin \dfrac{5\pi}{6}\right)$

95. Cube roots of $-27i$

96. Fourth roots of $625i$

97. Cube roots of $-\dfrac{125}{2}\left(1 + \sqrt{3}i\right)$

98. Cube roots of $-4\sqrt{2}(1 - i)$

99. Cube roots of 64

100. Fourth roots of i

101. Fifth roots of 1

102. Cube roots of 1000

103. Cube roots of -125

104. Fourth roots of -4

105. Fifth roots of $128(-1 + i)$

106. Sixth roots of $64i$

In Exercises 107–114, use the theorem on page 473 to find all the solutions of the equation and represent the solutions graphically.

107. $x^4 - i = 0$ **108.** $x^3 + 27 = 0$

109. $x^5 - 243 = 0$ **110.** $x^4 + 81 = 0$

111. $x^3 + 64i = 0$ **112.** $x^6 - 64i = 0$

113. $x^3 - (1 - i) = 0$ **114.** $x^4 + (1 + i) = 0$

Synthesis

True or False? **In Exercises 115 and 116, determine whether the statement is true or false. Justify your answer.**

115. $\frac{1}{2}(1 - \sqrt{3}i)$ is a ninth root of -1.

116. $\sqrt{3} + i$ is a solution of the equation $x^2 - 8i = 0$.

117. Given two complex numbers

$$z_1 = r_1(\cos \theta_1 + i \sin \theta_1)$$

and

$$z_2 = r_2(\cos \theta_2 + i \sin \theta_2), z_2 \neq 0,$$

prove that

$$\frac{z_1}{z_2} = \frac{r_1}{r_2}[\cos(\theta_1 - \theta_2) + i \sin(\theta_1 - \theta_2)].$$

118. Show that $\bar{z} = r[\cos(-\theta) + i \sin(-\theta)]$ is the complex conjugate of $z = r(\cos \theta + i \sin \theta)$.

119. Use the trigonometric forms of z and $\bar{z}$ in Exercise 118 to find (a) $z\bar{z}$ and (b) $z/\bar{z}, \bar{z} \neq 0$.

120. Show that the negative of $z = r(\cos \theta + i \sin \theta)$ is $-z = r[\cos(\theta + \pi) + i \sin(\theta + \pi)]$.

Review

In Exercises 121 and 122, find the altitude of the triangle.

121.   **122.**

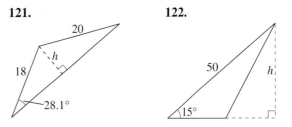

In Exercises 123–126, find a unit vector in the direction of the given vector.

123. $\mathbf{u} = \langle 10, 0 \rangle$ **124.** $\mathbf{v} = \langle -3, 7 \rangle$

125. $\mathbf{v} = 12\mathbf{i} - 5\mathbf{j}$ **126.** $\mathbf{w} = 8\mathbf{j}$

In Exercises 127–130, use the vectors $\mathbf{u} = \langle -1, 3 \rangle$ and $\mathbf{v} = \langle 5, -4 \rangle$ to find the indicated quantity.

127. $\mathbf{v} \cdot \mathbf{v}$ **128.** $\|\mathbf{u}\| + 6$

129. $3\mathbf{u} \cdot \mathbf{v}$ **130.** $\mathbf{u} \cdot 4\mathbf{v}$

In Exercises 131 and 132, determine whether $\mathbf{u}$ and $\mathbf{v}$ are orthogonal, parallel, or neither.

131. $\mathbf{u} = \langle -7, 35 \rangle$ **132.** $\mathbf{u} = 4\mathbf{i} - 5\mathbf{j}$

    $\mathbf{v} = \langle 3, -15 \rangle$ $\mathbf{v} = -10\mathbf{i} - 8\mathbf{j}$

What did you learn?

Section 6.1	Review Exercises
☐ How to use the Law of Sines to solve oblique triangles (AAS, ASA, or SSA)	1–10
☐ How to find areas of oblique triangles	11–14
☐ How to use the Law of Sines to model and solve real-life problems	15–18

Section 6.2

☐ How to use the Law of Cosines to solve oblique triangles (SSS or SAS)	19–28
☐ How to use the Law of Cosines to model and solve real-life problems	29–32
☐ How to use Heron's Area Formula to find areas of triangles	33–36

Section 6.3

☐ How to represent vectors as directed line segments	37, 38
☐ How to write the component form of vectors	39–44
☐ How to perform basic vector operations and represent vectors graphically	45–56
☐ How to write vectors as linear combinations of unit vectors	57–66
☐ How to find the direction angles of vectors	67–70
☐ How to use vectors to model and solve real-life problems	71–76

Section 6.4

☐ How to find the dot product of two vectors and use properties of the dot product	77–82
☐ How to find angles between vectors	83–90
☐ How to determine whether two vectors are orthogonal	91–94
☐ How to write vectors as sums of two vector components	95–98
☐ How to use vectors to find the work done by a force	99, 100

Section 6.5

☐ How to find absolute values of complex numbers	101–104
☐ How to write trigonometric forms of complex numbers	105–110
☐ How to multiply and divide complex numbers written in trigonometric form	111–116
☐ How to use DeMoivre's Theorem to find powers of complex numbers	117–120
☐ How to find nth roots of complex numbers	121–126

6 Review Exercises

6.1 **In Exercises 1–10, use the given information to solve the triangle. If two solutions exist, list both.**

1. $A = 12°$, $B = 58°$, $a = 8$
2. $B = 110°$, $C = 30°$, $c = 10.5$
3. $A = 75°$, $a = 2.5$, $b = 16.5$
4. $A = 130°$, $a = 60$, $b = 48$
5. $B = 115°$, $a = 9$, $b = 14.5$
6. $C = 50°$, $a = 25$, $c = 22$
7. $A = 15°$, $a = 5$, $b = 10$
8. $B = 150°$, $a = 64$, $b = 10$
9. $B = 25°$, $a = 6.2$, $b = 4$
10. $A = 74°$, $b = 12.8$, $a = 12.5$

In Exercises 11–14, find the area of the triangle having the indicated angle and sides.

11. $A = 27°$, $b = 5$, $c = 8$
12. $B = 80°$, $a = 4$, $c = 8$
13. $C = 122°$, $b = 18$, $a = 29$
14. $C = 100°$, $a = 120$, $b = 74$

15. **Height** From a certain distance, the angle of elevation to the top of a building is 17°. At a point 50 meters closer to the building, the angle of elevation is 31°. Approximate the height of the building.

16. **Altitude** The angles of elevation to an airplane from two points A and B on level ground are 51° and 68°, respectively. The points A and B are 2.5 miles apart, and the airplane is east of both points in the same vertical plane. Find the altitude of the plane.

17. **Height of a Tree** Find the height of a tree that stands on a hillside of slope 28° (from the horizontal) if, from a point 75 feet down the hill, the angle of elevation to the top of the tree is 45°.

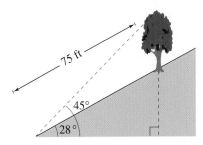

18. **River Width** Determine the width of a river that flows due east, if a surveyor finds that a tree on the opposite bank has a bearing of N 22° 30′ E from a certain point and a bearing of N 15° W from a point 400 feet downstream.

6.2 **In Exercises 19–28, solve the triangle having the indicated sides and angles.**

19. $a = 5$, $b = 8$, $c = 10$
20. $a = 13$, $b = 18$, $c = 26$
21. $a = 6$, $b = 9$, $C = 45°$
22. $B = 90°$, $a = 5$, $c = 12$
23. $B = 110°$, $a = 4$, $c = 4$
24. $B = 12°$, $a = 32$, $c = 36$
25. $a = 42$, $b = 25$, $c = 58$
26. $a = 80$, $b = 60$, $c = 100$
27. $B = 150°$, $a = 10$, $c = 20$
28. $a = 7.5$, $b = 15.0$, $c = 4.5$

29. **Geometry** The lengths of the diagonals of a parallelogram are 10 feet and 16 feet. Find the lengths of the sides of the parallelogram if the diagonals intersect at an angle of 28°.

30. **Geometry** The lengths of the diagonals of a parallelogram are 30 meters and 40 meters. Find the lengths of the sides of the parallelogram if the diagonals intersect at an angle of 34°.

31. **Surveying** To approximate the length of a marsh, a surveyor walks 425 meters from point A to point B. Then the surveyor turns 65° and walks 300 meters to point C. Approximate the length AC of the marsh.

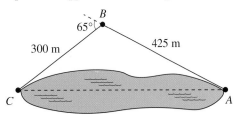

32. *Navigation* Two planes leave an airport at approximately the same time. One is flying at 425 miles per hour at a bearing of N 5° W, and the other is flying at 530 miles per hour at a bearing of N 67° E. Determine the distance between the planes after flying for 2 hours.

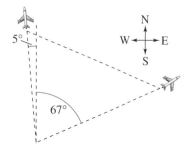

In Exercises 33–36, use Heron's Area Formula to find the area of the triangle with the given side lengths.

33. $a = 4$, $b = 5$, $c = 7$

34. $a = 15$, $b = 8$, $c = 10$

35. $a = 64.8$, $b = 49.2$, $c = 24.1$

36. $a = 8.55$, $b = 5.14$, $c = 12.73$

6.3 In Exercises 37 and 38, show that u = v.

37.

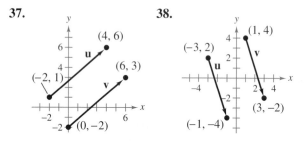

38.

In Exercises 39–44, find the component form of the vector v satisfying the given conditions.

39.

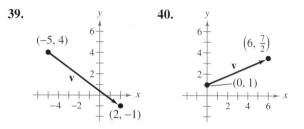

40.

41. Initial point: $(0, 10)$; Terminal point: $(7, 3)$

42. Initial point: $(1, 5)$; Terminal point: $(15, 9)$

43. $\|\mathbf{v}\| = 8$, $\theta = 120°$

44. $\|\mathbf{v}\| = \frac{1}{2}$, $\theta = 225°$

In Exercises 45–50, use the vectors $\mathbf{u} = \langle -1, -3 \rangle$, $\mathbf{v} = \langle -3, 6 \rangle$, and $\mathbf{w} = \langle 4, 5 \rangle$ to perform the given operations.

45. $-\mathbf{w}$

46. $\mathbf{w} - \mathbf{v}$

47. $\mathbf{w} + 2\mathbf{u}$

48. $2\mathbf{w} - 4\mathbf{u}$

49. $3\mathbf{v} + 3\mathbf{w}$

50. $\mathbf{u} - \frac{1}{2}\mathbf{v}$

In Exercises 51–56, find the component form of the specified vector and sketch its graph given that $\mathbf{u} = 6\mathbf{i} - 5\mathbf{j}$ and $\mathbf{v} = 10\mathbf{i} + 3\mathbf{j}$.

51. $\dfrac{1}{\|\mathbf{u}\|}\mathbf{u}$

52. $\dfrac{1}{\|\mathbf{v}\|}\mathbf{v}$

53. $3\mathbf{v}$

54. $\frac{1}{2}\mathbf{v}$

55. $4\mathbf{u} - 5\mathbf{v}$

56. $3\mathbf{v} - 2\mathbf{u}$

In Exercises 57–60, find a unit vector in the direction of the given vector.

57. $\mathbf{u} = \langle 0, -6 \rangle$

58. $\mathbf{v} = \langle -12, -5 \rangle$

59. $\mathbf{v} = 5\mathbf{i} - 2\mathbf{j}$

60. $\mathbf{w} = -7\mathbf{i}$

In Exercises 61–64, find a unit vector in the direction of $\overrightarrow{PQ}$.

61. $P(7, -4)$, $Q(-3, 2)$

62. $P(2, 5)$, $Q(-1, 6)$

63. $P(0, 3)$, $Q(5, -8)$

64. $P(-4, -9)$, $Q(0, -3)$

In Exercises 65 and 66, use the given information to write u as a linear combination of the standard unit vectors i and j.

65. Initial point: $(-8, 3)$
Terminal point: $(1, -5)$

66. Initial point: $(2, -3.2)$
Terminal point: $(-6.4, 10.8)$

In Exercises 67 and 68, write the vector v in the form $\|\mathbf{v}\|[(\cos \theta)\mathbf{i} + (\sin \theta)\mathbf{j}]$.

67. $\mathbf{v} = -10\mathbf{i} + 10\mathbf{j}$

68. $\mathbf{v} = 4\mathbf{i} - \mathbf{j}$

In Exercises 69 and 70, use a graphing utility to graph the vectors and the resultant of the vectors. Find the magnitude and direction of the resultant.

69.

70.

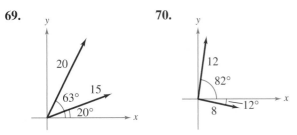

71. *Resultant Force*   Find the direction and magnitude of the resultant of the three forces shown in the graph.

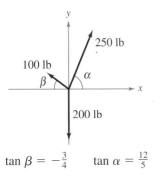

$$\tan \beta = -\tfrac{3}{4} \qquad \tan \alpha = \tfrac{12}{5}$$

72. *Resultant Force*   Forces of 85 pounds and 50 pounds act on a single point. The angle between the forces is 15°. Describe the resultant force.

73. *Rope Tension*   A 180-pound weight is supported by two ropes, as shown in the figure. Find the tension in each rope.

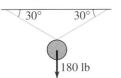

74. *Cable Tension*   In a manufacturing process, an electric hoist lifts 200-pound ingots. Find the tension in the supporting cables.

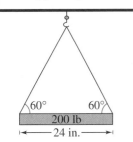

75. *Navigation*   An airplane has an airspeed of 430 miles per hour at a bearing of S 45° E. If the wind velocity is 35 miles per hour in the direction N 30° E, find the groundspeed and the direction of the plane.

76. *Navigation*   An airplane has an airspeed of 724 kilometers per hour at a bearing of N 30° E. If the wind velocity is from the west at 32 kilometers per hour, find the groundspeed and the direction of the plane.

6.4   In Exercises 77 and 78, find the dot product of u and v.

77. $\mathbf{u} = 6\mathbf{i} - \mathbf{j}$
   $\mathbf{v} = 2\mathbf{i} + 5\mathbf{j}$

78. $\mathbf{u} = 8\mathbf{i} - 7\mathbf{j}$
   $\mathbf{v} = 3\mathbf{i} - 4\mathbf{j}$

In Exercises 79–82, use the vectors $\mathbf{u} = \langle 6, -3 \rangle$ and $\mathbf{v} = \langle -2, -1 \rangle$ to find the indicated quantity.

79. $(\mathbf{u} \cdot \mathbf{v})\mathbf{u}$

80. $\|\mathbf{v}\| - 3$

81. $4\mathbf{u} \cdot \mathbf{v}$

82. $\|\mathbf{v}\|^2 + \|\mathbf{u}\|^2$

In Exercises 83–86, find the angle between u and v.

83. $\mathbf{u} = \cos \dfrac{7\pi}{4} \mathbf{i} + \sin \dfrac{7\pi}{4} \mathbf{j}$

   $\mathbf{v} = \cos \dfrac{5\pi}{6} \mathbf{i} + \sin \dfrac{5\pi}{6} \mathbf{j}$

84. $\mathbf{u} = \langle -6, -3 \rangle, \quad \mathbf{v} = \langle 4, 2 \rangle$

85. $\mathbf{u} = \langle 2\sqrt{2}, -4 \rangle, \quad \mathbf{v} = \langle -\sqrt{2}, 1 \rangle$

86. $\mathbf{u} = \langle 3, 1 \rangle, \quad \mathbf{v} = \langle 4, 5 \rangle$

In Exercises 87–90, use a graphing utility to sketch the vectors and find the degree measure of the angle between the vectors.

87. $\mathbf{u} = 4\mathbf{i} + \mathbf{j}$
   $\mathbf{v} = \mathbf{i} - 4\mathbf{j}$

88. $\mathbf{u} = 6\mathbf{i} + 2\mathbf{j}$
   $\mathbf{v} = -3\mathbf{i} - \mathbf{j}$

89. $\mathbf{u} = 7\mathbf{i} - 5\mathbf{j}$
   $\mathbf{v} = 10\mathbf{i} + 3\mathbf{j}$

90. $\mathbf{u} = -5.3\mathbf{i} + 2.8\mathbf{j}$
   $\mathbf{v} = -8.1\mathbf{i} - 4\mathbf{j}$

In Exercises 91–94, decide whether the vectors are orthogonal, parallel, or neither.

91. $\mathbf{u} = \langle 39, -12 \rangle$
   $\mathbf{v} = \langle -26, 8 \rangle$

92. $\mathbf{u} = \langle 8, -4 \rangle$
   $\mathbf{v} = \langle 5, 10 \rangle$

93. $\mathbf{u} = \langle 8, 5 \rangle$
   $\mathbf{v} = \langle -2, 4 \rangle$

94. $\mathbf{u} = \langle -15, 51 \rangle$
   $\mathbf{v} = \langle 20, -68 \rangle$

In Exercises 95–98, (a) find proj$_v$u and (b) write u as the sum of two vector components.

95. $\mathbf{u} = \langle -4, 3 \rangle$, $\mathbf{v} = \langle -8, -2 \rangle$

96. $\mathbf{u} = \langle 5, 6 \rangle$, $\mathbf{v} = \langle 10, 0 \rangle$

97. $\mathbf{u} = \langle 2, 7 \rangle$, $\mathbf{v} = \langle 1, -1 \rangle$

98. $\mathbf{u} = \langle -3, 5 \rangle$, $\mathbf{v} = \langle -5, 2 \rangle$

99. **Work**   Determine the work done by a crane lifting an 18,000-pound truck 48 inches.

100. **Braking Force**   A 500-pound motorcycle is headed up a hill inclined at 12°. What force is required to keep the motorcycle from rolling back down the hill when stopped at a red light?

6.5   **In Exercises 101–104, plot the complex number and find its absolute value.**

101. $-i$

102. $6i$

103. $7 - 5i$

104. $-3 - 10i$

In Exercises 105–110, find the trigonometric form of the complex number.

105.
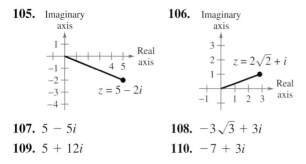

106.

107. $5 - 5i$

108. $-3\sqrt{3} + 3i$

109. $5 + 12i$

110. $-7 + 3i$

In Exercises 111 and 112, (a) express the two complex numbers in trigonometric form and (b) use the trigonometric form to find $z_1 z_2$ and z_1/z_2.

111. $z_1 = 2\sqrt{3} - 2i$, $z_2 = -10i$

112. $z_1 = -3(1 + i)$, $z_2 = 2(\sqrt{3} + i)$

In Exercises 113–116, perform the operation and simplify the result.

113. $\left[\dfrac{5}{2} \left(\cos \dfrac{\pi}{2} + i \sin \dfrac{\pi}{2} \right) \right] \left[4 \left(\cos \dfrac{\pi}{4} + i \sin \dfrac{\pi}{4} \right) \right]$

114. $\left[2 \left(\cos \dfrac{2\pi}{3} + i \sin \dfrac{2\pi}{3} \right) \right] \left[3 \left(\cos \dfrac{\pi}{6} + i \sin \dfrac{\pi}{6} \right) \right]$

115. $\dfrac{20(\cos 320° + i \sin 320°)}{5(\cos 80° + i \sin 80°)}$

116. $\dfrac{3(\cos 230° + i \sin 230°)}{9(\cos 95° + i \sin 95°)}$

In Exercises 117–120, find the indicated power of the complex number. Express the result in standard form.

117. $\left[5 \left(\cos \dfrac{\pi}{12} + i \sin \dfrac{\pi}{12} \right) \right]^4$

118. $\left[2 \left(\cos \dfrac{4\pi}{15} + i \sin \dfrac{4\pi}{15} \right) \right]^5$

119. $(2 + 3i)^6$ 120. $(1 - i)^8$

In Exercises 121 and 122, use the theorem on page 473 to find the roots of the complex number.

121. Sixth roots of $-729i$ 122. Fourth roots of 256

In Exercises 123–126, find all solutions of the equation and represent the solutions graphically.

123. $x^4 + 256 = 0$ 124. $x^5 - 32 = 0$

125. $x^3 + 8i = 0$ 126. $(x^3 - 1)(x^2 + 1) = 0$

Synthesis

In Exercises 127 and 128, determine whether the statement is true or false. Justify your answer.

127. The Law of Sines is true if one of the angles in the triangle is a right angle.

128. When the Law of Sines is used, the solution is always unique.

129. What characterizes a vector in the plane?

130. The figure shows z_1 and z_2. Describe $z_1 z_2$ and z_1/z_2.

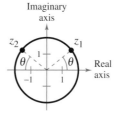

131. One of the fourth roots of a complex number z is shown in the graph.

(a) How many roots are not shown?

(b) Describe the other roots.

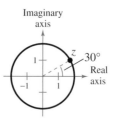

Chapter Project   *Adding Vectors Graphically*

The psuedocode at the right can be translated into a program for a graphing utility. (The program ADDVECT for several models of calculators can be found at our website, *college.hmco.com*.) The program sketches two vectors

$$\mathbf{u} = a\mathbf{i} + b\mathbf{j} \quad \text{and} \quad \mathbf{v} = c\mathbf{i} + d\mathbf{j}$$

in standard position. Then, using the parallelogram law for vector addition, the program also sketches the vector sum $\mathbf{u} + \mathbf{v}$. *Before* running the program, you should set values that produce an appropriate viewing window.

Program

- Input a
- Input b
- Input c
- Input d
- Draw a line from $(0, 0)$ to (a, b).
- Draw a line from $(0, 0)$ to (c, d).
- Add $a + c$ and store in e.
- Add $b + d$ and store in f.
- Draw a line from $(0, 0)$ to (e, f).
- Draw a line from (a, b) to (e, f).
- Draw a line from (c, d) to (e, f).
- Pause to view graph.
- End Program

a. Use the program to sketch the sum of the vectors $\mathbf{u} = 5\mathbf{i} + 2\mathbf{j}$ and $\mathbf{v} = -4\mathbf{i} + 3\mathbf{j}$. Set your viewing window as indicated in the graph below. Identify the vectors $\mathbf{u}$, $\mathbf{v}$, and $\mathbf{u} + \mathbf{v}$ in the graph.

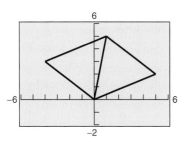

b. An airplane is headed N 60° W at a speed of 400 miles per hour. The airplane encounters wind of velocity 75 miles per hour in the direction N 40° E. Use the program to find the resultant speed and direction of the airplane.

Questions for Further Exploration

In Questions 1–4, use the program to sketch the sum of the vectors. Use the result to estimate graphically the components of the sum. Then check your result algebraically. (Use $-9 \le x \le 9$ and $-6 \le y \le 6$.)

1. $\mathbf{u} = 3\mathbf{i} + 4\mathbf{j}$, $\mathbf{v} = -5\mathbf{i} + \mathbf{j}$
2. $\mathbf{u} = 5\mathbf{i} - 4\mathbf{j}$, $\mathbf{v} = 3\mathbf{i} + 2\mathbf{j}$
3. $\mathbf{u} = -4\mathbf{i} + 4\mathbf{j}$, $\mathbf{v} = -2\mathbf{i} - 6\mathbf{j}$
4. $\mathbf{u} = 7\mathbf{i} + 3\mathbf{j}$, $\mathbf{v} = -2\mathbf{i} - 6\mathbf{j}$

5. After encountering the wind, is the airplane in part (b) above traveling at a higher speed or a lower speed? Explain.

6. Consider the airplane described in part (b), headed N 60° W at a speed of 400 miles per hour. What wind velocity, in the direction of N 40° E, will produce a resultant direction of N 50° W? Explain how to use the program listed above to obtain the answer *experimentally*. Then explain how to obtain the answer algebraically.

7. Consider the airplane described in part (b), headed N 60° W at a speed of 400 miles per hour. What wind direction, at a speed of 75 miles per hour, will produce a resultant direction of N 50° W? Explain how to use the program listed above to obtain the answer *experimentally*. Then explain how to obtain the answer algebraically.

6 Chapter Test

Take this test as you would take a test in class. After you are done, check your work against the answers given in the back of the book.

In Exercises 1–6, use the given information to solve the triangle. If two solutions exist, find both.

1. $c = 18.2$, $B = 98.4°$, $A = 37.6°$

2. $a = 14.9$, $b = 23.2$, $c = 10$

3. $A = 58°$, $b = 8.4$, $c = 11.2$

4. $A = 24.6°$, $b = 28$, $a = 15.6$

5. $B = 130°$, $c = 10.1$, $b = 5.2$

6. $A = 150°$, $b = 4.8$, $a = 9.4$

7. A plane flies 650 kilometers with a bearing of N 34° E from A to B. The plane then flies 1070 kilometers southeast from B to C. Find the bearing of the flight from B to C, as shown in the figure at the right.

8. Find the length of the pond shown at the right.

9. Find the component form and length of the vector **w** that has an initial point $(-8, -12)$ and terminal point $(4, 1)$.

10. Find a unit vector in the direction of the vector $\mathbf{v} = 7\mathbf{i} + 4\mathbf{j}$.

In Exercises 11–14, find (a) $2\mathbf{v} + \mathbf{u}$, (b) $\mathbf{u} - 3\mathbf{v}$, and (c) $5\mathbf{u} - \mathbf{v}$.

11. $\mathbf{u} = \langle 0, -4 \rangle$, $\mathbf{v} = \langle -2, 4 \rangle$

12. $\mathbf{u} = \langle -2, -3 \rangle$, $\mathbf{v} = \langle 1, 5 \rangle$

13. $\mathbf{u} = \mathbf{i} - \mathbf{j}$, $\mathbf{v} = 6\mathbf{i} + 9\mathbf{j}$

14. $\mathbf{u} = 2\mathbf{i} + 3\mathbf{j}$, $\mathbf{v} = -\mathbf{i} - 2\mathbf{j}$

15. Find the component form of the vector **v** with $\|\mathbf{v}\| = 12$, in the same direction as $\mathbf{u} = \langle 3, -5 \rangle$.

16. Forces with magnitudes of 250 pounds and 130 pounds act on an object at angles of 45° and −60°, respectively, with the x-axis. Find the direction and magnitude of the resultant of these forces.

17. Find the angle between the vectors $\mathbf{u} = 7\mathbf{i} + 2\mathbf{j}$ and $\mathbf{v} = -4\mathbf{j}$.

18. Are the vectors $\mathbf{u} = \langle 6, -4 \rangle$ and $\mathbf{v} = \langle 2, -3 \rangle$ orthogonal? Explain.

19. Find the projection of $\mathbf{u} = \langle 6, 7 \rangle$ onto $\mathbf{v} = \langle -5, -1 \rangle$. Then write **u** as the vector sum of two orthogonal vectors.

20. Write the complex number $z = -2 + 2i$ in trigonometric form.

In Exercises 21 and 22, write the complex number in standard form.

21. $100(\cos 240° + i \sin 240°)$

22. $24(\cos 330° + i \sin 330°)$

In Exercises 23 and 24, use DeMoivre's Theorem to find the indicated power of the complex number. Express the result in standard form.

23. $\left[3\left(\cos \dfrac{5\pi}{6} + i \sin \dfrac{5\pi}{6} \right) \right]^8$

24. $(3 - 3i)^6$

25. Find the fourth roots of $128\left(1 + \sqrt{3}i \right)$.

26. Find all solutions of the equation $x^4 - 625i = 0$ and represent the solutions graphically.

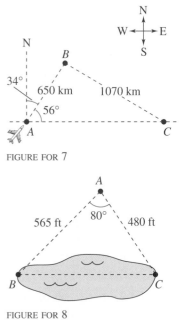

FIGURE FOR 7

FIGURE FOR 8

The *Interactive* CD-ROM and *Internet* versions of this text provide answers to the Chapter Tests and Cumulative Tests. They also offer Chapter Pre-Tests (which test key skills and concepts covered in previous chapters) and Chapter Post-Tests, both of which have randomly generated exercises with diagnostic capabilities.

Cumulative Test

Take this test to review the material from earlier chapters. After you are done, check your work against the answers given in the back of the book.

1. Consider the angle $\theta = -120°$.
   (a) Sketch the angle in standard position.
   (b) Determine a coterminal angle in the interval $[0°, 360°)$.
   (c) Convert the angle to radian measure.
   (d) Find its reference angle θ'.
   (e) Find the exact values of the six trigonometric functions of θ.

2. Convert the angle of magnitude 2.35 radians to degrees. Round the answer to one decimal place.

3. Find $\cos \theta$ if $\tan \theta = -\frac{4}{3}$ and $\sin \theta < 0$.

In Exercises 4–6, sketch the graph of the function by hand. Use a graphing utility to verify your graph.

4. $f(x) = 3 - 2 \sin \pi x$ 5. $f(x) = \dfrac{1}{2} \tan\left(x - \dfrac{\pi}{2}\right)$ 6. $f(x) = \dfrac{1}{2} \sec(x + \pi)$

7. Find a, b, and c such that the graph of the function $h(x) = a \cos(bx + c)$ matches the graph in the figure at the right.

FIGURE FOR 7

In Exercises 8 and 9, find the exact value without using a calculator.

8. $\tan(\arctan 6.7)$ 9. $\tan\left(\arcsin \frac{3}{5}\right)$

10. Write an algebraic expression equivalent to $\sin(\arccos 2x)$.

11. Find the altitude of the triangle shown at the right.

12. Subtract and simplify: $\dfrac{\sin \theta - 1}{\cos \theta} - \dfrac{\cos \theta}{\sin \theta - 1}$.

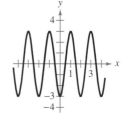

22 ft

74°

FIGURE FOR 11

In Exercises 13 and 14, verify the identity.

13. $\cot^2 \alpha(\sec^2 \alpha - 1) = 1$ 14. $\sin(x + y) \sin(x - y) = \sin^2 x - \sin^2 y$

In Exercises 15 and 16, find all solutions of the equation in the interval $[0, 2\pi)$.

15. $2 \cos^2 \beta - \cos \beta = 0$ 16. $3 \tan \theta - \cot \theta = 0$

17. Approximate the solutions to the equation $\cos^2 x - 5 \cos x - 1 = 0$ in the interval $[0, 2\pi)$.

In Exercises 18 and 19, use a graphing utility to graph the function and approximate its zeros in the interval $[0, 2\pi)$. If possible, find the exact values of the zeros algebraically.

18. $y = \dfrac{1 + \sin x}{\cos x} + \dfrac{\cos x}{1 + \sin x} - 4$ 19. $y = \tan^3 x - \tan^2 x + 3 \tan x - 3$

In Exercises 20 and 21, write the trigonometric expression as an algebraic expression.

20. $\cos(2 \arccos 2x)$ **21.** $\sin(2 \arctan x)$

22. Use the half-angle formulas to determine the exact values of the sine, cosine, and tangent of $\theta = 67° \, 30'$.

23. Write $\cos 8x + \cos 4x$ as a product.

In Exercises 24–27, verify the identity.

24. $\tan x(1 - \sin^2 x) = \frac{1}{2} \sin 2x$ **25.** $\sin 3\,\theta \sin \theta = \frac{1}{2}(\cos 2\theta - \cos 4\theta)$

26. $\sin 3x \cos 2x = \frac{1}{2}(\sin 5x + \sin x)$ **27.** $\dfrac{2 \cos 3x}{\sin 4x - \sin 2x} = \csc x$

In Exercises 28–31, use the given information to solve the triangle shown at the right.

28. $A = 46°, \; a = 14, \; b = 5$ **29.** $A = 30°, \; b = 8, \; c = 10$

30. $A = 24°, \; C = 101°, \; a = 10$ **31.** $a = 24, \; b = 30, \; c = 47$

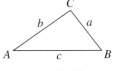

FIGURE FOR 28–31

32. Two sides of a triangle have lengths 14 and 19 inches. Their included angle measures 82°. Find the area of the triangle.

33. Find the area of a triangle with sides of length 11, 16, and 17 inches.

34. Write the vector $\mathbf{u} = \langle 3, 5 \rangle$ as a linear combination of the standard unit vectors $\mathbf{i}$ and $\mathbf{j}$.

35. Find a unit vector in the direction of $\mathbf{v} = \mathbf{i} + \mathbf{j}$.

36. Find $\mathbf{u} \cdot \mathbf{v}$ for $\mathbf{u} = 3\mathbf{i} + 4\mathbf{j}$ and $\mathbf{v} = \mathbf{i} - 2\mathbf{j}$.

37. Find the projection of $\mathbf{u} = \langle 8, -2 \rangle$ onto $\mathbf{v} = \langle 1, 5 \rangle$, and the vector component of $\mathbf{u}$ orthogonal to $\mathbf{v}$.

38. Find the trigonometric form of the complex number shown at the right.

39. Find the standard form of $8\left(\cos \dfrac{5\pi}{6} + i \sin \dfrac{5\pi}{6}\right)$.

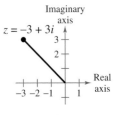

FIGURE FOR 38

40. Find the product of $[4(\cos 30° + i \sin 30°)][6(\cos 120° + i \sin 120°)]$. Write the answer in standard form.

41. Find the three cube roots of 1.

42. Write all the solutions of the equation $x^5 + 243 = 0$.

43. From a point 200 feet from a flagpole, the angles of elevation to the bottom and top of the flag are $16° \, 45'$ and 18°, respectively. Approximate the height of the flag to the nearest foot.

44. Write a model for a particle in simple harmonic motion with a displacement of 4 inches and a period of 8 seconds.

45. An airplane's velocity with respect to the air is 500 kilometers per hour, with a bearing of N 30° E. The wind at the altitude of the plane has a velocity of 50 kilometers per hour with a bearing of N 60° E. What is the true direction of the plane, and what is its speed relative to the ground?

46. Forces of 60 pounds and 100 pounds have a resultant force of 125 pounds. Find the angle between the two forces.

Systems of Equations and Inequalities

7.1 Solving Systems of Equations
7.2 Systems of Linear Equations in Two Variables
7.3 Multivariable Linear Systems
7.4 Systems of Inequalities
7.5 Linear Programming

The Big Picture

In this chapter you will learn how to

❑ solve systems of equations by substitution, by elimination, by Gaussian elimination, and graphically.
❑ recognize linear systems in row-echelon form and to use back substitution to solve the system.
❑ solve nonsquare systems of equations.
❑ sketch the graphs of inequalities in two variables and solve systems of inequalities.
❑ solve linear programming problems.
❑ use systems of equations and inequalities to model and solve real-life problems.

Ron Watts/CORBIS

In 1996, 57 million newspapers were printed daily in the United States. With a population of over 265 million, there were about 215 newspapers per 1000 people. (Source: U.S. Bureau of the Census and Editor & Publisher, Co.)

Important Vocabulary

As you encounter each new vocabulary term in this chapter, add the term and its definition to your notebook glossary.

Additional Resources Text-specific additional resources are available to help you do well in this course. See page xvi for details.

7.1 Solving Systems of Equations

The Method of Substitution

Up to this point in the text, most problems have involved either a function of one variable or a single equation in two variables. However, many problems in science, business, and engineering involve two or more equations in two or more variables. To solve such problems, you need to find solutions of **systems of equations.** Here is an example of a system of two equations in two unknowns, x and y.

$$\begin{cases} 2x + y = 5 & \text{Equation 1} \\ 3x - 2y = 4 & \text{Equation 2} \end{cases}$$

A **solution** of this system is an ordered pair that satisfies each equation in the system. Finding the set of all such solutions is called **solving the system of equations.** For instance, the ordered pair $(2, 1)$ is a solution of this system. To check this, you can substitute 2 for x and 1 for y in *each* equation.

Check $(2, 1)$ *in Equation 1:*

$$2x + y = 5 \qquad \text{Write Equation 1.}$$
$$2(2) + 1 \overset{?}{=} 5 \qquad \text{Substitute 2 for } x \text{ and 1 for } y.$$
$$4 + 1 = 5 \qquad \text{Solution checks in Equation 1.} \checkmark$$

Check $(2, 1)$ *in Equation 2:*

$$3x - 2y = 4 \qquad \text{Write Equation 2.}$$
$$3(2) - 2(1) \overset{?}{=} 4 \qquad \text{Substitute 2 for } x \text{ and 1 for } y.$$
$$6 - 2 = 4 \qquad \text{Solution checks in Equation 2.} \checkmark$$

In this chapter you will study four ways to solve equations, beginning with the **method of substitution.**

Method	Section	Type of System
1. Substitution	7.1	Linear or nonlinear, 2 variables
2. Graphical	7.1	Linear or nonlinear, 2 variables
3. Elimination	7.2	Linear, 2 variables
4. Gaussian Elimination	7.3	Linear, 3 or more variables

The Method of Substitution

1. Solve one of the equations for one variable in terms of the other.
2. Substitute the expression found in Step 1 into the other equation to obtain an equation in one variable.
3. Solve the equation obtained in Step 2.
4. Back-substitute the solution in Step 3 into the expression obtained in Step 1 to find the value of the other variable.
5. Check that the solution satisfies each of the original equations.

What You Should Learn:

- How to use the method of substitution to solve systems of equations in two variables
- How to solve systems of equations graphically
- How to use systems of equations to model and solve real-life problems

Why You Should Learn It:

You can use systems of equations in situations where the variables must satisfy two or more conditions. For instance, Exercise 73 on page 497 shows how to use a system of equations to compare two models for estimating the number of board feet in a 16-foot log.

Bruce Hands/Tony Stone Images

In the algebraic solution of Example 1, you use the method of substitution to solve the system of equations. In the graphical solution, note that the solution of the system corresponds to the **point of intersection** of the graphs.

EXAMPLE 1 Solving a System of Equations

Solve the system of equations.

$$\begin{cases} x + y = 4 & \text{Equation 1} \\ x - y = 2 & \text{Equation 2} \end{cases}$$

Algebraic Solution

Begin by solving for y in Equation 1.

$y = 4 - x$	Solve for y in Equation 1.

Next, substitute this expression for y into Equation 2 and solve the resulting single-variable equation for x.

$x - y = 2$	Write Equation 2.
$x - (4 - x) = 2$	Substitute $4 - x$ for y.
$x - 4 + x = 2$	Simplify.
$2x = 6$	Combine like terms.
$x = 3$	Divide each side by 2.

Finally, you can solve for y by *back-substituting* $x = 3$ into the equation $y = 4 - x$, to obtain

$y = 4 - x$	Write Equation 1.
$y = 4 - 3$	Substitute 3 for x.
$y = 1.$	Solve for y.

The solution is the ordered pair $(3, 1)$. Check this as follows.

Check $(3, 1)$ *in Equation 1:*

$x + y = 4$	Write Equation 1.
$3 + 1 \overset{?}{=} 4$	Substitute for x and y.
$4 = 4$	Solution checks in Equation 1. ✓

Check $(3, 1)$ *in Equation 2:*

$x - y = 2$	Write Equation 2.
$3 - 1 \overset{?}{=} 2$	Substitute for x and y.
$2 = 2$	Solution checks in Equation 2. ✓

Graphical Solution

Begin by solving both equations for y. Then use a graphing utility to graph the equations

$$y_1 = 4 - x \qquad \text{and} \qquad y_2 = x - 2$$

in the same viewing window. Use the *intersect* feature or the *zoom* and *trace* features of the graphing utility to approximate the point of intersection of the graphs. In Figure 7.1, the point of intersection is $(3, 1)$.

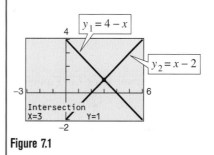

Figure 7.1

Check that $(3, 1)$ is the exact solution as follows.

Check $(3, 1)$ *in Equation 1:*

$x + y = 4$	Write Equation 1.
$3 + 1 \overset{?}{=} 4$	Substitute for x and y.
$4 = 4$	Solution checks in Equation 1. ✓

Check $(3, 1)$ *in Equation 2:*

$x - y = 2$	Write Equation 2.
$3 - 1 \overset{?}{=} 2$	Substitute for x and y.
$2 = 2$	Solution checks in Equation 2. ✓

In the algebraic solution of Example 1, note that the term *back-substitution* implies that you work *backward*. First you solve for one of the variables, and then you substitute that value *back* into one of the equations in the system to find the value of the other variable.

EXAMPLE 2 Solving a System by Substitution

A total of $12,000 is invested in two funds paying 9% and 11% simple interest. The yearly interest is $1180. How much is invested at each rate?

Solution

Verbal Model:

| 9% fund | + | 11% fund | = | Total investment |

| 9% interest | + | 11% interest | = | Total interest |

Labels: Amount in 9% fund = x (dollars)
Interest for 9% fund = $0.09x$ (dollars)
Amount in 11% fund = y (dollars)
Interest for 11% fund = $0.11y$ (dollars)
Total investment = $12,000$ (dollars)
Total interest = 1180 (dollars)

System:
$$\begin{cases} x + y = 12{,}000 & \text{Equation 1} \\ 0.09x + 0.11y = 1{,}180 & \text{Equation 2} \end{cases}$$

To begin, it is convenient to multiply both sides of Equation 2 by 100 to obtain $9x + 11y = 118{,}000$. This eliminates the need to work with decimals.

$$9x + 11y = 118{,}000 \qquad \text{Revised Equation 2}$$

To solve this system, you can solve for x in Equation 1.

$$x = 12{,}000 - y \qquad \text{Revised Equation 1}$$

Next, substitute this expression for x into Revised Equation 2 and solve the resulting equation for y.

$$9x + 11y = 118{,}000 \qquad \text{Write Revised Equation 2.}$$
$$9(12{,}000 - y) + 11y = 118{,}000 \qquad \text{Substitute } 12{,}000 - y \text{ for } x.$$
$$108{,}000 - 9y + 11y = 118{,}000 \qquad \text{Distributive Property}$$
$$2y = 10{,}000 \qquad \text{Combine like terms.}$$
$$y = 5000 \qquad \text{Divide each side by 2.}$$

Finally, back-substitute the value $y = 5000$ to solve for x.

$$x = 12{,}000 - y \qquad \text{Write Revised Equation 1.}$$
$$x = 12{,}000 - 5000 \qquad \text{Substitute 5000 for } y.$$
$$x = 7000 \qquad \text{Simplify.}$$

The solution is (7000, 5000). This means that $7000 is invested at 9% and $5000 is invested at 11% to yield yearly interest of $1180. Check this in the original system.

The equations in Examples 1 and 2 are linear. Substitution can also be used to solve systems in which one or both of the equations are nonlinear.

STUDY T!P

One way to check the answers you obtain in this section is to use a graphing utility. For instance, enter the two equations in Example 2

$$y_1 = 12{,}000 - x$$

$$y_2 = \frac{1180 - 0.09x}{0.11}$$

and find an appropriate viewing window that shows where the lines intersect. Then use the *intersect* feature or the *zoom* and *trace* features to find their point of intersection.

EXAMPLE 3   Substitution: Two-Solution Case

Solve the system of equations.

A computer animation of this concept appears in the *Interactive* CD-ROM and *Internet* versions of this text.

$$\begin{cases} x^2 + 4x - y = 7 & \text{Equation 1} \\ 2x - y = -1 & \text{Equation 2} \end{cases}$$

Algebraic Solution

Begin by solving for y in Equation 2 to obtain $y = 2x + 1$.
Next, substitute this expression for y into Equation 1 and solve for x.

$x^2 + 4x - y = 7$	Write Equation 1.
$x^2 + 4x - (2x + 1) = 7$	Substitute $2x + 1$ for y.
$x^2 + 2x - 8 = 0$	Write in general form.
$(x + 4)(x - 2) = 0$	Factor.
$x + 4 = 0 \implies x = -4$	Set 1st factor equal to 0.
$x - 2 = 0 \implies x = 2$	Set 2nd factor equal to 0.

Back-substituting these values of x to solve for the corresponding values of y produces the solutions $(-4, -7)$ and $(2, 5)$. Check these in the original system.

Graphical Solution

To graph each equation, first solve both equations for y. Then use a graphing utility to graph the equations in the same viewing window. Use the *intersect* feature or the *zoom* and *trace* features to approximate the points of intersection of the graphs. In Figure 7.2, the points of intersection are $(-4, -7)$ and $(2, 5)$. Check that $(-4, -7)$ and $(2, 5)$ are the exact solutions by substituting *both* ordered pairs into *both* equations.

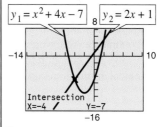

Figure 7.2

EXAMPLE 4   Substitution: No-Solution Case

Solve the system of equations.

$$\begin{cases} -x + y = 4 & \text{Equation 1} \\ x^2 + y = 3 & \text{Equation 2} \end{cases}$$

Solution

Begin by solving for y in Equation 1 to obtain $y = x + 4$. Next, substitute this expression for y into Equation 2 and solve for x.

$x^2 + y = 3$	Write Equation 2.
$x^2 + (x + 4) = 3$	Substitute $x + 4$ for y.
$x^2 + x + 1 = 0$	Simplify.
$x = \dfrac{-1 \pm \sqrt{1 - 4}}{2}$	Quadratic Formula

> **STUDY T!P**
>
> When using substitution, solve for the variable that is not raised to a power in either equation. For instance, in Example 4 it would not be practical to solve for x in Equation 2. Can you see why?

Because this yields two complex values, the equation $x^2 + x + 1 = 0$ has no *real* solution. So, the original system of equations has no *real* solution.

Try graphing the system of equations in Example 4. Do the graphs of the equations intersect? Why or why not?

From Examples 2, 3, and 4, you can see that a system of two equations in two unknowns can have exactly one solution, more than one solution, or no solution. For instance, in Figure 7.3(a), the two equations graph as two lines with a *single point* of intersection. The two equations in Example 3 graph as a parabola and a line with *two points* of intersection, as shown in Figure 7.3(b). The two equations in Example 4 graph as a line and a parabola that happen to have *no points* of intersection, as shown in Figure 7.3(c).

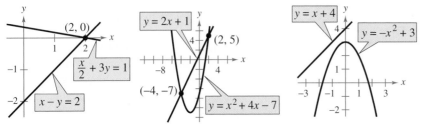

(a) One Intersection Point   **(b) Two Intersection Points**   **(c) No Intersection Points**
Figure 7.3

Example 5 shows the value of a graphical approach to solving systems of equations in two variables. Notice what would happen if you tried only the substitution method in Example 5. You would obtain the equation $x + \ln x = 1$. It would be difficult to solve this equation for x using standard algebraic techniques. In such cases, a graphical approach to solving a system of equations is more convenient.

EXAMPLE 5   Solving a System of Equations Graphically

Solve the system of equations.

$$\begin{cases} \quad y = \ln x & \text{Equation 1} \\ x + y = 1 & \text{Equation 2} \end{cases}$$

Solution

From the graphs of these equations, shown in Figure 7.4, it is clear that there is only one point of intersection. Use the *intersect* feature or the *zoom* and *trace* features of a graphing utility to approximate the solution point as $(1, 0)$. You can confirm this by substituting in *both* equations.

Check

$$0 = \ln 1 \qquad \text{Equation 1 checks. } \checkmark$$

$$1 + 0 = 1 \qquad \text{Equation 2 checks. } \checkmark$$

So, the solution is $(1, 0)$.

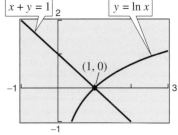

Figure 7.4

Applications

The total cost C of producing x units of a product typically has two components: the initial cost and the cost per unit. When enough units have been sold so that the total revenue R equals the total cost, the sales are said to have reached the **break-even point.** You will find that the break-even point corresponds to the point of intersection of the cost and revenue curves.

EXAMPLE 6 Break-Even Analysis

A small business invests $10,000 in equipment to produce a product. Each unit of the product costs $0.65 to produce and is sold for $1.20. How many items must be sold before the business breaks even?

Solution

The total cost of producing x units is

| Total cost | = | Cost per unit | . | Number of units | + | Initial cost |

$$C = 0.65x + 10,000. \qquad \text{Equation 1}$$

The revenue obtained by selling x units is

| Total revenue | = | Price per unit | . | Number of units |

$$R = 1.2x. \qquad \text{Equation 2}$$

Because the break-even point occurs when $R = C$, you have

$1.2x = 0.65x + 10,000$	Equate R and C.
$0.55x = 10,000$	Subtract 0.65x from each side.
$x = \dfrac{10,000}{0.55}$	Divide each side by 0.55.
$x \approx 18,182$ units.	Use a calculator.

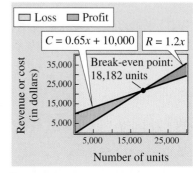

Figure 7.5

Note in Figure 7.5 that sales less than the break-even point correspond to an overall loss, whereas sales greater than the break-even point correspond to a profit. Verify the break-even point using the *intersect* feature or the *zoom* and *trace* features of a graphing utility.

Another way to view the solution in Example 6 is to consider the profit function

$$P = R - C.$$

The break-even point occurs when the profit is 0, which is the same as saying that $R = C$.

EXAMPLE 7   State Populations

From 1990 to 1997, the population of Arizona was increasing at a faster rate than the population of Alabama. Two models that approximate the populations P (in thousands) are

$$\begin{cases} P = 3631.2 + 131.7t \qquad \text{\small Arizona} \\ P = 4055.6 + 39.9t \qquad \text{\small Alabama} \end{cases}$$

where $t = 0$ represents 1990. According to these two models, when would you expect the population of Arizona to have exceeded the population of Alabama? (Source: U.S. Bureau of the Census)

Algebraic Solution

Because the first equation has already been solved for P in terms of t, you can substitute this value into the second equation and solve for t, as follows.

$$3631.2 + 131.7t = 4055.6 + 39.9t$$

$$131.7t - 39.9t = 4055.6 - 3631.2$$

$$91.8t = 424.4$$

$$t \approx 4.6$$

So, from the given models, you would expect that the population of Arizona exceeded the population of Alabama sometime during 1994.

Graphical Solution

Use a graphing utility to graph $y_1 = 3631.2 + 131.7x$ and $y_2 = 4055.6 + 39.9x$ in the same viewing window. Use the *intersect* feature or the *zoom* and *trace* features of the graphing utility to approximate the point of intersection of the graphs. In Figure 7.6, the point of intersection occurs when $x \approx 4.6$.

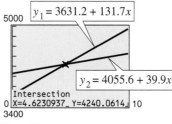

Figure 7.6

So, from the graph, it appears that the population of Arizona exceeded the population of Alabama sometime during 1994.

Writing About Math   *Points of Intersection*

In this section, you learned that the graphs of two equations can intersect at zero, one, or more points. Use a graphing utility to graph each of the following systems. Decide whether the system has no solution, one solution, two solutions, or more than two solutions.

a. $\begin{cases} -0.5x^2 + 0.25x + y = -1.5 \\ 1.75x + y = -2.625 \end{cases}$ b. $\begin{cases} -0.5x^2 + 0.25x + y = -1.5 \\ -1.25x + y = -4.5 \end{cases}$

c. $\begin{cases} -0.5x^2 + 0.25x + y = -1.5 \\ 3.5x + y = 2.25 \end{cases}$

Create three other systems of equations, one with no solution, one with one solution, and one with two solutions. Write a short paragraph that explains how you created these systems. Include a description of how you can check that the systems have the appropriate number of solutions.

7.1 Exercises

In Exercises 1–4, decide whether each ordered pair is a solution of the system of equations.

1. $\begin{cases} 4x - y = 1 \\ 6x + y = -6 \end{cases}$ (a) $(0, -3)$ (b) $(-1, -5)$
                     (c) $\left(-\frac{3}{2}, 3\right)$ (d) $\left(-\frac{1}{2}, -3\right)$

2. $\begin{cases} 4x^2 + y = 3 \\ -x - y = 11 \end{cases}$ (a) $(2, -13)$ (b) $(-2, -9)$
                     (c) $\left(-\frac{3}{2}, 6\right)$ (d) $\left(-\frac{7}{4}, -\frac{37}{4}\right)$

3. $\begin{cases} y = -2e^x \\ 3x - y = 2 \end{cases}$ (a) $(-2, 0)$ (b) $(0, -2)$
                     (c) $(0, -3)$ (d) $(-1, -5)$

4. $\begin{cases} -\log_{10} x + 3 = y \\ \frac{1}{9}x + y = \frac{28}{9} \end{cases}$ (a) $(100, 1)$ (b) $(10, 2)$
                     (c) $(1, 3)$ (d) $(1, 1)$

In Exercises 5–14, solve the system by the method of substitution. Check your solution graphically.

5. $\begin{cases} 2x + y = 6 \\ -x + y = 0 \end{cases}$ **6.** $\begin{cases} x - y = -4 \\ x + 2y = 5 \end{cases}$

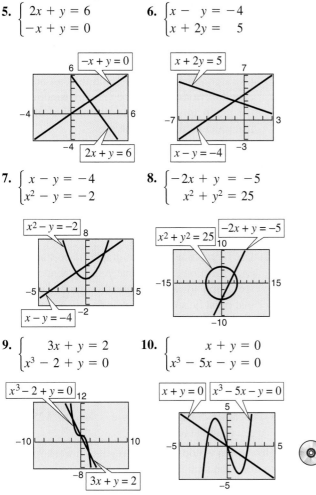

7. $\begin{cases} x - y = -4 \\ x^2 - y = -2 \end{cases}$ **8.** $\begin{cases} -2x + y = -5 \\ x^2 + y^2 = 25 \end{cases}$

9. $\begin{cases} 3x + y = 2 \\ x^3 - 2 + y = 0 \end{cases}$ **10.** $\begin{cases} x + y = 0 \\ x^3 - 5x - y = 0 \end{cases}$

11. $\begin{cases} x^2 + y = 0 \\ x^2 - 4x - y = 0 \end{cases}$ **12.** $\begin{cases} y = -2x^2 + 2 \\ y = 2(x^4 - 2x^2 + 1) \end{cases}$

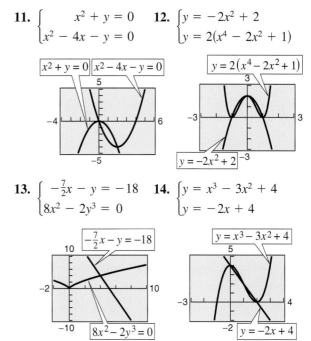

13. $\begin{cases} -\frac{7}{2}x - y = -18 \\ 8x^2 - 2y^3 = 0 \end{cases}$ **14.** $\begin{cases} y = x^3 - 3x^2 + 4 \\ y = -2x + 4 \end{cases}$

In Exercises 15–28, solve the system by substitution. Use a graphing utility to verify your result.

15. $\begin{cases} x - y = 0 \\ 5x - 3y = 10 \end{cases}$ **16.** $\begin{cases} x + 2y = 1 \\ 5x - 4y = -23 \end{cases}$

17. $\begin{cases} 2x - y + 2 = 0 \\ 4x + y - 5 = 0 \end{cases}$ **18.** $\begin{cases} 6x - 3y - 4 = 0 \\ x + 2y - 4 = 0 \end{cases}$

19. $\begin{cases} 1.5x + 0.8y = 2.3 \\ 0.3x - 0.2y = 0.1 \end{cases}$ **20.** $\begin{cases} 0.5x + 3.2y = 9.0 \\ 0.2x - 1.6y = -3.6 \end{cases}$

21. $\begin{cases} \frac{1}{5}x + \frac{1}{2}y = 8 \\ x + y = 20 \end{cases}$ **22.** $\begin{cases} \frac{1}{2}x + \frac{3}{4}y = 10 \\ \frac{3}{4}x - y = 4 \end{cases}$

23. $\begin{cases} 8x + 4y = 7 \\ 2x + y = 0 \end{cases}$ **24.** $\begin{cases} 2x - y = 4 \\ -4x + 2y = -12 \end{cases}$

25. $\begin{cases} -\frac{5}{3}x + y = 5 \\ -5x + 3y = 6 \end{cases}$ **26.** $\begin{cases} -\frac{2}{3}x + y = -2 \\ 2x - 3y = 6 \end{cases}$

27. $\begin{cases} x - y = 0 \\ 2x + y = 0 \end{cases}$ **28.** $\begin{cases} x - 2y = 0 \\ 3x - y = 0 \end{cases}$

The *Interactive* CD-ROM and *Internet* versions of this text contain step-by-step solutions to all odd-numbered Section and Review Exercises. They also provide Tutorial Exercises, which link to Guided Examples for additional help.

In Exercises 29–36, solve the system of equations graphically. Verify your solutions algebraically.

29. $\begin{cases} -x + 2y = 2 \\ 3x + y = 15 \end{cases}$ **30.** $\begin{cases} x + y = 0 \\ 3x - 2y = 10 \end{cases}$

31. $\begin{cases} x - 3y = -2 \\ 5x + 3y = 17 \end{cases}$ **32.** $\begin{cases} -x + 2y = 1 \\ x - y = 2 \end{cases}$

33. $\begin{cases} x + y = 4 \\ x^2 + y^2 - 4x = 0 \end{cases}$ **34.** $\begin{cases} -x + y = 3 \\ x^2 - 6x - 27 + y^2 = 0 \end{cases}$

35. $\begin{cases} x - y + 3 = 0 \\ x^2 - 4x + 7 = y \end{cases}$ **36.** $\begin{cases} y^2 - 4x + 11 = 0 \\ -\frac{1}{2}x + y = -\frac{1}{2} \end{cases}$

In Exercises 37–50, use a graphing utility to approximate all points of intersection of the graph of the system of equations. Verify your solutions by checking them in the original system.

37. $\begin{cases} 7x + 8y = 24 \\ x - 8y = 8 \end{cases}$ **38.** $\begin{cases} x - y = 0 \\ 5x - 2y = 6 \end{cases}$

39. $\begin{cases} 2x - y + 3 = 0 \\ x^2 + y^2 - 4x = 0 \end{cases}$ **40.** $\begin{cases} 3x - 2y = 0 \\ x^2 + y^2 = 4 \end{cases}$

41. $\begin{cases} x^2 + y^2 = 8 \\ y = x^2 \end{cases}$ **42.** $\begin{cases} x^2 + y^2 = 25 \\ (x - 8)^2 + y^2 = 41 \end{cases}$

43. $\begin{cases} y = e^x \\ x - y + 1 = 0 \end{cases}$ **44.** $\begin{cases} y = -4e^{-x} \\ y + 3x + 8 = 0 \end{cases}$

45. $\begin{cases} x + 2y = 8 \\ y = \log_2 x \end{cases}$

46. $\begin{cases} y = -2 + \ln(x - 1) \\ 3y + 2x = 9 \end{cases}$

47. $\begin{cases} y = \sqrt{x} \\ y = x \end{cases}$ **48.** $\begin{cases} x - y = 3 \\ x - y^2 = 1 \end{cases}$

49. $\begin{cases} x^2 + y^2 = 169 \\ x^2 - 8y = 104 \end{cases}$ **50.** $\begin{cases} x^2 + y^2 = 4 \\ 2x^2 - y = 2 \end{cases}$

In Exercises 51–62, solve the system graphically or algebraically. Explain your choice of method.

51. $\begin{cases} y = 2x \\ y = x^2 + 1 \end{cases}$ **52.** $\begin{cases} x + y = 4 \\ x^2 + y = 2 \end{cases}$

53. $\begin{cases} 3x - 7y + 6 = 0 \\ x^2 - y^2 = 4 \end{cases}$ **54.** $\begin{cases} x^2 + y^2 = 25 \\ 2x + y = 10 \end{cases}$

55. $\begin{cases} x - 2y = 4 \\ x^2 - y = 0 \end{cases}$ **56.** $\begin{cases} y = (x + 1)^3 \\ y = \sqrt{x - 1} \end{cases}$

57. $\begin{cases} y - e^{-x} = 1 \\ y - \ln x = 3 \end{cases}$ **58.** $\begin{cases} x^2 + y = 4 \\ e^x - y = 0 \end{cases}$

59. $\begin{cases} y = x^3 - 2x^2 + 1 \\ y = 1 - x^2 \end{cases}$ **60.** $\begin{cases} y = x^3 - 2x^2 + x - 1 \\ y = -x^2 + 3x - 1 \end{cases}$

61. $\begin{cases} xy - 1 = 0 \\ 2x - 4y + 7 = 0 \end{cases}$ **62.** $\begin{cases} x - 2y = 1 \\ y = \sqrt{x - 1} \end{cases}$

Break-Even Analysis In Exercises 63–66, use a graphing utility to graph the cost and revenue functions in the same viewing window. Find the sales x necessary to break even ($R = C$) and the corresponding revenue R obtained by selling x units. (Round to the nearest whole unit.)

	Cost	*Revenue*
63.	$C = 8650x + 250{,}000$	$R = 9950x$
64.	$C = 2.65x + 350{,}000$	$R = 4.15x$
65.	$C = 5.5\sqrt{x} + 10{,}000$	$R = 3.29x$
66.	$C = 7.8\sqrt{x} + 18{,}500$	$R = 12.84x$

67. *Break-Even Point* A small business invests $16,000 to produce an item that will sell for $5.95. Each unit can be produced for $3.45.

(a) Write cost and revenue functions for x units produced and sold.

(b) Use a graphing utility to graph the cost and revenue functions in the same viewing window. Use the graph to approximate the number of units that must be sold to break even.

(c) Verify the result of part (b) algebraically.

68. *Break-Even Point* A small business has an initial investment of $5000. The unit cost of the product is $21.60, and the selling price is $34.10.

(a) Write cost and revenue functions for x units produced and sold.

(b) Use a graphing utility to graph the cost and revenue functions in the same viewing window. Use the graph to approximate the number of units that must be sold to break even.

(c) Verify the result of part (b) algebraically.

69. *Choice of Two Jobs* You are offered two different jobs selling dental supplies.

• One company offers a straight commission of 6% of sales.

• The other company offers a salary of $250 per week plus 3% of sales.

How much would you have to sell in a week in order to make the straight commission offer better?

70. ***Choice of Two Jobs***   You are offered two different jobs selling college textbooks.

- One company offers an annual salary of $25,000 plus a year-end bonus of 1% of your total sales.
- The other company offers an annual salary of $20,000 plus a year-end bonus of 2% of your total sales.

Determine the annual sales that make the second offer better.

71. ***Finance***   A total of $20,000 is invested in two funds paying 6.5% and 8.5% simple interest. The 6.5% investment has a lower risk. The investor wants a yearly interest check of $1600 from the investment.

(a) Write a system of equations in which one equation represents the total amount invested and the other equation represents the $1600 required in interest. Let x and y represent the amounts invested at 6.5% and 8.5%, respectively.

(b) Use a graphing utility to graph the two equations in the same viewing window. As the amount invested at 6.5% increases, how do the amount invested at 8.5% and the amount of interest each change? Explain.

(c) What is the most that can be invested at 6.5% to meet the requirement of $1600 per year in interest?

72. ***Finance***   A total of $25,000 is invested in two funds paying 6% and 8.5% simple interest. The 6% investment has a lower risk. The investor wants a yearly interest income of $2000 from the two investments.

(a) Write a system of equations in which one equation represents the total amount invested and the other equation represents the $2000 required in interest. Let x and y represent the amounts invested at 6% and 8.5%, respectively.

(b) Use a graphing utility to graph the two equations in the same viewing window. As the amount invested at 6% increases, how do the amount invested at 8.5% and the amount of interest each change? Explain.

(c) What is the most that can be invested at 6% to meet the requirement of $2000 per year in interest?

73. ***Log Volume***   You are offered two different rules for estimating the number of board feet in a log that is 16 feet long. A board foot is a unit of measure for lumber equal to a board one foot square and one inch thick. One rule is the *Doyle Log Rule* and is modeled by

$$V = (D - 4)^2, \quad 5 \le D \le 40$$

and the other rule is the *Scribner Log Rule* and is modeled by

$$V = 0.79D^2 - 2D - 4, \quad 5 \le D \le 40$$

where D is the diameter of the log and V is its volume in board feet.

(a) Use a graphing utility to graph the log rules in the same viewing window.

(b) For what diameter do the two rules agree?

(c) If you were selling large logs, which rule would you use? Explain your reasoning.

Geometry   **In Exercises 74–77, find the dimensions of the rectangle meeting the specified conditions.**

74. The perimeter is 30 meters and the length is 3 meters greater than the width.

75. The perimeter is 280 centimeters and the width is 20 centimeters less than the length.

76. The perimeter is 42 inches and the width is three-fourths the length.

77. The perimeter is 210 feet and the length is $1\frac{1}{2}$ times the width.

78. ***Geometry***   What are the dimensions of a rectangular tract of land if its perimeter is 40 miles and its area is 96 square miles?

79. ***Geometry***   What are the dimensions of an isosceles right triangle with a 2-inch hypotenuse and an area of 1 square inch?

80. ***Data Analysis***   The table shows the amount y (in millions of short tons) of paperboard produced in the United States in the years 1993 through 1996. (Source: American Forest and Paper Association)

Year	1993	1994	1995	1996
y	43.1	45.7	46.6	47.9

(a) Use the regression capabilities of a graphing utility to find a linear model and a quadratic model that represent the data in the interval from 1993 through 1996. (Let $t = 3$ represent 1993.)

(b) Use the graphing utility to graph the data and the two models in the same viewing window.

(c) Approximate the points of intersection of the graphs of the models.

(d) Use the models to estimate newsprint production in 1998. Which model do you think gives the more accurate estimate? Explain.

81. ***Data Analysis***   The table shows the average hourly earnings E of production workers in manufacturing industries in the United States for the years 1992 to 1997.   (Source: U.S. Bureau of Labor Statistics)

Year	1992	1993	1994
E	$11.46	$11.74	$12.06

Year	1995	1996	1997
E	$12.37	$12.78	$13.17

Let t represent the time in years, with $t = 2$ corresponding to 1992.

(a) Use the regression capabilities of a graphing utility to fit a linear model and a quadratic model to the data.

(b) Use the graphing utility to graph the data and the two models in the same viewing window.

(c) Approximate the points of intersections of the graphs of the models.

(d) Use the models to estimate the average hourly earnings in 2000. Which model do you think gives the more accurate estimate? Explain.

Synthesis

True or False?   **In Exercises 82 and 83, determine whether the statement is true or false. Justify your answer.**

82. In order to solve a system of equations by substitution, you must always solve for y in one of the two equations and then back-substitute.

83. If a system consists of a parabola and a circle, then it can have at most two solutions.

84. ***Think About It***   When solving a system of equations by substitution, how do you recognize that the system has no solution?

85. ***Writing***   Write a brief paragraph describing any advantages of substitution over the graphical method of solving a system of equations.

86. ***Exploration***   Find an equation of a line whose graph intersects the graph of the parabola $y = x^2$ at the following numbers of points. (There are many correct answers.)

(a) Two points

(b) One point

(c) No points

87. ***Conjecture***

(a) Use a graphing utility to graph the system of equations
$$\begin{cases} y = b^x \\ y = x^b \end{cases}$$
for $b = 2$ and $b = 4$.

(b) For a fixed value of $b > 1$, make a conjecture about the number of points of intersection of the graphs in part (a).

Review

In Exercises 88–93, find the general form of the equation of the line through the two points.

88. $(-2, 7), (5, 5)$

89. $(3.5, 4), (10, 6)$

90. $(6, 3), (10, 3)$

91. $(4, -2), (4, 5)$

92. $\left(\frac{3}{5}, 0\right), (4, 6)$

93. $\left(-\frac{7}{3}, 8\right), \left(\frac{5}{2}, \frac{1}{2}\right)$

In Exercises 94–97, find the domain of the function and identify any horizontal or vertical asymptotes.

94. $f(x) = \dfrac{5}{x - 6}$

95. $f(x) = \dfrac{2x - 7}{3x + 2}$

96. $f(x) = \dfrac{x^2 + 2}{x^2 - 16}$

97. $f(x) = 3 - \dfrac{2}{x^2}$

In Exercises 98–101, solve the equation. Round your result to three decimal places.

98. $6^{7-x} = 1249$

99. $8(10^{2x}) = 28$

100. $e^{2x} - 9e^x + 18 = 0$

101. $\log_{10}(x + 3) - \log_{10} x = \log_{10}(x + 8)$

7.2   Systems of Linear Equations in Two Variables

The Method of Elimination

In Section 7.1, you studied two methods for solving a system of equations: substitution and graphing. Now you will study the **method of elimination.** The key step in this method is to obtain, for one of the variables, coefficients that differ only in sign so that *adding* the equations eliminates the variable.

$$\begin{cases} 3x + 5y = 7 \\ -3x - 2y = -1 \end{cases}$$

         Equation 1

         Equation 2

$$3y = 6$$ Add equations.

Note that by adding the two equations, you eliminate the variable x and obtain a single equation in y. Solving this equation for y produces $y = 2$, which you can then back-substitute into one of the original equations to solve for x.

EXAMPLE 1   The Method of Elimination

Solve the system of linear equations.

$$\begin{cases} 3x + 2y = 4 \\ 5x - 2y = 8 \end{cases}$$

     Equation 1

     Equation 2

Solution

You can eliminate y by adding the two equations.

$$3x + 2y = 4$$ Write Equation 1.

$$5x - 2y = 8$$ Write Equation 2.

$$8x \quad\quad = 12$$ Add equations.

So, $x = \frac{3}{2}$. By back-substituting, you can solve for y.

$$3x + 2y = 4$$ Write Equation 1.

$$3\left(\tfrac{3}{2}\right) + 2y = 4$$ Substitute $\frac{3}{2}$ for x.

$$y = -\tfrac{1}{4}$$ Solve for y.

The solution is $\left(\frac{3}{2}, -\frac{1}{4}\right)$. You can check the solution *algebraically* by substituting into the original system, or graphically as shown in Section 7.1.

Check

$$3\left(\tfrac{3}{2}\right) + 2\left(-\tfrac{1}{4}\right) \stackrel{?}{=} 4$$ Substitute into Equation 1.

$$\tfrac{9}{2} - \tfrac{1}{2} = 4$$ Equation 1 checks. ✓

$$5\left(\tfrac{3}{2}\right) - 2\left(-\tfrac{1}{4}\right) \stackrel{?}{=} 8$$ Substitute into Equation 2.

$$\tfrac{15}{2} + \tfrac{1}{2} = 8$$ Equation 2 checks. ✓

Try using substitution to solve the system given in Example 1. Which method is easier? Many people find that the method of elimination is more efficient.

What You Should Learn:

- How to use the method of elimination to solve systems of linear equations in two variables
- How to graphically interpret the number of solutions of systems of linear equations in two variables
- How to use systems of linear equations in two variables to model and solve real-life problems

Why You Should Learn It:

The method of elimination is one method of solving a system of linear equations. For instance, Exercise 72 on page 507 shows how to use a system of linear equations to recover information about types of shoes that were sold in a shoe store.

Frank Siteman/PhotoEdit

The Method of Elimination

To use the **method of elimination** to solve a system of two linear equations in x and y, perform the following steps.

1. Obtain coefficients for x (or y) that differ only in sign by multiplying all terms of one or both equations by suitably chosen constants.

2. Add the equations to eliminate one variable; solve the resulting equation.

3. Back-substitute the value obtained in Step 2 into either of the original equations and solve for the other variable.

4. Check your solution in both of the original equations.

EXAMPLE 2 The Method of Elimination

Solve the system of linear equations.

$$\begin{cases} 5x + 3y = 9 & \text{Equation 1} \\ 2x - 4y = 14 & \text{Equation 2} \end{cases}$$

Algebraic Solution

You can obtain coefficients that differ only in sign by multiplying Equation 1 by 4 and multiplying Equation 2 by 3.

$$5x + 3y = 9 \implies 20x + 12y = 36 \qquad \text{Multiply Equation 1 by 4.}$$

$$\underline{2x - 4y = 14} \implies \underline{6x - 12y = 42} \qquad \text{Multiply Equation 2 by 3.}$$

$$26x \qquad = 78 \qquad \text{Add equations.}$$

From this equation, you can see that $x = 3$. By back-substituting this value of x into Equation 2, you can solve for y.

$$2x - 4y = 14 \qquad \text{Write Equation 2.}$$

$$2(3) - 4y = 14 \qquad \text{Substitute 3 for } x.$$

$$-4y = 8 \qquad \text{Combine like terms.}$$

$$y = -2 \qquad \text{Solve for } y.$$

The solution is $(3, -2)$. You can check the solution algebraically by substituting into the original system.

Graphical Solution

Solve each equation for y. Then use a graphing utility to graph $y_1 = 3 - \frac{5}{3}x$ and $y_2 = -\frac{7}{2} + \frac{1}{2}x$ in the same viewing window. Use the *intersect* feature or the *zoom* and *trace* features to approximate the point of intersection of the graphs. From the graphs in Figure 7.7, the point of intersection is $(3, -2)$, so the approximate solution is $(3, -2)$. You can determine that this is the exact solution by checking $(3, -2)$ in both equations.

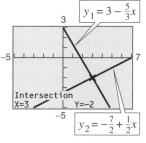

Figure 7.7

In Example 2, the two systems of linear equations

$$\begin{cases} 5x + 3y = 9 \\ 2x - 4y = 14 \end{cases} \quad \text{and} \quad \begin{cases} 20x + 12y = 36 \\ 6x - 12y = 42 \end{cases}$$

are called **equivalent systems** because they have precisely the same solution set. The operations that can be performed on a system of linear equations to produce an equivalent system are (1) interchanging any two equations, (2) multiplying an equation by a nonzero constant, and (3) adding a multiple of one equation to any other equation in the system.

Graphical Interpretation of Two-Variable Systems

It is possible for a *general* system of equations to have exactly one solution, two or more solutions, or no solution. If a system of *linear* equations has two different solutions, it must have an *infinite* number of solutions. To see why this is true, consider the following graphical interpretations of a system of two linear equations in two variables. (Remember that the graph of a linear equation in two variables is a straight line.)

Graphical Interpretation of Solutions

For a system of two linear equations in two variables, the number of solutions is one of the following.

Number of Solutions	Graphical Interpretation
1. Exactly one solution	The two lines intersect at one point.
2. Infinitely many solutions	The two lines are identical.
3. No solution	The two lines are parallel.

A system of linear equations is **consistent** if it has at least one solution. It is **inconsistent** if it has no solution.

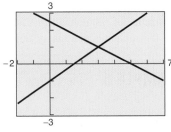

Exploration

Use a graphing utility to graph each system of equations.

a. $\begin{cases} y = 5x + 1 \\ y - x = -5 \end{cases}$

b. $\begin{cases} 3y = 4x - 1 \\ -8x + 2 = -6y \end{cases}$

c. $\begin{cases} 2y = -x + 3 \\ -4 = y + \frac{1}{2}x \end{cases}$

Determine the number of solutions each system has. Explain your reasoning.

EXAMPLE 3 Recognizing Graphs of Linear Systems

Match each system of linear equations (a, b, c) with its graph (i, ii, iii) in Figure 7.8. Describe the number of solutions. Then state whether the system is consistent or inconsistent.

a. $\begin{cases} 2x - 3y = 3 \\ -4x + 6y = 6 \end{cases}$ **b.** $\begin{cases} 2x - 3y = 3 \\ x + 2y = 5 \end{cases}$ **c.** $\begin{cases} 2x - 3y = 3 \\ -4x + 6y = -6 \end{cases}$

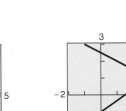

A computer simulation of this example appears in the *Interactive* CD-ROM and *Internet* versions of this text.

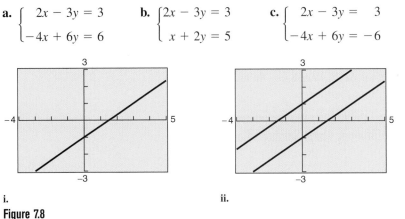

i.
Figure 7.8

ii.

iii.

Solution

a. The graph is a pair of parallel lines (ii). The lines have no point of intersection, so the system has no solution. The system is inconsistent.

b. The graph is a pair of intersecting lines (iii). The lines have one point of intersection, so the system has exactly one solution. The system is consistent.

c. The graph is a pair of lines that coincide (i). The lines have infinitely many points of intersection, so the system has infinitely many solutions. The system is consistent.

In Examples 4 and 5, you can use the method of elimination to determine that a system of linear equations has no solution or infinitely many solutions.

EXAMPLE 4   The Method of Elimination: No-Solution Case

Solve the system of linear equations.

$$\begin{cases} x - 2y = 3 & \text{Equation 1} \\ -2x + 4y = 1 & \text{Equation 2} \end{cases}$$

Algebraic Solution

To obtain coefficients that differ only in sign, multiply Equation 1 by 2.

By adding the equations, you obtain $0 = 7$. Because there are no values of x and y for which $0 = 7$, this is a false statement. So, you can conclude that the system is inconsistent and has no solution.

Graphical Solution

Solve each equation for y: $y_1 = -\frac{3}{2} + \frac{1}{2}x$ $y_2 = \frac{1}{4} + \frac{1}{2}x$

Notice that the lines have the same slope, so they are parallel. You can use a graphing utility to verify this by graphing both equations as shown in Figure 7.9. Then try using the *intersect* feature to find a point of intersection. Because the graphing utility cannot find a point of intersection, you will get an error. Use the *zoom* feature and then try using the *intersect* feature again. Because repeated zooming yields the same error message, you can conclude that the lines are parallel, and, therefore, the system has no solution.

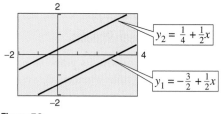

Figure 7.9

In Example 4, note that the occurrence of a false statement, such as $0 = 7$, indicates that the system has no solution. In the next example, note that the occurrence of a statement that is true for all values of the variables—in this case, $0 = 0$—indicates that the system has infinitely many solutions.

EXAMPLE 5   The Method of Elimination: Infinitely Many Solutions Case

Solve the system of linear equations: $\begin{cases} 2x - y = 1 & \text{Equation 1} \\ 4x - 2y = 2 & \text{Equation 2} \end{cases}$

Solution

To obtain coefficients that differ only in sign, multiply Equation 2 by $-\frac{1}{2}$.

$2x - y = 1$ ⟹	$2x - y = 1$	Write Equation 1.
$4x - 2y = 2$ ⟹	$-2x + y = -1$	Multiply Equation 2 by $-\frac{1}{2}$.
	$0 = 0$	Add equations.

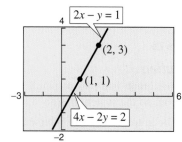

Figure 7.10

Because $0 = 0$ for all values of x and y, the two equations turn out to be equivalent (have the same solution set). You can conclude that the system has infinitely many solutions. The solution set consists of all points (x, y) lying on the line $2x - y = 1$, as shown in Figure 7.10.

Example 6 illustrates a strategy for solving a system of linear equations that has decimal coefficients.

EXAMPLE 6   A Linear System Having Decimal Coefficients

Solve the system of linear equations.

$$\begin{cases} 0.02x - 0.05y = -0.38 & \text{Equation 1} \\ 0.03x + 0.04y = 1.04 & \text{Equation 2} \end{cases}$$

Solution

Because the coefficients are two-place decimals, begin by multiplying each equation by 100 to produce a system with integer coefficients.

$$\begin{cases} 2x - 5y = -38 & \text{Revised Equation 1} \\ 3x + 4y = 104 & \text{Revised Equation 2} \end{cases}$$

Now, to obtain coefficients that differ only in sign, multiply Revised Equation 1 by 3 and multiply Revised Equation 2 by -2.

$$2x - 5y = -38 \quad \Longrightarrow \quad 6x - 15y = -114 \qquad \text{Multiply Revised Equation 1 by 3.}$$

$$\underline{3x + 4y = 104} \quad \Longrightarrow \quad \underline{-6x - 8y = -208} \qquad \text{Multiply Revised Equation 2 by } -2.$$

$$-23y = -322 \qquad \text{Add equations.}$$

So, you can conclude that

$$y = \frac{-322}{-23} = 14.$$

Back-substituting this value into Revised Equation 2 produces the following.

$$3x + 4y = 104 \qquad \text{Write revised Equation 2.}$$

$$3x + 4(14) = 104 \qquad \text{Substitute 14 for } y.$$

$$3x = 48 \qquad \text{Combine like terms.}$$

$$x = 16 \qquad \text{Solve for } x.$$

The solution is $(16, 14)$. Check this in the original system, as follows.

Check

$$0.02(16) - 0.05(14) \overset{?}{=} -0.38 \qquad \text{Substitute into Equation 1.}$$

$$0.32 - 0.70 = -0.38 \qquad \text{Equation 1 checks. } \checkmark$$

$$0.03(16) + 0.04(14) \overset{?}{=} 1.04 \qquad \text{Substitute into Equation 2.}$$

$$0.48 + 0.56 = 1.04 \qquad \text{Equation 2 checks. } \checkmark$$

STUDY T!P

The general solution of the linear system

$$\begin{cases} ax + by = c \\ dx + ey = f \end{cases}$$

is $x = (ce - bf)/(ae - db)$ and $y = (af - cd)/(ae - db)$. If $ae - db = 0$, the system does not have a unique solution. A program called SOLVE for solving such a system is given on our website, *college.hmco.com*. Try using this program to check the solution of the system in Example 6.

The *Interactive* CD-ROM and *Internet* versions of this text show every example with its solution; clicking on the *Try It!* button brings up similar problems. Guided Examples and Integrated Examples show step-by-step solutions to additional examples. Integrated Examples are related to several concepts in the section.

Applications

At this point, you may be asking the question "How can I tell which application problems can be solved using a system of linear equations?" The answer comes from the following considerations.

1. Does the problem involve more than one unknown quantity?

2. Are there two (or more) equations or conditions to be satisfied?

If one or both of these conditions occur, the appropriate mathematical model for the problem may be a system of linear equations. Example 7 shows how to construct such a model.

EXAMPLE 7   An Application of a Linear System

An airplane flying into a headwind travels the 2000-mile flying distance between two cities in 4 hours and 24 minutes. On the return flight, the same distance is traveled in 4 hours. Find the airspeed of the plane and the speed of the wind, assuming that both remain constant.

Solution

The two unknown quantities are the speeds of the wind and the plane. If r_1 is the speed of the plane and r_2 is the speed of the wind, then

$r_1 - r_2$ = speed of the plane *against* the wind

$r_1 + r_2$ = speed of the plane *with* the wind

as shown in Figure 7.11. Using the formula

Distance = (rate)(time)

for these two speeds, you obtain the following equations.

$$2000 = (r_1 - r_2)\left(4 + \frac{24}{60}\right)$$

$$2000 = (r_1 + r_2)(4)$$

These two equations simplify as follows.

$$\begin{cases} 5000 = 11r_1 - 11r_2 & \text{Equation 1} \\ 500 = r_1 + r_2 & \text{Equation 2} \end{cases}$$

By elimination, the solution is

$$r_1 = \frac{5250}{11} \approx 477.27 \text{ miles per hour} \qquad \text{Speed of plane}$$

$$r_2 = \frac{250}{11} \approx 22.73 \text{ miles per hour.} \qquad \text{Speed of wind}$$

Check this solution in the original statement of the problem.

Original flight

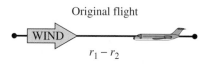

$r_1 - r_2$

Return flight

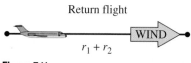

$r_1 + r_2$

Figure 7.11

7.2   E x e r c i s e s

In Exercises 1–10, solve the system of equations by the method of elimination. Label each graphed line with its equation.

1. $\begin{cases} 2x + y = 5 \\ x - y = 1 \end{cases}$ **2.** $\begin{cases} x + 3y = 1 \\ -x + 2y = 4 \end{cases}$

3. $\begin{cases} x + y = 0 \\ 3x + 2y = 1 \end{cases}$ **4.** $\begin{cases} 2x - y = 3 \\ 4x + 3y = 21 \end{cases}$

5. $\begin{cases} x - y = 2 \\ -2x + 2y = 5 \end{cases}$ **6.** $\begin{cases} 3x + 2y = 3 \\ 6x + 4y = 14 \end{cases}$

7. $\begin{cases} 3x - 2y = 5 \\ -6x + 4y = -10 \end{cases}$ **8.** $\begin{cases} 9x - 3y = -15 \\ -3x + y = 5 \end{cases}$

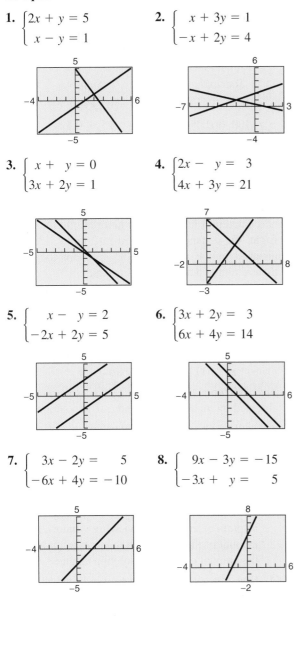

9. $\begin{cases} 9x + 3y = 1 \\ 3x - 6y = 5 \end{cases}$ **10.** $\begin{cases} 5x + 3y = -18 \\ 2x - 6y = 1 \end{cases}$

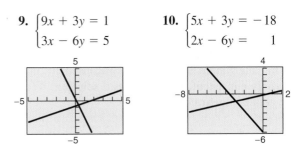

In Exercises 11–20, solve the system by the method of elimination and check any solution algebraically.

11. $\begin{cases} x + 2y = 4 \\ x - 2y = 1 \end{cases}$ **12.** $\begin{cases} 3x - 5y = 2 \\ 2x + 5y = 13 \end{cases}$

13. $\begin{cases} 2x + 3y = 18 \\ 5x - y = 11 \end{cases}$ **14.** $\begin{cases} x + 7y = 12 \\ 3x - 5y = 10 \end{cases}$

15. $\begin{cases} 3x + 2y = 10 \\ 2x + 5y = 3 \end{cases}$ **16.** $\begin{cases} 2r + 4s = 5 \\ 16r + 50s = 55 \end{cases}$

17. $\begin{cases} 5u + 6v = 24 \\ 3u + 5v = 18 \end{cases}$ **18.** $\begin{cases} 3x + 11y = 4 \\ -2x - 5y = 9 \end{cases}$

19. $\begin{cases} 1.8x + 1.2y = 4 \\ 9x + 6y = 3 \end{cases}$ **20.** $\begin{cases} 3.1x - 2.9y = -10.2 \\ 31x - 12y = 34 \end{cases}$

In Exercises 21–32, solve the system by the method of elimination and verify any solution using a graphing utility.

21. $\begin{cases} 4b + 3m = 3 \\ 3b + 11m = 13 \end{cases}$ **22.** $\begin{cases} 2x + 5y = 8 \\ 5x + 8y = 10 \end{cases}$

23. $\begin{cases} \dfrac{x}{4} + \dfrac{y}{6} = 1 \\ \dfrac{1}{2}x - \dfrac{1}{2}y = \dfrac{1}{6} \end{cases}$ **24.** $\begin{cases} \dfrac{2}{3}x + \dfrac{1}{6}y = \dfrac{2}{3} \\ \dfrac{10}{3}x - \dfrac{5}{6}y = \dfrac{10}{3} \end{cases}$

25. $\begin{cases} \frac{3}{4}x + y = \frac{1}{8} \\ \frac{9}{4}x + 3y = \frac{3}{8} \end{cases}$ **26.** $\begin{cases} \frac{2}{5}x - \frac{3}{2}y = 4 \\ \frac{1}{5}x - \frac{3}{4}y = -2 \end{cases}$

27. $\begin{cases} \dfrac{x+3}{4} + \dfrac{y-1}{3} = 1 \\ 2x - y = 12 \end{cases}$ **28.** $\begin{cases} \dfrac{x-1}{2} + \dfrac{y+2}{3} = 4 \\ x - 2y = 5 \end{cases}$

29. $\begin{cases} 2.5x - 3y = 1.5 \\ 2x - 2.4y = 1.2 \end{cases}$ **30.** $\begin{cases} 0.2x - 0.5y = -27.8 \\ 0.3x + 0.4y = 68.7 \end{cases}$

31. $\begin{cases} 0.02x - 0.05y = -0.19 \\ 0.03x + 0.04y = 0.52 \end{cases}$

32. $\begin{cases} 0.05x - 0.03y = 0.21 \\ 0.07x + 0.02y = 0.16 \end{cases}$

In Exercises 33–40, use a graphing utility to graph the lines in the system. Use the graphs to determine whether the system is consistent or inconsistent. If the system is consistent, determine the solution. Verify your results algebraically.

33. $\begin{cases} 2x - 5y = 0 \\ x - y = 3 \end{cases}$

34. $\begin{cases} 2x + y = 5 \\ x - 2y = -1 \end{cases}$

35. $\begin{cases} \frac{3}{5}x - y = 3 \\ -3x + 5y = 9 \end{cases}$

36. $\begin{cases} 4x - 6y = 9 \\ \frac{16}{3}x - 8y = 12 \end{cases}$

37. $\begin{cases} x + 7y = 2 \\ 4x - y = 9 \end{cases}$

38. $\begin{cases} 8x - 14y = 5 \\ 2x - 3.5y = 1.25 \end{cases}$

39. $\begin{cases} -x + 7y = 3 \\ -\frac{1}{7}x + y = 5 \end{cases}$

40. $\begin{cases} -7x + 6y = -4 \\ \frac{7}{6}x + y = -1 \end{cases}$

In Exercises 41–48, use a graphing utility to graph the two equations. Use the graphs to approximate the solution of the system.

41. $\begin{cases} 8x + 9y = 42 \\ 6x - y = 16 \end{cases}$

42. $\begin{cases} 4y = -8 \\ 7x - 2y = 25 \end{cases}$

43. $\begin{cases} \frac{3}{2}x - \frac{1}{5}y = 8 \\ -2x + 3y = 3 \end{cases}$

44. $\begin{cases} \frac{3}{4}x - \frac{5}{2}y = -9 \\ -x + 6y = 28 \end{cases}$

45. $\begin{cases} 0.5x + 2.2y = 9 \\ 6x + 0.4y = -22 \end{cases}$

46. $\begin{cases} 2.4x + 3.8y = -17.6 \\ 4x - 0.2y = -3.2 \end{cases}$

47. $\begin{cases} 7x - 2y = 24 \\ 5x + 6y = -20 \end{cases}$

48. $\begin{cases} 10x - 13y = -20 \\ 8x + 11y = -16 \end{cases}$

In Exercises 49–56, use any method to solve the system.

49. $\begin{cases} 3x - 5y = 7 \\ 2x + y = 9 \end{cases}$

50. $\begin{cases} -x + 3y = 17 \\ 4x + 3y = 7 \end{cases}$

51. $\begin{cases} y = 2x - 5 \\ y = 5x - 11 \end{cases}$

52. $\begin{cases} 7x + 3y = 16 \\ y = x + 2 \end{cases}$

53. $\begin{cases} x - 5y = 21 \\ 6x + 5y = 21 \end{cases}$

54. $\begin{cases} y = -3x - 8 \\ y = 15 - 2x \end{cases}$

55. $\begin{cases} -2x + 8y = 19 \\ y = x - 3 \end{cases}$

56. $\begin{cases} 4x - 3y = 6 \\ -5x + 7y = -1 \end{cases}$

Exploration In Exercises 57–60, find a system of linear equations that has the given solution. (There are many correct answers.)

57. $(6, 3)$

58. $(8, -2)$

59. $\left(3, \frac{5}{2}\right)$

60. $\left(-\frac{2}{3}, -10\right)$

Supply and Demand In Exercises 61–64, find the *point of equilibrium* of the demand and supply equations. The point of equilibrium is defined by the price *p* and the number of units *x* that satisfy both the demand and supply equations.

	Demand	Supply
61.	$p = 50 - 0.5x$	$p = 0.125x$
62.	$p = 100 - 0.05x$	$p = 25 + 0.1x$
63.	$p = 140 - 0.00002x$	$p = 80 + 0.00001x$
64.	$p = 400 - 0.0002x$	$p = 225 + 0.0005x$

65. *Airplane Speed* An airplane flying into a headwind travels the 1800-mile flying distance between two cities in 3 hours and 36 minutes. On the return flight, the distance is traveled in 3 hours. Find the airspeed of the plane and the speed of the wind, assuming that both remain constant.

66. *Airplane Speed* Two planes start from the same airport and fly in opposite directions. The second plane starts $\frac{1}{2}$ hour after the first plane, but its speed is 80 kilometers per hour faster. Find the airspeed of each plane if 2 hours after the first plane departs the planes are 3200 kilometers apart.

67. *Acid Mixture* Ten liters of a 30% acid solution is obtained by mixing a 20% and a 50% solution.

(a) Write a system of equations. One equation represents the amount of final mixture required and the other represents the amount of acid in the final mixture. Let *x* and *y* represent the amounts of 20% and 50% solutions, respectively.

(b) Use a graphing utility to graph the two equations in part (a) in the same viewing window. As the amount of the 20% solution increases, how does the amount of the 50% solution change?

(c) How much of each solution is required to obtain the specified concentration in the final mixture?

68. *Fuel Mixture* Five hundred gallons of 89-octane gasoline is obtained by mixing 87-octane gasoline with 92-octane gasoline.

(a) Write a system of equations. One equation represents the amount of final mixture required and the other represents the amount of 87- and

92-octane gasoline in the final mixture. Let x and y represent the number of 87-octane gallons and 92-octane gallons, respectively.

(b) Use a graphing utility to graph the two equations in part (a). As the amount of 87-octane gasoline increases, how does the amount of 92-octane gasoline change?

(c) How much of each type of gasoline is required to obtain the 500 gallons of 89-octane gasoline?

69. *Finance* A total of $12,000 is invested in two corporate bonds that pay 7.5% and 9% simple interest. The investor wants an annual interest income of $990 from the investments. What is the most that can be invested in the 7.5% bond?

70. *Finance* A total of $32,000 is invested in two municipal bonds that pay 5.75% and 6.25% simple interest. The investor wants an annual interest income of $1900 from the investments. What is the most that can be invested in the 5.75% bond?

71. *Ticket Sales* Five hundred tickets were sold for one performance of a play. The tickets for adults and children sold for $7.50 and $4.00, respectively, and the total receipts for the performance were $3312.50. How many of each kind of ticket were sold?

72. *Shoe Sales* On Saturday night, the manager of a shoe store evaluates the receipts of the previous week's sales. Two hundred and forty pairs of two different styles of tennis shoes were sold. One style sold for $66.95 and the other sold for $84.95. The total receipts were $17,652. The cash register that was supposed to record the number of each type of shoe sold malfunctioned. Can you recover the information? If so, how many shoes of each type were sold?

73. *Driving Distances* In a trip of 300 kilometers, two people drive. One person drives three times as far as the other. Find the distance that each person drives.

74. *Truck Scheduling* A contractor hires two trucking companies to haul 1600 tons of crushed stone for a highway construction project. The contracts state that one company is to haul four times as much as the other. Find the amount hauled by each company.

Fitting a Line to Data **In Exercises 75–78, find the least squares regression line $y = ax + b$ for the points $(x_1, y_1), (x_2, y_2), \ldots , (x_n, y_n)$ by solving the system for a and b. Then use the regression capabilities of a graphing utility to confirm your result. (For an explanation of how the coefficients of a and b in the system are obtained, see Appendix B.)**

75. $\begin{cases} 5b + 10a = 20.2 \\ 10b + 30a = 50.1 \end{cases}$ **76.** $\begin{cases} 5b + 10a = 11.7 \\ 10b + 30a = 25.6 \end{cases}$

77. $\begin{cases} 7b + 21a = 35.1 \\ 21b + 91a = 114.2 \end{cases}$ **78.** $\begin{cases} 6b + 15a = 23.6 \\ 15b + 55a = 48.8 \end{cases}$

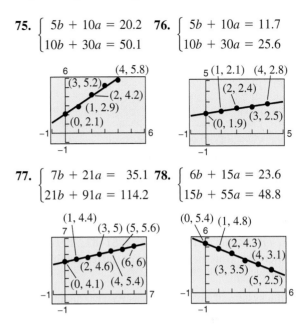

79. *Data Analysis* A farmer used four test plots to determine the relationship between wheat yield (in bushels per acre) and the amount of fertilizer (in hundreds of pounds per acre). The results are shown in the table.

Fertilizer, x	1.0	1.5	2.0	2.5
Yield, y	32	41	48	53

(a) Find the least squares regression line $y = ax + b$ for the data by solving the system for a and b.

$$\begin{cases} 4b + 7.0a = 174 \\ 7b + 13.5a = 322 \end{cases}$$

(b) Use the linear regression capabilities of a graphing utility to confirm the result.

(c) Plot the data and the linear regression model.

(d) Use the line to estimate the yield for a fertilizer application of 160 pounds per acre.

80. *Data Analysis* A store manager wants to know the demand for a certain product as a function of the price. The daily sales for the different prices of the product are shown in the table.

Price, x	$1.00	$1.25	$1.50
Demand, y	450	375	330

(a) Find the least squares regression line $y = ax + b$ for the data by solving the system for a and b.

$$\begin{cases} 3.00b + 3.7500a = 1155.00 \\ 3.75b + 4.8125a = 1413.75 \end{cases}$$

(b) Use the linear regression capabilities of a graphing utility to confirm the result.

(c) Plot the data and the linear regression model.

(d) Then, use the line to predict the demand when the price is $1.40.

Synthesis

True or False? **In Exercises 81–83, determine whether the statement is true or false. Justify your answer.**

81. If a system of linear equations has two distinct solutions, then it has an infinite number of solutions.

82. Solving a system of equations graphically will always give an exact solution.

83. If a system of linear equations has no solution, then the lines must be parallel.

Think About It **In Exercises 84 and 85, the graphs of the two equations appear to be parallel. Yet, when the system is solved algebraically, it is found that the system does have a solution. Find the solution and explain why it does not appear on the portion of the graph that is shown.**

84. $\begin{cases} 100y - x = 200 \\ 99y - x = -198 \end{cases}$ **85.** $\begin{cases} 21x - 20y = 0 \\ 13x - 12y = 120 \end{cases}$

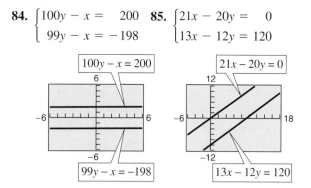

86. *Writing* Briefly explain whether or not it is possible for a consistent system of linear equations to have exactly two solutions.

87. *Think About It* Give examples of (a) a system of linear equations that has no solution and (b) a system that has an infinite number of solutions.

Exploration **In Exercises 88 and 89, find the value of k such that the system of equations is inconsistent.**

88. $\begin{cases} 4x - 8y = -3 \\ 2x + ky = 16 \end{cases}$ **89.** $\begin{cases} 15x + 3y = 6 \\ -10x + ky = 9 \end{cases}$

Advanced Applications **In Exercises 90–93, solve the system of equations for u and v. While solving for these variables, consider the transcendental functions as constants. (Systems of this type are found in a course in differential equations.)**

90. $\begin{cases} u \sin x + v \cos x = 0 \\ u \cos x - v \sin x = \sec x \end{cases}$

91. $\begin{cases} u \cos 2x + v \sin 2x = 0 \\ u(-2 \sin 2x) + v(2 \cos 2x) = \csc 2x \end{cases}$

92. $\begin{cases} ue^x + vxe^x = 0 \\ ue^x + v(x + 1)e^x = e^x \ln x \end{cases}$

93. $\begin{cases} ue^{2x} + vxe^{2x} = 0 \\ u(2e^{2x}) + v(2x + 1)e^{2x} = \dfrac{e^{2x}}{x} \end{cases}$

Review

In Exercises 94–97, use a graphing utility to graph the functions. Use the graph to determine the domain and range.

94. $f(x) = x^2 - 2x$ **95.** $g(x) = \sqrt[3]{x - 2}$

96. $h(x) = \sqrt{25 - x^2}$ **97.** $s(t) = 4/(9 - t^2)$

In Exercises 98–105, solve the inequality and graph the solution on a real number line.

98. $-11 - 6x \geq 33$ **99.** $2(x - 3) > -5x + 1$

100. $8x - 15 \leq -4(2x - 1)$

101. $-6 \leq 3x - 10 < 6$

102. $|x - 8| < 10$ **103.** $|x + 10| \geq -3$

104. $2x^2 + 3x - 35 < 0$ **105.** $3x^2 + 12x > 0$

In Exercises 106–109, write the expression as the logarithm of a single quantity.

106. $\ln x + \ln 6$ **107.** $\ln x - 5 \ln(x + 3)$

108. $\log_9 12 - \log_9 x$ **109.** $\frac{1}{4}\log_6 3x$

In Exercises 110 and 111, solve the system by the method of substitution.

110. $\begin{cases} 4x - y = -16 \\ -5x + 3y = 6 \end{cases}$ **111.** $\begin{cases} -12x + 9y = 51 \\ -x - 7y = -19 \end{cases}$

7.3 Multivariable Linear Systems

Row-Echelon Form and Back-Substitution

The method of elimination can be applied to a system of linear equations in more than two variables. When elimination is used to solve a system of linear equations, the goal is to rewrite the system in a form to which back-substitution can be applied. Consider the following two systems of linear equations.

System of 3 Linear Equations in 3 Variables (See Example 2):

$$\begin{cases} x - 2y + 3z = 9 \\ -x + 3y \quad\;\;\; = -4 \\ 2x - 5y + 5z = 17 \end{cases}$$

Equivalent System in Row-Echelon Form (See Example 1):

$$\begin{cases} x - 2y + 3z = 9 \\ \quad\;\; y + 3z = 5 \\ \quad\qquad z = 2 \end{cases}$$

The second system is said to be in **row-echelon form,** which means that it has a "stair-step" pattern with leading coefficients of 1. After comparing the two systems, it should be clear that it is easier to solve the system in row-echelon form.

EXAMPLE 1   Using Back-Substitution

Solve the system of linear equations.

$$\begin{cases} x - 2y + 3z = 9 & \text{Equation 1} \\ \quad\;\; y + 3z = 5 & \text{Equation 2} \\ \quad\qquad z = 2 & \text{Equation 3} \end{cases}$$

Solution

From Equation 3, you know the value of z. To solve for y, substitute $z = 2$ into Equation 2 to obtain

$$y + 3(2) = 5 \qquad\qquad \text{Substitute 2 for } z.$$

$$y = -1. \qquad\qquad \text{Solve for } y.$$

Finally, substitute $y = -1$ and $z = 2$ into Equation 1 to obtain

$$x - 2(-1) + 3(2) = 9 \qquad \text{Substitute } -1 \text{ for } y \text{ and 2 for } z.$$

$$x = 1. \qquad\qquad \text{Solve for } x.$$

The solution is $x = 1$, $y = -1$, and $z = 2$, which can be written as the **ordered triple** $(1, -1, 2)$. Check this in the original system of equations as follows.

Check $1 - 2(-1) + 3(2) = 9$ ✓

$-1 + 3(2) = 5$ ✓

$2 = 2$ ✓

What You Should Learn:

- How to recognize linear systems in row-echelon form and use back-substitution to solve the systems
- How to use Gaussian Elimination to solve systems of linear equations
- How to solve nonsquare systems of linear equations
- How to graphically interpret three-variable systems
- How to use systems of linear equations to write partial fraction decompositions of rational expressions
- How to use systems of linear equations in three or more variables to model and solve real-life problems

Why You Should Learn It:

Systems of linear equations in three or more variables can be used to model and solve real-life problems. For instance, Exercise 109 on page 523 shows how to use a system of linear equations to analyze an automobile's braking system.

Andy Sacks/Tony Stone Images

Gaussian Elimination

Two systems of equations are *equivalent* if they have the same solution set. To solve a system that is not in row-echelon form, first convert it to an *equivalent* system that *is* in row-echelon form by using one or more of the elementary row operations shown below. This process is called **Gaussian elimination,** after the German mathematician Carl Friedrich Gauss (1777–1855).

Elementary Row Operations

1. Interchange two equations.
2. Multiply one of the equations by a nonzero constant.
3. Add a multiple of one equation to another equation.

EXAMPLE 2 Using Gaussian Elimination to Solve a System

Solve the system of linear equations.

$$\begin{cases} x - 2y + 3z = 9 & \text{Equation 1} \\ -x + 3y = -4 & \text{Equation 2} \\ 2x - 5y + 5z = 17 & \text{Equation 3} \end{cases}$$

Solution

Because the leading coefficient of the first equation is 1, you can begin by saving the x in the upper left position and eliminating the other x's from the first column.

$$\begin{cases} x - 2y + 3z = 9 \\ y + 3z = 5 \\ 2x - 5y + 5z = 17 \end{cases}$$

> Adding the first equation to the second equation produces a new second equation.

$$\begin{cases} x - 2y + 3z = 9 \\ y + 3z = 5 \\ -y - z = -1 \end{cases}$$

> Adding -2 times the first equation to the third equation produces a new third equation.

Now that all but the first x have been eliminated from the first column, go to work on the second column. (You need to eliminate y from the third equation.)

$$\begin{cases} x - 2y + 3z = 9 \\ y + 3z = 5 \\ 2z = 4 \end{cases}$$

> Adding the second equation to the third equation produces a new third equation.

Finally, you need a coefficient of 1 for z in the third equation.

$$\begin{cases} x - 2y + 3z = 9 \\ y + 3z = 5 \\ z = 2 \end{cases}$$

> Multiplying the third equation by $\frac{1}{2}$ produces a new third equation.

This is the same system that was solved in Example 1. As in that example, you can conclude that the solution is $x = 1$, $y = -1$, and $z = 2$, written as $(1, -1, 2)$. Check this in the original system of the equations.

One of the most influential Chinese mathematics books was the *Chui-chang suan-shu* or *Nine Chapters on the Mathematical Art* (written in approximately 250 B.C.). Chapter Eight contained solutions of systems of linear equations using positive and negative numbers. One such system was

$$\begin{cases} 3x + 2y + 3z = 39 \\ 2x + 3y + 3z = 34. \\ 2x + 2y + 3z = 26 \end{cases}$$

This system was solved using column operations on a matrix. Matrices (plural for matrix) will be discussed in the next chapter.

STUDY T!P

The goal of Gaussian elimination is to use elementary row operations on a system in order to isolate one variable. You can then solve for the value of the variable and use back-substitution to find the values of the remaining variables.

Christopher Lui/China Stock

The next example involves an inconsistent system—one that has no solution. The key to recognizing an inconsistent system is that at some stage in the elimination process, you obtain a false statement such as $0 = -2$.

EXAMPLE 3   An Inconsistent System

Solve the system of linear equations.

$$\begin{cases} x - 3y + z = 1 & \text{Equation 1} \\ 2x - y - 2z = 2 & \text{Equation 2} \\ x + 2y - 3z = -1 & \text{Equation 3} \end{cases}$$

Solution

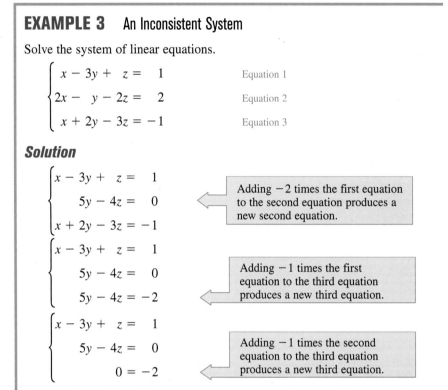

$$\begin{cases} x - 3y + z = 1 \\ 5y - 4z = 0 \\ x + 2y - 3z = -1 \end{cases}$$

Adding -2 times the first equation to the second equation produces a new second equation.

$$\begin{cases} x - 3y + z = 1 \\ 5y - 4z = 0 \\ 5y - 4z = -2 \end{cases}$$

Adding -1 times the first equation to the third equation produces a new third equation.

$$\begin{cases} x - 3y + z = 1 \\ 5y - 4z = 0 \\ 0 = -2 \end{cases}$$

Adding -1 times the second equation to the third equation produces a new third equation.

Because $0 = -2$ is a false statement, you can conclude that this system is inconsistent and therefore has no solution. Moreover, because this system is equivalent to the original system, you can conclude that the original system also has no solution.

As with a system of linear equations in two variables, the number of solutions of a system of linear equations in more than two variables must fall into one of three categories.

The Number of Solutions of a Linear System

For a system of linear equations, exactly one of the following is true.

1. There is exactly one solution.
2. There are infinitely many solutions.
3. There is no solution.

A system of linear equations is called *consistent* if it has at least one solution. A consistent system with exactly one solution is **independent.** A consistent system with infinitely many solutions is **dependent.** A system of linear equations is called *inconsistent* if it has no solution.

EXAMPLE 4   A System with Infinitely Many Solutions

Solve the system of linear equations.

$$\begin{cases} x + y - 3z = -1 & \text{Equation 1} \\ \quad\quad y - z = 0 & \text{Equation 2} \\ -x + 2y \quad\quad = 1 & \text{Equation 3} \end{cases}$$

Solution

$$\begin{cases} x + y - 3z = -1 \\ \quad y - z = 0 \\ \quad 3y - 3z = 0 \end{cases}$$

> Adding the first equation to the third equation produces a new third equation.

$$\begin{cases} x + y - 3z = -1 \\ \quad y - z = 0 \\ \quad\quad 0 = 0 \end{cases}$$

> Adding -3 times the second equation to the third equation produces a new third equation.

This means that Equation 3 depends on Equations 1 and 2 in the sense that it gives us no additional information about the variables. So, the original system is equivalent to the system

$$\begin{cases} x + y - 3z = -1 \\ \quad y - z = 0 \end{cases}.$$

In the last equation, solve for y in terms of z to obtain $y = z$. Back-substituting for y into the previous equation produces $x = 2z - 1$. Finally, letting $z = a$, where a is a real number, the solutions to the given system are all of the form

$$x = 2a - 1, \qquad y = a, \qquad \text{and} \qquad z = a.$$

So, every ordered triple of the form

$$(2a - 1, a, a), \qquad a \text{ is a real number}$$

is a solution of the system.

In Example 4, there are other ways to write the same infinite set of solutions. For instance, the solutions could have been written as

$$\left(b, \tfrac{1}{2}(b + 1), \tfrac{1}{2}(b + 1)\right), \qquad b \text{ is a real number}.$$

To convince yourself that this description produces the same set of solutions, consider the following.

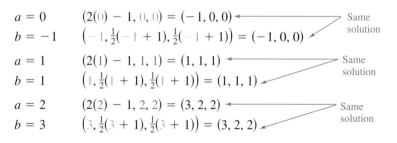

Substitution	Solution	
$a = 0$	$(2(0) - 1, 0, 0) = (-1, 0, 0)$	Same solution
$b = -1$	$\left(-1, \tfrac{1}{2}(-1 + 1), \tfrac{1}{2}(-1 + 1)\right) = (-1, 0, 0)$	
$a = 1$	$(2(1) - 1, 1, 1) = (1, 1, 1)$	Same solution
$b = 1$	$\left(1, \tfrac{1}{2}(1 + 1), \tfrac{1}{2}(1 + 1)\right) = (1, 1, 1)$	
$a = 2$	$(2(2) - 1, 2, 2) = (3, 2, 2)$	Same solution
$b = 3$	$\left(3, \tfrac{1}{2}(3 + 1), \tfrac{1}{2}(3 + 1)\right) = (3, 2, 2)$	

STUDY T!P

There are an infinite number of solutions to Example 4, but they all are of a specific form. By selecting, for example, a-values of 0, 1, and 3, you can verify that $(-1, 0, 0)$, $(1, 1, 1)$, and $(5, 3, 3)$ are specific solutions. It is incorrect to say simply that the solution to Example 4 is "infinite." You must also specify the form of the solutions.

Nonsquare Systems

So far, each system of linear equations has been *square*, which means that the number of equations is equal to the number of variables. In a **nonsquare system of equations,** the number of equations differs from the number of variables. A system of linear equations cannot have a unique solution unless there are at least as many equations as there are variables in the system.

EXAMPLE 5   A System with Fewer Equations Than Variables

Solve the system of linear equations.

$$\begin{cases} x - 2y + z = 2 & \text{Equation 1} \\ 2x - y - z = 1 & \text{Equation 2} \end{cases}$$

Solution

Begin by rewriting the system in row-echelon form.

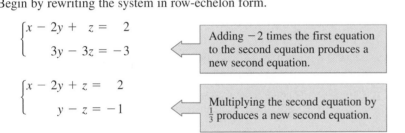

$$\begin{cases} x - 2y + z = 2 \\ 3y - 3z = -3 \end{cases}$$

Adding -2 times the first equation to the second equation produces a new second equation.

$$\begin{cases} x - 2y + z = 2 \\ y - z = -1 \end{cases}$$

Multiplying the second equation by $\frac{1}{3}$ produces a new second equation.

Solving for y in terms of z, you get $y = z - 1$, and back-substitution into Equation 1 yields

$$x - 2(z - 1) + z = 2$$

$$x - 2z + 2 + z = 2$$

$$x = z.$$

Finally, by letting $z = a$ where a is a real number, you have the solution

$$x = a, \qquad y = a - 1, \qquad \text{and} \qquad z = a.$$

So, every ordered triple of the form

$$(a, a - 1, a), \qquad a \text{ is a real number}$$

is a solution of the system.

In Example 5, try choosing some values of a to obtain different solutions of the system, such as $(1, 0, 1)$, $(2, 1, 2)$, and $(3, 2, 3)$. Then check each of the solutions in the original system as follows.

Check: $(1, 0, 1)$ *Check:* $(2, 1, 2)$ *Check:* $(3, 2, 3)$

$1 - 2(0) + 1 = 2$ $2 - 2(1) + 2 = 2$ $3 - 2(2) + 3 = 2$

$\qquad\qquad 2 = 2 \checkmark$ $\qquad\qquad 2 = 2 \checkmark$ $\qquad\qquad 2 = 2 \checkmark$

$2(1) - 0 - 1 = 1$ $2(2) - 1 - 2 = 1$ $2(3) - 2 - 3 = 1$

$\qquad\qquad 1 = 1 \checkmark$ $\qquad\qquad 1 = 1 \checkmark$ $\qquad\qquad 1 = 1 \checkmark$

Graphical Interpretation of Three-Variable Systems

Solutions of equations with three variables can be pictured with a **three-dimensional coordinate system.** To construct such a system, begin with the xy-coordinate plane in a horizontal position. Then draw the z-axis as a vertical line through the origin.

Every ordered triple (x, y, z) corresponds to a point on the three-dimensional coordinate system. For instance, the points corresponding to

$$(-2, 5, 4), \qquad (2, -5, 3), \qquad \text{and} \qquad (3, 3, -2)$$

are shown in Figure 7.12.

The **graph of an equation in three variables** consists of all points (x, y, z) that are solutions of the equation. The graph of a linear equation in three variables is a *plane*. Sketching graphs on a three-dimensional coordinate system is difficult because the sketch itself is only two-dimensional.

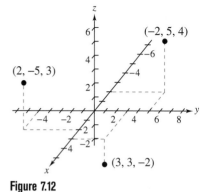

Figure 7.12

One technique for sketching a plane is to find the three points at which the plane intersects the axes. For instance, the plane

$$3x + 2y + 4z = 12$$

intersects the x-axis at the point $(4, 0, 0)$, the y-axis at the point $(0, 6, 0)$, and the z-axis at the point $(0, 0, 3)$.

By plotting these three points, connecting them with line segments, and shading the resulting triangular region, you can sketch a portion of the graph, as shown in Figure 7.13.

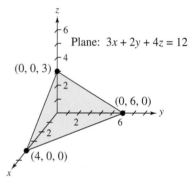

Figure 7.13

The graph of a system of three linear equations in three variables consists of *three* planes. When these planes intersect in a single point, the system has exactly one solution. [See Figure 7.14(a).] When the three planes have no point in common, the system has no solution. [See Figures 7.14(b) and (c).] When the three planes intersect in a line or a plane, the system has infinitely many solutions. [See Figures 7.14(d) and (e).]

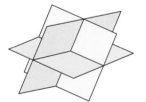

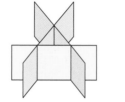

(a) **Solution: one point** (b) **Solution: none** (c) **Solution: none**

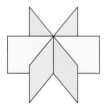

(d) **Solution: one line** (e) **Solution: one plane**

Figure 7.14

Partial Fraction Decomposition and Other Applications

A rational expression can be often written as the sum of two or more simpler rational expressions. For example, the rational expression

$$\frac{x + 7}{x^2 - x - 6}$$

can be written as the sum of two fractions with linear denominators. That is,

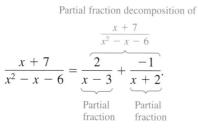

Partial fraction decomposition of

$$\frac{x + 7}{x^2 - x - 6}$$

$$\frac{x + 7}{x^2 - x - 6} = \frac{2}{x - 3} + \frac{-1}{x + 2}.$$

Partial   Partial
fraction   fraction

Each fraction on the right side of the equation is a **partial fraction,** and together they make up the **partial fraction decomposition** of the left side.

The method of partial fractions was introduced by **John Bernoulli (1667–1748),** a Swiss mathematician who was instrumental in the early development of calculus. Bernoulli was a professor at the University of Basel and taught many outstanding students, the most famous of whom was Leonhard Euler.

Decomposition of *N(x)/D(x)* into Partial Fractions

1. *Divide if improper:* If $N(x)/D(x)$ is an improper fraction (degree of $N(x) \geq$ degree of $D(x)$), divide the denominator into the numerator to obtain

$$\frac{N(x)}{D(x)} = (\text{polynomial}) + \frac{N_1(x)}{D(x)}$$

and apply Steps 2, 3, and 4 (below) to the proper rational expression $N_1(x)/D(x)$.

2. *Factor denominator:* Completely factor the denominator into factors of the form

$$(px + q)^m \quad \text{and} \quad (ax^2 + bx + c)^n$$

where $(ax^2 + bx + c)$ is irreducible.

3. *Linear factors:* For *each* factor of the form $(px + q)^m$, the partial fraction decomposition must include the following sum of *m* fractions.

$$\frac{A_1}{(px + q)} + \frac{A_2}{(px + q)^2} + \cdots + \frac{A_m}{(px + q)^m}$$

4. *Quadratic factors:* For *each* factor of the form $(ax^2 + bx + c)^n$, the partial fraction decomposition must include the following sum of *n* fractions.

$$\frac{B_1 x + C_1}{ax^2 + bx + c} + \frac{B_2 x + C_2}{(ax^2 + bx + c)^2} + \cdots + \frac{B_n x + C_n}{(ax^2 + bx + c)^n}$$

One of the most important applications of partial fractions is in calculus. If you go on to take a course in calculus, you will learn how partial fractions can be used in a calculus operation called antidifferentiation.

EXAMPLE 6   Partial Fraction Decomposition: Distinct Linear Factors

Write the partial fraction decomposition for

$$\frac{x + 7}{x^2 - x - 6}.$$

Solution

Because $x^2 - x - 6 = (x - 3)(x + 2)$, you should include one partial fraction with a constant numerator for each linear factor of the denominator and write

$$\frac{x + 7}{x^2 - x - 6} = \frac{A}{x - 3} + \frac{B}{x + 2}.$$

Multiplying both sides of this equation by the least common denominator, $(x - 3)(x + 2)$, leads to the **basic equation**

$$x + 7 = A(x + 2) + B(x - 3) \qquad \text{Basic equation}$$

$$= Ax + 2A + Bx - 3B$$

$$= (A + B)x + 2A - 3B. \qquad \text{Write in polynomial form.}$$

By equating coefficients of like terms on opposite sides of the equation, you obtain the following system of linear equations.

$$\begin{cases} A + B = 1 \\ 2A - 3B = 7 \end{cases}$$

You can solve the system of linear equations as follows.

$$A + B = 1 \quad \Longrightarrow \quad 3A + 3B = 3 \qquad \text{Multiply Equation 1 by 3.}$$

$$2A - 3B = 7 \quad \Longrightarrow \quad \underline{2A - 3B = 7} \qquad \text{Equation 2}$$

$$5A \quad\quad = 10 \qquad \text{Add equations}$$

From this equation, you can see that $A = 2$. By back-substituting this value of A into Equation 1, you can determine that $B = -1$.

Therefore, the partial fraction decomposition is

$$\frac{x + 7}{x^2 - x - 6} = \frac{2}{x - 3} - \frac{1}{x + 2}.$$

Check this result by combining the two partial fractions on the right side of the equation.

You can graphically check the decomposition found in Example 1. To do this, graph

$$y_1 = \frac{x + 7}{x^2 - x - 6} \quad \text{and} \quad y_2 = \frac{2}{x - 3} - \frac{1}{x + 2}$$

in the same viewing window. Their graphs should be identical, as shown in Figure 7.15.

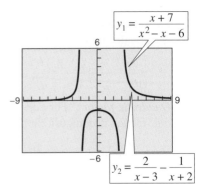

Figure 7.15

The next example shows how to find the partial fraction decomposition for a rational function whose denominator has a repeated linear factor.

EXAMPLE 7 Partial Fraction Decomposition: Repeated Linear Factors

Write the partial fraction decomposition for

$$\frac{5x^2 + 20x + 6}{x^3 + 2x^2 + x}.$$

Solution

Because the denominator factors as

$$x^3 + 2x^2 + x = x(x^2 + 2x + 1)$$
$$= x(x + 1)^2$$

you should include one partial fraction with a constant numerator for each power of x and $(x + 1)$ and write

$$\frac{5x^2 + 20x + 6}{x(x + 1)^2} = \frac{A}{x} + \frac{B}{x + 1} + \frac{C}{(x + 1)^2}.$$

Multiplying by the LCD, $x(x + 1)^2$, leads to the basic equation

$$5x^2 + 20x + 6 = A(x + 1)^2 + Bx(x + 1) + Cx \qquad \text{Basic equation}$$
$$= Ax^2 + 2Ax + A + Bx^2 + Bx + Cx$$
$$= (A + B)x^2 + (2A + B + C)x + A. \qquad \text{Polynomial form}$$

By equating coefficients of like terms on opposite sides of the equation, you obtain the following system of linear equations.

$$\begin{cases} A + B & = 5 \\ 2A + B + C = 20 \\ A & = 6 \end{cases}$$

Substituting 6 for A in the first equation produces

$$6 + B = 5$$
$$B = -1.$$

Substituting 6 for A and -1 for B in the second equation produces

$$2(6) + (-1) + C = 20$$
$$C = 9.$$

Therefore, the partial fraction decomposition is

$$\frac{5x^2 + 20x + 6}{x(x + 1)^2} = \frac{6}{x} - \frac{1}{x + 1} + \frac{9}{(x + 1)^2}.$$

Exploration

Partial fraction decomposition is practical only for rational functions whose denominators factor "nicely." For example, the factorization of the denominator $x^2 - x - 5$ is

$$\left(x - \frac{1 - \sqrt{21}}{2}\right)\left(x - \frac{1 + \sqrt{21}}{2}\right).$$

Write the basic equation and try to complete the decomposition for

$$\frac{x + 7}{x^2 - x - 5}.$$

What problems do you encounter?

EXAMPLE 8   Vertical Motion

The height at time t of an object that is moving in a (vertical) line with constant acceleration a is given by the *position equation* $s = \frac{1}{2}at^2 + v_0t + s_0$. The height s is measured in feet, t is measured in seconds, v_0 is the initial velocity (in feet per second) at $t = 0$, and s_0 is the initial height. Find the values of a, v_0, and s_0, if $s = 52$ at $t = 1$, $s = 52$ at $t = 2$, and $s = 20$ at $t = 3$, as shown in Figure 7.16.

Solution

You can obtain three linear equations in a, v_0, and s_0 as follows.

When $t = 1$: $\frac{1}{2}a(1)^2 + v_0(1) + s_0 = 52$ ⟹ $a + 2v_0 + 2s_0 = 104$

When $t = 2$: $\frac{1}{2}a(2)^2 + v_0(2) + s_0 = 52$ ⟹ $2a + 2v_0 + s_0 = 52$

When $t = 3$: $\frac{1}{2}a(3)^2 + v_0(3) + s_0 = 20$ ⟹ $9a + 6v_0 + 2s_0 = 40$

Solving this system yields $a = -32$, $v_0 = 48$, and $s_0 = 20$.

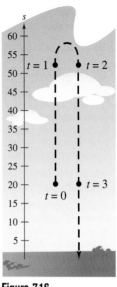

Figure 7.16

EXAMPLE 9   Data Analysis: Curve-Fitting

Find a quadratic equation, $y = ax^2 + bx + c$, whose graph passes through the points $(-1, 3)$, $(1, 1)$, and $(2, 6)$.

Solution

Because the graph of $y = ax^2 + bx + c$ passes through the points $(-1, 3)$, $(1, 1)$, and $(2, 6)$, you can write the following.

When $x = -1$, $y = 3$: $a(-1)^2 + b(-1) + c = 3$

When $x = 1$, $y = 1$: $a(1)^2 + b(1) + c = 1$

When $x = 2$, $y = 6$: $a(2)^2 + b(2) + c = 6$

This produces the following system of linear equations.

$$\begin{cases} a - b + c = 3 & \text{Equation 1} \\ a + b + c = 1 & \text{Equation 2} \\ 4a + 2b + c = 6 & \text{Equation 3} \end{cases}$$

The solution of this system is $a = 2$, $b = -1$, and $c = 0$. So, the equation of the parabola is $y = 2x^2 - x$, as shown in Figure 7.17. From the graph, you can check to see that the parabola passes through the given points $(-1, 3)$, $(1, 1)$, and $(2, 6)$.

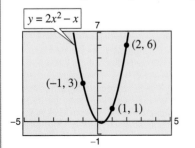

Figure 7.17

STUDY TIP

When you use a system of linear equations to solve an application, it is wise to interpret your solution in the context of the problem to see if it makes sense. For instance, in Example 8 the solution results in a position equation of

$$s = -16t^2 + 48t + 20$$

which implies that the object was thrown upward at a velocity of 48 feet per second from a height of 20 feet. The object undergoes a constant downward acceleration of 32 feet per second squared. (Physics will tell you that this is the value of the acceleration due to gravity.)

7.3 Exercises

In Exercises 1–4, decide whether each ordered triple is a solution of the system of equations.

1. $\begin{cases} 3x - y + z = 1 \\ 2x \qquad - 3z = -14 \\ \qquad 5y + 2z = 8 \end{cases}$

   (a) $(2, 5, 0)$ (b) $(-2, 0, 4)$
   (c) $(0, -1, 3)$ (d) $(-1, 0, 4)$

2. $\begin{cases} 3x + 4y - z = 17 \\ 5x - y + 2z = -2 \\ 2x - 3y + 7z = -21 \end{cases}$

   (a) $(3, 2, 0)$ (b) $(1, 3, -2)$
   (c) $(4, 1, -3)$ (d) $(1, 5, -1)$

3. $\begin{cases} 4x + y - z = 0 \\ -8x - 6y + z = -\frac{7}{4} \\ 3x - y = -\frac{9}{4} \end{cases}$

   (a) $(0, 1, 1)$ (b) $\left(-\frac{3}{2}, \frac{5}{4}, -\frac{5}{4}\right)$
   (c) $\left(-\frac{1}{2}, \frac{3}{4}, -\frac{5}{4}\right)$ (d) $\left(-\frac{1}{2}, 2, 0\right)$

4. $\begin{cases} -4x - y - 8z = -6 \\ \qquad y + z = 0 \\ 4x - 7y = 6 \end{cases}$

   (a) $(-2, -2, 2)$ (b) $\left(-\frac{33}{2}, -10, 10\right)$
   (c) $\left(\frac{1}{8}, -\frac{1}{2}, \frac{1}{2}\right)$ (d) $\left(-\frac{11}{2}, -4, 4\right)$

In Exercises 5–10, use back-substitution to solve the system of linear equations.

5. $\begin{cases} 2x - y + 5z = 24 \\ \quad y + 2z = 4 \\ \qquad z = 6 \end{cases}$ **6.** $\begin{cases} 4x - 3y - 2z = 21 \\ \quad 6y - 5z = -10 \\ \qquad z = -4 \end{cases}$

7. $\begin{cases} 2x + y - 3z = 10 \\ \quad y = -2 \\ \quad y - z = 4 \end{cases}$ **8.** $\begin{cases} x = 8 \\ 2x + 3y = 10 \\ x - y + 2z = 22 \end{cases}$

9. $\begin{cases} 4x - 2y + z = 8 \\ \qquad 2z = 4 \\ \quad -y + z = 4 \end{cases}$ **10.** $\begin{cases} 5x - 8z = 22 \\ 3y - 5z = 10 \\ \qquad z = -4 \end{cases}$

In Exercises 11 and 12, perform the row operation and write the equivalent system.

11. Add Equation 1 to Equation 2.

$\begin{cases} x - 2y + 3z = 5 & \text{Equation 1} \\ -x + 3y - 5z = 4 & \text{Equation 2} \\ 2x \qquad - 3z = 0 & \text{Equation 3} \end{cases}$

What did this operation accomplish?

12. Add -2 times Equation 1 to Equation 3.

$\begin{cases} x - 2y + 3z = 5 & \text{Equation 1} \\ -x + 3y - 5z = 4 & \text{Equation 2} \\ 2x \qquad - 3z = 0 & \text{Equation 3} \end{cases}$

What did this operation accomplish?

In Exercises 13–40, solve the system of linear equations and check any solution algebraically.

13. $\begin{cases} x + y + z = 6 \\ 2x - y + z = -1 \\ 3x \quad - z = -7 \end{cases}$ **14.** $\begin{cases} x + y + z = 2 \\ -x + 3y + 2z = 8 \\ 4x + y = 4 \end{cases}$

15. $\begin{cases} 2x + 2z = 6 \\ 5x + 3y = 11 \\ \quad 3y - 4z = 1 \end{cases}$ **16.** $\begin{cases} 4x + y - 3z = 11 \\ 2x - 3y + 2z = 9 \\ x + y + z = -3 \end{cases}$

17. $\begin{cases} 6y + 4z = -18 \\ 3x + 3y = 9 \\ 2x \quad - 3z = 12 \end{cases}$ **18.** $\begin{cases} 2x + 4y + z = -4 \\ 2x - 4y + 6z = 13 \\ 4x - 2y + z = 6 \end{cases}$

19. $\begin{cases} 3x - 2y + 4z = 1 \\ x + y - 2z = 3 \\ 2x - 3y + 6z = 8 \end{cases}$ **20.** $\begin{cases} 5x - 3y + 2z = 3 \\ 2x + 4y - z = 7 \\ x - 11y + 4z = 3 \end{cases}$

21. $\begin{cases} 3x + 3y + 5z = 1 \\ 3x + 5y + 9z = 0 \\ 5x + 9y + 17z = 0 \end{cases}$ **22.** $\begin{cases} 2x + y + 3z = 1 \\ 2x + 6y + 8z = 3 \\ 6x + 8y + 18z = 5 \end{cases}$

23. $\begin{cases} x + 2y - 7z = -4 \\ 2x + y + z = 13 \\ 3x + 9y - 36z = -33 \end{cases}$

24. $\begin{cases} 2x + y - 3z = 4 \\ 4x + 2z = 10 \\ -2x + 3y - 13z = -8 \end{cases}$

25. $\begin{cases} 3x - 3y + 6z = 6 \\ x + 2y - z = 5 \\ 5x - 8y + 13z = 7 \end{cases}$ **26.** $\begin{cases} x + 4z = 13 \\ 4x - 2y + z = 7 \\ 2x - 2y - 7z = -19 \end{cases}$

27. $\begin{cases} x - 2y + 3z = 4 \\ 3x - y + 2z = 0 \\ x + 3y - 4z = -2 \end{cases}$

28. $\begin{cases} -x + 3y + z = 4 \\ 4x - 2y - 5z = -7 \\ 2x + 4y - 3z = 12 \end{cases}$ **29.** $\begin{cases} x + 4z = 1 \\ x + y + 10z = 10 \\ 2x - y + 2z = -5 \end{cases}$

30. $\begin{cases} 3x - 2y - 6z = -4 \\ -3x + 2y + 6z = 1 \\ x - y - 5z = -3 \end{cases}$

31. $\begin{cases} 2x + 3y = 0 \\ 4x + 3y - z = 0 \\ 8x + 3y + 3z = 0 \end{cases}$ **32.** $\begin{cases} 4x + 3y + 17z = 0 \\ 5x + 4y + 22z = 0 \\ 4x + 2y + 19z = 0 \end{cases}$

33. $\begin{cases} x - 2y + 5z = 2 \\ 4x - z = 0 \end{cases}$

34. $\begin{cases} x - 3y + 2z = 18 \\ 5x - 13y + 12z = 80 \end{cases}$

35. $\begin{cases} 2x - 3y + z = -2 \\ -4x + 9y = 7 \end{cases}$

36. $\begin{cases} 2x + 3y + 3z = 7 \\ 4x + 18y + 15z = 44 \end{cases}$

37. $\begin{cases} 12x + 5y + z = 0 \\ 23x + 4y - z = 0 \end{cases}$ **38.** $\begin{cases} 10x - 3y + 2z = 0 \\ 19x - 5y - z = 0 \end{cases}$

39. $\begin{cases} x + 3w = 4 \\ 2y - z - w = 0 \\ 3y - 2w = 1 \\ 2x - y + 4z = 5 \end{cases}$

40. $\begin{cases} x + y + z + w = 6 \\ 2x + 3y - w = 0 \\ -3x + 4y + z + 2w = 4 \\ x + 2y - z + w = 0 \end{cases}$

Exploration **In Exercises 41–44, find a system of linear equations that has the given solution. (There are many correct answers.)**

41. $(4, -1, 2)$ **42.** $(-5, -2, 1)$

43. $\left(3, -\frac{1}{2}, \frac{7}{4}\right)$ **44.** $\left(-\frac{3}{2}, 4, -7\right)$

Three-Dimensional Graphics **In Exercises 45–48, sketch the plane represented by the linear equation. Then list four points that lie in the plane.**

45. $2x + 3y + 4z = 12$ **46.** $x + y + z = 6$

47. $2x + y + z = 4$ **48.** $x + 2y + 2z = 6$

In Exercises 49–54, write the form of the partial fraction decomposition of the rational expression. Do not solve for the constants.

49. $\dfrac{7}{x^2 - 14x}$ **50.** $\dfrac{x - 2}{x^2 + 4x + 3}$

51. $\dfrac{12}{x^3 - 10x^2}$ **52.** $\dfrac{x^2 - 3x + 2}{4x^3 + 11x^2}$

53. $\dfrac{4x^2 + 3}{(x - 5)^3}$ **54.** $\dfrac{6x + 5}{(x + 2)^4}$

In Exercises 55–72, write the partial fraction decomposition for the rational expression. Check your result algebraically by combining the fractions.

55. $\dfrac{1}{x^2 - 1}$ **56.** $\dfrac{1}{4x^2 - 9}$

57. $\dfrac{1}{x^2 + x}$ **58.** $\dfrac{3}{x^2 - 3x}$

59. $\dfrac{1}{2x^2 + x}$ **60.** $\dfrac{5}{x^2 + x - 6}$

61. $\dfrac{3}{x^2 + x - 2}$ **62.** $\dfrac{x + 1}{x^2 + 4x + 3}$

63. $\dfrac{x^2 + 12x + 12}{x^3 - 4x}$ **64.** $\dfrac{x + 2}{x(x - 4)}$

65. $\dfrac{4x^2 + 2x - 1}{x^2(x + 1)}$ **66.** $\dfrac{2x - 3}{(x - 1)^2}$

67. $\dfrac{3x}{(x - 3)^2}$ **68.** $\dfrac{6x^2 + 1}{x^2(x - 1)^3}$

69. $\dfrac{2x^3 - x^2 + x + 5}{x^2 + 3x + 2}$ **70.** $\dfrac{x^3 + 2x^2 - x + 1}{x^2 + 3x - 4}$

71. $\dfrac{x^4}{(x - 1)^3}$ **72.** $\dfrac{4x^4}{(2x - 1)^3}$

In Exercises 73–78, write the partial fraction decomposition for the rational expression. Use a graphing utility to check your result graphically.

73. $\dfrac{5 - x}{2x^2 + x - 1}$

74. $\dfrac{3x^2 - 7x - 2}{x^3 - x}$

75. $\dfrac{x - 1}{x^3 + x^2}$

76. $\dfrac{4x^2 - 1}{2x(x + 1)^2}$

77. $\dfrac{2x^3 - 4x^2 - 15x + 5}{x^2 - 2x - 8}$

78. $\dfrac{x^3 - x + 3}{x^2 + x - 2}$

Graphical Analysis   In Exercises 79 and 80, write the partial fraction decomposition for the rational function. Identify the graph of the rational function and the graphs of each term of its decomposition. State any relationship between the vertical asymptotes of the rational function and the vertical asymptotes of the terms of the decomposition.

79. $y = \dfrac{x - 12}{x(x - 4)}$

80. $y = \dfrac{2(4x - 3)}{x^2 - 9}$

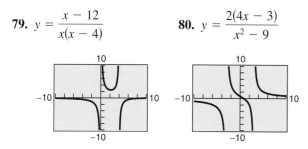

Vertical Motion   In Exercises 81–84, find the position equation $s = \frac{1}{2}at^2 + v_0 t + s_0$ for an object at the given heights moving vertically at the specified times.

81. At $t = 1$ second, $s = 128$ feet
At $t = 2$ seconds, $s = 80$ feet
At $t = 3$ seconds, $s = 0$ feet

82. At $t = 1$ second, $s = 48$ feet
At $t = 2$ seconds, $s = 64$ feet
At $t = 3$ seconds, $s = 48$ feet

83. At $t = 1$ second, $s = 452$ feet
At $t = 2$ seconds, $s = 260$ feet
At $t = 3$ seconds, $s = 116$ feet

84. At $t = 1$ second, $s = 132$ feet
At $t = 2$ seconds, $s = 100$ feet
At $t = 3$ seconds, $s = 36$ feet

In Exercises 85–88, find the equation of the parabola

$$y = ax^2 + bx + c$$

that passes through the given points. To verify your result, use a graphing utility to plot the points and graph the parabola.

85. $(0, 0), (2, -2), (4, 0)$

86. $(0, 3), (1, 4), (2, 3)$

87. $(2, 0), (3, -1), (4, 0)$

88. $(1, 3), (2, 2), (3, -3)$

In Exercises 89–92, find the equation of the circle

$$x^2 + y^2 + Dx + Ey + F = 0$$

that passes through the given points. To verify your result, use a graphing utility to plot the points and graph the circle.

89. $(0, 0), (2, 2), (4, 0)$

90. $(0, 0), (0, 6), (3, 3)$

91. $(-3, -1), (2, 4), (-6, 8)$

92. $(-6, -1), (-4, 3), (2, -5)$

93. *Investments*   An inheritance of \$16,000 was divided among three investments yielding a total of \$990 in interest per year. The interest rates were 5%, 6%, and 7%. Find the amount in each investment if the 5% and 6% investments were \$3000 and \$2000 less than the 7% investment.

94. *Investments*   A total of \$1520 a year is received in interest from three investments. The interest rates for the three investments are 5%, 7%, and 8%. The 5% investment is half of the 7% investment, and the 7% investment is \$1500 less than the 8% investment. Find the amount in each investment.

95. *Borrowing*   A small corporation borrowed \$775,000 to expand its product line. Some of the money was borrowed at 8%, some at 9%, and some at 10%. How much was borrowed at each rate if the annual interest was \$67,000 and the amount borrowed at 8% was four times the amount borrowed at 10%?

96. *Borrowing*   A small corporation borrowed \$800,000 to expand its product line. Some of the money was borrowed at 8%, some at 9%, and some at 10%. How much was borrowed at each rate if the annual interest was \$67,000 and the amount borrowed at 8% was five times the amount borrowed at 10%?

Investment Portfolio   **In Exercises 97 and 98, consider an investor with a portfolio totaling $500,000 that is invested in certificates of deposit, municipal bonds, blue-chip stocks, and growth or speculative stocks. How much is put in each type of investment?**

97. The certificates of deposit pay 10% annually, and the municipal bonds pay 8% annually. Over a 5-year period, the investor expects the blue-chip stocks to return 12% annually and the growth stocks to return 13% annually. The investor wants a combined annual return of 10% as well as only one-fourth of the portfolio invested in stocks.

98. The certificates of deposit pay 9% annually, and the municipal bonds pay 5% annually. Over a 5-year period, the investor expects the blue-chip stocks to return 12% annually and the growth stocks to return 14% annually. The investor wants a combined annual return of 10% and also wants to have only one-fourth of the portfolio invested in stocks.

99. *Crop Spraying*   A mixture of 12 liters of chemical A, 16 liters of chemical B, and 26 liters of chemical C is required to kill a certain destructive crop insect. Commercial spray X contains 1, 2, and 2 parts, respectively, of these chemicals. Commercial spray Y contains only chemical C. Commercial spray Z contains only chemicals A and B in equal amounts. How much of each type of commercial spray is needed to get the desired mixture?

100. *Chemistry*   A chemist needs 10 liters of a 25% acid solution. The solution is to be mixed from three solutions whose concentrations are 10%, 20%, and 50%. How many liters of each solution should the chemist use to satisfy the following?

 (a) Use as little as possible of the 50% solution.

 (b) Use as much as possible of the 50% solution.

 (c) Use 2 liters of the 50% solution.

101. *Truck Scheduling*   A small company that manufactures products A and B has an order for 15 units of product A and 16 units of product B. The company has trucks of three different sizes that can haul the products A and B in the amounts shown in the table. How many trucks of each size are needed to deliver the order? Give *two* possible solutions.

Truck	Large	Medium	Small
Product A	6	4	0
Product B	3	4	3

102. *Electrical Networks*   When Kirchhoff's Laws are applied to the electrical network in the figure, the currents I_1, I_2, and I_3 are the solution of the system

$$\begin{cases} I_1 - I_2 + I_3 = 0 \\ 3I_1 + 2I_2 \qquad = 7. \\ \qquad 2I_2 + 4I_3 = 8 \end{cases}$$

Find the currents.

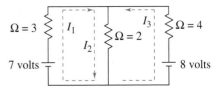

103. *Pulley System*   A system of pulleys is loaded with 128-pound and 32-pound weights. The tensions t_1 and t_2 in the ropes and the acceleration a of the 32-pound weight are found by solving the system

$$\begin{cases} t_1 - 2t_2 \qquad = 0 \\ t_1 \qquad - 2a = 128 \\ \qquad t_2 + a = 32 \end{cases}$$

where t_1 and t_2 are measured in pounds and a is in feet per second squared. Solve the system.

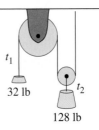

104. *Pulley System*   If the 32-pound weight is replaced with a 64-pound weight in the pulley system described in Exercise 103, the new pulley system is modeled by the following system of equations.

$$\begin{cases} t_1 - 2t_2 \qquad = 0 \\ t_1 \qquad - 2a = 128 \\ \qquad t_2 + 2a = 64 \end{cases}$$

Solve the system, and use your answer for the acceleration to describe what (if anything) is happening in the system.

Fitting a Parabola   **In Exercises 105–108, find the least squares regression parabola $y = ax^2 + bx + c$ for the points $(x_1, y_1), (x_2, y_2), \ldots, (x_n, y_n)$ by solv-**

ing the following system of linear equations for *a*, *b*, and *c*. Then use the regression capabilities of a graphing utility to confirm the result. (For an explanation of how the coefficients of *a*, *b*, and *c* in the system are obtained, see Appendix B.)

105.

$$\begin{cases} 4c & + 40a = 19 \\ 40b & = -12 \\ 40c & + 544a = 160 \end{cases}$$

106.

$$\begin{cases} 5c & + 10a = 8 \\ 10b & = 12 \\ 10c & + 34a = 22 \end{cases}$$

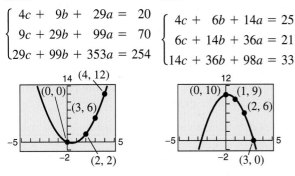

107.

$$\begin{cases} 4c + 9b + 29a = 20 \\ 9c + 29b + 99a = 70 \\ 29c + 99b + 353a = 254 \end{cases}$$

108.

$$\begin{cases} 4c + 6b + 14a = 25 \\ 6c + 14b + 36a = 21 \\ 14c + 36b + 98a = 33 \end{cases}$$

109. *Analyzing Data* During the testing of a new braking system on an automobile, the speeds (in miles per hour) and the stopping distances (in feet) were recorded as follows.

Speed, x	20	30	40	50	60
Stopping Distance, y	25	55	105	188	300

(a) Find the least squares regression parabola for the data. Use a graphing utility to graph the parabola and the data in the same viewing window.

(b) Use the model to estimate the stopping distance for a speed of 70 miles per hour.

110. *Analyzing Data* A wildlife management team studied the reproduction rates of deer in five 5-acre tracts of a wildlife preserve. For each tract, the number of females and the percent of females that had offspring the following year were recorded. The results are shown in the table.

Number, x	80	100	120	140	160
Percent, y	80	75	68	55	30

(a) Find the least squares regression parabola for the data.

(b) Use a graphing utility to graph the parabola and the data in the same viewing window.

(c) Use the model to predict the percent of females that would have offspring if $x = 170$.

111. *Exhaust Temperatures* The magnitude of the range of exhaust temperatures (in degrees Fahrenheit) in an experimental diesel engine is approximated by

$$R = \frac{2000(4 - 3x)}{(11 - 7x)(7 - 4x)}, \quad 0 \le x \le 1$$

where x is the relative load.

(a) Write the partial fraction decomposition for the rational function.

(b) The decomposition in part (a) is the difference of two fractions. The absolute values of the terms give the expected maximum and minimum temperatures of the exhaust gases. Use a graphing utility to graph each term.

112. *Cellular Phone Revenues* The amount of service revenue R (in billions of dollars) for cellular phones in the United States for the years 1991 through 1997 can be approximated by the model

$$R = \frac{-2617 + 4423t}{t(411 - 38t)}, \quad 1 \le t \le 7$$

where t represents the year with $t = 1$ corresponding to 1991. Write the partial fraction decomposition for the rational function. (Source: Cellular Telecommunications Industry Association)

Synthesis

True or False? In Exercises 113–115, determine whether the statement is true or false. Justify your answer.

113. The system $\begin{cases} x + 4y - 5z = 8 \\ 2y + z = 5 \\ z = 1 \end{cases}$ is in row-echelon form.

114. If a system of three linear equations is inconsistent, then its graph has no common point.

115. For the rational expression

$$\frac{x}{(x + 10)(x - 10)^2}$$

the partial fraction decomposition is of the form

$$\frac{A}{x + 10} + \frac{B}{(x - 10)^2}.$$

116. *Error Analysis* Suppose you are tutoring a student in algebra. In trying to find a partial fraction decomposition, your student writes the following.

$$\frac{x^2 + 1}{x(x - 1)} = \frac{A}{x} + \frac{B}{x - 1}$$

$$x^2 + 1 = A(x - 1) + Bx \qquad \text{Basic equation}$$

$$x^2 + 1 = (A + B)x - A$$

Your student then forms the following system of linear equations.

$$\begin{cases} A + B = 0 \\ -A \quad\ \ = 1 \end{cases}$$

Solve the system and check the partial fraction decomposition it yields. Has your student worked the problem correctly? If not, what went wrong?

In Exercises 117–120, write the partial fraction decomposition for the rational expression. Check your result algebraically. Then assign a value to the constant *a* and check the result graphically.

117. $\dfrac{1}{a^2 - x^2}$

118. $\dfrac{1}{(x + 1)(a - x)}$

119. $\dfrac{1}{y(a - y)}$

120. $\dfrac{1}{x(x + a)}$

121. *Think About It* Are the two systems of equations equivalent? Give reasons for your answer.

$$\begin{cases} x + 3y - \ z = 6 \\ 2x - \ y + 2z = 1 \\ 3x + 2y - \ z = 2 \end{cases} \qquad \begin{cases} x + 3y - \ z = \quad 6 \\ \quad\ \ -7y + 4z = \quad 1 \\ \quad\ \ -7y - 4z = -16 \end{cases}$$

122. *Think About It* When using Gaussian elimination to solve a system of linear equations, how can you recognize that it has no solution? Give an example that illustrates your answer.

Advanced Applications **In Exercises 123–126, find values of $x, y,$ and λ that satisfy the system. These systems arise in certain optimization problems in calculus; λ is called a Lagrange multiplier.**

123. $\begin{cases} y + \lambda = 0 \\ x + \lambda = 0 \\ x + y - 10 = 0 \end{cases}$ **124.** $\begin{cases} 2x + \lambda = 0 \\ 2y + \lambda = 0 \\ x + y - 4 = 0 \end{cases}$

125. $\begin{cases} 2x - 2x\lambda = 0 \\ -2y + \lambda = 0 \\ y - x^2 = 0 \end{cases}$ **126.** $\begin{cases} 2 + 2x + 2\lambda = 0 \\ 2x + 1 + \lambda = 0 \\ 2x + y - 100 = 0 \end{cases}$

Review

In Exercises 127–134, sketch the graph of the function.

127. $f(x) = -3x + 7$ **128.** $f(x) = 6 - x$

129. $f(x) = -2x^2$ **130.** $f(x) = \frac{1}{4}x^2 + 1$

131. $f(x) = x^2 - 9x + 18$ **132.** $f(x) = 2x^2 - 9x - 5$

133. $f(x) = -x^2(x - 3)$ **134.** $f(x) = \frac{1}{2}x^3 - 1$

In Exercises 135–138, (a) determine the real zeros of f and (b) sketch the graph of f.

135. $f(x) = x^3 + x^2 - 12x$

136. $f(x) = -8x^4 + 32x^2$

137. $f(x) = 2x^3 + 5x^2 - 21x - 36$

138. $f(x) = 6x^3 - 29x^2 - 6x + 5$

In Exercises 139–142, use a graphing utility to create a table of values. Then sketch the graph of the equation by hand.

139. $y = 4^{-x-4} - 5$ **140.** $y = \left(\frac{5}{2}\right)^{x-1} - 4$

141. $y = 2.9^{0.8x} - 3$ **142.** $y = -3.5^{x+2} - 6$

In Exercises 143–146, solve the system by elimination.

143. $\begin{cases} 3x + 3y = 7 \\ 3x + 5y = 3 \end{cases}$ **144.** $\begin{cases} 6x - \ 5y = 3 \\ 10x - 12y = 5 \end{cases}$

145. $\begin{cases} 2x + \ y = 120 \\ x + 2y = 120 \end{cases}$ **146.** $\begin{cases} 12x + 42y = -17 \\ 30x - 18y = \quad 19 \end{cases}$

7.4 Systems of Inequalities

The Graph of an Inequality

The statements $3x - 2y < 6$ and $2x^2 + 3y^2 \geq 6$ are inequalities in two variables. An ordered pair (a, b) is a **solution of an inequality** in x and y if the inequality is true when a and b are substituted for x and y, respectively. The **graph of an inequality** is the collection of all solutions of the inequality. To sketch the graph of an inequality, begin by sketching the graph of the *corresponding equation*. The graph of the equation will normally separate the plane into two or more regions. In each such region, one of the following must be true.

1. *All* points in the region are solutions of the inequality.

2. *No* point in the region is a solution of the inequality.

So, you can determine whether the points in an entire region satisfy the inequality by simply testing *one* point in the region.

Sketching the Graph of an Inequality in Two Variables

1. Replace the inequality sign with an equal sign, and sketch the graph of the corresponding equation. (Use a dashed line for $<$ or $>$ and a solid line for $\leq$ or $\geq$.)

2. Test one point in each of the regions formed by the graph in Step 1. If the point satisfies the inequality, shade the entire region to denote that every point in the region satisfies the inequality.

EXAMPLE 1 Sketching the Graph of an Inequality

To sketch the graph of $y \geq x^2 - 1$, begin by graphing the corresponding *equation* $y = x^2 - 1$, which is a parabola, as shown in Figure 7.18. By testing the point $(0, 0)$ *above* the parabola and the point $(0, -2)$ *below* the parabola, you can see that $(0, 0)$ satisfies the inequality because $0 \geq 0^2 - 1$ and that $(0, -2)$ does not satisfy the inequality because $-2 \ngeq 0^2 - 1$. So, the points that satisfy the inequality are those lying above (or on) the parabola.

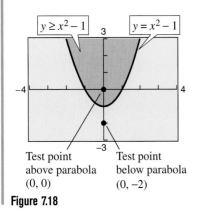

Test point above parabola $(0, 0)$
Test point below parabola $(0, -2)$

Figure 7.18

The inequality given in Example 1 is a nonlinear inequality in two variables. Most of the following examples involve **linear inequalities** such as $ax + by < c$. The graph of a linear inequality is a half-plane lying on one side of the line $ax + by = c$.

EXAMPLE 2   Sketching the Graphs of Linear Inequalities

Sketch the graph of each linear inequality.

a. $x > -2$ **b.** $y \leq 3$

Solution

a. The graph of the corresponding equation $x = -2$ is a vertical line. The points that satisfy the inequality $x > -2$ are those lying to the right of (but not on) this line, as shown in Figure 7.19.

b. The graph of the corresponding equation $y = 3$ is a horizontal line. The points that satisfy the inequality $y \leq 3$ are those lying below (or on) this line, as shown in Figure 7.20.

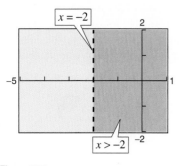

Figure 7.19

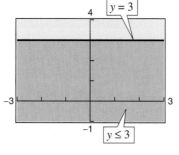

Figure 7.20

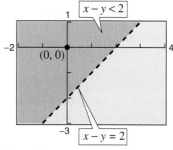

Figure 7.21

STUDY TIP

A graphing utility can be used to graph an inequality. For instance, to graph $y \geq x - 2$, enter $y = x - 2$ and use the *shade* feature of the graphing utility to shade the correct part of the graph. Consult your user's manual for instructions.

EXAMPLE 3   Sketching the Graph of a Linear Inequality

Sketch the graph of $x - y < 2$.

Solution

The graph of the corresponding equation $x - y = 2$ is a line, as shown in Figure 7.21. Because the origin $(0, 0)$ satisfies the inequality, the graph consists of the half-plane lying above the line. (Try checking a point below the line. Regardless of which point below the line you choose, you will see that it does not satisfy the inequality.)

To graph a linear inequality, it can help to write the inequality in slope-intercept form. For instance, by writing $x - y < 2$ in Example 3 in the form

$$y > x - 2$$

you can see that the solution points lie *above* the line $y = x - 2$ (or $x - y = 2$), as shown in Figure 7.21.

Systems of Inequalities

Many practical problems in business, science, and engineering involve systems of linear inequalities. A **solution of a system of inequalities** in x and y is a point (x, y) that satisfies each inequality in the system.

To sketch the graph of a system of inequalities in two variables, first sketch the graph of each individual inequality (on the same coordinate system) and then find the region that is *common* to every graph in the system. For systems of *linear* inequalities, it is helpful to find the vertices of the solution region.

EXAMPLE 4   Solving a System of Inequalities

Sketch the graph (and label the vertices) of the solution set of the system.

$$\begin{cases} x - y < & 2 & \text{Inequality 1} \\ x > -2 & & \text{Inequality 2} \\ y \le & 3 & \text{Inequality 3} \end{cases}$$

A computer animation of this example appears in the *Interactive* CD-ROM and *Internet* versions of this text.

Solution

The graphs of these inequalities are shown in Figures 7.19 through 7.21. The triangular region common to all three graphs can be found by superimposing the graphs on the same coordinate system, as shown in Figure 7.22. To find the vertices of the region, solve the three systems of corresponding equations obtained by taking pairs of equations representing the boundaries of the individual regions and solving these pairs of equations.

Vertex A: $(-2, -4)$ *Vertex B:* $(5, 3)$ *Vertex C:* $(-2, 3)$

$$\begin{cases} x - y = & 2 \\ x = -2 \end{cases} \qquad \begin{cases} x - y = 2 \\ y = 3 \end{cases} \qquad \begin{cases} x = -2 \\ y = & 3 \end{cases}$$

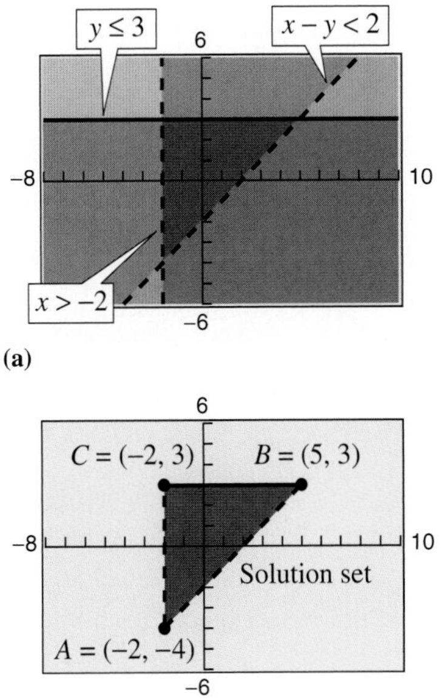

(a)

(b)

Figure 7.22

> ## STUDY T!P
>
> Using different colored pencils to shade the solution of each inequality in a system makes identifying the solution of the system of inequalities easier. The region common to every graph in the system is where all shaded regions overlap. This region represents the solution set of the system.

For the triangular region shown in Figure 7.22, each point of intersection of a pair of boundary lines corresponds to a vertex. With more complicated regions, two border lines can sometimes intersect at a point that is not a vertex of the region, as shown in Figure 7.23. To keep track of which points of intersection are actually vertices of the region, you should sketch the region and refer to your sketch as you find each point of intersection.

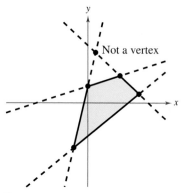

Figure 7.23

EXAMPLE 5   Solving a System of Inequalities

Sketch the region containing all points that satisfy the system.

$$\begin{cases} x^2 - y \le 1 & \text{Inequality 1} \\ -x + y \le 1 & \text{Inequality 2} \end{cases}$$

Solution

As shown in Figure 7.24, the points that satisfy the inequality $x^2 - y \le 1$ are the points lying above (or on) the parabola

$$y = x^2 - 1. \qquad \text{Parabola}$$

The points that satisfy the inequality $-x + y \le 1$ are the points lying below (or on) the line

$$y = x + 1. \qquad \text{Line}$$

To find the points of intersection of the parabola and the line, solve the system of corresponding equations.

$$\begin{cases} x^2 - y = 1 \\ -x + y = 1 \end{cases}$$

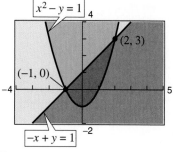

Figure 7.24

Using the method of substitution, you can find the solutions to be $(-1, 0)$ and $(2, 3)$. So, the region containing all points that satisfy the system are indicated by the purple shaded region in Figure 7.24.

When solving a system of inequalities, you should be aware that the system might have no solution, or it might be represented by an unbounded region in the plane. These two possibilities are shown in Examples 6 and 7.

EXAMPLE 6   A System with No Solution

Sketch the solution set of the system.

$$\begin{cases} x + y > 3 & \text{Inequality 1} \\ x + y < -1 & \text{Inequality 2} \end{cases}$$

Solution

From the way the system is written, it is clear that the system has no solutions, because the quantity $(x + y)$ cannot be both less than -1 and greater than 3. Graphically, the inequality $x + y > 3$ is represented by the half-plane lying above the line $x + y = 3$, and the inequality $x + y < -1$ is represented by the half-plane lying below the line $x + y = -1$, as shown in Figure 7.25. These two half-planes have no points in common. So the system of inequalities has no solution.

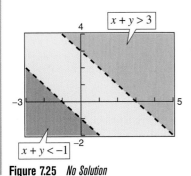

Figure 7.25   *No Solution*

EXAMPLE 7   An Unbounded Solution Set

Sketch the solution set of the system.

$$\begin{cases} x + y < 3 & \text{Inequality 1} \\ x + 2y > 3 & \text{Inequality 2} \end{cases}$$

Solution

The graph of the inequality $x + y < 3$ is the half-plane that lies below the line $x + y = 3$, as shown in Figure 7.26. The graph of the inequality $x + 2y > 3$ is the half-plane that lies above the line $x + 2y = 3$. The intersection of these two half-planes is an *infinite wedge* that has a vertex at $(3, 0)$. This unbounded region representing the solution set has an infinite number of points.

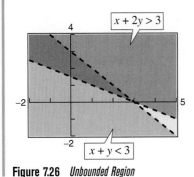

Figure 7.26   *Unbounded Region*

Applications

The next example discusses two concepts that economists call *consumer surplus* and *producer surplus*. As shown in Figure 7.27, the *point of equilibrium* is defined by the price p and the number of units x that satisfy both the demand and supply equations. Consumer surplus is defined as the area of the region that lies *below* the demand curve, *above* the horizontal line passing through the equilibrium point, and to the right of the p-axis. Similarly, the producer surplus is defined as the area of the region that lies *above* the supply curve, *below* the horizontal line passing through the equilibrium point, and to the right of the p-axis. The consumer surplus is a measure of the amount that consumers would have been willing to pay *above what they actually paid*, whereas the producer surplus is a measure of the amount that producers would have been willing to receive *below what they actually received*.

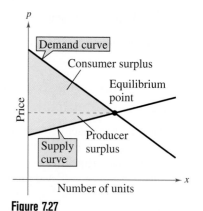

Figure 7.27

EXAMPLE 8 Consumer Surplus and Producer Surplus

The demand and supply functions for a certain type of calculator are

$$\begin{cases} p = 150 - 0.00001x & \text{Demand equation} \\ p = 60 + 0.00002x & \text{Supply equation} \end{cases}$$

where p is the price (in dollars) and x represents the number of units. Find the consumer surplus and producer surplus for these two equations.

Solution

Begin by finding the point of equilibrium by solving the equation

$$60 + 0.00002x = 150 - 0.00001x.$$

In Example 8 in Section 7.2, you saw that the solution is $x = 3,000,000$, which corresponds to an equilibrium price of $p = \$120$. So, the consumer surplus and producer surplus are the areas of the following triangular regions.

Consumer Surplus	*Producer Surplus*
$\begin{cases} p \le 150 - 0.00001x \\ p \ge 120 \\ x \ge 0 \end{cases}$	$\begin{cases} p \ge 60 + 0.00002x \\ p \le 120 \\ x \ge 0 \end{cases}$

In Figure 7.28, you can see that these regions are triangles. So, the consumer and producer surpluses are

Consumer surplus $= \frac{1}{2}$ (base)(height)

$= \frac{1}{2}(3,000,000)(30)$

$= \$45,000,000$

Producer surplus $= \frac{1}{2}$ (base)(height)

$= \frac{1}{2}(3,000,000)(60)$

$= \$90,000,000$

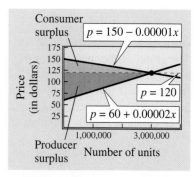

Figure 7.28

EXAMPLE 9   Nutrition

The liquid portion of a diet is to provide at least 300 calories, 36 units of vitamin A, and 90 units of vitamin C daily. A cup of dietary drink X provides 60 calories, 12 units of vitamin A, and 10 units of vitamin C. A cup of dietary drink Y provides 60 calories, 6 units of vitamin A, and 30 units of vitamin C. Set up a system of linear inequalities that must be satisfied in order to meet the minimum daily requirements for calories and vitamins.

Solution

Begin by letting x and y represent the following.

> x = number of cups of dietary drink X
>
> y = number of cups of dietary drink Y

To meet the minimum daily requirements, the following inequalities must be satisfied.

$$\begin{cases} 60x + 60y \geq 300 & \text{Calories} \\ 12x + 6y \geq 36 & \text{Vitamin A} \\ 10x + 30y \geq 90 & \text{Vitamin C} \\ x \geq 0 \\ y \geq 0 \end{cases}$$

The last two inequalities are included because x and y cannot be negative. The graph of this system of inequalities is shown in Figure 7.29. (More is said about this application in Example 6 in Section 7.5.)

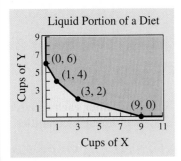

Liquid Portion of a Diet

Figure 7.29

> ### STUDY T!P
>
> When using a system of inequalities to represent a real-life application in which the variables cannot be negative, remember to include inequalities for this condition. For instance, in Example 9, x and y cannot be negative, so the inequalities $x \geq 0$ and $y \geq 0$ must be included in the system.

Writing About Math   *Creating a System of Inequalities*

Plot the points $(0, 0)$, $(4, 0)$, $(3, 2)$, and $(0, 2)$ in a coordinate plane. Draw the quadrilateral that has these four points as its vertices. Write a system of linear inequalities that has the quadrilateral as its solution. Explain how you found the system of inequalities.

In Exercises 1–8, match the inequality with its graph. [The graphs are labeled (a), (b), (c), (d), (e), (f), (g), and (h).]

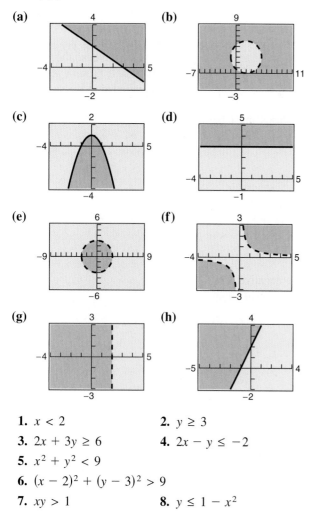

1. $x < 2$ **2.** $y \geq 3$

3. $2x + 3y \geq 6$ **4.** $2x - y \leq -2$

5. $x^2 + y^2 < 9$

6. $(x - 2)^2 + (y - 3)^2 > 9$

7. $xy > 1$ **8.** $y \leq 1 - x^2$

In Exercises 9–20, sketch the graph of the inequality.

9. $x \geq 2$ **10.** $x \leq -5$

11. $y \geq -1$ **12.** $y \leq 3$

13. $y < 2 - x$ **14.** $y > 2x - 4$

15. $2y - x \geq 4$ **16.** $5x + 3y \geq -15$

17. $y^2 - x < 0$ **18.** $4x + y^2 > 1$

19. $(x + 1)^2 + y^2 < 9$

20. $(x - 1)^2 + (y - 4)^2 > 9$

In Exercises 21–32, use a graphing utility to graph the inequality. Be sure to shade the region representing the solution.

21. $y \geq \frac{2}{3}x - 1$ **22.** $y \leq 6 - \frac{3}{2}x$

23. $y < -3.8x + 1.1$ **24.** $y \geq -20.74 + 2.66x$

25. $x^2 + 5y - 10 \leq 0$ **26.** $2x^2 - y - 3 > 0$

27. $y \leq \dfrac{1}{1 + x^2}$ **28.** $y > \dfrac{-10}{x^2 + x + 4}$

29. $y < \ln x$ **30.** $y \geq 4 - \ln(x + 5)$

31. $y > 3^{-x-4}$ **32.** $y \leq 2^{2x-1} - 3$

In Exercises 33–36, write an inequality for the shaded region shown in the graph.

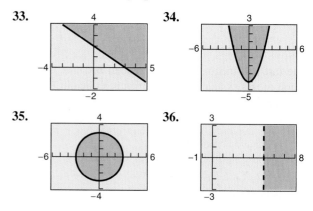

In Exercises 37 and 38, decide whether each ordered pair is a solution of the system of linear inequalities.

37. $\begin{cases} -2x + 5y \geq 3 \\ \qquad\quad y < 4 \\ -4x + 2y < 7 \end{cases}$ (a) $(0, 2)$ (b) $(-6, 4)$
    (c) $(-8, -2)$ (d) $(-3, 2)$

38. $\begin{cases} \quad 3x + y > 1 \\ -y - \frac{1}{2}x^2 \leq -4 \\ -15x + 4y > 0 \end{cases}$ (a) $(0, 10)$ (b) $(0, -1)$
    (c) $(2, 9)$ (d) $(-1, 6)$

In Exercises 39–50, sketch the graph of the solution of the system of inequalities. Verify your sketch by using a graphing utility to graph the system, shading the region that represents the solution of the system.

39. $\begin{cases} x + y \le 1 \\ -x + y \le 1 \\ \qquad y \ge 0 \end{cases}$ **40.** $\begin{cases} 3x + 2y < 6 \\ x \qquad > 0 \\ \qquad y > 0 \end{cases}$

41. $\begin{cases} -3x + 2y < \quad 6 \\ \quad x - 4y > -2 \\ \quad 2x + \ y < \quad 3 \end{cases}$ **42.** $\begin{cases} x - 7y > -36 \\ 5x + 2y > \quad 5 \\ 6x - 5y > \quad 6 \end{cases}$

43. $\begin{cases} 2x + \ y > 2 \\ 6x + 3y < 2 \end{cases}$ **44.** $\begin{cases} x - 2y < -6 \\ 5x - 3y > -9 \end{cases}$

45. $\begin{cases} -x^2 + y \ge 5 \\ \frac{1}{4}x + y < 3 \end{cases}$ **46.** $\begin{cases} y^2 - 3x \ge \quad 9 \\ x + \ y \ge -3 \end{cases}$

47. $\begin{cases} x > y^2 \\ x < y + 2 \end{cases}$ **48.** $\begin{cases} x - y^2 > 0 \\ x - y \ < 2 \end{cases}$

49. $\begin{cases} x^2 + y^2 \le 9 \\ x^2 + y^2 \ge 1 \end{cases}$ **50.** $\begin{cases} x < 2y - y^2 \\ 0 < x + y \end{cases}$

In Exercises 51–56, use a graphing utility to graph the inequalities. Be sure to shade the region representing the solution of the system.

51. $\begin{cases} y \le \sqrt{3x} + 1 \\ y \ge x^2 + 1 \end{cases}$ **52.** $\begin{cases} y < -x^2 + 2x + 3 \\ y > \quad x^2 - 4x + 3 \end{cases}$

53. $\begin{cases} y < x^3 - 2x + 1 \\ y > -2x \\ x \le 1 \end{cases}$ **54.** $\begin{cases} y \ge x^4 - 2x^2 + 1 \\ y \le 1 - x^2 \end{cases}$

55. $\begin{cases} \quad x^2 y \ge 1 \\ 0 < x \le 4 \\ \quad \ y \le 4 \end{cases}$ **56.** $\begin{cases} \quad y \le e^{-x^2/2} \\ \quad y \ge 0 \\ -2 \le x \le 2 \end{cases}$

In Exercises 57–64, find a set of inequalities to describe the region.

57. **58.**

59. **60.**

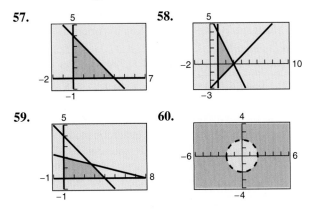

61. Rectangular region with vertices at $(2, 1)$, $(5, 1)$, $(5, 7)$, and $(2, 7)$

62. Parallelogrammatic region with vertices at $(0, 0)$, $(4, 0)$, $(1, 4)$, and $(5, 4)$

63. Triangular region with vertices at $(0, 0)$, $(5, 0)$, and $(2, 3)$

64. Triangular region with vertices at $(-1, 0)$, $(1, 0)$, and $(0, 1)$

In Exercises 65–69, (a) find a system of inequalities that models the problem and (b) use a graphing utility to graph the systems, shading the region that represents the solution of the system.

65. *Investment*   A person plans to invest some or all of $20,000 in two different interest-bearing accounts. Each account is to contain at least $5000, and one account should have at least twice the amount that is in the other account.

66. *Concert Ticket Sales*   One type of concert ticket costs $15 and another costs $25. The promoter of the concert must sell at least 15,000 tickets, including at least 8000 of the $15 tickets and at least 4000 of the $25 tickets, and the gross receipts must total at least $275,000 in order for the concert to be held.

67. *Furniture Production*   A furniture company can sell all the tables and chairs it produces. Each table requires 1 hour of assembly and $1\frac{1}{3}$ hours of finishing. Each chair requires $1\frac{1}{2}$ hours of assembly and $1\frac{1}{2}$ hours of finishing. The company's assembly center is available 12 hours per day, and its finishing center is available 15 hours per day.

68. *Computer Inventory*   A store sells two models of a certain brand of computer. Because of the demand, it is necessary to stock at least twice as many units of model A as units of model B. The costs to the store for the two models are $800 and $1200, respectively. The management does not want more than $20,000 in computer inventory at any one time, and it wants at least four model A computers and two model B computers in inventory at all times.

69. *Diet Supplement*   A dietitian is asked to design a special diet using two different foods. The minimum daily requirements in the new diet are 280 units of calcium, 160 units of iron, and 180 units of vitamin B. Each ounce of food X contains 20 units of calcium, 15 units of iron, and 10 units of vitamin B. Each ounce of food Y contains 10 units of calcium, 10 units of iron, and 20 units of vitamin B.

70. Physical Fitness Facility You plan an exercise facility that has an indoor running track with an exercise floor inside the track. The track must be at least 125 meters long, and the exercise floor must have an area of at least 500 square meters.

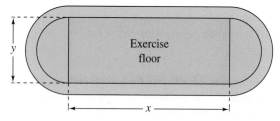

(a) Find a system of inequalities describing the requirements of the exercise facility.

(b) Sketch the graph of the system in part (a).

71. Graphical Reasoning Two concentric circles have radii of x and y meters, where $y > x$. The area between the boundaries of the circles must be at least 10 square meters.

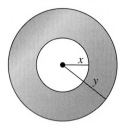

(a) Find an inequality describing the constraints on the circles.

(b) Use a graphing utility to graph the inequality in part (a). Graph the line $y = x$ in the same viewing window.

(c) Identify the graph of the line in relation to the boundary of the inequality. Explain its meaning in the context of the problem.

Consumer and Producer Surpluses In Exercises 72–75, use a graphing utility to graph the system representing the consumer surplus and producer surplus for the supply and demand equations. Be sure to shade the region representing the solution of the system. Find the consumer surplus and the producer surplus.

Demand	Supply
72. $p = 50 - 0.5x$	$p = 0.125x$
73. $p = 60 - x$	$p = 10 + \frac{7}{3}x$
74. $p = 300 - x$	$p = 100 + x$
75. $p = 140 - 0.00002x$	$p = 80 + 0.00001x$

Synthesis

True or False? In Exercises 76 and 77, determine whether the statement is true or false. Justify your answer.

76. The area of the figure defined by the system below is 99 square units.

$$\begin{cases} x \geq -3 \\ x \leq 6 \\ y \leq 5 \\ y \geq -6 \end{cases}$$

77. The graph below shows the solution of the system

$$\begin{cases} y \leq 6 \\ -4x - 9y > 6. \\ 3x + y^2 \geq 2 \end{cases}$$

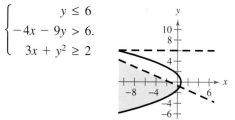

78. Think About It After graphing the boundary of an inequality in x and y, how do you decide on which side of the boundary the solution set of the inequality lies?

79. Writing Describe the difference between the solution set of a system of equations and the solution set of a system of inequalities.

Review

In Exercises 80–83, find the equation of the line passing through the two points.

80. $(-2, 6), (4, -4)$ **81.** $(-8, 0), (3, -1)$

82. $\left(\frac{3}{4}, -2\right), \left(-\frac{7}{2}, 5\right)$ **83.** $(3.4, -5.2), (-2.6, 0.8)$

84. Mortgage Loans The table shows the number of outstanding mortgage loans M (in millions) in the United States for the years 1992 to 1997. (Source: Mortgage Bankers Association of America)

Year	1992	1993	1994	1995	1996	1997
M	42.6	45.3	47.6	49.2	50.1	51.2

Use the regression capabilities of a graphing utility to find a linear model and a quadratic model that represent the data. Let $t = 0$ represent 1990. Use a graphing utility to plot the actual data and the models in the same viewing window. How closely do the models represent the data?

7.5 Linear Programming

Linear Programming: A Graphical Approach

Many applications in business and economics involve a process called **optimization,** in which you are asked to find the minimum or maximum value of a quantity. In this section you will study an optimization strategy called **linear programming.**

A two-dimensional linear programming problem consists of a linear **objective function** and a system of linear inequalities called **constraints.** The objective function gives the quantity that is to be maximized (or minimized), and the constraints determine the set of **feasible solutions.** For example, suppose you are asked to maximize the value of

$$z = ax + by \qquad \text{Objective function}$$

subject to a set of constraints that determines the region in Figure 7.30. Because every point in the region satisfies each constraint, it is not clear how you should go about finding the point that yields a maximum value of z. Fortunately, it can be shown that if there is an optimal solution, it must occur at one of the vertices. So, *you can find the maximum value by testing z at each of the vertices.*

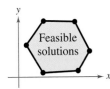

Figure 7.30

What You Should Learn:

- How to solve linear programming problems
- How to use linear programming to model and solve real-life problems

Why You Should Learn It:

Linear programming is a powerful tool used in business and industry to effectively manage resources in order to maximize profits or minimize costs. For instance, Exercise 41 on page 543 shows how to use linear programming to analyze the profitability of two models of compact disc players.

Jeff Greenberg/Unicorn Stock Photos

Optimal Solution of a Linear Programming Problem

If a linear programming problem has a solution, it must occur at a vertex of the set of feasible solutions. If there is more than one solution, at least one of them must occur at such a vertex. In either case, the value of the objective function is unique.

Here are some guidelines for solving a linear programming problem in two variables in which an objective function is to be maximized *or* minimized.

Solving a Linear Programming Problem

1. Sketch the region corresponding to the system of constraints. (The points inside or on the boundary of the region are *feasible solutions*.)

2. Find the vertices of the region.

3. Test the objective function at each of the vertices and select the values of the variables that optimize the objective function. For a bounded region, both a minimum and a maximum value will exist. (For an unbounded region, *if* an optimal solution exists, it will occur at a vertex.)

EXAMPLE 1   Solving a Linear Programming Problem

A computer animation of this example appears in the *Interactive* CD-ROM and *Internet* versions of this text.

Find the maximum value of

$$z = 3x + 2y \qquad \text{Objective function}$$

subject to the following constraints.

$$\left.\begin{array}{r} x \geq 0 \\ y \geq 0 \\ x + 2y \leq 4 \\ x - y \leq 1 \end{array}\right\} \qquad \text{Constraints}$$

Solution

The constraints form the region shown in Figure 7.31. At the four vertices of this region, the objective function has the following values.

At $(0, 0)$: $z = 3(0) + 2(0) = 0$

At $(1, 0)$: $z = 3(1) + 2(0) = 3$

At $(2, 1)$: $z = 3(2) + 2(1) = 8$ Maximum value of z

At $(0, 2)$: $z = 3(0) + 2(2) = 4$

So, the maximum value of z is 8, and this occurs when $x = 2$ and $y = 1$.

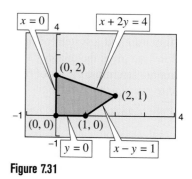

Figure 7.31

In Example 1, try testing some of the *interior* points in the region. You will see that the corresponding values of z are less than 8. Here are some examples.

At $(1, 1)$: $z = 3(1) + 2(1) = 5$

At $\left(1, \frac{1}{2}\right)$: $z = 3(1) + 2\left(\frac{1}{2}\right) = 4$

At $\left(\frac{1}{2}, \frac{3}{2}\right)$: $z = 3\left(\frac{1}{2}\right) + 2\left(\frac{3}{2}\right) = \frac{9}{2}$

To see why the maximum value of the objective function in Example 1 must occur at a vertex, consider writing the objective function in the form

$$y = -\frac{3}{2}x + \frac{z}{2} \qquad \text{Family of lines}$$

where $z/2$ is the y-intercept of the objective function. This equation represents a family of lines, each of slope $-\frac{3}{2}$. Of these infinitely many lines, you want the one that has the largest z-value while still intersecting the region determined by the constraints. In other words, of all the lines with a slope of $-\frac{3}{2}$, you want the one that has the largest y-intercept *and* intersects the given region, as shown in Figure 7.32. It should be clear that such a line will pass through one (or more) of the vertices of the region.

STUDY T!P

Remember that a vertex of a region can be found using a system of linear equations. The system will consist of the equations of the lines passing through the vertex.

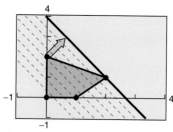

Figure 7.32

The next example shows that the same basic procedure can be used to solve a problem in which the objective function is to be *minimized*.

EXAMPLE 2   Minimizing an Objective Function

Find the minimum value of

$$z = 5x + 7y \qquad \text{Objective function}$$

where $x \geq 0$ and $y \geq 0$, subject to the following constraints.

$$\left. \begin{array}{r} 2x + 3y \geq 6 \\ 3x - y \leq 15 \\ -x + y \leq 4 \\ 2x + 5y \leq 27 \end{array} \right\} \qquad \text{Constraints}$$

Solution

The region bounded by the constraints is shown in Figure 7.33. By testing the objective function at each vertex, you obtain the following.

At $(0, 2)$: $z = 5(0) + 7(2) = 14$ Minimum value of z

At $(0, 4)$: $z = 5(0) + 7(4) = 28$

At $(1, 5)$: $z = 5(1) + 7(5) = 40$

At $(6, 3)$: $z = 5(6) + 7(3) = 51$

At $(5, 0)$: $z = 5(5) + 7(0) = 25$

At $(3, 0)$: $z = 5(3) + 7(0) = 15$

So, the minimum value of z is 14, and this occurs when $x = 0$ and $y = 2$.

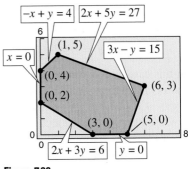

Figure 7.33

EXAMPLE 3   Maximizing an Objective Function

Find the maximum value of

$$z = 5x + 7y \qquad \text{Objective function}$$

where $x \geq 0$ and $y \geq 0$, subject to the following constraints.

$$\left. \begin{array}{r} 2x + 3y \geq 6 \\ 3x - y \leq 15 \\ -x + y \leq 4 \\ 2x + 5y \leq 27 \end{array} \right\} \qquad \text{Constraints}$$

Solution

This linear programming problem is identical to that given in Example 2 above, except that the objective function is *maximized* instead of minimized. Using the values of z at the vertices shown above, you can conclude that the maximum value of z is 51, and that this value occurs when $x = 6$ and $y = 3$.

It is possible for the maximum (or minimum) value in a linear programming problem to occur at *two* different vertices. For instance, at the vertices of the region shown in Figure 7.34, the objective function

$$z = 2x + 2y \qquad \text{Objective function}$$

has the following values.

At $(0, 0)$: $\quad z = 2(0) + 2(0) = \quad 0$

At $(0, 4)$: $\quad z = 2(0) + 2(4) = \quad 8$

At $(2, 4)$: $\quad z = 2(2) + 2(4) = 12 \qquad$ Maximum value of z

At $(5, 1)$: $\quad z = 2(5) + 2(1) = 12 \qquad$ Maximum value of z

At $(5, 0)$: $\quad z = 2(5) + 2(0) = 10$

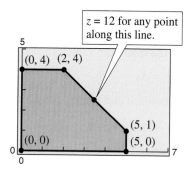

Figure 7.34

In this case, you can conclude that the objective function has a maximum value (of 12) not only at the vertices $(2, 4)$ and $(5, 1)$, but also at *any point on the line segment connecting these two vertices*, as shown in Figure 7.34. Note that by rewriting the objective function as

$$y = -x + \frac{1}{2}z$$

you can see that it has the same slope as the line through the vertices $(2, 4)$ and $(5, 1)$.

Some linear programming problems have no optimal solution. This can occur if the region determined by the constraints is *unbounded*.

EXAMPLE 4   An Unbounded Region

Find the maximum value of

$$z = 4x + 2y \qquad \text{Objective function}$$

where $x \geq 0$ and $y \geq 0$, subject to the following constraints.

$$\left.\begin{array}{r} x + 2y \geq 4 \\ 3x + y \geq 7 \\ -x + 2y \leq 7 \end{array}\right\} \qquad \text{Constraints}$$

Solution

The region determined by the constraints is shown in Figure 7.35. For this unbounded region, there is no maximum value of z. To see this, note that the point $(x, 0)$ lies in the region for all values of $x \geq 4$. By choosing large values of x, you can obtain values of $z = 4(x) + 2(0) = 4x$ that are as large as you want. So, there is no maximum value of z. For the vertices of the region, the objective function has the following values. So, there *is* a minimum value of $z = 10$, which occurs at the vertex $(2, 1)$.

Figure 7.35

At $(1, 4)$: $\quad z = 4(1) + 2(4) = 12$

At $(2, 1)$: $\quad z = 4(2) + 2(1) = 10 \qquad$ Minimum value of z

At $(4, 0)$: $\quad z = 4(4) + 2(0) = 16$

Applications

Example 5 shows how linear programming can be used to find the maximum profit in a business application.

EXAMPLE 5   Maximum Profit

A manufacturer wants to maximize the profit for two products. Product I yields a profit of $1.50 per unit, and product II yields a profit of $2.00 per unit. Market tests and available resources have indicated the following constraints.

1. The combined production level should not exceed 1200 units per month.

2. The demand for product II is no more than half the demand for product I.

3. The production level of product I is less than or equal to 600 units plus three times the production level of product II.

Solution

If you let x be the number of units of product I and y be the number of units of product II, the objective function (for the combined profit) is

$$P = 1.5x + 2y. \qquad \text{Objective function}$$

The three constraints translate into the following linear inequalities.

1. $x + y \leq 1200$ $x + y \leq 1200$

2. $y \leq \frac{1}{2}x$ $-x + 2y \leq 0$

3. $x \leq 3y + 600$ $x - 3y \leq 600$

Because neither x nor y can be negative, you also have the two additional constraints of $x \geq 0$ and $y \geq 0$. Figure 7.36 shows the region determined by the constraints. To find the maximum profit, test the value of P at the vertices of the region.

At $(0, 0)$: $P = 1.5(0) + 2(0)$

$ = 0$

At $(800, 400)$: $P = 1.5(800) + 2(400)$

$ = 2000 \qquad \text{Maximum profit}$

At $(1050, 150)$: $P = 1.5(1050) + 2(150)$

$ = 1875$

At $(600, 0)$: $P = 1.5(600) + 2(0)$

$ = 900$

So, the maximum profit is $2000, and it occurs when the monthly production consists of 800 units of product I and 400 units of product II.

In Example 5, suppose the manufacturer improved the production of product I so that it yielded a profit of $2.50 per unit. This would change the vertex at which the maximum profit would occur. In this case, a maximum profit of $2925 would result from producing 1050 units of product I and 150 units of product II.

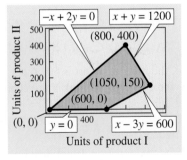

Figure 7.36

EXAMPLE 6   Minimum Cost

The liquid portion of a diet is to provide at least 300 calories, 36 units of vitamin A, and 90 units of vitamin C daily. A cup of dietary drink X costs $0.12 and provides 60 calories, 12 units of vitamin A, and 10 units of vitamin C. A cup of dietary drink Y costs $0.15 and provides 60 calories, 6 units of vitamin A, and 30 units of vitamin C. How many cups of each drink should be consumed each day to minimize the cost and still meet the daily requirements?

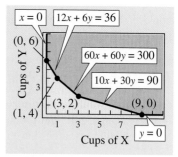

Figure 7.37

Solution

As in Example 9 on page 531, let x be the number of cups of dietary drink X and let y be the number of cups of dietary drink Y.

$$
\left.\begin{array}{lrr}
\textit{For Calories:} & 60x + 60y \geq & 300 \\
\textit{For Vitamin A:} & 12x + 6y \geq & 36 \\
\textit{For Vitamin C:} & 10x + 30y \geq & 90 \\
& x \geq & 0 \\
& y \geq & 0
\end{array}\right\} \quad \text{Constraints}
$$

The cost C is given by $C = 0.12x + 0.15y$.   Objective function

The graph of the region determined by the constraints is shown in Figure 7.37. To determine the minimum cost, test C at each vertex of the region.

At $(0, 6)$: $C = 0.12(0) + 0.15(6) = 0.90$

At $(1, 4)$: $C = 0.12(1) + 0.15(4) = 0.72$

At $(3, 2)$: $C = 0.12(3) + 0.15(2) = 0.66$ Minimum value of C

At $(9, 0)$: $C = 0.12(9) + 0.15(0) = 1.08$

So, the minimum cost is $0.66 per day, and this occurs when three cups of drink X and two cups of drink Y are consumed each day.

Writing About Math   *Creating a Linear Programming Problem*

Sketch the region determined by the constraints.

$$
\left.\begin{array}{rr}
x \geq & 0 \\
y \geq & 0 \\
x + 2y \leq & 10 \\
x + y \leq & 7
\end{array}\right\} \quad \text{Constraints}
$$

Find, if possible, an objective function of the form $z = ax + by$ that has a maximum at the indicated vertex of the region.

a. Maximum at $(0, 5)$ **b.** Maximum at $(4, 3)$

c. Maximum at $(7, 0)$ **d.** Maximum at $(0, 0)$

Write a short paragraph explaining how you found each objective function.

7.5   E x e r c i s e s

In Exercises 1–12, find the minimum and maximum values of the objective function and where they occur, subject to the indicated constraints. (For each exercise, the graph of the region determined by the constraints is provided.)

1. Objective function:

$z = 3x + 5y$

Constraints:

$x \geq 0$

$y \geq 0$

$x + y \leq 6$

2. Objective function:

$z = 2x + 8y$

Constraints:

$x \geq 0$

$y \geq 0$

$2x + y \leq 4$

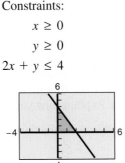

3. Objective function:

$z = 10x + 7y$

Constraints:
(See Exercise 1.)

4. Objective function:

$z = 7x + 3y$

Constraints:
(See Exercise 2.)

5. Objective function:

$z = 5x + 2y$

Constraints:

$x \geq 0$

$y \geq 0$

$x + 3y \leq 15$

$4x + y \leq 16$

6. Objective function:

$z = 4x + 3y$

Constraints:

$x \geq 0$

$2x + 3y \geq 6$

$3x - 2y \leq 9$

$x + 5y \leq 20$

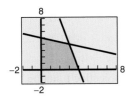

7. Objective function:

$z = 5x + 0.5y$

Constraints:
(See Exercise 5.)

8. Objective function:

$z = x + 6y$

Constraints:
(See Exercise 6.)

9. Objective function:

$z = 10x + 7y$

Constraints:

$0 \leq x \leq 60$

$0 \leq y \leq 45$

$5x + 6y \leq 420$

10. Objective function:

$z = 50x + 35y$

Constraints:

$x \geq 0$

$y \geq 0$

$8x + 9y \leq 7200$

$8x + 9y \geq 5400$

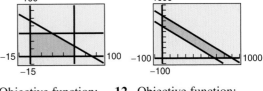

11. Objective function:

$z = 25x + 30y$

Constraints:
(See Exercise 9.)

12. Objective function:

$z = 16x + 18y$

Constraints:
(See Exercise 10.)

In Exercises 13–20, sketch the constraint region. Then find the minimum and maximum values of the objective function and where they occur, subject to the constraints.

13. Objective function:

$z = 6x + 10y$

Constraints:

$x \geq 0$

$y \geq 0$

$2x + 5y \leq 10$

14. Objective function:

$z = 4x + 6y$

Constraints:

$-x + y \leq 11$

$x + y \leq 27$

$2x + 5y \leq 90$

$x \geq 0$

$y \geq 0$

15. Objective function:

$z = 9x + 24y$

Constraints:
(See Exercise 13.)

16. Objective function:

$z = 7x + 2y$

Constraints:
(See Exercise 14.)

17. Objective function:

$z = 4x + 3y$

Constraints:

$x \geq 0$

$2x + 3y \geq 6$

$3x - 2y \leq 9$

$x + 5y \leq 20$

18. Objective function:

$z = 4x + 5y$

Constraints:

$x \geq 0$

$y \geq 0$

$2x + 2y \leq 10$

$x + 2y \leq 6$

19. Objective function:

$z = 3x + 7y$

Constraints:

(See Exercise 17.)

20. Objective function:

$z = 2x - y$

Constraints:

(See Exercise 18.)

In Exercises 21–26, use a graphing utility to sketch the region determined by the constraints. Then find the minimum and maximum values of the objective function and where they occur, subject to the constraints.

21. Objective function:

$z = 4x + y$

Constraints:

$$x \geq 0$$
$$y \geq 0$$
$$x + 2y \leq 40$$
$$2x + 3y \geq 72$$

22. Objective function:

$z = x$

Constraints:

$$x \geq 0$$
$$y \geq 0$$
$$2x + 3y \leq 60$$
$$2x + y \leq 28$$
$$4x + y \leq 48$$

23. Objective function:

$z = x + 4y$

Constraints:

(See Exercise 21.)

24. Objective function:

$z = y$

Constraints:

(See Exercise 22.)

25. Objective function:

$z = 2x + 3y$

Constraints:

(See Exercise 21.)

26. Objective function:

$z = 3x + 2y$

Constraints:

(See Exercise 22.)

Exploration In Exercises 27–30, perform the following: (a) use a graphing utility to graph the region bounded by the following constraints.

$$3x + y \leq 15$$
$$4x + 3y \leq 30$$
$$x \geq 0$$
$$y \geq 0$$

(b) Graph the objective function for the given maximum value of z in the same viewing window as the graph of the constraints. (c) Use the graph to determine the feasible point or points that yield the maximum. Explain how you arrived at your answer.

Objective Function	Maximum
27. $z = 2x + y$	$z = 12$
28. $z = 5x + y$	$z = 25$

Objective Function	Maximum
29. $z = x + y$	$z = 10$
30. $z = 3x + y$	$z = 15$

Exploration In Exercises 31–34, perform the following: (a) use a graphing utility to graph the region bounded by the following constraints.

$$x + 4y \leq 20$$
$$x + y \leq 8$$
$$3x + 2y \leq 21$$
$$x \geq 0$$
$$y \geq 0$$

(b) Graph the objective function for the given maximum value of z in the same viewing window as the graph of the constraints. (c) Use the graph to determine the feasible point or points that yield the maximum. Explain how you arrived at your answer.

Objective Function	Maximum
31. $z = x + 5y$	$z = 25$
32. $z = 2x + 4y$	$z = 24$
33. $z = 4x + 5y$	$z = 36$
34. $z = 4x + y$	$z = 28$

In Exercises 35–40, the linear programming problem has an unusual characteristic. Sketch a graph of the solution region for the problem and describe the unusual characteristic. The objective function is to be maximized in each case.

35. Objective function:

$z = 2.5x + y$

Constraints:

$$x \geq 0$$
$$y \geq 0$$
$$3x + 5y \leq 15$$
$$5x + 2y \leq 10$$

36. Objective function:

$z = x + y$

Constraints:

$$x \geq 0$$
$$y \geq 0$$
$$-x + y \leq 1$$
$$-x + 2y \leq 4$$

37. Objective function:

$z = -x + 2y$

Constraints:

$$x \geq 0$$
$$y \geq 0$$
$$x \leq 10$$
$$x + y \leq 7$$

38. Objective function:

$z = x + y$

Constraints:

$$x \geq 0$$
$$y \geq 0$$
$$-x + y \leq 0$$
$$-3x + y \geq 3$$

39. Objective function:
$$z = 3x + 4y$$
Constraints:
$$x \geq 0$$
$$y \geq 0$$
$$x + y \leq 1$$
$$2x + y \leq 4$$

40. Objective function:
$$z = x + 2y$$
Constraints:
$$x \geq 0$$
$$y \geq 0$$
$$x + 2y \leq 4$$
$$2x + y \leq 4$$

41. ***Maximum Profit*** A merchant plans to sell two models of compact disc players at costs of $250 and $400. The $250 model yields a profit of $45, and the $400 model yields a profit of $50. The merchant estimates that the total monthly demand will not exceed 250 units. The merchant does not want to invest more than $70,000 in inventory for these products. Find the number of units of each model that should be stocked in order to maximize profit. What is the maximum profit?

42. ***Maximum Profit*** A fruit grower has 150 acres of land available to raise two crops, A and B. It takes 1 day to trim an acre of crop A and 2 days to trim an acre of crop B, and there are 240 days per year available for trimming. It takes 0.3 day to pick an acre of crop A and 0.1 day to pick an acre of crop B, and there are 30 days available for picking. The profits are $140 per acre for crop A and $235 per acre for crop B. Find the number of acres of each fruit that should be planted to maximize profit. What is the maximum profit?

43. ***Minimum Cost*** Two gasolines, type A ($1.13 per gallon) and type B ($1.28 per gallon), have octane ratings of 80 and 92, respectively. Determine the blend of minimum cost with an octane rating of at least 90. What is the minimum cost? (*Hint:* Let x be the fraction of each gallon that is type A and let y be the fraction that is type B.)

44. ***Maximum Revenue*** An accounting firm has 900 hours of staff time and 100 hours of reviewing time available each week. The firm charges $2000 for an audit and $300 for a tax return. Each audit requires 100 hours of staff time and 10 hours of review time. Each tax return requires 12.5 hours of staff time and 2.5 hours of review time. What numbers of audits and tax returns will yield the maximum revenue? What is the maximum revenue?

45. ***Maximum Revenue*** The accounting firm in Exercise 44 lowers its charge for an audit to $1000. What numbers of audits and tax returns will yield the maximum revenue? What is the maximum revenue?

46. ***Maximum Profit*** A manufacturer produces two models of bicycles. The amounts of time (in hours) required for assembling, painting, and packaging the two models are as follows.

	Model A	Model B
Assembling	2	2.5
Painting	4	1
Packaging	1	0.75

The total amounts of time available for assembling, painting, and packaging are 4000, 4800, and 1500 hours, respectively. The profits per unit are $45 for model A and $50 for model B. How many of each model should be produced to maximize profit? What is the maximum profit?

47. ***Maximum Profit*** A manufacturer produces two models of snowboards. The amounts of time (in hours) required for assembling, painting, and packaging the two models are as follows.

	Model A	Model B
Assembling	2.5	3
Painting	2	1
Packaging	0.75	1.25

The total amounts of time available for assembling, painting, and packaging are 4000, 2500, and 1500 hours, respectively. The profits per unit are $50 for model A and $52 for model B. How many of each model should be produced to maximize profit? What is the maximum profit?

48. ***Minimum Cost*** A farming cooperative mixes two brands of cattle feed. Brand X costs $25 per bag and contains 2 units of nutritional element A, 2 units of element B, and 2 units of element C. Brand Y costs $20 per bag and contains 1 unit of nutritional element A, 9 units of element B, and 3 units of element C. The minimum requirements for nutrients A, B, and C are 12 units, 36 units, and 24 units, respectively. Find the number of bags of each brand that should be mixed to produce a mixture having a minimum cost per bag. What is the minimum cost per mixed bag?

Synthesis

True or False? **In Exercises 49 and 50, determine whether the statement is true or false. Justify your answer.**

49. If an objective function has a maximum value at the adjacent vertices $(4, 7)$ and $(8, 3)$, you can conclude that it also has a maximum value at the points $(4.5, 6.5)$ and $(7.8, 3.2)$.

50. When solving a linear programming problem, if the objective function has a maximum value at two adjacent vertices, you can assume that there is an infinite number of points that will produce the maximum value.

Think About It **In Exercises 51–54, find an objective function that has a maximum or minimum value at the indicated vertex of the constraint region shown below. (There are many correct answers.)**

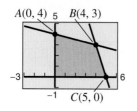

51. The maximum occurs at vertex A.

52. The maximum occurs at vertex B.

53. The maximum occurs at vertex C.

54. The minimum occurs at vertex C.

In Exercises 55 and 56, determine values of t such that the objective function has a maximum value at each indicated vertex.

55. Objective function:

$z = 3x + ty$

Constraints:

$$x \geq 0$$
$$y \geq 0$$
$$x + 3y \leq 15$$
$$4x + y \leq 16$$

(a) $(0, 5)$

(b) $(3, 4)$

56. Objective function:

$z = 3x + ty$

Constraints:

$$x \geq 0$$
$$y \geq 0$$
$$x + 2y \leq 4$$
$$x - y \leq 1$$

(a) $(2, 1)$

(b) $(0, 2)$

Review

In Exercises 57–62, solve the equation algebraically. Round your result to three decimal places.

57. $e^{2x} + 2e^x - 15 = 0$

58. $e^{2x} - 10e^x + 24 = 0$

59. $8(62 - e^{x/4}) = 192$

60. $\dfrac{150}{e^{-x} - 4} = 75$

61. $7 \ln 3x = 12$

62. $\ln(x + 9)^2 = 2$

In Exercises 63–66, use any method to solve the system.

63. $\begin{cases} -3x + 5y = -9 \\ x - 4y = 10 \end{cases}$

64. $\begin{cases} 5x - y = -13 \\ -x + 2y = 17 \end{cases}$

65. $\begin{cases} 4x + 5y = 6 \\ 5x - 11y = -27 \end{cases}$

66. $\begin{cases} -\frac{3}{2}x + 2y = 9 \\ 5x - \frac{1}{2}y = 44 \end{cases}$

In Exercises 67–70, sketch the graph of the solution of the system of inequalities. Use a graphing utility to verify your solution.

67. $\begin{cases} 2x + y \leq 7 \\ y \leq 4 \\ x \geq -1 \\ y \geq 0 \end{cases}$

68. $\begin{cases} x + y \leq 6 \\ y \geq -2 \\ x \geq -5 \end{cases}$

69. $\begin{cases} x^2 + y \leq 1 \\ y - 3x \geq -3 \\ y \geq -3 \end{cases}$

70. $\begin{cases} x^2 + y^2 \leq 9 \\ x + y \leq 3 \\ x \geq 0 \end{cases}$

7 Chapter Summary

What did you learn?

Section 7.1	**Review Exercises**
☐ How to use the method of substitution to solve systems of equations in two variables	1–6
☐ How to solve systems of equations graphically	7–12
☐ How to use systems of equations to model and solve real-life problems	13–16

Section 7.2	
☐ How to use the method of elimination to solve systems of linear equations in two variables	17–24
☐ How to graphically interpret the number of solutions of systems of linear equations in two variables	25–30
☐ How to use systems of linear equations in two variables to model and solve real-life problems	31–34

Section 7.3	
☐ How to recognize linear systems in row-echelon form and use back-substitution to solve the systems	35, 36
☐ How to use Gaussian elimination to solve systems of linear equations	37–42
☐ How to solve nonsquare systems of linear equations	43, 44
☐ How to graphically interpret three-variable systems	45, 46
☐ How to use systems of linear equations to write partial fraction decompositions of rational expressions	47–52
☐ How to use systems of linear equations in three or more variables to model and solve real-life problems	53, 54

Section 7.4	
☐ How to sketch graphs of inequalities in two variables	55–60
☐ How to solve systems of inequalities	61–70
☐ How to use systems of inequalities in two variables to model and solve real-life problems	71–76

Section 7.5	
☐ How to solve linear programming problems	77–80
☐ How to use linear programming to model and solve real-life problems	81, 82

7 Review Exercises

7.1 In Exercises 1–6, solve the system by the method of substitution.

1. $\begin{cases} x + y = 2 \\ x - y = 0 \end{cases}$

2. $\begin{cases} 2x = 3(y - 1) \\ y = x \end{cases}$

3. $\begin{cases} x^2 - y^2 = 9 \\ x - y = 1 \end{cases}$

4. $\begin{cases} x^2 + y^2 = 169 \\ 3x + 2y = 39 \end{cases}$

5. $\begin{cases} y = 2x^2 \\ y = x^4 - 2x^2 \end{cases}$

6. $\begin{cases} x = y + 3 \\ x = y^2 + 1 \end{cases}$

In Exercises 7–12, use a graphing utility to solve the system of equations. If you cannot identify the exact solution, find the solution accurate to two decimal places.

7. $\begin{cases} 5x + 6y = 7 \\ -x - 4y = 0 \end{cases}$

8. $\begin{cases} 8x - 3y = -3 \\ 2x + 5y = 28 \end{cases}$

9. $\begin{cases} y^2 - 2y + x = 0 \\ x + y = 0 \end{cases}$

10. $\begin{cases} y = 2x^2 - 4x + 1 \\ y = x^2 - 4x + 3 \end{cases}$

11. $\begin{cases} y = 2(6 - x) \\ y = 2^{x-2} \end{cases}$

12. $\begin{cases} y = \ln(x - 1) - 3 \\ y = 4 - \frac{1}{2}x \end{cases}$

13. **Break-Even Point** You set up a business and make an initial investment of $10,000. The unit cost of the product is $2.85 and the selling price is $4.95. How many units must you sell to break even?

14. **Choice of Two Jobs** You are offered two sales jobs. One company offers an annual salary of $22,500 plus a year-end bonus of 1.5% of your total sales. The other company offers a salary of $20,000 plus a year-end bonus of 2% of total sales. What amount of sales will make the second offer better? Explain.

15. **Geometry** The perimeter of a rectangle is 480 meters and its length is 1.5 times its width. Find the dimensions of the rectangle.

16. **Geometry** The perimeter of a rectangle is 68 feet and its width is $\frac{8}{9}$ times its length. Find the dimensions of the rectangle.

7.2 In Exercises 17–24, solve by elimination.

17. $\begin{cases} 2x - y = 2 \\ 6x + 8y = 39 \end{cases}$

18. $\begin{cases} 40x + 30y = 24 \\ 20x - 50y = -14 \end{cases}$

19. $\begin{cases} \frac{1}{5}x + \frac{3}{10}y = \frac{7}{50} \\ \frac{2}{5}x + \frac{1}{2}y = \frac{1}{5} \end{cases}$

20. $\begin{cases} \frac{5}{12}x - \frac{3}{4}y = \frac{25}{4} \\ -x + \frac{7}{8}y = -\frac{38}{5} \end{cases}$

21. $\begin{cases} 3x - 2y = 0 \\ 3x + 2(y + 5) = 10 \end{cases}$

22. $\begin{cases} 7x + 12y = 63 \\ 2x + 3y = 15 \end{cases}$

23. $\begin{cases} 1.25x - 2y = 3.5 \\ 5x - 8y = 14 \end{cases}$

24. $\begin{cases} 1.5x + 2.5y = 8.5 \\ 6x + 10y = 24 \end{cases}$

In Exercises 25–30, use a graphing utility to graph the system. Use the graph to determine whether the system is consistent or inconsistent. If the system is consistent, determine the solution. Verify your results algebraically.

25. $\begin{cases} 3x + 2y = 0 \\ x - y = 4 \end{cases}$

26. $\begin{cases} x + y = 6 \\ -2y = -12 + 2x \end{cases}$

27. $\begin{cases} \frac{1}{4}x - \frac{1}{5}y = 2 \\ -5x + 4y = 8 \end{cases}$

28. $\begin{cases} \frac{7}{2}x - 7y = -1 \\ -x + 2y = 4 \end{cases}$

29. $\begin{cases} 8x - 2y = 11 \\ -4x + y = -5.5 \end{cases}$

30. $\begin{cases} -x + 3.2y = 10.4 \\ 3x - 9.6y = 6.4 \end{cases}$

31. **Flying Speeds** Two planes leave Pittsburgh and Philadelphia at the same time, each going to the other city. One plane flies 25 miles per hour faster than the other. Find the air speed of each plane if the cities are 275 miles apart and the planes pass each other after 40 minutes of flying time.

32. **Finance** A total of $46,000 is invested in two corporate bonds that pay 6.75% and 7.25% simple interest. The investor wants an annual interest income of $3245 from the investments. What is the most that can be invested in the 6.75% bond?

Supply and Demand In Exercises 33 and 34, find the point of equilibrium.

Demand Function	Supply Function
33. $p = 37 - 0.0002x$	$p = 22 + 0.00001x$
34. $p = 120 - 0.0001x$	$p = 45 + 0.0002x$

7.3 In Exercises 35 and 36, use back-substitution to solve the system of linear equations.

35. $\begin{cases} x - 4y + 3z = 3 \\ -y + z = -1 \\ z = -5 \end{cases}$

36. $\begin{cases} x - 7y + 8z = 85 \\ y - 9z = -35 \\ z = 3 \end{cases}$

In Exercises 37–40, solve the system of linear equations and check any solution algebraically.

37. $\begin{cases} x + 3y - z = 13 \\ 2x \quad\quad - 5z = 23 \\ 4x - y - 2z = 14 \end{cases}$

38. $\begin{cases} x + 2y + 6z = 4 \\ -3x + 2y - z = -4 \\ 4x + \quad\quad 2z = 16 \end{cases}$

39. $\begin{cases} x - 2y + z = -6 \\ 2x - 3y \quad\quad = -7 \\ -x + 3y - 3z = 11 \end{cases}$

40. $\begin{cases} 2x + \quad\quad 6z = -9 \\ 3x - 2y + 11z = -16 \\ 3x - y + 7z = -11 \end{cases}$

In Exercises 41 and 42, find the equation of the parabola $y = ax^2 + bx + c$ that passes through the given points. Use a graphing utility to verify your result.

41.   **42.**

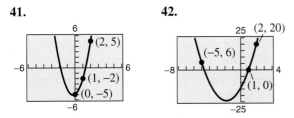

In Exercises 43 and 44, solve the nonsquare system of equations.

43. $\begin{cases} 5x - 12y + 7z = 16 \\ 3x - 7y + 4z = 9 \end{cases}$ **44.** $\begin{cases} 2x + 5y - 19z = 34 \\ 3x + 8y - 31z = 54 \end{cases}$

In Exercises 45 and 46, sketch the plane represented by the linear equation. Then list four points that lie in the plane.

45. $2x - 4y + z = 8$ **46.** $3x + 3y - z = 9$

In Exercises 47–52, write the partial fraction decomposition for the rational expression.

47. $\dfrac{4 - x}{x^2 + 6x + 8}$ **48.** $\dfrac{-x}{x^2 + 3x + 2}$

49. $\dfrac{x^2}{x^2 + 2x - 15}$ **50.** $\dfrac{9}{x^2 - 9}$

51. $\dfrac{x^2 + 2x}{x^3 - x^2 + x - 1}$ **52.** $\dfrac{3x^3 + 4x}{(x^2 + 1)^2}$

53. *Agriculture*   A mixture of 6 gallons of chemical A, 8 gallons of chemical B, and 13 gallons of chemical C is required to kill a certain destructive crop insect. Commercial spray X contains 1, 2, and 2 parts, respectively, of these chemicals. Commercial spray Y contains only chemical C. Commercial spray Z contains chemicals A, B, and C in equal amounts. How much of each type of commercial spray is needed to get the desired mixture?

54. *Investments*   An inheritance of $20,000 was divided among three investments yielding $1780 in interest per year. The interest rates for the three investments were 7%, 9%, and 11%. Find the amount placed in each investment if the second and third were $3000 and $1000 less than the first, respectively.

7.4   In Exercises 55–60, sketch the graph of the inequality.

55. $x \le 6$ **56.** $y \ge -10$

57. $y \le 5 - \frac{1}{2}x$ **58.** $3y - x \ge 7$

59. $y - 4x^2 > -1$ **60.** $y \ge 2e^x - 6$

In Exercises 61–64, match the system of inequalities with the graph of its solution. [The graphs are labeled (a), (b), (c), and (d).]

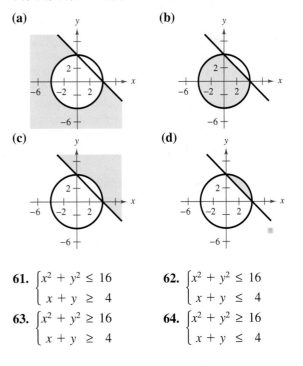

61. $\begin{cases} x^2 + y^2 \le 16 \\ x + y \ge 4 \end{cases}$ **62.** $\begin{cases} x^2 + y^2 \le 16 \\ x + y \le 4 \end{cases}$

63. $\begin{cases} x^2 + y^2 \ge 16 \\ x + y \ge 4 \end{cases}$ **64.** $\begin{cases} x^2 + y^2 \ge 16 \\ x + y \le 4 \end{cases}$

In Exercises 65–70, sketch a graph of the solution set of the system of inequalities. Use a graphing utility to verify your result.

65. $\begin{cases} x + 2y \leq 160 \\ 3x + y \leq 180 \\ x \geq 0 \\ y \geq 0 \end{cases}$

66. $\begin{cases} 2x + 3y \leq 24 \\ 2x + y \leq 16 \\ x \geq 0 \\ y \geq 0 \end{cases}$

67. $\begin{cases} y < x + 1 \\ y > x^2 - 1 \end{cases}$

68. $\begin{cases} y \leq 6 - 2x - x^2 \\ y \geq x + 6 \end{cases}$

69. $\begin{cases} 2x - 3y \geq 0 \\ 2x - y \leq 8 \\ y \geq 0 \end{cases}$

70. $\begin{cases} x^2 + y^2 \leq 9 \\ (x - 3)^2 + y^2 \leq 9 \end{cases}$

In Exercises 71 and 72, derive a set of inequalities to describe the region.

71. *Parallelogram:* Vertices at $(1, 5)$, $(3, 1)$, $(6, 10)$, $(8, 6)$

72. *Triangle:* Vertices at $(1, 2)$, $(6, 7)$, $(8, 1)$

In Exercises 73 and 74, find a system of inequalities that models the description. Use a graphing utility to graph and shade the solution of the system.

73. *Fruit Distribution* A Pennsylvania fruit grower has at most 1500 bushels of apples that are to be divided between markets in Harrisburg and philadelphia. These two markets need at least 400 bushels and 600 bushels, respectively.

74. *Inventory Costs* A warehouse operator has 24,000 square feet of floor space in which to store two products. Each unit of product I requires 20 square feet of floor space and costs $12 per day to store. Each unit of product II requires 30 square feet of floor space and costs $8 per day to store. The total storage cost per day cannot exceed $12,400.

In Exercises 75 and 76, use a graphing utility to shade the regions representing the consumer and producer surpluses for the equations. Find the surpluses.

Demand	*Supply*
75. $p = 160 - 0.0001x$	$p = 70 + 0.0002x$
76. $p = 130 - 0.0002x$	$p = 30 + 0.0003x$

7.5 In Exercises 77–80, use a graphing utility to graph the region determined by the constraints. Then find the minimum or maximum values of the objective function and where they occur, subject to the constraints.

77. Maximize:
$z = 3x + 4y$
Constraints:
$x \geq 0$
$y \geq 0$
$2x + 5y \leq 50$
$4x + y \leq 28$

78. Minimize:
$z = 10x + 7y$
Constraints:
$x \geq 0$
$y \geq 0$
$2x + y \geq 100$
$x + y \geq 75$

79. Minimize:
$z = 1.75x + 2.25y$
Constraints:
$x \geq 0$
$y \geq 0$
$2x + y \geq 25$
$3x + 2y \geq 45$

80. Maximize:
$z = 50x + 70y$
Constraints:
$x \geq 0$
$y \geq 0$
$x + 2y \leq 1500$
$5x + 2y \leq 3500$

81. *Maximum Revenue* A student is working part time as a cosmetologist to pay college expenses. The student may work no more than 24 hours per week. Haircuts cost $17 and require an average of 20 minutes, and permanents cost $60 and require an average of 1 hour and 10 minutes. What combination of haircuts and/or permanent will yield a maximum revenue? What is the maximum revenue?

82. *Minimum Cost* Two gasolines, type A and type B, have octane ratings of 80 and 92, respectively. Type A costs $1.25 per gallon and type B costs $1.55 per gallon. Determine the blend of minimum cost with an octane rating of at least 88. (*Hint:* Let x be the fraction of each gallon that is type A and let y be the fraction that is type B.) What is the minimum cost?

Synthesis

True or False? In Exercises 83 and 84, determine whether the statement is true or false. Justify your answer.

83. The system $\begin{cases} y \leq 5 \\ y \geq -2 \\ y \geq \frac{7}{2}x - 9 \\ y \geq -\frac{7}{2}x + 26 \end{cases}$ represents a region shaped like an isosceles trapezoid.

84. It is possible for an objective function of a linear programming problem to have exactly ten maximum value points.

Chapter Project *Fitting Models to Data*

In this project, you will find and use models relating to newspaper circulation in the United States by using *least squares regression analysis*. This procedure can be performed with a computer or graphing utility.

The numbers of morning and evening newspapers published in the United States from 1990 through 1997 are shown in the table.

a. Use the regression capabilities of a graphing utility to find linear models for both sets of data. Then use the models to estimate the numbers of morning and evening newspapers published in 2003. In the table, $t = 0$ represents 1990. (Source: Editor and Publisher Company)

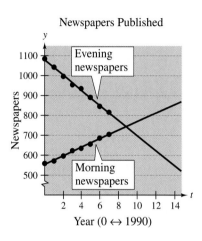

Newspapers Published

Year, t	0	1	2	3	4	5	6	7
Morning	559	571	596	623	635	656	686	705
Evening	1084	1042	996	954	935	891	846	816

b. Use the models you found in part (a) to predict the year when the number of morning papers published is equal to the number of evening papers published. Use the graph of the models shown to verify your result.

Questions for Further Exploration

1. The numbers (in millions) of morning newspapers sold each day in the United States from 1990 to 1997 are shown in the table below. In the table $t = 0$ represents 1990. Find a linear model that represents the data. Use your model to estimate the number of morning papers sold each day in 2003.

Year, t	0	1	2	3
Morning	41.3	41.5	42.4	43.1

Year, t	4	5	6	7
Morning	43.4	44.3	44.8	45.4

2. From 1990 through 1997, the number of morning newspapers published and sold increased. Did the average circulation per morning paper (number sold ÷ number published) increase or decrease? Explain.

3. If you had the opportunity to invest in a company that published only one type of newspaper (morning or evening), which would you choose? Explain your reasoning.

7 Chapter Test

Take this test as you would take a test in class. After you are done, check your work against the answers in the back of the book.

In Exercises 1–3, solve the system by the method of substitution.

1. $\begin{cases} x - y = 6 \\ 3x + 5y = 2 \end{cases}$
 2. $\begin{cases} y = x - 1 \\ y = (x - 1)^3 \end{cases}$
 3. $\begin{cases} 4x - y^2 = 7 \\ x - y = 3 \end{cases}$

In Exercises 4–6, solve the system graphically.

4. $\begin{cases} 4x - 3y = -15 \\ 4x + 3y = -9 \end{cases}$
 5. $\begin{cases} y = 16 - x^2 \\ y = x + 4 \end{cases}$
 6. $\begin{cases} y - \ln x = 8 \\ 3x + y + 10 = 21 \end{cases}$

In Exercises 7–9, solve the linear system by elimination.

7. $\begin{cases} 2x + 5y = -11 \\ 5x - y = 19 \end{cases}$
 8. $\begin{cases} x - 2y + 3z = -5 \\ 2x \quad - z = -4 \\ 3y + z = 17 \end{cases}$
 9. $\begin{cases} 5x + 5y - z = 0 \\ 10x + 5y + 2z = 0 \\ 5x + 15y - 9z = 0 \end{cases}$

10. Find a system of linear equations that has the solution $\left(\frac{4}{3}, -8\right)$.

11. Find a system of linear equations that has the solution $\left(-\frac{1}{2}, 5, -\frac{9}{4}\right)$.

12. Find the equation of the parabola $y = ax^2 + bx + c$ passing through the points $(0, 6)$, $(-2, 2)$, and $\left(3, \frac{9}{2}\right)$.

13. Write the partial fraction decomposition for the rational expression $\dfrac{5x - 2}{(x - 1)^2}$.

In Exercises 14–16, sketch the graph of the inequality.

14. $x \geq -7$
 15. $-4x + 6y < -11$
 16. $(x - 4)^2 + y^2 < 16$

In Exercises 17–19, graph the inequalities and shade the region representing the solution.

17. $\begin{cases} 2x + y \leq 6 \\ 2x - y \leq 0 \\ x \geq -5 \end{cases}$
 18. $\begin{cases} y < -x^4 + x^2 + 4 \\ y > 4x \end{cases}$
 19. $\begin{cases} x^2 + y^2 \leq 16 \\ x \geq 1 \\ y \geq -3 \end{cases}$

20. Derive a set of inequalities to describe the region in the figure at the right.

21. Find the maximum value of the objective function $z = 20x + 12y$ subject to the constraints $x \geq 0$, $y \geq 0$, $x + 4y \leq 32$, and $3x + 2y \leq 36$.

22. A merchant plans to sell two models of compact disc players at costs of $275 and $400. The $275 model yields a profit of $55 and the $400 model yields a profit of $75. The merchant estimates that the total monthly demand will not exceed 300 units. The merchant does not want to invest more than $100,000 in inventory for these products. Find the number of units of each model that should be stocked in order to maximize profit.

The *Interactive* CD-ROM and *Internet* versions of this text provide answers to the Chapter Tests and Cumulative Tests. They also offer Chapter Pre-Tests (that test key skills and concepts covered in previous chapters) and Chapter Post-Tests, both of which have randomly generated exercises with diagnostic capabilities.

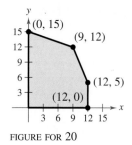

FIGURE FOR 20

Matrices and Determinants

8.1 Matrices and Systems of Equations

8.2 Operations with Matrices

8.3 The Inverse of a Square Matrix

8.4 The Determinant of a Square Matrix

8.5 Applications of Matrices and Determinants

Superstock

The Big Picture

In this chapter you will learn how to

❏ write matrices, identify their order, and perform elementary row operations.

❏ use Gaussian elimination and Gauss-Jordan elimination with matrices to solve systems of linear equations.

❏ add, subtract, and multiply two matrices, and multiply a matrix by a real number.

❏ verify that two matrices are inverses of each other and find inverses of matrices.

❏ use inverse matrices to solve systems of linear equations.

❏ find the determinants of square matrices.

❏ use Cramer's Rule to solve systems of linear equations.

Heating and shaping iron changes its physical properties. Molten iron in a steelmaking furnace reaches temperatures of 2800°F.

Important Vocabulary

As you encounter each new vocabulary term in this chapter, add the term and its definition to your notebook glossary.

- matrix (p. 552)
- entry of a matrix (p. 552)
- order of a matrix (p. 552)
- square matrix (p. 552)
- main diagonal (p. 552)
- row matrix (p. 552)
- column matrix (p. 552)
- augmented matrix (p. 553)
- coefficient matrix (p. 553)
- elementary row operations (p. 554)
- row-equivalent matrices (p. 554)
- row-echelon form (p. 556)

- reduced row-echelon form (p. 556)
- Gauss-Jordan elimination (p. 559)
- scalar multiple (p. 568)
- zero matrix (p. 571)
- additive identity (p. 571)
- matrix multiplication (p. 572)
- identity matrix of order n (p. 574)
- inverse of a matrix (p. 582)
- invertible or nonsingular matrix (p. 583)
- singular matrix (p. 583)
- determinant (p. 591)

- minors (p. 593)
- cofactors (p. 593)
- expanding by cofactors (p. 594)
- triangular matrix (p. 595)
- upper triangular matrix (p. 595)
- lower triangular matrix (p. 595)
- diagonal matrix (p. 595)
- collinear points (p. 600)
- Cramer's Rule (p. 601)
- uncoded row matrices (p. 604)
- coded row matrices (p. 604)

Additional Resources Text-specific additional resources are available to help you do well in this course. See page xvi for details.

8.1   Matrices and Systems of Equations

Matrices

In this section you will study a streamlined technique for solving systems of linear equations. This technique involves the use of a rectangular array of real numbers called a **matrix.** The plural of matrix is *matrices.*

What You Should Learn:

- How to write matrices and identify their orders
- How to perform elementary row operations on matrices
- How to use matrices and Gaussian elimination to solve systems of linear equations
- How to use matrices and Gauss-Jordan elimination to solve systems of linear equations

Definition of Matrix

If m and n are positive integers, an $m \times n$ (read "m by n") matrix is a rectangular array

$$\begin{bmatrix} a_{11} & a_{12} & a_{13} & \cdots & a_{1n} \\ a_{21} & a_{22} & a_{23} & \cdots & a_{2n} \\ a_{31} & a_{32} & a_{33} & \cdots & a_{3n} \\ \vdots & \vdots & \vdots & & \vdots \\ a_{m1} & a_{m2} & a_{m3} & \cdots & a_{mn} \end{bmatrix} \Bigg\} \; m \text{ rows}$$

$\underbrace{\hphantom{aaaaaaaaaaa}}_{n \text{ columns}}$

in which each **entry** a_{ij} of the matrix is a real number. An $m \times n$ matrix has m rows (horizontal lines) and n columns (vertical lines).

The entry in the ith row and jth column is denoted by the *double subscript* notation a_{ij}. For instance, the entry a_{23} is the entry in the second row and third column. A matrix having m rows and n columns is said to be of **order** $m \times n$. If $m = n$, the matrix is **square** of order n. For a square matrix, the entries $a_{11}, a_{22}, a_{33}, \ldots$ are the **main diagonal** entries.

Why You Should Learn It:

Matrices can be used to solve systems of linear equations in two or more variables. For instance, Exercise 82 on page 565 shows how a matrix can be used to help find a model for the parabolic path of a baseball.

Amwell/Tony Stone Images

EXAMPLE 1   Order of Matrices

Determine the order of each matrix.

a. $\begin{bmatrix} 2 \end{bmatrix}$ **b.** $\begin{bmatrix} 1 & -3 & 0 & \frac{1}{2} \end{bmatrix}$ **c.** $\begin{bmatrix} 0 & 0 \\ 0 & 0 \end{bmatrix}$ **d.** $\begin{bmatrix} 5 & 0 \\ 2 & -2 \\ -7 & 4 \end{bmatrix}$

Solution

a. *Order:* 1×1

$\begin{bmatrix} 2 \end{bmatrix}$ 1 row, 1 column

b. *Order:* 1×4

$\begin{bmatrix} 1 & -3 & 0 & \frac{1}{2} \end{bmatrix}$ 1 row, 4 columns

c. *Order:* 2×2

$\begin{bmatrix} 0 & 0 \\ 0 & 0 \end{bmatrix}$ 2 rows, 2 columns

d. *Order:* 3×2

$\begin{bmatrix} 5 & 0 \\ 2 & -2 \\ -7 & 4 \end{bmatrix}$ 3 rows, 2 columns

A matrix that has only one row [such as the matrix in Example 1(b)] is called a **row matrix,** and a matrix that has only one column is called a **column matrix.**

A matrix derived from a system of linear equations (each written in standard form with the constant term on the right) is the **augmented matrix** of the system. Moreover, the matrix derived from the coefficients of the system (but not including the constant terms) is the **coefficient matrix** of the system.

System
$$\begin{cases} x - 4y + 3z = 5 \\ -x + 3y - z = -3 \\ 2x - 4z = 6 \end{cases}$$

Augmented Matrix
$$\begin{bmatrix} 1 & -4 & 3 & \vdots & 5 \\ -1 & 3 & -1 & \vdots & -3 \\ 2 & 0 & -4 & \vdots & 6 \end{bmatrix}$$

Coefficient Matrix
$$\begin{bmatrix} 1 & -4 & 3 \\ -1 & 3 & -1 \\ 2 & 0 & -4 \end{bmatrix}$$

Note the use of 0 for the missing y-variable in the third equation, and also note the fourth column (of constant terms) in the augmented matrix. The optional dotted line in the augmented matrix helps to identify the constant terms.

When forming either the coefficient matrix or the augmented matrix of a system, you should begin by vertically aligning the variables in the equations and using 0's for the missing variables.

EXAMPLE 2 Writing an Augmented Matrix

Write the augmented matrix for the system of linear equations.

$$\begin{cases} x + 3y = 9 \\ -y + 4z = -2 \\ x - 5z = 0 \end{cases}$$

What is the order of the augmented matrix?

Solution

Begin by writing the linear system and aligning the variables.

$$\begin{cases} x + 3y = 9 \\ -y + 4z = -2 \\ x - 5z = 0 \end{cases}$$

Next, use the coefficients and constant terms as the matrix entries. Include zeros for each missing coefficient.

$$\begin{bmatrix} 1 & 3 & 0 & \vdots & 9 \\ 0 & -1 & 4 & \vdots & -2 \\ 1 & 0 & -5 & \vdots & 0 \end{bmatrix}$$

The augmented matrix has 3 rows and 4 columns, so it is a 3×4 matrix.

Elementary Row Operations

In Section 7.3, you studied three operations that can be used on a system of linear equations to produce an equivalent system.

1. Interchange two equations.

2. Multiply an equation by a nonzero constant.

3. Add a multiple of an equation to another equation.

In matrix terminology these three operations correspond to **elementary row operations.** An elementary row operation on an augmented matrix corresponds to a new (but equivalent) system of linear equations. Two matrices are **row-equivalent** if one can be obtained from the other by a sequence of elementary row operations.

The *Interactive* CD-ROM and *Internet* versions of this text show every example with its solution; clicking on the *Try It!* button brings up similar problems. Guided Examples and Integrated Examples show step-by-step solutions to additional examples. Integrated Examples are related to several concepts in the section.

Elementary Row Operations

1. Interchange two rows.

2. Multiply a row by a nonzero constant.

3. Add a multiple of a row to another row.

Although elementary row operations are simple to perform, they involve a lot of arithmetic. Because it is easy to make a mistake, you should get in the habit of noting the elementary row operations performed in each step so that you can go back and check your work.

EXAMPLE 3   Elementary Row Operations

a. Interchange the first and second rows.

Original Matrix

$$\begin{bmatrix} 0 & 1 & 3 & 4 \\ -1 & 2 & 0 & 3 \\ 2 & -3 & 4 & 1 \end{bmatrix}$$

New Row-Equivalent Matrix

$$\begin{matrix} R_2 \\ R_1 \end{matrix} \begin{bmatrix} -1 & 2 & 0 & 3 \\ 0 & 1 & 3 & 4 \\ 2 & -3 & 4 & 1 \end{bmatrix}$$

b. Multiply the first row by $\frac{1}{2}$.

Original Matrix

$$\begin{bmatrix} 2 & -4 & 6 & -2 \\ 1 & 3 & -3 & 0 \\ 5 & -2 & 1 & 2 \end{bmatrix}$$

New Row-Equivalent Matrix

$$\frac{1}{2}R_1 \rightarrow \begin{bmatrix} 1 & -2 & 3 & -1 \\ 1 & 3 & -3 & 0 \\ 5 & -2 & 1 & 2 \end{bmatrix}$$

c. Add -2 times the first row to the third row.

Original Matrix

$$\begin{bmatrix} 1 & 2 & -4 & 3 \\ 0 & 3 & -2 & -1 \\ 2 & 1 & 5 & -2 \end{bmatrix}$$

New Row-Equivalent Matrix

$$\begin{bmatrix} 1 & 2 & -4 & 3 \\ 0 & 3 & -2 & -1 \\ -2R_1 + R_3 \rightarrow 0 & -3 & 13 & -8 \end{bmatrix}$$

Note that the elementary row operation is written beside the row that is *changed*.

STUDY TIP

Most graphing utilities can perform elementary row operations on matrices. Consult your user's manual to determine how to enter a matrix and perform elementary row operations. The screen in the top figure below shows how one graphing utility displays the original matrix in Example 3(a). The screen in the bottom figure below shows the new row-equivalent matrix in Example 3(a).

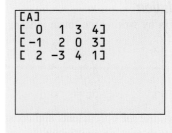

```
[A]
[  0   1   3   4]
[ -1   2   0   3]
[  2  -3   4   1]
```

```
[A]
[ -1   2   0   3]
[  0   1   3   4]
[  2  -3   4   1]
```

Gaussian Elimination with Back Substitution

In Example 2 of Section 7.3, you used Gaussian elimination with back-substitution to solve a system of linear equations. The next example demonstrates the matrix version of Gaussian elimination. The two methods are essentially the same. The basic difference is that with matrices you do not need to keep writing the variables.

EXAMPLE 4   Using Elementary Row Operations

Linear System	*Associated Augmented Matrix*

$$\begin{cases} x - 2y + 3z = 9 \\ -x + 3y = -4 \\ 2x - 5y + 5z = 17 \end{cases}$$

$$\begin{bmatrix} 1 & -2 & 3 & \vdots & 9 \\ -1 & 3 & 0 & \vdots & -4 \\ 2 & -5 & 5 & \vdots & 17 \end{bmatrix}$$

The *Interactive* CD-ROM and *Internet* versions of this text offer a built-in graphing calculator, which can be used with the Examples, Explorations, and Exercises.

Add the first equation to the second equation.

Add the first row to the second row $(R_1 + R_2)$.

$$\begin{cases} x - 2y + 3z = 9 \\ y + 3z = 5 \\ 2x - 5y + 5z = 17 \end{cases}$$

$$R_1 + R_2 \rightarrow \begin{bmatrix} 1 & -2 & 3 & \vdots & 9 \\ 0 & 1 & 3 & \vdots & 5 \\ 2 & -5 & 5 & \vdots & 17 \end{bmatrix}$$

Add -2 times the first equation to the third equation.

Add -2 times the first row to the third row $(-2R_1 + R_3)$.

$$\begin{cases} x - 2y + 3z = 9 \\ y + 3z = 5 \\ -y - z = -1 \end{cases}$$

$$-2R_1 + R_3 \rightarrow \begin{bmatrix} 1 & -2 & 3 & \vdots & 9 \\ 0 & 1 & 3 & \vdots & 5 \\ 0 & -1 & -1 & \vdots & -1 \end{bmatrix}$$

Add the second equation to the third equation.

Add the second row to the third row $(R_2 + R_3)$.

$$\begin{cases} x - 2y + 3z = 9 \\ y + 3z = 5 \\ 2z = 4 \end{cases}$$

$$R_2 + R_3 \rightarrow \begin{bmatrix} 1 & -2 & 3 & \vdots & 9 \\ 0 & 1 & 3 & \vdots & 5 \\ 0 & 0 & 2 & \vdots & 4 \end{bmatrix}$$

Multiply the third equation by $\frac{1}{2}$.

Multiply the third row by $\frac{1}{2}$.

$$\begin{cases} x - 2y + 3z = 9 \\ y + 3z = 5 \\ z = 2 \end{cases}$$

$$\tfrac{1}{2}R_3 \rightarrow \begin{bmatrix} 1 & -2 & 3 & \vdots & 9 \\ 0 & 1 & 3 & \vdots & 5 \\ 0 & 0 & 1 & \vdots & 2 \end{bmatrix}$$

At this point, you can use back-substitution to find that the solution is $x = 1$, $y = -1$, and $z = 2$, as was done in Example 2 of Section 7.3.

Remember that you should check a solution by substituting the values of x, y, and z into each equation in the original system.

The last matrix in Example 4 is said to be in **row-echelon form.** The term *echelon* refers to the stair-step pattern formed by the nonzero elements of the matrix. To be in this form, a matrix must have the properties listed on page 556.

Row-Echelon Form and Reduced Row-Echelon Form

A matrix in **row-echelon form** has the following properties.

1. All rows consisting entirely of zeros occur at the bottom of the matrix.

2. For each row that does not consist entirely of zeros, the first nonzero entry is 1 (called a **leading 1**).

3. For two successive (nonzero) rows, the leading 1 in the higher row is farther to the left than the leading 1 in the lower row.

A matrix in *row-echelon form* is in **reduced row-echelon form** if every column that has a leading 1 has zeros in every position above and below its leading 1.

STUDY T!P

Some graphing utilities can automatically transform a matrix to row-echelon form and reduced row-echelon form. Consult your user's manual to see whether your graphing utility has this capability. If so, use it to verify the results in this section.

It is worth mentioning that the row-echelon form of a matrix is not unique. That is, two different sequences of elementary row operations may yield different row-echelon forms.

EXAMPLE 5   Row-Echelon Form

Decide whether each matrix is in row-echelon form. If it is, decide whether the matrix is in reduced row-echelon form.

a. $\begin{bmatrix} 1 & 2 & -1 & 4 \\ 0 & 1 & 0 & 3 \\ 0 & 0 & 1 & -2 \end{bmatrix}$ b. $\begin{bmatrix} 1 & 2 & -1 & 2 \\ 0 & 0 & 0 & 0 \\ 0 & 1 & 2 & -4 \end{bmatrix}$

c. $\begin{bmatrix} 1 & -5 & 2 & -1 & 3 \\ 0 & 0 & 1 & 3 & -2 \\ 0 & 0 & 0 & 1 & 4 \\ 0 & 0 & 0 & 0 & 1 \end{bmatrix}$ d. $\begin{bmatrix} 1 & 0 & 0 & -1 \\ 0 & 1 & 0 & 2 \\ 0 & 0 & 1 & 3 \\ 0 & 0 & 0 & 0 \end{bmatrix}$

e. $\begin{bmatrix} 1 & 2 & -3 & 4 \\ 0 & 2 & 1 & -1 \\ 0 & 0 & 1 & -3 \end{bmatrix}$ f. $\begin{bmatrix} 0 & 1 & 0 & 5 \\ 0 & 0 & 1 & 3 \\ 0 & 0 & 0 & 0 \end{bmatrix}$

Solution

The matrices in (a), (c), (d), and (f) are in row-echelon form. The matrices in (d) and (f) are in reduced form because every column that has a leading 1 has zeros in every position above and below its leading 1. The matrix in (b) is not in row-echelon form because a row of all zeros does not occur at the bottom of the matrix. The matrix in (e) is not in row-echelon form because the first nonzero entry in row 2 is not a leading 1.

Every matrix has a row-equivalent matrix that is in row-echelon form. For instance, in Example 5, you can change the matrix in part (e) to row-echelon form by multiplying its second row by $\frac{1}{2}$. What elementary row operation could you perform on the matrix in part (b) so that it would be in row-echelon form?

You have seen that the row-echelon form of a given matrix is *not* unique; however, the *reduced* row-echelon form of a given matrix *is* unique.

Gaussian elimination with back-substitution works well for solving systems of linear equations by hand or with a computer. For this algorithm, the order in which the elementary row operations are performed is important. You should operate from *left to right by columns*, using elementary row operations to obtain zeros in all entries directly below the leading 1's.

EXAMPLE 6   Gaussian Elimination with Back-Substitution

Solve the system $\begin{cases} y + z - 2w = -3 \\ x + 2y - z = 2 \\ 2x + 4y + z - 3w = -2 \\ x - 4y - 7z - w = -19 \end{cases}$.

A computer animation of this example appears in the *Interactive* CD-ROM and *Internet* versions of this text.

Solution

Interchange rows 1 and 2 in order to have a 1 in the upper left corner.

$$\begin{matrix} \overset{\curvearrowright R_2}{\underset{R_1}{}} \end{matrix} \begin{bmatrix} 1 & 2 & -1 & 0 & \vdots & 2 \\ 0 & 1 & 1 & -2 & \vdots & -3 \\ 2 & 4 & 1 & -3 & \vdots & -2 \\ 1 & -4 & -7 & -1 & \vdots & -19 \end{bmatrix}$$ First column has leading 1 in upper left corner.

$$\begin{matrix} \\ \\ -2R_1 + R_3 \rightarrow \\ -R_1 + R_4 \rightarrow \end{matrix} \begin{bmatrix} 1 & 2 & -1 & 0 & \vdots & 2 \\ 0 & 1 & 1 & -2 & \vdots & -3 \\ 0 & 0 & 3 & -3 & \vdots & -6 \\ 0 & -6 & -6 & -1 & \vdots & -21 \end{bmatrix}$$ First column has zeros below its leading 1.

$$\begin{matrix} \\ \\ \\ 6R_2 + R_4 \rightarrow \end{matrix} \begin{bmatrix} 1 & 2 & -1 & 0 & \vdots & 2 \\ 0 & 1 & 1 & -2 & \vdots & -3 \\ 0 & 0 & 3 & -3 & \vdots & -6 \\ 0 & 0 & 0 & -13 & \vdots & -39 \end{bmatrix}$$ Second column has zeros below its leading 1.

$$\begin{matrix} \\ \\ \tfrac{1}{3}R_3 \rightarrow \\ \\ \end{matrix} \begin{bmatrix} 1 & 2 & -1 & 0 & \vdots & 2 \\ 0 & 1 & 1 & -2 & \vdots & -3 \\ 0 & 0 & 1 & -1 & \vdots & -2 \\ 0 & 0 & 0 & -13 & \vdots & -39 \end{bmatrix}$$ Third column has zeros below its leading 1.

$$\begin{matrix} \\ \\ \\ -\tfrac{1}{13}R_4 \rightarrow \end{matrix} \begin{bmatrix} 1 & 2 & -1 & 0 & \vdots & 2 \\ 0 & 1 & 1 & -2 & \vdots & -3 \\ 0 & 0 & 1 & -1 & \vdots & -2 \\ 0 & 0 & 0 & 1 & \vdots & 3 \end{bmatrix}$$ Fourth column has a leading 1.

The matrix is now in row-echelon form, and the corresponding system is

$$\begin{cases} x + 2y - z = 2 \\ y + z - 2w = -3. \\ z - w = -2 \\ w = 3 \end{cases}$$

Using back-substitution, you can determine that the solution is $x = -1$, $y = 2$, $z = 1$, and $w = 3$. Check this in the original system of equations.

The following steps summarize the procedure used in Example 6.

Gaussian Elimination with Back-Substitution

1. Write the augmented matrix of the system of linear equations.
2. Use elementary row operations to rewrite the augmented matrix in row-echelon form.
3. Write the system of linear equations corresponding to the matrix in row-echelon form and use back-substitution to find the solution.

When solving a system of linear equations, remember that it is possible for the system to have no solution. If, in the elimination process, you obtain a row with zeros except for the last entry, it is unnecessary to continue the elimination process. You can simply conclude that the system is inconsistent.

EXAMPLE 7   A System with No Solution

Solve the system.

$$\begin{cases} x - y + 2z = 4 \\ x \qquad + z = 6 \\ 2x - 3y + 5z = 4 \\ 3x + 2y - z = 1 \end{cases}$$

Solution

$$\begin{bmatrix} 1 & -1 & 2 & \vdots & 4 \\ 1 & 0 & 1 & \vdots & 6 \\ 2 & -3 & 5 & \vdots & 4 \\ 3 & 2 & -1 & \vdots & 1 \end{bmatrix} \quad -R_1 + R_2 \rightarrow \begin{bmatrix} 1 & -1 & 2 & \vdots & 4 \\ 0 & 1 & -1 & \vdots & 2 \\ 2 & -3 & 5 & \vdots & 4 \\ 3 & 2 & -1 & \vdots & 1 \end{bmatrix}$$

$$-2R_1 + R_3 \rightarrow \begin{bmatrix} 1 & -1 & 2 & \vdots & 4 \\ 0 & 1 & -1 & \vdots & 2 \\ 0 & -1 & 1 & \vdots & -4 \\ 3 & 2 & -1 & \vdots & 1 \end{bmatrix}$$

$$-3R_1 + R_4 \rightarrow \begin{bmatrix} 1 & -1 & 2 & \vdots & 4 \\ 0 & 1 & -1 & \vdots & 2 \\ 0 & -1 & 1 & \vdots & -4 \\ 0 & 5 & -7 & \vdots & -11 \end{bmatrix}$$

$$R_2 + R_3 \rightarrow \begin{bmatrix} 1 & -1 & 2 & \vdots & 4 \\ 0 & 1 & -1 & \vdots & 2 \\ 0 & 0 & 0 & \vdots & -2 \\ 0 & 5 & -7 & \vdots & -11 \end{bmatrix}$$

Note that the third row of this matrix consists of zeros except for the last entry. Converting this row back to a linear equation yields $0 = -2$, which is a false statement. This means that the original system of linear equations is *inconsistent* and the system has no solution.

Gauss-Jordan Elimination

With Gaussian elimination, elementary row operations are applied to a matrix to obtain a (row-equivalent) row-echelon form. A second method of elimination, called **Gauss-Jordan elimination,** after Carl Friedrich Gauss (1777–1855) and Wilhelm Jordan (1842–1899), continues the reduction process until a reduced row-echelon form is obtained. This procedure is demonstrated in Example 8.

EXAMPLE 8   Gauss-Jordan Elimination

Use Gauss-Jordan elimination to solve the system.

$$\begin{cases} x - 2y + 3z = 9 \\ -x + 3y = -4 \\ 2x - 5y + 5z = 17 \end{cases}$$

Solution

In Example 4, Gaussian elimination was used to obtain the row-echelon form

$$\begin{bmatrix} 1 & -2 & 3 & \vdots & 9 \\ 0 & 1 & 3 & \vdots & 5 \\ 0 & 0 & 1 & \vdots & 2 \end{bmatrix}.$$

Now, rather than using back-substitution, apply additional elementary row operations until you obtain a matrix in *reduced* row-echelon form. To do this, you must produce zeros above each of the leading 1's, as follows.

$$\begin{matrix} 2R_2 + R_1 \rightarrow \end{matrix} \begin{bmatrix} 1 & 0 & 9 & \vdots & 19 \\ 0 & 1 & 3 & \vdots & 5 \\ 0 & 0 & 1 & \vdots & 2 \end{bmatrix}$$

Second column has zeros above its leading 1.

$$\begin{matrix} -9R_3 + R_1 \rightarrow \\ -3R_3 + R_2 \rightarrow \end{matrix} \begin{bmatrix} 1 & 0 & 0 & \vdots & 1 \\ 0 & 1 & 0 & \vdots & -1 \\ 0 & 0 & 1 & \vdots & 2 \end{bmatrix}$$

Third column has zeros above its leading 1.

Now, converting back to a system of linear equations, you have

$$\begin{cases} x = 1 \\ y = -1 \\ z = 2 \end{cases}$$

which is the same solution that was obtained using Gaussian elimination.

The beauty of Gauss-Jordan elimination is that, from the reduced row-echelon form, you can simply read the solution. Which technique do you prefer: Gaussian elimination or Gauss-Jordan elimination?

STUDY TIP

For a demonstration of a graphical approach to Gauss-Jordan elimination on a 2×3 matrix, see the graphing calculator program ROWOPS, available for several models of graphing calculators on our website at *college.hmco.com*.

The elimination procedures described in this section employ an algorithmic approach that is easily adapted to computer use. However, the procedure makes no effort to avoid fractional coefficients. For instance, if the system given in Example 8 had been listed as

$$\begin{cases} 2x - 5y + 5z = 17 \\ x - 2y + 3z = 9 \\ -x + 3y = -4 \end{cases}$$

the procedure would have required multiplication of the first row by $\frac{1}{2}$, which would have introduced fractions in the first row. For hand computations, fractions can sometimes be avoided by judiciously choosing the order in which the elementary row operations are applied.

EXAMPLE 9   A System with an Infinite Number of Solutions

Solve the system.

$$\begin{cases} 2x + 4y - 2z = 0 \\ 3x + 5y = 1 \end{cases}$$

Solution

$$\begin{bmatrix} 2 & 4 & -2 & \vdots & 0 \\ 3 & 5 & 0 & \vdots & 1 \end{bmatrix} \qquad \frac{1}{2}R_1 \rightarrow \begin{bmatrix} 1 & 2 & -1 & \vdots & 0 \\ 3 & 5 & 0 & \vdots & 1 \end{bmatrix}$$

$$-3R_1 + R_2 \rightarrow \begin{bmatrix} 1 & 2 & -1 & \vdots & 0 \\ 0 & -1 & 3 & \vdots & 1 \end{bmatrix}$$

$$-R_2 \rightarrow \begin{bmatrix} 1 & 2 & -1 & \vdots & 0 \\ 0 & 1 & -3 & \vdots & -1 \end{bmatrix}$$

$$-2R_2 + R_1 \rightarrow \begin{bmatrix} 1 & 0 & 5 & \vdots & 2 \\ 0 & 1 & -3 & \vdots & -1 \end{bmatrix}$$

The corresponding system of equations is

$$\begin{cases} x + 5z = 2 \\ y - 3z = -1 \end{cases}.$$

Solving for x and y in terms of z, you have $x = -5z + 2$ and $y = 3z - 1$. Then, letting $z = a$, where a is a real number, the solution set has the form

$$(-5a + 2, 3a - 1, a).$$

Recall from Section 7.3 that a solution set of this form represents an infinite number of solutions. Try substituting values for a to obtain a few solutions. Then check each solution in the original system of equations.

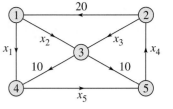

Figure 8.1

EXAMPLE 10   Analysis of a Network

Set up a system of linear equations representing the network shown in Figure 8.1.

Solution

In a network, it is assumed that the total flow into a junction is equal to the total flow out of the junction. For example, because Junction 1 in Figure 8.1 has 20 units flowing into it, there must be 20 units flowing out of it. This is represented by the linear equation $x_1 + x_2 = 20$. A linear equation can be written for each of the network's five junctions, so the network is modeled by the following system.

$$\begin{cases} x_1 + x_2 && = & 20 & \text{Junction 1} \\ & x_3 - x_4 & = & -20 & \text{Junction 2} \\ x_2 + x_3 && = & 20. & \text{Junction 3} \\ x_1 && - x_5 = & -10 & \text{Junction 4} \\ && - x_4 + x_5 = & -10 & \text{Junction 5} \end{cases}$$

Using Gauss-Jordan elimination on the augmented matrix produces the matrix in reduced row-echelon form.

Augmented Matrix

$$\begin{bmatrix} 1 & 1 & 0 & 0 & 0 & \vdots & 20 \\ 0 & 0 & 1 & -1 & 0 & \vdots & -20 \\ 0 & 1 & 1 & 0 & 0 & \vdots & 20 \\ 1 & 0 & 0 & 0 & -1 & \vdots & -10 \\ 0 & 0 & 0 & -1 & 1 & \vdots & -10 \end{bmatrix}$$

Matrix in Reduced Row-Echelon Form

$$\begin{bmatrix} 1 & 0 & 0 & 0 & -1 & \vdots & -10 \\ 0 & 1 & 0 & 0 & 1 & \vdots & 30 \\ 0 & 0 & 1 & 0 & -1 & \vdots & -10 \\ 0 & 0 & 0 & 1 & -1 & \vdots & 10 \\ 0 & 0 & 0 & 0 & 0 & \vdots & 0 \end{bmatrix}$$

Letting $x_5 = t$, where t is a real number, you have $x_1 = t - 10$, $x_2 = -t + 30$, $x_3 = t - 10$, and $x_4 = t + 10$. So, this system has an infinite number of solutions.

Writing About Math   *Error Analysis*

One of your classmates has submitted the following steps for a solution of a system by Gauss-Jordan elimination. Find the error(s) in the solution. Write a short paragraph explaining the error(s) to your classmate.

$$\begin{bmatrix} 1 & 1 & \vdots & 4 \\ 2 & 3 & \vdots & 5 \end{bmatrix}$$

$$-2R_1 + R_2 \rightarrow \begin{bmatrix} 1 & 1 & \vdots & 4 \\ 0 & 1 & \vdots & 5 \end{bmatrix}$$

$$-1R_2 + R_1 \rightarrow \begin{bmatrix} 1 & 0 & \vdots & 4 \\ 0 & 1 & \vdots & 5 \end{bmatrix}$$

8.1 Exercises

In Exercises 1–6, determine the order of the matrix.

1. $\begin{bmatrix} 3 & -2 \\ 7 & 0 \\ 0 & 9 \end{bmatrix}$

2. $\begin{bmatrix} 6 & -3 & 10 & 8 \end{bmatrix}$

3. $\begin{bmatrix} 4 \\ 32 \\ 3 \end{bmatrix}$

4. $\begin{bmatrix} -3 & 7 & 15 & 0 \\ 0 & 0 & 3 & 3 \\ 1 & 1 & 6 & 7 \end{bmatrix}$

5. $\begin{bmatrix} 33 & 45 \\ -9 & 20 \end{bmatrix}$

6. $\begin{bmatrix} 9 \end{bmatrix}$

In Exercises 7–10, write the augmented matrix for the system of linear equations.

7. $\begin{cases} 4x - 5y = 33 \\ -x + 5y = -27 \end{cases}$

8. $\begin{cases} 7x + 4y = 22 \\ 5x - 9y = 15 \end{cases}$

9. $\begin{cases} x + 10y - 2z = 2 \\ 5x - 3y + 4z = 0 \\ 2x + y = 6 \end{cases}$

10. $\begin{cases} -x - 8y + 5z = 17 \\ -6x + 12z = -24 \\ 3x + y - 8z = 11 \end{cases}$

In Exercises 11–14, write the system of linear equations represented by the augmented matrix. (Use variables x, y, z, and w.)

11. $\begin{bmatrix} 1 & 2 & \vdots & 7 \\ 2 & -3 & \vdots & 4 \end{bmatrix}$

12. $\begin{bmatrix} 7 & -5 & \vdots & 0 \\ 8 & 3 & \vdots & -2 \end{bmatrix}$

13. $\begin{bmatrix} 9 & 12 & 3 & 0 & \vdots & 0 \\ -2 & 18 & 5 & 2 & \vdots & 10 \\ 1 & 7 & -8 & 0 & \vdots & -4 \end{bmatrix}$

14. $\begin{bmatrix} 6 & 2 & -1 & -5 & \vdots & -25 \\ -1 & 0 & 7 & 3 & \vdots & 7 \\ 4 & -1 & -10 & 6 & \vdots & 23 \\ 0 & 8 & 1 & -11 & \vdots & -21 \end{bmatrix}$

In Exercises 15–18, determine whether the matrix is in row-echelon form. If it is, determine if it is also in reduced row-echelon form.

15. $\begin{bmatrix} 1 & 0 & 0 & 0 \\ 0 & 1 & 1 & 5 \\ 0 & 0 & 0 & 0 \end{bmatrix}$

16. $\begin{bmatrix} 1 & 3 & 0 & 0 \\ 0 & 0 & 1 & 8 \\ 0 & 0 & 0 & 0 \end{bmatrix}$

17. $\begin{bmatrix} 2 & 0 & 4 & 0 \\ 0 & -1 & 3 & 6 \\ 0 & 0 & 1 & 5 \end{bmatrix}$

18. $\begin{bmatrix} 1 & 0 & 2 & 1 \\ 0 & 1 & -3 & 10 \\ 0 & 0 & 1 & 0 \end{bmatrix}$

In Exercises 19–22, fill in the blanks using elementary row operations to form a row-equivalent matrix.

19. $\begin{bmatrix} 1 & 4 & 3 \\ 2 & 10 & 5 \end{bmatrix}$
$\begin{bmatrix} 1 & 4 & 3 \\ 0 & & -1 \end{bmatrix}$

20. $\begin{bmatrix} 3 & 6 & 8 \\ 4 & -3 & 6 \end{bmatrix}$
$\begin{bmatrix} 1 & & \frac{8}{3} \\ 4 & -3 & 6 \end{bmatrix}$

21. $\begin{bmatrix} 1 & 1 & 4 & -1 \\ 3 & 8 & 10 & 3 \\ -2 & 1 & 12 & 6 \end{bmatrix}$
$\begin{bmatrix} 1 & 1 & 4 & -1 \\ 0 & 5 & & \\ 0 & 3 & & \end{bmatrix}$
$\begin{bmatrix} 1 & 1 & 4 & -1 \\ 0 & 1 & -\frac{2}{5} & \frac{6}{5} \\ 0 & 3 & & \end{bmatrix}$

22. $\begin{bmatrix} 2 & 4 & 8 & 3 \\ 1 & -1 & -3 & 2 \\ 2 & 6 & 4 & 9 \end{bmatrix}$
$\begin{bmatrix} 1 & & & \\ 1 & -1 & -3 & 2 \\ 2 & 6 & 4 & 9 \end{bmatrix}$
$\begin{bmatrix} 1 & 2 & 4 & \frac{3}{2} \\ 0 & & -7 & \frac{1}{2} \\ 0 & 2 & & \end{bmatrix}$

In Exercises 23–26, identify the elementary row operation being performed to obtain the new row-equivalent matrix.

23. *Original Matrix*   *New Row-Equivalent Matrix*
$\begin{bmatrix} -2 & 5 & 1 \\ 3 & -1 & -8 \end{bmatrix}$ $\begin{bmatrix} 13 & 0 & -39 \\ 3 & -1 & -8 \end{bmatrix}$

24. *Original Matrix*   *New Row-Equivalent Matrix*
$\begin{bmatrix} 3 & -1 & -4 \\ -4 & 3 & 7 \end{bmatrix}$ $\begin{bmatrix} 3 & -1 & -4 \\ 5 & 0 & -5 \end{bmatrix}$

25. *Original Matrix*   *New Row-Equivalent Matrix*
$\begin{bmatrix} 0 & -1 & -5 & 5 \\ -1 & 3 & -7 & 6 \\ 4 & -5 & 1 & 3 \end{bmatrix}$ $\begin{bmatrix} -1 & 3 & -7 & 6 \\ 0 & -1 & -5 & 5 \\ 4 & -5 & 1 & 3 \end{bmatrix}$

26. *Original Matrix*   *New Row-Equivalent Matrix*
$\begin{bmatrix} -1 & -2 & 3 & -2 \\ 2 & -5 & 1 & -7 \\ 5 & 4 & -7 & 6 \end{bmatrix}$ $\begin{bmatrix} -1 & -2 & 3 & -2 \\ 2 & -5 & 1 & -7 \\ 0 & -6 & 8 & -4 \end{bmatrix}$

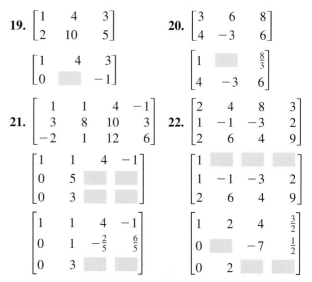

27. Perform the sequence of row operations on the matrix. What did the operations accomplish?

$$\begin{bmatrix} 1 & 2 & 3 \\ 2 & -1 & -4 \\ 3 & 1 & -1 \end{bmatrix}$$

(a) Add -2 times Row 1 to Row 2.

(b) Add -3 times Row 1 to Row 3.

(c) Add -1 times Row 2 to Row 3.

(d) Multiply Row 2 by $-\frac{1}{5}$.

(e) Add -2 times Row 2 to Row 1.

28. Perform the sequence of row operations on the matrix. What did the operations accomplish?

$$\begin{bmatrix} 7 & 1 \\ 0 & 2 \\ -3 & 4 \\ 4 & 1 \end{bmatrix}$$

(a) Add Row 3 to Row 4.

(b) Interchange Rows 1 and 4.

(c) Add 3 times Row 1 to Row 3.

(d) Add -7 times Row 1 to Row 4.

(e) Multiply Row 2 by $\frac{1}{2}$.

(f) Add the appropriate multiples of Row 2 to Rows 1, 3, and 4 so that each entry not on the main diagonal is zero.

29. Repeat steps (a) through (e) in Exercise 27 using a graphing utility.

30. Repeat steps (a) through (f) in Exercise 28 using a graphing utility.

In Exercises 31–34, write the matrix in row-echelon form. Remember that the row-echelon form of a matrix is not unique.

31. $\begin{bmatrix} 1 & 1 & 0 & 5 \\ -2 & -1 & 2 & -10 \\ 3 & 6 & 7 & 14 \end{bmatrix}$

32. $\begin{bmatrix} 1 & 2 & -1 & 3 \\ 3 & 7 & -5 & 14 \\ -2 & -1 & -3 & 8 \end{bmatrix}$

33. $\begin{bmatrix} 1 & -1 & -1 & 1 \\ 5 & -4 & 1 & 8 \\ -6 & 8 & 18 & 0 \end{bmatrix}$

34. $\begin{bmatrix} 1 & -3 & 0 & -7 \\ -3 & 10 & 1 & 23 \\ 4 & -10 & 2 & -24 \end{bmatrix}$

In Exercises 35–38, use the matrix capabilities of a graphing utility to write the matrix in reduced row-echelon form.

35. $\begin{bmatrix} 3 & 3 & 3 \\ -1 & 0 & -4 \\ 2 & 4 & -2 \end{bmatrix}$ **36.** $\begin{bmatrix} 1 & 3 & 2 \\ 5 & 15 & 9 \\ 2 & 6 & 10 \end{bmatrix}$

37. $\begin{bmatrix} -3 & 5 & 1 & 12 \\ 1 & -1 & 1 & 4 \end{bmatrix}$

38. $\begin{bmatrix} 5 & 1 & 2 & 4 \\ -1 & 5 & 10 & -32 \end{bmatrix}$

In Exercises 39–42, write the system of linear equations represented by the augmented matrix. Then use back-substitution to find the solution. (Use variables x, y, and z.)

39. $\begin{bmatrix} 1 & -2 & \vdots & 4 \\ 0 & 1 & \vdots & -3 \end{bmatrix}$ **40.** $\begin{bmatrix} 1 & 5 & \vdots & 0 \\ 0 & 1 & \vdots & -1 \end{bmatrix}$

41. $\begin{bmatrix} 1 & -1 & 2 & \vdots & 4 \\ 0 & 1 & -1 & \vdots & 2 \\ 0 & 0 & 1 & \vdots & -2 \end{bmatrix}$

42. $\begin{bmatrix} 1 & 2 & -2 & \vdots & -1 \\ 0 & 1 & 1 & \vdots & 9 \\ 0 & 0 & 1 & \vdots & -3 \end{bmatrix}$

In Exercises 43–46, an augmented matrix that represents a system of linear equations (in variables x, y, and z) has been reduced using Gauss-Jordan elimination. Write the solution represented by the augmented matrix.

43. $\begin{bmatrix} 1 & 0 & \vdots & 7 \\ 0 & 1 & \vdots & -5 \end{bmatrix}$ **44.** $\begin{bmatrix} 1 & 0 & \vdots & -2 \\ 0 & 1 & \vdots & 4 \end{bmatrix}$

45. $\begin{bmatrix} 1 & 0 & 0 & \vdots & -4 \\ 0 & 1 & 0 & \vdots & -8 \\ 0 & 0 & 1 & \vdots & 2 \end{bmatrix}$

46. $\begin{bmatrix} 1 & 0 & 0 & \vdots & 3 \\ 0 & 1 & 0 & \vdots & -1 \\ 0 & 0 & 1 & \vdots & 0 \end{bmatrix}$

In Exercises 47–60, use matrices to solve the system of equations if possible. Use Gaussian elimination with back-substitution or Gauss-Jordan elimination.

47. $\begin{cases} x + 2y = 7 \\ 2x + y = 8 \end{cases}$ **48.** $\begin{cases} 2x + 6y = 14 \\ 2x + 3y = 2 \end{cases}$

49. $\begin{cases} -3x + 5y = -28 \\ 3x + 4y = 10 \\ 4x - 8y = 40 \end{cases}$ **50.** $\begin{cases} x + 2y = 0 \\ x + y = 6 \\ 3x - 2y = 8 \end{cases}$

51. $\begin{cases} 8x - 4y = 13 \\ 5x + 2y = 7 \end{cases}$ **52.** $\begin{cases} x - 3y = 5 \\ -2x + 6y = -10 \end{cases}$

53. $\begin{cases} -x + 2y = 1.5 \\ 2x - 4y = 3 \end{cases}$ **54.** $\begin{cases} 2x - y = -0.1 \\ 3x + 2y = 1.6 \end{cases}$

55. $\begin{cases} x \quad\quad - 3z = -2 \\ 3x + y - 2z = 5 \\ 2x + 2y + z = 4 \end{cases}$ **56.** $\begin{cases} 2x - y + 3z = 26 \\ 2y - z = 12 \\ 7x - 5y = -8 \end{cases}$

57. $\begin{cases} x + y - 5z = 3 \\ x \quad\quad - 2z = 1 \\ 2x - y - z = 0 \end{cases}$ **58.** $\begin{cases} 2x \quad\quad + 3z = 3 \\ 4x - 3y + 7z = 5 \\ 8x - 9y + 15z = 9 \end{cases}$

59. $\begin{cases} x + 2y = 0 \\ -x - y = 0 \end{cases}$ **60.** $\begin{cases} x + 2y = 0 \\ 2x + 4y = 0 \end{cases}$

In Exercises 61–66, use the matrix capabilities of a graphing utility to reduce the augmented matrix and solve the system of equations.

61. $\begin{cases} 3x + 3y + 12z = 6 \\ x + y + 4z = 2 \\ 2x + 5y + 20z = 10 \\ -x + 2y + 8z = 4 \end{cases}$

62. $\begin{cases} 2x + 10y + 2z = 6 \\ x + 5y + 2z = 6 \\ x + 5y + z = 3 \\ -3x - 15y - 3z = -9 \end{cases}$

63. $\begin{cases} 2x + y - z + 2w = -6 \\ 3x + 4y + w = 1 \\ x + 5y + 2z + 6w = -3 \\ 5x + 2y - z - w = 3 \end{cases}$

64. $\begin{cases} x + 2y + 2z + 4w = 11 \\ 3x + 6y + 5z + 12w = 30 \end{cases}$

65. $\begin{cases} x + y + z = 0 \\ 2x + 3y + z = 0 \\ 3x + 5y + z = 0 \end{cases}$ **66.** $\begin{cases} x + 2y + z + 3w = 0 \\ x - y + w = 0 \\ y - z + 2w = 0 \end{cases}$

In Exercises 67–70, determine whether the two systems of linear equations yield the same solutions. If so, find the solutions.

67. (a) $\begin{cases} x - 2y + z = -6 \\ y - 5z = 16 \\ z = -3 \end{cases}$ (b) $\begin{cases} x + y - 2z = 6 \\ y + 3z = -8 \\ z = -3 \end{cases}$

68. (a) $\begin{cases} x - 3y + 4z = -11 \\ y - z = -4 \\ z = 2 \end{cases}$ (b) $\begin{cases} x + 4y = -11 \\ y + 3z = 4 \\ z = 2 \end{cases}$

69. (a) $\begin{cases} x - 4y + 5z = 27 \\ y - 7z = -54 \\ z = 8 \end{cases}$ (b) $\begin{cases} x - 6y + z = 15 \\ y + 5z = 42 \\ z = 8 \end{cases}$

70. (a) $\begin{cases} x + 3y - z = 19 \\ y + 6z = -18 \\ z = -4 \end{cases}$ (b) $\begin{cases} x - y + 3z = -15 \\ y - 2z = 14 \\ z = -4 \end{cases}$

In Exercises 71–76, find the specified equation that passes through the points. Use a graphing utility to verify your result.

71. Parabola:
$$y = ax^2 + bx + c$$

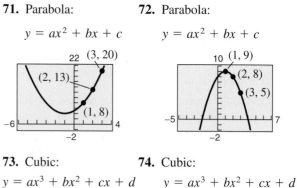

72. Parabola:
$$y = ax^2 + bx + c$$

73. Cubic:
$$y = ax^3 + bx^2 + cx + d$$

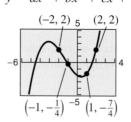

74. Cubic:
$$y = ax^3 + bx^2 + cx + d$$

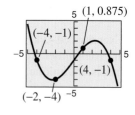

75. Quartic:
$$y = ax^4 + bx^3 + cx^2 + dx + e$$

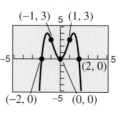

76. Quartic:
$$y = ax^4 + bx^3 + cx^2 + dx + e$$

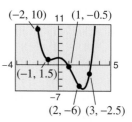

77. ***Borrowing Money*** A small corporation borrowed $1,500,000 to expand its product line. Some of the money was borrowed at 8%, some at 9%, and some at 12%. How much was borrowed at each rate if the annual interest was $133,000 and the amount borrowed at 8% was 4 times the amount borrowed at 12%?

78. ***Borrowing Money*** A small corporation borrowed $500,000 to build a new office building. Some of the money was borrowed at 9%, some at 10%, and some at 12%. How much was borrowed at each rate if the annual interest was $52,000 and the amount borrowed at 10% was $2\frac{1}{2}$ times the amount borrowed at 9%?

79. ***Electrical Network*** The currents in an electrical network are given by the solution of the system

$$\begin{cases} I_1 - I_2 + I_3 = 0 \\ 2I_1 + 2I_2 \qquad = 7 \\ \qquad 2I_2 + 4I_3 = 8 \end{cases}$$

where I_1, I_2, and I_3 are measured in amperes. Solve the system of equations.

80. Redo Exercise 79 for the system

$$\begin{cases} 3I_1 - 4I_2 + I_3 = -\frac{9}{8} \\ I_1 + 6I_2 \qquad = 12. \\ \qquad 3I_2 + 2I_3 = 8 \end{cases}$$

81. ***Data Analysis*** The bar graph gives the value y (in millions of dollars) for new orders of civil jet transport aircraft built by U.S. companies for the years 1995 through 1997. (Source: Aerospace Industries Association of America)

(a) Find the equation of the parabola that passes through the points. Let $t = 5$ represent 1995.

(b) Use a graphing utility to graph the parabola.

(c) Use the equation in part (a) to estimate y in the year 1998.

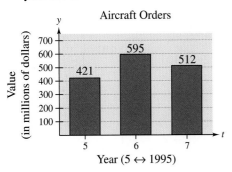

Aircraft Orders

82. ***Mathematical Modeling*** After the path of a ball thrown by a baseball player is videotaped, it is analyzed on a television set with a grid covering the screen. The tape is paused three times, and the position of the ball is measured each time. The coordinates are approximately (0, 5.0), (15, 9.6), and (30, 12.4). The x-coordinate measures the horizontal distance from the player (in feet), and the y-coordinate is the height of the ball (in feet).

(a) Find the equation of the parabola $y = ax^2 + bx + c$ that passes through the three points.

(b) Use a graphing utility to graph the parabola. Approximate the maximum height of the ball and the point at which the ball strikes the ground.

(c) Find algebraically the maximum height of the ball and the point at which it strikes the ground.

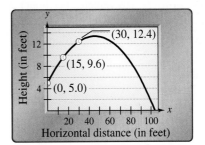

Network Analysis **In Exercises 83–85, answer the questions about the specified network.**

83. Water flowing through a network of pipes (in thousands of cubic meters per hour) is shown in the figure.

(a) Solve this system for the water flow represented by x_i, $i = 1, 2, 3, 4, 5, 6$, and 7.

(b) Find the network flow pattern when
$$x_6 = x_7 = 0.$$

(c) Find the network flow pattern when $x_5 = 1000$ and $x_6 = 0$.

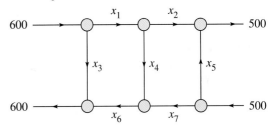

84. The flow of traffic (in vehicles per hour) through a network of streets is shown in the figure.

(a) Solve this system for the traffic flow represented by x_i, $i = 1, 2, 3, 4$, and 5.

(b) Find the traffic flow when $x_2 = 200$ and $x_3 = 50$.

(c) Find the traffic flow when $x_2 = 150$ and $x_3 = 0$.

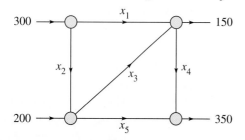

85. The flow of traffic (in vehicles per hour) through a network of streets is shown in the figure.

(a) Solve this system for the traffic flow represented by x_i, $i = 1, 2, 3$, and 4.

(b) Find the traffic flow when $x_4 = 0$.

(c) Find the traffic flow when $x_4 = 100$.

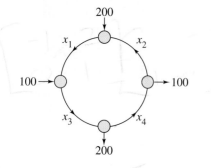

Synthesis

True or False? **In Exercises 86–88, determine whether the statement is true or false. Justify your answer.**

86. $\begin{bmatrix} 6 & 0 & -3 & 10 \\ -2 & 5 & -6 & 2 \end{bmatrix}$ is a 4 × 2 matrix.

87. The matrix $\begin{bmatrix} 0 & 0 & 0 & 0 \\ 0 & 0 & 1 & -4 \\ 0 & 1 & 0 & 2 \\ 1 & 0 & 0 & 5 \end{bmatrix}$ is in reduced row-echelon form.

88. The difference between Gaussian elimination and Gauss-Jordan elimination is that Gaussian elimination reduces a matrix until a reduced row-echelon form is obtained.

89. *Think About It* The augmented matrix represents a system of linear equations (in variables x, y, and z) that has been reduced using Gauss-Jordan elimination. Write a system of equations with *nonzero* coefficients that is represented by the reduced matrix. (There are many correct answers.)

$$\begin{bmatrix} 1 & 0 & 3 & \vdots & -2 \\ 0 & 1 & 4 & \vdots & 1 \\ 0 & 0 & 0 & \vdots & 0 \end{bmatrix}$$

90. Use the system $\begin{cases} x + 3y + z = 3 \\ x + 5y + 5z = 1 \\ 2x + 6y + 3z = 8 \end{cases}$ to write two different row-echelon forms that yield the same solutions.

91. *Think About It*

(a) Describe the row-echelon form of an augmented matrix that corresponds to a system of linear equations that is inconsistent.

(b) Describe the row-echelon form of an augmented matrix that corresponds to a system of linear equations that has an infinite number of solutions.

Review

In Exercises 92–95, sketch the graph of the function. Identify any asymptotes.

92. $f(x) = \dfrac{7}{-x - 1}$

93. $f(x) = \dfrac{4x}{5x^2 + 2}$

94. $f(x) = \dfrac{x^2 - 2x - 3}{x - 4}$

95. $f(x) = \dfrac{x^2 - 36}{x + 1}$

In Exercises 96–99, sketch the graph of the function by hand. Use a graphing utility to verify your sketch.

96. $f(x) = 2^{x-1}$

97. $g(x) = 3^{-x+2}$

98. $h(x) = \ln(x - 1)$

99. $f(x) = 3 + \ln x$

In Exercises 100–103, use any method to solve the system of equations.

100. $\begin{cases} x - 2y = 4 \\ 6x + 2y = 10 \end{cases}$

101. $\begin{cases} 3x - 8y = 47 \\ -5x + 9y = -48 \end{cases}$

102. $\begin{cases} -x + 4y - 5z = 19 \\ 3x - 2y + z = -9 \\ -4x + 3y - 7z = 24 \end{cases}$

103. $\begin{cases} 6x - y + 4z = -11 \\ 5x + 4y - 2z = 0 \\ -8x - y + 5z = 18 \end{cases}$

8.2   Operations with Matrices

Equality of Matrices

What You Should Learn:

- How to decide whether two matrices are equal
- How to add and subtract matrices and multiply matrices by a real number
- How to solve matrix equations
- How to multiply two matrices
- How to use matrix operations to model and solve real-life problems

In Section 8.1, you used matrices to solve systems of linear equations. Matrices, however, can do much more than this. There is a rich mathematical theory of matrices, and its applications are numerous. This section and the next two introduce some fundamentals of matrix theory. It is standard mathematical convention to represent matrices in any of the following three ways.

Representations of Matrices

1. A matrix can be denoted by an uppercase letter such as A, B, or C.

2. A matrix can be denoted by a representative element enclosed in brackets, such as $[a_{ij}]$, $[b_{ij}]$, or $[c_{ij}]$.

3. A matrix can be denoted by a rectangular array of numbers such as

$$A = [a_{ij}] = \begin{bmatrix} a_{11} & a_{12} & a_{13} & \cdots & a_{1n} \\ a_{21} & a_{22} & a_{23} & \cdots & a_{2n} \\ a_{31} & a_{32} & a_{33} & \cdots & a_{3n} \\ \vdots & \vdots & \vdots & & \vdots \\ a_{m1} & a_{m2} & a_{m3} & \cdots & a_{mn} \end{bmatrix}.$$

Two matrices $A = [a_{ij}]$ and $B = [b_{ij}]$ are **equal** if they have the same order $(m \times n)$ and $a_{ij} = b_{ij}$ for $1 \le i \le m$ and $1 \le j \le n$. In other words, two matrices are equal if all of their corresponding entries are equal.

Why You Should Learn It:

Matrix algebra provides a systematic way of performing mathematical operations on large arrays of numbers. In Exercise 84 on page 580, you will use matrix multiplication to help analyze the labor and wage requirements for a boat manufacturer.

Jim Brown/The Stock Market

EXAMPLE 1   Equality of Matrices

Solve for a_{11}, a_{12}, a_{21}, and a_{22} in the following matrix equation.

$$\begin{bmatrix} a_{11} & a_{12} \\ a_{21} & a_{22} \end{bmatrix} = \begin{bmatrix} 2 & -1 \\ -3 & 0 \end{bmatrix}$$

Solution

Because two matrices are equal only if their corresponding entries are equal, you can conclude the following.

$a_{11} = \quad 2$

$a_{12} = -1$

$a_{21} = -3$

$a_{22} = \quad 0$

Matrix Addition and Scalar Multiplication

You can add two matrices (of the same order) by adding their corresponding entries.

Definition of Matrix Addition

If $A = [a_{ij}]$ and $B = [b_{ij}]$ are matrices of order $m \times n$, their sum is the $m \times n$ matrix

$$A + B = [a_{ij} + b_{ij}].$$

The sum of two matrices of different orders is undefined.

The Granger Collection

Arthur Cayley (1821–1895), a British mathematician, invented matrices around 1858. Cayley was a Cambridge University graduate and a lawyer by profession. His groundbreaking work on matrices was begun as he studied the theory of transformations. Cayley also was instrumental in the development of determinants. Cayley and two American mathematicians, **Benjamin Peirce (1809–1880)** and his son **Charles S. Peirce (1839–1914),** are credited with developing "matrix algebra."

EXAMPLE 2   Addition of Matrices

a. $\begin{bmatrix} -1 & 2 \\ 0 & 1 \end{bmatrix} + \begin{bmatrix} 1 & 3 \\ -1 & 2 \end{bmatrix} = \begin{bmatrix} -1+1 & 2+3 \\ 0-1 & 1+2 \end{bmatrix} = \begin{bmatrix} 0 & 5 \\ -1 & 3 \end{bmatrix}$

b. $\begin{bmatrix} 0 & 1 & -2 \\ 1 & 2 & 3 \end{bmatrix} + \begin{bmatrix} 0 & 0 & 0 \\ 0 & 0 & 0 \end{bmatrix} = \begin{bmatrix} 0 & 1 & -2 \\ 1 & 2 & 3 \end{bmatrix}$

c. $\begin{bmatrix} 1 \\ -3 \\ -2 \end{bmatrix} + \begin{bmatrix} -1 \\ 3 \\ 2 \end{bmatrix} = \begin{bmatrix} 0 \\ 0 \\ 0 \end{bmatrix}$

d. The sum of

$$A = \begin{bmatrix} 2 & 1 & 0 \\ 4 & 0 & -1 \\ 3 & -2 & 2 \end{bmatrix} \quad \text{and} \quad B = \begin{bmatrix} 0 & 1 \\ -1 & 3 \\ 2 & 4 \end{bmatrix}$$

is undefined because A is of order 3×3 and B is of order 3×2.

In work with matrices, numbers are usually referred to as **scalars.** In this text, scalars will always be real numbers. You can multiply a matrix A by a scalar c by multiplying each entry in A by c.

Definition of Scalar Multiplication

If $A = [a_{ij}]$ is an $m \times n$ matrix and c is a scalar, the **scalar multiple** of A by c is the $m \times n$ matrix

$$cA = [ca_{ij}].$$

Most graphing utilities can perform matrix addition and scalar multiplication. If you have such a graphing utility, duplicate the matrix operations in Examples 2 and 3. Try adding two matrices of different orders such as

$$A = \begin{bmatrix} 1 & 2 \\ 3 & 4 \end{bmatrix} \quad \text{and} \quad B = \begin{bmatrix} 5 \\ 6 \end{bmatrix}.$$

What error message does your utility display?

The symbol $-A$ represents the scalar product $(-1)A$. Moreover, if A and B are of the same order, $A - B$ represents the sum of A and $(-1)B$. That is,

$$A - B = A + (-1)B.$$ Subtraction of matrices

EXAMPLE 3   Scalar Multiplication and Matrix Subtraction

For the following matrices, find (a) $3A$, (b) $-B$, and (c) $3A - B$.

$$A = \begin{bmatrix} 2 & 2 & 4 \\ -3 & 0 & -1 \\ 2 & 1 & 2 \end{bmatrix} \quad \text{and} \quad B = \begin{bmatrix} 2 & 0 & 0 \\ 1 & -4 & 3 \\ -1 & 3 & 2 \end{bmatrix}$$

Solution

a. $3A = 3 \begin{bmatrix} 2 & 2 & 4 \\ -3 & 0 & -1 \\ 2 & 1 & 2 \end{bmatrix}$ Scalar multiplication

$$= \begin{bmatrix} 3(2) & 3(2) & 3(4) \\ 3(-3) & 3(0) & 3(-1) \\ 3(2) & 3(1) & 3(2) \end{bmatrix}$$ Multiply each entry by 3.

$$= \begin{bmatrix} 6 & 6 & 12 \\ -9 & 0 & -3 \\ 6 & 3 & 6 \end{bmatrix}$$ Simplify.

b. $-B = (-1) \begin{bmatrix} 2 & 0 & 0 \\ 1 & -4 & 3 \\ -1 & 3 & 2 \end{bmatrix}$ Definition of negation

$$= \begin{bmatrix} -2 & 0 & 0 \\ -1 & 4 & -3 \\ 1 & -3 & -2 \end{bmatrix}$$ Multiply each entry by -1.

c. $3A - B = \begin{bmatrix} 6 & 6 & 12 \\ -9 & 0 & -3 \\ 6 & 3 & 6 \end{bmatrix} - \begin{bmatrix} 2 & 0 & 0 \\ 1 & -4 & 3 \\ -1 & 3 & 2 \end{bmatrix}$ Matrix subtraction

$$= \begin{bmatrix} 4 & 6 & 12 \\ -10 & 4 & -6 \\ 7 & 0 & 4 \end{bmatrix}$$ Subtract corresponding entries.

It is often convenient to rewrite the scalar multiple cA by factoring c out of every entry in the matrix. For instance, in the following example, the scalar $\frac{1}{2}$ has been factored out of the matrix.

$$\begin{bmatrix} \frac{1}{2} & -\frac{3}{2} \\ \frac{5}{2} & \frac{1}{2} \end{bmatrix} = \begin{bmatrix} \frac{1}{2}(1) & \frac{1}{2}(-3) \\ \frac{1}{2}(5) & \frac{1}{2}(1) \end{bmatrix}$$

$$= \frac{1}{2} \begin{bmatrix} 1 & -3 \\ 5 & 1 \end{bmatrix}$$

STUDY T!P

The order of operations for matrix expressions is similar to that for real numbers. In particular, you perform scalar multiplication before matrix addition and subtraction, as shown in Example 3(c).

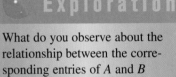

Exploration

What do you observe about the relationship between the corresponding entries of A and B below? Use a graphing utility to find $A + B$. What conclusion can you make about the entries of A and B and the sum of $A + B$?

$$A = \begin{bmatrix} -1 & 5 \\ 2 & -6 \end{bmatrix}$$

$$B = \begin{bmatrix} 1 & -5 \\ -2 & 6 \end{bmatrix}$$

The properties of matrix addition and scalar multiplication are similar to those of addition and multiplication of real numbers.

Properties of Matrix Addition and Scalar Multiplication

Let A, B, and C be $m \times n$ matrices and let c and d be scalars.

1. $A + B = B + A$ Commutative Property of Matrix Addition

2. $A + (B + C) = (A + B) + C$ Associative Property of Matrix Addition

3. $(cd)A = c(dA)$ Associative Property of Scalar Multiplication

4. $1A = A$ Scalar Identity

5. $c(A + B) = cA + cB$ Distributive Property

6. $(c + d)A = cA + dA$ Distributive Property

Note that the Associative Property of Matrix Addition allows you to write expressions such as $A + B + C$ without ambiguity because the same sum occurs no matter how the matrices are grouped. In other words, you obtain the same sum whether you group $A + B + C$ as $(A + B) + C$ or as $A + (B + C)$. This same reasoning applies to sums of four or more matrices.

EXAMPLE 4   Addition of More Than Two Matrices

Find the sum.

$$\begin{bmatrix} 1 \\ 2 \\ -3 \end{bmatrix} + \begin{bmatrix} -1 \\ -1 \\ 2 \end{bmatrix} + \begin{bmatrix} 0 \\ 1 \\ 4 \end{bmatrix} + \begin{bmatrix} 2 \\ -3 \\ -2 \end{bmatrix}$$

Solution

By adding corresponding entries, you obtain the following sum of four matrices.

$$\begin{bmatrix} 1 \\ 2 \\ -3 \end{bmatrix} + \begin{bmatrix} -1 \\ -1 \\ 2 \end{bmatrix} + \begin{bmatrix} 0 \\ 1 \\ 4 \end{bmatrix} + \begin{bmatrix} 2 \\ -3 \\ -2 \end{bmatrix} = \begin{bmatrix} 2 \\ -1 \\ 1 \end{bmatrix}$$

EXAMPLE 5   Using the Distributive Property

Evaluate the expression $3\left(\begin{bmatrix} -2 & 0 \\ 4 & 1 \end{bmatrix} + \begin{bmatrix} 4 & -2 \\ 3 & 7 \end{bmatrix} \right)$.

Solution

$$3\left(\begin{bmatrix} -2 & 0 \\ 4 & 1 \end{bmatrix} + \begin{bmatrix} 4 & -2 \\ 3 & 7 \end{bmatrix} \right) = 3\begin{bmatrix} -2 & 0 \\ 4 & 1 \end{bmatrix} + 3\begin{bmatrix} 4 & -2 \\ 3 & 7 \end{bmatrix}$$

$$= \begin{bmatrix} -6 & 0 \\ 12 & 3 \end{bmatrix} + \begin{bmatrix} 12 & -6 \\ 9 & 21 \end{bmatrix}$$

$$= \begin{bmatrix} 6 & -6 \\ 21 & 24 \end{bmatrix}$$

STUDY TIP

In Example 5, you could have added the two matrices first and then multiplied the matrix by 3. The result would have been the same.

One important property of addition of real numbers is that the number 0 is the additive identity. That is, $c + 0 = c$ for any real number c. For matrices, a similar property holds. That is, if A is an $m \times n$ matrix and O is the $m \times n$ **zero matrix** consisting entirely of zeros, then $A + O = A$.

In other words, O is the **additive identity** for the set of all $m \times n$ matrices. For example, the following matrices are the additive identities for the set of all 2×3 and 2×2 matrices.

$$O = \begin{bmatrix} 0 & 0 & 0 \\ 0 & 0 & 0 \end{bmatrix} \quad \text{and} \quad O = \begin{bmatrix} 0 & 0 \\ 0 & 0 \end{bmatrix}.$$

$\underbrace{\qquad\qquad}_{\text{Zero } 2 \times 3 \text{ matrix}}$ $\underbrace{\qquad\qquad}_{\text{Zero } 2 \times 2 \text{ matrix}}$

The algebra of real numbers and the algebra of matrices have many similarities. For example, compare the following solutions.

Real Numbers (Solve for x.)	$m \times n$ Matrices (Solve for X.)
$x + a = b$	$X + A = B$
$x + a + (-a) = b + (-a)$	$X + A + (-A) = B + (-A)$
$x + 0 = b - a$	$X + O = B - A$
$x = b - a$	$X = B - A$

The algebra of real numbers and the algebra of matrices also have important differences, which will be discussed later.

Exploration

Use a graphing utility to find the sum of A and B below. Use your result to predict the sum of C and D below. Then check your prediction using a graphing utility.

$$A = \begin{bmatrix} -2 & 1 \\ 0 & 3 \end{bmatrix} \quad B = \begin{bmatrix} 0 & 0 \\ 0 & 0 \end{bmatrix}$$

$$C = \begin{bmatrix} 1 & -1 \\ 2 & -1 \\ 0 & 2 \end{bmatrix} \quad D = \begin{bmatrix} 0 & 0 \\ 0 & 0 \\ 0 & 0 \end{bmatrix}$$

EXAMPLE 6   Solving a Matrix Equation

Solve for X in the equation $3X + A = B$, where

$$A = \begin{bmatrix} 1 & -2 \\ 0 & 3 \end{bmatrix} \quad \text{and} \quad B = \begin{bmatrix} -3 & 4 \\ 2 & 1 \end{bmatrix}.$$

Solution

Begin by solving the equation for X to obtain

$$3X = B - A$$

$$X = \tfrac{1}{3}(B - A).$$

Now, using the matrices A and B, you have

$$X = \tfrac{1}{3}\left(\begin{bmatrix} -3 & 4 \\ 2 & 1 \end{bmatrix} - \begin{bmatrix} 1 & -2 \\ 0 & 3 \end{bmatrix} \right)$$

$$= \tfrac{1}{3}\begin{bmatrix} -4 & 6 \\ 2 & -2 \end{bmatrix}$$

$$= \begin{bmatrix} -\tfrac{4}{3} & 2 \\ \tfrac{2}{3} & -\tfrac{2}{3} \end{bmatrix}.$$

Matrix Multiplication

The third basic matrix operation is **matrix multiplication.** At first glance, the following definition may seem unusual. You will see later, however, that this definition of the product of two matrices has many practical applications.

> ## Definition of Matrix Multiplication
>
> If $A = [a_{ij}]$ is an $m \times n$ matrix and $B = [b_{ij}]$ is an $n \times p$ matrix, the product AB is an $m \times p$ matrix
>
> $$AB = [c_{ij}]$$
>
> where $c_{ij} = a_{i1}b_{1j} + a_{i2}b_{2j} + a_{i3}b_{3j} + \cdots + a_{in}b_{nj}.$

The definition of matrix multiplication indicates a *row-by-column* multiplication, where the entry in the ith row and jth column of the product AB is obtained by multiplying the entries in the ith row of A by the corresponding entries in the jth column of B and then adding the results. Example 7 illustrates this process.

EXAMPLE 7 Finding the Product of Two Matrices

Find the product AB where

$$A = \begin{bmatrix} -1 & 3 \\ 4 & -2 \\ 5 & 0 \end{bmatrix} \quad \text{and} \quad B = \begin{bmatrix} -3 & 2 \\ -4 & 1 \end{bmatrix}.$$

Solution

First, note that the product AB is defined because the number of columns of A is equal to the number of rows of B. Moreover, the product AB has order 3×2, and is of the form

$$\begin{bmatrix} -1 & 3 \\ 4 & -2 \\ 5 & 0 \end{bmatrix} \begin{bmatrix} -3 & 2 \\ -4 & 1 \end{bmatrix} = \begin{bmatrix} c_{11} & c_{12} \\ c_{21} & c_{22} \\ c_{31} & c_{32} \end{bmatrix}.$$

To find the entries of the product, multiply each row of A by each column of B, as follows. Use a graphing utility to check this result.

$$AB = \begin{bmatrix} -1 & 3 \\ 4 & -2 \\ 5 & 0 \end{bmatrix} \begin{bmatrix} -3 & 2 \\ -4 & 1 \end{bmatrix}$$

$$= \begin{bmatrix} (-1)(-3) + (3)(-4) & (-1)(2) + (3)(1) \\ (4)(-3) + (-2)(-4) & (4)(2) + (-2)(1) \\ (5)(-3) + (0)(-4) & (5)(2) + (0)(1) \end{bmatrix}$$

$$= \begin{bmatrix} -9 & 1 \\ -4 & 6 \\ -15 & 10 \end{bmatrix}$$

A computer animation of this example appears in the *Interactive* CD-ROM and *Internet* versions of this text.

STUDY T!P

Some graphing utilities are able to multiply matrices. If you have such a graphing utility, enter the matrices

$$A = \begin{bmatrix} 1 & 2 & 3 \\ 2 & -5 & 1 \end{bmatrix} \quad \text{and}$$

$$B = \begin{bmatrix} -3 & 2 & 1 \\ 4 & -2 & 0 \\ 1 & 2 & 3 \end{bmatrix}$$

to find the product AB. You should get:

$$\begin{bmatrix} 8 & 4 & 10 \\ -25 & 16 & 5 \end{bmatrix}.$$

Be sure you understand that for the product of two matrices to be defined, the number of *columns* of the first matrix must equal the number of *rows* of the second matrix. That is, the middle two indices must be the same and the outside two indices give the order of the product, as shown in the following diagram.

$$A \quad \times \quad B \quad = \quad AB$$

$m \times n \qquad n \times p \qquad\qquad m \times p$

equal

order of *AB*

EXAMPLE 8   Matrix Multiplication

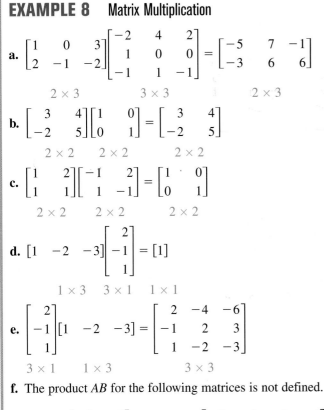

a. $\begin{bmatrix} 1 & 0 & 3 \\ 2 & -1 & -2 \end{bmatrix} \begin{bmatrix} -2 & 4 & 2 \\ 1 & 0 & 0 \\ -1 & 1 & -1 \end{bmatrix} = \begin{bmatrix} -5 & 7 & -1 \\ -3 & 6 & 6 \end{bmatrix}$

$\quad 2 \times 3 \qquad\qquad 3 \times 3 \qquad\qquad 2 \times 3$

b. $\begin{bmatrix} 3 & 4 \\ -2 & 5 \end{bmatrix} \begin{bmatrix} 1 & 0 \\ 0 & 1 \end{bmatrix} = \begin{bmatrix} 3 & 4 \\ -2 & 5 \end{bmatrix}$

$\quad 2 \times 2 \qquad 2 \times 2 \qquad\quad 2 \times 2$

c. $\begin{bmatrix} 1 & 2 \\ 1 & 1 \end{bmatrix} \begin{bmatrix} -1 & 2 \\ 1 & -1 \end{bmatrix} = \begin{bmatrix} 1 & 0 \\ 0 & 1 \end{bmatrix}$

$\quad 2 \times 2 \qquad 2 \times 2 \qquad\quad 2 \times 2$

d. $\begin{bmatrix} 1 & -2 & -3 \end{bmatrix} \begin{bmatrix} 2 \\ -1 \\ 1 \end{bmatrix} = \begin{bmatrix} 1 \end{bmatrix}$

$\quad\quad 1 \times 3 \qquad 3 \times 1 \quad\ 1 \times 1$

e. $\begin{bmatrix} 2 \\ -1 \\ 1 \end{bmatrix} \begin{bmatrix} 1 & -2 & -3 \end{bmatrix} = \begin{bmatrix} 2 & -4 & -6 \\ -1 & 2 & 3 \\ 1 & -2 & -3 \end{bmatrix}$

$\quad 3 \times 1 \qquad 1 \times 3 \qquad\qquad 3 \times 3$

f. The product *AB* for the following matrices is not defined.

$$A = \begin{bmatrix} -2 & 1 \\ 1 & -3 \\ 1 & 4 \end{bmatrix} \quad \text{and} \quad B = \begin{bmatrix} -2 & 3 & 1 & 4 \\ 0 & 1 & -1 & 2 \\ 2 & -1 & 0 & 1 \end{bmatrix}$$

$\qquad\quad 3 \times 2 \qquad\qquad\qquad\quad\ 3 \times 4$

In parts (d) and (e) of Example 8, note that the two products are different. Matrix multiplication is not, in general, commutative. That is, for most matrices, $AB \neq BA$. This is one way in which the algebra of real numbers and the algebra of matrices differ.

Exploration

Use a graphing utility to multiply the matrices

$$A = \begin{bmatrix} 1 & 2 \\ 3 & 4 \end{bmatrix} \quad \text{and}$$

$$B = \begin{bmatrix} 0 & 1 \\ 2 & 3 \end{bmatrix}.$$

Do you obtain the same result for the product *AB* as for the product *BA*? What does this tell you about matrix multiplication and commutativity?

In summary, the general pattern for matrix multiplication is as follows. To obtain the entry in the ith row and the jth column of the product AB, use the ith row of A and the jth column of B.

$$\begin{bmatrix} a_{11} & a_{12} & a_{13} & \cdots & a_{1n} \\ a_{21} & a_{22} & a_{23} & \cdots & a_{2n} \\ a_{31} & a_{32} & a_{33} & \cdots & a_{3n} \\ \vdots & \vdots & \vdots & & \vdots \\ a_{i1} & a_{i2} & a_{i3} & \cdots & a_{in} \\ \vdots & \vdots & \vdots & & \vdots \\ a_{m1} & a_{m2} & a_{m3} & \cdots & a_{mn} \end{bmatrix} \begin{bmatrix} b_{11} & b_{12} & \cdots & b_{1j} & \cdots & b_{1p} \\ b_{21} & b_{22} & \cdots & b_{2j} & \cdots & b_{2p} \\ b_{31} & b_{32} & \cdots & b_{3j} & \cdots & b_{3p} \\ \vdots & \vdots & & \vdots & & \vdots \\ b_{n1} & b_{n2} & \cdots & b_{nj} & \cdots & b_{np} \end{bmatrix} = \begin{bmatrix} c_{11} & c_{12} & \cdots & c_{1j} & \cdots & c_{1p} \\ c_{21} & c_{22} & \cdots & c_{2j} & \cdots & c_{2p} \\ \vdots & \vdots & & \vdots & & \vdots \\ c_{i1} & c_{i2} & \cdots & c_{ij} & \cdots & c_{ip} \\ \vdots & \vdots & & \vdots & & \vdots \\ c_{m1} & c_{m2} & \cdots & c_{mj} & \cdots & c_{mp} \end{bmatrix}$$

$$a_{i1}b_{1j} + a_{i2}b_{2j} + a_{i3}b_{3j} + \cdots + a_{in}b_{nj} = c_{ij}$$

Properties of Matrix Multiplication

Let A, B, and C be matrices and let c be a scalar.

1. $A(BC) = (AB)C$ Associative Property of Matrix Multiplication

2. $A(B + C) = AB + AC$ Distributive Property

3. $(A + B)C = AC + BC$ Distributive Property

4. $c(AB) = (cA)B = A(cB)$ Associative Property of Scalar Multiplication

Definition of Identity Multiplication

The $n \times n$ matrix that consists of 1's on its main diagonal and 0's elsewhere is called the **identity matrix of order n** and is denoted by

$$I_n = \begin{bmatrix} 1 & 0 & 0 & \cdots & 0 \\ 0 & 1 & 0 & \cdots & 0 \\ 0 & 0 & 1 & \cdots & 0 \\ \vdots & \vdots & \vdots & & \vdots \\ 0 & 0 & 0 & \cdots & 1 \end{bmatrix}. \quad \text{Identity matrix}$$

Note that an identity matrix must be *square*. When the order is understood to be n, you can denote I_n simply by I.

If A is an $n \times n$ matrix, the identity matrix has the property that $AI_n = A$ and $I_n A = A$. For example,

$$\begin{bmatrix} 3 & -2 & 5 \\ 1 & 0 & 4 \\ -1 & 2 & -3 \end{bmatrix} \begin{bmatrix} 1 & 0 & 0 \\ 0 & 1 & 0 \\ 0 & 0 & 1 \end{bmatrix} = \begin{bmatrix} 3 & -2 & 5 \\ 1 & 0 & 4 \\ -1 & 2 & -3 \end{bmatrix} \quad AI = A$$

and

$$\begin{bmatrix} 1 & 0 & 0 \\ 0 & 1 & 0 \\ 0 & 0 & 1 \end{bmatrix} \begin{bmatrix} 3 & -2 & 5 \\ 1 & 0 & 4 \\ -1 & 2 & -3 \end{bmatrix} = \begin{bmatrix} 3 & -2 & 5 \\ 1 & 0 & 4 \\ -1 & 2 & -3 \end{bmatrix}. \quad IA = A$$

Applications

Matrix multiplication can be used to represent a system of linear equations. Note how the system

$$\begin{cases} a_{11}x_1 + a_{12}x_2 + a_{13}x_3 = b_1 \\ a_{21}x_1 + a_{22}x_2 + a_{23}x_3 = b_2 \\ a_{31}x_1 + a_{32}x_2 + a_{33}x_3 = b_3 \end{cases}$$

can be written as the matrix equation $AX = B$, where A is the *coefficient matrix* of the system, and X and B are column matrices. The column matrix B is also called a *constant* matrix. Its entries are the constant terms in the system of equations.

$$\underset{A}{\begin{bmatrix} a_{11} & a_{12} & a_{13} \\ a_{21} & a_{22} & a_{23} \\ a_{31} & a_{32} & a_{33} \end{bmatrix}} \times \underset{X}{\begin{bmatrix} x_1 \\ x_2 \\ x_3 \end{bmatrix}} = \underset{B}{\begin{bmatrix} b_1 \\ b_2 \\ b_3 \end{bmatrix}}$$

EXAMPLE 9 Solving a System of Linear Equations

Consider the system of linear equations $\begin{cases} x_1 - 2x_2 + x_3 = -4 \\ x_2 + 2x_3 = 4. \\ 2x_1 + 3x_2 - 2x_3 = 2 \end{cases}$

A computer animation of this example appears in the *Interactive* CD-ROM and *Internet* versions of this text.

a. Write this system as a matrix equation, $AX = B$.

b. Use Gauss-Jordan elimination on the augmented matrix $[A \vdots B]$ to solve for the matrix X.

Solution

a. In matrix form $AX = B$, the system can be written as follows.

$$\begin{bmatrix} 1 & -2 & 1 \\ 0 & 1 & 2 \\ 2 & 3 & -2 \end{bmatrix} \begin{bmatrix} x_1 \\ x_2 \\ x_3 \end{bmatrix} = \begin{bmatrix} -4 \\ 4 \\ 2 \end{bmatrix}$$

b. The augmented matrix is

$$[A \vdots B] = \begin{bmatrix} 1 & -2 & 1 & \vdots & -4 \\ 0 & 1 & 2 & \vdots & 4 \\ 2 & 3 & -2 & \vdots & 2 \end{bmatrix}.$$

Using Gauss-Jordan elimination, you can rewrite this equation as

$$[I \vdots X] = \begin{bmatrix} 1 & 0 & 0 & \vdots & -1 \\ 0 & 1 & 0 & \vdots & 2 \\ 0 & 0 & 1 & \vdots & 1 \end{bmatrix}.$$

So, the solution of the system of linear equations is $x_1 = -1$, $x_2 = 2$, and $x_3 = 1$. The solution of the matrix equation is

$$X = \begin{bmatrix} x_1 \\ x_2 \\ x_3 \end{bmatrix} = \begin{bmatrix} -1 \\ 2 \\ 1 \end{bmatrix}.$$

STUDY T!P

You can use some graphing utilities to obtain the reduced row-echelon form of a matrix. Consult your user's manual to determine how to perform this matrix operation. The screen in the figure below shows how one graphing utility displays the reduced row-echelon form of the augmented matrix in Example 9.

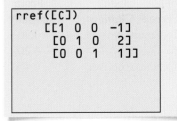

```
rref([C])
      [[1 0 0 -1]
       [0 1 0  2]
       [0 0 1  1]]
```

EXAMPLE 10 Softball Team Expenses

Two softball teams submit equipment lists to their sponsors.

	Women's Team	Men's Team
Bats	12	15
Balls	45	38
Gloves	15	17

Each bat costs $48, each ball costs $4, and each glove costs $42. Use matrices to find the total cost of equipment for each team.

Solution

The equipment lists and the costs per item can be written in matrix form as

$$E = \begin{bmatrix} 12 & 15 \\ 45 & 38 \\ 15 & 17 \end{bmatrix} \quad \text{and} \quad C = \begin{bmatrix} 48 & 4 & 42 \end{bmatrix}.$$

To find the total cost of the equipment for each team, you need to find the product. Note that you cannot find the total cost using the product EC because EC is not defined. That is, the number of columns of E (2 columns) does not match the number of rows of C (one row). You can find the total cost using the product CE because the number of columns of C (3 columns) matches the number of rows of E (3 rows). Therefore, the total cost of equipment for each team is given by the product

$$CE = \begin{bmatrix} 48 & 4 & 42 \end{bmatrix} \begin{bmatrix} 12 & 15 \\ 45 & 38 \\ 15 & 17 \end{bmatrix}$$

$$= \begin{bmatrix} 1386 & 1586 \end{bmatrix}.$$

So, the total cost of equipment for the women's team is $1386, and the total cost of equipment for the men's team is $1586.

Writing About Math *Number of Cable Subscribers*

Two competing companies offer cable television to a city with 100,000 households. Company A has 25,000 subscribers, and Company B has 30,000. (The other 45,000 households do not subscribe.) The percent changes in cable subscriptions each year are shown below. Write a short paragraph explaining how matrix multiplication can be used to find the number of subscribers each company will have in one year.

Percent Changes	From Company A	From Company B	From Nonsubscribers
To Company A	0.70	0.15	0.15
To Company B	0.20	0.80	0.15
To Nonsubscriber	0.10	0.05	0.70

8.2 Exercises

In Exercises 1–4, find x, y, and z.

1. $\begin{bmatrix} x & -2 \\ 7 & y \end{bmatrix} = \begin{bmatrix} -4 & -2 \\ 7 & 22 \end{bmatrix}$

2. $\begin{bmatrix} -5 & x \\ y & 8 \end{bmatrix} = \begin{bmatrix} -5 & 13 \\ 12 & 8 \end{bmatrix}$

3. $\begin{bmatrix} 16 & 4 & 5 & 4 \\ -3 & 13 & 15 & 12 \\ 0 & 2 & 4 & 0 \end{bmatrix} = \begin{bmatrix} 16 & 4 & 2x+7 & 4 \\ -3 & 13 & & 15 & 3y \\ 0 & 2 & 3z-14 & 0 \end{bmatrix}$

4. $\begin{bmatrix} x+4 & 8 & -3 \\ 1 & 22 & 2y \\ 7 & -2 & z+2 \end{bmatrix} = \begin{bmatrix} 2x+9 & 8 & -3 \\ 1 & 22 & -8 \\ 7 & -2 & 11 \end{bmatrix}$

In Exercises 5–12, if possible, find (a) $A + B$, (b) $A - B$, (c) $3A$, and (d) $3A - 2B$. Use the matrix capabilities of a graphing utility to verify your results.

5. $A = \begin{bmatrix} 1 & -1 \\ 2 & -1 \end{bmatrix}$, $B = \begin{bmatrix} 2 & -1 \\ -1 & 8 \end{bmatrix}$

6. $A = \begin{bmatrix} 1 & 2 \\ 2 & 1 \end{bmatrix}$, $B = \begin{bmatrix} -3 & -2 \\ 4 & 2 \end{bmatrix}$

7. $A = \begin{bmatrix} 8 & -1 \\ 2 & 3 \\ -4 & 5 \end{bmatrix}$, $B = \begin{bmatrix} 1 & 6 \\ -1 & -5 \\ 1 & 10 \end{bmatrix}$

8. $A = \begin{bmatrix} 2 & 1 & 1 \\ -1 & -1 & 5 \end{bmatrix}$, $B = \begin{bmatrix} 6 & -3 & 2 \\ -4 & 1 & -2 \end{bmatrix}$

9. $A = \begin{bmatrix} 2 & 2 & -1 & 0 & 1 \\ 1 & 1 & -2 & 0 & -1 \end{bmatrix}$,

$B = \begin{bmatrix} 1 & 1 & -1 & 1 & 0 \\ -3 & 4 & 9 & -6 & -7 \end{bmatrix}$

10. $A = \begin{bmatrix} -1 & 4 & 0 \\ 3 & -2 & 2 \\ 5 & 4 & -1 \\ 0 & 8 & -6 \\ -4 & -1 & 0 \end{bmatrix}$, $B = \begin{bmatrix} -3 & 5 & 1 \\ 2 & -4 & -7 \\ 10 & -9 & -1 \\ 3 & 2 & -4 \\ 0 & 1 & -2 \end{bmatrix}$

11. $A = \begin{bmatrix} 6 & 0 & 3 \\ -1 & -4 & 0 \end{bmatrix}$, $B = \begin{bmatrix} 8 & -1 \\ 4 & -3 \end{bmatrix}$

12. $A = \begin{bmatrix} 3 \\ 2 \\ -1 \end{bmatrix}$, $B = \begin{bmatrix} -4 & 6 & 2 \end{bmatrix}$

In Exercises 13–18, evaluate the expression.

13. $\begin{bmatrix} -4 & 0 \\ 5 & -6 \end{bmatrix} + \begin{bmatrix} 9 & 1 \\ -2 & -2 \end{bmatrix} + \begin{bmatrix} -12 & -10 \\ 14 & 6 \end{bmatrix}$

14. $\begin{bmatrix} 6 & 9 \\ -1 & 0 \\ 7 & 1 \\ 0 & -10 \end{bmatrix} + \begin{bmatrix} 0 & 5 \\ -2 & -1 \\ 3 & -6 \\ -4 & 2 \end{bmatrix} + \begin{bmatrix} -13 & -7 \\ 4 & -1 \\ -6 & 0 \\ -1 & 25 \end{bmatrix}$

15. $4\left(\begin{bmatrix} -4 & 0 & 1 \\ 0 & 2 & 3 \end{bmatrix} - \begin{bmatrix} 2 & 1 & -2 \\ 3 & -6 & 0 \end{bmatrix} \right)$

16. $\frac{1}{2}([5 \quad -2 \quad 4 \quad 0] + [14 \quad 6 \quad -18 \quad 9])$

17. $-3\left(\begin{bmatrix} 0 & -3 \\ 7 & 2 \end{bmatrix} + \begin{bmatrix} -6 & 3 \\ 8 & 1 \end{bmatrix} \right) - \begin{bmatrix} 4 & -4 \\ 7 & -9 \end{bmatrix}$

18. $-1\begin{bmatrix} 4 & 11 \\ -2 & -1 \\ 9 & 3 \end{bmatrix} + \frac{1}{6}\left(\begin{bmatrix} -5 & -1 \\ 3 & 4 \\ 0 & 13 \end{bmatrix} + \begin{bmatrix} 7 & 5 \\ -9 & -1 \\ 6 & -1 \end{bmatrix} \right)$

In Exercises 19–22, use the matrix capabilities of a graphing utility to evaluate the expression. Round your results to three decimal places, if necessary.

19. $\frac{3}{7}\begin{bmatrix} 2 & 5 \\ -1 & -4 \end{bmatrix}$

20. $55\left(\begin{bmatrix} 14 & -11 \\ -22 & 19 \end{bmatrix} + \begin{bmatrix} -22 & 20 \\ 13 & 6 \end{bmatrix} \right)$

21. $-\begin{bmatrix} 3.211 & 6.829 \\ -1.004 & 4.914 \\ 0.055 & -3.889 \end{bmatrix} - \begin{bmatrix} -1.630 & -3.090 \\ 5.256 & 8.335 \\ -9.768 & 4.251 \end{bmatrix}$

22. $-12\left(\begin{bmatrix} 6 & 20 \\ 1 & -9 \\ -2 & 5 \end{bmatrix} + \begin{bmatrix} 14 & -15 \\ -8 & -6 \\ 7 & 0 \end{bmatrix} + \begin{bmatrix} -31 & -19 \\ 16 & 10 \\ 24 & 10 \end{bmatrix} \right)$

In Exercises 23–26, solve for X given

$$A = \begin{bmatrix} -2 & -1 \\ 1 & 0 \\ 3 & -4 \end{bmatrix} \quad \text{and} \quad B = \begin{bmatrix} 0 & 3 \\ 2 & 0 \\ -4 & -1 \end{bmatrix}.$$

23. $X = 3A - 2B$ **24.** $2X = 2A - B$

25. $2X + 3A = B$ **26.** $2A + 4B = -2X$

In Exercises 27–32, if possible, find (a) AB, (b) BA, and (c) A^2. (*Note:* $A^2 = AA$.) Use the matrix capabilities of a graphing utility to verify your results.

27. $A = \begin{bmatrix} 1 & 2 \\ 5 & 2 \end{bmatrix}$, $B = \begin{bmatrix} 2 & -1 \\ -1 & 8 \end{bmatrix}$

28. $A = \begin{bmatrix} 2 & -1 \\ 1 & -4 \end{bmatrix}$, $B = \begin{bmatrix} -2 & 0 \\ 3 & -3 \end{bmatrix}$

29. $A = \begin{bmatrix} 3 & -1 \\ 1 & 3 \end{bmatrix}, \quad B = \begin{bmatrix} 1 & -3 \\ 3 & 1 \end{bmatrix}$

30. $A = \begin{bmatrix} 1 & -1 \\ 1 & 1 \end{bmatrix}, \quad B = \begin{bmatrix} 1 & 3 \\ -3 & 1 \end{bmatrix}$

31. $A = \begin{bmatrix} 1 & -1 & 7 \\ 2 & -1 & 8 \\ 3 & 1 & -1 \end{bmatrix}, \quad B = \begin{bmatrix} 1 & 1 & 2 \\ 2 & 1 & 1 \\ 1 & -3 & 2 \end{bmatrix}$

32. $A = \begin{bmatrix} 3 & 2 & 1 \end{bmatrix}, \quad B = \begin{bmatrix} 2 \\ 3 \\ 0 \end{bmatrix}$

In Exercises 33–40, find *AB*, if possible.

33. $A = \begin{bmatrix} 2 & 1 \\ -5 & 4 \\ -1 & -6 \end{bmatrix}, \quad B = \begin{bmatrix} 0 & -3 & 3 \\ 4 & 0 & 2 \\ 8 & -2 & 7 \end{bmatrix}$

34. $A = \begin{bmatrix} 0 & -1 & 2 \\ 6 & 0 & 3 \\ 7 & -1 & 8 \end{bmatrix}, \quad B = \begin{bmatrix} 2 & -1 \\ 4 & -5 \\ 1 & 6 \end{bmatrix}$

35. $A = \begin{bmatrix} -1 & 6 \\ -4 & 5 \\ 0 & 3 \end{bmatrix}, \quad B = \begin{bmatrix} 2 & 3 \\ 0 & 9 \end{bmatrix}$

36. $A = \begin{bmatrix} 1 & 0 & 0 \\ 0 & 4 & 0 \\ 0 & 0 & -2 \end{bmatrix}, \quad B = \begin{bmatrix} 3 & 0 & 0 \\ 0 & -1 & 0 \\ 0 & 0 & 5 \end{bmatrix}$

37. $A = \begin{bmatrix} 5 & 0 & 0 \\ 0 & -8 & 0 \\ 0 & 0 & 7 \end{bmatrix}, \quad B = \begin{bmatrix} \frac{1}{5} & 0 & 0 \\ 0 & -\frac{1}{8} & 0 \\ 0 & 0 & \frac{1}{2} \end{bmatrix}$

38. $A = \begin{bmatrix} 0 & 0 & 5 \\ 0 & 0 & -3 \\ 0 & 0 & 4 \end{bmatrix}, \quad B = \begin{bmatrix} 6 & -11 & 4 \\ 8 & 16 & 4 \\ 0 & 0 & 0 \end{bmatrix}$

39. $A = \begin{bmatrix} 10 \\ 11 \end{bmatrix}, \quad B = \begin{bmatrix} 4 & -2 & -1 & 8 \end{bmatrix}$

40. $A = \begin{bmatrix} 1 & 0 & 3 & -2 \\ 6 & 13 & 8 & -17 \end{bmatrix}, \quad B = \begin{bmatrix} 1 & 6 \\ 4 & 2 \end{bmatrix}$

In Exercises 41–46, use the matrix capabilities of a graphing utility to find *AB*.

41. $A = \begin{bmatrix} 7 & 5 & -4 \\ -2 & 5 & 1 \\ 10 & -4 & -7 \end{bmatrix}, \quad B = \begin{bmatrix} 2 & -2 & 3 \\ 8 & 1 & 4 \\ -4 & 2 & -8 \end{bmatrix}$

42. $A = \begin{bmatrix} 11 & -12 & 4 \\ 14 & 10 & 12 \\ 6 & -2 & 9 \end{bmatrix}, \quad B = \begin{bmatrix} 12 & 10 \\ -5 & 12 \\ 15 & 16 \end{bmatrix}$

43. $A = \begin{bmatrix} -3 & 8 & -6 & 8 \\ -12 & 15 & 9 & 6 \\ 5 & -1 & 1 & 5 \end{bmatrix},$

$B = \begin{bmatrix} 3 & 1 & 6 \\ 24 & 15 & 14 \\ 16 & 10 & 21 \\ 8 & -4 & 10 \end{bmatrix}$

44. $A = \begin{bmatrix} -2 & 6 & 12 \\ 21 & -5 & 6 \\ 13 & -2 & 9 \end{bmatrix},$

$B = \begin{bmatrix} 3 & 0 \\ -7 & 18 \\ 34 & 14 \\ 0.5 & 1.4 \end{bmatrix}$

45. $A = \begin{bmatrix} 9 & 10 & -38 & 18 \\ 100 & -50 & 250 & 75 \end{bmatrix},$

$B = \begin{bmatrix} 52 & -85 & 27 & 45 \\ 40 & -35 & 60 & 82 \end{bmatrix}$

46. $A = \begin{bmatrix} 16 & -18 \\ -4 & 13 \\ -9 & 21 \end{bmatrix},$

$B = \begin{bmatrix} -7 & 20 & -1 \\ 7 & 15 & 26 \end{bmatrix}$

In Exercises 47–50, use the matrix capabilities of a graphing utility to evaluate the expression.

47. $\begin{bmatrix} 3 & 1 \\ 0 & -2 \end{bmatrix} \begin{bmatrix} 1 & 0 \\ -2 & 2 \end{bmatrix} \begin{bmatrix} 1 & 0 \\ 2 & 4 \end{bmatrix}$

48. $-3 \left(\begin{bmatrix} 6 & 5 & -1 \\ 1 & -2 & 0 \end{bmatrix} \begin{bmatrix} 0 & 3 \\ -1 & -3 \\ 4 & 1 \end{bmatrix} \right)$

49. $\begin{bmatrix} 0 & 2 & -2 \\ 4 & 1 & 2 \end{bmatrix} \left(\begin{bmatrix} 4 & 0 \\ 0 & -1 \\ -1 & 2 \end{bmatrix} + \begin{bmatrix} -2 & 3 \\ -3 & 5 \\ 0 & -3 \end{bmatrix} \right)$

50. $\begin{bmatrix} 3 \\ -1 \\ 5 \\ 7 \end{bmatrix} \left(\begin{bmatrix} 5 & -6 \end{bmatrix} + \begin{bmatrix} 7 & -1 \end{bmatrix} + \begin{bmatrix} -8 & 9 \end{bmatrix} \right)$

In Exercises 51–56, use matrix multiplication to determine which of the given matrices is a solution of the system of equations. Use a graphing utility to verify your result.

51. $\begin{cases} x + 2y = 4 \\ 3x + 2y = 0 \end{cases}$

(a) $\begin{bmatrix} 2 \\ 1 \end{bmatrix}$ (b) $\begin{bmatrix} -2 \\ 3 \end{bmatrix}$

(c) $\begin{bmatrix} -4 \\ 4 \end{bmatrix}$ (d) $\begin{bmatrix} 2 \\ -3 \end{bmatrix}$

52. $\begin{cases} 6x + 2y = 0 \\ -x + 5y = 16 \end{cases}$

(a) $\begin{bmatrix} -1 \\ 3 \end{bmatrix}$ (b) $\begin{bmatrix} 2 \\ -6 \end{bmatrix}$

(c) $\begin{bmatrix} 3 \\ -9 \end{bmatrix}$ (d) $\begin{bmatrix} -3 \\ 9 \end{bmatrix}$

53. $\begin{cases} -2x - 3y = -6 \\ 4x + 2y = 20 \end{cases}$

(a) $\begin{bmatrix} 3 \\ 0 \end{bmatrix}$ (b) $\begin{bmatrix} 6 \\ -2 \end{bmatrix}$

(c) $\begin{bmatrix} -6 \\ 6 \end{bmatrix}$ (d) $\begin{bmatrix} 4 \\ 2 \end{bmatrix}$

54. $\begin{cases} 5x - 7y = -15 \\ 3x + y = 17 \end{cases}$

(a) $\begin{bmatrix} 4 \\ 5 \end{bmatrix}$ (b) $\begin{bmatrix} 5 \\ 2 \end{bmatrix}$

(c) $\begin{bmatrix} -4 \\ -5 \end{bmatrix}$ (d) $\begin{bmatrix} 2 \\ 11 \end{bmatrix}$

55. $\begin{cases} x - 3y + 4z = 3 \\ 2x - 5y = 19 \\ -2x + 4y - 5z = -6 \end{cases}$

(a) $\begin{bmatrix} 3 \\ 2 \\ -2 \end{bmatrix}$ (b) $\begin{bmatrix} 2 \\ -3 \\ -2 \end{bmatrix}$ (c) $\begin{bmatrix} 1 \\ 3 \\ -2 \end{bmatrix}$ (d) $\begin{bmatrix} 3 \\ -3 \\ 2 \end{bmatrix}$

56. $\begin{cases} -3x + y - 6z = 12 \\ x - 4y + 2z = -26 \\ 4x + y + z = -23 \end{cases}$

(a) $\begin{bmatrix} 6 \\ -8 \\ 3 \end{bmatrix}$ (b) $\begin{bmatrix} 5 \\ 4 \\ -1 \end{bmatrix}$ (c) $\begin{bmatrix} -2 \\ 6 \\ 0 \end{bmatrix}$ (d) $\begin{bmatrix} -8 \\ 6 \\ 3 \end{bmatrix}$

In Exercises 57–64, (a) write each system of equations as a matrix equation $AX = B$ and (b) use Gauss-Jordan elimination on the augmented matrix $[A \vdots B]$ to solve for the matrix X. Use a graphing utility to check your solution.

57. $\begin{cases} -x_1 + x_2 = 4 \\ -2x_1 + x_2 = 0 \end{cases}$

58. $\begin{cases} 2x_1 + 3x_2 = 5 \\ x_1 + 4x_2 = 10 \end{cases}$

59. $\begin{cases} -2x_1 - 3x_2 = -4 \\ 6x_1 + x_2 = -36 \end{cases}$

60. $\begin{cases} -4x_1 + 9x_2 = -13 \\ x_1 - 3x_2 = 12 \end{cases}$

61. $\begin{cases} x_1 - 2x_2 + 3x_3 = 9 \\ -x_1 + 3x_2 - x_3 = -6 \\ 2x_1 - 5x_2 + 5x_3 = 17 \end{cases}$

62. $\begin{cases} x_1 + x_2 - 3x_3 = -1 \\ -x_1 + 2x_2 = 1 \\ -x_2 + x_3 = 0 \end{cases}$

63. $\begin{cases} x_1 - 5x_2 + 2x_3 = -20 \\ -3x_1 + x_2 - x_3 = 8 \\ -2x_2 + 5x_3 = -16 \end{cases}$

64. $\begin{cases} x_1 - x_2 + 4x_3 = 17 \\ x_1 + 3x_2 = -11 \\ -6x_2 + 5x_3 = 40 \end{cases}$

In Exercises 65–68, use the matrix capabilities of a graphing utility to find

$$f(A) = a_0 I_n + a_1 A + a_2 A^2 + \cdots + a_n A^n.$$

65. $f(x) = x^2 - 5x + 2, \quad A = \begin{bmatrix} 2 & 0 \\ 4 & 5 \end{bmatrix}$

66. $f(x) = x^2 - 7x + 6, \quad A = \begin{bmatrix} 5 & 4 \\ 1 & 2 \end{bmatrix}$

67. $f(x) = x^3 - 10x^2 + 31x - 30, \quad A = \begin{bmatrix} 3 & 1 & 4 \\ 0 & 2 & 6 \\ 0 & 0 & 5 \end{bmatrix}$

68. $f(x) = x^2 - 10x + 24, \quad A = \begin{bmatrix} 8 & -4 \\ 2 & 2 \end{bmatrix}$

Think About It In Exercises 69–78, use matrices A and B each of order 2×3, C of order 3×2, and D of order 2×2. Determine whether the matrices are of proper order to perform the operation(s). If so, give the order of the answer.

69. $A + 2C$ **70.** $B - 3C$

71. AB **72.** BC

73. $BC - D$ **74.** $CB - D$

75. $(CA)D$ **76.** $(BC)D$

77. $D(A - 3B)$ **78.** $(BC - D)A$

79. *Manufacturing* A certain corporation has three factories, each of which manufactures two products. The number of units of product i produced at factory j in one day is represented by a_{ij} in the matrix

$$A = \begin{bmatrix} 60 & 40 & 20 \\ 30 & 90 & 60 \end{bmatrix}.$$

Find the production levels if production is increased by 20%. (*Hint:* Because an increase of 20% corresponds to $100\% + 20\%$, multiply the given matrix by 1.2.)

80. *Manufacturing*   A certain corporation has four factories, each of which manufactures two products. The number of units of product i produced at factory j in one day is represented by a_{ij} in the matrix

$$A = \begin{bmatrix} 100 & 90 & 70 & 30 \\ 40 & 20 & 60 & 60 \end{bmatrix}.$$

Find the production levels if production is increased by 10%.

81. *Crop Production*   A fruit grower raises two crops, which are shipped to three outlets. The number of units of crop i that are shipped to outlet j is represented by a_{ij} in the matrix

$$A = \begin{bmatrix} 100 & 75 & 75 \\ 125 & 150 & 100 \end{bmatrix}.$$

The profit per unit is represented by the matrix

$$B = [\$3.75 \quad \$7.00].$$

Find the product BA and state what each entry of the product represents.

82. *Revenue*   A manufacturer produces three models of a product that are shipped to two warehouses. The number of units of model i that are shipped to warehouse j is represented by a_{ij} in the matrix

$$A = \begin{bmatrix} 5{,}000 & 4{,}000 \\ 6{,}000 & 10{,}000 \\ 8{,}000 & 5{,}000 \end{bmatrix}.$$

The price per unit is represented by the matrix

$$B = [\$20.50 \quad \$26.50 \quad \$29.50].$$

Compute BA and state what each entry of the product represents.

83. *Inventory Levels*   Five models of computers are sold through three retail outlets. The inventories are given by S. The wholesale and retail prices are given by T. Compute ST and interpret the result.

Model

$$S = \begin{matrix} & A \;\; B \;\; C \;\; D \;\; E \\ \begin{bmatrix} 3 & 2 & 2 & 3 & 0 \\ 0 & 2 & 3 & 4 & 3 \\ 4 & 2 & 1 & 3 & 2 \end{bmatrix} & \begin{matrix} 1 \\ 2 \\ 3 \end{matrix} \end{matrix} \right\} \text{Outlet}$$

Price

$$T = \begin{matrix} \text{Wholesale} \;\; \text{Retail} \\ \begin{bmatrix} \$840 & \$1100 \\ \$1200 & \$1350 \\ \$1450 & \$1650 \\ \$2650 & \$3000 \\ \$3050 & \$3200 \end{bmatrix} \begin{matrix} A \\ B \\ C \\ D \\ E \end{matrix} \right\} \text{Model}$$

84. *Labor/Wage Requirements*   A company that manufactures boats has the following labor-hour and wage requirements. Compute ST and interpret the result.

Labor per Boat

Department

$$S = \begin{matrix} \text{Cutting} \;\; \text{Assembly} \;\; \text{Packaging} \\ \begin{bmatrix} 1.0\,\text{hr} & 0.5\,\text{hr} & 0.2\,\text{hr} \\ 1.6\,\text{hr} & 1.0\,\text{hr} & 0.2\,\text{hr} \\ 2.5\,\text{hr} & 2.0\,\text{hr} & 0.4\,\text{hr} \end{bmatrix} \begin{matrix} \text{Small} \\ \text{Medium} \\ \text{Large} \end{matrix} \right\} \text{Boat size}$$

Wages per Hour

Plant

$$T = \begin{matrix} \text{A} \quad\;\; \text{B} \\ \begin{bmatrix} \$12 & \$10 \\ \$9 & \$8 \\ \$6 & \$5 \end{bmatrix} \begin{matrix} \text{Cutting} \\ \text{Assembly} \\ \text{Packaging} \end{matrix} \right\} \text{Department}$$

85. *Voting Preference*   The matrix

From

$$P = \begin{matrix} \text{R} \quad\;\; \text{D} \quad\;\; \text{I} \\ \begin{bmatrix} 0.6 & 0.1 & 0.1 \\ 0.2 & 0.7 & 0.1 \\ 0.2 & 0.2 & 0.8 \end{bmatrix} \begin{matrix} \text{R} \\ \text{D} \\ \text{I} \end{matrix} \right\} \text{To}$$

is called a stochastic matrix. Each entry $p_{ij}(i \neq j)$ represents the proportion of the voting population that changes from party i to party j, and p_{ii} represents the proportion that remains loyal to the party from one election to the next. Compute and interpret P^2.

86. *Voting Preference*   Use a graphing utility to find P^3, P^4, P^5, P^6, P^7, and P^8 for the matrix given in Exercise 85. Can you detect a pattern as P is raised to higher powers?

Synthesis

True or False?   **In Exercises 87–89, determine whether the statement is true or false. Justify your answer.**

87. Two matrices can be added only if they have the same order.

88. $\begin{bmatrix} -6 & -2 \\ 2 & -6 \end{bmatrix} \begin{bmatrix} 4 & 0 \\ 0 & -1 \end{bmatrix} = \begin{bmatrix} 4 & 0 \\ 0 & -1 \end{bmatrix} \begin{bmatrix} -6 & -2 \\ 2 & -6 \end{bmatrix}$

89. $\begin{bmatrix} -2 & 4 \\ -3 & 0 \\ 6 & 1 \end{bmatrix} \begin{bmatrix} 1 & 1 \\ 1 & 1 \end{bmatrix} = \begin{bmatrix} -2 & 4 \\ -3 & 0 \\ 6 & 1 \end{bmatrix}$

90. *Think About It* If a, b, and c are real numbers such that $c \neq 0$ and $ac = bc$, then $a = b$. However, if A, B, and C are nonzero matrices such that $AC = BC$, then A is *not necessarily* equal to B. Illustrate this using the following matrices.

$$A = \begin{bmatrix} 0 & 1 \\ 0 & 1 \end{bmatrix}, \quad B = \begin{bmatrix} 1 & 0 \\ 1 & 0 \end{bmatrix}, \quad C = \begin{bmatrix} 2 & 3 \\ 2 & 3 \end{bmatrix}$$

91. *Think About It* If a and b are real numbers such that $ab = 0$, then $a = 0$ or $b = 0$. However, if A and B are matrices such that $AB = 0$, it is *not necessarily* true that $A = 0$ or $B = 0$. Illustrate this using the following matrices.

$$A = \begin{bmatrix} 3 & 3 \\ 4 & 4 \end{bmatrix}, \quad B = \begin{bmatrix} 1 & -1 \\ -1 & 1 \end{bmatrix}$$

Exploration **In Exercises 92 and 93, let $i = \sqrt{-1}$.**

92. Consider the matrix

$$A = \begin{bmatrix} i & 0 \\ 0 & i \end{bmatrix}.$$

Find A^2, A^3, and A^4. Identify any similarities with i^2, i^3, and i^4.

93. Find and identify A^2 for the matrix

$$A = \begin{bmatrix} 0 & -i \\ i & 0 \end{bmatrix}.$$

94. *Exploration* Let A and B be unequal diagonal matrices of the same order. (A *diagonal matrix* is a square matrix in which each entry not on the main diagonal is zero.) Determine the products AB for several pairs of such matrices. Make a conjecture about a quick rule for such products.

95. *Exploration* Consider matrices of the form

$$A = \begin{bmatrix} 0 & a_{12} & a_{13} & a_{14} & \cdots & a_{1n} \\ 0 & 0 & a_{23} & a_{24} & \cdots & a_{2n} \\ 0 & 0 & 0 & a_{34} & \cdots & a_{3n} \\ \vdots & \vdots & \vdots & \vdots & \cdots & \vdots \\ 0 & 0 & 0 & 0 & \cdots & a_{(n-1)n} \\ 0 & 0 & 0 & 0 & \cdots & 0 \end{bmatrix}.$$

(a) Write a 2×2 matrix and a 3×3 matrix in the form of A.

(b) Use a graphing utility to raise each of the matrices to higher powers. Describe the result.

(c) Use the result of part (b) to make a conjecture about powers of A if A is a 4×4 matrix. Use a graphing utility to test your conjecture.

(d) Use the results of parts (b) and (c) to make a conjecture about powers of A if A is an $n \times n$ matrix.

Review

In Exercises 96–101, solve the equation.

96. $3x^2 + 20x - 32 = 0$ **97.** $8x^2 - 10x - 3 = 0$
98. $4x^2 + 10x - 3 = 0$ **99.** $3x^3 + 22x^2 - 45x = 0$
100. $3x^3 - 12x^2 + 5x - 20 = 0$
101. $2x^3 - 5x^2 - 12x + 30 = 0$

In Exercises 102–105, expand the logarithmic expression.

102. $\ln\left(\dfrac{100}{e^2}\right)$ **103.** $\log_2\left(\dfrac{12}{x}\right)$

104. $\ln\left[x^2(x-2)^3\right]$ **105.** $\ln\left(\dfrac{x^2-9}{x^4}\right)$

In Exercises 106–109, condense the logarithmic expression.

106. $3\ln 4 - \frac{1}{3}\ln(x^2+3)$
107. $\ln x - 3[\ln(x+6) + \ln(x-6)]$
108. $\frac{1}{2}[2\ln(x+5) + \ln x - \ln(x-8)]$
109. $\frac{3}{2}\ln 7t^4 - \frac{3}{5}\ln t^5$

In Exercises 110 and 111, solve the system of equations. Check your answer algebraically.

110. $\begin{cases} -x + 4y - 2z = 22 \\ 2x + 5y - 3z = 28 \\ 6x - y - z = -4 \end{cases}$

111. $\begin{cases} 3x - 8y + z = -8 \\ -2x - 4y - 3z = -11 \\ 4x + 9y - 5z = -10 \end{cases}$

8.3 The Inverse of a Square Matrix

The Inverse of a Matrix

This section further develops the algebra of matrices. To begin, consider the real number equation $ax = b$. To solve this equation for x, multiply each side of the equation by a^{-1} (provided that $a \neq 0$).

$$ax = b$$
$$(a^{-1}a)x = a^{-1}b$$
$$(1)x = a^{-1}b$$
$$x = a^{-1}b$$

The number a^{-1} is called the *multiplicative inverse of a* because $a^{-1}a = 1$. The definition of the multiplicative **inverse of a matrix** is similar.

Definition of the Inverse of a Square Matrix

Let A be an $n \times n$ matrix. If there exists a matrix A^{-1} such that

$$AA^{-1} = I_n = A^{-1}A$$

then A^{-1} is called the **inverse** of A. The symbol A^{-1} is read "*A* inverse."

EXAMPLE 1 The Inverse of a Matrix

Show that B is the inverse of A, where $A = \begin{bmatrix} -1 & 2 \\ -1 & 1 \end{bmatrix}$ and $B = \begin{bmatrix} 1 & -2 \\ 1 & -1 \end{bmatrix}$.

Solution

To show that B is the inverse of A, show that $AB = I = BA$, as follows.

$$AB = \begin{bmatrix} -1 & 2 \\ -1 & 1 \end{bmatrix}\begin{bmatrix} 1 & -2 \\ 1 & -1 \end{bmatrix}$$

$$= \begin{bmatrix} -1+2 & 2-2 \\ -1+1 & 2-1 \end{bmatrix}$$

$$= \begin{bmatrix} 1 & 0 \\ 0 & 1 \end{bmatrix}$$

$$BA = \begin{bmatrix} 1 & -2 \\ 1 & -1 \end{bmatrix}\begin{bmatrix} -1 & 2 \\ -1 & 1 \end{bmatrix}$$

$$= \begin{bmatrix} -1+2 & 2-2 \\ -1+1 & 2-1 \end{bmatrix}$$

$$= \begin{bmatrix} 1 & 0 \\ 0 & 1 \end{bmatrix}$$

Recall that it is not always true that $AB = BA$, even if both products are defined. However, if A and B are both square matrices and $AB = I_n$, it can be shown that $BA = I_n$. So, in Example 1, you need only to check that $AB = I_2$.

What You Should Learn:

- How to verify that two matrices are inverses of each other
- How to use Gauss-Jordan elimination to find inverses of matrices
- How to use a formula to find inverses of 2×2 matrices
- How to use inverse matrices to solve systems of linear equations

Why You Should Learn It:

Systems of equations can be solved using the inverse of the coefficient matrix, which is particularly useful when the coefficients are the same for several systems, but the constants are different. Exercises 69 and 70 on page 590 show how to use an inverse matrix to find unknown currents in electrical circuits.

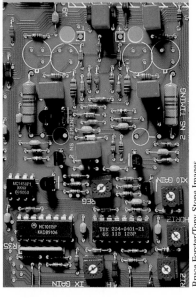

If a matrix A has an inverse, A is called **invertible** (or **nonsingular**); otherwise, A is called **singular.** A nonsquare matrix cannot have an inverse. To see this, note that if A is of order $m \times n$ and B is of order $n \times m$ (where $m \neq n$), the products AB and BA are of different orders and so cannot be equal to each other. Not all square matrices possess inverses, as you will see at the bottom of page 585. If, however, a matrix does have an inverse, that inverse is unique. Example 2 shows how to use systems of equations to find the inverse of a matrix.

EXAMPLE 2   Finding the Inverse of a Matrix

Find the inverse of

$$A = \begin{bmatrix} 1 & 4 \\ -1 & -3 \end{bmatrix}.$$

Solution

To find the inverse of A, try to solve the matrix equation $AX = I$ for X.

$$\underset{A}{\begin{bmatrix} 1 & 4 \\ -1 & -3 \end{bmatrix}} \underset{X}{\begin{bmatrix} x_{11} & x_{12} \\ x_{21} & x_{22} \end{bmatrix}} = \underset{I}{\begin{bmatrix} 1 & 0 \\ 0 & 1 \end{bmatrix}}$$

$$\begin{bmatrix} x_{11} + 4x_{21} & x_{12} + 4x_{22} \\ -x_{11} - 3x_{21} & -x_{12} - 3x_{22} \end{bmatrix} = \begin{bmatrix} 1 & 0 \\ 0 & 1 \end{bmatrix}$$

Equating corresponding entries, you obtain the following two systems of linear equations.

$$\begin{cases} x_{11} + 4x_{21} = 1 \\ -x_{11} - 3x_{21} = 0 \end{cases} \qquad \begin{cases} x_{12} + 4x_{22} = 0 \\ -x_{12} - 3x_{22} = 1 \end{cases}$$

Solve the first system using elementary row operations to determine that $x_{11} = -3$ and $x_{21} = 1$. From the second system you can determine that $x_{12} = -4$ and $x_{22} = 1$. Therefore, the inverse of A is

$$X = A^{-1}$$

$$= \begin{bmatrix} -3 & -4 \\ 1 & 1 \end{bmatrix}.$$

You can use matrix multiplication to check this result.

Check

$$AA^{-1} = \begin{bmatrix} 1 & 4 \\ -1 & -3 \end{bmatrix} \begin{bmatrix} -3 & -4 \\ 1 & 1 \end{bmatrix} = \begin{bmatrix} 1 & 0 \\ 0 & 1 \end{bmatrix} \checkmark$$

$$A^{-1}A = \begin{bmatrix} -3 & -4 \\ 1 & 1 \end{bmatrix} \begin{bmatrix} 1 & 4 \\ -1 & -3 \end{bmatrix} = \begin{bmatrix} 1 & 0 \\ 0 & 1 \end{bmatrix} \checkmark$$

Exploration

Most graphing utilities have the capability of finding the inverse of a square matrix. Try using a graphing utility to find the inverse of the matrix

$$A = \begin{bmatrix} 2 & -3 & 1 \\ -1 & 2 & -1 \\ -2 & 0 & 1 \end{bmatrix}.$$

After you find A^{-1}, store it as $[B]$ and use the graphing utility to find $[A] \times [B]$ and $[B] \times [A]$. What can you conclude?

Finding Inverse Matrices

In Example 2, note that the two systems of linear equations have the *same coefficient matrix A*. Rather than solve the two systems represented by

$$\begin{bmatrix} 1 & 4 & \vdots & 1 \\ -1 & -3 & \vdots & 0 \end{bmatrix}$$

and

$$\begin{bmatrix} 1 & 4 & \vdots & 0 \\ -1 & -3 & \vdots & 1 \end{bmatrix}$$

separately, you can solve them *simultaneously* by *adjoining* the identity matrix to the coefficient matrix to obtain

$$\begin{matrix} A & & I \\ \begin{bmatrix} 1 & 4 & \vdots & 1 & 0 \\ -1 & -3 & \vdots & 0 & 1 \end{bmatrix} \end{matrix}.$$

Then, applying Gauss-Jordan elimination to this matrix, you can solve *both* systems with a single elimination process.

$$\begin{bmatrix} 1 & 4 & \vdots & 1 & 0 \\ -1 & -3 & \vdots & 0 & 1 \end{bmatrix}$$

$$R_1 + R_2 \rightarrow \begin{bmatrix} 1 & 4 & \vdots & 1 & 0 \\ 0 & 1 & \vdots & 1 & 1 \end{bmatrix}$$

$$-4R_2 + R_1 \rightarrow \begin{bmatrix} 1 & 0 & \vdots & -3 & -4 \\ 0 & 1 & \vdots & 1 & 1 \end{bmatrix}$$

So, from the "doubly augmented" matrix $[A \vdots I]$, you obtained the matrix $[I \vdots A^{-1}]$.

$$\begin{matrix} A & I & I & A^{-1} \\ \begin{bmatrix} 1 & 4 & \vdots & 1 & 0 \\ -1 & -3 & \vdots & 0 & 1 \end{bmatrix} & \Longrightarrow & \begin{bmatrix} 1 & 0 & \vdots & -3 & -4 \\ 0 & 1 & \vdots & 1 & 1 \end{bmatrix} \end{matrix}$$

This procedure (or algorithm) works for any square matrix that has an inverse.

Exploration

Select two 2×2 matrices A and B that have inverses. Enter them into your graphing utility and calculate $(AB)^{-1}$. Then calculate $B^{-1}A^{-1}$ and $A^{-1}B^{-1}$. Make a conjecture about the inverse of a product of two invertible matrices.

Finding an Inverse Matrix

Let A be a square matrix of order n.

1. Write the $n \times 2n$ matrix that consists of the given matrix A on the left and the $n \times n$ identity matrix I on the right to obtain $[A \vdots I]$.

2. If possible, row reduce A to I using elementary row operations on the *entire* matrix $[A \vdots I]$. The result will be the matrix $[I \vdots A^{-1}]$. If this is not possible, A is not invertible.

3. Check your work by multiplying to see that $AA^{-1} = I = A^{-1}A$.

EXAMPLE 3   Finding the Inverse of a Matrix

Find the inverse of $A = \begin{bmatrix} 1 & -1 & 0 \\ 1 & 0 & -1 \\ 6 & -2 & -3 \end{bmatrix}$.

Solution

Begin by adjoining the identity matrix to A to form the matrix

$$[A \; \vdots \; I] = \begin{bmatrix} 1 & -1 & 0 & \vdots & 1 & 0 & 0 \\ 1 & 0 & -1 & \vdots & 0 & 1 & 0 \\ 6 & -2 & -3 & \vdots & 0 & 0 & 1 \end{bmatrix}.$$

Using elementary row operations to obtain the form $[I \; \vdots \; A^{-1}]$ results in

$$\begin{bmatrix} 1 & 0 & 0 & \vdots & -2 & -3 & 1 \\ 0 & 1 & 0 & \vdots & -3 & -3 & 1 \\ 0 & 0 & 1 & \vdots & -2 & -4 & 1 \end{bmatrix}.$$

Therefore, the matrix A is invertible and its inverse is

$$A^{-1} = \begin{bmatrix} -2 & -3 & 1 \\ -3 & -3 & 1 \\ -2 & -4 & 1 \end{bmatrix}.$$

Try using a graphing utility to confirm this result by multiplying A by A^{-1} to obtain I.

The process shown in Example 3 applies to any $n \times n$ matrix A. If A has an inverse, this process will find it. If A does not have an inverse, the process will tell you so. That is, matrix A will not reduce to the identity matrix. For instance, the following matrix has no inverse.

$$A = \begin{bmatrix} 1 & 2 & 0 \\ 3 & -1 & 2 \\ -2 & 3 & -2 \end{bmatrix}$$

To confirm that matrix A above has no inverse, begin by adjoining the identity matrix to A to form

$$[A \; \vdots \; I] = \begin{bmatrix} 1 & 2 & 0 & \vdots & 1 & 0 & 0 \\ 3 & -1 & 2 & \vdots & 0 & 1 & 0 \\ -2 & 3 & -2 & \vdots & 0 & 0 & 1 \end{bmatrix}.$$

Then use elementary row operations to obtain

$$\begin{bmatrix} 1 & 2 & 0 & \vdots & 1 & 0 & 0 \\ 0 & -7 & 2 & \vdots & -3 & 1 & 0 \\ 0 & 0 & 0 & \vdots & -1 & 1 & 1 \end{bmatrix}.$$

At this point in the elimination process you can see that it is impossible to obtain the identity matrix I on the left. Therefore, A is not invertible.

The Inverse of a 2 × 2 Matrix

Using Gauss-Jordan elimination to find the inverse of a matrix works well (even as a computer technique) for matrices of order 3 × 3 or greater. For 2 × 2 matrices, however, many people prefer to use a formula for the inverse rather than Gauss-Jordan elimination. This simple formula, which works *only* for 2 × 2 matrices, is explained as follows. If A is the 2 × 2 matrix

$$A = \begin{bmatrix} a & b \\ c & d \end{bmatrix}$$

then A is invertible if and only if $ad - bc \neq 0$. If $ad - bc \neq 0$, the inverse is

$$A^{-1} = \frac{1}{ad - bc} \begin{bmatrix} d & -b \\ -c & a \end{bmatrix}.$$

Try verifying this inverse by multiplication. The denominator $ad - bc$ is called the *determinant* of the 2 × 2 matrix A. You will study determinants in the next section.

> ### Exploration
>
> Use a graphing utility with matrix operations to find the inverse of the matrix
>
> $$A = \begin{bmatrix} 1 & -3 \\ -2 & 6 \end{bmatrix}.$$
>
> What message appears on the screen? Why does the graphing utility display this message?

EXAMPLE 4   Finding the Inverse of a 2 × 2 Matrix

If possible, find the inverse of the matrix.

a. $A = \begin{bmatrix} 3 & -1 \\ -2 & 2 \end{bmatrix}$

b. $B = \begin{bmatrix} 3 & -1 \\ -6 & 2 \end{bmatrix}$

Solution

a. For the matrix A, apply the formula for the inverse of a 2 × 2 matrix to obtain

$$ad - bc = (3)(2) - (-1)(-2)$$
$$= 4.$$

Because this quantity is not zero, the inverse is formed by interchanging the entries on the main diagonal, changing the signs of the other two entries, and multiplying by the scalar $\frac{1}{4}$, as follows.

$$A^{-1} = \frac{1}{4} \begin{bmatrix} 2 & 1 \\ 2 & 3 \end{bmatrix}$$

$$= \begin{bmatrix} \frac{1}{2} & \frac{1}{4} \\ \frac{1}{2} & \frac{3}{4} \end{bmatrix}$$

b. For the matrix B, you have

$$ad - bc = (3)(2) - (-1)(-6)$$
$$= 0$$

which means that B is not invertible.

Systems of Linear Equations

You know that a system of linear equations can have exactly one solution, infinitely many solutions, or no solution. If the coefficient matrix A of a *square* system (a system that has the same number of equations as variables) is invertible, the system has a unique solution, which is given as follows.

A System of Equations with a Unique Solution

If A is an invertible matrix, the system of linear equations represented by $AX = B$ has a unique solution

$$X = A^{-1}B.$$

The formula $X = A^{-1}B$ is used on most graphing utilities to solve linear systems that have invertible coefficient matrices. That is, you enter the $n \times n$ coefficient matrix $[A]$ and the $n \times 1$ column matrix $[B]$. The solution X is given by $[A]^{-1}[B]$.

EXAMPLE 5   Solving a System of Equations Using an Inverse

Use an inverse matrix to solve the system.

$$\begin{cases} 2x + 3y + z = -1 \\ 3x + 3y + z = 1 \\ 2x + 4y + z = -2 \end{cases}$$

Solution

Begin by writing the system as $AX = B$.

$$\begin{bmatrix} 2 & 3 & 1 \\ 3 & 3 & 1 \\ 2 & 4 & 1 \end{bmatrix} \begin{bmatrix} x \\ y \\ z \end{bmatrix} = \begin{bmatrix} -1 \\ 1 \\ -2 \end{bmatrix}$$

Then, use Gauss-Jordan elimination to find A^{-1}.

$$A^{-1} = \begin{bmatrix} -1 & 1 & 0 \\ -1 & 0 & 1 \\ 6 & -2 & -3 \end{bmatrix}$$

Finally, multiply B by A^{-1} on the left to obtain the solution.

$$X = A^{-1}B$$

$$= \begin{bmatrix} -1 & 1 & 0 \\ -1 & 0 & 1 \\ 6 & -2 & -3 \end{bmatrix} \begin{bmatrix} -1 \\ 1 \\ -2 \end{bmatrix}$$

$$= \begin{bmatrix} 2 \\ -1 \\ -2 \end{bmatrix}$$

So, the solution is $x = 2$, $y = -1$, and $z = -2$.

Use a graphing utility to verify A^{-1} for the system of equations in Example 5.

> ## STUDY T!P
>
> Remember that matrix multiplication is not commutative. So, you must multiply matrices in the correct order. For instance, in Example 5, you must multiply B by A^{-1} on the left.

8.3 Exercises

In Exercises 1–8, show that B is the inverse of A.

1. $A = \begin{bmatrix} 2 & 1 \\ 5 & 3 \end{bmatrix}, \quad B = \begin{bmatrix} 3 & -1 \\ -5 & 2 \end{bmatrix}$

2. $A = \begin{bmatrix} 1 & -1 \\ -1 & 2 \end{bmatrix}, \quad B = \begin{bmatrix} 2 & 1 \\ 1 & 1 \end{bmatrix}$

3. $A = \begin{bmatrix} 1 & 2 \\ 3 & 4 \end{bmatrix}, \quad B = \begin{bmatrix} -2 & 1 \\ \frac{3}{2} & -\frac{1}{2} \end{bmatrix}$

4. $A = \begin{bmatrix} 1 & -1 \\ 2 & 3 \end{bmatrix}, \quad B = \begin{bmatrix} \frac{3}{5} & \frac{1}{5} \\ -\frac{2}{5} & \frac{1}{5} \end{bmatrix}$

5. $A = \begin{bmatrix} 2 & -17 & 11 \\ -1 & 11 & -7 \\ 0 & 3 & -2 \end{bmatrix}, \quad B = \begin{bmatrix} 1 & 1 & 2 \\ 2 & 4 & -3 \\ 3 & 6 & -5 \end{bmatrix}$

6. $A = \begin{bmatrix} -4 & 1 & 5 \\ -1 & 2 & 4 \\ 0 & -1 & -1 \end{bmatrix}, \quad B = \begin{bmatrix} -\frac{1}{2} & 1 & \frac{3}{2} \\ \frac{1}{4} & -1 & -\frac{11}{4} \\ -\frac{1}{4} & 1 & \frac{7}{4} \end{bmatrix}$

7. $A = \begin{bmatrix} -2 & 2 & 3 \\ 1 & -1 & 0 \\ 0 & 1 & 4 \end{bmatrix}, \quad B = \frac{1}{3}\begin{bmatrix} -4 & -5 & 3 \\ -4 & -8 & 3 \\ 1 & 2 & 0 \end{bmatrix}$

8. $A = \begin{bmatrix} -1 & 1 & 0 & -1 \\ 1 & -1 & 1 & 0 \\ -1 & 1 & 2 & 0 \\ 0 & -1 & 1 & 1 \end{bmatrix}$,

$B = \frac{1}{3}\begin{bmatrix} -3 & 1 & 1 & -3 \\ -3 & -1 & 2 & -3 \\ 0 & 1 & 1 & 0 \\ -3 & -2 & 1 & 0 \end{bmatrix}$

In Exercises 9–12, use the matrix capabilities of a graphing utility to show that B is the inverse of A.

9. $A = \begin{bmatrix} -1 & -4 \\ 1 & 2 \end{bmatrix}, \quad B = \begin{bmatrix} 1 & 2 \\ -\frac{1}{2} & -\frac{1}{2} \end{bmatrix}$

10. $A = \begin{bmatrix} 11 & -12 \\ 2 & -2 \end{bmatrix}, \quad B = \begin{bmatrix} -1 & 6 \\ -1 & \frac{11}{2} \end{bmatrix}$

11. $A = \begin{bmatrix} 1.6 & 2 \\ -3.5 & -4.5 \end{bmatrix}, \quad B = \begin{bmatrix} 22.5 & 10 \\ -17.5 & -8 \end{bmatrix}$

12. $A = \begin{bmatrix} 4 & 0 & -2 \\ 1 & 2 & -4 \\ 0 & 3 & 1 \end{bmatrix}, \quad B = \begin{bmatrix} 0.28 & -0.12 & 0.08 \\ -0.02 & 0.08 & 0.28 \\ 0.06 & -0.24 & 0.16 \end{bmatrix}$

In Exercises 13–26, find the inverse of the matrix (if it exists).

13. $\begin{bmatrix} 2 & 0 \\ 0 & 3 \end{bmatrix}$

14. $\begin{bmatrix} 1 & 2 \\ 3 & 7 \end{bmatrix}$

15. $\begin{bmatrix} 1 & -2 \\ 2 & -3 \end{bmatrix}$

16. $\begin{bmatrix} -7 & 33 \\ 4 & -19 \end{bmatrix}$

17. $\begin{bmatrix} -1 & 1 \\ -2 & 1 \end{bmatrix}$

18. $\begin{bmatrix} 2 & 4 \\ 4 & 8 \end{bmatrix}$

19. $\begin{bmatrix} 2 & 7 & 1 \\ -3 & -9 & 2 \end{bmatrix}$

20. $\begin{bmatrix} -2 & 5 \\ 6 & -15 \\ 0 & 1 \end{bmatrix}$

21. $\begin{bmatrix} 1 & 1 & 1 \\ 3 & 5 & 4 \\ 3 & 6 & 5 \end{bmatrix}$

22. $\begin{bmatrix} 1 & 2 & 2 \\ 3 & 7 & 9 \\ -1 & -4 & -7 \end{bmatrix}$

23. $\begin{bmatrix} 1 & 0 & 0 \\ 3 & 4 & 0 \\ 2 & 5 & 5 \end{bmatrix}$

24. $\begin{bmatrix} 1 & 0 & 0 \\ 3 & 0 & 0 \\ 2 & 5 & 5 \end{bmatrix}$

25. $\begin{bmatrix} -8 & 0 & 0 & 0 \\ 0 & 1 & 0 & 0 \\ 0 & 0 & 4 & 0 \\ 0 & 0 & 0 & -5 \end{bmatrix}$

26. $\begin{bmatrix} 1 & 3 & -2 & 0 \\ 0 & 2 & 4 & 6 \\ 0 & 0 & -2 & 1 \\ 0 & 0 & 0 & 5 \end{bmatrix}$

In Exercises 27–38, use the matrix capabilities of a graphing utility to find the inverse of the matrix (if it exists).

27. $\begin{bmatrix} 1 & 2 & -1 \\ 3 & 7 & -10 \\ -5 & -7 & -15 \end{bmatrix}$

28. $\begin{bmatrix} 10 & 5 & -7 \\ -5 & 1 & 4 \\ 3 & 2 & -2 \end{bmatrix}$

29. $\begin{bmatrix} 1 & 1 & 2 \\ 3 & 1 & 0 \\ -2 & 0 & 3 \end{bmatrix}$

30. $\begin{bmatrix} 3 & 2 & 2 \\ 2 & 2 & 2 \\ -4 & 4 & 3 \end{bmatrix}$

31. $\begin{bmatrix} -\frac{1}{2} & \frac{3}{4} & \frac{1}{4} \\ 1 & 0 & -\frac{3}{2} \\ 0 & -1 & \frac{1}{2} \end{bmatrix}$

32. $\begin{bmatrix} -\frac{5}{6} & \frac{1}{3} & \frac{11}{6} \\ 0 & \frac{2}{3} & 2 \\ 1 & -\frac{1}{2} & -\frac{5}{2} \end{bmatrix}$

33. $\begin{bmatrix} 0.1 & 0.2 & 0.3 \\ -0.3 & 0.2 & 0.2 \\ 0.5 & 0.4 & 0.4 \end{bmatrix}$

34. $\begin{bmatrix} 0.6 & 0 & -0.3 \\ 0.7 & -1 & 0.2 \\ 1 & 0 & -0.9 \end{bmatrix}$

35. $\begin{bmatrix} 1 & 0 & 3 & 0 \\ 0 & 2 & 0 & 4 \\ 1 & 0 & 3 & 0 \\ 0 & 2 & 0 & 4 \end{bmatrix}$

36. $\begin{bmatrix} 4 & 8 & -7 & 14 \\ 2 & 5 & -4 & 6 \\ 0 & 2 & 1 & -7 \\ 3 & 6 & -5 & 10 \end{bmatrix}$

37. $\begin{bmatrix} -1 & 0 & 1 & 0 \\ 0 & 2 & 0 & -1 \\ 2 & 0 & -1 & 0 \\ 0 & -1 & 0 & 1 \end{bmatrix}$

38. $\begin{bmatrix} 1 & -2 & -1 & -2 \\ 3 & -5 & -2 & -3 \\ 2 & -5 & -2 & -5 \\ -1 & 4 & 4 & 11 \end{bmatrix}$

In Exercises 39–42, use the formula on page 586 to find the inverse of the 2 × 2 matrix.

39. $\begin{bmatrix} -4 & -6 \\ 2 & 3 \end{bmatrix}$ **40.** $\begin{bmatrix} -12 & 3 \\ 5 & -2 \end{bmatrix}$

41. $\begin{bmatrix} \frac{7}{2} & -\frac{3}{4} \\ \frac{1}{5} & \frac{4}{5} \end{bmatrix}$ **42.** $\begin{bmatrix} -\frac{1}{4} & \frac{9}{4} \\ \frac{5}{3} & \frac{8}{9} \end{bmatrix}$

In Exercises 43–46, use an inverse matrix to solve the system of linear equations. (Use the inverse matrix found in Exercise 15.)

43. $\begin{cases} x - 2y = 5 \\ 2x - 3y = 10 \end{cases}$ **44.** $\begin{cases} x - 2y = 0 \\ 2x - 3y = 3 \end{cases}$

45. $\begin{cases} x - 2y = 4 \\ 2x - 3y = 2 \end{cases}$ **46.** $\begin{cases} x - 2y = 1 \\ 2x - 3y = -2 \end{cases}$

In Exercises 47 and 48, use an inverse matrix to solve the system of linear equations. (Use the inverse matrix found in Exercise 21.)

47. $\begin{cases} x + y + z = 0 \\ 3x + 5y + 4z = 5 \\ 3x + 6y + 5z = 2 \end{cases}$ **48.** $\begin{cases} x + y + z = -1 \\ 3x + 5y + 4z = 2 \\ 3x + 6y + 5z = 0 \end{cases}$

In Exercises 49 and 50, use an inverse matrix and the matrix capabilities of a graphing utility to solve the system of linear equations. (Use the inverse matrix found in Exercise 38.)

49. $\begin{cases} x_1 - 2x_2 - x_3 - 2x_4 = 0 \\ 3x_1 - 5x_2 - 2x_3 - 3x_4 = 1 \\ 2x_1 - 5x_2 - 2x_3 - 5x_4 = -1 \\ -x_1 + 4x_2 + 4x_3 + 11x_4 = 2 \end{cases}$

50. $\begin{cases} x_1 - 2x_2 - x_3 - 2x_4 = 1 \\ 3x_1 - 5x_2 - 2x_3 - 3x_4 = -2 \\ 2x_1 - 5x_2 - 2x_3 - 5x_4 = 0 \\ -x_1 + 4x_2 + 4x_3 + 11x_4 = -3 \end{cases}$

In Exercises 51–58, use an inverse matrix (if it exists) to solve the system of linear equations.

51. $\begin{cases} 3x + 4y = -2 \\ 5x + 3y = 4 \end{cases}$ **52.** $\begin{cases} 18x + 12y = 13 \\ 30x + 24y = 23 \end{cases}$

53. $\begin{cases} -0.4x + 0.8y = 1.6 \\ 2x - 4y = 5 \end{cases}$ **54.** $\begin{cases} 0.2x - 0.6y = 2.4 \\ -x + 1.4y = -8.8 \end{cases}$

55. $\begin{cases} -\frac{1}{4}x + \frac{3}{8}y = -2 \\ \frac{3}{2}x + \frac{3}{4}y = -12 \end{cases}$ **56.** $\begin{cases} \frac{5}{6}x - y = -20 \\ \frac{4}{3}x - \frac{7}{2}y = -51 \end{cases}$

57. $\begin{cases} 4x - y + z = -5 \\ 2x + 2y + 3z = 10 \\ 5x - 2y + 6z = 1 \end{cases}$ **58.** $\begin{cases} 4x - 2y + 3z = -2 \\ 2x + 2y + 5z = 16 \\ 8x - 5y - 2z = 4 \end{cases}$

In Exercises 59–64, use the matrix capabilities of a graphing utility to solve (if possible) the system of linear equations.

59. $\begin{cases} 5x - 3y + 2z = 2 \\ 2x + 2y - 3z = 3 \\ -x + 7y - 8z = 4 \end{cases}$ **60.** $\begin{cases} 2x + 3y + 5z = 4 \\ 3x + 5y - 9z = 7 \\ 5x + 9y + 17z = 13 \end{cases}$

61. $\begin{cases} 3x - 2y + z = -29 \\ -4x + y - 3z = 37 \\ x - 5y + z = -24 \end{cases}$

62. $\begin{cases} -8x + 7y - 10z = -151 \\ 12x + 3y - 5z = 86 \\ 15x - 9y + 2z = 187 \end{cases}$

63. $\begin{cases} 7x - 3y + 2w = 41 \\ -2x + y - w = -13 \\ 4x + z - 2w = 12 \\ -x + y - w = -8 \end{cases}$

64. $\begin{cases} 2x + 5y + w = 11 \\ x + 4y + 2z - 2w = -7 \\ 2x - 2y + 5z + w = 3 \\ x - 3w = -1 \end{cases}$

Bond Investments In Exercises 65–68, consider a person who invests in AAA-rated bonds, A-rated bonds, and B-rated bonds. The average yields are 6.5% on AAA-bonds, 7% on A-bonds, and 9% on B-bonds. The person invests twice as much in B-bonds as in A-bonds. Let x, y, and z represent the amounts invested in AAA-, A-, and B-bonds, respectively.

$\begin{cases} x + y + z = \text{(total investment)} \\ 0.065x + 0.07y + 0.09z = \text{(annual return)} \\ 2y - z = 0 \end{cases}$

Use the inverse of the coefficient matrix of this system to find the amount invested in each type of bond.

65. Total investment = $25,000
Annual return = $1900

66. Total investment = $45,000
Annual return = $3750

67. Total investment = $12,000
Annual return = $835

68. Total investment = $500,000
Annual return = $38,000

Circuit Analysis   **In Exercises 69 and 70, consider the circuit in the figure below. The currents I_1, I_2, and I_3, in amperes, are given by the solution of the system of linear equations**

$$\begin{cases} 2I_1 \qquad + 4I_3 = E_1 \\ \quad\; I_2 + 4I_3 = E_2 \\ I_1 + I_2 - \; I_3 = \; 0 \end{cases}$$

where E_1 and E_2 are voltages. Use the inverse of the coefficient matrix of this system to find the unknown currents for the given voltages.

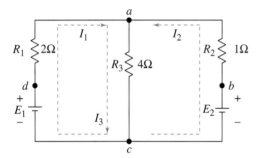

69. $E_1 = 14$ V, $E_2 = 28$ V

70. $E_1 = 10$ V, $E_2 = 10$ V

Synthesis

True or False?   **In Exercises 71–73, determine whether the statement is true or false. Justify your answer.**

71. Multiplication of an invertible matrix and its inverse is commutative.

72. If you multiply two matrices and obtain an identity matrix, you can assume the matrices are inverses of each other.

73. No nonsquare matrices have inverses.

74. *Writing*   Write a brief paragraph explaining the advantage of using an inverse matrix to solve the systems of linear equations in Exercises 43–50.

75. If A is a 2 × 2 matrix given by $A = \begin{bmatrix} a & b \\ c & d \end{bmatrix}$, then A is invertible if and only if $ad - bc \neq 0$. If $ad - bc \neq 0$, verify that the inverse is

$$A^{-1} = \frac{1}{ad - bc} \begin{bmatrix} d & -b \\ -c & a \end{bmatrix}.$$

Then use the result to find the inverse of each matrix.

(a) $\begin{bmatrix} 5 & -2 \\ 2 & 3 \end{bmatrix}$ (b) $\begin{bmatrix} 7 & 12 \\ -8 & -5 \end{bmatrix}$

76. *Exploration*   Consider the matrices of the form

$$A = \begin{bmatrix} a_{11} & 0 & 0 & 0 & \cdots & 0 \\ 0 & a_{22} & 0 & 0 & \cdots & 0 \\ 0 & 0 & a_{33} & 0 & \cdots & 0 \\ \vdots & \vdots & \vdots & \vdots & \cdots & \vdots \\ 0 & 0 & 0 & 0 & \cdots & a_{nn} \end{bmatrix}.$$

(a) Write a 2 × 2 matrix and a 3 × 3 matrix in the form of A. Find the inverse of each.

(b) Use the result of part (a) to make a conjecture about the inverse of a matrix of the form of A.

Review

In Exercises 77–80, find all the real solutions of the polynomial equation.

77. $x^3 + 5x^2 - 6x - 30 = 0$

78. $x^3 - 9x^2 + 14x + 24 = 0$

79. $x^4 - x^3 - 10x^2 - 2x - 24 = 0$

80. $x^4 + 2x^3 - 11x^2 - 12x = 0$

In Exercises 81–84, use a graphing utility to create a table of values for the function. Then sketch its graph by hand.

81. $f(x) = -3^{x-5} + 3$ **82.** $f(x) = 4^{-x-1} + 8$

83. $f(x) = 4 + e^{x-4}$ **84.** $f(x) = -2e^{0.6x} + 4$

In Exercises 85–90, perform the matrix operation.

85. $-3 \begin{bmatrix} -4 & 6 \\ 2 & -8 \\ 1 & 12 \end{bmatrix}$ **86.** $-\frac{1}{12} \begin{bmatrix} -6 & 2 & 0 \\ -4 & -24 & 18 \end{bmatrix}$

87. $\begin{bmatrix} 2 & -7 \\ -3 & -1 \end{bmatrix} - 5 \begin{bmatrix} -1 & 4 \\ 8 & -5 \end{bmatrix}$

88. $10 \begin{bmatrix} 2 & -3 \\ 1 & 0 \end{bmatrix} + \begin{bmatrix} 16 & 17 \\ -7 & 8 \end{bmatrix}$

89. $\begin{bmatrix} -2 & 0 \\ -1 & 4 \end{bmatrix} \begin{bmatrix} -3 & 1 \\ 3 & 2 \end{bmatrix}$

90. $\begin{bmatrix} 6 & -8 \\ 5 & 2 \\ -1 & 0 \end{bmatrix} \begin{bmatrix} 0 & -1 & 2 \\ 4 & -5 & -3 \end{bmatrix}$

8.4 The Determinant of a Square Matrix

The Determinant of a Matrix

Every *square* matrix can be associated with a real number called its **determinant.** Determinants have many uses, and several will be discussed in this and the next section. Historically, the use of determinants arose from special number patterns that occur when systems of linear equations are solved. For instance, the system

$$\begin{cases} a_1 x + b_1 y = c_1 \\ a_2 x + b_2 y = c_2 \end{cases}$$

has a solution

$$x = \frac{c_1 b_2 - c_2 b_1}{a_1 b_2 - a_2 b_1}$$

and

$$y = \frac{a_1 c_2 - a_2 c_1}{a_1 b_2 - a_2 b_1}$$

provided that $a_1 b_2 - a_2 b_1 \neq 0$. Note that the denominator of each fraction is the same. This denominator is called the *determinant* of the coefficient matrix of the system.

Why You Should Learn It:
Determinants are often used in other branches of mathematics. For instance, Exercises 53–58 on page 597 show some types of determinants that are useful in calculus.

What You Should Learn:
- How to find determinants of 2×2 matrices
- How to find minors and cofactors of square matrices
- How to find determinants of square matrices
- How to find determinants of triangular matrices

Coefficient Matrix   *Determinant*

$$A = \begin{bmatrix} a_1 & b_1 \\ a_2 & b_2 \end{bmatrix} \qquad \det(A) = a_1 b_2 - a_2 b_1$$

The determinant of the matrix A can also be denoted by vertical bars on both sides of the matrix, as indicated in the following definition.

Definition of the Determinant of a 2×2 Matrix

The **determinant** of the matrix

$$A = \begin{bmatrix} a_1 & b_1 \\ a_2 & b_2 \end{bmatrix}$$

is

$$\det(A) = |A| = \begin{vmatrix} a_1 & b_1 \\ a_2 & b_2 \end{vmatrix} = a_1 b_2 - a_2 b_1.$$

In this text, $\det(A)$ and $|A|$ are used interchangeably to represent the determinant of A. Although vertical bars are also used to denote the absolute value of a real number, the context will show which use is intended.

A convenient method for remembering the formula for the determinant of a 2×2 matrix is shown in the diagram.

$$\det(A) = \begin{vmatrix} a_1 & b_1 \\ a_2 & b_2 \end{vmatrix} = a_1 b_2 - a_2 b_1$$

Note that the determinant is the difference of the products of the two diagonals of the matrix.

EXAMPLE 1   The Determinant of a 2 × 2 Matrix

Find the determinant of each matrix.

a. $A = \begin{bmatrix} 2 & -3 \\ 1 & 2 \end{bmatrix}$ **b.** $B = \begin{bmatrix} 2 & 1 \\ 4 & 2 \end{bmatrix}$ **c.** $C = \begin{bmatrix} 0 & \frac{3}{2} \\ 2 & 4 \end{bmatrix}$

Solution

a. $\det(A) = \begin{vmatrix} 2 & -3 \\ 1 & 2 \end{vmatrix} = 2(2) - 1(-3)$

$$= 4 + 3$$

$$= 7$$

b. $\det(B) = \begin{vmatrix} 2 & 1 \\ 4 & 2 \end{vmatrix} = 2(2) - 4(1)$

$$= 4 - 4$$

$$= 0$$

c. $\det(C) = \begin{vmatrix} 0 & \frac{3}{2} \\ 2 & 4 \end{vmatrix} = 0(4) - 2\left(\frac{3}{2}\right)$

$$= 0 - 3$$

$$= -3$$

Notice in Example 1 that the determinant of a matrix can be positive, zero, or negative.

The determinant of a matrix of order 1×1 is defined simply as the entry of the matrix. For instance, if $A = [-2]$, $\det(A) = -2$.

Most graphing utilities can evaluate the determinant of a matrix. Consult your user's guide to find how to evaluate a determinant. For instance, you can evaluate the determinant of

$$A = \begin{bmatrix} 2 & -3 \\ 1 & 2 \end{bmatrix}$$

by entering the matrix as $[A]$ and then choosing the determinant feature. The result should be 7, as in Example 1(a). Try evaluating the determinant of

$$B = \begin{bmatrix} 3 & -1 & 1 \\ 0 & 2 & 1 \end{bmatrix}.$$

What happens when you try to evaluate the determinant of a nonsquare matrix?

Minors and Cofactors

To define the determinant of a square matrix of order 3×3 or higher, it is helpful to introduce the concepts of **minors** and **cofactors.**

> ### Minors and Cofactors of a Square Matrix
>
> If A is a square matrix, the **minor** M_{ij} of the entry a_{ij} is the determinant of the matrix obtained by deleting the ith row and jth column of A. The **cofactor** C_{ij} of the entry a_{ij} is
>
> $$C_{ij} = (-1)^{i+j}M_{ij}.$$

Sign Pattern for Cofactors

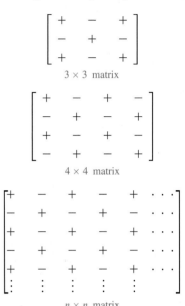

EXAMPLE 2   Finding the Minors and Cofactors of a Matrix

Find all the minors and cofactors of

$$A = \begin{bmatrix} 0 & 2 & 1 \\ 3 & -1 & 2 \\ 4 & 0 & 1 \end{bmatrix}.$$

Solution

To find the minor M_{11}, delete the first row and first column of A and evaluate the determinant of the resulting matrix.

$$\begin{bmatrix} 0 & 2 & 1 \\ 3 & -1 & 2 \\ 4 & 0 & 1 \end{bmatrix}$$

$$M_{11} = \begin{vmatrix} -1 & 2 \\ 0 & 1 \end{vmatrix} = -1(1) - 0(2) = -1$$

Similarly, to find M_{12}, delete the first row and second column.

$$\begin{bmatrix} 0 & 2 & 1 \\ 3 & -1 & 2 \\ 4 & 0 & 1 \end{bmatrix}$$

$$M_{12} = \begin{vmatrix} 3 & 2 \\ 4 & 1 \end{vmatrix} = 3(1) - 4(2) = -5$$

Continuing this pattern, you obtain the minors.

$$\begin{array}{lll} M_{11} = -1 & M_{12} = -5 & M_{13} = 4 \\ M_{21} = 2 & M_{22} = -4 & M_{23} = -8 \\ M_{31} = 5 & M_{32} = -3 & M_{33} = -6 \end{array}$$

Now, to find the cofactors, combine these minors with the checkerboard pattern of signs for a 3×3 matrix shown at the upper right.

$$\begin{array}{lll} C_{11} = -1 & C_{12} = 5 & C_{13} = 4 \\ C_{21} = -2 & C_{22} = -4 & C_{23} = 8 \\ C_{31} = 5 & C_{32} = 3 & C_{33} = -6 \end{array}$$

STUDY T!P

In the sign pattern for cofactors above, notice that *odd* positions (where $i + j$ is odd) have negative signs and *even* positions (where $i + j$ is even) have positive signs.

The Determinant of a Square Matrix

The following definition is called *inductive* because it uses determinants of matrices of order $n - 1$ to define the determinant of a matrix of order n.

Determinant of a Square Matrix

If A is a square matrix (of order 2×2 or greater), the determinant of A is the sum of the entries in any row (or column) of A multiplied by their respective cofactors. For instance, expanding along the first row yields

$$|A| = a_{11}C_{11} + a_{12}C_{12} + \cdots + a_{1n}C_{1n}.$$

Applying this definition to find a determinant is called **expanding by cofactors.**

Try checking that for a 2×2 matrix this definition yields

$$|A| = a_{11}a_{22} - a_{21}a_{12}$$

as previously defined.

EXAMPLE 3   The Determinant of a Matrix of Order 3 × 3

Find the determinant of $A = \begin{bmatrix} 0 & 2 & 1 \\ 3 & -1 & 2 \\ 4 & 0 & 1 \end{bmatrix}$.

Solution

Note that this is the same matrix that was in Example 2. There you found the cofactors of the entries in the first row to be

$$C_{11} = -1, \qquad C_{12} = 5, \qquad \text{and} \qquad C_{13} = 4.$$

So, by the definition of the determinant of a square matrix, you have

$$|A| = a_{11}C_{11} + a_{12}C_{12} + a_{13}C_{13} \qquad \text{First-row expansion}$$

$$= 0(-1) + 2(5) + 1(4)$$

$$= 14.$$

In Example 3, the determinant was found by expanding by the cofactors in the first row. You could have used any row or column. For instance, you could have expanded along the second row to obtain

$$|A| = a_{21}C_{21} + a_{22}C_{22} + a_{23}C_{23} \qquad \text{Second-row expansion}$$

$$= 3(-2) + (-1)(-4) + 2(8)$$

$$= 14.$$

When expanding by cofactors, you do not need to find cofactors of zero entries, because zero times its cofactor is zero.

$$a_{ij}C_{ij} = (0)C_{ij} = 0$$

So, the row (or column) containing the most zeros is usually the best choice for expansion by cofactors.

Triangular Matrices

Evaluating determinants of matrices of order 4 or higher can be tedious. There is, however, an important exception: the determinant of a **triangular** matrix. A triangular matrix is a square matrix with all zero entries either below or above its main diagonal. A square matrix is **upper triangular** if it has all zero entries below its main diagonal and **lower triangular** if it has all zero entries above its main diagonal. A matrix that is both upper and lower triangular is called **diagonal.** That is, a diagonal matrix is a square matrix in which all entries above and below the main diagonal are zero.

Upper Triangular Matrix

$$\begin{bmatrix} a_{11} & a_{12} & a_{13} & \cdots & a_{1n} \\ 0 & a_{22} & a_{23} & \cdots & a_{2n} \\ 0 & 0 & a_{33} & \cdots & a_{3n} \\ \vdots & \vdots & \vdots & & \vdots \\ 0 & 0 & 0 & \cdots & a_{nn} \end{bmatrix}$$

Lower Triangular Matrix

$$\begin{bmatrix} a_{11} & 0 & 0 & \cdots & 0 \\ a_{21} & a_{22} & 0 & \cdots & 0 \\ a_{31} & a_{32} & a_{33} & \cdots & 0 \\ \vdots & \vdots & \vdots & & \vdots \\ a_{n1} & a_{n2} & a_{n3} & \cdots & a_{nn} \end{bmatrix}$$

Diagonal Matrix

$$\begin{bmatrix} a_{11} & 0 & 0 & \cdots & 0 \\ 0 & a_{22} & 0 & \cdots & 0 \\ 0 & 0 & a_{33} & \cdots & 0 \\ \vdots & \vdots & \vdots & & \vdots \\ 0 & 0 & 0 & \cdots & a_{nn} \end{bmatrix}$$

To find the determinant of a triangular matrix of any order, simply form the product of the entries on the main diagonal.

> **Exploration**
>
> The formula for the determinant of a triangular matrix (discussed at the left) is only one of many properties of matrices. You can use a computer or calculator to discover other properties. For instance, how is $|cA|$ related to $|A|$? How are $|A|$ and $|B|$ related to $|AB|$?

EXAMPLE 4 The Determinant of a Triangular Matrix

a.
$$\begin{vmatrix} 2 & 0 & 0 & 0 \\ 4 & -2 & 0 & 0 \\ -5 & 6 & 1 & 0 \\ 1 & 5 & 3 & 3 \end{vmatrix} = (2)(-2)(1)(3) = -12$$

b.
$$\begin{vmatrix} -1 & 0 & 0 & 0 & 0 \\ 0 & 3 & 0 & 0 & 0 \\ 0 & 0 & 2 & 0 & 0 \\ 0 & 0 & 0 & 4 & 0 \\ 0 & 0 & 0 & 0 & -2 \end{vmatrix} = (-1)(3)(2)(4)(-2) = 48$$

Writing About Math *The Determinant of a Triangular Matrix*

Write an argument that explains why the determinant of a 3×3 triangular matrix is the product of its main-diagonal entries.

$$\begin{bmatrix} a_{11} & a_{12} & a_{13} \\ 0 & a_{22} & a_{23} \\ 0 & 0 & a_{33} \end{bmatrix} = a_{11} a_{22} a_{33}$$

8.4   E x e r c i s e s

In Exercises 1–12, find the determinant of the matrix.

1. $[7]$

2. $[-3]$

3. $\begin{bmatrix} 1 & 4 \\ 7 & 3 \end{bmatrix}$

4. $\begin{bmatrix} -3 & 1 \\ 5 & 2 \end{bmatrix}$

5. $\begin{bmatrix} 6 & 2 \\ -5 & 3 \end{bmatrix}$

6. $\begin{bmatrix} 3 & -3 \\ 4 & -8 \end{bmatrix}$

7. $\begin{bmatrix} -7 & 6 \\ \frac{1}{2} & 3 \end{bmatrix}$

8. $\begin{bmatrix} 4 & -3 \\ 0 & 0 \end{bmatrix}$

9. $\begin{bmatrix} 2 & -1 & 0 \\ 4 & 2 & 1 \\ 4 & 2 & 1 \end{bmatrix}$

10. $\begin{bmatrix} -2 & 2 & 3 \\ 1 & -1 & 0 \\ 0 & 1 & 4 \end{bmatrix}$

11. $\begin{bmatrix} -1 & 2 & -5 \\ 0 & 3 & 4 \\ 0 & 0 & 3 \end{bmatrix}$

12. $\begin{bmatrix} 1 & 0 & 0 \\ -4 & -1 & 0 \\ 5 & 1 & 5 \end{bmatrix}$

In Exercises 13 and 14, use the matrix capabilities of a graphing utility to find the determinant of the matrix.

13. $\begin{bmatrix} 0.3 & 0.2 & 0.2 \\ 0.2 & 0.2 & 0.2 \\ -0.4 & 0.4 & 0.3 \end{bmatrix}$

14. $\begin{bmatrix} 0.1 & 0.2 & 0.3 \\ -0.3 & 0.2 & 0.2 \\ 0.5 & 0.4 & 0.4 \end{bmatrix}$

In Exercises 15–18, find (a) all minors and (b) all cofactors of the matrix.

15. $\begin{bmatrix} 3 & 4 \\ 2 & -5 \end{bmatrix}$

16. $\begin{bmatrix} 11 & 0 \\ -3 & 2 \end{bmatrix}$

17. $\begin{bmatrix} 3 & -2 & 8 \\ 3 & 2 & -6 \\ -1 & 3 & 6 \end{bmatrix}$

18. $\begin{bmatrix} -2 & 9 & 4 \\ 7 & -6 & 0 \\ 6 & 7 & -6 \end{bmatrix}$

In Exercises 19–22, find the determinant of the matrix by the method of expansion by cofactors. Expand using the indicated row or column.

19. $\begin{bmatrix} -3 & 2 & 1 \\ 4 & 5 & 6 \\ 2 & -3 & 1 \end{bmatrix}$ (a) Row 1
(b) Column 2

20. $\begin{bmatrix} -3 & 4 & 2 \\ 6 & 3 & 1 \\ 4 & -7 & -8 \end{bmatrix}$ (a) Row 2
(b) Column 3

21. $\begin{bmatrix} 6 & 0 & -3 & 5 \\ 4 & 13 & 6 & -8 \\ -1 & 0 & 7 & 4 \\ 8 & 6 & 0 & 2 \end{bmatrix}$ (a) Row 2
(b) Column 2

22. $\begin{bmatrix} 10 & 8 & 3 & -7 \\ 4 & 0 & 5 & -6 \\ 0 & 3 & 2 & 7 \\ 1 & 0 & -3 & 2 \end{bmatrix}$ (a) Row 3
(b) Column 1

In Exercises 23–30, find the determinant of the matrix. Expand by cofactors on the row or column that appears to make the computations easiest.

23. $\begin{bmatrix} 1 & 4 & -2 \\ 3 & 2 & 0 \\ -1 & 4 & 3 \end{bmatrix}$

24. $\begin{bmatrix} 2 & -1 & 3 \\ 1 & 4 & 4 \\ 1 & 0 & 2 \end{bmatrix}$

25. $\begin{bmatrix} 2 & 4 & 6 \\ 0 & 3 & 1 \\ 0 & 0 & -5 \end{bmatrix}$

26. $\begin{bmatrix} -3 & 0 & 0 \\ 7 & 11 & 0 \\ 1 & 2 & 2 \end{bmatrix}$

27. $\begin{bmatrix} 2 & 6 & 6 & 2 \\ 2 & 7 & 3 & 6 \\ 1 & 5 & 0 & 1 \\ 3 & 7 & 0 & 7 \end{bmatrix}$

28. $\begin{bmatrix} 3 & 6 & -5 & 4 \\ -2 & 0 & 6 & 0 \\ 1 & 1 & 2 & 2 \\ 0 & 3 & -1 & -1 \end{bmatrix}$

29. $\begin{bmatrix} 3 & 2 & 4 & -1 & 5 \\ -2 & 0 & 1 & 3 & 2 \\ 1 & 0 & 0 & 4 & 0 \\ 6 & 0 & 2 & -1 & 0 \\ 3 & 0 & 5 & 1 & 0 \end{bmatrix}$

30. $\begin{bmatrix} 5 & 2 & 0 & 0 & -2 \\ 0 & 1 & 4 & 3 & 2 \\ 0 & 0 & 2 & 6 & 3 \\ 0 & 0 & 3 & 4 & 1 \\ 0 & 0 & 0 & 0 & 2 \end{bmatrix}$

In Exercises 31–34, evaluate the determinant. Do not use a graphing utility.

31. $\begin{vmatrix} 4 & 0 & 0 & 0 \\ 6 & -5 & 0 & 0 \\ 1 & 3 & 1 & 0 \\ 1 & -2 & 7 & 3 \end{vmatrix}$

32. $\begin{vmatrix} -1 & 3 & 5 & 0 \\ 0 & -10 & 9 & -6 \\ 0 & 0 & 7 & -1 \\ 0 & 0 & 0 & -2 \end{vmatrix}$

33. $\begin{vmatrix} 8 & 0 & 4 & 12 & -1 \\ 0 & -3 & -6 & -10 & 2 \\ 0 & 0 & -1 & 5 & -3 \\ 0 & 0 & 0 & 9 & 7 \\ 0 & 0 & 0 & 0 & 1 \end{vmatrix}$

34. $\begin{vmatrix} -2 & 0 & 0 & 0 & 0 \\ -1 & 4 & 0 & 0 & 0 \\ 3 & 5 & 1 & 0 & 0 \\ 6 & -11 & 8 & 10 & 0 \\ 0 & 13 & -9 & 0 & -3 \end{vmatrix}$

In Exercises 35–38, use the matrix capabilities of a graphing utility to evaluate the determinant.

35. $\begin{vmatrix} 1 & -1 & 8 & 4 \\ 2 & 6 & 0 & -4 \\ 2 & 0 & 2 & 6 \\ 0 & 2 & 8 & 0 \end{vmatrix}$ **36.** $\begin{vmatrix} 0 & -3 & 8 & 2 \\ 8 & 1 & -1 & 6 \\ -4 & 6 & 0 & 9 \\ -7 & 0 & 0 & 14 \end{vmatrix}$

37. $\begin{vmatrix} 3 & -2 & 4 & 3 & 1 \\ -1 & 0 & 2 & 1 & 0 \\ 5 & -1 & 0 & 3 & 2 \\ 4 & 7 & -8 & 0 & 0 \\ 1 & 2 & 3 & 0 & 2 \end{vmatrix}$

38. $\begin{vmatrix} -2 & 0 & 0 & 0 & 0 \\ 0 & 3 & 0 & 0 & 0 \\ 0 & 0 & -1 & 0 & 0 \\ 0 & 0 & 0 & 2 & 0 \\ 0 & 0 & 0 & 0 & -4 \end{vmatrix}$

In Exercises 39–42, find (a) $|A|$, **(b)** $|B|$, **(c)** AB, **and (d)** $|AB|$.

39. $A = \begin{bmatrix} -1 & 0 \\ 0 & 3 \end{bmatrix}, \quad B = \begin{bmatrix} 2 & 0 \\ 0 & -1 \end{bmatrix}$

40. $A = \begin{bmatrix} -2 & 1 \\ 4 & -2 \end{bmatrix}, \quad B = \begin{bmatrix} 1 & 2 \\ 0 & -1 \end{bmatrix}$

41. $A = \begin{bmatrix} -1 & 2 & 1 \\ 1 & 0 & 1 \\ 0 & 1 & 0 \end{bmatrix}, \quad B = \begin{bmatrix} -1 & 0 & 0 \\ 0 & 2 & 0 \\ 0 & 0 & 3 \end{bmatrix}$

42. $A = \begin{bmatrix} 2 & 0 & 1 \\ 1 & -1 & 2 \\ 3 & 1 & 0 \end{bmatrix}, \quad B = \begin{bmatrix} 2 & -1 & 4 \\ 0 & 1 & 3 \\ 3 & -2 & 1 \end{bmatrix}$

In Exercises 43 and 44, use the matrix capabilities of a graphing utility to find (a) $|A|$, **(b)** $|B|$, **(c)** AB, **and (d)** $|AB|$.

43. $A = \begin{bmatrix} 6 & 4 & 0 & 1 \\ 2 & -3 & -2 & -4 \\ 0 & 1 & 5 & 0 \\ -1 & 0 & -1 & 1 \end{bmatrix}$

$B = \begin{bmatrix} 0 & -5 & 0 & -2 \\ -2 & 4 & -1 & -4 \\ 3 & 0 & 1 & 0 \\ 1 & -2 & 3 & 0 \end{bmatrix}$

44. $A = \begin{bmatrix} -1 & 5 & 2 & 0 \\ 0 & 0 & 1 & 1 \\ 3 & -3 & -1 & 0 \\ 4 & 2 & 4 & -1 \end{bmatrix}, B = \begin{bmatrix} 1 & 5 & 0 & 0 \\ 10 & -1 & 2 & 4 \\ 2 & 0 & 0 & 1 \\ -3 & 2 & 5 & 0 \end{bmatrix}$

In Exercises 45–48, evaluate the determinants to verify the equation.

45. $\begin{vmatrix} w & x \\ y & z \end{vmatrix} = -\begin{vmatrix} y & z \\ w & x \end{vmatrix}$

46. $\begin{vmatrix} w & cx \\ y & cz \end{vmatrix} = c\begin{vmatrix} w & x \\ y & z \end{vmatrix}$

47. $\begin{vmatrix} w & x \\ y & z \end{vmatrix} = \begin{vmatrix} w & x + cw \\ y & z + cy \end{vmatrix}$

48. $\begin{vmatrix} w & x \\ cw & cx \end{vmatrix} = 0$

In Exercises 49 and 50, evaluate the determinant to verify the equation.

49. $\begin{vmatrix} 1 & x & x^2 \\ 1 & y & y^2 \\ 1 & z & z^2 \end{vmatrix} = (y - x)(z - x)(z - y)$

50. $\begin{vmatrix} a + b & a & a \\ a & a + b & a \\ a & a & a + b \end{vmatrix} = b^2(3a + b)$

In Exercises 51 and 52, solve for x.

51. $\begin{vmatrix} x - 1 & 2 \\ 3 & x - 2 \end{vmatrix} = 0$ **52.** $\begin{vmatrix} x - 2 & -1 \\ -3 & x \end{vmatrix} = 0$

In Exercises 53–58, evaluate the determinant, where the entries are functions. Determinants of this type occur in calculus.

53. $\begin{vmatrix} 4u & -1 \\ -1 & 2v \end{vmatrix}$ **54.** $\begin{vmatrix} 3x^2 & -3y^2 \\ 1 & 1 \end{vmatrix}$

55. $\begin{vmatrix} e^{2x} & e^{3x} \\ 2e^{2x} & 3e^{3x} \end{vmatrix}$ **56.** $\begin{vmatrix} e^{-x} & xe^{-x} \\ -e^{-x} & (1 - x)e^{-x} \end{vmatrix}$

57. $\begin{vmatrix} x & \ln x \\ 1 & 1/x \end{vmatrix}$ **58.** $\begin{vmatrix} x & x \ln x \\ 1 & 1 + \ln x \end{vmatrix}$

Synthesis

True or False? **In Exercises 59 and 60, determine whether the statement is true or false. Justify your answer.**

59. If a square matrix has an entire row of zeros, the determinant will always be zero.

60. If two columns of a square matrix are the same, then the determinant of the matrix will be zero.

61. ***Exploration*** Find square matrices A and B to demonstrate that

$$|A + B| \neq |A| + |B|.$$

62. ***Exploration*** Consider square matrices in which the entries are consecutive integers. An example of such a matrix is

$$\begin{bmatrix} 4 & 5 & 6 \\ 7 & 8 & 9 \\ 10 & 11 & 12 \end{bmatrix}.$$

Use a graphing utility to evaluate four determinants of this type. Make a conjecture based on the results. Then verify your conjecture.

63. ***Writing*** Write a brief paragraph explaining the difference between a square matrix and its determinant.

64. ***Think About It*** If A is a matrix of order 3×3 such that $|A| = 5$, is it possible to find $|2A|$? Explain.

Properties of Determinants **In Exercises 65–67, a property of determinants is given. State how the property has been applied to the given determinants and use a graphing utility to verify the results.**

65. If A and B are square matrices and B is obtained from A by interchanging two rows of A or interchanging two columns of A, then $|B| = -|A|$.

(a) $\begin{vmatrix} 1 & 3 & 4 \\ -7 & 2 & -5 \\ 6 & 1 & 2 \end{vmatrix} = - \begin{vmatrix} 1 & 4 & 3 \\ -7 & -5 & 2 \\ 6 & 2 & 1 \end{vmatrix}$

(b) $\begin{vmatrix} 1 & 3 & 4 \\ -2 & 2 & 0 \\ 1 & 6 & 2 \end{vmatrix} = - \begin{vmatrix} 1 & 6 & 2 \\ -2 & 2 & 0 \\ 1 & 3 & 4 \end{vmatrix}$

66. If A and B are square matrices and B is obtained from A by adding a multiple of a row of A to another row of A or by adding a multiple of a column of A to another column of A, then $|B| = |A|$.

(a) $\begin{vmatrix} 1 & -3 \\ 5 & 2 \end{vmatrix} = \begin{vmatrix} 1 & -3 \\ 0 & 17 \end{vmatrix}$

(b) $\begin{vmatrix} 5 & 4 & 2 \\ 2 & -3 & 4 \\ 7 & 6 & 3 \end{vmatrix} = \begin{vmatrix} 1 & 10 & -6 \\ 2 & -3 & 4 \\ 7 & 6 & 3 \end{vmatrix}$

67. If A and B are square matrices and B is obtained from A by multiplying a row of A by a nonzero constant c or multiplying a column of A by a nonzero constant c, then $|B| = c|A|$.

(a) $\begin{vmatrix} 5 & 10 & 15 \\ 2 & -3 & 4 \\ 2 & -7 & 1 \end{vmatrix} = 5 \begin{vmatrix} 1 & 2 & 3 \\ 2 & -3 & 4 \\ 2 & -7 & 1 \end{vmatrix}$

(b) $\begin{vmatrix} 1 & 8 & -3 \\ 3 & -12 & 6 \\ 7 & 4 & 9 \end{vmatrix} = 12 \begin{vmatrix} 1 & 2 & -1 \\ 3 & -3 & 2 \\ 7 & 1 & 3 \end{vmatrix}$

Review

In Exercises 68–71, solve the system of equations using substitution or elimination. Verify graphically.

68. $\begin{cases} 3x - 10y = 46 \\ x + y = -2 \end{cases}$

69. $\begin{cases} 5x + 7y = 23 \\ -4x - 2y = -4 \end{cases}$

70. $\begin{cases} -x + 9y = 11 \\ 6x - 3y = -47 \end{cases}$

71. $\begin{cases} 12x - 15y = 3 \\ -13x + 5y = -1 \end{cases}$

In Exercises 72 and 73, sketch the graph of the system of inequalities.

72. $\begin{cases} x + y \leq 8 \\ x \geq -3 \\ 2x - y \leq 4 \end{cases}$

73. $\begin{cases} -x - y \geq -4 \\ y \leq 1 \\ x + y \geq -4 \end{cases}$

In Exercises 74–77, use a graphing utility to find the inverse of the matrix (if it exists).

74. $\begin{bmatrix} -4 & 1 \\ 8 & -1 \end{bmatrix}$

75. $\begin{bmatrix} -5 & -8 \\ 3 & 6 \end{bmatrix}$

76. $\begin{bmatrix} -7 & 2 & 9 \\ 2 & -4 & -6 \\ 3 & 5 & 2 \end{bmatrix}$

77. $\begin{bmatrix} -6 & 2 & 0 \\ 1 & 3 & -2 \\ -2 & 0 & 1 \end{bmatrix}$

8.5 Applications of Matrices and Determinants

What You Should Learn:

- How to use determinants to find areas of triangles
- How to use determinants to decide whether points are collinear
- How to use Cramer's Rule to solve systems of linear equations
- How to use matrices to encode and decode messages

Why You Should Learn It:

A determinant can be used to find the area of a region of forest infected with gypsy moths, as shown in Exercise 27 on page 607.

Area of a Triangle

In this section, you will study some additional applications of matrices and determinants. The first involves a formula for finding the area of a triangle whose vertices are given by three points on a rectangular coordinate system.

Area of a Triangle

The area of a triangle with vertices (x_1, y_1), (x_2, y_2), and (x_3, y_3) is

$$\text{Area} = \pm\frac{1}{2}\begin{vmatrix} x_1 & y_1 & 1 \\ x_2 & y_2 & 1 \\ x_3 & y_3 & 1 \end{vmatrix}$$

where the symbol $(\pm)$ indicates that the appropriate sign should be chosen to yield a positive area.

EXAMPLE 1   Finding the Area of a Triangle

Find the area of a triangle whose vertices are $(1, 0)$, $(2, 2)$, and $(4, 3)$, as shown in Figure 8.2.

Solution

Let $(x_1, y_1) = (1, 0)$, $(x_2, y_2) = (2, 2)$, and $(x_3, y_3) = (4, 3)$. Then, to find the area of a triangle, evaluate the determinant

$$\begin{vmatrix} x_1 & y_1 & 1 \\ x_2 & y_2 & 1 \\ x_3 & y_3 & 1 \end{vmatrix} = \begin{vmatrix} 1 & 0 & 1 \\ 2 & 2 & 1 \\ 4 & 3 & 1 \end{vmatrix}$$

$$= 1(-1)^2\begin{vmatrix} 2 & 1 \\ 3 & 1 \end{vmatrix} + 0(-1)^3\begin{vmatrix} 2 & 1 \\ 4 & 1 \end{vmatrix} + 1(-1)^4\begin{vmatrix} 2 & 2 \\ 4 & 3 \end{vmatrix}$$

$$= 1(-1) + 0 + 1(-2)$$

$$= -3.$$

Using this value, you can conclude that the area of the triangle is

$$\text{Area} = -\frac{1}{2}\begin{vmatrix} 1 & 0 & 1 \\ 2 & 2 & 1 \\ 4 & 3 & 1 \end{vmatrix}$$

$$= -\frac{1}{2}(-3)$$

$$= \frac{3}{2} \text{ square units.}$$

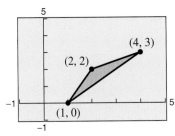

Figure 8.2

Collinear Points

What if the three points in Example 1 had been on the same line? What would have happened had the area formula been applied to three such points? The answer is that the determinant would have been zero. Consider, for instance, the three collinear points $(0, 1)$, $(2, 2)$, and $(4, 3)$, as shown in Figure 8.3. The area of the "triangle" that has these three points as vertices is

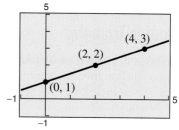

$$\frac{1}{2}\begin{vmatrix} 0 & 1 & 1 \\ 2 & 2 & 1 \\ 4 & 3 & 1 \end{vmatrix} = \frac{1}{2}\left[0(-1)^2\begin{vmatrix} 2 & 1 \\ 3 & 1 \end{vmatrix} + 1(-1)^3\begin{vmatrix} 2 & 1 \\ 4 & 1 \end{vmatrix} + 1(-1)^4\begin{vmatrix} 2 & 2 \\ 4 & 3 \end{vmatrix} \right]$$

$$= \frac{1}{2}[0 + 1(2) + 1(-2)]$$

$$= 0$$

Figure 8.3

This result is generalized as follows.

Test for Collinear Points

Three points (x_1, y_1), (x_2, y_2), and (x_3, y_3) are **collinear** (lie on the same line) if and only if

$$\begin{vmatrix} x_1 & y_1 & 1 \\ x_2 & y_2 & 1 \\ x_3 & y_3 & 1 \end{vmatrix} = 0.$$

EXAMPLE 2   Testing for Collinear Points

Determine whether the points $(-2, -2)$, $(1, 1)$, and $(7, 5)$ lie on the same line. (See Figure 8.4.)

Solution

Letting $(x_1, y_1) = (-2, -2)$, $(x_2, y_2) = (1, 1)$, and $(x_3, y_3) = (7, 5)$, you have

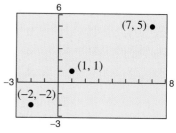

$$\begin{vmatrix} x_1 & y_1 & 1 \\ x_2 & y_2 & 1 \\ x_3 & y_3 & 1 \end{vmatrix} = \begin{vmatrix} -2 & -2 & 1 \\ 1 & 1 & 1 \\ 7 & 5 & 1 \end{vmatrix}$$

Figure 8.4

$$= -2(-1)^2\begin{vmatrix} 1 & 1 \\ 5 & 1 \end{vmatrix} + (-2)(-1)^3\begin{vmatrix} 1 & 1 \\ 7 & 1 \end{vmatrix} + 1(-1)^4\begin{vmatrix} 1 & 1 \\ 7 & 5 \end{vmatrix}$$

$$= -2(-4) + (-2)(6) + 1(-2)$$

$$= -6.$$

Because the value of this determinant is not zero, you can conclude that the three points do not lie on the same line.

Cramer's Rule

So far, you have studied three methods for solving a system of linear equations: substitution, elimination (with equations), and elimination (with matrices). You will now study one more method, **Cramer's Rule,** named after Gabriel Cramer (1704–1752). This rule uses determinants to write the solution of a system of linear equations. To see how Cramer's Rule works, take another look at the solution described at the beginning of Section 8.4. There, it was pointed out that the system

$$\begin{cases} a_1 x + b_1 y = c_1 \\ a_2 x + b_2 y = c_2 \end{cases}$$

has a solution

$$x = \frac{c_1 b_2 - c_2 b_1}{a_1 b_2 - a_2 b_1}$$

and

$$y = \frac{a_1 c_2 - a_2 c_1}{a_1 b_2 - a_2 b_1}$$

provided that $a_1 b_2 - a_2 b_1 \neq 0$. Each numerator and denominator in this solution can be expressed as a determinant, as follows.

$$x = \frac{c_1 b_2 - c_2 b_1}{a_1 b_2 - a_2 b_1} = \frac{\begin{vmatrix} c_1 & b_1 \\ c_2 & b_2 \end{vmatrix}}{\begin{vmatrix} a_1 & b_1 \\ a_2 & b_2 \end{vmatrix}}$$

$$y = \frac{a_1 c_2 - a_2 c_1}{a_1 b_2 - a_2 b_1} = \frac{\begin{vmatrix} a_1 & c_1 \\ a_2 & c_2 \end{vmatrix}}{\begin{vmatrix} a_1 & b_1 \\ a_2 & b_2 \end{vmatrix}}$$

Relative to the original system, the denominator for x and y is simply the determinant of the *coefficient* matrix of the system. This determinant is denoted by D. The numerators for x and y are denoted by D_x and D_y, respectively. They are formed by using the column of constants as replacements for the coefficients of x and y, as follows.

Coefficient Matrix	D	D_x	D_y
$\begin{bmatrix} a_1 & b_1 \\ a_2 & b_2 \end{bmatrix}$	$\begin{vmatrix} a_1 & b_1 \\ a_2 & b_2 \end{vmatrix}$	$\begin{vmatrix} c_1 & b_1 \\ c_2 & b_2 \end{vmatrix}$	$\begin{vmatrix} a_1 & c_1 \\ a_2 & c_2 \end{vmatrix}$

EXAMPLE 3   Using Cramer's Rule for a 2×2 System

Use Cramer's Rule to solve $\begin{cases} 4x - 2y = 10 \\ 3x - 5y = 11 \end{cases}$.

Solution

To begin, find the determinant of the coefficient matrix.

$$D = \begin{vmatrix} 4 & -2 \\ 3 & -5 \end{vmatrix} = -20 - (-6) = -14$$

Because this determinant is not zero, apply Cramer's Rule.

$$x = \frac{D_x}{D} = \frac{\begin{vmatrix} 10 & -2 \\ 11 & -5 \end{vmatrix}}{-14} = \frac{(-50) - (-22)}{-14} = \frac{-28}{-14} = 2$$

$$y = \frac{D_y}{D} = \frac{\begin{vmatrix} 4 & 10 \\ 3 & 11 \end{vmatrix}}{-14} = \frac{44 - 30}{-14} = \frac{14}{-14} = -1$$

Therefore, the solution is $x = 2$ and $y = -1$. Check this in the original system.

Cramer's Rule generalizes easily to systems of n equations in n variables. The value of each variable is given as the quotient of two determinants. The denominator is the determinant of the coefficient matrix, and the numerator is the determinant of the matrix formed by replacing the column corresponding to the variable (being solved for) with the column representing the constants. For instance, the solution for x_3 in the following system is shown.

$$\begin{array}{l} a_{11}x_1 + a_{12}x_2 + a_{13}x_3 = b_1 \\ a_{21}x_1 + a_{22}x_2 + a_{23}x_3 = b_2 \\ a_{31}x_1 + a_{32}x_2 + a_{33}x_3 = b_3 \end{array} \qquad x_3 = \frac{|A_3|}{|A|} = \frac{\begin{vmatrix} a_{11} & a_{12} & b_1 \\ a_{21} & a_{22} & b_2 \\ a_{31} & a_{32} & b_3 \end{vmatrix}}{\begin{vmatrix} a_{11} & a_{12} & a_{13} \\ a_{21} & a_{22} & a_{23} \\ a_{31} & a_{32} & a_{33} \end{vmatrix}}$$

Cramer's Rule

If a system of n linear equations in n variables has a coefficient matrix A with a *nonzero* determinant $|A|$, the solution is

$$x_1 = \frac{|A_1|}{|A|}, \quad x_2 = \frac{|A_2|}{|A|}, \quad \dots, \quad x_n = \frac{|A_n|}{|A|}$$

where the ith column of A_i is the column of constants in the system of equations. If the determinant of the coefficient matrix is zero, the system has either no solution or infinitely many solutions.

EXAMPLE 4　Using Cramer's Rule for a 3 × 3 System

Use Cramer's Rule, if possible, to solve the following system of linear equations.

$$\begin{cases} -x & + z = & 4 \\ 2x - y + & z = & -3 \\ & y - 3z = & 1 \end{cases}$$

Solution

Using the matrix capabilities of a graphing utility to evaluate the determinant of the coefficient matrix A, you find that Cramer's Rule cannot be applied because $|A| = 0$.

EXAMPLE 5　Using Cramer's Rule for a 3 × 3 System

Use Cramer's Rule, if possible, to solve the system of linear equations.

$$\begin{cases} -x + 2y - 3z = 1 \\ 2x \quad + z = 0 \\ 3x - 4y + 4z = 2 \end{cases}$$

A computer animation of this example appears in the *Interactive* CD-ROM and *Internet* versions of this text.

Solution

The coefficient matrix

$$\begin{bmatrix} -1 & 2 & -3 \\ 2 & 0 & 1 \\ 3 & -4 & 4 \end{bmatrix}$$

can be expanded along the second row, as follows.

$$D = 2(-1)^3 \begin{vmatrix} 2 & -3 \\ -4 & 4 \end{vmatrix} + 0(-1)^4 \begin{vmatrix} -1 & -3 \\ 3 & 4 \end{vmatrix} + 1(-1)^5 \begin{vmatrix} -1 & 2 \\ 3 & -4 \end{vmatrix}$$

$$= -2(-4) + 0 - 1(-2) = 10$$

Because this determinant is not zero, you can apply Cramer's Rule.

$$x = \frac{D_x}{D} = \frac{\begin{vmatrix} 1 & 2 & -3 \\ 0 & 0 & 1 \\ 2 & -4 & 4 \end{vmatrix}}{10} = \frac{8}{10} = \frac{4}{5}$$

$$y = \frac{D_y}{D} = \frac{\begin{vmatrix} -1 & 1 & -3 \\ 2 & 0 & 1 \\ 3 & 2 & 4 \end{vmatrix}}{10} = \frac{-15}{10} = -\frac{3}{2}$$

$$z = \frac{D_z}{D} = \frac{\begin{vmatrix} -1 & 2 & 1 \\ 2 & 0 & 0 \\ 3 & -4 & 2 \end{vmatrix}}{10} = \frac{-16}{10} = -\frac{8}{5}$$

The solution is $\left(\frac{4}{5}, -\frac{3}{2}, -\frac{8}{5} \right)$. Check this in the original system.

Cryptography

A **cryptogram** is a message written according to a secret code. (The Greek word *kryptos* means "hidden.") Matrix multiplication can be used to encode and decode messages. To begin, you need to assign a number to each letter in the alphabet (with 0 assigned to a blank space), as follows.

0 = _	9 = I	18 = R
1 = A	10 = J	19 = S
2 = B	11 = K	20 = T
3 = C	12 = L	21 = U
4 = D	13 = M	22 = V
5 = E	14 = N	23 = W
6 = F	15 = O	24 = X
7 = G	16 = P	25 = Y
8 = H	17 = Q	26 = Z

Then the message is converted to numbers and partitioned into **uncoded row matrices**, each having n entries, as demonstrated in Example 6.

EXAMPLE 6   Forming Uncoded Row Matrices

Write the uncoded row matrices of order 1×3 for the message
    MEET ME MONDAY.

Solution

Partitioning the message (including blank spaces, but ignoring punctuation) into groups of three produces the following uncoded row matrices.

$$\begin{bmatrix} 13 & 5 & 5 \end{bmatrix} \quad \begin{bmatrix} 20 & 0 & 13 \end{bmatrix} \quad \begin{bmatrix} 5 & 0 & 13 \end{bmatrix} \quad \begin{bmatrix} 15 & 14 & 4 \end{bmatrix} \quad \begin{bmatrix} 1 & 25 & 0 \end{bmatrix}$$
    M   E   E   T   M   E   M   O   N   D   A   Y

Note that a blank space is used to fill out the last uncoded row matrix.

To encode a message, choose an $n \times n$ invertible matrix A and multiply the uncoded row matrices by A (on the right) to obtain **coded row matrices**. Here is an example.

    Uncoded Matrix   Encoding Matrix A   Coded Matrix

$$\begin{bmatrix} 13 & 5 & 5 \end{bmatrix} \begin{bmatrix} 1 & -2 & 2 \\ -1 & 1 & 3 \\ 1 & -1 & -4 \end{bmatrix} = \begin{bmatrix} 13 & -26 & 21 \end{bmatrix}$$

This technique is further illustrated in Example 7.

EXAMPLE 7   Encoding a Message

Use the following matrix to encode the message MEET ME MONDAY.

$$A = \begin{bmatrix} 1 & -2 & 2 \\ -1 & 1 & 3 \\ 1 & -1 & -4 \end{bmatrix}$$

Solution

The coded row matrices are obtained by multiplying each of the uncoded row matrices found in Example 6 by the matrix A, as follows.

Uncoded Matrix   *Encoding Matrix A*   *Coded Matrix*

$$\begin{bmatrix} 13 & 5 & 5 \end{bmatrix} \begin{bmatrix} 1 & -2 & 2 \\ -1 & 1 & 3 \\ 1 & -1 & -4 \end{bmatrix} = \begin{bmatrix} 13 & -26 & 21 \end{bmatrix}$$

$$\begin{bmatrix} 20 & 0 & 13 \end{bmatrix} \begin{bmatrix} 1 & -2 & 2 \\ -1 & 1 & 3 \\ 1 & -1 & -4 \end{bmatrix} = \begin{bmatrix} 33 & -53 & -12 \end{bmatrix}$$

$$\begin{bmatrix} 5 & 0 & 13 \end{bmatrix} \begin{bmatrix} 1 & -2 & 2 \\ -1 & 1 & 3 \\ 1 & -1 & -4 \end{bmatrix} = \begin{bmatrix} 18 & -23 & -42 \end{bmatrix}$$

$$\begin{bmatrix} 15 & 14 & 4 \end{bmatrix} \begin{bmatrix} 1 & -2 & 2 \\ -1 & 1 & 3 \\ 1 & -1 & -4 \end{bmatrix} = \begin{bmatrix} 5 & -20 & 56 \end{bmatrix}$$

$$\begin{bmatrix} 1 & 25 & 0 \end{bmatrix} \begin{bmatrix} 1 & -2 & 2 \\ -1 & 1 & 3 \\ 1 & -1 & -4 \end{bmatrix} = \begin{bmatrix} -24 & 23 & 77 \end{bmatrix}$$

So, the sequence of coded row matrices is

$$\begin{bmatrix} 13 & -26 & 21 \end{bmatrix}\begin{bmatrix} 33 & -53 & -12 \end{bmatrix}\begin{bmatrix} 18 & -23 & -42 \end{bmatrix}\begin{bmatrix} 5 & -20 & 56 \end{bmatrix}\begin{bmatrix} -24 & 23 & 77 \end{bmatrix}.$$

Finally, removing the matrix notation produces the following cryptogram.

$$13 \ -26 \ 21 \ 33 \ -53 \ -12 \ 18 \ -23 \ -42 \ 5 \ -20 \ 56 \ -24 \ 23 \ 77$$

For those who do not know the matrix A, decoding the cryptogram found in Example 7 is difficult. But for an authorized receiver who knows the matrix A, decoding is simple. The receiver need only multiply the coded row matrices by A^{-1} (on the right) to retrieve the uncoded row matrices. Here is an example.

$$\underbrace{\begin{bmatrix} 13 & -26 & 21 \end{bmatrix}}_{\text{Coded}} \underbrace{\begin{bmatrix} -1 & -10 & -8 \\ -1 & -6 & -5 \\ 0 & -1 & -1 \end{bmatrix}}_{A^{-1}} = \underbrace{\begin{bmatrix} 13 & 5 & 5 \end{bmatrix}}_{\text{Uncoded}}$$

EXAMPLE 8   Decoding a Message

Use the inverse of the matrix

$$A = \begin{bmatrix} 1 & -2 & 2 \\ -1 & 1 & 3 \\ 1 & -1 & -4 \end{bmatrix}$$

to decode the cryptogram

13 −26 21 33 −53 −12 18 −23 −42 5 −20 56 −24 23 77.

Solution

Partition the message into groups of three to form the coded row matrices. Then multiply each coded row matrix by A^{-1} (on the right).

Coded Matrix   Decoding Matrix A^{-1} Decoded Matrix

$$\begin{bmatrix} 13 & -26 & 21 \end{bmatrix} \begin{bmatrix} -1 & -10 & -8 \\ -1 & -6 & -5 \\ 0 & -1 & -1 \end{bmatrix} = \begin{bmatrix} 13 & 5 & 5 \end{bmatrix}$$

$$\begin{bmatrix} 33 & -53 & -12 \end{bmatrix} \begin{bmatrix} -1 & -10 & -8 \\ -1 & -6 & -5 \\ 0 & -1 & -1 \end{bmatrix} = \begin{bmatrix} 20 & 0 & 13 \end{bmatrix}$$

$$\begin{bmatrix} 18 & -23 & -42 \end{bmatrix} \begin{bmatrix} -1 & -10 & -8 \\ -1 & -6 & -5 \\ 0 & -1 & -1 \end{bmatrix} = \begin{bmatrix} 5 & 0 & 13 \end{bmatrix}$$

$$\begin{bmatrix} 5 & -20 & 56 \end{bmatrix} \begin{bmatrix} -1 & -10 & -8 \\ -1 & -6 & -5 \\ 0 & -1 & -1 \end{bmatrix} = \begin{bmatrix} 15 & 14 & 4 \end{bmatrix}$$

$$\begin{bmatrix} -24 & 23 & 77 \end{bmatrix} \begin{bmatrix} -1 & -10 & -8 \\ -1 & -6 & -5 \\ 0 & -1 & -1 \end{bmatrix} = \begin{bmatrix} 1 & 25 & 0 \end{bmatrix}$$

So, the message is as follows.

$$\begin{bmatrix} 13 & 5 & 5 \end{bmatrix} \quad \begin{bmatrix} 20 & 0 & 13 \end{bmatrix} \quad \begin{bmatrix} 5 & 0 & 13 \end{bmatrix} \quad \begin{bmatrix} 15 & 14 & 4 \end{bmatrix} \quad \begin{bmatrix} 1 & 25 & 0 \end{bmatrix}$$

M E E T   M E   M O N D A Y

Writing About Math   *Cryptography*

Use your school's library, the Internet, or some other reference source to research a few current real-life uses of cryptography. Write a short summary of these uses. Include a description of how messages are encoded and decoded in each case.

8 Chapter Summary

What did you learn?

Section 8.1

	Review Exercises
☐ How to write matrices and identify their orders	1–12
☐ How to perform elementary row operations on matrices	13–18
☐ How to use matrices and Gaussian elimination to solve systems of linear equations	19–26
☐ How to use matrices and Gauss-Jordan elimination to solve systems of linear equations	27–36

Section 8.2

☐ How to decide whether two matrices are equal	37–40
☐ How to add and subtract matrices and multiply matrices by a real number	41–48
☐ How to solve matrix equations	49–52
☐ How to multiply two matrices	53–62
☐ How to use matrix operations to model and solve real-life problems	63, 64

Section 8.3

☐ How to verify that two matrices are inverses of each other	65–68
☐ How to use Gauss-Jordan elimination to find inverses of matrices	69–76
☐ How to use a formula to find inverses of 2×2 matrices	77–82
☐ How to use inverse matrices to solve systems of linear equations	83–96

Section 8.4

☐ How to find determinants of 2×2 matrices	97–100
☐ How to find minors and cofactors of square matrices	101–104
☐ How to find determinants of square matrices	105–114
☐ How to find determinants of triangular matrices	115, 116

Section 8.5

☐ How to use determinants to find areas of triangles	117–120
☐ How to use determinants to decide whether points are collinear	121–124
☐ How to use Cramer's Rule to solve systems of linear equations	125–140
☐ How to use matrices to encode and decode messages	141–144

8 Review Exercises

8.1 In Exercises 1–6, determine the order of the matrix.

1. $\begin{bmatrix} -3 \\ 1 \\ 10 \end{bmatrix}$

2. $\begin{bmatrix} 3 & -1 & 0 & 6 \\ -2 & 7 & 1 & 4 \end{bmatrix}$

3. $\begin{bmatrix} 14 \end{bmatrix}$

4. $\begin{bmatrix} 6 & 7 & -5 & 0 & -8 \end{bmatrix}$

5. $\begin{bmatrix} 8 & 4 \\ -1 & 6 \\ 0 & 11 \\ 12 & 9 \end{bmatrix}$

6. $\begin{bmatrix} -1 & 0 & 5 \\ 2 & 3 & 20 \end{bmatrix}$

In Exercises 7–10, form the augmented matrix for the system of linear equations.

7. $\begin{cases} 3x - 10y = 15 \\ 5x + 4y = 22 \end{cases}$

8. $\begin{cases} -x + y = 12 \\ 10x - 4y = -90 \end{cases}$

9. $\begin{cases} 8x - 7y + 4z = 12 \\ 3x - 5y + 2z = 20 \\ 5x + 3y - 3z = 26 \end{cases}$

10. $\begin{cases} 3x - 5y + z = 25 \\ -4x - 2z = -14 \\ 6x + y = 15 \end{cases}$

In Exercises 11 and 12, write the system of linear equations represented by the augmented matrix. (Use variables x, y, z, and w.)

11. $\begin{bmatrix} 5 & 1 & 7 & \vdots & -9 \\ 4 & 2 & 0 & \vdots & 10 \\ 9 & 4 & 2 & \vdots & 3 \end{bmatrix}$

12. $\begin{bmatrix} 13 & 16 & 7 & 3 & \vdots & 2 \\ 1 & 21 & 8 & 5 & \vdots & 12 \\ 4 & 10 & -4 & 3 & \vdots & -1 \end{bmatrix}$

In Exercises 13 and 14, write the matrix in reduced row-echelon form.

13. $\begin{bmatrix} 0 & 1 & 1 \\ 1 & 2 & 3 \\ 2 & 2 & 2 \end{bmatrix}$

14. $\begin{bmatrix} 1 & 1 & 1 & 0 \\ 1 & 1 & 0 & 1 \\ 1 & 0 & 1 & 1 \\ 0 & 1 & 1 & 1 \end{bmatrix}$

In Exercises 15–18, use the matrix capabilities of a graphing utility to write the matrix in reduced row-echelon form.

15. $\begin{bmatrix} 3 & -2 & 1 & 0 \\ 4 & -3 & 0 & 1 \end{bmatrix}$

16. $\begin{bmatrix} 1 & 1 & 2 & 1 & 0 & 0 \\ -1 & 0 & 3 & 0 & 1 & 0 \\ 1 & 2 & 8 & 0 & 0 & 1 \end{bmatrix}$

17. $\begin{bmatrix} 1 & 3 & 4 \\ 0 & 1 & 1 \\ 2 & 4 & 6 \end{bmatrix}$

18. $\begin{bmatrix} 4 & 8 & 16 \\ 3 & -1 & 2 \\ -2 & 10 & 12 \end{bmatrix}$

In Exercises 19–26, use Gaussian elimination with back-substitution to solve the system of equations.

19. $\begin{cases} 5x + 4y = 2 \\ -x + y = -22 \end{cases}$

20. $\begin{cases} 2x - 5y = 2 \\ 3x - 7y = 1 \end{cases}$

21. $\begin{cases} 2x + y = 0.3 \\ 3x - y = -1.3 \end{cases}$

22. $\begin{cases} 0.2x - 0.1y = 0.07 \\ 0.4x - 0.5y = -0.01 \end{cases}$

23. $\begin{cases} 2x + 3y + 3z = 3 \\ 6x + 6y + 12z = 13 \\ 12x + 9y - z = 2 \end{cases}$

24. $\begin{cases} x + 2y + 6z = 1 \\ 2x + 5y + 15z = 4 \\ 3x + y + 3z = -6 \end{cases}$

25. $\begin{cases} 3x + 21y - 29z = -1 \\ 2x + 15y - 21z = 0 \end{cases}$

26. $\begin{cases} x + 2y + w = 3 \\ -3y + 3z = 0 \\ 4x + 4y + z + 2w = 0 \\ 2x + z = 3 \end{cases}$

In Exercises 27–30, use Gauss-Jordan elimination to solve the system of equations.

27. $\begin{cases} -x + y + 2z = 1 \\ 2x + 3y + z = -2 \\ 5x + 4y + 2z = 4 \end{cases}$

28. $\begin{cases} 4x + 4y + 4z = 5 \\ 4x - 2y - 8z = 1 \\ 5x + 3y + 8z = 6 \end{cases}$

29. $\begin{cases} 2x - y + 9z = -8 \\ -x - 3y + 4z = -15 \\ 5x + 2y - z = 17 \end{cases}$

30. $\begin{cases} -3x + y + 7z = -20 \\ 5x - 2y - z = 34 \\ -x + y + 4z = -8 \end{cases}$

In Exercises 31–36, use the matrix capabilities of a graphing utility to reduce the augmented matrix and solve the system of equations.

31. $\begin{cases} x - 3y = -2 \\ x + y = 2 \end{cases}$ **32.** $\begin{cases} -x + 3y = 5 \\ 4x - y = 2 \end{cases}$

33. $\begin{cases} x + 2y - z = 7 \\ - y - z = 4 \\ 4x - z = 16 \end{cases}$ **34.** $\begin{cases} 3x + 6z = 0 \\ -2x + y = 5 \\ y + 2z = 3 \end{cases}$

35. $\begin{cases} 3x - y + 5z - 2w = -44 \\ x + 6y + 4z - w = 1 \\ 5x - y + z + 3w = -15 \\ 4y - z - 8w = 58 \end{cases}$

36. $\begin{cases} 4x + 12y + 2z = 20 \\ x + 6y + 4z = 12 \\ x + 6y + z = 8 \\ -2x - 10y - 2z = -10 \end{cases}$

8.2 In Exercises 37–40, find *x* and *y*.

37. $\begin{bmatrix} -1 & x \\ y & 9 \end{bmatrix} = \begin{bmatrix} -1 & 12 \\ -7 & 9 \end{bmatrix}$

38. $\begin{bmatrix} -1 & 0 \\ x & 5 \\ -4 & y \end{bmatrix} = \begin{bmatrix} -1 & 0 \\ 8 & 5 \\ -4 & 0 \end{bmatrix}$

39. $\begin{bmatrix} x+3 & 4 & -4y \\ 0 & -3 & 2 \\ -2 & y+5 & 6x \end{bmatrix} = \begin{bmatrix} 5x-1 & 4 & -44 \\ 0 & -3 & 2 \\ -2 & 16 & 6 \end{bmatrix}$

40. $\begin{bmatrix} -9 & 4 & 2 & -5 \\ 0 & -3 & 7 & -4 \\ 6 & -1 & 1 & 0 \end{bmatrix} = \begin{bmatrix} -9 & 4 & x-10 & -5 \\ 0 & -3 & 7 & 2y \\ \frac{1}{2}x & -1 & 1 & 0 \end{bmatrix}$

In Exercises 41–46, perform the matrix operations. If it is not possible, explain why.

41. $\begin{bmatrix} 7 & 3 \\ -1 & 5 \end{bmatrix} + \begin{bmatrix} 10 & -20 \\ 14 & -3 \end{bmatrix}$

42. $\begin{bmatrix} -11 & 16 & 19 \\ -7 & -2 & 1 \end{bmatrix} - \begin{bmatrix} 6 & 0 \\ 8 & -4 \\ -2 & 10 \end{bmatrix}$

43. $\begin{bmatrix} 2 & 1 & 0 \\ 0 & 5 & -4 \end{bmatrix} - 3\begin{bmatrix} 5 & 3 & -6 \\ 0 & -2 & 5 \end{bmatrix}$

44. $-2\begin{bmatrix} 1 & 2 \\ 5 & -4 \\ 6 & 0 \end{bmatrix} + 8\begin{bmatrix} 7 & 1 \\ 1 & 2 \\ 1 & 4 \end{bmatrix}$

45. $-\begin{bmatrix} 8 & -1 & 8 \\ -2 & 4 & 12 \\ 0 & -6 & 0 \end{bmatrix} - 5\begin{bmatrix} -2 & 0 & -4 \\ 3 & -1 & 1 \\ 6 & 12 & -8 \end{bmatrix}$

46. $6\begin{bmatrix} -4 & -1 & -3 & 4 \\ 2 & -5 & 7 & -10 \end{bmatrix} + 2\begin{bmatrix} -1 & 1 & 13 & -7 \\ 14 & -3 & 8 & -1 \end{bmatrix}$

In Exercises 47 and 48, use a graphing utility to perform the matrix operations.

47. $3\begin{bmatrix} 8 & -2 & 5 \\ 1 & 3 & -1 \end{bmatrix} + 6\begin{bmatrix} 4 & -2 & -3 \\ 2 & 7 & 6 \end{bmatrix}$

48. $-5\begin{bmatrix} 2 & 0 \\ 7 & -2 \\ 8 & 2 \end{bmatrix} + 4\begin{bmatrix} 4 & -2 \\ 6 & 11 \\ -1 & 3 \end{bmatrix}$

In Exercises 49–52, solve for *X* given

$$A = \begin{bmatrix} -4 & 0 \\ 1 & -5 \\ -3 & 2 \end{bmatrix} \quad \text{and} \quad B = \begin{bmatrix} 1 & 2 \\ -2 & 1 \\ 4 & 4 \end{bmatrix}.$$

49. $X = 3A - 2B$ **50.** $6X = 4A + 3B$

51. $3X + 2A = B$ **52.** $2A - 5B = 3X$

In Exercises 53–58, multiply the matrices. If it is not possible, explain why.

53. $\begin{bmatrix} 1 & 2 \\ 5 & -4 \\ 6 & 0 \end{bmatrix}\begin{bmatrix} 6 & -2 & 8 \\ 4 & 0 & 0 \end{bmatrix}$

54. $\begin{bmatrix} 1 & 5 & 6 \\ 2 & -4 & 0 \end{bmatrix}\begin{bmatrix} 6 & -2 & 8 \\ 4 & 0 & 0 \end{bmatrix}$

55. $\begin{bmatrix} 1 & 5 & 6 \\ 2 & -4 & 0 \end{bmatrix}\begin{bmatrix} 6 & 4 \\ -2 & 0 \\ 8 & 0 \end{bmatrix}$

56. $\begin{bmatrix} 1 & 3 & 2 \\ 0 & 2 & -4 \\ 0 & 0 & 3 \end{bmatrix}\begin{bmatrix} 4 & -3 & 2 \\ 0 & 3 & -1 \\ 0 & 0 & 2 \end{bmatrix}$

57. $\begin{bmatrix} 2 & 1 \\ 6 & 0 \end{bmatrix}\left(\begin{bmatrix} 4 & 2 \\ -3 & 1 \end{bmatrix} + \begin{bmatrix} -2 & 4 \\ 0 & 4 \end{bmatrix}\right)$

58. $\begin{bmatrix} 1 & -1 \\ 4 & 2 \end{bmatrix}\left(\begin{bmatrix} 0 & 3 \\ 1 & 2 \end{bmatrix}\begin{bmatrix} 1 & 0 \\ 5 & -3 \end{bmatrix}\right)$

In Exercises 59 and 60, use a graphing utility to multiply the matrices.

59. $\begin{bmatrix} 4 & 1 \\ 11 & -7 \\ 12 & 3 \end{bmatrix} \begin{bmatrix} 3 & -5 & 6 \\ 2 & -2 & -2 \end{bmatrix}$

60. $\begin{bmatrix} -2 & 3 & 10 \\ 4 & -2 & 2 \end{bmatrix} \begin{bmatrix} 1 & 1 \\ -5 & 2 \\ 3 & 2 \end{bmatrix}$

61. Write the system of linear equations represented by the matrix equation

$$\begin{bmatrix} 5 & 4 \\ -1 & 1 \end{bmatrix} \begin{bmatrix} x \\ y \end{bmatrix} = \begin{bmatrix} 2 \\ -22 \end{bmatrix}.$$

62. Write the matrix equation $AX = B$ for the following system of linear equations.

$$\begin{cases} 2x + 3y + z = 10 \\ 2x - 3y - 3z = 22 \\ 4x - 2y + 3z = -2 \end{cases}$$

63. *Manufacturing* A manufacturing company produces three models of a product that are shipped to two warehouses. The number of units of model i that are shipped to warehouse j is represented by a_{ij} in the matrix

$$A = \begin{bmatrix} 8200 & 7400 \\ 6500 & 9800 \\ 5400 & 4800 \end{bmatrix}.$$

The price per unit is represented by the matrix $B = [\$10.25 \quad \$14.50 \quad \$17.75]$.

Use a graphing utility to compute BA and interpret the result.

64. Suppose in Exercise 63 that the number of units of each model shipped to the warehouse is increased by 25% for the following shipment. Use your graphing utility to find a new matrix A_n representing the number of units being shipped to the warehouses. Then calculate BA_n.

8.3 **In Exercises 65–68, show that B is the inverse of A.**

65. $A = \begin{bmatrix} -4 & -1 \\ 7 & 2 \end{bmatrix}, \quad B = \begin{bmatrix} -2 & -1 \\ 7 & 4 \end{bmatrix}$

66. $A = \begin{bmatrix} 5 & -1 \\ 11 & -2 \end{bmatrix}, \quad B = \begin{bmatrix} -2 & 1 \\ -11 & 5 \end{bmatrix}$

67. $A = \begin{bmatrix} 1 & 1 & 0 \\ 1 & 0 & 1 \\ 6 & 2 & 3 \end{bmatrix}, \quad B = \begin{bmatrix} -2 & -3 & 1 \\ 3 & 3 & -1 \\ 2 & 4 & -1 \end{bmatrix}$

68. $A = \begin{bmatrix} 1 & -1 & 0 \\ -1 & 0 & -1 \\ 8 & -4 & 2 \end{bmatrix}, \quad B = \begin{bmatrix} -2 & 1 & \frac{1}{2} \\ -3 & 1 & \frac{1}{2} \\ 2 & -2 & -\frac{1}{2} \end{bmatrix}$

In Exercises 69–72, use Gauss-Jordan elimination to find the inverse of the matrix (if it exists).

69. $\begin{bmatrix} -6 & 5 \\ -5 & 4 \end{bmatrix}$ **70.** $\begin{bmatrix} -3 & -5 \\ 2 & 3 \end{bmatrix}$

71. $\begin{bmatrix} -1 & -2 & -2 \\ 3 & 7 & 9 \\ 1 & 4 & 7 \end{bmatrix}$ **72.** $\begin{bmatrix} 0 & -2 & 1 \\ -5 & -2 & -3 \\ 7 & 3 & 4 \end{bmatrix}$

In Exercises 73–76, use a graphing utility to find the inverse of the matrix (if it exists).

73. $\begin{bmatrix} 2 & 6 \\ 3 & -6 \end{bmatrix}$ **74.** $\begin{bmatrix} 3 & -10 \\ 4 & 2 \end{bmatrix}$

75. $\begin{bmatrix} 2 & 0 & 3 \\ -1 & 1 & 1 \\ 2 & -2 & 1 \end{bmatrix}$ **76.** $\begin{bmatrix} 1 & 4 & 6 \\ 2 & -3 & 1 \\ -1 & 18 & 16 \end{bmatrix}$

In Exercises 77–82, use the formula on page 586 to find the inverse of the matrix (if it exists).

77. $\begin{bmatrix} -7 & 2 \\ -8 & 2 \end{bmatrix}$ **78.** $\begin{bmatrix} 10 & 4 \\ 7 & 3 \end{bmatrix}$

79. $\begin{bmatrix} -6 & -5 \\ 3 & 3 \end{bmatrix}$ **80.** $\begin{bmatrix} 10 & 45 \\ 2 & 9 \end{bmatrix}$

81. $\begin{bmatrix} -1 & 20 \\ \frac{3}{10} & -6 \end{bmatrix}$ **82.** $\begin{bmatrix} -\frac{3}{4} & \frac{5}{2} \\ -\frac{4}{5} & -\frac{8}{3} \end{bmatrix}$

In Exercises 83–90, use an inverse matrix (if it exists) to solve the system of linear equations.

83. $\begin{cases} -x + 4y = 8 \\ 2x - 7y = -5 \end{cases}$ **84.** $\begin{cases} 5x - y = 13 \\ -9x + 2y = -24 \end{cases}$

85. $\begin{cases} -3x + 10y = 8 \\ 5x - 17y = -13 \end{cases}$ **86.** $\begin{cases} 4x - 2y = -10 \\ -19x + 9y = 47 \end{cases}$

87. $\begin{cases} 3x + 2y - z = 6 \\ x - y + 2z = -1 \\ 5x + y + z = 7 \end{cases}$ **88.** $\begin{cases} -x + 4y - 2z = 12 \\ 2x - 9y + 5z = -25 \\ -x + 5y - 4z = 10 \end{cases}$

89. $\begin{cases} -2x + y + 2z = -13 \\ -x - 4y + z = -11 \\ -y - z = 0 \end{cases}$

90. $\begin{cases} 3x - y + 5z = -14 \\ -x + y + 6z = 8 \\ -8x + 4y - z = 44 \end{cases}$

In Exercises 91–96, use a graphing utility to solve the system of linear equations using the inverse of the coefficient matrix (if it exists).

91. $\begin{cases} x + 2y = -1 \\ 3x + 4y = -5 \end{cases}$ **92.** $\begin{cases} x + 3y = 23 \\ -6x + 2y = -18 \end{cases}$

93. $\begin{cases} -3x - 3y - 4z = 2 \\ \qquad y + z = -1 \\ 4x + 3y + 4z = -1 \end{cases}$

94. $\begin{cases} x + 3y + 2z = 2 \\ -2x - 5y - z = 10 \\ 2x + 4y = -12 \end{cases}$

95. $\begin{cases} -x + y + z = 6 \\ 4x - 3y + z = 20 \\ 2x - y + 3z = 8 \end{cases}$

96. $\begin{cases} 2x + 3y - 4z = 1 \\ x - y + 2z = -4 \\ 3x + 7y - 10z = 0 \end{cases}$

8.4   **In Exercises 97–100, find the determinant of the matrix.**

97. $\begin{bmatrix} 8 & 5 \\ 2 & -4 \end{bmatrix}$ **98.** $\begin{bmatrix} -9 & 11 \\ 7 & -4 \end{bmatrix}$

99. $\begin{bmatrix} 50 & -30 \\ 10 & 5 \end{bmatrix}$ **100.** $\begin{bmatrix} 14 & -24 \\ 12 & -15 \end{bmatrix}$

In Exercises 101–104, find all (a) minors and (b) cofactors of the matrix.

101. $\begin{bmatrix} 2 & -1 \\ 7 & 4 \end{bmatrix}$ **102.** $\begin{bmatrix} 3 & 6 \\ 5 & -4 \end{bmatrix}$

103. $\begin{bmatrix} 3 & 2 & -1 \\ -2 & 5 & 0 \\ 1 & 8 & 6 \end{bmatrix}$ **104.** $\begin{bmatrix} 8 & 3 & 4 \\ 6 & 5 & -9 \\ -4 & 1 & 2 \end{bmatrix}$

In Exercises 105 and 106, find the determinant of the matrix by the method of expansion by cofactors. Expand using the indicated row or column.

105. $\begin{bmatrix} 5 & 0 & -3 \\ 0 & 12 & 4 \\ 1 & 6 & 3 \end{bmatrix}$ (a) Row 2   (b) Column 2

106. $\begin{bmatrix} 10 & -5 & 5 \\ 30 & 0 & 10 \\ 0 & 10 & 1 \end{bmatrix}$ (a) Row 3   (b) Column 1

In Exercises 107–114, evaluate the determinant. Expand by cofactors on the row or column that appears to make the computations easiest.

107. $\begin{vmatrix} -2 & 4 & 1 \\ -6 & 0 & 2 \\ 5 & 3 & 4 \end{vmatrix}$ **108.** $\begin{vmatrix} 4 & 7 & -1 \\ 2 & -3 & 4 \\ -5 & 1 & -1 \end{vmatrix}$

109. $\begin{vmatrix} 1 & 0 & -2 \\ 0 & 1 & 0 \\ -2 & 0 & 1 \end{vmatrix}$ **110.** $\begin{vmatrix} 0 & 3 & 1 \\ 5 & -2 & 1 \\ 1 & 6 & 1 \end{vmatrix}$

111. $\begin{vmatrix} 3 & 0 & -4 & 0 \\ 0 & 8 & 1 & 2 \\ 6 & 1 & 8 & 2 \\ 0 & 3 & -4 & 1 \end{vmatrix}$ **112.** $\begin{vmatrix} -5 & 6 & 0 & 0 \\ 0 & 1 & -1 & 2 \\ -3 & 4 & -5 & 1 \\ 1 & 6 & 0 & 3 \end{vmatrix}$

113. $\begin{vmatrix} 5 & 3 & 0 & 6 \\ 4 & 6 & 4 & 12 \\ 0 & 2 & -3 & 4 \\ 0 & 1 & -2 & 2 \end{vmatrix}$ **114.** $\begin{vmatrix} 1 & 4 & 3 & 2 \\ -5 & 6 & 2 & 1 \\ 0 & 0 & 0 & 0 \\ 3 & -2 & 1 & 5 \end{vmatrix}$

In Exercises 115 and 116, evaluate the determinant.

115. $\begin{vmatrix} 8 & 6 & 0 & 2 \\ 0 & -1 & 1 & -4 \\ 0 & 0 & 4 & 5 \\ 0 & 0 & 0 & 3 \end{vmatrix}$ **116.** $\begin{vmatrix} -5 & 0 & 0 & 0 \\ 7 & -2 & 0 & 0 \\ 11 & 21 & 2 & 0 \\ -6 & 9 & 12 & 14 \end{vmatrix}$

8.5   **In Exercises 117–120, use a determinant to find the area of the triangle with the given vertices.**

117. $(1, 0), (5, 0), (5, 8)$ **118.** $(-4, 0), (4, 0), (0, 6)$

119. $\left(\frac{1}{2}, 1\right), \left(2, -\frac{5}{2}\right), \left(\frac{3}{2}, 1\right)$ **120.** $\left(\frac{3}{2}, 1\right), \left(4, -\frac{1}{2}\right), (4, 2)$

In Exercises 121–124, use a determinant to decide whether the points are collinear.

121. $(-1, 7), (3, -9), (-3, 15)$

122. $(0, -5), (-2, -6), (8, -1)$

123. $(9, -10), (4, -1), (1, 5)$

124. $(-2, -8), (3, 7), (7, 19)$

In Exercises 125–130, use Cramer's Rule (if possible) to solve the system of equations.

125. $\begin{cases} x + 2y = 5 \\ -x + y = 1 \end{cases}$ **126.** $\begin{cases} 2x - y = -10 \\ 3x + 2y = -1 \end{cases}$

127. $\begin{cases} 5x - 2y = 6 \\ -11x + 3y = -23 \end{cases}$ **128.** $\begin{cases} 3x + 8y = -7 \\ 9x - 5y = 37 \end{cases}$

129. $\begin{cases} -2x + 3y - 5z = -11 \\ 4x - y + z = -3 \\ -x - 4y + 6z = 15 \end{cases}$

130. $\begin{cases} 5x - 2y + z = 15 \\ 3x - 3y - z = -7 \\ 2x - y - 7z = -3 \end{cases}$

In Exercises 131–134, use a graphing utility and Cramer's Rule (if possible) to solve the system of equations.

131. $\begin{cases} 3x + 6y = 5 \\ 6x + 14y = 11 \end{cases}$ **132.** $\begin{cases} -0.4x + 0.8y = 1.6 \\ 0.2x + 0.3y = 2.2 \end{cases}$

133. $\begin{cases} 5x - 3y + 2z = 2 \\ 2x + 2y - 3z = 3 \\ x - 7y + 8z = -4 \end{cases}$

134. $\begin{cases} 14x - 21y - 7z = 10 \\ -4x + 2y - 2z = 4 \\ 56x - 21y + 7z = 5 \end{cases}$

In Exercises 135–140, use Cramer's Rule to solve.

135. *Mixture Problem*   A florist wants to arrange a dozen flowers consisting of two varieties: carnations and roses. Carnations cost $0.75 each and roses cost $1.50 each. How many of each should the florist use so that the arrangement will cost $12.00?

136. *Mixture Problem*   One hundred liters of a 60% acid solution is obtained by mixing a 75% solution with a 50% solution. How many liters of each must be used to obtain the desired mixture?

137. *Fitting a Parabola to Three Points*   Find an equation of the parabola $y = ax^2 + bx + c$ that passes through the points $(-1, 2)$, $(0, 3)$, and $(1, 6)$.

138. *Fitting a Parabola to Three Points*   Find an equation of the parabola $y = ax^2 + bx + c$ that passes through the points $(0, 1)$, $(2, -11)$, and $(-3, 4)$.

139. *Break-Even Point*   A business invests $25,000 in equipment to produce a product. Each unit of the product costs $3.75 to produce and is sold for $5.25. How many items must be sold before the business breaks even?

140. *Break-Even Point*   A business invests $48,000 in equipment to produce a product. Each unit of the product costs $2.50 to produce and is sold for $8.75. How many items must be sold before the business breaks even?

In Exercises 141 and 142, find the uncoded 1×3 row matrices for the message. Then encode the message using the matrix.

Message	*Matrix*

141. LOOK OUT BELOW $\begin{bmatrix} 2 & -2 & 0 \\ 3 & 0 & -3 \\ -6 & 2 & 3 \end{bmatrix}$

142. RETURN TO BASE $\begin{bmatrix} 2 & 1 & 0 \\ -6 & -6 & -2 \\ 3 & 2 & 1 \end{bmatrix}$

In Exercises 143 and 144, decode the cryptogram by using the inverse of the matrix.

$$A = \begin{bmatrix} -5 & 4 & -3 \\ 10 & -7 & 6 \\ 8 & -6 & 5 \end{bmatrix}.$$

143. $-5\ 11\ -2\ 370\ -265\ 225\ -57\ 48\ -33\ 32$ $-15\ 20\ 245\ -171\ 147$

144. $145\ -105\ 92\ 264\ -188\ 160\ 23\ -16\ 15\ 129\ -84$ $78\ -9\ 8\ -5\ 159\ -118\ 100\ 219\ -152\ 133\ 370$ $-265\ 225\ -105\ 84\ -63$

Synthesis

True or False?   **In Exercises 145 and 146, determine whether the statement is true or false. Justify your answer.**

145. It is possible to find the determinant of a 4×5 matrix.

146. $\begin{vmatrix} a_{11} & a_{12} & a_{13} \\ a_{21} & a_{22} & a_{23} \\ a_{31} + c_1 & a_{32} + c_2 & a_{33} + c_3 \end{vmatrix} =$

$\begin{vmatrix} a_{11} & a_{12} & a_{13} \\ a_{21} & a_{22} & a_{23} \\ a_{31} & a_{32} & a_{33} \end{vmatrix} + \begin{vmatrix} a_{11} & a_{12} & a_{13} \\ a_{21} & a_{22} & a_{23} \\ c_1 & c_2 & c_3 \end{vmatrix}$

147. What is the relationship between the three elementary row operations on an augmented matrix and the operations that lead to equivalent systems of equations?

Chapter Project *Solving Systems of Equations*

Matrices have always been a powerful mathematical tool—especially for dealing with problems that involve a lot of data. With technology available to perform the calculations, matrices have also become a practical mathematical tool. In this project, you will use matrices to represent linear systems that model real-life data.

Three iron alloys contain different percents of carbon, chromium, and iron, shown in the matrix at the right. Alloy *X* is a type of wrought iron, alloy *Y* is a type of stainless steel, and alloy *Z* is a type of cast iron. How much of each of the three alloys can you make with 15 tons of carbon, 39 tons of chromium, and 546 tons of iron?

	Alloy X	Alloy Y	Alloy Z
Carbon	1%	1%	4%
Chromium	0%	15%	3%
Iron	99%	84%	93%

a. Let *x*, *y*, and *z* represent the amounts of the three iron alloys. Write a linear system to model the situation.

b. Write the matrix equation $AX = B$ that represents this system.

c. With a graphing utility or computer, solve the equation in part (b) for *X* by finding $X = A^{-1}B$.

Chapter Project Investigations

1. Three different gold alloys contain the percents of gold, copper, and silver shown in the following matrix. You have 20,144 grams of gold, 766 grams of copper, and 1990 grams of silver. How much of each alloy can you make?

Percent by Weight

	Alloy X	Alloy Y	Alloy Z
Gold	94%	92%	80%
Copper	4%	2%	4%
Silver	2%	6%	16%

2. The percent (by age group) of the total amounts spent on three types of shoes in 1996 is shown in the following matrix. The total amounts (in millions) spent by each age group on the three types of shoes were $518.97 (14–17 age group), $336.16 (18–24 age group), and $753.37 (25–34 age group). How many dollars worth of gym shoes, jogging shoes, and walking shoes were sold in 1996? (Source: National Sporting Goods Association)

		Gym shoes	Jogging shoes	Walking shoes
Age Group	14–17	0.14	0.13	0.03
	18–24	0.05	0.10	0.04
	25–34	0.10	0.19	0.11

3. The percent (by household size) of the total number of households owning dogs, cats, and birds in 1996 is shown in the following matrix. The total number (in millions) owned by each household size was 923.86 (1-person households), 1975.74 (2-person households), and 3380.4 (3-or-more-person households). How many households owned dogs, cats, and birds in 1996?

		Dogs	Cats	Birds
Family size	1 person	13.2	16.8	12.7
	2 persons	31.0	32.6	27.9
	3 or more persons	55.8	50.6	59.4

Take this test as you would take a test in class. After you are done, check your work against the answers given in the back of the book.

The *Interactive* CD-ROM and *Internet* versions of this text provide answers to the Chapter Tests and Cumulative Tests. They also offer Chapter Pre-Tests (that test key skills and concepts covered in previous chapters) and Chapter Post-Tests, both of which have randomly generated exercises with diagnostic capabilities.

In Exercises 1 and 2, write the matrix in reduced row-echelon form. Use a graphing utility to verify your result.

1. $\begin{bmatrix} 1 & -1 & 5 \\ 6 & 2 & 3 \\ 5 & 3 & -3 \end{bmatrix}$

2. $\begin{bmatrix} 1 & 0 & -1 & 2 \\ -1 & 1 & 1 & -3 \\ 1 & 1 & -1 & 1 \\ 3 & 2 & -3 & 4 \end{bmatrix}$

In Exercises 3 and 4, use matrices and elementary row operations to solve (if possible) the system of equations.

3. $\begin{cases} 2x + y + 2z = 4 \\ 2x + 2y = 5 \\ 2x - y + 6z = 2 \end{cases}$

4. $\begin{cases} 2x + 3y + z = 10 \\ 2x - 3y - 3z = 22 \\ 4x - 2y + 3z = -2 \end{cases}$

5. Find (a) $A - B$, (b) $3A$, and (c) $3A - 2B$.

$A = \begin{bmatrix} 5 & 4 & 4 \\ -4 & -4 & 0 \end{bmatrix}$, $B = \begin{bmatrix} 4 & -1 & 6 \\ -4 & 0 & -3 \end{bmatrix}$

6. Find AB, if possible, when $A = \begin{bmatrix} 2 & -2 & 6 \\ 3 & -1 & 7 \\ 2 & 0 & -2 \end{bmatrix}$ and $B = \begin{bmatrix} 4 & 4 \\ 3 & 2 \\ 1 & -2 \end{bmatrix}$.

7. Find A^{-1} for $A = \begin{bmatrix} -6 & 4 \\ 10 & -5 \end{bmatrix}$ and use A^{-1} to solve the system

$\begin{cases} -6x + 4y = 10 \\ 10x - 5y = 20 \end{cases}$.

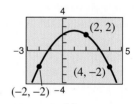

FIGURE FOR 11

In Exercises 8–10, evaluate the determinant.

8. $\begin{vmatrix} -25 & 18 \\ 6 & -7 \end{vmatrix}$

9. $\begin{vmatrix} 4 & 0 & 3 \\ 1 & -8 & 2 \\ 3 & 2 & 2 \end{vmatrix}$

10. $\begin{vmatrix} -10 & 4 & 12 & 21 \\ 0 & 2 & -16 & 0 \\ 0 & 0 & 5 & -9 \\ 0 & 0 & 0 & -3 \end{vmatrix}$

11. Use a determinant to find the area of the triangle in the figure at the right.

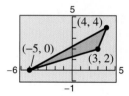

FIGURE FOR 14

In Exercises 12 and 13, use Cramer's Rule to solve (if possible) the system of equations.

12. $\begin{cases} 20x + 8y = 11 \\ 12x - 24y = 21 \end{cases}$

13. $\begin{cases} 13x - 6y = 17 \\ 26x - 12y = 8 \end{cases}$

14. Find the equation of the parabola $y = ax^2 + bx + c$ that passes through the points in the graph at the right. Use a graphing utility to verify your result.

15. The flow of traffic (in vehicles per hour) through a network of streets is shown in the figure at the right. Solve the system for the traffic flow represented by x_i, $i = 1, 2, 3, 4,$ and 5.

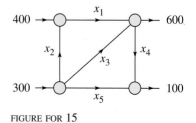

FIGURE FOR 15

Sequences, Series, and Probability

9.1 Sequences and Series
9.2 Arithmetic Sequences and Partial Sums
9.3 Geometric Sequences and Series
9.4 Mathematical Induction

9.5 The Binomial Theorem
9.6 Counting Principles
9.7 Probability

Superstock

In 1998, the U.S. Postal Service handled 41 percent of the world's mail volume, approximately 630 million pieces every day. (Source: U.S. Postal Service)

Important Vocabulary

As you encounter each new vocabulary term in this chapter, add the term and its definition to your notebook glossary.

- infinite sequence (p. 618)
- finite sequence (p. 618)
- recursive (p. 620)
- factorial (p. 620)
- summation or sigma notation (p. 622)
- infinite series (p. 623)
- finite series or nth partial sum (p. 623)
- arithmetic sequence (p. 629)
- geometric sequence (p. 638)

- infinite geometric series (p. 642)
- mathematical induction (p. 648)
- binomial coefficients (p. 656)
- Binomial Theorem (p. 656)
- Pascal's Triangle (p. 658)
- expanding a binomial (p. 659)
- Fundamental Counting Principle (p. 665)
- permutation (p. 666)

- permutation of n elements taken r at a time (p. 666)
- distinguishable permutation (p. 668)
- combination of n elements taken r at a time (p. 669)
- probability (p. 675)
- independent events (p. 680)
- complement of an event (p. 681)

Additional Resources Text-specific additional resources are available to help you do well in this course. See page xvi for details.

9.1 Sequences and Series

Sequences

In mathematics, the word *sequence* is used in much the same way as in ordinary English. Saying that a collection is listed in *sequence* means that it is ordered so that it has a first member, a second member, a third member, and so on.

Mathematically, you can think of a sequence as a *function* whose domain is the set of positive integers. Instead of using function notation, sequences are usually written using subscript notation, as shown in the following definition.

> ### Definition of Sequence
>
> An **infinite sequence** is a function whose domain is the set of positive integers. The function values
>
> $$a_1, a_2, a_3, a_4, \ldots, a_n, \ldots$$
>
> are the **terms** of the sequence. If the domain of the function consists of the first n positive integers only, the sequence is a **finite sequence.**

On occasion, it is convenient to begin subscripting a sequence with 0 instead of 1 so that the terms of the sequence become $a_0, a_1, a_2, a_3, \ldots$.

EXAMPLE 1   Finding the Terms of a Sequence

Find the first four terms of the sequences given by

a. $a_n = 3n - 2$ **b.** $a_n = 3 + (-1)^n$.

Solution

a. The first four terms of the sequence given by $a_n = 3n - 2$ are

$\qquad a_1 = 3(1) - 2 = 1$ 1st term

$\qquad a_2 = 3(2) - 2 = 4$ 2nd term

$\qquad a_3 = 3(3) - 2 = 7$ 3rd term

$\qquad a_4 = 3(4) - 2 = 10.$ 4th term

b. The first four terms of the sequence given by $a_n = 3 + (-1)^n$ are

$\qquad a_1 = 3 + (-1)^1 = 3 - 1 = 2$ 1st term

$\qquad a_2 = 3 + (-1)^2 = 3 + 1 = 4$ 2nd term

$\qquad a_3 = 3 + (-1)^3 = 3 - 1 = 2$ 3rd term

$\qquad a_4 = 3 + (-1)^4 = 3 + 1 = 4.$ 4th term

To graph a sequence using a graphing utility, set the mode to *dot* and *sequence* and enter the sequence. Consult your user's manual for instructions. Try graphing the sequences in Example 1 and using the *value* or *trace* feature to identify the terms.

What You Should Learn:

- How to use sequence notation to write the terms of sequences
- How to use factorial notation
- How to use summation notation to write sums
- How to find sums of infinite series
- How to use sequences and series to model and solve real-life problems

Why You Should Learn It:

Sequences and series are useful in modeling sets of values in order to identify a pattern. For instance, Exercise 115 on page 627 shows how a sequence can be used to model the average daily cost to community hospitals per patient from 1989 to 1996.

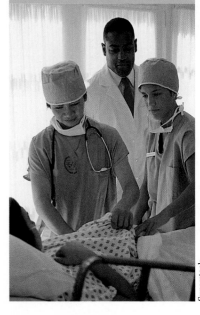
Superstock

EXAMPLE 2   Finding the Terms of a Sequence

Find the first five terms of the sequence given by $a_n = \dfrac{(-1)^n}{2n - 1}$.

Algebraic Solution

The first five terms of the sequence given by

$$a_n = \frac{(-1)^n}{2n - 1} \text{ are as follows.}$$

$a_1 = \dfrac{(-1)^1}{2(1) - 1} = \dfrac{-1}{2 - 1} = -1$ 1st term

$a_2 = \dfrac{(-1)^2}{2(2) - 1} = \dfrac{1}{4 - 1} = \dfrac{1}{3}$ 2nd term

$a_3 = \dfrac{(-1)^3}{2(3) - 1} = \dfrac{-1}{6 - 1} = -\dfrac{1}{5}$ 3rd term

$a_4 = \dfrac{(-1)^4}{2(4) - 1} = \dfrac{1}{8 - 1} = \dfrac{1}{7}$ 4th term

$a_5 = \dfrac{(-1)^5}{2(5) - 1} = \dfrac{-1}{10 - 1} = -\dfrac{1}{9}$ 5th term

Numerical Solution

Use the *table* feature of a graphing utility to create a table showing the terms of the sequence $u_n = (-1)^n/(2n - 1)$ for $n = 1, 2, 3, 4,$ and 5. From the table in Figure 9.1, you can estimate the first five terms of the sequence as follows.

$u_1 = -1, \qquad u_2 = 0.33333 \approx \dfrac{1}{3}, \qquad u_3 = -0.2 = -\dfrac{1}{5},$

$u_4 = 0.14286 \approx \dfrac{1}{7} \quad$ and $\quad u_5 = -0.1111 \approx -\dfrac{1}{9}$

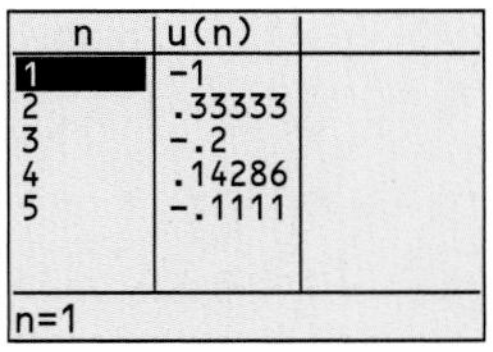

n	u(n)
1	-1
2	.33333
3	-.2
4	.14286
5	-.1111

n=1

Figure 9.1

Simply listing the first few terms is not sufficient to define a unique sequence—the nth term *must be given*. To see this, consider the following sequences, both of which have the same first three terms.

$$\frac{1}{2}, \frac{1}{4}, \frac{1}{8}, \frac{1}{16}, \ldots \cdots \frac{1}{2^n}, \cdots$$

$$\frac{1}{2}, \frac{1}{4}, \frac{1}{8}, \frac{1}{15}, \ldots \cdots \frac{6}{(n + 1)(n^2 - n + 6)}, \cdots$$

EXAMPLE 3   Finding the nth Term of a Sequence

Write an expression for the apparent nth term (a_n) of each sequence.

a. $1, 3, 5, 7, \ldots$ **b.** $2, 5, 10, 17, \ldots$

Solution

a. n: 1 2 3 4 . . . n

    Terms: 1 3 5 7 . . . a_n

    Apparent Pattern: Each term is 1 less than twice n, which implies that

$$a_n = 2n - 1.$$

b. n: 1 2   3   4 . . . n

    Terms: 2 5 10 17 . . . a_n

    Apparent Pattern: Each term is 1 more than the square of n, which implies that

$$a_n = n^2 + 1.$$

Some sequences are defined **recursively.** To define a sequence recursively, you need to be given one or more of the first few terms. All other terms of the sequence are then defined using previous terms. A well-known example is the Fibonacci sequence shown in Example 4.

EXAMPLE 4   The Fibonacci Sequence: A Recursive Sequence

The Fibonacci sequence is defined recursively as follows.

$$a_0 = 1, \ a_1 = 1, \ a_k = a_{k-2} + a_{k-1}, \qquad \text{where } k \geq 2$$

Write the first six terms of this sequence.

Solution

$a_0 = 1$	0th term is given.
$a_1 = 1$	1st term is given.
$a_2 = a_0 + a_1 = 1 + 1 = 2$	Use recursive formula.
$a_3 = a_1 + a_2 = 1 + 2 = 3$	Use recursive formula.
$a_4 = a_2 + a_3 = 2 + 3 = 5$	Use recursive formula.
$a_5 = a_3 + a_4 = 3 + 5 = 8$	Use recursive formula.

The *Interactive* CD-ROM and *Internet* versions if this text show every example with its solution; clicking on the *Try It!* button brings up similar problems. Guided Examples and Integrated Examples show step-by-step solutions to additional examples. Integrated Examples are related to several concepts in the section.

Factorial Notation

Some very important sequences in mathematics involve terms that are defined with special types of products called **factorials.**

Definition of Factorial

If n is a positive integer, **n factorial** is defined by

$$n! = 1 \cdot 2 \cdot 3 \cdot 4 \cdots (n-1) \cdot n.$$

As a special case, zero factorial is defined as $0! = 1$.

Here are some values of $n!$ for the first several nonnegative integers. Notice that $0! = 1$ by definition.

$$0! = 1$$
$$1! = 1$$
$$2! = 1 \cdot 2 = 2$$
$$3! = 1 \cdot 2 \cdot 3 = 6$$
$$4! = 1 \cdot 2 \cdot 3 \cdot 4 = 24$$
$$5! = 1 \cdot 2 \cdot 3 \cdot 4 \cdot 5 = 120$$

The value of n does not have to be very large before the value of $n!$ becomes huge. For instance, $10! = 3,628,800$.

STUDY T!P

Most graphing utilities have the capability to compute $n!$. Use your utility to compare $3 \cdot 5!$ and $(3 \cdot 5)!$. How do they differ? How large a value of $n!$ will your graphing utility allow you to compute?

Factorials follow the same conventions for order of operations as do exponents. For instance,

$$2n! = 2(n!) = 2(1 \cdot 2 \cdot 3 \cdot 4 \cdots n)$$

whereas $(2n)! = 1 \cdot 2 \cdot 3 \cdot 4 \cdots 2n.$

EXAMPLE 5   Finding the Terms of a Sequence Involving Factorials

List the first five terms of the sequence given by $a_n = \dfrac{2^n}{n!}$. Begin with $n = 0$.

Algebraic Solution

$$a_0 = \frac{2^0}{0!} = \frac{1}{1} = 1 \qquad \text{0th term}$$

$$a_1 = \frac{2^1}{1!} = \frac{2}{1} = 2 \qquad \text{1st term}$$

$$a_2 = \frac{2^2}{2!} = \frac{4}{2} = 2 \qquad \text{2nd term}$$

$$a_3 = \frac{2^3}{3!} = \frac{8}{6} = \frac{4}{3} \qquad \text{3rd term}$$

$$a_4 = \frac{2^4}{4!} = \frac{16}{24} = \frac{2}{3} \qquad \text{4th term}$$

Graphical Solution

Using a graphing utility set to *dot* and *sequence* modes, enter the sequence $u_n = 2^n/n!$.

Set the viewing window to $0 \le n \le 4$, $0 \le x \le 6$, and $0 \le y \le 4$. Then graph the sequence as shown in Figure 9.2. Use the *value* or *trace* feature to approximate the first five terms as follows.

$$u_0 = 1, \quad u_1 = 2, \quad u_2 = 2, \quad u_3 \approx 1.333 \approx \frac{4}{3}, \quad u_4 \approx 0.666 \approx \frac{2}{3}$$

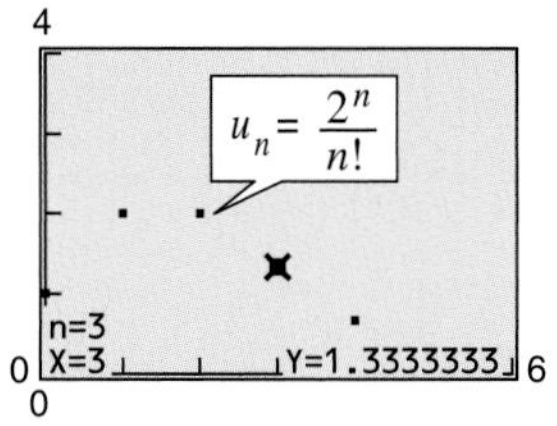

Figure 9.2

When working with fractions involving factorials, you will often find that the fractions can be reduced to simplify the computations.

EXAMPLE 6   Evaluating Factorial Expressions

Evaluate each factorial expression.

a. $\dfrac{8!}{2! \cdot 6!}$ **b.** $\dfrac{2! \cdot 6!}{3! \cdot 5!}$ **c.** $\dfrac{n!}{(n-1)!}$

Solution

a. $\dfrac{8!}{2! \cdot 6!} = \dfrac{1 \cdot 2 \cdot 3 \cdot 4 \cdot 5 \cdot 6 \cdot 7 \cdot 8}{1 \cdot 2 \cdot 1 \cdot 2 \cdot 3 \cdot 4 \cdot 5 \cdot 6} = \dfrac{7 \cdot 8}{2} = 28$

b. $\dfrac{2! \cdot 6!}{3! \cdot 5!} = \dfrac{1 \cdot 2 \cdot 1 \cdot 2 \cdot 3 \cdot 4 \cdot 5 \cdot 6}{1 \cdot 2 \cdot 3 \cdot 1 \cdot 2 \cdot 3 \cdot 4 \cdot 5} = \dfrac{6}{3} = 2$

c. $\dfrac{n!}{(n-1)!} = \dfrac{1 \cdot 2 \cdot 3 \cdots (n-1) \cdot n}{1 \cdot 2 \cdot 3 \cdots (n-1)} = n$

Note in Example 6(a) that you can simplify the computation as follows.

$$\frac{8!}{2!\,6!} = \frac{8 \cdot 7 \cdot 6!}{2!\,6!} = \frac{8 \cdot 7}{2 \cdot 1} = 28$$

Summation Notation

There is a convenient notation for the sum of the terms of a finite sequence. It is called **summation notation** or **sigma notation** because it involves the use of the uppercase Greek letter sigma, written as Σ.

Definition of Summation Notation

The sum of the first n terms of a sequence is represented by

$$\sum_{i=1}^{n} a_i = a_1 + a_2 + a_3 + a_4 + \cdots + a_n$$

where i is called the **index of summation**, n is the **upper limit of summation**, and 1 is the **lower limit of summation**.

EXAMPLE 7   Sigma Notation for Sums

a. $\displaystyle\sum_{i=1}^{5} 3i = 3(1) + 3(2) + 3(3) + 3(4) + 3(5)$

$$= 3(1 + 2 + 3 + 4 + 5)$$
$$= 3(15) = 45$$

b. $\displaystyle\sum_{k=3}^{6} (1 + k^2) = (1 + 3^2) + (1 + 4^2) + (1 + 5^2) + (1 + 6^2)$

$$= 10 + 17 + 26 + 37 = 90$$

c. $\displaystyle\sum_{n=0}^{8} \frac{1}{n!} = \frac{1}{0!} + \frac{1}{1!} + \frac{1}{2!} + \frac{1}{3!} + \frac{1}{4!} + \frac{1}{5!} + \frac{1}{6!} + \frac{1}{7!} + \frac{1}{8!}$

$$= 1 + 1 + \frac{1}{2} + \frac{1}{6} + \frac{1}{24} + \frac{1}{120} + \frac{1}{720} + \frac{1}{5040} + \frac{1}{40,320}$$

$$\approx 2.71828$$

For this summation, note that the sum is very close to the irrational number $e \approx 2.718281828$. It can be shown that as more terms of the sequence whose nth term is $1/n!$ are added, the sum becomes closer and closer to e.

In Example 7, note that the lower limit of a summation does not have to be 1. Also note that the index of summation does not have to be the letter i. For instance, in part (b) the letter k is the index of summation.

Most graphing utilities are able to sum the first n terms of a sequence. Check your user's manual for a *sum sequence* feature or a *series* feature. Figure 9.3 is an example of how one graphing utility displays the sum of the terms of the sequence

$$a_n = \frac{1}{n!} \quad \text{from} \quad n = 0 \quad \text{to} \quad n = 8.$$

In Example 7(a), note that $\displaystyle\sum_{i=1}^{5} 3i = 3(1 + 2 + 3 + 4 + 5) = 3\sum_{i=1}^{5} i.$

This is an example of one of the *properties of sums* listed on page 623.

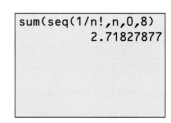

Figure 9.3

Properties of Sums

1. $\displaystyle\sum_{i=1}^{n} ca_i = c\sum_{i=1}^{n} a_i,$ c is any constant.

2. $\displaystyle\sum_{i=1}^{n} (a_i + b_i) = \sum_{i=1}^{n} a_i + \sum_{i=1}^{n} b_i$

3. $\displaystyle\sum_{i=1}^{n} (a_i - b_i) = \sum_{i=1}^{n} a_i - \sum_{i=1}^{n} b_i$

A proof of Property 1 is given in Appendix A.

Series

Many applications involve the sum of the terms of an infinite sequence. Such a sum is called an **infinite series** or simply a **series.**

Definition of a Series

Consider the infinite sequence $a_1, a_2, a_3, \ldots, a_i, \ldots$

1. The sum of all terms of the infinite sequence is called an **infinite series** and is denoted by

$$a_1 + a_2 + a_3 + \cdots + a_i + \cdots = \sum_{i=1}^{\infty} a_i.$$

2. The sum of the first n terms of the sequence is called a **finite series** or the **nth partial sum** of the sequence and is denoted by

$$a_1 + a_2 + a_3 + \cdots + a_n = \sum_{i=1}^{n} a_i.$$

STUDY TIP

Variations in the upper and lower limits of summation can produce quite different-looking summation notations for *the same sum.* For example, the following two sums have identical terms.

$$\sum_{i=1}^{5} 3(2^i) \quad \text{and} \quad \sum_{i=0}^{4} 3(2^{i+1})$$

EXAMPLE 8   Finding the Sum of a Series

For the series $\displaystyle\sum_{i=1}^{\infty} \frac{3}{10^i}$, find (a) the 3rd partial sum and (b) the sum.

Solution

a. The 3rd partial sum is

$$\sum_{i=1}^{3} \frac{3}{10^i} = \frac{3}{10^1} + \frac{3}{10^2} + \frac{3}{10^3} = 0.3 + 0.03 + 0.003 = 0.333.$$

b. The sum of the series is

$$\sum_{i=1}^{\infty} \frac{3}{10^i} = \frac{3}{10^1} + \frac{3}{10^2} + \frac{3}{10^3} + \frac{3}{10^4} + \frac{3}{10^5} + \cdots$$

$$= 0.3 + 0.03 + 0.003 + 0.0003 + 0.00003 + \cdots$$

$$= 0.33333 \ldots \approx \frac{1}{3}.$$

Notice in Example 8(b) that the sum of an infinite series can be a finite number.

Application

Sequences have many applications in situations that involve a recognizable pattern. One is illustrated in Example 9.

EXAMPLE 9 Population of the United States

From 1960 to 1997, the resident population of the United States can be approximated by the model

$$a_n = \sqrt{33{,}282 + 801.3n + 6.12n^2}, \qquad n = 0, 1, \ldots, 37$$

where a_n is the population in millions and n represents the calendar year, with $n = 0$ corresponding to 1960. Find the last five terms of this finite sequence. (Source: U.S. Bureau of the Census)

Algebraic Solution

The last five terms of this finite sequence are as follows.

$a_{33} = \sqrt{33{,}282 + 801.3(33) + 6.12(33)^2}$

$\qquad \approx 257.7$ 1993 population

$a_{34} = \sqrt{33{,}282 + 801.3(34) + 6.12(34)^2}$

$\qquad \approx 260.0$ 1994 population

$a_{35} = \sqrt{33{,}282 + 801.3(35) + 6.12(35)^2}$

$\qquad \approx 262.3$ 1995 population

$a_{36} = \sqrt{33{,}282 + 801.3(36) + 6.12(36)^2}$

$\qquad \approx 264.7$ 1996 population

$a_{37} = \sqrt{33{,}282 + 801.3(37) + 6.12(37)^2}$

$\qquad \approx 267.0$ 1997 population

Graphical Solution

Using a graphing utility set to *dot* and *sequence* modes, enter the sequence

$$u_n = \sqrt{33{,}282 + 801.3n + 6.12n^2}.$$

Set the viewing window to $0 \le n \le 40$, $0 \le x \le 40$, and $140 \le y \le 280$. Then graph the sequence, as shown in Figure 9.4. Use the *value* or *trace* feature to approximate the last five terms.

$u_{33} \approx 257.7 \qquad u_{34} \approx 260.0 \qquad u_{35} \approx 262.3$

$u_{36} \approx 264.7 \qquad u_{37} \approx 267.0$

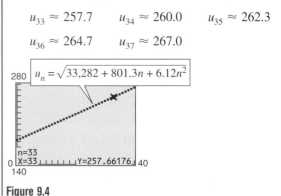

Figure 9.4

Writing About Math

The stars in the figure at the right are formed by placing n equally spaced points on a circle and connecting each point with the third point from it. For these stars, the measure of the angle (in degrees) of each point is

$$d_n = \frac{180(n-6)}{n}, \qquad n \ge 7.$$

a. Write the first four terms of the sequence.

b. If you form the stars by connecting each point with the fourth point from it, you obtain stars with the following number of points and angle measures: 9 points $(20°)$, 10 points $(36°)$, 11 points $\left(49\frac{1}{11}°\right)$, 12 points $(60°)$. Find a formula for the measure of the angle (in degrees) of each point of an n-pointed star. Explain how you found the formula.

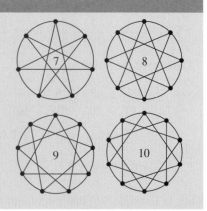

9.1 EXERCISES

In Exercises 1–22, write the first five terms of the sequence. (Assume *n* begins with 1.) Use the *table* feature of a graphing utility to verify your results.

1. $a_n = 2n + 5$ **2.** $a_n = 4n - 7$

3. $a_n = 2^n$ **4.** $a_n = \left(\frac{1}{2}\right)^n$

5. $a_n = (-2)^n$ **6.** $a_n = \left(-\frac{1}{2}\right)^n$

7. $a_n = \frac{n+1}{n}$ **8.** $a_n = \frac{n}{n+1}$

9. $a_n = \frac{6n}{3n^2 - 1}$ **10.** $a_n = \frac{3n^2 - n + 4}{2n^2 + 1}$

11. $a_n = \frac{1 + (-1)^n}{n}$ **12.** $a_n = \frac{1 + (-1)^n}{2n}$

13. $a_n = 3 - \frac{1}{2^n}$ **14.** $a_n = \frac{3^n}{4^n}$

15. $a_n = \frac{1}{n^{3/2}}$ **16.** $a_n = \frac{10}{n^{2/3}}$

17. $a_n = \frac{3^n}{n!}$ **18.** $a_n = \frac{n!}{2^n}$

19. $a_n = \frac{(-1)^n}{n^2}$ **20.** $a_n = (-1)^n\left(\frac{n}{n+1}\right)$

21. $a_n = (2n - 1)(2n + 1)$ **22.** $a_n = n(n - 1)(n - 2)$

In Exercises 23–28, find the indicated term of the sequence.

23. $a_n = (-1)^n(3n - 2)$

     $a_{25} = $ ▇

24. $a_n = (-1)^{n-1}[n(n - 1)]$

     $a_{16} = $ ▇

25. $a_n = \frac{2^n}{n!}$ **26.** $a_n = \frac{n!}{2n}$

     $a_{10} = $ ▇ $a_8 = $ ▇

27. $a_n = \frac{4n}{2n^2 - 3}$ **28.** $a_n = \frac{4n^2 - n + 3}{n(n - 1)(n + 2)}$

     $a_{12} = $ ▇ $a_{15} = $ ▇

In Exercises 29–34, write the first five terms of the sequence defined recursively.

29. $a_1 = 28$, $a_{k+1} = a_k - 4$

30. $a_1 = 15$, $a_{k+1} = a_k + 3$

31. $a_1 = 3$, $a_{k+1} = 2(a_k - 1)$

32. $a_1 = 32$, $a_{k+1} = \frac{1}{2}a_k$

33. $a_1 = 2, a_2 = 6$, $a_{k+2} = a_{k+1} + 2a_k$

34. $a_1 = 52, a_2 = 40$, $a_{k+2} = \frac{1}{2}a_{k+1} - a_k$

In Exercises 35–40, use a graphing utility to graph the first ten terms of the sequence. (Assume *n* begins with 1.)

35. $a_n = \frac{2}{3}n$ **36.** $a_n = 2 - \frac{4}{n}$

37. $a_n = 16(-0.5)^{n-1}$ **38.** $a_n = 8(0.75)^{n-1}$

39. $a_n = \frac{2n}{n+1}$ **40.** $a_n = \frac{3n^2}{n^2 + 1}$

In Exercises 41–46, use the *table* feature of a graphing utility to find the first ten terms of the sequence. (Assume *n* begins with 1.)

41. $a_n = 2(3n - 1) + 5$

42. $a_n = 2n(n + 1)(n + 2)$

43. $a_n = \frac{6^n}{n!}$ **44.** $a_n = \frac{n!}{(n^2 - 10)}$

45. $a_n = 1 + \frac{n+1}{n}$ **46.** $a_n = \frac{4n^2}{n + 2}$

In Exercises 47–50, match the sequence with its graph. [The graphs are labeled (a), (b), (c), and (d).]

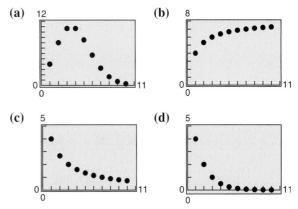

The *Interactive* CD-ROM and *Internet* versions of this text contain step-by-step solutions to all odd-numbered Section and Review Exercises. They also provide Tutorial Exercises, which link to Guided Examples for additional help.

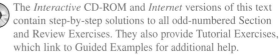

47. $a_n = \dfrac{8}{n+1}$ **48.** $a_n = \dfrac{8n}{n+1}$

49. $a_n = 4(0.5)^{n-1}$ **50.** $a_n = \dfrac{4^n}{n!}$

In Exercises 51–64, write an expression for the apparent nth term of the sequence. (Assume n begins with 1.)

51. $1, 4, 7, 10, 13, \ldots$ **52.** $3, 7, 11, 15, 19, \ldots$

53. $0, 3, 8, 15, 24, \ldots$ **54.** $1, \frac{1}{4}, \frac{1}{9}, \frac{1}{16}, \frac{1}{25}, \ldots$

55. $\frac{2}{3}, \frac{3}{4}, \frac{4}{5}, \frac{5}{6}, \frac{6}{7}, \ldots$ **56.** $\frac{2}{1}, \frac{3}{3}, \frac{4}{5}, \frac{5}{7}, \frac{6}{9}, \ldots$

57. $\frac{1}{2}, \frac{-1}{4}, \frac{1}{8}, \frac{-1}{16}, \ldots$ **58.** $\frac{1}{3}, -\frac{2}{9}, \frac{4}{27}, -\frac{8}{81}, \ldots$

59. $1 + \frac{1}{1}, 1 + \frac{1}{2}, 1 + \frac{1}{3}, 1 + \frac{1}{4}, 1 + \frac{1}{5}, \ldots$

60. $1 + \frac{1}{2}, 1 + \frac{3}{4}, 1 + \frac{7}{8}, 1 + \frac{15}{16}, 1 + \frac{31}{32}, \ldots$

61. $1, \frac{1}{2}, \frac{1}{6}, \frac{1}{24}, \frac{1}{120}, \ldots$

62. $1, 2, \dfrac{2^2}{2}, \dfrac{2^3}{6}, \dfrac{2^4}{24}, \dfrac{2^5}{120}, \ldots$

63. $1, 3, 1, 3, 1, \ldots$

64. $1, -1, 1, -1, 1, \ldots$

In Exercises 65–68, write the first five terms of the sequence defined recursively. Use the pattern to write the nth term of the sequence as a function of n. (Assume n begins with 1.)

65. $a_1 = 6, \quad a_{k+1} = a_k + 2$

66. $a_1 = 25, \quad a_{k+1} = a_k - 5$

67. $a_1 = 81, \quad a_{k+1} = \frac{1}{3}a_k$

68. $a_1 = 14, \quad a_{k+1} = -2a_k$

In Exercises 69–78, simplify the ratio of factorials.

69. $\dfrac{3!}{6!}$ **70.** $\dfrac{4!}{7!}$

71. $\dfrac{10!}{8!}$ **72.** $\dfrac{25!}{23!}$

73. $\dfrac{12!}{4! \cdot 8!}$ **74.** $\dfrac{10! \cdot 3!}{4! \cdot 6!}$

75. $\dfrac{(n+1)!}{n!}$ **76.** $\dfrac{(n+2)!}{n!}$

77. $\dfrac{(2n-1)!}{(2n+1)!}$ **78.** $\dfrac{(2n+2)!}{(2n)!}$

In Exercises 79–90, find the sum.

79. $\displaystyle\sum_{i=1}^{5} (2i + 1)$ **80.** $\displaystyle\sum_{i=1}^{6} (3i - 1)$

81. $\displaystyle\sum_{k=1}^{4} 10$ **82.** $\displaystyle\sum_{k=1}^{5} 6$

83. $\displaystyle\sum_{i=0}^{4} i^2$ **84.** $\displaystyle\sum_{i=0}^{5} 3i^2$

85. $\displaystyle\sum_{k=0}^{3} \dfrac{1}{k^2 + 1}$ **86.** $\displaystyle\sum_{j=3}^{5} \dfrac{1}{j}$

87. $\displaystyle\sum_{i=1}^{4} [(i-1)^2 + (i+1)^3]$ **88.** $\displaystyle\sum_{k=2}^{5} (k+1)(k-3)$

89. $\displaystyle\sum_{i=1}^{4} 2^i$ **90.** $\displaystyle\sum_{j=0}^{4} (-2)^j$

In Exercises 91–94, use a graphing utility to find the sum.

91. $\displaystyle\sum_{j=1}^{6} (24 - 3j)$ **92.** $\displaystyle\sum_{j=1}^{10} \dfrac{3}{j+1}$

93. $\displaystyle\sum_{k=0}^{4} \dfrac{(-1)^k}{k+1}$ **94.** $\displaystyle\sum_{k=0}^{4} \dfrac{(-1)^k}{k!}$

In Exercises 95–104, use sigma notation to write the sum. Then use a graphing utility to find the sum.

95. $\dfrac{1}{3(1)} + \dfrac{1}{3(2)} + \dfrac{1}{3(3)} + \cdots + \dfrac{1}{3(9)}$

96. $\dfrac{5}{1+1} + \dfrac{5}{1+2} + \dfrac{5}{1+3} + \cdots + \dfrac{5}{1+15}$

97. $\left[2\left(\frac{1}{8}\right) + 3\right] + \left[2\left(\frac{2}{8}\right) + 3\right] + \cdots + \left[2\left(\frac{8}{8}\right) + 3\right]$

98. $\left[1 - \left(\frac{1}{6}\right)^2\right] + \left[1 - \left(\frac{2}{6}\right)^2\right] + \cdots + \left[1 - \left(\frac{6}{6}\right)^2\right]$

99. $3 - 9 + 27 - 81 + 243 - 729$

100. $1 - \dfrac{1}{2} + \dfrac{1}{4} - \dfrac{1}{8} + \cdots - \dfrac{1}{128}$

101. $\dfrac{1}{1^2} - \dfrac{1}{2^2} + \dfrac{1}{3^2} - \dfrac{1}{4^2} + \cdots - \dfrac{1}{20^2}$

102. $\dfrac{1}{1 \cdot 3} + \dfrac{1}{2 \cdot 4} + \dfrac{1}{3 \cdot 5} + \cdots + \dfrac{1}{10 \cdot 12}$

103. $\dfrac{1}{4} + \dfrac{3}{8} + \dfrac{7}{16} + \dfrac{15}{32} + \dfrac{31}{64}$

104. $\dfrac{1}{2} + \dfrac{2}{4} + \dfrac{6}{8} + \dfrac{24}{16} + \dfrac{120}{32} + \dfrac{720}{64}$

In Exercises 105–108, find the indicated partial sum of the series.

105. $\displaystyle\sum_{i=1}^{\infty} 5\left(\frac{1}{2}\right)^i$,

4th partial sum

106. $\displaystyle\sum_{i=1}^{\infty} 2\left(\frac{1}{3}\right)^i$,

5th partial sum

107. $\displaystyle\sum_{n=1}^{\infty} 4\left(-\frac{1}{2}\right)^n$,

3rd partial sum

108. $\displaystyle\sum_{n=1}^{\infty} 8\left(-\frac{1}{4}\right)^n$,

4th partial sum

In Exercises 109–112, find the sum of the infinite series.

109. $\displaystyle\sum_{i=1}^{\infty} 6\left(\frac{1}{10}\right)^i$

110. $\displaystyle\sum_{k=1}^{\infty} 4\left(\frac{1}{10}\right)^k$

111. $\displaystyle\sum_{k=1}^{\infty} \left(\frac{1}{10}\right)^k$

112. $\displaystyle\sum_{i=1}^{\infty} 2\left(\frac{1}{10}\right)^i$

113. *Compound Interest*   A deposit of \$5000 is made in an account that earns 8% interest compounded quarterly. The balance in the account after n quarters is

$$A_n = 5000\left(1 + \frac{0.08}{4}\right)^n, \quad n = 1, 2, 3, \ldots.$$

(a) Compute the first eight terms of this sequence.

(b) Find the balance in this account after 10 years by computing the 40th term of the sequence.

114. *Compound Interest*   A deposit of \$100 is made *each month* in an account that earns 12% interest compounded monthly. The balance in the account after n months is

$$A_n = 100(101)[(1.01)^n - 1], \quad n = 1, 2, 3, \ldots.$$

(a) Compute the first six terms of this sequence.

(b) Find the balance in this account after 5 years by computing the 60th term of the sequence.

(c) Find the balance in this account after 20 years by computing the 240th term of the sequence.

115. *Per Capita Hospital Care*   The average cost to community hospitals per patient per day from 1989 to 1996 can be approximated by the model $a_n = 696.39 + 66.44n - 2.37n^2$, $n = -1, \ldots, 6$ where a_n is the cost (in dollars) and n is the year, with $n = 0$ corresponding to 1990. Find the terms of this finite sequence and use a graphing utility to construct a bar graph that represents the sequence.

What does the pattern of the bar graph say about the future of hospital costs?   (Source: American Hospital Association)

116. *Federal Debt*   From 1987 to 1998, the federal debt rose from just over \$2 trillion dollars to over \$5 trillion dollars. The federal debt from 1987 to 1998 is approximated by the model

$$a_n = \sqrt{11.7 + 2.4n}, \quad n = -3, \ldots, 8$$

where a_n is the debt (in trillions of dollars) and n is the year, with $n = 0$ corresponding to 1990. Find the terms of this finite sequence and use a graphing utility to construct a bar graph that represents the sequence. What does the pattern in the bar graph say about the future of the federal debt?   (Source: Bureau of the Public Debt)

117. *Corporate Income*   The net income a_n (in millions of dollars) of Wal-Mart for the years 1990 through 1998 are shown in the graph. The income can be approximated by the model

$$a_n = 1215.16 + 608.19n - 114.83n^2 + 11n^3,$$
$$n = 0, \ldots, 8$$

where $n = 0$ represents 1990. Use this model to approximate the total net income from 1990 through 1998. Compare this sum with the result of adding the incomes shown in the graph.   (Source: Wal-Mart Stores, Inc.)

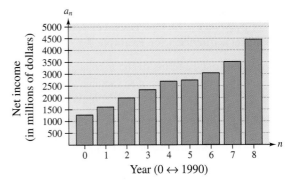

118. *Corporate Dividends*   The dividends a_n (in dollars) declared per share of common stock of Procter & Gamble Company for the years 1990 through 1998 are shown in the graph. These dividends can be approximated by the model

$$a_n = 4.27 + 0.29n - 2.93 \ln n,$$
$$n = 10, \ldots, 18$$

where $n = 10$ represents 1990. Use this model to approximate the total dividends per share of common stock from 1990 through 1998. Compare this sum with the result of adding the dividends shown in the graph.   (Source: Procter & Gamble Company)

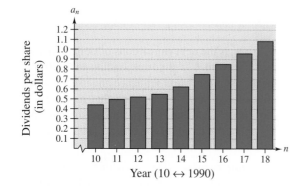

Synthesis

True or False? **In Exercises 119 and 120, determine whether the statement is true or false. Justify your answer.**

119. $\displaystyle\sum_{i=1}^{4}(i^2 + 2i) = \sum_{i=1}^{4}i^2 + 2\sum_{i=1}^{4}i$

120. $\displaystyle\sum_{j=1}^{4}2^j = \sum_{j=3}^{6}2^{j-2}$

Fibonacci Sequence **In Exercises 121 and 122, use the Fibonacci sequence. (See Example 4.)**

121. Write the first 12 terms of the Fibonacci sequence a_n and the first 10 terms of the sequence given by

$$b_n = \frac{a_{n+1}}{a_n}, \quad n > 1.$$

122. Using the definition for b_n in Exercise 121, show that b_n can be defined recursively by

$$b_n = 1 + \frac{1}{b_{n-1}}.$$

123. Find the first few terms of $a_n = n^2 - n + 11$. Describe any pattern or make an observation about the terms of the sequence.

In Exercises 124–127, find the first five terms of the sequence.

124. $a_n = \dfrac{x^n}{n!}$

125. $a_n = \dfrac{(-1)^n x^{2n+1}}{2n + 1}$

126. $a_n = \dfrac{(-1)^n x^{2n}}{(2n)!}$

127. $a_n = \dfrac{(-1)^n x^{2n+1}}{(2n + 1)!}$

Review

In Exercises 128 and 129, write the augmented matrix for the system of linear equations.

128. $\begin{cases} -4x + y = -7 \\ 6x - 9y = 3 \end{cases}$ **129.** $\begin{cases} 2x + y + 3z = -3 \\ -x + 5y = 14 \\ -3x - 6y - 7z = -7 \end{cases}$

In Exercises 130–133, find (a) $A - B$, (b) $2B - 3A$, (c) AB, (d) BA.

130. $A = \begin{bmatrix} 6 & 5 \\ 3 & 4 \end{bmatrix}$, $B = \begin{bmatrix} -2 & 4 \\ 6 & -3 \end{bmatrix}$

131. $A = \begin{bmatrix} 10 & 7 \\ -4 & 6 \end{bmatrix}$, $B = \begin{bmatrix} 0 & -12 \\ 8 & 11 \end{bmatrix}$

132. $A = \begin{bmatrix} -2 & -3 & 6 \\ 4 & 5 & 7 \\ 1 & 7 & 4 \end{bmatrix}$, $B = \begin{bmatrix} 1 & 4 & 2 \\ 0 & 1 & 6 \\ 0 & 3 & 1 \end{bmatrix}$

133. $A = \begin{bmatrix} -1 & 4 & 0 \\ 5 & 1 & 2 \\ 0 & -1 & 3 \end{bmatrix}$, $B = \begin{bmatrix} 0 & 4 & 0 \\ 3 & 1 & -2 \\ -1 & 0 & 2 \end{bmatrix}$

In Exercises 134–137, find the determinant of the matrix.

134. $A = \begin{bmatrix} 3 & 7 \\ -2 & 9 \end{bmatrix}$ **135.** $A = \begin{bmatrix} -4 & 11 \\ 13 & 20 \end{bmatrix}$

136. $A = \begin{bmatrix} 4 & 0 & 5 \\ 0 & -7 & 2 \\ 9 & 1 & -1 \end{bmatrix}$

137. $A = \begin{bmatrix} 10 & 9 & 12 & 2 \\ -2 & 5 & 8 & 7 \\ -2 & -1 & 0 & 3 \\ -4 & 6 & 2 & 1 \end{bmatrix}$

9.2 Arithmetic Sequences and Partial Sums

Arithmetic Sequences

A sequence whose consecutive terms have a common difference is called an **arithmetic sequence.**

Definition of Arithmetic Sequence

A sequence is **arithmetic** if the differences between consecutive terms are the same. So, the sequence

$$a_1, a_2, a_3, a_4, \ldots, a_n, \ldots$$

is arithmetic if there is a number d such that

$$a_2 - a_1 = a_3 - a_2 = a_4 - a_3 = \cdots = d.$$

The number d is the **common difference** of the arithmetic sequence.

What You Should Learn:

- How to recognize, write, and find the nth terms of arithmetic sequences
- How to find nth partial sums of arithmetic sequences
- How to use arithmetic sequences to model and solve real-life problems

Why You Should Learn It:

Arithmetic sequences can reduce the amount of time it takes to find the sum of a sequence of numbers with a common difference. In Exercise 85 on page 636, you will use an arithmetic sequence to find the number of bricks needed to lay a brick patio.

Index Stock

EXAMPLE 1 Examples of Arithmetic Sequences

a. The sequence whose nth term is $4n + 3$ is arithmetic. For this sequence, the common difference between consecutive terms is 4.

$$\underbrace{7, 11,}_{11 - 7 = 4} 15, 19, \ldots, 4n + 3, \ldots$$

b. The sequence whose nth term is $7 - 5n$ is arithmetic. For this sequence, the common difference between consecutive terms is -5.

$$\underbrace{2, -3,}_{-3 - 2 = -5} -8, -13, \ldots, 7 - 5n, \ldots$$

c. The sequence whose nth term is $\frac{1}{4}(n + 3)$ is arithmetic. For this sequence, the common difference between consecutive terms is $\frac{1}{4}$.

$$\underbrace{1, \frac{5}{4},}_{\frac{5}{4} - 1 = \frac{1}{4}} \frac{3}{2}, \frac{7}{4}, \ldots, \frac{n + 3}{4}, \ldots$$

d. The sequence $1, 4, 9, 16, \ldots$, whose nth term is n^2 is *not* arithmetic. The difference between the first two terms is

$$a_2 - a_1 = 4 - 1 = 3$$

but the difference between the second and third terms is

$$a_3 - a_2 = 9 - 4 = 5.$$

In Example 1, notice that each of the arithmetic sequences in parts (a), (b), and (c) has an nth term that is of the form $dn + c$, where the common difference of the sequence is d. This result is summarized as follows.

The nth Term of an Arithmetic Sequence

The nth term of an arithmetic sequence has the form

$$a_n = dn + c$$

where d is the common difference between consecutive terms of the sequence and $c = a_1 - d$.

An arithmetic sequence $a_n = dn + c$ can be thought of as "counting by d's" after a shift of c units from d. For instance, the sequence

$$2, 6, 10, 14, 18, \ldots$$

has a common difference of 4, so you are counting by 4's after a shift of 2 units below 4 (beginning with $a_1 = 2$). So, the nth term is $4n - 2$. Similarly, the nth term of the sequence

$$6, 11, 16, 21, \ldots$$

is $5n + 1$ because you are counting by 5's after a shift of 1 unit above 5 (beginning with $a_1 = 6$).

EXAMPLE 2   Finding the nth Term of an Arithmetic Sequence

Find a formula for the nth term of the arithmetic sequence whose common difference is 3 and whose first term is 2.

Solution

Because the sequence is arithmetic, you know that the formula for the nth term is of the form $a_n = dn + c$. Moreover, because the common difference is $d = 3$, the formula must have the form $a_n = 3n + c$. Because $a_1 = 2$, it follows that

$$c = a_1 - d = 2 - 3 = -1.$$

So, the formula for the nth term is $a_n = 3n - 1$. The sequence therefore has the following form.

$$2, 5, 8, 11, 14, \ldots, 3n - 1, \ldots$$

A graph of the first 15 terms of the sequence is shown in Figure 9.5. Notice that the points lie on a line. This makes sense because a_n is a linear function of n. In other words, the terms "arithmetic" and "linear" are closely connected.

Another way to find a formula for the nth term of the sequence in Example 2 is to begin by writing the terms of the sequence.

a_1	a_2	a_3	a_4	a_5	a_6	a_7	
2	$2 + 3$	$5 + 3$	$8 + 3$	$11 + 3$	$14 + 3$	$17 + 3$	$\ldots$
2	5	8	11	14	17	20	$\ldots$

From these terms, you can reason that the nth term is of the form

$$a_n = dn + c = 3n - 1.$$

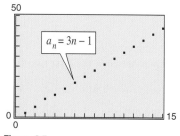

Figure 9.5

EXAMPLE 3   Writing the Terms of an Arithmetic Sequence

The fourth term of an arithmetic sequence is 20, and the 13th term is 65. Write the first several terms of this sequence.

Solution

The fourth and 13th terms of the sequence are related by

$$a_{13} = a_4 + 9d.$$

Using $a_4 = 20$ and $a_{13} = 65$, you can conclude that $d = 5$, which implies that the sequence is as follows.

$$a_1 \quad a_2 \quad a_3 \quad a_4 \quad a_5 \quad a_6 \quad a_7 \quad a_8 \quad a_9 \quad a_{10} \quad a_{11} \quad a_{12} \quad a_{13} \; \cdots$$
$$5, \quad 10, \quad 15, \quad 20, \quad 25, \quad 30, \quad 35, \quad 40, \quad 45, \quad 50, \quad 55, \quad 60, \quad 65, \; \cdots$$

If you know the nth term of an arithmetic sequence *and* you know the common difference of the sequence, you can find the $(n + 1)$th term by using the *recursive formula*

$$a_{n+1} = a_n + d. \qquad \text{Recursive formula}$$

With this formula, you can find any term of an arithmetic sequence, *provided* that you know the previous term. For instance, if you know the first term, you can find the second term. Then, knowing the second term, you can find the third term, and so on.

If you substitute $a_1 - d$ for c in the formula $a_n = dn + c$, the nth term of an arithmetic sequence has the alternative recursive formula

$$a_n = a_1 + (n - 1)d. \qquad \text{Alternative recursive formula}$$

Use this formula to solve Example 4. You should get the same answer.

EXAMPLE 4   Using a Recursive Formula

Find the seventh term of the arithmetic sequence whose first two terms are 2 and 9.

Algebraic Solution

To find the seventh term, first find a formula for the nth term. Because the first term is 2, it follows that

$$c = a_1 - d = 2 - 7 = -5.$$

Therefore, a formula for the nth term is

$$a_n = dn + c$$
$$= 7n - 5$$

which implies that the seventh term is

$$a_7 = 7(7) - 5$$
$$= 44.$$

Numerical Solution

For this sequence, the common difference is $d = 9 - 2 = 7$. Use the *sequence* and *list* features of a graphing utility to create a table that begins at 2 and increases by 7 in each row, as shown in Figure 9.6. The number in the seventh row of the table is 44, so 44 is the seventh term of the arithmetic sequence.

L1	L2	L3
1	2	-----
2	9	
3	16	
4	23	
5	30	
6	37	
7	44	

L2(7)=44

Figure 9.6

The Sum of a Finite Arithmetic Sequence

There is a simple formula for the *sum* of a finite arithmetic sequence. A proof of the formula is given in Appendix A.

The Sum of a Finite Arithmetic Sequence

The sum of a finite arithmetic sequence with n terms is

$$S_n = \frac{n}{2}(a_1 + a_n).$$

Be sure you see that this formula works only for *arithmetic* sequences. Using this formula reduces the amount of time it takes to find the sum of an arithmetic sequence, as you will see in the following example.

EXAMPLE 5 Finding the Sum of a Finite Arithmetic Sequence

Find the sum: $1 + 3 + 5 + 7 + 9 + 11 + 13 + 15 + 17 + 19$.

Solution

To begin, notice that the sequence is arithmetic (with a common difference of 2). Moreover, the sequence has 10 terms. So, the sum of the sequence is

$$S_n = 1 + 3 + 5 + 7 + 9 + 11 + 13 + 15 + 17 + 19$$

$$= \frac{n}{2}(a_1 + a_n)$$

$$= \frac{10}{2}(1 + 19) \qquad n = 10,\ a_1 = 1,\ a_{10} = 19$$

$$= 5(20)$$

$$= 100.$$

When **Carl Friedrich Gauss (1777–1855)** was ten years old, his teacher asked him to add all the integers from 1 to 100. When Gauss returned with the correct answer after only a few moments the teacher could only look at him in astounded silence. This is what Gauss did:

$$\begin{array}{r} 1 + \quad 2 + \quad 3 + \cdots + 100 \\ \underline{100 + \quad 99 + \quad 98 + \cdots + \quad 1} \\ 101 + 101 + 101 + \cdots + 101 \end{array}$$

$$\frac{100 \times 101}{2} = 5050$$

EXAMPLE 6 Finding the Sum of a Finite Arithmetic Sequence

Find the sum of the integers from 1 to 100.

Solution

The integers from 1 to 100 form an arithmetic sequence that has 100 terms. So, you can use the formula for the sum of an arithmetic sequence, as follows.

$$S_n = 1 + 2 + 3 + 4 + 5 + 6 + \cdots + 99 + 100$$

$$= \frac{n}{2}(a_1 + a_n)$$

$$= \frac{100}{2}(1 + 100) \qquad n = 100,\ a_1 = 1,\ a_{100} = 100$$

$$= 50(101)$$

$$= 5050$$

The sum of the first n terms of an infinite sequence is called the **nth partial sum.**

EXAMPLE 7   Finding a Partial Sum of an Arithmetic Sequence

Find the 150th partial sum of the arithmetic sequence

$$5, 16, 27, 38, 49, \ldots .$$

Solution

For this arithmetic sequence, you have $a_1 = 5$ and $d = 16 - 5 = 11$. So,

$$c = a_1 - d = 5 - 11 = -6$$

and the nth term is

$$a_n = 11n - 6.$$

Therefore, $a_{150} = 11(150) - 6 = 1644$, and the sum of the first 150 terms is

$$S_n = \frac{n}{2}(a_1 + a_n)$$

$$= \frac{150}{2}(5 + 1644) \qquad n = 150, a_1 = 5, a_{150} = 1644$$

$$= 75(1649)$$

$$= 123,675.$$

Applications

EXAMPLE 8   Seating Capacity

An auditorium has 20 rows of seats. There are 20 seats in the first row, 21 seats in the second row, 22 seats in the third row, and so on. (See Figure 9.7.) How many seats are there in all 20 rows?

Solution

The numbers of seats in the 20 rows form an arithmetic sequence in which the common difference is $d = 1$. Because

$$c = a_1 - d = 20 - 1 = 19$$

you can determine that the formula for the nth term of the sequence is $a_n = n + 19$. So, the 20th term in the sequence is $a_{20} = 20 + 19 = 39$, and the total number of seats is

$$S_n = 20 + 21 + 22 + \cdots + 39$$

$$= \frac{n}{2}(a_1 + a_{20})$$

$$= \frac{20}{2}(20 + 39) \qquad n = 20, a_1 = 20, a_{20} = 39$$

$$= 10(59)$$

$$= 590.$$

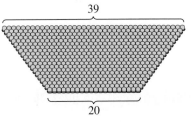

Figure 9.7

EXAMPLE 9 Total Sales

A small business sells \$10,000 worth of products during its first year. The owner of the business has set a goal of increasing annual sales by \$7500 each year for 19 years. Assuming that this goal is met, find the total sales during the first 20 years this business is in operation.

Algebraic Solution

The annual sales form an arithmetic sequence in which $a_1 = 10{,}000$ and $d = 7500$. So,

$$c = a_1 - d$$
$$= 10{,}000 - 7500$$
$$= 2500$$

and the nth term of the sequence is

$$a_n = 7500n + 2500.$$

This implies that the 20th term of the sequence is

$$a_{20} = 7500(20) + 2500$$
$$= 152{,}500.$$

The sum of the first 20 terms of the sequence is

$$S_n = \frac{n}{2}(a_1 + a_{20})$$

$$S_{20} = \frac{20}{2}(10{,}000 + 152{,}500) \quad n = 20, \ a_1 = 10{,}000, \ a_{20} = 152{,}500$$

$$= 10(162{,}500)$$

$$= 1{,}625{,}000.$$

So, the total sales for the first 20 years is \$1,625,500.

Numerical Solution

The annual sales form an arithmetic sequence in which $a_1 = 10{,}000$ and $d = 7500$. So, $c = a_1 - d = 10{,}000 - 7500 = 2500$. Use a graphing utility to create a table that shows the sales $u_n = 7500n + 2500$ for each of the 20 years, as shown in Figure 9.8. Then use the graphing utility to find that the sum of the data in the table is 1,625,000. So, the total sales for the first 20 years is \$1,625,000.

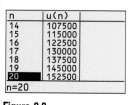

Figure 9.8

Writing About Math *Numerical Relationships*

Decide whether it is possible to fill in the blanks in each of the sequences such that the resulting sequence is arithmetic. If so, find a recursive formula for the sequence. Write a short paragraph explaining how you made your decisions.

a. -7, ___, ___, ___, ___, ___, 11

b. 17, ___, ___, ___, ___, ___, ___, ___, 71

c. $2, 6$, ___, ___, 162

d. $4, 7.5$, ___, ___, ___, ___, ___, ___, ___, 39

e. $8, 12$, ___, ___, ___, 60.75

9.2 Exercises

In Exercises 1–8, determine whether the sequence is arithmetic. If it is, find the common difference.

1. 10, 8, 6, 4, 2, . . .

2. 4, 9, 14, 19, 24, . . .

3. $3, \frac{5}{2}, 2, \frac{3}{2}, 1, \ldots$

4. $\frac{1}{3}, \frac{2}{3}, \frac{4}{3}, \frac{8}{3}, \frac{16}{3}, \ldots$

5. $-24, -16, -8, 0, 8, \ldots$

6. ln 1, ln 2, ln 3, ln 4, ln 5, . . .

7. 3.7, 4.3, 4.9, 5.5, 6.1, . . .

8. $1^2, 2^2, 3^2, 4^2, 5^2, \ldots$

In Exercises 9–16, write the first five terms of the sequence. Determine whether the sequence is arithmetic. If it is, find the common difference.

9. $a_n = 8 + 13n$

10. $a_n = (2^n)n$

11. $a_n = \dfrac{1}{n + 1}$

12. $a_n = 1 + (n - 1)4$

13. $a_n = 150 - 7n$

14. $a_n = 2^{n-1}$

15. $a_n = 3 + \dfrac{(-1)^n 2}{n}$

16. $a_n = (-1)^n$

In Exercises 17–20, write the first five terms of the arithmetic sequence. Find the common difference and write the nth term of the sequence as a function of n.

17. $a_1 = 15, \quad a_{k+1} = a_k + 9$

18. $a_1 = 200, \quad a_{k+1} = a_k - 20$

19. $a_1 = \frac{7}{2}, \quad a_{k+1} = a_k - \frac{1}{4}$

20. $a_1 = 0.375, \quad a_{k+1} = a_k + 0.25$

In Exercises 21–28, write the first five terms of the arithmetic sequence. Use a graphing utility to verify your results numerically.

21. $a_1 = 5, d = 6$

22. $a_1 = 5, d = -\frac{3}{4}$

23. $a_1 = -2.6, d = -0.4$

24. $a_4 = 16, a_{10} = 46$

25. $a_8 = 26, a_{12} = 42$

26. $a_6 = -38, a_{11} = -73$

27. $a_3 = 19, a_{15} = -1.7$

28. $a_5 = 16, a_{14} = 38.5$

In Exercises 29–34, the first two terms of the arithmetic sequences are given. Find the missing term. Use a graphing utility to verify the result numerically.

29. $a_1 = 5, \quad a_2 = 11, \quad a_{10} =$

30. $a_1 = 3, \quad a_2 = 13, \quad a_9 =$

31. $a_1 = 2, \quad a_2 = -2, \quad a_{14} =$

32. $a_1 = -1, \quad a_2 = -10, \quad a_{25} =$

33. $a_1 = 4.2, \quad a_2 = 6.6, \quad a_7 =$

34. $a_1 = -0.7, \quad a_2 = -13.8, \quad a_8 =$

In Exercises 35–44, find a formula for a_n for the arithmetic sequence.

35. $a_1 = 1, d = 3$

36. $a_1 = 15, d = 4$

37. $a_1 = 100, d = -8$

38. $a_1 = 0, d = -\frac{2}{3}$

39. $4, \frac{3}{2}, -1, -\frac{7}{2}, \ldots$

40. $10, 5, 0, -5, -10, \ldots$

41. $a_1 = 5, a_4 = 15$

42. $a_1 = -4, a_5 = 16$

43. $a_3 = 94, a_6 = 85$

44. $a_5 = 190, a_{10} = 115$

In Exercises 45–48, match the sequence with its graph. [The graphs are labeled (a), (b), (c), and (d).]

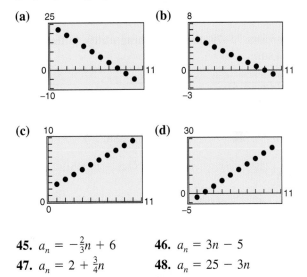

45. $a_n = -\frac{2}{3}n + 6$

46. $a_n = 3n - 5$

47. $a_n = 2 + \frac{3}{4}n$

48. $a_n = 25 - 3n$

In Exercises 49–52, use a graphing utility in *dot* mode to graph the first ten terms of the sequence.

49. $a_n = 15 - \frac{3}{2}n$

50. $a_n = -5 + 2n$

51. $a_n = 0.2n + 3$

52. $a_n = -0.3n + 8$

In Exercises 53–58, use the *table* feature of a graphing utility to find the first ten terms of the sequence.

53. $a_n = 4n - 5$

54. $a_n = 17 + 3n$

55. $a_n = 20 - \frac{3}{4}n$

56. $a_n = \frac{4}{5}n + 12$

57. $a_n = 1.5 + 0.005n$

58. $a_n = -12.4n + 9$

In Exercises 59–64, find the indicated *n*th partial sum of the arithmetic sequence.

59. $8, 26, 44, 62, \ldots, \quad n = 10$

60. $-6, -2, 2, 6, \ldots, \quad n = 50$

61. $0.5, 1.3, 2.1, 2.9, \ldots, \quad n = 10$

62. $40, 29, 18, 7, \ldots, \quad n = 10$

63. $a_1 = 100, \quad a_{25} = 220, \quad n = 25$

64. $a_1 = 15, \quad a_{100} = 307, \quad n = 100$

In Exercises 65–72, find the partial sum without using a graphing utility.

65. $\displaystyle\sum_{n=1}^{50} n$

66. $\displaystyle\sum_{n=1}^{100} 2n$

67. $\displaystyle\sum_{n=1}^{100} 5n$

68. $\displaystyle\sum_{n=51}^{100} 7n$

69. $\displaystyle\sum_{n=11}^{30} n - \sum_{n=1}^{10} n$

70. $\displaystyle\sum_{n=51}^{100} n - \sum_{n=1}^{50} n$

71. $\displaystyle\sum_{n=1}^{500} (n + 3)$

72. $\displaystyle\sum_{n=1}^{250} (1000 - n)$

In Exercises 73–78, use a graphing utility to find the partial sum.

73. $\displaystyle\sum_{n=1}^{20} (2n + 5)$

74. $\displaystyle\sum_{n=1}^{100} \frac{n + 4}{2}$

75. $\displaystyle\sum_{n=0}^{50} (1000 - 5n)$

76. $\displaystyle\sum_{n=0}^{100} \frac{8 - 3n}{16}$

77. $\displaystyle\sum_{i=1}^{60} \left(250 - \frac{8}{3}i\right)$

78. $\displaystyle\sum_{j=1}^{200} (4.5 + 0.025j)$

79. Find the sum of the first 100 positive odd integers.

80. Find the sum of the integers from -10 to 50.

Job Offer **In Exercises 81 and 82, consider a job offer with the given starting salary and guaranteed salary increase for the first 5 years of employment.**

(a) Determine the person's salary during the sixth year of employment.

(b) Determine the person's total compensation from the company through 6 full years of employment.

(c) Verify your results in parts (a) and (b) numerically.

	Starting Salary	Annual Raise
81.	$32,500	$1500
82.	$36,800	$1750

83. *Seating Capacity* Determine the seating capacity of an auditorium with 30 rows of seats if there are 20 seats in the first row, 24 seats in the second row, 28 seats in the third row, and so on.

84. *Seating Capacity* Determine the seating capacity of an auditorium with 36 rows of seats if there are 15 seats in the first row, 18 seats in the second row, 21 seats in the third row, and so on.

85. *Brick Pattern* A brick patio has the approximate shape of a trapezoid, as shown in the figure. The patio has 18 rows of bricks. The first row has 14 bricks and the 18th row has 31 bricks. How many bricks are in the patio?

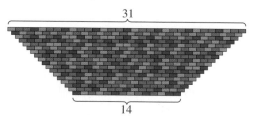

86. *Number of Logs* Logs are stacked in a pile, as shown in the figure. The top row has 15 logs and the bottom row has 24 logs. How many logs are in the stack?

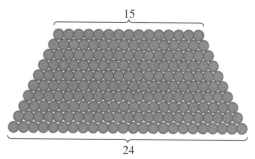

87. *Auditorium Seating* Each row in a small auditorium has two more seats than the preceding row, as shown in the figure. Find the seating capacity of the auditorium if the front row seats 25 people and there are 15 rows of seats.

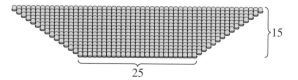

88. *Baling Hay* In the first two trips around a field baling hay, a farmer makes 93 bales and 89 bales, respectively, as shown in the figure. Because each trip is shorter than the preceding trip, the farmer estimates that the same pattern will continue. Estimate the total number of bales made if there are another six trips around the field.

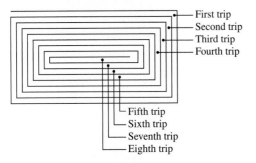

89. *Grandfather Clock* Each hour, a grandfather clock strikes the number of times corresponding to the hour of the day. How many times does the clock strike in a day?

90. *Falling Object* An object (with negligible air resistance) is dropped from an airplane. During the first second of fall, the object falls 4.9 meters; during the second second of fall, it falls 14.7 meters; during the third second, it falls 24.5 meters; and during the fourth second, it falls 34.3 meters. If this arithmetic pattern continues, how many meters will the object have fallen after 10 seconds?

Synthesis

True or False? **In Exercises 91 and 92, determine whether the statement is true or false. Justify your answer.**

91. Given an arithmetic sequence for which only the first and second terms are known, it is possible to find the nth term.

92. If the only known information about a finite arithmetic sequence is its first term and its last term, then it is possible to find the sum of the sequence.

In Exercises 93 and 94, find the first ten terms of the sequence.

93. $a_1 = x, d = 2x$ **94.** $a_1 = -y, d = 5y$

95. *Pattern Recognition*

(a) Compute the following five sums of positive odd integers.

$1 + 3 = $

$1 + 3 + 5 = $

$1 + 3 + 5 + 7 = $

$1 + 3 + 5 + 7 + 9 = $

$1 + 3 + 5 + 7 + 9 + 11 = $

(b) Use the sums in part (a) to make a conjecture about the sums of positive odd integers. Check your conjecture for the sum

$1 + 3 + 5 + 7 + 9 + 11 + 13 = $.

(c) Verify your conjecture algebraically.

96. *Think About It* The sum of the first 20 terms of an arithmetic sequence with a common difference of 3 is 650. Find the first term.

97. *Think About It* The sum of the first n terms of an arithmetic sequence with first term a_1 and common difference d is S_n. Determine the sum if the first term is increased by 5. Explain.

Review

In Exercises 98 and 99, use Gauss-Jordan elimination to solve the system of equations.

98. $\begin{cases} 2x - y + 7z = -10 \\ 3x + 2y - 4z = 17 \\ 6x - 5y + z = -20 \end{cases}$

99. $\begin{cases} -x + 4y + 10z = 4 \\ 5x - 3y + z = 31 \\ 8x + 2y - 3z = -5 \end{cases}$

In Exercises 100 and 101, use a determinant to find the area of the triangle with the given vertices.

100. $(0, 0), (4, -3), (2, 6)$ **101.** $(-1, 2), (5, 1), (3, 8)$

In Exercises 102 and 103, simplify the ratio of factorials.

102. $\dfrac{6!}{5! \cdot 2!}$ **103.** $\dfrac{6! \cdot 8!}{14!}$

9.3 Geometric Sequences and Series

Geometric Sequences

In Section 9.2, you learned that a sequence whose consecutive terms have a common *difference* is an arithmetic sequence. In this section, you will study another important type of sequence called a **geometric sequence.** Consecutive terms of a geometric sequence have a common *ratio*.

Definition of Geometric Sequence

A sequence is a **geometric sequence** if the ratios of consecutive terms are the same.

$$\frac{a_2}{a_1} = \frac{a_3}{a_2} = \frac{a_4}{a_3} = \cdots = r, \qquad r \neq 0$$

The number r is the **common ratio** of the sequence.

EXAMPLE 1   Examples of Geometric Sequences

a. The sequence whose nth term is 2^n is geometric. For this sequence, the common ratio between consecutive terms is 2.

$$\underbrace{2, 4, 8, 16, \ldots, 2^n, \ldots}_{}$$
$$\tfrac{4}{2} = 2$$

b. The sequence whose nth term is $4(3^n)$ is geometric. For this sequence, the common ratio between consecutive terms is 3.

$$\underbrace{12, 36, 108, 324, \ldots, 4(3^n), \ldots}_{}$$
$$\tfrac{36}{12} = 3$$

c. The sequence whose nth term is $\left(-\frac{1}{3}\right)^n$ is geometric. For this sequence, the common ratio between consecutive terms is $-\frac{1}{3}$.

$$\underbrace{-\frac{1}{3}, \frac{1}{9}, -\frac{1}{27}, \frac{1}{81}, \ldots, \left(-\frac{1}{3}\right)^n, \ldots}_{}$$
$$\frac{\frac{1}{9}}{-\frac{1}{3}} = -\frac{1}{3}$$

d. The sequence 1, 4, 9, 16, . . . , whose nth term is n^2 is *not* geometric. The ratio of the second term to first term is

$$\frac{a_2}{a_1} = \frac{4}{1} = 4$$

but the ratio of the third term to the second term is

$$\frac{a_3}{a_2} = \frac{9}{4}.$$

What You Should Learn:

- How to recognize, write, and find the nth terms of geometric sequences
- How to find nth partial sums of geometric sequences
- How to find sums of infinite geometric series
- How to use geometric sequences to model and solve real-life problems

Why You Should Learn It:

Geometric sequences can reduce the amount of time it takes to find the sum of a sequence of numbers with a common ratio. For instance, Exercise 92 on page 645 shows how to use a geometric sequence to estimate the population growth of a city.

In Example 1, each of the geometric sequences in parts (a), (b), and (c) has an *n*th term in the form ar^n, where the common ratio of the sequence is *r*.

The *n*th Term of a Geometric Sequence

The *n*th term of a geometric sequence has the form

$$a_n = a_1 r^{n-1}$$

where *r* is the common ratio of consecutive terms of the sequence. So, every geometric sequence can be written in the following form.

$$a_1, \quad a_2, \quad a_3, \quad a_4, \quad a_5, \quad \cdots \cdots, \quad a_n, \quad \cdots \cdots$$

$$a_1, a_1 r, a_1 r^2, a_1 r^3, a_1 r^4, \ldots, a_1 r^{n-1}, \ldots$$

If you know the *n*th term of a geometric sequence, you can find the $(n + 1)$th term by multiplying by *r*. That is, $a_{n+1} = ra_n$.

EXAMPLE 2   Finding the Terms of a Geometric Sequence

Write the first five terms of the geometric sequence whose first term is $a_1 = 3$ and whose common ratio is $r = 2$.

Solution
Starting with 3, repeatedly multiply by 2 to obtain the following.

$a_1 = 3$	1st term	
$a_2 = 3(2^1) = 6$	2nd term	
$a_3 = 3(2^2) = 12$	3rd term	
$a_4 = 3(2^3) = 24$	4th term	
$a_5 = 3(2^4) = 48$	5th term	

> **STUDY T!P**
>
> You can use a graphing utility to generate the geometric sequence in Example 2 using the following steps.
>
> 3 (enter key)
> 2 ⊠ (previous answer key)
>
> Now press the enter key repeatedly to generate the terms of the sequence.
>   Most graphing utilities have a built-in function that will display the terms of a geometric sequence. Consult your user's manual for instructions.

EXAMPLE 3   Finding a Term of a Geometric Sequence

Find the 15th term of the geometric sequence whose first term is 20 and whose common ratio is 1.05.

Algebraic Solution

$$a_n = a_1 r^{n-1} \qquad \text{Formula for a geometric sequence}$$

$$a_{15} = 20(1.05)^{15-1} \qquad \text{Substitute for } a_1, r, \text{ and } n.$$

$$\approx 39.599 \qquad \text{Use a calculator.}$$

Numerical Solution

For this sequence, $r = 1.05$ and $a_1 = 20$. Use the *table* feature of a graphing utility to create a table that shows the value of $u_n = 20(1.05)^{n-1}$ for $n = 1$ through $n = 15$. From Figure 9.9, the number in the fifteenth row is approximately 39.599, so the 15th term of the geometric sequence is about 39.599.

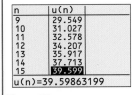

n	u(n)
9	29.549
10	31.027
11	32.578
12	34.207
13	35.917
14	37.713
15	39.599

u(n)=39.59863199

Figure 9.9

EXAMPLE 4   Finding a Term of a Geometric Sequence

Find a formula for the nth term of the following geometric sequence. What is the 9th term of the sequence?

$$5, 15, 45, \ldots$$

Solution

The common ratio of this sequence is

$$r = \frac{15}{5} = 3.$$

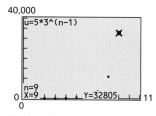

Figure 9.10

Because the first term is $a_1 = 5$, the formula must have the form

$$a_n = a_1 r^{n-1} = 5(3)^{n-1}.$$

You can determine the 9th term $(n = 9)$ to be

$$a_9 = 5(3)^{9-1} \qquad \text{Substitute 9 for } n.$$

$$= 5(6561) \qquad \text{Use a calculator.}$$

$$= 32{,}805. \qquad \text{Simplify.}$$

A graph of the first 9 terms of the sequence is shown in Figure 9.10. Notice that the points lie on an exponential curve. This makes sense because a_n is an exponential function of n.

If you know *any* two terms of a geometric sequence, you can use that information to find a formula for the nth term of the sequence.

EXAMPLE 5   Finding a Term of a Geometric Sequence

The fourth term of a geometric sequence is 125, and the 10th term is 125/64. Find the 14th term. (Assume that the terms of the sequence are positive.)

Solution

The 10th term is related to the fourth term by the equation

$$a_{10} = a_4 r^6. \qquad \text{Multiply 4th term by } r^{10-4}.$$

Because $a_{10} = 125/64$ and $a_4 = 125$, you can solve for r as follows.

$$\frac{125}{64} = 125r^6$$

$$\frac{1}{64} = r^6 \qquad \Longrightarrow \qquad \frac{1}{2} = r$$

You can obtain the 14th term by multiplying the 10th term by r^4.

$$a_{14} = a_{10}r^4$$

$$= \frac{125}{64}\left(\frac{1}{2}\right)^4 = \frac{125}{1024}$$

STUDY T!P

Remember that r is the common ratio of consecutive terms of a geometric sequence. So, in Example 5,

$$a_{10} = a_1 r^9$$

$$= a_1 \cdot r \cdot r \cdot r \cdot r^6$$

$$= a_1 \cdot \frac{a_2}{a_1} \cdot \frac{a_3}{a_2} \cdot \frac{a_4}{a_3} \cdot r^6$$

$$= a_4 r^6.$$

The Sum of a Finite Geometric Sequence

The formula for the sum of a *finite* geometric sequence is as follows. A proof of the formula is given in Appendix A.

The Sum of a Finite Geometric Sequence

The sum of the geometric sequence

$$a_1, \ a_1 r, \ a_1 r^2, \ a_1 r^3, \ a_1 r^4, \ . \ . \ ., a_1 r^{n-1}$$

with common ratio $r \neq 1$ is

$$S_n = a_1 \left(\frac{1 - r^n}{1 - r} \right).$$

EXAMPLE 6   Finding the Sum of a Finite Geometric Sequence

Find the sum $\displaystyle\sum_{n=1}^{12} 4(0.3)^n$.

Solution
By writing out a few terms, you have

$$\sum_{n=1}^{12} 4(0.3)^n = 4(0.3) + 4(0.3)^2 + 4(0.3)^3 + \cdots + 4(0.3)^{12}.$$

Now, because $a_1 = 4(0.3)$, $r = 0.3$, and $n = 12$, you can apply the formula for the sum of a finite geometric sequence to obtain

$$\sum_{n=1}^{12} 4(0.3)^n = a_1 \left(\frac{1 - r^n}{1 - r} \right) \qquad \text{Formula for sum of a finite geometric sequence}$$

$$= 4(0.3) \left[\frac{1 - (0.3)^{12}}{1 - 0.3} \right] \qquad \text{Substitute for } a_1, r, \text{ and } n.$$

$$\approx 1.714.$$

When using the formula for the sum of a geometric sequence, be careful to check that the index begins at $i = 1$. If the index begins at $i = 0$, you must adjust the formula for the nth partial sum. For instance, if the index in Example 6 had begun with $n = 0$, the sum would have been

$$\sum_{n=0}^{12} 4(0.3)^n = 4 + \sum_{n=1}^{12} 4(0.3)^n$$

$$\approx 4 + 1.714$$

$$= 5.714.$$

STUDY T!P

Using a graphing calculator, you can calculate the sum of the sequence in Example 6 to be 1.7142848.

Calculate the sum beginning at $n = 0$. You should obtain a sum of 5.7142848.

Geometric Series

The summation of the terms of an infinite geometric sequence is called an **infinite geometric series** or simply a **geometric series.**

The formula for the sum of a *finite* geometric sequence can, depending on the value of r, be extended to produce a formula for the sum of an *infinite* geometric series. Specifically, if the common ratio r has the property that $|r| < 1$, it can be shown that r^n becomes arbitrarily close to zero as n increases without bound. Consequently,

$$a_1\left(\frac{1 - r^n}{1 - r}\right) \longrightarrow a_1\left(\frac{1 - 0}{1 - r}\right) \quad \text{as} \quad n \longrightarrow \infty.$$

This result is summarized as follows.

The Sum of an Infinite Geometric Series

If $|r| < 1$, then the infinite geometric series

$$a_1, a_1r, a_1r^2, a_1r^3, \dots, a_1r^{n-1}, \dots$$

has the sum

$$S = \frac{a_1}{1 - r}.$$

EXAMPLE 7   Finding the Sum of an Infinite Geometric Series

Find each sum.

a. $\displaystyle\sum_{n=1}^{\infty} 4(0.6)^{n-1}$ **b.** $3 + 0.3 + 0.03 + 0.003 + \cdots$

Solution

a. $\displaystyle\sum_{n=1}^{\infty} 4(0.6)^{n-1} = 4(1) + 4(0.6) + 4(0.6)^2 + 4(0.6)^3 + \cdots + 4(0.6)^{n-1} + \cdots$

$$= \frac{4}{1 - (0.6)} \qquad \frac{a_1}{1 - r}$$

$$= 10$$

b. $3 + 0.3 + 0.03 + 0.003 + \cdots = 3 + 3(0.1) + 3(0.1)^2 + 3(0.1)^3 + \cdots$

$$= \frac{3}{1 - (0.1)} \qquad \frac{a_1}{1 - r}$$

$$= \frac{10}{3}$$

$$\approx 3.33$$

Application

EXAMPLE 8   Compound Interest

A deposit of $50 is made on the first day of each month in a savings account that pays 6% compounded monthly. What is the balance of this annuity at the end of 2 years?

Solution

The first deposit will gain interest for 24 months, and its balance will be

$$A_{24} = 50\left(1 + \frac{0.06}{12}\right)^{24} = 50(1.005)^{24}.$$

The second deposit will gain interest for 23 months, and its balance will be

$$A_{23} = 50\left(1 + \frac{0.06}{12}\right)^{23} = 50(1.005)^{23}.$$

The last deposit will gain interest for only 1 month, and its balance will be

$$A_{1} = 50\left(1 + \frac{0.06}{12}\right)^{1} = 50(1.005).$$

The total balance in the account will be the sum of the balances of the 24 deposits. Using the formula for the sum of a finite geometric sequence, with $A_1 = 50(1.005)$ and $r = 1.005$, you have

$$S_n = a_1\left(\frac{1 - r^n}{1 - r}\right) \qquad \text{Formula for sum of a finite geometric sequence}$$

$$S_{24} = 50(1.005)\left[\frac{1 - (1.005)^{24}}{1 - 1.005}\right] \qquad \text{Substitute values for } a_1, r, \text{ and } n.$$

$$= \$1277.96. \qquad \text{Simplify.}$$

Writing About Math   *An Experiment*

You will need a piece of string or yarn, a pair of scissors, and a tape measure. Measure out any length of string at least 5 feet long. Double over the string and cut it in half. Take one of the resulting halves, double it over, and cut it in half. Continue this process until you are no longer able to cut a length of string in half. How many cuts were you able to make? Construct a sequence of the resulting string lengths after each cut, starting with the original length of the string. Find a formula for the *n*th term of this sequence. How many cuts could you theoretically make? Write a short paragraph discussing why you were not able to make that many cuts.

9.3 E x e r c i s e s

In Exercises 1–10, determine whether the sequence is geometric. If it is, find the common ratio.

1. 5, 15, 45, 135, . . . **2.** 3, 15, 75, 375, . . .

3. 6, 18, 30, 42, . . . **4.** 1, -2, 4, -8, . . .

5. 1, $-\frac{1}{2}$, $\frac{1}{4}$, $-\frac{1}{8}$, . . . **6.** 3, 0.6, 0.12, 0.024, . . .

7. $\frac{1}{2}$, $\frac{2}{3}$, $\frac{3}{4}$, $\frac{4}{5}$, . . . **8.** 9, -6, 4, $-\frac{8}{3}$, . . .

9. 1, $\frac{1}{2}$, $\frac{1}{3}$, $\frac{1}{4}$, . . . **10.** $\frac{1}{5}$, $\frac{2}{7}$, $\frac{3}{9}$, $\frac{4}{11}$, . . .

In Exercises 11–20, write the first five terms of the geometric sequence.

11. $a_1 = 8$, $r = 3$ **12.** $a_1 = 10$, $r = 2$

13. $a_1 = 1$, $r = \frac{1}{2}$ **14.** $a_1 = 2$, $r = \frac{1}{3}$

15. $a_1 = 5$, $r = -\frac{1}{10}$ **16.** $a_1 = 6$, $r = -\frac{1}{4}$

17. $a_1 = 3.5$, $r = 5$ **18.** $a_1 = 0.4$, $r = \frac{5}{2}$

19. $a_1 = 1$, $r = e$ **20.** $a_1 = 4$, $r = \sqrt{3}$

In Exercises 21–26, write the first five terms of the geometric sequence. Determine the common ratio and write the nth term of the sequence as a function of n.

21. $a_1 = 64$, $a_{k+1} = \frac{1}{2}a_k$

22. $a_1 = 81$, $a_{k+1} = \frac{1}{3}a_k$

23. $a_1 = 4$, $a_{k+1} = 3a_k$

24. $a_1 = 5$, $a_{k+1} = -2a_k$

25. $a_1 = 6$, $a_{k+1} = -\frac{3}{2}a_k$

26. $a_1 = 36$, $a_{k+1} = -\frac{2}{3}a_k$

In Exercises 27–36, find the nth term of the geometric sequence. Use a graphing utility to verify your answer numerically.

27. $a_1 = 4$, $r = \frac{1}{2}$, $n = 10$

28. $a_1 = 5$, $r = \frac{3}{2}$, $n = 8$

29. $a_1 = 6$, $r = -\frac{1}{3}$, $n = 12$

30. $a_1 = 8$, $r = \sqrt{5}$, $n = 9$

31. $a_1 = 500$, $r = 1.02$, $n = 14$

32. $a_1 = 1000$, $r = 1.005$, $n = 11$

33. $a_1 = 16$, $a_4 = \frac{27}{4}$, $n = 3$

34. $a_2 = 3$, $a_5 = \frac{3}{64}$, $n = 1$

35. $a_2 = -18$, $a_5 = \frac{2}{3}$, $n = 6$

36. $a_3 = \frac{16}{3}$, $a_5 = \frac{64}{27}$, $n = 7$

In Exercises 37–42, find the indicated nth term of the geometric sequence.

37. 9th term: 7, 21, 63, . . .

38. 7th term: 3, 36, 432, . . .

39. 10th term: 5, 30, 180, . . .

40. 22nd term: 4, 8, 16, . . .

41. 12th term: $\frac{3}{16}$, $\frac{3}{4}$, 3, . . .

42. 8th term: $\frac{1}{2}$, 8, 128, . . .

In Exercises 43–46, match the sequence with its graph. [The graphs are labeled (a), (b), (c), and (d).]

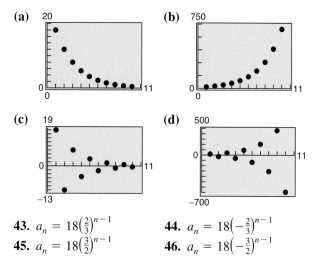

(a)

(b)

(c)

(d)

43. $a_n = 18\left(\frac{2}{3}\right)^{n-1}$ **44.** $a_n = 18\left(-\frac{2}{3}\right)^{n-1}$

45. $a_n = 18\left(\frac{3}{2}\right)^{n-1}$ **46.** $a_n = 18\left(-\frac{3}{2}\right)^{n-1}$

In Exercises 47–50, use a graphing utility to graph the first ten terms of the sequence.

47. $a_n = 12(-0.75)^{n-1}$ **48.** $a_n = 12(-0.4)^{n-1}$

49. $a_n = 2(1.3)^{n-1}$ **50.** $a_n = 2(-1.4)^{n-1}$

In Exercises 51 and 52, find the first four terms of the sequence of partial sums of the geometric series. In a sequence of partial sums, the term S_n is the sum of the first n terms of the sequence. For instance, S_2 is the sum of the first two terms.

51. 8, -4, 2, -1, $\frac{1}{2}$, . . . **52.** 8, 12, 18, 27, $\frac{81}{2}$, . . .

In Exercises 53 and 54, use a graphing utility to create a table showing the sequence of partial sums of the first ten terms of the series.

53. $\displaystyle\sum_{n=1}^{\infty} 16\left(\tfrac{1}{2}\right)^{n-1}$ **54.** $\displaystyle\sum_{n=1}^{\infty} 4(0.2)^{n-1}$

In Exercises 55–64, find the sum. Use a graphing utility to verify your result.

55. $\displaystyle\sum_{n=1}^{9} 2^{n-1}$ **56.** $\displaystyle\sum_{n=1}^{9} (-2)^{n-1}$

57. $\displaystyle\sum_{i=1}^{7} 64\left(-\tfrac{1}{2}\right)^{i-1}$ **58.** $\displaystyle\sum_{i=1}^{6} 32\left(\tfrac{1}{4}\right)^{i-1}$

59. $\displaystyle\sum_{n=0}^{20} 3\left(\tfrac{3}{2}\right)^{n}$ **60.** $\displaystyle\sum_{n=0}^{15} 2\left(\tfrac{4}{3}\right)^{n}$

61. $\displaystyle\sum_{i=1}^{10} 8\left(-\tfrac{1}{4}\right)^{i-1}$ **62.** $\displaystyle\sum_{i=1}^{10} 5\left(-\tfrac{1}{3}\right)^{i-1}$

63. $\displaystyle\sum_{n=0}^{5} 300(1.06)^{n}$ **64.** $\displaystyle\sum_{n=0}^{6} 500(1.04)^{n}$

In Exercises 65–68, use summation notation to express the sum.

65. $5 + 15 + 45 + \cdots + 3645$

66. $7 + 14 + 28 + \cdots + 896$

67. $2 - \tfrac{1}{2} + \tfrac{1}{8} - \cdots + \tfrac{1}{2048}$

68. $15 - 3 + \tfrac{3}{5} - \cdots - \tfrac{3}{625}$

In Exercises 69–84, find the sum of the infinite geometric series, if possible.

69. $\displaystyle\sum_{n=0}^{\infty} \left(\tfrac{1}{2}\right)^{n}$ **70.** $\displaystyle\sum_{n=0}^{\infty} 2\left(\tfrac{2}{3}\right)^{n}$

71. $\displaystyle\sum_{n=0}^{\infty} \left(-\tfrac{1}{2}\right)^{n}$ **72.** $\displaystyle\sum_{n=0}^{\infty} 2\left(-\tfrac{2}{3}\right)^{n}$

73. $\displaystyle\sum_{n=0}^{\infty} 4\left(\tfrac{1}{4}\right)^{n}$ **74.** $\displaystyle\sum_{n=0}^{\infty} \left(\tfrac{1}{10}\right)^{n}$

75. $\displaystyle\sum_{n=1}^{\infty} 2\left(\tfrac{7}{3}\right)^{n-1}$ **76.** $\displaystyle\sum_{n=1}^{\infty} \tfrac{1}{2}(2)^{n}$

77. $\displaystyle\sum_{n=0}^{\infty} (0.4)^{n}$ **78.** $\displaystyle\sum_{n=0}^{\infty} 4(0.2)^{n}$

79. $\displaystyle\sum_{n=0}^{\infty} -3(0.9)^{n}$ **80.** $\displaystyle\sum_{n=0}^{\infty} -10(0.2)^{n}$

81. $8 + 6 + \tfrac{9}{2} + \tfrac{27}{8} + \cdots$

82. $9 + 6 + 4 + \tfrac{8}{3} + \cdots$

83. $3 - 1 + \tfrac{1}{3} - \tfrac{1}{9} + \cdots$

84. $-6 + 5 - \tfrac{25}{6} + \tfrac{125}{36} - \cdots$

In Exercises 85–88, find the rational number representation of the repeating decimal.

85. $0.\overline{36}$ **86.** $0.\overline{297}$

87. $0.3\overline{18}$ **88.** $1.3\overline{8}$

89. *Compound Interest* A principal of $1000 is invested at 8% interest. Find the amount after 10 years if the interest is compounded (a) annually, (b) semiannually, (c) quarterly, (d) monthly, and (e) daily.

90. *Compound Interest* A principal of $2500 is invested at 7% interest. Find the amount after 20 years if the interest is compounded (a) annually, (b) semiannually, (c) quarterly, (d) monthly, and (e) daily.

91. *Depreciation* A company buys a machine for $155,000 and it depreciates at a rate of 30% per year. (In other words, at the end of each year the depreciated value is 70% of what it was at the beginning of the year.) Find the depreciated value of the machine after 5 full years.

92. *Population Growth* A city of 350,000 people is growing at a rate of 1.3% per year. Estimate the population of the city 30 years from now.

93. *Annuities* A deposit of $100 is made at the beginning of each month in an account that pays 6% interest, compounded monthly. The balance A in the account at the end of 5 years is

$$A = 100\left(1 + \frac{0.06}{12}\right)^{1} + \cdots + 100\left(1 + \frac{0.06}{12}\right)^{60}.$$

Find A.

94. *Annuities* A deposit of $50 is made at the beginning of each month in an account that pays 8% interest, compounded monthly. The balance A in the account at the end of 5 years is

$$A = 50\left(1 + \frac{0.08}{12}\right)^{1} + \cdots + 50\left(1 + \frac{0.08}{12}\right)^{60}.$$

Find A.

95. *Annuities* A deposit of P dollars is made at the beginning of each month in an account earning an annual interest rate r, compounded monthly. The balance A after t years is

$$A = P\left(1 + \frac{r}{12}\right) + P\left(1 + \frac{r}{12}\right)^{2} + \cdots$$
$$+ P\left(1 + \frac{r}{12}\right)^{12t}.$$

Show that the balance is

$$A = P\left[\left(1 + \frac{r}{12}\right)^{12t} - 1\right]\left(1 + \frac{12}{r}\right).$$

96. *Annuities* A deposit of P dollars is made at the beginning of each month in an account earning an annual interest rate r, compounded continuously. The balance A after t years is

$$A = Pe^{r/12} + Pe^{2r/12} + \cdots + Pe^{12tr/12}.$$

Show that the balance is $A = \dfrac{Pe^{r/12}(e^{rt} - 1)}{e^{r/12} - 1}$.

Annuities **In Exercises 97–100, consider making monthly deposits of P dollars in a savings account earning an annual interest rate r. Use the results of Exercises 95 and 96 to find the balance A after t years if the interest is compounded (a) monthly and (b) continuously.**

97. $P = \$50$, $r = 7\%$, $t = 20$ years

98. $P = \$75$, $r = 9\%$, $t = 25$ years

99. $P = \$100$, $r = 10\%$, $t = 40$ years

100. $P = \$20$, $r = 6\%$, $t = 50$ years

101. *Annuities* Consider an initial deposit of P dollars in an account earning an annual interest rate r, compounded monthly. At the end of each month, a withdrawal of W dollars will occur and the account will be depleted in t years. The amount of the initial deposit required is

$$P = W\left(1 + \frac{r}{12}\right)^{-1} + W\left(1 + \frac{r}{12}\right)^{-2} + \cdots$$
$$+ W\left(1 + \frac{r}{12}\right)^{-12t}.$$

Show that the initial deposit is

$$P = W\left(\frac{12}{r}\right)\left[1 - \left(1 + \frac{r}{12}\right)^{-12t}\right].$$

102. *Annuities* Determine the amount required in an individual retirement account for an individual who retires at age 65 and wants an income of $2000 from the account each month for 20 years. Use the result of Exercise 101, and assume that the account earns 9% compounded monthly.

103. *Geometry* The sides of a square are 16 inches in length. A new square is formed by connecting the midpoints of the sides of the original square, and two of the triangles are shaded. If this process is repeated five more times, determine the total area of the shaded region.

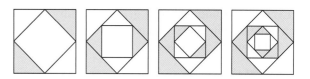

FIGURE FOR 103

104. *Geometry* The sides of a square are 27 inches in length. New squares are formed by dividing the original square into nine squares. The center square is then shaded. If this process is repeated three more times, determine the total area of the shaded region.

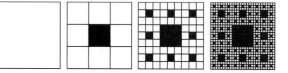

105. *Corporate Revenue* The annual revenues a_n (in billions of dollars) for the Coca-Cola Company for 1990 through 1996 can be approximated by the model

$$a_n = 3.978e^{0.11n}, \quad n = 0, 1, \ldots, 6$$

where $n = 0$ represents 1990. Use this model and the formula for the sum of a finite geometric sequence to approximate the total revenue earned during this 7-year period. (Source: Coca-Cola Enterprises, Inc.)

106. *Distance* A ball is dropped from a height of 16 feet. Each time it drops h feet, it rebounds $0.81h$ feet.

(a) Find the total distance traveled by the ball.

(b) The ball takes the following time for each fall.

$$s_1 = -16t^2 + 16, \qquad s_1 = 0 \text{ if } t = 1$$
$$s_2 = -16t^2 + 16(0.81), \qquad s_2 = 0 \text{ if } t = 0.9$$
$$s_3 = -16t^2 + 16(0.81)^2, \qquad s_3 = 0 \text{ if } t = (0.9)^2$$
$$s_4 = -16t^2 + 16(0.81)^3, \qquad s_4 = 0 \text{ if } t = (0.9)^3$$
$$\vdots \qquad\qquad \vdots$$
$$s_n = -16t^2 + 16(0.81)^{n-1}, \quad s_n = 0 \text{ if } t = (0.9)^{n-1}$$

Beginning with s_2, the ball takes the same amount of time to bounce up as it does to fall, and so the total time elapsed before it comes to rest is

$$t = 1 + 2\sum_{n=1}^{\infty} (0.9)^n.$$

Find this total.

107. *Salary* Suppose you go to work for a company that pays $0.01 the first day, $0.02 the second day, $0.04 the third day, and so on. If the daily wage keeps doubling, what will your total income be for working (a) 29 days? (b) 30 days? (c) 31 days?

108. *Salary* A company is offering a job with a salary of $30,000 for the first year. Suppose that during the next 39 years, there is a 5% raise each year. Determine the total compensation over the 40-year period.

Synthesis

True or False? **In Exercises 109–111, determine whether the statement is true or false. Justify your answer.**

109. A sequence is geometric if the ratios of consecutive differences of consecutive terms are the same.

110. You can find the nth term of a geometric sequence by multiplying its common ratio by the first term of the sequence raised to the $(n - 1)$ power.

111. A geometric sequence with a common ratio of 1 is also an arithmetic sequence.

In Exercises 112–115, write the first five terms of the geometric sequence.

112. $a_1 = 3, r = \dfrac{x}{2}$ **113.** $a_1 = 8, r = \dfrac{2x}{3}$

114. $a_1 = 5, r = 2x$ **115.** $a_1 = \frac{1}{2}, r = 7x$

In Exercises 116–119, find the nth term of the geometric sequence.

116. $a_1 = 100, r = e^x, n = 9$

117. $a_1 = 6, r = 3e^x, n = 8$

118. $a_1 = 1, r = -\dfrac{x}{3}, n = 7$

119. $a_1 = 4, r = \dfrac{4x}{3}, n = 6$

Graphical Reasoning **In Exercises 120 and 121, use a graphing utility to graph the function. Identify the horizontal asymptote of the graph and determine its relationship to the sum.**

120. $f(x) = 6\left[\dfrac{1 - (0.5)^x}{1 - (0.5)}\right], \quad \sum_{n=0}^{\infty} 6\left(\dfrac{1}{2}\right)^n$

121. $f(x) = 2\left[\dfrac{1 - (0.8)^x}{1 - (0.8)}\right], \quad \sum_{n=0}^{\infty} 2\left(\dfrac{4}{5}\right)^n$

122. *Writing* Write a brief paragraph explaining why the terms of a geometric sequence decrease in magnitude when $-1 < r < 1$.

123. *Writing* Write a brief paragraph explaining how to use the first two terms of a geometric sequence to find the nth term.

Review

124. The ratio of cement to sand in a 90-pound bag of dry mix is 1 to 4. Find the number of pounds of sand in the bag.

125. A truck traveled at an average speed of 50 miles per hour on a 200-mile trip. On the return trip, the average speed was 42 miles per hour. Find the average speed for the round trip.

126. Find two consecutive positive even integers whose product is 624.

127. Suppose your friend can mow a lawn in 4 hours and you can mow it in 6 hours. How long will it take both of you to mow the lawn?

In Exercises 128–131, perform the matrix operation.

128. $\begin{bmatrix} 4 & -1 \\ 6 & 2 \end{bmatrix}\begin{bmatrix} 1 & 3 \\ -2 & 5 \end{bmatrix}$

129. $-4\begin{bmatrix} 7 & 2 \\ -6 & 0 \end{bmatrix} + \begin{bmatrix} -5 & 5 \\ 3 & 1 \end{bmatrix}$

130. $\begin{bmatrix} -1 & 3 & 4 \\ -2 & 8 & 0 \\ 2 & 5 & -1 \end{bmatrix}\begin{bmatrix} -1 & 0 & 4 \\ -4 & 3 & 5 \\ 0 & 2 & -3 \end{bmatrix}$

131. $2\begin{bmatrix} -1 & 3 & 4 \\ -2 & 8 & 0 \\ 2 & 5 & -1 \end{bmatrix} - 4\begin{bmatrix} -1 & 0 & 4 \\ -4 & 3 & 5 \\ 0 & 2 & -3 \end{bmatrix}$

In Exercises 132–135, find the sum.

132. $\sum_{i=1}^{4} (3i + 4)$ **133.** $\sum_{i=0}^{6} 4i^2$

134. $\sum_{k=1}^{5} 12$ **135.** $\sum_{k=0}^{4} \dfrac{2}{k^2 + 2}$

9.4 Mathematical Induction

Introduction

In this section you will study a form of mathematical proof called **mathematical induction.** It is important that you clearly see the logical need for it, so let's take a closer look at a problem discussed in Example 5 on page 632.

$$S_1 = 1 = 1^2$$
$$S_2 = 1 + 3 = 2^2$$
$$S_3 = 1 + 3 + 5 = 3^2$$
$$S_4 = 1 + 3 + 5 + 7 = 4^2$$
$$S_5 = 1 + 3 + 5 + 7 + 9 = 5^2$$

Judging from the pattern formed by these first five sums, it appears that the sum of the first n odd integers is

$$S_n = 1 + 3 + 5 + 7 + 9 + \cdots + (2n - 1) = n^2.$$

Although this particular formula is valid, it is important for you to see that recognizing a pattern and then simply *jumping to the conclusion* that the pattern must be true for all values of n is *not* a logically valid method of proof. There are many examples in which a pattern appears to be developing for small values of n but then fails at some point. One of the most famous cases of this was the conjecture by the French mathematician Pierre de Fermat (1601–1665), who speculated that all numbers of the form

$$F_n = 2^{2^n} + 1, \quad n = 0, 1, 2, \ldots$$

are prime. For $n = 0, 1, 2, 3,$ and 4, the conjecture is true.

$$F_0 = 3$$
$$F_1 = 5$$
$$F_2 = 17$$
$$F_3 = 257$$
$$F_4 = 65,537$$

The size of the next Fermat number ($F_5 = 4,294,967,297$) is so great that it was difficult for Fermat to determine whether or not it was prime. However, another well-known mathematician, Leonhard Euler (1707–1783), later found a factorization

$$F_5 = 4,294,967,297$$
$$= 641(6,700,417)$$

which proved that F_5 is not prime. Therefore Fermat's conjecture was false.

Just because a rule, pattern, or formula seems to work for several values of n, you cannot simply decide that it is valid for *all* values of n without going through a *legitimate proof.*

What You Should Learn:

- How to use mathematical induction to prove statements involving a positive integer n
- How to find sums of powers of integers
- How to find finite differences of sequences

Why You Should Learn It:

Mathematical induction can be used to prove statements involving positive integers. For instance, in Exercises 27–35 on page 654, you are asked to use mathematical induction to prove properties of positive integers.

The Principle of Mathematical Induction

Let P_n be a statement involving the positive integer n. If

1. P_1 is true, and
2. the truth of P_k implies the truth of P_{k+1}, for every positive k,

then P_n must be true for all positive integers n.

To apply the Principle of Mathematical Induction, you need to be able to determine the statement P_{k+1} for a given statement P_k.

EXAMPLE 1   A Preliminary Example

Find P_{k+1} for the following.

a. $P_k : S_k = \dfrac{k^2(k+1)^2}{4}$

b. $P_k : S_k = 1 + 5 + 9 + \cdots + [4(k-1) - 3] + (4k - 3)$

c. $P_k : 3^k \geq 2k + 1$

Solution

a. $P_{k+1} : S_{k+1} = \dfrac{(k+1)^2(k+1+1)^2}{4}$ Substitute $k+1$ for k.

$\qquad\qquad = \dfrac{(k+1)^2(k+2)^2}{4}$ Simplify.

b. $P_{k+1} : S_{k+1} = 1 + 5 + 9 + \cdots + \{4[(k+1) - 1] - 3\} + [4(k+1) - 3]$

$\qquad\qquad = 1 + 5 + 9 + \cdots + (4k - 3) + (4k + 1)$

c. $P_{k+1} : 3^{k+1} \geq 2(k+1) + 1$

$\qquad\qquad 3^{k+1} \geq 2k + 3$

A well-known illustration used to explain why the Principle of Mathematical Induction works is the unending line of dominoes represented by Figure 9.11. If the line actually contains infinitely many dominoes, it is clear that you could not knock down the entire line by knocking down only *one domino* at a time. However, suppose it were true that each domino would knock down the next one as it fell. Then you could knock them all down simply by pushing the first one and starting a chain reaction. Mathematical induction works in the same way. If the truth of P_k implies the truth of P_{k+1} and if P_1 is true, the chain reaction proceeds as follows: P_1 implies P_2, P_2 implies P_3, P_3 implies P_4, and so on.

Figure 9.11

When using mathematical induction to prove a *summation* formula (such as the one in Example 2), it is helpful to think of S_{k+1} as

$$S_{k+1} = S_k + a_{k+1}$$

where a_{k+1} is the $(k + 1)$ term of the original sum.

EXAMPLE 2   Using Mathematical Induction

Use mathematical induction to prove the following formula.

$$S_n = 1 + 3 + 5 + 7 + \cdots + (2n - 1)$$
$$= n^2$$

Solution

Mathematical induction consists of two distinct parts. First, you must show that the formula is true when $n = 1$.

1. When $n = 1$, the formula is valid, because

$$S_1 = 1 = 1^2.$$

The second part of mathematical induction has two steps. The first step is to assume that the formula is valid for *some* integer k. The second step is to use this assumption to prove that the formula is valid for the next integer, $k + 1$.

2. Assuming that the formula

$$S_k = 1 + 3 + 5 + 7 + \cdots + (2k - 1)$$
$$= k^2$$

is true, you must show that the formula $S_{k+1} = (k + 1)^2$ is true.

$$
\begin{aligned}
S_{k+1} &= 1 + 3 + 5 + 7 + \cdots + (2k - 1) + [2(k + 1) - 1] \\
&= [1 + 3 + 5 + 7 + \cdots + (2k - 1)] + (2k + 2 - 1) \\
&= S_k + (2k + 1) &&\text{Group terms to form } S_k. \\
&= k^2 + 2k + 1 &&\text{Substitute } k^2 \text{ for } S_k. \\
&= (k + 1)^2
\end{aligned}
$$

Combining the results of parts (1) and (2), you can conclude by mathematical induction that the formula is valid for *all* positive integer values of n.

It occasionally happens that a statement involving natural numbers is not true for the first $k - 1$ positive integers but is true for all values of $n \geq k$. In these instances, you use a slight variation of the Principle of Mathematical Induction in which you verify P_k rather than P_1. This variation is called the *extended principle of mathematical induction*. To see the validity of this, note from Figure 9.11 that all but the first $k - 1$ dominoes can be knocked down by knocking over the kth domino. This suggests that you can prove a statement P_n to be true for $n \geq k$ by showing that P_k is true and that P_k implies P_{k+1}. In Exercises 21–26 in this section, you are asked to apply this extension of mathematical induction.

EXAMPLE 3   Using Mathematical Induction

Use mathematical induction to prove the formula

$$S_n = 1^2 + 2^2 + 3^2 + 4^2 + \cdots + n^2$$

$$= \frac{n(n + 1)(2n + 1)}{6}$$

for all $n \geq 1$.

Solution

1. When $n = 1$, the formula is valid, because

$$S_1 = 1^2 = \frac{1(2)(3)}{6}.$$

2. Assuming that

$$S_k = 1^2 + 2^2 + 3^2 + 4^2 + \cdots + k^2$$

$$= \frac{k(k + 1)(2k + 1)}{6}$$

you must show that

$$S_{k+1} = \frac{(k + 1)(k + 1 + 1)[2(k + 1) + 1]}{6} = \frac{(k + 1)(k + 2)(2k + 3)}{6}.$$

To do this, write the following.

$$S_{k+1} = S_k + a_{k+1}$$

$$= (1^2 + 2^2 + 3^2 + 4^2 + \cdots + k^2) + (k + 1)^2$$

$$= \frac{k(k + 1)(2k + 1)}{6} + (k + 1)^2$$

$$= \frac{k(k + 1)(2k + 1) + 6(k + 1)^2}{6}$$

$$= \frac{(k + 1)[k(2k + 1) + 6(k + 1)]}{6}$$

$$= \frac{(k + 1)(2k^2 + 7k + 6)}{6}$$

$$= \frac{(k + 1)(k + 2)(2k + 3)}{6}$$

Combining the results of parts (1) and (2), you can conclude by mathematical induction that the formula is valid for *all* $n \geq 1$.

When proving a formula with mathematical induction, the only statement that you *need* to verify is P_1. As a check, however, it is a good idea to try verifying some of the other statements. For instance, in Example 3, try verifying P_2 and P_3.

Sums of Powers of Integers

The formula in Example 3 is one of a collection of useful summation formulas. This and other formulas dealing with the sums of various powers of the first n positive integers are summarized as follows.

Sums of Powers of Integers

1. $\displaystyle\sum_{i=1}^{n} i = 1 + 2 + 3 + 4 + \cdots + n = \dfrac{n(n+1)}{2}$

2. $\displaystyle\sum_{i=1}^{n} i^2 = 1^2 + 2^2 + 3^2 + 4^2 + \cdots + n^2 = \dfrac{n(n+1)(2n+1)}{6}$

3. $\displaystyle\sum_{i=1}^{n} i^3 = 1^3 + 2^3 + 3^3 + 4^3 + \cdots + n^3 = \dfrac{n^2(n+1)^2}{4}$

4. $\displaystyle\sum_{i=1}^{n} i^4 = 1^4 + 2^4 + 3^4 + 4^4 + \cdots + n^4 = \dfrac{n(n+1)(2n+1)(3n^2+3n-1)}{30}$

5. $\displaystyle\sum_{i=1}^{n} i^5 = 1^5 + 2^5 + 3^5 + 4^5 + \cdots + n^5 = \dfrac{n^2(n+1)^2(2n^2+2n-1)}{12}$

Each of these formulas for sums can be proven by mathematical induction. (See Exercises 13, 14, 17, and 18 in this section.)

EXAMPLE 4   Proving an Inequality by Mathematical Induction

Prove that $n < 2^n$ for all positive integers n.

Solution

1. For $n = 1$ and $n = 2$, the formula is true, because

$$1 < 2^1 \text{ and } 2 < 2^2.$$

2. Assuming that

$$k < 2^k$$

you need to show that $k + 1 < 2^{k+1}$. Note first that

$$2^{k+1} = 2(2^k) > 2(k) = 2k. \qquad \text{By assumption } 2^k > k.$$

Because $2k = k + k > k + 1$ for all $k > 1$, it follows that

$$2^{k+1} > 2k > k + 1$$

or

$$k + 1 < 2^{k+1}.$$

Therefore, $n < 2^n$ for all integers $n \geq 1$.

Finite Differences

The **first differences** of a sequence are found by subtracting consecutive terms. The **second differences** are found by subtracting consecutive first differences. The first and second differences of the sequence 3, 5, 8, 12, 17, 23, . . . are as follows.

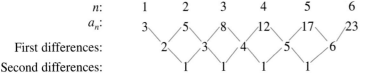

For this sequence, the second differences are all the same. When this happens, and the second differences are nonzero, the sequence has a perfect quadratic model. If the first differences are all the same nonzero number, the sequence has a linear model—that is, it is arithmetic.

EXAMPLE 5   Finding a Quadratic Model

Find the quadratic model for the sequence 3, 5, 8, 12, 17, 23, ...

Solution

You know from the second differences shown above the model is quadratic and has the form

$$a_n = an^2 + bn + c.$$

By substituting 1, 2, and 3 for n, you can obtain a system of three linear equations in three variables.

$a_1 = a(1)^2 + b(1) + c = 3$ Substitute 1 for n.

$a_2 = a(2)^2 + b(2) + c = 5$ Substitute 2 for n.

$a_3 = a(3)^2 + b(3) + c = 8$ Substitute 3 for n.

You now have a system of three equations in a, b, and c.

$$\begin{cases} a + b + c = 3 & \text{Equation 1} \\ 4a + 2b + c = 5 & \text{Equation 2} \\ 9a + 3b + c = 8 & \text{Equation 3} \end{cases}$$

Solving this system of equations using techniques discussed in Chapter 7, you can find the solution to be $a = \frac{1}{2}$, $b = \frac{1}{2}$, and $c = 2$. So, the quadratic model is

$$a_n = \frac{1}{2}n^2 + \frac{1}{2}n + 2.$$

Check the values of $a_1, a_2,$ and a_3 as follows.

Check

$a_1 = \frac{1}{2}(1)^2 + \frac{1}{2}(1) + 2 = 3$ Solution checks. ✓

$a_2 = \frac{1}{2}(2)^2 + \frac{1}{2}(2) + 2 = 5$ Solution checks. ✓

$a_3 = \frac{1}{2}(3)^2 + \frac{1}{2}(3) + 2 = 8$ Solution checks. ✓

9.4 Exercises

In Exercises 1–6, find P_{k+1} for the given P_k.

1. $P_k = \dfrac{5}{k(k+1)}$

2. $P_k = \dfrac{4}{(k+2)(k+3)}$

3. $P_k = \dfrac{k^2(k+3)^2}{6}$

4. $P_k = \dfrac{k}{2}(5k-3)$

5. $P_k = 1 + 6 + 11 + \cdots + [5(k-1) - 4] + (5k - 4)$

6. $P_k = 7 + 13 + 19 + \cdots + [6(k-1) + 1] + (6k + 1)$

In Exercises 7–20, use mathematical induction to prove the formula for every positive integer n.

7. $2 + 4 + 6 + 8 + \cdots + 2n = n(n+1)$

8. $1 + 5 + 9 + 13 + \cdots + (4n - 3) = n(2n - 1)$

9. $3 + 8 + 13 + 18 + \cdots + (5n - 2) = \dfrac{n}{2}(5n + 1)$

10. $1 + 4 + 7 + 10 + \cdots + (3n - 2) = \dfrac{n}{2}(3n - 1)$

11. $1 + 2 + 2^2 + 2^3 + \cdots + 2^{n-1} = 2^n - 1$

12. $2(1 + 3 + 3^2 + 3^3 + \cdots + 3^{n-1}) = 3^n - 1$

13. $1 + 2 + 3 + 4 + \cdots + n = \dfrac{n(n+1)}{2}$

14. $1^3 + 2^3 + 3^3 + 4^3 + \cdots + n^3 = \dfrac{n^2(n+1)^2}{4}$

15. $1^2 + 3^2 + 5^2 + \cdots + (2n-1)^2 = \dfrac{n(2n-1)(2n+1)}{3}$

16. $\left(1 + \dfrac{1}{1}\right)\left(1 + \dfrac{1}{2}\right)\left(1 + \dfrac{1}{3}\right) \cdots \left(1 + \dfrac{1}{n}\right) = n + 1$

17. $\displaystyle\sum_{i=1}^{n} i^4 = \dfrac{n(n+1)(2n+1)(3n^2 + 3n - 1)}{30}$

18. $\displaystyle\sum_{i=1}^{n} i^5 = \dfrac{n^2(n+1)^2(2n^2 + 2n - 1)}{12}$

19. $\displaystyle\sum_{i=1}^{n} i(i+1) = \dfrac{n(n+1)(n+2)}{3}$

20. $\displaystyle\sum_{i=1}^{n} \dfrac{1}{(2i-1)(2i+1)} = \dfrac{n}{2n+1}$

In Exercises 21–26, prove the inequality for the indicated integer values of n.

21. $n! > 2^n, \quad n \geq 4$

22. $\left(\tfrac{4}{3}\right)^n > n, \quad n \geq 7$

23. $\dfrac{1}{\sqrt{1}} + \dfrac{1}{\sqrt{2}} + \dfrac{1}{\sqrt{3}} + \cdots + \dfrac{1}{\sqrt{n}} > \sqrt{n}, \quad n \geq 2$

24. $\left(\dfrac{x}{y}\right)^{n+1} < \left(\dfrac{x}{y}\right)^{n}, \quad n \geq 1 \text{ and } 0 < x < y$

25. $(1 + a)^n \geq na, \quad n \geq 1, \quad a > 1$

26. $3^n > n \, 2^n, \quad n \geq 1$

In Exercises 27–35, use mathematical induction to prove the given property for all positive integers n.

27. $(ab)^n = a^n b^n$

28. $\left(\dfrac{a}{b}\right)^n = \dfrac{a^n}{b^n}$

29. If $x_1 \neq 0, \, x_2 \neq 0, \, \ldots, x_n \neq 0$, then
$(x_1 x_2 x_3 \cdots x_n)^{-1} = x_1^{-1} x_2^{-1} x_3^{-1} \cdots x_n^{-1}$.

30. If $x_1 > 0, \, x_2 > 0, \, \ldots, x_n > 0$, then
$\ln(x_1 x_2 x_3 \cdots x_n) = \ln x_1 + \ln x_2 + \ln x_3$
$\qquad\qquad\qquad\quad + \cdots + \ln x_n.$

31. Generalized Distributive Law:
$x(y_1 + y_2 + \cdots + y_n) = xy_1 + xy_2 + \cdots + xy_n$

32. $(a + bi)^n$ and $(a - bi)^n$ are complex conjugates for all $n \geq 1$.

33. A factor of $(n^3 + 3n^2 + 2n)$ is 3.

34. A factor of $(2^{2n-1} + 3^{2n-1})$ is 5.

35. A factor of $(9^n - 8n - 1)$ is 64 for all $n \geq 2$.

In Exercises 36–39, write the first five terms of the sequence.

36. $a_0 = 1$
$a_n = a_{n-1} + 2$

37. $a_0 = 10$
$a_n = 4a_{n-1}$

38. $a_0 = 4$
$a_1 = 2$
$a_n = a_{n-1} - a_{n-2}$

39. $a_0 = 0$
$a_1 = 2$
$a_n = a_{n-1} + 2a_{n-2}$

In Exercises 40–49, write the first five terms of the sequence beginning with the given term. Then calculate the first and second differences of the sequence. Does the sequence have a linear model, a quadratic model, or neither?

40. $a_1 = 0$
$a_n = a_{n-1} + 3$

41. $a_1 = 2$
$a_n = n - a_{n-1}$

42. $a_1 = 3$

$a_n = a_{n-1} - n$

43. $a_2 = -3$

$a_n = -2a_{n-1}$

44. $a_0 = 0$

$a_n = a_{n-1} + n$

45. $a_0 = 2$

$a_n = (a_{n-1})^2$

46. $a_1 = 2$

$a_n = a_{n-1} + 2$

47. $a_1 = 0$

$a_n = a_{n-1} + 2n$

48. $a_0 = 1$

$a_n = a_{n-1} + n^2$

49. $a_0 = 0$

$a_n = a_{n-1} - 1$

In Exercises 50–53, find a quadratic model for the sequence with the indicated terms.

50. $a_0 = 3$, $a_1 = 3$, $a_4 = 15$

51. $a_0 = 7$, $a_1 = 6$, $a_3 = 10$

52. $a_0 = -3$, $a_2 = 1$, $a_4 = 9$

53. $a_0 = 3$, $a_2 = 0$, $a_6 = 36$

Synthesis

True or False? **In Exercises 54–56, determine whether the statement is true or false. Justify your answer.**

54. If the statement P_k is true and P_k implies P_{k+1}, then P_1 is also true.

55. If a sequence is arithmetic, then the first differences of the sequence are all zero.

56. A sequence with n terms has $n - 1$ second differences.

57. ***Writing*** In your own words, explain what is meant by a proof by mathematical induction.

58. ***Think About It*** What conclusion can be drawn from the given information about the sequence of statements P_n?

(a) P_3 is true and P_k implies P_{k+1}.

(b) $P_1, P_2, P_3, \ldots, P_{50}$ are all true.

(c) P_1, P_2, and P_3 are all true, but the truth of P_k does not imply that P_{k+1} is true.

(d) P_2 is true and P_{2k} implies P_{2k+2}.

Review

In Exercises 59–62, solve the system of equations.

59. $\begin{cases} x - y = 2 \\ -4x + 5y = -3 \end{cases}$

60. $\begin{cases} x - 3y = 1 \\ 7x - 6y = -38 \end{cases}$

61. $\begin{cases} y = x^2 \\ -3x + 2y = 2 \end{cases}$

62. $\begin{cases} x - y^3 = 0 \\ x - 2y^2 = 0 \end{cases}$

In Exercises 63–66, use Gauss-Jordan elimination to solve the system.

63. $\begin{cases} x - y = -1 \\ x + 2y - 2z = 3 \\ 3x - y + 2z = 3 \end{cases}$

64. $\begin{cases} 2x + y - 2z = 1 \\ x - z = 1 \\ 3x + 3y + z = 12 \end{cases}$

65. $\begin{cases} -3x + y + 5z = 25 \\ x - 2y + 3z = 7 \\ 2x + 3y - z = 0 \end{cases}$

66. $\begin{cases} 2x - y + 4z = 21 \\ -4x + 3y + z = -14 \\ -x - 4y + 7z = 12 \end{cases}$

In Exercises 67 and 68, find the determinant of the matrix.

67. $\begin{bmatrix} 7 & 6 \\ -4 & 2 \end{bmatrix}$

68. $\begin{bmatrix} 2 & 4 & 8 \\ 0 & 6 & -9 \\ 4 & -3 & 8 \end{bmatrix}$

In Exercises 69–72, expand the expression.

69. $(2x^2 - 1)^2$

70. $(2x - y)^2$

71. $(5 - 4x)^3$

72. $(2x - 4y)^3$

9.5 The Binomial Theorem

Binomial Coefficients

Recall that a *binomial* is a polynomial that has two terms. In this section, you will study a formula that provides a quick method of raising a binomial to a power. To begin, look at the expansion of

$$(x + y)^n$$

for several values of n.

$$(x + y)^0 = 1$$

$$(x + y)^1 = x + y$$

$$(x + y)^2 = x^2 + 2xy + y^2$$

$$(x + y)^3 = x^3 + 3x^2y + 3xy^2 + y^3$$

$$(x + y)^4 = x^4 + 4x^3y + 6x^2y^2 + 4xy^3 + y^4$$

$$(x + y)^5 = x^5 + 5x^4y + 10x^3y^2 + 10x^2y^3 + 5xy^4 + y^5$$

There are several observations you can make about these expansions.

1. In each expansion, there are $n + 1$ terms.

2. In each expansion, x and y have symmetric roles. The powers of x decrease by 1 in successive terms, whereas the powers of y increase by 1.

3. The sum of the powers of each term is n. For instance, in the expansion of $(x + y)^5$, the sum of the powers of each term is 5.

$$\overset{4 + 1 = 5 \quad 3 + 2 = 5}{}$$
$$(x + y)^5 = x^5 + 5\overbrace{x^4y^1} + 10\overbrace{x^3y^2} + 10x^2y^3 + 5x^1y^4 + y^5$$

4. The coefficients increase and then decrease in a symmetric pattern.

The coefficients of a binomial expansion are called **binomial coefficients.** To find them, you can use the **Binomial Theorem.** A proof of this theorem is given in Appendix A.

What You Should Learn:

- How to use the Binomial Theorem to calculate binomial coefficients
- How to use Pascal's Triangle to calculate binomial coefficients
- How to use binomial coefficients to write binomial expansions

Why You Should Learn It:

You can use binomial coefficients to predict future behavior. For instance, in Exercise 84 on page 662, you are asked to use binomial coefficients to find the probability that a baseball player gets 3 hits during the next 10 times at bat.

Jonathan Daniel/Allsport

The Binomial Theorem

In the expansion of $(x + y)^n$

$$(x + y)^n = x^n + nx^{n-1}y + \cdots +{}_nC_r\, x^{n-r}y^r + \cdots + nxy^{n-1} + y^n$$

the coefficient of $x^{n-r}y^r$ is

$${}_nC_r = \frac{n!}{(n - r)!\,r!}.$$

The symbol $\binom{n}{r}$ is often used in place of ${}_nC_r$ to denote binomial coefficients.

EXAMPLE 1   Finding Binomial Coefficients

Find the binomial coefficients.

a. $_8C_2$ **b.** $\binom{10}{3}$ **c.** $_7C_0$ **d.** $\binom{8}{8}$

Solution

a. $_8C_2 = \dfrac{8!}{6! \cdot 2!} = \dfrac{(8 \cdot 7) \cdot 6!}{6! \cdot 2!} = \dfrac{8 \cdot 7}{2 \cdot 1} = 28$

b. $\binom{10}{3} = \dfrac{10!}{7! \cdot 3!} = \dfrac{(10 \cdot 9 \cdot 8) \cdot 7!}{7! \cdot 3!} = \dfrac{10 \cdot 9 \cdot 8}{3 \cdot 2 \cdot 1} = 120$

c. $_7C_0 = \dfrac{7!}{7! \cdot 0!} = 1$

d. $\binom{8}{8} = \dfrac{8!}{0! \cdot 8!} = 1$

When $r \neq 0$ and $r \neq n$, as in parts (a) and (b) above, there is a simple pattern for evaluating binomial coefficients.

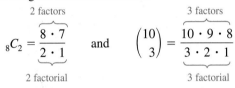

$$_8C_2 = \overbrace{\dfrac{8 \cdot 7}{\underbrace{2 \cdot 1}_{\text{2 factorial}}}}^{\text{2 factors}} \quad \text{and} \quad \binom{10}{3} = \overbrace{\dfrac{10 \cdot 9 \cdot 8}{\underbrace{3 \cdot 2 \cdot 1}_{\text{3 factorial}}}}^{\text{3 factors}}$$

EXAMPLE 2   Finding Binomial Coefficients

Find the binomial coefficients.

a. $_7C_3$ **b.** $_7C_4$ **c.** $_{12}C_1$ **d.** $_{12}C_{11}$

Solution

a. $_7C_3 = \dfrac{7 \cdot 6 \cdot 5}{3 \cdot 2 \cdot 1} = 35$

b. $_7C_4 = \dfrac{7 \cdot 6 \cdot 5 \cdot 4}{4 \cdot 3 \cdot 2 \cdot 1} = 35$

c. $_{12}C_1 = \dfrac{12}{1} = 12$

d. $_{12}C_{11} = \dfrac{12!}{1! \cdot 11!} = \dfrac{(12) \cdot 11!}{1! \cdot 11!} = \dfrac{12}{1} = 12$

It is not a coincidence that the results in parts (a) and (b) of Example 2 are the same and that the results in parts (c) and (d) are the same. In general, it is true that

$$_nC_r = {}_nC_{n-r}.$$

This shows the symmetric property of binomial coefficients that was identified earlier.

STUDY T!P

Most graphing calculators are programmed to evaluate $_nC_r$. Consult your user's manual to evaluate the binomial coefficients in Example 1.

Exploration

Find the following pairs of binomial coefficients.

a. $_7C_0, {}_7C_7$

b. $_8C_0, {}_8C_8$

c. $_{10}C_0, {}_{10}C_{10}$

d. $_7C_1, {}_7C_6$

e. $_8C_1, {}_8C_7$

f. $_{10}C_1, {}_{10}C_9$

What do you observe about the pairs in (a), (b), and (c)? What do you observe about the pairs in (d), (e), and (f)? Write two conjectures from your observations. Develop a convincing argument for your two conjectures.

Pascal's Triangle

There is a convenient way to remember a pattern for binomial coefficients. By arranging the coefficients in a triangular pattern, you obtain the following array, which is called **Pascal's Triangle.** This triangle is named after the famous French mathematician Blaise Pascal (1623–1662).

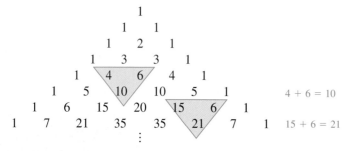

4 + 6 = 10

15 + 6 = 21

The first and last number in each row of Pascal's Triangle is 1. Every other number in each row is formed by adding the two numbers immediately above the number. Pascal noticed that numbers in this triangle are precisely the same numbers as the coefficients of binomial expansions.

$$(x + y)^0 = 1$$
$$(x + y)^1 = 1x + 1y$$
$$(x + y)^2 = 1x^2 + 2xy + 1y^2$$
$$(x + y)^3 = 1x^3 + 3x^2y + 3xy^2 + 1y^3$$
$$(x + y)^4 = 1x^4 + 4x^3y + 6x^2y^2 + 4xy^3 + 1y^4$$
$$(x + y)^5 = 1x^5 + 5x^4y + 10x^3y^2 + 10x^2y^3 + 5xy^4 + 1y^5$$
$$(x + y)^6 = 1x^6 + 6x^5y + 15x^4y^2 + 20x^3y^3 + 15x^2y^4 + 6xy^5 + 1y^6$$
$$(x + y)^7 = 1x^7 + 7x^6y + 21x^5y^2 + 35x^4y^3 + 35x^3y^4 + 21x^2y^5 + 7xy^6 + 1y^7$$
$$\vdots$$

The top row in Pascal's Triangle is called the *zero row* because it corresponds to the binomial expansion $(x + y)^0 = 1$.

Similarly, the next row is called the *first row* because it corresponds to the binomial expansion $(x + y)^1 = 1(x) + 1(y)$.

In general, the *nth row* in Pascal's Triangle gives the coefficients of $(x + y)^n$.

EXAMPLE 3   Using Pascal's Triangle

Use the seventh row of Pascal's Triangle to find the binomial coefficients.

$$_8C_0, \, _8C_1, \, _8C_2, \, _8C_3, \, _8C_4, \, _8C_5, \, _8C_6, \, _8C_7, \, _8C_8$$

Solution

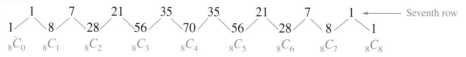

Seventh row

Exploration

Complete the following table and describe the result.

n	r	$_nC_r$	$_nC_{n-r}$
9	5		
7	1		
12	4		
6	0		
10	7		

What characteristics of Pascal's Triangle are illustrated by the table?

Binomial Expansions

As mentioned at the beginning of this section, when you write out the coefficients for a binomial that is raised to a power, you are **expanding a binomial.** The formulas for binomial coefficients give you an easy way to expand binomials, as demonstrated in the next four examples.

EXAMPLE 4   Expanding a Binomial

Write the expansion for the expression $(x + 1)^3$.

Solution
The binomial coefficients from the third row of Pascal's Triangle are

$1, 3, 3, 1.$

Therefore, the expansion is as follows.

$$(x + 1)^3 = (1)x^3 + (3)x^2(1) + (3)x(1^2) + (1)(1^3)$$

$$= x^3 + 3x^2 + 3x + 1$$

To expand binomials representing *differences*, rather than sums, you alternate signs. Here are two examples.

$$(x - 1)^3 = x^3 - 3x^2 + 3x - 1$$

$$(x - 1)^4 = x^4 - 4x^3 + 6x^2 - 4x + 1$$

EXAMPLE 5   Expanding a Binomial

Write the expansion for the expression

$(2x - 3)^4.$

Solution
The binomial coefficients from the fourth row of Pascal's Triangle are

$1, 4, 6, 4, 1.$

Therefore, the expansion is as follows.

$$(2x - 3)^4 = (1)(2x)^4 - (4)(2x)^3(3) + (6)(2x)^2(3^2) - (4)(2x)(3^3) + (1)(3^4)$$

$$= 16x^4 - 96x^3 + 216x^2 - 216x + 81$$

You can use a graphing utility to check the expansion by graphing the original binomial expression and the expansion in the same viewing window. The graphs should coincide, as shown in Figure 9.12.

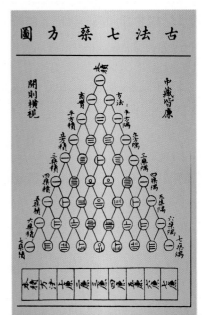

"Pascal's" Triangle and forms of the Binomial Theorem were known in Eastern cultures prior to the Western "discovery" of the theorem. The Chinese text *Precious Mirror* contains a triangle of binomial expansions through the eighth power.

A computer animation of this concept appears in the *Interactive* CD-ROM and *Internet* versions of this text.

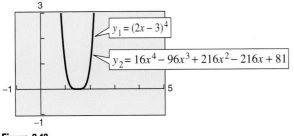

Figure 9.12

EXAMPLE 6   Expanding a Binomial

Write the expansion for $(x - 2y)^4$.

Solution

Use the fourth row of Pascal's Triangle, as follows.

$$(x - 2y)^4 = (1)x^4 - (4)x^3(2y) + (6)x^2(2y)^2 - (4)x(2y)^3 + (1)(2y)^4$$
$$= x^4 - 8x^3y + 24x^2y^2 - 32xy^3 + 16y^4$$

EXAMPLE 7   Expanding a Binomial

Write the expansion for $(x^2 + 4)^3$.

Solution

Use the third row of Pascal's Triangle, as follows.

$$(x^2 + 4)^3 = (1)(x^2)^3 + (3)(x^2)^2(4) + (3)x^2(4^2) + (1)(4^3)$$
$$= x^6 + 12x^4 + 48x^2 + 64$$

To find a specific term in a binomial expansion, use the fact that from the Binomial Theorem, the $(r + 1)$st term is

$$_nC_r x^{n-r} y^r.$$

EXAMPLE 8   Finding a Term in a Binomial Expansion

Find the sixth term of $(a + 2b)^8$.

Solution

For the sixth term in this binomial expansion, use $n = 8$ and $r = 5$ to get

$$_8C_5 a^{8-5}(2b)^5 = 56 \cdot a^3 \cdot (2b)^5$$
$$= 56(2^5)a^3b^5$$
$$= 1792a^3b^5.$$

Writing About Math   *Error Analysis*

Suppose you are a math instructor and receive the following solutions from one of your students on a quiz. Find the error(s) in each solution and write a short paragraph discussing ways that your student could avoid the error(s) in the future.

a. Find the second term in the expansion of $(2x - 3y)^5$.

$$5(2x)^4(3y)^2 = 720x^4y^2 \ \times$$

b. Find the fourth term in the expansion of $\left(\frac{1}{2}x + 7y\right)^6$.

$$_6C_4\left(\tfrac{1}{2}x\right)^2(7y)^4 = 9003.75x^2y^4 \ \times$$

9.5   E x e r c i s e s

In Exercises 1–12, evaluate $_nC_r$.

1. $_7C_5$

2. $_9C_6$

3. $\binom{12}{0}$

4. $\binom{20}{20}$

5. $_{20}C_{15}$

6. $_{12}C_3$

7. $_{14}C_1$

8. $_{18}C_{17}$

9. $\binom{100}{98}$

10. $\binom{10}{7}$

11. $_{100}C_2$

12. $_{13}C_8$

In Exercises 13–18, use a graphing utility to evaluate $_nC_r$.

13. $_{32}C_{28}$

14. $_{17}C_4$

15. $_{22}C_9$

16. $_{52}C_{47}$

17. $_{41}C_{36}$

18. $_{34}C_4$

In Exercises 19–22, evaluate using Pascal's Triangle.

19. $_7C_2$

20. $_6C_4$

21. $_8C_5$

22. $_8C_6$

In Exercises 23–44, use the Binomial Theorem to expand and simplify the expression.

23. $(x + 1)^4$

24. $(x + 1)^6$

25. $(a + 3)^3$

26. $(a + 2)^4$

27. $(y - 2)^4$

28. $(y - 2)^5$

29. $(x + y)^5$

30. $(x + y)^6$

31. $(r + 2s)^6$

32. $(x + 3y)^4$

33. $(x - y)^5$

34. $(2x - y)^5$

35. $(1 - 4x)^3$

36. $(5 - 2y)^3$

37. $(x^2 + 5)^4$

38. $(x^2 + y^2)^6$

39. $\left(\dfrac{1}{x} + y\right)^5$

40. $\left(\dfrac{1}{x} + 2y\right)^6$

41. $2(x - 3)^4 + 5(x - 3)^2$

42. $3(x + 1)^5 - 4(x + 1)^3$

43. $-3(x - 2)^3 - 4(x + 1)^6$

44. $6(x + 2)^5 - 2(x - 1)^2$

In Exercises 45–48, expand the binomial using Pascal's Triangle to determine the coefficients.

45. $(3t - s)^5$

46. $(x + 2y)^5$

47. $(3 - 2z)^4$

48. $(3y + 2)^5$

In Exercises 49–56, find the coefficient a of the given term in the expansion of the binomial.

Binomial	*Term*
49. $(x + 3)^{12}$	ax^4
50. $(x^2 + 3)^{12}$	ax^{10}
51. $(x - 2y)^{10}$	ax^8y^2
52. $(4x - y)^{10}$	ax^2y^8
53. $(3x - 2y)^9$	ax^6y^3
54. $(2x - 3y)^8$	ax^4y^4
55. $(x^2 + y)^{10}$	ax^8y^6
56. $(z^2 - 1)^{12}$	az^6

In Exercises 57–60, use the Binomial Theorem to expand and simplify the expression.

57. $\left(\sqrt{x} + 5\right)^4$

58. $\left(4\sqrt{t} - 1\right)^3$

59. $(x^{2/3} - y^{1/3})^3$

60. $(u^{3/5} + 2)^5$

In Exercises 61–64, expand the binomial in the difference quotient and simplify.

$$\dfrac{f(x + h) - f(x)}{h}$$

61. $f(x) = x^3$

62. $f(x) = x^4$

63. $f(x) = \sqrt{x}$

64. $f(x) = \dfrac{1}{x}$

In Exercises 65–70, use the Binomial Theorem to expand the complex number. Simplify your result.

65. $(1 + i)^4$

66. $(4 - i)^5$

67. $(2 - 3i)^6$

68. $\left(5 + \sqrt{-9}\right)^3$

69. $\left(-\dfrac{1}{2} + \dfrac{\sqrt{3}}{2}i\right)^3$

70. $\left(5 - \sqrt{3}i\right)^4$

Approximation   In Exercises 71–74, use the Binomial Theorem to approximate the given quantity accurate to three decimal places. For example, in Exercise 71, use the expansion

$$(1.02)^8 = (1 + 0.02)^8 = 1 + 8(0.02) + 28(0.02)^2 + \cdots.$$

71. $(1.02)^8$

72. $(2.005)^{10}$

73. $(2.99)^{12}$

74. $(1.98)^9$

Graphical Reasoning   In Exercises 75–78, use a graphing utility to obtain the graph of f and g in the same viewing window. What is the relationship between the two graphs? Use the Binomial Theorem to write the polynomial function g in standard form.

75. $f(x) = x^3 - 4x, \quad g(x) = f(x + 6)$

76. $f(x) = -x^4 + 4x^2 - 1, \quad g(x) = f(x - 4)$

77. $f(x) = -x^2 + 3x + 2, \quad g(x) = f(x - 2)$

78. $f(x) = 2x^2 - 4x + 1, \quad g(x) = f(x + 3)$

Exploration   In Exercises 79 and 80, use a graphing utility to evaluate and determine which two are equal.

79. (a) $_{12}C_5$ (b) $(_6C_5)^2$

(c) $_{11}C_5 + _{11}C_4$ (d) $_6C_5 + _6C_5$

80. (a) $_{25}C_6$ (b) $2(_{25}C_2 + _{25}C_4)$

(c) $\displaystyle\sum_{k=0}^{5} [(_{10}C_k)(_8C_{5-k})]$ (d) $_{18}C_5$

Graphical Reasoning   In Exercises 81 and 82, use a graphing utility to obtain the graphs of the functions in the given order and in the same viewing window. Compare the graphs. Which two functions have identical graphs and why?

81. (a) $f(x) = (1 - x)^3$

(b) $g(x) = 1 - 3x$

(c) $h(x) = 1 - 3x + 3x^2$

(d) $p(x) = 1 - 3x + 3x^2 - x^3$

82. (a) $f(x) = \left(1 - \frac{1}{2}x\right)^4$

(b) $g(x) = 1 - 2x + \frac{3}{2}x^2$

(c) $h(x) = 1 - 2x + \frac{3}{2}x^2 - \frac{1}{2}x^3$

(d) $p(x) = 1 - 2x + \frac{3}{2}x^2 - \frac{1}{2}x^3 + \frac{1}{16}x^4$

Probability   In Exercises 83–86, consider n independent trials of an experiment where each trial has two possible outcomes, called a success and a failure, respectively. The probability of a success on each trial is p and the probability of a failure is $q = 1 - p$. In this context, the term $_nC_k\, p^k q^{n-k}$ in the expansion of $(p + q)^n$ gives the probability of k successes in the n trials of the experiment.

83. A fair coin is tossed seven times. To find the probability of obtaining 4 heads, evaluate the term

$$_7C_4\left(\tfrac{1}{2}\right)^4\left(\tfrac{1}{2}\right)^3$$

in the expansion $\left(\tfrac{1}{2} + \tfrac{1}{2}\right)^7$.

84. The probability of a baseball player getting a hit on any given time at bat is $\frac{1}{4}$. To find the probability that the player gets 3 hits during the next 10 times at bat, evaluate the term

$$_{10}C_3\left(\tfrac{1}{4}\right)^3\left(\tfrac{3}{4}\right)^7$$

in the expansion $\left(\tfrac{1}{4} + \tfrac{3}{4}\right)^{10}$.

85. The probability of a sales representative making a sale with any one customer is $\frac{1}{3}$. The sales representative makes 8 contacts a day. To find the probability of making 4 sales, evaluate the term

$$_8C_4\left(\tfrac{1}{3}\right)^4\left(\tfrac{2}{3}\right)^4$$

in the expansion $\left(\tfrac{1}{3} + \tfrac{2}{3}\right)^8$.

86. To find the probability that the sales representative in Exercise 85 makes 4 sales if the probability of a sale with any one customer is $\frac{1}{2}$, evaluate the term

$$_8C_4\left(\tfrac{1}{2}\right)^4\left(\tfrac{1}{2}\right)^4$$

in the expansion $\left(\tfrac{1}{2} + \tfrac{1}{2}\right)^8$.

87. *Life Insurance*   The average amount of life insurance per household f (in thousands of dollars) from 1980 through 1996 can be approximated by

$$f(t) = 0.0348t^2 + 5.1083t + 41.0250, \quad 0 \le t \le 16$$

where $t = 0$ represents 1980. You want to adjust this model so that $t = 0$ corresponds to 1990 rather than 1980. To do this, you shift the graph of f 10 units *to the left* and obtain

$$g(t) = f(t + 10).$$

(Source: American Council of Life Insurance)

(a) Write $g(t)$ in standard form.

(b) Use a graphing utility to graph f and g in the same viewing window.

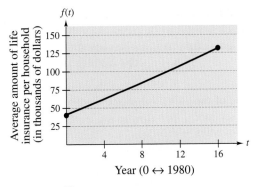

FIGURE FOR 87

88. *Health Maintenance Organizations*   The number of people f (in millions) enrolled in health maintenance organizations in the United States from 1976 through 1997 can be approximated by the model

$$f(t) = 0.0834t^2 + 0.7657t + 5.3680, \quad 0 \le t \le 21$$

where $t = 0$ represents 1976. You want to adjust this model so that $t = 0$ corresponds to 1980 rather than 1976. To do this, you shift the graph of f 4 units *to the left* and obtain $g(t) = f(t + 4)$.   (Source: Group Health Association of America, Interstudy)

(a) Write $g(t)$ in standard form.

(b) Use a graphing utility to graph f and g in the same viewing window.

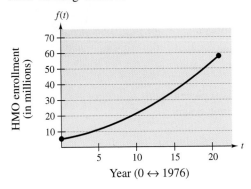

Synthesis

True or False?   **In Exercises 89 and 90, determine whether the statement is true or false. Justify your answer.**

89. One of the terms in the expansion of $(x - 2y)^{12}$ is $7920x^4y^8$.

90. The x^{10}-term and the x^{14}-term of the expansion of $(x^2 + 3)^{12}$ have identical coefficients.

91. *Writing*   In your own words, explain how to form the rows of Pascal's Triangle.

92. Form the first nine rows of Pascal's Triangle.

93. *Think About It*   How many terms are in the expansion of $(x + y)^n$?

94. *Think About It*   How do the expansions of $(x + y)^n$ and $(x - y)^n$ differ?

In Exercises 95–98, prove the given property for all integers r and n where $0 \le r \le n$.

95. $_nC_r = {}_nC_{n-r}$

96. $_nC_0 - {}_nC_1 + {}_nC_2 - \cdots \pm {}_nC_n = 0$

97. $_{n+1}C_r = {}_nC_r + {}_nC_{r-1}$

98. The sum of the numbers in the nth row of Pascal's Triangle is 2^n.

Review

In Exercises 99–102, describe the relationship between the graphs of f and g.

99. $g(x) = f(x) + 8$ **100.** $g(x) = f(x - 3)$

101. $g(x) = f(-x)$ **102.** $g(x) = -f(x)$

In Exercises 103–108, perform the matrix operation using the given matrices.

$$A = \begin{bmatrix} -2 & 3 & 0 \\ 5 & -1 & 2 \\ 4 & 0 & 1 \end{bmatrix} \quad B = \begin{bmatrix} 0 & -4 & -3 \\ 3 & 1 & 2 \\ -5 & -2 & 6 \end{bmatrix}$$

103. $A + B$ **104.** $4A - B$

105. $-3A - 5B$ **106.** $6A + 10B$

107. AB **108.** BA

In Exercises 109 and 110, find the inverse of the matrix.

109. $\begin{bmatrix} -6 & 5 \\ -5 & 4 \end{bmatrix}$ **110.** $\begin{bmatrix} 1.2 & -2.3 \\ -2 & 4 \end{bmatrix}$

9.6 Counting Principles

Simple Counting Problems

The last two sections of this chapter present a brief introduction to some of the basic counting principles and their application to probability. In the next section, you will see that much of probability has to do with counting the number of ways an event can occur.

EXAMPLE 1   Selecting Pairs of Numbers at Random

Eight pieces of paper are numbered from 1 to 8 and placed in a box. One piece of paper is drawn from the box, its number is written down, and the piece of paper is replaced in the box. Then, a piece of paper is again drawn from the box, and its number is written down. Finally, the two numbers are added together. How many different ways can a total of 12 be obtained?

Solution

To solve this problem, count the different ways that a total of 12 can be obtained using two numbers from 1 to 8.

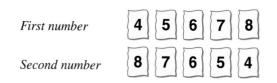

First number   4 5 6 7 8

Second number   8 7 6 5 4

From this list, you can see that a total of 12 can occur in five different ways.

EXAMPLE 2   Selecting Pairs of Numbers at Random

Eight pieces of paper are numbered from 1 to 8 and placed in a box. Two pieces of paper are drawn from the box, and the numbers on the paper are written down and totaled. How many different ways can a total of 12 be obtained?

Solution

To solve this problem, count the different ways that a total of 12 can be obtained using two *different* numbers from 1 to 8.

First number   4 5 7 8

Second number   8 7 5 4

So, a total of 12 can be obtained in four different ways.

The difference between the counting problems in Examples 1 and 2 can be distinguished by saying that the random selection in Example 1 occurs *with replacement,* whereas the random selection in Example 2 occurs *without replacement,* which eliminates the possibility of choosing two 6's.

The Fundamental Counting Principle

Examples 1 and 2 describe simple counting problems in which you can *list* each possible way that an event can occur. When it is possible, this is always the best way to solve a counting problem. However, some events can occur in so many different ways that it is not feasible to write out the entire list. In such cases, you must rely on formulas and counting principles. The most important of these is the **Fundamental Counting Principle.**

> ## Fundamental Counting Principle
>
> Let E_1 and E_2 be two events. The first event E_1 can occur in m_1 different ways. After E_1 has occurred, E_2 can occur in m_2 different ways. The number of ways that the two events can occur is
>
> $m_1 \cdot m_2.$

The Fundamental Counting Principle can be extended to three or more events. For instance, the number of ways that three events E_1, E_2, and E_3 can occur is

$m_1 \cdot m_2 \cdot m_3.$

EXAMPLE 3   Using the Fundamental Counting Principle

How many different pairs of letters from the English alphabet are possible?

Solution
This experiment has two events. The first event is the choice of the first letter, and the second event is the choice of the second letter. Because the English alphabet contains 26 letters, it follows that the number of letter pairs is

$26 \cdot 26 = 676.$

EXAMPLE 4   Counting Telephone Numbers

Telephone numbers in the United States have 10 digits. The first three are the *area code* and the next seven are the *local telephone number*. How many different telephone numbers are possible within each area code? (Note that a local telephone number cannot begin with 0 or 1.)

Solution
Because the first digit cannot be 0 or 1, there are only eight choices for the first digit. For each of the other six digits, there are 10 choices.

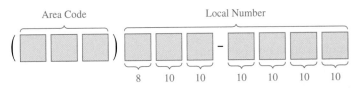

So, the number of local telephone numbers that are possible within each area code is $8 \cdot 10 \cdot 10 \cdot 10 \cdot 10 \cdot 10 \cdot 10 = 8,000,000.$

Permutations

One important application of the Fundamental Counting Principle is in determining the number of ways that n elements can be arranged (in order). An ordering of n elements is called a **permutation** of the elements.

Definition of Permutation

A **permutation** of n different elements is an ordering of the elements such that one element is first, one is second, one is third, and so on.

EXAMPLE 5   Finding the Number of Permutations of n Elements

How many permutations are possible for the letters A, B, C, D, E, and F?

Solution

Consider the following reasoning.

First position:	Any of the *six* letters
Second position:	Any of the remaining *five* letters
Third position:	Any of the remaining *four* letters
Fourth position:	Any of the remaining *three* letters
Fifth position:	Either of the remaining *two* letters
Sixth position:	The *one* remaining letter

So, the number of choices for the six positions are as follows.

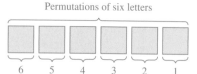

Permutations of six letters

6   5   4   3   2   1

The total number of permutations of the six letters is

$$6! = 6 \cdot 5 \cdot 4 \cdot 3 \cdot 2 \cdot 1$$
$$= 720.$$

Number of Permutations of n Elements

The number of permutations of n elements is

$$n \cdot (n - 1) \cdots 4 \cdot 3 \cdot 2 \cdot 1 = n!.$$

In other words, there are $n!$ different ways that n elements can be ordered.

Occasionally, you are interested in ordering a *subset* of a collection of elements rather than the entire collection. For example, you might want to choose (and order) r elements out of a collection of n elements. Such an ordering is called a **permutation of n elements taken r at a time.**

EXAMPLE 6   Counting Horse Race Finishes

Eight horses are running in a race. In how many different ways can these horses come in first, second, and third? (Assume that there are no ties.)

Solution

Here are the different possibilities.

Win (first position):	*Eight* choices
Place (second position):	*Seven* choices
Show (third position):	*Six* choices

The number of choices for the three positions are as follows.

Different orders of horses

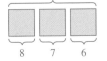

8   7   6

So, there are

$$8 \cdot 7 \cdot 6 = 336$$

different ways the eight horses can come in first, second, and third.

Permutations of *n* Elements Taken *r* at a Time

The number of **permutations of *n* elements taken *r* at a time** is

$$_nP_r = \frac{n!}{(n-r)!}$$

$$= n(n-1)(n-2) \cdots (n-r+1).$$

Using this formula, you can rework Example 6 to find that the number of permutations of eight horses taken three at a time is

$$_8P_3 = \frac{8!}{5!}$$

$$= \frac{8 \cdot 7 \cdot 6 \cdot 5!}{5!}$$

$$= 336$$

which is the same answer.

Most graphing utilities are programmed to evaluate $_nP_r$. Consult your user's manual to learn how to evaluate the permutation $_8P_3$. You should get 336, as shown in Example 6.

Remember that for permutations, order is important. So, if you are looking at the possible permutations of the letters A, B, C, and D taken three at a time, the permutations (A,B,D) and (B,A,D) would be counted as different because the *order* of the elements is different.

Suppose, however, that you are asked to find the possible permutations of the letters A, A, B, and C. The total number of permutations of the four letters would be $_4P_4 = 4!$. However, not all of these arrangements would be *distinguishable* because there are two A's in the list. To find the number of distinguishable permutations, you can use the following formula.

Distinguishable Permutations

Suppose a set of n objects has n_1 of one kind of object, n_2 of a second kind, n_3 of a third kind, and so on, with

$$n = n_1 + n_2 + n_3 + \cdots + n_k.$$

The number of **distinguishable permutations** of the n objects is

$$\frac{n!}{n_1! \cdot n_2! \cdot n_3! \cdots n_k!}.$$

EXAMPLE 7 Distinguishable Permutations

In how many distinguishable ways can the letters in BANANA be written?

Solution

This word has six letters, of which three are A's, two are N's, and one is a B. So, the number of distinguishable ways the letters can be written is

$$\frac{6!}{3! \cdot 2! \cdot 1!} = \frac{6 \cdot 5 \cdot 4 \cdot 3!}{3! \cdot 2!}$$

$$= 60.$$

The 60 different arrangements are as follows.

AAABNN	AAANBN	AAANNB	AABANN
AABNAN	AABNNA	AANABN	AANANB
AANBAN	AANBNA	AANNAB	AANNBA
ABAANN	ABANAN	ABANNA	ABNAAN
ABNANA	ABNNAA	ANAABN	ANAANB
ANABAN	ANABNA	ANANAB	ANANBA
ANBAAN	ANBANA	ANBNAA	ANNAAB
ANNABA	ANNBAA	BAAANN	BAANAN
BAANNA	BANAAN	BANANA	BANNAA
BNAAAN	BNAANA	BNANAA	BNNAAA
NAAABN	NAAANB	NAABAN	NAABNA
NAANAB	NAANBA	NABAAN	NABANA
NABNAA	NANAAB	NANABA	NANBAA
NBAAAN	NBAANA	NBANAA	NBNAAA
NNAAAB	NNAABA	NNABAA	NNBAAA

Combinations

When you count the number of possible permutations of a set of elements, order is important. As a final topic in this section, you will look at a method of selecting subsets of a larger set in which order *is not* important. Such subsets are called **combinations of *n* elements taken *r* at a time.** For instance, the combinations

$$\{A, B, C\} \quad \text{and} \quad \{B, A, C\}$$

are equivalent because both sets contain the same three elements, and the order in which the elements are listed is not important. So, you would count only one of the two sets. A common example of how a combination occurs is a card game in which the player is free to reorder the cards after they have been dealt.

EXAMPLE 8   Combinations of *n* Elements Taken *r* at a Time

In how many different ways can three letters be chosen from the letters A, B, C, D, and E? (The order of the three letters is not important.)

Solution
The following subsets represent the different combinations of three letters that can be chosen from five letters.

$$\{A, B, C\} \quad \{A, B, D\}$$

$$\{A, B, E\} \quad \{A, C, D\}$$

$$\{A, C, E\} \quad \{A, D, E\}$$

$$\{B, C, D\} \quad \{B, C, E\}$$

$$\{B, D, E\} \quad \{C, D, E\}$$

From this list, you can conclude that there are 10 different ways that three letters can be chosen from five letters.

Combinations of *n* Elements Taken *r* at a Time

The number of **combinations of *n* elements taken *r* at a time** is

$$_nC_r = \frac{n!}{(n-r)!r!}.$$

Note that the formula for $_nC_r$ is the same one given for binomial coefficients. To see how this formula is used, solve the counting problem in Example 8. In that problem, you are asked to find the number of combinations of five elements taken three at a time. So, $n = 5$, $r = 3$, and the number of combinations is

$$_5C_3 = \frac{5!}{2!3!} = \frac{5 \cdot 4 \cdot \overset{2}{3!}}{2 \cdot 1 \cdot 3!} = 10$$

which is the same answer obtained in Example 8.

EXAMPLE 9 Counting Card Hands

A standard poker hand consists of five cards dealt from a deck of 52. How many different poker hands are possible?

Solution

Note that after the cards are dealt, the players may reorder them, and therefore order is not important. So, you can find the number of different poker hands by using the formula for the number of combinations of 52 elements taken five at a time, as follows.

$$_{52}C_5 = \frac{52!}{47!5!} = \frac{52 \cdot 51 \cdot 50 \cdot 49 \cdot 48 \cdot 47!}{5 \cdot 4 \cdot 3 \cdot 2 \cdot 1 \cdot 47!} = 2{,}598{,}960$$

EXAMPLE 10 The Number of Subsets of a Set

Find the total number of subsets of a set that has 10 elements.

Solution

Begin by considering the number of subsets with 0 elements, the number with 1 element, the number with 2 elements, and so on.

Number of subsets	=	Subsets with 0 elements	+	Subsets with 1 element	+	Subsets with 2 elements	+ · · · +	Subsets with 10 elements

$$= \binom{10}{0} + \binom{10}{1} + \binom{10}{2} + \cdots + \binom{10}{10}$$

Compare this expression to the binomial expansion of $(1 + 1)^{10}$ to see that they are the same.

$$(1 + 1)^{10} = \binom{10}{0}1^{10}1^0 + \binom{10}{1}1^91^1 + \binom{10}{2}1^81^2 + \cdots + \binom{10}{10}1^01^{10}$$

$$= \binom{10}{0} + \binom{10}{1} + \binom{10}{2} + \cdots + \binom{10}{10}$$

This implies that the total number of subsets of a set of 10 elements is

$$(1 + 1)^{10} = 2^{10} = 1024.$$

The result of Example 10 can be generalized to conclude that the total number of subsets of a set of n elements is 2^n.

Writing About Math *Problem Posing*

According to NASA, each space shuttle astronaut consumes an average of 3000 calories per day. An evening meal normally consists of a main dish, a vegetable dish, and two different desserts. The space shuttle food list contains ten items classified as main dishes, eight vegetable dishes, and 13 desserts. How many different evening meal menus are possible? Write a short paragraph explaining how you determined this number. Create two other problems that could be asked about the evening meal menus and solve them. (Source: NASA)

9.6   E x e r c i s e s

Random Selection   **In Exercises 1–8, determine the number of ways a computer can randomly generate one or more such integers from 1 through 12.**

1. An odd integer

2. An even integer

3. A prime integer

4. An integer that is greater than 7

5. An integer that is divisible by 4

6. An integer that is divisible by 3

7. Two integers whose sum is 10

8. Two *distinct* integers whose sum is 10

9. ***Entertainment Systems***   A customer can choose one of four amplifiers, one of six compact disc players, and one of eight speaker models for an entertainment system. Determine the number of possible system configurations.

10. ***Computer Systems***   A customer in a computer store can choose one of three monitors, one of two keyboards, and one of seven computers. If all the choices are compatible, determine the number of possible system configurations.

11. ***Job Applicants***   A college needs two additional faculty members: a chemist and a statistician. In how many ways can these positions be filled if there are three applicants for the chemistry position and six applicants for the statistics position?

12. ***Course Schedule***   A college student is preparing a course schedule for the next semester. The student must select one of two mathematics courses, one of three science courses, and one of five courses from the social sciences and humanities. How many schedules are possible?

13. ***Periodic Table***   You are taking a chemistry test and are asked to list the first ten elements *in order* as they appear in the periodic table of elements. Suppose you have no idea of the correct order and simply guess. In how many different orders could you list the elements?

14. ***True-False Exam***   In how many ways can a ten-question true-false exam be answered? (Assume that no questions are omitted.)

15. ***Toboggan Ride***   Four people are lining up for a ride on a toboggan, but only two of the four are willing to take the first position. With that constraint, in how many ways can the four people be seated on the toboggan?

16. ***Taking a Trip***   Four people are taking a long trip in a four-seat car. Three of the people agree to share the driving. In how many different arrangements can the four people sit?

17. ***License Plate Numbers***   In a certain state the automobile license plates consist of two letters followed by a four-digit number. How many distinct license plate numbers can be formed?

18. ***License Plate Numbers***   In a certain state the automobile license plates consist of two letters followed by a four-digit number. To avoid confusion between "O" and "zero" and "I" and "one," the letters "O" and "I" are not used. How many distinct license plate numbers can be formed?

19. ***Three-Digit Numbers***   How many three-digit numbers can be formed under the following conditions?

    (a) The leading digit cannot be zero.

    (b) The leading digit cannot be zero and no repetition of digits is allowed.

    (c) The leading digit cannot be zero and the number must be a multiple of 5.

    (d) The number is at least 400.

20. ***Four-Digit Numbers***   How many four-digit numbers can be formed under the following conditions?

    (a) The leading digit cannot be zero.

    (b) The leading digit cannot be zero and no repetition of digits is allowed.

    (c) The leading digit cannot be zero and the number must be less than 5000.

    (d) The leading digit cannot be zero and the number must be even.

21. ***Combination Lock***   A combination lock will open when the right choice of three numbers (from 1 to 40, inclusive) is selected. How many different lock combinations are possible?

22. Telephone Numbers In 1997, the Commonwealth of Massachusetts had three area codes: one for the Boston metropolitan area, one for the rest of eastern Massachusetts, and one for western Massachusetts. Using the information about telephone numbers in Example 4, how many telephone numbers could the phone system have accommodated in the state of Massachusetts?

23. Concert Seats Three couples have reserved seats in a given row for a concert. In how many different ways can they be seated if

(a) there are no seating restrictions?

(b) the two members of each couple wish to sit together?

24. Single File In how many orders can five girls and three boys walk through a doorway single file if

(a) there are no restrictions?

(b) the girls walk through before the boys?

In Exercises 25–30, evaluate $_nP_r$ using the formula.

25. $_4P_4$

26. $_5P_5$

27. $_8P_3$

28. $_{20}P_2$

29. $_5P_4$

30. $_7P_4$

In Exercises 31 and 32, solve for n.

31. $14 \cdot {}_nP_3 = {}_{n+2}P_4$

32. $_nP_5 = 18 \cdot {}_{n-2}P_4$

In Exercises 33–38, evaluate using a graphing utility.

33. $_{20}P_6$

34. $_{100}P_5$

35. $_{120}P_4$

36. $_{10}P_8$

37. $_{20}C_4$

38. $_{10}C_7$

39. Posing for a Photograph In how many ways can five children line up in a row?

40. Riding in a Car In how many ways can four people sit in a four-passenger car?

41. Morse Code In Morse code, all characters are transmitted using a sequence of *dots* and *dashes*.

(a) How many different characters can be formed with a sequence of four symbols, each of which is a dot or dash?

(b) How many can be formed with a sequence of one, two, or three symbols?

42. Assembly Line Production Four processes are involved in assembling a certain product, and they can be performed in any order. The management wants to test each order to determine which is the least time consuming. How many different orders will have to be tested?

In Exercises 43–46, find the number of distinguishable permutations of the group of letters.

43. A, A, G, E, E, E, M

44. B, B, B, T, T, T, T, T

45. A, L, G, E, B, R, A

46. M, I, S, S, I, S, S, I, P, P, I

47. Write all permutations of the letters A, B, C, and D.

48. Write all the permutations of the letters A, B, C, and D if the letters B and C must remain between the letters A and D.

49. Write all the possible selections of two letters that can be formed from the letters A, B, C, D, E, and F. (The order of the two letters is not important.)

50. Write all the possible selections of three letters that can be formed from the letters A, B, C, D, E, and F. (The order of the three letters is not important.)

51. Forming an Experimental Group In order to conduct a certain experiment, four students are randomly selected from a class of 20. How many different groups of four students are possible?

52. Test Questions You can answer any 12 questions from a total of 14 questions on an exam. In how many different ways can you select the questions?

53. Lottery Choices There are 40 numbers in a particular state lottery. In how many ways can a player select six of the numbers?

54. Lottery Choices There are 50 numbers in a particular state lottery. In how many ways can a player select six of the numbers?

55. Number of Subsets How many subsets of five elements can be formed from a set of 100 elements?

56. Number of Subsets How many subsets of six elements can be formed from a set of 80 elements?

57. Geometry Three points that are not on a line determine three lines. How many lines are determined by nine points, no three of which are on a line?

58. Defective Units A shipment of 25 television sets contains three defective units. In how many ways can a vending company purchase four of these units and receive (a) all good units, (b) two good units, and (c) at least two good units?

59. *Job Applicants* An employer interviews 12 people for four openings in the company. Five of the 12 people are women. If all 12 are qualified, in how many ways can the employer fill the four positions if (a) the selection is random and (b) exactly two women are selected?

60. *Poker Hand* Five cards are selected from an ordinary deck of 52 playing cards. In how many ways can you get a full house? (A full house consists of three of one kind and two of another. For example, 8-8-8-5-5 and K-K-K-10-10 are full houses.)

61. *Forming a Committee* Four people are to be selected at random from a group of four couples. In how many ways can this be done, given the following conditions?

(a) There are no restrictions.

(b) The group must have at least one couple.

(c) Each couple must be represented in the group.

62. *Interpersonal Relationships* The complexity of the interpersonal relationships increases dramatically as the size of a group increases. Determine the number of different two-person relationships in a group of people of size (a) 3, (b) 8, (c) 12, and (d) 20.

In Exercises 63–66, find the number of diagonals of the polygon. (A line segment connecting any two non-adjacent vertices is called a *diagonal* of the polygon.)

63. Pentagon

64. Hexagon

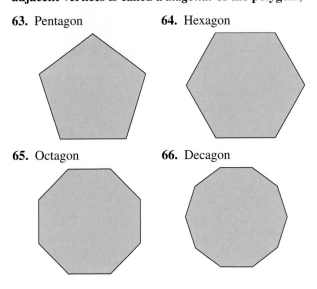

65. Octagon

66. Decagon

Synthesis

True or False? **In Exercises 67–69, determine whether the statement is true or false. Justify your answer.**

67. The number of pairs of letters that can be formed from any of the first 13 letters in the alphabet (A–M) is an example of a permutation.

68. The number of permutations of n elements can be derived by using the Fundamental Counting Principle.

69. $_nP_r > {_nC_r}$ always.

70. *Think About It* Can your calculator evaluate $_{100}P_{80}$? If not, explain why.

71. *Writing* Explain in words the meaning of $_nP_r$.

72. What is the relationship between $_nC_r$ and $_nC_{n-r}$?

73. Without calculating the numbers, determine which of the following is greater. Explain.

(a) The combinations of 10 elements taken 6 at a time

(b) The permutations of 10 elements taken 6 at a time

In Exercises 74–77, prove the identity.

74. $_nP_{n-1} = {_nP_n}$

75. $_nC_n = {_nC_0}$

76. $_nC_{n-1} = {_nC_1}$

77. $_nC_r = \dfrac{_nP_r}{r!}$

Review

In Exercises 78–81, solve the equation. Round your answer to two decimal places, if necessary.

78. $\sqrt{x - 3} = x - 6$

79. $\dfrac{4}{t} + \dfrac{3}{2t} = 1$

80. $\log_2(x - 3) = 5$

81. $e^{x/3} = 16$

In Exercises 82–85, use Cramer's Rule to solve the system of equations.

82. $\begin{cases} -5x + 3y = -14 \\ 7x - 2y = 2 \end{cases}$

83. $\begin{cases} 8x + y = 35 \\ 6x + 2y = 10 \end{cases}$

84. $\begin{cases} -3x - 4y = -1 \\ 9x + 5y = -4 \end{cases}$

85. $\begin{cases} 10x - 11y = -74 \\ -8x - 4y = 8 \end{cases}$

In Exercises 86–89, use the Binomial Theorem to expand and simplify the expression.

86. $(x - 4)^3$

87. $(x - 1)^6$

88. $(x^2 + 4)^5$

89. $(3x - y)^4$

9.7 Probability

The Probability of an Event

Any happening whose result is uncertain is called an **experiment.** The possible results of the experiment are **outcomes,** the set of all possible outcomes of the experiment is the **sample space** of the experiment, and any subcollection of a sample space is an **event.**

For instance, when a six-sided die is tossed, the sample space can be represented by the numbers from 1 through 6. For this experiment, each of the outcomes is *equally likely.*

To describe sample spaces in such a way that each outcome is equally likely, you must sometimes distinguish between various outcomes in ways that appear artificial. Example 1 illustrates such a situation.

EXAMPLE 1   Finding the Sample Space

Find the sample space for each of the following.

a. One coin is tossed.

b. Two coins are tossed.

c. Three coins are tossed.

Solution

a. Because the coin will land either heads up (denoted by H) or tails up (denoted by T), the sample space is

$$S = \{H, T\}.$$

b. Because either coin can land heads up or tails up, the possible outcomes are as follows.

  HH = heads up on both coins

  HT = heads up on first coin and tails up on second coin

  TH = tails up on first coin and heads up on second coin

  TT = tails up on both coins

So, the sample space is

$$S = \{HH, HT, TH, TT\}.$$

Note that this list distinguishes between the two cases HT and TH, even though these two outcomes appear to be similar.

c. Following the notation of part (b), the sample space is

$$S = \{HHH, HHT, HTH, HTT, THH, THT, TTH, TTT\}.$$

Note that this list distinguishes between the cases HHT, HTH, and THH, and between the cases HTT, THT, and TTH.

Telegraph Colour Library/FPG

To calculate the probability of an event, count the number of outcomes in the event and in the sample space. The *number of outcomes* in event E is denoted by $n(E)$ and the number of outcomes in the sample space S is denoted by $n(S)$. The probability that event E will occur is given by $n(E)/n(S)$.

The Probability of an Event

If an event E has $n(E)$ equally likely outcomes and its sample space S has $n(S)$ equally likely outcomes, the **probability** of event E is

$$P(E) = \frac{n(E)}{n(S)}.$$

Because the number of outcomes in an event must be less than or equal to the number of outcomes in the sample space, the probability of an event must be a number from 0 to 1, inclusive. That is,

$$0 \le P(E) \le 1,$$

as indicated in Figure 9.13. If $P(E) = 0$, event E *cannot occur*, and E is called an **impossible event.** If $P(E) = 1$, event E *must occur*, and E is called a **certain event.**

EXAMPLE 2 Finding the Probability of an Event

a. Two coins are tossed. What is the probability that both land heads up?

b. A card is drawn from a standard deck of playing cards. What is the probability that it is an ace?

Solution

a. Following the procedure in Example 1(b), let

$$E = \{HH\}$$

and

$$S = \{HH, HT, TH, TT\}.$$

The probability of getting two heads is

$$P(E) = \frac{n(E)}{n(S)}$$

$$= \frac{1}{4}.$$

b. Because there are 52 cards in a standard deck of playing cards and there are four aces (one in each suit), the probability of drawing an ace is

$$P(E) = \frac{n(E)}{n(S)}$$

$$= \frac{4}{52} = \frac{1}{13}.$$

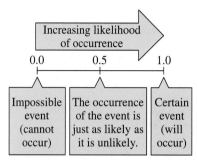

Exploration

Toss two coins 40 times and write down the number of heads that occur on each toss (0, 1, or 2). How many times did two heads occur? How many times would you expect two heads to occur if you did the experiment 1000 times?

Increasing likelihood of occurrence

0.0　　　　0.5　　　　1.0

| Impossible event (cannot occur) | The occurrence of the event is just as likely as it is unlikely. | Certain event (will occur) |

Figure 9.13

Blaise Pascal (1623–1662) was a French mathematician and scientist. He, along with Pierre de Fermat, laid the foundation for the mathematical theory of probability.

EXAMPLE 3 Finding the Probability of an Event

Two six-sided dice are tossed. What is the probability that the total of the two dice is 7? (See Figure 9.14.)

Solution

Because there are six possible outcomes on each die, you can use the Fundamental Counting Principle to conclude that there are 6 · 6 or 36 different outcomes when two dice are tossed. To find the probability of rolling a total of 7, you must first count the number of ways this can occur.

Figure 9.14

First Die	1	2	3	4	5	6
Second Die	6	5	4	3	2	1

A computer simulation of this example appears in the *Interactive* CD-ROM and *Internet* versions of this text.

So, a total of 7 can be rolled in six ways, which means that the probability of rolling a 7 is

$$P(E) = \frac{n(E)}{n(S)}$$

$$= \frac{6}{36}$$

$$= \frac{1}{6}.$$

You could have written out each sample space in Examples 2 and 3 and simply counted the outcomes in the desired events. For larger sample spaces, however, using the counting principles discussed in Section 9.6 should save you time.

EXAMPLE 4 Finding the Probability of an Event

Twelve-sided dice, as shown in Figure 9.15, can be constructed (in the shape of regular dodecahedrons) so that each of the numbers from 1 to 6 appears twice on each die. Prove that these dice can be used in any game requiring ordinary six-sided dice without changing the probabilities of different outcomes.

Solution

For an ordinary six-sided die, each of the numbers 1, 2, 3, 4, 5, and 6 occurs only once, so the probability of any particular number coming up is

$$P(E) = \frac{n(E)}{n(S)} = \frac{1}{6}.$$

For one of the 12-sided dice, each number occurs twice, so the probability of any particular number coming up is

$$P(E) = \frac{n(E)}{n(S)} = \frac{2}{12} = \frac{1}{6}.$$

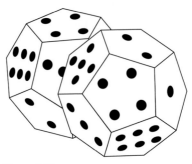

Figure 9.15

EXAMPLE 5   The Probability of Winning a Lottery

In a state lottery, a player chooses six different numbers from 1 to 40. If these six numbers match the six numbers drawn by the lottery commission, the player wins (or shares) the top prize. What is the probability of winning the top prize?

Solution

To find the number of elements in the sample space, use the formula for the number of combinations of 40 elements taken six at a time.

$$n(S) = {}_{40}C_6$$

$$= \frac{40 \cdot 39 \cdot 38 \cdot 37 \cdot 36 \cdot 35}{6 \cdot 5 \cdot 4 \cdot 3 \cdot 2 \cdot 1}$$

$$= 3{,}838{,}380$$

If a person buys only one ticket, the probability of winning is

$$P(E) = \frac{n(E)}{n(S)}$$

$$= \frac{1}{3{,}838{,}380}.$$

EXAMPLE 6   Random Selection

The numbers of colleges and universities in various regions of the United States in 1996 are shown in Figure 9.16. One institution is selected at random. What is the probability that the institution is in one of the two north central regions?   (Source: U.S. National Center for Education Statistics)

Solution

From the figure, the total number of colleges and universities is 3696. Because there are 397 + 578 = 975 colleges and universities in the two north central regions, the probability that the institution is in one of these regions is

$$P(E) = \frac{n(E)}{n(S)} = \frac{975}{3696} \approx 0.264.$$

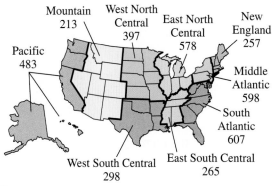

Figure 9.16

Mutually Exclusive Events

Two events A and B (from the same sample space) are **mutually exclusive** if A and B have no outcomes in common. In the terminology of sets, the intersection of A and B is the empty set and

$P(A \cap B) = 0.$

For instance, if two dice are tossed, the event A of rolling a total of 6 and the event B of rolling a total of 9 are mutually exclusive. To find the probability that one or the other of two mutually exclusive events will occur, you can *add* their individual probabilities.

Probability of the Union of Two Events

If A and B are events in the same sample space, the probability of A *or* B occurring is

$P(A \cup B) = P(A) + P(B) - P(A \cap B).$

If A and B are mutually exclusive, then

$P(A \cup B) = P(A) + P(B).$

EXAMPLE 7   The Probability of a Union

One card is selected from a standard deck of 52 playing cards. What is the probability that the card is either a heart or a face card?

Solution

Because the deck has 13 hearts, the probability of selecting a heart (event A) is

$P(A) = \dfrac{13}{52}.$

Similarly, because the deck has 12 face cards, the probability of selecting a face card (event B) is

$P(B) = \dfrac{12}{52}.$

Because three of the cards are hearts and face cards (see Figure 9.17), it follows that

$P(A \cap B) = \dfrac{3}{52}.$

Finally, applying the formula for the probability of the union of two events, you can conclude that the probability of selecting a heart or a face card is

$P(A \cup B) = P(A) + P(B) - P(A \cap B)$

$= \dfrac{13}{52} + \dfrac{12}{52} - \dfrac{3}{52}$

$= \dfrac{22}{52} \approx 0.423.$

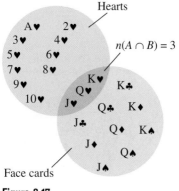

Figure 9.17

EXAMPLE 8   Probability of Mutually Exclusive Events

The personnel department of a company has compiled data on employees' numbers of years of service. The results are shown in the table.

Years of Service	Number of Employees
0–4	157
5–9	89
10–14	74
15–19	63
20–24	42
25–29	38
30–34	37
35–39	21
40–44	8

If an employee is chosen at random, what is the probability that the employee has had nine or fewer years of service?

Solution

To begin, add the number of employees and find that the total is 529. Next, let event A represent choosing an employee with 0 to 4 years of service and let event B represent choosing an employee with 5 to 9 years of service. Then

$$P(A) = \frac{157}{529}$$

and

$$P(B) = \frac{89}{529}.$$

Because A and B have no outcomes in common, you can conclude that these two events are mutually exclusive and that

$$P(A \cup B) = P(A) + P(B)$$

$$= \frac{157}{529} + \frac{89}{529}$$

$$= \frac{246}{529}$$

$$\approx 0.465.$$

So, the probability of choosing an employee who has nine or fewer years of service is about 0.465.

Independent Events

Two events are **independent** if the occurrence of one has no effect on the occurrence of the other. To find the probability that two independent events will occur, *multiply* the probabilities of each. For instance, rolling a total of 12 with two six-sided dice has no effect on the outcome of future rolls of the dice.

Probability of Independent Events

If A and B are **independent events,** the probability that both A and B will occur is

$$P(A \text{ and } B) = P(A) \cdot P(B).$$

EXAMPLE 9 Probability of Independent Events

A random number generator on a computer selects three integers from 1 to 20. What is the probability that all three numbers are less than or equal to 5?

Solution

The probability of selecting a number from 1 to 5 is

$$P(A) = \frac{5}{20} = \frac{1}{4}.$$

So, the probability that all three numbers are less than or equal to 5 is

$$P(A) \cdot P(A) \cdot P(A) = \left(\frac{1}{4}\right)\left(\frac{1}{4}\right)\left(\frac{1}{4}\right)$$

$$= \frac{1}{64}.$$

EXAMPLE 10 Probability of Independent Events

In 1997, 58% of the population of the United States were 30 years old or older. Suppose that in a survey, ten people were chosen at random from the population. What is the probability that all ten were 30 years old or older?   (Source: U.S. Bureau of the Census)

Solution

Let A represent choosing a person who was 30 years old or older. Because the probability of choosing a person who was 30 years old or older was 0.58, you can conclude that the probability that all ten people were 30 years old or older is

$$[P(A)]^{10} = (0.58)^{10}$$

$$\approx 0.0043.$$

The Complement of an Event

The **complement of an event** *A* is the collection of all outcomes in the sample space that are *not* in *A*. The complement of event *A* is denoted by *A′*. Because $P(A$ or $A′) = 1$ and because *A* and *A′* are mutually exclusive, it follows that $P(A) + P(A′) = 1$. Therefore, the probability of *A′* is

$$P(A′) = 1 - P(A).$$

For instance, if the probability of *winning* a certain game is

$$P(A) = \frac{1}{4}$$

then the probability of *losing* the game is

$$P(A′) = 1 - \frac{1}{4}$$

$$= \frac{3}{4}.$$

Exploration

Suppose you are in a class with 22 other people. What is the probability that the birthdays of at least two of the 23 people fall on the same day of the year? What if you know the probability of everyone having the same birthday? Do you think that would help you to find the answer?

Probability of a Complement

Let *A* be an event and let *A′* be its complement. If the probability of *A* is $P(A)$, the probability of the complement is

$$P(A′) = 1 - P(A).$$

EXAMPLE 11   Finding the Probability of a Complement

A manufacturer has determined that a certain machine averages one faulty unit for every 1000 it produces. What is the probability that an order of 200 units will have one or more faulty units?

Solution

To solve this problem as stated, you would need to find the probabilities of having exactly one faulty unit, exactly two faulty units, exactly three faulty units, and so on. However, using complements, you can simply find the probability that all units are perfect and then subtract this value from 1. Because the probability that any given unit is perfect is 999/1000, the probability that all 200 units are perfect is

$$P(A) = \left(\frac{999}{1000}\right)^{200}$$

$$\approx 0.8186.$$

Therefore, the probability that at least one unit is faulty is

$$P(A′) = 1 - P(A)$$

$$\approx 0.1814.$$

9.7 Exercises

In Exercises 1–6, determine the sample space for the given experiment.

1. A coin and a six-sided die are tossed.

2. A six-sided die is tossed twice and the sum of the points is recorded.

3. A taste tester has to rank three varieties of orange juice, A, B, and C, according to preference.

4. Two marbles are selected (without replacement) from a sack containing two red marbles, two blue marbles, and one black marble. The color of each marble is recorded.

5. Two county supervisors are selected from five supervisors, A, B, C, D, and E, to study a recycling plan.

6. A sales representative makes presentations about a product in three homes per day. In each home there may be a sale (denote by S) or there may be no sale (denote by F).

Heads or Tails **In Exercises 7–10, find the probability in the experiment of tossing a coin three times. Use the sample space $S = \{HHH, HHT, HTH, HTT, THH, THT, TTH, TTT\}$.**

7. The probability of getting exactly two tails

8. The probability of getting a head on the first toss

9. The probability of getting at least one head

10. The probability of getting at least two heads

Drawing a Card **In Exercises 11–14, find the probability in the experiment of selecting one card from a standard deck of 52 playing cards.**

11. The card is a face card.

12. The card is not a face card.

13. The card is a black face card.

14. The card is an 8 or lower. (Aces are low.)

Tossing a Die **In Exercises 15–20, find the probability in the experiment of tossing a six-sided die twice.**

15. The sum is 5.

16. The sum is at least 8.

17. The sum is less than 12.

18. The sum is 2, 3, or 12.

19. The sum is odd and no more than 5.

20. The sum is odd or prime.

Drawing Marbles **In Exercises 21–24, find the probability in the experiment of drawing two marbles (without replacement) from a bag containing one green, two yellow, and three red marbles.**

21. Both marbles are red.

22. Both marbles are yellow.

23. Neither marble is yellow.

24. The marbles are of different colors.

In Exercises 25–28, you are given the probability that an event *will* happen. Find the probability that the event *will not* happen.

25. $P(E) = 0.7$ 26. $P(E) = 0.36$

27. $P(E) = \frac{1}{3}$ 28. $P(E) = \frac{5}{6}$

In Exercises 29–32, you are given the probability that an event *will not* happen. Find the probability that the event *will* happen.

29. $P(E') = 0.15$ 30. $P(E') = 0.84$

31. $P(E') = \frac{13}{20}$ 32. $P(E') = \frac{59}{100}$

33. *Graphical Reasoning* In 1997 there were approximately 1.3 million temporary help agency workers in the United States. The circle graph shows the age profile of these workers. (Source: U.S. Bureau of Labor Statistics)

(a) Estimate the number of temporary help agency workers in the age group 16–19.

(b) What is the probability that a person selected at random from the population of temporary agency workers is in the 25–34 age group?

(c) What is the probability that a person selected at random from the population of temporary agency workers is in the 35–54 age group?

(d) What is the probability that a person selected at random from the population of temporary agency workers is 55 or over?

Age of U.S. Workers

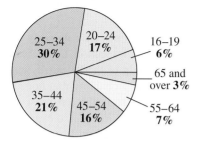

FIGURE FOR 33

34. Graphical Reasoning In 1997 there were approximately 111 million workers in the civilian labor force in the United States. The circle graph shows the educational attainment of these workers. (Source: U.S. Bureau of Labor Statistics)

(a) Estimate the number of workers whose highest education level was a high school education.

(b) A person is selected at random from the civilian work force. What is the probability that the person's highest education level is 1 to 3 years of college?

(c) A person is selected at random from the civilian work force. What is the probability that the person has had more than a high school education?

Educational Attainment
of Workers

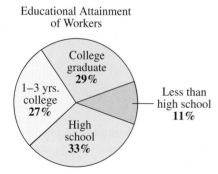

35. Data Analysis A study of the effectiveness of a flu vaccine was conducted with a sample of 500 people. Some in the study were given no vaccine, some were given one injection, and some were given two injections. The results of the study are shown below.

	No Vaccine	One Injection	Two Injections	Total
Flu	7	2	13	22
No flu	149	52	277	478
Total	156	54	290	500

A person is selected at random from the sample. Find the specified probability.

(a) The person had two injections.

(b) The person did not get the flu.

(c) The person got the flu and had one injection.

36. Data Analysis One hundred college students were interviewed to determine their political-party affiliations and whether they favored a balanced-budget amendment to the Constitution. The results of the study are listed in the table.

	Favor	Not favor	Unsure	Total
Democrat	23	25	7	55
Republican	32	9	4	45
Total	55	34	11	100

A person is selected at random from the sample. Find the probability that the described person is selected.

(a) A person who doesn't favor the amendment

(b) A Republican

(c) A Democrat who favors the amendment

37. Alumni Association A college sends a survey to selected members of the class of 2000. Of the 1254 people who graduated that year, 672 are women, of whom 124 went on to graduate school. Of the 582 male graduates, 198 went on to graduate school. If an alumni member is selected at random, what is the probability that the person is (a) female, (b) male, and (c) female and did not attend graduate school?

38. Post–High School Education In a high school graduating class of 128 students, 52 are on the honor roll. Of these, 48 are going on to college; and of the other 76 students, 56 are going on to college. If a student is selected at random from the class, what is the probability that the person chosen is (a) going to college, (b) not going to college, and (c) not going to college and on the honor roll?

39. Winning an Election Taylor, Moore, and Perez are candidates for public office. It is estimated that Moore and Perez have about the same probability of winning, and Taylor is believed to be twice as likely to win as either of the others. Find the probability of each candidate's winning the election.

40. *Payroll Error*   The employees of a company are in six departments: 31 are in sales, 54 are in research, 42 are in marketing, 20 are in engineering, 47 are in finance, and 58 are in production. If one employee's paycheck is lost, what is the probability that the employee is in the research department?

In Exercises 41–52, the sample spaces are large and you should use the counting principles discussed in Section 9.6.

41. *Preparing for a Test*   A class is given a list of 20 study problems from which ten will be part of an upcoming exam. If a given student knows how to solve 15 of the problems, find the probability that the student will be able to answer (a) all ten questions on the exam, (b) exactly eight questions on the exam, and (c) at least nine questions on the exam.

42. *Preparing for a Test*   A class is given a list of eight study problems from which five will be part of an upcoming exam. If a given student knows how to solve six of the problems, find the probability that the student will be able to answer (a) all five questions on the exam, (b) exactly four questions on the exam, and (c) at least four questions on the exam.

43. *Letter Mix-Up*   Four letters and envelopes are addressed to four different people. If the letters are randomly inserted into the envelopes, what is the probability that (a) exactly one is inserted in the correct envelope and (b) at least one is inserted in the correct envelope?

44. *Payroll Mix-Up*   Five paychecks and envelopes are addressed to five different people. If the paychecks are randomly inserted into the envelopes, what is the probability that (a) exactly one is inserted in the correct envelope and (b) at least one is inserted in the correct envelope?

45. *Game Show*   On a game show you are given five digits to arrange in the proper order to form the price of a car. If you are correct, you win the car. What is the probability of winning, given the following conditions?

(a) You guess the position of each digit.

(b) You know the first digit and guess the others.

46. *Card Game*   The deck of a card game is made up of 108 cards. Twenty-five each are red, yellow, blue, and green, and eight are wild cards. Each player is randomly dealt a seven-card hand. (a) What is the probability that a hand will have exactly two wild cards? (b) What is the probability that a hand will have two wild cards, two red cards, and three blue cards?

47. *Hospital Inspection*   As part of a monthly inspection at a hospital, the inspection team randomly selects reports from eight of the 84 nurses who are on duty. (a) What is the probability that none of the reports selected will be from the ten most experienced nurses on duty? (b) What is the probability that all of the reports selected will be from the 20 least experienced nurses on duty?

48. *Poker Hand*   Five cards are drawn from an ordinary deck of 52 playing cards. What is the probability of getting a full house? (A full house consists of three of one kind and two of another. For example, 3-3-3-7-7 is a full house.)

49. *Defective Units*   A shipment of 12 microwave ovens contains three defective units. A vending company has ordered four of these units, and because all are packaged identically, the selection will be random. What is the probability that

(a) all four units are good?

(b) exactly two units are good?

(c) at least two units are good?

50. *Defective Units*   A shipment of 20 compact disc players contains four defective units. A retail outlet has ordered five of these units, and because all are packaged identically, the selection will be random. What is the probability that

(a) all five units are good?

(b) exactly four units are good?

(c) at least one unit is good?

51. *Random Number Generator*   Two integers (from 1 through 30) are chosen by a random number generator. What is the probability that (a) the numbers are both even, (b) one number is even and one is odd, (c) both numbers are less than 10, and (d) the same number is chosen twice?

52. *Random Number Generator*   Two integers (from 1 through 40) are chosen by a random number generator. What is the probability that (a) the numbers are both even, (b) one number is even and one is odd, (c) both numbers are less than 30, and (d) the same number is chosen twice?

53. *Backup System*   A space vehicle has an independent backup system for one of its communication networks. The probability that either system will function satisfactorily for the duration of a flight is

0.985. What is the probability that during a given flight (a) both systems function satisfactorily, (b) at least one system functions satisfactorily, and (c) both systems fail?

54. *Backup Vehicle* A fire company keeps two rescue vehicles to serve the community. Because of the demand on the company's time and the chance of mechanical failure, the probability that a specific vehicle is available when needed is 90%. If the availability of one vehicle is *independent* of the other, find the probability that (a) both vehicles are available at a given time, (b) neither vehicle is available at a given time, and (c) at least one vehicle is available at a given time.

55. *Making a Sale* A sales representative makes sales at approximately one-fifth of all calls. If, on a given day, the representative contacts six potential clients, what is the probability that a sale will be made with (a) all six contacts, (b) none of the contacts, and (c) at least one contact?

56. *A Boy or a Girl?* Assume that the probability of the birth of a child of a particular sex is 50%. In a family with four children, what is the probability that (a) all the children are boys, (b) all the children are the same sex, and (c) there is at least one boy?

57. *Is That Cash or Charge?* Suppose that the methods used by shoppers to pay for merchandise are as shown in the circle graph. If two shoppers are chosen at random, what is the probability that both shoppers paid for their purchases only in cash?

How Shoppers Pay
for Merchandise

58. *Flexible Work Hours* In a survey, people were asked if they would prefer to work flexible hours—even if it meant slower career advancement—so they could spend more time with their families. The results of the survey are shown in the circle graph. Suppose that three people from the survey were chosen at random. What is the probability that all three people would prefer flexible work hours?

Flexible Work Hours

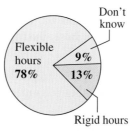

FIGURE FOR 58

59. *Geometry* You and a friend agree to meet at your favorite fast-food restaurant between 5:00 and 6:00 P.M. The one who arrives first will wait 15 minutes for the other, after which the first person will leave. What is the probability that the two of you will actually meet, assuming that your arrival times are random within the hour?

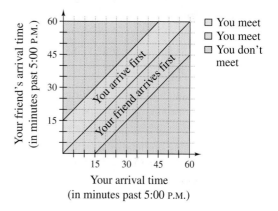

60. *Estimating* π A coin of diameter d is dropped onto a paper that contains a grid of squares d units on a side.

(a) Find the probability that the coin covers a vertex of one of the squares on the grid.

(b) Perform the experiment 100 times and use the results to approximate π.

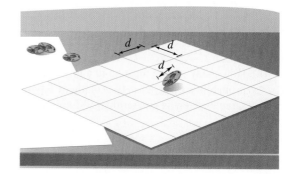

Synthesis

True or False?   **In Exercises 61 and 62, determine whether the statement is true or false. Justify your answer.**

61. If the probability of an outcome in a sample space is 1, then the probability of the other outcomes in the sample space is 0.

62. Rolling a number less than 3 on a normal 6-sided die has a probability of $\frac{1}{3}$. The complement of this event is to roll a number greater than 3 and its probability is $\frac{1}{2}$.

63. *Pattern Recognition and Exploration*   Consider a group of n people.

(a) Explain why the following gives the probability that the n people have distinct birthdays.

$$n = 2: \quad \frac{365}{365} \cdot \frac{364}{365} = \frac{365 \cdot 364}{365^2}$$

$$n = 3: \quad \frac{365}{365} \cdot \frac{364}{365} \cdot \frac{363}{365} = \frac{365 \cdot 364 \cdot 363}{365^3}$$

(b) Use the pattern in part (a) to write an expression for the probability that four people ($n = 4$) have distinct birthdays.

(c) Let P_n be the probability that the n people have distinct birthdays. Verify that this probability can be obtained recursively by

$$P_1 = 1 \quad \text{and} \quad P_n = \frac{365 - (n - 1)}{365} P_{n-1}.$$

(d) Explain why $Q_n = 1 - P_n$ gives the probability that at least two people in a group of n people have the same birthday.

(e) Use the results of parts (c) and (d) to complete the table.

n	10	15	20	23	30	40	50
P_n							
Q_n							

(f) How many people must be in a group before the probability of at least two of them having the same birthday is greater than $\frac{1}{2}$? Explain.

64. *Think About It*   The weather forecast indicates that the probability of rain is 40%. Explain what this means.

Review

In Exercises 65–68, find all real solutions of the rational equation.

65. $\dfrac{2}{x - 5} = 4$

66. $\dfrac{3}{2x + 3} - 4 = \dfrac{-1}{2x + 3}$

67. $\dfrac{3}{x - 2} + \dfrac{x}{x + 2} = 1$

68. $\dfrac{2}{x} - \dfrac{5}{x - 2} = \dfrac{-13}{x^2 - 2x}$

In Exercises 69–72, find all real solutions of the logarithmic equation.

69. $\ln x = 8$

70. $3 - 4 \ln x = 6$

71. $4 \ln 6x = 16$

72. $5 \ln 2x - 4 = 11$

In Exercises 73–76, sketch the graph of the solution of the system of inequalities.

73. $\begin{cases} y \geq -3 \\ x \geq -1 \\ -x - y \geq -8 \end{cases}$

74. $\begin{cases} x \leq 3 \\ y \leq 6 \\ 5x + 2y \geq 10 \end{cases}$

75. $\begin{cases} x^2 + y \geq -2 \\ y \geq x - 4 \end{cases}$

76. $\begin{cases} x^2 + y^2 \leq 4 \\ x + y \geq -2 \end{cases}$

In Exercises 77–80, evaluate $_nC_r$. Verify your result using a graphing utility.

77. $_6C_2$

78. $_9C_5$

79. $_{11}C_8$

80. $_{16}C_{13}$

9 Chapter Summary

What did you learn?

Section 9.1	Review Exercises
☐ How to use sequence notation to write the terms of sequences	1–10
☐ How to use factorial notation	11–14
☐ How to use summation notation to write sums	15–26
☐ How to find sums of infinite series	27–30
☐ How to use sequences and series to model and solve real-life problems	31, 32

Section 9.2	
☐ How to recognize, write, and find the nth terms of arithmetic sequences	33–46
☐ How to find nth partial sums of arithmetic sequences	47–52
☐ How to use arithmetic sequences to model and solve real-life problems	53, 54

Section 9.3	
☐ How to recognize, write, and find the nth terms of geometric sequences	55–66
☐ How to find nth partial sums of geometric sequences	67–76
☐ How to find sums of infinite geometric series	77–80
☐ How to use geometric sequences to model and solve real-life problems	81–84

Section 9.4	
☐ How to use mathematical induction to prove statements involving a positive integer n	85–88
☐ How to find sums of powers of integers	89–92
☐ How to find finite differences of sequences	93–96

Section 9.5	
☐ How to use the Binomial Theorem to calculate binomial coefficients	97–100
☐ How to use Pascal's Triangle to calculate binomial coefficients	101–104
☐ How to use binomial coefficients to write binomial expansions	105–110

Section 9.6	
☐ How to solve simple counting problems	111, 112
☐ How to use the Fundamental Counting Principle to solve more complicated counting problems	113, 114
☐ How to use permutations to solve counting problems	115–117
☐ How to use combinations to solve counting problems	118–120

Section 9.7	
☐ How to find probabilities of events	121–124
☐ How to find probabilities of mutually exclusive events	125
☐ How to find probabilities of independent events	126, 127
☐ How to find probabilities of complements of events	128, 129

9 Review Exercises

9.1 **In Exercises 1–6, write the first five terms of the sequence. (Assume n begins with 1.)**

1. $a_n = 2 + \dfrac{6}{n}$

2. $a_n = 8 - \dfrac{1}{2^n}$

3. $a_n = \dfrac{5n}{2n - 1}$

4. $a_n = n(n - 1)$

5. $a_n = \dfrac{72}{n!}$

6. $a_n = \dfrac{1}{(n + 1)!}$

In Exercises 7–10, use a graphing utility to graph the first ten terms of the sequence (Assume n begins with 1.)

7. $a_n = \frac{3}{2}n$

8. $a_n = \dfrac{3n}{n + 2}$

9. $a_n = 4(0.4)^{n - 1}$

10. $a_n = -8(0.5)^{n - 1}$

In Exercises 11–14, evaluate the factorial expression.

11. $\dfrac{18!}{20!}$

12. $3! \cdot 2!$

13. $\dfrac{3! \cdot 5!}{6!}$

14. $\dfrac{7! \cdot 6!}{6! \cdot 8!}$

In Exercises 15–22, find the sum. Use a graphing utility to verify your result.

15. $\displaystyle\sum_{i=1}^{6} 5$

16. $\displaystyle\sum_{k=2}^{5} 4k$

17. $\displaystyle\sum_{j=1}^{4} \dfrac{6}{j^2}$

18. $\displaystyle\sum_{i=1}^{8} \dfrac{i}{i + 1}$

19. $\displaystyle\sum_{k=1}^{10} 2k^3$

20. $\displaystyle\sum_{j=0}^{4} (j^2 + 1)$

21. $\displaystyle\sum_{n=0}^{10} (n^2 + 3)$

22. $\displaystyle\sum_{n=1}^{100} \left(\dfrac{1}{n} - \dfrac{1}{n + 1}\right)$

In Exercises 23–26, use sigma notation to write the sum.

23. $\dfrac{1}{2(1)} + \dfrac{1}{2(2)} + \dfrac{1}{2(3)} + \cdots + \dfrac{1}{2(20)}$

24. $2(1^2) + 2(2^2) + 2(3^2) + \cdots + 2(9^2)$

25. $\dfrac{1}{2} + \dfrac{2}{3} + \dfrac{3}{4} + \cdots + \dfrac{9}{10}$

26. $1 - \dfrac{1}{3} + \dfrac{1}{9} - \dfrac{1}{27} + \cdots$

In Exercises 27–30, find (a) the 4th partial sum and (b) the sum of the infinite series.

27. $\displaystyle\sum_{k=1}^{\infty} \dfrac{5}{10^k}$

28. $\displaystyle\sum_{k=1}^{\infty} 8\left(\tfrac{1}{10}\right)^k$

29. $\displaystyle\sum_{k=1}^{\infty} 2\left(\tfrac{1}{100}\right)^k$

30. $\displaystyle\sum_{k=1}^{\infty} 7\left(\tfrac{1}{10}\right)^k$

31. *Compound Interest* A deposit of $2500 is made in an account that earns 8% interest compounded quarterly. The balance in the account after n quarters is

$$a_n = 2500\left(1 + \dfrac{0.08}{4}\right)^n, \qquad n = 1, 2, 3, \ldots$$

   (a) Compute the first eight terms of this sequence.

   (b) Find the balance in this account after 10 years by computing the 40th term of the sequence.

32. *Corporate Sales* The sales for the Whirlpool Corporation from 1990 to 1998 can be approximated by the model

$$a_n = 29.65n^2 + 186.19n + 6474.70$$

where a_n is the sales (in millions of dollars) and n is the year, with $n = 0$ corresponding to 1990. Find the terms of this finite sequence and use a graphing utility to construct a bar graph that represents the sequence. (Source: Whirlpool Corporation)

9.2 **In Exercises 33–36, determine whether the sequence is arithmetic. If it is, find the common difference.**

33. $5, -2, -9, -16, -23, \ldots$

34. $0, 1, 3, 6, 10, \ldots$

35. $\frac{1}{2}, 1, \frac{3}{2}, 2, \frac{5}{2}, \ldots$

36. $\frac{9}{9}, \frac{8}{9}, \frac{7}{9}, \frac{6}{9}, \frac{5}{9}, \ldots$

In Exercises 37–40, write the first five terms of the arithmetic sequence.

37. $a_1 = 3, \ d = 4$

38. $a_1 = 8, \ d = -2$

39. $a_4 = 10, \ a_{10} = 28$

40. $a_2 = 14, \ a_6 = 22$

In Exercises 41–44, write the first five terms of the arithmetic sequence defined recursively. Determine the common difference and write the *n*th term of the sequence as a function of *n*.

41. $a_1 = 35,\quad a_{k+1} = a_k - 3$

42. $a_1 = 15,\quad a_{k+1} = a_k + \frac{5}{2}$

43. $a_1 = 9,\quad a_{k+1} = a_k + 7$

44. $a_1 = 100,\quad a_{k+1} = a_k - 5$

In Exercises 45 and 46, write an expression for the *n*th term of the arithmetic sequence and find the sum of the first 20 terms of the sequence.

45. $a_1 = 100,\quad d = -3$ **46.** $a_1 = 10,\quad a_3 = 28$

In Exercises 47–50, find the sum. Use a graphing utility to verify your result.

47. $\displaystyle\sum_{j=1}^{10}(2j - 3)$ **48.** $\displaystyle\sum_{j=1}^{8}(20 - 3j)$

49. $\displaystyle\sum_{k=1}^{11}\left(\tfrac{2}{3}k + 4\right)$ **50.** $\displaystyle\sum_{k=1}^{25}\left(\frac{3k + 1}{4}\right)$

51. Find the sum of the first 100 positive multiples of 5.

52. Find the sum of the integers from 20 to 80 (inclusive).

53. *Job Offer* A job has a starting salary of \$34,000 and a guaranteed salary increase of \$2250 per year for the first 4 years of employment. Determine (a) the salary during the fifth year and (b) the total compensation through 5 full years of employment.

54. *Baling Hay* In his first trip baling hay around a field, a farmer makes 123 bales. In his second trip he makes 11 fewer bales. Because each trip is shorter than the preceding trip, the farmer estimates that the same pattern will continue. Estimate the total number of bales made if there are another six trips around the field.

9.3 **In Exercises 55–58, write the first five terms of the geometric sequence.**

55. $a_1 = 4,\ r = -\frac{1}{4}$ **56.** $a_1 = 2,\ r = 2$

57. $a_1 = 9,\ a_3 = 4$ **58.** $a_1 = 2,\ a_3 = 12$

In Exercises 59–62, write the first five terms of the geometric sequence defined recursively. Determine the common ratio and write the *n*th term of the sequence as a function of *n*.

59. $a_1 = 120,\quad a_{k+1} = \frac{1}{3}a_k$

60. $a_1 = 200,\quad a_{k+1} = 0.1a_k$

61. $a_1 = 25,\quad a_{k+1} = -\frac{3}{5}a_k$

62. $a_1 = 18,\quad a_{k+1} = \frac{5}{3}a_k$

In Exercises 63–66, write an expression for the *n*th term of the geometric sequence and find the sum of the first 20 terms of the sequence.

63. $a_1 = 16,\quad a_2 = -8$ **64.** $a_3 = 6,\quad a_4 = 1$

65. $a_1 = 100,\quad r = 1.05$ **66.** $a_1 = 5,\quad r = 0.2$

In Exercises 67–72, find the sum.

67. $\displaystyle\sum_{i=1}^{7}2^{i-1}$ **68.** $\displaystyle\sum_{i=1}^{5}3^{i-1}$

69. $\displaystyle\sum_{n=1}^{7}(-4)^{n-1}$ **70.** $\displaystyle\sum_{n=1}^{4}12\left(-\tfrac{1}{2}\right)^{n-1}$

71. $\displaystyle\sum_{n=0}^{4}250(1.02)^n$ **72.** $\displaystyle\sum_{n=0}^{5}400(1.08)^n$

In Exercises 73–76, use a graphing utility to find the sum.

73. $\displaystyle\sum_{i=1}^{10}10\left(\tfrac{3}{5}\right)^{i-1}$ **74.** $\displaystyle\sum_{i=1}^{15}20(0.2)^{i-1}$

75. $\displaystyle\sum_{i=1}^{25}100(1.06)^{i-1}$ **76.** $\displaystyle\sum_{i=1}^{20}8\left(\tfrac{6}{5}\right)^{i-1}$

In Exercises 77–80, find the sum of the infinite series.

77. $\displaystyle\sum_{i=1}^{\infty}\left(\tfrac{7}{8}\right)^{i-1}$ **78.** $\displaystyle\sum_{i=1}^{\infty}\left(\tfrac{1}{3}\right)^{i-1}$

79. $\displaystyle\sum_{k=1}^{\infty}4\left(\tfrac{2}{3}\right)^{k-1}$ **80.** $\displaystyle\sum_{k=1}^{\infty}1.3\left(\tfrac{1}{10}\right)^{k-1}$

81. *Depreciation* A company buys a fleet of six vans for \$120,000. During the next 5 years, the fleet will depreciate at a rate of 30% per year. (That is, at the end of each year, the depreciated value is 70% of what it was at the beginning of the year.)

(a) Find the formula for the *n*th term of a geometric sequence that gives the value of the fleet *t* full years after it was purchased.

(b) Find the depreciated value of the fleet at the end of 5 full years.

82. *Total Compensation* A job pays a salary of $32,000 the first year. During the next 39 years, there is a 5.5% raise each year. What is the total salary over the 40-year period?

83. *Annuity* A deposit of $75 is made at the beginning of each month in an account that pays 4% interest, compounded monthly. The balance A in the account at the end of 4 years is

$$A = 75\left(1 + \frac{0.04}{12}\right)^1 + \cdots + 75\left(1 + \frac{0.04}{12}\right)^{48}.$$

Find A.

84. *Compound Interest* A deposit of $100 is made at the beginning of each month for 10 years in an account that pays 6.5%, compounded monthly. What is the balance in the account at the end of the 10 years?

9.4 **In Exercises 85–88, use mathematical induction to prove the formula for every positive integer n.**

85. $2 + 7 + \cdots + (5n - 3) = \dfrac{n}{2}(5n - 1)$

86. $1 + \dfrac{3}{2} + 2 + \dfrac{5}{2} + \cdots + \dfrac{1}{2}(n + 1) = \dfrac{n}{4}(n + 3)$

87. $\displaystyle\sum_{i=0}^{n-1} ar^i = \dfrac{a(1 - r^n)}{1 - r}$

88. $\displaystyle\sum_{k=0}^{n-1} (a + kd) = \dfrac{n}{2}[2a + (n - 1)d]$

In Exercises 89–92, find the sum using the formulas for the sums of powers of integers.

89. $\displaystyle\sum_{n=1}^{30} n$

90. $\displaystyle\sum_{n=1}^{10} n^2$

91. $\displaystyle\sum_{n=1}^{7} n^4$

92. $\displaystyle\sum_{n=1}^{6} n^5$

In Exercises 93–96, find the first five terms of the sequence beginning with a_1. Then calculate the first and second differences of the sequence. Does the sequence have a linear model, a quadratic model, or neither?

93. $a_1 = 5$
$a_n = a_{n-1} + 5$

94. $a_1 = -3$
$a_n = a_{n-1} - 2n$

95. $a_1 = 16$
$a_n = a_{n-1} - 1$

96. $a_1 = 1$
$a_n = n - a_{n-1}$

9.5 **In Exercises 97–100, evaluate $_nC_r$. Use a graphing utility to verify your answer.**

97. $_{10}C_8$

98. $_{12}C_5$

99. $\dbinom{9}{4}$

100. $\dbinom{14}{12}$

In Exercises 101–104, evaluate using Pascal's Triangle.

101. $_6C_3$

102. $_9C_7$

103. $\dbinom{8}{4}$

104. $\dbinom{10}{5}$

In Exercises 105–110, use the Binomial Theorem to expand the binomial. Simplify your answer. (Remember that $i = \sqrt{-1}$.)

105. $(a - 3b)^5$

106. $(3x + y^2)^7$

107. $\left(\dfrac{x}{2} + y\right)^4$

108. $\left(\dfrac{2}{x} - 3x\right)^6$

109. $(5 + 2i)^4$

110. $(4 - 5i)^3$

9.6 **111.** *Numbers in a Hat* If slips of paper numbered 1 through 14 are placed in a hat, in how many ways could two numbers be drawn so that the sum of the numbers was 12? Assume the random selection is without replacement.

112. *Morse Code* In Morse code, each character is transmitted using a sequence of dits and dahs. How many different characters can be formed by a sequence of three dits and dahs? (These can be repeated. For example, dit-dit-dit represents the letter *s*.)

113. *Course Schedule* A college student is preparing a course schedule for the next semester. The student must select one of four mathematics courses, one of six biology courses, and one of two art courses. How many schedules are possible?

114. *Amateur Radio* A novice amateur radio license consists of two letters, one digit, and then three more letters. How many different licenses can be issued if no restrictions are placed on the letters or digits?

In Exercises 115 and 116, find the number of distinguishable permutations of the group of letters.

115. C, A, L, C, U, L, U, S

116. I, N, V, E, R, T, E, B, R, A, T, E

117. *Bike Race* There are ten bicyclists entered in a race. In how many different orders could the ten bicyclists finish?

118. *Team Captains* From a pool of 11 players on a football team, two will be chosen as co-captains. In how many different ways could these positions be filled?

119. *Test Questions* A student can answer any 15 questions from a total of 20 questions on an exam. In how many different ways could the student select the questions?

120. *Magic Act* A magician is performing in front of an audience of 68 people. In order to do a demonstration, six people are randomly selected from the audience. How many different groups of six people are possible? Use a graphing utility to find your answer.

9.7 121. *Matching Socks* A man has five pairs of socks (no two pairs are the same color). If he randomly selects two socks from a drawer, what is the probability that he gets a matched pair?

122. *Bookshelf Order* A child returns a five-volume set of books to a bookshelf. The child is not able to read and so cannot distinguish one volume from another. What is the probability that the books are shelved in the correct order?

123. *Roll of the Dice* Are the chances of rolling a 3 with one die the same as rolling a total of 6 with two dice? If not, which has the higher probability?

124. *Roll of the Dice* A six-sided die is rolled six times. What is the probability that each side appears exactly once?

125. *Data Analysis* A sample of college students, faculty members, and administrators were asked whether they favored a proposed increase in the annual activity fee to enhance student life on campus. The results of the study are shown in the table.

	Students	Faculty	Admin.	Total
Favor	237	37	18	292
Oppose	163	38	7	208
Total	400	75	25	500

A person is selected at random from the sample. Find the specified probability.

(a) The person is not in favor of the proposal.

(b) The person is a student.

(c) The person is a faculty member and is in favor of the proposal.

126. *Card Game* Five cards are drawn from an ordinary deck of 52 playing cards. Find the probability of getting two pairs. (For example, the hand could be A-A-5-5-Q or 4-4-7-7-K.)

127. *Parental Independence* Suppose that in a survey, senior citizens were asked if they would live with their children when they reached the point of not being able to live alone. The results are shown in the figure. If three senior citizens who could not live alone are randomly selected, what is the probability that all three are not living with their children?

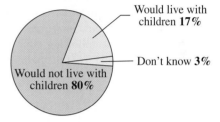

Would live with children **17%**

Don't know **3%**

Would not live with children **80%**

128. *Picking a Card* You randomly select a card from a 52-card deck. What is the probability that the card is *not* a club?

129. *Tossing a Coin* Find the probability of obtaining at least one tail when a coin is tossed five times.

Synthesis

True or False? In Exercises 130 and 131, determine whether the statement is true or false. Justify your answer.

130. $\dfrac{(n+2)!}{n!} = (n+2)(n+1)$

131. $\displaystyle\sum_{k=1}^{8} 3k = 3\sum_{k=1}^{8} k$

132. An infinite sequence is a function. What is the domain of the function?

133. How do the two sequences differ?

(a) $a_n = \dfrac{(-1)^n}{n}$ (b) $a_n = \dfrac{(-1)^{n+1}}{n}$

134. In your own words, explain what makes a sequence (a) arithmetic and (b) geometric.

135. The graphs of two sequences are shown below. Identify each sequence as arithmetic or geometric. Explain your reasoning.

(a)   (b)

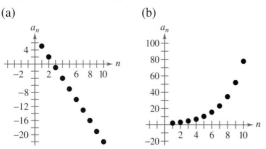

136. *Population Growth*   Consider an idealized population with the characteristic that each member of the population produces one offspring at the end of every time period. If each member has a life span of three time periods and the population begins with ten newborn members, then the following table shows the population during the first five time periods.

	Time Period				
Age Bracket	1	2	3	4	5
0–1	10	10	20	40	70
1–2		10	10	20	40
2–3			10	10	20
Total	10	20	40	70	130

The sequence for the total population has the property that

$$S_n = S_{n-1} + S_{n-2} + S_{n-3}, \quad n > 3.$$

Find the total population during the next five time periods.

137. Explain what a recursive formula is.

138. Explain why the terms of a geometric sequence of positive terms decrease when $0 < r < 1$.

In Exercises 139–142, match the sequence or sum of a sequence with its graph without doing any calculations. Explain your reasoning. [The graphs are labeled (a), (b), (c), and (d).]

(a)   (b)

(c)   (d)

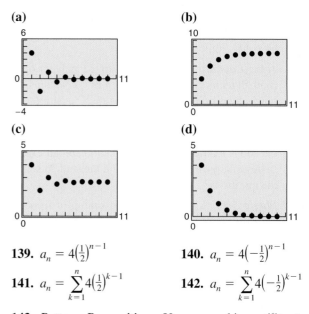

139. $a_n = 4\left(\frac{1}{2}\right)^{n-1}$

140. $a_n = 4\left(-\frac{1}{2}\right)^{n-1}$

141. $a_n = \sum_{k=1}^{n} 4\left(\frac{1}{2}\right)^{k-1}$

142. $a_n = \sum_{k=1}^{n} 4\left(-\frac{1}{2}\right)^{k-1}$

143. *Pattern Recognition*   Use a graphing utility to complete the table. What number is the nth term of the sequence approaching as n gets larger?

$$a_n = \left(1 + \frac{1}{n}\right)^n$$

n	100	500	1000	5000
a_n				

n	10,000	15,000	20,000	25,000
a_n				

144. How do the expansions of $(-x + y)^n$ and $(x - y)^n$ differ?

145. The probability of an event must be a real number in what interval? Is the interval open or closed?

146. The probability of an event is $\frac{2}{3}$. What is the probability that the event does not occur? Explain.

147. The weather forecast indicates that the probability of rain is 60%. Explain what this means.

Chapter Project *Fitting Models to Data*

In this project, you will explore difference quotients and see how they can be used to find average rates.

Suppose you are driving from Atlanta to Miami. The trip is about 700 miles and takes you 12 hours. Your average speed is

$$\text{Average speed} = \frac{\text{distance}}{\text{time}} = \frac{700}{12} \approx 58.3 \text{ miles per hour.}$$

This concept can be generalized as follows. Let f be a function defined on the interval $[a, b]$. The **average rate of change of** f from a to b is

$$\text{Average rate of change} = \frac{f(b) - f(a)}{b - a}.$$

This expression is called a *difference quotient*.

a. Calculate the average rate of change of $f(x) = 3x + 4$ on the interval $[2, 6]$. Select any other interval and show that you obtain the same average rate of change.

b. Calculate the average rates of change of $f(x) = x^2$ on the intervals $[1, 3]$ and $[4, 6]$. Are they equal? Explain.

c. During a one-hour trip, your average speed is 50 miles per hour. Discuss the relationship between this speed and the speeds shown on your speedometer.

Questions for Further Exploration

1. Let $f(x) = x^2 - 2$. Calculate the average rates of change of f on the intervals $[1, 3], [1, 2], [1, 1.5]$, and $[1, 1.1]$.

(a) Find an expression for the average rate of change of f on the interval $[1, 1 + h]$, where h is any real number.

(b) Create a table that shows the average rate of change of f for several values of h. Choose values of h that get closer and closer to 0. What value does the average rate of change of f approach as h approaches 0?

(c) The answer to part (b) is the slope of the tangent line to the graph of f at the point $(1, f(1))$. Graph this line and f in the same viewing window. Describe the behavior of the graphs near the point $(1, f(1))$.

2. The data in the table gives the costs of first-class postage in the United States for selected years from 1971 to 1999, where $t = 0$ corresponds to 1900. (Source: U.S. Postal Service)

t	71	74	75	78	81
Postage	$0.08	$0.10	$0.13	$0.15	$0.20

t	85	88	91	95	99
Postage	$0.22	$0.25	$0.29	$0.32	$0.33

(a) Find the average rate of change of the cost of postage between each two adjacent time intervals.

(b) When was the average rate of change largest? smallest?

(c) Are there intervals over which the average rate of change was zero? What does this mean?

9 Chapter Test

Take this test as you would take a test in class. After you are done, check your work against the answers given in the back of the book.

In Exercises 1 and 2, write the first five terms of the sequence.

1. $a_n = \left(-\frac{2}{3}\right)^{n-1}$. (Begin with $n = 1$.)

2. $a_1 = 12$ and $a_{k+1} = a_k + 4$

3. Simplify $\dfrac{11! \, 4!}{4! \, 7!}$.

In Exercises 4 and 5, find a formula for the nth term of the sequence.

4. Arithmetic: $a_1 = 5000$, $d = -100$

5. Geometric: $a_1 = 4$, $a_{k+1} = \frac{1}{2}a_k$

6. Use sigma notation to write: $\dfrac{2}{3(1) + 1} + \dfrac{2}{3(2) + 1} + \cdots + \dfrac{2}{3(12) + 1}$.

7. Find the sum of the first 50 positive multiples of 3.

In Exercises 8–10, find the sum.

8. $\displaystyle\sum_{n=1}^{7} (8n - 5)$

9. $\displaystyle\sum_{n=1}^{8} 24\left(\frac{1}{6}\right)^{n-1}$

10. $\displaystyle\sum_{n=1}^{\infty} 5\left(\frac{1}{10}\right)^{n-1}$

11. Find the balance in an increasing annuity in which a principal of $50 is deposited at the beginning of each month for 25 years. Assume that the amount in the fund is compounded monthly at 8%.

12. Use mathematical induction to prove the formula

$$3 + 6 + 9 + \cdots + 3n = \frac{3n(n + 1)}{2}.$$

In Exercises 13–16, evaluate $_nC_r$.

13. $_9C_3$

14. $_{20}C_3$

15. $_{18}C_5$

16. $_{40}C_{38}$

17. Find the coefficient of the term x^3y^5 in the expansion of $(x + y)^8$.

18. How many distinct license plates can be issued with one letter followed by a three-digit number?

19. Four students are randomly selected from a class of 25 to answer questions from a reading assignment. In how many ways can the four be selected?

20. A card is drawn from a standard deck of 52 playing cards. Find the probability that it is a red face card.

21. Suppose that two spark plugs require replacement in a four-cylinder engine. If the mechanic randomly removes two plugs, find the probability that they are the two defective plugs.

22. Two integers (from 1 to 60) are chosen by a random number generator. What is the probability that (a) both numbers are odd, (b) both numbers are less than 12, and (c) the same number is chosen twice?

The *Interactive* CD-ROM and *Internet* versions of this text provide answers to the Chapter Tests and Cumulative Tests. They also offer Chapter Pre-Tests (which test key skills and concepts covered in previous chapters) and Chapter Post-Tests, both of which have randomly generated exercises with diagnostic capabilities.

Topics in Analytic Geometry

10

Derke O'Hara/Tony Stone Images

Ice in the nucleus of a comet is heated and vaporized by the sun. The escaping gas collects dust particles, forming a tail which always points away from the sun.

The Big Picture

In this chapter you will learn how to

❑ write the standard equations of parabolas, ellipses, and hyperbolas.
❑ analyze and sketch the graphs of parabolas, ellipses, and hyperbolas.
❑ rotate the coordinate axis to eliminate the *xy*-term in equations of conics and use the discriminant to classify conics.
❑ solve systems of quadratic equations.
❑ rewrite sets of parametric equations as rectangular equations and find sets of parametric equations for graphs.
❑ write equations in polar form.
❑ graph polar equations and recognize special polar graphs.
❑ write equations of conics in polar form.

Important Vocabulary

As you encounter each new vocabulary term in this chapter, add the term and its definition to your notebook glossary.

- conic section or conic (pp. 696, 754)
- parabola (pp. 697, 754)
- directrix (p. 697)
- focus or foci (pp. 697, 704, 713)
- tangent (p. 699)
- ellipse (pp. 704, 754)
- vertices (pp. 704, 713)
- major axis (p. 704)
- minor axis (p. 704)
- center (pp. 704, 713)
- eccentricity (pp. 708, 754)
- hyperbola (pp. 713, 754)
- transverse axis (p. 713)
- asymptotes (p. 715)
- conjugate axis (p. 715)
- discriminant (p. 726)
- parameter (p. 731)
- parametric equations (p. 731)
- plane curve (p. 731)
- orientation (p. 732)
- polar coordinate system (p. 739)
- pole or origin (p. 739)
- polar axis (p. 739)
- polar coordinates (p. 739)

Additional Resources   Text-specific additional resources are available to help you do well in this course. See page xvi for details.

10.1 Introduction to Conics: Parabolas

Conic sections were discovered during the classical Greek period, 600 to 300 B.C. The early Greeks were concerned largely with the geometric properties of conics. It was not until the early 17th century that the broad applicability of conics became apparent, and they then played a prominent role in the early development of calculus.

Each **conic section** (or simply **conic**) is the intersection of a plane and a double-napped cone. Notice in Figure 10.1(a) that in the formation of the four basic conics, the intersecting plane does not pass through the vertex of the cone. When the plane does pass through the vertex, the resulting figure is a *degenerate conic*, as shown in Figure 10.1(b).

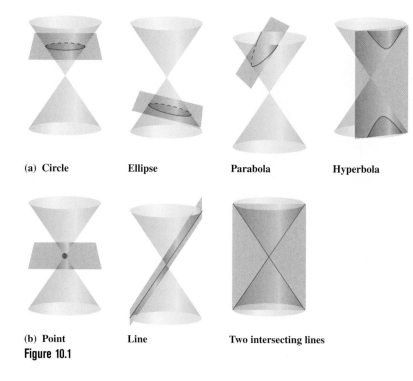

(a) Circle　　　**Ellipse**　　　**Parabola**　　　**Hyperbola**

(b) Point　　　**Line**　　　**Two intersecting lines**
Figure 10.1

What You Should Learn:

- How to recognize a conic as the intersection of a plane and a double-napped cone
- How to write equations of parabolas in standard form
- How to use the reflective property of parabolas to solve real-life problems

Why You Should Learn It:

Parabolas can be used to model and solve many types of real-life problems. For instance, in Exercise 58 on page 702, a parabola is used to model the cables of a suspension bridge.

Adam Woolfitt/CORBIS

A computer animation of this concept appears in the *Interactive* CD-ROM and *Internet* versions of this text.

There are several ways to approach the study of conics. You could begin by defining conics in terms of the intersections of planes and cones, as the Greeks did, or you could define them algebraically, in terms of the general second-degree equation

$$Ax^2 + Bxy + Cy^2 + Dx + Ey + F = 0.$$

However, you will study a third approach, in which each of the conics is defined as a **locus** (collection) of points satisfying a geometric property. For example, the definition of a circle as the collection of all points (x, y) that are equidistant from a fixed point (h, k) leads to the standard equation of a circle

$$(x - h)^2 + (y - k)^2 = r^2. \qquad \text{Equation of circle}$$

Parabolas

The first type of conic is called a **parabola,** and it is defined as follows.

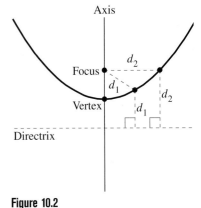

Figure 10.2

Definition of a Parabola

A **parabola** is the set of all points (x, y) that are equidistant from a fixed line (**directrix**) and a fixed point (**focus**) not on the line.

The midpoint between the focus and the directrix is called the **vertex,** and the line passing through the focus and the vertex is called the **axis** of the parabola. Note in Figure 10.2 that a parabola is symmetric with respect to its axis. Using the definition of a parabola, you can derive the following **standard form of the equation of a parabola** whose directrix is parallel to the x-axis or to the y-axis. See Appendix A for a proof.

Standard Equation of a Parabola

The **standard form of the equation of a parabola** with vertex at (h, k) is as follows.

$(x - h)^2 = 4p(y - k), \; p \neq 0$ Vertical axis; directrix: $y = k - p$

$(y - k)^2 = 4p(x - h), \; p \neq 0$ Horizontal axis; directrix: $x = h - p$

The focus lies on the axis p units (*directed distance*) from the vertex. If the vertex is at the origin $(0, 0)$, the equation takes one of the following forms.

$x^2 = 4py$ Vertical axis

$y^2 = 4px$ Horizontal axis

See Figure 10.3.

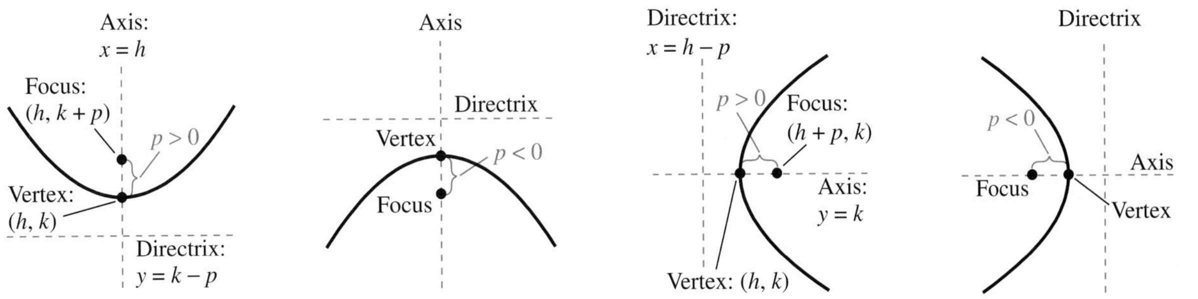

(a) Vertical axis: $p > 0$
$$(x - h)^2 = 4p(y - k)$$

(b) Vertical axis: $p < 0$

(c) Horizontal axis: $p > 0$

(d) Horizontal axis: $p < 0$
$$(y - k)^2 = 4p(x - h)$$

Figure 10.3

EXAMPLE 1 Finding the Standard Equation of a Parabola

Find the standard form of the equation of the parabola with vertex $(2, 1)$ and focus $(2, 4)$.

Solution

Because the axis of the parabola is vertical, consider the equation

$$(x - h)^2 = 4p(y - k)$$

where $h = 2$, $k = 1$, and $p = 4 - 1 = 3$. So, the standard form is

$$(x - 2)^2 = 4(3)(y - 1) = 12(y - 1).$$

You can obtain the more common quadratic form as follows.

$(x - 2)^2 = 12(y - 1)$	Write original equation.
$x^2 - 4x + 4 = 12y - 12$	Multiply.
$x^2 - 4x + 16 = 12y$	Add 12 to each side.
$y = \frac{1}{12}(x^2 - 4x + 16)$	Divide each side by 12.

Try using a graphing utility to confirm the graph of this parabola, as shown in Figure 10.4.

The *Interactive* CD-ROM and *Internet* versions of this text show every example with its solution; clicking on the *Try It!* button brings up similar problems. Guided Examples and Integrated Examples show step-by-step solutions to additional examples. Integrated Examples are related to several concepts in the section.

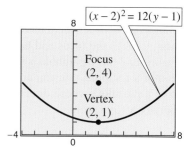

Figure 10.4

EXAMPLE 2 Finding the Focus of a Parabola

Find the focus of the parabola

$$y = -\frac{1}{2}x^2 - x + \frac{1}{2}.$$

Solution

To find the focus, convert to standard form by completing the square.

$y = -\frac{1}{2}x^2 - x + \frac{1}{2}$	Write original equation.
$-2y = x^2 + 2x - 1$	Multiply each side by –2.
$1 - 2y = x^2 + 2x$	Group terms.
$1 + 1 - 2y = x^2 + 2x + 1$	Complete the square.
$2 - 2y = x^2 + 2x + 1$	Combine like terms.
$-2(y - 1) = (x + 1)^2$	Write in standard form.

Comparing this equation with

$$(x - h)^2 = 4p(y - k)$$

you can conclude that $h = -1$, $k = 1$, and $p = -\frac{1}{2}$. Because p is negative, the parabola opens downward, as shown in Figure 10.5. Therefore, the focus of the parabola is

$$(h, k + p) = \left(-1, \frac{1}{2}\right). \qquad \text{Focus}$$

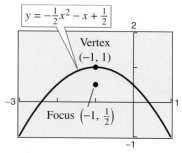

Figure 10.5

EXAMPLE 3 Vertex at the Origin

Find the standard equation of the parabola with vertex at the origin and focus $(2, 0)$.

Solution

The axis of the parabola is horizontal, passing through $(0, 0)$ and $(2, 0)$, as shown in Figure 10.6.

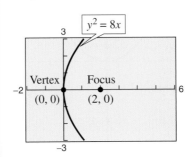

Figure 10.6

So, the standard form is

$$y^2 = 4px$$

where $h = k = 0$ and $p = 2$. Therefore, the equation is

$$y^2 = 8x.$$

Applications

A line segment that passes through the focus of a parabola and has endpoints on the parabola is called a **focal chord.** The specific focal chord perpendicular to the axis of the parabola is called the **latus rectum.**

Parabolas occur in a wide variety of applications. For instance, a parabolic reflector can be formed by revolving a parabola around its axis. The resulting surface has the property that all incoming rays parallel to the axis are reflected through the focus of the parabola; this is the principle behind the construction of the parabolic mirrors used in reflecting telescopes. Conversely, the light rays emanating from the focus of a parabolic reflector used in a flashlight are all parallel to one another, as shown in Figure 10.7.

A line is **tangent** to a parabola at a point on the parabola if the line intersects, but does not cross, the parabola at the point. Tangent lines to parabolas have special properties that are related to the use of parabolas in constructing reflective surfaces.

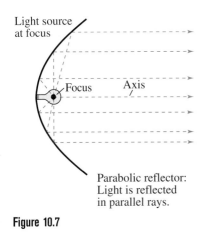

Parabolic reflector: Light is reflected in parallel rays.

Figure 10.7

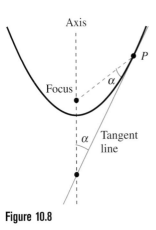

Figure 10.8

Reflective Property of a Parabola

The tangent line to a parabola at a point P makes equal angles with the following two lines (see Figure 10.8).

1. The line passing through P and the focus
2. The axis of the parabola

EXAMPLE 4   Finding the Tangent Line at a Point on a Parabola

Find the equation of the tangent line to the parabola given by $y = x^2$ at the point $(1, 1)$.

Solution

For this parabola, $p = \frac{1}{4}$ and the focus is $\left(0, \frac{1}{4}\right)$, as shown in Figure 10.9. You can find the y-intercept $(0, b)$ of the tangent line by equating the lengths of the two sides of the isosceles triangle

$$d_1 = \frac{1}{4} - b$$

and

$$d_2 = \sqrt{(1 - 0)^2 + \left[1 - \left(\frac{1}{4}\right)\right]^2} = \frac{5}{4}$$

shown in Figure 10.9. Setting $d_1 = d_2$ produces

$$\frac{1}{4} - b = \frac{5}{4}$$

$$b = -1.$$

So, the slope of the tangent line is

$$m = \frac{1 - (-1)}{1 - 0} = 2$$

and its slope-intercept equation is

$$y = 2x - 1.$$

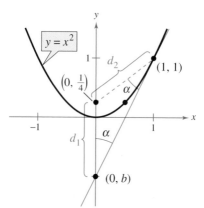

Figure 10.9

Try using a graphing utility to confirm the result of Example 4. By graphing

$$y = x^2 \qquad \text{and} \qquad y = 2x - 1$$

in the same viewing window, you should be able to see that the line touches the parabola at the point $(1, 1)$.

Writing About Math   *Television Antenna Dishes*

Cross sections of television antenna dishes are parabolic in shape. Write a paragraph describing why these dishes are parabolic. Include a graphical representation of your description.

10.1 E x e r c i s e s

In Exercises 1–6, match the equation with its graph. [The graphs are labeled (a), (b), (c), (d), (e), and (f).]

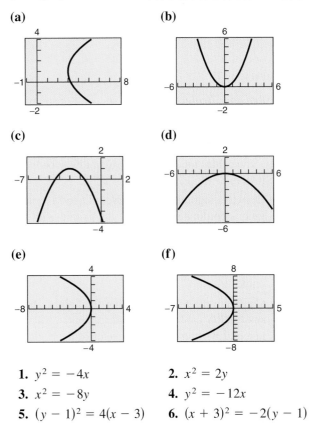

(a)

(b)

(c)

(d)

(e)

(f)

1. $y^2 = -4x$

2. $x^2 = 2y$

3. $x^2 = -8y$

4. $y^2 = -12x$

5. $(y - 1)^2 = 4(x - 3)$

6. $(x + 3)^2 = -2(y - 1)$

In Exercises 7–20, find the vertex, focus, and directrix of the parabola and sketch its graph.

7. $y = \frac{1}{2}x^2$

8. $y = -2x^2$

9. $y^2 = -6x$

10. $y^2 = 3x$

11. $x^2 + 6y = 0$

12. $x + y^2 = 0$

13. $(x - 1)^2 + 8(y + 2) = 0$

14. $(x + 5) + (y - 1)^2 = 0$

15. $\left(x + \frac{3}{2}\right)^2 = 4(y - 2)$

16. $\left(x + \frac{1}{2}\right)^2 = 4(y - 1)$

17. $y = \frac{1}{4}(x^2 - 2x + 5)$

18. $x = \frac{1}{4}(y^2 + 2y + 33)$

19. $y^2 + 6y + 8x + 25 = 0$

20. $y^2 - 4y - 4x = 0$

The *Interactive* CD-ROM and *Internet* versions of this text contain step-by-step solutions to all odd-numbered Section and Review Exercises. They also provide Tutorial Exercises, which link to Guided Examples for additional help.

In Exercises 21–24, find the vertex, focus, and directrix of the parabola and sketch its graph. Use a graphing utility to verify your graph.

21. $x^2 + 4x + 6y - 2 = 0$

22. $x^2 - 2x + 8y + 9 = 0$

23. $y^2 + x + y = 0$

24. $y^2 - 4x - 4 = 0$

In Exercises 25 and 26, change the equation so that its graph matches the given graph.

25. $y^2 = -6x$

26. $y^2 = 9x$

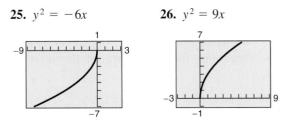

In Exercises 27–38, find the standard form of the equation of the parabola with its vertex at the origin.

27.

28.

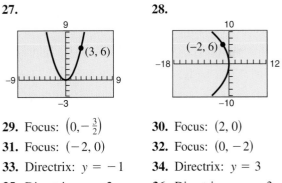

(3, 6)

(−2, 6)

29. Focus: $\left(0, -\frac{3}{2}\right)$

30. Focus: $(2, 0)$

31. Focus: $(-2, 0)$

32. Focus: $(0, -2)$

33. Directrix: $y = -1$

34. Directrix: $y = 3$

35. Directrix: $x = 2$

36. Directrix: $x = -3$

37. Horizontal axis and passes through the point $(4, 6)$

38. Vertical axis and passes through the point $(-3, -3)$

In Exercises 39–48, find the standard form of the equation of the parabola.

39.

40.

(3, 1)

(4, 0)

(2, 0)

(4.5, 4)

(5, 3)

41. **42.**

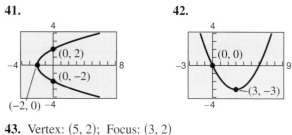

43. Vertex: $(5, 2)$; Focus: $(3, 2)$

44. Vertex: $(-1, 2)$; Focus: $(-1, 0)$

45. Vertex: $(0, 4)$; Directrix: $y = 2$

46. Vertex: $(-2, 1)$; Directrix: $x = 1$

47. Focus: $(2, 2)$; Directrix: $x = -2$

48. Focus: $(0, 0)$; Directrix: $y = 8$

In Exercises 49 and 50, the equations of a parabola and a tangent line to the parabola are given. Use a graphing utility to graph both in the same viewing window. Determine the coordinates of the point of tangency.

Parabola	*Tangent Line*
49. $y^2 - 8x = 0$	$x - y + 2 = 0$
50. $x^2 + 12y = 0$	$x + y - 3 = 0$

In Exercises 51–54, find an equation of the tangent line to the parabola at the given point and find the x-intercept of the line.

51. $x^2 = 2y$, $(4, 8)$ **52.** $x^2 = 2y$, $\left(-3, \frac{9}{2}\right)$

53. $y = -2x^2$, $(-1, -2)$ **54.** $y = -2x^2$, $(2, -8)$

55. *Revenue*   The revenue R generated by the sale of x units of a product is

$$R = 265x - \frac{5}{4}x^2.$$

Use a graphing utility to graph the function and approximate the number of sales that will maximize revenue.

56. *Revenue*   The revenue R generated by the sale of x units of a product is

$$R = 378x - \frac{7}{5}x^2.$$

Use a graphing utility to graph the function and approximate the number of sales that will maximize revenue.

57. *Satellite Antenna*   The receiver in a parabolic television dish antenna is 3.5 feet from the vertex and is located at the focus. Find an equation of a cross

section of the reflector. (Assume that the dish is directed upward and that the vertex is at the origin.)

FIGURE FOR 57

58. *Suspension Bridge*   Each cable of a suspension bridge is suspended (in the shape of a parabola) between two towers that are 120 meters apart, and the top of each tower is 20 meters above the roadway. The cables touch the roadway midway between the towers.

(a) Draw a diagram for the bridge. Draw a rectangular coordinate system on the bridge with the center of the bridge at the origin. Identify the coordinates of the known points.

(b) Find an equation for the parabolic shape of each cable.

(c) Complete the table by finding the height of the suspension cables above the roadway at a distance of x meters from the center of the bridge.

x	0	20	40	60
y				

59. *Road Design*   Roads are often designed with parabolic surfaces to allow rain to drain off. A particular road that is 32 feet wide is 0.4 foot higher in the center than it is on the sides.

(a) Find an equation of the parabola. (Assume that the origin is at the center of the road.)

(b) How far from the center of the road is the road surface 0.1 foot lower than in the middle?

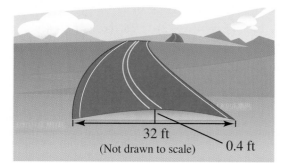

32 ft
(Not drawn to scale)   0.4 ft

60. *Highway Design*   Highway engineers design a parabolic curve for an entrance ramp from a straight street to an interstate highway. Find an equation of the parabola.

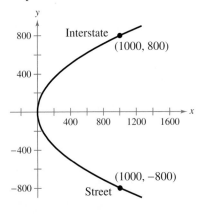

61. *Satellite Orbit*   An earth satellite in a 100-mile-high circular orbit around the earth has a velocity of approximately 17,500 miles per hour. If this velocity is multiplied by $\sqrt{2}$, the satellite will have the minimum velocity necessary to escape the earth's gravity and it will follow a parabolic path with the center of the earth as the focus.

(a) Find the escape velocity of the satellite.

(b) Find an equation of its path (assume the radius of the earth is 4000 miles).

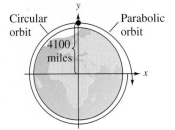

62. *Path of a Projectile*   The path of a softball is given by the equation

$$y = -0.08x^2 + x + 4.$$

The coordinates x and y are measured in feet, with $x = 0$ corresponding to the position where the ball was thrown.

(a) Use a graphing utility to graph the trajectory of the softball.

(b) Move the cursor along the path to approximate the highest point and the range of the trajectory.

Projectile Motion   **In Exercises 63 and 64, consider the path of a projectile projected horizontally with a velocity of v feet per second at a height of s feet, where the model for the path is**

$$y = -\frac{16}{v^2}t^2 + s.$$

In this model, air resistance is disregarded and y is the height (in feet) of the projectile t seconds after its release.

63. A ball is thrown from the top of a 75-foot tower with a velocity of 32 feet per second.

(a) Find the equation of the parabolic path.

(b) How far does the ball travel horizontally before striking the ground?

64. A bomber flying due east at 550 miles per hour at an altitude of 42,000 feet releases a bomb. Determine the distance the bomb travels horizontally before striking the ground.

Synthesis

True or False?   **In Exercises 65 and 66, determine whether the statement is true or false. Justify your answer.**

65. It is possible for a parabola to intersect its directrix.

66. If the vertex and focus of a parabola are on a horizontal line, then the directrix of the parabola is vertical.

Review

In Exercises 67–70, list the possible rational zeros given by the Rational Zero Test.

67. $f(x) = x^3 - 2x^2 + 2x - 4$

68. $f(x) = 2x^3 + 4x^2 - 3x + 10$

69. $f(x) = 2x^5 + x^2 + 16$

70. $f(x) = 3x^3 - 12x + 22$

10.2 Ellipses

Introduction

The second type of conic is called an **ellipse.** It is defined as follows.

Definition of an Ellipse

An **ellipse** is the set of all points (x, y) the sum of whose distances from two distinct fixed points (**foci**) is constant. (See Figure 10.10.)

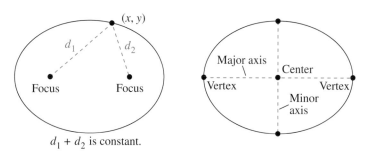

$d_1 + d_2$ is constant.

Figure 10.10

The line through the foci intersects the ellipse at two points called **vertices.** The chord joining the vertices is the **major axis,** and its midpoint is the **center** of the ellipse. The chord perpendicular to the major axis at the center is the **minor axis** of the ellipse.

To derive the standard form of the equation of an ellipse, consider the ellipse in Figure 10.11 with the following points: center, (h, k); vertices, $(h \pm a, k)$; foci, $(h \pm c, k)$.

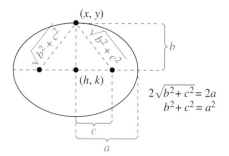

$$2\sqrt{b^2 + c^2} = 2a$$
$$b^2 + c^2 = a^2$$

Figure 10.11

The sum of the distances from any point on the ellipse to the two foci is constant. Using a vertex point, this constant sum is

$$(a + c) + (a - c) = 2a \qquad \text{Length of major axis}$$

or simply the length of the major axis.

© Royal Observatory, Edinburgh

Now, if you let (x, y) be *any* point on the ellipse, the sum of the distances between (x, y) and the two foci must also be $2a$. That is,

$$\sqrt{[x - (h - c)]^2 + (y - k)^2} + \sqrt{[x - (h + c)]^2 + (y - k)^2} = 2a.$$

Finally, using Figure 10.11 and $b^2 = a^2 - c^2$, you obtain the following equation of the ellipse.

$$b^2(x - h)^2 + a^2(y - k)^2 = a^2b^2$$

$$\frac{(x - h)^2}{a^2} + \frac{(y - k)^2}{b^2} = 1$$

In the development above, you would obtain a similar equation by starting with a vertical major axis. Both results are summarized as follows.

Standard Equation of an Ellipse

The **standard form of the equation of an ellipse,** with center (h, k) and major and minor axes of lengths $2a$ and $2b$, respectively, where $0 < b < a$, is

$$\frac{(x - h)^2}{a^2} + \frac{(y - k)^2}{b^2} = 1 \qquad \text{Major axis is horizontal.}$$

$$\frac{(x - h)^2}{b^2} + \frac{(y - k)^2}{a^2} = 1. \qquad \text{Major axis is vertical.}$$

The foci lie on the major axis, c units from the center, with $c^2 = a^2 - b^2$. If the center is at the origin $(0, 0)$, the equation takes one of the following forms.

$$\frac{x^2}{a^2} + \frac{y^2}{b^2} = 1 \qquad \text{Major axis is horizontal.}$$

$$\frac{x^2}{b^2} + \frac{y^2}{a^2} = 1 \qquad \text{Major axis is vertical.}$$

STUDY TIP

Don't confuse the equation

$$c^2 = a^2 - b^2$$

with the Pythagorean Theorem—there is a sign difference.

Figure 10.12 shows both the vertical and horizontal orientations for an ellipse.

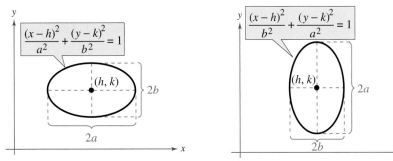

Figure 10.12

Figure 10.13

You can visualize the definition of an ellipse by imagining two thumbtacks placed at the foci, as shown in Figure 10.13. If the ends of a fixed length of string are fastened to the thumbtacks and the string is drawn taut with a pencil, the path traced by the pencil will be an ellipse.

A computer animation of this concept appears in the *Interactive* CD-ROM and *Internet* versions of this text.

EXAMPLE 1   Finding the Standard Equation of an Ellipse

Find the standard form of the equation of the ellipse having foci at $(0, 1)$ and $(4, 1)$, and a major axis of length 6, as shown in Figure 10.14.

Solution

Because the foci occur at $(0, 1)$ and $(4, 1)$, the center of the ellipse is $(2, 1)$ and the distance from the center to one of the foci is $c = 2$. Because $2a = 6$ you know that $a = 3$. Now, from $c^2 = a^2 - b^2$, you have

$$b = \sqrt{a^2 - c^2}$$
$$= \sqrt{9 - 4}$$
$$= \sqrt{5}.$$

Because the major axis is horizontal, the standard equation is

$$\frac{(x - 2)^2}{3^2} + \frac{(y - 1)^2}{(\sqrt{5})^2} = 1.$$

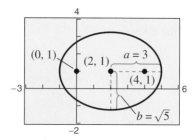

Figure 10.14

EXAMPLE 2   Sketching an Ellipse

Sketch the ellipse given by $4x^2 + y^2 = 36$ and identify the vertices.

Algebraic Solution

$$4x^2 + y^2 = 36 \qquad \text{Write original equation.}$$

$$\frac{4x^2}{36} + \frac{y^2}{36} = \frac{36}{36} \qquad \text{Divide each side by 36.}$$

$$\frac{x^2}{3^2} + \frac{y^2}{6^2} = 1 \qquad \text{Write in standard form.}$$

Because the denominator of the y^2-term is larger than the denominator to the x^2-term, you can conclude that the major axis is vertical. Moreover, because $a = 6$, the vertices are $(0, -6)$ and $(0, 6)$. Finally, because $b = 3$, the endpoints of the minor axis are $(-3, 0)$ and $(3, 0)$, as shown in Figure 10.15.

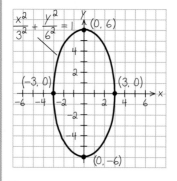

Figure 10.15

Graphical Solution

Solve the equation of the ellipse for y as follows.

$$4x^2 + y^2 = 36$$
$$y^2 = 36 - 4x^2$$
$$y = \pm\sqrt{36 - 4x^2}$$

Then use a graphing utility to graph both $y_1 = \sqrt{36 - 4x^2}$ and $y_2 = -\sqrt{36 - 4x^2}$ in the same viewing window. Be sure to use a square setting. From the graph in Figure 10.16, you can see that the major axis is vertical. You can use the *zoom* and *trace* features to approximate the vertices to be $(0, 6)$ and $(0, -6)$.

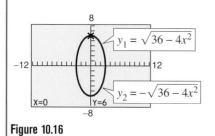

Figure 10.16

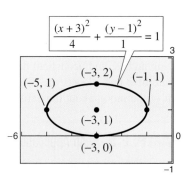

Figure 10.17

EXAMPLE 3 Writing an Equation in Standard Form

Write the equation of the ellipse in standard form and sketch the graph of the ellipse.

$$x^2 + 4y^2 + 6x - 8y + 9 = 0.$$

Solution

To write the given equation in standard form you must complete the square twice. In the fourth step, note that 9 and 4 are added to *both* sides of the equation.

$$x^2 + 4y^2 + 6x - 8y + 9 = 0 \quad \text{Write original equation.}$$

$$\left(x^2 + 6x + \quad\right) + \left(4y^2 - 8y + \quad\right) = -9 \quad \text{Group terms.}$$

$$\left(x^2 + 6x + \quad\right) + 4\left(y^2 - 2y + \quad\right) = -9 \quad \text{Factor 4 out of } y\text{-terms.}$$

$$(x^2 + 6x + 9) + 4(y^2 - 2y + 1) = -9 + 9 + 4(1)$$

$$(x + 3)^2 + 4(y - 1)^2 = 4 \quad \text{Write in completed square form.}$$

$$\frac{(x + 3)^2}{2^2} + \frac{(y - 1)^2}{1^2} = 1 \quad \text{Write in standard form.}$$

Now you see that the center is at $(h, k) = (-3, 1)$. Because the denominator of the x-term is $a^2 = 2^2$, you can locate the endpoints of the major axis two units to the right and left of the center. Similarly, because the denominator of the y-term is $b^2 = 1^2$, you can locate the endpoints of the minor axis one unit up and down from the center. The graph of this ellipse is shown in Figure 10.17.

EXAMPLE 4 Analyzing an Ellipse

Find the center, vertices, and foci of the ellipse $4x^2 + y^2 - 8x + 4y - 8 = 0$.

Solution

By completing the square, you can write the given equation in standard form.

$$4x^2 + y^2 - 8x + 4y - 8 = 0$$

$$\left(4x^2 - 8x + \quad\right) + \left(y^2 + 4y + \quad\right) = 8$$

$$4\left(x^2 - 2x + \quad\right) + \left(y^2 + 4y + \quad\right) = 8$$

$$4(x^2 - 2x + 1) + (y^2 + 4y + 4) = 8 + 4(1) + 4$$

$$4(x - 1)^2 + (y + 2)^2 = 16$$

$$\frac{(x - 1)^2}{2^2} + \frac{(y + 2)^2}{4^2} = 1$$

So, the major axis is vertical, where $h = 1, k = -2, a = 4, b = 2$, and

$$c = \sqrt{16 - 4} = 2\sqrt{3}.$$

Therefore, you have the following.

Center: $(1, -2)$ Vertices: $(1, -6)$ Foci: $\left(1, -2 - 2\sqrt{3}\right)$
                               $(1, 2)$ $\left(1, -2 + 2\sqrt{3}\right)$

The graph of the ellipse is shown in Figure 10.18.

You can use a graphing utility to graph an ellipse by graphing the upper and lower portions in the same viewing window. For instance, to graph the ellipse in Example 3, first solve for y to get

$$y_1 = 1 + \sqrt{1 - \frac{(x + 3)^2}{4}}$$

and

$$y_2 = 1 - \sqrt{1 - \frac{(x + 3)^2}{4}}.$$

Use a viewing window in which $-6 \le x \le 0$ and $-1 \le y \le 3$. You should obtain the graph shown in Figure 10.17.

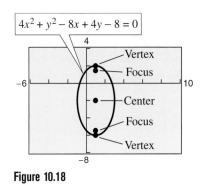

Figure 10.18

Application

Ellipses have many practical and aesthetic uses. For instance, machine gears, supporting arches, and acoustical designs often involve elliptical shapes. The orbits of satellites and planets are also ellipses. Example 5 investigates the elliptical orbit of the moon about the earth.

EXAMPLE 5   An Application Involving an Elliptical Orbit

The moon travels about the earth in an elliptical orbit with the earth at one focus, as shown in Figure 10.19. The major and minor axes of the orbit have lengths of 768,806 kilometers and 767,746 kilometers, respectively. Find the greatest and least distances (the apogee and perigee) from the earth's center to the moon's center.

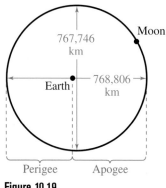

Figure 10.19

Solution

Because $2a = 768,806$ and $2b = 767,746$, you have $a = 384,403$ and $b = 383,873$, which implies that

$$c = \sqrt{a^2 - b^2}$$

$$= \sqrt{384,403^2 - 383,873^2}$$

$$\approx 20,179.$$

Therefore, the greatest distance between the center of the earth and the center of the moon is

$$a + c \approx 384,403 + 20,179$$

$$= 404,582 \text{ km}$$

and the least distance is

$$a - c \approx 384,403 - 20,179$$

$$= 364,224 \text{ km}.$$

Eccentricity

One of the reasons it was difficult for early astronomers to detect that the orbits of the planets are ellipses is that the foci of the planetary orbits are relatively close to their centers, and so the orbits are nearly circular. To measure the ovalness of an ellipse, you can use the concept of **eccentricity.**

Definition of Eccentricity

The **eccentricity** e of an ellipse is given by the ratio

$$e = \frac{c}{a}.$$

Note that $0 < e < 1$ for *every* ellipse.

Exploration

Use a graphing utility to graph the following ellipses in the same viewing window.

$$\frac{x^2}{9} + \frac{y^2}{8} = 1$$

$$\frac{x^2}{25} + \frac{y^2}{9} = 1$$

$$\frac{x^2}{49} + \frac{y^2}{9} = 1$$

Calculate the eccentricity $e = c/a$ for each ellipse. How does the value of e relate to the shape of the ellipse?

To see how this ratio is used to describe the shape of an ellipse, note that because the foci of an ellipse are located along the major axis between the vertices and the center, it follows that

$$0 < c < a.$$

For an ellipse that is nearly circular, the foci are close to the center and the ratio c/a is small [see Figure 10.20(a)]. On the other hand, for an elongated ellipse, the foci are close to the vertices and the ratio c/a is close to 1 [see Figure 10.20(b)].

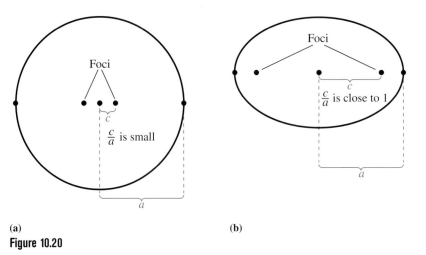

(a)   **(b)**

Figure 10.20

The orbit of the moon has an eccentricity of $e = 0.0549$, and the eccentricities of the nine planetary orbits are as follows.

Mercury:	$e = 0.2056$	Saturn:	$e = 0.0543$
Venus:	$e = 0.0068$	Uranus:	$e = 0.0460$
Earth:	$e = 0.0167$	Neptune:	$e = 0.0082$
Mars:	$e = 0.0934$	Pluto:	$e = 0.2481$
Jupiter:	$e = 0.0484$		

W r i t i n g   A b o u t   M a t h   *Graphing Ellipses*

Write an equation of an ellipse in standard form and graph it on graph paper. Do not write the equation on your graph. Exchange graphs with another student. Use the graph you receive to reconstruct the equation of the ellipse it represents. Find the eccentricity of the ellipse. Compare your results and write a short paragraph that discusses your findings.

10.2 Exercises

In Exercises 1–6, match the equation with its graph. [The graphs are labeled (a), (b), (c), (d), (e), and (f).]

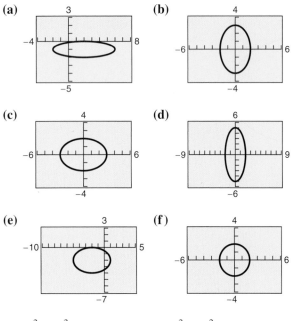

(a)

(b)

(c)

(d)

(e)

(f)

1. $\dfrac{x^2}{4} + \dfrac{y^2}{9} = 1$

2. $\dfrac{x^2}{9} + \dfrac{y^2}{4} = 1$

3. $\dfrac{x^2}{4} + \dfrac{y^2}{25} = 1$

4. $\dfrac{y^2}{4} + \dfrac{x^2}{4} = 1$

5. $\dfrac{(x-2)^2}{16} + (y+1)^2 = 1$

6. $\dfrac{(x+2)^2}{9} + \dfrac{(y+2)^2}{4} = 1$

In Exercises 7–22, find the center, vertices, foci, and eccentricity of the ellipse, and sketch its graph. Use a graphing utility to verify your graph.

7. $\dfrac{x^2}{25} + \dfrac{y^2}{16} = 1$

8. $\dfrac{x^2}{81} + \dfrac{y^2}{144} = 1$

9. $\dfrac{x^2}{5} + \dfrac{y^2}{9} = 1$

10. $\dfrac{x^2}{64} + \dfrac{y^2}{28} = 1$

11. $\dfrac{(x+3)^2}{16} + \dfrac{(y-5)^2}{25} = 1$

12. $\dfrac{(x-4)^2}{12} + \dfrac{(y+3)^2}{16} = 1$

13. $\dfrac{(x+5)^2}{9/4} + (y-1)^2 = 1$

14. $(x+2)^2 + \dfrac{(y+4)^2}{1/4} = 1$

15. $9x^2 + 4y^2 + 36x - 24y + 36 = 0$

16. $9x^2 + 4y^2 - 54x + 40y + 37 = 0$

17. $x^2 + 5y^2 - 8x - 30y - 39 = 0$

18. $3x^2 + y^2 + 18x - 2y - 8 = 0$

19. $6x^2 + 2y^2 + 18x - 10y + 2 = 0$

20. $x^2 + 4y^2 - 6x + 20y - 2 = 0$

21. $16x^2 + 25y^2 - 32x + 50y + 16 = 0$

22. $9x^2 + 25y^2 - 36x - 50y + 61 = 0$

In Exercises 23–26, use a graphing utility to graph the ellipse. Find the center, foci, and vertices. (*Hint:* Use two equations.)

23. $5x^2 + 3y^2 = 15$

24. $3x^2 + 4y^2 = 12$

25. $12x^2 + 20y^2 - 12x + 40y - 37 = 0$

26. $36x^2 + 9y^2 + 48x - 36y + 43 = 0$

In Exercises 27–34, find the standard form of the equation of the ellipse with its center at the origin.

27. **28.**

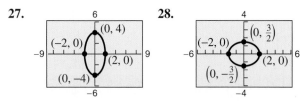

29. Vertices: $(\pm 6, 0)$; Foci: $(\pm 2, 0)$

30. Vertices: $(0, \pm 8)$; Foci: $(0, \pm 4)$

31. Foci: $(\pm 5, 0)$; Major axis of length 12

32. Foci: $(\pm 2, 0)$; Major axis of length 8

33. Vertices: $(0, \pm 5)$; Passes through the point $(4, 2)$

34. Major axis is vertical; Passes through the points $(0, 4)$ and $(2, 0)$

In Exercises 35–46, find the standard form of the equation of the specified ellipse.

35.

36.

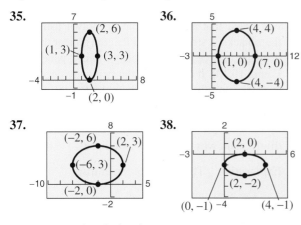

37.

38.

39. Vertices: $(0, 4)$, $(4, 4)$; Minor axis of length 2

40. Foci: $(0, 0)$, $(4, 0)$; Major axis of length 8

41. Foci: $(0, 0)$, $(0, 8)$; Major axis of length 16

42. Center: $(2, -1)$; Vertex: $\left(2, \frac{1}{2}\right)$; Minor axis of length 2

43. Vertices: $(3, 1)$, $(3, 9)$; Minor axis of length 6

44. Center: $(3, 2)$; $a = 3c$; Foci: $(1, 2)$, $(5, 2)$

45. Center: $(0, 4)$; $a = 2c$; Vertices: $(-4, 4)$, $(4, 4)$

46. Vertices: $(5, 0)$, $(5, 12)$; Endpoints of the minor axis: $(0, 6)$, $(10, 6)$

47. Find an equation of the ellipse with vertices $(\pm 5, 0)$ and eccentricity $e = \frac{3}{5}$.

48. Find an equation of the ellipse with vertices $(0, \pm 8)$ and eccentricity $e = \frac{1}{2}$.

49. *Fireplace Arch* A fireplace arch is to be built in the shape of a semiellipse. The opening is to have a height of 2 feet at the center and a width of 6 feet along the base. The contractor draws the outline of the ellipse using the method described on page 705. Give the required positions of the tacks and the length of the string.

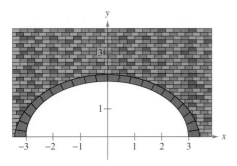

50. *Mountain Travel* A semielliptical arch over a tunnel for a road through a mountain has a major axis of 80 feet and a height at the center of 30 feet.

    (a) Draw a rectangular coordinate system on a sketch of the tunnel with the center of the road entering the tunnel at the origin. Identify the coordinates of the known points.

    (b) Find an equation of the elliptical tunnel.

    (c) Determine the height of the arch 5 feet from the edge of the tunnel.

51. *Geometry* The area of the ellipse in the figure is twice the area of the circle. What is the length of the major axis? (*Hint:* $A = \pi ab$ for an ellipse.)

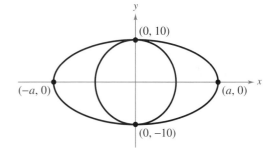

52. *Comet Orbit* Halley's comet has an elliptical orbit with the sun at one focus. The eccentricity of the orbit is approximately 0.97. The length of the major axis of the orbit is about 36.23 astronomical units. (An astronomical unit is about 93 million miles.) Find an equation for the orbit. Place the center of the orbit at the origin and place the major axis on the *x*-axis.

53. *Comet Orbit* The comet Encke has an elliptical orbit with the sun at one focus. Encke ranges from 0.34 to 4.08 astronomical units from the sun. Find an equation of the orbit. Place the center of the orbit at the origin and place the major axis on the *x*-axis.

54. *Satellite Orbit*   The first artificial satellite to orbit earth was *Sputnik I* (launched by Russia in 1957). Its highest point above earth's surface was 938 kilometers, and its lowest point was 212 kilometers. The radius of earth is 6378 kilometers. Find the eccentricity of the orbit.

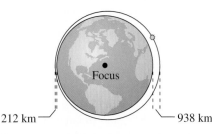

212 km — Focus — 938 km

55. *Geometry*   A line segment through a focus with endpoints on the ellipse and perpendicular to the major axis is called a **latus rectum** of the ellipse. Therefore, an ellipse has two latera recta. Knowing the length of the latera recta is helpful in sketching an ellipse because it yields other points on the curve. Show that the length of each latus rectum is $2b^2/a$.

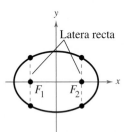

Latera recta

F_1 F_2

In Exercises 56–59, sketch the graph of the ellipse, making use of the latera recta (see Exercise 55).

56. $\dfrac{x^2}{4} + \dfrac{y^2}{1} = 1$ **57.** $\dfrac{x^2}{9} + \dfrac{y^2}{16} = 1$

58. $9x^2 + 4y^2 = 36$ **59.** $5x^2 + 3y^2 = 15$

Synthesis

True or False?   In Exercises 60–63, determine whether the statement is true or false. Justify your answer.

60. The graph of $(x^2/4) + y^4 = 1$ is an ellipse.

61. It is easier to distinguish the graph of an ellipse from the graph of a circle if the eccentricity of the ellipse is large (close to 1).

62. The area of a circle with diameter $d = 2r = 8$ is greater than the area of an ellipse with major axis $2a = 8$.

63. It is possible for the foci of an ellipse to occur outside the ellipse.

64. *Think About It*   At the beginning of this section it was noted that an ellipse can be drawn using two thumbtacks, a string of fixed length (greater than the distance between the two tacks), and a pencil (see Figure 10.13). If the ends of the string are fastened at the tacks and the string is drawn taut with a pencil, the path traced by the pencil is an ellipse.

(a) What is the length of the string in terms of a?

(b) Explain why the path is an ellipse.

65. *Exploration*   The area A of the ellipse

$$\frac{x^2}{a^2} + \frac{y^2}{b^2} = 1 \text{ is } A = \pi ab.$$

For a particular application, $a + b = 20$.

(a) Write the area of the ellipse as a function of a.

(b) Find the equation of an ellipse with an area of 264 square centimeters.

(c) Complete the table and make a conjecture about the shape of the ellipse with a maximum area.

a	8	9	10	11	12	13
A						

(d) Use a graphing utility to graph the area function, and use the graph to make a conjecture about the shape of the ellipse that yields a maximum area.

Review

In Exercises 66–69, determine whether the sequence is arithmetic, geometric, or neither.

66. 66, 55, 44, 33, 22, . . .   **67.** 80, 40, 20, 10, 5, . . .

68. $\frac{1}{4}, \frac{1}{2}, 1, 2, 4, \ldots$ **69.** $-\frac{1}{2}, \frac{1}{2}, \frac{3}{2}, \frac{5}{2}, \frac{7}{2}, \ldots$

In Exercises 70–73, find a formula for a_n for the arithmetic sequence.

70. $a_1 = 13,\ d = 3$ **71.** $a_1 = 0,\ d = -\frac{1}{4}$

72. $a_1 = 5,\ a_4 = 9.5$ **73.** $a_3 = 27,\ a_8 = 72$

In Exercises 74–77, find the sum.

74. $\displaystyle\sum_{n=0}^{6} 3^n$ **75.** $\displaystyle\sum_{n=0}^{6} (-3)^n$

76. $\displaystyle\sum_{n=1}^{10} 4\left(\frac{3}{4}\right)^{n-1}$ **77.** $\displaystyle\sum_{n=0}^{10} 5\left(\frac{4}{3}\right)^n$

10.3 Hyperbolas

Introduction

The definition of a hyperbola parallels that of an ellipse. The difference is that for an ellipse, the *sum* of the distances between the foci and a point on the ellipse is fixed; whereas for a hyperbola, the *difference* of these distances is fixed.

What You Should Learn:

- How to write equations of hyperbolas in standard form
- How to find asymptotes of hyperbolas
- How to use properties of hyperbolas to solve real-life problems
- How to classify conics from their general equations

Definition of a Hyperbola

A **hyperbola** is the set of all points (x, y) the difference of whose distances from two distinct fixed points (**foci**) is a positive constant. (See Figure 10.21.)

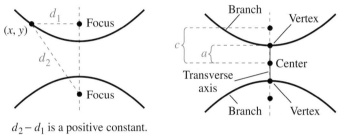

$d_2 - d_1$ is a positive constant.

Figure 10.21

Why You Should Learn It:

Hyperbolas can be used to model and solve many types of real-life problems. For instance, in Exercise 41 on page 721, hyperbolas are used to locate the position of an explosion that was recorded by three listening stations.

James Foote/Photo Researchers, Inc.

Every hyperbola has two disconnected **branches.** The line through the two foci intersects a hyperbola at its two **vertices.** The line segment connecting the vertices is called the **transverse axis,** and the midpoint of the transverse axis is called the **center** of the hyperbola. The development of the **standard form of the equation of a hyperbola** is similar to that of an ellipse. Note that a, b, and c are related differently for hyperbolas than for ellipses.

Standard Equation of a Hyperbola

The **standard form of the equation of a hyperbola** with center at (h, k) is

$$\frac{(x - h)^2}{a^2} - \frac{(y - k)^2}{b^2} = 1 \qquad \text{Transverse axis is horizontal.}$$

$$\frac{(y - k)^2}{a^2} - \frac{(x - h)^2}{b^2} = 1. \qquad \text{Transverse axis is vertical.}$$

The vertices are a units from the center, and the foci are c units from the center. Moreover, $c^2 = a^2 + b^2$. If the center of the hyperbola is at the origin $(0, 0)$, the equation takes one of the following forms.

$$\frac{x^2}{a^2} - \frac{y^2}{b^2} = 1 \qquad \text{Transverse axis is horizontal.} \qquad \frac{y^2}{a^2} - \frac{x^2}{b^2} = 1 \qquad \text{Transverse axis is vertical.}$$

Figure 10.22 shows both the horizontal and vertical orientations for a hyperbola.

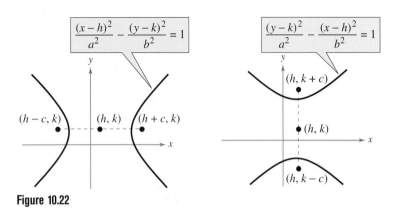

Figure 10.22

EXAMPLE 1   Finding the Standard Equation of a Hyperbola

Find the standard form of the equation of the hyperbola with foci at $(-1, 2)$ and $(5, 2)$, and vertices at $(0, 2)$ and $(4, 2)$.

Solution

By the Midpoint Formula, the center of the hyperbola occurs at the point $(2, 2)$. Furthermore, $c = 3$ and $a = 2$, and it follows that

$$b = \sqrt{c^2 - a^2}$$
$$= \sqrt{3^2 - 2^2}$$
$$= \sqrt{9 - 4}$$
$$= \sqrt{5}.$$

So, the equation of the hyperbola is

$$\frac{(x - 2)^2}{2^2} - \frac{(y - 2)^2}{(\sqrt{5})^2} = 1.$$

Figure 10.23 shows the hyperbola.

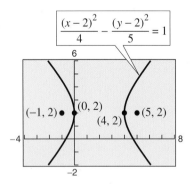

Figure 10.23

Asymptotes of a Hyperbola

Each hyperbola has two **asymptotes** that intersect at the center of the hyperbola, as shown in Figure 10.24. The asymptotes pass through the vertices of a rectangle of dimensions $2a$ by $2b$, with its center at (h, k).

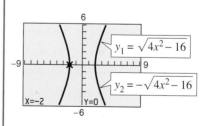

Asymptotes of a Hyperbola

$$y = k \pm \frac{b}{a}(x - h)$$ Asymptotes for horizontal transverse axis $$y = k \pm \frac{a}{b}(x - h)$$ Asymptotes for vertical transverse axis

The **conjugate axis** of a hyperbola is the line segment of length $2b$ joining $(h, k + b)$ and $(h, k - b)$ if the transverse axis is horizontal, and the line segment of length $2b$ joining $(h + b, k)$ and $(h - b, k)$ if the transverse axis is vertical.

Figure 10.24

EXAMPLE 2   Sketching the Graph of a Hyperbola

Sketch the hyperbola whose equation is $4x^2 - y^2 = 16$.

Algebraic Solution

$$4x^2 - y^2 = 16$$ Write original equation.

$$\frac{4x^2}{16} - \frac{y^2}{16} = \frac{16}{16}$$ Divide each side by 16.

$$\frac{x^2}{2^2} - \frac{y^2}{4^2} = 1$$ Write in standard form.

Because the x^2-term is positive, you can conclude that the transverse axis is horizontal. So, the vertices occur at $(-2, 0)$ and $(2, 0)$, and the endpoints of the conjugate axis occur at $(0, -4)$ and $(0, 4)$. Using these four points, you can sketch the rectangle shown in Figure 10.25(a). Finally, by drawing the asymptotes through the corners of this rectangle, you can complete the sketch, as shown in Figure 10.25(b).

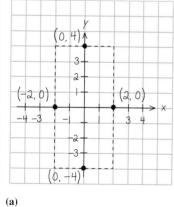

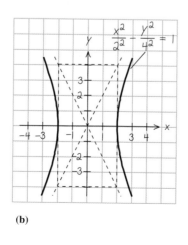

(a)   **(b)**

Figure 10.25

Graphical Solution

Solve the equation of the hyperbola for y as follows.

$$4x^2 - y^2 = 16$$

$$4x^2 - 16 = y^2$$

$$\pm\sqrt{4x^2 - 16} = y$$

Then use a graphing utility to graph $y_1 = \sqrt{4x^2 - 16}$ and $y_2 = -\sqrt{4x^2 - 16}$ in the same viewing window. Be sure to use a square setting. From the graph in Figure 10.26, you can see that the transverse axis is horizontal. You can use the *zoom* and *trace* features to approximate the vertices to be $(-2, 0)$ and $(2, 0)$.

Figure 10.26

EXAMPLE 3   Finding the Asymptotes of a Hyperbola

Sketch the hyperbola given by $4x^2 - 3y^2 + 8x + 16 = 0$ and find the equations of its asymptotes.

Solution

$$4x^2 - 3y^2 + 8x + 16 = 0 \qquad \text{Write original equation.}$$

$$4(x^2 + 2x) - 3y^2 = -16 \qquad \text{Subtract 16 from each side; factor.}$$

$$4(x^2 + 2x + 1) - 3y^2 = -16 + 4 \qquad \text{Complete the square.}$$

$$4(x + 1)^2 - 3y^2 = -12 \qquad \text{Write in completed square form.}$$

$$\frac{y^2}{2^2} - \frac{(x + 1)^2}{\left(\sqrt{3}\right)^2} = 1 \qquad \text{Write in standard form.}$$

From this equation you can conclude that the hyperbola is centered at $(-1, 0)$ and has vertices at $(-1, 2)$ and $(-1, -2)$, and that the ends of the conjugate axis occur at $\left(-1 - \sqrt{3}, 0\right)$ and $\left(-1 + \sqrt{3}, 0\right)$. To sketch the hyperbola, draw a rectangle through these four points. The asymptotes are the lines passing through the corners of the rectangle, as shown in Figure 10.27. Finally, using $a = 2$ and $b = \sqrt{3}$, you can conclude that the equations of the asymptotes are

$$y = \frac{2}{\sqrt{3}}(x + 1) \qquad \text{and} \qquad y = -\frac{2}{\sqrt{3}}(x + 1).$$

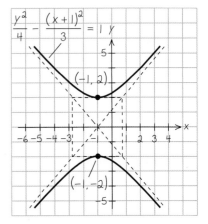

Figure 10.27

If the constant term F in the equation in Example 3 had been $F = 4$ instead of 16, you would have obtained the following degenerate case.

$$\text{Two Intersecting Lines:} \quad \frac{y^2}{4} - \frac{(x + 1)^2}{3} = 0$$

STUDY T!P

You can use a graphing utility to graph a hyperbola by graphing the upper and lower portions in the same viewing window. For instance, to graph the hyperbola in Example 3, first solve for y to get

$$y_1 = 2\sqrt{1 + \frac{(x + 1)^2}{3}}$$

and

$$y_2 = -2\sqrt{1 + \frac{(x + 1)^2}{3}}.$$

Use a viewing window in which $-8 \le x \le 6$ and $-6 \le y \le 6$. You should obtain the graph shown in Figure 10.28.

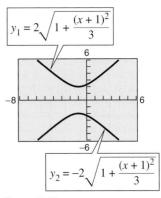

Figure 10.28

EXAMPLE 4   Using Asymptotes to Find the Standard Equation

Find the standard form of the equation of the hyperbola having vertices at $(3, -5)$ and $(3, 1)$ and with asymptotes

$$y = 2x - 8 \quad \text{and} \quad y = -2x + 4$$

as shown in Figure 10.29.

Solution

By the Midpoint Formula, the center of the hyperbola is at $(3, -2)$. Furthermore, the hyperbola has a vertical transverse axis with $a = 3$. From the given equations, you can determine the slopes of the asymptotes to be

$$m_1 = 2 = \frac{a}{b}$$

and

$$m_2 = -2 = -\frac{a}{b}$$

and because $a = 3$, you can conclude that $b = \frac{3}{2}$. So, the standard equation is

$$\frac{(y + 2)^2}{3^2} - \frac{(x - 3)^2}{(3/2)^2} = 1.$$

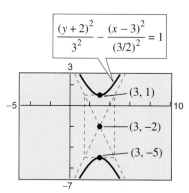

Figure 10.29

As with ellipses, the *eccentricity* of a hyperbola is

$$e = \frac{c}{a} \qquad \text{Eccentricity}$$

and because $c > a$ it follows that $e > 1$. If the eccentricity is large, the branches of the hyperbola are nearly flat. If the eccentricity is close to 1, the branches of the hyperbola are more pointed. See Figure 10.30.

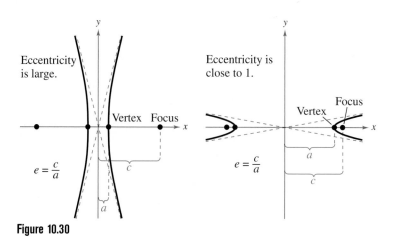

Figure 10.30

Applications

The following application was developed during World War II. It shows how the properties of hyperbolas can be used in radar and other detection systems.

EXAMPLE 5   An Application Involving Hyperbolas

Two microphones, 1 mile apart, record an explosion. Microphone A received the sound 2 seconds before microphone B. Where was the explosion?

Solution

Assuming sound travels at 1100 feet per second, you know that the explosion took place 2200 feet further from B than from A, as shown in Figure 10.31. The locus of all points that are 2200 feet closer to A than to B is one branch of the hyperbola

$$\frac{x^2}{a^2} + \frac{y^2}{b^2} = 1$$

where

$$c = \frac{5280}{2} = 2640$$

and

$$a = \frac{2200}{2} = 1100.$$

So, $b^2 = c^2 - a^2 = 5{,}759{,}600$, and you can conclude that the explosion occurred somewhere on the right branch of the hyperbola

$$\frac{x^2}{1{,}210{,}000} - \frac{y^2}{5{,}759{,}600} = 1.$$

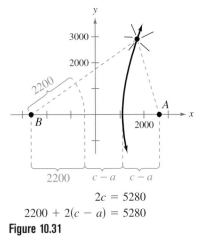

$$2c = 5280$$
$$2200 + 2(c - a) = 5280$$

Figure 10.31

Another interesting application of conic sections involves the orbits of comets in our solar system. Of the 610 comets identified prior to 1970, 245 have elliptical orbits, 295 have parabolic orbits, and 70 have hyperbolic orbits. The center of the sun is a focus of each of these orbits, and each orbit has a vertex at the point where the comet is closest to the sun, as shown in Figure 10.32. Undoubtedly, there have been many comets with parabolic or hyperbolic orbits that have not been identified. You get to see such comets only *once*. Comets with elliptical orbits, such as Halley's comet, are the only ones that remain in our solar system.

If p is the distance between the vertex and the focus in meters, and v is the velocity of the comet at the vertex in meters per second, the type of orbit is determined as follows.

1. Ellipse: $v < \sqrt{2GM/p}$
2. Parabola: $v = \sqrt{2GM/p}$
3. Hyperbola: $v > \sqrt{2GM/p}$

In each of these equations, $M \approx 1.991 \times 10^{30}$ kilograms (the mass of the sun) and $G \approx 6.67 \times 10^{-11}$ cubic meters per kilogram-second squared (the universal gravitational constant).

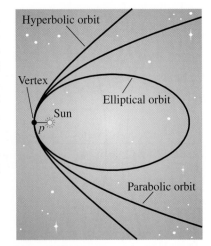

Figure 10.32

Classifying a Conic from Its General Equation

The graph of $Ax^2 + Cy^2 + Dx + Ey + F = 0$ is one of the following.

1. Circle: $A = C$ $A \neq 0$

2. Parabola: $AC = 0$ $A = 0$ or $C = 0$, but not both.

3. Ellipse: $AC > 0$ A and C have like signs.

4. Hyperbola: $AC < 0$ A and C have unlike signs.

The test above is valid *if* the graph is a conic. The test does not apply to equations such as $x^2 + y^2 = -1$, which is not a conic.

EXAMPLE 6   Classifying Conics from General Equations

Classify each graph.

a. $4x^2 - 9x + y - 5 = 0$

b. $4x^2 - y^2 + 8x - 6y + 4 = 0$

c. $2x^2 + 4y^2 - 4x + 12y = 0$

d. $2x^2 + 2y^2 - 8x + 12y + 2 = 0$

Solution

a. For the equation $4x^2 - 9x + y - 5 = 0$, you have

$$AC = 4(0) = 0. \qquad \text{Parabola}$$

So, the graph is a parabola.

b. For the equation $4x^2 - y^2 + 8x - 6y + 4 = 0$, you have

$$AC = 4(-1) < 0. \qquad \text{Hyperbola}$$

So, the graph is a hyperbola.

c. For the equation $2x^2 + 4y^2 - 4x + 12y = 0$, you have

$$AC = 2(4) > 0. \qquad \text{Ellipse}$$

So, the graph is an ellipse.

d. For the equation $2x^2 + 2y^2 - 8x + 12y + 2 = 0$, you have

$$A = C = 2. \qquad \text{Circle}$$

So, the graph is a circle.

The first woman to be credited with detecting a new comet was the English astronomer **Caroline Herschel (1750–1848).** During her long life, Caroline Herschel discovered a total of eight new comets.

The Granger Collection

Writing About Math   *Identifying Equations of Conics*

Use the Internet to research information about the orbits of comets in our solar system. What can you find about the orbits of comets that have been identified since 1970? Write a summary of your results. Identify your source. Does it seem reliable?

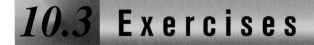

10.3 Exercises

In Exercises 1–4, match the equation with its graph. [The graphs are labeled (a), (b), (c), and (d).]

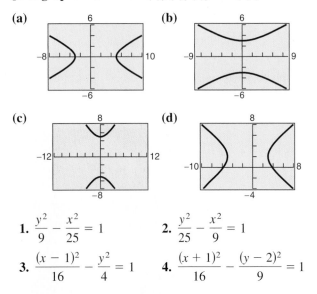

(a)

(b)

(c)

(d)

1. $\dfrac{y^2}{9} - \dfrac{x^2}{25} = 1$ **2.** $\dfrac{y^2}{25} - \dfrac{x^2}{9} = 1$

3. $\dfrac{(x-1)^2}{16} - \dfrac{y^2}{4} = 1$ **4.** $\dfrac{(x+1)^2}{16} - \dfrac{(y-2)^2}{9} = 1$

In Exercises 5–18, find the center, vertices, foci, and asymptotes of the hyperbola, and sketch its graph, using the asymptotes as an aid. Use a graphing utility to verify your graph.

5. $x^2 - y^2 = 1$ **6.** $\dfrac{x^2}{9} - \dfrac{y^2}{25} = 1$

7. $\dfrac{y^2}{1} - \dfrac{x^2}{4} = 1$ **8.** $\dfrac{y^2}{9} - \dfrac{x^2}{1} = 1$

9. $\dfrac{y^2}{25} - \dfrac{x^2}{81} = 1$ **10.** $\dfrac{x^2}{36} - \dfrac{y^2}{4} = 1$

11. $\dfrac{(x-1)^2}{4} - \dfrac{(y+2)^2}{1} = 1$

12. $\dfrac{(x+3)^2}{144} - \dfrac{(y-2)^2}{25} = 1$

13. $(y+6)^2 - (x-2)^2 = 1$

14. $\dfrac{(y-1)^2}{1/4} - \dfrac{(x+3)^2}{1/16} = 1$

15. $9x^2 - y^2 - 36x - 6y + 18 = 0$

16. $x^2 - 9y^2 + 36y - 72 = 0$

17. $x^2 - 9y^2 + 2x - 54y - 80 = 0$

18. $16y^2 - x^2 + 2x + 64y + 63 = 0$

In Exercises 19–22, find the center, vertices, foci, and the equations of the asymptotes of the hyperbola. Use a graphing utility to graph the hyperbola and its asymptotes.

19. $2x^2 - 3y^2 = 6$

20. $6y^2 - 3x^2 = 18$

21. $9y^2 - x^2 + 2x + 54y + 62 = 0$

22. $9x^2 - y^2 + 54x + 10y + 55 = 0$

In Exercises 23–40, find the standard form of the equation of the specified hyperbola.

23. Vertices: $(0, \pm 2)$; Foci: $(0, \pm 4)$

24. Vertices: $(\pm 2, 0)$; Foci: $(\pm 5, 0)$

25. Vertices: $(\pm 1, 0)$; Asymptotes: $y = \pm 5x$

26. Vertices: $(0, \pm 3)$; Asymptotes: $y = \pm 3x$

27. Foci: $(0, \pm 8)$; Asymptotes: $y = \pm 4x$

28. Foci: $(\pm 10, 0)$; Asymptotes: $y = \pm \frac{3}{4}x$

29. Vertices: $(2, 0), (6, 0)$; Foci: $(0, 0), (8, 0)$

30. Vertices: $(2, 3), (2, -3)$; Foci: $(2, 5), (2, -5)$

31. Vertices: $(4, 1), (4, 9)$; Foci: $(4, 0), (4, 10)$

32. Vertices: $(-2, 1), (2, 1)$; Foci: $(-3, 1), (3, 1)$

33. Vertices: $(2, 3), (2, -3)$;
Passes through the point $(0, 5)$

34. Vertices: $(-2, 1), (2, 1)$;
Passes through the point $(5, 4)$

35. Vertices: $(0, 4), (0, 0)$;
Passes through the point $\left(\sqrt{5}, 5\right)$

36. Vertices: $(1, 2), (1, -2)$;
Passes through the point $\left(0, \sqrt{5}\right)$

37. Vertices: $(1, 2), (3, 2)$;
Asymptotes: $y = x$, $y = 4 - x$

38. Vertices: $(3, 0), (3, -6)$;
Asymptotes: $y = x - 6$, $y = -x$

39. Vertices: $(0, 2), (6, 2)$;
Asymptotes: $y = \frac{2}{3}x$, $y = 4 - \frac{2}{3}x$

40. Vertices: $(3, 0), (3, 4)$;
Asymptotes: $y = \frac{2}{3}x$, $y = 4 - \frac{2}{3}x$

41. *Sound Location*   Three listening stations located at $(3300, 0)$, $(3300, 1100)$, and $(-3300, 0)$ monitor an explosion. If the last two stations detect the explosion 1 second and 4 seconds after the first, respectively, determine the coordinates of the explosion. (Assume that the coordinate system is measured in feet and that sound travels at 1100 feet per second.)

42. *Navigation*   Long distance radio navigation for aircraft and ships uses synchronized pulses transmitted by widely separated transmitting stations. These pulses travel at the speed of light (186,000 miles per second). The difference in the times of arrival of these pulses at an aircraft or ship is constant on a hyperbola having the transmitting stations as foci. Assume that two stations, 300 miles apart, are positioned on the rectangular coordinate system at points with coordinates $(-150, 0)$ and $(150, 0)$, and that a ship is traveling on a path with coordinates $(x, 75)$. Find the x-coordinate of the position of the ship if the time difference between the pulses from the transmitting stations is 1000 microseconds (0.001 second).

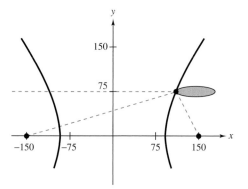

43. *Hyperbolic Mirror*   A hyperbolic mirror (used in some telescopes) has the property that a light ray directed at a focus will be reflected to the other focus. The focus of a hyperbolic mirror has coordinates $(24, 0)$. Find the vertex of the mirror if its mount has coordinates $(24, 24)$.

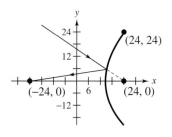

In Exercises 44–51, classify the graph of the equation as a circle, a parabola, an ellipse, or a hyperbola.

44. $x^2 + y^2 - 6x + 4y + 9 = 0$

45. $x^2 + 4y^2 - 6x + 16y + 21 = 0$

46. $4x^2 - y^2 - 4x - 3 = 0$

47. $y^2 - 4y - 4x = 0$

48. $4x^2 + 3y^2 + 8x - 24y + 51 = 0$

49. $4y^2 - 2x^2 - 4y - 8x - 15 = 0$

50. $25x^2 - 10x - 200y - 119 = 0$

51. $4x^2 + 4y^2 - 16y + 15 = 0$

Synthesis

***True or False?*   In Exercises 52 and 53, determine whether the statement is true or false. Justify your answer.**

52. In the standard form of the equation of a hyperbola, the larger the ratio of b to a, the larger the eccentricity of the hyperbola.

53. In the standard form of the equation of a hyperbola, the trivial solution of two intersecting lines occurs when $b = 0$.

54. *Think About It*   Consider a hyperbola centered at the origin with a horizontal transverse axis. Use the definition of a hyperbola to derive its standard form.

55. *Think About It*   Explain how the central rectangle of a hyperbola can be used to sketch its asymptotes.

Review

In Exercises 56–59, perform the indicated polynomial operation.

56. Subtract: $(x^3 - 3x^2) - (6 - 2x - 4x^2)$

57. Multiply: $\left(3x - \frac{1}{2}\right)(x + 4)$

58. Divide: $\dfrac{x^3 - 3x + 4}{x + 2}$

59. Expand: $[(x + y) + 3]^2$

In Exercises 60–65, factor the polynomial.

60. $x^3 - 16x$

61. $x^2 + 14x + 49$

62. $2x^3 - 24x^2 + 72x$

63. $6x^3 - 11x^2 - 10x$

64. $16x^3 + 54$

65. $4 - x + 4x^2 - x^3$

10.4 Rotation and Systems of Quadratic Equations

Rotation

In the previous section you learned that the equation of a conic with axes parallel to the coordinate axes has a standard form that can be written in the general form

$$Ax^2 + Cy^2 + Dx + Ey + F = 0. \qquad \text{Horizontal or vertical axes}$$

In this section you will study the equations of conics whose axes are rotated so that they are not parallel to either the *x*-axis or the *y*-axis. The general equation for such conics contains an *xy*-term.

$$Ax^2 + Bxy + Cy^2 + Dx + Ey + F = 0 \qquad \text{Equation in } xy\text{-plane}$$

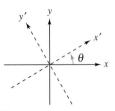

Figure 10.33

To eliminate this *xy*-term, you can use a procedure called rotation of axes. The objective is to rotate the *x*- and *y*-axes until they are parallel to the axes of the conic. The rotated axes are denoted as the *x′*-axis and the *y′*-axis, as shown in

Figure 10.33. After the rotation, the equation of the conic in the new *x′y′*-plane will have the form

$$A'(x')^2 + C'(y')^2 + D'x' + E'y' + F' = 0. \qquad \text{Equation in } x'y'\text{-plane}$$

Because this equation has no *xy*-term, you can obtain a standard form by completing the square. The following theorem identifies how much to rotate the axes to eliminate the *xy*-term and also the equations for determining the new coefficients A', C', D', E', and F'. For a proof of this theorem, see Appendix A.

Rotation of Axes to Eliminate an *xy*-Term

The general second-degree equation $Ax^2 + Bxy + Cy^2 + Dx + Ey + F = 0$ can be rewritten as

$$A'(x')^2 + C'(y')^2 + D'x' + E'y' + F' = 0$$

by rotating the coordinate axes through an angle θ, where $\cot 2\theta = \dfrac{A - C}{B}$.

The coefficients of the new equation are obtained by making the substitutions

$$x = x'\cos\theta - y'\sin\theta \qquad \text{and} \qquad y = x'\sin\theta + y'\cos\theta.$$

EXAMPLE 1   Rotation of Axes for a Hyperbola

Rotate the axes to eliminate the xy-term in the equations $xy - 1 = 0$. Then write the equation in standard form.

Solution
Because $A = 0$, $B = 1$, and $C = 0$, you have

$$\cot 2\theta = \frac{A - C}{B} = 0 \quad \Longrightarrow \quad 2\theta = \frac{\pi}{2} \quad \Longrightarrow \quad \theta = \frac{\pi}{4}$$

which implies that

$$x = x' \cos \frac{\pi}{4} - y' \sin \frac{\pi}{4}$$

$$= x'\left(\frac{\sqrt{2}}{2}\right) - y'\left(\frac{\sqrt{2}}{2}\right)$$

$$= \frac{x' - y'}{\sqrt{2}}$$

and

$$y = x' \sin \frac{\pi}{4} + y' \cos \frac{\pi}{4}$$

$$= x'\left(\frac{\sqrt{2}}{2}\right) + y'\left(\frac{\sqrt{2}}{2}\right)$$

$$= \frac{x' + y'}{\sqrt{2}}.$$

The equation in the $x'y'$-system is obtained by substituting these expressions into the equation $xy - 1 = 0$.

$$\left(\frac{x' - y'}{\sqrt{2}}\right)\left(\frac{x' + y'}{\sqrt{2}}\right) - 1 = 0$$

$$\frac{(x')^2 - (y')^2}{2} - 1 = 0$$

$$\frac{(x')^2}{(\sqrt{2})^2} - \frac{(y')^2}{(\sqrt{2})^2} = 1 \qquad \text{Write in standard form.}$$

In the $x'y'$-system, this is a hyperbola centered at the origin with vertices at $\left(\pm\sqrt{2}, 0\right)$, as shown in Figure 10.34. To find the coordinates of the vertices in the xy-system, substitute the coordinates $\left(\pm\sqrt{2}, 0\right)$ into the equations

$$x = \frac{x' - y'}{\sqrt{2}} \qquad \text{and} \qquad y = \frac{x' + y'}{\sqrt{2}}.$$

This substitution yields the vertices $(1, 1)$ and $(-1, -1)$ in the xy-system. Note also that the asymptotes of the hyperbola have equations $y' = \pm x'$, which correspond to the original x- and y-axes.

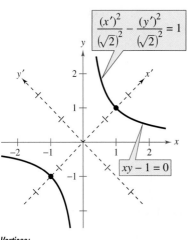

Vertices:
In x′y′-system: $\left(\sqrt{2}, 0\right), \left(-\sqrt{2}, 0\right)$
In xy-system: $(1, 1), (-1, -1)$
Figure 10.34

EXAMPLE 2 Rotation of Axes for an Ellipse

Rotate the axes to eliminate the xy-term in the equation

$$7x^2 - 6\sqrt{3}\,xy + 13y^2 - 16 = 0.$$

Then write the equation in standard form and sketch its graph.

Solution

Because $A = 7$, $B = -6\sqrt{3}$, and $C = 13$, you have

$$\cot 2\theta = \frac{A - C}{B}$$

$$= \frac{7 - 13}{-6\sqrt{3}}$$

$$= \frac{1}{\sqrt{3}}$$

which implies that $\theta = \pi/6$. The equation in the $x'y'$-system is obtained by making the substitutions

$$x = x' \cos \frac{\pi}{6} - y' \sin \frac{\pi}{6}$$

$$= x'\!\left(\frac{\sqrt{3}}{2}\right) - y'\!\left(\frac{1}{2}\right)$$

$$= \frac{\sqrt{3}x' - y'}{2}$$

and

$$y = x' \sin \frac{\pi}{6} + y' \cos \frac{\pi}{6}$$

$$= x'\!\left(\frac{1}{2}\right) + y'\!\left(\frac{\sqrt{3}}{2}\right)$$

$$= \frac{x' + \sqrt{3}y'}{2}$$

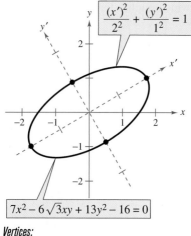

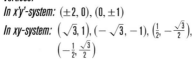

Vertices:
In x'y'-system: $(\pm 2, 0)$, $(0, \pm 1)$

In xy-system: $\left(\sqrt{3}, 1\right), \left(-\sqrt{3}, -1\right), \left(\frac{1}{2}, -\frac{\sqrt{3}}{2}\right),$ $\left(-\frac{1}{2}, \frac{\sqrt{3}}{2}\right)$

Figure 10.35

into the original equation. So, you have

$$7x^2 - 6\sqrt{3}\,xy + 13y^2 - 16 = 0$$

$$7\!\left(\frac{\sqrt{3}x' - y'}{2}\right)^{2} - 6\sqrt{3}\!\left(\frac{\sqrt{3}x' - y'}{2}\right)\!\left(\frac{x' + \sqrt{3}y'}{2}\right) + 13\!\left(\frac{x' + \sqrt{3}y'}{2}\right)^{2} - 16 = 0$$

which simplifies to

$$4(x')^2 + 16(y')^2 - 16 = 0$$

$$4(x')^2 + 16(y')^2 = 16$$

$$\frac{(x')^2}{2^2} + \frac{(y')^2}{1^2} = 1. \qquad \text{Write in standard form.}$$

This is the equation of an ellipse centered at the origin with vertices $(\pm 2, 0)$ in the $x'y'$-system, as shown in Figure 10.35.

EXAMPLE 3   Rotation of Axes for a Parabola

Rotate the axes to eliminate the xy-term in the equation

$$x^2 - 4xy + 4y^2 + 5\sqrt{5}\,y + 1 = 0.$$

Then write the equation in standard form and sketch its graph.

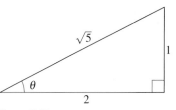

Figure 10.36

Solution

Because $A = 1$, $B = -4$, and $C = 4$, you have

$$\cot 2\theta = \frac{A - C}{B} = \frac{1 - 4}{-4} = \frac{3}{4}.$$

Using the identity $\cot 2\theta = (\cot^2 \theta - 1)/(2 \cot \theta)$ produces

$$\cot 2\theta = \frac{3}{4} = \frac{\cot^2 \theta - 1}{2 \cot \theta}$$

from which you obtain the equation

$$4 \cot^2 \theta - 4 = 6 \cot \theta$$

$$4 \cot^2 \theta - 6 \cot \theta - 4 = 0$$

$$(2 \cot \theta - 4)(2 \cot \theta + 1) = 0.$$

Considering $0 < \theta < \pi/2$, you have $2 \cot \theta = 4$. So,

$$\cot \theta = 2 \quad \Longrightarrow \quad \theta \approx 26.6°.$$

From the triangle in Figure 10.36, you obtain $\sin \theta = 1/\sqrt{5}$ and $\cos \theta = 2/\sqrt{5}$. So, you use the substitutions

$$x = x' \cos \theta - y' \sin \theta = x'\!\left(\frac{2}{\sqrt{5}}\right) - y'\!\left(\frac{1}{\sqrt{5}}\right) = \frac{2x' - y'}{\sqrt{5}}$$

$$y = x' \sin \theta + y' \cos \theta = x'\!\left(\frac{1}{\sqrt{5}}\right) + y'\!\left(\frac{2}{\sqrt{5}}\right) = \frac{x' + 2y'}{\sqrt{5}}.$$

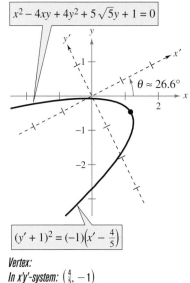

$$x^2 - 4xy + 4y^2 + 5\sqrt{5}\,y + 1 = 0$$

$$\theta \approx 26.6°$$

$$(y' + 1)^2 = (-1)\!\left(x' - \frac{4}{5}\right)$$

Vertex:
In x'y'-system: $\left(\frac{4}{5}, -1\right)$
In xy-system: $\left(\frac{13}{5\sqrt{5}}, -\frac{6}{5\sqrt{5}}\right)$

Figure 10.37

Substituting these expressions into the original equation, you have

$$x^2 - 4xy + 4y^2 + 5\sqrt{5}\,y + 1 = 0$$

$$\left(\frac{2x' - y'}{\sqrt{5}}\right)^2 - 4\left(\frac{2x' - y'}{\sqrt{5}}\right)\left(\frac{x' + 2y'}{\sqrt{5}}\right) + 4\left(\frac{x' + 2y'}{\sqrt{5}}\right)^2 + 5\sqrt{5}\left(\frac{x' + 2y'}{\sqrt{5}}\right) + 1 = 0$$

which simplifies as follows.

$$5(y')^2 + 5x' + 10y' + 1 = 0$$

$$5[(y')^2 + 2y'] = -5x' - 1 \qquad \text{Group terms.}$$

$$5[(y')^2 + 2y' + 1] = -5x' - 1 + 5 \qquad \text{Complete the square.}$$

$$5(y' + 1)^2 = -5x' + 4 \qquad \text{Write in completed square form.}$$

$$(y' + 1)^2 = (-1)\!\left(x' - \frac{4}{5}\right) \qquad \text{Write in standard form.}$$

The graph of this equation is a parabola with vertex at $\left(\frac{4}{5}, -1\right)$. Its axis is parallel to the x'-axis in the $x'y'$-system, as shown in Figure 10.37.

Invariants Under Rotation

In the rotation of axes theorem listed at the beginning of this section, note that the constant term is the same in both equations—that is, $F' = F$. Such quantities are **invariant under rotation.** The next theorem lists some other rotation invariants.

Rotation Invariants

The rotation of the coordinate axes through an angle θ that transforms the equation $Ax^2 + Bxy + Cy^2 + Dx + Ey + F = 0$ into the form

$$A'(x')^2 + C'(y')^2 + D'x' + E'y' + F' = 0$$

has the following rotation invariants.

1. $F = F'$

2. $A + C = A' + C'$

3. $B^2 - 4AC = (B')^2 - 4A'C'$

You can use the results of this theorem to classify the graph of a second-degree equation *with* an xy-term in much the same way you do for a second-degree equation *without* an xy-term. Note that because $B' = 0$, the invariant $B^2 - 4AC$ reduces to

$$B^2 - 4AC = -4A'C'. \qquad \text{Discriminant}$$

This quantity is called the **discriminant** of the equation

$$Ax^2 + Bxy + Cy^2 + Dx + Ey + F = 0.$$

Now, from the classification procedure given in Section 10.3, you know that the sign of $A'C'$ determines the type of graph for the equation

$$A'(x')^2 + C'(y')^2 + D'x' + E'y' + F' = 0.$$

Consequently, the sign of $B^2 - 4AC$ will determine the type of graph for the original equation, as given in the following classification.

Classification of Conics by the Discriminant

The graph of the equation $Ax^2 + Bxy + Cy^2 + Dx + Ey + F = 0$ is, except in degenerate cases, determined by its discriminant as follows.

1. Ellipse or circle: $B^2 - 4AC < 0$

2. Parabola: $B^2 - 4AC = 0$

3. Hyperbola: $B^2 - 4AC > 0$

For example, in the general equation

$$3x^2 + 7xy + 5y^2 - 6x - 7y + 15 = 0$$

you have $A = 3$, $B = 7$, and $C = 5$. So, the discriminant is

$$B^2 - 4AC = 7^2 - 4(3)(5) = 49 - 60 = -11.$$

Because $-11 < 0$, the graph of the equation is an ellipse or a circle.

EXAMPLE 4   Rotations and Graphing Utilities

For each of the following, classify the graph, use the quadratic formula to solve for y, and then use a graphing utility to graph the equation.

a. $2x^2 - 3xy + 2y^2 - 2x = 0$

b. $x^2 - 6xy + 9y^2 - 2y + 1 = 0$

c. $3x^2 + 8xy + 4y^2 - 7 = 0$

Solution

a. Because $B^2 - 4AC = 9 - 16 < 0$, the graph is a circle or an ellipse. Solve for y as follows.

$$2x^2 - 3xy + 2y^2 - 2x = 0 \qquad \text{Write original equation.}$$

$$2y^2 - 3xy + (2x^2 - 2x) = 0 \qquad \text{Quadratic form } ay^2 + by + c = 0$$

$$y = \frac{-(-3x) \pm \sqrt{(-3x)^2 - 4(2)(2x^2 - 2x)}}{2(2)}$$

$$y = \frac{3x \pm \sqrt{x(16 - 7x)}}{4}$$

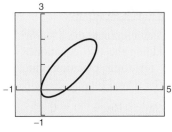

Figure 10.38

Graph both of the equations to obtain the ellipse in Figure 10.38.

$$y = \frac{3x + \sqrt{x(16 - 7x)}}{4} \qquad \text{Top half of ellipse}$$

$$y = \frac{3x - \sqrt{x(16 - 7x)}}{4} \qquad \text{Bottom half of ellipse}$$

b. Because $B^2 - 4AC = 36 - 36 = 0$, the graph is a parabola.

$$x^2 - 6xy + 9y^2 - 2y + 1 = 0 \qquad \text{Write original equation.}$$

$$9y^2 - (6x + 2)y + (x^2 + 1) = 0 \qquad \text{Quadratic form } ay^2 + by + c = 0$$

$$y = \frac{(6x + 2) \pm \sqrt{(6x + 2)^2 - 4(9)(x^2 + 1)}}{18}$$

$$y = \frac{3x + 1 \pm \sqrt{2(3x - 4)}}{9}$$

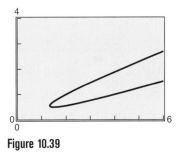

Figure 10.39

Graphing the resulting two equations gives the parabola in Figure 10.39.

c. Because $B^2 - 4AC = 64 - 48 > 0$, the graph is a hyperbola.

$$3x^2 + 8xy + 4y^2 - 7 = 0 \qquad \text{Write original equation.}$$

$$4y^2 + 8xy + (3x^2 - 7) = 0 \qquad \text{Quadratic form } ay^2 + by + c = 0$$

$$y = \frac{-8x \pm \sqrt{(8x)^2 - 4(4)(3x^2 - 7)}}{8}$$

$$y = \frac{-2x \pm \sqrt{x^2 + 7}}{2}$$

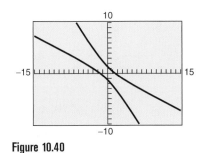

Figure 10.40

The graph of the resulting two equations yields the hyperbola in Figure 10.40.

Systems of Quadratic Equations

To find the points of intersection of two conics, you can use elimination or substitution, as demonstrated in Examples 5 and 6.

EXAMPLE 5   Solving a Quadratic System

Solve the system of quadratic equations.

$$\begin{cases} x^2 + y^2 - 16x + 39 = 0 & \text{Equation 1} \\ x^2 - y^2 - 9 = 0 & \text{Equation 2} \end{cases}$$

Algebraic Solution

You can eliminate the y^2-term by adding the two equations. The resulting equation can then be solved for x.

$$2x^2 - 16x + 30 = 0$$

$$2(x - 3)(x - 5) = 0$$

There are two real solutions: $x = 3$ and $x = 5$. The corresponding y-values are $y = 0$ and $y = \pm 4$. So, the graphs have three points of intersection:

$$(3, 0), (5, 4), \text{ and } (5, -4).$$

Graphical Solution

Begin by solving each equation for y as follows.

$$y = \pm\sqrt{-x^2 + 16x - 39} \qquad y = \pm\sqrt{x^2 - 9}$$

Use a graphing utility to graph all four equations $y_1 = \sqrt{-x^2 + 16x - 39}$, $y_2 = -\sqrt{-x^2 + 16x - 39}$, $y_3 = \sqrt{x^2 - 9}$, and $y_4 = -\sqrt{x^2 - 9}$ in the same viewing window. In Figure 10.41, you can see that the graphs appear to intersect at the points $(3, 0)$, $(5, 4)$, and $(5, -4)$. Use the *intersect* feature to confirm this.

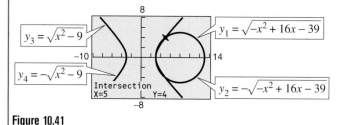

Figure 10.41

EXAMPLE 6   Solving a Quadratic System by Substitution

Solve the system of quadratic equations.

$$\begin{cases} x^2 + 4y^2 - 4x - 8y + 4 = 0 & \text{Equation 1} \\ x^2 + 4y - 4 = 0 & \text{Equation 2} \end{cases}$$

Solution

Because Equation 2 has no y^2-term, solve the equation for y to obtain $y = 1 - (1/4)x^2$. Next, substitute this into Equation 1 and solve for x.

$$x^2 + 4\left(1 - \frac{1}{4}x^2\right)^2 - 4x - 8\left(1 - \frac{1}{4}x^2\right) + 4 = 0$$

$$x^2 + 4 - 2x^2 + \frac{1}{4}x^4 - 4x - 8 + 2x^2 + 4 = 0$$

$$x^4 + 4x^2 - 16x = 0$$

$$x(x - 2)(x^2 + 2x + 8) = 0$$

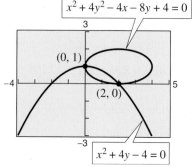

Figure 10.42

In factored form, you can see that the equation has two real solutions: $x = 0$ and $x = 2$. The corresponding values of y are $y = 1$ and $y = 0$. This implies that the solutions of the system of equations are $(0, 1)$ and $(2, 0)$, as shown in Figure 10.42.

10.4 Exercises

In Exercises 1–4, the $x'y'$-coordinate system has been rotated θ degrees from the xy-coordinate system. The coordinates of a point on the xy-coordinate system are given. Find the coordinates of the point on the rotated coordinate system.

1. $\theta = 90°, (0, 4)$ 2. $\theta = 45°, (3, 3)$

3. $\theta = 30°, (1, 6)$ 4. $\theta = 60°, (5, 1)$

In Exercises 5–16, rotate the axes to eliminate the xy-term in the equation. Then write the equation in standard form. Sketch the graph of the equation, showing both sets of axes.

5. $xy + 1 = 0$

6. $xy - 2 = 0$

7. $x^2 - 8xy + y^2 + 1 = 0$

8. $xy + x - 2y + 3 = 0$

9. $xy - 2y - 4x = 0$

10. $13x^2 + 6\sqrt{3}xy + 7y^2 - 16 = 0$

11. $5x^2 - 6xy + 5y^2 - 12 = 0$

12. $2x^2 - 3xy - 2y^2 + 10 = 0$

13. $3x^2 - 2\sqrt{3}xy + y^2 + 2x + 2\sqrt{3}y = 0$

14. $16x^2 - 24xy + 9y^2 - 60x - 80y + 100 = 0$

15. $9x^2 + 24xy + 16y^2 + 90x - 130y = 0$

16. $9x^2 + 24xy + 16y^2 + 80x - 60y = 0$

In Exercises 17–22, use a graphing utility to graph the conic. Determine the angle θ through which the axes are rotated. Explain how you used the graphing utility to obtain the graph.

17. $x^2 + xy + y^2 = 12$

18. $x^2 - 4xy + 2y^2 = 10$

19. $17x^2 + 32xy - 7y^2 = 75$

20. $40x^2 + 36xy + 25y^2 = 52$

21. $32x^2 + 50xy + 7y^2 = 52$

22. $4x^2 - 12xy + 9y^2 + \left(4\sqrt{13} - 12\right)x - \left(6\sqrt{13} + 8\right)y = 91$

In Exercises 23–28, match the graph with its equation. [The graphs are labeled (a), (b), (c), (d), (e), and (f).]

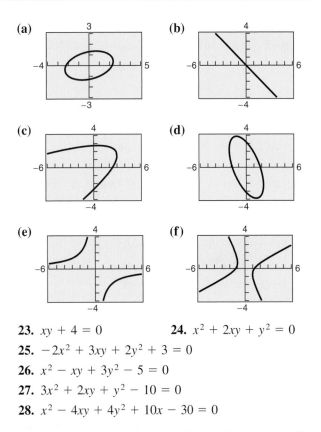

(a)   (b)   (c)   (d)   (e)   (f)

23. $xy + 4 = 0$ 24. $x^2 + 2xy + y^2 = 0$

25. $-2x^2 + 3xy + 2y^2 + 3 = 0$

26. $x^2 - xy + 3y^2 - 5 = 0$

27. $3x^2 + 2xy + y^2 - 10 = 0$

28. $x^2 - 4xy + 4y^2 + 10x - 30 = 0$

In Exercises 29–36, (a) use the discriminant to classify the graph, (b) use the quadratic formula to solve for y, and (c) use a graphing utility to graph the equation.

29. $16x^2 - 24xy + 9y^2 - 30x - 40y = 0$

30. $x^2 - 8xy - 2y^2 - 6 = 0$

31. $15x^2 - 8xy + 7y^2 - 45 = 0$

32. $2x^2 + 4xy + 5y^2 + 3x - 4y - 20 = 0$

33. $x^2 - 6xy - 5y^2 + 4x - 22 = 0$

34. $36x^2 - 60xy + 25y^2 + 9y = 0$

35. $x^2 + 4xy + 4y^2 - 5x - y - 3 = 0$

36. $x^2 + xy + 4y^2 + x + y - 4 = 0$

In Exercises 37–40, sketch (if possible) the graph of the degenerate conic.

37. $y^2 - 16x^2 = 0$

38. $x^2 + y^2 - 2x + 6y + 10 = 0$

39. $x^2 + 2xy + y^2 - 4 = 0$

40. $x^2 - 10xy + y^2 = 0$

In Exercises 41–48, solve the system of quadratic equations algebraically by the method of elimination. Then verify your results by using a graphing utility to graph the equations and find any points of intersection of the graphs.

41. $\begin{cases} -x^2 + y^2 + 4x - 6y + 4 = 0 \\ x^2 + y^2 - 4x - 6y + 12 = 0 \end{cases}$

42. $\begin{cases} -x^2 - y^2 - 8x + 20y - 7 = 0 \\ x^2 + 9y^2 + 8x + 4y + 7 = 0 \end{cases}$

43. $\begin{cases} -4x^2 - y^2 - 16x + 24y - 16 = 0 \\ 4x^2 + y^2 + 40x - 24y + 208 = 0 \end{cases}$

44. $\begin{cases} x^2 - 4y^2 - 20x - 64y - 172 = 0 \\ 16x^2 + 4y^2 - 320x + 64y + 1600 = 0 \end{cases}$

45. $\begin{cases} x^2 - y^2 - 12x + 16y - 64 = 0 \\ x^2 + y^2 - 12x - 16y + 64 = 0 \end{cases}$

46. $\begin{cases} x^2 + 4y^2 - 2x - 8y + 1 = 0 \\ -x^2 + 2x - 4y - 1 = 0 \end{cases}$

47. $\begin{cases} -16x^2 - y^2 + 24y - 80 = 0 \\ 16x^2 + 25y^2 - 400 = 0 \end{cases}$

48. $\begin{cases} 16x^2 - y^2 + 16y - 128 = 0 \\ y^2 - 48x - 16y - 32 = 0 \end{cases}$

In Exercises 49–54, solve the system of quadratic equations algebraically by the method of substitution. Then verify your results by using a graphing utility to graph the equations and find any points of intersection of the graphs.

49. $\begin{cases} x^2 + y^2 - 4 = 0 \\ 3x - y^2 = 0 \end{cases}$

50. $\begin{cases} 4x^2 + 9y^2 - 36y = 0 \\ x^2 + 9y - 27 = 0 \end{cases}$

51. $\begin{cases} x^2 + 2y^2 - 4x + 6y - 5 = 0 \\ -x + y - 4 = 0 \end{cases}$

52. $\begin{cases} x^2 + 2y^2 - 4x + 6y - 5 = 0 \\ x^2 - 4x - y + 4 = 0 \end{cases}$

53. $\begin{cases} xy + x - 2y + 3 = 0 \\ x^2 + 4y^2 - 9 = 0 \end{cases}$

54. $\begin{cases} 5x^2 - 2xy + 5y^2 - 12 = 0 \\ x + y - 1 = 0 \end{cases}$

Synthesis

True or False? In Exercises 55 and 56, determine whether the statement is true or false. Justify your answer.

55. The graph of $x^2 + xy + ky^2 + 6x + 10 = 0$, where k is any constant less than $\frac{1}{4}$, is a hyperbola.

56. After using a rotation of axes to eliminate the xy-term from an equation of the form

$$Ax^2 + Bxy + Cy^2 + Dx + Ey + F = 0$$

the coefficients of the x^2- and y^2-terms remain A and B, respectively.

57. Show that the equation $x^2 + y^2 = r^2$ is invariant under rotation of axes.

58. Find the lengths of the major and minor axes of the ellipse in Exercise 10.

Review

In Exercises 59–62, sketch the graph of the rational function. Identify all intercepts and asymptotes.

59. $g(x) = \dfrac{2}{2 - x}$ **60.** $f(x) = \dfrac{2x}{2 - x}$

61. $h(t) = \dfrac{t^2}{2 - t}$ **62.** $g(s) = \dfrac{2}{4 - s^2}$

In Exercises 63–66, find (a) AB, (b) BA, and, if possible, (c) A^2.

63. $A = \begin{bmatrix} 1 & -3 \\ 2 & 5 \end{bmatrix}$, $B = \begin{bmatrix} 0 & 6 \\ 5 & -1 \end{bmatrix}$

64. $A = \begin{bmatrix} 1 & 5 \\ 0 & -2 \end{bmatrix}$, $B = \begin{bmatrix} 3 & 2 \\ -3 & 8 \end{bmatrix}$

65. $A = \begin{bmatrix} 4 & -2 & 5 \end{bmatrix}$, $B = \begin{bmatrix} 3 \\ -4 \\ 5 \end{bmatrix}$

66. $A = \begin{bmatrix} 0 & -2 & 0 \\ 1 & 1 & 5 \\ 3 & 4 & 0 \end{bmatrix}$, $B = \begin{bmatrix} 1 & 0 & -3 \\ -4 & 5 & -1 \\ 6 & 3 & 2 \end{bmatrix}$

In Exercises 67–70, find the coefficient a of the given term in the expansion of the binomial.

Binomial	Term
67. $(x + 8)^7$	ax^2
68. $(3x - y)^6$	ax^3y^3
69. $(x - 4y)^{10}$	ax^6y^4
70. $(3x + 2y)^8$	ax^2y^6

10.5 Parametric Equations

Plane Curves

Up to this point, you have been representing a graph by a single equation involving *two* variables such as x and y. In this section, you will study situations in which it is useful to introduce a *third* variable to represent a curve in the plane.

To see the usefulness of this procedure, consider the path followed by an object that is propelled into the air at an angle of 45°. If the initial velocity of the object is 48 feet per second, it can be shown that the object follows the parabolic path

$$y = -\frac{x^2}{72} + x \qquad \text{Rectangular equation}$$

as shown in Figure 10.43. However, this equation does not tell the whole story. Although it does tell us *where* the object has been, it doesn't tell us *when* the object was at a given point (x, y) on the path. To determine this time, you can introduce a third variable t, which is called a **parameter.** It is possible to write both x and y as functions of t to obtain the **parametric equations**

$$x = 24\sqrt{2}\,t \qquad \text{Parametric equation for } x$$

$$y = -16t^2 + 24\sqrt{2}\,t. \qquad \text{Parametric equation for } y$$

From this set of equations you can determine that at time $t = 0$, the object is at the point $(0, 0)$. Similarly, at time $t = 1$, the object is at the point $(24\sqrt{2}, 24\sqrt{2} - 16)$, and so on.

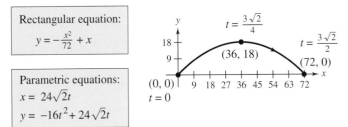

Rectangular equation:
$$y = -\frac{x^2}{72} + x$$

Parametric equations:
$$x = 24\sqrt{2}\,t$$
$$y = -16t^2 + 24\sqrt{2}\,t$$

Curvilinear motion: two variables for position, one variable for time
Figure 10.43

For this particular motion problem, x and y are continuous functions of t, and the resulting path is a **plane curve.** (Recall that a *continuous function* is one whose graph can be traced without lifting the pencil from the paper.)

Definition of a Plane Curve

If f and g are continuous functions of t on an interval I, the set of ordered pairs $(f(t), g(t))$ is a **plane curve** C. The equations

$$x = f(t) \qquad \text{and} \qquad y = g(t)$$

are **parametric equations** for C, and t is the **parameter.**

What You Should Learn:

- How to evaluate sets of parametric equations for given values of the parameter
- How to graph curves that are represented by sets of parametric equations
- How to rewrite sets of parametric equations as single rectangular equations by eliminating the parameter
- How to find sets of parametric equations for graphs

Why You Should Learn It:

Parametric equations are useful for modeling the path of an object. For instance, in Exercise 59 on page 738, a set of parametric equations is used to model the path of a baseball.

Jonathan Daniel/Allsport

A computer animation of this concept appears in th *Interactive* CD-ROM and *Internet* versions of this text.

Sketching a Plane Curve

One way to sketch a curve represented by a pair of parametric equations is to plot points in the xy-plane. Each set of coordinates (x, y) is determined from a value chosen for the parameter t. By plotting the resulting points in the order of *increasing* values of t, you trace the curve in a specific direction. This is called the **orientation** of the curve.

Library of Functions

Parametric equations consist of a pair of functions $x = f(t)$ and $y = g(t)$, each of which is a function of the parameter t. These equations define a plane curve, which might not be the graph of a function, as in Example 1. Most graphing utilities have a *parametric* mode.

EXAMPLE 1   Sketching a Plane Curve

Sketch the curve given by the parametric equations

$$x = t^2 - 4 \quad \text{and} \quad y = \frac{t}{2}, \quad -2 \le t \le 3.$$

Describe the orientation of the curve.

A computer animation of this example appears in the *Interactive* CD-ROM and *Internet* versions of this text.

Solution

Using values of t in the given interval, the parametric equations yield the points (x, y) shown in the table.

t	-2	-1	0	1	2	3
x	0	-3	-4	-3	0	5
y	-1	$-\frac{1}{2}$	0	$\frac{1}{2}$	1	$\frac{3}{2}$

By plotting these points in the order of increasing t, you obtain the curve shown in Figure 10.44(a). The arrows on the curve indicate its orientation as t increases from -2 to 3. So, if a particle were moving on this curve, it would start at $(0, -1)$ and then move along the curve to the point $\left(5, \frac{3}{2}\right)$.

The graph shown in Figure 10.44(a) does not define y as a function of x. This points out one benefit of parametric equations—they can be used to represent graphs that are more general than graphs of functions.

Two different sets of parametric equations can have the same graph. For example, the set of parametric equations

$$x = 4t^2 - 4 \quad \text{and} \quad y = t, \quad -1 \le t \le \frac{3}{2}$$

has the same graph as the set given in Example 1. [See Figure 10.44(b).] However, by comparing the values of t in Figures 10.44(a) and (b), you can see that this second graph is traced out more *rapidly* (considering t as time) than the first graph. So, in applications, different parametric representations can be used to represent various *speeds* at which objects travel along a given path.

Another way to display a curve represented by a pair of parametric equations is to use a graphing utility, as shown in Example 2.

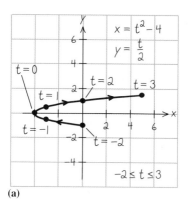

(a)

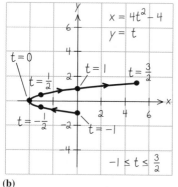

(b)

Figure 10.44

EXAMPLE 2   Using a Graphing Utility in Parametric Mode

Use a graphing utility to graph the curves represented by the parametric equations. For which curve is y a function of x? (Use $-4 \le t \le 4$.)

a. $x = t^2$ **b.** $x = t$ **c.** $x = t^2$

$\quad y = t^3$ $\quad\ \ y = t^3$ $\quad\ \ y = t$

Solution

Begin by setting the graphing utility to *parametric* mode. When choosing a viewing window, you must set not only minimum and maximum values of x and y but also minimum and maximum values of t.

a. Enter the parametric equations for x and y.

$\quad$ X1T $= $ T^2,$\quad$ Y1T $= $ T^3

The curve is shown in Figure 10.45(a). From the graph, you can see that y *is not* a function of x.

b. Enter the parametric equations for x and y.

$\quad$ X1T $= $ T,$\quad$ Y1T $= $ T^3

The curve is shown in Figure 10.45(b). From the graph, you can see that y *is* a function of x.

c. Enter the parametric equations for x and y.

$\quad$ X1T $= $ T^2,$\quad$ Y1T $= $ T

The curve is shown in Figure 10.45(c). From the graph, you can see that y *is not* a function of x.

The *Interactive* CD-ROM and *Internet* versions of this text offer a built-in graphing calculator, which can be used with the Examples, Explorations, and Exercises.

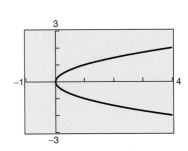

(a)$\qquad\qquad\qquad\qquad\qquad$ (b)$\qquad\qquad\qquad\qquad\qquad$ (c)

Figure 10.45

Exploration

Use a graphing utility set in *parametric* mode to graph the curve

$\quad$ X1T $= $ T and Y1T $= 1 - $ T^2.

Set the viewing window so that $-4 \le x \le 4$ and $-12 \le y \le 2$. Now, graph the curve with various settings for t. Use the following.

a. $0 \le t \le 3$$\qquad\qquad$ **b.** $-3 \le t \le 0$$\qquad\qquad$ **c.** $-3 \le t \le 3$

Compare the curves given by the different t settings. Repeat this experiment using X1T $= -$T. How does this change the results?

Eliminating the Parameter

Many curves that are represented by sets of parametric equations have graphs that can also be represented by rectangular equations (in x and y). The process of finding the rectangular equation is called **eliminating the parameter.**

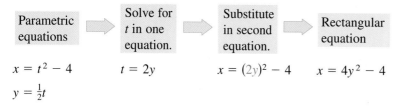

$$x = t^2 - 4 \qquad t = 2y \qquad x = (2y)^2 - 4 \qquad x = 4y^2 - 4$$

$$y = \tfrac{1}{2}t$$

After eliminating the parameter, you can recognize that the curve is a parabola with a horizontal axis and vertex at $(-4, 0)$.

Converting equations from parametric to rectangular form can change the ranges of x and y. In such cases, you should restrict x and y in the rectangular equation so that its graph matches the graph of the parametric equations.

EXAMPLE 3 Eliminating the Parameter

Identify the curve represented by the equations

$$x = \frac{1}{\sqrt{t + 1}} \quad \text{and} \quad y = \frac{t}{t + 1}.$$

Solution

Solving for t in the equation for x produces

$$x^2 = \frac{1}{t + 1} \quad \text{or} \quad \frac{1}{x^2} = t + 1$$

which implies that $t = (1/x^2) - 1$. Substituting in the equation for y, you obtain

$$
\begin{aligned}
y &= \frac{t}{t + 1} \\[1mm]
&= \frac{\left(\frac{1}{x^2}\right) - 1}{\left(\frac{1}{x^2}\right) - 1 + 1} \\[1mm]
&= \frac{\dfrac{1 - x^2}{x^2}}{\left(\dfrac{1}{x^2}\right)} \cdot \frac{x^2}{x^2} \\[1mm]
&= 1 - x^2.
\end{aligned}
$$

From the rectangular equation, you can recognize the curve to be a parabola that opens downward and has its vertex at $(0, 1)$, as shown in Figure 10.46(a). The rectangular equation is defined for all values of x. The parametric equation for x, however, is defined only when $t > -1$. From the graph of the parametric equation, you can see that x is always positive, as shown in Figure 10.46(b). So, you should restrict the domain of x to positive values, as shown in Figure 10.46(c).

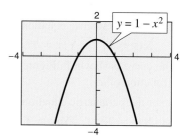

(a)

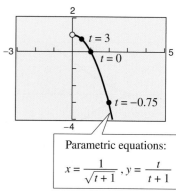

(b)

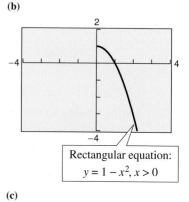

(c)

Figure 10.46

It is not necessary for the parameter in a set of parametric equations to represent time. The next example uses an *angle* as the parameter.

EXAMPLE 4   Eliminating the Parameter

Sketch the curve represented by $x = 3 \cos \theta$ and $y = 4 \sin \theta$, $0 \le \theta \le 2\pi$, by eliminating the parameter.

Solution

Begin by solving for $\cos \theta$ and $\sin \theta$ in the given equations.

$$\cos \theta = \frac{x}{3} \quad \text{and} \quad \sin \theta = \frac{y}{4} \qquad \text{Solve for } \cos \theta \text{ and } \sin \theta.$$

Make use of the identity $\sin^2 \theta + \cos^2 \theta = 1$ to form an equation involving only x and y.

$$\cos^2 \theta + \sin^2 \theta = 1 \qquad \text{Pythagorean identity}$$

$$\left(\frac{x}{3}\right)^2 + \left(\frac{y}{4}\right)^2 = 1 \qquad \text{Substitute.}$$

$$\frac{x^2}{9} + \frac{y^2}{16} = 1 \qquad \text{Rectangular equation}$$

From this rectangular equation, you can see that the graph is an ellipse centered at $(0, 0)$, with vertices at $(0, 4)$ and $(0, -4)$, and minor axis of length $2b = 6$, as shown in Figure 10.47. Note that the elliptic curve is traced out *counterclockwise* as θ varies from 0 to 2π.

Finding Parametric Equations for a Graph

How can you determine a set of parametric equations for a given graph or a given physical description? From the discussion following Example 1, you know that such a representation is not unique. This is further demonstrated in Example 5.

EXAMPLE 5   Finding Parametric Equations for a Given Graph

Find a set of parametric equations to represent the graph of $y = 1 - x^2$ using the following parameters. **a.** $t = x$ **b.** $t = 1 - x$

Solution

a. Letting $t = x$, you obtain the parametric equations $x = t$ and $y = 1 - t^2$.

     The graph of these equations is shown in Figure 10.48(a).

b. Letting $t = 1 - x$, you obtain the following parametric equations.

$$x = 1 - t \qquad \text{Parametric equation for } x$$

$$y = 1 - (1 - t)^2 \qquad \text{Substitute } 1 - t \text{ for } x.$$

$$= 2t - t^2 \qquad \text{Parametric equation for } y$$

The graph of these equations is shown in Figure 10.48(b). In this figure, note how the resulting curve is oriented by the increasing values of t. In Figure 10.48(a), the curve has the opposite orientation.

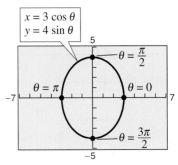

Figure 10.47

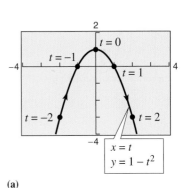

(a)

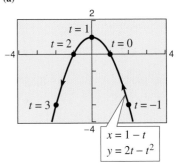

(b)

Figure 10.48

10.5 Exercises

**In Exercises 1–8, match the equation with its graph.
[The graphs are labeled (a), (b), (c), (d), (e), (f), (g), and (h).]**

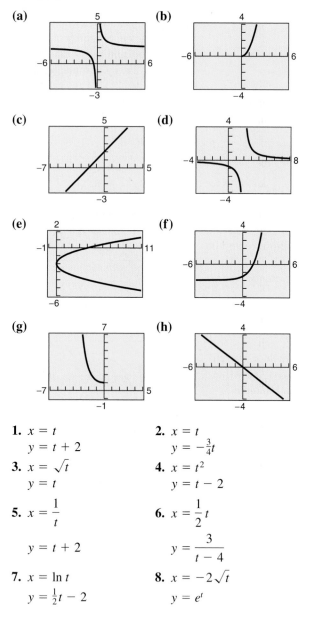

(a)

(b)

(c)

(d)

(e)

(f)

(g)

(h)

1. $x = t$
   $y = t + 2$

2. $x = t$
   $y = -\frac{3}{4}t$

3. $x = \sqrt{t}$
   $y = t$

4. $x = t^2$
   $y = t - 2$

5. $x = \dfrac{1}{t}$

   $y = t + 2$

6. $x = \dfrac{1}{2}t$

   $y = \dfrac{3}{t - 4}$

7. $x = \ln t$
   $y = \frac{1}{2}t - 2$

8. $x = -2\sqrt{t}$
   $y = e^t$

9. Consider the parametric equations
   $x = \sqrt{t}$ and $y = 2 - t.$

   (a) Complete the table.

t	0	1	2	3	4
x					
y					

   (b) Plot the points (x, y) generated in part (a) and sketch a graph of the parametric equations.

   (c) Use a graphing utility to graph the curve represented by the parametric equations.

   (d) Find the rectangular equation by eliminating the parameter. Sketch its graph. How do the graphs differ from those in parts (b) and (c)?

10. Consider the parametric equations
   $x = 4\cos^2\theta$ and $y = 2\sin\theta.$

   (a) Complete the table.

θ	$-\dfrac{\pi}{2}$	$-\dfrac{\pi}{4}$	0	$\dfrac{\pi}{4}$	$\dfrac{\pi}{2}$
x					
y					

   (b) Plot the points (x, y) generated in part (a) and sketch a graph of the parametric equations.

   (c) Use a graphing utility to graph the curve represented by the parametric equations.

   (d) Find the rectangular equation by eliminating the parameter. Sketch its graph. How do the graphs differ from those in parts (b) and (c)?

In Exercises 11–26, sketch the curve represented by the parametric equations (indicate the direction of the curve). Use a graphing utility to confirm your result. Then eliminate the parameter and write the corresponding rectangular equation whose graph represents the curve.

11. $x = t$
   $y = -4t$

12. $x = t$
   $y = \frac{1}{2}t$

13. $x = 3t + 1$
$y = 2t - 1$

14. $x = 3 - 2t$
$y = 2 + 3t$

15. $x = \frac{1}{4}t$
$y = t^2$

16. $x = t$
$y = t^3$

17. $x = t + 5$
$y = t^2$

18. $x = \sqrt{t}$
$y = 1 - t$

19. $x = 2t$
$y = |t - 2|$

20. $x = |t - 1|$
$y = t + 2$

21. $x = 3 \cos \theta$
$y = 3 \sin \theta$

22. $x = \cos \theta$
$y = 3 \sin \theta$

23. $x = e^{-t}$
$y = e^{3t}$

24. $x = e^{2t}$
$y = e^t$

25. $x = t^3$
$y = 3 \ln t$

26. $x = \ln 2t$
$y = 2t^2$

In Exercises 27–34, use a graphing utility to graph the curve represented by the parametric equations.

27. $x = 4 \sin 2\theta$
$y = 2 \cos 2\theta$

28. $x = \cos \theta$
$y = 2 \sin 2\theta$

29. $x = 4 + 2 \cos \theta$
$y = -1 + \sin \theta$

30. $x = 4 + 2 \cos \theta$
$y = -1 + 2 \sin \theta$

31. $x = 4 \sec \theta$
$y = 3 \tan \theta$

32. $x = \sec \theta$
$y = \tan \theta$

33. $x = t/2$
$y = \ln(t^2 + 1)$

34. $x = 10 - 0.01e^t$
$y = 0.4t^2$

In Exercises 35 and 36, determine how the plane curves differ from each other.

35. (a) $x = t$
$y = 2t + 1$

(b) $x = \cos \theta$
$y = 2 \cos \theta + 1$

(c) $x = e^{-t}$
$y = 2e^{-t} + 1$

(d) $x = e^t$
$y = 2e^t + 1$

36. (a) $x = t$
$y = t$

(b) $x = t^2$
$y = t^2$

(c) $x = -t$
$y = -t$

(d) $x = t^3$
$y = t^3$

In Exercises 37–40, eliminate the parameter and obtain the standard form of the rectangular equation.

37. Line through (x_1, y_1) and (x_2, y_2):
$x = x_1 + t(x_2 - x_1)$
$y = y_1 + t(y_2 - y_1)$

38. Circle: $x = h + r \cos \theta$
$y = k + r \sin \theta$

39. Ellipse: $x = h + a \cos \theta$
$y = k + b \sin \theta$

40. Hyperbola: $x = h + a \sec \theta$
$y = k + b \tan \theta$

In Exercises 41–46, use the results of Exercises 37–40 to find a set of parametric equations for the line or conic.

41. Line: Passes through $(0, 0)$ and $(5, -2)$
42. Line: Passes through $(1, 4)$ and $(5, -2)$
43. Circle: Center: $(2, 1)$; Radius: 4
44. Circle: Center: $(-3, 1)$; Radius: 3
45. Ellipse: Vertices: $(\pm 5, 0)$; Foci: $(\pm 4, 0)$
46. Hyperbola: Vertices: $(0, \pm 1)$; Foci: $(0, \pm 2)$

In Exercises 47 and 48, find two different sets of parametric equations for the given rectangular equation.

47. $y = 3x - 2$

48. $y = x^2$

In Exercises 49–54, use a graphing utility to obtain a graph of the curve represented by the parametric equations.

49. Cycloid: $x = 2(\theta - \sin \theta)$
$y = 2(1 - \cos \theta)$

50. Prolate cycloid: $x = 2\theta - 4 \sin \theta$
$y = 2 - 4 \cos \theta$

51. Hypocycloid: $x = 3 \cos^3 \theta$
$y = 3 \sin^3 \theta$

52. Curtate cycloid: $x = 2\theta - \sin \theta$
$y = 2 - \cos \theta$

53. Witch of Agnesi: $x = 2 \cot \theta$
$y = 2 \sin^2 \theta$

54. Folium of Descartes: $x = \dfrac{3t}{1 + t^3}$
$y = \dfrac{3t^2}{1 + t^3}$

In Exercises 55–58, match the parametric equations with the correct graph. [The graphs are labeled (a), (b), (c), and (d).]

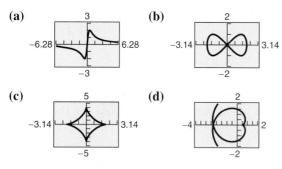

(a)

3

−6.28 6.28

−3

(b)

2

−3.14 3.14

−2

(c)

5

−3.14 3.14

−5

(d)

2

−4 2

−2

55. Lissajous curve: $x = 2 \cos \theta$, $y = \sin 2\theta$

56. Evolute of ellipse: $x = 2 \cos^3 \theta$, $y = 4 \sin^3 \theta$

57. Involute of circle: $x = \frac{1}{2}(\cos \theta + \theta \sin \theta)$
$$y = \frac{1}{2}(\sin \theta - \theta \cos \theta)$$

58. Serpentine curve: $x = \frac{1}{2} \cot \theta$, $y = 4 \sin \theta \cos \theta$

Projectile Motion **In Exercises 59 and 60, consider a projectile launched at a height *h* feet above the ground at an angle θ with the horizontal. If the initial velocity is v_0 feet per second, the path of the projectile is modeled by the parametric equations**

$$x = (v_0 \cos \theta)t \quad \text{and} \quad y = h + (v_0 \sin \theta)t - 16t^2.$$

59. *Baseball* The center-field fence in a ballpark is 10 feet high and 400 feet from home plate. The baseball is hit 3 feet above the ground. It leaves the bat at an angle of θ degrees with the horizontal at a speed of 100 miles per hour.

(a) Write a set of parametric equations for the path of the baseball.

(b) Use a graphing utility to sketch the path of the baseball for $\theta = 15°$. Is the hit a home run?

(c) Use a graphing utility to sketch the path of the baseball for $\theta = 23°$. Is the hit a home run?

(d) Find the minimum angle required for the hit to be a home run.

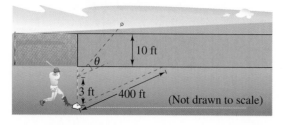

10 ft

θ

3 ft 400 ft

(Not drawn to scale)

60. *Football* The quarterback of a football team releases a pass at a height of 7 feet above the playing field, and the football is caught by a receiver at a height of 4 feet, 30 yards directly downfield. The pass is released at an angle of 35° with the horizontal.

(a) Write a set of parametric equations for the path of the football.

(b) Find the speed of the football when it is released.

(c) Use a graphing utility to graph the path of the football and approximate its maximum height.

(d) Find the time the receiver has to position himself after the quarterback releases the football.

Synthesis

True or False? **In Exercises 61 and 62, determine whether the statement is true or false. Justify your answer.**

61. The two sets of parametric equations $x = t$, $y = t^2 + 1$ and $x = 3t$, $y = 9t^2 + 1$ correspond to the same rectangular equation.

62. The graph of the parametric equations $x = t^2$ and $y = t^2$ is the line $y = x$.

63. *Think About It* The graph of the parametric equations $x = t^3$ and $y = t - 1$ is shown below. Would the graph change for the equations $x = (-t^3)$ and $y = -t - 1$? If so, how would it change?

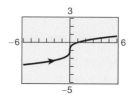

3

−6 6

−5

Review

In Exercises 64–67, find all solutions of the equation.

64. $5x^2 + 8 = 0$

65. $x^2 - 6x + 4 = 0$

66. $4x^2 + 4x - 11 = 0$

67. $x^4 - 18x^2 + 18 = 0$

In Exercises 68–73, find the sum. Use a graphing utility to verify your result.

68. $\sum_{n=1}^{50} 8n$

69. $\sum_{n=1}^{200} (n - 8)$

70. $\sum_{n=1}^{40} \left(300 - \frac{1}{2}n\right)$

71. $\sum_{n=1}^{70} \frac{7 - 5n}{12}$

72. $\sum_{n=0}^{18} 8\left(\frac{1}{2}\right)^n$

73. $\sum_{n=0}^{10} 10\left(\frac{2}{3}\right)^n$

10.6 Polar Coordinates

Introduction

So far, you have been representing graphs of equations as collections of points (x, y) on the rectangular coordinate system, where x and y represent the directed distances from the coordinate axes to the point (x, y). In this section, you will study a second system called the **polar coordinate system.**

To form the polar coordinate system in the plane, fix a point O, called the **pole** (or **origin**), and construct from O an initial ray called the **polar axis,** as shown in Figure 10.49. Then each point P in the plane can be assigned **polar coordinates** (r, θ) as follows.

1. $r = directed\ distance$ from O to P
2. $\theta = directed\ angle$, counterclockwise from polar axis to segment $\overline{OP}$

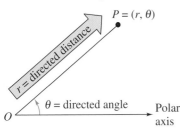

Figure 10.49

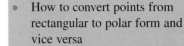
EXAMPLE 1   Plotting Points in the Polar Coordinate System

a. The point $(r, \theta) = (2, \pi/3)$ lies two units from the pole on the terminal side of the angle $\theta = \pi/3$, as shown in Figure 10.50(a).

b. The point $(r, \theta) = (3, -\pi/6)$ lies three units from the pole on the terminal side of the angle $\theta = -\pi/6$, as shown in Figure 10.50(b).

c. The point $(r, \theta) = (3, 11\pi/6)$ coincides with the point $(3, -\pi/6)$, as shown in Figure 10.50(c).

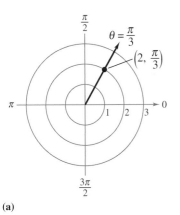

(a)

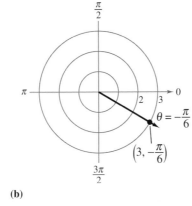

(b)

(c)

Figure 10.50

In rectangular coordinates, each point (x, y) has a unique representation. This is not true for polar coordinates. For instance, the coordinates (r, θ) and $(r, 2\pi + \theta)$ represent the same point, as illustrated in Example 1. Another way to obtain multiple representations of a point is to use negative values for r. Because r is a *directed distance*, the coordinates (r, θ) and $(-r, \theta + \pi)$ represent the same point. In general, the point (r, θ) can be represented as

$$(r, \theta) = (r, \theta \pm 2n\pi) \qquad \text{or} \qquad (r, \theta) = (-r, \theta \pm (2n + 1)\pi)$$

where n is any integer. Moreover, the pole is represented by $(0, \theta)$, where θ is any angle.

EXAMPLE 2 Multiple Representation of Points

Plot the point $(3, -3\pi/4)$ and find three additional polar representations of this point, using $-2\pi < \theta < 2\pi$.

Solution

The point is shown in Figure 10.51. Three other representations are as follows.

$$\left(3, -\frac{3\pi}{4} + 2\pi\right) = \left(3, \frac{5\pi}{4}\right) \qquad \text{Add } 2\pi \text{ to } \theta.$$

$$\left(-3, -\frac{3\pi}{4} - \pi\right) = \left(-3, -\frac{7\pi}{4}\right) \qquad \text{Replace } r \text{ by } -r; \text{ subtract } \pi \text{ from } \theta.$$

$$\left(-3, -\frac{3\pi}{4} + \pi\right) = \left(-3, \frac{\pi}{4}\right) \qquad \text{Replace } r \text{ by } -r; \text{ add } \pi \text{ to } \theta.$$

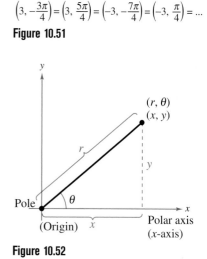

$$\left(3, -\tfrac{3\pi}{4}\right) = \left(3, \tfrac{5\pi}{4}\right) = \left(-3, -\tfrac{7\pi}{4}\right) = \left(-3, \tfrac{\pi}{4}\right) = \dots$$

Figure 10.51

Coordinate Conversion

To establish the relationship between polar and rectangular coordinates, let the polar axis coincide with the positive x-axis and the pole with the origin, as shown in Figure 10.52. Because (x, y) lies on a circle of radius r, it follows that $r^2 = x^2 + y^2$. Moreover, for $r > 0$, the definitions of the trigonometric functions imply that

$$\tan \theta = \frac{y}{x}, \qquad \cos \theta = \frac{x}{r}, \qquad \text{and} \qquad \sin \theta = \frac{y}{r}.$$

You can show that the same relationships hold for $r < 0$.

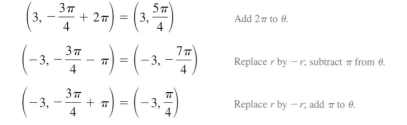

Figure 10.52

Coordinate Conversion

The polar coordinates (r, θ) are related to the rectangular coordinates (x, y) as follows.

$$x = r \cos \theta \qquad \text{and} \qquad \tan \theta = \frac{y}{x}$$

$$y = r \sin \theta \qquad \qquad r^2 = x^2 + y^2$$

EXAMPLE 3 Polar-to-Rectangular Conversion

Convert each point to rectangular coordinates. (See Figure 10.53.)

a. $(2, \pi)$ **b.** $\left(\sqrt{3}, \dfrac{\pi}{6} \right)$

Solution
a. For the point $(r, \theta) = (2, \pi)$, you have

$$x = r \cos \theta = 2 \cos \pi = -2$$

and

$$y = r \sin \theta = 2 \sin \pi = 0.$$

The rectangular coordinates are $(x, y) = (-2, 0)$.

b. For the point $(r, \theta) = \left(\sqrt{3}, \pi/6 \right)$, you have

$$x = \sqrt{3} \cos \frac{\pi}{6} = \sqrt{3} \left(\frac{\sqrt{3}}{2} \right) = \frac{3}{2}$$

and

$$y = \sqrt{3} \sin \frac{\pi}{6} = \sqrt{3} \left(\frac{1}{2} \right) = \frac{\sqrt{3}}{2}.$$

The rectangular coordinates are $(x, y) = \left(3/2, \sqrt{3}/2 \right)$.

EXAMPLE 4 Rectangular-to-Polar Conversion

Convert each point to polar coordinates.

a. $(-1, 1)$ **b.** $(0, 2)$

Solution
a. For the second-quadrant point $(x, y) = (-1, 1)$, you have

$$\tan \theta = \frac{y}{x} = \frac{1}{-1} = -1$$

$$\theta = \frac{3\pi}{4}.$$

Because θ lies in the same quadrant as (x, y), use positive r.

$$r = \sqrt{x^2 + y^2} = \sqrt{(-1)^2 + (1)^2} = \sqrt{2}$$

So, *one* set of polar coordinates is $(r, \theta) = \left(\sqrt{2}, 3\pi/4 \right)$, as shown in Figure 10.54(a).

b. Because the point $(x, y) = (0, 2)$ lies on the positive y-axis, choose

$$\theta = \pi/2 \quad \text{and} \quad r = 2.$$

This implies that one set of polar coordinates is $(r, \theta) = (2, \pi/2)$, as shown in Figure 10.54(b).

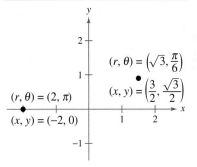

Figure 10.53

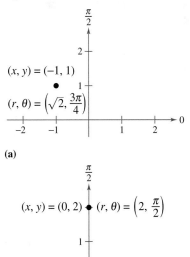

(a)

(b)
Figure 10.54

Equation Conversion

By comparing Examples 3 and 4, you see that point conversion from the polar to the rectangular system is straightforward, whereas point conversion from the rectangular to the polar system is more involved. For equations, the opposite is true. To convert a rectangular equation to polar form, you simply replace x by $r \cos \theta$ and y by $r \sin \theta$. For instance, the rectangular equation $y = x^2$ can be written in polar form as follows.

$$y = x^2 \qquad \text{Rectangular equation}$$
$$r \sin \theta = (r \cos \theta)^2 \qquad \text{Polar equation}$$
$$r = \sec \theta \tan \theta \qquad \text{Simplest form}$$

On the other hand, converting a polar equation to rectangular form requires considerable ingenuity.

Example 5 demonstrates several polar-to-rectangular conversions that enable you to sketch the graphs of some polar equations.

EXAMPLE 5 Converting Polar Equations to Rectangular Form

Describe the graph of each polar equation and find the corresponding rectangular equation.

a. $r = 2$ **b.** $\theta = \dfrac{\pi}{3}$ **c.** $r = \sec \theta$

Solution

a. The graph of the polar equation $r = 2$ consists of all points that are two units from the pole. In other words, this graph is a circle centered at the origin with a radius of 2, as shown in Figure 10.55(a). You can confirm this by converting to rectangular form, using the relationship $r^2 = x^2 + y^2$.

$$r = 2 \quad \Longrightarrow \quad r^2 = 2^2 \quad \Longrightarrow \quad x^2 + y^2 = 2^2$$
$$\underbrace{\hphantom{r = 2}}_{\text{Polar equation}} \qquad\qquad \underbrace{\hphantom{x^2 + y^2 = 2^2}}_{\text{Rectangular equation}}$$

b. The graph of the polar equation $\theta = \pi/3$ consists of all points on the line that make an angle of $\pi/3$ with the positive x-axis, as shown in Figure 10.55(b). To convert to rectangular form, you make use of the relationship $\tan \theta = y/x$.

$$\theta = \frac{\pi}{3} \quad \Longrightarrow \quad \tan \theta = \sqrt{3} \quad \Longrightarrow \quad y = \sqrt{3}x$$
$$\underbrace{\hphantom{\theta = \frac{\pi}{3}}}_{\text{Polar equation}} \qquad\qquad \underbrace{\hphantom{y = \sqrt{3}x}}_{\text{Rectangular equation}}$$

c. The graph of the polar equation $r = \sec \theta$ is not evident by simple inspection, so you convert to rectangular form by using the relationship $r \cos \theta = x$.

$$r = \sec \theta \quad \Longrightarrow \quad r \cos \theta = 1 \quad \Longrightarrow \quad x = 1$$
$$\underbrace{\hphantom{r = \sec \theta}}_{\text{Polar equation}} \qquad\qquad \underbrace{\hphantom{x = 1}}_{\text{Rectangular equation}}$$

Now you see that the graph is a vertical line, as shown in Figure 10.55(c).

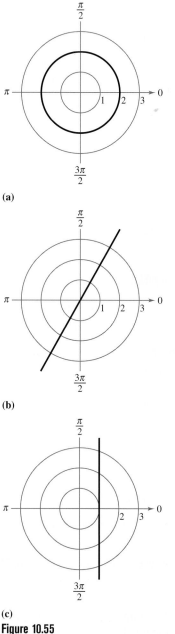

(a)

(b)

(c)

Figure 10.55

10.6 Exercises

In Exercises 1–4, a point in polar coordinates is given. Find the corresponding rectangular coordinates for the point.

1. $\left(4, \dfrac{\pi}{2}\right)$　　　**2.** $\left(4, \dfrac{3\pi}{2}\right)$

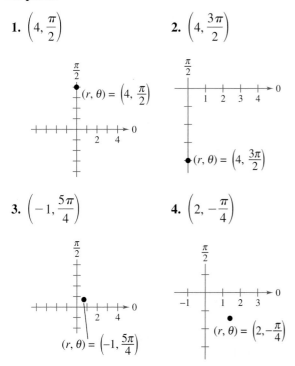

3. $\left(-1, \dfrac{5\pi}{4}\right)$　　**4.** $\left(2, -\dfrac{\pi}{4}\right)$

In Exercises 5–12, plot the point given in polar coordinates and find three additional polar representations of the point, using $-2\pi < \theta < 2\pi$.

5. $\left(4, \dfrac{2\pi}{3}\right)$　　　**6.** $\left(1, \dfrac{7\pi}{4}\right)$

7. $\left(5, -\dfrac{5\pi}{3}\right)$　　**8.** $\left(-3, -\dfrac{7\pi}{6}\right)$

9. $\left(\sqrt{3}, \dfrac{5\pi}{6}\right)$　　**10.** $\left(5\sqrt{2}, -\dfrac{11\pi}{6}\right)$

11. $\left(\dfrac{3}{2}, -\dfrac{3\pi}{2}\right)$　　**12.** $\left(-\dfrac{7}{8}, -\dfrac{\pi}{6}\right)$

In Exercises 13–22, plot the point given in polar coordinates and find the corresponding rectangular coordinates for the point.

13. $\left(4, -\dfrac{\pi}{3}\right)$　　　**14.** $\left(2, \dfrac{7\pi}{6}\right)$

15. $\left(-1, -\dfrac{3\pi}{4}\right)$　　**16.** $\left(-3, -\dfrac{2\pi}{3}\right)$

17. $\left(0, -\dfrac{7\pi}{6}\right)$　　**18.** $\left(0, \dfrac{5\pi}{4}\right)$

19. $\left(32, \dfrac{5\pi}{2}\right)$　　**20.** $\left(18, -\dfrac{3\pi}{2}\right)$

21. $\left(\sqrt{2}, 2.36\right)$　　**22.** $(-3, -1.57)$

In Exercises 23–26, use a graphing utility to find the rectangular coordinates for the point given in polar coordinates.

23. $\left(2, \dfrac{3\pi}{4}\right)$　　**24.** $\left(-2, \dfrac{7\pi}{6}\right)$

25. $(-4.5, 1.3)$　　**26.** $(8.25, 3.5)$

In Exercises 27–36, the rectangular coordinates of a point are given. Plot the point and find *two* sets of polar coordinates for the point for $0 \le \theta < 2\pi$.

27. $(-7, 0)$　　　**28.** $(0, -5)$

29. $(1, 1)$　　　　**30.** $(-3, -3)$

31. $(-3, 4)$　　　**32.** $(3, -1)$

33. $\left(-\sqrt{3}, -\sqrt{3}\right)$　　**34.** $(2, -2)$

35. $(4, 6)$　　　　**36.** $(5, 12)$

In Exercises 37–44, use a graphing utility to find one set of polar coordinates for the point given in rectangular coordinates.

37. $(3, -2)$　　　**38.** $(-4, 1)$

39. $\left(\sqrt{3}, 2\right)$　　**40.** $\left(3\sqrt{2}, 3\sqrt{2}\right)$

41. $\left(\dfrac{5}{2}, \dfrac{4}{3}\right)$　　**42.** $\left(\dfrac{11}{4}, -\dfrac{5}{8}\right)$

43. $(0, -5)$　　　**44.** $(-8, 0)$

In Exercises 45–54, convert the rectangular equation to polar form. Assume $a \ge 0$.

45. (a) $x^2 + y^2 = 49$　　(b) $x^2 + y^2 = a^2$

46. (a) $x^2 + y^2 - 6x = 0$　(b) $x^2 + y^2 - 8y = 0$

47. (a) $x^2 + y^2 - 2ax = 0$　(b) $x^2 + y^2 - 2ay = 0$

48. (a) $y = 4$　　　　(b) $y = b$

49. (a) $x = 12$　　　(b) $x = a$

50. (a) $3x - 6y + 2 = 0$　(b) $4x + 7y - 2 = 0$

51. (a) $xy = 4$ (b) $2xy = 1$

52. (a) $y = x$ (b) $y^2 = 2x$

53. (a) $y^2 = x^3$ (b) $x^2 = y^3$

54. (a) $(x^2 + y^2)^2 - 9(x^2 - y^2) = 0$

    (b) $y^2 - 8x - 16 = 0$

In Exercises 55–70, convert the polar equation to rectangular form.

55. $r = 4 \sin \theta$ **56.** $r = 4 \cos \theta$

57. $\theta = \dfrac{\pi}{6}$ **58.** $\theta = \dfrac{5\pi}{3}$

59. $r = 4$ **60.** $r = 10$

61. $r = -3 \csc \theta$ **62.** $r = 2 \sec \theta$

63. $r^2 = \cos \theta$ **64.** $r^2 = \sin 2\theta$

65. $r = 2 \sin 3\theta$ **66.** $r = 3 \cos 2\theta$

67. $r = \dfrac{1}{1 - \cos \theta}$ **68.** $r = \dfrac{2}{1 + \sin \theta}$

69. $r = \dfrac{6}{2 - 3 \sin \theta}$ **70.** $r = \dfrac{6}{2 \cos \theta - 3 \sin \theta}$

In Exercises 71–76, describe the graph of the polar equation and find the corresponding rectangular equation. Sketch its graph.

71. $r = 3$ **72.** $r = 8$

73. $\theta = \dfrac{\pi}{4}$ **74.** $\theta = \dfrac{5\pi}{6}$

75. $r = 3 \sec \theta$ **76.** $r = 2 \csc \theta$

Synthesis

True or False? **In Exercises 77 and 78, determine whether the statement is true or false. Justify your answer.**

77. If (r_1, θ_1) and (r_2, θ_2) represent the same point in the polar coordinate system, then $|r_1| = |r_2|$.

78. If (r, θ_1) and (r, θ_2) represent the same point in the polar coordinate system, then $\theta_1 = \theta_2 + 2\pi n$ for some integer n.

79. *Think About It*

(a) Show that the distance between the points (r_1, θ_1) and (r_2, θ_2) is

$$\sqrt{r_1^{\,2} + r_2^{\,2} - 2r_1 r_2 \cos(\theta_1 - \theta_2)}.$$

(b) Describe the position of the points relative to

each other if $\theta_1 = \theta_2$. Simplify the distance formula for this case. Is the simplification what you expected? Explain.

(c) Simplify the distance formula if $\theta_1 - \theta_2 = 90°$. Is the simplification what you expected? Explain.

(d) Choose two points on the polar coordinate system and find the distance between them. Then choose different polar representations of the same two points and apply the distance formula again. Discuss the result.

80. *Exploration*

(a) Set the viewing window of your graphing utility to rectangular coordinates and locate the cursor at any position off the coordinate axes. Move the cursor horizontally and observe any changes in the displayed coordinates of the points. Explain the changes. Now repeat the process moving the cursor vertically.

(b) Set the viewing window of your graphing utility to polar coordinates and locate the cursor at any position off the coordinate axes. Move the cursor horizontally and observe any changes in the displayed coordinates of the points. Explain the changes. Now repeat the process moving the cursor vertically.

(c) Explain why the results of parts (a) and (b) are not the same.

Review

In Exercises 81–84 use determinants to solve the system of equations.

81. $\begin{cases} 5x - 7y = -11 \\ -3x + y = -3 \end{cases}$ **82.** $\begin{cases} 3x + 5y = 10 \\ 4x - 2y = -5 \end{cases}$

83. $\begin{cases} 3a - 2b + c = 0 \\ 2a + b - 3c = 0 \\ a - 3b + 9c = 0 \end{cases}$ **84.** $\begin{cases} 5u + 7v + 9w = 15 \\ u - 2v - 3w = 7 \\ 8u - 2v + w = 0 \end{cases}$

In Exercises 85–88, find the coefficient a of the given term in the expansion of the binomial.

Binomial	Term
85. $(x + 5)^8$	ax^3
86. $(x^2 - 3)^{10}$	ax^8
87. $(2x - y)^{12}$	ax^7y^5
88. $(3x - 2y)^7$	ax^3y^4

10.7 Graphs of Polar Equations

Introduction

In previous chapters you spent a lot of time learning how to sketch graphs in rectangular coordinates. You began with the basic point-plotting method. Then you used sketching aids such as a graphing utility, symmetry, intercepts, asymptotes, periods, and shifts to further investigate the nature of the graph. This section approaches curve sketching in the polar coordinate system similarly.

EXAMPLE 1   Graphing a Polar Equation by Point Plotting

Sketch the graph of the polar equation $r = 4 \sin \theta$.

Solution

The sine function is periodic, so you can get a full range of r-values by considering values of θ in the interval $0 \le \theta \le 2\pi$, as shown in the table.

θ	0	$\dfrac{\pi}{6}$	$\dfrac{\pi}{3}$	$\dfrac{\pi}{2}$	$\dfrac{2\pi}{3}$	$\dfrac{5\pi}{6}$	π	$\dfrac{7\pi}{6}$	$\dfrac{3\pi}{2}$	$\dfrac{11\pi}{6}$	2π
r	0	2	$2\sqrt{3}$	4	$2\sqrt{3}$	2	0	-2	-4	-2	0

If you plot these points as shown in Figure 10.56, it appears that the graph is a circle of radius 2 whose center is at the point $(x, y) = (0, 2)$.

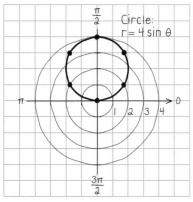

Figure 10.56

You can confirm the graph found in Example 1 in three ways.

1. *Convert to Rectangular Form*   Multiply both sides of the polar equation by r and convert the result to rectangular form.
2. *Use a Polar Coordinate Mode*   Set your graphing utility to *polar* mode and graph the polar equation. (Use $0 \le \theta \le 2\pi$, $-6 \le x \le 6$, and $-4 \le y \le 4$.)
3. *Use a Parametric Mode*   Set your graphing utility to *parametric* mode and graph $x = (4 \sin t) \cos t$ and $y = (4 \sin t) \sin t$.

Most graphing utilities have a *polar-coordinate* graphing mode. If yours doesn't, you can use the following parametric conversion to graph a polar equation.

Polar Equations in Parametric Form

The graph of the polar equation $r = f(\theta)$ can be written in parametric form, using t as a parameter, as follows.

$$x = f(t) \cos t \qquad \text{and} \qquad y = f(t) \sin t.$$

Symmetry

In Figure 10.56, note that as θ increases from 0 to 2π the graph is traced out twice. Moreover, note that the graph is *symmetric with respect to the line* $\theta = \pi/2$. Had you known about this symmetry and retracing ahead of time, you could have used fewer points.

Symmetry with respect to the line $\theta = \pi/2$ is one of three important types of symmetry to consider in polar curve sketching. (See Figure 10.57.)

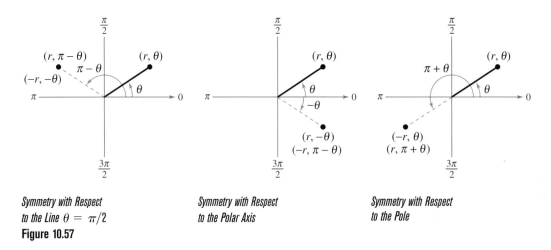

Symmetry with Respect
to the Line $\theta = \pi/2$
Figure 10.57

Symmetry with Respect
to the Polar Axis

Symmetry with Respect
to the Pole

Tests for Symmetry on Polar Coordinates

The graph of a polar equation is symmetric with respect to the following if the given substitution yields an equivalent equation.

1. The line $\theta = \pi/2$:   Replace (r, θ) by $(r, \pi - \theta)$ or $(-r, -\theta)$.
2. The polar axis: Replace (r, θ) by $(r, -\theta)$ or $(-r, \pi - \theta)$.
3. The pole: Replace (r, θ) by $(r, \pi + \theta)$ or $(-r, \theta)$.

EXAMPLE 2   Using Symmetry to Sketch a Polar Graph

Use symmetry to sketch the graph of $r = 3 + 2 \cos \theta$.

Solution

Replacing (r, θ) by $(r, -\theta)$ produces

$$r = 3 + 2 \cos(-\theta)$$

$$= 3 + 2 \cos \theta.$$

So, you can conclude that the curve is symmetric with respect to the polar axis. Plotting the points in the table and using polar axis symmetry, you obtain the graph shown in Figure 10.58. This graph is called a **limaçon.**

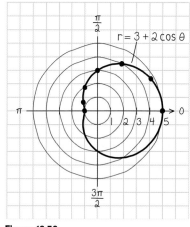

Figure 10.58

θ	0	$\dfrac{\pi}{6}$	$\dfrac{\pi}{3}$	$\dfrac{\pi}{2}$	$\dfrac{2\pi}{3}$	$\dfrac{5\pi}{6}$	π
r	5	$3 + \sqrt{3}$	4	3	2	$3 - \sqrt{3}$	1

Use a graphing utility to confirm this graph.

The three tests for symmetry in polar coordinates on page 746 are sufficient to guarantee symmetry, but they are not necessary. For instance, Figure 10.59 shows the graph of

$$r = \theta + 2\pi. \qquad \text{Spiral of Archimedes}$$

From the figure, you can see that the graph is symmetric with respect to the line $\theta = \pi/2$. Yet the tests on page 746 fail to indicate symmetry because neither of the following replacements yields an equivalent equation.

Original Equation	*Replacement*	*New Equation*
$r = \theta + 2\pi$	(r, θ) by $(-r, -\theta)$	$-r = -\theta + 2\pi$
$r = \theta + 2\pi$	(r, θ) by $(r, \pi - \theta)$	$r = -\theta + 3\pi$

The equations discussed in Examples 1 and 2 are of the form

$$r = 4 \sin \theta = f(\sin \theta)$$

and

$$r = 3 + 2 \cos \theta = g(\cos \theta).$$

The graph of the first equation is symmetric with respect to the line $\theta = \pi/2$, and the graph of the second equation is symmetric with respect to the polar axis. This observation can be generalized to yield the following *quick test for symmetry.*

1. The graph of $r = f(\sin \theta)$ is symmetric with respect to the line $\theta = \pi/2$.
2. The graph of $r = g(\cos \theta)$ is symmetric with respect to the polar axis.

STUDY T!P

The *table* feature of a graphing utility is very useful in constructing tables of values for polar equations. Set your graphing utility to *polar* mode and enter the polar equation in Example 2. You can verify the table of values in Example 2 by starting the table at $\theta = 0$ and incrementing the value of θ by $\pi/6$.

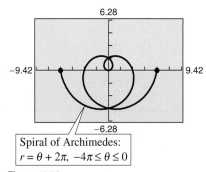

Spiral of Archimedes:
$r = \theta + 2\pi, \ -4\pi \le \theta \le 0$

Figure 10.59

Zeros and Maximum *r*-Values

Two additional aids to sketching graphs of polar equations involve knowing the θ-values for which $|r|$ is maximum and knowing the θ-values for which $r = 0$. In Example 1, the maximum value of $|r|$ for $r = 4 \sin \theta$ is $|r| = 4$, and this occurs when $\theta = \pi/2$ (see Figure 10.56). Moreover, $r = 0$ when $\theta = 0$.

EXAMPLE 3   Finding Maximum *r*-Values of a Polar Graph

Find the maximum value of r for the graph of $r = 1 - 2 \cos \theta$.

Graphical Solution

Because the polar equation is of the form

$$r = 1 - 2 \cos \theta = g(\cos \theta)$$

you know the graph is symmetric with respect to the polar axis. You can confirm this by graphing the polar equation, as shown in Figure 10.60. (In the graph, θ varies from 0 to 2π.) To find the maximum r-value for the graph, use your graphing utility's *trace* feature. When you do this, you should find that the graph has a maximum r-value of 3. This value of r occurs when $\theta = \pi$. In the graph, note that the point $(3, \pi)$ is farthest from the pole.

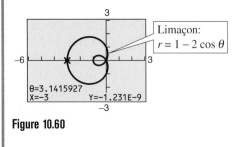

Figure 10.60

Numerical Solution

To approximate the maximum value of r for the graph of $r = 1 - 2 \cos \theta$, use the *table* feature of a graphing utility to create a table that begins at $\theta = 0$ and increments by $\pi/12$, as shown in Figure 10.61. From the table, the maximum value of r appears to be 3 when $\theta = 3.1416 \approx \pi$.

θ	r1	
2.0944	2	
2.3562	2.4142	
2.618	2.7321	
2.8798	2.9319	
3.1416	**3**	
3.4034	2.9319	
3.6652	2.7321	
θ=3.14159265359		

Figure 10.61

By creating a second table that begins at $\theta = \pi/2$ and increments by $\pi/24$, as shown in Figure 10.62, the maximum value of r still appears to be 3 when $\theta = 3.1416 \approx \pi$.

θ	r1	
2.7489	2.8478	
2.8798	2.9319	
3.0107	2.9829	
3.1416	**3**	
3.2725	2.9829	
3.4034	2.9319	
3.5343	2.8478	
θ=3.14159265359		

Figure 10.62

Note how the negative r-values determine the *inner loop* of the graph in Figure 10.60. This type of graph is a limaçon.

Exploration

The graph of the polar equation $r = e^{\cos\theta} - 2 \cos 4\theta + \sin^5(\theta/12)$ is called *the butterfly curve*, as shown in Figure 10.63.

a. The graph at the right was produced using $0 \le \theta \le 2\pi$. Does this show the entire graph? Explain your reasoning.

b. Use the *trace* feature of your graphing calculator to approximate the maximum r-value of the graph. Does this value change if you use $0 \le \theta \le 4\pi$ instead of $0 \le \theta \le 2\pi$? Explain.

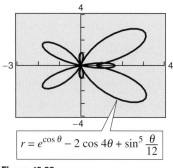

$$r = e^{\cos \theta} - 2 \cos 4\theta + \sin^5 \frac{\theta}{12}$$

Figure 10.63

Some curves reach their zeros and maximum *r*-values at more than one point. Example 4 shows how to handle this situation.

EXAMPLE 4   Analyzing a Polar Graph

Analyze the graph of $r = 2 \cos 3\theta$.

Solution

Symmetry	With respect to the polar axis
Maximum value of $\|r\|$	$\|r\| = 2$ when $3\theta = 0, \pi, 2\pi, 3\pi$
	or $\quad \theta = 0, \pi/3, 2\pi/3, \pi$
Zeros of r	$r = 0$ when $3\theta = \pi/2, 3\pi/2, 5\pi/2$
	or $\quad \theta = \pi/6, \pi/2, 5\pi/6$

θ	0	$\dfrac{\pi}{12}$	$\dfrac{\pi}{6}$	$\dfrac{\pi}{4}$	$\dfrac{\pi}{3}$	$\dfrac{5\pi}{12}$	$\dfrac{\pi}{2}$
r	2	$\sqrt{2}$	0	$-\sqrt{2}$	-2	$-\sqrt{2}$	0

By plotting these points and using the specified symmetry, zeros, and maximum values, you can obtain the graph shown in Figure 10.64. This graph is called a **rose curve,** and each loop on the graph is called a *petal*. Note how the entire curve is generated as θ increases from 0 to π.

A computer animation of this example appears in the *Interactive* CD-ROM and *Internet* versions of this text.

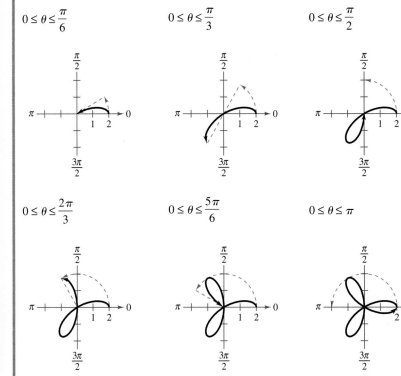

$0 \leq \theta \leq \dfrac{\pi}{6}$ $\qquad$ $0 \leq \theta \leq \dfrac{\pi}{3}$ $\qquad$ $0 \leq \theta \leq \dfrac{\pi}{2}$

$0 \leq \theta \leq \dfrac{2\pi}{3}$ $\qquad$ $0 \leq \theta \leq \dfrac{5\pi}{6}$ $\qquad$ $0 \leq \theta \leq \pi$

Figure 10.64

Exploration

Notice that the rose curve in Example 4 has three petals. How many petals does the rose curve $r = 2 \cos 4\theta$ have? Experiment with other rose curves and determine the number of petals for the curves $r = 2 \cos n\theta$ and $r = 2 \sin n\theta$, where *n* is a positive integer.

Special Polar Graphs

Several important types of graphs have equations that are simpler in polar form than in rectangular form. For example, the circle

$r = 4 \sin \theta$

in Example 1 has the more complicated rectangular equation

$x^2 + (y - 2)^2 = 4.$

The following list gives several other types of graphs that have simple polar equations.

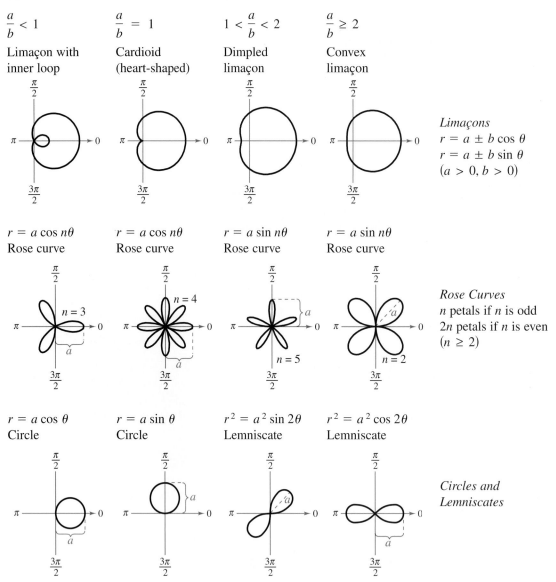

$\dfrac{a}{b} < 1$	$\dfrac{a}{b} = 1$	$1 < \dfrac{a}{b} < 2$	$\dfrac{a}{b} \geq 2$	
Limaçon with inner loop	Cardioid (heart-shaped)	Dimpled limaçon	Convex limaçon	*Limaçons* $r = a \pm b \cos \theta$ $r = a \pm b \sin \theta$ $(a > 0, b > 0)$
$r = a \cos n\theta$ Rose curve	$r = a \cos n\theta$ Rose curve	$r = a \sin n\theta$ Rose curve	$r = a \sin n\theta$ Rose curve	*Rose Curves* n petals if n is odd $2n$ petals if n is even $(n \geq 2)$
$n = 3$	$n = 4$	$n = 5$	$n = 2$	
$r = a \cos \theta$ Circle	$r = a \sin \theta$ Circle	$r^2 = a^2 \sin 2\theta$ Lemniscate	$r^2 = a^2 \cos 2\theta$ Lemniscate	*Circles and Lemniscates*

EXAMPLE 5   Analyzing a Rose Curve

Analyze the graph of

$r = 3 \cos 2\theta.$

Solution

Begin with an analysis of the basic features of the graph.

Type of curve	Rose curve with $2n = 4$ petals
Symmetry	With respect to polar axis, the line $\theta = \pi/2$, and the pole
Maximum value of $|r|$	$|r| = 3$ when $\theta = 0, \pi/2, \pi, 3\pi/2$
Zeros of r	$r = 0$ when $\theta = \pi/4, 3\pi/4$

Using a graphing utility (with $0 \le \theta \le 2\pi$), you can obtain the graph shown in Figure 10.65.

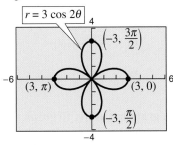

Figure 10.65

EXAMPLE 6   Analyzing a Lemniscate

Analyze the graph

$r^2 = 9 \sin 2\theta.$

Solution

Begin with an analysis of the basic features of the graph.

Type of curve	Lemniscate
Symmetry	With respect to pole
Maximum value of $|r|$	$|r| = 3$ when $\theta = \pi/4$
Zeros of r	$r = 0$ when $\theta = 0, \pi/2$

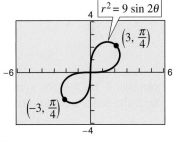

Figure 10.66

Using a graphing utility (with $r = \sqrt{9 \sin 2\theta}$ and $0 \le \theta \le 2\pi$), you can obtain the graph shown in Figure 10.66.

Writing About Math   *Heart to Bell*

Use a graphing utility to graph the polar equation

$r = \cos 5\theta + n \cos \theta$

for $0 \le \theta < \pi$ for the integers $n = -5$ to $n = 5$. As you graph these equations, you should see the graph change shape from a heart to a bell.
Write a short paragraph explaining what values of n produce the heart portion of the curve and what values of n produce the bell.

10.7 Exercises

In Exercises 1–6, identify the type of polar graph.

1.

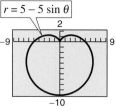

$r = 3 \cos 2\theta$

2.

$r = 5 - 5 \sin \theta$

3.

$r = 3 \cos \theta$

4.

$r^2 = 9 \cos 2\theta$

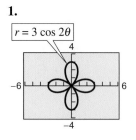

5.

$r = 6 \sin 2\theta$

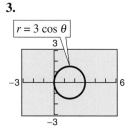

6.

$r = 1 + 4 \cos \theta$

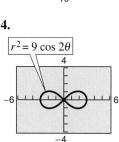

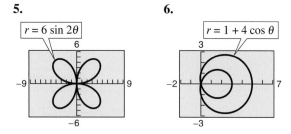

In Exercises 7–16, test for symmetry with respect to $\theta = \pi/2$, the polar axis, and the pole.

7. $r = 10 + 4 \cos \theta$

8. $r = 16 \cos 3\theta$

9. $r = \dfrac{6}{1 + \sin \theta}$

10. $r = \dfrac{4}{1 - \cos \theta}$

11. $r = 6 \sin \theta$

12. $r = 4 - \sin \theta$

13. $r = 4 \sec \theta \csc \theta$

14. $r = 2 \csc \theta \cos \theta$

15. $r^2 = 25 \sin 2\theta$

16. $r^2 = 25 \cos 4\theta$

In Exercises 17–20, find the maximum value of $|r|$ and any zeros of r. Verify your answers numerically.

17. $r = 10 - 10 \sin \theta$

18. $r = 6 + 12 \cos \theta$

19. $r = 4 \cos 3\theta$

20. $r = 5 \sin 2\theta$

In Exercises 21–38, sketch the graph of the polar equation. Use a graphing utility to confirm your graph.

21. $r = 5$

22. $r = 2$

23. $\theta = \dfrac{\pi}{6}$

24. $\theta = -\dfrac{5\pi}{3}$

25. $r = 3 \sin \theta$

26. $r = 3 \cos \theta$

27. $r = 3(1 - \cos \theta)$

28. $r = 2(1 - \sin \theta)$

29. $r = 3 - 4 \cos \theta$

30. $r = 5 - 4 \sin \theta$

31. $r = 6 + \sin \theta$

32. $r = 4 + 5 \cos \theta$

33. $r = 5 \cos 3\theta$

34. $r = -\sin 5\theta$

35. $r = 7 \sin 2\theta$

36. $r = 3 \cos 5\theta$

37. $r = \dfrac{\theta}{2}$

38. $r = \theta$

In Exercises 39–54, use a graphing utility to graph the polar equation. Describe your viewing window.

39. $r = \dfrac{\theta}{4}$

40. $r = -\dfrac{\theta}{3}$

41. $r = 6 \cos \theta$

42. $r = \cos 2\theta$

43. $r = 2(3 - \sin \theta)$

44. $r = 6 - 4 \sin \theta$

45. $r = 3 - 6 \cos \theta$

46. $r = 2(3 - 2 \sin \theta)$

47. $r = \dfrac{3}{\sin \theta - 2 \cos \theta}$

48. $r = \dfrac{6}{2 \sin \theta - 3 \cos \theta}$

49. $r^2 = 4 \cos 2\theta$

50. $r^2 = 4 \sin \theta$

51. $r = 4 \sin \theta \cos^2 \theta$

52. $r = 2 \cos(3\theta - 2)$

53. $r = 4 \csc \theta + 5$

54. $r = 4 - \sec \theta$

In Exercises 55–62, use a graphing utility to graph the polar equation. Find an interval for θ for which the graph is traced *only once*.

55. $r = 3 - 2 \cos \theta$

56. $r = 2(1 - 2 \sin \theta)$

57. $r = 2 + \sin \theta$

58. $r = 4 + 3 \cos \theta$

59. $r = 2 \cos\left(\dfrac{3\theta}{2}\right)$

60. $r = 3 \sin\left(\dfrac{5\theta}{2}\right)$

61. $r^2 = 4 \sin 2\theta$

62. $r^2 = \dfrac{1}{\theta}$

In Exercises 63–66, use a graphing utility to graph the polar equation and show that the indicated line is an asymptote of the graph.

Name of Graph	*Polar Equation*	*Asymptote*
63. Conchoid	$r = 2 - \sec\theta$	$x = -1$
64. Conchoid	$r = 2 + \csc\theta$	$y = 1$
65. Hyperbolic spiral	$r = \dfrac{2}{\theta}$	$y = 2$
66. Strophoid	$r = 2\cos 2\theta \sec\theta$	$x = -2$

Synthesis

True or False? **In Exercises 67–70, determine whether the statement is true or false. Justify your answer.**

67. The point with polar coordinate $\left(6, \dfrac{11\pi}{6}\right)$ lies on the graph of $r = 2\sin\theta + 5$.

68. The graph of $r = 4\cos 8\theta$ is a rose curve with 8 petals.

69. The graph of $r = 10\sin 5\theta$ is a rose curve with 10 petals.

70. A rose curve will always have symmetry with respect to the line $\theta = \pi/2$.

71. *Graphical Reasoning* Use a graphing utility to graph the polar equation

$$r = 6[1 + \cos(\theta - \phi)]$$

for (a) $\phi = 0$, (b) $\phi = \pi/4$, and (c) $\phi = \pi/2$. Use the graphs to describe the effect of the angle ϕ. Write the equation as a function of $\sin\theta$ for part (c).

72. The graph of $r = f(\theta)$ is rotated about the pole through an angle ϕ. Show that the equation of the rotated graph is $r = f(\theta - \phi)$.

73. Consider the graph of $r = f(\sin\theta)$.

(a) Show that if the graph is rotated counterclockwise $\pi/2$ radians about the pole, the equation of the rotated graph is $r = f(-\cos\theta)$.

(b) Show that if the graph is rotated counterclockwise π radians about the pole, the equation of the rotated graph is $r = f(-\sin\theta)$.

(c) Show that if the graph is rotated counterclockwise $3\pi/2$ radians about the pole, the equation of the rotated graph is $r = f(\cos\theta)$.

In Exercises 74–76, use the results of Exercise 72 and 73.

74. Write an equation for the limaçon $r = 2 - \sin\theta$ after it has been rotated through the given angle.

(a) $\dfrac{\pi}{4}$ (b) $\dfrac{\pi}{2}$ (c) π (d) $\dfrac{3\pi}{2}$

75. Write an equation for the rose curve $r = 2\sin 2\theta$ after it has been rotated through the given angle.

(a) $\dfrac{\pi}{6}$ (b) $\dfrac{\pi}{2}$ (c) $\dfrac{2\pi}{3}$ (d) π

76. Sketch the graph of each equation.

(a) $r = 1 - \sin\theta$ (b) $r = 1 - \sin\left(\theta - \dfrac{\pi}{4}\right)$

77. *Exploration* Use a graphing utility to graph the polar equation $r = 2 + k\cos\theta$ for $k = 0$, $k = 1$, $k = 2$, and $k = 3$. Identify each graph.

78. *Exploration* Consider the polar equation $r = 3\sin k\theta$.

(a) Use a graphing utility to graph the equation for $k = 1.5$. Find the interval for θ for which the graph is traced only once.

(b) Use a graphing utility to graph the equation for $k = 2.5$. Find the interval for θ for which the graph is traced only once.

(c) Is it possible to find an interval for θ for which the graph is traced only once for any rational number k? Explain.

Review

In Exercises 79–82, write the first five terms of the arithmetic sequence. Find the common difference and write the nth term of the sequence as a function of n.

79. $a_1 = 2$, $a_8 = 23$

80. $a_1 = \frac{5}{2}$, $a_3 = \frac{11}{6}$

81. $a_1 = 150$, $a_{k+1} = a_k - 18$

82. $a_1 = 0.525$, $a_{k+1} = a_k + 0.75$

In Exercises 83–88, find the sum.

83. $\displaystyle\sum_{n=1}^{20} 4n$

84. $\displaystyle\sum_{n=1}^{50} 8n$

85. $\displaystyle\sum_{n=1}^{120} (n + 5)$

86. $\displaystyle\sum_{n=1}^{200} (300 - n)$

87. $\displaystyle\sum_{n=0}^{\infty} \left(\tfrac{1}{8}\right)^n$

88. $\displaystyle\sum_{n=1}^{\infty} 6(0.4)^{n-1}$

10.8 Polar Equations of Conics

Alternative Definition of Conics

In Sections 10.2 and 10.3, you learned that the rectangular equations of ellipses and hyperbolas take simple forms when the origin lies at the *center*. As it happens, there are many important applications of conics in which it is more convenient to use one of the *foci* as the origin for the coordinate system. For example, the sun lies at one focus of the earth's orbit. Similarly the light source of a parabolic reflector lies at its focus. In this section you will learn that polar equations of conics take simple forms if one of the foci lies at the pole.

To begin, consider the following alternative definition of a conic that uses the concept of eccentricity.

Alternative Definition of a Conic

The locus of a point in the plane which moves so that its distance from a fixed point (focus) is in constant ratio to its distance from a fixed line (directrix) is a **conic.** The constant ratio is the **eccentricity** of the conic and is denoted by e. Moreover, the conic is an **ellipse** if $e < 1$, a **parabola** if $e = 1$, and a **hyperbola** if $e > 1$.

NASA

In Figure 10.67, note that for each type of conic, the pole corresponds to the fixed point (focus) given in the definition.

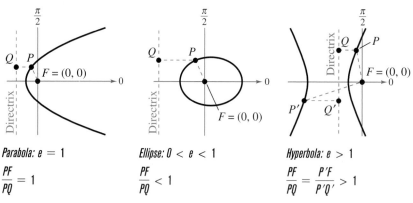

Parabola: $e = 1$
$$\frac{PF}{PQ} = 1$$

Ellipse: $0 < e < 1$
$$\frac{PF}{PQ} < 1$$

Hyperbola: $e > 1$
$$\frac{PF}{PQ} = \frac{P'F}{P'Q'} > 1$$

Figure 10.67

Polar Equations of Conics

The benefit of locating a focus of a conic at the pole is that the equation of the conic takes on a simpler form. A proof of the polar form is given in Appendix A.

Polar Equations of Conics

The graph of a polar equation of the form

1. $r = \dfrac{ep}{1 \pm e \cos \theta}$ **2.** $r = \dfrac{ep}{1 \pm e \sin \theta}$

is a conic, where $e > 0$ is the eccentricity and $|p|$ is the distance between the focus (pole) and the directrix.

Equations of the form $r = \dfrac{ep}{1 \pm e \cos \theta}$ Vertical directrix

correspond to conics with vertical directrices and equations of the form

$$r = \frac{ep}{1 \pm e \sin \theta}$$ Horizontal directrix

correspond to conics with horizontal directrices. Moreover, the converse is also true—that is, any conic with a focus at the pole and having a horizontal or vertical directrix can be represented by one of the given equations.

A computer simulation of this concept appears in the *Interactive* CD-ROM and *Internet* versions of this text.

EXAMPLE 1   Determining a Conic from Its Equation

Determine the type of conic represented by the equation $r = \dfrac{15}{3 - 2 \cos \theta}$.

Algebraic Solution

To determine the type of conic, rewrite the equation in the form $r = ep/(1 \pm e \cos \theta)$.

$$r = \frac{15}{3 - 2 \cos \theta}$$

$$= \frac{5}{1 - (2/3) \cos \theta}$$ Divide numerator and denominator by 3.

From this form you can conclude that the graph is an ellipse with $e = \frac{2}{3}$.

Graphical Solution

Use a graphing utility in *polar* mode to graph $r = \dfrac{15}{3 - 2 \cos \theta}$.

Be sure to use a square setting. From the graph in Figure 10.68, you can see that the conic appears to be an ellipse.

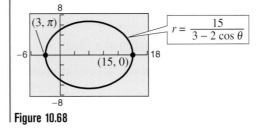

$r = \dfrac{15}{3 - 2 \cos \theta}$

Figure 10.68

For the ellipse in Figure 10.68, the major axis is horizontal and the vertices lie at $(r, \theta) = (15, 0)$ and $(r, \theta) = (3, \pi)$. So, the length of the *major* axis is $2a = 18$. To find the length of the *minor* axis, you can use the equations $e = c/a$ and $b^2 = a^2 - c^2$ to conclude that

$$b^2 = a^2 - c^2 = a^2 - (ea)^2 = a^2(1 - e^2).$$ Ellipse

Because $e = \frac{2}{3}$, you have $b^2 = 9^2[1 - (2/3)^2] = 45$, which implies that $b = \sqrt{45} = 3\sqrt{5}$. So, the length of the minor axis is $2b = 6\sqrt{5}$. A similar analysis for hyperbolas yields

$$b^2 = c^2 - a^2 = (ea)^2 - a^2 = a^2(e^2 - 1).$$ Hyperbola

EXAMPLE 2   Analyzing the Graph of a Polar Equation

Analyze the graph of the polar equation

$$r = \frac{32}{3 + 5 \sin \theta}.$$

Solution

Dividing the numerator and denominator by 3 produces

$$r = \frac{32/3}{1 + (5/3) \sin \theta}.$$

Because $e = 5/3 > 1$, the graph is a hyperbola. The transverse axis of the hyperbola lies on the line $\theta = \pi/2$ and the vertices occur at $(r, \theta) = (4, \pi/2)$ and $(r, \theta) = (-16, 3\pi/2)$. Because the length of the transverse axis is 12, you can see that $a = 6$. To find b, write

$$b^2 = a^2(e^2 - 1) = 6^2\left[\left(\frac{5}{3}\right)^2 - 1\right] = 64.$$

Therefore, $b = 8$. The asymptotes of the hyperbola are $y = 10 \pm \frac{3}{4}x$, as shown in Figure 10.69.

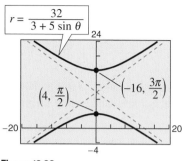

Figure 10.69

In the next example, you are asked to find a polar equation for a specified conic. To do this, let p be the distance between the pole and the directrix.

1. Horizontal directrix above the pole: $r = \dfrac{ep}{1 + e \sin \theta}$

2. Horizontal directrix below the pole: $r = \dfrac{ep}{1 - e \sin \theta}$

3. Vertical directrix to the right of the pole: $r = \dfrac{ep}{1 + e \cos \theta}$

4. Vertical directrix to the left of the pole: $r = \dfrac{ep}{1 - e \cos \theta}$

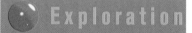

Exploration

Try using a graphing utility in *polar* mode to verify the four orientations shown at the left. Remember that e must be positive, but p can be positive or negative.

EXAMPLE 3   Finding the Polar Equation of a Conic

Find the polar equation of the parabola whose focus is the pole and whose directrix is the line $y = 3$.

Solution

From Figure 10.70, you can see that the directrix is horizontal. So, you can choose an equation of the form

$$r = \frac{ep}{1 + e \sin \theta}.$$

Moreover, because the eccentricity of a parabola is $e = 1$ and the distance between the pole and the directrix is $p = 3$, you have the equation

$$r = \frac{3}{1 + \sin \theta}.$$

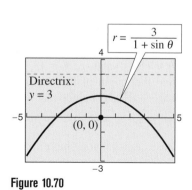

Figure 10.70

Application

Kepler's Laws (listed below), named after the German astronomer Johannes Kepler (1571–1630), can be used to describe the orbits of the planets about the sun.

1. Each planet moves in an elliptical orbit with the sun as a focus.
2. A ray from the sun to the planet sweeps out equal areas of the ellipse in equal times.
3. The square of the period is proportional to the cube of the mean distance between the planet and the sun.

Although Kepler simply stated these laws on the basis of observation, they were later validated by Isaac Newton (1642–1727). In fact, Newton was able to show that each law can be deduced from a set of universal laws of motion and gravitation that govern the movement of all heavenly bodies, including comets and satellites. This is illustrated in the next example, which involves the comet named after the English mathematician and physicist Edmund Halley (1656–1742).

If you use earth as a reference with a period of 1 year and a distance of 1 astronomical unit, the proportionality constant in Kepler's third law is 1. For example, because Mars has a mean distance to the sun of $d = 1.523$ AU, its period P is given by $d^3 = P^2$. So, the period for Mars is $P = 1.88$ years.

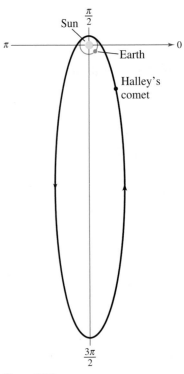

Figure 10.71

EXAMPLE 4   Halley's Comet

Halley's comet has an elliptical orbit with an eccentricity of $e \approx 0.97$. The length of the major axis of the orbit is approximately 36.18 astronomical units. (An *astronomical unit* is defined as the mean distance between earth and the sun, 93 million miles.) Find a polar equation for the orbit. How close does Halley's comet come to the sun?

Solution

Using a vertical axis, as shown in Figure 10.71, choose an equation of the form $r = ep/(1 + e \sin \theta)$. Because the vertices of the ellipse occur when $\theta = \pi/2$ and $\theta = 3\pi/2$, you can determine the length of the major axis to be the sum of the r-values of the vertices. That is,

$$2a = \frac{0.97p}{1 + 0.97} + \frac{0.97p}{1 - 0.97} \approx 32.83p \approx 36.18.$$

So, $p \approx 1.102$ and $ep \approx (0.97)(1.102) \approx 1.069$. Using this value in the equation, you have

$$r = \frac{1.069}{1 + 0.97 \sin \theta}$$

where r is measured in astronomical units. To find the closest point to the sun (the focus), substitute $\theta = \pi/2$ into this equation to obtain

$$r = \frac{1.069}{1 + 0.97 \sin(\pi/2)} \approx 0.54 \text{ astronomical units} \approx 50,000,000 \text{ miles.}$$

10.8 Exercises

Graphical Reasoning In Exercises 1–4, use a graphing utility to graph the polar equation when (a) $e = 1$, (b) $e = 0.5$, and (c) $e = 1.5$.

1. $r = \dfrac{2e}{1 + e \cos \theta}$ **2.** $r = \dfrac{2e}{1 - e \cos \theta}$

3. $r = \dfrac{2e}{1 - e \sin \theta}$ **4.** $r = \dfrac{2e}{1 + e \sin \theta}$

In Exercises 5–8, match the polar equation with the correct graph. [The graphs are labeled (a), (b), (c), and (d).]

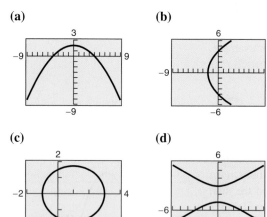

(a) **(b)**

(c) **(d)**

5. $r = \dfrac{4}{1 - \cos \theta}$ **6.** $r = \dfrac{3}{2 - \cos \theta}$

7. $r = \dfrac{3}{1 + 2 \sin \theta}$ **8.** $r = \dfrac{4}{1 + \sin \theta}$

In Exercises 9–18, determine the type of conic represented by the equation algebraically. Use a graphing utility to confirm your result graphically.

9. $r = \dfrac{6}{1 - \cos \theta}$ **10.** $r = \dfrac{6}{1 + \sin \theta}$

11. $r = \dfrac{4}{4 - \cos \theta}$ **12.** $r = \dfrac{7}{7 + \sin \theta}$

13. $r = \dfrac{8}{4 + 3 \sin \theta}$ **14.** $r = \dfrac{6}{3 - 2 \cos \theta}$

15. $r = \dfrac{4}{2 + 3 \sin \theta}$ **16.** $r = \dfrac{5}{-1 + 2 \cos \theta}$

17. $r = \dfrac{3}{4 - 8 \cos \theta}$ **18.** $r = \dfrac{10}{3 + 9 \sin \theta}$

In Exercises 19–22, use a graphing utility to graph the polar equation. Identify the graph.

19. $r = \dfrac{-5}{1 - \sin \theta}$ **20.** $r = \dfrac{-3}{2 + 4 \sin \theta}$

21. $r = \dfrac{12}{2 - \cos \theta}$ **22.** $r = \dfrac{14}{14 + 17 \sin \theta}$

In Exercises 23–26, use a graphing utility to graph the rotated conic.

23. $r = \dfrac{6}{1 - \cos(\theta - \pi/4)}$ (See Exercise 9.)

24. $r = \dfrac{7}{7 + \sin(\theta - \pi/3)}$ (See Exercise 12.)

25. $r = \dfrac{8}{4 + 3 \sin(\theta + \pi/6)}$ (See Exercise 13.)

26. $r = \dfrac{5}{-1 + 2 \cos(\theta + 2\pi/3)}$ (See Exercise 16.)

In Exercises 27–38, find a polar equation of the conic with its focus at the pole.

Conic	Eccentricity	Directrix
27. Parabola	$e = 1$	$x = -1$
28. Parabola	$e = 1$	$y = -4$
29. Ellipse	$e = \frac{1}{2}$	$y = 1$
30. Ellipse	$e = \frac{3}{4}$	$y = -4$
31. Hyperbola	$e = 2$	$x = 1$
32. Hyperbola	$e = \frac{3}{2}$	$x = -1$

Conic	Vertex or Vertices
33. Parabola	$\left(1, -\dfrac{\pi}{2}\right)$
34. Parabola	$\left(10, \dfrac{\pi}{2}\right)$
35. Ellipse	$(2, 0), (8, \pi)$
36. Ellipse	$\left(2, \dfrac{\pi}{2}\right), \left(4, \dfrac{3\pi}{2}\right)$

Conic	*Vertex or Vertices*
37. Hyperbola	$\left(1, \dfrac{3\pi}{2}\right), \left(9, \dfrac{3\pi}{2}\right)$
38. Hyperbola	$\left(4, \dfrac{\pi}{2}\right), \left(-1, \dfrac{3\pi}{2}\right)$

39. *Planetary Motion*   The planets travel in elliptical orbits with the sun as a focus. Assume that the focus is at the pole, the major axis lies on the polar axis, and the length of the major axis is $2a$. Show that the polar equation of the orbit is

$$r = \frac{(1 - e^2)a}{1 - e \cos \theta},$$ where e is the eccentricity.

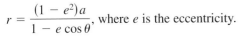
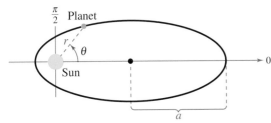

40. *Planetary Motion*   Use the result of Exercise 39 to show that the minimum distance (*perihelion*) from the sun to the planet is $r = a(1 - e)$ and that the maximum distance (*aphelion*) is $r = a(1 + e)$.

In Exercises 41 and 42, use the results of Exercises 39 and 40 to find the polar equation of the planet and the perihelion and aphelion distances.

41. Earth $a = 92.957 \times 10^6$ miles
$\qquad\qquad\; e = 0.0167$

42. Pluto $a = 5.900 \times 10^9$ kilometers
$\qquad\qquad\; e = 0.2444$

43. *Explorer 18*   On November 27, 1963, the United States launched *Explorer 18*. Its low and high points over the surface of earth were 119 miles and 122,800 miles, respectively. The center of earth is the focus of the orbit. Find the polar equation for the orbit and then find the distance between the surface of earth (assume a radius of 4000 miles) and the satellite when $\theta = 60°$.

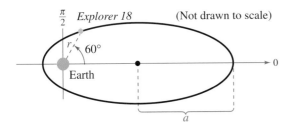

44. *Explorer 18*   In Exercise 43, find the distance between the surface of earth and the satellite when $\theta = 30°$.

Synthesis

True or False?   **In Exercises 45 and 46, determine whether the statement is true or false. Justify your answer.**

45. The graph of $r = 4/(-3 - 3 \sin \theta)$ has a horizontal directrix above the pole.

46. The conic represented by the following equation is an ellipse.

$$r^2 = \frac{16}{9 - 4 \cos \left(\theta + \dfrac{\pi}{4}\right)}$$

47. Show that the polar equation for the ellipse

$$\frac{x^2}{a^2} + \frac{y^2}{b^2} = 1 \quad \text{is} \quad r^2 = \frac{b^2}{1 - e^2 \cos^2 \theta}.$$

48. Show that the polar equation for the hyperbola

$$\frac{x^2}{a^2} - \frac{y^2}{b^2} = 1 \quad \text{is} \quad r^2 = \frac{-b^2}{1 - e^2 \cos^2 \theta}.$$

In Exercises 49 and 50, use the results of Exercises 47 and 48 to write the polar form of the equation of the conic.

49. $\dfrac{x^2}{169} + \dfrac{y^2}{144} = 1$ **50.** $\dfrac{x^2}{9} - \dfrac{y^2}{16} = 1$

Review

In Exercises 51–54, find the number of distinguishable permutations of the group of letters.

51. M, A, M, M, A, L   **52.** B, A, R, B, E, C, U, E
53. C, L, E, V, E, L, A, N, D
54. C, I, N, C, I, N, N, A, T, I

In Exercises 55–58, evaluate the expression. Do not use a calculator.

55. $_{12}C_9$ **56.** $_{18}C_{16}$
57. $_{10}P_3$ **58.** $_{29}P_2$

What did you learn?

Section 10.1	Review Exercises
☐ How to recognize a conic as the intersection of a plane and a double-napped cone	1, 2
☐ How to write equations of parabolas in standard form	3–6
☐ How to use the reflective property of parabolas to solve real-life problems	7–10

Section 10.2	
☐ How to write equations of ellipses in standard form	11–14
☐ How to use properties of ellipses to model and solve real-life problems	15, 16
☐ How to find eccentricities of ellipses	17–20

Section 10.3	
☐ How to write equations of hyperbolas in standard form	21–24
☐ How to find asymptotes of hyperbolas	25–28
☐ How to use properties of hyperbolas to solve real-life problems	29, 30
☐ How to classify conics from their general equations	31, 32

Section 10.4	
☐ How to rotate the coordinate axes to eliminate the xy-term in equations of conics	33–36
☐ How to use the discriminant to classify conics	37–40
☐ How to solve systems of quadratic equations	41, 42

Section 10.5	
☐ How to evaluate sets of parametric equations for given values of the parameter	43–46
☐ How to graph curves that are represented by sets of parametric equations and rewrite sets of parametric equations as single rectangular equations	47–54, 57, 58
☐ How to find sets of parametric equations for graphs	55, 56

Section 10.6	
☐ How to plot points in the polar coordinate system	59–64
☐ How to convert points from rectangular to polar form and vice versa	65–72
☐ How to convert equations from rectangular to polar form and vice versa	73–88

Section 10.7	
☐ How to graph a polar equation by point plotting	89–100
☐ How to use symmetry, zeros, and maximum r-values as graphing aids	101–106
☐ How to recognize special polar graphs	107–114

Section 10.8	
☐ How to define conics in terms of eccentricities	115–120
☐ How to write equations of conics in polar form	121–124
☐ How to use equations of conics in polar form to model real-life problems	125

10 Review Exercises

10.1 **In Exercises 1 and 2, state what type of conic is formed by the intersection of the plane and the double-napped cone.**

1.   **2.**

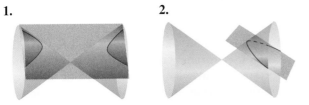

In Exercises 3–6, find the standard form of the equation of the parabola.

3. Vertex: $(4, 2)$

Focus: $(4, 0)$

4. Vertex: $(2, 0)$

Focus: $(0, 0)$

5. Vertex: $(0, 2)$

Directrix: $x = -3$

6. Vertex: $(2, 2)$

Directrix: $y = 0$

In Exercises 7 and 8, find an equation of a tangent line to the parabola at the given point and find the x-intercept of the line.

7. $x^2 = -2y$, $(2, -2)$

8. $x^2 = -2y$, $(-4, -8)$

9. *Parabolic Archway*   A parabolic archway is 12 meters high at the vertex. At a height of 10 meters, the width of the archway is 8 meters. How wide is the archway at ground level?

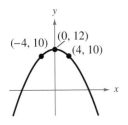

$(-4, 10)$ $(0, 12)$

$(4, 10)$

10. *Flashlight*   The light bulb in a flashlight is at the focus of its parabolic reflector, 1.5 centimeters from the vertex of the reflector. Write an equation for a cross section of the flashlight's reflector with its focus on the positive x-axis and its vertex at the origin.

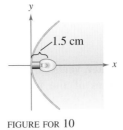

1.5 cm

FIGURE FOR 10

10.2 **In Exercises 11–14, find the standard form of the equation of the ellipse.**

11. Vertices: $(-3, 0), (7, 0)$; Foci: $(0, 0), (4, 0)$

12. Vertices: $(2, 0), (2, 4)$; Foci: $(2, 1), (2, 3)$

13. Vertices: $(0, 1), (4, 1)$;

Endpoints of the minor axis: $(2, 0), (2, 2)$

14. Vertices: $(-4, -1), (-4, 11)$;

Endpoints of the minor axis: $(-6, 5), (-2, 5)$

15. *Semielliptical Archway*   A semielliptical archway is set on pillars that are 10 feet apart. Its height (atop the pillars) is 4 feet. Where should the foci be placed in order to sketch the semielliptical arch?

16. *Wading Pool*   You are building a wading pool that is in the shape of an ellipse. Your plans give an equation for the elliptical shape of the pool measured in feet as

$$\frac{x^2}{324} + \frac{y^2}{196} = 1.$$

Find the longest distance across the pool, the shortest distance, and the distance between the foci.

In Exercises 17–20, find the center, vertices, foci, and eccentricity of the ellipse.

17. $16x^2 + 9y^2 - 32x + 72y + 16 = 0$

18. $4x^2 + 25y^2 + 16x - 150y + 141 = 0$

19. $\dfrac{(x + 2)^2}{81} + \dfrac{(y - 1)^3}{100} = 1$

20. $\dfrac{(x - 5)^2}{1} + \dfrac{(y + 3)^2}{36} = 1$

10.3 **In Exercises 21–24, find the standard form of the equation of the hyperbola.**

21. Vertices: $(-10, 3), (6, 3)$; Foci: $(-12, 3), (8, 3)$

22. Vertices: $(2, 2), (-2, 2)$; Foci: $(4, 2), (-4, 2)$

23. Foci: $(0, 0), (8, 0)$; Asymptotes: $y = \pm 2(x - 4)$

24. Foci: $(3, \pm 2)$; Asymptotes: $y = \pm 2(x - 3)$

In Exercises 25–28, find the center, vertices, foci, and the equations of the asymptotes of the hyperbola. Then sketch its graph.

25. $9x^2 - 16y^2 - 18x - 32y - 151 = 0$

26. $-4x^2 + 25y^2 - 8x + 150y + 121 = 0$

27. $\dfrac{(x - 3)^2}{16} - \dfrac{(y + 5)^2}{4} = 1$

28. $\dfrac{(y - 1)^2}{4} - x^2 = 1$

29. *Loran* A radio transmitting station A is located 200 miles east of transmitting station B. A ship is in an area to the north and 40 miles west of the station A. Synchronized radio pulses transmitted at 186,000 miles per second by the two stations are received 0.0005 seconds sooner from station A than from station B. How far north is the ship?

30. *Locating an Explosion* Two of your friends live 4 miles apart and on the same "east-west" street, and you live halfway between them. You are having a three-way phone conversation when you hear an explosion. Six seconds later your friend to the east hears the explosion, and your friend to the west hears it 8 seconds after you do. Find equations of two hyperbolas that would locate the explosion. (Sound travels at a rate of 1100 feet per second.)

In Exercises 31 and 32, classify the conic from its general equation.

31. $3x^2 + 2y^2 - 12x + 12y + 29 = 0$

32. $4x^2 - 4y^2 - 4x + 8y - 11 = 0$

10.4 **In Exercises 33–36, rotate the axes to eliminate the xy-term in the equation. Then write the equation in standard form. Sketch the graph of the equation, showing both sets of axes.**

33. $xy - 4 = 0$

34. $x^2 - 10xy + y^2 + 1 = 0$

35. $5x^2 - 2xy + 5y^2 - 12 = 0$

36. $4x^2 + 8xy + 4y^2 + 7\sqrt{2}x + 9\sqrt{2}y = 0$

In Exercises 37–40, (a) use the discriminant to classify the graph, (b) use the quadratic formula to solve for y, and then (c) use a graphing utility to graph the equation.

37. $16x^2 - 8xy + y^2 - 10x + 5y = 0$

38. $13x^2 - 8xy + 7y^2 - 45 = 0$

39. $x^2 + y^2 + 2xy + 2\sqrt{2}x - 2\sqrt{2}y + 2 = 0$

40. $x^2 - 4xy - 2y^2 - 6 = 0$

In Exercises 41 and 42, use any method to solve the system of quadratic equations algebraically. Then verify your results by using a graphing utility to graph the equations and find any points of intersection of the graphs.

41. $\begin{cases} -4x^2 - y^2 - 32x + 24y - 64 = 0 \\ 4x^2 + y^2 + 56x - 24y + 304 = 0 \end{cases}$

42. $\begin{cases} x^2 + y^2 - 25 = 0 \\ 9x - 4y^2 = 0 \end{cases}$

10.5 **In Exercises 43–46, evaluate the parametric equations $x = 3 \cos \theta$ and $y = 2 \sin^2 \theta$ for the given value of θ.**

43. $\theta = 0$

44. $\theta = \dfrac{\pi}{3}$

45. $\theta = \dfrac{\pi}{6}$

46. $\theta = -\dfrac{\pi}{4}$

In Exercises 47–54, sketch the curve represented by the parametric equations and, where possible, write the corresponding rectangular equation by eliminating the parameter. Verify your result with a graphing utility.

47. $x = 1 + 4t$
$y = 2 - 3t$

48. $x = t + 4$
$y = t^2$

49. $x = \dfrac{1}{t}$
$y = t^2$

50. $x = \dfrac{1}{t}$
$y = 2t + 3$

51. $x = 6 \cos \theta$
$y = 6 \sin \theta$

52. $x = 3 + 3 \cos \theta$
$y = 2 + 5 \sin \theta$

53. $x = \sec \theta$
$y = \tan \theta$

54. $x = 2\theta - \sin \theta$
$y = 2 - \cos \theta$

55. Find a parametric representation of the ellipse with center at $(-3, 4)$, major axis horizontal and eight units in length, and minor axis six units in length.

56. Find a parametric representation of the hyperbola with vertices $(0, \pm 4)$ and foci $(0, \pm 5)$.

57. *Rotary Engine* The rotary engine was developed by Felix Wankel in the 1950s. It features a rotor that is basically a modified equilateral triangle. The rotor moves in a chamber that, in two dimensions, is an epitrochoid. Use a graphing utility to graph the chamber modeled by the parametric equations $x = \cos 3\theta + 5 \cos \theta$ and $y = \sin 3\theta + 5 \sin \theta$.

58. *Involute of a Circle* The involute of a circle is described by the endpoint P of a string that is held taut as it is unwound from a spool. The spool does not rotate. Use a graphing utility to graph the involute of a circle modeled by the parametric equations. Use $r = 2$. Describe your viewing window.

$$x = r(\cos \theta + \theta \sin \theta)$$
$$y = r(\sin \theta - \theta \cos \theta)$$

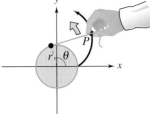

10.6 In Exercises 59–64, plot the point in the polar coordinate system and find three additional polar representations of the point, using $-2\pi < \theta < 2\pi$.

59. $\left(1, \dfrac{5\pi}{4}\right)$ **60.** $\left(-2, \dfrac{7\pi}{6}\right)$ **61.** $\left(-\dfrac{5}{2}, \dfrac{\pi}{6}\right)$

62. $\left(\dfrac{3}{4}, -\dfrac{\pi}{3}\right)$ **63.** $\left(\sqrt{5}, \dfrac{4\pi}{3}\right)$ **64.** $\left(\sqrt{10}, \dfrac{3\pi}{4}\right)$

In Exercises 65–68, plot the points given in polar coordinates and find the corresponding rectangular coordinates for the point.

65. $\left(5, -\dfrac{7\pi}{6}\right)$ **66.** $\left(-3, \dfrac{2\pi}{3}\right)$

67. $\left(12, -\dfrac{\pi}{2}\right)$ **68.** $\left(\sqrt{3}, 1.78\right)$

In Exercises 69–72, the rectangular coordinates of a point are given. Find two sets of polar coordinates for the point for $0 \le \theta < 2\pi$.

69. $(0, -9)$ **70.** $(-6, 8)$

71. $(5, -5)$ **72.** $\left(-3, -\sqrt{3}\right)$

In Exercises 73–80, convert the polar equation to rectangular form.

73. $r = 5$ **74.** $r = 12$

75. $r = 3 \cos \theta$ **76.** $r = 8 \sin \theta$

77. $r^2 = \cos 2\theta$ **78.** $r^2 = \sin \theta$

79. $r = \dfrac{2}{2 - \sin \theta}$ **80.** $r = \dfrac{10}{4 - 7 \cos \theta}$

In Exercises 81–88, convert the rectangular equation to polar form.

81. $x^2 + y^2 = 9$ **82.** $x^2 + y^2 = 20$

83. $y = 6$ **84.** $x = 14$

85. $x^2 + y^2 - 4x = 0$ **86.** $x^2 + y^2 - 6y = 0$

87. $xy = 5$ **88.** $xy = -2$

10.7 In Exercises 89–96, sketch the graph of the polar equation. Use a graphing utility to verify your graph.

89. $r = 5$ **90.** $r = 11$

91. $\theta = \dfrac{\pi}{2}$ **92.** $\theta = -\dfrac{5\pi}{6}$

93. $r = 5 \cos \theta$ **94.** $r = 2 \sin \theta$

95. $r = -2(1 + \cos \theta)$ **96.** $r = 4 - 3 \cos \theta$

In Exercises 97–100, use a graphing utility to graph the polar equation. Describe the graph.

97. $r^2 = 4 \sin^2 2\theta$ **98.** $r^2 = 9 \cos^2 2\theta$

99. $r = \dfrac{3}{\cos\left(\theta - \dfrac{\pi}{4}\right)}$ **100.** $r = \dfrac{-4}{\sin\left(\theta + \dfrac{2\pi}{3}\right)}$

In Exercises 101–106, sketch the graph of the polar equation. Identify any symmetry, maximum r-values, and zeros of r. Use a graphing utility to verify your graph.

101. $r = 5 + 4 \cos \theta$ **102.** $r = 3 - 5 \sin \theta$

103. $r = -3 \cos 2\theta$ **104.** $r = \cos 5\theta$

105. $r^2 = \cos 2\theta$ **106.** $r^2 = 5 \sin 2\theta$

In Exercises 107–114, identify and sketch the graph of the polar equation. Use a graphing utility to verify your graph.

107. $r = 6 - \cos \theta$

108. $r = 7(1 - \sin \theta)$

109. $r = 10 + 15 \sin \theta$

110. $r = 8 + 3 \cos \theta$

111. $r = 4 \sin 5\theta$

112. $r = 2 \cos 3\theta$

113. $r^2 = 8 \cos 2\theta$

114. $r^2 = 36 \sin 2\theta$

10.7 **In Exercises 115–120, identify and sketch the graph of the polar equation. Use a graphing utility to verify your graph.**

115. $r = \dfrac{2}{1 - \sin \theta}$

116. $r = \dfrac{1}{1 + 2 \sin \theta}$

117. $r = \dfrac{4}{5 - 3 \cos \theta}$

118. $r = \dfrac{6}{-1 + 4 \cos \theta}$

119. $r = \dfrac{5}{6 + 2 \sin \theta}$

120. $r = \dfrac{3}{4 - 4 \cos \theta}$

In Exercises 121–125, find a polar equation for the conic.

121. Circle — Center: $(8, \pi/2)$; Solution point: $(0, 0)$

122. Parabola — Vertex: $(2, \pi/2)$; Focus: $(0, 0)$

123. Ellipse — Vertices: $(5, 0), (1, \pi)$; One focus: $(0, 0)$

124. Hyperbola — Vertices: $(1, 0), (7, 0)$; One focus: $(0, 0)$

125. *Planetary Motion* The planet Mars has an elliptical orbit with an eccentricity of $e \approx 0.092$. The length of the major axis of the orbit is approximately 3.05 astronomical units. Find a polar equation for the orbit and its perihelion and aphelion distances.

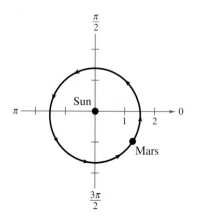

Synthesis

True or False? **In Exercises 126–128, determine whether the statement is true or false. Justify your answer.**

126. The graph of $x^2/4 - y^4 = 1$ is a hyperbola.

127. There is only one set of parametric equations that represents the line $y = 3 - 2x$.

128. There is a unique polar coordinate representation of each point in the plane.

Writing **In Exercises 129 and 130, an equation and four variations are given. In your own words, describe how the graph of each of the variations differs from the graph of the first equation.**

129. $y^2 = 8x$

(a) $(y - 2)^2 = 8x$ (b) $y^2 = 8(x + 1)$

(c) $y^2 = -8x$ (d) $y^2 = 4x$

130. $\dfrac{x^2}{4} + \dfrac{y^2}{9} = 1$

(a) $\dfrac{x^2}{9} + \dfrac{y^2}{4} = 1$ (b) $\dfrac{x^2}{4} + \dfrac{y^2}{4} = 1$

(c) $\dfrac{x^2}{4} + \dfrac{y^2}{25} = 1$ (d) $\dfrac{(x - 3)^2}{4} + \dfrac{y^2}{9} = 1$

131. Consider an ellipse whose major axis is horizontal and 10 units in length. The number b in the standard form of the ellipse must be less than what real number? Explain the change in the shape of the ellipse as b approaches this number.

132. The curve represented by the parametric equations $x = 2 \sec t$ and $y = 3 \tan t$ is shown in the graph. Would the curve change for the equations $x = 2 \sec(-t)$ and $y = 3 \tan(-t)$? If so, how would the graph change?

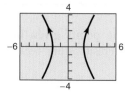

Chapter Project *Polar, Rectangular, and Parametric Forms*

In this project, you will compare the polar, rectangular, and parametric equations for conics.

a. Consider the polar equation

$$r = \frac{15}{3 - 2\cos\theta}.$$

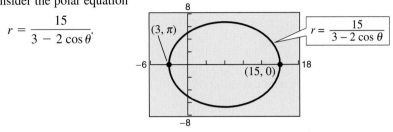

In Section 10.8, you learned that the graph of this polar equation is an ellipse and that one of the ellipse's foci is at the pole, as shown in the graph above. To find the rectangular equation for this ellipse, begin by rewriting the equation as

$$3r - 2r\cos\theta = 15.$$

Then use the substitutions $r = \sqrt{x^2 + y^2}$ and $r\cos\theta = x$ to find the rectangular equation. After you find the rectangular equation, write it in the standard form

$$\frac{(x - h)^2}{a^2} + \frac{(y - k)^2}{b^2} = 1.$$

b. Use the standard form of the ellipse in part (a) to find the center, foci, and eccentricity of the ellipse. Compare your results with those obtained in Example 1 in Section 10.8. Use the function mode of a graphing utility to graph the ellipse.

c. Use the standard form of the ellipse and the identity $\sin^2\theta + \cos^2\theta = 1$ to write parametric equations for the ellipse. Then use the parametric mode of a graphing utility to graph the ellipse.

Questions for Further Exploration

1. Sketch the graph of the polar equation

$$r = \frac{32}{3 + 5\sin\theta}.$$

Which type of conic is this?

2. Write the standard form of the rectangular equation of the conic in Question 1. Then sketch the graph using the function mode of a graphing utility. Does your graph agree with the graph obtained in Question 1? Which of the two graphs is easier to obtain? Explain your reasoning.

3. Write parametric equations for the conic given in Question 1. Then use the parametric mode of a graphing utility to graph the conic. Does your graph agree with the graph obtained in Question 1?

4. Consider the parametric equations

$$x = 4 + 5\cos t \qquad \text{and} \qquad y = 3\sin t.$$

With appropriate scaling of the coordinate system, could these equations represent the path of a comet about the sun? Explain your reasoning.

10   Chapter Test

Take this test as you would take a test in class. After you are done, check your work against the answers given in the back of the book.

The *Interactive* CD-ROM and *Internet* versions of this text provide answers to the Chapter Tests and Cumulative Tests. They also offer Chapter Pre-Tests (which test key skills and concepts covered in previous chapters) and Chapter Post Tests, both of which have randomly generated exercises with diagnostic capabilities.

In Exercises 1–4, graph the conic and identify any vertices and foci.

1. $y^2 - 8x = 0$

2. $\dfrac{(x-3)^2}{81} + \dfrac{(y-2)^2}{100} = 1$

3. $x^2 - 6x + 2y + 9 = 0$

4. $x^2 - 4y^2 - 4x = 0$

5. Find the equation and sketch the graph of a parabola with a focus at $(8, -2)$ and directrix at $x = 4$.

6. Find an equation of the ellipse shown at the right.

7. Find an equation of the hyperbola with vertices $(0, \pm 3)$ and asymptotes $y = \pm \frac{3}{2}x$.

8. (a) Determine the number of degrees the axis must be rotated to eliminate the xy-term of the conic $x^2 + 6xy + y^2 - 6 = 0$.

    (b) Graph the conic and use a graphing utility to confirm your result.

9. Solve the system of quadratic equations $x^2 + 2y^2 - 4x + 6y - 5 = 0$ and $x + y + 5 = 0$ algebraically by the method of substitution. Then verify your results by using a graphing utility to graph the equations and find any points of intersection of the graphs.

10. Sketch the curve represented by the parametric equations $x = \sqrt{t^2 + 2}$ and $y = t/4$. Eliminate the parameter and write the corresponding rectangular equation.

11. Find two different sets of parametric equations for the rectangular equation $y = \frac{1}{4}x - 5$.

12. Convert the polar coordinate $\left(-14, \dfrac{5\pi}{3}\right)$ to rectangular form.

13. Convert the rectangular coordinate $(-2, -2)$ to polar form and find two additional representations for this point.

14. Convert the rectangular equation $x^2 + y^2 - 12y = 0$ to polar form.

In Exercises 15–17, sketch the graph of the polar equation and identify the type of graph.

15. $r = 2 + 3\sin\theta$ 16. $r = 8\cos 3\theta$ 17. $r = \dfrac{8}{4 + 6\sin\theta}$

18. Find a polar equation of an ellipse with its focus at the pole, an eccentricity of $e = \frac{1}{4}$, and directrix at $y = 4$.

19. Mercury orbits the sun in an elliptical path with the center of the sun at one focus. The eccentricity of Mercury's orbit is $e = 0.2056$. The length of the major axis of the orbit is 72 million miles. Find a polar equation for the orbit and its perihelion and aphelion distances.

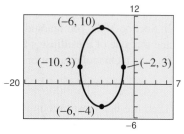

FIGURE FOR 6

7–10 Cumulative Test

Take this test to review the material from earlier chapters. After you are done, check your work against the answers in the back of the book.

In Exercises 1 and 2, solve the system of equations. Use Gaussian elimination with back substitution or Gauss-Jordan elimination.

1. $\begin{cases} -x - 3y = 5 \\ 4x + 2y = 10 \end{cases}$

2. $\begin{cases} 2x - 3y + z = 13 \\ -4x + y - 2z = -6 \\ x - 3y + 3z = 12 \end{cases}$

In Exercises 3–6, perform the matrix operations given

$$A = \begin{bmatrix} -3 & 0 & -4 \\ 2 & 4 & 5 \\ -4 & 8 & 1 \end{bmatrix} \quad \text{and} \quad B = \begin{bmatrix} -1 & 5 & 2 \\ 6 & -3 & 3 \\ 0 & 4 & -2 \end{bmatrix}.$$

3. $3A - 2B$ **4.** $5A + 3B$ **5.** AB **6.** BA

7. Find (a) the inverse of A (if it exists), and (b) the determinant of A. Use a graphing utility to check your answers.

$$A = \begin{bmatrix} 1 & 2 & -1 \\ 3 & 7 & -10 \\ -5 & -7 & -15 \end{bmatrix}$$

8. Use an inverse matrix to solve (if possible) the system on the right.

$\begin{cases} x - 3y - 2z = 8 \\ -2x + 7y + 3z = -19 \\ x - y - 3z = 3 \end{cases}$

FOR EXERCISE 8

9. Use a determinant to find the area of the triangle with vertices $(0, 0)$, $(6, 2)$, and $(8, 10)$.

10. Find the sum of the first 20 terms of the arithmetic sequence 8, 13, 18, 23, ...

In Exercises 11–14, find the sum. Use a graphing utility to verify your answer.

11. $\sum_{k=1}^{6} (7k - 2)$ **12.** $\sum_{k=1}^{4} \frac{2}{k^2 + 4}$ **13.** $\sum_{n=0}^{10} 9\left(\frac{3}{4}\right)^n$ **14.** $\sum_{n=1}^{\infty} 8(0.9)^{n-1}$

15. Find the sum of the integers from 40 to 94 (inclusive).

16. Use mathematical induction to prove the formula
$3 + 7 + 11 + 15 + \ldots + (4n - 1) = n(2n + 1)$.

17. Use the Binomial Theorem to expand the expression $(x - 2y)^6$.

18. Find the number of distinguishable permutations of the group of letters: A, N, T, A, R, C, T, I, C, A.

In Exercises 19–22, identify the conic and sketch its graph.

19. $\dfrac{(y + 3)^2}{36} - \dfrac{(x - 5)^2}{121} = 1$

20. $\dfrac{(x - 2)^2}{4} + \dfrac{(y + 1)^2}{9} = 1$

21. $5y^2 - 4x^2 = 20$ **22.** $x^2 + y^2 - 2x - 4y - 4 = 0$

In Exercises 23–25, find the standard form of the equation of the graph of the conic.

23.   **24.**   **25.**

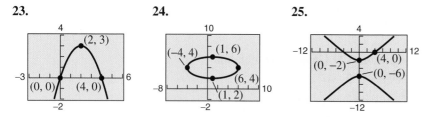

26. Find the standard form of the equation of the hyperbola with foci $(0, 0)$ and $(0, 4)$, and asymptotes $y = \pm\frac{1}{2}x + 2$.

27. Use a graphing utility to graph $x^2 - 4xy + 2y^2 = 6$. Determine the angle θ through which the axes are rotated.

28. Solve the system of quadratic equations algebraically by the method of elimination. Then verify your results by using a graphing utility to graph the equations and find any points of intersection of the graphs.

$$\begin{cases} x^2 - y^2 - 12x + 12y - 36 = 0 \\ x^2 + y^2 - 12x - 12y + 36 = 0 \end{cases}$$

29. Sketch the curve represented by the parametric equations $x = 4 \ln t$ and $y = \frac{1}{2}t^2$. Then eliminate the parameter and write the corresponding rectangular equation whose graph represents the curve.

In Exercises 30–33, plot the point in polar coordinates and find three additional polar representations for $-2\pi < \theta < 2\pi$.

30. $\left(8, \dfrac{5\pi}{6}\right)$ **31.** $\left(5, -\dfrac{3\pi}{4}\right)$ **32.** $\left(-2, \dfrac{5\pi}{4}\right)$ **33.** $\left(-3, -\dfrac{11\pi}{6}\right)$

34. Convert the rectangular equation $-8x - 3y + 5 = 0$ to polar form.

35. Convert the polar equation $r = \dfrac{2}{4 - 5 \cos \theta}$ to rectangular form.

In Exercises 36–38, sketch and identify the graph of the polar equation.

36. $\theta = -\dfrac{\pi}{6}$ **37.** $r = 3 - 2 \sin \theta$ **38.** $r = 2 + 5 \cos \theta$

39. Find a polar equation of the ellipse with vertices at $(14, 0)$ and $(2, \pi)$ and one focus at the pole.

40. A personnel manager has 15 applicants to fill four different positions. In how many ways can this be done, assuming that all the applicants are qualified for any of the four positions?

41. Use a graphing utility to determine the number of subsets of six elements that can be formed from a set of 72 elements.

42. On a game show, the digits 3, 4, and 5 must be arranged in the proper order to form the price of an appliance. If they are arranged correctly, the contestant wins the appliance. What is the probability of winning if the contestant knows that the price is at least $400?

Appendix A Proofs of Selected Theorems

Section P.1, page 57

The Midpoint Formula

The midpoint of the segment joining the points (x_1, y_1) and (x_2, y_2) is given by the Midpoint Formula

$$\text{Midpoint} = \left(\frac{x_1 + x_2}{2}, \frac{y_1 + y_2}{2}\right).$$

Proof

Using the figure, you must show that

$$d_1 = d_2 \quad \text{and} \quad d_1 + d_2 = d_3.$$

By the Distance Formula, you obtain

$$d_1 = \sqrt{\left(\frac{x_1 + x_2}{2} - x_1\right)^2 + \left(\frac{y_1 + y_2}{2} - y_1\right)^2} = \frac{1}{2}\sqrt{(x_2 - x_1)^2 + (y_2 - y_1)^2}$$

$$d_2 = \sqrt{\left(x_2 - \frac{x_1 + x_2}{2}\right)^2 + \left(y_2 - \frac{y_1 + y_2}{2}\right)^2} = \frac{1}{2}\sqrt{(x_2 - x_1)^2 + (y_2 - y_1)^2}$$

$$d_3 = \sqrt{(x_2 - x_1)^2 + (y_2 - y_1)^2}.$$

So, it follows that $d_1 = d_2$ and $d_1 + d_2 = d_3$.

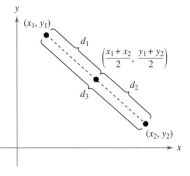

Midpoint Formula

Section 2.3, page 164

The Remainder Theorem

If a polynomial $f(x)$ is divided by $x - k$, the remainder is

$$r = f(k).$$

Proof

From the Division Algorithm, you have

$$f(x) = (x - k)q(x) + r(x)$$

and because either $r(x) = 0$ or the degree of $r(x)$ is less than the degree of $x - k$, you know that $r(x)$ must be a constant. That is, $r(x) = r$. Now, by evaluating $f(x)$ at $x = k$, you have

$$f(k) = (k - k)q(k) + r = (0)q(k) + r = r.$$

Section 2.3, page 164

The Factor Theorem

A polynomial $f(x)$ has a factor $(x - k)$ if and only if $f(k) = 0$.

Proof

Using the Division Algorithm with the factor $(x - k)$, you have

$$f(x) = (x - k)q(x) + r(x).$$

By the Remainder Theorem, $r(x) = r = f(k)$, and you have

$$f(x) = (x - k)q(x) + f(k)$$

where $q(x)$ is a polynomial of lesser degree than $f(x)$. If $f(k) = 0$, then

$$f(x) = (x - k)q(x)$$

and you see that $(x - k)$ is a factor of $f(x)$. Conversely, if $(x - k)$ is a factor of $f(x)$, division of $f(x)$ by $(x - k)$ yields a remainder of 0. So, by the Remainder Theorem, you have $f(k) = 0$.

Section 2.5, page 182

Linear Factorization Theorem

If $f(x)$ is a polynomial of degree n, where $n > 0$, then f has precisely n linear factors

$$f(x) = a_n(x - c_1)(x - c_2) \cdots (x - c_n)$$

where $c_1, c_2, \ldots, c_n$ are complex numbers.

Proof

Using the Fundamental Theorem of Algebra, you know that f must have at least one zero, c_1. Consequently, $(x - c_1)$ is a factor of $f(x)$, and you have

$$f(x) = (x - c_1)f_1(x).$$

If the degree of $f_1(x)$ is greater than zero, you again apply the Fundamental Theorem to conclude that f_1 must have a zero c_2, which implies that

$$f(x) = (x - c_1)(x - c_2)f_2(x).$$

It is clear that the degree of $f_1(x)$ is $n - 1$, that the degree of $f_2(x)$ is $n - 2$, and that you can repeatedly apply the Fundamental Theorem n times until you obtain

$$f(x) = a_n(x - c_1)(x - c_2) \cdots (x - c_n)$$

where a_n is the leading coefficient of the polynomial $f(x)$.

Section 2.5, page 184

Factors of a Polynomial

Every polynomial of degree $n > 0$ with real coefficients can be written as the product of linear and quadratic factors with real coefficients, where the quadratic factors have no real zeros.

Proof

To begin, use the Linear Factorization Theorem to conclude that $f(x)$ can be completely factored in the form

$$f(x) = d(x - c_1)(x - c_2)(x - c_3) \cdots (x - c_n).$$

If each c_i is real, there is nothing more to prove. If any c_i is complex ($c_i = a + bi$, $b \neq 0$), then, because the coefficients of $f(x)$ are real, you know that the conjugate $c_j = a - bi$ is also a zero. By multiplying the corresponding factors, you obtain

$$(x - c_i)(x - c_j) = [x - (a + bi)][x - (a - bi)]$$
$$= x^2 - 2ax + (a^2 + b^2)$$

where each coefficient is real.

Section 3.3, page 241

Properties of Logarithms

Let a be a positive number such that $a \neq 1$, and let n be a real number. If u and v are positive real numbers, the following properties are true.

1. $\log_a(uv) = \log_a u + \log_a v$ **1.** $\ln(uv) = \ln u + \ln v$

2. $\log_a \dfrac{u}{v} = \log_a u - \log_a v$ **2.** $\ln \dfrac{u}{v} = \ln u - \ln v$

3. $\log_a u^n = n \log_a u$ **3.** $\ln u^n = n \ln u$

Proof

To prove Property 1, let

$$x = \log_a u \quad \text{and} \quad y = \log_a v.$$

The corresponding exponential forms of these two equations are

$$a^x = u \quad \text{and} \quad a^y = v.$$

Multiplying u and v produces $uv = a^x a^y = a^{x+y}$. The corresponding logarithmic form of $uv = a^{x+y}$ is $\log_a(uv) = x + y$. So, $\log_a(uv) = \log_a u + \log_a v$.

Section 5.4, page 404

Sum and Difference Formulas

$\sin(u + v) = \sin u \cos v + \cos u \sin v$

$\sin(u - v) = \sin u \cos v - \cos u \sin v$

$\cos(u + v) = \cos u \cos v - \sin u \sin v$

$\cos(u - v) = \cos u \cos v + \sin u \sin v$

$\tan(u + v) = \dfrac{\tan u + \tan v}{1 - \tan u \tan v}$

$\tan(u - v) = \dfrac{\tan u - \tan v}{1 + \tan u \tan v}$

Proof

Here are proofs for the formulas for $\cos(u \pm v)$. In the top figure, let A be the point $(1, 0)$ and then use u and v to locate the points $B = (x_1, y_1)$, $C = (x_2, y_2)$, and $D = (x_3, y_3)$ on the unit circle. So, $x_i^2 + y_i^2 = 1$ for $i = 1, 2$, and 3. For convenience, assume that $0 < v < u < 2\pi$. From the bottom figure, note that arcs AC and BD have the same length. So, line segments AC and BD are also equal in length, which implies that

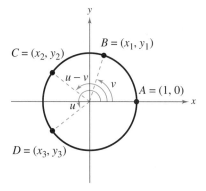

$$\sqrt{(x_2 - 1)^2 + (y_2 - 0)^2} = \sqrt{(x_3 - x_1)^2 + (y_3 - y_1)^2}$$

$$x_2^2 - 2x_2 + 1 + y_2^2 = x_3^2 - 2x_1x_3 + x_1^2 + y_3^2 - 2y_1y_3 + y_1^2$$

$$(x_2^2 + y_2^2) + 1 - 2x_2 = (x_3^2 + y_3^2) + (x_1^2 + y_1^2) - 2x_1x_3 - 2y_1y_3$$

$$1 + 1 - 2x_2 = 1 + 1 - 2x_1x_3 - 2y_1y_3$$

$$x_2 = x_3x_1 + y_3y_1.$$

Finally, by substituting the values $x_2 = \cos(u - v)$, $x_3 = \cos u$, $x_1 = \cos v$, $y_3 = \sin u$, and $y_1 = \sin v$, you obtain

$$\cos(u - v) = \cos u \cos v + \sin u \sin v.$$

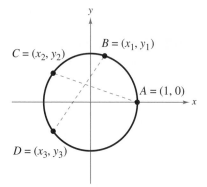

The formula for $\cos(u + v)$ can be established by considering $u + v = u - (-v)$ and using the formula just derived to obtain

$$\cos(u + v) = \cos[u - (-v)]$$

$$= \cos u \cos(-v) + \sin u \sin(-v)$$

$$= \cos u \cos v - \sin u \sin v.$$

Section 5.5, page 411

Double-Angle Formulas

$\sin 2u = 2 \sin u \cos u$

$\cos 2u = \cos^2 u - \sin^2 u$

$\quad\quad = 2 \cos^2 u - 1$

$\quad\quad = 1 - 2 \sin^2 u$

$\tan 2u = \dfrac{2 \tan u}{1 - \tan^2 u}$

Proof

To prove the first formula, let $v = u$ in the formula for $\sin(u + v)$.

$$\sin 2u = \sin(u + u)$$
$$= \sin u \cos u + \cos u \sin u$$
$$= 2 \sin u \cos u$$

To prove the second formula, let $v = u$ in the formula for $\cos(u + v)$.

$$\cos 2u = \cos(u + u)$$
$$= \cos u \cos u - \sin u \sin u$$
$$= \cos^2 u - \sin^2 u$$

The tangent double-angle formula can be proven in a similar way.

Section 5.5, page 413

Power-Reducing Formulas

$$\sin^2 u = \frac{1 - \cos 2u}{2} \qquad \cos^2 u = \frac{1 + \cos 2u}{2} \qquad \tan^2 u = \frac{1 - \cos 2u}{1 + \cos 2u}$$

Proof

The first two formulas can be verified by solving for $\sin^2 u$ and $\cos^2 u$, respectively, in the double-angle formulas

$$\cos 2u = 1 - 2 \sin^2 u \qquad \text{and} \qquad \cos 2u = 2 \cos^2 u - 1.$$

The third formula can be verified using the fact that

$$\tan^2 u = \frac{\sin^2 u}{\cos^2 u}.$$

Section 5.5, page 416

Sum-to-Product Formulas

$$\sin x + \sin y = 2 \sin\left(\frac{x + y}{2}\right) \cos\left(\frac{x - y}{2}\right)$$

$$\sin x - \sin y = 2 \cos\left(\frac{x + y}{2}\right) \sin\left(\frac{x - y}{2}\right)$$

$$\cos x + \cos y = 2 \cos\left(\frac{x + y}{2}\right) \cos\left(\frac{x - y}{2}\right)$$

$$\cos x - \cos y = -2 \sin\left(\frac{x + y}{2}\right) \sin\left(\frac{x - y}{2}\right)$$

Proof

To prove the first formula, let $x = u + v$ and $y = u - v$. Then substitute $u = (x + y)/2$ and $v = (x - y)/2$ in the product-to-sum formula.

$$\sin u \cos v = \frac{1}{2}[\sin(u + v) + \sin(u - v)]$$

$$\sin\left(\frac{x + y}{2}\right)\cos\left(\frac{x - y}{2}\right) = \frac{1}{2}(\sin x + \sin y)$$

$$2\sin\left(\frac{x + y}{2}\right)\cos\left(\frac{x - y}{2}\right) = \sin x + \sin y$$

Section 6.1, page 428

Law of Sines

If ABC is a triangle with sides a, b, and c, then

$$\frac{a}{\sin A} = \frac{b}{\sin B} = \frac{c}{\sin C}.$$

Oblique Triangles

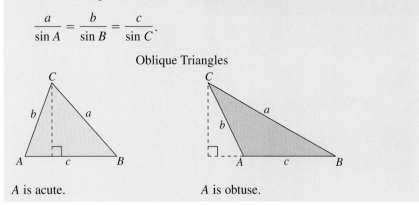

A is acute.   A is obtuse.

Proof

Let h be the altitude of either triangle found in the figure above. Then you have

$$\sin A = \frac{h}{b} \qquad \text{or} \qquad h = b \sin A$$

$$\sin B = \frac{h}{a} \qquad \text{or} \qquad h = a \sin B.$$

Equating these two values of h, you have

$$a \sin B = b \sin A \qquad \text{or} \qquad \frac{a}{\sin A} = \frac{b}{\sin B}.$$

Note that $\sin A \neq 0$ and $\sin B \neq 0$ because no angle of a triangle can have a measure of $0°$ or $180°$. In a similar manner, by constructing an altitude from vertex B to side AC (extended), you can show that

$$\frac{a}{\sin A} = \frac{c}{\sin C}.$$

So, the Law of Sines is established.

Section 6.2, page 437

Law of Cosines

Standard Form

$a^2 = b^2 + c^2 - 2bc \cos A$

$b^2 = a^2 + c^2 - 2ac \cos B$

$c^2 = a^2 + b^2 - 2ab \cos C$

Alternative Form

$\cos A = \dfrac{b^2 + c^2 - a^2}{2bc}$

$\cos B = \dfrac{a^2 + c^2 - b^2}{2ac}$

$\cos C = \dfrac{a^2 + b^2 - c^2}{2ab}$

Proof

Consider a triangle that has three acute angles, as shown in the figure. Note that vertex B has coordinates $(c, 0)$. Furthermore, C has coordinates (x, y), where $x = b \cos A$ and $y = b \sin A$. Because a is the distance from vertex C to vertex B, it follows that

$a = \sqrt{(x - c)^2 + (y - 0)^2}$

$a^2 = (b \cos A - c)^2 + (b \sin A)^2$

$a^2 = b^2 \cos^2 A - 2bc \cos A + c^2 + b^2 \sin^2 A$

$a^2 = b^2(\sin^2 A + \cos^2 A) + c^2 - 2ab \cos A$

$a^2 = b^2 + c^2 - 2bc \cos A.$ $\qquad \sin^2 A + \cos^2 A = 1$

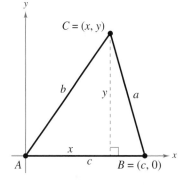

A similar argument can be used for a triangle having an obtuse angle.

Section 6.2, page 440

Heron's Area Formula

Given any triangle with sides of lengths a, b, and c, the area of the triangle is

$\text{Area} = \sqrt{s(s - a)(s - b)(s - c)}$

where $s = (a + b + c)/2$.

Proof

You know that

$$\text{Area} = \frac{1}{2} bc \sin A$$

$$= \sqrt{\frac{1}{4} b^2 c^2 \sin^2 A}$$

$$= \sqrt{\frac{1}{4} b^2 c^2 (1 - \cos^2 A)}$$

$$= \sqrt{\left[\frac{1}{2} bc(1 + \cos A) \right] \left[\frac{1}{2} bc(1 - \cos A) \right]}.$$

Using the Law of Cosines, you can show that

$$\frac{1}{2} bc(1 + \cos A) = \frac{a + b + c}{2} \cdot \frac{-a + b + c}{2}$$

and

$$\frac{1}{2} bc(1 - \cos A) = \frac{a - b + c}{2} \cdot \frac{a + b - c}{2}.$$

Letting $s = (a + b + c)/2$, these two equations can be rewritten as

$$\frac{1}{2} bc(1 + \cos A) = s(s - a)$$

and

$$\frac{1}{2} bc(1 - \cos A) = (s - b)(s - c).$$

So, you can conclude that

$$\text{Area} = \sqrt{s(s - a)(s - b)(s - c)}.$$

Section 6.4, page 458

Properties of the Dot Product

Let $\mathbf{u}$, $\mathbf{v}$, and $\mathbf{w}$ be vectors in the plane or in space and let c be a scalar.

1. $\mathbf{u} \cdot \mathbf{v} = \mathbf{v} \cdot \mathbf{u}$

2. $\mathbf{0} \cdot \mathbf{v} = 0$

3. $\mathbf{u} \cdot (\mathbf{v} + \mathbf{w}) = \mathbf{u} \cdot \mathbf{v} + \mathbf{u} \cdot \mathbf{w}$

4. $\mathbf{v} \cdot \mathbf{v} = \|\mathbf{v}\|^2$

5. $c(\mathbf{u} \cdot \mathbf{v}) = c\mathbf{u} \cdot \mathbf{v} = \mathbf{u} \cdot c\mathbf{v}$

Proof

To prove Property 1, let $\mathbf{u} = \langle u_1, u_2 \rangle$ and $\mathbf{v} = \langle v_1, v_2 \rangle$. Then

$$\mathbf{u} \cdot \mathbf{v} = u_1 v_1 + u_2 v_2 = v_1 u_1 + v_2 u_2 = \mathbf{v} \cdot \mathbf{u}.$$

To prove Property 4, let $\mathbf{v} = \langle v_1, v_2 \rangle$. Then

$$\mathbf{v} \cdot \mathbf{v} = v_1{}^2 + v_2{}^2 = \left(\sqrt{v_1{}^2 + v_2{}^2} \right)^2 = \|\mathbf{v}\|^2.$$

To prove the fifth property, let $\mathbf{u} = \langle u_1, u_2 \rangle$ and $\mathbf{v} = \langle v_1, v_2 \rangle$ and let c be a scalar. Then,

$$\begin{aligned}
c(\mathbf{u} \cdot \mathbf{v}) &= c(\langle u_1, u_2 \rangle \cdot \langle v_1, v_2 \rangle) \\
&= c(u_1 v_1 + u_2 v_2) \\
&= (cu_1)v_1 + (cu_2)v_2 \\
&= \langle cu_1, cu_2 \rangle \cdot \langle v_1, v_2 \rangle = c\mathbf{u} \cdot \mathbf{v} \\
&= u_1(cv_1) + u_2(cv_2) \\
&= \langle u_1, u_2 \rangle \cdot \langle cv_1, cv_2 \rangle = \mathbf{u} \cdot c\mathbf{v}.
\end{aligned}$$

Section 6.4, page 459

Angle Between Two Vectors

If θ is the angle between two nonzero vectors $\mathbf{u}$ and $\mathbf{v}$, then

$$\cos \theta = \frac{\mathbf{u} \cdot \mathbf{v}}{\|\mathbf{u}\| \, \|\mathbf{v}\|}.$$

Proof

Consider the triangle determined by vectors $\mathbf{u}$, $\mathbf{v}$, and $\mathbf{v} - \mathbf{u}$, as shown in the figure. By the Law of Cosines, you can write

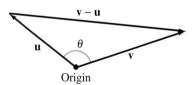

$$\|\mathbf{v} - \mathbf{u}\|^2 = \|\mathbf{u}\|^2 + \|\mathbf{v}\|^2 - 2\|\mathbf{u}\| \, \|\mathbf{v}\| \cos \theta$$

$$(\mathbf{v} - \mathbf{u}) \cdot (\mathbf{v} - \mathbf{u}) = \|\mathbf{u}\|^2 + \|\mathbf{v}\|^2 - 2\|\mathbf{u}\| \, \|\mathbf{v}\| \cos \theta$$

$$(\mathbf{v} - \mathbf{u}) \cdot \mathbf{v} - (\mathbf{v} - \mathbf{u}) \cdot \mathbf{u} = \|\mathbf{u}\|^2 + \|\mathbf{v}\|^2 - 2\|\mathbf{u}\| \, \|\mathbf{v}\| \cos \theta$$

$$\mathbf{v} \cdot \mathbf{v} - \mathbf{u} \cdot \mathbf{v} - \mathbf{v} \cdot \mathbf{u} + \mathbf{u} \cdot \mathbf{u} = \|\mathbf{u}\|^2 + \|\mathbf{v}\|^2 - 2\|\mathbf{u}\| \, \|\mathbf{v}\| \cos \theta$$

$$\|\mathbf{v}\|^2 - 2\mathbf{u} \cdot \mathbf{v} + \|\mathbf{u}\|^2 = \|\mathbf{u}\|^2 + \|\mathbf{v}\|^2 - 2\|\mathbf{u}\| \, \|\mathbf{v}\| \cos \theta$$

$$-2\mathbf{u} \cdot \mathbf{v} = -2\|\mathbf{u}\| \, \|\mathbf{v}\| \cos \theta$$

$$\cos \theta = \frac{\mathbf{u} \cdot \mathbf{v}}{\|\mathbf{u}\| \, \|\mathbf{v}\|}.$$

Section 9.1, page 623

Properties of Sums

1. $\displaystyle\sum_{i=1}^{n} ca_i = c\sum_{i=1}^{n} a_i,$ c is any constant.

2. $\displaystyle\sum_{i=1}^{n} (a_i + b_i) = \sum_{i=1}^{n} a_i + \sum_{i=1}^{n} b_i$

3. $\displaystyle\sum_{i=1}^{n} (a_i - b_i) = \sum_{i=1}^{n} a_i - \sum_{i=1}^{n} b_i$

Proof

Each of these properties follows directly from the Associative Property of Addition, the Commutative Property of Addition, and the Distributive Property of multiplication over addition. For example, note the use of the Distributive Property in the proof of Property 1.

$$\sum_{i=1}^{n} ca_i = ca_1 + ca_2 + ca_3 + \cdots + ca_n$$

$$= c(a_1 + a_2 + a_3 + \cdots + a_n) = c\sum_{i=1}^{n} a_i$$

Section 9.2, page 632

The Sum of a Finite Arithmetic Sequence

The sum of a finite arithmetic sequence with n terms is

$$S_n = \frac{n}{2}(a_1 + a_n).$$

Proof

Begin by generating the terms of the arithmetic sequence in two ways. In the first way, repeatedly add d to the first term to obtain

$$S_n = a_1 + a_2 + a_3 + \cdots + a_{n-2} + a_{n-1} + a_n$$

$$= a_1 + [a_1 + d] + [a_1 + 2d] + \cdots + [a_1 + (n-1)d].$$

In the second way, repeatedly subtract d from the nth term to obtain

$$S_n = a_n + a_{n-1} + a_{n-2} + \cdots + a_3 + a_2 + a_1$$

$$= a_n + [a_n - d] + [a_n - 2d] + \cdots + [a_n - (n-1)d].$$

If you add these two versions of S_n, the multiples of d cancel and you obtain

$$2S_n = \overbrace{(a_1 + a_n) + (a_1 + a_n) + (a_1 + a_n) + \cdots + (a_1 + a_n)}^{n \text{ terms}}$$

$$= n(a_1 + a_n).$$

So, you have $S_n = \dfrac{n}{2}(a_1 + a_n).$

Section 9.3, page 641

The Sum of a Finite Geometric Sequence

The sum of the geometric sequence

$$a_1, \ a_1r, \ a_1r^2, \ a_1r^3, \ a_1r^4, \ \ldots, a_1r^{n-1}$$

with common ratio $r \neq 1$ is $S_n = a_1\left(\dfrac{1 - r^n}{1 - r}\right)$.

Proof

Begin by writing out the nth partial sum.

$$S_n = a_1 + a_1r + a_1r^2 + \cdots + a_1r^{n-2} + a_1r^{n-1}$$

Multiplication by r yields

$$rS_n = a_1r + a_1r^2 + a_1r^3 + \cdots + a_1r^{n-1} + a_1r^n.$$

Subtracting the second equation from the first yields

$$S_n - rS_n = a_1 - a_1r^n.$$

So, $S_n(1 - r) = a_1(1 - r^n)$, and, because $r \neq 1$, you have

$$S_n = a_1\left(\frac{1 - r^n}{1 - r}\right).$$

Section 9.5, page 656

The Binomial Theorem

In the expansion of $(x + y)^n$

$$(x + y)^n = x^n + nx^{n-1}y + \cdots + {}_nC_r \, x^{n-r}y^r + \cdots + nxy^{n-1} + y^n$$

the coefficient of $x^{n-r}y^r$ is ${}_nC_r = \dfrac{n!}{(n - r)!r!}$.

Proof

The Binomial Theorem can be proved quite nicely using mathematical induction. The steps are straightforward but look a little messy, so only an outline of the proof is presented.

1. If $n = 1$, you have

$$(x + y)^1 = x^1 + y^1 = {}_1C_0x + {}_1C_1y$$

and the formula is valid.

2. Assuming that the formula is true for $n = k$, the coefficient of $x^{k-r}y^r$ is

$${}_kC_r = \frac{k!}{(k - r)!r!} = \frac{k(k - 1)(k - 2) \cdots (k - r + 1)}{r!}.$$

To show that the formula is true for $n = k + 1$, look at the coefficient of $x^{k+1-r}y^r$ in the expansion of

$$(x + y)^{k+1} = (x + y)^k(x + y).$$

From the right-hand side, you can determine that the term involving $x^{k+1-r}y^r$ is the sum of two products.

$$({}_kC_r x^{k-r}y^r)(x) + ({}_kC_{r-1}x^{k+1-r}y^{r-1})(y)$$

$$= \left[\frac{k!}{(k-r)!r!} + \frac{k!}{(k-r+1)!(r-1)!}\right]x^{k+1-r}y^r$$

$$= \left[\frac{(k+1-r)k!}{(k+1-r)!r!} + \frac{k!r}{(k+1-r)!r!}\right]x^{k+1-r}y^r$$

$$= \left[\frac{k!(k+1-r+r)}{(k+1-r)!r!}\right]x^{k+1-r}y^r$$

$$= \left[\frac{(k+1)!}{(k+1-r)!r!}\right]x^{k+1-r}y^r$$

$$= {}_{k+1}C_r x^{k+1-r}y^r$$

So, by mathematical induction, the Binomial Theorem is valid for all positive integers n.

Section 10.1, page 697

Standard Equation of a Parabola (Vertex at Origin)

The standard form of the equation of a parabola with vertex at $(0, 0)$ and directrix $y = -p$ is

$$x^2 = 4py, \qquad p \neq 0. \qquad \text{Vertical axis}$$

For directrix $x = -p$, the equation is

$$y^2 = 4px, \qquad p \neq 0. \qquad \text{Horizontal axis}$$

The focus is on the axis p units (directed distance) from the vertex.

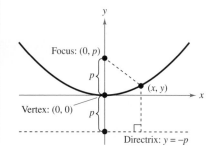

Proof

Because the two cases are similar, a proof will be given for the first case only. Suppose the directrix $(y = -p)$ is parallel to the x-axis. In the figure, you assume that $p > 0$, and because p is the directed distance from the vertex to the focus, the focus must lie above the vertex. Because the point (x, y) is equidistant from $(0, p)$ and $y = -p$, you can apply the Distance Formula to obtain

$$\sqrt{(x - 0)^2 + (y - p)^2} = y + p$$

$$x^2 + (y - p)^2 = (y + p)^2$$

$$x^2 + y^2 - 2py + p^2 = y^2 + 2py + p^2$$

$$x^2 = 4py.$$

Section 10.4, page 722

Rotation of Axes to Eliminate an *xy*-term

The general second-degree equation $Ax^2 + Bxy + Cy^2 + Dx + Ey + F = 0$ can be rewritten as

$$A'(x')^2 + C'(y')^2 + D'x' + E'y' + F' = 0$$

by rotating the coordinate axes through an angle θ, where

$$\cot 2\theta = \frac{A - C}{B}.$$

The coefficients of the new equation are obtained by making the substitutions

$$x = x'\cos\theta - y'\sin\theta \quad \text{and} \quad y = x'\sin\theta + y'\cos\theta.$$

Proof

You need to discover how the coordinates in the *xy*-system are related to the coordinates in the *x'y'*-system. To do this, choose a point $P = (x, y)$ in the original system and attempt to find its coordinates (x', y') in the rotated system. In either system, the distance r between the point P and the origin is the same. So, the equations for x, y, x', and y' are those given in the figure. Using the formulas for the sine and cosine of the difference of two angles, you have the following.

$$x' = r\cos(\alpha - \theta)$$

$$= r(\cos\alpha\cos\theta + \sin\alpha\sin\theta)$$

$$= r\cos\alpha\cos\theta + r\sin\alpha\sin\theta$$

$$= x\cos\theta + y\sin\theta$$

$$y' = r\sin(\alpha - \theta)$$

$$= r(\sin\alpha\cos\theta - \cos\alpha\sin\theta)$$

$$= r\sin\alpha\cos\theta - r\cos\alpha\sin\theta$$

$$= y\cos\theta - x\sin\theta$$

Solving this system for x and y yields

$$x = x'\cos\theta - y'\sin\theta$$

$$y = x'\sin\theta + y'\cos\theta.$$

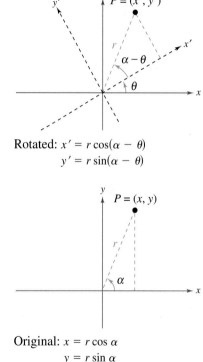

Rotated: $x' = r\cos(\alpha - \theta)$
$y' = r\sin(\alpha - \theta)$

Original: $x = r\cos\alpha$
$y = r\sin\alpha$

Finally, by substituting these values for x and y into the original equation and collecting terms, you obtain

$$A' = A\cos^2\theta + B\cos\theta\sin\theta + C\sin^2\theta$$

$$C' = A\sin^2\theta - B\cos\theta\sin\theta + C\cos^2\theta$$

$$D' = D\cos\theta + E\sin\theta$$

$$E' = -D\sin\theta + E\cos\theta$$

$$F' = F.$$

To eliminate the $x'y'$-term, you must select θ so that $B' = 0$.

$$B' = 2(C - A) \sin \theta \cos \theta + B(\cos^2\theta - \sin^2\theta)$$

$$= (C - A) \sin 2\theta + B \cos 2\theta$$

$$= B(\sin 2\theta)\left(\frac{C - A}{B} + \cot 2\theta\right)$$

$$= 0, \quad \sin 2\theta \neq 0$$

If $B = 0$, no rotation is necessary because the xy-term is not present in the original equation. If $B \neq 0$, the only way to make $B' = 0$ is to let

$$\cot 2\theta = \frac{A - C}{B}, \quad B \neq 0.$$

So, you have established the desired results.

Section 10.8, page 754

Polar Equations of Conics

The graph of a polar equation of the form

1. $r = \dfrac{ep}{1 \pm e \cos \theta}$ **2.** $r = \dfrac{ep}{1 \pm e \sin \theta}$

is a conic, where $e > 0$ is the eccentricity and $|p|$ is the distance between the focus (pole) and the directrix.

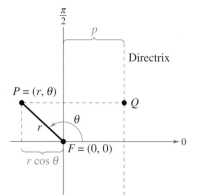

Proof

A proof for $r = ep/(1 + e \cos \theta)$ with $p > 0$ is listed here. The proofs of the other cases are similar. In the figure, consider a vertical directrix, p units to the right of the focus $F = (0, 0)$. If $P = (r, \theta)$ is a point on the graph of $r = ep/(1 + e \cos \theta)$ the distance between P and the directrix is

$$PQ = |p - x|$$

$$= |p - r \cos \theta|$$

$$= \left| p - \left(\frac{ep}{1 + e \cos\theta}\right)\cos \theta \right|$$

$$= \left| p\left(1 - \frac{e \cos\theta}{1 + e \cos \theta}\right) \right|$$

$$= \left| \frac{p}{1 + e \cos \theta} \right| = \left| \frac{r}{e} \right|.$$

Moreover, because the distance between P and the pole is simply $PF = |r|$, the ratio of PF to PQ is

$$\frac{PF}{PQ} = \frac{|r|}{|r/e|} = |e| = e$$

and by definition, the graph of the equation must be a conic.

Appendix B Concepts in Statistics

Appendix B.1 Representing Data

Line Plots

Statistics is the branch of mathematics that studies techniques for collecting, organizing, and interpreting data. In this section, you will study several ways to organize data. The first is a **line plot,** which uses a portion of a real number line to order numbers. Line plots are especially useful for ordering small sets of numbers (about 50 or less) by hand.

EXAMPLE 1   Constructing a Line Plot

Use a line plot to organize the following test scores. Which number occurs with the greatest frequency?

93, 70, 76, 67, 86, 93, 82, 78, 83, 86, 64, 78, 76, 66, 83,
83, 96, 74, 69, 76, 64, 74, 79, 76, 88, 76, 81, 82, 74, 70

Solution

Begin by scanning the data to find the smallest and largest numbers. For this data, the smallest number is 64 and the largest is 96. Next, draw a portion of a real number line that includes the interval [64, 96]. To create the line plot, start with the first number, 93, and enter an × above 93 on the number line. Continue recording ×'s for each number in the list until you obtain the line plot shown in Figure B.1. From the line plot, you can see that 76 occurs with the greatest frequency.

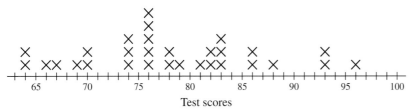

Figure B.1

Stem-and-Leaf Plots

Another type of plot that can be used to organize sets of numbers by hand is a stem-and-leaf plot. A set of test scores and the corresponding stem-and-leaf plot are shown at the right.

Test Scores

93, 70, 76, 58, 86, 93, 82, 78,
83, 86, 64, 78, 76, 66, 83, 83,
96, 74, 69, 76, 64, 74, 79, 76,
88, 76, 81, 82, 74, 70

Stems	Leaves
5	8
6	4 4 6 9
7	0 0 4 4 4 6 6 6 6 8 8 9
8	1 2 2 3 3 3 6 6 8
9	3 3 6

Note that the *leaves* represent the units digits of the numbers and the *stems* represent the tens digits. Stem-and-leaf plots can also be used to compare two sets of data, as shown in the following example.

EXAMPLE 2 Comparing Two Sets of Data

Use a stem-and-leaf plot to compare the test scores given on page A15 with the following test scores. Which set of test scores is better?

90, 81, 70, 62, 64, 73, 81, 92, 73, 81, 92, 93, 83, 75, 76,
83, 94, 96, 86, 77, 77, 86, 96, 86, 77, 86, 87, 87, 79, 88

Solution
Begin by ordering the second set of scores.

62, 64, 70, 73, 73, 75, 76, 77, 77, 77, 79, 81, 81, 81, 83,
83, 86, 86, 86, 86, 87, 87, 88, 90, 92, 92, 93, 94, 96, 96

Now that the data has been ordered, you can construct a *double* stem-and-leaf plot by letting the leaves to the right of the stems represent the units digits for the first group of test scores and letting the leaves to the left of the stems represent the units digits for the second group of test scores.

Leaves (2nd Group)	Stems	Leaves (1st Group)
	5	8
4 2	6	4 4 6 9
9 7 7 7 6 5 3 3 0	7	0 0 4 4 4 6 6 6 6 6 6 8 8 9
8 7 7 6 6 6 6 3 3 1 1 1	8	1 2 2 3 3 3 6 6 8
6 6 4 3 2 2 0	9	3 3 6

By comparing the two sets of leaves, you can see that the second group of test scores is better than the first group.

EXAMPLE 3 Using a Stem-and-Leaf Plot

The table at the right above shows the percent of the population of each state and the District of Columbia that was at least 65 years old in 1997. Use a stem-and-leaf plot to organize the data. (Source: U.S. Bureau of the Census)

Solution
Begin by ordering the numbers, as shown below.

5.3, 8.7, 9.9, 10.1, 10.1, 11.1, 11.2, 11.2, 11.3, 11.3, 11.4, 11.5, 11.5, 11.5,
12.1, 12.1, 12.2, 12.3, 12.3, 12.4, 12.5, 12.5, 12.5, 12.5, 12.5, 12.9, 13.0,
13.2, 13.2, 13.2, 13.2, 13.3, 13.4, 13.4, 13.4, 13.5, 13.7, 13.7, 13.7, 13.9,
13.9, 14.1, 14.3, 14.3, 14.4, 14.4, 15.0, 15.1, 15.8, 15.8, 18.5

Next construct the stem-and-leaf plot using the leaves to represent the digits to the right of the decimal points, as shown at the right. From the stem-and-leaf plot, you can see that Alaska has the lowest percent and Florida has the highest percent.

AK	5.3	MT	13.2
AL	13.0	NC	12.5
AR	14.3	ND	14.4
AZ	13.2	NE	13.7
CA	11.1	NH	12.1
CO	10.1	NJ	13.7
CT	14.4	NM	11.2
DC	13.9	NV	11.5
DE	12.9	NY	13.4
FL	18.5	OH	13.4
GA	9.9	OK	13.4
HI	13.2	OR	13.3
IA	15.0	PA	15.8
ID	11.3	RI	15.8
IL	12.5	SC	12.1
IN	12.5	SD	14.3
KS	13.5	TN	12.5
KY	12.5	TX	10.1
LA	11.4	UT	8.7
MA	14.1	VA	11.2
MD	11.5	VT	12.3
ME	13.9	WA	11.5
MI	12.4	WI	13.2
MN	12.3	WV	15.1
MO	13.7	WY	11.3
MS	12.2		

Stems	Leaves
5.	3
6.	
7.	
8.	7
9.	9
10.	1 1
11.	1 2 2 3 3 4 5 5 5
12.	1 1 2 3 3 4 5 5 5 5 5 9
13.	0 2 2 2 2 3 4 4 4 5 7 7 7 9 9
14.	1 3 3 4 4
15.	0 1 8 8
16.	
17.	
18.	5

Histograms and Frequency Distributions

With data such as that given in Example 3, it is useful to group the numbers into intervals and plot the frequency of the data in each interval. For instance, the frequency distribution and the histogram shown in Figure B.2 represent the data given in Example 3.

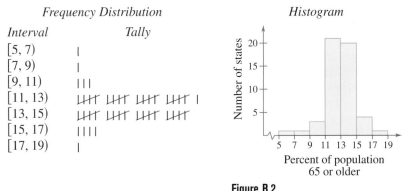

Figure B.2

A histogram has a portion of a real number line as its horizontal axis. A histogram is similar to a bar graph, except that the rectangles (bars) in a bar graph can be either horizontal or vertical and the labels of the bars are not necessarily numbers. Another difference between a bar graph and a histogram is that the bars in a bar graph are usually separated by spaces, whereas the bars in a histogram are not.

EXAMPLE 4   Constructing a Histogram

A company has 48 sales representatives who sold the following numbers of units during the first quarter of 2000. Construct a grouped frequency distribution for this data.

107	162	184	170	177	102	145	141
105	193	167	149	195	127	193	191
150	153	164	167	171	163	141	129
109	171	150	138	100	164	147	153
171	163	118	142	107	144	100	132
153	107	124	162	192	134	187	177

Interval	Tally
100–109	LHT III
110–119	I
120–129	III
130–139	III
140–149	LHT II
150–159	LHT
160–169	LHT III
170–179	LHT I
180–189	II
190–199	LHT

Solution

To begin constructing a grouped frequency distribution, you must first decide on the number of groups. There are several ways to group this data. However, because the smallest number is 100 and the largest is 195, it seems that 10 groups of 10 each would be appropriate. The first group would be 100–109, the second group would be 110–119, and so on. By tallying the data into the 10 groups, you obtain the distribution shown at the right above. A histogram for the distribution is shown in Figure B.3.

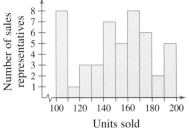

Figure B.3

Fitting a Line to Data

Many real-life situations involve finding relationships between two variables, such as the year and the number of people in the labor force. In a typical situation, data is collected and written as a set of ordered pairs. The graph of such a set is called a *scatter plot.*

EXAMPLE 5   Constructing a Scatter Plot

The data in the table shows the number of people P (in millions) in the United States who were part of the labor force from 1987 through 1997. In the table, t represents the year, with $t = 7$ corresponding to 1987. Sketch a scatter plot of the data.   (Source: U.S. Bureau of Labor Statistics)

t	7	8	9	10	11	12	13	14	15	16	17
P	120	122	124	126	126	128	129	131	132	134	136

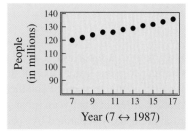

Figure B.4

Solution

Begin by representing the data with a set of ordered pairs.

$(7, 120), (8, 122), (9, 124), (10, 126), (11, 126), (12, 128),$

$(13, 129), (14, 131), (15, 132), (16, 134), (17, 136)$

Then plot each point in a coordinate plane, as shown in Figure B.4.

Most graphing utilities have built-in statistical programs that can create scatter plots. Try using your graphing utility to plot the points shown in the table in Example 5.

From the scatter plot in Figure B.4, it appears that the points describe a relationship that is nearly linear. The relationship is not *exactly* linear because the labor force did not increase by precisely the same amount each year.

A mathematical equation that approximates the relationship between t and P is a *mathematical model.* When developing a mathematical model to describe a set of data, you strive for two (often conflicting) goals—accuracy and simplicity. For the data above, a linear model of the form $P = at + b$ appears to be best. It is simple and relatively accurate.

Finding a linear model to represent the relationship described by a scatter plot is called **fitting a line to data.** You can do this graphically by simply sketching the line that appears to fit the points, finding two points on the line, and then finding the equation of the line that passes through the two points.

EXAMPLE 6   Fitting a Line to Data

Find a linear model that relates the year with the number of people in the United States labor force. (See Example 5.)

t	7	8	9	10	11	12	13	14	15	16	17
P	120	122	124	126	126	128	129	131	132	134	136

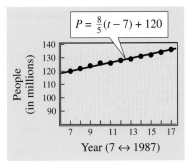

Figure B.5

Solution

After plotting the data in the table, draw the line that you think best represents the data, as shown in Figure B.5. Two points that lie on this line are $(7, 120)$ and $(17, 136)$. Using the point-slope form, you can find the equation of the line to be

$$P = \frac{8}{5}(t - 7) + 120. \qquad \text{Linear model}$$

Once you have found a model, you can measure how well the model fits the data by comparing the actual values with the values given by the model, as shown in the following table.

	t	7	8	9	10	11	12	13	14	15	16	17
Actual ➡	P	120	122	124	126	126	128	129	131	132	134	136
Model ➡	P	120	121.6	123.2	124.8	126.4	128	129.6	131.2	132.8	134.4	136

The model in Example 6 is based on the two data points chosen. If different points were chosen, the model may change somewhat. For instance, if you choose $(10, 126)$ and $(15, 132)$, the new model is $P = \frac{6}{5}(t - 10) + 126$.

If you use the regression capabilities of a graphing calculator or computer program to find a linear model for the data in Example 6, you will notice that the program may also output a value of $r \approx 0.995$. This number is the **correlation coefficient** of the data. Correlation coefficients vary between -1 and 1. Basically, the closer $|r|$ is to 1, the better the points can be described by a line. Three examples are shown in Figure B.6.

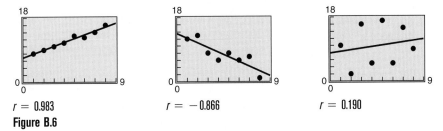

$r = 0.983$ $r = -0.866$ $r = 0.190$
Figure B.6

B.1 Exercises

1. *Gasoline Prices*   The line plot shows a sample of prices of unleaded regular gasoline from 25 different cities.

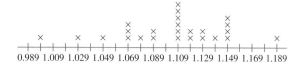

(a) What price occurred with the greatest frequency?

(b) What is the range of prices?

2. *Livestock Weights*   The line plot shows the weights (to the nearest hundred pounds) of 30 head of cattle sold by a rancher.

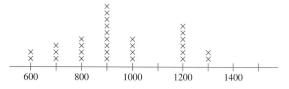

(a) What weight occurred with the greatest frequency?

(b) What is the range of weights?

Quiz and Exam Scores   **In Exercises 3–6, use the following scores from a math class of 30 students. The scores are for two 25-point quizzes and two 100-point exams.**

Quiz #1   20, 15, 14, 20, 16, 19, 10, 21, 24, 15, 15, 14, 15, 21, 19, 15, 20, 18, 18, 22, 18, 16, 18, 19, 21, 19, 16, 20, 14, 12

Quiz #2   22, 22, 23, 22, 21, 24, 22, 19, 21, 23, 23, 25, 24, 22, 22, 23, 23, 23, 23, 22, 24, 23, 22, 24, 21, 24, 16, 21, 16, 14

Exam #1   77, 100, 77, 70, 83, 89, 87, 85, 81, 84, 81, 78, 89, 78, 88, 85, 90, 92, 75, 81, 85, 100, 98, 81, 78, 75, 85, 89, 82, 75

Exam #2   76, 78, 73, 59, 70, 81, 71, 66, 66, 73, 68, 67, 63, 67, 77, 84, 87, 71, 78, 78, 90, 80, 77, 70, 80, 64, 74, 68, 68, 68

3. Construct a line plot for each quiz. For each quiz, which score occurred with the greatest frequency?

4. Construct a line plot for each exam. For each exam, which score occurred with the greatest frequency?

5. Construct a stem-and-leaf plot for Exam #1.

6. Construct a double stem-and-leaf plot to compare the scores for Exam #1 and Exam #2. Which set of scores is higher?

7. *Insurance Coverage*   The table shows the total numbers of persons (in thousands) without health insurance coverage in the 50 states and the District of Columbia in 1996. Use a stem-and-leaf plot to organize the data.   (Source: U.S. Bureau of the Census)

AK	89	AL	550	AR	566	AZ	1159	CA	6514
CO	644	CT	368	DC	80	DE	98	FL	2722
GA	1319	HI	101	IA	335	ID	196	IL	1337
IN	600	KS	292	KY	601	LA	890	MA	766
MD	581	ME	146	MI	857	MN	480	MO	700
MS	518	MT	124	NC	1160	ND	62	NE	190
NH	109	NJ	1317	NM	412	NV	255	NY	3132
OH	1292	OK	570	OR	496	PA	1133	RI	93
SC	634	SD	67	TN	841	TX	4680	UT	240
VA	811	VT	65	WA	761	WI	438	WV	261
WY	66								

8. *Snowfall*   The data below shows the seasonal snowfall (in inches) at Lincoln, Nebraska for the years 1968 through 1997 (the amounts are listed in order by year). How would you organize this data? Explain your reasoning.   (Source: University of Nebraska–Lincoln)

39.8, 26.2, 49.0, 21.6, 29.2, 33.6, 42.1, 21.1, 21.8, 31.0, 34.4, 23.3, 13.0, 32.3, 38.0, 47.5, 21.5, 18.9, 15.7, 13.0, 19.1, 18.7, 25.8, 23.8, 32.1, 21.3, 21.8, 30.7, 29.0, 44.6

9. *Retirement Contributions*   The employees of a company must contribute 7% of their monthly salaries to a company-sponsored retirement plan. The contributed amounts (in dollars) for the company's 35 employees are as follows.

100, 200, 130, 136, 161, 156, 209, 126, 135, 98, 114, 117, 168, 133, 140, 124, 172, 127, 143, 157, 124, 152, 104, 126, 155, 92, 194, 115, 120, 136, 148, 112, 116, 146, 96

(a) Construct a frequency distribution using groups of 20. The first group should be 90–109.

(b) Construct a histogram for this frequency distribution.

10. *Quiz Scores*   The following ordered pairs give the scores of two consecutive 15-point quizzes for a class of 18 students.

(7, 13), (9, 7), (14, 14), (15, 15), (10, 15),

(9, 7), (14, 11), (14, 15), (8, 10), (9, 10),

(15, 9), (10, 11), (11, 14), (7, 14),

(11, 10), (14, 11), (10, 15), (9, 6)

(a) Create a scatter plot for the data.

(b) Does the relationship between consecutive quiz scores appear to be approximately linear? If not, give some possible explanations.

11. *Dentists*   The data in the table shows the number of dentists D (in thousands) in the United States from 1985 through 1996. In the table, t represents the year, with $t = 5$ representing 1985.

t	5	6	7	8	9	10
D	156	158	161	164	168	173

t	11	12	13	14	15	16
D	179	183	187	191	194	196

(a) Sketch a scatter plot of the data.

(b) Find the equation of the line that seems to best fit the data.

(c) Use the regression capabilities of a graphing utility to find a linear model for the data. Compare with your answer to part (b).

(d) Use the model from part (c) to estimate the number of dentists in the United States in 2000, 2002, and 2005.

12. *Consumerism*   The data in the table shows the per capita consumption W (in gallons) of bottled water in the United States from 1988 through 1996. In the table, t represents the year, with $t = 0$ representing 1990.   (Source: U.S. Department of Agriculture)

t	-2	-1	0	1	2
W	6.5	7.4	8	8	8.2

t	3	4	5	6
W	9.4	10.7	11.6	12.4

(a) Sketch a scatter plot of the data.

(b) Find the equation of a line that seems to best fit the data.

(c) Use the regression capabilities of a graphing utility to find a linear model for the data. Compare with your answer to part (b).

(d) Use the model from part (c) to estimate the per capita consumption in 2000, 2004, and 2007.

13. *World Food Production*   The data in the table shows the world production P (in millions of metric tons) of vegetables (including melons) for 1990 through 1997.   (Source: U.S. Department of Agriculture)

Year	1990	1991	1992	1993
P	461.4	462.4	478.6	509.5

Year	1994	1995	1996	1997
P	532.9	559.9	589.1	595.6

(a) Use the regression capabilities of a graphing utility to find a linear model for the data.

(b) Use a graphing utility to graph the actual data and the model in the same viewing window. How closely does the model represent the data?

(c) Create a table showing the actual values of P and the values of P given by the model.

Appendix B.2   Measures of Central Tendency and Dispersion

Mean, Median, and Mode

In many real-life situations, it is helpful to describe data by a single number that is most representative of the entire collection of numbers. Such a number is called a **measure of central tendency.** The most commonly used measures are as follows.

1. The **mean,** or **average,** of n numbers is the sum of the numbers divided by n.

2. The **median** of n numbers is the middle number when the numbers are written in order. If n is even, the median is the average of the two middle numbers.

3. The **mode** of n numbers is the number that occurs most frequently. If two numbers tie for most frequent occurrence, the collection has two modes and is called **bimodal.**

What You Should Learn:

- How to find and interpret the mean, median, and mode of a set of data
- How to determine the measure of central tendency that best represents a set of data
- How to find the standard deviations of a set of data
- How to use box-and-whisker plots

Why You Should Learn It:

Measures of central tendency and dispersion provide a convenient way to describe and compare sets of data. For instance, in Exercise 36 on page A30, the mean and standard deviation are used to analyze the price of gold for the years 1978 through 1997.

EXAMPLE 1   Comparing Measures of Central Tendency

On an interview for a job, the interviewer tells you that the average annual income of the company's 25 employees is $60,849. The actual annual incomes of the 25 employees are shown below. What are the mean, median, and mode of the incomes? Was the person telling you the truth?

$17,305,	$478,320,	$45,678,	$18,980,	$17,408,
$25,676,	$28,906,	$12,500,	$24,540,	$33,450,
$12,500,	$33,855,	$37,450,	$20,432,	$28,956,
$34,983,	$36,540,	$250,921,	$36,853,	$16,430,
$32,654,	$98,213,	$48,980,	$94,024,	$35,671

Solution

The mean of the incomes is

$$\text{Mean} = \frac{17,305 + 478,320 + 45,678 + 18,980 + \cdots + 35,671}{25}$$

$$= \frac{1,521,225}{25} = \$60,849.$$

To find the median, order the incomes as follows.

$12,500,	$12,500,	$16,430,	$17,305,	$17,408,
$18,980,	$20,432,	$24,540,	$25,676,	$28,906,
$28,956,	$32,654,	$33,450,	$33,855,	$34,983,
$35,671,	$36,540,	$36,853,	$37,450,	$45,678,
$48,980,	$94,024,	$98,213,	$250,921,	$478,320

From this list, you can see that the median (the middle number) is $33,450. From the same list, you can see that $12,500 is the only income that occurs more than once. So, the mode is $12,500. Technically, the person was telling the truth because the average is (generally) defined to be the mean. However, of the three measures of central tendency—*Mean:* $60,849 *Median:* $33,450 *Mode:* $12,500—it seems clear that the median is most representative. The mean is inflated by the two highest salaries.

Choosing a Measure of Central Tendency

Which of the three measures of central tendency is the most representative? The answer is that it depends on the distribution of the data *and* the way in which you plan to use the data.

For instance, in Example 1, the mean salary of $60,849 does not seem very representative to a potential employee. To a city income tax collector who wants to estimate 1% of the total income of the 25 employees, however, the mean is precisely the right measure.

EXAMPLE 2   Choosing a Measure of Central Tendency

Which measure of central tendency is the most representative of the data given in each of the following frequency distributions?

a. *Number*	*Tally*	b. *Number*	*Tally*	c. *Number*	*Tally*
1	7	1	9	1	6
2	20	2	8	2	1
3	15	3	7	3	2
4	11	4	6	4	3
5	8	5	5	5	5
6	3	6	6	6	5
7	2	7	7	7	4
8	0	8	8	8	3
9	15	9	9	9	0

Solution
a. For this data, the mean is 4.23, the median is 3, and the mode is 2. Of these, the mode is probably the most representative.
b. For this data, the mean and median are each 5 and the modes are 1 and 9 (the distribution is bimodal). Of these, the mean or median is the most representative.
c. For this data, the mean is 4.59, the median is 5, and the mode is 1. Of these, the mean or median is the most representative.

Variance and Standard Deviation

Very different sets of numbers can have the same mean. You will now study two **measures of dispersion,** which give you an idea of how much the numbers in a set differ from the mean of the set. These two measures are called the *variance* of the set and the *standard deviation* of the set.

Definitions of Variance and Standard Deviation

Consider a set of numbers $\{x_1, x_2, \ldots, x_n\}$ with a mean of $\bar{x}$. The **variance** of the set is

$$v = \frac{(x_1 - \bar{x})^2 + (x_2 - \bar{x})^2 + \cdots + (x_n - \bar{x})^2}{n}$$

and the **standard deviation** of the set is $\sigma = \sqrt{v}$ (σ is the lowercase Greek letter *sigma*).

The standard deviation of a set is a measure of how much a typical number in the set differs from the mean. The greater the standard deviation, the more the numbers in the set vary from the mean. For instance, each of the following sets has a mean of 5.

$$\{5, 5, 5, 5\}, \quad \{4, 4, 6, 6\}, \quad \text{and} \quad \{3, 3, 7, 7\}$$

The standard deviations of the sets are 0, 1, and 2.

$$\sigma_1 = \sqrt{\frac{(5-5)^2 + (5-5)^2 + (5-5)^2 + (5-5)^2}{4}}$$

$$= 0$$

$$\sigma_2 = \sqrt{\frac{(4-5)^2 + (4-5)^2 + (6-5)^2 + (6-5)^2}{4}}$$

$$= 1$$

$$\sigma_3 = \sqrt{\frac{(3-5)^2 + (3-5)^2 + (7-5)^2 + (7-5)^2}{4}}$$

$$= 2$$

EXAMPLE 3   Estimations of Standard Deviation

Consider the three sets of data represented by the bar graphs in Figure B.7. Which set has the smallest standard deviation? Which has the largest?

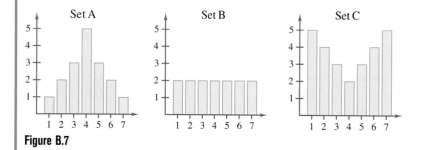

Figure B.7

Solution

Of the three sets, the numbers in set *A* are grouped most closely to the center and the numbers in set *C* are the most dispersed. So, set *A* has the smallest standard deviation and set *C* has the largest standard deviation.

EXAMPLE 4   Find Standard Deviation

Find the standard deviation of each set shown in Example 3.

Solution

Because of the symmetry of each bar graph, you can conclude that each has a mean of $\bar{x} = 4$. The standard deviation of set A is

$$\sigma = \sqrt{\frac{(-3)^2 + 2(-2)^2 + 3(-1)^2 + 5(0)^2 + 3(1)^2 + 2(2)^2 + (3)^2}{17}}$$

$$\approx 1.53.$$

The standard deviation of set B is

$$\sigma = \sqrt{\frac{2(-3)^2 + 2(-2)^2 + 2(-1)^2 + 2(0)^2 + 2(1)^2 + 2(2)^2 + 2(3)^2}{14}}$$

$$= 2.$$

The standard deviation of set C is

$$\sigma = \sqrt{\frac{5(-3)^2 + 4(-2)^2 + 3(-1)^2 + 2(0)^2 + 3(1)^2 + 4(2)^2 + 5(3)^2}{26}}$$

$$\approx 2.22.$$

These values confirm the results of Example 3. That is, set A has the smallest standard deviation and set C has the largest.

The following alternative formula provides a more efficient way to compute the standard deviation.

Alternative Formula for Standard Deviation

The standard deviation of $\{x_1, x_2, \ldots, x_n\}$ is

$$\sigma = \sqrt{\frac{x_1^2 + x_2^2 + \cdots + x_n^2}{n} - \bar{x}^2}.$$

Because of messy computations, this formula is difficult to verify. Conceptually, however, the process is straightforward. It consists of showing that the expressions

$$\sqrt{\frac{(x_1 - \bar{x})^2 + (x_2 - \bar{x})^2 + \cdots + (x_n - \bar{x})^2}{n}}$$

and

$$\sqrt{\frac{x_1^2 + x_2^2 + \cdots + x_n^2}{n} - \bar{x}^2}$$

are equivalent. Try verifying this equivalence for the set $\{x_1, x_2, x_3\}$ with $\bar{x} = (x_1 + x_2 + x_3)/3$.

EXAMPLE 5 Using the Alternative Formula

Use the alternative formula for standard deviation to find the standard deviation of the following set of numbers.

5, 6, 6, 7, 7, 8, 8, 8, 9, 10

Solution

Begin by finding the mean of the set, which is 7.4. So, the standard deviation is

$$\sigma = \sqrt{\frac{5^2 + 2(6^2) + 2(7^2) + 3(8^2) + 9^2 + 10^2}{10} - (7.4)^2}$$

$$= \sqrt{\frac{568}{10} - 54.76}$$

$$= \sqrt{2.04}$$

$$\approx 1.43.$$

You can use the statistical features of a graphing utility to check this result.

A well-known theorem in statistics, called *Chebychev's Theorem*, states that at least

$$1 - \frac{1}{k^2}$$

of the numbers in a distribution must lie within k standard deviations of the mean. So, 75% of the numbers in a collection must lie within two standard deviations of the mean, and at least 88.9% of the numbers must lie within three standard deviations of the mean. For most distributions, these percentages are low. For instance, in all three distributions shown in Example 3, 100% of the numbers lie within two standard deviations of the mean.

AK	66	MT	62
AL	40	NC	42
AR	39	ND	47
AZ	51	NE	63
CA	62	NH	59
CO	69	NJ	77
CT	80	NM	45
DC	94	NV	49
DE	44	NY	73
FL	50	OH	55
GA	46	OK	47
HI	80	OR	70
IA	55	PA	61
ID	53	RI	56
IL	61	SC	41
IN	47	SD	49
KS	51	TN	53
KY	53	TX	47
LA	45	UT	66
MA	74	VA	54
MD	68	VT	57
ME	47	WA	68
MI	62	WI	65
MN	67	WV	43
MO	53	WY	52
MS	37		

EXAMPLE 6 Describing a Distribution

The table at the right above shows the number of dentists (per 100,000 people) in each state and the District of Columbia. Find the mean and standard deviation of the numbers. What percent of the numbers lie within two standard deviations of the mean? (Source: American Dental Association)

Solution

Begin by entering the numbers into a graphing utility that has a standard deviation program. After running the program, you should obtain

$$\bar{x} \approx 56.76 \qquad \text{and} \qquad \sigma = 12.14.$$

The interval that contains all numbers that lie within two standard deviations of the mean is

$$[56.76 - 2(12.14), 56.76 + 2(12.14)] \qquad \text{or} \qquad [32.48, 81.04].$$

From the histogram in Figure B.8, you can see that all but one of the numbers (98%) lie in this interval—all but the number that corresponds to the number of dentists (per 100,000 people) in Washington, D.C.

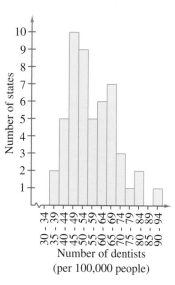

Figure B.8

Box-and-Whisker Plots

Standard deviation is the measure of dispersion that is associated with the mean. Quartiles measure dispersion associated with the median.

Definition of Quartiles

Consider an ordered set of numbers whose median is m. The **lower quartile** is the median of the numbers that occur before m. The **upper quartile** is the median of the numbers that occur after m.

EXAMPLE 7   Finding Quartiles of a Set

Find the lower and upper quartiles for the following set.

34, 14, 24, 16, 12, 18, 20, 24, 16, 26, 13, 27

Solution

Begin by ordering the set.

12, 13, 14,	16, 16, 18,	20, 24, 24,	26, 27, 34
1st 25%	2nd 25%	3rd 25%	4th 25%

The median of the entire set is 19. The median of the six numbers that are less than 19 is 15. So, the lower quartile is 15. The median of the six numbers that are greater than 19 is 25. So, the upper quartile is 25.

Quartiles are represented graphically by a **box-and-whisker plot,** as shown in Figure B.9. In the plot, notice that five numbers are listed: the smallest number, the lower quartile, the median, the upper quartile, and the largest number. Also notice that the numbers are spaced proportionally, as though they were on a real number line.

```
●————┬──┬──────┬─────┬─────────────────●
    12   15    19    25                34
```

Figure B.9

The next example shows how to find quartiles when the number of elements in a set is not divisible by 4.

EXAMPLE 8 Sketching Box-and-Whisker Plots

Sketch a box-and-whisker plot for each of the given sets.

a. 27, 28, 30, 42, 45, 50, 50, 61, 62, 64, 66

b. 82, 82, 83, 85, 87, 89, 90, 94, 95, 95, 96, 98, 99

c. 11, 13, 13, 15, 17, 18, 20, 24, 24, 27

Solution

a. This set has 11 numbers. The median is 50 (the sixth number). The lower quartile is 30 (the median of the first five numbers). The upper quartile is 62 (the median of the last five numbers).

b. This set has 13 numbers. The median is 90 (the seventh number). The lower quartile is 84 (the median of the first six numbers). The upper quartile is 95.5 (the median of the last six numbers).

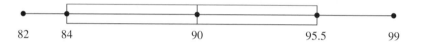

c. This set has 10 numbers. The median is 17.5 (the average of the fifth and sixth numbers). The lower quartile is 13 (the median of the first five numbers). The upper quartile is 24 (the median of the last five numbers).

B.2 Exercises

In Exercises 1–6, find the mean, median, and mode of the set of measurements.

1. 5, 12, 7, 14, 8, 9, 7

2. 30, 37, 32, 39, 33, 34, 32

3. 5, 12, 7, 24, 8, 9, 7

4. 20, 37, 32, 39, 33, 34, 32

5. 5, 12, 7, 14, 9, 7

6. 30, 37, 32, 39, 34, 32

7. *Reasoning* Compare your answers for Exercises 1 and 3 with those for Exercises 2 and 4. Which of the measures of central tendency is sensitive to extreme measurements? Explain your reasoning.

8. *Reasoning*

(a) Add 6 to each measurement in Exercise 1 and calculate the mean, median, and mode of the revised measurements. How are the measures of central tendency changed?

(b) If a constant k is added to each measurement in a set of data, how will the measures of central tendency change?

9. *Electric Bills*   A person had the following monthly bills for electricity. What are the mean and median of the collection of bills?

January	$67.92	February	$59.84
March	$52.00	April	$52.50
May	$57.99	June	$65.35
July	$81.76	August	$74.98
September	$87.82	October	$83.18
November	$65.35	December	$57.00

10. *Car Rental*   A car rental company kept the following record of the numbers of miles a rental car was driven. What are the mean, median, and mode of this data?

Monday	410	Tuesday	260
Wednesday	320	Thursday	320
Friday	460	Saturday	150

11. *Six-Child Families*   A study was done on families having six children. The table gives the numbers of families in the study with the indicated numbers of girls. Determine the mean, median, and mode of this set of data.

Number of girls	0	1	2	3	4	5	6
Frequency	1	24	45	54	50	19	7

12. *Baseball*   A baseball fan examined the records of a favorite baseball player's performance during his last 50 games. The numbers of games in which the player had 0, 1, 2, 3, and 4 hits are recorded in the table.

Number of hits	0	1	2	3	4
Frequency	14	26	7	2	1

(a) Determine the average number of hits per game.

(b) Determine the player's batting average if he had 200 at-bats during the 50-game series.

13. *Think About It*   Construct a collection of numbers that has the following properties. If this is not possible, explain why it is not.

Mean = 6, median = 4, mode = 4

14. *Think About It*   Construct a collection of numbers that has the following properties. If this is not possible, explain why it is not.

Mean = 6, median = 6, mode = 4

15. *Test Scores*   A professor records the following scores for a 100-point exam.

99, 64, 80, 77, 59, 72, 87, 79, 92, 88,
90, 42, 20, 89, 42, 100, 98, 84, 78, 91

Which measure of central tendency best describes these test scores?

16. *Shoe Sales*   A salesman sold eight pairs of a certain style of men's shoes. The sizes of the eight pairs were as follows: $10\frac{1}{2}$, 8, 12, $10\frac{1}{2}$, 10, $9\frac{1}{2}$, 11, and $10\frac{1}{2}$. Which measure (or measures) of central tendency best describes the typical shoe size for this data?

In Exercises 17–24, find the mean ($\bar{x}$), variance (v), and standard deviation (σ) of the numbers.

17. 4, 10, 8, 2

18. 3, 15, 6, 9, 2

19. 0, 1, 1, 2, 2, 2, 3, 3, 4

20. 2, 2, 2, 2, 2, 2

21. 1, 2, 3, 4, 5, 6, 7

22. 1, 1, 1, 5, 5, 5

23. 49, 62, 40, 29, 32, 70

24. 1.5, 0.4, 2.1, 0.7, 0.8

In Exercises 25–30, use the alternative formula to find the standard deviation of the numbers.

25. 2, 4, 6, 6, 13, 5

26. 10, 25, 50, 26, 15, 33, 29, 4

27. 246, 336, 473, 167, 219, 359

28. 6.0, 9.1, 4.4, 8.7, 10.4

29. 8.1, 6.9, 3.7, 4.2, 6.1

30. 9.0, 7.5, 3.3, 7.4, 6.0

In Exercises 31 and 32, line plots of sets of data are given. Determine the mean and standard deviation of each set.

31. (a)

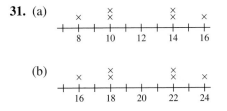

(b)

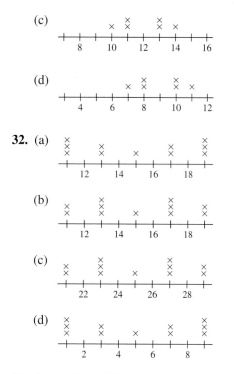

(c)

(d)

32. (a)

(b)

(c)

(d)

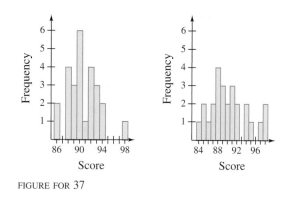

FIGURE FOR 37

33. *Reasoning* Without calculating the standard deviation, explain why the set {4, 4, 20, 20} has a standard deviation of 8.

34. *Reasoning* If the standard deviation of a set of numbers is 0, what does this imply about the set?

35. *Test Scores* An instructor adds five points to each student's exam score. Will this change the mean or standard deviation of the exam scores? Explain.

36. *Price of Gold* The following data represents the average prices of gold (in dollars per fine ounce) for the years 1978 to 1997. Use a computer or calculator to find the mean, variance, and standard deviation of the data. What percent of the data lies within two standard deviations of the mean? (Source: U.S. Bureau of Mines)

194,	308,	613,	460,	376,
424,	361,	318,	368,	448,
438,	383,	385,	363,	345,
361,	385,	386,	389,	333

37. *Think About It* The histograms represent the test scores of two classes of a college course in mathematics. Which histogram has the smaller standard deviation?

In Exercises 38–41, sketch a box-and-whisker plot for the data without the aid of a graphing utility.

38. 23, 15, 14, 23, 13, 14, 13, 20, 12

39. 11, 10, 11, 14, 17, 16, 14, 11, 8, 14, 20

40. 46, 48, 48, 50, 52, 47, 51, 47, 49, 53

41. 25, 20, 22, 28, 24, 28, 25, 19, 27, 29, 28, 21

In Exercises 42–45, use a graphing utility to create a box-and-whisker plot for the data.

42. 19, 12, 14, 9, 14, 15, 17, 13, 19, 11, 10, 19

43. 9, 5, 5, 5, 6, 5, 4, 12, 7, 10, 7, 11, 8, 9, 9

44. 20.1, 43.4, 34.9, 23.9, 33.5, 24.1, 22.5, 42.4, 25.7, 17.4, 23.8, 33.3, 17.3, 36.4, 21.8

45. 78.4, 76.3, 107.5, 78.5, 93.2, 90.3, 77.8, 37.1, 97.1, 75.5, 58.8, 65.6

46. *Product Lifetime* A company has redesigned a product in an attempt to increase the lifetime of the product. The two sets of data list the lifetimes (in months) of 20 units with the original design and 20 units with the new design. Create a box-and-whisker plot for each set of data, and then comment on the differences between the plots.

Original Design

15.1	78.3	56.3	68.9	30.6
27.2	12.5	42.7	72.7	20.2
53.0	13.5	11.0	18.4	85.2
10.8	38.3	85.1	10.0	12.6

New Design

55.8	71.5	25.6	19.0	23.1
37.2	60.0	35.3	18.9	80.5
46.7	31.1	67.9	23.5	99.5
54.0	23.2	45.5	24.8	87.8

Appendix B.3 Least Squares Regression

In many of the examples and exercises in this text, you have been asked to use the regression capabilities of a graphing utility to find mathematical models for sets of data. Another way to find a mathematical model for a set of data is to use the **method of least squares.** As a measure of how well a model fits a set of data points

$$\{(x_1, y_1), (x_2, y_2), (x_3, y_3), . . . ,(x_n, y_n)\}$$

you can add the squares of the differences between the actual y-values and the values given by the model to obtain the **sum of the squared differences.** For instance, the table shows the heights x (in feet) and the diameters y (in inches) of eight trees. The table also shows the values of a linear model $y^* = 0.69x - 40$ for each x-value. The sum of squared differences for the model is 53.3.

x	70	72	75	76	85	78	77	80
y	8.3	10.5	11.0	11.4	12.9	14.0	16.3	18.0
y^*	8.3	9.68	11.75	12.44	18.65	13.82	13.13	15.2

The model that has the least sum of squared differences is the **least squares regression line** for the data. The least squares regression line for the data in the table is $y \approx 0.43x - 20.3$. The sum of squared differences is 43.32.

To find the least squares regression line $y = ax + b$ for the points $\{(x_1, y_1), (x_2, y_2), (x_3, y_3), . . . , (x_n, y_n)\}$ algebraically, you need to solve the following system for a and b.

$$\begin{cases} nb + \left(\sum_{i=1}^{n} x_i \right) a = \sum_{i=1}^{n} y_i \\ \left(\sum_{i=1}^{n} x_i \right) b + \left(\sum_{i=1}^{n} x_i^2 \right) a = \sum_{i=1}^{n} x_i y_i \end{cases}$$

In the system,

$$\sum_{i=1}^{n} x_i = x_1 + x_2 + \cdots + x_n$$

$$\sum_{i=1}^{n} y_i = y_1 + y_2 + \cdots + y_n$$

$$\sum_{i=1}^{n} x_i^2 = x_1^2 + x_2^2 + \cdots + x_n^2$$

$$\sum_{i=1}^{n} x_i y_i = x_1 y_1 + x_2 y_2 + \cdots + x_n y_n.$$

EXAMPLE 1   Finding a Least Squares Regression Line

Find the least squares regression line for the points $(-3, 0)$, $(-1, 1)$, $(0, 2)$, and $(2, 3)$.

Solution

Begin by constructing a table like that shown below.

x	y	xy	x^2
-3	0	0	9
-1	1	-1	1
0	2	0	0
2	3	6	4
$\displaystyle\sum_{i=1}^{n} x_i = -2$	$\displaystyle\sum_{i=1}^{n} y_i = 6$	$\displaystyle\sum_{i=1}^{n} x_i y_i = 5$	$\displaystyle\sum_{i=1}^{n} x_i^2 = 14$

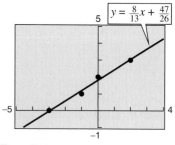

Figure B.10

Applying the system for the least squares regression line with $n = 4$ produces

$$\begin{cases} nb + \left(\displaystyle\sum_{i=1}^{n} x_i\right) a = \displaystyle\sum_{i=1}^{n} y_i \\ \left(\displaystyle\sum_{i=1}^{n} x_i\right) b + \left(\displaystyle\sum_{i=1}^{n} x_i^2\right) a = \displaystyle\sum_{i=1}^{n} x_i y_i \end{cases}$$ $$\begin{cases} 4b - 2a = 6 \\ -2b + 14a = 5 \end{cases}.$$

Solving this system of equations produces $a = \frac{8}{13}$ and $b = \frac{47}{26}$. So, the least squares regression line is $y = \frac{8}{13}x + \frac{47}{26}$, as shown in Figure B.10.

The least squares regression parabola $y = ax^2 + bx + c$ for the points

$$\{(x_1, y_1), (x_2, y_2), (x_3, y_3), \ldots, (x_n, y_n)\}$$

is obtained in a similar manner by solving the following system of three equations in three unknowns for a, b, and c.

$$\begin{cases} nc + \left(\displaystyle\sum_{i=1}^{n} x_i\right) b + \left(\displaystyle\sum_{i=1}^{n} x_i^2\right) a = \displaystyle\sum_{i=1}^{n} y_i \\ \left(\displaystyle\sum_{i=1}^{n} x_i\right) c + \left(\displaystyle\sum_{i=1}^{n} x_i^2\right) b + \left(\displaystyle\sum_{i=1}^{n} x_i^3\right) a = \displaystyle\sum_{i=1}^{n} x_i y_i \\ \left(\displaystyle\sum_{i=1}^{n} x_i^2\right) c + \left(\displaystyle\sum_{i=1}^{n} x_i^3\right) b + \left(\displaystyle\sum_{i=1}^{n} x_i^4\right) a = \displaystyle\sum_{i=1}^{n} x_i^2 y_i \end{cases}$$

Fortunately, graphing utilities have built-in least squares regression capabilities.

Appendix C   Solving Linear Equations and Inequalities

Linear Equations

A *linear equation* in one variable x is an equation that can be written in the standard form $ax + b = 0$, where a and b are real numbers with $a \neq 0$.

A linear equation has exactly one solution. To see this, consider the following steps. (Remember that $a \neq 0$.)

$ax + b = 0$	Original equation
$ax = -b$	Subtract b from both sides.
$x = -\dfrac{b}{a}$	Divide both sides by a.

To solve a linear equation in x, isolate x on one side of the equation by a sequence of *equivalent* (and usually simpler) equations, each having the same solution(s) as the original equation. The operations that yield equivalent equations come from the Substitution Principle and the simplification techniques discussed in Chapter P.

Generating Equivalent Equations

An equation can be transformed into an *equivalent equation* by one or more of the following steps.

	Original Equation	*Equivalent Equation*
1. Remove symbols of grouping, combine like terms, or reduce fractions on one or both sides of the equation.	$2x - x = 4$	$x = 4$
2. Add (or subtract) the same quantity to (from) *each* side of the equation.	$x + 1 = 6$	$x = 5$
3. Multiply (or divide) *each* side of the equation by the same *nonzero* quantity.	$2x = 6$	$x = 3$
4. Interchange the two sides of the equation.	$2 = x$	$x = 2$

After solving an equation, you should check each solution in the original equation.

EXAMPLE 1   Solving a Linear Equation

a. $3x - 6 = 0$ Original equation

$3x = 6$ Add 6 to each side.

$x = 2$ Divide each side by 3.

b. $4(2x + 3) = 6$ Original equation

$8x + 12 = 6$ Distributive Property

$8x = -6$ Subtract 12 from each side.

$x = -\dfrac{6}{8}$ Divide each side by 8.

$x = -\dfrac{3}{4}$ Simplify.

Linear Inequalities

Solving a linear inequality in one variable is much like solving a linear equation in one variable. To solve the inequality, you isolate the variable on one side using transformations that produce *equivalent inequalities*, which have the same solution(s).

Generating Equivalent Inequalities

An inequality can be transformed into an *equivalent inequality* by one or more of the following steps.

		Original Inequality	*Equivalent Inequality*
1.	Remove symbols of grouping, combine like terms, or reduce fractions on one or both sides of the inequality.	$4x + x \geq 2$	$5x \geq 2$
2.	Add (or subtract) the same number to (from) *each* side of the inequality.	$x - 3 < 5$	$x < 8$
3.	Multiply (or divide) each side of the inequality by the same *positive* number.	$\frac{1}{2}x > 3$	$x > 6$
4.	Multiply (or divide) each side of the inequality by the same *negative* number and *reverse* the inequality symbol.	$-2x \leq 6$	$x \geq -3$

EXAMPLE 2   Solving Linear Inequalities

a. $x + 5 \geq 3$ Original inequality

$x + 5 - 5 \geq 3 - 5$ Subtract 5 from each side.

$x \geq -2$ Simplify.

The solution is all real numbers greater than or equal to -2. Check several numbers that are greater than or equal to -2 in the original inequality.

b. $\dfrac{a}{3} \leq 12$ Original inequality

$3 \cdot \dfrac{a}{3} \leq 3 \cdot 12$ Multiply each side by 3.

$a \leq 36$ Simplify.

The solution is all real numbers less than or equal to 36. Check several numbers that are less than or equal to 36 in the original inequality.

c. $-4.2m > 6.3$ Original inequality

$\dfrac{-4.2m}{-4.2} < \dfrac{6.3}{-4.2}$ Divide each side by -4.2 and reverse inequality symbol.

$m < -1.5$ Simplify.

The solution is all real numbers less than -1.5. Check several numbers that are less than -1.5 in the original inequality.

C   E x e r c i s e s

In Exercises 1–18, solve the equation and check your solution.

1. $x + 4 = 7$

2. $x + 6 = 11$

3. $x - 2 = 5$

4. $x - 5 = 1$

5. $3x = 12$

6. $2x = 6$

7. $\dfrac{x}{5} = 4$

8. $\dfrac{x}{4} = 5$

9. $8x + 7 = 39$

10. $12x - 5 = 43$

11. $24 - 7x = 3$

12. $13 + 6x = 61$

13. $15x - 4 = 16$

14. $3x - 8 = 2$

15. $3x - 2(x + 5) = 10$

16. $4x + 2(7 - x) = 5$

17. $2x + 3 = 2x - 2$

18. $8(x - 2) = 4(2x - 4)$

In Exercises 19–36, solve the inequality and check your solution.

19. $x + 5 < 7$

20. $4 + x > -12$

21. $-x - 8 > -17$

22. $-3 + x < 19$

23. $6 + x \leq -8$

24. $x - 10 \geq -6$

25. $\frac{4}{5}x > 8$

26. $\frac{2}{3}x < -4$

27. $-\frac{3}{4}x > -3$

28. $-\frac{1}{6}x < -2$

29. $4x < 12$

30. $2x > -3$

31. $-11x \leq -22$

32. $-7x \geq 21$

33. $x - 3(x + 1) \geq 7$

34. $2(4x - 5) - 3x \leq -15$

35. $7x - 12 < 4x + 6$

36. $11 - 6x \leq 2x + 7$

Looking for solutions to your math problems? These study aids offer more than just the answers!

Two technology options offer complete solutions to every odd-numbered problem, help you practice and assess your skills, and provide problem-solving tools.

Interactive Precalculus Functions and Graphs: A Graphing Approach 2.0 CD-ROM

- Enhance your understanding with step-by-step solutions to all odd-numbered exercises in the text.
- Assess your skill levels with diagnostic pre- and post-tests for each chapter.
- Strengthen your skills with tutorial exercises accompanied by examples and diagnostics.
- Visualize and graph with a built-in Meridian Graphing Calculator Emulator.
- Enjoy learning with additional interactive features.

Internet Precalculus Functions and Graphs: A Graphing Approach 1.0

A subscription to this web site offers all of the CD-ROM features listed above, plus:
- Chat rooms for peer support
- Bulletin boards for sharing ideas and insights on each chapter

To purchase Interactive Precalculus Functions and Graphs: A Graphing Approach 2.0 CD-ROM

- Visit Houghton Mifflin's College Store at **college.hmco.com** or contact your campus bookstore.

To subscribe online to Internet Precalculus Functions and Graphs: A Graphing Approach 1.0

- Visit Houghton Mifflin's College Division web site at **college.hmco.com** and select mathematics.

Answers to Odd-Numbered Exercises and Tests

Chapter P

Section P.1 *(page 9)*

1.

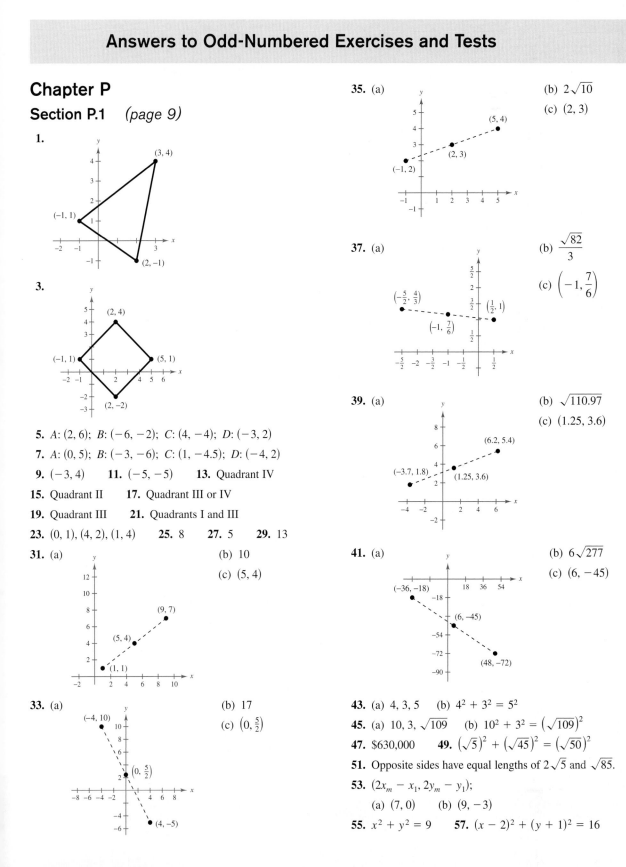

3.

5. $A: (2, 6)$; $B: (-6, -2)$; $C: (4, -4)$; $D: (-3, 2)$

7. $A: (0, 5)$; $B: (-3, -6)$; $C: (1, -4.5)$; $D: (-4, 2)$

9. $(-3, 4)$ **11.** $(-5, -5)$ **13.** Quadrant IV

15. Quadrant II   **17.** Quadrant III or IV

19. Quadrant III   **21.** Quadrants I and III

23. $(0, 1), (4, 2), (1, 4)$ **25.** 8   **27.** 5   **29.** 13

31. (a)   (b) 10

(c) $(5, 4)$

33. (a)   (b) 17

(c) $\left(0, \frac{5}{2}\right)$

35. (a)   (b) $2\sqrt{10}$

(c) $(2, 3)$

37. (a)   (b) $\dfrac{\sqrt{82}}{3}$

(c) $\left(-1, \dfrac{7}{6}\right)$

39. (a)   (b) $\sqrt{110.97}$

(c) $(1.25, 3.6)$

41. (a)   (b) $6\sqrt{277}$

(c) $(6, -45)$

43. (a) $4, 3, 5$ (b) $4^2 + 3^2 = 5^2$

45. (a) $10, 3, \sqrt{109}$ (b) $10^2 + 3^2 = \left(\sqrt{109}\right)^2$

47. \$630,000   **49.** $\left(\sqrt{5}\right)^2 + \left(\sqrt{45}\right)^2 = \left(\sqrt{50}\right)^2$

51. Opposite sides have equal lengths of $2\sqrt{5}$ and $\sqrt{85}$.

53. $(2x_m - x_1, 2y_m - y_1)$;

(a) $(7, 0)$ (b) $(9, -3)$

55. $x^2 + y^2 = 9$ **57.** $(x - 2)^2 + (y + 1)^2 = 16$

59. $(x + 1)^2 + (y - 2)^2 = 5$

61. $(x - 3)^2 + (y - 4)^2 = 25$

63.
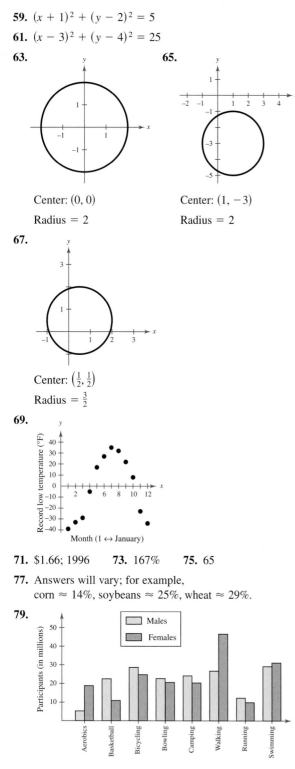

Center: $(0, 0)$

Radius $= 2$

65.

Center: $(1, -3)$

Radius $= 2$

67.

Center: $\left(\frac{1}{2}, \frac{1}{2}\right)$

Radius $= \frac{3}{2}$

69.

71. $1.66; 1996   **73.** 167%   **75.** 65

77. Answers will vary; for example,
corn $\approx 14\%$, soybeans $\approx 25\%$, wheat $\approx 29\%$.

79.

81. (a) $\approx 52\%$

(b) No. The trend limits the amount of funds available for capital improvements in industry. (Answers will vary.)

83. (a) During 1993

(b)
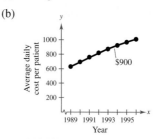

85. ≈ 180.28 kilometers

87. (a) Answers will vary; from 1991 on, the data are nearly in a straight line. Estimate: 7 new members in 2001.

(b) Answers will vary; elections for inclusion in the Hall of Fame began in 1986.

89. $1245.05 million

91. True. The lengths of the sides from $(-8, 4)$ to $(2, 11)$ and from $(2, 11)$ to $(-5, 1)$ are both $\sqrt{149}$.

93. 0; 0

Section P.2 *(page 21)*

1. (a) Yes   (b) Yes   **3.** (a) No   (b) Yes

5. (a) No   (b) Yes   **7.** (a) Yes   (b) Yes

9.

x	-1	0	1	$\frac{3}{2}$	2
y	5	3	1	0	-1

11.

x	-1	0	1	2	3
y	3	0	-1	0	3

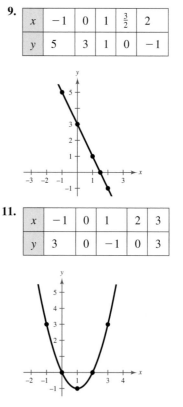

13.

x	0	1	2	3	4
y	1	2	3	2	1

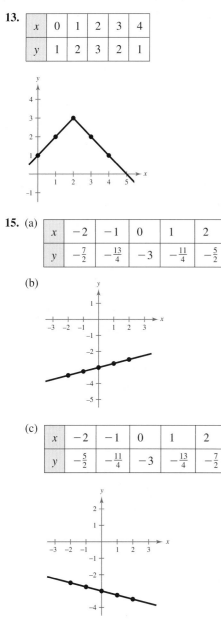

15. (a)

x	-2	-1	0	1	2
y	$-\frac{7}{2}$	$-\frac{13}{4}$	-3	$-\frac{11}{4}$	$-\frac{5}{2}$

(b)

(c)

x	-2	-1	0	1	2
y	$-\frac{5}{2}$	$-\frac{11}{4}$	-3	$-\frac{13}{4}$	$-\frac{7}{2}$

The lines have opposite slopes but the same y-intercept.

17. (d) **19.** (f) **21.** (a)

23. **25.**

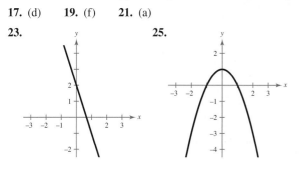

27. **29.**

31. **33.**

35. **37.**

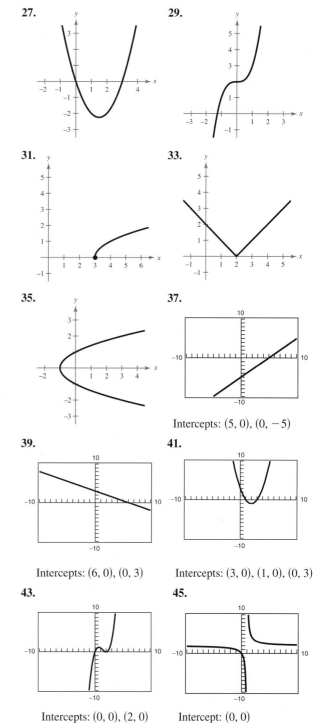

Intercepts: $(5, 0)$, $(0, -5)$

39. **41.**

Intercepts: $(6, 0)$, $(0, 3)$ Intercepts: $(3, 0)$, $(1, 0)$, $(0, 3)$

43. **45.**

Intercepts: $(0, 0)$, $(2, 0)$ Intercept: $(0, 0)$

47.

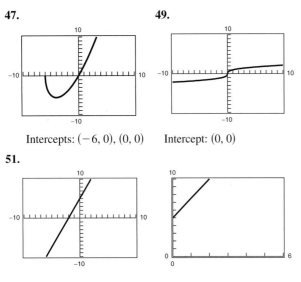

Intercepts: $(-6, 0)$, $(0, 0)$

49.

Intercept: $(0, 0)$

51.

The standard setting gives a more complete graph.

53.

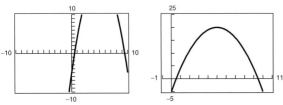

The specified setting gives a more complete graph.

55.

Xmin = -5
Xmax = 5
Xscl = 1
Ymin = -30
Ymax = 10
Yscl = 5

57.

Xmin = -30
Xmax = 30
Xscl = 5
Ymin = -10
Ymax = 50
Yscl = 5

59. $y_1 = \sqrt{64 - x^2}$
$y_2 = -\sqrt{64 - x^2}$

61. $y_1 = \sqrt{49 - x^2}$
$y_2 = -\sqrt{49 - x^2}$

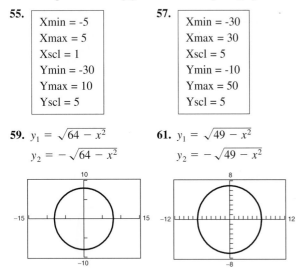

63. The graphs are identical. Distributive Property

65. The graphs are identical. Associative Property of Multiplication

67.

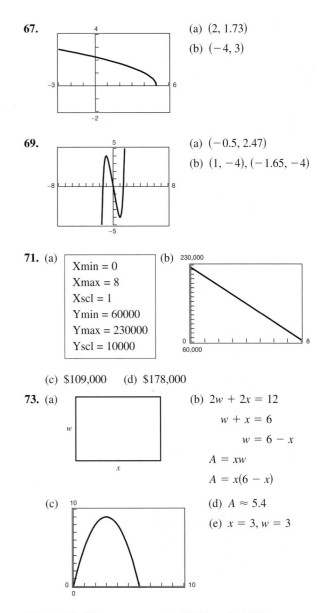

(a) $(2, 1.73)$

(b) $(-4, 3)$

69.

(a) $(-0.5, 2.47)$

(b) $(1, -4)$, $(-1.65, -4)$

71. (a)

Xmin = 0
Xmax = 8
Xscl = 1
Ymin = 60000
Ymax = 230000
Yscl = 10000

(b)

(c) \$109,000   (d) \$178,000

73. (a)

(b) $2w + 2x = 12$
$w + x = 6$
$w = 6 - x$
$A = xw$
$A = x(6 - x)$

(c)

(d) $A \approx 5.4$

(e) $x = 3, w = 3$

75. (a) The life expectancy of a child born in 1950

(b) 1973   (c) ≈ 66.3 years   (d) ≈ 78.7 years

77. (a)

x	10	20	30	40	50
y	107.33	26.56	11.60	6.36	3.94

x	60	70	80	90	100
y	2.62	1.83	1.31	0.96	0.71

(b) $x = 45.64$ (c) $y = 1.103$

(d) They are inversely related.

79. False. $x = 0$ has an infinite number of x-intercepts.

81. Answers will vary.

Section P.3 *(page 33)*

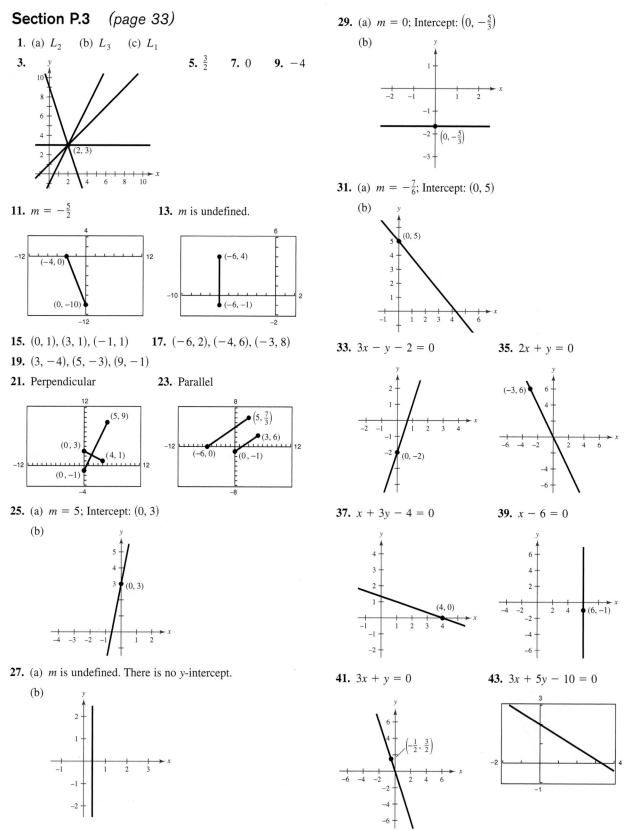

1. (a) L_2 (b) L_3 (c) L_1

3.

5. $\frac{3}{2}$ **7.** 0 **9.** -4

11. $m = -\frac{5}{2}$

13. m is undefined.

15. $(0, 1), (3, 1), (-1, 1)$ **17.** $(-6, 2), (-4, 6), (-3, 8)$

19. $(3, -4), (5, -3), (9, -1)$

21. Perpendicular

23. Parallel

25. (a) $m = 5$; Intercept: $(0, 3)$

(b)

27. (a) m is undefined. There is no y-intercept.

(b)

29. (a) $m = 0$; Intercept: $\left(0, -\frac{5}{3}\right)$

(b)

31. (a) $m = -\frac{7}{6}$; Intercept: $(0, 5)$

(b)

33. $3x - y - 2 = 0$ **35.** $2x + y = 0$

37. $x + 3y - 4 = 0$ **39.** $x - 6 = 0$

41. $3x + y = 0$ **43.** $3x + 5y - 10 = 0$

45. $x + 8 = 0$

47. $x + 2y - 3 = 0$

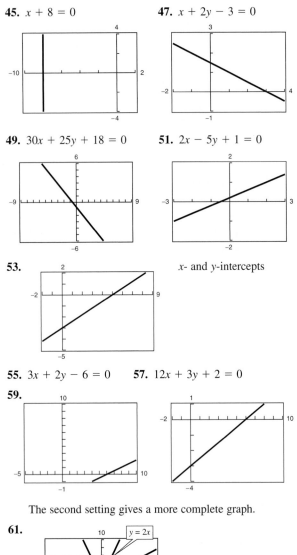

49. $30x + 25y + 18 = 0$

51. $2x - 5y + 1 = 0$

53.

x- and y-intercepts

55. $3x + 2y - 6 = 0$ **57.** $12x + 3y + 2 = 0$

59.

The second setting gives a more complete graph.

61.

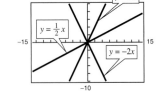

$y = \frac{1}{2}x$ and $y = -2x$ are perpendicular.

63.

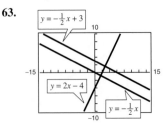

$y = -\frac{1}{2}x$ and $y = -\frac{1}{2}x + 3$ are parallel. Both are perpendicular to $y = 2x - 4$.

65. (a) $2x - y - 3 = 0$ (b) $x + 2y - 4 = 0$

67. (a) $-6x - 8y + 3 = 0$ (b) $96x - 72y + 127 = 0$

69. (a) $10x - 10y + 43 = 0$ (b) $10x + 10y - 93 = 0$

71. $3x - 2y - 1 = 0$

73. (a) Sales increase of $135

(b) No sales increase

(c) Sales decrease of $40

75. (a) Greatest increases per share: 1990 and 1996

Greatest decrease per share: 1997

(b) $37x - 1000y + 943 = 0$

(c) Slope is the average increase per share per year.

(d) $1.46. Answers will vary.

77. 16,667 feet   **79.** $125t - V + 2415 = 0$

81. $2000t + V - 22,400 = 0$

83. (b); slope $= -10$; the amount owed decreases by $10 per week.

85. (a); slope $= 0.25$; expenses increase by $0.25 per mile.

87. $F = \frac{9}{5}C + 32$ **89.** $39,500

91. (a) $V = -175t + 875$

(b)

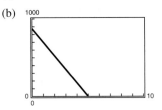

t	0	1	2	3	4	5
V	875	700	525	350	175	0

93. (a) $C = 16.75t + 36,500$ (b) $R = 27t$

(c) $P = 10.25t - 36,500$ (d) $t \approx 3561$ hours

95. (a) $y = 92.84t + 487.82$ (b) $1,601,000

(c) Average increase per year

97. False. The line through $(10, -3)$ and $(2, -9)$ is $y = \frac{3}{4}x + \frac{9}{4}$, and $\left(-12, -\frac{37}{2}\right)$ is not on this line.

99. -4

101. No. The slopes of two perpendicular lines have opposite signs (assuming that neither line is vertical nor horizontal).

Section P.4   *(page 50)*

1. (a) Yes   (b) No   (c) No   (d) No

3. (a) Yes   (b) No   (c) No   (d) No

5. (a) No   (b) No   (c) No   (d) Yes

7. Identity   **9.** Identity   **11.** Conditional   **13.** $-\frac{96}{23}$

15. -4 **17.** $-\frac{6}{5}$ **19.** 10   **21.** 4   **23.** 5

25. $\frac{11}{6}$ **27.** $\frac{5}{3}$ **29.** No solution

31. $(5, 0), (0, -5)$ **33.** $(-2, 0), (1, 0), (0, -2)$

35. $(-2, 0), (0, 0)$ **37.** $(-2, 0), (6, 0), (0, -2)$

39. $(1, 0), \left(0, \frac{1}{2}\right)$

41.

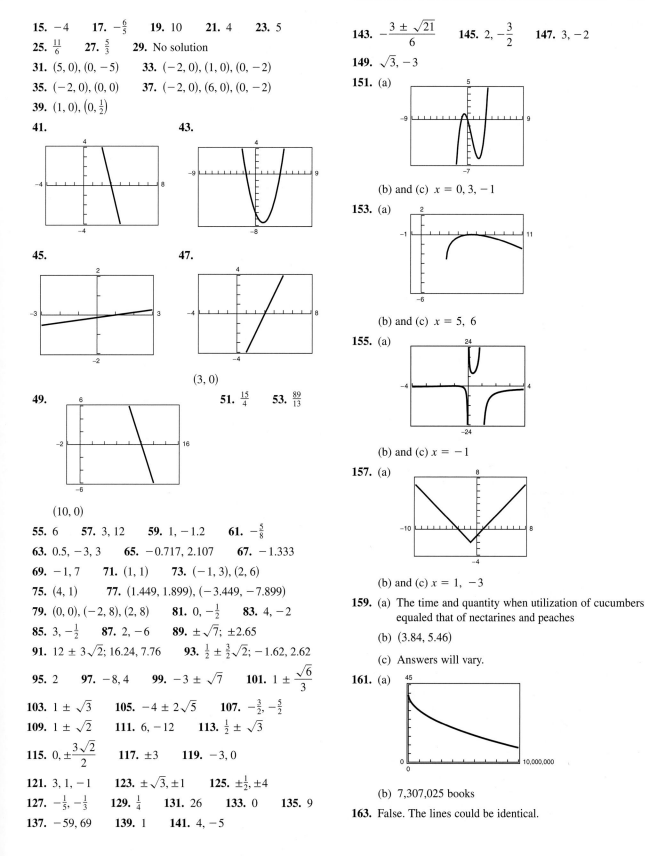

43.

45.

47.

$(3, 0)$

49.

51. $\frac{15}{4}$ **53.** $\frac{89}{13}$

$(10, 0)$

55. 6   **57.** 3, 12   **59.** 1, -1.2 **61.** $-\frac{5}{8}$

63. $0.5, -3, 3$ **65.** $-0.717, 2.107$ **67.** -1.333

69. $-1, 7$ **71.** $(1, 1)$ **73.** $(-1, 3), (2, 6)$

75. $(4, 1)$ **77.** $(1.449, 1.899), (-3.449, -7.899)$

79. $(0, 0), (-2, 8), (2, 8)$ **81.** $0, -\frac{1}{2}$ **83.** $4, -2$

85. $3, -\frac{1}{2}$ **87.** $2, -6$ **89.** $\pm\sqrt{7}; \pm2.65$

91. $12 \pm 3\sqrt{2}; 16.24, 7.76$ **93.** $\frac{1}{2} \pm \frac{3}{2}\sqrt{2}; -1.62, 2.62$

95. 2   **97.** $-8, 4$ **99.** $-3 \pm \sqrt{7}$ **101.** $1 \pm \frac{\sqrt{6}}{3}$

103. $1 \pm \sqrt{3}$ **105.** $-4 \pm 2\sqrt{5}$ **107.** $-\frac{3}{2}, -\frac{5}{2}$

109. $1 \pm \sqrt{2}$ **111.** $6, -12$ **113.** $\frac{1}{2} \pm \sqrt{3}$

115. $0, \pm\frac{3\sqrt{2}}{2}$ **117.** ±3 **119.** $-3, 0$

121. $3, 1, -1$ **123.** $\pm\sqrt{3}, \pm1$ **125.** $\pm\frac{1}{2}, \pm4$

127. $-\frac{1}{5}, -\frac{1}{3}$ **129.** $\frac{1}{4}$ **131.** 26   **133.** 0   **135.** 9

137. $-59, 69$ **139.** 1   **141.** $4, -5$

143. $-\dfrac{3 \pm \sqrt{21}}{6}$ **145.** $2, -\dfrac{3}{2}$ **147.** $3, -2$

149. $\sqrt{3}, -3$

151. (a)

(b) and (c) $x = 0, 3, -1$

153. (a)

(b) and (c) $x = 5, 6$

155. (a)

(b) and (c) $x = -1$

157. (a)

(b) and (c) $x = 1, -3$

159. (a) The time and quantity when utilization of cucumbers equaled that of nectarines and peaches

(b) $(3.84, 5.46)$

(c) Answers will vary.

161. (a)

(b) 7,307,025 books

163. False. The lines could be identical.

Section P.5 *(page 63)*

1. (d)   **3.** (c)

5. (a) Yes   (b) No   (c) Yes   (d) No

7. (a) No   (b) Yes   (c) Yes   (d) No

9. $x > -4$

11. $x < -\frac{1}{2}$

13. $-1 < x < 3$

15. $-2 < x \leq 5$

17. $-\frac{9}{2} < x < \frac{15}{2}$

19.

$x > 2$

21.

$x \leq 2$

23.

$-\frac{4}{3} < x < \frac{1}{3}$

25.

(a) $x \geq 2$

(b) $x \leq \frac{3}{2}$

27.

(a) $-2 \leq x \leq 4$

(b) $x \leq 4$

29. $x < -2,\ x > 2$

31. $1 < x < 13$

33. $x < -28,\ x > 0$

35. $-2 < x < 3$

37.

(a) $1 \leq x \leq 5$

(b) $x \leq -1,\ x \geq 7$

39. $|x| \leq 3$ **41.** $|x - 7| \geq 3$ **43.** $|x - 12| \leq 10$

45. $(-7, 3)$ **47.** $(-\infty, -5], [1, \infty)$

49. $[-2, 0], [2, \infty)$

51.

(a) $x \leq -1,\ x \geq 3$

(b) $0 \leq x \leq 2$

53.

(a) $-2 \leq x \leq 0,$
$2 \leq x < \infty$

(b) $x \leq 4$

55. $(-\infty, -1), (0, 1)$ **57.** $(-\infty, -1), (4, \infty)$

59.

(a) $0 \leq x < 2$

(b) $2 < x \leq 4$

61.

(a) $2 \leq |x|$

(b) $-\infty < x < \infty$

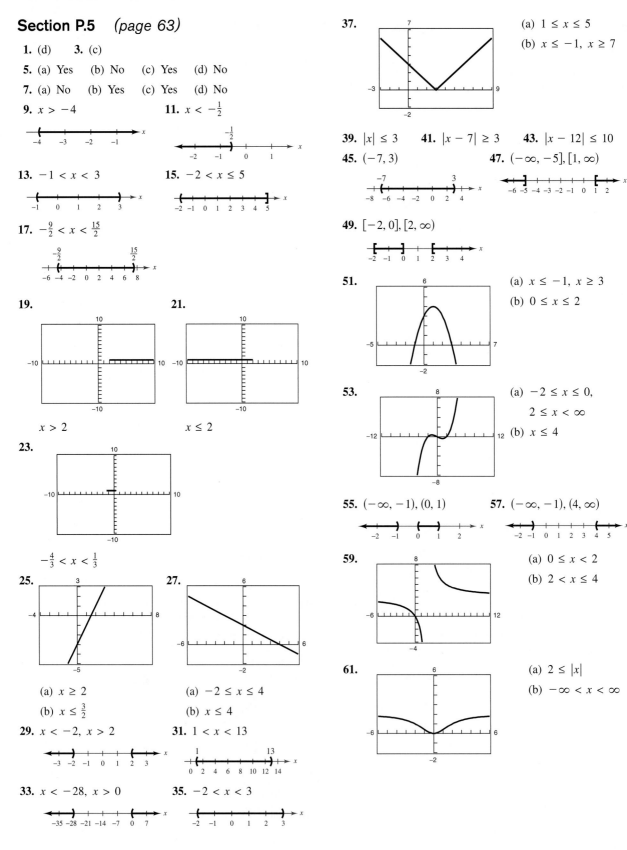

63. $[5, \infty)$ **65.** $(-\infty, \infty)$ **67.** $\left[-\frac{5}{2}, \infty\right)$

69. (a) and (b)

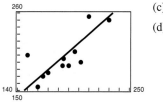

(c) 186.23 pounds

(d) Answers will vary.

71. $[65.8, 71.2]$

73. (a) $333\frac{1}{3}$ vibrations per second (b) 3.6 millimeters

(c) $1.2 < t < 2.4$ (d) $0 < v < 500$

75. False. $10 \geq -x$

77. (a) a, b

(b)

Interval	Expression	Sign
$(-\infty, a)$	$(x - a)$	$-$
	$(x - b)$	$-$
	$(x - a)(x - b)$	$+$
(a, b)	$(x - a)$	$+$
	$(x - b)$	$-$
	$(x - a)(x - b)$	$-$
(b, ∞)	$(x - a)$	$+$
	$(x - b)$	$+$
	$(x - a)(x - b)$	$+$

(c) x-values that are zeros of the function

Review Exercises *(page 67)*

1.

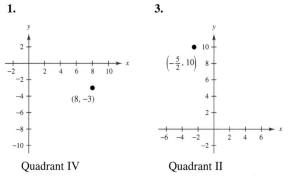

Quadrant IV

3.

$\left(-\frac{5}{2}, 10\right)$

Quadrant II

5. Quadrant IV

7. (a)

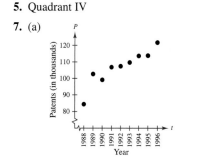

(b) The number of patents issued has increased fairly steadily since 1988.

9.

Distance: 5

11.

$2\left(\sqrt{185}\right)^2 = \left(\sqrt{370}\right)^2$

13.

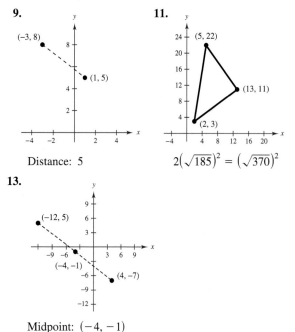

Midpoint: $(-4, -1)$

15. (a) \$352.35 million

(b) The estimate is \$2.95 million over the actual revenue, which is less than a 1% error.

17. $(x - 3)^2 + (y - 2)^2 = 65$

19.

x	-1	0	1	2	3
y	4	0	-2	-2	0

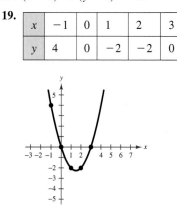

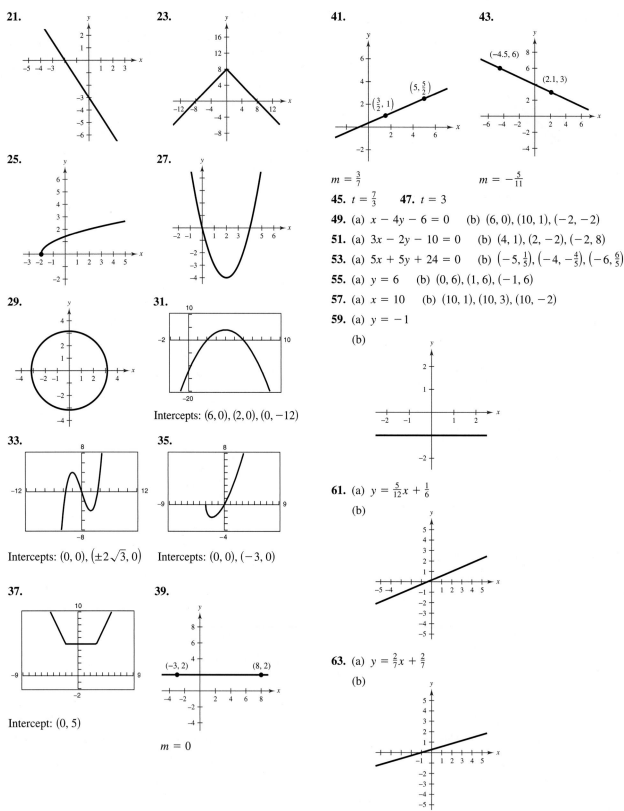

21.

23.

25.

27.

29.

31.

Intercepts: $(6, 0), (2, 0), (0, -12)$

33.

Intercepts: $(0, 0), (\pm 2\sqrt{3}, 0)$

35.

Intercepts: $(0, 0), (-3, 0)$

37.

Intercept: $(0, 5)$

39.

$m = 0$

41.

$m = \frac{3}{7}$

43.

$m = -\frac{5}{11}$

45. $t = \frac{7}{3}$ **47.** $t = 3$

49. (a) $x - 4y - 6 = 0$ (b) $(6, 0), (10, 1), (-2, -2)$

51. (a) $3x - 2y - 10 = 0$ (b) $(4, 1), (2, -2), (-2, 8)$

53. (a) $5x + 5y + 24 = 0$ (b) $\left(-5, \frac{1}{5}\right), \left(-4, -\frac{4}{5}\right), \left(-6, \frac{6}{5}\right)$

55. (a) $y = 6$ (b) $(0, 6), (1, 6), (-1, 6)$

57. (a) $x = 10$ (b) $(10, 1), (10, 3), (10, -2)$

59. (a) $y = -1$

(b)

61. (a) $y = \frac{5}{12}x + \frac{1}{6}$

(b)

63. (a) $y = \frac{2}{7}x + \frac{2}{7}$

(b)

65. (a) $5x - 4y - 23 = 0$

(b) $4x + 5y - 2 = 0$

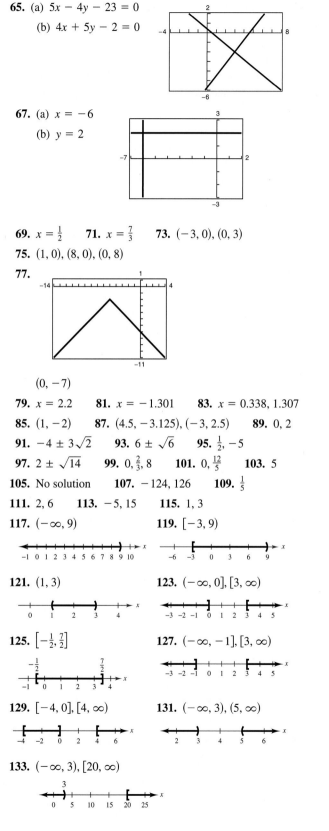

67. (a) $x = -6$

(b) $y = 2$

69. $x = \frac{1}{2}$ **71.** $x = \frac{7}{3}$ **73.** $(-3, 0), (0, 3)$

75. $(1, 0), (8, 0), (0, 8)$

77.

$(0, -7)$

79. $x = 2.2$ **81.** $x = -1.301$ **83.** $x = 0.338, 1.307$

85. $(1, -2)$ **87.** $(4.5, -3.125), (-3, 2.5)$ **89.** $0, 2$

91. $-4 \pm 3\sqrt{2}$ **93.** $6 \pm \sqrt{6}$ **95.** $\frac{1}{2}, -5$

97. $2 \pm \sqrt{14}$ **99.** $0, \frac{2}{3}, 8$ **101.** $0, \frac{12}{5}$ **103.** 5

105. No solution   **107.** $-124, 126$ **109.** $\frac{1}{5}$

111. $2, 6$ **113.** $-5, 15$ **115.** $1, 3$

117. $(-\infty, 9)$ **119.** $[-3, 9)$

121. $(1, 3)$ **123.** $(-\infty, 0], [3, \infty)$

125. $\left[-\frac{1}{2}, \frac{7}{2}\right]$ **127.** $(-\infty, -1], [3, \infty)$

129. $[-4, 0], [4, \infty)$ **131.** $(-\infty, 3), (5, \infty)$

133. $(-\infty, 3), [20, \infty)$

135. $[430.044, 435.244]$ **137.** True

139. The real zeros of a function are the values of x for which the graph of the function crosses the x-axis (the x-intercepts). They are also the values of x which satisfy the equation $f(x) = 0$.

Chapter Test   *(page 71)*

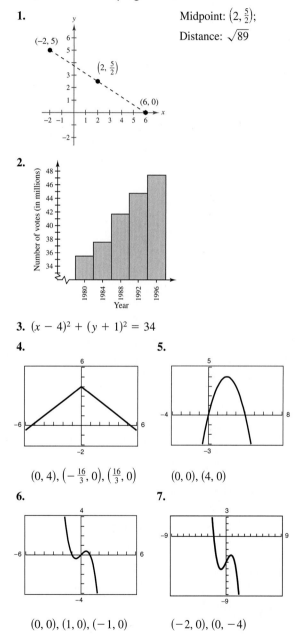

1. Midpoint: $\left(2, \frac{5}{2}\right)$; Distance: $\sqrt{89}$

2.

3. $(x - 4)^2 + (y + 1)^2 = 34$

4.

$(0, 4), \left(-\frac{16}{3}, 0\right), \left(\frac{16}{3}, 0\right)$

5.

$(0, 0), (4, 0)$

6.

$(0, 0), (1, 0), (-1, 0)$

7.

$(-2, 0), (0, -4)$

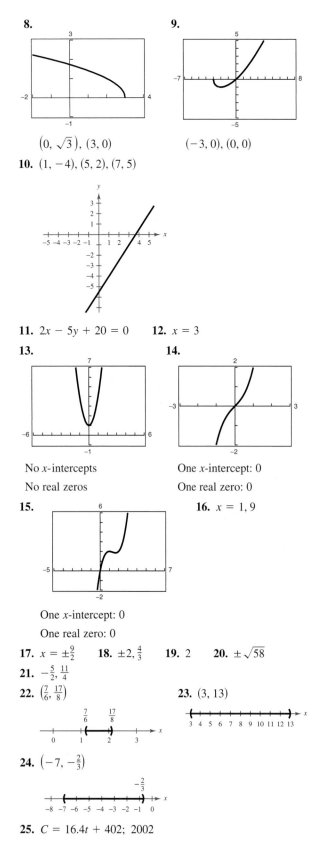

8.

$\left(0, \sqrt{3}\right), (3, 0)$

9.

$(-3, 0), (0, 0)$

10. $(1, -4), (5, 2), (7, 5)$

11. $2x - 5y + 20 = 0$ **12.** $x = 3$

13.

No x-intercepts

No real zeros

14.

One x-intercept: 0

One real zero: 0

15.

One x-intercept: 0

One real zero: 0

16. $x = 1, 9$

17. $x = \pm \frac{9}{2}$ **18.** $\pm 2, \frac{4}{3}$ **19.** 2   **20.** $\pm \sqrt{58}$

21. $-\frac{5}{2}, \frac{11}{4}$

22. $\left(\frac{7}{6}, \frac{17}{8}\right)$ **23.** $(3, 13)$

24. $\left(-7, -\frac{2}{3}\right)$

25. $C = 16.4t + 402;\ 2002$

Chapter 1
Section 1.1   *(page 82)*

1. Yes   **3.** No

5. Yes. Each input value is matched with one output value.

7. No. The same input value is matched with two different output values.

9. (a) Function

(b) Not a function because the element 1 in A corresponds to two elements, -2 and 1, in B.

(c) Function   (d) Not a function

11. Each is a function. To each year there corresponds one and only one circulation.

13. Not a function   **15.** Function   **17.** Function

19. Not a function   **21.** Function   **23.** Not a function

25. (a) $\dfrac{1}{5}$ (b) 1   (c) $\dfrac{1}{4t + 1}$ (d) $\dfrac{1}{x + c + 1}$

27. (a) -1 (b) -9 (c) $2x - 5$

29. (a) 0   (b) -0.75 (c) $x^2 + 2x$

31. (a) 1   (b) 2.5   (c) $3 - 2|x|$

33. (a) $-\dfrac{1}{9}$ (b) Undefined   (c) $\dfrac{1}{y^2 + 6y}$

35. (a) 1   (b) -1 (c) 1   **37.** (a) -1 (b) 2   (c) 6

39.

x	-2	-1	0	1	2
$f(x)$	1	-2	-3	-2	1

41.

t	-5	-4	-3	-2	-1
$h(t)$	1	$\frac{1}{2}$	0	$\frac{1}{2}$	1

43.

x	-2	-1	0	1	2
$f(x)$	5	$\frac{9}{2}$	4	1	0

45. 5   **47.** $\frac{4}{3}$ **49.** ± 3 **51.** ± 4 **53.** $2, -1$

55. 0, 3   **57.** All real numbers x

59. All real numbers t except $t = 0$

61. All real numbers y such that $y \geq 10$

63. All real numbers x such that $-1 \leq x \leq 1$

65. All real numbers x except $x = 0, -2$

67. All real numbers s such that $s \neq 4$ and $s \geq 1$

69. All real numbers x except $x = 0$

71. $\{(-2, 4), (-1, 1), (0, 0), (1, 1), (2, 4)\}$

73. $\left\{(-2, 0), (-1, 1), \left(0, \sqrt{2}\right), \left(1, \sqrt{3}\right), (2, 2)\right\}$

75. $g(x) = -2x^2;\ c = -2$ **77.** $r(x) = \dfrac{32}{x};\ c = 32$

79. 2 **81.** $3 + h$, $h \neq 0$ **83.** $3x^2 + 3xc + c^2$, $c \neq 0$

85. $-\dfrac{1}{t}$, $t \neq 1$ **87.** $A = \dfrac{C^2}{4\pi}$ **89.** $A = \dfrac{s^2}{2}$

91. (a)

Height, x	Width	Volume, V
1	$24 - 2(1)$	$1[24 - 2(1)]^2 = 484$
2	$24 - 2(2)$	$2[24 - 2(2)]^2 = 800$
3	$24 - 2(3)$	$3[24 - 2(3)]^2 = 972$
4	$24 - 2(4)$	$4[24 - 2(4)]^2 = 1024$
5	$24 - 2(5)$	$5[24 - 2(5)]^2 = 980$
6	$24 - 2(6)$	$6[24 - 2(6)]^2 = 864$

Maximum when $x = 4$

(b) $V = x(24 - 2x)^2$, $0 < x < 12$

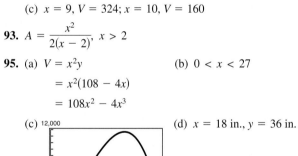

(c) $x = 9$, $V = 324$; $x = 10$, $V = 160$

93. $A = \dfrac{x^2}{2(x - 2)}$, $x > 2$

95. (a) $V = x^2 y$ (b) $0 < x < 27$

$\quad = x^2(108 - 4x)$

$\quad = 108x^2 - 4x^3$

(c)  (d) $x = 18$ in., $y = 36$ in.

97. (a) $C = 12.30x + 98,000$ (b) $R = 17.98x$

(c) $P = 5.68x - 98,000$

99. (a) $R = \dfrac{240n - n^2}{20}$

(b)

n	90	100	110	120	130	140	150
$R(n)$	\$675	\$700	\$715	\$720	\$715	\$700	\$675

The revenue is maximum when $n = 120$.

(c)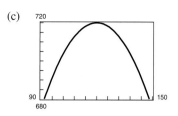

The revenue is maximum when $n = 120$.

101. (a)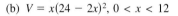

(b) $h = \sqrt{d^2 - 3000^2}$; $d \geq 3000$

(c)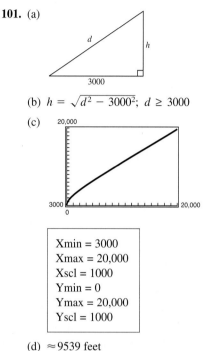

Xmin = 3000
Xmax = 20,000
Xscl = 1000
Ymin = 0
Ymax = 20,000
Yscl = 1000

(d) ≈ 9539 feet

103. False. Range is $[-1, \infty)$.

105. No. 3 corresponds to both u and v.

107. Function notation is a convenient way of referencing the value of a function for a specific domain value.

109. 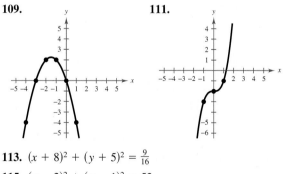 **111.**

113. $(x + 8)^2 + (y + 5)^2 = \dfrac{9}{16}$

115. $(x - 2)^2 + (y - 1)^2 = 52$

Section 1.2 *(page 96)*

1. Domain: $(-\infty, \infty)$; Range: $(-\infty, 1]$

3. Domain: $(-\infty, -1]$, $[1, \infty)$; Range: $[0, \infty)$

5. Domain: $(-\infty, \infty)$; Range: $[0, \infty)$

7.

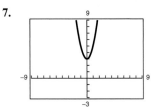

Domain: $(-\infty, \infty)$; Range: $[3, \infty)$

9.

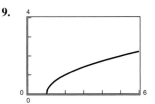

Domain: $[1, \infty)$; Range: $[0, \infty)$

11.

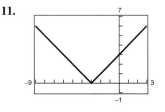

Domain: $(-\infty, \infty)$; Range: $[0, \infty)$

13. Function. Graph the given function over the window shown in the figure.

15. Not a function. Solve for y and graph the resulting two functions.

17. Function. Solve for y and graph the resulting function.

19. (a) Increasing on $(-\infty, \infty)$ (b) Odd function

21. (a) Increasing on $(-\infty, 0)$, $(2, \infty)$

Decreasing on $(0, 2)$

(b) Neither even nor odd

23. (a)

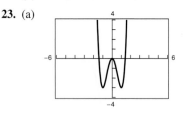

(b) Increasing on $(-1, 0)$, $(1, \infty)$

Decreasing on $(-\infty, -1)$, $(0, 1)$

(c) Even function

25. (a)

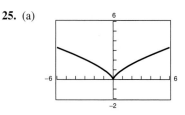

(b) Decreasing on $(-\infty, 0)$; Increasing on $(0, \infty)$

(c) Even function

27. (a)

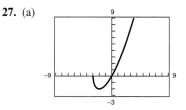

(b) Increasing on $(-2, \infty)$; Decreasing on $(-3, -2)$

(c) Neither even nor odd

29. (a)

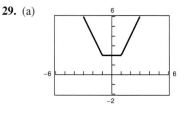

(b) Decreasing on $(-\infty, -1)$; Constant on $(-1, 1)$; Increasing on $(1, \infty)$

(c) Even function

31.

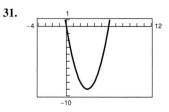

Relative minimum: $(3, -9)$

33.

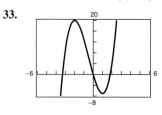

Relative minimum: $(1, -7)$

Relative maximum: $(-2, 20)$

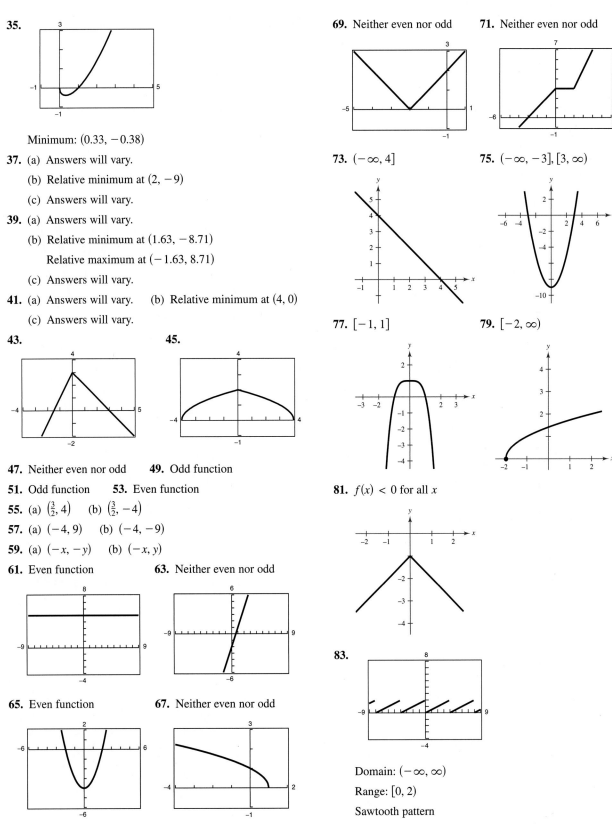

35.

Minimum: $(0.33, -0.38)$

37. (a) Answers will vary.

    (b) Relative minimum at $(2, -9)$

    (c) Answers will vary.

39. (a) Answers will vary.

    (b) Relative minimum at $(1.63, -8.71)$

        Relative maximum at $(-1.63, 8.71)$

    (c) Answers will vary.

41. (a) Answers will vary.   (b) Relative minimum at $(4, 0)$

    (c) Answers will vary.

43.   **45.**

47. Neither even nor odd   **49.** Odd function

51. Odd function   **53.** Even function

55. (a) $\left(\frac{3}{2}, 4\right)$ (b) $\left(\frac{3}{2}, -4\right)$

57. (a) $(-4, 9)$ (b) $(-4, -9)$

59. (a) $(-x, -y)$ (b) $(-x, y)$

61. Even function   **63.** Neither even nor odd

65. Even function   **67.** Neither even nor odd

69. Neither even nor odd   **71.** Neither even nor odd

73. $(-\infty, 4]$ **75.** $(-\infty, -3], [3, \infty)$

77. $[-1, 1]$ **79.** $[-2, \infty)$

81. $f(x) < 0$ for all x

83.

Domain: $(-\infty, \infty)$

Range: $[0, 2)$

Sawtooth pattern

85. (a) Answers will vary.

(b)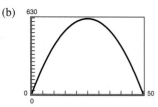

(c) 625 square meters; 25 × 25 meters

87. (a) C_2 is the appropriate model. The cost of the first minute is $1.05 and the cost increases $0.38 when the next minute begins, etc.

(b)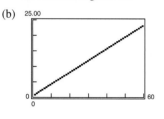

$7.89

89. $h = -x^2 + 4x - 3, \ 1 \le x \le 3$

91. $h = 2x - x^2, \ 0 \le x \le 2$ **93.** $L = \frac{1}{2}y^2, \ 0 \le y \le 4$

95. (a) $y = 1.473x^3 - 16.411x^2 + 31.24x - 95.2$

(b) Domain: $[0, 7]$

(c)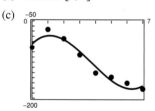

(d) Most accurate: 1992; Least accurate: 1991

(e) Yes. The cubic curve is starting an upswing for future years.

97. False. Counterexample: $f(x) = \sqrt{1 + x^2}$

99. If $y = a_{2n+1}x^{2n+1} + a_{2n-1}x^{2n-1} + \cdots + a_3x^3 + a_1x$, each exponent is odd. Then

$$f(-x) = -a_{2n+1}x^{2n+1} - a_{2n-1}x^{2n-1} -$$
$$\cdots - a_3x^3 - a_1x,$$

which is equal to $-f(x)$. Therefore, by definition, the original function is odd.

101. (a) Even. g is a reflection in the x-axis.

(b) Even. g is a reflection in the y-axis.

(c) Even. g is a vertical shift downward.

(d) Neither even nor odd. g is shifted to the right and reflected in the x-axis.

103. No. x is not a function of y because horizontal lines can be drawn to intersect the graph twice, so each y-value corresponds to two distinct x-values if $-5 < y < 5$.

105. (a) $d = 10$ (b) Midpoint: $(-1, 3)$

107. (a) $d = \dfrac{\sqrt{733}}{4}$ (b) Midpoint: $\left(-\dfrac{21}{8}, \dfrac{5}{12}\right)$

109. (a) -17 (b) 1 (c) $-x^2 + 3x + 1$

111. (a) 6 (b) -55 (c) $\frac{1}{9}$ **113.** $-h - 6, \ h \neq 0$

Section 1.3 *(page 106)*

1. 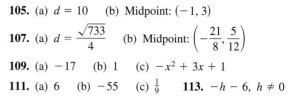 **3.**

5. **7.**

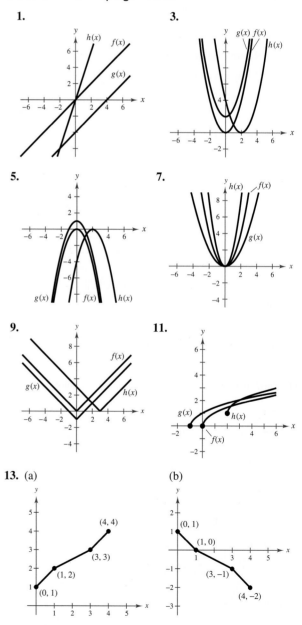

9. **11.**

13. (a) (b)

(c)

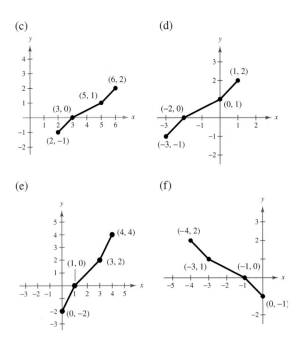

(d)

(e)

(f)

15. Vertical shrink of $y = x$ **17.** Constant function
$y = \frac{1}{2}x$ $y = 7$

19. Reflection in the x-axis and a vertical shift of $y = \sqrt{x}$
$y = 1 - \sqrt{x}$

21. Horizontal shift of $y = |x|$ **23.** Vertical shift of $y = x^2$
$y = |x + 2|$ $y = x^2 - 1$

25. Reflection in the x-axis of $y = x^3$ followed by a vertical shift
$y = 1 - x^3$

27. Vertical shift 2 units upward

29. Horizontal shift 2 units to the right

31. Vertical stretch   **33.** Horizontal shift 2 units to the left

35. Reflection in the x-axis   **37.** Vertical shrink

39. Reflection in the x-axis and vertical shift 4 units upward

41. Horizontal shift 2 units to the left and vertical shrink

43. Vertical shrink and vertical shift 2 units upward

45.

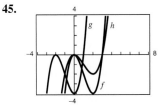

g is a horizontal shift and h is a vertical shrink.

47.

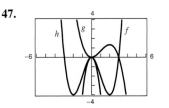

g is a vertical shrink and a reflection in the x-axis and h is a reflection in the y-axis.

49. $g(x) = -(x^3 - 3x^2) + 1$

51. (a) $f(x) = x^2$

(b) Reflection in the x-axis and vertical shift 12 units upward

(c)   (d) $g(x) = 12 - f(x)$

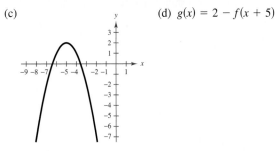

53. (a) $f(x) = x^2$

(b) Horizontal shift 5 units to the left, reflection in the x-axis, and vertical shift 2 units upward

(c)   (d) $g(x) = 2 - f(x + 5)$

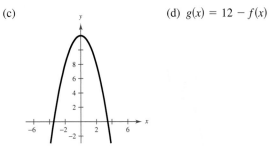

55. (a) $f(x) = x^2$

(b) Horizontal shift 4 units to the right, vertical stretch, and vertical shift 3 units upward

(c)   (d) $g(x) = 3 + 2f(x - 4)$

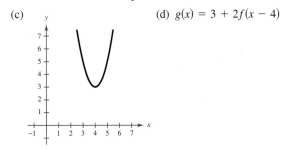

57. (a) $f(x) = x^3$

(b) Vertical shift 7 units upward

(c)

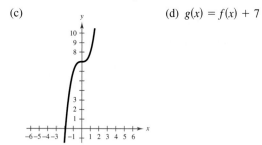

(d) $g(x) = f(x) + 7$

59. (a) $f(x) = x^3$

(b) Horizontal shift 1 unit to the right and vertical shift 2 units upward

(c)

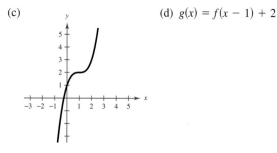

(d) $g(x) = f(x - 1) + 2$

61. (a) $f(x) = x^3$

(b) Horizontal shift 2 units to the right and vertical stretch

(c)

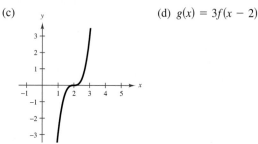

(d) $g(x) = 3f(x - 2)$

63. (a) $f(x) = |x|$

(b) Reflection in the x-axis and vertical shift 2 units downward

(c)

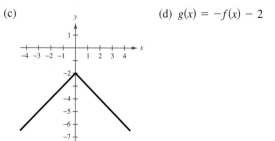

(d) $g(x) = -f(x) - 2$

65. (a) $f(x) = |x|$

(b) Horizontal shift 4 units to the left, reflection in the x-axis, and vertical shift 8 units upward

(c)

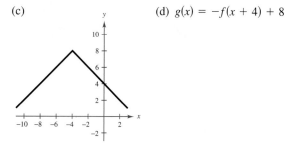

(d) $g(x) = -f(x + 4) + 8$

67. (a) $f(x) = |x|$

(b) Horizontal shift 1 unit to the right, reflection in the x-axis, and vertical stretch

(c)

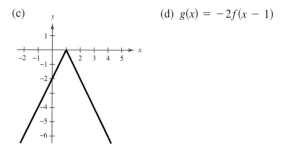

(d) $g(x) = -2f(x - 1)$

69. (a) $f(x) = \sqrt{x}$

(b) Horizontal shift 9 units to the right

(c)

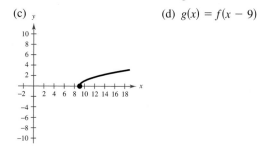

(d) $g(x) = f(x - 9)$

71. (a) $f(x) = \sqrt{x}$

(b) Reflection in the y-axis, horizontal shift 7 units to the right, and vertical shift 2 units downward

(c)

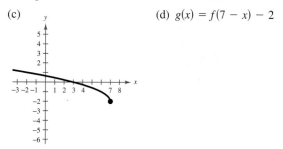

(d) $g(x) = f(7 - x) - 2$

73. (a) $f(x) = \sqrt{x}$

(b) Horizontal shift 1 unit to the right and vertical stretch 4 units

(c)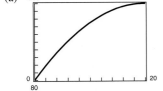

(d) $g(x) = 4f(x - 1)$

75. (a)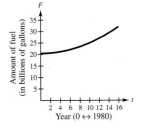

(b) $P(x) = 55 + 20x - 0.5x^2$; vertical shift

(c) $P(x) = 80 + \dfrac{1}{5}x - \dfrac{x^2}{20,000}$; horizontal stretch

77. (a) Vertical shrink and vertical shift

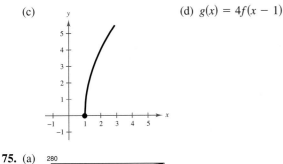

(b) $G(t) = 0.04t^2 + 0.8t + 24.46$; $G(t) = F(t + 10)$

79. (a) To each time t there corresponds one and only one temperature T.

(b) $60°, 72°$

(c) All the temperature changes would be 1 hour later.

(d) The temperature would be decreased by 1 degree.

81. False. The point $(-1, 28)$ does not lie on the graph of $g(x) = -(x - 6)^2 + 3$.

83. $y = x^7$ resembles the cubic graph and $y = x^8$ resembles the quadratic graph. Both are steeper on $(-\infty, -1)$ and $(1, \infty)$ and both are closer to 0 on $-1 < x < 1$.

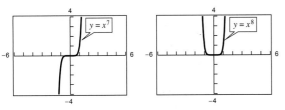

85.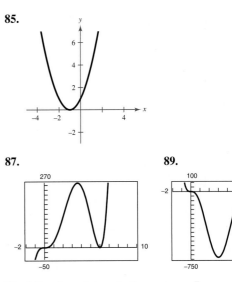

87.

89.

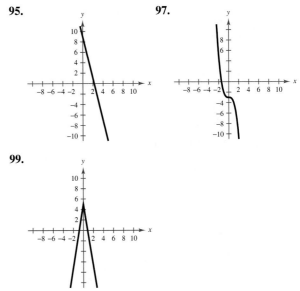

91. All real numbers $x \geq 5$ except $x = 7$

93. All real numbers

95.

97.

99.

Section 1.4 *(page 116)*

1. (a) $2x$ (b) 2 (c) $x^2 - 1$ (d) $\dfrac{x + 1}{x - 1}$, $x \neq 1$

(e) All $x \neq 1$

3. (a) $x^2 - x + 1$ (b) $x^2 + x - 1$ (c) $x^2 - x^3$

(d) $\dfrac{x^2}{1 - x}$, $x \neq 1$ (e) All $x \neq 1$

5. (a) $x^2 + 5 + \sqrt{1 - x}$ (b) $x^2 + 5 - \sqrt{1 - x}$

(c) $(x^2 + 5)\sqrt{1 - x}$ (d) $\dfrac{x^2 + 5}{\sqrt{1 - x}}$, $x < 1$

(e) $x < 1$

7. (a) $\dfrac{x+1}{x^2}$ (b) $\dfrac{x-1}{x^2}$ (c) $\dfrac{1}{x^3}$ (d) $x,\ x \neq 0$

 (e) $x \neq 0$

9. 9   **11.** 5   **13.** 0   **15.** 26   **17.** $4t^2 - 2t + 5$

19. $-125t^3 - 100t^2 - 5t - 4$ **21.** $\dfrac{t^2+1}{-t-4}$

23.

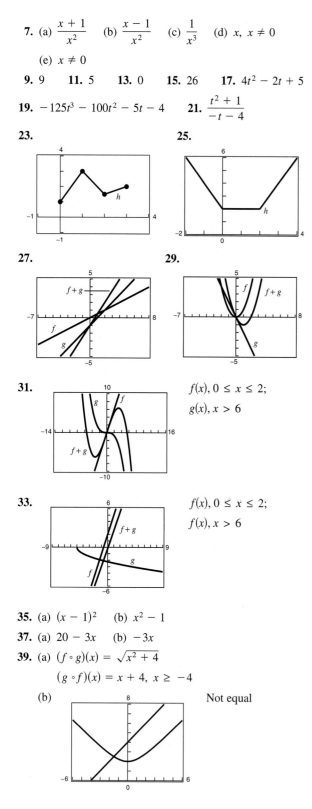

25.

27.

29.

31.

$f(x),\ 0 \leq x \leq 2;$

$g(x),\ x > 6$

33.

$f(x),\ 0 \leq x \leq 2;$

$f(x),\ x > 6$

35. (a) $(x-1)^2$ (b) $x^2 - 1$

37. (a) $20 - 3x$ (b) $-3x$

39. (a) $(f \circ g)(x) = \sqrt{x^2 + 4}$

 $(g \circ f)(x) = x + 4,\ x \geq -4$

 (b)   Not equal

41. (a) $(f \circ g)(x) = x - \dfrac{8}{3};\ (g \circ f)(x) = x - 8$

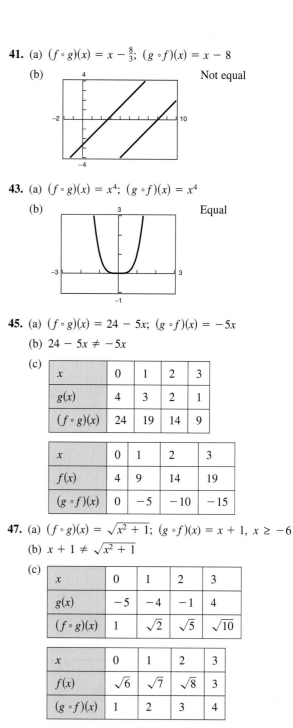

 (b)   Not equal

43. (a) $(f \circ g)(x) = x^4;\ (g \circ f)(x) = x^4$

 (b)   Equal

45. (a) $(f \circ g)(x) = 24 - 5x;\ (g \circ f)(x) = -5x$

 (b) $24 - 5x \neq -5x$

 (c)

x	0	1	2	3
$g(x)$	4	3	2	1
$(f \circ g)(x)$	24	19	14	9

x	0	1	2	3
$f(x)$	4	9	14	19
$(g \circ f)(x)$	0	-5	-10	-15

47. (a) $(f \circ g)(x) = \sqrt{x^2 + 1};\ (g \circ f)(x) = x + 1,\ x \geq -6$

 (b) $x + 1 \neq \sqrt{x^2 + 1}$

 (c)

x	0	1	2	3
$g(x)$	-5	-4	-1	4
$(f \circ g)(x)$	1	$\sqrt{2}$	$\sqrt{5}$	$\sqrt{10}$

x	0	1	2	3
$f(x)$	$\sqrt{6}$	$\sqrt{7}$	$\sqrt{8}$	3
$(g \circ f)(x)$	1	2	3	4

49. (a) $(f \circ g)(x) = |2x + 2|;\ (g \circ f)(x) = 2|x + 3| - 1$

 (b) $(f \circ g)(x) = \begin{cases} 2x + 2, & x \geq -1 \\ -2x - 2, & x < -1 \end{cases}$

 $(g \circ f)(x) = \begin{cases} 2x + 5, & x \geq -3 \\ -2x - 7, & x < -3 \end{cases}$

 $(f \circ g)(x) \neq (g \circ f)(x)$

(c)

x	0	-1	-3	-5
$g(x)$	-1	-3	-7	-11
$(f \circ g)(x)$	2	0	4	8

x	0	-1	-3	-5
$f(x)$	3	2	0	2
$(g \circ f)(x)$	5	3	-1	3

51. (a) 3   (b) 0   **53.** (a) 0   (b) 4

55. (a) 0   (b) 4   **57.** $f(x) = x^2$, $g(x) = 2x + 1$

59. $f(x) = \sqrt[3]{x}$, $g(x) = x^2 - 4$

61. $f(x) = \dfrac{1}{x}$, $g(x) = x + 2$

63. $f(x) = x^2 + 2x$, $g(x) = x + 4$

65. (a) $x \geq 0$ (b) All real numbers   (c) All real numbers

67. (a) All real numbers except $x = 0$ (b) All real numbers
   (c) All real numbers except $x = -3$

69. (a) All real numbers except $x = 0$ (b) All real numbers
   (c) All real numbers except $x = 1$

71. $3, h \neq 0$ **73.** $-2x - h, h \neq 0$

75. $\dfrac{-4}{x(x + h)}, h \neq 0$

77. $\dfrac{2}{\sqrt{2(x + h) + 1} + \sqrt{2x + 1}}, h \neq 0$

79. (a) $T = \frac{3}{4}x + \frac{1}{15}x^2$

(b)

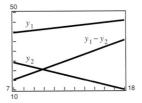

(c) B. For example, $B(60) = 240$ while $R(60)$ is only 45.

81. $y_1 = -0.59t^2 + 7.66t + 144.9$

     $y_2 = 16.58t + 245.06$

     $y_3 = 1.84t + 21.92$

83. $(A \circ r)(t) = 0.36\pi t^2$

   $A \circ r$ represents the area of the circle at time t.

85. (a) $(C \circ x)(t) = 3000t + 750$

     $C \circ x$ represents the cost after t production hours.

(b)

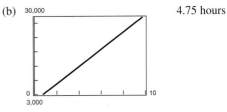

4.75 hours

87. $g(f(x))$ represents 3 percent of an amount over \$500,000.

89.

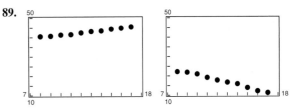

Both data sets appear to be linear.

$y_1 = 0.57x + 35.6$; $y_2 = -1.29x + 33.3$

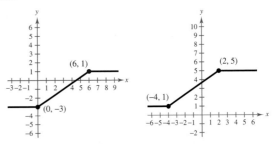

The difference between the morning and evening news-paper circulations is increasing.

91. True   **93.** Odd (Proofs will vary.)

95. $\frac{1}{2}[f(x) + f(-x)] + \frac{1}{2}[f(x) - f(-x)]$

$\qquad = \frac{1}{2}[f(x) + f(-x) + f(x) - f(-x)]$

$\qquad = \frac{1}{2}[2f(x)]$

$\qquad = f(x)$

97. $(0, -5), (1, -5), (2, -7)$ (Answers will vary.)

99. $\left(0, 2\sqrt{6}\right), \left(1, \sqrt{23}\right), \left(2, 2\sqrt{5}\right)$ (Answers will vary.)

101. $10x - y + 38 = 0$ **103.** $30x + 11y - 34 = 0$

105.   **107.**

109.

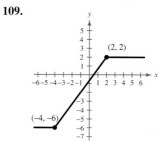

Section 1.5 *(page 127)*

1. (c) **3.** (a) **5.** $f^{-1}(x) = \frac{1}{8}x$ **7.** $f^{-1}(x) = x - 10$

9. $f^{-1}(x) = \frac{1}{2}(x - 1)$ **11.** $f^{-1}(x) = x^3$

13. (a) $f(g(x)) = f\left(\dfrac{x}{2}\right) = 2\left(\dfrac{x}{2}\right) = x$

$$g(f(x)) = g(2x) = \dfrac{(2x)}{2} = x$$

(b)

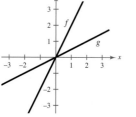

15. (a) $f(g(x)) = f\left(\dfrac{x-1}{5}\right) = 5\left(\dfrac{x-1}{5}\right) + 1 = x$

$$g(f(x)) = g(5x + 1) = \dfrac{(5x+1) - 1}{5} = x$$

(b)

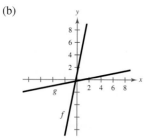

17. $f(g(x)) = f(\sqrt[3]{x}) = (\sqrt[3]{x})^3 = x$

$g(f(x)) = g(x^3) = \sqrt[3]{x^3} = x$

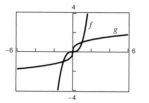

Reflections in the line $y = x$

19. $f(g(x)) = f(x^2 + 4), \quad x \geq 0$

$$= \sqrt{(x^2 + 4) - 4} = x$$

$g(f(x)) = g(\sqrt{x - 4})$

$$= (\sqrt{x - 4})^2 + 4 = x$$

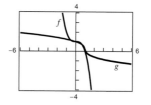

Reflections in the line $y = x$

21. $f(g(x)) = f(\sqrt[3]{1 - x}) = 1 - (\sqrt[3]{1 - x})^3 = x$

$g(f(x)) = g(1 - x^3) = \sqrt[3]{1 - (1 - x^3)} = x$

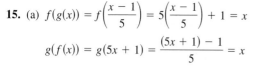

Reflections in the line $y = x$

23. (a) $f(g(x)) = f\left(-\dfrac{2x + 6}{7}\right)$

$$= -\dfrac{7}{2}\left(-\dfrac{2x + 6}{7}\right) - 3 = x$$

$g(f(x)) = g\left(-\dfrac{7}{2}x - 3\right)$

$$= -\dfrac{2\left(-\frac{7}{2}x - 3\right) + 6}{7} = x$$

(b)

x	0	2	-2	6
$f(x)$	-3	-10	4	-24

x	-3	-10	4	-24
$g(x)$	0	2	-2	6

25. (a) $f(g(x)) = f(\sqrt[3]{x - 5}) = (\sqrt[3]{x - 5})^3 + 5 = x$

$g(f(x)) = g(x^3 + 5) = \sqrt[3]{(x^3 + 5) - 5} = x$

(b)

x	0	1	-1	-2	4
$f(x)$	5	6	4	-3	69

x	5	6	4	-3	69
$g(x)$	0	1	-1	-2	4

27. (a) $f(g(x)) = f(8 + x^2)$
$$= -\sqrt{(8 + x^2) - 8}$$
$$= -\sqrt{x^2} = -(-x) = x, \; x \le 0$$
$$g(f(x)) = g\left(-\sqrt{x - 8}\right)$$
$$= 8 + \left(-\sqrt{x - 8}\right)^2$$
$$= 8 + (x - 8) = x, \; x \ge 8$$

(b)

x	8	9	12	15
$f(x)$	0	-1	-2	$-\sqrt{7}$

x	0	-1	-2	$-\sqrt{7}$
$g(x)$	8	9	12	15

29.

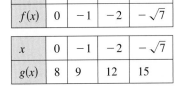

One-to-one

31.

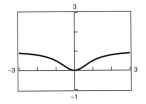

Not one-to-one

33.

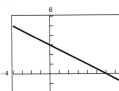

Not one-to-one

35.

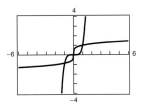

One-to-one

37.

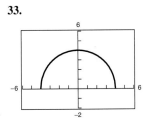

Not one-to-one

39.

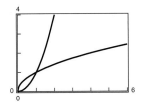

One-to-one

41.

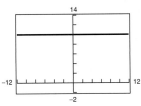

Not one-to-one

43. $f^{-1}(x) = \dfrac{x + 3}{2}$

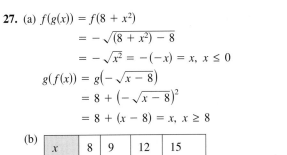

Reflections in the line $y = x$

45. $f^{-1}(x) = \sqrt[5]{x}$

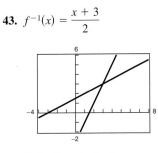

Reflections in the line $y = x$

47. $f^{-1}(x) = x^2, \; x \ge 0$

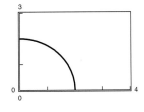

Reflections in the line $y = x$

49. $f^{-1}(x) = \sqrt{4 - x^2}, \; 0 \le x \le 2$

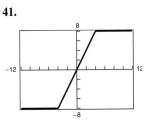

Reflections in the line $y = x$

51. $f^{-1}(x) = x^3 + 1$

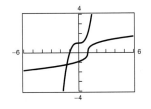

Reflections in the line $y = x$

53. $f^{-1}(x) = \dfrac{4}{x}$

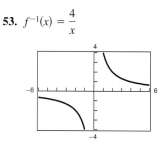

Reflections in the line $y = x$

55. Not one-to-one   **57.** $f^{-1}(x) = \dfrac{5x - 4}{3}$

59. $f^{-1}(x) = \sqrt{x} - 3,\ x \ge 0$

61. Not one-to-one

63. $f^{-1}(x) = \dfrac{x^2 - 3}{2},\ x \ge 0$ **65.** Not one-to-one

67. $f^{-1}(x) = \dfrac{x - b}{a}$

69. $y = \sqrt{x} + 2,\ x \ge 0$ **71.** $y = x - 2,\ x \ge 0$

73.

x	-4	-2	2	3
$f^{-1}(x)$	-2	-1	1	3

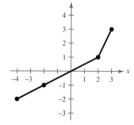

75. (a) and (b)

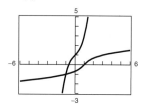

(c) Inverse function because it satisfies the Vertical Line Test

77. (a) and (b)

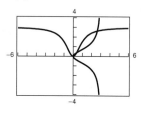

(c) Not an inverse function because it does not satisfy the Vertical Line Test

79. 32   **81.** 600   **83.** $2\sqrt[3]{x + 3}$

85. $\dfrac{x + 1}{2}$ **87.** $\dfrac{x + 1}{2}$

89. (a) $y = \dfrac{x - 8}{0.75}$

   $y = $ number of units produced

   $x = $ hourly wage

(b) 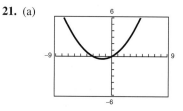   (c) \$15.50

   (d) 19 units

91. (a) Yes

(b) $f^{-1}(t)$ represents the year new car sales totaled \$$t$ billion.

(c) 5, or 1995

(d) No. The inverse is not a function because f is not one-to-one.

93. True   **95.** Answers will vary.   **97.** 2   **99.** 0

101. $f^{-1}(x) = \frac{1}{12}x$ **103.** $f^{-1}(x) = \sqrt[3]{x - 7}$

Review Exercises *(page 131)*

1. (a) Not a function because element 20 in A corresponds to two elements, 4 and 6, in B.

(b) Function   (c) Function

(d) Not a function because 30 in A corresponds to no element in B.

3. Not a function   **5.** Function

7. (a) 5   (b) 17   (c) $t^4 + 1$ (d) $-x^2 - 1$

9. All real numbers   **11.** $[-5, 5]$

13. All real numbers except $s = 3$

15. (a) $C = 5.35x + 16,000$ (b) $P = 2.85x - 16,000$

17. Domain: $(-\infty, \infty)$; Range: $(-\infty, 3]$

19. Domain: $[-6, 6]$; Range: $[0, 6]$

21. (a)

(b) Function

23. (a)

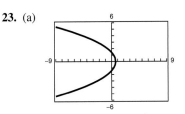

(b) Not a function

25. Increasing on $(-\infty, -1)$, $(1, \infty)$;
Decreasing on $(-1, 1)$

27. Increasing on $(6, \infty)$

29. Relative maximum: $(0, 16)$
Relative minimums: $(-2, 0)$, $(2, 0)$

31. Relative maximum: $(3, 27)$

33.

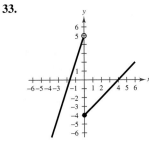

35. Even

37. Absolute value function $f(x) = |x|$
Vertical shift 3 units upward
$g(x) = |x| + 3$

39.

41.

43.

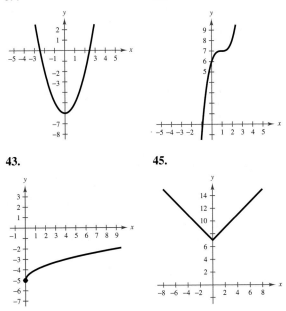

45.

47.

49.

51.

53.

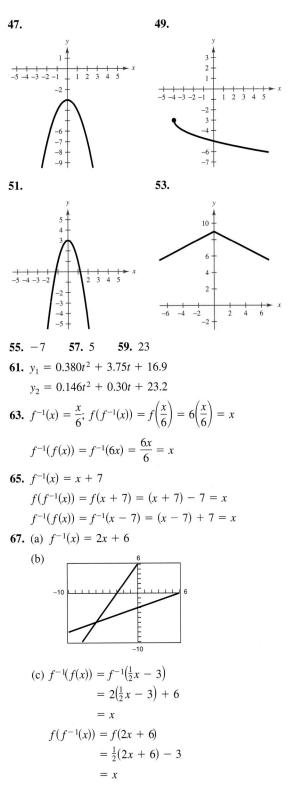

55. -7 **57.** 5   **59.** 23

61. $y_1 = 0.380t^2 + 3.75t + 16.9$
$y_2 = 0.146t^2 + 0.30t + 23.2$

63. $f^{-1}(x) = \dfrac{x}{6}$; $f(f^{-1}(x)) = f\left(\dfrac{x}{6}\right) = 6\left(\dfrac{x}{6}\right) = x$

$f^{-1}(f(x)) = f^{-1}(6x) = \dfrac{6x}{6} = x$

65. $f^{-1}(x) = x + 7$
$f(f^{-1}(x)) = f(x + 7) = (x + 7) - 7 = x$
$f^{-1}(f(x)) = f^{-1}(x - 7) = (x - 7) + 7 = x$

67. (a) $f^{-1}(x) = 2x + 6$
(b)

(c) $f^{-1}(f(x)) = f^{-1}\left(\tfrac{1}{2}x - 3\right)$
$= 2\left(\tfrac{1}{2}x - 3\right) + 6$
$= x$

$f(f^{-1}(x)) = f(2x + 6)$
$= \tfrac{1}{2}(2x + 6) - 3$
$= x$

69. (a) $f^{-1}(x) = x^2 - 1$, $x \geq 0$

(b)

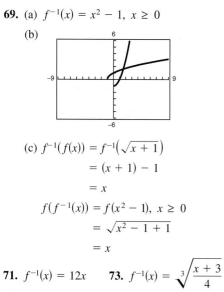

(c) $f^{-1}(f(x)) = f^{-1}(\sqrt{x+1})$

$\qquad = (x+1) - 1$

$\qquad = x$

$f(f^{-1}(x)) = f(x^2 - 1)$, $x \geq 0$

$\qquad = \sqrt{x^2 - 1 + 1}$

$\qquad = x$

71. $f^{-1}(x) = 12x$ **73.** $f^{-1}(x) = \sqrt[3]{\dfrac{x+3}{4}}$

75. $f^{-1}(x) = x^2 - 10$, $x \geq 0$

77. True. For odd n, $f(x) = x^n$ is a one-to-one function.

79. Vertical lines of the form $x = c$ are not functions because an infinite number of y-values correspond with the value of $x = c$.

Chapter Test   *(page 134)*

1. No. To some x there correspond more than one value of y.

2. 7   **3.** $10 - \sqrt{6-t}$ **4.** $\dfrac{\sqrt{3-x} - 1}{2 - x}$

5. $(-\infty, 3]$ **6.** $C = 5.60x + 24{,}000$

$\qquad\qquad\qquad P = 3.60x - 24{,}000$

7. (a)

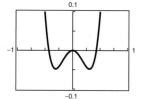

(b) Increasing on $(-0.31, 0)$, $(0.31, \infty)$

Decreasing on $(-\infty, -0.31)$, $(0, 0.31)$

(c) Even

8. (a)

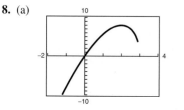

(b) Increasing on $(-\infty, 2)$

Decreasing on $(2, 3)$

(c) Neither even nor odd

9. (a)

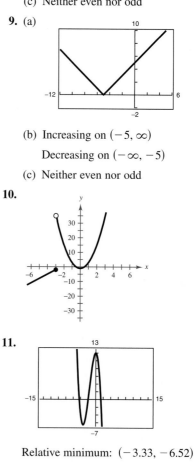

(b) Increasing on $(-5, \infty)$

Decreasing on $(-\infty, -5)$

(c) Neither even nor odd

10.

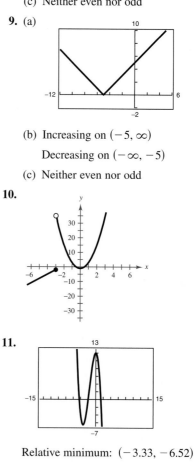

11.

Relative minimum: $(-3.33, -6.52)$

Relative maximum: $(0, 12)$

12.

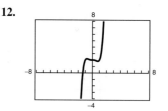

Relative maximum: $(-0.77, 2.19)$

Relative minimum: $(0.77, 1.81)$

13. (a) $f(x) = x^3$

(b) Horizontal shift 5 units to the right, reflection in x-axis, vertical stretch, and then a vertical shift 3 units upward

(c)

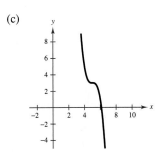

14. (a) $f(x) = \sqrt{x}$

(b) Reflection in y-axis and a horizontal shift 7 units to the left

(c)

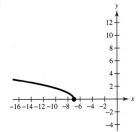

15. (a) $f(x) = |x|$

(b) Reflection in y-axis (no effect), vertical stretch, and vertical shift 7 units downward

(c)

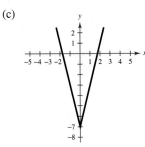

16. (a) $f(x) = x^2$

(b) Horizontal shift 3 units to the right and a reflection in the x-axis, followed by a vertical shift 5 units upward

(c)

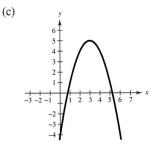

17. (a) $x^2 - \sqrt{2 - x}$, $(-\infty, 2]$ (b) $\dfrac{x^2}{\sqrt{2 - x}}$, $(-\infty, 2)$

(c) $2 - x$, $(-\infty, 2]$ (d) $2 - x^2$, $[0, \infty)$

18. $f^{-1}(x) = \sqrt[3]{x - 8}$ **19.** No inverse

20. $f^{-1}(x) = \left(\frac{8}{3}x\right)^{2/3}$, $x \geq 0$

21. $A = -x^2 + 50x$, $0 < x < 50$

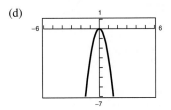

Maximum area = 625 square inches

Chapter 2

Section 2.1 *(page 143)*

1. (g)   **3.** (b)   **5.** (f)   **7.** (e)

9. (a)

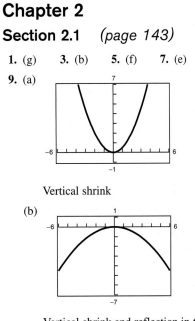

Vertical shrink

(b)

Vertical shrink and reflection in the x-axis

(c)

Vertical stretch

(d)

Vertical stretch and reflection in the x-axis

11. (a)

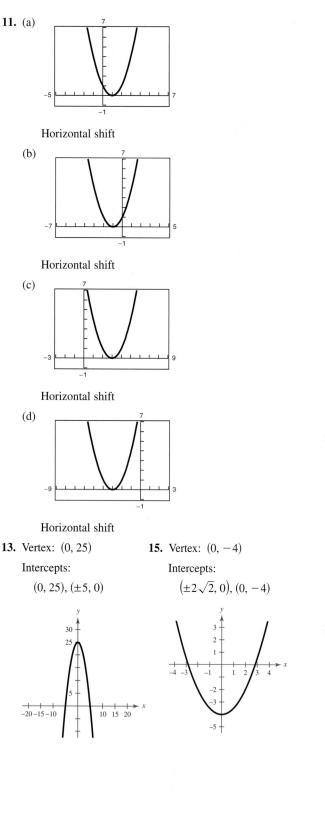

Horizontal shift

(b)

Horizontal shift

(c)

Horizontal shift

(d)

Horizontal shift

13. Vertex: $(0, 25)$
Intercepts:
$(0, 25), (\pm 5, 0)$

15. Vertex: $(0, -4)$
Intercepts:
$(\pm 2\sqrt{2}, 0), (0, -4)$

17. Vertex: $(-4, -3)$
Intercepts:
$(\pm\sqrt{3} - 4, 0), (0, 13)$

19. Vertex: $(4, 0)$
Intercepts:
$(4, 0), (0, 16)$

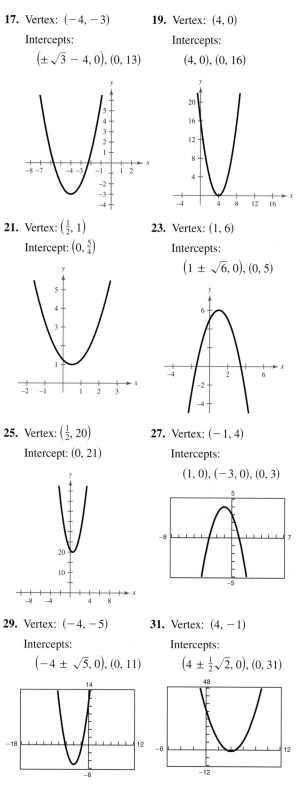

21. Vertex: $\left(\frac{1}{2}, 1\right)$
Intercept: $\left(0, \frac{5}{4}\right)$

23. Vertex: $(1, 6)$
Intercepts:
$\left(1 \pm \sqrt{6}, 0\right), (0, 5)$

25. Vertex: $\left(\frac{1}{2}, 20\right)$
Intercept: $(0, 21)$

27. Vertex: $(-1, 4)$
Intercepts:
$(1, 0), (-3, 0), (0, 3)$

29. Vertex: $(-4, -5)$
Intercepts:
$\left(-4 \pm \sqrt{5}, 0\right), (0, 11)$

31. Vertex: $(4, -1)$
Intercepts:
$\left(4 \pm \frac{1}{2}\sqrt{2}, 0\right), (0, 31)$

33. Vertex: $(-2, -3)$

Intercepts: $\left(-2 \pm \sqrt{6}, 0\right), (0, -1)$

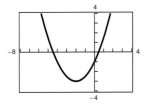

35. $y = (x - 1)^2$ **37.** $y = -(x + 1)^2 + 4$

39. $f(x) = (x + 2)^2 + 5$ **41.** $f(x) = -\frac{1}{2}(x - 3)^2 + 4$

43. $f(x) = 2(x + 2)^2 - 2$ **45.** $f(x) = \frac{19}{81}\left(x - \frac{5}{2}\right)^2 - \frac{3}{4}$

47. $(\pm 4, 0)$; They are the same.

49. $(5, 0), (-1, 0)$; They are the same.

51.

53.

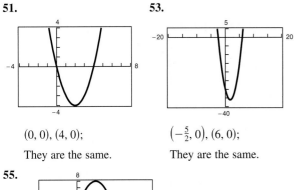

$(0, 0), (4, 0);$

$\left(-\frac{5}{2}, 0\right), (6, 0);$

They are the same.

They are the same.

55.

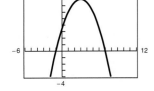

$(7, 0), (-1, 0)$; They are the same.

57. $f(x) = x^2 - 2x - 3$ **59.** $f(x) = 2x^2 + 7x + 3$

$g(x) = -x^2 + 2x + 3$ $g(x) = -2x^2 - 7x - 3$

61. 55, 55   **63.** 12, 6

65. (a) $A = x(50 - x), \ 0 < x < 50$

(b)

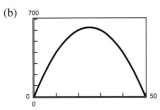

(c) 25 feet × 25 feet

67. (a)

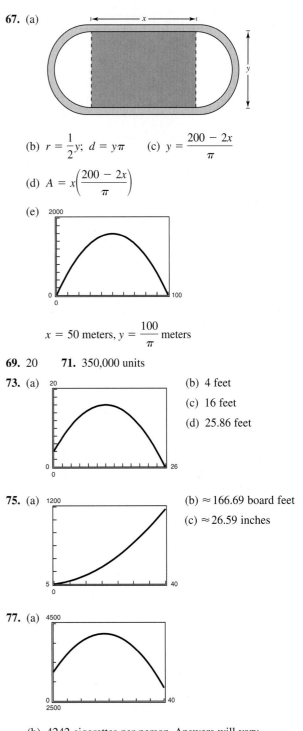

(b) $r = \frac{1}{2}y; \ d = y\pi$ (c) $y = \dfrac{200 - 2x}{\pi}$

(d) $A = x\left(\dfrac{200 - 2x}{\pi}\right)$

(e)

$x = 50$ meters, $y = \dfrac{100}{\pi}$ meters

69. 20   **71.** 350,000 units

73. (a)

(b) 4 feet

(c) 16 feet

(d) 25.86 feet

75. (a)

(b) ≈ 166.69 board feet

(c) ≈ 26.59 inches

77. (a)

(b) 4242 cigarettes per person. Answers will vary.

(c) 9703 cigarettes per smoker per year; 27 cigarettes per smoker per day

79. True   **81.** (a). The profits are positive and rising.

83. $(4, 2)$ **85.** $(2, 11)$ **87.** No   **89.** No

Section 2.2 *(page 156)*

1. (f)   **3.** (c)   **5.** (e)   **7.** (g)

9. (a)   (b)

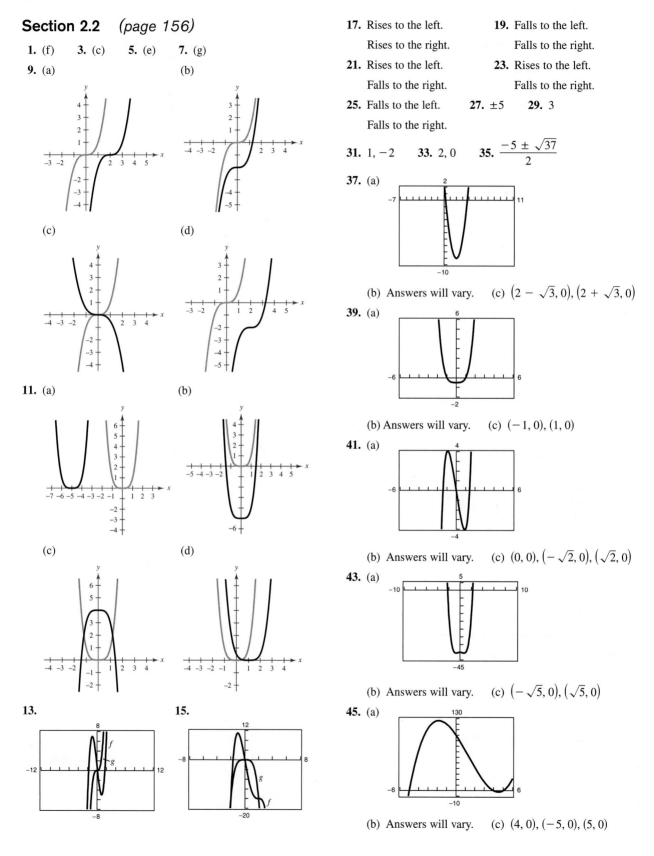

(c)   (d)

11. (a)   (b)

(c)   (d)

13.   **15.**

17. Rises to the left.   **19.** Falls to the left.
Rises to the right.   Falls to the right.

21. Rises to the left.   **23.** Rises to the left.
Falls to the right.   Falls to the right.

25. Falls to the left.   **27.** ±5   **29.** 3
Falls to the right.

31. 1, −2   **33.** 2, 0   **35.** $\dfrac{-5 \pm \sqrt{37}}{2}$

37. (a)

(b) Answers will vary.   (c) $\left(2 - \sqrt{3}, 0\right), \left(2 + \sqrt{3}, 0\right)$

39. (a)

(b) Answers will vary.   (c) $(-1, 0), (1, 0)$

41. (a)

(b) Answers will vary.   (c) $(0, 0), \left(-\sqrt{2}, 0\right), \left(\sqrt{2}, 0\right)$

43. (a)

(b) Answers will vary.   (c) $\left(-\sqrt{5}, 0\right), \left(\sqrt{5}, 0\right)$

45. (a)

(b) Answers will vary.   (c) $(4, 0), (-5, 0), (5, 0)$

47. (a)

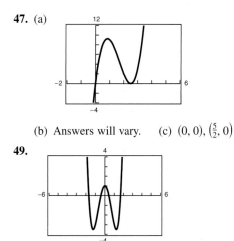

(b) Answers will vary. (c) $(0, 0), \left(\frac{5}{2}, 0\right)$

49.

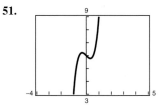

Relative maximum: $(0, 1)$

Relative minimums: $(\pm 1.225, -3.500)$

51.

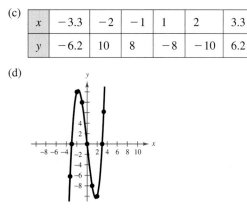

Relative maximum: $(-0.324, 6.218)$

Relative minimum: $(0.324, 5.782)$

53. $f(x) = x^2 - 12x$ **55.** $f(x) = x^2 + 4x - 12$

57. $f(x) = x^3 + 7x^2 + 12x$

59. $f(x) = x^4 - 4x^3 - 9x^2 + 36x$

61. $f(x) = x^2 - 2x - 2$

63. $f(x) = x^3 - 10x^2 + 27x - 22$

65. (a) Falls to the left, rises to the right

(b) $(0, 0), (3, 0), (-3, 0)$

(c)

x	-3.3	-2	-1	1	2	3.3
y	-6.2	10	8	-8	-10	6.2

(d)

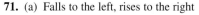

67. (a) Rises to the left and the right

(b) No zeros

(c)

t	-3	-2	-1	0	1	2	3
y	7.5	5.8	4.5	3.8	3.5	3.8	4.5

(d)

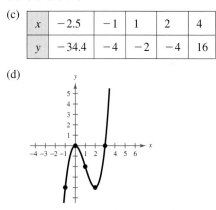

69. (a) Falls to the left, rises to the right

(b) $(0, 0), (3, 0)$

(c)

x	-2.5	-1	1	2	4
y	-34.4	-4	-2	-4	16

(d)

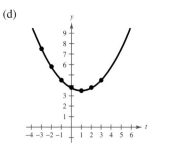

71. (a) Falls to the left, rises to the right

(b) $(0, 0), (2, 0), (3, 0)$

(c)

x	-0.5	0.5	1	1.5	2.5	3.5
y	-13	5.6	6	3.4	-1.9	7.9

(d)

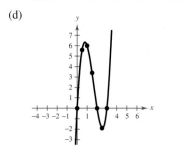

73. (a) Rises to the left, falls to the right

(b) $(0, 0), (-5, 0)$

(c)

x	-6	-4	-3	-2	1	2
y	36	-16	-18	-12	-6	-28

(d)

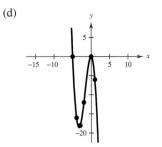

75. (a) Falls to the left, rises to the right

(b) $(0, 0), (4, 0)$

(c)

x	-2	-1	1	2	3	5
y	-24	-5	-3	-8	-9	25

(d)

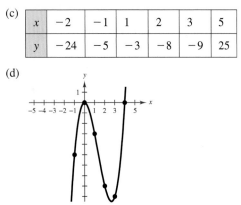

77. (a) Falls to the left and right

(b) $(-2, 0), (2, 0)$

(c)

t	-4	-3	-1	0	1	3
y	-36	-6.3	-2.3	-4	-2.3	-6.3

(d)

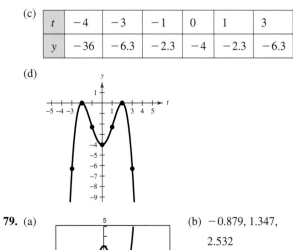

79. (a)   (b) $-0.879, 1.347,$
                     2.532

$(-1, 0), (1, 2), (2, 3)$

81. (a)   (b) $-1.585, 0.779$

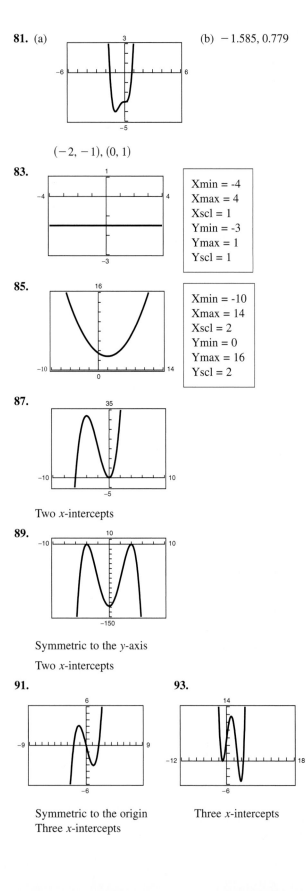

$(-2, -1), (0, 1)$

83.

| Xmin = -4 |
| Xmax = 4 |
| Xscl = 1 |
| Ymin = -3 |
| Ymax = 1 |
| Yscl = 1 |

85.

| Xmin = -10 |
| Xmax = 14 |
| Xscl = 2 |
| Ymin = 0 |
| Ymax = 16 |
| Yscl = 2 |

87.

Two x-intercepts

89.

Symmetric to the y-axis

Two x-intercepts

91.   **93.**

Symmetric to the origin   Three x-intercepts
Three x-intercepts

95. (a) and (b)

Height	Width	Volume
1	$36 - 2(1)$	$1[36 - 2(1)]^2 = 1156$
2	$36 - 2(2)$	$2[36 - 2(2)]^2 = 2048$
3	$36 - 2(3)$	$3[36 - 2(3)]^2 = 2700$
4	$36 - 2(4)$	$4[36 - 2(4)]^2 = 3136$
5	$36 - 2(5)$	$5[36 - 2(5)]^2 = 3380$
6	$36 - 2(6)$	$6[36 - 2(6)]^2 = 3456$
7	$36 - 2(7)$	$7[36 - 2(7)]^2 = 3388$

$5 < x < 7$

(c) Answers will vary. Domain: $0 < x < 18$

(d) 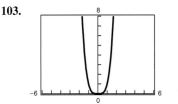 $x = 6$

97. $(200, 160)$

99. (a) $0.002983x^4 - 0.02386x^3 + 0.0194x^2 + 0.113x$

(b) 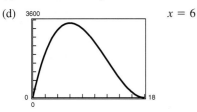   (c) 0. Yes

The model fits well.

101. False. It can have at most three turning points.

103.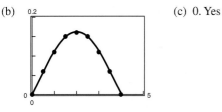

(a) Vertical shift of two units; even

(b) Horizontal shift of two units; neither even nor odd

(c) Reflection in the y-axis; even

(d) Reflection in the x-axis; even

(e) Vertical shrink; even   (f) Vertical shrink; even

(g) $g(x) = x^3$; odd   (h) $g(x) = x^{16}$; even

105. 33   **107.** $-\frac{4}{3} \approx -1.3$ **109.** 72

111. $x \le -\frac{1}{2}$, $x \ge 1$ **113.** $x \le -24$, $x \ge 8$

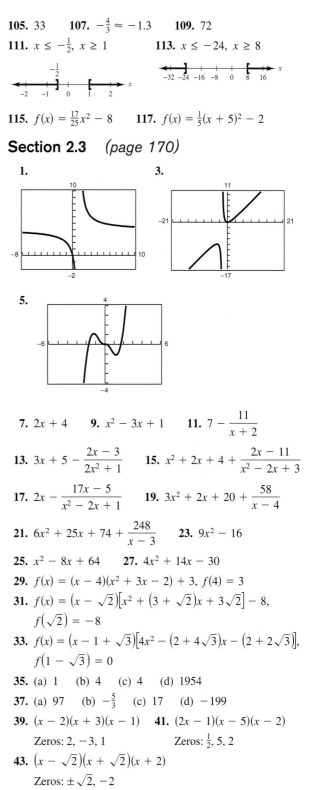

115. $f(x) = \frac{17}{25}x^2 - 8$ **117.** $f(x) = \frac{1}{5}(x + 5)^2 - 2$

Section 2.3   *(page 170)*

1.   **3.**

5.

7. $2x + 4$ **9.** $x^2 - 3x + 1$ **11.** $7 - \dfrac{11}{x + 2}$

13. $3x + 5 - \dfrac{2x - 3}{2x^2 + 1}$ **15.** $x^2 + 2x + 4 + \dfrac{2x - 11}{x^2 - 2x + 3}$

17. $2x - \dfrac{17x - 5}{x^2 - 2x + 1}$ **19.** $3x^2 + 2x + 20 + \dfrac{58}{x - 4}$

21. $6x^2 + 25x + 74 + \dfrac{248}{x - 3}$ **23.** $9x^2 - 16$

25. $x^2 - 8x + 64$ **27.** $4x^2 + 14x - 30$

29. $f(x) = (x - 4)(x^2 + 3x - 2) + 3$, $f(4) = 3$

31. $f(x) = \left(x - \sqrt{2}\right)\left[x^2 + \left(3 + \sqrt{2}\right)x + 3\sqrt{2}\right] - 8$,
$f\left(\sqrt{2}\right) = -8$

33. $f(x) = \left(x - 1 + \sqrt{3}\right)\left[4x^2 - \left(2 + 4\sqrt{3}\right)x - \left(2 + 2\sqrt{3}\right)\right]$,
$f\left(1 - \sqrt{3}\right) = 0$

35. (a) 1   (b) 4   (c) 4   (d) 1954

37. (a) 97   (b) $-\frac{5}{3}$ (c) 17   (d) -199

39. $(x - 2)(x + 3)(x - 1)$ **41.** $(2x - 1)(x - 5)(x - 2)$
Zeros: $2, -3, 1$ Zeros: $\frac{1}{2}, 5, 2$

43. $\left(x - \sqrt{2}\right)\left(x + \sqrt{2}\right)(x + 2)$
Zeros: $\pm\sqrt{2}, -2$

45. (a) Answers will vary.   (b) $(2x - 1)$
(c) $(x + 2)(x - 1)(2x - 1)$ (d) $-2, 1, \frac{1}{2}$

47. (a) Answers will vary.   (b) $(x - 1), (x - 2)$

   (c) $(x - 5)(x + 4)(x - 1)(x - 2)$ (d) $-4, 1, 2, 5$

49. (a) Answers will vary.   (b) $(x + 7)$

   (c) $(2x + 1)(3x - 2)(x + 7)$ (d) $-7, -\frac{1}{2}, \frac{2}{3}$

51. $\pm 1, \pm 3$

53. $\pm 1, \pm 3, \pm 5, \pm 9, \pm 15, \pm 45, \pm\frac{1}{2}, \pm\frac{3}{2}, \pm\frac{5}{2}, \pm\frac{9}{2}, \pm\frac{15}{2}, \pm\frac{45}{2}$

55. (a) $\pm 1, \pm 2, \pm 4$

  (b) 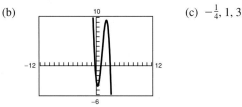   (c) $-2, -1, 2$

57. (a) $\pm 1, \pm 3, \pm\frac{1}{2}, \pm\frac{3}{2}, \pm\frac{1}{4}, \pm\frac{3}{4}$

  (b) 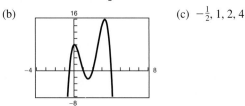   (c) $-\frac{1}{4}, 1, 3$

59. (a) $\pm 1, \pm 2, \pm 4, \pm 8, \pm\frac{1}{2}$

  (b) 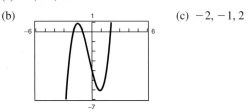   (c) $-\frac{1}{2}, 1, 2, 4$

61. (a) $\pm 1, \pm 2, \pm 4, \pm 8, \pm\frac{1}{2}, \pm\frac{1}{3}, \pm\frac{2}{3}, \pm\frac{4}{3}, \pm\frac{8}{3}, \pm\frac{1}{6}$

  (b) 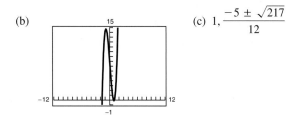   (c) $1, \dfrac{-5 \pm \sqrt{217}}{12}$

63. $-1, 2$ **65.** $0, -1, -3, 4$ **67.** $-2, -\frac{1}{2}, 4$

69. (a) $-2.236, 2, 2.236$ (b) 2

   (c) $f(x) = (x - 2)\left(x - \sqrt{5}\right)\left(x + \sqrt{5}\right)$

71. (a) $-2, 0.268, 3.732$ (b) -2

   (c) $h(t) = (t + 2)\left(t - 2 + \sqrt{3}\right)\left(t - 2 - \sqrt{3}\right)$

73. (a) $0, 3, 4, -1.414, 1.414$ (b) $0, 3, 4$

   (c) $h(x) = x(x - 3)(x - 4)\left(x + \sqrt{2}\right)\left(x - \sqrt{2}\right)$

75. and 77. Answers will vary.   **79.** $\pm 2, \pm\frac{3}{2}$ **81.** $\pm 1, \frac{1}{4}$

83. (d)   **85.** (b)

87. (a) and (b)

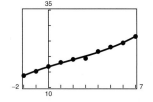

$$R = 0.01326t^3 - 0.0677t^2 + 1.231t + 16.68$$

The model is a good fit.

(c)

t	-2	-1	0	1	2
R	13.84	15.37	16.68	17.86	18.98

t	3	4	5	6	7
R	20.12	21.37	22.8	24.49	26.53

(d) $R(12) = 44.62$. No; the model will turn sharply upward.

89. (a)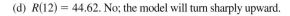

(b) $V = x(9 - 2x)(15 - 2x)$; Domain: $0 < x < \frac{9}{2}$

(c)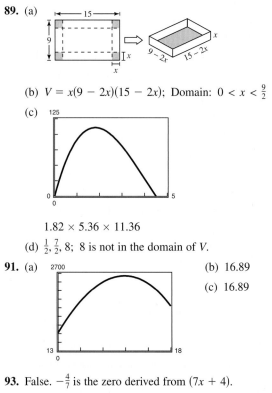

  $1.82 \times 5.36 \times 11.36$

(d) $\frac{1}{2}, \frac{7}{2}, 8$; 8 is not in the domain of V.

91. (a)   (b) 16.89

                                (c) 16.89

93. False. $-\frac{4}{7}$ is the zero derived from $(7x + 4)$.

95. True   **97.** $x^{2n} - x^n + 3$

99. Multiply the divisor and the quotient to obtain the dividend.

101. (a) $2x^2 - 3$ (b) $4x^2 - 20x + 26$

103. (a) $\dfrac{1}{4x + x^2}$ (b) $\dfrac{4}{x} + \dfrac{1}{x^2}$

105. $f(x) = x^3 - 6x^2 - 19x + 24$ **107.** $f(x) = x^2 - 4x + 1$

Section 2.4   *(page 180)*

1. $a = -9, b = 4$ **3.** $a = 6, b = 5$ **5.** $4 + 5i$

7. 12 **9.** $-1 - 5i$ **11.** -75 **13.** $0.3i$

15. $11 - i$ **17.** $7 - 3\sqrt{2}\,i$ **19.** $-14 + 20i$

21. $\frac{1}{6} + \frac{7}{6}i$ **23.** $-4.2 + 7.5i$ **25.** $-2\sqrt{3}$

27. -10 **29.** $5 + i$ **31.** $12 + 30i$

33. 24 **35.** $-9 + 40i$

37. $\sqrt{-6}\sqrt{-6} = \sqrt{6}\,i\sqrt{6}\,i = 6i^2 = -6$ **39.** 25

41. 41 **43.** 484 **45.** 11 **47.** $-6i$

49. $\frac{16}{41} + \frac{20}{41}i$ **51.** $\frac{3}{5} + \frac{4}{5}i$ **53.** $-7 - 6i$

55. $-\frac{9}{1681} + \frac{40}{1681}i$ **57.** $-\frac{1}{2} - \frac{5}{2}i$ **59.** $-1 + 6i$

61. $-5i$ **63.** $-375\sqrt{3}\,i$ **65.** i **67.** $4 + 3i$

69. $6i$

71.

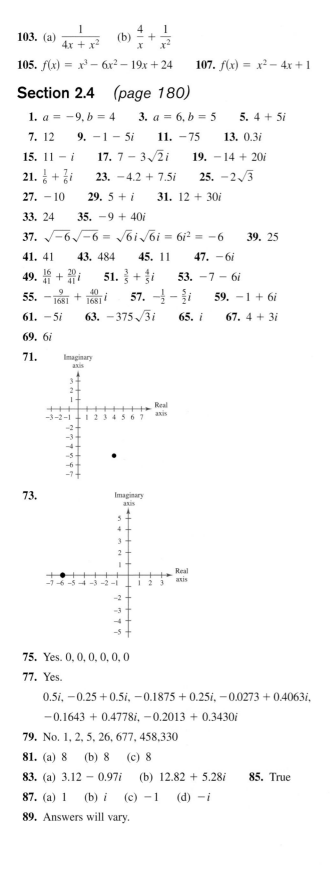

73.

75. Yes. $0, 0, 0, 0, 0, 0$

77. Yes.

$0.5i, -0.25 + 0.5i, -0.1875 + 0.25i, -0.0273 + 0.4063i,$
$-0.1643 + 0.4778i, -0.2013 + 0.3430i$

79. No. $1, 2, 5, 26, 677, 458{,}330$

81. (a) 8 (b) 8 (c) 8

83. (a) $3.12 - 0.97i$ (b) $12.82 + 5.28i$ **85.** True

87. (a) 1 (b) i (c) -1 (d) $-i$

89. Answers will vary.

91. (a) $2x + 3y + 7 = 0$ (b) $3x - 2y + 30 = 0$

93. $(-4, 0), (2, 0), (0, -8)$ **95.** $(-1, 0), (1, 0), (0, -1)$

Section 2.5   *(page 187)*

1. $0, 6, 6$ **3.** $2, -4, -4, -4$ **5.** $-6, i, -i$

7. $2, -3 + 5i, -3 - 5i$

9. Zeros: $4, -i, i$. One real zero; same

11. Zeros: $\sqrt{2}\,i, \sqrt{2}\,i, -\sqrt{2}\,i, -\sqrt{2}\,i$. No real zeros; same

13. $2 \pm \sqrt{3}$

$\left(x - 2 - \sqrt{3}\right)\!\left(x - 2 + \sqrt{3}\right)$

15. $6 \pm \sqrt{10}$

$\left(x - 6 - \sqrt{10}\right)\!\left(x - 6 + \sqrt{10}\right)$

17. $\pm 5i$ **19.** $\pm 3, \pm 3i$

$(x + 5i)(x - 5i)$ $(x + 3)(x - 3)(x + 3i)(x - 3i)$

21. $1 \pm i$

$(z - 1 + i)(z - 1 - i)$

23. $-5, 4 \pm 3i$

$(t + 5)(t - 4 + 3i)(t - 4 - 3i)$

25. $-\frac{3}{4}, 1 \pm \frac{1}{2}i$

$(4x + 3)\!\left(x - 1 - \frac{1}{2}i\right)\!\left(x - 1 + \frac{1}{2}i\right)$

27. $\pm i, \pm 3i$

$(x + i)(x - i)(x + 3i)(x - 3i)$

29. $2, 2, \pm 2i$

$(x - 2)^2(x + 2i)(x - 2i)$

31. $-2, -\frac{1}{2}, \pm i$

$(x + 2)(2x + 1)(x - i)(x + i)$

33. (a) $7 \pm \sqrt{3}$ (b) $\left(x - 7 - \sqrt{3}\right)\!\left(x - 7 + \sqrt{3}\right)$
  (c) $\left(7 \pm \sqrt{3}, 0\right)$

35. (a) $-7 \pm \sqrt{5}$ (b) $\left(x + 7 - \sqrt{5}\right)\!\left(x + 7 + \sqrt{5}\right)$
  (c) $\left(-7 \pm \sqrt{5}, 0\right)$

37. (a) $-6, 3 \pm 4i$ (b) $(x + 6)(x - 3 - 4i)(x - 3 + 4i)$
  (c) $(-6, 0)$

39. (a) $\pm 4i, \pm 3i$ (b) $(x + 4i)(x - 4i)(x + 3i)(x - 3i)$
  (c) None

41. $x^3 - x^2 + 25x - 25$ **43.** $x^3 - 10x^2 + 33x - 34$

45. $x^4 + 37x^2 + 36$ **47.** $x^4 + 8x^3 + 9x^2 - 10x + 100$

49. (a) $(x^2 + 1)(x^2 - 7)$ (b) $(x^2 + 1)\!\left(x + \sqrt{7}\right)\!\left(x - \sqrt{7}\right)$
  (c) $(x + i)(x - i)\!\left(x + \sqrt{7}\right)\!\left(x - \sqrt{7}\right)$

51. (a) $(x^2 - 6)(x^2 - 2x + 3)$
  (b) $\left(x + \sqrt{6}\right)\!\left(x - \sqrt{6}\right)(x^2 - 2x + 3)$
  (c) $\left(x + \sqrt{6}\right)\!\left(x - \sqrt{6}\right)\!\left(x - 1 - \sqrt{2}\,i\right)\!\left(x - 1 + \sqrt{2}\,i\right)$

53. $-\frac{3}{2}, \pm 5i$ **55.** $-3, 5 \pm 2i$ **57.** $-\frac{2}{3}, 1 \pm \sqrt{3}i$

59. $\frac{3}{4}, \frac{1}{2}(1 \pm \sqrt{5}i)$

61. (a) $1.000, 2.000$ (b) $1, 2$ (c) $1, 2, -3 \pm \sqrt{2}i$

63. (a) 0.750 (b) $\frac{3}{4}$ (c) $\frac{3}{4}, \frac{1}{2} \pm \frac{\sqrt{5}}{2}i$

65. No. Setting $h = 64$ and solving the resulting equation yields imaginary roots.

67. False. A polynomial can only have an even number of complex zeros, so one of the zeros of a third-degree polynomial must be real.

69. (a) $0 < k < 4$ (b) $k = 4$ (c) $k < 0$ (d) $k > 4$

71. $f(x) = x^2 + b$

73.   **75.**

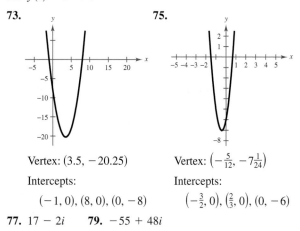

Vertex: $(3.5, -20.25)$ Vertex: $\left(-\frac{5}{12}, -7\frac{1}{24}\right)$

Intercepts: Intercepts:

   $(-1, 0), (8, 0), (0, -8)$ $\left(-\frac{3}{2}, 0\right), \left(\frac{2}{3}, 0\right), (0, -6)$

77. $17 - 2i$ **79.** $-55 + 48i$

Section 2.6   *(page 195)*

1. (a)

x	$f(x)$	x	$f(x)$
0.5	-2	1.5	2
0.9	-10	1.1	10
0.99	-100	1.01	100
0.999	-1000	1.001	1000

x	$f(x)$	x	$f(x)$
5	0.25	-5	$-0.\overline{16}$
10	$0.\overline{1}$	-10	$-0.\overline{09}$
100	$0.\overline{01}$	-100	$-0.\overline{0099}$
1000	$0.\overline{001}$	-1000	$-0.\overline{000999}$

(b) Vertical asymptote: $x = 1$
   Horizontal asymptote: $y = 0$

(c) Domain: all $x \neq 1$

3. (a)

x	$f(x)$	x	$f(x)$
0.5	3	1.5	9
0.9	27	1.1	33
0.99	297	1.01	303
0.999	2997	1.001	3003

x	$f(x)$	x	$f(x)$
5	3.75	-5	-2.5
10	$3.\overline{33}$	-10	-2.727
100	$3.\overline{03}$	-100	-2.97
1000	$3.\overline{003}$	-1000	-2.997

(b) Vertical asymptote: $x = 1$
   Horizontal asymptotes: $y = \pm 3$

(c) Domain: all $x \neq 1$

5. (a)

x	$f(x)$	x	$f(x)$
0.5	-1	1.5	5.4
0.9	-12.79	1.1	17.29
0.99	-147.8	1.01	152.3
0.999	-1498	1.001	1502.3

x	$f(x)$	x	$f(x)$
5	3.125	-5	3.125
10	$3.\overline{03}$	-10	$3.\overline{03}$
100	$3.\overline{0003}$	-100	$3.\overline{0003}$
1000	3	-1000	3.000003

(b) Vertical asymptotes: $x = \pm 1$
   Horizontal asymptote: $y = 3$

(c) Domain: all $x \neq \pm 1$

7. (a)   **9.** (c)   **11.** (b)

13. (a) Domain: all $x \neq 0$
   (b) Vertical asymptote: $x = 0$
      Horizontal asymptote: $y = 0$

15. (a) Domain: all $x \neq 3$
   (b) Vertical asymptote: $x = 3$
      Horizontal asymptote: $y = -1$

17. (a) Domain: all $x \neq \pm 1$
   (b) Vertical asymptotes: $x = \pm 1$

19. (a) Domain of f: all $x \neq -2$

Domain of g: all real numbers

(b) Vertical asymptote: None

(c)

x	-4	-3	-2.5	-2	-1.5	-1	0
$f(x)$	-6	-5	-4.5	Undef.	-3.5	-3	-2
$g(x)$	-6	-5	-4.5	-4	-3.5	-3	-2

(d) Values differ only where f is undefined.

21. (a) Domain of f: all $x \neq 0, 3$; Domain of g: all $x \neq 0$

(b) Vertical asymptote: $x = 0$

(c)

x	-1	-0.5	0	0.5	2	3	4
$f(x)$	-1	-2	Undef.	2	$\frac{1}{2}$	Undef.	$\frac{1}{4}$
$g(x)$	-1	-2	Undef.	2	$\frac{1}{2}$	$\frac{1}{3}$	$\frac{1}{4}$

(d) Values differ only where f is undefined and g is defined.

23. 4; Less than; Greater than

25. 2; Greater than; Less than

27. ± 3 **29.** 7

31. (a) \$28.33 million   (b) \$170 million

(c) \$765 million

(d)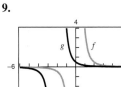

Answers will vary.

(e) No. The function is undefined at the 100% level.

33. (a)

M	200	400	600	800	1000
t	0.472	0.596	0.710	0.817	0.916

M	1200	1400	1600	1800	2000
t	1.009	1.096	1.178	1.255	1.328

The greater the mass, the more time required per oscillation.

(b) $M \approx 1306$ grams

35. (a) 333 deer, 500 deer, 800 deer

(b) 1500. Because the degrees of the numerator and the denominator are equal, the limiting size is the ratio of the leading coefficients, $60/0.4 = 1500$.

37. (a) 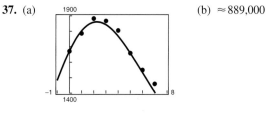   (b) $\approx 889{,}000$

(c) Not likely. The graph of the model dips quite low in the future.

39. True. Can be written as

$$\frac{N(x)}{D(x)} = \frac{x^3 - 2x^2 - 5x + 6}{1}.$$

41. $f(x) = \dfrac{1}{1 + x^2}$ **43.** $f(x) = \dfrac{-3(x + 1)^2}{x\left(x - \frac{5}{2}\right)}$

45. $x = 5$ **47.** $t = 0, \pm 5\sqrt{2}$

49. $x = \pm\sqrt{15}, \pm\sqrt{15}\,i$ **51.** $x - 7 - \dfrac{6}{x - 3}$

53. $4x - 21 + \dfrac{116}{x + 6}$ **55.** $x^3 + 2x^2 + 36x + 72$

57. $x^3 + 5x^2 + 7x - 13$

Section 2.7   *(page 204)*

1.

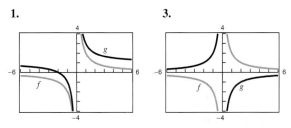

Vertical shift

3.

Reflection in the x-axis

5.

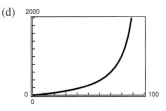

Vertical shift

7.

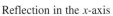

Horizontal shift

9.

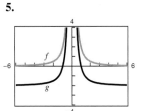

Horizontal shift

11.

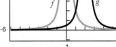

Reflection in the x-axis

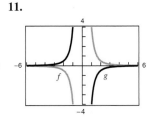

13.

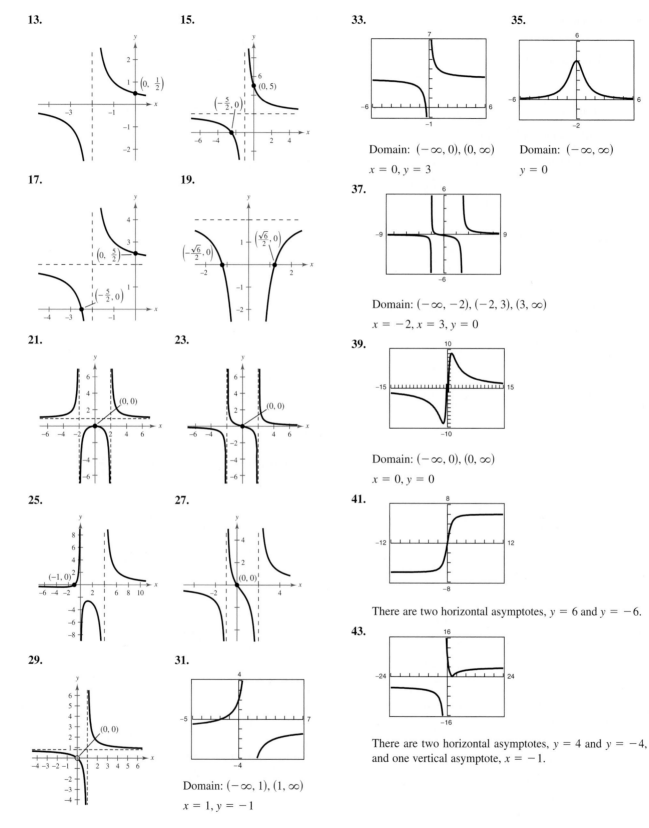

15.

17.

19.

21.

23.

25.

27.

29.

31.

Domain: $(-\infty, 1), (1, \infty)$

$x = 1, y = -1$

33.

Domain: $(-\infty, 0), (0, \infty)$

$x = 0, y = 3$

35.

Domain: $(-\infty, \infty)$

$y = 0$

37.

Domain: $(-\infty, -2), (-2, 3), (3, \infty)$

$x = -2, x = 3, y = 0$

39.

Domain: $(-\infty, 0), (0, \infty)$

$x = 0, y = 0$

41.

There are two horizontal asymptotes, $y = 6$ and $y = -6$.

43.

There are two horizontal asymptotes, $y = 4$ and $y = -4$, and one vertical asymptote, $x = -1$.

45.

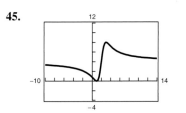

The graph crosses the horizontal asymptote, $y = 4$.

47.

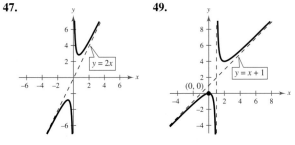

49.

51.

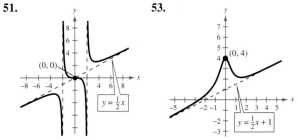

53.

55. $(-1, 0)$ **57.** $(1, 0), (-1, 0)$

59.

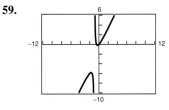

Domain: $(-\infty, -1), (-1, \infty)$

Vertical asymptote: $x = -1$

$y = 2x - 1$

61.

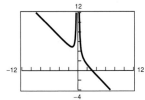

Domain: $(-\infty, 0), (0, \infty)$

Vertical asymptote: $x = 0$

$y = -x + 3$

63.

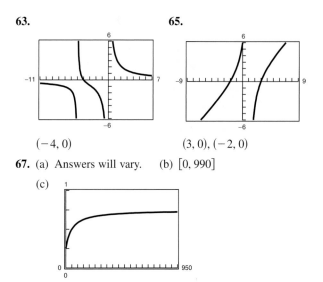

$(-4, 0)$

65.

$(3, 0), (-2, 0)$

67. (a) Answers will vary.   (b) $[0, 990]$

(c)

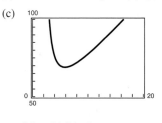

Increases more slowly; the concentration reaches 74.5% when the tank is full.

69. (a) Answers will vary.   (b) $(2, \infty)$

(c)

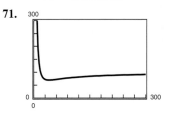

5.9×11.8 inches

71.

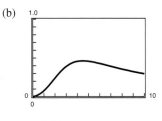

$x \approx 40$

73. (a) $C = 0$. The chemical will eventually dissipate.

(b)

$t \approx 4.5$ hours

(c) Before ≈ 2.6 hours and after ≈ 8.3 hours

75. (a)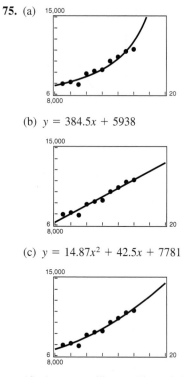

(b) $y = 384.5x + 5938$

(c) $y = 14.87x^2 + 42.5x + 7781$

(d) Answers will vary. The rational model seems to start climbing too fast. The quadratic model may fit the data a little better than the linear model, is easy to use, and would be most useful.

77. False. A graph with a vertical asymptote is not continuous.

79.

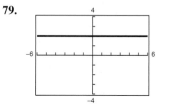

The denominator is a factor of the numerator.

81. No. (Discussions will vary.)

83. $f(x) = \dfrac{x^2 + 2x - 15}{x + 4}$ **85.** $f(x) = \dfrac{-2x - 12}{x - 3}$

87. 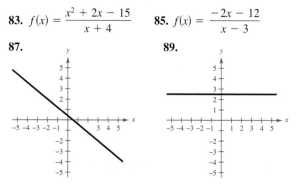   **89.**

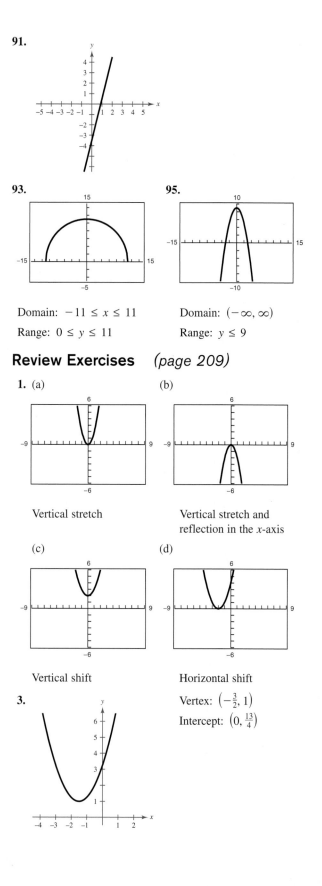

91.

93.

95.

Domain: $-11 \le x \le 11$

Range: $0 \le y \le 11$

Domain: $(-\infty, \infty)$

Range: $y \le 9$

Review Exercises   *(page 209)*

1. (a)   (b)

Vertical stretch   Vertical stretch and reflection in the x-axis

(c)   (d)

Vertical shift   Horizontal shift

Vertex: $\left(-\frac{3}{2}, 1\right)$

Intercept: $\left(0, \frac{13}{4}\right)$

3.

5.

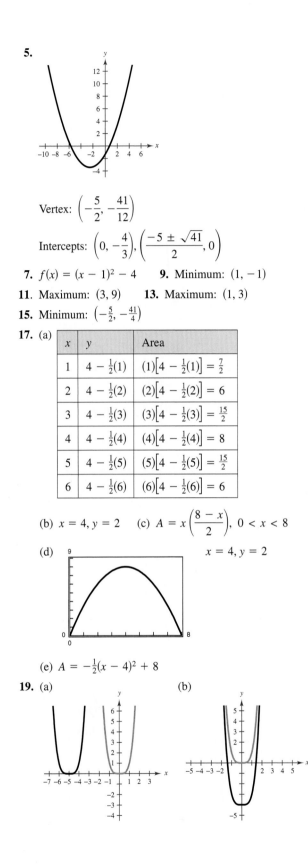

Vertex: $\left(-\dfrac{5}{2}, -\dfrac{41}{12}\right)$

Intercepts: $\left(0, -\dfrac{4}{3}\right), \left(\dfrac{-5 \pm \sqrt{41}}{2}, 0\right)$

7. $f(x) = (x - 1)^2 - 4$ **9.** Minimum: $(1, -1)$

11. Maximum: $(3, 9)$ **13.** Maximum: $(1, 3)$

15. Minimum: $\left(-\dfrac{5}{2}, -\dfrac{41}{4}\right)$

17. (a)

x	y	Area
1	$4 - \frac{1}{2}(1)$	$(1)\left[4 - \frac{1}{2}(1)\right] = \frac{7}{2}$
2	$4 - \frac{1}{2}(2)$	$(2)\left[4 - \frac{1}{2}(2)\right] = 6$
3	$4 - \frac{1}{2}(3)$	$(3)\left[4 - \frac{1}{2}(3)\right] = \frac{15}{2}$
4	$4 - \frac{1}{2}(4)$	$(4)\left[4 - \frac{1}{2}(4)\right] = 8$
5	$4 - \frac{1}{2}(5)$	$(5)\left[4 - \frac{1}{2}(5)\right] = \frac{15}{2}$
6	$4 - \frac{1}{2}(6)$	$(6)\left[4 - \frac{1}{2}(6)\right] = 6$

(b) $x = 4, y = 2$ (c) $A = x\left(\dfrac{8 - x}{2}\right), \ 0 < x < 8$

(d) $x = 4, y = 2$

(e) $A = -\frac{1}{2}(x - 4)^2 + 8$

19. (a)   (b)

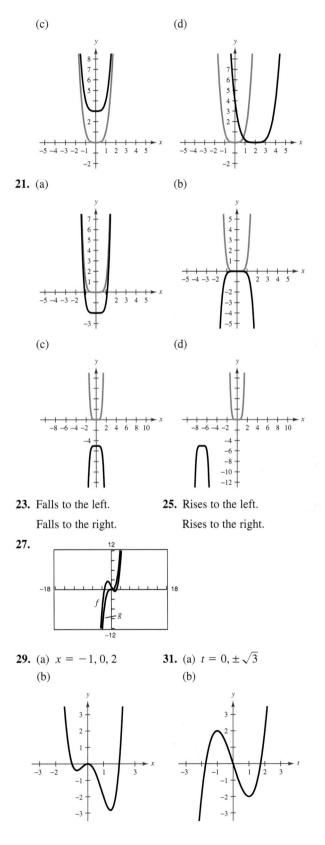

(c)   (d)

21. (a)   (b)

(c)   (d)

23. Falls to the left.   **25.** Rises to the left.
Falls to the right.   Rises to the right.

27.

29. (a) $x = -1, 0, 2$ **31.** (a) $t = 0, \pm\sqrt{3}$
(b)   (b)

33. (a) $x = -3, 0$

(b)

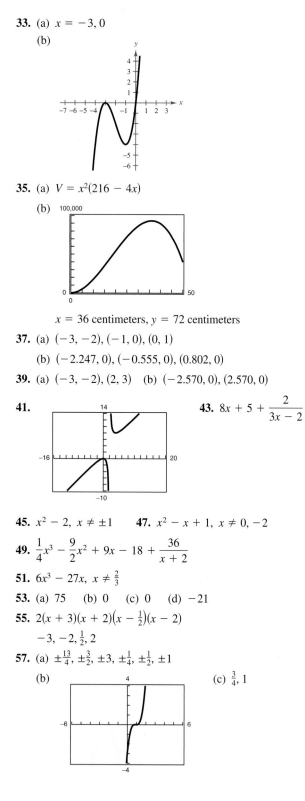

35. (a) $V = x^2(216 - 4x)$

(b)

$x = 36$ centimeters, $y = 72$ centimeters

37. (a) $(-3, -2), (-1, 0), (0, 1)$

(b) $(-2.247, 0), (-0.555, 0), (0.802, 0)$

39. (a) $(-3, -2), (2, 3)$ (b) $(-2.570, 0), (2.570, 0)$

41.

43. $8x + 5 + \dfrac{2}{3x - 2}$

45. $x^2 - 2, \ x \neq \pm 1$ **47.** $x^2 - x + 1, \ x \neq 0, -2$

49. $\dfrac{1}{4}x^3 - \dfrac{9}{2}x^2 + 9x - 18 + \dfrac{36}{x + 2}$

51. $6x^3 - 27x, \ x \neq \dfrac{2}{3}$

53. (a) 75   (b) 0   (c) 0   (d) -21

55. $2(x + 3)(x + 2)\left(x - \dfrac{1}{2}\right)(x - 2)$

$-3, -2, \dfrac{1}{2}, 2$

57. (a) $\pm\dfrac{13}{4}, \pm\dfrac{3}{2}, \pm 3, \pm\dfrac{1}{4}, \pm\dfrac{1}{2}, \pm 1$

(b)

(c) $\dfrac{3}{4}, 1$

59. (a) $\pm\dfrac{10}{3}, \pm\dfrac{20}{3}, \pm 10, \pm 20, \pm\dfrac{5}{3}, \pm 5, \pm\dfrac{5}{6}, \pm\dfrac{5}{2}, \pm\dfrac{2}{3}, \pm\dfrac{4}{3}, \pm 2,$

$\pm 4, \pm\dfrac{1}{3}, \pm\dfrac{1}{6}, \pm\dfrac{1}{2}, \pm 1$

(b)

(c) $\dfrac{5}{6}$

61. (a) $\pm 3, \pm 6, \pm 9, \pm 18, \pm\dfrac{3}{2}, \pm\dfrac{9}{2}, \pm 2, \pm\dfrac{1}{2}, \pm\dfrac{1}{3}, \pm\dfrac{2}{3}, \pm\dfrac{1}{6}, \pm 1$

(b)

(c) $-1, \dfrac{2}{3}, \dfrac{3}{2}, 3$

63. Answers will vary.   **65.** $6 + 5i$ **67.** $2 + 7i$

69. $3 + 7i$ **71.** $40 + 65i$ **73.** $-4 - 46i$

75. -80 **77.** $1 - 6i$ **79.** $\dfrac{4}{3}i$

81.

83.

85. (a)

(b) Two

(c) $-1, -0.54$

87. (a)

(b) One

(c) 3.26

89. $2, 1 \pm i$

$(x - 2)(x - 1 - i)(x - 1 + i)$

91. $-4, -1 \pm \sqrt{2}i$

$(x + 4)(x + 1 + \sqrt{2}i)(x + 1 - \sqrt{2}i)$

93. $\pm 3i, \pm 5i$

$(x - 3i)(x + 3i)(x - 5i)(x + 5i)$

95. $f(x) = x^4 + 4x^3 + 29x^2 + 100x + 100$

97. $f(x) = 3x^4 - 13x^3 + 5x^2 + 43x + 22$

99. (a) $(x^2 - 5)(x^2 - x + 4)$

(b) $(x - \sqrt{5})(x + \sqrt{5})(x^2 - x + 4)$

(c) $(x - \sqrt{5})(x + \sqrt{5})\left(x - \dfrac{1 \pm \sqrt{15}i}{2}\right)\left(x - \dfrac{1 - \sqrt{15}i}{2}\right)$

101. Domain: all $x \neq 1$

Vertical asymptote: $x = 1$

Horizontal asymptote: $y = -1$

103. Domain: all $x \neq 6, -3$

Vertical asymptotes: $x = 6, x = -3$

Horizontal asymptote: $y = 0$

105. $y = -1$ **107.** $y = 2$ **109.** $y = \pm 1$

111. (a)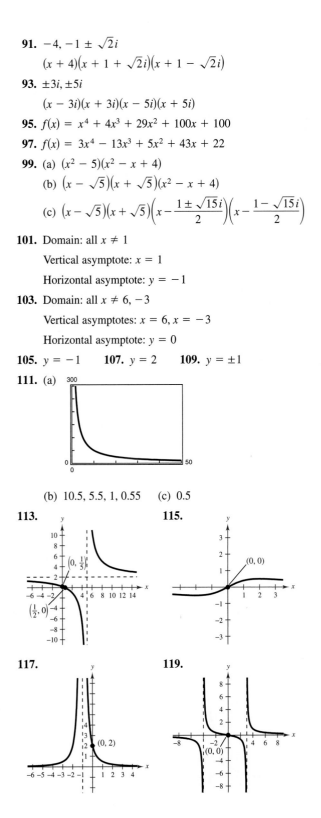

(b) 10.5, 5.5, 1, 0.55   (c) 0.5

113.

115.

117.

119.

121.

123.

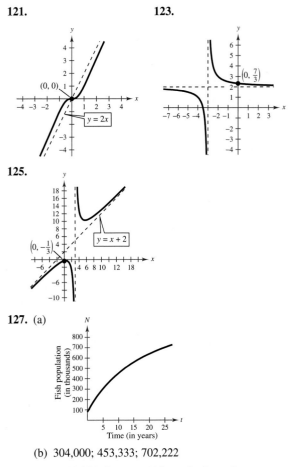

125.

127. (a)

(b) 304,000; 453,333; 702,222

(c) 1,200,000, because N has a horizontal asymptote at $y = 1,200,000$.

129. True

Chapter Test   *(page 214)*

1. (a) Reflection in the x-axis followed by a vertical shift

(b) Horizontal shift

2. Vertex: $(-2, -1)$

Intercepts: $(0, 3), (-3, 0), (-1, 0)$

3. $y = (x - 3)^2 - 6$

4. (a) 50 feet

(b) 5. Changing the constant term results in a vertical shift of the graph and therefore changes the maximum height.

5. $3x + \dfrac{x - 1}{x^2 + 1}$ **6.** $2x^3 + 4x^2 + 3x + 6 + \dfrac{9}{x - 2}$

7. $\pm 1, \pm 2, \pm 3, \pm 4, \pm 6, \pm 8, \pm 12, \pm 24, \pm \frac{1}{2}, \pm \frac{3}{2}$

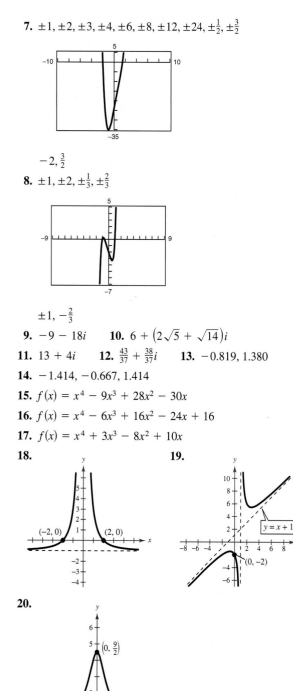

$-2, \frac{3}{2}$

8. $\pm 1, \pm 2, \pm \frac{1}{3}, \pm \frac{2}{3}$

$\pm 1, -\frac{2}{3}$

9. $-9 - 18i$ **10.** $6 + (2\sqrt{5} + \sqrt{14})i$

11. $13 + 4i$ **12.** $\frac{43}{37} + \frac{38}{37}i$ **13.** $-0.819, 1.380$

14. $-1.414, -0.667, 1.414$

15. $f(x) = x^4 - 9x^3 + 28x^2 - 30x$

16. $f(x) = x^4 - 6x^3 + 16x^2 - 24x + 16$

17. $f(x) = x^4 + 3x^3 - 8x^2 + 10x$

18.   **19.**

20.

21. 24.8 years

Chapter 3

Section 3.1 *(page 225)*

1. 4112.033   **3.** 77,494.076   **5.** 19.568   **7.** 1.649

9. 9897.129   **11.** $f(x) = h(x)$

13. $f(x) = g(x) = h(x)$ **15.** (c)   **17.** (e)   **19.** (g)

21. (a)   **23.** Right shift of 5 units

25. Left shift of 4 units and reflection in x-axis

27.   **29.**

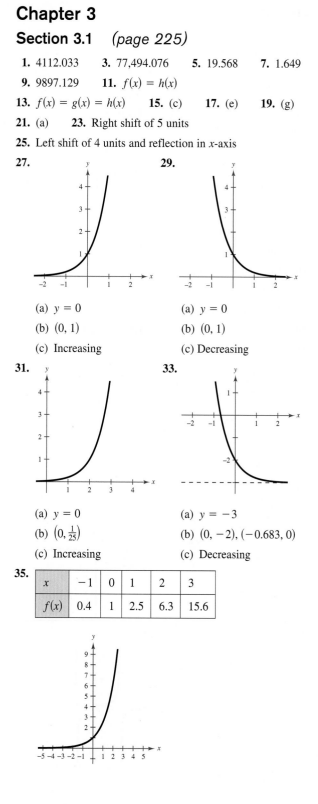

(a) $y = 0$ (a) $y = 0$

(b) $(0, 1)$ (b) $(0, 1)$

(c) Increasing   (c) Decreasing

31.   **33.**

(a) $y = 0$ (a) $y = -3$

(b) $\left(0, \frac{1}{25}\right)$ (b) $(0, -2), (-0.683, 0)$

(c) Increasing   (c) Decreasing

35.

x	-1	0	1	2	3
$f(x)$	0.4	1	2.5	6.3	15.6

37.

x	-1	0	1	2
$f(x)$	0.2	1	6	36

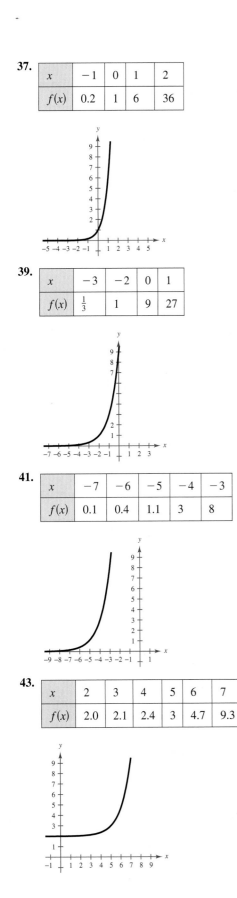

39.

x	-3	-2	0	1
$f(x)$	$\frac{1}{3}$	1	9	27

41.

x	-7	-6	-5	-4	-3
$f(x)$	0.1	0.4	1.1	3	8

43.

x	2	3	4	5	6	7
$f(x)$	2.0	2.1	2.4	3	4.7	9.3

45.

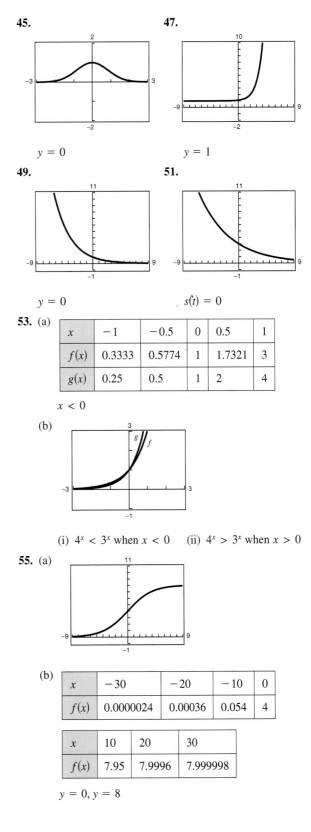

$y = 0$

47.

$y = 1$

49.

$y = 0$

51.

$s(t) = 0$

53. (a)

x	-1	-0.5	0	0.5	1
$f(x)$	0.3333	0.5774	1	1.7321	3
$g(x)$	0.25	0.5	1	2	4

$x < 0$

(b)

(i) $4^x < 3^x$ when $x < 0$ (ii) $4^x > 3^x$ when $x > 0$

55. (a)

(b)

x	-30	-20	-10	0
$f(x)$	0.0000024	0.00036	0.054	4

x	10	20	30
$f(x)$	7.95	7.9996	7.999998

$y = 0, y = 8$

57. (a)

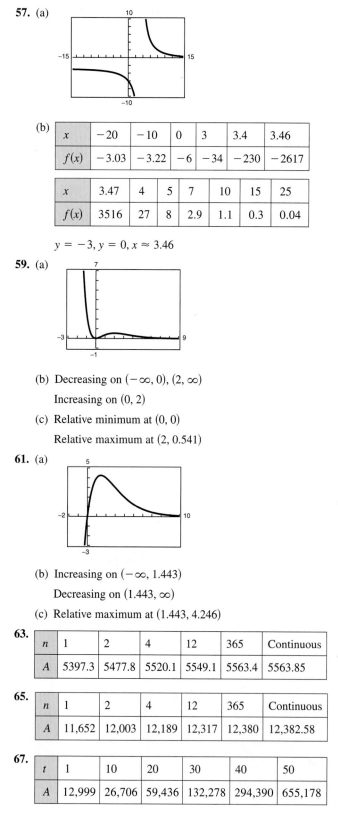

(b)

x	-20	-10	0	3	3.4	3.46
$f(x)$	-3.03	-3.22	-6	-34	-230	-2617

x	3.47	4	5	7	10	15	25
$f(x)$	3516	27	8	2.9	1.1	0.3	0.04

$y = -3, y = 0, x \approx 3.46$

59. (a)

(b) Decreasing on $(-\infty, 0), (2, \infty)$

Increasing on $(0, 2)$

(c) Relative minimum at $(0, 0)$

Relative maximum at $(2, 0.541)$

61. (a)

(b) Increasing on $(-\infty, 1.443)$

Decreasing on $(1.443, \infty)$

(c) Relative maximum at $(1.443, 4.246)$

63.

n	1	2	4	12	365	Continuous
A	5397.3	5477.8	5520.1	5549.1	5563.4	5563.85

65.

n	1	2	4	12	365	Continuous
A	11,652	12,003	12,189	12,317	12,380	12,382.58

67.

t	1	10	20	30	40	50
A	12,999	26,706	59,436	132,278	294,390	655,178

69.

t	1	10	20	30	40	50
A	12,804	22,946	43,877	83,902	160,435	306,782

71. (a)

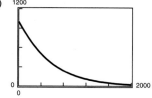

(b) $421.12

(c) $350.13

(d)

x	100	200	300	400
P	849.53	717.64	603.25	504.94

x	500	600	700
P	421.12	350.13	290.35

73. (a)

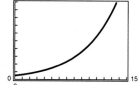

(b) $P(0) = 100; P(5) \approx 300; P(10) \approx 900$

75. (a) 25 units   (b) 16.30 units

(c)

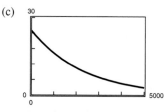

(d) Never. The graph has a horizontal asymptote at $Q = 0$.

77. (a) and (b)

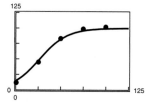

The model fits the data well.

(c)

x	0	25	50	75	100
y	15	47	82	96	99

(d) 64.7% (e) 37.4

79. (a) (b) $35.45

81. True

83.

(a) $y = e^x$

(b) The exponential function increases at a faster rate.

(c) It usually implies rapid growth.

85. $1 < \sqrt{2} < 2$, so $2^1 < 2^{\sqrt{2}} < 2^2$

87.

Answers will vary.

89. $f^{-1}(x) = -\frac{3}{2}\left(x - \frac{5}{2}\right)$ **91.** No inverse function

93.   **95.**

Section 3.2 *(page 236)*

1. $4^3 = 64$ **3.** $7^{-2} = \frac{1}{49}$ **5.** $32^{2/5} = 4$ **7.** $e^0 = 1$

9. $\log_5 125 = 3$ **11.** $\log_{81} 3 = \frac{1}{4}$ **13.** $\log_6 \frac{1}{36} = -2$

15. $\ln 20.0855\ldots = 3$ **17.** $\ln 13.463\ldots = 2.6$

19. 4   **21.** $-\frac{1}{2}$ **23.** -2 **25.** 9   **27.** 8   **29.** 2

31. 2.538   **33.** -0.097 **35.** 1.746   **37.** 1.869

39. 7.022   **41.** 22.276

43.   **45.**

Reflections in the   Reflections in the
line $y = x$ line $y = x$
$g = f^{-1}$ $g = f^{-1}$

47. (c)   **49.** (d)   **51.** (b)

53. Domain: $(0, \infty)$

Vertical asymptote: $x = 0$

Intercept: $(1, 0)$

55. Domain: $(3, \infty)$

Vertical asymptote: $x = 3$

Intercept: $(4, 0)$

57. Domain: $(0, \infty)$

Vertical asymptote: $x = 0$

Intercept: $(9, 0)$

59. Domain: $(3, \infty)$

Vertical asymptote: $x = 3$

Intercept:
$(3 + 6^{-6}, 0) \approx (3, 0)$

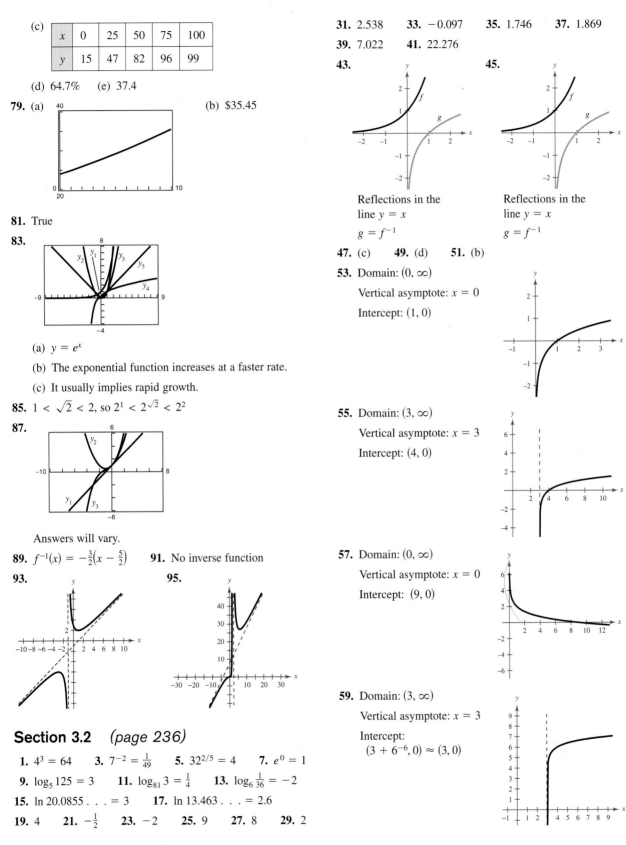

61. Domain: $(0, \infty)$

Vertical asymptote: $x = 0$

Intercept: $(5, 0)$

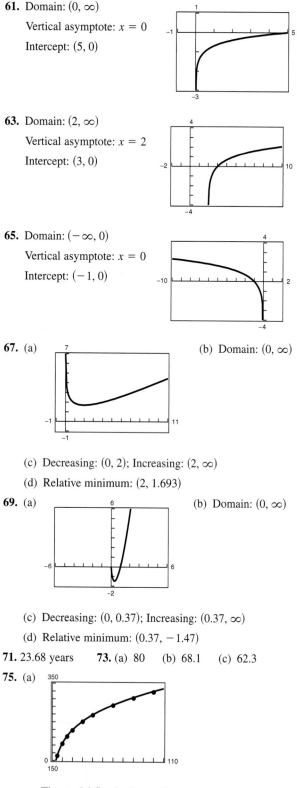

63. Domain: $(2, \infty)$

Vertical asymptote: $x = 2$

Intercept: $(3, 0)$

65. Domain: $(-\infty, 0)$

Vertical asymptote: $x = 0$

Intercept: $(-1, 0)$

67. (a)   (b) Domain: $(0, \infty)$

(c) Decreasing: $(0, 2)$; Increasing: $(2, \infty)$

(d) Relative minimum: $(2, 1.693)$

69. (a)   (b) Domain: $(0, \infty)$

(c) Decreasing: $(0, 0.37)$; Increasing: $(0.37, \infty)$

(d) Relative minimum: $(0.37, -1.47)$

71. 23.68 years   **73.** (a) 80   (b) 68.1   (c) 62.3

75. (a)

The model fits the data well.

(b) 67.3 pounds per square inch   (c) 306.5°F

77.

r	0.005	0.010	0.015
t	138.6	69.3	46.2

r	0.020	0.025	0.030
t	34.7	27.7	23.1

As the growth rate of the population increases, the time for the population to double decreases.

79. (a) 120 decibels   (b) 100 decibels   (c) No

81. (a) 15 cubic feet per minute   (b) 382 cubic feet

(c) 382 square feet

83. 10 years

85. Total amount: $199,108.80; Interest: $49,108.80

87. True   **89.**

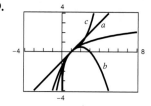

91. (a)

x	1	5	10	10^2	10^4	10^6
$f(x)$	0	0.322	0.230	0.046	0.00092	0.0000138

(b) 0

(c)

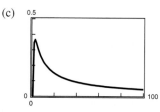

93. Vertical asymptote: $x = 0$

95. Vertical asymptote: $x = 7$ **97.** 33.115   **99.** 0.002

Section 3.3   *(page 244)*

1.

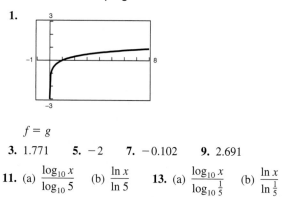

$f = g$

3. 1.771   **5.** -2 **7.** -0.102 **9.** 2.691

11. (a) $\dfrac{\log_{10} x}{\log_{10} 5}$ (b) $\dfrac{\ln x}{\ln 5}$ **13.** (a) $\dfrac{\log_{10} x}{\log_{10} \frac{1}{5}}$ (b) $\dfrac{\ln x}{\ln \frac{1}{5}}$

15. (a) $\dfrac{\log_{10} \frac{3}{10}}{\log_{10} x}$ (b) $\dfrac{\ln \frac{3}{10}}{\ln x}$ **17.** (a) $\dfrac{\log_{10} x}{\log_{10} 2.6}$ (b) $\dfrac{\ln x}{\ln 2.6}$

19. $\dfrac{\log_{10} x}{\log_{10} 2}$ **21.** $\dfrac{\log_{10} x}{\log_{10} \frac{1}{2}}$

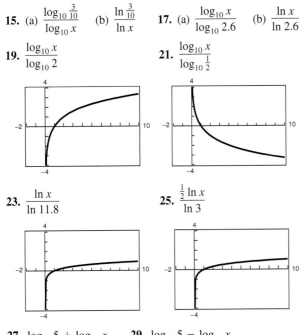

23. $\dfrac{\ln x}{\ln 11.8}$ **25.** $\dfrac{\frac{1}{2} \ln x}{\ln 3}$

27. $\log_{10} 5 + \log_{10} x$ **29.** $\log_{10} 5 - \log_{10} x$
31. $4 \log_8 x$ **33.** $\frac{1}{2} \ln z$ **35.** $\ln x + \ln y + \ln z$
37. $\frac{1}{2} \ln(a - 1)$ **39.** $\ln z + 2 \ln(z - 1)$
41. $\frac{1}{3} \ln x - \frac{1}{3} \ln y$ **43.** $4 \ln x + \frac{1}{2} \ln y - 5 \ln z$
45. $2 \log_b x - 2 \log_b y - 3 \log_b z$
47.

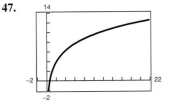

$y_1 = y_2$ for positive values of x.

49. $\ln 4x$ **51.** $\log_4 \dfrac{z}{y}$ **53.** $\log_2(x + 3)^2$

55. $\log_3 \sqrt[3]{7x}$ **57.** $\ln \dfrac{x}{(x + 1)^3}$ **59.** $\ln \dfrac{x - 2}{x + 2}$

61. $\ln \dfrac{x}{(x^2 - 4)^2}$ **63.** $\ln \sqrt[3]{\dfrac{x(x + 3)^2}{x^2 - 1}}$

65. $\ln \dfrac{\sqrt[3]{y(y + 4)^2}}{y - 1}$ **67.** $\ln \dfrac{9}{\sqrt{x^2 + 1}}$

69. **71.**

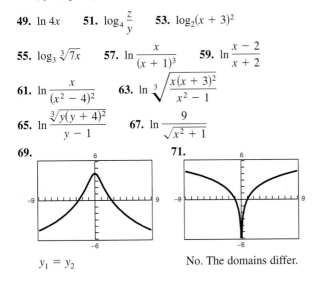

$y_1 = y_2$ No. The domains differ.

73. 2 **75.** 6.8 **77.** -4 is not in the domain of $\log_2 x$.
79. 2 **81.** -4 **83.** 0 is not in the domain of $\log_{10} x$.
85. 8.5 **87.** $\frac{3}{2}$ **89.** $\frac{1}{2}(1 + \log_7 10)$
91. $-3 - \log_5 2$ **93.** $6 + \ln 5$
95. (a) $120 + 10 \log_{10} I$

(b)

I	10^{-4}	10^{-6}	10^{-8}	10^{-10}	10^{-12}	10^{-14}
B	80	60	40	20	0	-20

97. (a)

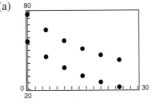

(b) $T = 54.438(0.964)^x + 21$

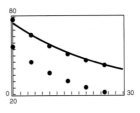

(c) $\ln(T - 21) = -0.037t + 3.997$

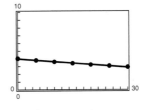

$T = e^{(-0.037t + 3.997)} + 21$

(d)

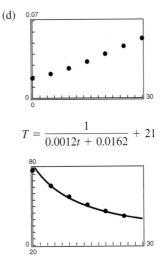

$T = \dfrac{1}{0.0012t + 0.0162} + 21$

99. False. ln 1 = 0   **101.** False. $f(x) - f(2) = \ln \frac{x}{2}$

103. False. $u = v^2$

105.

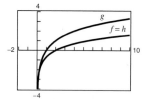

$f(x) = h(x)$. The graphs are indentical because for each positive value of x, Property 2 of logarithms holds.

107. Answers will vary.

109.   **111.**

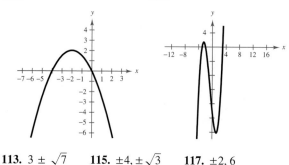

113. $3 \pm \sqrt{7}$ **115.** $\pm 4, \pm \sqrt{3}$ **117.** $\pm 2, 6$

119. 0.052   **121.** 15,235.494   **123.** 2.342

125. 0.697

Section 3.4   *(page 254)*

1. (a) Yes   (b) No

3. (a) No   (b) Yes   (c) Yes, approximate

5. (a) Yes, approximate   (b) No   (c) Yes

7. (a) Yes   (b) Yes, approximate   (c) No

9.   **11.**

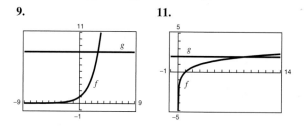

(3, 8)   (9, 2)

13.

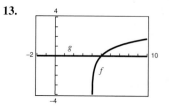

(5, 0)

15. 2   **17.** 4   **19.** $\frac{2}{3}$ **21.** −3   **23.** 4   **25.** 5

27. ln 4 ≈ 1.386   **29.** e^{-7} ≈ 0.00091   **31.** 5

33. 0.1   **35.** 1   **37.** x^2 **39.** $5x + 2$ **41.** x^2

43. 2.756   **45.** ln 10 ≈ 2.303   **47.** 2   **49.** −6.142

51. 0.511   **53.** 0   **55.** ln 5 ≈ 1.609

57. 2 ln 108 ≈ 9.364   **59.** 6.960

61.

x	0.6	0.7	0.8	0.9	1.0
$f(x)$	6.05	8.17	11.02	14.88	20.09

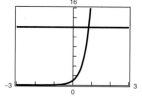

0.828

63.

x	5	6	7	8	9
$f(x)$	1756	1598	1338	908	200

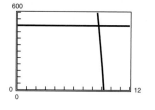

8.635

65. 1.881   **67.** 0.051   **69.** 6.146   **71.** 21.330

73. 3.656

75.   **77.**

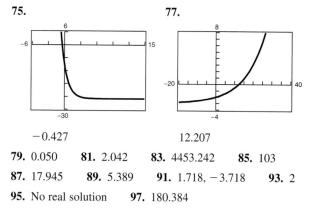

−0.427   12.207

79. 0.050   **81.** 2.042   **83.** 4453.242   **85.** 103

87. 17.945   **89.** 5.389   **91.** 1.718, −3.718   **93.** 2

95. No real solution   **97.** 180.384

99.

x	2	3	4	5	6
f(x)	1.39	1.79	2.08	2.30	2.48

5.512

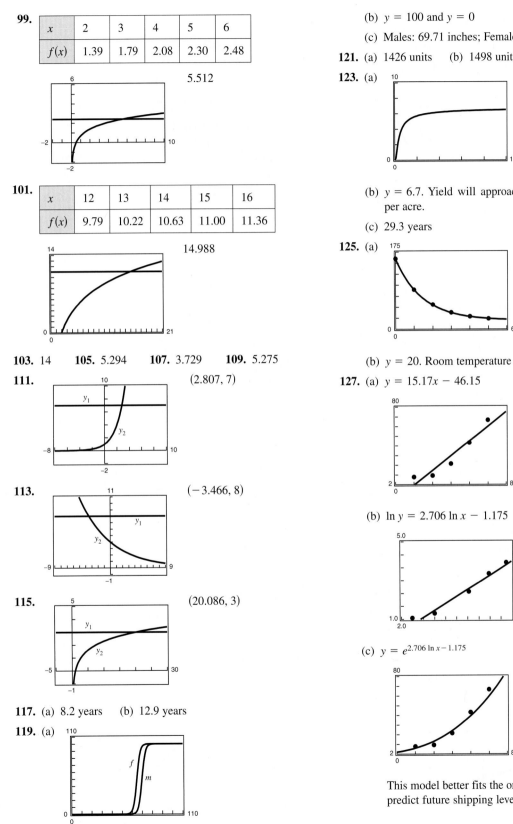

101.

x	12	13	14	15	16
f(x)	9.79	10.22	10.63	11.00	11.36

14.988

103. 14   **105.** 5.294   **107.** 3.729   **109.** 5.275

111.   (2.807, 7)

113.   (−3.466, 8)

115.   (20.086, 3)

117. (a) 8.2 years   (b) 12.9 years

119. (a)

(b) $y = 100$ and $y = 0$

(c) Males: 69.71 inches; Females: 64.51 inches

121. (a) 1426 units   (b) 1498 units

123. (a)

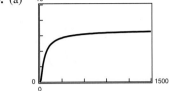

(b) $y = 6.7$. Yield will approach 6.7 million cubic feet per acre.

(c) 29.3 years

125. (a)

(b) $y = 20$. Room temperature   (c) 0.81 hour

127. (a) $y = 15.17x - 46.15$

1999

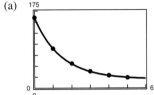

(b) $\ln y = 2.706 \ln x - 1.175$

1998

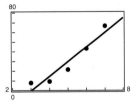

(c) $y = e^{2.706 \ln x - 1.175}$

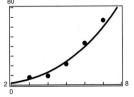

This model better fits the original data and will better predict future shipping levels.

129. False.

For example, $\ln(2x - 1) + \ln(x + 2) = \ln(x^2 - x + 1)$ has 2 extraneous solutions, $x = -1$ and $x = -3$.

131. No. It is dependent on the interest rate.

133.

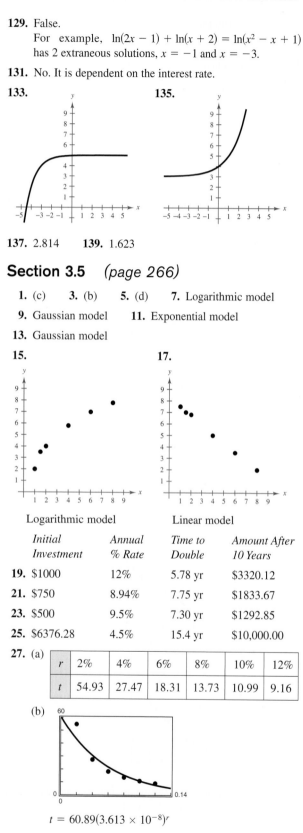

135.

137. 2.814   **139.** 1.623

Section 3.5 *(page 266)*

1. (c)   **3.** (b)   **5.** (d)   **7.** Logarithmic model

9. Gaussian model   **11.** Exponential model

13. Gaussian model

15.

17.

Logarithmic model

Linear model

	Initial Investment	Annual % Rate	Time to Double	Amount After 10 Years
19.	$1000	12%	5.78 yr	$3320.12
21.	$750	8.94%	7.75 yr	$1833.67
23.	$500	9.5%	7.30 yr	$1292.85
25.	$6376.28	4.5%	15.4 yr	$10,000.00

27. (a)

r	2%	4%	6%	8%	10%	12%
t	54.93	27.47	18.31	13.73	10.99	9.16

(b)

$t = 60.89(3.613 \times 10^{-8})^r$

29.

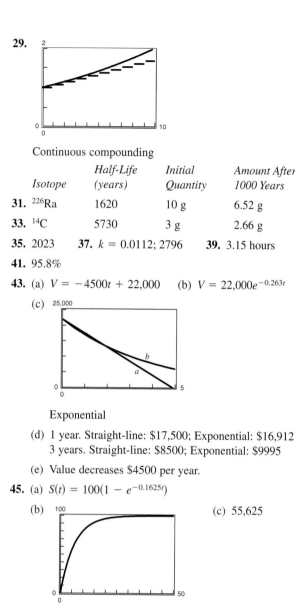

Continuous compounding

Isotope	Half-Life (years)	Initial Quantity	Amount After 1000 Years
31. ^{226}Ra	1620	10 g	6.52 g
33. ^{14}C	5730	3 g	2.66 g

35. 2023   **37.** $k = 0.0112$; 2796   **39.** 3.15 hours

41. 95.8%

43. (a) $V = -4500t + 22,000$ (b) $V = 22,000e^{-0.263t}$

(c)

Exponential

(d) 1 year. Straight-line: $17,500; Exponential: $16,912
3 years. Straight-line: $8500; Exponential: $9995

(e) Value decreases $4500 per year.

45. (a) $S(t) = 100(1 - e^{-0.1625t})$

(b)

(c) 55,625

47. (a) $N = 30(1 - e^{-0.050t})$ (b) 36 days

(c) No. It is not a linear function.

49. (a) 7.6   (b) 7.1

51. (a) 20 decibels   (b) 70 decibels   (c) 120 decibels

53. 95%   **55.** 4.64   **57.** 10,000,000 times

59. (a)

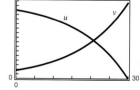

(b) Interest. $t \approx 20.7$ years

(c)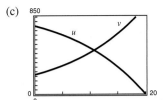

Interest. $t \approx 10.7$ years

Less interest is paid over the life of the loan.

61. $y = e^{0.768x}$ **63.** $y = \frac{1}{2}e^{0.576x}$

65. (a) $t_3 = 0.2729s - 6.0143$; $t_4 = 1.5385e^{0.0291s}$

(b)

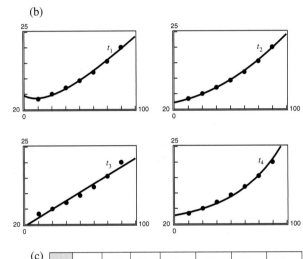

(c)

s	30	40	50	60	70	80	90
t_1	3.6	4.7	6.7	9.4	12.5	15.9	19.6
t_2	3.3	4.9	7.0	9.5	12.5	15.9	19.9
t_3	2.2	4.9	7.6	10.4	13.1	15.8	18.5
t_4	3.7	4.9	6.6	8.8	11.8	15.8	21.1

(d) Model: t_1; Sum ≈ 1.9

Model: t_2; Sum ≈ 1.1

Model: t_3; Sum ≈ 5.6

Model: t_4; Sum ≈ 2.6

Quadratic model fits best.

67. 7:30 A.M.

69. (a) $y = 0.082x + 4.45$

(b) $y = 4.536(1.015)^x$

(c) The linear model fits the data better; Answers will vary.

(d) Linear: 6.50 billion; Exponential: 6.58 billion

71. (a) $y = 298.794(1.085)^x$

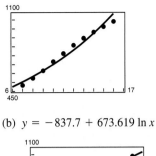

(b) $y = -837.7 + 673.619 \ln x$

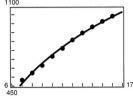

(c) The logarithmic model is better. It will continue to be better if growth of healthcare costs is slowed.

73. (a) $y_1 = -1.81x^3 + 14.58x^2 + 16.39x + 10.00$

$y_2 = 23.07 + 121.08 \ln x$

$y_3 = 38.38(1.4227)^x$

(b)

Cubic model

(c)

x	y	$y - y_1$	$(y - y_1)^2$	$y - y_2$	$(y - y_2)^2$
1	40	0.84	0.71	16.93	286.62
2	85	-1.62	2.62	-22.00	483.84
3	140	-1.52	2.31	-16.09	258.89
4	200	7.00	49.00	9.08	82.40
5	225	-5.20	27.04	7.06	49.83
6	245	2.74	7.51	4.98	24.84

x	$y - y_3$	$(y - y_3)^2$
1	-14.60	213.25
2	7.32	53.52
3	29.48	869.01
4	42.76	1828.56
5	1.30	1.68
6	-73.26	5367.34

(d) y_1: 89.19; y_2: 1186.42; y_3: 8333.36

Cubic model

(e) Sum of the squares of the errors

75. False. For example, $y = \dfrac{3}{1 + 2e^{-0.5x}}$ does not have an x-intercept.

77. True   **79.** (b); $(0, -3)$, $\left(\frac{9}{4}, 0\right)$ **81.** (f); $(0, 25)$, $\left(\frac{100}{9}, 0\right)$

83. (d); $(0, 3)$ **85.** $4x^2 - 12x + 9, x \neq -4$

87. $2x^2 + 3 + \dfrac{3}{x - 4}$

89.

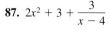

91.

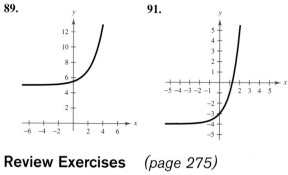

Review Exercises *(page 275)*

1. 10.325   **3.** 0.201   **5.** (e)   **7.** (b)   **9.** (d)

11.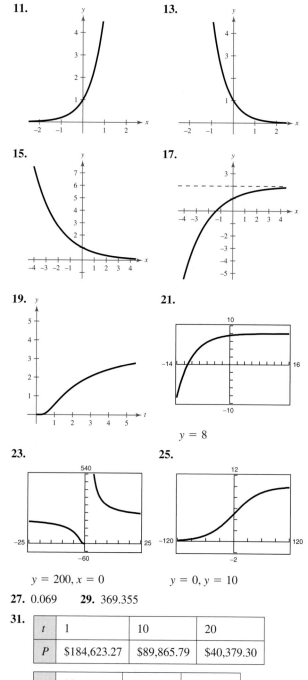

13.

15.

17.

19.

21.

$y = 8$

23.

25.

$y = 200, x = 0$ $y = 0, y = 10$

27. 0.069   **29.** 369.355

31.

t	1	10	20
P	\$184,623.27	\$89,865.79	\$40,379.30

t	30	40	50
P	\$18,143.59	\$8152.44	\$3663.13

33. (a)

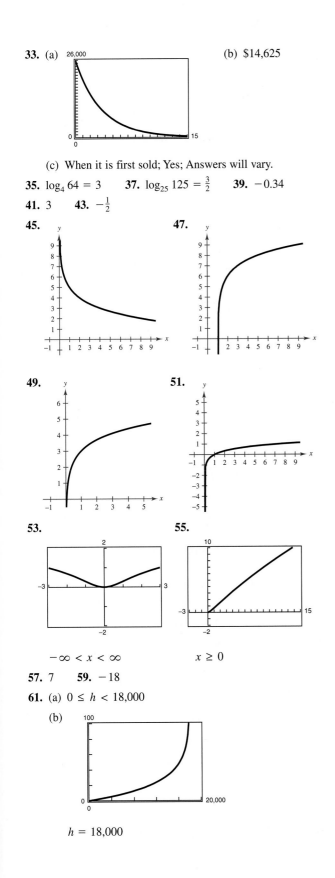

(b) $14,625

(c) When it is first sold; Yes; Answers will vary.

35. $\log_4 64 = 3$ **37.** $\log_{25} 125 = \frac{3}{2}$ **39.** -0.34

41. 3   **43.** $-\frac{1}{2}$

45.

47.

49.

51.

53.

55.

$-\infty < x < \infty$ $x \ge 0$

57. 7   **59.** -18

61. (a) $0 \le h < 18,000$

(b)

$h = 18,000$

(c) The time required to further increase its altitude increases.

(d) 5.46 minutes

63. 1.585   **65.** 2.132   **67.** $\ln 4 + \ln 5$

69. $\ln 5 - 3 \ln 4$ **71.** 1.6542   **73.** 0.2823

75. $1 + 2 \log_5 x$ **77.** $\log_{10} 5 + \frac{1}{2} \log_{10} y - 2 \log_{10} x$

79. $\ln(x^2 + 1) + \ln(x - 1)$ **81.** $\log_2 5x$

83. $\ln \dfrac{\sqrt{|2x - 1|}}{(x + 1)^2}$ **85.** $\ln \dfrac{3 \sqrt[3]{4 - x^2}}{x}$

87. (a)

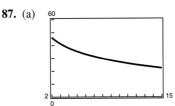

(b)

h	4	6	8	10	12	14
s	38	33	30	27	25	23

(c) The decrease in productivity starts to level off.

89. 3   **91.** -3 **93.** 2401   **95.** $\ln 12 \approx 2.485$

97. $-\dfrac{\ln 44}{5} \approx -0.757$ **99.** $\ln 22 \approx 3.091$

101. $\log_5 17 \approx 1.760$ **103.** $\ln 5 \approx 1.609, \ln 2 \approx 0.693$

105. $\frac{1}{3} e^{8.2} \approx 1213.650$ **107.** $\frac{1}{4} e^{15/2} \approx 452.011$

109. $3e^2 \approx 22.167$ **111.** $e^4 - 1 \approx 53.598$

113. No solution   **115.** $\frac{9}{10}$ **117.** ≈ 15.2 years

119. (e)   **121.** (f)   **123.** (a)   **125.** 2025

127. (a) 5.78%   (b) $10,595.03   (c) 5.95%

129. (a) 7.7 weeks   (b) 13.3 weeks

131. $y = 2e^{0.1014x}$ **133.** $y = \frac{1}{2} e^{0.4605x}$

135. $y = 234.684(0.8746)^x$

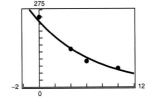

137. True   **139.** False. $\ln(xy) = \ln x + \ln y$ **141.** True

Chapter Test *(page 280)*

1.

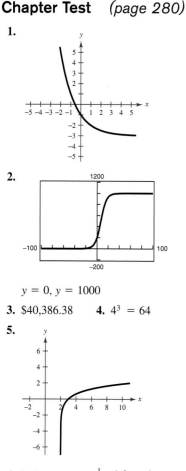

2.

$y = 0,\ y = 1000$

3. $40,386.38$ **4.** $4^3 = 64$

5.

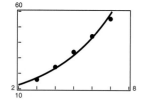

6. $\ln 6 + 2 \ln x - \frac{1}{2} \ln(x^2 + 1)$ **7.** 2   **8.** -3

9. 14.955   **10.** 18.447   **11.** -3.975

12. $10,204   **13.** (a) 0.154   (b) 0.487   (c) 0.811

14. (a) 300   (b) 570   (c) At the end of the 8th year

15. (c); it passes through the point $(0, 0)$. Symmetric to the y-axis, and $y = 6$ is a horizontal asymptote.

16. $y = 6.775(1.361)^x$

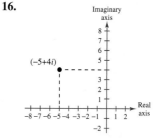

Cumulative Test for Chapters 1–3 *(page 281)*

1. The graph does not represent y as a function of x because for many values of x there correspond two values of y.

2. (a) $\frac{3}{2}$ (b) No solution   (c) $1 + \dfrac{2}{s}$

3. (a) -13 (b) -14 (c) -8

4.

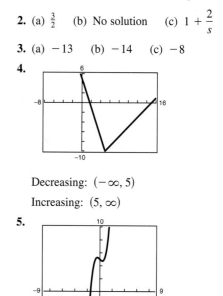

Decreasing: $(-\infty, 5)$
Increasing: $(5, \infty)$

5.

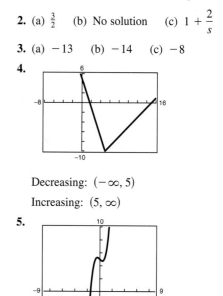

Relative maximum: $(-0.41, 5.27)$
Relative minimum: $(0.41, 4.73)$

6. (a) The graph of r is a vertical shrink of the graph of y.

   (b) The graph of h can be obtained by shifting the graph of y 2 units up.

   (c) The graph of g can be obtained by shifting the graph of y 2 units left.

7. -53 **8.** $\frac{197}{16}$ **9.** -79 **10.** 42

11. $(-\infty, 1) \cup (1, \infty)$ **12.** $f^{-1}(x) = x^2 + 2,\ x \ge 0$

13. Vertex: $\left(\frac{3}{4}, -\frac{49}{8}\right)$; Intercepts: $(-1, 0), \left(\frac{5}{2}, 0\right), (0, -5)$

14. $y = \frac{3}{2}x^2 + 6x + 9$

15. $-3\ \big|\ \begin{array}{ccccc} 2 & 5 & 0 & 7 & -6 \\ & -6 & 3 & -9 & 6 \\ \hline 2 & -1 & 3 & -2 & 0 \end{array}$

A remainder of 0 indicates that $(x + 3)$ is a factor.

16.

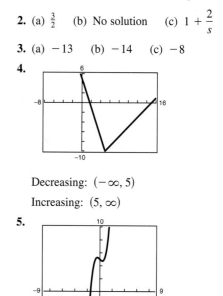

17. $4 - 5i$ **18.** $4\sqrt{15} + 6\sqrt{10}\,i$ **19.** $\dfrac{-1 + 12i}{5}$

20. $29 + 26i$

21. $f(x) = x^5 - 5x^4 + 19x^3 - 71x^2 + 48x + 144$

22. $\frac{1}{2}, -3, \pm 2i$; $f(x) = (2x - 1)(x + 3)(x + 2i)(x - 2i)$

23.

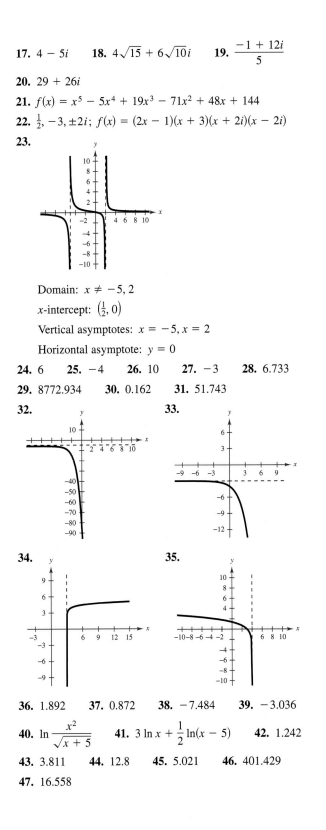

Domain: $x \neq -5, 2$

x-intercept: $\left(\frac{1}{2}, 0\right)$

Vertical asymptotes: $x = -5, x = 2$

Horizontal asymptote: $y = 0$

24. 6   **25.** -4 **26.** 10   **27.** -3 **28.** 6.733

29. 8772.934   **30.** 0.162   **31.** 51.743

32.   **33.**

34.   **35.**

36. 1.892   **37.** 0.872   **38.** -7.484 **39.** -3.036

40. $\ln \dfrac{x^2}{\sqrt{x + 5}}$ **41.** $3 \ln x + \dfrac{1}{2}\ln(x - 5)$ **42.** 1.242

43. 3.811   **44.** 12.8   **45.** 5.021   **46.** 401.429

47. 16.558

48. (a) $A(x) = 273x - x^2$

(b)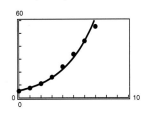

Domain: $0 < x < 273$

(c) 76 ft by 197 ft

49. $5.549(1.412)^x$

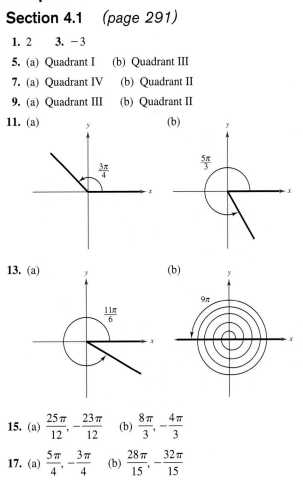

Chapter 4

Section 4.1   *(page 291)*

1. 2   **3.** -3

5. (a) Quadrant I   (b) Quadrant III

7. (a) Quadrant IV   (b) Quadrant II

9. (a) Quadrant III   (b) Quadrant II

11. (a)   (b)

13. (a)   (b)

15. (a) $\dfrac{25\pi}{12}, -\dfrac{23\pi}{12}$ (b) $\dfrac{8\pi}{3}, -\dfrac{4\pi}{3}$

17. (a) $\dfrac{5\pi}{4}, -\dfrac{3\pi}{4}$ (b) $\dfrac{28\pi}{15}, -\dfrac{32\pi}{15}$

19. (a) Complement: $\dfrac{\pi}{6}$; Supplement: $\dfrac{2\pi}{3}$

   (b) Complement: none; Supplement: $\dfrac{\pi}{4}$

21. (a) Complement: none; Supplement: none

   (b) Complement: none; Supplement: $\dfrac{\pi}{2}$

23. $210°$ **25.** $-45°$

27. (a) Quadrant II   (b) Quadrant IV

29. (a) Quadrant III   (b) Quadrant I

31. (a)   (b)

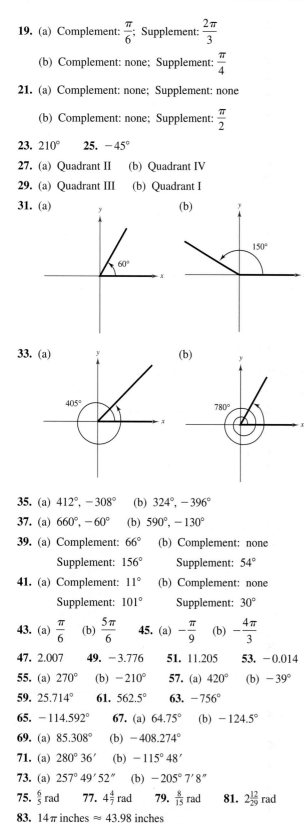

33. (a)   (b)

35. (a) $412°,\ -308°$ (b) $324°,\ -396°$

37. (a) $660°,\ -60°$ (b) $590°,\ -130°$

39. (a) Complement: $66°$ (b) Complement: none

   Supplement: $156°$ Supplement: $54°$

41. (a) Complement: $11°$ (b) Complement: none

   Supplement: $101°$ Supplement: $30°$

43. (a) $\dfrac{\pi}{6}$ (b) $\dfrac{5\pi}{6}$ **45.** (a) $-\dfrac{\pi}{9}$ (b) $-\dfrac{4\pi}{3}$

47. 2.007 **49.** -3.776 **51.** 11.205 **53.** -0.014

55. (a) $270°$ (b) $-210°$ **57.** (a) $420°$ (b) $-39°$

59. $25.714°$ **61.** $562.5°$ **63.** $-756°$

65. $-114.592°$ **67.** (a) $64.75°$ (b) $-124.5°$

69. (a) $85.308°$ (b) $-408.274°$

71. (a) $280°\,36'$ (b) $-115°\,48'$

73. (a) $257°\,49'\,52''$ (b) $-205°\,7'\,8''$

75. $\dfrac{6}{5}$ rad   **77.** $4\dfrac{4}{7}$ rad   **79.** $\dfrac{8}{15}$ rad   **81.** $2\dfrac{12}{29}$ rad

83. 14π inches ≈ 43.98 inches

85. 4π meters ≈ 12.57 meters   **87.** 1141.02 miles

89. 0.094 rad $\approx 5.39°$ **91.** $\dfrac{5}{12}$ rad $\approx 23.87°$

93. (a) $540° \approx 9.42$ rad   (b) $900° \approx 15.71$ rad

   (c) $1260° \approx 21.99$ rad

95. (a) 448.2 rev/min

   (b) 2816 rad/min

97. 20.16π in./sec

99. False. A radian is larger: 1 rad $\approx 57.3°$.

101. True. The sum of the angles is π, or $180°$.

103. Two angles in standard position are coterminal angles if they have the same initial and terminal sides.

105. $\dfrac{50\pi}{3}$ square meters

107. (a) $A = 0.4r^2,\ r > 0;\ s = 0.8r,\ r > 0$

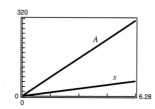

The area function changes more rapidly for $r > 1$ because it is quadratic and the arc length function is linear.

   (b) $A = 50\theta,\ 0 < \theta < 2\pi;\ s = 10\theta,\ 0 < \theta < 2\pi$

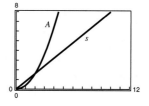

109.   **111.**

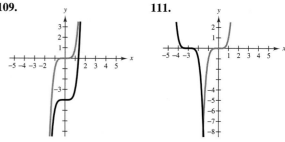

113. **115.**

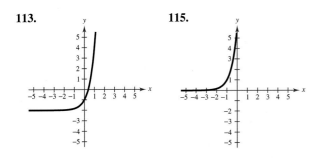

Section 4.2 *(page 300)*

1. $\sin\theta = \frac{15}{17}$

$\cos\theta = -\frac{8}{17}$

$\tan\theta = -\frac{15}{8}$

$\csc\theta = \frac{17}{15}$

$\sec\theta = -\frac{17}{8}$

$\cot\theta = -\frac{8}{15}$

3. $\sin\theta = -\frac{5}{13}$

$\cos\theta = \frac{12}{13}$

$\tan\theta = -\frac{5}{12}$

$\csc\theta = -\frac{13}{5}$

$\sec\theta = \frac{13}{12}$

$\cot\theta = -\frac{12}{5}$

5. $\left(\frac{\sqrt{2}}{2}, \frac{\sqrt{2}}{2}\right)$ **7.** $\left(-\frac{\sqrt{3}}{2}, -\frac{1}{2}\right)$ **9.** $\left(-\frac{1}{2}, -\frac{\sqrt{3}}{2}\right)$

11. $(0, -1)$

13. $\sin\frac{\pi}{4} = \frac{\sqrt{2}}{2}$

$\cos\frac{\pi}{4} = \frac{\sqrt{2}}{2}$

$\tan\frac{\pi}{4} = 1$

15. $\sin\left(-\frac{\pi}{6}\right) = -\frac{1}{2}$

$\cos\left(-\frac{\pi}{6}\right) = \frac{\sqrt{3}}{2}$

$\tan\left(-\frac{\pi}{6}\right) = -\frac{\sqrt{3}}{3}$

17. $\sin\left(-\frac{7\pi}{4}\right) = \frac{\sqrt{2}}{2}$

$\cos\left(-\frac{7\pi}{4}\right) = \frac{\sqrt{2}}{2}$

$\tan\left(-\frac{7\pi}{4}\right) = 1$

19. $\sin\frac{11\pi}{6} = -\frac{1}{2}$

$\cos\frac{11\pi}{6} = \frac{\sqrt{3}}{2}$

$\tan\frac{11\pi}{6} = -\frac{\sqrt{3}}{3}$

21. $\sin\left(-\frac{3\pi}{2}\right) = 1$

$\cos\left(-\frac{3\pi}{2}\right) = 0$

$\tan\left(-\frac{3\pi}{2}\right)$ is undefined.

23. $\sin\frac{3\pi}{4} = \frac{\sqrt{2}}{2}$ $\quad\csc\frac{3\pi}{4} = \sqrt{2}$

$\cos\frac{3\pi}{4} = -\frac{\sqrt{2}}{2}$ $\quad\sec\frac{3\pi}{4} = -\sqrt{2}$

$\tan\frac{3\pi}{4} = -1$ $\quad\cot\frac{3\pi}{4} = -1$

25. $\sin\frac{\pi}{2} = 1$ $\qquad\csc\frac{\pi}{2} = 1$

$\cos\frac{\pi}{2} = 0$ $\qquad\sec\frac{\pi}{2}$ is undefined.

$\tan\frac{\pi}{2}$ is undefined. $\qquad\cot\frac{\pi}{2} = 0$

27. $\sin\left(-\frac{\pi}{3}\right) = -\frac{\sqrt{3}}{2}$ $\quad\csc\left(-\frac{\pi}{3}\right) = -\frac{2\sqrt{3}}{3}$

$\cos\left(-\frac{\pi}{3}\right) = \frac{1}{2}$ $\qquad\sec\left(-\frac{\pi}{3}\right) = 2$

$\tan\left(-\frac{\pi}{3}\right) = -\sqrt{3}$ $\quad\cot\left(-\frac{\pi}{3}\right) = -\frac{\sqrt{3}}{3}$

29. $\sin 5\pi = \sin\pi = 0$ **31.** $\cos\frac{8\pi}{3} = \cos\frac{2\pi}{3} = -\frac{1}{2}$

33. $\cos(-3\pi) = \cos(-\pi) = -1$

35. $\sin\left(-\frac{9\pi}{4}\right) = \sin\frac{7\pi}{4} = -\frac{\sqrt{2}}{2}$ **37.** (a) $-\frac{1}{3}$ (b) -3

39. (a) $-\frac{1}{5}$ (b) -5 **41.** (a) $\frac{4}{5}$ (b) $-\frac{4}{5}$

43. 0.7071 **45.** 1.0378 **47.** -0.1288

49. 1.3940 **51.** -1.4486 **53.** (a) -1 (b) -0.4

55. (a) $0.25, 2.89$ (b) $1.82, 4.46$ **57.** 0.7943

59. (a) 0.2500 foot (b) 0.0177 foot (c) -0.2475 foot

61. $0.0707 = \cos 1.5 \neq 2\cos 0.75 = 1.4634$

63. False. $\sin(-t) = -\sin t$ means that the function is odd, not that the sine of a negative angle is a negative number.

65. (a) y-axis (b) $\sin t_1 = \sin(\pi - t_1)$

(c) $\cos(\pi - t_1) = -\cos t_1$

67. Answers will vary. **69.** It is an even function.

71. $f^{-1}(x) = \sqrt[3]{4(x - 1)}$ **73.** $f^{-1}(x) = \frac{x}{2 - x}, x < 2$

75. **77.**

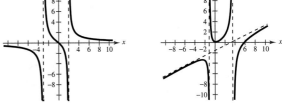

79. \$108.63 **81.** (a) 300 (b) 570 (c) 9

Section 4.3 *(page 310)*

1. $\sin\theta = \frac{3}{5}$ $\qquad\csc\theta = \frac{5}{3}$

$\cos\theta = \frac{4}{5}$ $\qquad\sec\theta = \frac{5}{4}$

$\tan\theta = \frac{3}{4}$ $\qquad\cot\theta = \frac{4}{3}$

3. $\sin \theta = \dfrac{8}{17}$

$\cos \theta = \dfrac{15}{17}$

$\tan \theta = \dfrac{8}{15}$

$\csc \theta = \dfrac{17}{8}$

$\sec \theta = \dfrac{17}{15}$

$\cot \theta = \dfrac{15}{8}$

5. $\sin \theta = \dfrac{3\sqrt{13}}{13}$

$\cos \theta = \dfrac{2\sqrt{13}}{13}$

$\tan \theta = \dfrac{3}{2}$

$\csc \theta = \dfrac{\sqrt{13}}{3}$

$\sec \theta = \dfrac{\sqrt{13}}{2}$

$\cot \theta = \dfrac{2}{3}$

7. $\sin \theta = \dfrac{1}{3}$

$\cos \theta = \dfrac{2\sqrt{2}}{3}$

$\tan \theta = \dfrac{\sqrt{2}}{4}$

$\csc \theta = 3$

$\sec \theta = \dfrac{3\sqrt{2}}{4}$

$\cot \theta = 2\sqrt{2}$

The triangles are similar and corresponding sides are proportional.

9. $\sin \theta = \dfrac{3}{5}$

$\cos \theta = \dfrac{4}{5}$

$\tan \theta = \dfrac{3}{4}$

$\csc \theta = \dfrac{5}{3}$

$\sec \theta = \dfrac{5}{4}$

$\cot \theta = \dfrac{4}{3}$

The triangles are similar and corresponding sides are proportional.

11. $\cos \theta = \dfrac{\sqrt{11}}{6}$

$\tan \theta = \dfrac{5\sqrt{11}}{11}$

$\csc \theta = \dfrac{6}{5}$

$\sec \theta = \dfrac{6\sqrt{11}}{11}$

$\cot \theta = \dfrac{\sqrt{11}}{5}$

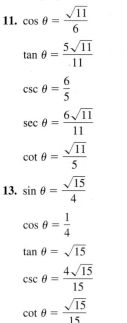

13. $\sin \theta = \dfrac{\sqrt{15}}{4}$

$\cos \theta = \dfrac{1}{4}$

$\tan \theta = \sqrt{15}$

$\csc \theta = \dfrac{4\sqrt{15}}{15}$

$\cot \theta = \dfrac{\sqrt{15}}{15}$

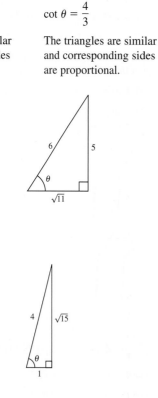

15. $\sin \theta = \dfrac{3\sqrt{10}}{10}$

$\cos \theta = \dfrac{\sqrt{10}}{10}$

$\csc \theta = \dfrac{\sqrt{10}}{3}$

$\sec \theta = \sqrt{10}$

$\cot \theta = \dfrac{1}{3}$

17. $\sin \theta = \dfrac{4\sqrt{97}}{97}$

$\cos \theta = \dfrac{9\sqrt{97}}{97}$

$\tan \theta = \dfrac{4}{9}$

$\csc \theta = \dfrac{\sqrt{97}}{4}$

$\sec \theta = \dfrac{\sqrt{97}}{9}$

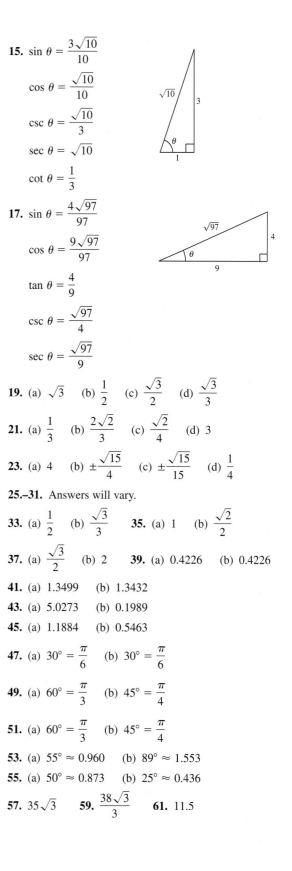

19. (a) $\sqrt{3}$ (b) $\dfrac{1}{2}$ (c) $\dfrac{\sqrt{3}}{2}$ (d) $\dfrac{\sqrt{3}}{3}$

21. (a) $\dfrac{1}{3}$ (b) $\dfrac{2\sqrt{2}}{3}$ (c) $\dfrac{\sqrt{2}}{4}$ (d) 3

23. (a) 4 (b) $\pm\dfrac{\sqrt{15}}{4}$ (c) $\pm\dfrac{\sqrt{15}}{15}$ (d) $\dfrac{1}{4}$

25.–31. Answers will vary.

33. (a) $\dfrac{1}{2}$ (b) $\dfrac{\sqrt{3}}{3}$ **35.** (a) 1 (b) $\dfrac{\sqrt{2}}{2}$

37. (a) $\dfrac{\sqrt{3}}{2}$ (b) 2 **39.** (a) 0.4226 (b) 0.4226

41. (a) 1.3499 (b) 1.3432

43. (a) 5.0273 (b) 0.1989

45. (a) 1.1884 (b) 0.5463

47. (a) $30° = \dfrac{\pi}{6}$ (b) $30° = \dfrac{\pi}{6}$

49. (a) $60° = \dfrac{\pi}{3}$ (b) $45° = \dfrac{\pi}{4}$

51. (a) $60° = \dfrac{\pi}{3}$ (b) $45° = \dfrac{\pi}{4}$

53. (a) $55° \approx 0.960$ (b) $89° \approx 1.553$

55. (a) $50° \approx 0.873$ (b) $25° \approx 0.436$

57. $35\sqrt{3}$ **59.** $\dfrac{38\sqrt{3}}{3}$ **61.** 11.5

63. (a)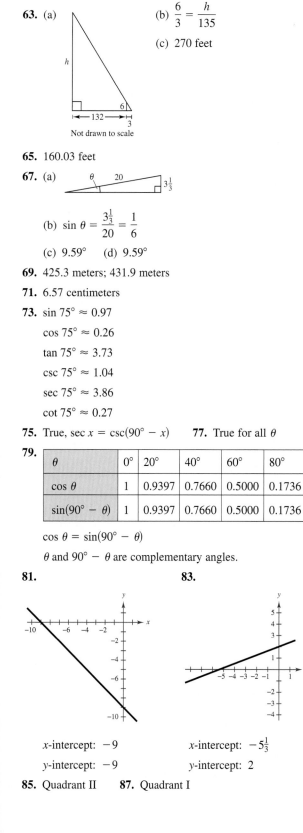

(b) $\dfrac{6}{3} = \dfrac{h}{135}$

(c) 270 feet

h

$\leftarrow 132 \rightarrow$ $\dfrac{6}{3}$

Not drawn to scale

65. 160.03 feet

67. (a)

θ 20 $3\frac{1}{3}$

(b) $\sin \theta = \dfrac{3\frac{1}{3}}{20} = \dfrac{1}{6}$

(c) 9.59°   (d) 9.59°

69. 425.3 meters; 431.9 meters

71. 6.57 centimeters

73. $\sin 75° \approx 0.97$

$\cos 75° \approx 0.26$

$\tan 75° \approx 3.73$

$\csc 75° \approx 1.04$

$\sec 75° \approx 3.86$

$\cot 75° \approx 0.27$

75. True, $\sec x = \csc(90° - x)$ **77.** True for all θ

79.

θ	0°	20°	40°	60°	80°
$\cos \theta$	1	0.9397	0.7660	0.5000	0.1736
$\sin(90° - \theta)$	1	0.9397	0.7660	0.5000	0.1736

$\cos \theta = \sin(90° - \theta)$

θ and $90° - \theta$ are complementary angles.

81.

x-intercept: -9

y-intercept: -9

83.

x-intercept: $-5\frac{1}{3}$

y-intercept: 2

85. Quadrant II   **87.** Quadrant I

Section 4.4   *(page 320)*

1. (a) $\sin \theta = \frac{3}{5}$

$\cos \theta = \frac{4}{5}$

$\tan \theta = \frac{3}{4}$

$\csc \theta = \frac{5}{3}$

$\sec \theta = \frac{5}{4}$

$\cot \theta = \frac{4}{3}$

(b) $\sin \theta = -\frac{15}{17}$

$\cos \theta = -\frac{8}{17}$

$\tan \theta = \frac{15}{8}$

$\csc \theta = -\frac{17}{15}$

$\sec \theta = -\frac{17}{8}$

$\cot \theta = \frac{8}{15}$

3. (a) $\sin \theta = -\dfrac{1}{2}$

$\cos \theta = -\dfrac{\sqrt{3}}{2}$

$\tan \theta = \dfrac{\sqrt{3}}{3}$

$\csc \theta = -2$

$\sec \theta = -\dfrac{2\sqrt{3}}{3}$

$\cot \theta = \sqrt{3}$

(b) $\sin \theta = \dfrac{\sqrt{2}}{2}$

$\cos \theta = -\dfrac{\sqrt{2}}{2}$

$\tan \theta = -1$

$\csc \theta = \sqrt{2}$

$\sec \theta = -\sqrt{2}$

$\cot \theta = -1$

5. $\sin \theta = \frac{24}{25}$

$\cos \theta = \frac{7}{25}$

$\tan \theta = \frac{24}{7}$

$\csc \theta = \frac{25}{24}$

$\sec \theta = \frac{25}{7}$

$\cot \theta = \frac{7}{24}$

7. $\sin \theta = -\frac{12}{13}$

$\cos \theta = \frac{5}{13}$

$\tan \theta = -\frac{12}{5}$

$\csc \theta = -\frac{13}{12}$

$\sec \theta = \frac{13}{5}$

$\cot \theta = -\frac{5}{12}$

9. $\sin \theta = \dfrac{5\sqrt{29}}{29}$

$\cos \theta = -\dfrac{2\sqrt{29}}{29}$

$\tan \theta = -\dfrac{5}{2}$

$\csc \theta = \dfrac{\sqrt{29}}{5}$

$\sec \theta = -\dfrac{\sqrt{29}}{2}$

$\cot \theta = -\dfrac{2}{5}$

11. $\sin \theta = \dfrac{9\sqrt{85}}{85}$

$\cos \theta = -\dfrac{2\sqrt{85}}{85}$

$\tan \theta = -\dfrac{9}{2}$

$\csc \theta = \dfrac{\sqrt{85}}{9}$

$\sec \theta = -\dfrac{\sqrt{85}}{2}$

$\cot \theta = -\dfrac{2}{9}$

13. Quadrant III   **15.** Quadrant II   **17.** Quadrant I

19. $\sin \theta = \frac{3}{5}$

$\cos \theta = -\frac{4}{5}$

$\tan \theta = -\frac{3}{4}$

$\csc \theta = \frac{5}{3}$

$\sec \theta = -\frac{5}{4}$

$\cot \theta = -\frac{4}{3}$

21. $\sin \theta = -\frac{15}{17}$

$\cos \theta = \frac{8}{17}$

$\tan \theta = -\frac{15}{8}$

$\csc \theta = -\frac{17}{15}$

$\sec \theta = \frac{17}{8}$

$\cot \theta = -\frac{8}{15}$

23. $\sin \theta = \dfrac{\sqrt{3}}{2}$

$\cos \theta = -\dfrac{1}{2}$

$\tan \theta = -\sqrt{3}$

$\csc \theta = \dfrac{2\sqrt{3}}{3}$

$\sec \theta = -2$

$\cot \theta = -\dfrac{\sqrt{3}}{3}$

25. $\sin \theta = 0$

$\cos \theta = -1$

$\tan \theta = 0$

$\csc \theta$ is undefined.

$\sec \theta = -1$

$\cot \theta$ is undefined.

27. $\sin \theta = \dfrac{\sqrt{2}}{2}$

$\cos \theta = -\dfrac{\sqrt{2}}{2}$

$\tan \theta = -1$

$\csc \theta = \sqrt{2}$

$\sec \theta = -\sqrt{2}$

$\cot \theta = -1$

29. $\sin \theta = -\dfrac{2\sqrt{5}}{5}$

$\cos \theta = -\dfrac{\sqrt{5}}{5}$

$\tan \theta = 2$

$\csc \theta = -\dfrac{\sqrt{5}}{2}$

$\sec \theta = -\sqrt{5}$

$\cot \theta = \dfrac{1}{2}$

31. -1 **33.** 0 **35.** 1 **37.** Undefined

39. $\theta' = 28°$ **41.** $\theta' = 65°$

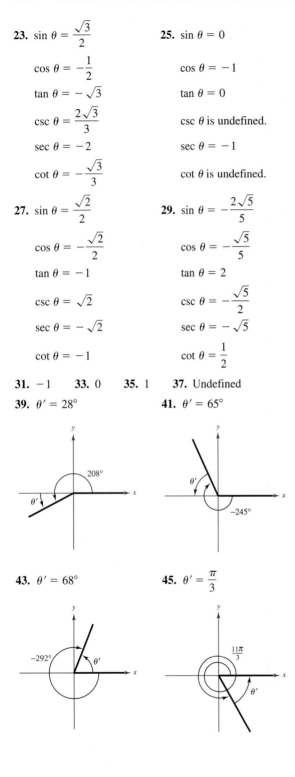

43. $\theta' = 68°$ **45.** $\theta' = \dfrac{\pi}{3}$

47. $\theta' = 3.5 - \pi$ **49.** $\theta' = 3.68 - \pi \approx 0.5384$

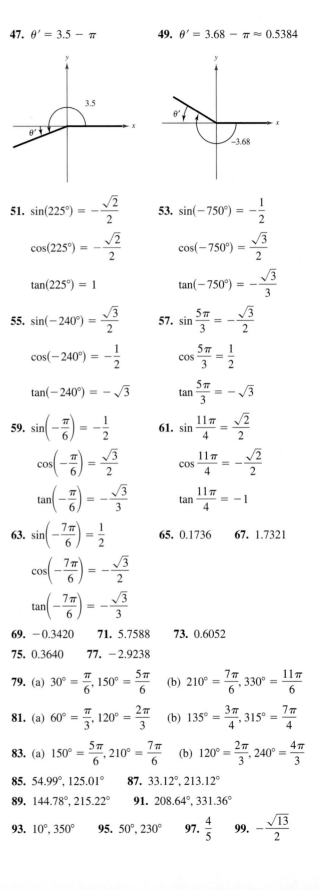

51. $\sin(225°) = -\dfrac{\sqrt{2}}{2}$

$\cos(225°) = -\dfrac{\sqrt{2}}{2}$

$\tan(225°) = 1$

53. $\sin(-750°) = -\dfrac{1}{2}$

$\cos(-750°) = \dfrac{\sqrt{3}}{2}$

$\tan(-750°) = -\dfrac{\sqrt{3}}{3}$

55. $\sin(-240°) = \dfrac{\sqrt{3}}{2}$

$\cos(-240°) = -\dfrac{1}{2}$

$\tan(-240°) = -\sqrt{3}$

57. $\sin \dfrac{5\pi}{3} = -\dfrac{\sqrt{3}}{2}$

$\cos \dfrac{5\pi}{3} = \dfrac{1}{2}$

$\tan \dfrac{5\pi}{3} = -\sqrt{3}$

59. $\sin\left(-\dfrac{\pi}{6}\right) = -\dfrac{1}{2}$

$\cos\left(-\dfrac{\pi}{6}\right) = \dfrac{\sqrt{3}}{2}$

$\tan\left(-\dfrac{\pi}{6}\right) = -\dfrac{\sqrt{3}}{3}$

61. $\sin \dfrac{11\pi}{4} = \dfrac{\sqrt{2}}{2}$

$\cos \dfrac{11\pi}{4} = -\dfrac{\sqrt{2}}{2}$

$\tan \dfrac{11\pi}{4} = -1$

63. $\sin\left(-\dfrac{7\pi}{6}\right) = \dfrac{1}{2}$

$\cos\left(-\dfrac{7\pi}{6}\right) = -\dfrac{\sqrt{3}}{2}$

$\tan\left(-\dfrac{7\pi}{6}\right) = -\dfrac{\sqrt{3}}{3}$

65. 0.1736 **67.** 1.7321

69. -0.3420 **71.** 5.7588 **73.** 0.6052

75. 0.3640 **77.** -2.9238

79. (a) $30° = \dfrac{\pi}{6}, 150° = \dfrac{5\pi}{6}$ (b) $210° = \dfrac{7\pi}{6}, 330° = \dfrac{11\pi}{6}$

81. (a) $60° = \dfrac{\pi}{3}, 120° = \dfrac{2\pi}{3}$ (b) $135° = \dfrac{3\pi}{4}, 315° = \dfrac{7\pi}{4}$

83. (a) $150° = \dfrac{5\pi}{6}, 210° = \dfrac{7\pi}{6}$ (b) $120° = \dfrac{2\pi}{3}, 240° = \dfrac{4\pi}{3}$

85. $54.99°, 125.01°$ **87.** $33.12°, 213.12°$

89. $144.78°, 215.22°$ **91.** $208.64°, 331.36°$

93. $10°, 350°$ **95.** $50°, 230°$ **97.** $\dfrac{4}{5}$ **99.** $-\dfrac{\sqrt{13}}{2}$

101. $\frac{8}{5}$

103. (a) 25.2°F   (b) 65.1°F   (c) 50.8°F

105. (a) 12 miles   (b) 6 miles   (c) 6.9 miles

107. False. $\theta' \neq 180° - \theta$

109. (a)

θ	0°	20°	40°	60°	80°
$\sin \theta$	0	0.3420	0.6428	0.8660	0.9848
$\sin(180° - \theta)$	0	0.3420	0.6428	0.8660	0.9848

(b) $\sin \theta = \sin(180° - \theta)$

111. Answers will vary.

113.   **115.**

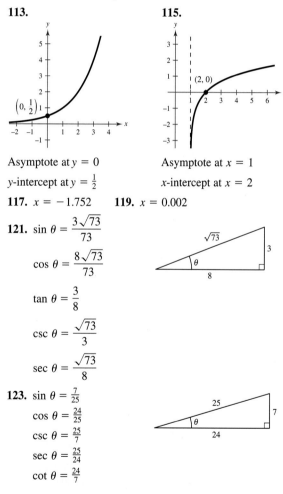

Asymptote at $y = 0$

y-intercept at $y = \frac{1}{2}$

Asymptote at $x = 1$

x-intercept at $x = 2$

117. $x = -1.752$ **119.** $x = 0.002$

121. $\sin \theta = \dfrac{3\sqrt{73}}{73}$

$\cos \theta = \dfrac{8\sqrt{73}}{73}$

$\tan \theta = \dfrac{3}{8}$

$\csc \theta = \dfrac{\sqrt{73}}{3}$

$\sec \theta = \dfrac{\sqrt{73}}{8}$

123. $\sin \theta = \frac{7}{25}$

$\cos \theta = \frac{24}{25}$

$\csc \theta = \frac{25}{7}$

$\sec \theta = \frac{25}{24}$

$\cot \theta = \frac{24}{7}$

Section 4.5   *(page 330)*

1. Period: π; Amplitude: 3

3. Period: 4π; Amplitude: $\frac{5}{2}$

5. Period: 2; Amplitude: $\frac{2}{3}$

7. Period: 2π; Amplitude: 2

9. Period: $\dfrac{\pi}{3}$; Amplitude: 3

11. Period: 3π; Amplitude: $\frac{1}{4}$

13. Period: $\frac{1}{2}$; Amplitude: 3

15. g is a shift of f π units to the right.

17. g is a reflection of f about the x-axis.

19. g is a reflection in the x-axis and five times the amplitude of f.

21. g is the graph of f shifted 4 units up.

23. g has twice the amplitude of f.

25. g is a horizontal shift of f π units to the right.

27.   **29.**

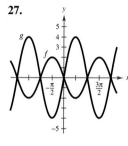
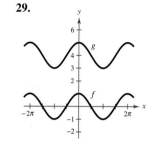

31.   **33.**

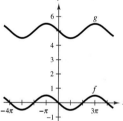
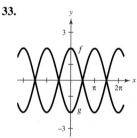

35.

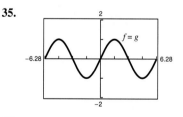

37.

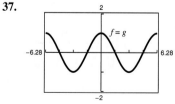

39.

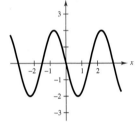

41.

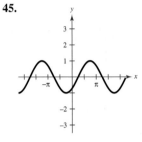

59.

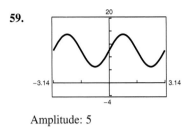

Amplitude: 5

Period: π

43.

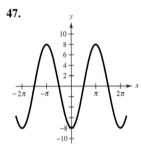

45.

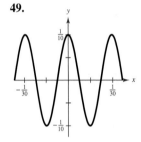

61.

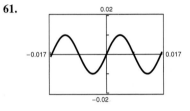

Amplitude: $\frac{1}{100}$

Period: $\frac{1}{60}$

63. $f(x) = 4 - 4 \cos x$ **65.** $f(x) = 1 - 6 \cos x$

47.

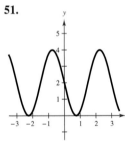

49.

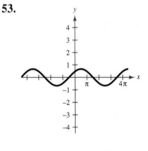

67. $f(x) = -3 \sin(2x)$ **69.** $f(x) = \sin\left(x - \frac{\pi}{4}\right)$

71.

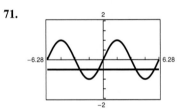

$$x = -\frac{\pi}{6}, -\frac{5\pi}{6}, \frac{7\pi}{6}, \frac{11\pi}{6}$$

51.

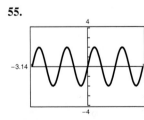

53.

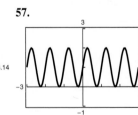

73.

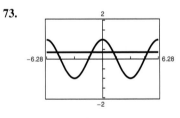

$$x = \pm\frac{\pi}{4}, \pm\frac{7\pi}{4}$$

55.

Amplitude: 2

Period: $\frac{\pi}{2}$

57.

Amplitude: 1

Period: 1

75. (a)

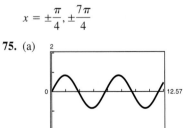

(b) 6 seconds

(c) 10 cycles per minute

77.

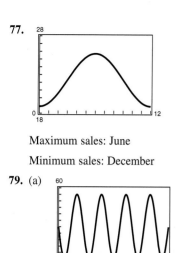

Maximum sales: June

Minimum sales: December

79. (a)

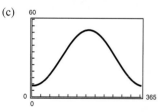

(b) Minimum height: 5 feet

Maximum height: 55 feet

81. (a) 365 days. The cycle is one year.

(b) 30.35 gallons per day; Both terms of the model; Average of the minimum and maximum of the model

(c)

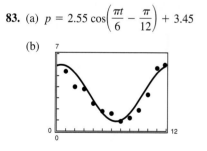

Consumption exceeds 40 gallons per day from the beginning of May through part of September.

83. (a) $p = 2.55 \cos\left(\dfrac{\pi t}{6} - \dfrac{\pi}{12}\right) + 3.45$

(b)

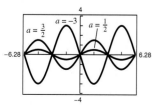

The model fits the data fairly well.

85. True

87.

Amplitude changes

89.

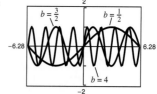

Period changes

91.

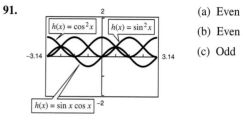

(a) Even

(b) Even

(c) Odd

93. g is the periodic function and f is the parabola.

95. **97.**

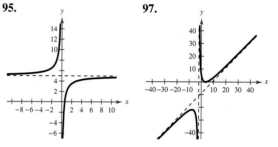

Asymptotes: $x = 0, y = 5$ Asymptotes: $x = -2, y = x - 9$

99. $-20°$ **101.** $-27.502°$

Section 4.6 *(page 341)*

1. (g), 4π **3.** (f), $\dfrac{\pi}{2}$ **5.** (b), 2   **7.** (e), 2π

9. **11.**

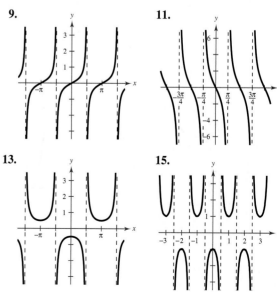

13. **15.**

17.

19.

21.

23.

25.

27.

29.

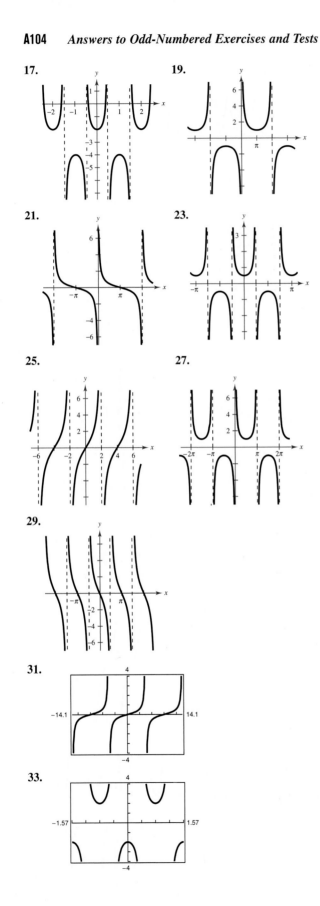

31.

33.

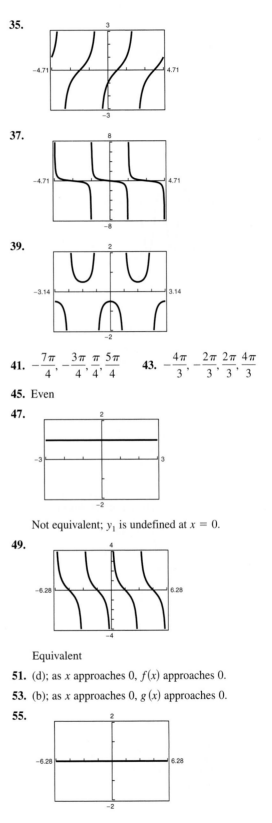

35.

37.

39.

41. $-\dfrac{7\pi}{4}, -\dfrac{3\pi}{4}, \dfrac{\pi}{4}, \dfrac{5\pi}{4}$ **43.** $-\dfrac{4\pi}{3}, -\dfrac{2\pi}{3}, \dfrac{2\pi}{3}, \dfrac{4\pi}{3}$

45. Even

47.

Not equivalent; y_1 is undefined at $x = 0$.

49.

Equivalent

51. (d); as x approaches 0, $f(x)$ approaches 0.

53. (b); as x approaches 0, $g(x)$ approaches 0.

55.

Equal

57.

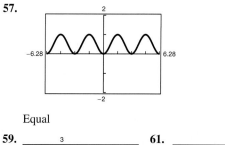

Equal

59.

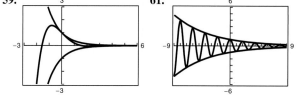

$f \to 0$ as $x \to \infty$ $f \to 0$ as $x \to \infty$

61.

63.

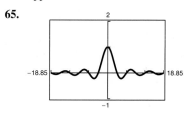

Approaches $\pm\infty$

65.

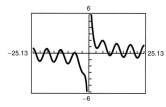

Approaches 1

67.

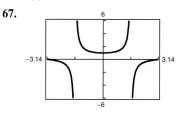

Approaches 1

69. $d = 5 \cot x$

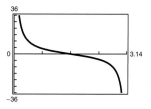

71. As the predator population increases, the number of prey decreases. When the number of prey is small, the number of predators decreases.

73. (a)

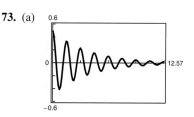

(b) Not periodic and damped; goes to 0 as t increases.

75. (a)

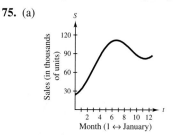

(b) According to the graph, the greatest number of sales occurs when $t = 6$, or in June. The least number of sales is in January.

77. (a) $0 < c < 8$ (b) $0.22 < b < 1$

79. True   **81.** True

83. As x approaches π from the left, f approaches ∞. As x approaches π from the right, f approaches $-\infty$.

85. (a)

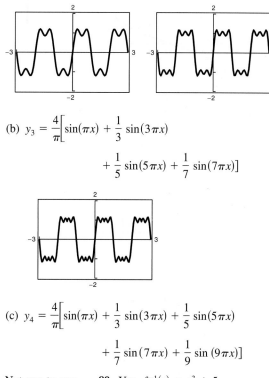

(b) $y_3 = \dfrac{4}{\pi}\Big[\sin(\pi x) + \dfrac{1}{3}\sin(3\pi x)$

$\qquad\qquad + \dfrac{1}{5}\sin(5\pi x) + \dfrac{1}{7}\sin(7\pi x)\Big]$

(c) $y_4 = \dfrac{4}{\pi}\Big[\sin(\pi x) + \dfrac{1}{3}\sin(3\pi x) + \dfrac{1}{5}\sin(5\pi x)$

$\qquad\qquad + \dfrac{1}{7}\sin(7\pi x) + \dfrac{1}{9}\sin(9\pi x)\Big]$

87. Not one-to-one   **89.** Yes. $f^{-1}(x) = x^3 + 5$

91. $\sin \theta = \dfrac{9\sqrt{277}}{277}$ $\qquad$ $\csc \theta = \dfrac{\sqrt{277}}{9}$

$\cos \theta = \dfrac{14\sqrt{277}}{277}$ $\qquad$ $\sec \theta = \dfrac{\sqrt{277}}{14}$

$\tan \theta = \dfrac{9}{14}$ $\qquad$ $\cot \theta = \dfrac{14}{9}$

Section 4.7 *(page 351)*

1. (a)

x	−1	−0.8	−0.6	−0.4	−0.2
y	−1.5708	−0.9273	−0.6435	−0.4115	−0.2014

x	0	0.2	0.4	0.6	0.8	1
y	0	0.2014	0.4115	0.6435	0.9273	1.5708

(b)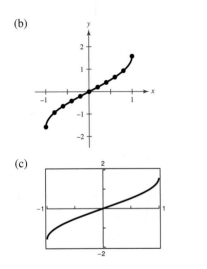

(c)

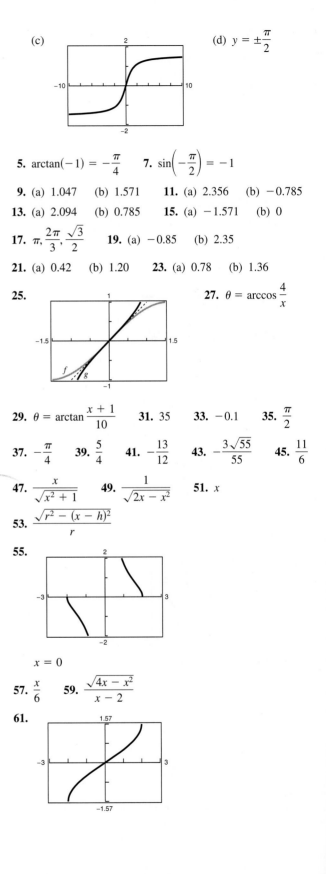

(d) Intercept: $(0, 0)$; Symmetric about the origin

3. (a)

x	−10	−8	−6	−4	−2
y	−1.4711	−1.4464	−1.4056	−1.3258	−1.1071

x	0	2	4	6	8	10
y	0	1.1071	1.3258	1.4056	1.4464	1.4711

(b)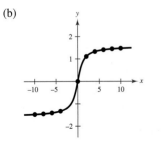

(c)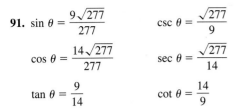

(d) $y = \pm\dfrac{\pi}{2}$

5. $\arctan(-1) = -\dfrac{\pi}{4}$ $\qquad$ **7.** $\sin\left(-\dfrac{\pi}{2}\right) = -1$

9. (a) 1.047 $\quad$ (b) 1.571 $\qquad$ **11.** (a) 2.356 $\quad$ (b) −0.785

13. (a) 2.094 $\quad$ (b) 0.785 $\qquad$ **15.** (a) −1.571 $\quad$ (b) 0

17. $\pi, \dfrac{2\pi}{3}, \dfrac{\sqrt{3}}{2}$ $\qquad$ **19.** (a) −0.85 $\quad$ (b) 2.35

21. (a) 0.42 $\quad$ (b) 1.20 $\qquad$ **23.** (a) 0.78 $\quad$ (b) 1.36

25.

27. $\theta = \arccos \dfrac{4}{x}$

29. $\theta = \arctan \dfrac{x+1}{10}$ $\qquad$ **31.** 35 $\qquad$ **33.** −0.1 $\qquad$ **35.** $\dfrac{\pi}{2}$

37. $-\dfrac{\pi}{4}$ $\qquad$ **39.** $\dfrac{5}{4}$ $\qquad$ **41.** $-\dfrac{13}{12}$ $\qquad$ **43.** $-\dfrac{3\sqrt{55}}{55}$ $\qquad$ **45.** $\dfrac{11}{6}$

47. $\dfrac{x}{\sqrt{x^2+1}}$ $\qquad$ **49.** $\dfrac{1}{\sqrt{2x-x^2}}$ $\qquad$ **51.** x

53. $\dfrac{\sqrt{r^2-(x-h)^2}}{r}$

55.

$x = 0$

57. $\dfrac{x}{6}$ $\qquad$ **59.** $\dfrac{\sqrt{4x-x^2}}{x-2}$

61.

63.

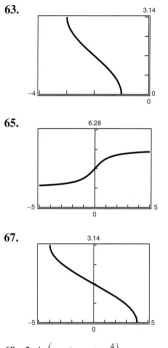

65.

67.

69. $5 \sin\left(\pi t + \arctan \frac{4}{3}\right)$

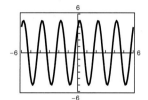

The two forms are equivalent.

71. (a) $\theta = \arctan \frac{s}{750}$ (b) 0.49 rad, 1.13 rad

73. (a) 0.45 rad   (b) 24.39 feet

75. (a) $\frac{\pi}{4}$ (b) $\frac{\pi}{2}$ (c) 1.25   (d) 2.03

77. (a) $\theta = \arctan \frac{x}{20}$ (b) 0.24 rad, 0.54 rad

79. False. Inverse trigonometric functions do not have the same relationships that trigonometric functions have.

81.

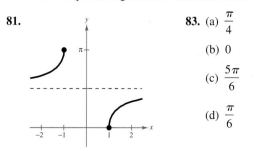

83. (a) $\frac{\pi}{4}$

(b) 0

(c) $\frac{5\pi}{6}$

(d) $\frac{\pi}{6}$

85.–89. Answers will vary.

91. $\sin(-585°) = \dfrac{\sqrt{2}}{2}$

$\cos(-585°) = -\dfrac{\sqrt{2}}{2}$

$\tan(-585°) = -1$

93. $\sin\left(-\dfrac{19\pi}{4}\right) = -\dfrac{\sqrt{2}}{2}$

$\cos\left(-\dfrac{19\pi}{4}\right) = -\dfrac{\sqrt{2}}{2}$

$\tan\left(-\dfrac{19\pi}{4}\right) = 1$

95.

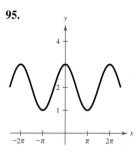

97.

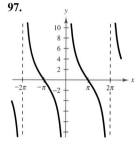

Section 4.8   *(page 361)*

1. $a \approx 4.66$
$c \approx 11.03$
$B = 65°$

3. $a \approx 8.26$
$c \approx 25.38$
$A = 19°$

5. $c \approx 17.09$
$A \approx 20.56°$
$B \approx 69.44°$

7. $a \approx 37.95$
$A \approx 64.62°$
$B \approx 25.38°$

9. $a \approx 91.34$
$b \approx 420.70$
$B = 77° 45'$

11. 2.56 inches   **13.** 6.30 feet

15. (a) $L = 60 \cot \theta$

(b)

θ	10°	20°	30°	40°	50°
L	340	165	104	72	50

(c) No. Cotangent is not a linear function.

17. (a) $h = 20 \sin \theta$

(b)

θ	60°	65°	70°	75°	80°
h	17.3	18.1	18.8	19.3	19.7

19. (a)

(b) $h = 50(\tan 47° 40' - \tan 35°)$ (c) 19.9 feet

21. ≈ 2090 feet   **23.** $\approx 38.3°$ **25.** $\approx 15.5°$

27. ≈ 5099 feet   **29.** ≈ 0.66 miles

31. ≈437 miles north; ≈700 miles east

33. (a) N 58° E   (b) ≈68.82 meters   **35.** N 55° W

37. ≈1933.3 feet   **39.** 17,054 feet ≈ 3.23 miles

41. 78.69°   **43.** ≈35.3°   **45.** $y = \sqrt{3}\,r$

47. ≈28.2 inches   **49.** $a \approx 12.2, b \approx 7$

51. (a) 4   (b) 4   (c) $\frac{1}{16}$

53. (a) $\frac{1}{16}$ (b) 70   (c) $\frac{1}{140}$ **55.** $y = 8 \sin \pi t$

57. $y = 3 \cos\left(\dfrac{10\pi t}{9}\right)$ **59.** $\omega = 528\pi$

61. (a)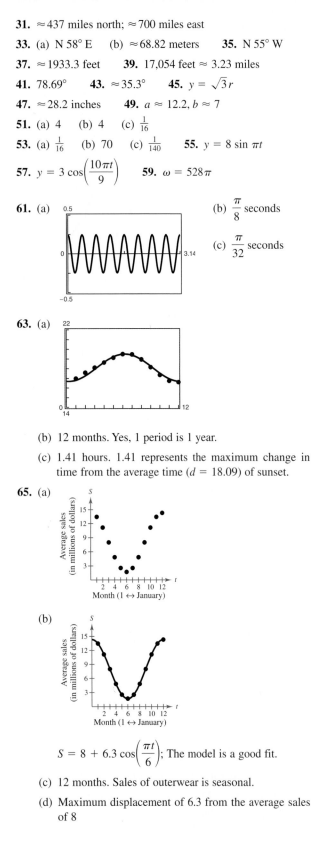
(b) $\dfrac{\pi}{8}$ seconds

(c) $\dfrac{\pi}{32}$ seconds

63. (a)

(b) 12 months. Yes, 1 period is 1 year.

(c) 1.41 hours. 1.41 represents the maximum change in time from the average time ($d = 18.09$) of sunset.

65. (a)

(b)

$S = 8 + 6.3 \cos\left(\dfrac{\pi t}{6}\right)$; The model is a good fit.

(c) 12 months. Sales of outerwear is seasonal.

(d) Maximum displacement of 6.3 from the average sales of 8

67. False. $a = \dfrac{22.56}{\tan(48.1°)}$ **69.** $x = 1.994$

71. $x = \pm 4.851$ **73.** 1.44 radians or 82.53°

75. −0.11 radian or −6.32°

Review Exercises   *(page 368)*

1. 0.5 rad   **3.** 4.5 rad

5. (a)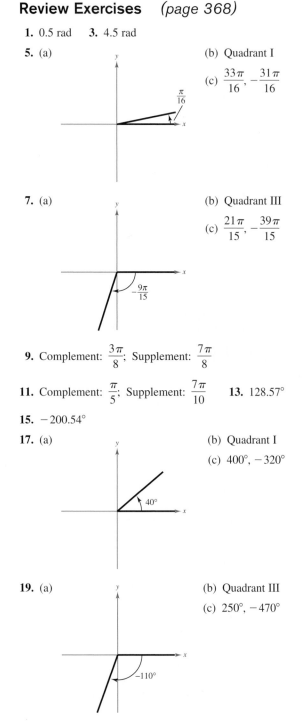
(b) Quadrant I

(c) $\dfrac{33\pi}{16}, -\dfrac{31\pi}{16}$

7. (a)
(b) Quadrant III

(c) $\dfrac{21\pi}{15}, -\dfrac{39\pi}{15}$

9. Complement: $\dfrac{3\pi}{8}$; Supplement: $\dfrac{7\pi}{8}$

11. Complement: $\dfrac{\pi}{5}$; Supplement: $\dfrac{7\pi}{10}$ **13.** 128.57°

15. −200.54°

17. (a)
(b) Quadrant I

(c) 400°, −320°

19. (a)
(b) Quadrant III

(c) 250°, −470°

21. Complement: 82°; Supplement: 172°

23. Complement: none; Supplement: 9° **25.** 135.28°

27. 5.38° **29.** 135° 17′ 24″ **31.** −85° 21′ 36″

33. 8.3776 **35.** −0.5890 **37.** 2.083 rad

39. 48.171 meters **41.** 0.094 rad ≈ 5.39°

43. $\left(-\dfrac{1}{2}, \dfrac{\sqrt{3}}{2}\right)$ **45.** $\left(-\dfrac{\sqrt{3}}{2}, \dfrac{1}{2}\right)$

47. $\sin \dfrac{7\pi}{6} = -\dfrac{1}{2}$

$\cos \dfrac{7\pi}{6} = -\dfrac{\sqrt{3}}{2}$

$\tan \dfrac{7\pi}{6} = \dfrac{\sqrt{3}}{3}$

$\csc \dfrac{7\pi}{6} = -2$

$\sec \dfrac{7\pi}{6} = -\dfrac{2\sqrt{3}}{3}$

$\cot \dfrac{7\pi}{6} = \sqrt{3}$

49. $\sin\left(-\dfrac{2\pi}{3}\right) = -\dfrac{\sqrt{3}}{2}$

$\cos\left(-\dfrac{2\pi}{3}\right) = -\dfrac{1}{2}$

$\tan\left(-\dfrac{2\pi}{3}\right) = \sqrt{3}$

$\csc\left(-\dfrac{2\pi}{3}\right) = -\dfrac{2\sqrt{3}}{2}$

$\sec\left(-\dfrac{2\pi}{3}\right) = -2$

$\cot\left(-\dfrac{2\pi}{3}\right) = \dfrac{\sqrt{3}}{3}$

51. $\sin \dfrac{11\pi}{4} = \sin \dfrac{3\pi}{4} = \dfrac{\sqrt{2}}{2}$

53. $\sin\left(-\dfrac{17\pi}{6}\right) = \sin\left(-\dfrac{5\pi}{6}\right) = -\dfrac{1}{2}$ **55.** −0.89 **57.** $\dfrac{1}{2}$

59. $\sin \theta = \dfrac{5\sqrt{61}}{61}$

$\cos \theta = \dfrac{6\sqrt{61}}{61}$

$\tan \theta = \dfrac{5}{6}$

$\csc \theta = \dfrac{\sqrt{61}}{5}$

$\sec \theta = \dfrac{\sqrt{61}}{6}$

$\cot \theta = \dfrac{6}{5}$

61. $\sin \theta = \dfrac{\sqrt{65}}{9}$

$\cos \theta = \dfrac{4}{9}$

$\tan \theta = \dfrac{\sqrt{65}}{4}$

$\csc \theta = \dfrac{9\sqrt{65}}{65}$

$\sec \theta = \dfrac{9}{4}$

$\cot \theta = \dfrac{4\sqrt{65}}{65}$

63. Answers will vary. **65.** (a) 0.1045 (b) 0.1045

67. (a) 0.7071 (b) 1.4142 **69.** 9.2 meters

71. $\sin \theta = \frac{4}{5}$ $\csc \theta = \frac{5}{4}$

$\cos \theta = \frac{3}{5}$ $\sec \theta = \frac{5}{3}$

$\tan \theta = \frac{4}{3}$ $\cot \theta = \frac{3}{4}$

73. $\sin \theta = \dfrac{2\sqrt{53}}{53}$ $\csc \theta = \dfrac{\sqrt{53}}{2}$

$\cos \theta = -\dfrac{7\sqrt{53}}{53}$ $\sec \theta = -\dfrac{\sqrt{53}}{7}$

$\tan \theta = -\dfrac{2}{7}$ $\cot \theta = -\dfrac{7}{2}$

75. $\sin \theta = \dfrac{15\sqrt{241}}{241}$

$\cos \theta = \dfrac{4\sqrt{241}}{241}$

$\tan \theta = \dfrac{15}{4}$

$\csc \theta = \dfrac{\sqrt{241}}{15}$

$\sec \theta = \dfrac{\sqrt{241}}{4}$

$\cot \theta = \dfrac{4}{15}$

77. $\sin \theta = -\dfrac{\sqrt{11}}{6}$

$\cos \theta = \dfrac{5}{6}$

$\tan \theta = -\dfrac{\sqrt{11}}{5}$

$\csc \theta = -\dfrac{6\sqrt{11}}{11}$

$\cot \theta = -\dfrac{5\sqrt{11}}{11}$

79. $\cos \theta = -\dfrac{\sqrt{55}}{8}$

$\tan \theta = -\dfrac{3\sqrt{55}}{55}$

$\csc \theta = \dfrac{8}{3}$

$\sec \theta = -\dfrac{8\sqrt{55}}{55}$

$\cot \theta = -\dfrac{\sqrt{55}}{3}$

81. $\sqrt{3}$ **83.** $\dfrac{\sqrt{3}}{2}$ **85.** $-\dfrac{\sqrt{2}}{2}$ **87.** 0.65

89. 3.24 **91.** 84° **93.** $\dfrac{\pi}{5}$

95. $\sin 240° = -\dfrac{\sqrt{3}}{2}$

$\cos 240° = -\dfrac{1}{2}$

$\tan 240° = \sqrt{3}$

97. $\sin(-210°) = \dfrac{1}{2}$

$\cos(-210°) = -\dfrac{\sqrt{3}}{2}$

$\tan(-210°) = -\dfrac{\sqrt{3}}{3}$

99. $\sin\left(-\dfrac{9\pi}{4}\right) = -\dfrac{\sqrt{2}}{2}$

$\cos\left(-\dfrac{9\pi}{4}\right) = \dfrac{\sqrt{2}}{2}$

$\tan\left(-\dfrac{9\pi}{4}\right) = -1$

101. $\sin\left(-\dfrac{\pi}{2}\right) = -1$

$\cos\left(-\dfrac{\pi}{2}\right) = 0$

$\tan\left(-\dfrac{\pi}{2}\right)$ is undefined.

103. $\left(-\dfrac{1}{2}, \dfrac{\sqrt{3}}{2}\right)$

$\sin \dfrac{2\pi}{3} = \dfrac{\sqrt{3}}{2}$

$\cos \dfrac{2\pi}{3} = -\dfrac{1}{2}$

$\tan \dfrac{2\pi}{3} = -\sqrt{3}$

105. $\left(-\dfrac{\sqrt{3}}{2}, -\dfrac{1}{2}\right)$

$\sin \dfrac{7\pi}{6} = -\dfrac{1}{2}$

$\cos \dfrac{7\pi}{6} = -\dfrac{\sqrt{3}}{2}$

$\tan \dfrac{7\pi}{6} = \dfrac{\sqrt{3}}{3}$

107.

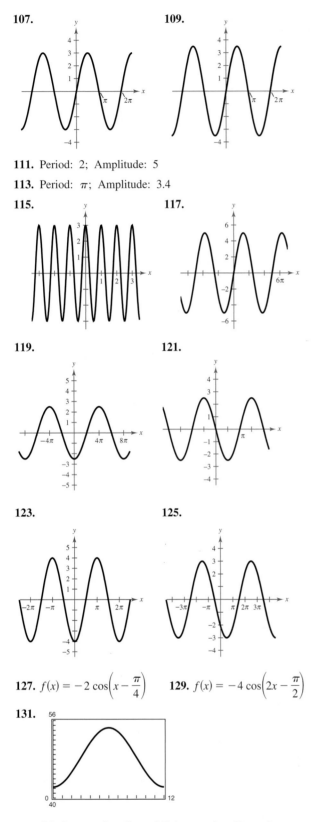

109.

111. Period: 2; Amplitude: 5

113. Period: π; Amplitude: 3.4

115.

117.

119.

121.

123.

125.

127. $f(x) = -2\cos\left(x - \dfrac{\pi}{4}\right)$ **129.** $f(x) = -4\cos\left(2x - \dfrac{\pi}{2}\right)$

131.

Maximum sales: June; Minimum sales: December

133. $f(x) = \frac{1}{2}\tan\left(\frac{1}{2}x\right)$

135.

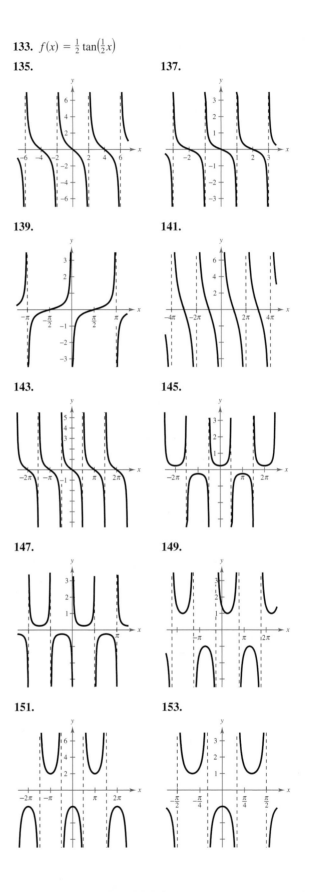

137.

139.

141.

143.

145.

147.

149.

151.

153.

155. 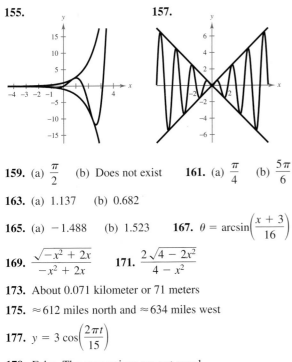   **157.**

159. (a) $\dfrac{\pi}{2}$ (b) Does not exist   **161.** (a) $\dfrac{\pi}{4}$ (b) $\dfrac{5\pi}{6}$

163. (a) 1.137   (b) 0.682

165. (a) -1.488 (b) 1.523   **167.** $\theta = \arcsin\!\left(\dfrac{x+3}{16}\right)$

169. $\dfrac{\sqrt{-x^2 + 2x}}{-x^2 + 2x}$ **171.** $\dfrac{2\sqrt{4 - 2x^2}}{4 - x^2}$

173. About 0.071 kilometer or 71 meters

175. ≈ 612 miles north and ≈ 634 miles west

177. $y = 3\cos\!\left(\dfrac{2\pi t}{15}\right)$

179. False. The expressions are not equal.

181. False. $3\pi/4$ is not in the range of the arctangent function.

183. Increases. The linear velocity is proportional to the radius.

Chapter Test   *(page 374)*

1. (a) 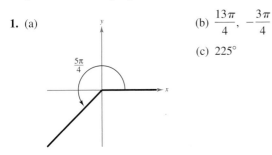   (b) $\dfrac{13\pi}{4}, \; -\dfrac{3\pi}{4}$

(c) $225°$

2. 3000 radians per minute

3. $\sin\theta = \dfrac{4\sqrt{17}}{17}$

$\cos\theta = -\dfrac{\sqrt{17}}{17}$

$\tan\theta = -4$

$\csc\theta = \dfrac{\sqrt{17}}{4}$

$\sec\theta = -\sqrt{17}$

$\cot\theta = -\dfrac{1}{4}$

4. $\sin\theta = \pm\dfrac{11\sqrt{157}}{157}$

$\cos\theta = \pm\dfrac{6\sqrt{157}}{157}$

$\csc\theta = \pm\dfrac{\sqrt{157}}{11}$

$\sec\theta = \pm\dfrac{\sqrt{157}}{6}$

$\cot\theta = \dfrac{6}{11}$

5. $\theta' = 70°$

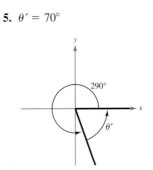

6. Quadrant III   **7.** $150°, 210°$ **8.** 1.33, 1.81

9. $\sin\theta = -\dfrac{\sqrt{11}}{6}$

$\cos\theta = \dfrac{5}{6}$

$\tan\theta = -\dfrac{\sqrt{11}}{5}$

$\csc\theta = -\dfrac{6\sqrt{11}}{11}$

$\cot\theta = -\dfrac{5\sqrt{11}}{11}$

10. 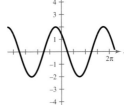   **11.**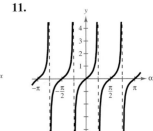

12.   **13.**

14. 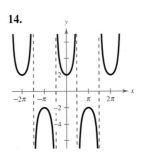   **15.**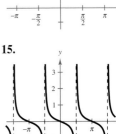

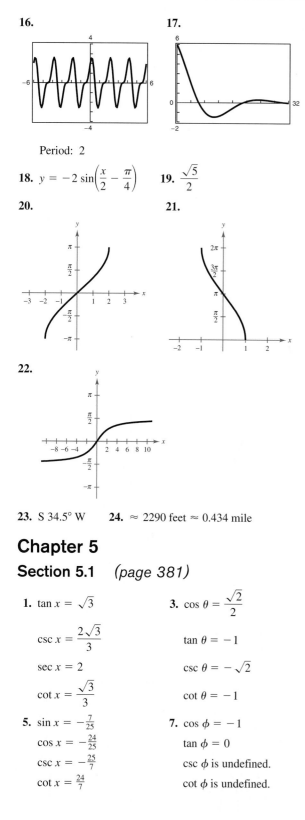

16.

17.

Period: 2

18. $y = -2 \sin\left(\dfrac{x}{2} - \dfrac{\pi}{4}\right)$ **19.** $\dfrac{\sqrt{5}}{2}$

20.

21.

22.

23. S 34.5° W **24.** ≈ 2290 feet ≈ 0.434 mile

Chapter 5

Section 5.1 *(page 381)*

1. $\tan x = \sqrt{3}$

$\csc x = \dfrac{2\sqrt{3}}{3}$

$\sec x = 2$

$\cot x = \dfrac{\sqrt{3}}{3}$

3. $\cos \theta = \dfrac{\sqrt{2}}{2}$

$\tan \theta = -1$

$\csc \theta = -\sqrt{2}$

$\cot \theta = -1$

5. $\sin x = -\dfrac{7}{25}$

$\cos x = -\dfrac{24}{25}$

$\csc x = -\dfrac{25}{7}$

$\cot x = \dfrac{24}{7}$

7. $\cos \phi = -1$

$\tan \phi = 0$

$\csc \phi$ is undefined.

$\cot \phi$ is undefined.

9. $\sin x = \dfrac{2}{3}$

$\cos x = -\dfrac{\sqrt{5}}{3}$

$\csc x = \dfrac{3}{2}$

$\sec x = -\dfrac{3\sqrt{5}}{5}$

$\cot x = -\dfrac{\sqrt{5}}{2}$

11. $\sin \theta = -\dfrac{4\sqrt{17}}{17}$

$\cos \theta = -\dfrac{\sqrt{17}}{17}$

$\csc \theta = -\dfrac{\sqrt{17}}{4}$

$\sec \theta = -\sqrt{17}$

$\cot \theta = \dfrac{1}{4}$

13. $\cos \theta = 0$

$\tan \theta$ is undefined.

$\csc \theta = -1$

$\sec \theta$ is undefined.

15. $1, 1$ **17.** $\infty, 0$ **19.** (d) **21.** (a)

23. (e) **25.** (b) **27.** (f) **29.** (e)

31. $\cos x$ **33.** $\cos^2 \phi$ **35.** $\cos x$

37. 1 **39.** $-\tan x$ **41.** $\cot x$ **43.** $1 + \sin y$

45.–61. Answers will vary.

63. $\cos^2 x$ **65.** $\sin^2 x \tan^2 x$ **67.** $\sec^4 x$

69. $\sin^2 x - \cos^2 x$ **71.** $1 + 2 \sin x \cos x$

73. $\tan^2 x$ **75.** $2 \csc^2 x$ **77.** $2 \sec x$

79. $1 + \cos y$ **81.** $3(\sec x + \tan x)$

83.

x	0.2	0.4	0.6	0.8
y_1	0.1987	0.3894	0.5646	0.7174
y_2	0.1987	0.3894	0.5646	0.7174

x	1.0	1.2	1.4
y_1	0.8415	0.9320	0.9854
y_2	0.8415	0.9320	0.9854

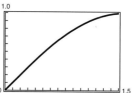

$y_1 = y_2$

85.

x	0.2	0.4	0.6	0.8
y_1	1.2230	1.5085	1.8958	2.4650
y_2	1.2230	1.5085	1.8958	2.4650

x	1.0	1.2	1.4
y_1	3.4082	5.3319	11.6814
y_2	3.4082	5.3319	11.6814

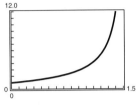

$y_1 = y_2$

87. $\csc x$ **89.** $\tan x$ **91.** $5 \cos \theta$ **93.** $3 \tan \theta$

95. $5 \sec \theta$ **97.** $0 \le \theta \le \pi$

99. $0 \le \theta < \dfrac{\pi}{2}, \dfrac{3\pi}{2} < \theta < 2\pi$ **101.** $\ln|\cot \theta|$

103. $\ln[(\cos x)(1 + \sin x)]$

105. (a) $\csc^2(132°) - \cot^2(132°) \approx 1.8107 - 0.8107 = 1$

    (b) $\csc^2\left(\dfrac{2\pi}{7}\right) - \cot^2\left(\dfrac{2\pi}{7}\right) \approx 1.63596 - 0.63596 = 1$

107. (a) $\cos(90° - 80) = \cos(10°) \approx 0.9848;$

    $\sin(80°) \approx 0.9848$

    (b) $\cos\left(\dfrac{\pi}{2} - 0.8\right) \approx 0.7174;\ \sin(0.8) \approx 0.7174$

109. Answers will vary.

111. False. $5 \sec \theta = \dfrac{5}{\cos \theta} \ne \dfrac{1}{5 \cos \theta}$

113. False. $\sin \theta \csc \phi = \dfrac{\sin \theta}{\sin \phi} \ne 1$ when $\theta \ne \phi$.

115. $\sin \theta = \pm\sqrt{1 - \cos^2 \theta}$

    $\tan \theta = \pm\dfrac{\sqrt{1 - \cos^2 \theta}}{\cos \theta}$

    $\csc \theta = \pm\dfrac{1}{\sqrt{1 - \cos^2 \theta}}$

    $\sec \theta = \dfrac{1}{\cos \theta}$

    $\cot \theta = \pm\dfrac{\cos \theta}{\sqrt{1 - \cos^2 \theta}}$

The sign depends on the choice of θ.

117. $\theta = 341°, \theta' = 19°$ **119.** $\theta = -212°, \theta' = 32°$

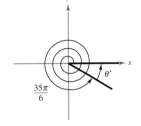

121. $\theta = \dfrac{35\pi}{6}, \theta' = \dfrac{\pi}{6}$

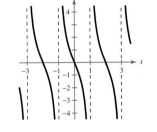

123.

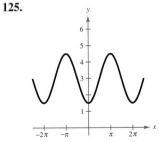

125.

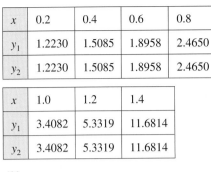

127. $B = 62°$ **129.** $A = 60.37°$

    $C = 90°$ $B = 29.63°$

    $a = 9.39$ $C = 90°$

    $b = 17.66$ $a = 10.9$

Section 5.2   *(page 389)*

1.–9. Answers will vary.

11.

x	0.2	0.4	0.6	0.8
y_1	4.835	2.1785	1.2064	0.6767
y_2	4.835	2.1785	1.2064	0.6767

x	1.0	1.2	1.4
y_1	0.3469	0.1409	0.0293
y_2	0.3469	0.1409	0.0293

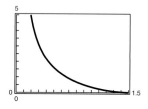

13.

x	0.2	0.4	0.6	0.8
y_1	4.835	2.1785	1.2064	0.6767
y_2	4.835	2.1785	1.2064	0.6767

x	1.0	1.2	1.4
y_1	0.3469	0.1409	0.0293
y_2	0.3469	0.1409	0.0293

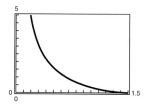

15.

x	0.2	0.4	0.6	0.8
y_1	5.0335	2.5679	1.771	1.394
y_2	5.0335	2.5679	1.771	1.394

x	1.0	1.2	1.4
y_1	1.1884	1.0729	1.0148
y_2	1.1884	1.0729	1.0148

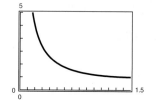

17.

x	0.2	0.4	0.6	0.8
y_1	5.1359	2.7880	2.1458	2.0009
y_2	5.1359	2.7880	2.1458	2.0009

x	1.0	1.2	1.4
y_1	2.1995	2.9609	5.9704
y_2	2.1995	2.9609	5.9704

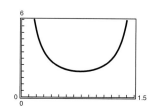

19. $\cot(-x) = -\cot(x)$ so,
$$(1 + \tan)[1 + \cos(-x)] = \tan x - \cot x.$$

21. $(\tan^2 x)^2$

23.–51. Answers will vary.

53.

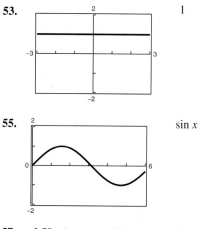

1

55.

$\sin x$

57. and 59. Answers will vary. **61.** 1 **63.** 2

65 and 67. Answers will vary. **69.** $\mu = \tan \theta, W \neq 0$

71. Answers will vary.

73. True. Cosine and secant are even functions.

75. False. $\sin^2(x) = (\sin x)^2 \neq \sin(x^2)$

77. $|\sin \theta| = \sqrt{1 - \cos^2 \theta}; \dfrac{7\pi}{4}$

79. $\sqrt{\sin^2 x + \cos^2 x} \neq \sin x + \cos x$. The left side is 1 for any x, but the right side is not necessarily 1; $\pi/4$.

81. Answers will vary.

83. Answer is not unique. Sample answer: $y = x^4 + 17x + 16$

85. Answer is not unique. Sample answer:
$$y = x^5 - 4x^4 + 6x^3 - 4x^2$$

87. 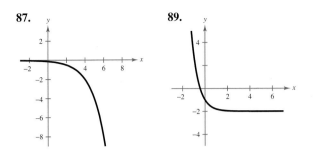 **89.**

91. $\theta \approx 2.36$ rad   **93.** Quadrant III   **95.** Quadrant III

Section 5.3   *(page 400)*

1.–5. Answers will vary.   **7.** $-1, 3$ **9.** ± 2

11. $\dfrac{2\pi}{3}, \dfrac{4\pi}{3}$ **13.** $\dfrac{\pi}{6}, \dfrac{11\pi}{6}$ **15.** $\dfrac{\pi}{3}, \dfrac{2\pi}{3}, \dfrac{4\pi}{3}, \dfrac{5\pi}{3}$

17. $\dfrac{\pi}{8}, \dfrac{3\pi}{8}, \dfrac{5\pi}{8}, \dfrac{7\pi}{8}, \dfrac{9\pi}{8}, \dfrac{11\pi}{8}, \dfrac{13\pi}{8}, \dfrac{15\pi}{8}$

19. $\dfrac{\pi}{6}, \dfrac{5\pi}{6}, \dfrac{7\pi}{6}, \dfrac{11\pi}{6}$ **21.** $\dfrac{\pi}{3}, \dfrac{2\pi}{3}, \dfrac{4\pi}{3}, \dfrac{5\pi}{3}$

23. $\dfrac{\pi}{6}, \dfrac{\pi}{3}, \dfrac{2\pi}{3}, \dfrac{5\pi}{6}, \dfrac{7\pi}{6}, \dfrac{4\pi}{3}, \dfrac{5\pi}{3}, \dfrac{11\pi}{6}$ **25.** $0, \dfrac{\pi}{2}, \pi, \dfrac{3\pi}{2}$

27. $0, \dfrac{\pi}{6}, \dfrac{5\pi}{6}, \pi, \dfrac{7\pi}{6}, \dfrac{11\pi}{6}$ **29.** $\dfrac{\pi}{3}, \pi, \dfrac{5\pi}{3}$

31. No solution   **33.** $\dfrac{\pi}{2}$ **35.** $\dfrac{\pi}{2}$ **37.** π

39. $\dfrac{\pi}{6}, \dfrac{5\pi}{6}, \dfrac{7\pi}{6}, \dfrac{11\pi}{6}$ **41.** $2.0344, 5.1760, \dfrac{\pi}{4}, \dfrac{5\pi}{4}$

43. $1.4595, 4.8237$ **45.** $0.8614, 5.4218$

47. $0.7854, 2.3562, 3.6652, 3.9270, 5.4978, 5.7596$

49. $0.5236, 2.6180$ **51.** 4.9172

53. $0, 2.6779, 3.1416, 5.8195$

55. $0.3398, 0.8481, 2.2935, 2.8018$

57. $-1.154, 0.534$ **59.** 1.110

61. (a)

x	0	1	2	3
$f(x)$	Undef.	0.83	-1.36	-2.93

x	4	5	6
$f(x)$	-4.46	-6.34	-13.02

The zero is in the interval $(1, 2)$ because f changes signs in that interval.

(b)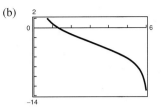

The interval is the same as in part (a).

(c) 1.3065

63. (a)

x	0	1	2	3
$f(x)$	-1	1.39	1.65	-0.70

x	4	5	6
$f(x)$	-1.94	-2.00	-1.48

The zeros are in the intervals $(0, 1)$ and $(2, 3)$ because f changes signs in those intervals.

(b)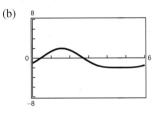

The intervals are the same as in part (a).

(c) $0.4271, 2.7145$

65. (a)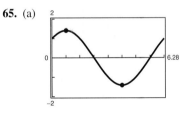

Maximum: $(0.785, 1.41)$

Minimum: $(3.93, -1.41)$

(b) $\dfrac{\pi}{4}, \dfrac{5\pi}{4}$

67. (a)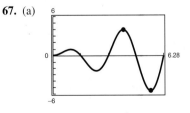

Maximum: $(3.99, 3.96)$

Minimum: $(5.54, -5.52)$

(b) $0, 1.01, 2.46, 3.99, 5.44$

69. 1

71. (a) All real numbers x except $x = 0$

    (b) y-axis symmetry; Horizontal asymptote: $y = 1$

    (c) Oscillates   (d) Infinite number of solutions

    (e) Yes, 0.6366

73. 0.04 second, 0.43 second , 0.83 second

75. $123 \le t \le 223$ days

77. 1.91°   **79.** 1.96 seconds

81. (a) 3.33

    (b)

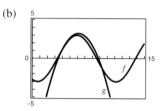

    (c) 3.46, 8.81. The zero of $g(x)$ on $[0, 6]$ is 3.46, which is closest to the zero of $f(x)$.

83. False. Some trigonometric equations don't have any solutions in the interval $[0, 2\pi)$.

85. 2.164 rad   **87.** -0.007 rad

89. 24.249   **91.** 24.892

93.   **95.**

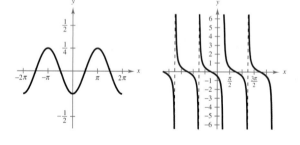

Section 5.4 *(page 408)*

1. (a) 0 (b) $\dfrac{1 + \sqrt{3}}{2}$ **3.** (a) $\dfrac{1}{2}$ (b) $\dfrac{-1 - \sqrt{3}}{2}$

5. (a) $-\dfrac{\sqrt{2}}{2}$ (b) $\dfrac{2 - \sqrt{2}}{2}$

7. (a) $\dfrac{-\sqrt{2} - \sqrt{6}}{4}$ (b) $\dfrac{-\sqrt{2} - \sqrt{3}}{2}$

9. $\sin(75°) = \dfrac{\sqrt{2} + \sqrt{6}}{4}$ **11.** $\sin(105°) = \dfrac{\sqrt{6} + \sqrt{2}}{4}$

    $\cos(75°) = \dfrac{\sqrt{6} - \sqrt{2}}{4}$ $\cos(105°) = \dfrac{\sqrt{2} - \sqrt{6}}{4}$

    $\tan(75°) = 2 + \sqrt{3}$ $\tan(105°) = -2 - \sqrt{3}$

13. $\sin(195°) = \dfrac{\sqrt{2} - \sqrt{6}}{4}$

    $\cos(195°) = \dfrac{-\sqrt{2} - \sqrt{6}}{4}$

    $\tan(195°) = 2 - \sqrt{3}$

15. $\sin\left(\dfrac{11\pi}{12}\right) = \dfrac{\sqrt{6} - \sqrt{2}}{4}$ **17.** $\sin\left(-\dfrac{\pi}{12}\right) = \dfrac{\sqrt{2} - \sqrt{6}}{4}$

    $\cos\left(\dfrac{11\pi}{12}\right) = \dfrac{-\sqrt{6} - \sqrt{2}}{4}$ $\cos\left(-\dfrac{\pi}{12}\right) = \dfrac{\sqrt{6} + \sqrt{2}}{4}$

    $\tan\left(\dfrac{11\pi}{12}\right) = -2 + \sqrt{3}$ $\tan\left(-\dfrac{\pi}{12}\right) = -2 + \sqrt{3}$

19. $\cos 55°$ **21.** $\sin 290°$ **23.** $\tan 239°$

25. $\sin 1.8$ **27.** $\cos \dfrac{12\pi}{35}$

29.

x	0.2	0.4	0.6	0.8
y_1	0.9801	0.9211	0.8253	0.6967
y_2	0.9801	0.9211	0.8253	0.6967

x	1.0	1.2	1.4
y_1	0.5403	0.3624	0.1700
y_2	0.5403	0.3624	0.1700

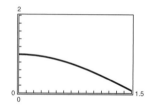

31.

x	0.2	0.4	0.6	0.8
y_1	0.6621	0.7978	0.9017	0.9696
y_2	0.6621	0.7978	0.9017	0.9696

x	1.0	1.2	1.4
y_1	0.9989	0.9883	0.9384
y_2	0.9989	0.9883	0.9384

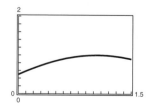

33.

x	0.2	0.4	0.6	0.8
y_1	0.9605	0.8484	0.6812	0.4854
y_2	0.9605	0.8484	0.6812	0.4854

x	1.0	1.2	1.4
y_1	0.2919	0.1313	0.0289
y_2	0.2919	0.1313	0.0289

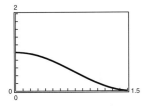

35. $\frac{33}{65}$ **37.** $-\frac{56}{65}$ **39.** $-\frac{3}{5}$ **41.** $\frac{44}{125}$

43.–49. Answers will vary. **51.** 1

53. $\dfrac{2x^2 - \sqrt{1 - x^2}}{\sqrt{4x^2 + 1}}$ **55.** $\dfrac{\pi}{2}$ **57.** $\dfrac{5\pi}{4}, \dfrac{7\pi}{4}$

59. $0, \dfrac{\pi}{3}, \pi, \dfrac{5\pi}{3}$ **61.** 0.7854, 5.4978 **63.** 0, 3.1416

65. Answers will vary.

67. False. $\sin(u \pm v) = \sin u \cos v \pm \cos u \sin v$

69. False. $\sin(75°) = \dfrac{\sqrt{6} + \sqrt{2}}{4}$

71. and 73. Answers will vary.

75. (a) $\sqrt{2} \sin\left(\theta + \dfrac{\pi}{4}\right)$

(b) $\sqrt{2} \cos\left(\theta - \dfrac{\pi}{4}\right)$

77. (a) $13 \sin(3\theta + 0.3948)$

(b) $13 \cos(3\theta - 1.1760)$

79. $2 \cos \theta$ **81.** $u + v = \omega$

83.

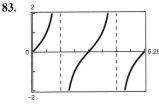

$h(\theta) = \tan \theta$

85. (a) $(-\infty, 0), (0, \infty)$

(b)

h	0.01	0.02	0.05
$f(h)$	-0.5043	-0.5086	-0.5214
$g(h)$	-0.5043	-0.5086	-0.5214

h	0.1	0.2	0.5
$f(h)$	-0.5424	-0.5830	-0.6915
$g(h)$	-0.5424	-0.5830	-0.6915

(c) 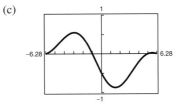 (d) $-\dfrac{1}{2}$

87. $(38, 0), (0, 19)$ **89.** $(0, 4), (2, 0), (7, 0)$

91. $\dfrac{\pi}{6}$ **93.** $\dfrac{\pi}{2}$

Section 5.5 *(page 418)*

1. $\dfrac{3}{5}$ **3.** $\dfrac{7}{25}$ **5.** $\dfrac{24}{7}$ **7.** $\dfrac{25}{24}$ **9.** $0, \dfrac{\pi}{3}, \pi, \dfrac{5\pi}{3}$

11. 0.1263, 1.4445, 3.2679, 4.5860 **13.** $\dfrac{\pi}{3}, \pi, \dfrac{5\pi}{3}$

15. $0, \dfrac{\pi}{2}, \pi, \dfrac{3\pi}{2}$ **17.** $4 \sin 2x$ **19.** $5 \cos 2x$

21. $\sin 2u = \frac{24}{25}$ **23.** $\sin 2u = \frac{4}{5}$

$\cos 2u = \frac{7}{25}$ $\cos 2u = \frac{3}{5}$

$\tan 2u = \frac{24}{7}$ $\tan 2u = \frac{4}{3}$

25. $\frac{1}{8}(3 + 4 \cos 2x + \cos 4x)$ **27.** $\frac{1}{8}(1 - \cos 4x)$

29. $\frac{1}{32}(2 + \cos 2x - 2 \cos 4x - \cos 6x)$

31. $\dfrac{5\sqrt{26}}{26}$ **33.** $\dfrac{1}{5}$ **35.** $\sqrt{26}$ **37.** $\dfrac{5}{13}$

39. $\sin 15° = \dfrac{\sqrt{2 - \sqrt{3}}}{2}$ **41.** $\sin 112° 30' = \dfrac{\sqrt{2 + \sqrt{2}}}{2}$

$\cos 15° = \dfrac{\sqrt{2 + \sqrt{3}}}{2}$ $\cos 112° 30' = -\dfrac{\sqrt{2 - \sqrt{2}}}{2}$

$\tan 15° = 2 - \sqrt{3}$ $\tan 112° 30' = -1 - \sqrt{2}$

43. $\sin \dfrac{\pi}{8} = \dfrac{\sqrt{2 - \sqrt{2}}}{2}$ **45.** $\sin \dfrac{3\pi}{8} = \dfrac{\sqrt{2 + \sqrt{2}}}{2}$

$\cos \dfrac{\pi}{8} = \dfrac{\sqrt{2 + \sqrt{2}}}{2}$ $\cos \dfrac{3\pi}{8} = \dfrac{\sqrt{2 - \sqrt{2}}}{2}$

$\tan \dfrac{\pi}{8} = \sqrt{2} - 1$ $\tan \dfrac{3\pi}{8} = \sqrt{2} + 1$

47. $\sin\dfrac{u}{2} = \dfrac{5\sqrt{26}}{26}$

$\cos\dfrac{u}{2} = \dfrac{\sqrt{26}}{26}$

$\tan\dfrac{u}{2} = 5$

49. $\sin\dfrac{u}{2} = \sqrt{\dfrac{89 - 5\sqrt{89}}{178}}$

$\cos\dfrac{u}{2} = -\sqrt{\dfrac{89 + 5\sqrt{89}}{178}}$

$\tan\dfrac{u}{2} = \dfrac{5 - \sqrt{89}}{8}$

51. $|\sin 3x|$ **53.** $-|\tan 4x|$ **55.** $\dfrac{\pi}{3}, \dfrac{5\pi}{3}$

57. $\dfrac{\pi}{3}, \pi, \dfrac{5\pi}{3}$ **59.** $3\sin\dfrac{2\pi}{3}$ **61.** $\dfrac{1}{2}(\sin 8\theta + \sin 2\theta)$

63. $\dfrac{5}{2}(\cos 8\beta + \cos 2\beta)$ **65.** $2\sin 45° \cos 15°$

67. $2\cos 3\theta \sin 2\theta$ **69.** $2\cos\alpha \sin\beta$

71. $-2\sin\theta \sin\dfrac{\pi}{2} = -2\sin\theta$

73. $0, \dfrac{\pi}{4}, \dfrac{\pi}{2}, \dfrac{3\pi}{4}, \pi, \dfrac{5\pi}{4}, \dfrac{3\pi}{2}, \dfrac{7\pi}{4}$ **75.** $\dfrac{\pi}{6}, \dfrac{5\pi}{6}$ **77.** $\dfrac{25}{169}$

79. $\dfrac{4}{13}$ **81.–91.** Answers will vary.

93.

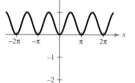

95. (a)

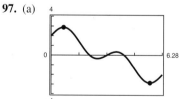

Maximum: $(\pi, 3)$

(b) π

97. (a)

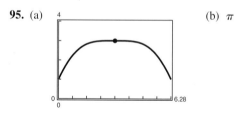

Minimum: $(5.5839, -2.8642)$

Maximum: $(0.6993, 2.8642)$

(b) $0.6693, 2.6078, 3.6754, 5.5839$

99. $2x\sqrt{1 - x^2}$ **101.** $1 - 2x^2$ **103.** $\dfrac{1}{16}v_0 \sin\theta \cos\theta$

105. $\theta = 0.4482$

107. False. $\sin\dfrac{x}{2} = -\sqrt{\dfrac{1 - \cos x}{2}}$ for $\pi \le \dfrac{x}{2} \le 2\pi$.

109. (a)

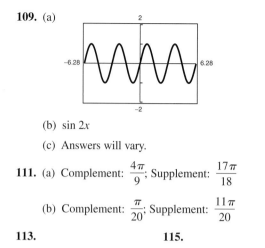

(b) $\sin 2x$

(c) Answers will vary.

111. (a) Complement: $\dfrac{4\pi}{9}$; Supplement: $\dfrac{17\pi}{18}$

(b) Complement: $\dfrac{\pi}{20}$; Supplement: $\dfrac{11\pi}{20}$

113.   **115.**

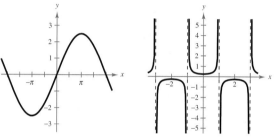

Review Exercises *(page 422)*

1. $\sec x$ **3.** $\cos x$ **5.** $\cot x$ **7.** $\sec x$ **9.** $\sec x$

11. $\tan x = \dfrac{3}{4}$

$\csc x = \dfrac{5}{3}$

$\sec x = \dfrac{5}{4}$

$\cot x = \dfrac{4}{3}$

13. $\cos x = \dfrac{\sqrt{2}}{2}$

$\tan x = -1$

$\csc x = -\sqrt{2}$

$\sec x = \sqrt{2}$

$\cot x = -1$

15. $\sin^2 x$ **17.** $1 + \cot\alpha$ **19.** 1 **21.** $\csc x$

23.–37. Answers will vary.

39. $\dfrac{\pi}{6}, \dfrac{5\pi}{6}$ **41.** $\dfrac{\pi}{3}, \dfrac{2\pi}{3}$ **43.** $\dfrac{\pi}{6}, \dfrac{7\pi}{6}$

45. $\dfrac{\pi}{3}, \dfrac{2\pi}{3}, \dfrac{4\pi}{3}, \dfrac{5\pi}{3}$ **47.** $\dfrac{\pi}{6}, \dfrac{5\pi}{6}, \dfrac{7\pi}{6}, \dfrac{11\pi}{6}$ **49.** $0, \pi$

51. $0, \dfrac{2\pi}{3}, \dfrac{4\pi}{3}$ **53.** $0, \dfrac{\pi}{2}, \pi$ **55.** $\dfrac{\pi}{8}, \dfrac{3\pi}{8}, \dfrac{9\pi}{8}, \dfrac{11\pi}{8}$

57. $0, \dfrac{\pi}{8}, \dfrac{3\pi}{8}, \dfrac{5\pi}{8}, \dfrac{7\pi}{8}, \dfrac{9\pi}{8}, \dfrac{11\pi}{8}, \dfrac{13\pi}{8}, \dfrac{15\pi}{8}$ **59.** $\dfrac{\pi}{2}, \dfrac{3\pi}{2}$

61. $0, \pi$ **63.** $1.2490, 1.8158, 4.3906, 4.9574$

65. $\sin 285° = \dfrac{-\sqrt{2} - \sqrt{6}}{4}$ **67.** $\sin \dfrac{5\pi}{12} = \dfrac{\sqrt{6} + \sqrt{2}}{4}$

$\cos 285° = \dfrac{\sqrt{6} - \sqrt{2}}{4}$ $\cos \dfrac{5\pi}{12} = \dfrac{\sqrt{6} - \sqrt{2}}{4}$

$\tan 285° = -2 - \sqrt{3}$ $\tan \dfrac{5\pi}{12} = 2 + \sqrt{3}$

69. $\sin 190°$ **71.** $\tan 35°$ **73.** $\dfrac{-15 - 12\sqrt{7}}{52}$

75. $\dfrac{5\sqrt{7} + 36}{52}$ **77.** $\dfrac{5\sqrt{7} - 36}{52}$

79.–83. Answers will vary.   **85.** $\dfrac{\pi}{6}, \dfrac{11\pi}{6}$

87.–91. Answers will vary.

93. $\sin 2u = \dfrac{20\sqrt{6}}{49}$

$\cos 2u = -\dfrac{1}{49}$

$\tan 2u = -20\sqrt{6}$

95. $15°, 75°$ **97.** $\frac{1}{128}(3 - 4\cos 4x + \cos 8x)$

99. $\frac{1}{8}(3 - 4\cos 4x + \cos 8x)$

101. $\sin 67° 30' = \dfrac{\sqrt{2 + \sqrt{2}}}{2}$

$\cos 67° 30' = \dfrac{\sqrt{2 - \sqrt{2}}}{2}$

$\tan 67° 30' = \sqrt{2} + 1$

103. $\sin \dfrac{11\pi}{12} = \dfrac{\sqrt{2 - \sqrt{3}}}{2}$

$\cos \dfrac{11\pi}{12} = -\dfrac{\sqrt{2 + \sqrt{3}}}{2}$

$\tan \dfrac{11\pi}{12} = -2 + \sqrt{3}$

105. $\tan 3x$ **107.** Answers will vary.

109. $2 \cos \dfrac{5\theta}{2} \cos \dfrac{\theta}{2}$ **111.** $-2 \sin \dfrac{\pi}{2} \sin \dfrac{\pi}{4}$

113. $\dfrac{1}{2}(\cos \alpha - \cos 5\alpha)$ **115.** $3 \sin \dfrac{\pi}{2}$

117. (a) $y = \dfrac{1}{2}\sqrt{10} \sin\left(8t - \arctan \dfrac{1}{3}\right)$

(b)

(c) $\dfrac{1}{2}\sqrt{10}$ **(d)** $\dfrac{4}{\pi}$

119. False. $\sin(x \pm y) = \sin x \cos y \pm \cos x \sin y$

121. True   **123.** No. Answers will vary.

125. $y_3: 1 - y_2$

Chapter Test   *(page 426)*

1. $\sin \theta = -\dfrac{6\sqrt{61}}{61}$

$\cos \theta = -\dfrac{5\sqrt{61}}{61}$

$\csc \theta = -\dfrac{\sqrt{61}}{6}$

$\sec \theta = -\dfrac{\sqrt{61}}{5}$

$\cot \theta = \dfrac{5}{6}$

2. 1 **3.** 1 **4.** $\csc \theta \sec \theta$

5. $\dfrac{\pi}{2} < \theta \le \pi, \dfrac{3\pi}{2} < \theta < 2\pi$

6.

$y_1 = y_2$

7.–12. Answers will vary.

13. $\sin\left(-\dfrac{7\pi}{12}\right) = \dfrac{-\sqrt{6} - \sqrt{2}}{4}$

$\cos\left(-\dfrac{7\pi}{12}\right) = \dfrac{-\sqrt{6} + \sqrt{2}}{4}$

$\tan\left(-\dfrac{7\pi}{12}\right) = 2 + \sqrt{3}$

14. $\frac{1}{8}(1 - \cos 4x)$ **15.** $\frac{3}{2}(\cos 4\theta - \cos 8\theta)$

16. $2 \cos 4\theta \cos \theta$ **17.** $0, \dfrac{3\pi}{4}, \pi, \dfrac{7\pi}{4}$ **18.** $\dfrac{\pi}{6}, \dfrac{\pi}{2}, \dfrac{5\pi}{6}, \dfrac{3\pi}{2}$

19. $\dfrac{\pi}{6}, \dfrac{5\pi}{6}, \dfrac{7\pi}{6}, \dfrac{11\pi}{6}$ **20.** $\dfrac{\pi}{6}, \dfrac{5\pi}{6}, \dfrac{3\pi}{2}$ **21.** 1.306

22. $\sin 2u = \frac{4}{5}$ **23.** Answers will vary.

$\tan 2u = -\frac{4}{3}$

24. $-2 - \sqrt{3}$ **25.** $76.52°$

Chapter 6

Section 6.1 *(page 434)*

1. $C = 105°$, $b \approx 16.97$, $c \approx 23.18$

3. $C = 110°$, $b \approx 22.44$, $c \approx 24.35$

5. $B = 13.6°$, $C \approx 130.4°$, $c \approx 12.96$

7. $B = 10°$, $b \approx 69.46$, $c \approx 136.81$

9. $B = 42°4'$, $a \approx 22.05$, $b \approx 14.88$

11. $A \approx 10°11'$, $C \approx 154°19'$, $c \approx 11.03$

13. $B \approx 18°13'$, $C \approx 51°32'$, $c \approx 40.05$

15. No solution

17. Two solutions

$B \approx 72.2°$, $C \approx 49.8°$, $c \approx 10.27$

$B \approx 107.8°$, $C \approx 14.2°$, $c \approx 48.23$

19. $B \approx 48.74°$, $C \approx 21.26°$, $c \approx 48.23$

21. (a) $b \le 5$, $b = \dfrac{5}{\sin 36°}$ (b) $5 < b < \dfrac{5}{\sin 36°}$

(c) $b > \dfrac{5}{\sin 36°}$

23. ≈ 28.2 square units **25.** ≈ 1782.3 square units

27. ≈ 1057.1 square units

29. (a) 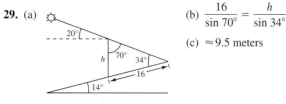 (b) $\dfrac{16}{\sin 70°} = \dfrac{h}{\sin 34°}$

(c) ≈ 9.5 meters

31. S 65° W

33. (a) 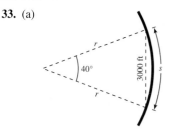 (b) 4385.71 feet

(c) 3061.80 feet

35. Distance from A: 42.3 kilometers

Distance from B: 25.8 kilometers

37. ≈ 4.55 miles

39. (a)

θ	0°	45°	90°	135°	180°
d	0	0.5338	1.6905	2.6552	3

(b) 0.0071 inches

41. $d = \dfrac{2 \sin \theta}{\sin(\phi - \theta)}$

43. False. The triangle cannot be solved if only three angles are known.

45. False. For the case SSA there could be two possible triangles, one triangle, or no triangle.

47. $\tan \theta = -\frac{12}{5}$; $\csc \theta = -\frac{13}{12}$; $\sec \theta = \frac{13}{5}$; $\cot \theta = -\frac{5}{12}$

49. $\sin \theta = \dfrac{\sqrt{122}}{122}$; $\cos \theta = -\dfrac{11\sqrt{122}}{122}$; $\tan \theta = -\dfrac{1}{11}$;

$\csc \theta = \sqrt{122}$

51. Answers will vary. **53.** $3(\sin 11\theta + \sin 5\theta)$

55. $\dfrac{3}{2}\left[\sin \dfrac{11\pi}{6} - \sin\left(-\dfrac{9\pi}{6} \right) \right]$

Section 6.2 *(page 441)*

1. $A \approx 26.4°$, $B \approx 36.3°$, $C \approx 117.3°$

3. $B \approx 29.4°$, $C \approx 100.6°$, $a \approx 23.4$

5. $A \approx 36.9°$, $B \approx 53.1°$, $C = 90°$

7. $A \approx 103.5°$, $B \approx 38.2°$, $C \approx 38.2°$

9. $b \approx 8.6$, $A \approx 153.9°$, $C \approx 17.7°$

	a	b	c	d	θ	ϕ
11.	4	8	11.6	5	30°	150°
13.	10	14	20	13.86	68.2°	111.8°
15.	10	11.58	18	12	67.1°	112.9°

17. ≈ 22.45 square units **19.** ≈ 15.52 square units

21. 96.82 square units

23.

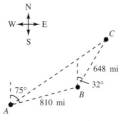

1357.8 miles, S 56° W

25. 483.4 meters **27.** 43.3 miles

29. (a) N 58.4° W (b) S 81.5° W **31.** ≈ 63.7 feet

33. $PQ \approx 9.4$, $QS \approx 5.0$, $RS \approx 12.8$

35. (a) $49 = 2.25 + x^2 - 3x \cos \theta$

(b) $x = \frac{1}{2}\left(3 \cos \theta + \sqrt{9 \cos^2\theta + 187} \right)$

(c)

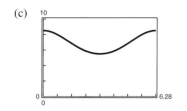

(d) 6 inches

37. ≈ 2.76 feet

39. False. A triangle cannot be formed with sides of length 10 cm, 16 cm, and 5 cm.

41. (a) and (b) Answers will vary.

43. $r \approx 64.5$ feet, track length ≈ 405 feet   **45.** $\dfrac{\pi}{3}, \dfrac{5\pi}{3}$

47. $\dfrac{\pi}{6}, \dfrac{11\pi}{6}$ **49.** $-2 \sin\left(\dfrac{7\pi}{12}\right) \sin\left(\dfrac{\pi}{4}\right)$

Section 6.3 *(page 453)*

1. $\|\mathbf{u}\| = \|\mathbf{v}\| = \sqrt{17}$, $\text{slope}_\mathbf{u} = \text{slope}_\mathbf{v} = \frac{1}{4}$

 $\mathbf{u}$ and $\mathbf{v}$ have the same magnitude and direction so they are equal.

3. $\langle 4, 3 \rangle$, $\|\mathbf{v}\| = 5$ **5.** $\langle -3, 2 \rangle$, $\|\mathbf{v}\| = \sqrt{13}$

7. $\langle 0, 5 \rangle$, $\|\mathbf{v}\| = 5$ **9.** $\langle -\frac{9}{2}, -\frac{5}{2} \rangle$, $\|\mathbf{v}\| = \frac{1}{2}\sqrt{106}$

11. $\langle 8, 6 \rangle$, $\|\mathbf{v}\| = 10$

13. $\langle 7.9, -17.9 \rangle$, $\|\mathbf{v}\| = \sqrt{382.82} \approx 19.6$

15.   **17.**

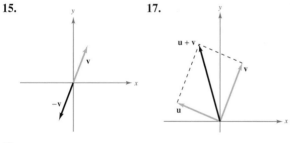

19.

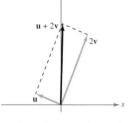

21. (a) $\langle 11, 3 \rangle$ (b) $\langle -3, 1 \rangle$ (c) $\langle -13, 1 \rangle$ (d) $\langle 23, 9 \rangle$

23. (a) $\langle -4, -5 \rangle$ (b) $\langle -6, 1 \rangle$ (c) $\langle -13, 5 \rangle$
 (d) $\langle -19, -11 \rangle$

25. (a) $3\mathbf{i} - 2\mathbf{j}$ (b) $-\mathbf{i} + 4\mathbf{j}$ (c) $-4\mathbf{i} + 11\mathbf{j}$
 (d) $6\mathbf{i} + \mathbf{j}$

27. $\langle 1, 0 \rangle$ **29.** $\left\langle -\dfrac{\sqrt{2}}{2}, \dfrac{\sqrt{2}}{2} \right\rangle$ **31.** $\left\langle -\dfrac{24}{25}, -\dfrac{7}{25} \right\rangle$

33. $\dfrac{4}{5}\mathbf{i} - \dfrac{3}{5}\mathbf{j}$ **35.** $\mathbf{j}$ **37.** $\left\langle \dfrac{5\sqrt{2}}{2}, \dfrac{5\sqrt{2}}{2} \right\rangle$

39. $\frac{21}{5}\mathbf{i} + \frac{28}{5}\mathbf{j}$ **41.** $-8\mathbf{i}$

43. $\mathbf{v} = \left\langle 3, -\frac{3}{2} \right\rangle$ **45.** $\mathbf{v} = \langle 4, 3 \rangle$

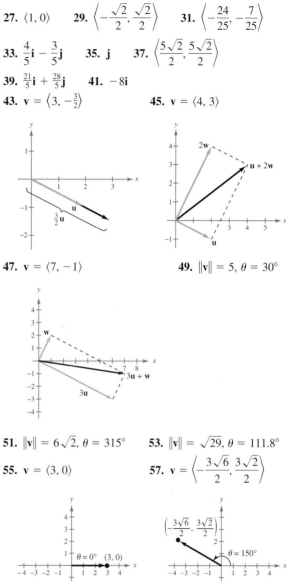

47. $\mathbf{v} = \langle 7, -1 \rangle$ **49.** $\|\mathbf{v}\| = 5$, $\theta = 30°$

51. $\|\mathbf{v}\| = 6\sqrt{2}$, $\theta = 315°$ **53.** $\|\mathbf{v}\| = \sqrt{29}$, $\theta = 111.8°$

55. $\mathbf{v} = \langle 3, 0 \rangle$ **57.** $\mathbf{v} = \left\langle -\dfrac{3\sqrt{6}}{2}, \dfrac{3\sqrt{2}}{2} \right\rangle$

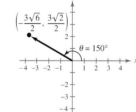

59. $\mathbf{v} = \left\langle \dfrac{\sqrt{10}}{5}, \dfrac{3\sqrt{10}}{5} \right\rangle$

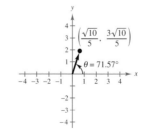

61. $\left\langle \dfrac{5}{2}, \dfrac{5\sqrt{3}+10}{2} \right\rangle$ **63.** $\left\langle 10\sqrt{2}-25\sqrt{3},\ 10\sqrt{2}+25 \right\rangle$

65. $90°$ **67.** $63.4°$

69.

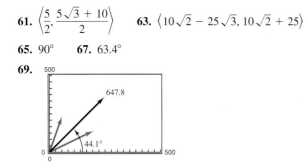

71. $62.7°$ **73.** $8.7°$, 2396.19 newtons

75. Horizontal component: $70 \cos 40° \approx 53.62$ feet per second

     Vertical component: $70 \sin 40° \approx 45.00$ feet per second

77. $T_{AC} \approx 3611.1$ pounds, $T_{BC} \approx 2169.5$ pounds

79. (a) 3192.5 pounds

     (b) $T = 3000 \sec \theta$; Domain: $0° \le \theta < 90°$

     (c)

θ	10°	20°	30°	40°	50°	60°
T	3046.3	3192.5	3464.1	3916.2	4667.2	6000.0

     (d)

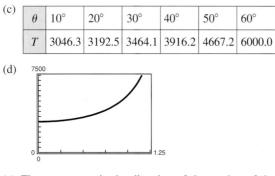

     (e) The component in the direction of the motion of the barge decreases.

81. N 26.7° E, 130.35 kilometers per hour

83. (a) $12.1°$, 357.85 newtons

     (b) $M = 10\sqrt{660 \cos \theta + 709}$,

$$\alpha = \arctan \dfrac{15 \sin \theta}{15 \cos \theta + 22}$$

     (c)

θ	0°	30°	60°	90°
M	370.0	357.9	322.3	266.3
α	0°	12.1°	23.8°	34.3°

θ	120°	150°	180°
M	194.7	117.2	70.0
α	41.9°	39.8°	0°

(d)

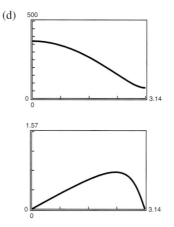

     (e) For increasing θ, the two vectors tend to work against each other, resulting in a decrease in the magnitude of the resultant.

85. True   **87.** True. $a = b = 0$

89. (a) $0°$ (b) $180°$

     (c) No. The magnitude is equal to the sum when the angle between the vectors is $0°$.

91. Answers will vary.   **93.** $\langle 1, 3 \rangle$ or $\langle -1, -3 \rangle$

95. $7 \cos \theta$ **97.** $10 \csc \theta$

99. $C = 87°$, $b = 18.16$, $c = 20.73$

101. $A = 24.15°$, $B = 30.75°$, $C = 125.10°$

Section 6.4 *(page 464)*

1. -18 **3.** 11 **5.** 8, scalar   **7.** $\langle 2, -8 \rangle$, vector

9. 4, scalar   **11.** 13 **13.** $5\sqrt{41}$ **15.** 6

17. $90°$ **19.** $70.56°$ **21.** $90°$ **23.** $\dfrac{5\pi}{12}$

25.   **27.**

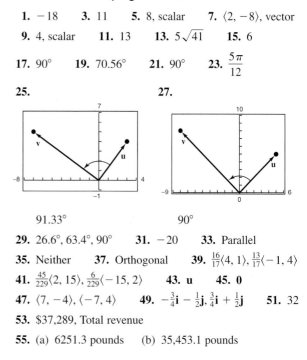

     $91.33°$ $90°$

29. $26.6°, 63.4°, 90°$ **31.** -20 **33.** Parallel

35. Neither   **37.** Orthogonal   **39.** $\dfrac{16}{17}\langle 4, 1 \rangle, \dfrac{13}{17}\langle -1, 4 \rangle$

41. $\dfrac{45}{229}\langle 2, 15 \rangle, \dfrac{6}{229}\langle -15, 2 \rangle$ **43.** $\mathbf{u}$ **45.** $\mathbf{0}$

47. $\langle 7, -4 \rangle, \langle -7, 4 \rangle$ **49.** $-\dfrac{3}{4}\mathbf{i} - \dfrac{1}{2}\mathbf{j}, \dfrac{3}{4}\mathbf{i} + \dfrac{1}{2}\mathbf{j}$ **51.** 32

53. \$37,289, Total revenue

55. (a) 6251.3 pounds   (b) 35,453.1 pounds

57. 735 newton-meters   **59.** 779.4 foot-pounds

61. 725.0 foot-pounds

63. True. The zero vector is orthogonal to every vector.

65. Orthogonal. $\mathbf{u} \cdot \mathbf{v} = 0$

67. (a) $\mathbf{u}$ and $\mathbf{v}$ are parallel.   (b) $\mathbf{u}$ and $\mathbf{v}$ are orthogonal.

69. Answers will vary.   **71.** (a) and (b) Answers will vary.

73. $\dfrac{\pi}{2}, \dfrac{3\pi}{2}$ **75.** $\dfrac{5\pi}{6}, \dfrac{7\pi}{6}$ **77.** 59.43 square units

79. (a) $\langle 7, -1 \rangle$ (b) $\langle 5, -2 \rangle$ (c) $\langle -11, 5 \rangle$

81. (a) $\langle -6, 3 \rangle$ (b) $\langle -6, 12 \rangle$ (c) $\langle 14, -31 \rangle$

83. Buy now.   **85.** 8

Section 6.5   *(page 475)*

1. 6

3. 4

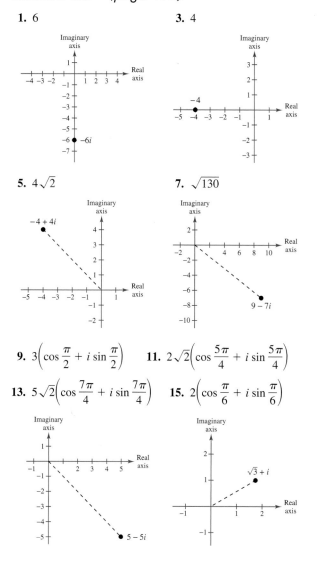

5. $4\sqrt{2}$

7. $\sqrt{130}$

9. $3\left(\cos \dfrac{\pi}{2} + i \sin \dfrac{\pi}{2}\right)$ **11.** $2\sqrt{2}\left(\cos \dfrac{5\pi}{4} + i \sin \dfrac{5\pi}{4}\right)$

13. $5\sqrt{2}\left(\cos \dfrac{7\pi}{4} + i \sin \dfrac{7\pi}{4}\right)$ **15.** $2\left(\cos \dfrac{\pi}{6} + i \sin \dfrac{\pi}{6}\right)$

17. $4\left(\cos \dfrac{4\pi}{3} + i \sin \dfrac{4\pi}{3}\right)$ **19.** $8\left(\cos \dfrac{\pi}{2} + i \sin \dfrac{\pi}{2}\right)$

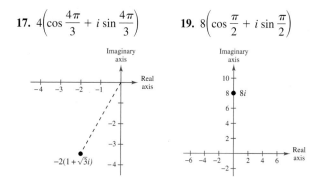

21. $\sqrt{65}(\cos 150.26° + i \sin 150.26°)$

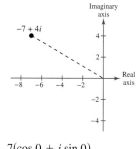

23. $7(\cos 0 + i \sin 0)$

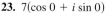

25. $\sqrt{101}(\cos 84.29° + i \sin 84.29°)$

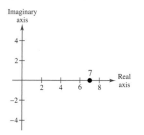

27. $5\sqrt{2}(\cos 188.13° + i \sin 188.13°)$

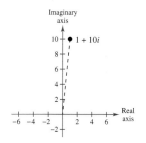

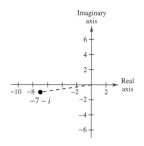

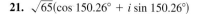

29. $5.385(\cos 21.80° + i \sin 21.80°)$

31. $8.185(\cos 301.22° + i \sin 301.22°)$

33. $-1 + \sqrt{3}i$

35. $\dfrac{3\sqrt{3}}{4} - \dfrac{3}{4}i$

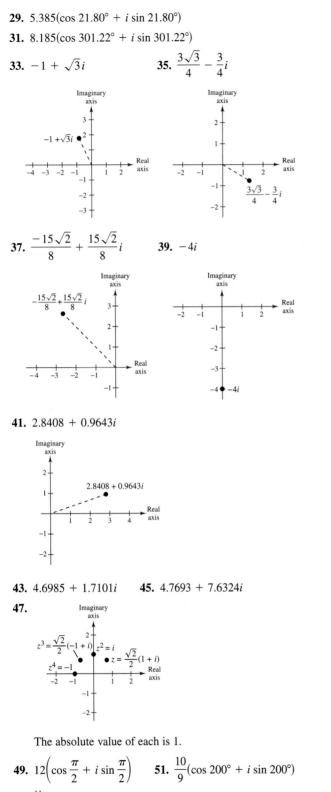

37. $\dfrac{-15\sqrt{2}}{8} + \dfrac{15\sqrt{2}}{8}i$

39. $-4i$

41. $2.8408 + 0.9643i$

43. $4.6985 + 1.7101i$

45. $4.7693 + 7.6324i$

47.

The absolute value of each is 1.

49. $12\left(\cos \dfrac{\pi}{2} + i \sin \dfrac{\pi}{2}\right)$

51. $\dfrac{10}{9}(\cos 200° + i \sin 200°)$

53. $\dfrac{11}{50}(\cos 130° + i \sin 130°)$

55. $\cos 30° + i \sin 30°$

57. $\dfrac{1}{2}(\cos 80° + i \sin 80°)$

59. $6[\cos(-48°) + i \sin(-48°)]$

61. (a) $2\sqrt{2}\left(\cos \dfrac{\pi}{4} + i \sin \dfrac{\pi}{4}\right)$

$\sqrt{2}\left[\cos\left(-\dfrac{\pi}{4}\right) + i \sin\left(-\dfrac{\pi}{4}\right)\right]$

(b) $4(\cos 0 + i \sin 0) = 4$ (c) 4

63. (a) $2\left[\cos\left(-\dfrac{\pi}{2}\right) + i \sin\left(-\dfrac{\pi}{2}\right)\right]$

$\sqrt{2}\left(\cos \dfrac{\pi}{4} + i \sin \dfrac{\pi}{4}\right)$

(b) $2\sqrt{2}\left[\cos\left(-\dfrac{\pi}{4}\right) + i \sin\left(-\dfrac{\pi}{4}\right)\right] = 2 - 2i$

(c) $2 - 2i$

65. (a) $5(\cos 0 + i \sin 0)$

$\sqrt{13}(\cos 56.31° + i \sin 56.31°)$

(b) $\dfrac{5\sqrt{13}}{13}[\cos(-56.31°) + i \sin(-56.31°)]$

$\approx 0.7692 - 1.1538i$

(c) $\dfrac{10}{13} - \dfrac{15}{13}i \approx 0.7692 - 1.1538i$

67.

69.

71. $-2 + 2i$ **73.** $-32i$ **75.** $-32\sqrt{3} + 32i$

77. $\dfrac{125}{2} + \dfrac{125\sqrt{3}}{2}i$ **79.** i

81. $140.0189 + 1014.3819i$ **83.** $-597.0295 - 121.9895i$

85. $\dfrac{81}{2} + \dfrac{81\sqrt{3}}{2}i$

87. $\sqrt[6]{1}\left[\cos \dfrac{0 + 2\pi(4)}{6} + i \sin \dfrac{0 + 2\pi(4)}{6}\right] = -\dfrac{1}{2}(1 + \sqrt{3}i)$

89. (a) $2(\cos 30° + i \sin 30°)$

$2(\cos 150° + i \sin 150°)$

$2(\cos 270° + i \sin 270°)$

(b) and (c) $8i$

91. (a) $\sqrt{5}(\cos 60° + i \sin 60°)$

$\sqrt{5}(\cos 240° + i \sin 240°)$

(b)

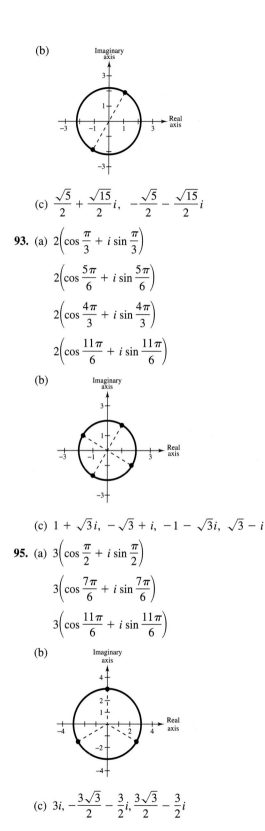

(c) $\dfrac{\sqrt{5}}{2} + \dfrac{\sqrt{15}}{2}i, \quad -\dfrac{\sqrt{5}}{2} - \dfrac{\sqrt{15}}{2}i$

93. (a) $2\left(\cos\dfrac{\pi}{3} + i\sin\dfrac{\pi}{3}\right)$

$2\left(\cos\dfrac{5\pi}{6} + i\sin\dfrac{5\pi}{6}\right)$

$2\left(\cos\dfrac{4\pi}{3} + i\sin\dfrac{4\pi}{3}\right)$

$2\left(\cos\dfrac{11\pi}{6} + i\sin\dfrac{11\pi}{6}\right)$

(b)

(c) $1 + \sqrt{3}\,i, \ -\sqrt{3} + i, \ -1 - \sqrt{3}\,i, \ \sqrt{3} - i$

95. (a) $3\left(\cos\dfrac{\pi}{2} + i\sin\dfrac{\pi}{2}\right)$

$3\left(\cos\dfrac{7\pi}{6} + i\sin\dfrac{7\pi}{6}\right)$

$3\left(\cos\dfrac{11\pi}{6} + i\sin\dfrac{11\pi}{6}\right)$

(b)

(c) $3i, \ -\dfrac{3\sqrt{3}}{2} - \dfrac{3}{2}i, \ \dfrac{3\sqrt{3}}{2} - \dfrac{3}{2}i$

97. (a) $5\left(\cos\dfrac{4\pi}{9} + i\sin\dfrac{4\pi}{9}\right)$

$5\left(\cos\dfrac{10\pi}{9} + i\sin\dfrac{10\pi}{9}\right)$

$5\left(\cos\dfrac{16\pi}{9} + i\sin\dfrac{16\pi}{9}\right)$

(b)

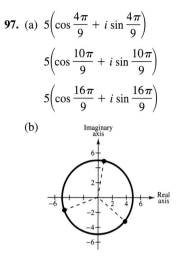

(c) $0.8682 + 4.9240i, \ -4.6985 - 1.7101i,$

$3.8302 - 3.2139i$

99. (a) $4(\cos 0 + i\sin 0)$

$4\left(\cos\dfrac{2\pi}{3} + i\sin\dfrac{2\pi}{3}\right)$

$4\left(\cos\dfrac{4\pi}{3} + i\sin\dfrac{4\pi}{3}\right)$

(b)

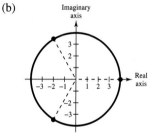

(c) $4, \ -2 + 2\sqrt{3}\,i, \ -2 - 2\sqrt{3}\,i$

101. (a) $\cos 0 + i\sin 0$

$\cos\dfrac{2\pi}{5} + i\sin\dfrac{2\pi}{5}$

$\cos\dfrac{4\pi}{5} + i\sin\dfrac{4\pi}{5}$

$\cos\dfrac{6\pi}{5} + i\sin\dfrac{6\pi}{5}$

$\cos\dfrac{8\pi}{5} + i\sin\dfrac{8\pi}{5}$

(b)

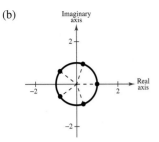

(c) $1, 0.3090 + 0.9511i, -0.8090 + 0.5878i,$
$-0.8090 - 0.5878i, 0.3090 - 0.9511i$

103. (a) $5\left(\cos\dfrac{\pi}{3} + i\sin\dfrac{\pi}{3}\right)$

$5(\cos\pi + i\sin\pi)$

$5\left(\cos\dfrac{5\pi}{3} + i\sin\dfrac{5\pi}{3}\right)$

(b)

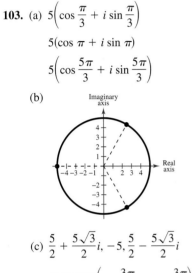

(c) $\dfrac{5}{2} + \dfrac{5\sqrt{3}}{2}i, -5, \dfrac{5}{2} - \dfrac{5\sqrt{3}}{2}i$

105. (a) $\sqrt[10]{32{,}768}\left(\cos\dfrac{3\pi}{20} + i\sin\dfrac{3\pi}{20}\right)$

$\sqrt[10]{32{,}768}\left(\cos\dfrac{11\pi}{20} + i\sin\dfrac{11\pi}{20}\right)$

$\sqrt[10]{32{,}768}\left(\cos\dfrac{19\pi}{20} + i\sin\dfrac{19\pi}{20}\right)$

$\sqrt[10]{32{,}768}\left(\cos\dfrac{27\pi}{20} + i\sin\dfrac{27\pi}{20}\right)$

$\sqrt[10]{32{,}768}\left(\cos\dfrac{7\pi}{4} + i\sin\dfrac{7\pi}{4}\right)$

(b)

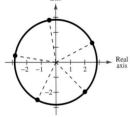

(c) $2.5201 + 1.2841i, -0.4425 + 2.7936i,$
$-2.7936 + 0.4425i, -1.2841 - 2.5201i,$
$2.0000 - 2.0000i$

107. $\cos\dfrac{\pi}{8} + i\sin\dfrac{\pi}{8}$

$\cos\dfrac{5\pi}{8} + i\sin\dfrac{5\pi}{8}$

$\cos\dfrac{9\pi}{8} + i\sin\dfrac{9\pi}{8}$

$\cos\dfrac{13\pi}{8} + i\sin\dfrac{13\pi}{8}$

109. $3(\cos 0 + i\sin 0)$

$3\left(\cos\dfrac{2\pi}{5} + i\sin\dfrac{2\pi}{5}\right)$

$3\left(\cos\dfrac{4\pi}{5} + i\sin\dfrac{4\pi}{5}\right)$

$3\left(\cos\dfrac{6\pi}{5} + i\sin\dfrac{6\pi}{5}\right)$

$3\left(\cos\dfrac{8\pi}{5} + i\sin\dfrac{8\pi}{5}\right)$

111. $4\left(\cos\dfrac{\pi}{2} + i\sin\dfrac{\pi}{2}\right)$

$4\left(\cos\dfrac{7\pi}{6} + i\sin\dfrac{7\pi}{6}\right)$

$4\left(\cos\dfrac{11\pi}{6} + i\sin\dfrac{11\pi}{6}\right)$

113. $\sqrt[6]{2}\left(\cos\dfrac{7\pi}{12} + i\sin\dfrac{7\pi}{12}\right)$

$\sqrt[6]{2}\left(\cos\dfrac{5\pi}{4} + i\sin\dfrac{5\pi}{4}\right)$

$\sqrt[6]{2}\left(\cos\dfrac{23\pi}{12} + i\sin\dfrac{23\pi}{12}\right)$

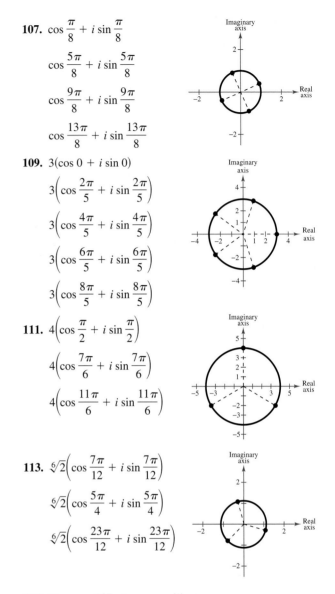

115. True   **117.** Answers will vary.

119. $z\bar{z} = r^2; z/\bar{z} = \cos 2\theta + i\sin 2\theta$ **121.** 8.4782

123. $\langle 1, 0\rangle$ **125.** $\frac{12}{13}\mathbf{i} - \frac{5}{13}\mathbf{j}$ **127.** 41

129. -51 **131.** Parallel

Review Exercises   *(page 479)*

1. $C = 110°, b \approx 32.6, c \approx 36.2$ **3.** No solution

5. $A \approx 34.2°, C \approx 30.8°, c \approx 8.2$

7. $B \approx 31.2°, C \approx 133.8°, c \approx 13.9$

$B \approx 148.8°, C \approx 16.2°, c \approx 5.4$

9. $A \approx 40.9°, C \approx 114.1°, c \approx 8.6$

$A \approx 139.1°, C \approx 15.9°, c \approx 2.6$

11. ≈ 9.08 square units   **13.** ≈ 221.3 square units

15. ≈ 31.1 meters   **17.** ≈ 31.0 feet

19. $A \approx 29.7°, B \approx 52.4°, C \approx 97.9°$

21. $c \approx 6.37, A \approx 41.7°, B \approx 93.3°$

23. $A = 35°, C = 35°, b \approx 6.6$

25. $A \approx 39.9°, B \approx 22.4°, C \approx 117.7°$

27. $b \approx 29.1, A \approx 9.9°, C \approx 20.1°$

29. ≈ 4.3 feet, ≈ 12.6 feet   **31.** ≈ 615.1 meters

33. ≈ 9.8 square units   **35.** ≈ 511.7 square units

37. $\|\mathbf{u}\| = \|\mathbf{v}\| = \sqrt{61},$ slope$_{\mathbf{u}}$ = slope$_{\mathbf{v}}$ = $\frac{5}{6}$

39. $\langle 7, -5 \rangle$ **41.** $\langle 7, -7 \rangle$ **43.** $\langle -4, 4\sqrt{3} \rangle$

45. $\langle -4, -5 \rangle$ **47.** $\langle 2, -1 \rangle$ **49.** $\langle 3, 33 \rangle$

51. $\left\langle \dfrac{6}{\sqrt{61}}, -\dfrac{5}{\sqrt{61}} \right\rangle$ **53.** $\langle 30, 9 \rangle$

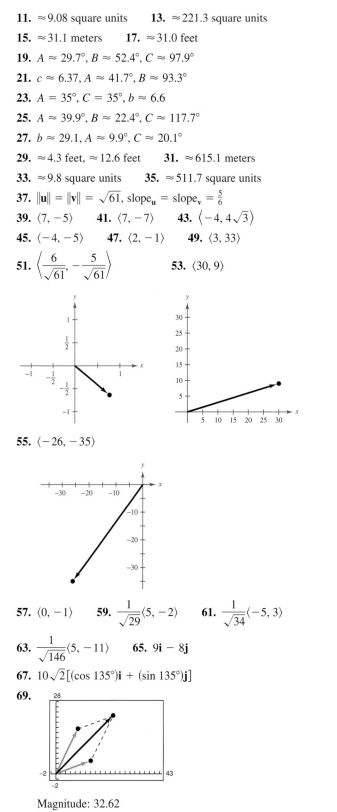

55. $\langle -26, -35 \rangle$

57. $\langle 0, -1 \rangle$ **59.** $\dfrac{1}{\sqrt{29}}\langle 5, -2 \rangle$ **61.** $\dfrac{1}{\sqrt{34}}\langle -5, 3 \rangle$

63. $\dfrac{1}{\sqrt{146}}\langle 5, -11 \rangle$ **65.** $9\mathbf{i} - 8\mathbf{j}$

67. $10\sqrt{2}[(\cos 135°)\mathbf{i} + (\sin 135°)\mathbf{j}]$

69.

Magnitude: 32.62

Direction: 44.72°

71. 79.9°, 92.2 pounds   **73.** 180 pounds

75. 422.3 miles per hour, S 49.6° E   **77.** 7

79. $\langle -54, 27 \rangle$ **81.** -36 **83.** $\dfrac{11\pi}{12}$ **85.** 160.5°

87.   **89.**

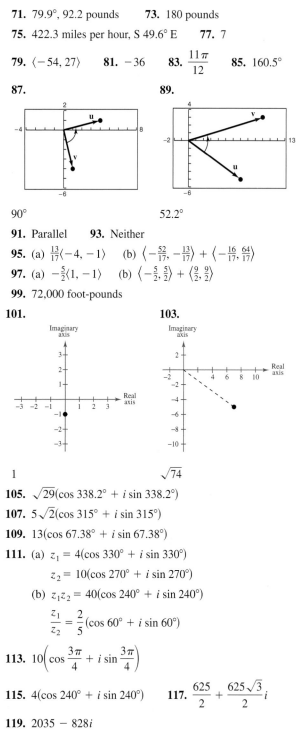

90°   52.2°

91. Parallel   **93.** Neither

95. (a) $\frac{13}{17}\langle -4, -1 \rangle$ (b) $\left\langle -\frac{52}{17}, -\frac{13}{17} \right\rangle + \left\langle -\frac{16}{17}, \frac{64}{17} \right\rangle$

97. (a) $-\frac{5}{2}\langle 1, -1 \rangle$ (b) $\left\langle -\frac{5}{2}, \frac{5}{2} \right\rangle + \left\langle \frac{9}{2}, \frac{9}{2} \right\rangle$

99. 72,000 foot-pounds

101.   **103.**

1                          $\sqrt{74}$

105. $\sqrt{29}(\cos 338.2° + i \sin 338.2°)$

107. $5\sqrt{2}(\cos 315° + i \sin 315°)$

109. $13(\cos 67.38° + i \sin 67.38°)$

111. (a) $z_1 = 4(\cos 330° + i \sin 330°)$

   $z_2 = 10(\cos 270° + i \sin 270°)$

   (b) $z_1 z_2 = 40(\cos 240° + i \sin 240°)$

   $\dfrac{z_1}{z_2} = \dfrac{2}{5}(\cos 60° + i \sin 60°)$

113. $10\left(\cos \dfrac{3\pi}{4} + i \sin \dfrac{3\pi}{4} \right)$

115. $4(\cos 240° + i \sin 240°)$ **117.** $\dfrac{625}{2} + \dfrac{625\sqrt{3}}{2}i$

119. $2035 - 828i$

121. $3\left(\cos \dfrac{\pi}{4} + i \sin \dfrac{\pi}{4}\right)$

$3\left(\cos \dfrac{7\pi}{12} + i \sin \dfrac{7\pi}{12}\right)$

$3\left(\cos \dfrac{11\pi}{12} + i \sin \dfrac{11\pi}{12}\right)$

$3\left(\cos \dfrac{5\pi}{4} + i \sin \dfrac{5\pi}{4}\right)$

$3\left(\cos \dfrac{19\pi}{12} + i \sin \dfrac{19\pi}{12}\right)$

$3\left(\cos \dfrac{23\pi}{12} + i \sin \dfrac{23\pi}{12}\right)$

123. $4\left(\cos \dfrac{\pi}{4} + i \sin \dfrac{\pi}{4}\right) = 2\sqrt{2} + 2\sqrt{2}i$

$4\left(\cos \dfrac{3\pi}{4} + i \sin \dfrac{3\pi}{4}\right) = -2\sqrt{2} + 2\sqrt{2}i$

$4\left(\cos \dfrac{5\pi}{4} + i \sin \dfrac{5\pi}{4}\right) = -2\sqrt{2} - 2\sqrt{2}i$

$4\left(\cos \dfrac{7\pi}{4} + i \sin \dfrac{7\pi}{4}\right) = 2\sqrt{2} - 2\sqrt{2}i$

125. $2\left(\cos \dfrac{\pi}{2} + i \sin \dfrac{\pi}{2}\right) = 2i$

$2\left(\cos \dfrac{7\pi}{6} + i \sin \dfrac{7\pi}{6}\right) = -\sqrt{3} - i$

$2\left(\cos \dfrac{11\pi}{6} + i \sin \dfrac{11\pi}{6}\right) = \sqrt{3} - i$

127. True **129.** Direction and magnitude

131. (a) 3

(b) The modulus of each is 2, and the arguments are 120°, 210°, and 300°.

Chapter Test *(page 484)*

1. $C = 44°$, $a \approx 16.0$, $b \approx 25.9$

2. $A \approx 26.2°$, $B \approx 136.5°$, $C \approx 17.3°$

3. $a \approx 9.8$, $B \approx 46.6°$, $C \approx 75.5°$

4. $B \approx 48.3°$, $C \approx 107.1°$, $c \approx 35.8$

$B \approx 131.7°$, $C \approx 23.7°$, $c \approx 15.1$

5. No solution **6.** $B \approx 14.7°$, $C \approx 15.3°$, $c \approx 4.9$

7. S 59.8° E **8.** ≈ 675 feet

9. $\mathbf{w} = \langle 12, 13 \rangle$, $\|\mathbf{w}\| \approx 17.69$ **10.** $\left\langle \dfrac{7}{\sqrt{65}}, \dfrac{4}{\sqrt{65}} \right\rangle$

11. (a) $\langle -4, 4 \rangle$ (b) $\langle 6, -16 \rangle$ (c) $\langle 2, -24 \rangle$

12. (a) $\langle 0, 7 \rangle$ (b) $\langle -5, -18 \rangle$ (c) $\langle -11, -20 \rangle$

13. (a) $\langle 13, 17 \rangle$ (b) $\langle -17, -28 \rangle$ (c) $\langle -1, -14 \rangle$

14. (a) $\langle 0, -1 \rangle$ (b) $\langle 5, 9 \rangle$ (c) $\langle 11, 17 \rangle$

15. $\left\langle \dfrac{36}{\sqrt{34}}, -\dfrac{60}{\sqrt{34}} \right\rangle$ **16.** $\theta \approx 14.9°$, ≈ 250.2 pounds

17. $\approx 105.9°$ **18.** No, because $\mathbf{u} \cdot \mathbf{v} = 24$, not 0.

19. $\left\langle \dfrac{185}{26}, \dfrac{37}{26} \right\rangle$, $\mathbf{u} = \left\langle \dfrac{185}{26}, \dfrac{37}{26} \right\rangle + \left\langle -\dfrac{29}{26}, \dfrac{145}{26} \right\rangle$

20. $z = 2\sqrt{2}\left(\cos \dfrac{3\pi}{4} + i \sin \dfrac{3\pi}{4}\right)$

21. $-50 - 50\sqrt{3}i$ **22.** $12\sqrt{3} - 12i$

23. $-\dfrac{6561}{2} + \dfrac{6561\sqrt{3}}{2}i$ **24.** $5832i$

25. $4\left(\cos \dfrac{\pi}{12} + i \sin \dfrac{\pi}{12}\right) \approx 3.8637 + 1.0353i$

$4\left(\cos \dfrac{7\pi}{12} + i \sin \dfrac{7\pi}{12}\right) \approx -1.0353 + 3.8637i$

$4\left(\cos \dfrac{13\pi}{12} + i \sin \dfrac{13\pi}{12}\right) \approx -3.8637 - 1.0353i$

$4\left(\cos \dfrac{19\pi}{12} + i \sin \dfrac{19\pi}{12}\right) \approx 1.0353 - 3.8637i$

26. $5\left(\cos \dfrac{\pi}{8} + i \sin \dfrac{\pi}{8}\right)$

$5\left(\cos \dfrac{5\pi}{8} + i \sin \dfrac{5\pi}{8}\right)$

$5\left(\cos \dfrac{9\pi}{8} + i \sin \dfrac{9\pi}{8}\right)$

$5\left(\cos \dfrac{13\pi}{8} + i \sin \dfrac{13\pi}{8}\right)$

Cumulative Test for Chapters 4–6
(page 485)

1. (a)

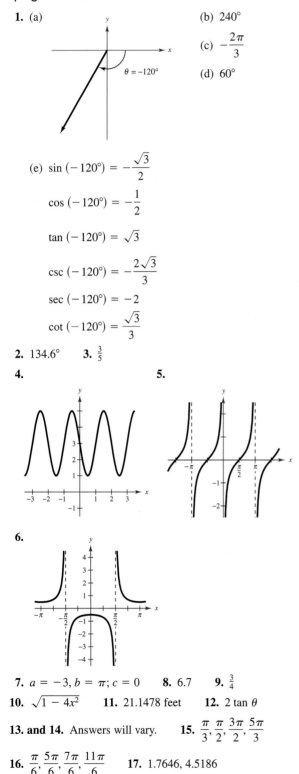

$\theta = -120°$

(b) $240°$

(c) $-\dfrac{2\pi}{3}$

(d) $60°$

(e) $\sin(-120°) = -\dfrac{\sqrt{3}}{2}$

$\cos(-120°) = -\dfrac{1}{2}$

$\tan(-120°) = \sqrt{3}$

$\csc(-120°) = -\dfrac{2\sqrt{3}}{3}$

$\sec(-120°) = -2$

$\cot(-120°) = \dfrac{\sqrt{3}}{3}$

2. $134.6°$ **3.** $\frac{3}{5}$

4. **5.**

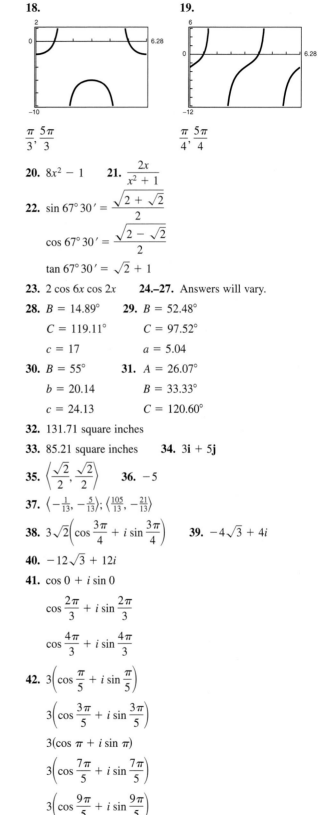

6.

7. $a = -3, b = \pi; c = 0$ **8.** 6.7 **9.** $\frac{3}{4}$

10. $\sqrt{1 - 4x^2}$ **11.** 21.1478 feet **12.** $2\tan\theta$

13. and 14. Answers will vary. **15.** $\dfrac{\pi}{3}, \dfrac{\pi}{2}, \dfrac{3\pi}{2}, \dfrac{5\pi}{3}$

16. $\dfrac{\pi}{6}, \dfrac{5\pi}{6}, \dfrac{7\pi}{6}, \dfrac{11\pi}{6}$ **17.** $1.7646, 4.5186$

18.

19.

$\dfrac{\pi}{3}, \dfrac{5\pi}{3}$

$\dfrac{\pi}{4}, \dfrac{5\pi}{4}$

20. $8x^2 - 1$ **21.** $\dfrac{2x}{x^2 + 1}$

22. $\sin 67° 30' = \dfrac{\sqrt{2 + \sqrt{2}}}{2}$

$\cos 67° 30' = \dfrac{\sqrt{2 - \sqrt{2}}}{2}$

$\tan 67° 30' = \sqrt{2} + 1$

23. $2\cos 6x \cos 2x$ **24.–27.** Answers will vary.

28. $B = 14.89°$ **29.** $B = 52.48°$
  $C = 119.11°$ $C = 97.52°$
  $c = 17$ $a = 5.04$

30. $B = 55°$ **31.** $A = 26.07°$
  $b = 20.14$ $B = 33.33°$
  $c = 24.13$ $C = 120.60°$

32. 131.71 square inches

33. 85.21 square inches **34.** $3\mathbf{i} + 5\mathbf{j}$

35. $\left\langle \dfrac{\sqrt{2}}{2}, \dfrac{\sqrt{2}}{2} \right\rangle$ **36.** -5

37. $\left\langle -\dfrac{1}{13}, -\dfrac{5}{13} \right\rangle; \left\langle \dfrac{105}{13}, -\dfrac{21}{13} \right\rangle$

38. $3\sqrt{2}\left(\cos\dfrac{3\pi}{4} + i\sin\dfrac{3\pi}{4} \right)$ **39.** $-4\sqrt{3} + 4i$

40. $-12\sqrt{3} + 12i$

41. $\cos 0 + i\sin 0$

$\cos\dfrac{2\pi}{3} + i\sin\dfrac{2\pi}{3}$

$\cos\dfrac{4\pi}{3} + i\sin\dfrac{4\pi}{3}$

42. $3\left(\cos\dfrac{\pi}{5} + i\sin\dfrac{\pi}{5} \right)$

$3\left(\cos\dfrac{3\pi}{5} + i\sin\dfrac{3\pi}{5} \right)$

$3(\cos\pi + i\sin\pi)$

$3\left(\cos\dfrac{7\pi}{5} + i\sin\dfrac{7\pi}{5} \right)$

$3\left(\cos\dfrac{9\pi}{5} + i\sin\dfrac{9\pi}{5} \right)$

43. ≈ 4.79 feet   **44.** $d = 4 \sin \dfrac{\pi}{4} t$

45. N 32.63° E; 543.88 kilometers per hour

46. $\approx 80.28°$

Chapter 7

Section 7.1   *(page 495)*

1. (a) No   (b) No   (c) No   (d) Yes

3. (a) No   (b) Yes   (c) No   (d) No

5. $(2, 2)$ **7.** $(2, 6), (-1, 3)$

9. $(0, 2), (\sqrt{3}, 2 - 3\sqrt{3}), (-\sqrt{3}, 2 + 3\sqrt{3})$

11. $(0, 0), (2, -4)$ **13.** $(4, 4)$ **15.** $(5, 5)$

17. $(\frac{1}{2}, 3)$ **19.** $(1, 1)$ **21.** $(\frac{20}{3}, \frac{40}{3})$ **23.** No solution

25. No solution   **27.** $(0, 0)$ **29.** $(4, 3)$ **31.** $(\frac{5}{2}, \frac{3}{2})$

33. $(2, 2), (4, 0)$ **35.** $(1, 4), (4, 7)$

37.

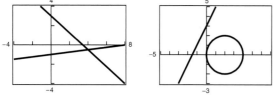

39.

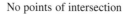

$(4, -0.5)$ No points of intersection

41.

43.

$(\pm 1.540, 2.372)$ $(0, 1)$

45.

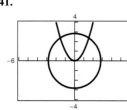

47.

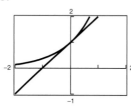

$(4, 2)$ $(0, 0), (1, 1)$

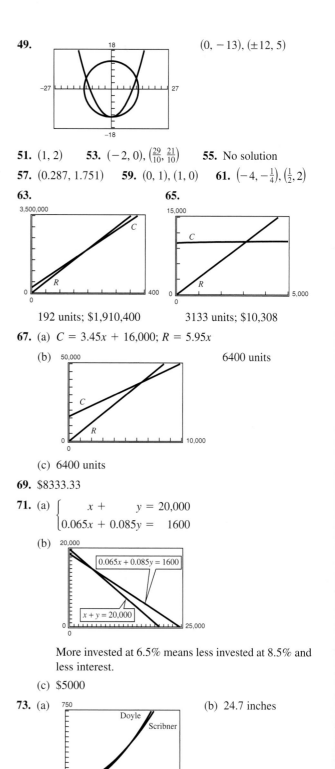

49. $(0, -13), (\pm 12, 5)$

51. $(1, 2)$ **53.** $(-2, 0), (\frac{29}{10}, \frac{21}{10})$ **55.** No solution

57. $(0.287, 1.751)$ **59.** $(0, 1), (1, 0)$ **61.** $(-4, -\frac{1}{4}), (\frac{1}{2}, 2)$

63.   **65.**

192 units; \$1,910,400   3133 units; \$10,308

67. (a) $C = 3.45x + 16,000; R = 5.95x$

(b)   6400 units

(c) 6400 units

69. \$8333.33

71. (a) $\begin{cases} x + y = 20,000 \\ 0.065x + 0.085y = 1600 \end{cases}$

(b)

More invested at 6.5% means less invested at 8.5% and less interest.

(c) \$5000

73. (a)   (b) 24.7 inches

(c) Doyle Log Rule; for logs of large diameter, V is larger using the Doyle Log Rule.

75. 60×80 centimeters   **77.** 42×63 feet

79. $2 \times \sqrt{2} \times \sqrt{2}$ inches

81. (a) $E = 0.342t + 10.72$

$E = 0.0163t^2 + 0.196t + 11.00$

(b)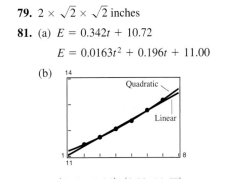

(c) $(6.18, 12.83), (2.78, 11.67)$

(d) Linear: \$14.14; Quadratic: \$14.59. Answers will vary.

83. False. Example: a parabola and a circle can have 4 intersection points.

85. Graphical solutions may be approximate.

87. (a)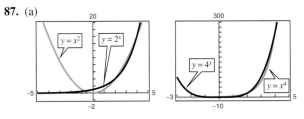

(b) Three points of intersection when b is even

89. $4x - 13y + 38 = 0$ **91.** $x = 4$

93. $45x + 29y - 127 = 0$

95. Domain: $-\infty < x < -\frac{2}{3}, -\frac{2}{3} < x < \infty$

Asymptotes: $x = -\frac{2}{3}, y = \frac{2}{3}$

97. Domain: $-\infty < x < 0, 0 < x < \infty$

Asymptotes: $x = 0, y = 3$

99. 0.272 **101.** 0.405

Section 7.2   *(page 505)*

1. $(2, 1)$ **3.** $(1, -1)$ **5.** Inconsistent

7. All points on $3x - 2y - 5 = 0$ **9.** $\left(\frac{1}{3}, -\frac{2}{3}\right)$

11. $\left(\frac{5}{2}, \frac{3}{4}\right)$ **13.** $(3, 4)$ **15.** $(4, -1)$ **17.** $\left(\frac{12}{7}, \frac{18}{7}\right)$

19. Inconsistent   **21.** $\left(-\frac{6}{35}, \frac{43}{35}\right)$ **23.** $\left(\frac{38}{15}, \frac{11}{5}\right)$

25. All points on $6x + 8y - 1 = 0$ **27.** $(5, -2)$

29. All points on $5x - 6y - 3 = 0$ **31.** $(8, 7)$

33.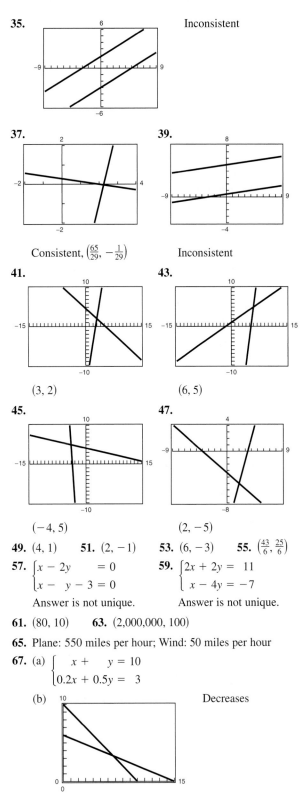

Consistent, $(5, 2)$

35.

Inconsistent

37.   **39.**

Consistent, $\left(\frac{65}{29}, -\frac{1}{29}\right)$ Inconsistent

41.   **43.**

$(3, 2)$ $(6, 5)$

45.   **47.**

$(-4, 5)$ $(2, -5)$

49. $(4, 1)$ **51.** $(2, -1)$ **53.** $(6, -3)$ **55.** $\left(\frac{43}{6}, \frac{25}{6}\right)$

57. $\begin{cases} x - 2y = 0 \\ x - y - 3 = 0 \end{cases}$ **59.** $\begin{cases} 2x + 2y = 11 \\ x - 4y = -7 \end{cases}$

Answer is not unique.   Answer is not unique.

61. $(80, 10)$ **63.** $(2{,}000{,}000, 100)$

65. Plane: 550 miles per hour; Wind: 50 miles per hour

67. (a) $\begin{cases} x + y = 10 \\ 0.2x + 0.5y = 3 \end{cases}$

(b)

Decreases

(c) 20% solution: $6\frac{2}{3}$ liters; 50% solution: $3\frac{1}{3}$ liters

69. $6000 **71.** 375 adults, 125 children

73. 225 kilometers and 75 kilometers

75. $y = 0.97x + 2.10$ **77.** $y = 0.318x + 4.061$

79. (a) and (b) $y = 14x + 19$

(c)

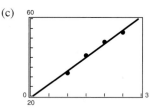

(d) 41.4 bushels

81. True **83.** True

85. $(300, 315)$. It is necessary to change the scale on the axes to see the point of intersection.

87. (a) $\begin{cases} x + y = 10 \\ x + y = 20 \end{cases}$ (b) $\begin{cases} x + y = 4 \\ 3x + 3y = 12 \end{cases}$

89. $k = -2$ **91.** $u = -\dfrac{1}{2}, v = \dfrac{\cot 2x}{2}$

93. $u = -1, v = \dfrac{1}{x}$

95.

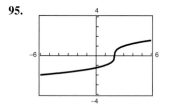

Domain: All real numbers
Range: All real numbers

97.

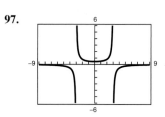

Domain: $-\infty < x < -3, -3 < x < 3, 3 < x < \infty$
Range: $-\infty < y < 0, 0 < y < \infty$

99. $x > 1$ **101.** $\dfrac{4}{3} \le x < \dfrac{16}{3}$

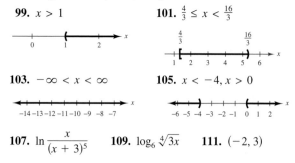

103. $-\infty < x < \infty$ **105.** $x < -4, x > 0$

107. $\ln \dfrac{x}{(x + 3)^5}$ **109.** $\log_6 \sqrt[4]{3x}$ **111.** $(-2, 3)$

Section 7.3 *(page 519)*

1. (a) No (b) No (c) No (d) Yes

3. (a) No (b) No (c) Yes (d) No

5. $(-7, -8, 6)$ **7.** $(-3, -2, -6)$ **9.** $\left(\frac{1}{2}, -2, 2\right)$

11. $\begin{cases} x - 2y + 3z = 5 \\ y - 2z = 9 \\ 2x \quad\quad - 3z = 0 \end{cases}$

First step in putting the system in row-echelon form

13. $(-1, 3, 4)$ **15.** $(-2, 7, 5)$ **17.** $(6, -3, 0)$

19. Inconsistent **21.** $\left(1, -\frac{3}{2}, \frac{1}{2}\right)$

23. $(-3a + 10, 5a - 7, a)$ **25.** $(-a + 3, a + 1, a)$

27. Inconsistent **29.** Inconsistent **31.** $(0, 0, 0)$

33. $(2a, 21a - 1, 8a)$ **35.** $\left(\frac{1}{2} - \frac{3}{2}a, 1 - \frac{2}{3}a, a\right)$

37. $(9a, -35a, 67a)$ **39.** $(1, 1, 1, 1)$

41. $\begin{cases} 3x + y - z = 9 \\ x + 2y - z = 0 \\ -x + y + 3z = 1 \end{cases}$ **43.** $\begin{cases} x + 6y + 4z = 7 \\ 2x - 2y - 4z = 0 \\ -x + y + z = -\frac{7}{4} \end{cases}$

Answer is not unique. Answer is not unique.

45.

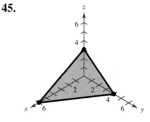

$(6, 0, 0), (0, 4, 0), (0, 0, 3), (4, 0, 1)$

47.

$(2, 0, 0), (0, 4, 0), (0, 0, 4), (0, 2, 2)$

49. $\dfrac{A}{x} + \dfrac{B}{x - 14}$ **51.** $\dfrac{A}{x} + \dfrac{B}{x^2} + \dfrac{C}{x - 10}$

53. $\dfrac{A}{x - 5} + \dfrac{B}{(x - 5)^2} + \dfrac{C}{(x - 5)^3}$

55. $\dfrac{1}{2}\left(\dfrac{1}{x - 1} - \dfrac{1}{x + 1}\right)$ **57.** $\dfrac{1}{x} - \dfrac{1}{x + 1}$

59. $\dfrac{1}{x} - \dfrac{2}{2x + 1}$ **61.** $\dfrac{1}{x - 1} - \dfrac{1}{x + 2}$

63. $-\dfrac{3}{x} - \dfrac{1}{x + 2} + \dfrac{5}{x - 2}$ **65.** $\dfrac{3}{x} - \dfrac{1}{x^2} + \dfrac{1}{x + 1}$

67. $\dfrac{3}{x - 3} + \dfrac{9}{(x - 3)^2}$ **69.** $2x - 7 + \dfrac{17}{x + 2} + \dfrac{1}{x + 1}$

71. $x + 3 + \dfrac{6}{x - 1} + \dfrac{4}{(x - 1)^2} + \dfrac{1}{(x - 1)^3}$

73. $\dfrac{3}{2x - 1} - \dfrac{2}{x + 1}$ **75.** $\dfrac{2}{x} - \dfrac{1}{x^2} - \dfrac{2}{x + 1}$

77. $2x + \dfrac{1}{2}\left(\dfrac{3}{x - 4} - \dfrac{1}{x + 2} \right)$

79. $\dfrac{3}{x} - \dfrac{2}{x - 4}$

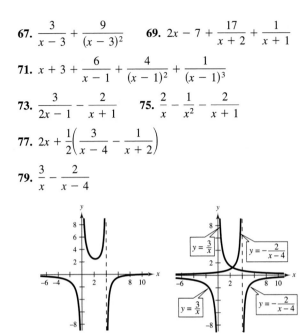

Vertical asymptotes are the same.

81. $s = -16t^2 + 144$ **83.** $s = 24t^2 - 264t + 692$

85. $y = \dfrac{1}{2}x^2 - 2x$ **87.** $y = x^2 - 6x + 8$

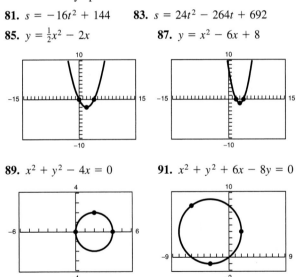

89. $x^2 + y^2 - 4x = 0$ **91.** $x^2 + y^2 + 6x - 8y = 0$

93. $4000 at 5%, $5000 at 6%, $7000 at 7%

95. $366,666.67 at 8%, $316,666.67 at 9%, $91,666.67 at 10%

97. $250,000 - \dfrac{1}{2}s$ in certificates of deposit

$125,000 + \dfrac{1}{2}s$ in municipal bonds

$125,000 - s$ in blue-chip stocks

s in growth stocks

99. 20 liters of spray X, 18 liters of spray Y, 16 liters of spray Z

101. Use four medium trucks or two large trucks, one medium truck, and two small trucks. (Other answers possible.)

103. $t_1 = 96$ pounds

$t_2 = 48$ pounds

$a = -16$ feet per second squared

105. $y = -\dfrac{5}{24}x^2 - \dfrac{3}{10}x + \dfrac{41}{6}$ **107.** $y = x^2 - x$

109. (a) $y = 0.14x^2 - 4.4x + 58$ (b) 436 feet

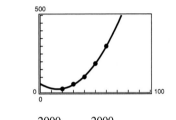

111. (a) $\dfrac{2000}{7 - 4x} - \dfrac{2000}{11 - 7x}$, $0 \le x \le 1$

(b)

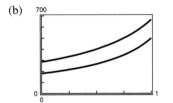

113. False. Leading coefficients are not all 1.

115. False. $\dfrac{A}{x + 10} + \dfrac{B}{x - 10} + \dfrac{C}{(x - 10)^2}$

117. $\dfrac{1}{2a}\left(\dfrac{1}{a + x} + \dfrac{1}{a - x} \right)$ **119.** $\dfrac{1}{a}\left(\dfrac{1}{y} + \dfrac{1}{a - y} \right)$

Answers will vary.   Answers will vary.

121. No. There are two arithmetic errors. They are the constant in the second equation and the coefficient of z in the third equation.

123. $x = 5, y = 5, \lambda = -5$

125. $x = \dfrac{\sqrt{2}}{2}, y = \dfrac{1}{2}, \lambda = 1$

$x = -\dfrac{\sqrt{2}}{2}, y = \dfrac{1}{2}, \lambda = 1$

$x = 0, y = 0, \lambda = 0$

127.   **129.**

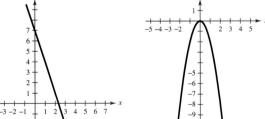

131.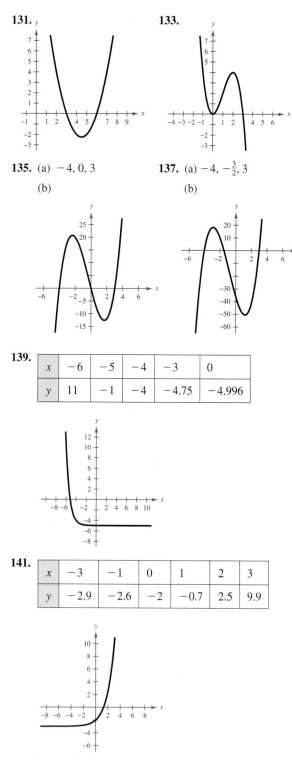

133.

135. (a) $-4, 0, 3$

(b)

137. (a) $-4, -\frac{3}{2}, 3$

(b)

139.

x	-6	-5	-4	-3	0
y	11	-1	-4	-4.75	-4.996

141.

x	-3	-1	0	1	2	3
y	-2.9	-2.6	-2	-0.7	2.5	9.9

143. $\left(4\frac{1}{3}, -2\right)$ **145.** $(40, 40)$

Section 7.4 *(page 532)*

1. (g) **3.** (a) **5.** (e) **7.** (f)

9. **11.**

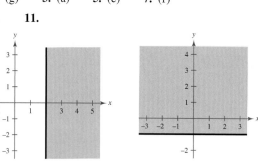

13. **15.**

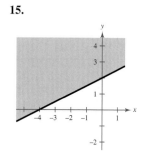

17. **19.**

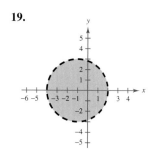

21. **23.**

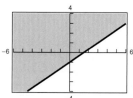

25. **27.**

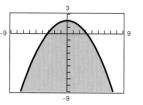

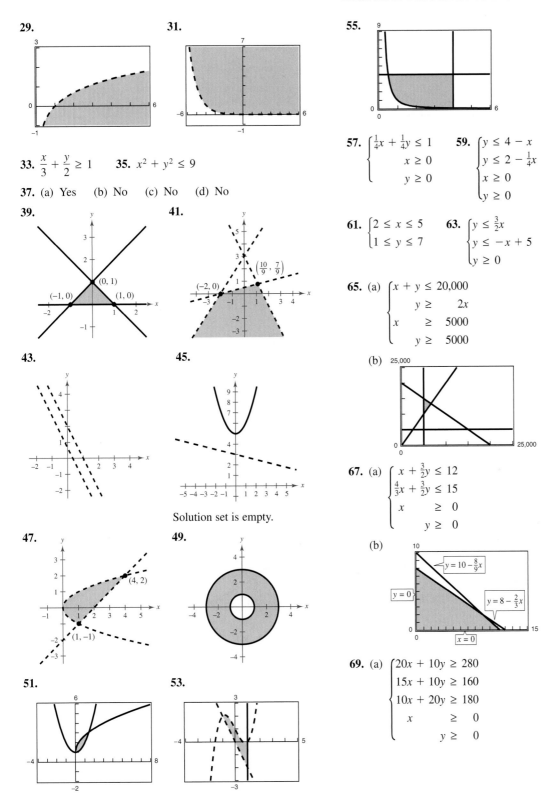

29.

31.

33. $\dfrac{x}{3} + \dfrac{y}{2} \geq 1$ **35.** $x^2 + y^2 \leq 9$

37. (a) Yes   (b) No   (c) No   (d) No

39.

41.

43.

45.

Solution set is empty.

47.

49.

51.

53.

55.

57. $\begin{cases} \frac{1}{4}x + \frac{1}{4}y \leq 1 \\ \quad\; x \geq 0 \\ \quad\; y \geq 0 \end{cases}$ **59.** $\begin{cases} y \leq 4 - x \\ y \leq 2 - \frac{1}{4}x \\ x \geq 0 \\ y \geq 0 \end{cases}$

61. $\begin{cases} 2 \leq x \leq 5 \\ 1 \leq y \leq 7 \end{cases}$ **63.** $\begin{cases} y \leq \frac{3}{2}x \\ y \leq -x + 5 \\ y \geq 0 \end{cases}$

65. (a) $\begin{cases} x + y \leq 20{,}000 \\ \quad\quad y \geq 2x \\ x \quad\quad \geq 5000 \\ \quad\quad y \geq 5000 \end{cases}$

(b)

67. (a) $\begin{cases} x + \frac{3}{2}y \leq 12 \\ \frac{4}{3}x + \frac{3}{2}y \leq 15 \\ x \quad\quad \geq 0 \\ \quad\quad y \geq 0 \end{cases}$

(b)

69. (a) $\begin{cases} 20x + 10y \geq 280 \\ 15x + 10y \geq 160 \\ 10x + 20y \geq 180 \\ \quad x \quad\quad \geq 0 \\ \quad\quad y \geq 0 \end{cases}$

(b)

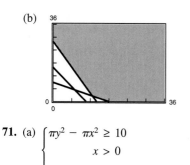

71. (a) $\begin{cases} \pi y^2 - \pi x^2 \geq 10 \\ \quad\quad\quad x > 0 \\ \quad\quad\quad y > x \end{cases}$

(b)

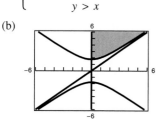

(c) The line is an asymptote to the boundary. The larger the circles, the closer the radii can be and the constraint still be satisfied.

73.

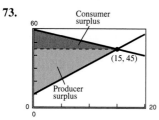

Consumer surplus: 112.5
Producer surplus: 262.5

75.

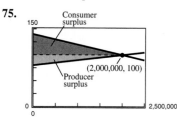

Consumer surplus: 40,000,000
Producer surplus: 20,000,000

77. False. $3x + y^2 \geq 2$ is outside the parabola.

79. Answers will vary.   **81.** $x + 11y + 8 = 0$

83. $x + y + 1.8 = 0$

Section 7.5   *(page 541)*

1. Minimum at $(0, 0)$: 0   **3.** Minimum at $(0, 0)$: 0
    Maximum at $(0, 6)$: 30   Maximum at $(6, 0)$: 60

5. Minimum at $(0, 0)$: 0   **7.** Minimum at $(0, 0)$: 0
    Maximum at $(3, 4)$: 23   Maximum at $(4, 0)$: 20

9. Minimum at $(0, 0)$: 0
    Maximum at $(60, 20)$: 740

11. Minimum at $(0, 0)$: 0
    Maximum at any point on the line segment connecting $(60, 20)$ and $(30, 45)$: 2100

13.

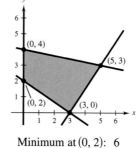

15.

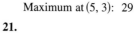

Minimum at $(0, 0)$: 0   Minimum at $(0, 0)$: 0
Maximum at $(5, 0)$: 30   Maximum at $(0, 2)$: 48

17.

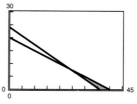

19.

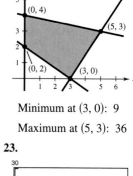

Minimum at $(0, 2)$: 6   Minimum at $(3, 0)$: 9
Maximum at $(5, 3)$: 29   Maximum at $(5, 3)$: 36

21.

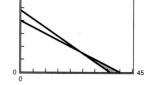

23.

Minimum at $(24, 8)$: 104   Minimum at $(36, 0)$: 36
Maximum at $(40, 0)$: 160   Maximum at $(24, 8)$: 56

25.

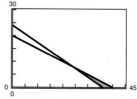

Minimum at any point on the line segment connecting $(24, 8)$ and $(36, 0)$: 72

Maximum at $(40, 0)$: 80

27. (a) and (b)   (c) $(3, 6)$

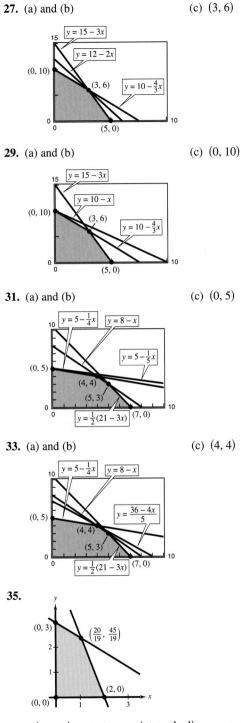

29. (a) and (b)   (c) $(0, 10)$

31. (a) and (b)   (c) $(0, 5)$

33. (a) and (b)   (c) $(4, 4)$

35.

z is maximum at any point on the line segment connecting $(2, 0)$ and $\left(\frac{20}{19}, \frac{45}{19}\right)$.

37.

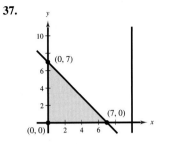

The constraint $x \leq 10$ is extraneous.
Maximum at $(0, 7)$: 14

39.

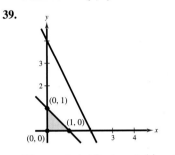

The constraint $2x + y \leq 4$ is extraneous.
Maximum at $(0, 1)$: 4

41. 200 units of the \$250 model; 50 units of the \$400 model
Maximum profit: \$11,500

43. Type A: $\frac{1}{6}$ gallon; Type B: $\frac{5}{6}$ gallon
Minimum cost: \$1.255 per gallon

45. No audits; 40 tax returns
Maximum revenue: \$12,000

47. 1000 units of Model A; 500 units of Model B
Maximum profit: \$76,000

49. True   **51.** $z = x + 5y$ (Answer is not unique.)

53. $z = 4x + y$ (Answer is not unique.)

55. (a) $t > 9$ (b) $\frac{3}{4} < t < 9$ **57.** 1.099

59. 14.550   **61.** 1.851   **63.** $(-2, -3)$ **65.** $(-1, 2)$

67.   **69.**

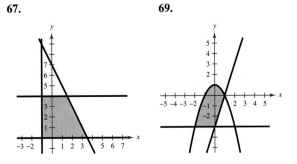

Review Exercises *(page 546)*

1. $(1, 1)$ **3.** $(5, 4)$ **5.** $(0, 0), (2, 8), (-2, 8)$

7. $\left(2, -\frac{1}{2}\right)$ **9.** $(0, 0), (-3, 3)$ **11.** $(4, 4)$

13. 4762 units   **15.** 96×144 meters   **17.** $\left(\frac{5}{2}, 3\right)$

19. $(-0.5, 0.8)$ **21.** $(0, 0)$ **23.** $\left(\frac{14}{5} + \frac{8}{5}a, a\right)$

25.

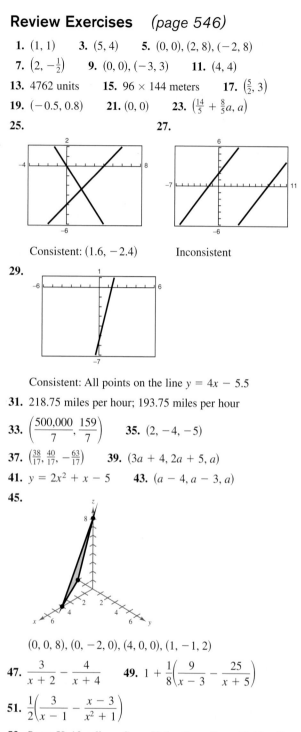

Consistent: $(1.6, -2.4)$

27.

Inconsistent

29.

Consistent: All points on the line $y = 4x - 5.5$

31. 218.75 miles per hour; 193.75 miles per hour

33. $\left(\dfrac{500{,}000}{7}, \dfrac{159}{7}\right)$ **35.** $(2, -4, -5)$

37. $\left(\dfrac{38}{17}, \dfrac{40}{17}, -\dfrac{63}{17}\right)$ **39.** $(3a + 4, 2a + 5, a)$

41. $y = 2x^2 + x - 5$ **43.** $(a - 4, a - 3, a)$

45.

$(0, 0, 8), (0, -2, 0), (4, 0, 0), (1, -1, 2)$

47. $\dfrac{3}{x + 2} - \dfrac{4}{x + 4}$ **49.** $1 + \dfrac{1}{8}\left(\dfrac{9}{x - 3} - \dfrac{25}{x + 5}\right)$

51. $\dfrac{1}{2}\left(\dfrac{3}{x - 1} - \dfrac{x - 3}{x^2 + 1}\right)$

53. Spray X: 10 gallons; Spray Y: 5 gallons; Spray Z: 12 gallons

55.

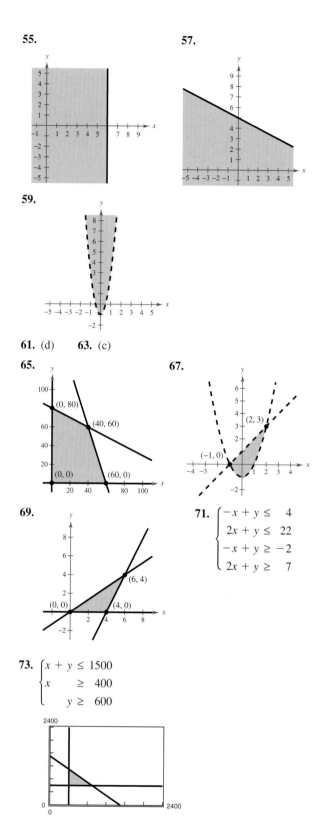

57.

59.

61. (d)   **63.** (c)

65.

67.

69.

71. $\begin{cases} -x + y \le 4 \\ 2x + y \le 22 \\ -x + y \ge -2 \\ 2x + y \ge 7 \end{cases}$

73. $\begin{cases} x + y \le 1500 \\ x \ge 400 \\ y \ge 600 \end{cases}$

不

75.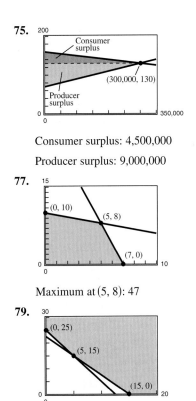

Consumer surplus: 4,500,000

Producer surplus: 9,000,000

77.

Maximum at $(5, 8)$: 47

79.

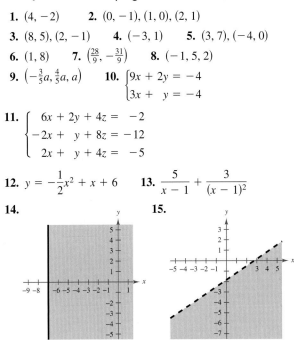

Minimum at $(15, 0)$: 26.25

81. 20 perms, 2 haircuts; Maximum revenue: $1234

83. False. There is no solution for the system.

Chapter Test *(page 550)*

1. $(4, -2)$ **2.** $(0, -1), (1, 0), (2, 1)$

3. $(8, 5), (2, -1)$ **4.** $(-3, 1)$ **5.** $(3, 7), (-4, 0)$

6. $(1, 8)$ **7.** $\left(\frac{28}{9}, -\frac{31}{9}\right)$ **8.** $(-1, 5, 2)$

9. $\left(-\frac{3}{5}a, \frac{4}{5}a, a\right)$ **10.** $\begin{cases} 9x + 2y = -4 \\ 3x + y = -4 \end{cases}$

11. $\begin{cases} 6x + 2y + 4z = -2 \\ -2x + y + 8z = -12 \\ 2x + y + 4z = -5 \end{cases}$

12. $y = -\frac{1}{2}x^2 + x + 6$ **13.** $\dfrac{5}{x-1} + \dfrac{3}{(x-1)^2}$

14.   **15.**

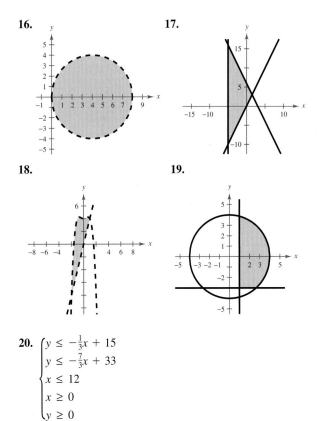

16.   **17.**

18.   **19.**

20. $\begin{cases} y \le -\frac{1}{3}x + 15 \\ y \le -\frac{7}{3}x + 33 \\ x \le 12 \\ x \ge 0 \\ y \ge 0 \end{cases}$

21. Maximum at $(12, 0)$: 240

22. Stock 160 units of the $275 model and 140 units of the $400 model.

Chapter 8

Section 8.1 *(page 562)*

1. 3×2 **3.** 3×1 **5.** 2×2

7. $\begin{bmatrix} 4 & -5 & \vdots & 33 \\ -1 & 5 & \vdots & -27 \end{bmatrix}$ **9.** $\begin{bmatrix} 1 & 10 & -2 & \vdots & 2 \\ 5 & -3 & 4 & \vdots & 0 \\ 2 & 1 & 0 & \vdots & 6 \end{bmatrix}$

11. $\begin{cases} x + 2y = 7 \\ 2x - 3y = 4 \end{cases}$ **13.** $\begin{cases} 9x + 12y + 3z = 0 \\ -2x + 18y + 5z + 2w = 10 \\ x + 7y - 8z = -4 \end{cases}$

15. Reduced row-echelon form

17. Not in row-echelon form   **19.** $\begin{bmatrix} 1 & 4 & 3 \\ 0 & 2 & -1 \end{bmatrix}$

21. $\begin{bmatrix} 1 & 1 & 4 & -1 \\ 0 & 5 & -2 & 6 \\ 0 & 3 & 20 & 4 \end{bmatrix}, \begin{bmatrix} 1 & 1 & 4 & -1 \\ 0 & 1 & -\frac{2}{5} & \frac{6}{5} \\ 0 & 3 & 20 & 4 \end{bmatrix}$

23. Add 5 times R_2 to R_1.   **25.** Interchange R_1 and R_2.

27. (a) $\begin{bmatrix} 1 & 2 & 3 \\ 0 & -5 & -10 \\ 3 & 1 & -1 \end{bmatrix}$ (b) $\begin{bmatrix} 1 & 2 & 3 \\ 0 & -5 & -10 \\ 0 & -5 & -10 \end{bmatrix}$

(c) $\begin{bmatrix} 1 & 2 & 3 \\ 0 & -5 & -10 \\ 0 & 0 & 0 \end{bmatrix}$ (d) $\begin{bmatrix} 1 & 2 & 3 \\ 0 & 1 & 2 \\ 0 & 0 & 0 \end{bmatrix}$

(e) $\begin{bmatrix} 1 & 0 & -1 \\ 0 & 1 & 2 \\ 0 & 0 & 0 \end{bmatrix}$

The matrix is in reduced row-echelon form.

29. (a)   (b)

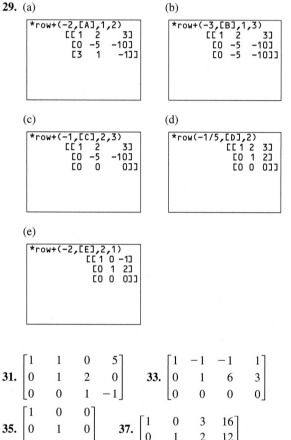

(c)   (d)

(e)

31. $\begin{bmatrix} 1 & 1 & 0 & 5 \\ 0 & 1 & 2 & 0 \\ 0 & 0 & 1 & -1 \end{bmatrix}$ **33.** $\begin{bmatrix} 1 & -1 & -1 & 1 \\ 0 & 1 & 6 & 3 \\ 0 & 0 & 0 & 0 \end{bmatrix}$

35. $\begin{bmatrix} 1 & 0 & 0 \\ 0 & 1 & 0 \\ 0 & 0 & 1 \end{bmatrix}$ **37.** $\begin{bmatrix} 1 & 0 & 3 & 16 \\ 0 & 1 & 2 & 12 \end{bmatrix}$

39. $\begin{cases} x - 2y = 4 \\ y = -3 \end{cases}$ **41.** $\begin{cases} x - y + 2z = 4 \\ y - z = 2 \\ z = -2 \end{cases}$

$(-2, -3)$ $(8, 0, -2)$

43. $(7, -5)$ **45.** $(-4, -8, 2)$ **47.** $(3, 2)$

49. $(6, -2)$ **51.** $\left(\frac{3}{2}, -\frac{1}{4}\right)$ **53.** Inconsistent

55. $(4, -3, 2)$ **57.** $(2a + 1, 3a + 2, a)$ **59.** $(0, 0)$

61. $(0, 2 - 4a, a)$ **63.** $(1, 0, 4, -2)$ **65.** $(-2a, a, a)$

67. Yes; $(-1, 1, -3)$ **69.** No **71.** $y = x^2 + 2x + 5$

73. $\frac{1}{4}x^3 + x^2 - x - 2$ **75.** $y = -x^4 + 4x^2$

77. \$800,000 at 8%, \$500,000 at 9%, \$200,000 at 12%

79. $I_1 = \frac{13}{10}, I_2 = \frac{11}{5}, I_3 = \frac{9}{10}$

81. (a) $y = -128.5t^2 + 1587.5t - 4304$

(b)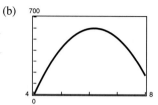

(c) \$172 million

83. (a) $x_1 = s, x_2 = t, x_3 = 600 - s,$

$x_4 = s - t, x_5 = 500 - t, x_6 = s, x_7 = t$

(b) $x_1 = 0, x_2 = 0, x_3 = 600, x_4 = 0, x_5 = 500,$

$x_6 = 0, x_7 = 0$

(c) $x_1 = 0, x_2 = -500, x_3 = 600, x_4 = 500,$

$x_5 = 1000, x_6 = 0, x_7 = -500$

85. (a) $x_1 = 100 + t, x_2 = -100 + t, x_3 = 200 + t, x_4 = t$

(b) $x_1 = 100, x_2 = -100, x_3 = 200, x_4 = 0$

(c) $x_1 = 200, x_2 = 0, x_3 = 300, x_4 = 100$

87. False. Rows are not arranged.

89. $\begin{cases} x + y + 7z = -1 \\ x + 2y + 11z = 0 \\ 2x + y + 10z = -3 \end{cases}$

(Answer is not unique.)

91. (a) There exists a row with all zeros except for the entry in the last column.

(b) There are fewer rows with nonzero entries than variables.

93.   **95.**

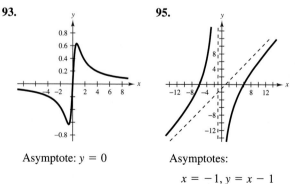

Asymptote: $y = 0$ Asymptotes:

$x = -1, y = x - 1$

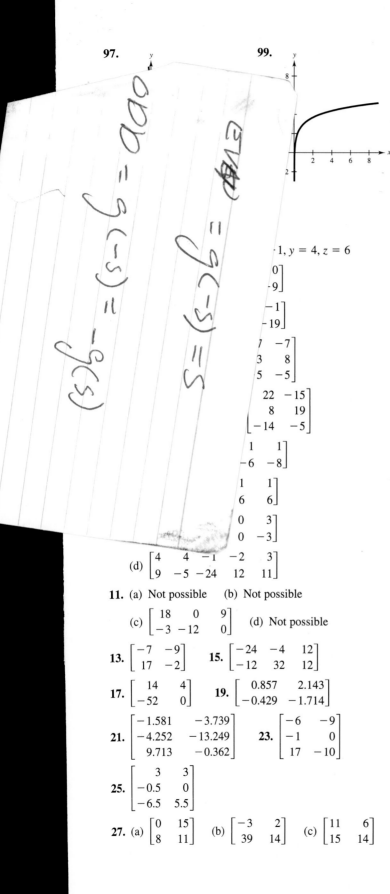

97.

99.

$x = 1, y = 4, z = 6$

0
-9

-1
-19

7 -7
3   8
5 -5

22 -15
8   19
-14 -5

1   1
-6 -8

1   1
6   6

0   3
0 -3

(d) $\begin{bmatrix} 4 & 4 & -1 & -2 & 3 \\ 9 & -5 & -24 & 12 & 11 \end{bmatrix}$

11. (a) Not possible (b) Not possible

(c) $\begin{bmatrix} 18 & 0 & 9 \\ -3 & -12 & 0 \end{bmatrix}$ (d) Not possible

13. $\begin{bmatrix} -7 & -9 \\ 17 & -2 \end{bmatrix}$ **15.** $\begin{bmatrix} -24 & -4 & 12 \\ -12 & 32 & 12 \end{bmatrix}$

17. $\begin{bmatrix} 14 & 4 \\ -52 & 0 \end{bmatrix}$ **19.** $\begin{bmatrix} 0.857 & 2.143 \\ -0.429 & -1.714 \end{bmatrix}$

21. $\begin{bmatrix} -1.581 & -3.739 \\ -4.252 & -13.249 \\ 9.713 & -0.362 \end{bmatrix}$ **23.** $\begin{bmatrix} -6 & -9 \\ -1 & 0 \\ 17 & -10 \end{bmatrix}$

25. $\begin{bmatrix} 3 & 3 \\ -0.5 & 0 \\ -6.5 & 5.5 \end{bmatrix}$

27. (a) $\begin{bmatrix} 0 & 15 \\ 8 & 11 \end{bmatrix}$ (b) $\begin{bmatrix} -3 & 2 \\ 39 & 14 \end{bmatrix}$ (c) $\begin{bmatrix} 11 & 6 \\ 15 & 14 \end{bmatrix}$

29. (a) $\begin{bmatrix} 0 & -10 \\ 10 & 0 \end{bmatrix}$ (b) $\begin{bmatrix} 0 & -10 \\ 10 & 0 \end{bmatrix}$ (c) $\begin{bmatrix} 8 & -6 \\ 6 & 8 \end{bmatrix}$

31. (a) $\begin{bmatrix} 6 & -21 & 15 \\ 8 & -23 & 19 \\ 4 & 7 & 5 \end{bmatrix}$ (b) $\begin{bmatrix} 9 & 0 & 13 \\ 7 & -2 & 21 \\ 1 & 4 & -19 \end{bmatrix}$

(c) $\begin{bmatrix} 20 & 7 & -8 \\ 24 & 7 & -2 \\ 2 & -5 & 30 \end{bmatrix}$

33. Not possible **35.** $\begin{bmatrix} -2 & 51 \\ -8 & 33 \\ 0 & 27 \end{bmatrix}$ **37.** $\begin{bmatrix} 1 & 0 & 0 \\ 0 & 1 & 0 \\ 0 & 0 & \frac{7}{2} \end{bmatrix}$

39. $\begin{bmatrix} 40 & -20 & -10 & 80 \\ 44 & -22 & -11 & 88 \end{bmatrix}$ **41.** $\begin{bmatrix} 70 & -17 & 73 \\ 32 & 11 & 6 \\ 16 & -38 & 70 \end{bmatrix}$

43. $\begin{bmatrix} 151 & 25 & 48 \\ 516 & 279 & 387 \\ 47 & -20 & 87 \end{bmatrix}$ **45.** Not possible

47. $\begin{bmatrix} 5 & 8 \\ -4 & -16 \end{bmatrix}$ **49.** $\begin{bmatrix} -4 & 10 \\ 3 & 14 \end{bmatrix}$

51. (b) **53.** (b) **55.** (b)

57. (a) $\begin{bmatrix} -1 & 1 \\ -2 & 1 \end{bmatrix}\begin{bmatrix} x_1 \\ x_2 \end{bmatrix} = \begin{bmatrix} 4 \\ 0 \end{bmatrix}$ (b) $\begin{bmatrix} 4 \\ 8 \end{bmatrix}$

59. (a) $\begin{bmatrix} -2 & -3 \\ 6 & 1 \end{bmatrix}\begin{bmatrix} x_1 \\ x_2 \end{bmatrix} = \begin{bmatrix} -4 \\ -36 \end{bmatrix}$ (b) $\begin{bmatrix} -7 \\ 6 \end{bmatrix}$

61. (a) $\begin{bmatrix} 1 & -2 & 3 \\ -1 & 3 & -1 \\ 2 & -5 & 5 \end{bmatrix}\begin{bmatrix} x_1 \\ x_2 \\ x_3 \end{bmatrix} = \begin{bmatrix} 9 \\ -6 \\ 17 \end{bmatrix}$ (b) $\begin{bmatrix} 1 \\ -1 \\ 2 \end{bmatrix}$

63. (a) $\begin{bmatrix} 1 & -5 & 2 \\ -3 & 1 & -1 \\ 0 & -2 & 5 \end{bmatrix}\begin{bmatrix} x_1 \\ x_2 \\ x_3 \end{bmatrix} = \begin{bmatrix} -20 \\ 8 \\ -16 \end{bmatrix}$ (b) $\begin{bmatrix} -1 \\ 3 \\ -2 \end{bmatrix}$

65. $\begin{bmatrix} -4 & 0 \\ 8 & 2 \end{bmatrix}$ **67.** $\begin{bmatrix} 0 & 0 & 0 \\ 0 & 0 & 0 \\ 0 & 0 & 0 \end{bmatrix}$

69. Not possible **71.** Not possible **73.** 2×2

75. Not possible **77.** 2×3 **79.** $\begin{bmatrix} 72 & 48 & 24 \\ 36 & 108 & 72 \end{bmatrix}$

81. $BA = \begin{bmatrix} \$1250 & \$1331.25 & \$981.25 \end{bmatrix}$

Each entry represents the total profit from the two products at one of the three outlets.

83. $\begin{bmatrix} \$15,770 & \$18,300 \\ \$26,500 & \$29,250 \\ \$21,260 & \$24,150 \end{bmatrix}$

The entries are the total wholesale and retail prices of the inventory at each outlet.

85. $\begin{bmatrix} 0.40 & 0.15 & 0.15 \\ 0.28 & 0.53 & 0.17 \\ 0.32 & 0.32 & 0.68 \end{bmatrix}$

P^2 represents the changes in party affiliations after two elections.

87. True. To add two matrices, you add corresponding entries.

89. False. The product is $\begin{bmatrix} 2 & 2 \\ -3 & -3 \\ 7 & 3 \end{bmatrix}$.

91. $AB = 0$, but $A \neq 0$ and $B \neq 0$.   **93.** $A^2 = \begin{bmatrix} 1 & 0 \\ 0 & 1 \end{bmatrix}$

95. (a) $A = \begin{bmatrix} 0 & 2 \\ 0 & 0 \end{bmatrix}$, $B = \begin{bmatrix} 0 & 2 & 3 \\ 0 & 0 & 4 \\ 0 & 0 & 0 \end{bmatrix}$

(Answers will vary.)

(b) A^2 and B^3 are zero matrices.

(c) $A = \begin{bmatrix} 0 & 2 & 3 & 4 \\ 0 & 0 & 5 & 6 \\ 0 & 0 & 0 & 7 \\ 0 & 0 & 0 & 0 \end{bmatrix}$

A^4 is the zero matrix.

(d) A^n is the zero matrix.

97. $\frac{3}{2}, -\frac{1}{4}$ **99.** $-9, 0, \frac{5}{3}$ **101.** $\frac{5}{2}, \pm\sqrt{6}$

103. $2 + \log_2 3 - \log_2 x$

105. $\ln(x + 3) + \ln(x - 3) - 4 \ln x$ **107.** $\ln \dfrac{x}{(x^2 - 36)^3}$

109. $\ln(7^{3/2} t^3)$ **111.** $(-1, 1, 3)$

Section 8.3 *(page 588)*

1.–11. Answers will vary.

13. $\begin{bmatrix} \frac{1}{2} & 0 \\ 0 & \frac{1}{3} \end{bmatrix}$ **15.** $\begin{bmatrix} -3 & 2 \\ -2 & 1 \end{bmatrix}$ **17.** $\begin{bmatrix} 1 & -1 \\ 2 & -1 \end{bmatrix}$

19. Does not exist

21. $\begin{bmatrix} 1 & 1 & -1 \\ -3 & 2 & -1 \\ 3 & -3 & 2 \end{bmatrix}$ **23.** $\begin{bmatrix} 1 & 0 & 0 \\ -0.75 & 0.25 & 0 \\ 0.35 & -0.25 & 0.2 \end{bmatrix}$

25. $\begin{bmatrix} -\frac{1}{8} & 0 & 0 & 0 \\ 0 & 1 & 0 & 0 \\ 0 & 0 & \frac{1}{4} & 0 \\ 0 & 0 & 0 & -\frac{1}{5} \end{bmatrix}$ **27.** $\begin{bmatrix} -175 & 37 & -13 \\ 95 & -20 & 7 \\ 14 & -3 & 1 \end{bmatrix}$

29. $\frac{1}{2}\begin{bmatrix} -3 & 3 & 2 \\ 9 & -7 & -6 \\ -2 & 2 & 2 \end{bmatrix}$ **31.** $\begin{bmatrix} -12 & -5 & -9 \\ -4 & -2 & -4 \\ -8 & -4 & -6 \end{bmatrix}$

33. $\frac{5}{11}\begin{bmatrix} 0 & -4 & 2 \\ -22 & 11 & 11 \\ 22 & -6 & -8 \end{bmatrix}$ **35.** Does not exist

37. $\begin{bmatrix} 1 & 0 & 1 & 0 \\ 0 & 1 & 0 & 1 \\ 2 & 0 & 1 & 0 \\ 0 & 1 & 0 & 2 \end{bmatrix}$ **39.** Does not exist

41. $\frac{1}{59}\begin{bmatrix} 16 & 15 \\ -4 & 70 \end{bmatrix}$ **43.** $(5, 0)$ **45.** $(-8, -6)$

47. $(3, 8, -11)$ **49.** $(2, 1, 0, 0)$ **51.** $(2, -2)$

53. No solution   **55.** $(-4, -8)$ **57.** $(-1, 3, 2)$

59. $(0.3125s + 0.8125, 1.1875s + 0.6875, s)$

61. $(-7, 3, -2)$ **63.** $(5, 0, -2, 3)$

65. \$10,000 in AAA-rated bonds, \$5000 in A-rated bonds, \$10,000 in B-rated bonds

67. \$9000 in AAA-rated bonds, \$1000 in A-rated bonds, \$2000 in B-rated bonds

69. $I_1 = -3$ amperes, $I_2 = 8$ amperes, $I_3 = 5$ amperes

71. True   **73.** True

75. Answers will vary.

(a) $\frac{1}{19}\begin{bmatrix} 3 & 2 \\ -2 & 5 \end{bmatrix}$ (b) $\frac{1}{61}\begin{bmatrix} -5 & -12 \\ 8 & 7 \end{bmatrix}$

77. $-5, \pm\sqrt{6}$ **79.** $-3, 4$

81.

x	1	2	3	5	6	7	8
$f(x)$	2.99	2.96	2.89	2.0	0	-6	-24

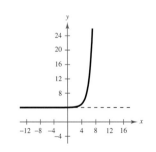

83.

x	0	2	3	5	6	7
$f(x)$	4.02	4.14	4.37	6.7	11.4	24

85. $\begin{bmatrix} 12 & -18 \\ -6 & 24 \\ -3 & -36 \end{bmatrix}$ **87.** $\begin{bmatrix} 7 & -27 \\ -43 & 24 \end{bmatrix}$ **89.** $\begin{bmatrix} 6 & -2 \\ 15 & 7 \end{bmatrix}$

Section 8.4   *(page 596)*

1. 7   **3.** -25 **5.** 28   **7.** -24 **9.** 0

11. -9 **13.** -0.002

15. (a) $M_{11} = -5, M_{12} = 2, M_{21} = 4, M_{22} = 3$

(b) $C_{11} = -5, C_{12} = -2, C_{21} = -4, C_{22} = 3$

17. (a) $M_{11} = 30, M_{12} = 12, M_{13} = 11, M_{21} = -36,$
$M_{22} = 26, M_{23} = 7, M_{31} = -4, M_{32} = -42,$
$M_{33} = 12$

(b) $C_{11} = 30, C_{12} = -12, C_{13} = 11, C_{21} = 36, C_{22} = 26,$
$C_{23} = -7, C_{31} = -4, C_{32} = 42, C_{33} = 12$

19. (a) -75 (b) -75 **21.** (a) 170   (b) 170

23. -58 **25.** -30 **27.** -168 **29.** 412

31. -60 **33.** 216   **35.** -336 **37.** 410

39. (a) -3 (b) -2 (c) $\begin{bmatrix} -2 & 0 \\ 0 & -3 \end{bmatrix}$ (d) 6

41. (a) 2   (b) -6 (c) $\begin{bmatrix} 1 & 4 & 3 \\ -1 & 0 & 3 \\ 0 & 2 & 0 \end{bmatrix}$ (d) -12

43. (a) -25 (b) -220

(c) $\begin{bmatrix} -7 & -16 & -1 & -28 \\ -4 & -14 & -11 & 8 \\ 13 & 4 & 4 & -4 \\ -2 & 3 & 2 & 2 \end{bmatrix}$ (d) 5500

45.–49. Answers will vary.   **51.** $-1, 4$ **53.** $8uv - 1$

55. e^{5x} **57.** $1 - \ln x$ **59.** True

61. $A = \begin{bmatrix} 1 & 3 \\ -2 & 4 \end{bmatrix}, B = \begin{bmatrix} -4 & 0 \\ 3 & 5 \end{bmatrix}$

$|A + B| = -30, |A| + |B| = -10$
(Answers will vary.)

63. A square matrix is a square array of numbers. A determinant of a square matrix is a real number.

65. (a) Columns 2 and 3 are interchanged.

(b) Rows 1 and 3 are interchanged.

67. (a) 5 is factored from the first row of the matrix.

(b) 4 and 3 are factored from the second and third columns.

69. $(-1, 4)$ **71.** $(0, -0.2)$

73.

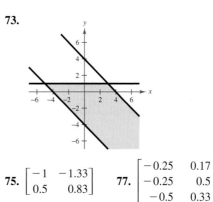

75. $\begin{bmatrix} -1 & -1.33 \\ 0.5 & 0.83 \end{bmatrix}$ **77.** $\begin{bmatrix} -0.25 & 0.17 & 0.33 \\ -0.25 & 0.5 & 1 \\ -0.5 & 0.33 & 1.67 \end{bmatrix}$

Section 8.5   *(page 607)*

1. 14   **3.** 7   **5.** $\frac{33}{8}$ **7.** 10   **9.** $x = \frac{16}{5}, 0$

11. Collinear   **13.** Not collinear   **15.** $x = -3$

17. $(2, -2)$ **19.** $(-3, -2)$ **21.** $\left(\frac{32}{7}, \frac{30}{7}\right)$

23. $(-1, 3, 2)$ **25.** $\left(0, -\frac{1}{2}, \frac{1}{2}\right)$ **27.** 250 square miles

29. Uncoded: $[20\ 18\ 15], [21\ 2\ 12], [5\ 0\ 9], [14\ 0\ 18],$
$[9\ 22\ 5], [18\ 0\ 3], [9\ 20\ 25]$

Encoded: $-52\ 10\ 27\ -49\ 3\ 34\ -49\ 13\ 27\ -94\ 22$
$54\ 1\ 1\ -7\ 0\ -12\ 9\ -121\ 41\ 55$

31. $28\ 34\ 5\ 10\ -6\ -45\ -15\ -70\ -130\ 23\ 51\ 61\ 44$
$75\ 63\ 80\ 185\ 235$

33. $-5\ -41\ -87\ 91\ 207\ 257\ 11\ -5\ -41\ 40\ 80\ 84\ 76$
$177\ 227$

35. HAPPY NEW YEAR   **37.** SEND PLANES

39. MEET ME TONIGHT RON   **41.** True

43. Answers will vary.   **45.** $8x + y + 6 = 0$

47. $5x + 4y - 28 = 0$

49.

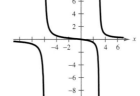

51. $(2, -2, 5)$

Review Exercises   *(page 610)*

1. 3×1 **3.** 1×1 **5.** 4×2

7. $\begin{bmatrix} 3 & -10 & \vdots & 15 \\ 5 & 4 & \vdots & 22 \end{bmatrix}$ **9.** $\begin{bmatrix} 8 & -7 & 4 & \vdots & 12 \\ 3 & -5 & 2 & \vdots & 20 \\ 5 & 3 & -3 & \vdots & 26 \end{bmatrix}$

11. $\begin{cases} 5x + y + 7z = -9 \\ 4x + 2y = 10 \\ 9x + 4y + 2z = 3 \end{cases}$

13. $\begin{bmatrix} 1 & 0 & 0 \\ 0 & 1 & 0 \\ 0 & 0 & 1 \end{bmatrix}$　**15.** $\begin{bmatrix} 1 & 0 & 3 & -2 \\ 0 & 1 & 4 & -3 \end{bmatrix}$

17. $\begin{bmatrix} 1 & 0 & 1 \\ 0 & 1 & 1 \\ 0 & 0 & 0 \end{bmatrix}$　**19.** $(10, -12)$　**21.** $(-0.2, 0.7)$

23. $\left(\frac{1}{2}, -\frac{1}{3}, 1\right)$　**25.** $\left(-2a - 5, \frac{5}{3}a + \frac{2}{3}, a\right)$

27. $(2, -3, 3)$　**29.** $(2, 3, -1)$　**31.** $(1, 1)$

33. $(3, 0, -4)$　**35.** $(2, 6, -10, -3)$

37. $x = 12, y = -7$　**39.** $x = 1, y = 11$

41. $\begin{bmatrix} 17 & -17 \\ 13 & 2 \end{bmatrix}$　**43.** $\begin{bmatrix} -13 & -8 & 18 \\ 0 & 11 & -19 \end{bmatrix}$

45. $\begin{bmatrix} 2 & 1 & 12 \\ -13 & 1 & -17 \\ -30 & -54 & 40 \end{bmatrix}$　**47.** $\begin{bmatrix} 48 & -18 & -3 \\ 15 & 51 & 33 \end{bmatrix}$

49. $\begin{bmatrix} -14 & -4 \\ 7 & -17 \\ -17 & -2 \end{bmatrix}$　**51.** $\frac{1}{3}\begin{bmatrix} 9 & 2 \\ -4 & 11 \\ 10 & 0 \end{bmatrix}$

53. $\begin{bmatrix} 14 & -2 & 8 \\ 14 & -10 & 40 \\ 36 & -12 & 48 \end{bmatrix}$　**55.** $\begin{bmatrix} 44 & 4 \\ 20 & 8 \end{bmatrix}$

57. $\begin{bmatrix} 1 & 17 \\ 12 & 36 \end{bmatrix}$　**59.** $\begin{bmatrix} 14 & -22 & 22 \\ 19 & -41 & 80 \\ 42 & -66 & 66 \end{bmatrix}$

61. $\begin{cases} 5x + 4y = 2 \\ -x + y = -22 \end{cases}$

63. $BA = [\$274,150 \quad \$303,150]$

The entries represent the total value of the units shipped to each warehouse.

65. and 67. Answers will vary.

69. $\begin{bmatrix} 4 & -5 \\ 5 & -6 \end{bmatrix}$　**71.** $\begin{bmatrix} 13 & 6 & -4 \\ -12 & -5 & 3 \\ 5 & 2 & -1 \end{bmatrix}$　**73.** $\begin{bmatrix} \frac{1}{5} & \frac{1}{5} \\ \frac{1}{10} & -\frac{1}{15} \end{bmatrix}$

75. $\begin{bmatrix} \frac{1}{2} & -1 & -\frac{1}{2} \\ \frac{1}{2} & -\frac{2}{3} & -\frac{5}{6} \\ 0 & \frac{2}{3} & \frac{1}{3} \end{bmatrix}$　**77.** $\begin{bmatrix} 1 & -1 \\ 4 & -\frac{7}{2} \end{bmatrix}$　**79.** $\begin{bmatrix} -1 & -\frac{5}{3} \\ 1 & 2 \end{bmatrix}$

81. Does not exist　**83.** $(36, 11)$　**85.** $(-6, -1)$

87. $(2, -1, -2)$　**89.** $(6, 1, -1)$　**91.** $(-3, 1)$

93. $(1, 1, -2)$

95. No inverse. The system is inconsistent.

97. -42　**99.** 550

101. (a) $M_{11} = 4, M_{12} = 7, M_{21} = -1, M_{22} = 2$

(b) $C_{11} = 4, C_{12} = -7, C_{21} = 1, C_{22} = 2$

103. (a) $M_{11} = 30, M_{12} = -12, M_{13} = -21, M_{21} = 20,$
$M_{22} = 19, M_{23} = 22, M_{31} = 5, M_{32} = -2,$
$M_{33} = 19$

(b) $C_{11} = 30, C_{12} = 12, C_{13} = -21, C_{21} = -20,$
$C_{22} = 19, C_{23} = -22, C_{31} = 5, C_{32} = 2, C_{33} = 19$

105. (a) 96　(b) 96　**107.** 130　**109.** -3

111. 279　**113.** 0　**115.** -96　**117.** 16

119. 1.75　**121.** Collinear　**123.** Not collinear

125. $(1, 2)$　**127.** $(4, 7)$　**129.** $(-1, 4, 5)$

131. $\left(\frac{2}{3}, \frac{1}{2}\right)$　**133.** Cramer's rule does not apply.

135. 8 carnations, 4 roses　**137.** $y = x^2 + 2x + 3$

139. 16,667 units

141. Uncoded: $[12 \ 15 \ 15], [11 \ 0 \ 15], [21 \ 20 \ 0],$
$[2 \ 5 \ 12], [15 \ 23 \ 0]$

Encoded: $-21 \ 6 \ 0 \ -68 \ 8 \ 45 \ 102 \ -42 \ -60 \ -53$
$20 \ 21 \ 99 \ -30 \ -69$

143. SEE YOU FRIDAY

145. False. Only square matrices have determinants.

147. Elementary row operations correspond to the operations on a system of equations.

Chapter Test　*(page 616)*

1. $\begin{bmatrix} 1 & 0 & 0 \\ 0 & 1 & 0 \\ 0 & 0 & 1 \end{bmatrix}$　**2.** $\begin{bmatrix} 1 & 0 & -1 & 2 \\ 0 & 1 & 0 & -1 \\ 0 & 0 & 0 & 0 \\ 0 & 0 & 0 & 0 \end{bmatrix}$

3. $(-2a + 1.5, 2a + 1, a)$　**4.** $(5, 2, -6)$

5. (a) $\begin{bmatrix} 1 & 5 & -2 \\ 0 & -4 & 3 \end{bmatrix}$　(b) $\begin{bmatrix} 15 & 12 & 12 \\ -12 & -12 & 0 \end{bmatrix}$

(c) $\begin{bmatrix} 7 & 14 & 0 \\ -4 & -12 & 6 \end{bmatrix}$

6. $\begin{bmatrix} 8 & -8 \\ 16 & -4 \\ 6 & 12 \end{bmatrix}$　**7.** $A^{-1} = \begin{bmatrix} \frac{1}{2} & \frac{2}{5} \\ 1 & \frac{3}{5} \end{bmatrix}$; $(13, 22)$

8. 67　**9.** -2　**10.** 300　**11.** 7

12. $\left(\frac{3}{4}, -\frac{1}{2}\right)$

13. Cramer's rule does not apply. System is inconsistent.

14. $y = -\frac{1}{2}x^2 + x + 2$

15. $x_1 = 700 - s - t, \ x_2 = 300 - s - t, \ x_3 = s,$
$x_4 = 100 - t, \ x_5 = t$

Chapter 9

Section 9.1 *(page 625)*

1. 7, 9, 11, 13, 15　　**3.** 2, 4, 8, 16, 32

5. $-2, 4, -8, 16, -32$　　**7.** $2, \frac{3}{2}, \frac{4}{3}, \frac{5}{4}, \frac{6}{5}$

9. $3, \frac{12}{11}, \frac{9}{13}, \frac{24}{47}, \frac{15}{37}$　　**11.** $0, 1, 0, \frac{1}{2}, 0$

13. $\frac{5}{2}, \frac{11}{4}, \frac{23}{8}, \frac{47}{16}, \frac{95}{32}$　　**15.** $1, \frac{1}{2^{3/2}}, \frac{1}{3^{3/2}}, \frac{1}{4^{3/2}}, \frac{1}{5^{3/2}}$

17. $3, \frac{9}{2}, \frac{9}{2}, \frac{27}{8}, \frac{81}{40}$　　**19.** $-1, \frac{1}{4}, -\frac{1}{9}, \frac{1}{16}, -\frac{1}{25}$

21. 3, 15, 35, 63, 99　　**23.** -73　　**25.** $\frac{4}{14,175}$

27. $\frac{48}{285} = \frac{16}{95}$　　**29.** 28, 24, 20, 16, 12

31. 3, 4, 6, 10, 18　　**33.** 2, 6, 10, 22, 42

35.

37.

39.

41. 9, 15, 21, 27, 33, 39, 45, 51, 57, 63

43. $6, 18, 36, 54, \frac{324}{5}, \frac{324}{5}, \frac{1944}{35}, \frac{1458}{35}, \frac{972}{35}, \frac{2916}{175}$

45. $3, \frac{5}{2}, \frac{7}{3}, \frac{9}{4}, \frac{11}{5}, \frac{13}{6}, \frac{15}{7}, \frac{17}{8}, \frac{19}{9}, \frac{21}{10}$　　**47.** (c)　　**49.** (d)

51. $a_n = 3n - 2$　　**53.** $a_n = n^2 - 1$　　**55.** $a_n = \frac{n+1}{n+2}$

57. $a_n = \frac{(-1)^{n+1}}{2^n}$　　**59.** $a_n = 1 + \frac{1}{n}$　　**61.** $a_n = \frac{1}{n!}$

63. $(-1)^n + 2(1)^n = (-1)^n + 2$

65. $6, 8, 10, 12, 14; a_n = 2n + 4$

67. $81, 27, 9, 3, 1; a_n = \frac{243}{3^n}$　　**69.** $\frac{1}{120}$　　**71.** 90

73. 495　　**75.** $n + 1$　　**77.** $\frac{1}{2n(2n+1)}$　　**79.** 35

81. 40　　**83.** 30　　**85.** $\frac{9}{5}$　　**87.** 238　　**89.** 30

91. 81　　**93.** $\frac{47}{60}$　　**95.** $\sum_{i=1}^{9} \frac{1}{3i} = 0.94299$

97. $\sum_{i=1}^{8} \left[2\left(\frac{i}{8}\right) + 3 \right] = 33$　　**99.** $\sum_{i=1}^{6} (-1)^{i+1} 3^i = -546$

101. $\sum_{i=1}^{20} \frac{(-1)^{i+1}}{i^2} = 0.821$　　**103.** $\sum_{i=1}^{5} \frac{2^i - 1}{2^{i+1}} = \frac{129}{64}$

105. $\frac{75}{16}$　　**107.** $-\frac{3}{2}$　　**109.** $\frac{2}{3}$　　**111.** $\frac{1}{9}$

113. (a) $A_1 = \$5100.00$, $A_2 = \$5202.00$, $A_3 = \$5306.04$,
$A_4 = \$5412.16$, $A_5 = \$5520.40$, $A_6 = \$5630.81$,
$A_7 = \$5743.43$, $A_8 = \$5858.30$

(b) $\$11,040.20$

115.

According to the graph, hospital costs are increasing.

117. $\$23,661.96$ million; the sums are approximately the same.

119. True

121. 1, 1, 2, 3, 5, 8, 13, 21, 34, 55, 89, 144;
$1, 2, \frac{3}{2}, \frac{5}{3}, \frac{8}{5}, \frac{13}{8}, \frac{21}{13}, \frac{34}{21}, \frac{55}{34}, \frac{89}{55}$

123. a_n increases by $2(n - 1)$. The terms seem to be prime numbers, however, $a_{11} = 121$ is not prime.

125. $-\frac{x^3}{3}, \frac{x^5}{5}, -\frac{x^7}{7}, \frac{x^9}{9}, -\frac{x^{11}}{11}$

127. $-\frac{x^3}{6}, \frac{x^5}{120}, -\frac{x^7}{5040}, \frac{x^9}{362,880}, -\frac{x^{11}}{39,916,800}$

129. $\begin{bmatrix} 2 & 1 & 3 & \vdots & -3 \\ -1 & 5 & 0 & \vdots & 14 \\ -3 & -6 & -7 & \vdots & -7 \end{bmatrix}$

131. (a) $\begin{bmatrix} 10 & 19 \\ -12 & -5 \end{bmatrix}$　(b) $\begin{bmatrix} -30 & -45 \\ 28 & 4 \end{bmatrix}$

(c) $\begin{bmatrix} 56 & -43 \\ 48 & 114 \end{bmatrix}$　(d) $\begin{bmatrix} 48 & -72 \\ 36 & 122 \end{bmatrix}$

133. (a) $\begin{bmatrix} -1 & 0 & 0 \\ 2 & 0 & 4 \\ 1 & -1 & 1 \end{bmatrix}$　(b) $\begin{bmatrix} 3 & -4 & 0 \\ -9 & -1 & -10 \\ -2 & 3 & -5 \end{bmatrix}$

(c) $\begin{bmatrix} 12 & 0 & -8 \\ 1 & 21 & 2 \\ -6 & -1 & 8 \end{bmatrix}$　(d) $\begin{bmatrix} 20 & 4 & 8 \\ 2 & 15 & -4 \\ 1 & -6 & 6 \end{bmatrix}$

135. -223　　**137.** 664

Section 9.2 *(page 635)*

1. Arithmetic sequence, $d = -2$

3. Arithmetic sequence, $d = -\frac{1}{2}$

5. Arithmetic sequence, $d = 8$

7. Arithmetic sequence, $d = 0.6$

9. 21, 34, 47, 60, 73

Arithmetic sequence, $d = 13$

11. $\frac{1}{2}, \frac{1}{3}, \frac{1}{4}, \frac{1}{5}, \frac{1}{6}$

Not an arithmetic sequence

13. 143, 136, 129, 122, 115

Arithmetic sequence, $d = -7$

15. $1, 4, \frac{7}{3}, \frac{7}{2}, \frac{13}{5}$

Not an arithmetic sequence

17. 15, 24, 33, 42, 51; $d = 9$; $a_n = 9n + 6$

19. $\frac{7}{2}, \frac{13}{4}, 3, \frac{11}{4}, \frac{5}{2}$; $d = -\frac{1}{4}$; $a_n = -\frac{1}{4}n + \frac{15}{4}$

21. 5, 11, 17, 23, 29   **23.** $-2.6, -3, -3.4, -3.8, -4.2$

25. $-2, 2, 6, 10, 14$ **27.** 22.45, 20.725, 19, 17.275, 15.55

29. 59   **31.** -50 **33.** 18.6   **35.** $a_n = -2 + 3n$

37. $a_n = 108 - 8n$ **39.** $a_n = \frac{13}{2} - \frac{5}{2}n$

41. $a_n = \frac{10}{3}n + \frac{5}{3}$ **43.** $a_n = 103 - 3n$

45. (b)   **47.** (c)

49.   **51.**

53. $-1, 3, 7, 11, 15, 19, 23, 27, 31, 35$

55. 19.25, 18.5, 17.75, 17, 16.25, 15.5, 14.75, 14, 13.25, 12.5

57. 1.505, 1.51, 1.515, 1.52, 1.525, 1.53, 1.535, 1.54, 1.545, 1.55

59. 890   **61.** 41   **63.** 4000   **65.** 1275

67. 25,250   **69.** 355   **71.** 126,750   **73.** 520

75. 44,625   **77.** 10,120   **79.** 10,000

81. (a) \$40,000   (b) \$217,500   (c) \$40,000; \$217,500

83. 2340 seats   **85.** 405 bricks   **87.** 585 seats

89. 156 times   **91.** True

93. $x, 3x, 5x, 7x, 9x, 11x, 13x, 15x, 17x, 19x$

95. (a) 4, 9, 16, 25, 36

(b) Sum of first n positive odd integers is n^2; 49

(c) $\frac{n}{2}[1 + (2n - 1)] = n^2$

97. $S_n + 5n$ **99.** $(2, -6, 3)$ **101.** 20 square units

103. $\frac{1}{3003}$

Section 9.3   *(page 644)*

1. Geometric sequence, $r = 3$

3. Not a geometric sequence

5. Geometric sequence, $r = -\frac{1}{2}$

7. Not a geometric sequence

9. Not a geometric sequence   **11.** 8, 24, 72, 216, 648

13. $1, \frac{1}{2}, \frac{1}{4}, \frac{1}{8}, \frac{1}{16}$ **15.** $5, -\frac{1}{2}, \frac{1}{20}, -\frac{1}{200}, \frac{1}{2000}$

17. 3.5, 17.5, 87.5, 437.5, 2187.5   **19.** $1, e, e^2, e^3, e^4$

21. $64, 32, 16, 8, 4; \frac{1}{2}$ **23.** 4, 12, 36, 108, 324; 3

$a_n = 128\left(\frac{1}{2}\right)^n$ $a_n = \frac{4}{3}(3)^n$

25. $6, -9, \frac{27}{2}, -\frac{81}{4}, \frac{243}{8}; -\frac{3}{2}$

$a_n = 6\left(-\frac{3}{2}\right)^{n-1}$

27. $\left(\frac{1}{2}\right)^7$ **29.** $-\frac{2}{3^{10}}$ **31.** $500(1.02)^{13}$ **33.** 9

35. $-\frac{2}{9}$ **37.** 45,927   **39.** 50,388,480

41. 786,432   **43.** (a)   **45.** (b)

47.   **49.**

51. 8, 4, 6, 5

53.

n	S_n
1	16
2	24
3	28
4	30
5	31
6	31.5
7	31.75
8	31.875
9	31.9375
10	31.96875

55. 511   **57.** 43

59. 29,921.31   **61.** 6.4   **63.** 2092.60

65. $\displaystyle\sum_{n=1}^{7} 5(3)^{n-1}$ **67.** $\displaystyle\sum_{n=1}^{7} 2\left(-\frac{1}{4}\right)^{n-1}$ **69.** 2   **71.** $\frac{2}{3}$

73. $\frac{16}{3}$ **75.** Series does not have a finite sum.   **77.** $\frac{5}{3}$

79. -30 **81.** 32   **83.** $\frac{9}{4}$ **85.** $\frac{4}{11}$ **87.** $\frac{7}{22}$

89. (a) \$2158.92   (b) \$2191.12   (c) \$2208.04

(d) \$2219.64   (e) \$2225.35

91. $26,050.85   **93.** $7011.89   **95.** Answers will vary.

97. (a) $26,198.27   (b) $26,263.88

99. (a) $637,678.02   (b) $645,861.43

101. Answers will vary.   **103.** 126 square inches

105. $39.7 billion

107. (a) $5,368,709.11   (b) $10,737,418.23

    (c) $21,474,836.47

109. False. Any arithmetic sequence can be used as a counterexample.

111. True. It is an arithmetic sequence with $d = 0$.

113. $8, \dfrac{16x}{3}, \dfrac{32x^2}{9}, \dfrac{64x^3}{27}, \dfrac{128x^4}{81}$

115. $\dfrac{1}{2}, \dfrac{7x}{2}, \dfrac{49x^2}{2}, \dfrac{343x^3}{2}, \dfrac{2401x^4}{2}$

117. $13,122e^{7x}$ **119.** $\dfrac{4096x^5}{243}$

121.

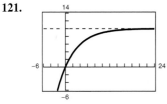

Horizontal asymptote: $y = 10$

Corresponds to the sum of the series

123. Divide the second term by the first to obtain the common ratio. The nth term is the first term times the common ratio raised to the $n - 1$ power.

125. 45.65 miles per hour   **127.** 2.4 hours

129. $\begin{bmatrix} -33 & -3 \\ 27 & 1 \end{bmatrix}$ **131.** $\begin{bmatrix} 2 & 6 & -8 \\ 12 & 4 & -20 \\ 4 & 2 & 10 \end{bmatrix}$

133. 364   **135.** $\dfrac{227}{99}$

Section 9.4 *(page 654)*

1. $\dfrac{5}{(k + 1)(k + 2)}$ **3.** $\dfrac{(k + 1)^2(k + 4)^2}{6}$

5. $1 + 6 + 11 + \cdots + (5k - 4) + (5k + 1)$

7.–35. Answers will vary.   **37.** 10, 40, 160, 640, 2560

39. 0, 2, 2, 6, 10

41. 2, 0, 3, 1, 4

    First differences: $-2, 3, -2, 3$

    Second differences: $5, -5, 5$

    Neither

43. $-3, 6, -12, 24, -48$

    First differences: $9, -18, 36, -72$

    Second differences: $-27, 54, -108$

    Neither

45. 2, 4, 16, 256, 65,536

    First differences: 2, 12, 240, 65,280

    Second differences: 10, 288, 65,040

    Neither

47. 0, 4, 10, 18, 28

    First differences: 4, 6, 8, 10

    Second differences: 2, 2, 2

    Quadratic

49. $0, -1, -2, -3, -4$

    First differences: $-1, -1, -1, -1$

    Second differences: 0, 0, 0

    Linear

51. $a_n = n^2 - 2n + 7$ **53.** $a_n = \frac{7}{4}n^2 - 5n + 3$

55. False. The first differences are all the same.

57. Answers will vary.   **59.** $(7, 5)$ **61.** $\left(-\frac{1}{2}, \frac{1}{4}\right), (2, 4)$

63. $(1, 2, 1)$ **65.** $(-1, 2, 4)$ **67.** 38

69. $4x^4 - 4x^2 + 1$ **71.** $-64x^3 + 240x^2 - 300x + 125$

Section 9.5 *(page 661)*

1. 21   **3.** 1   **5.** 15,504   **7.** 14   **9.** 4950

11. 4950   **13.** 35,960   **15.** 497,420   **17.** 749,398

19. 21   **21.** 56   **23.** $x^4 + 4x^3 + 6x^2 + 4x + 1$

25. $a^3 + 9a^2 + 27a + 27$

27. $y^4 - 8y^3 + 24y^2 - 32y + 16$

29. $x^5 + 5x^4y + 10x^3y^2 + 10x^2y^3 + 5xy^4 + y^5$

31. $r^6 + 12r^5s + 60r^4s^2 + 160r^3s^3 + 240r^2s^4 + 192rs^5$

    $+ 64s^6$

33. $x^5 - 5x^4y + 10x^3y^2 - 10x^2y^3 + 5xy^4 - y^5$

35. $1 - 12x + 48x^2 - 64x^3$

37. $x^8 + 20x^6 + 150x^4 + 500x^2 + 625$

39. $\dfrac{1}{x^5} + \dfrac{5y}{x^4} + \dfrac{10y^2}{x^3} + \dfrac{10y^3}{x^2} + \dfrac{5y^4}{x} + y^5$

41. $2x^4 - 24x^3 + 113x^2 - 246x + 207$

43. $-4x^6 - 24x^5 - 60x^4 - 83x^3 - 42x^2 - 60x + 20$

45. $243t^5 - 405t^4s + 270t^3s^2 - 90t^2s^3 + 15ts^4 - s^5$

47. $81 - 216z + 216z^2 - 96z^3 + 16z^4$ **49.** 3,247,695

51. 180   **53.** $-489,888$ **55.** 210

57. $x^2 + 20x^{3/2} + 150x + 500x^{1/2} + 625$

59. $x^2 - 3x^{4/3}y^{1/3} + 3x^{2/3}y^{2/3} - y$

61. $3x^2 + 3xh + h^2$, $h \neq 0$

63. $\dfrac{\sqrt{x + h} - \sqrt{x}}{h} = \dfrac{1}{\sqrt{x + h} + \sqrt{x}}$, $h \neq 0$

65. -4 **67.** $2035 + 828i$ **69.** 1 **71.** 1.172

73. 510,568.785

75.

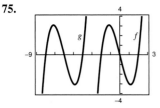

g is shifted 6 units left of f.
$g(x) = x^3 + 18x^2 + 104x + 192$

77.

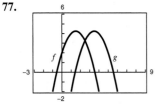

g is shifted 2 units right of f.
$g(x) = -x^2 + 7x - 8$

79. (a) 792   (b) 36   (c) 792   (d) 12

81.

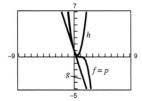

$p(x)$ is the expansion of $f(x)$.

83. 0.273   **85.** 0.171

87. (a) $g(t) = 0.0348t^2 + 5.8043t + 95.588$

(b)

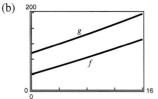

89. False. The correct term is $126{,}720x^4y^8$.

91. The first and last numbers in each row are 1. Every other number in each row is formed by adding the two numbers immediately above the number.

93. $n + 1$ terms   **95. and 97.** Answers will vary.

99. $g(x)$ is shifted 8 units up from $f(x)$.

101. $g(x)$ is the reflection of $f(x)$ in the y-axis.

103. $\begin{bmatrix} -2 & -1 & -3 \\ 8 & 0 & 4 \\ -1 & -2 & 7 \end{bmatrix}$ **105.** $\begin{bmatrix} 6 & 11 & 15 \\ -30 & -2 & -16 \\ 13 & 10 & -33 \end{bmatrix}$

107. $\begin{bmatrix} 9 & 11 & 12 \\ -13 & -25 & -5 \\ -5 & -18 & -6 \end{bmatrix}$ **109.** $\begin{bmatrix} 4 & -5 \\ 5 & -6 \end{bmatrix}$

Section 9.6   *(page 671)*

1. 6   **3.** 5   **5.** 3   **7.** 9   **9.** 192   **11.** 18

13. 3,628,800   **15.** 12   **17.** 6,760,000

19. (a) 900   (b) 648   (c) 180   (d) 600

21. 64,000   **23.** (a) 720   (b) 48   **25.** 24

27. 336   **29.** 120   **31.** $n = 5$ or $n = 6$

33. 27,907,200   **35.** 197,149,680   **37.** 4845

39. 120   **41.** (a) 16   (b) 14   **43.** 420   **45.** 2520

47. ABCD, ABDC, ACBD, ACDB, ADBC, ADCB, BACD, BADC, CABD, CADB, DABC, DACB, BCAD, BDAC, CBAD, CDAB, DBAC, DCAB, BCDA, BDCA, CBDA, CDBA, DBCA, DCBA

49. AB, AC, AD, AE, AF, BC, BD, BE, BF, CD, CE, CF, DE, DF, EF

51. 4845   **53.** 3,838,380   **55.** 75,287,520   **57.** 36

59. (a) 495   (b) 210   **61.** (a) 70   (b) 54   (c) 16

63. 5   **65.** 20

67. False. This is an example of a combination.

69. False. $_nP_r = {}_nC_r$ if $r = 1$ or 0.

71. $_nP_r$ represents the number of ways to choose and order r elements out of a collection of n elements.

73. (b). Numerous permutations can be made from each combination.

75. and 77. Answers will vary.   **79.** $\frac{11}{2}$ **81.** 8.32

83. $(6, -13)$ **85.** $(-3, 4)$

87. $x^6 - 6x^5 + 15x^4 - 20x^3 + 15x^2 - 6x + 1$

89. $81x^4 - 108x^3y + 54x^2y^2 - 12xy^3 + y^4$

Section 9.7   *(page 682)*

1. $\{(H, 1), (H, 2), (H, 3), (H, 4), (H, 5), (H, 6),$
$(T, 1), (T, 2), (T, 3), (T, 4), (T, 5), (T, 6)\}$

3. $\{ABC, ACB, BAC, BCA, CAB, CBA\}$

5. $\{(A, B), (A, C), (A, D), (A, E), (B, C), (B, D),$
$(B, E), (C, D), (C, E), (D, E)\}$

7. $\frac{3}{8}$ **9.** $\frac{7}{8}$ **11.** $\frac{3}{13}$ **13.** $\frac{3}{26}$ **15.** $\frac{1}{9}$ **17.** $\frac{35}{36}$

19. $\frac{1}{6}$ **21.** $\frac{1}{5}$ **23.** $\frac{2}{5}$ **25.** 0.3   **27.** $\frac{2}{3}$

29. 0.85   **31.** $\frac{7}{20}$

33. (a) 78,000   (b) 0.3   (c) 0.37   (d) 0.1

35. (a) 0.58   (b) 0.956   (c) 0.004

37. (a) $\frac{672}{1254}$ (b) $\frac{582}{1254}$ (c) $\frac{548}{1254}$

39. $P(\{\text{Taylor wins}\}) = \frac{1}{2}$

$P(\{\text{Moore wins}\}) = P(\{\text{Perez wins}\}) = \frac{1}{4}$

41. (a) $\frac{21}{1292} \approx 0.016$ (b) $\frac{225}{646} \approx 0.348$ (c) $\frac{49}{323} \approx 0.152$

43. (a) $\frac{1}{3}$ (b) $\frac{5}{8}$ **45.** (a) $\frac{1}{120}$ (b) $\frac{1}{24}$

47. (a) 0.346   (b) 0.0000029

49. (a) $\frac{14}{55}$ (b) $\frac{12}{55}$ (c) $\frac{54}{55}$

51. (a) $\frac{1}{4}$ (b) $\frac{1}{2}$ (c) $\frac{9}{100}$ (d) $\frac{1}{30}$

53. (a) 0.9702   (b) 0.9998   (c) 0.0002

55. (a) $\dfrac{1}{15,625}$ (b) $\dfrac{4096}{15,625}$ (c) $\dfrac{11,529}{15,625}$

57. 0.1024   **59.** $\frac{7}{16}$ **61.** True

63. (a) As you consider successive people with distinct birthdays, the probabilities must decrease to take into account the birth dates already used. Because the birth dates of people are independent events, multiply the respective probabilities of distinct birthdays.

(b) $\frac{365}{365} \cdot \frac{364}{365} \cdot \frac{363}{365} \cdot \frac{362}{365}$

(c) Answers will vary.

(d) Q_n is the probability that the birthdays are *not* distinct, which is equivalent to at least two people having the same birthday.

(e)

n	10	15	20	23	30	40	50
P_n	0.88	0.75	0.59	0.49	0.29	0.11	0.03
Q_n	0.12	0.25	0.41	0.51	0.71	0.89	0.97

(f) 23

65. $x = \frac{22}{4} = \frac{11}{2}$ **67.** $x = -10$ **69.** $x = e^8$

71. $x = \frac{1}{6}e^4$

73.   **75.**

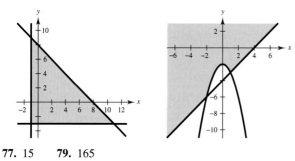

77. 15   **79.** 165

Review Exercises   *(page 688)*

1. $8, 5, 4, \frac{7}{2}, \frac{16}{5}$ **3.** $5, \frac{10}{3}, 3, \frac{20}{7}, \frac{25}{9}$ **5.** $72, 36, 12, 3, \frac{3}{5}$

7.   **9.**

11. $\frac{1}{380}$ **13.** 1   **15.** 30   **17.** $\frac{205}{24}$ **19.** 6050

21. 418   **23.** $\displaystyle\sum_{k=1}^{20} \frac{1}{2k}$ **25.** $\displaystyle\sum_{k=1}^{9} \frac{k}{k+1}$

27. (a) $\frac{1111}{2000}$ (b) $\frac{5}{9}$ **29.** (a) $\frac{2,020,202}{100,000,000}$ (b) $\frac{2}{99}$

31. (a) $a_1 = 2550, a_2 = 2601, a_3 = 2653.02, a_4 = 2706.08,$

$a_5 = 2760.20, a_6 = 2815.41, a_7 = 2871.71,$

$a_8 = 2929.15$

(b) $5520.10

33. Arithmetic sequence, $d = -7$

35. Arithmetic sequence, $d = \frac{1}{2}$ **37.** 3, 7, 11, 15, 19

39. 1, 4, 7, 10, 13

41. $35, 32, 29, 26, 23; -3$ **43.** $9, 16, 23, 30, 37; 7$

$a_n = 38 - 3n$ $\qquad\qquad\qquad a_n = 2 + 7n$

45. $a_n = 103 - 3n; 1430$ **47.** 80   **49.** 88

51. 25,250   **53.** (a) $43,000   (b) $192,500

55. $4, -1, \frac{1}{4}, -\frac{1}{16}, \frac{1}{64}$ **57.** $9, 6, 4, \frac{8}{3}, \frac{16}{9}$ or $9, -6, 4, -\frac{8}{3}, \frac{16}{9}$

59. $120, 40, \frac{40}{3}, \frac{40}{9}, \frac{40}{27}; \frac{1}{3}$ **61.** $25, -15, 9, -\frac{27}{5}, \frac{81}{25}; -\frac{3}{5}$

$a_n = 120\left(\frac{1}{3}\right)^{n-1}$ $\qquad\qquad a_n = 25\left(-\frac{3}{5}\right)^{n-1}$

63. $a_n = 16\left(-\frac{1}{2}\right)^{n-1}; 10.67$

65. $a_n = 100(1.05)^{n-1}; 3306.60$ **67.** 127   **69.** 3277

71. 1301.01   **73.** 24.85   **75.** 5486.45   **77.** 8

79. 12   **81.** (a) $a_t = 120,000(0.7)^t$ (b) $20,168.40

83. $3909.96   **85. and 87.** Answers will vary.

89. 465   **91.** 4676

93. 5, 10, 15, 20, 25

First differences: 5, 5, 5, 5

Second differences: 0, 0, 0

Linear model

95. 16, 15, 14, 13, 12

First differences: $-1, -1, -1, -1$

Second differences: 0, 0, 0

Linear model

97. 45   **99.** 126   **101.** 20   **103.** 70

105. $a^5 - 15a^4b + 90a^3b^2 - 270a^2b^3 + 405ab^4 - 243b^5$

107. $\dfrac{x^4}{16} + \dfrac{x^3y}{2} + \dfrac{3x^2y^2}{2} + 2xy^3 + y^4$ **109.** $41 + 840i$

111. 10   **113.** 48   **115.** 5040   **117.** 3,628,800

119. 15,504   **121.** $\frac{1}{9}$

123. $P(\{3\}) = \frac{1}{6}$

$P(\{(1, 5), (5, 1), (2, 4), (4, 2), (3, 3)\}) = \frac{5}{36}$

There is a higher probability of rolling a 3 with one die.

125. (a) 0.416   (b) 0.8   (c) 0.074   **127.** 0.512

129. $\frac{31}{32}$ **131.** True

133. (a) Odd-numbered terms are negative.

(b) Even-numbered terms are negative.

135. (a) Arithmetic. There is a constant difference between consecutive terms.

(b) Geometric. Each term is a constant multiple of the preceding term. In this case the common ratio is greater than 1.

137. Each term of the sequence is defined using a previous term or terms.

139. (d)   **141.** (b)

143.

n	100	500	1000	5000
a_n	2.704	2.7156	2.7169	2.7180

n	10,000	15,000	20,000	25,000
a_n	2.71814	2.71819	2.71821	2.71823

a_n approaches e.

145. $0 \le P \le 1$; closed

147. Meteorological records gathered over an extended period of time indicate that under similar weather conditions it will rain 60% of the time.

Chapter Test   *(page 694)*

1. $1, -\frac{2}{3}, \frac{4}{9}, -\frac{8}{27}, \frac{16}{81}$ **2.** 12, 16, 20, 24, 28   **3.** 7920

4. $a_n = 5100 - 100n$ **5.** $a_n = 4\left(\frac{1}{2}\right)^{n-1}$

6. $\displaystyle\sum_{n=1}^{12} \dfrac{2}{3n + 1}$ **7.** 3825   **8.** 189   **9.** 28.80

10. $\frac{50}{9}$ **11.** $47,868.33   **12.** Answers will vary.

13. 84   **14.** 1140   **15.** 8568   **16.** 780

17. 56   **18.** 26,000   **19.** 12,650   **20.** $\frac{3}{26}$ **21.** $\frac{1}{6}$

22. (a) $\frac{1}{4}$ (b) $\frac{121}{3600}$ (c) $\frac{1}{60}$

Chapter 10

Section 10.1   *(page 701)*

1. (e)   **3.** (d)   **5.** (a)

7. Vertex: $(0, 0)$ **9.** Vertex: $(0, 0)$

Focus: $\left(0, \frac{1}{2}\right)$ Focus: $\left(-\frac{3}{2}, 0\right)$

Directrix: $y = -\frac{1}{2}$ Directrix: $x = \frac{3}{2}$

11. Vertex: $(0, 0)$ **13.** Vertex: $(1, -2)$

Focus: $\left(0, -\frac{3}{2}\right)$ Focus: $(1, -4)$

Directrix: $y = \frac{3}{2}$ Directrix: $y = 0$

15. Vertex: $\left(-\frac{3}{2}, 2\right)$ **17.** Vertex: $(1, 1)$

Focus: $\left(-\frac{3}{2}, 3\right)$ Focus: $(1, 2)$

Directrix: $y = 1$ Directrix: $y = 0$

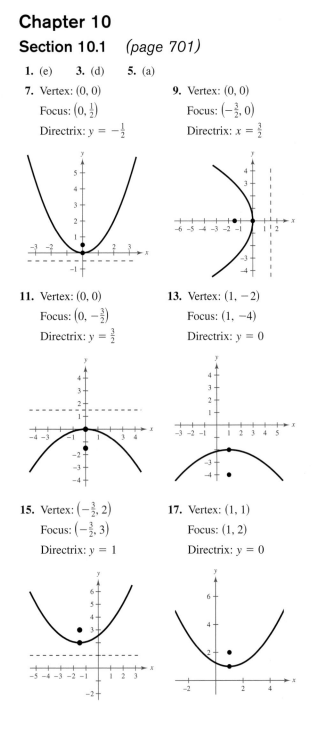

19. Vertex: $(-2, -3)$
  Focus: $(-4, -3)$
  Directrix: $x = 0$

21. Vertex: $(-2, 1)$
  Focus: $\left(-2, -\frac{1}{2}\right)$
  Directrix: $y = \frac{5}{2}$

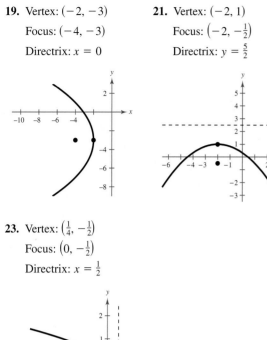

23. Vertex: $\left(\frac{1}{4}, -\frac{1}{2}\right)$
  Focus: $\left(0, -\frac{1}{2}\right)$
  Directrix: $x = \frac{1}{2}$

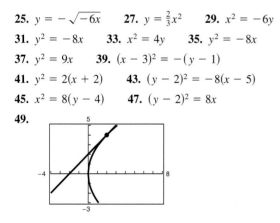

25. $y = -\sqrt{-6x}$ **27.** $y = \frac{2}{3}x^2$ **29.** $x^2 = -6y$

31. $y^2 = -8x$ **33.** $x^2 = 4y$ **35.** $y^2 = -8x$

37. $y^2 = 9x$ **39.** $(x - 3)^2 = -(y - 1)$

41. $y^2 = 2(x + 2)$ **43.** $(y - 2)^2 = -8(x - 5)$

45. $x^2 = 8(y - 4)$ **47.** $(y - 2)^2 = 8x$

49.

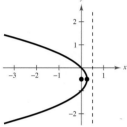

  $(2, 4)$

51. $4x - y - 8 = 0;\ (2, 0)$ **53.** $4x - y + 2 = 0;\ \left(-\frac{1}{2}, 0\right)$

55.

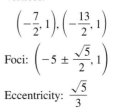

  About 106 sales

57. $y = \frac{1}{14}x^2$ **59.** (a) $x^2 = -640y$ (b) 8 feet

61. (a) $17{,}500\sqrt{2}$ miles per hour
  (b) $x^2 = -16{,}400(y - 4100)$

63. (a) $y = -\frac{1}{64}x^2 + 75$ (b) 69.3 feet

65. False. If the graph intersected the directrix, there would exist points nearer the directrix then the focus.

67. $\pm 1, \pm 2, \pm 4$ **69.** $\pm\frac{1}{2}, \pm 1, \pm 2, \pm 4, \pm 8, \pm 16$

Section 10.2   *(page 710)*

1. (b)   **3.** (d)   **5.** (a)

7. Center: $(0, 0)$
  Vertices: $(\pm 5, 0)$
  Foci: $(\pm 3, 0)$
  Eccentricity: $\frac{3}{5}$

9. Center: $(0, 0)$
  Vertices: $(0, \pm 3)$
  Foci: $(0, \pm 2)$
  Eccentricity: $\frac{2}{3}$

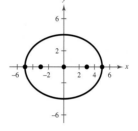
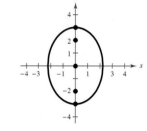

11. Center: $(-3, 5)$
  Vertices: $(-3, 0), (-3, 10)$
  Foci: $(-3, 8), (-3, 2)$
  Eccentricity: $\frac{3}{5}$

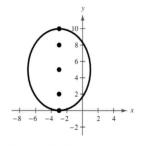

13. Center: $(-5, 1)$
  Vertices:
  $$\left(-\frac{7}{2}, 1\right), \left(-\frac{13}{2}, 1\right)$$
  Foci: $\left(-5 \pm \frac{\sqrt{5}}{2}, 1\right)$
  Eccentricity: $\frac{\sqrt{5}}{3}$

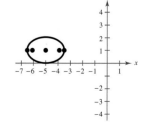

15. Center: $(-2, 3)$

Vertices:

$(-2, 6), (-2, 0)$

Foci: $\left(-2, 3 \pm \sqrt{5}\right)$

Eccentricity: $\dfrac{\sqrt{5}}{3}$

17. Center: $(4, 3)$

Vertices:

$(-6, 3), (14, 3)$

Foci: $\left(4 \pm 4\sqrt{5}, 3\right)$

Eccentricity: $\dfrac{2\sqrt{5}}{5}$

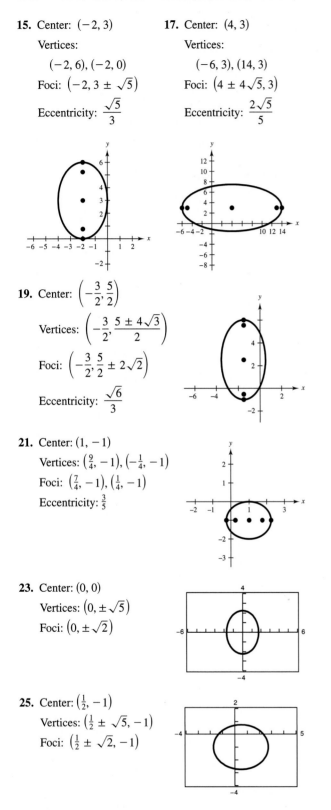

19. Center: $\left(-\dfrac{3}{2}, \dfrac{5}{2}\right)$

Vertices: $\left(-\dfrac{3}{2}, \dfrac{5 \pm 4\sqrt{3}}{2}\right)$

Foci: $\left(-\dfrac{3}{2}, \dfrac{5}{2} \pm 2\sqrt{2}\right)$

Eccentricity: $\dfrac{\sqrt{6}}{3}$

21. Center: $(1, -1)$

Vertices: $\left(\dfrac{9}{4}, -1\right), \left(-\dfrac{1}{4}, -1\right)$

Foci: $\left(\dfrac{7}{4}, -1\right), \left(\dfrac{1}{4}, -1\right)$

Eccentricity: $\dfrac{3}{5}$

23. Center: $(0, 0)$

Vertices: $\left(0, \pm\sqrt{5}\right)$

Foci: $\left(0, \pm\sqrt{2}\right)$

25. Center: $\left(\dfrac{1}{2}, -1\right)$

Vertices: $\left(\dfrac{1}{2} \pm \sqrt{5}, -1\right)$

Foci: $\left(\dfrac{1}{2} \pm \sqrt{2}, -1\right)$

27. $\dfrac{x^2}{4} + \dfrac{y^2}{16} = 1$ **29.** $\dfrac{x^2}{36} + \dfrac{y^2}{32} = 1$

31. $\dfrac{x^2}{36} + \dfrac{y^2}{11} = 1$ **33.** $\dfrac{21x^2}{400} + \dfrac{y^2}{25} = 1$

35. $\dfrac{(x-2)^2}{1} + \dfrac{(y-3)^2}{9} = 1$

37. $\dfrac{(x+2)^2}{16} + \dfrac{(y-3)^2}{9} = 1$

39. $\dfrac{(x-2)^2}{4} + \dfrac{(y-4)^2}{1} = 1$ **41.** $\dfrac{x^2}{48} + \dfrac{(y-4)^2}{64} = 1$

43. $\dfrac{(x-3)^2}{9} + \dfrac{(y-5)^2}{16} = 1$ **45.** $\dfrac{x^2}{16} + \dfrac{(y-4)^2}{12} = 1$

47. $\dfrac{x^2}{25} + \dfrac{y^2}{16} = 1$ **49.** $\left(\pm\sqrt{5}, 0\right); 6$ feet   **51.** 40

53. $\dfrac{x^2}{4.88} + \dfrac{y^2}{1.39} = 1$ **55.** Answers will vary.

57.

$\left(-\dfrac{9}{4}, \sqrt{7}\right)$ $\left(\dfrac{9}{4}, \sqrt{7}\right)$

$\left(-\dfrac{9}{4}, -\sqrt{7}\right)$ $\left(\dfrac{9}{4}, -\sqrt{7}\right)$

59.

$\left(-\dfrac{3\sqrt{5}}{5}, \sqrt{2}\right)$ $\left(\dfrac{3\sqrt{5}}{5}, \sqrt{2}\right)$

$\left(-\dfrac{3\sqrt{5}}{5}, -\sqrt{2}\right)$ $\left(\dfrac{3\sqrt{5}}{5}, -\sqrt{2}\right)$

61. True   **63.** False

65. (a) $A = \pi a(20 - a)$

(b) $\dfrac{x^2}{196} + \dfrac{y^2}{36} = 1$

(c)

a	8	9	10	11	12	13
A	301.6	311.0	314.2	311.0	301.6	285.9

$a = 10$. Circle

(d)

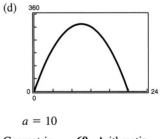

$a = 10$

67. Geometric   **69.** Arithmetic   **71.** $a_n = -\dfrac{1}{4}n + \dfrac{1}{4}$

73. $a_n = 9n$ **75.** 547 **77.** 340.1545

Section 10.3 *(page 720)*

1. (b) **3.** (a)

5. Center: $(0, 0)$

Vertices: $(\pm 1, 0)$

Foci: $\left(\pm \sqrt{2}, 0\right)$

Asymptotes: $y = \pm x$

7. Center: $(0, 0)$

Vertices: $(0, \pm 1)$

Foci: $\left(0, \pm \sqrt{5}\right)$

Asymptotes: $y = \pm \frac{1}{2}x$

9. Center: $(0, 0)$

Vertices: $(0, \pm 5)$

Foci: $\left(0, \pm \sqrt{106}\right)$

Asymptotes: $y = \pm \frac{5}{9}x$

11. Center: $(1, -2)$

Vertices: $(3, -2), (-1, -2)$

Foci: $\left(1 \pm \sqrt{5}, -2\right)$

Asymptotes:

$y = -2 \pm \frac{1}{2}(x - 1)$

13. Center: $(2, -6)$

Vertices: $(2, -5), (2, -7)$

Foci: $\left(2, -6 \pm \sqrt{2}\right)$

Asymptotes:

$y = -6 \pm (x - 2)$

15. Center: $(2, -3)$

Vertices: $(3, -3), (1, -3)$

Foci: $\left(2 \pm \sqrt{10}, -3\right)$

Asymptotes:

$y = -3 \pm 3(x - 2)$

17. The graph of this equation is two lines intersecting at $(-1, -3)$.

19. Center: $(0, 0)$

Vertices: $\left(\pm \sqrt{3}, 0\right)$

Foci: $\left(\pm \sqrt{5}, 0\right)$

21. Center: $(1, -3)$

Vertices: $\left(1, -3 \pm \sqrt{2}\right)$

Foci: $\left(1, -3 \pm 2\sqrt{5}\right)$

23. $\dfrac{y^2}{4} - \dfrac{x^2}{12} = 1$ **25.** $\dfrac{x^2}{1} - \dfrac{y^2}{25} = 1$

27. $\dfrac{17y^2}{1024} - \dfrac{17x^2}{64} = 1$ **29.** $\dfrac{(x - 4)^2}{4} - \dfrac{y^2}{12} = 1$

31. $\dfrac{(y - 5)^2}{16} - \dfrac{(x - 4)^2}{9} = 1$ **33.** $\dfrac{y^2}{9} - \dfrac{4(x - 2)^2}{9} = 1$

35. $\dfrac{(y - 2)^2}{4} - \dfrac{x^2}{4} = 1$ **37.** $\dfrac{(x - 2)^2}{1} - \dfrac{(y - 2)^2}{1} = 1$

39. $\dfrac{(x - 3)^2}{9} - \dfrac{(y - 2)^2}{4} = 1$ **41.** $(3300, -2750)$

43. $\left(12\left(\sqrt{5} - 1\right), 0\right) \approx (14.83, 0)$ **45.** Ellipse

47. Parabola **49.** Hyperbola **51.** Circle

53. False. $b = 0$ yields an undefined term. For the trivial solution of two intersecting lines to occur, the standard form of the equation of the hyperbola would be equal to zero.

$$\frac{(x - h)^2}{a^2} - \frac{(y - k)^2}{b^2} = 0 \text{ or } \frac{(y - k)^2}{a^2} - \frac{(x - h)^2}{b^2} = 0.$$

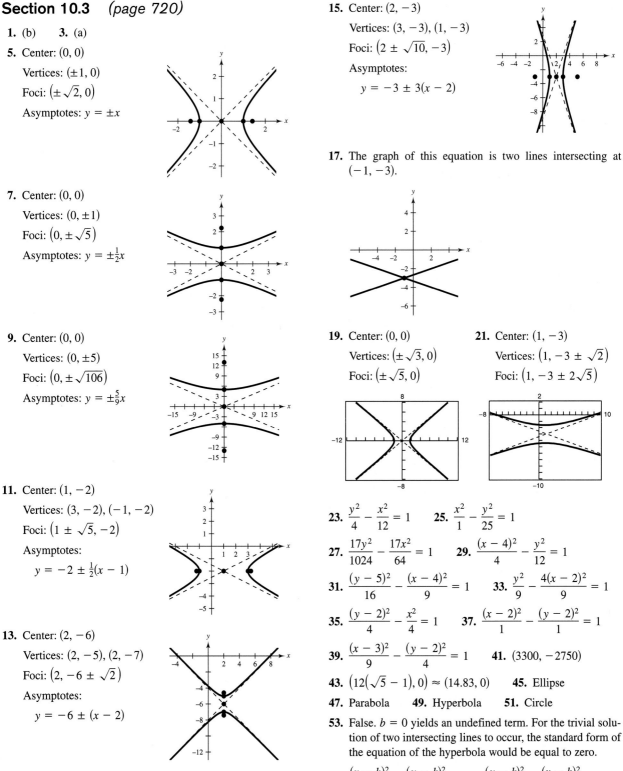

55. Answers will vary. **57.** $3x^2 + \frac{23}{2}x - 2$

59. $x^2 + 2xy + y^2 + 6x + 6y + 9$ **61.** $(x + 7)^2$

63. $x(3x + 2)(2x - 5)$ **65.** $(4 - x)(x^2 + 1)$

Section 10.4 *(page 729)*

1. $(4, 0)$ **3.** $\left(\dfrac{6 + \sqrt{3}}{2}, \dfrac{6\sqrt{3} - 1}{2}\right)$

5. $\dfrac{(y')^2}{2} - \dfrac{(x')^2}{2} = 1$ **7.** $\dfrac{(x')^2}{1/3} - \dfrac{(y')^2}{1/5} = 1$

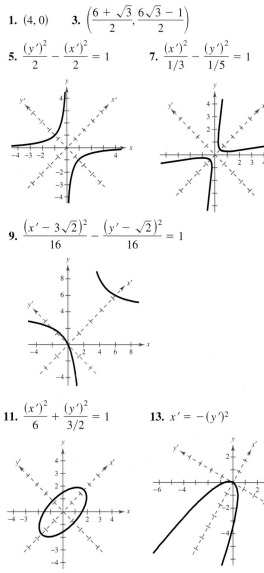

9. $\dfrac{\left(x' - 3\sqrt{2}\right)^2}{16} - \dfrac{\left(y' - \sqrt{2}\right)^2}{16} = 1$

11. $\dfrac{(x')^2}{6} + \dfrac{(y')^2}{3/2} = 1$ **13.** $x' = -(y')^2$

15. $y' = \frac{1}{6}(x')^2 - \frac{1}{3}x'$ **17.**

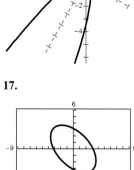

$\theta = 45°$

19.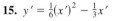

$\theta = 26.57°$ **21.** $\theta = 31.72°$

23. (e) **25.** (b) **27.** (d)

29. (a) Parabola

(b) $y = \dfrac{24x + 40 \pm \sqrt{3000x + 1600}}{18}$

(c)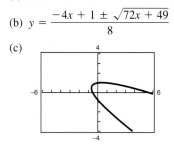

31. (a) Ellipse

(b) $y = \dfrac{8x \pm \sqrt{-356x^2 + 1260}}{14}$

(c)

33. (a) Hyperbola

(b) $y = \dfrac{6x \pm \sqrt{56x^2 + 80x - 440}}{-10}$

(c)

35. (a) Parabola

(b) $y = \dfrac{-4x + 1 \pm \sqrt{72x + 49}}{8}$

(c)

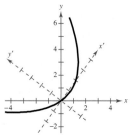

37. **39.**

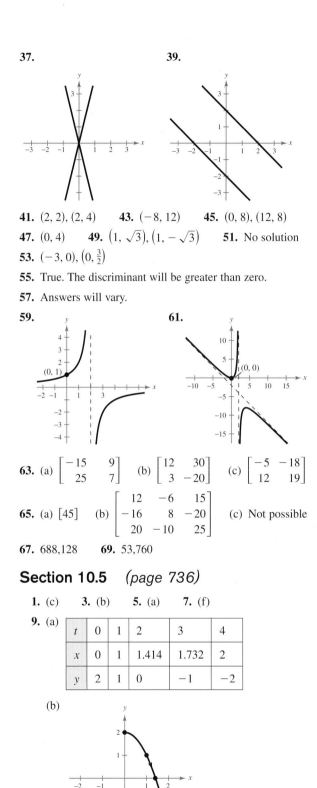

41. $(2, 2), (2, 4)$ **43.** $(-8, 12)$ **45.** $(0, 8), (12, 8)$
47. $(0, 4)$ **49.** $\left(1, \sqrt{3}\right), \left(1, -\sqrt{3}\right)$ **51.** No solution
53. $(-3, 0), \left(0, \frac{3}{2}\right)$
55. True. The discriminant will be greater than zero.
57. Answers will vary.
59. **61.**

63. (a) $\begin{bmatrix} -15 & 9 \\ 25 & 7 \end{bmatrix}$ (b) $\begin{bmatrix} 12 & 30 \\ 3 & -20 \end{bmatrix}$ (c) $\begin{bmatrix} -5 & -18 \\ 12 & 19 \end{bmatrix}$

65. (a) $[45]$ (b) $\begin{bmatrix} 12 & -6 & 15 \\ -16 & 8 & -20 \\ 20 & -10 & 25 \end{bmatrix}$ (c) Not possible

67. $688{,}128$ **69.** $53{,}760$

Section 10.5   *(page 736)*

1. (c)   **3.** (b)   **5.** (a)   **7.** (f)
9. (a)

t	0	1	2	3	4
x	0	1	1.414	1.732	2
y	2	1	0	-1	-2

(b)

(c)

(d) $y = 2 - x^2$

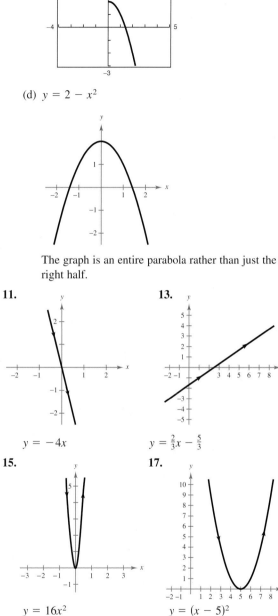

The graph is an entire parabola rather than just the right half.

11. **13.**

$y = -4x$ $y = \frac{2}{3}x - \frac{5}{3}$

15. **17.**

$y = 16x^2$ $y = (x - 5)^2$

19. **21.**

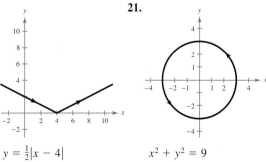

$y = \frac{1}{2}|x - 4|$ $x^2 + y^2 = 9$

23.

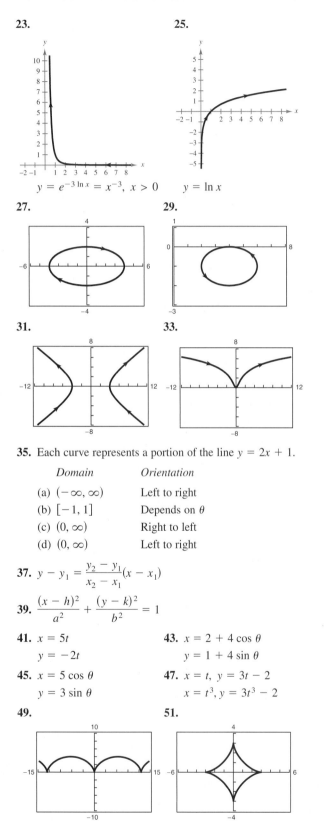

$$y = e^{-3 \ln x} = x^{-3}, \ x > 0$$

25.

$$y = \ln x$$

27.

29.

31.

33.

35. Each curve represents a portion of the line $y = 2x + 1$.

Domain	Orientation
(a) $(-\infty, \infty)$	Left to right
(b) $[-1, 1]$	Depends on θ
(c) $(0, \infty)$	Right to left
(d) $(0, \infty)$	Left to right

37. $y - y_1 = \dfrac{y_2 - y_1}{x_2 - x_1}(x - x_1)$

39. $\dfrac{(x - h)^2}{a^2} + \dfrac{(y - k)^2}{b^2} = 1$

41. $x = 5t$
$y = -2t$

43. $x = 2 + 4 \cos \theta$
$y = 1 + 4 \sin \theta$

45. $x = 5 \cos \theta$
$y = 3 \sin \theta$

47. $x = t, \ y = 3t - 2$
$x = t^3, \ y = 3t^3 - 2$

49.

51.

53.

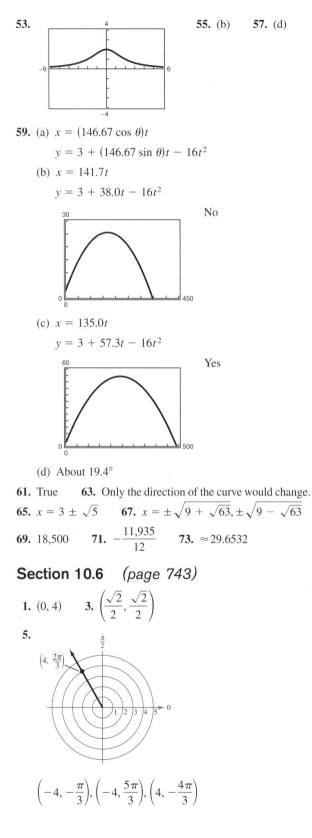

55. (b)   **57.** (d)

59. (a) $x = (146.67 \cos \theta)t$
$y = 3 + (146.67 \sin \theta)t - 16t^2$

(b) $x = 141.7t$
$y = 3 + 38.0t - 16t^2$

No

(c) $x = 135.0t$
$y = 3 + 57.3t - 16t^2$

Yes

(d) About $19.4°$

61. True   **63.** Only the direction of the curve would change.

65. $x = 3 \pm \sqrt{5}$ **67.** $x = \pm\sqrt{9 + \sqrt{63}}, \pm\sqrt{9 - \sqrt{63}}$

69. 18,500   **71.** $-\dfrac{11{,}935}{12}$ **73.** ≈ 29.6532

Section 10.6 *(page 743)*

1. $(0, 4)$ **3.** $\left(\dfrac{\sqrt{2}}{2}, \dfrac{\sqrt{2}}{2}\right)$

5.

$$\left(-4, -\dfrac{\pi}{3}\right), \left(-4, \dfrac{5\pi}{3}\right), \left(4, -\dfrac{4\pi}{3}\right)$$

7.

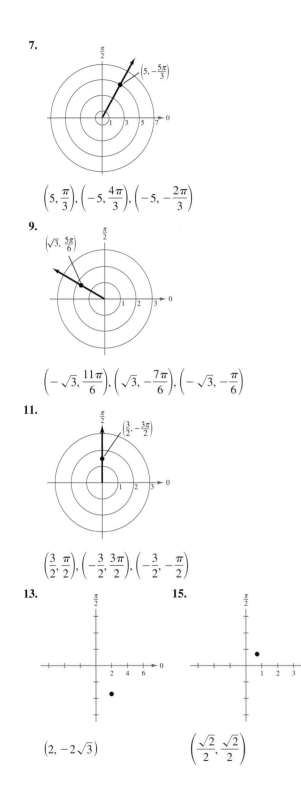

$\left(5, \dfrac{\pi}{3}\right), \left(-5, \dfrac{4\pi}{3}\right), \left(-5, -\dfrac{2\pi}{3}\right)$

9.

$\left(-\sqrt{3}, \dfrac{11\pi}{6}\right), \left(\sqrt{3}, -\dfrac{7\pi}{6}\right), \left(-\sqrt{3}, -\dfrac{\pi}{6}\right)$

11.

$\left(\dfrac{3}{2}, \dfrac{\pi}{2}\right), \left(-\dfrac{3}{2}, \dfrac{3\pi}{2}\right), \left(-\dfrac{3}{2}, -\dfrac{\pi}{2}\right)$

13.   **15.**

$\left(2, -2\sqrt{3}\right)$ $\left(\dfrac{\sqrt{2}}{2}, \dfrac{\sqrt{2}}{2}\right)$

17.   **19.**

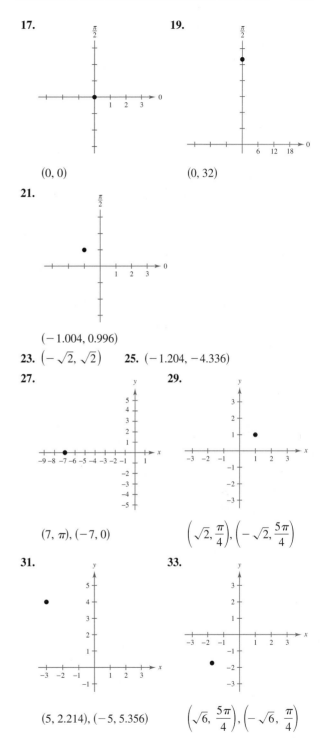

$(0, 0)$ $(0, 32)$

21.

$(-1.004, 0.996)$

23. $\left(-\sqrt{2}, \sqrt{2}\right)$ **25.** $(-1.204, -4.336)$

27.   **29.**

$(7, \pi), (-7, 0)$ $\left(\sqrt{2}, \dfrac{\pi}{4}\right), \left(-\sqrt{2}, \dfrac{5\pi}{4}\right)$

31.   **33.**

$(5, 2.214), (-5, 5.356)$ $\left(\sqrt{6}, \dfrac{5\pi}{4}\right), \left(-\sqrt{6}, \dfrac{\pi}{4}\right)$

35.

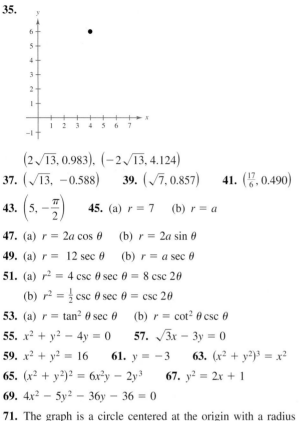

$\left(2\sqrt{13}, 0.983\right), \left(-2\sqrt{13}, 4.124\right)$

37. $\left(\sqrt{13}, -0.588\right)$ **39.** $\left(\sqrt{7}, 0.857\right)$ **41.** $\left(\frac{17}{6}, 0.490\right)$

43. $\left(5, -\dfrac{\pi}{2}\right)$ **45.** (a) $r = 7$ (b) $r = a$

47. (a) $r = 2a\cos\theta$ (b) $r = 2a\sin\theta$

49. (a) $r = 12\sec\theta$ (b) $r = a\sec\theta$

51. (a) $r^2 = 4\csc\theta\sec\theta = 8\csc 2\theta$

(b) $r^2 = \frac{1}{2}\csc\theta\sec\theta = \csc 2\theta$

53. (a) $r = \tan^2\theta\sec\theta$ (b) $r = \cot^2\theta\csc\theta$

55. $x^2 + y^2 - 4y = 0$ **57.** $\sqrt{3}x - 3y = 0$

59. $x^2 + y^2 = 16$ **61.** $y = -3$ **63.** $(x^2 + y^2)^3 = x^2$

65. $(x^2 + y^2)^2 = 6x^2y - 2y^3$ **67.** $y^2 = 2x + 1$

69. $4x^2 - 5y^2 - 36y - 36 = 0$

71. The graph is a circle centered at the origin with a radius of 3; $x^2 + y^2 = 9$.

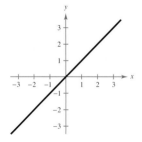

73. The graph consists of all points on the line that make an angle of $\pi/4$ with the positive x-axis; $x - y = 0$.

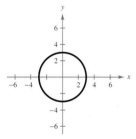

75. The graph is not evident by simple inspection; $x - 3 = 0$.

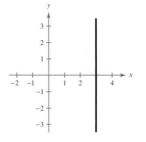

77. True. Because r is a directed distance, (r, θ) can be represented by $(-r, \theta \pm (2n + 1)\pi)$, so $|r| = |-r|$.

79. (a) Answers will vary.

(b) The points lie on a line.

$$d = \sqrt{r_1{}^2 + r_2{}^2 - 2r_1r_2} = |r_1 - r_2|$$

(c) $d = \sqrt{r_1{}^2 + r_2{}^2}$ (Pythagorean Theorem)

(d) Answers will vary. The distance formulas should give the same results.

81. $(2, 3)$ **83.** $\left(\frac{8}{7}, \frac{88}{35}, \frac{8}{5}\right)$ **85.** 175,000

87. $-101,376$

Section 10.7 *(page 752)*

1. Rose curve **3.** Circle **5.** Rose curve

7. Polar axis **9.** $\theta = \dfrac{\pi}{2}$ **11.** $\theta = \dfrac{\pi}{2}$

13. $\theta = \dfrac{\pi}{2}$, polar axis, pole **15.** Pole

17. Maximum: $|r| = 20$ when $\theta = \dfrac{3\pi}{2}$

Zero: $r = 0$ when $\theta = \dfrac{\pi}{2}$

19. Maximum: $|r| = 4$ when $\theta = 0, \dfrac{\pi}{3}, \dfrac{2\pi}{3}, \pi$

Zeros: $r = 0$ when $\theta = \dfrac{\pi}{6}, \dfrac{\pi}{2}, \dfrac{5\pi}{6}$

21.   **23.**

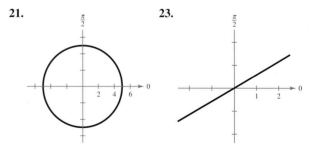

25.

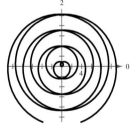

27.

29.

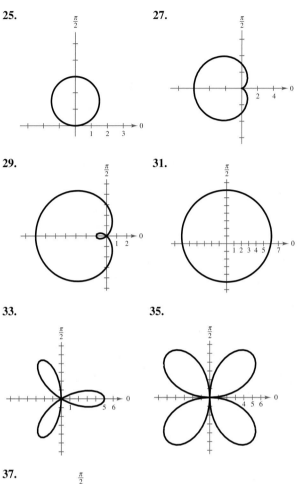

31.

33.

35.

37.

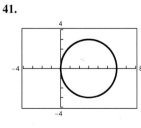

39.

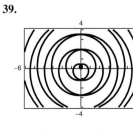

$-10\pi \leq \theta < 10\pi$

41.

$0 \leq \theta < \pi$

43.

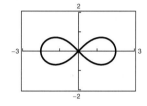

$0 \leq \theta < 2\pi$

45.

$0 \leq \theta < 2\pi$

47.

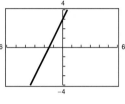

$0 \leq \theta < \dfrac{\pi}{2}$

49.

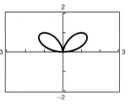

$-2\pi \leq \theta < 2\pi$

51.

$0 \leq \theta < \pi$

53.

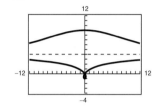

$0 \leq \theta < 2\pi$

55.

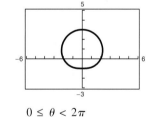

$0 \leq \theta < 2\pi$

57.

$0 \leq \theta < 2\pi$

59.

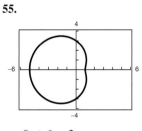

$0 \leq \theta < 4\pi$

61.

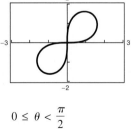

$0 \leq \theta < \dfrac{\pi}{2}$

63.

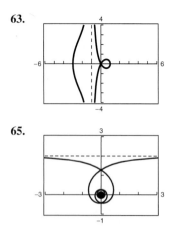

65.

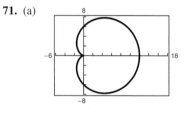

67. False. When the coordinates are substituted into the equation you get $6 = 4$, which is a false statement.

69. False. The rose curve will have 5 petals.

71. (a)

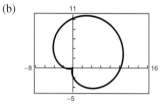

(b)

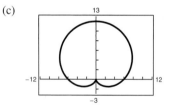

(c)

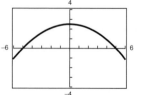

$r = 6(1 + \sin \theta)$

ϕ rotates the graph around the pole.

73. (a), (b), and (c) Answers will vary.

75. (a) $r = 4 \sin\left(\theta - \dfrac{\pi}{6}\right) \cos\left(\theta - \dfrac{\pi}{6}\right)$

(b) $r = -4 \sin \theta \cos \theta$

(c) $r = 4 \sin\left(\theta - \dfrac{2\pi}{3}\right) \cos\left(\theta - \dfrac{2\pi}{3}\right)$

(d) $r = 4 \sin \theta \cos \theta$

77.

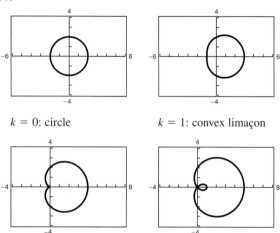

$k = 0$: circle $k = 1$: convex limaçon

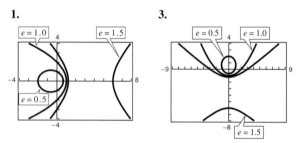

$k = 2$: cardioid $k = 3$: limaçon with inner loop

79. 2, 5, 8, 11, 14; $d = 3$; $a_n = -1 + 3n$

81. 150, 132, 114, 96, 78; $d = -18$; $a_n = 168 - 18n$

83. 840   **85.** 7860   **87.** $\frac{8}{7}$

Section 10.8   *(page 758)*

1.

3.

5. (b)   **7.** (d)   **9.** Parabola   **11.** Ellipse

13. Ellipse   **15.** Hyperbola   **17.** Hyperbola

19.

21.

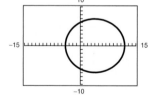

Parabola   Ellipse

23.

25.

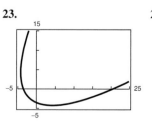

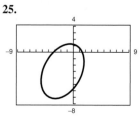

27. $r = \dfrac{1}{1 - \cos\theta}$ **29.** $r = \dfrac{1}{2 + \sin\theta}$

31. $r = \dfrac{2}{1 + 2\cos\theta}$ **33.** $r = \dfrac{2}{1 - \sin\theta}$

35. $r = \dfrac{16}{5 + 3\cos\theta}$ **37.** $r = \dfrac{9}{4 - 5\sin\theta}$

39. Answers will vary.

41. $r = \dfrac{9.2931 \times 10^7}{1 - 0.0167\cos\theta}$

Perihelion: 9.1405×10^7 miles

Aphelion: 9.4509×10^7 miles

43. $r = \dfrac{7977.2}{1 - 0.937\cos\theta}$

11,008 miles

45. False. The equation can be rewritten as

$r = \dfrac{-4/3}{1 + \sin\theta}$.

Because ep is negative, you know that e is positive and p is negative and that p represents the distance between the pole and the directrix, so the directrix has to be below the pole.

47. Answers will vary.   **49.** $r^2 = \dfrac{24{,}336}{169 - 25\cos^2\theta}$

51. 60   **53.** 90,720   **55.** 220   **57.** 720

Review Exercises *(page 761)*

1. Hyperbola   **3.** $(x - 4)^2 = -8(y - 2)$

5. $(y - 2)^2 = 12x$ **7.** $2x + y - 2 = 0;\ (1, 0)$

9. About 19.6 meters   **11.** $\dfrac{(x - 2)^2}{25} + \dfrac{y^2}{21} = 1$

13. $\dfrac{(x - 2)^2}{4} + (y - 1)^2 = 1$ **15.** 3 feet atop the pillars

17. Center: $(1, -4)$ **19.** Center: $(-2, 1)$

Vertices: Vertices:

$(1, 0), (1, -8)$ $(-2, 11), (-2, -9)$

Foci: $\left(1, -4 \pm \sqrt{7}\right)$ Foci: $\left(-2, 1 \pm \sqrt{19}\right)$

Eccentricity: $\dfrac{\sqrt{7}}{4}$ Eccentricity: $\dfrac{\sqrt{19}}{10}$

21. $\dfrac{(x + 2)^2}{64} - \dfrac{(y - 3)^2}{36} = 1$ **23.** $\dfrac{5(x - 4)^2}{16} - \dfrac{5y^2}{64} = 1$

25. Center: $(1, -1)$

Vertices: $(5, -1), (-3, -1)$

Foci: $(6, -1), (-4, -1)$

Asymptotes:

$y = -1 \pm \frac{3}{4}(x - 1)$

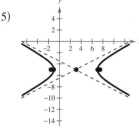

27. Center: $(3, -5)$

Vertices: $(7, -5), (-1, -5)$

Foci: $\left(3 \pm 2\sqrt{5}, -5\right)$

Asymptotes:

$y = -5 \pm \frac{1}{2}(x - 3)$

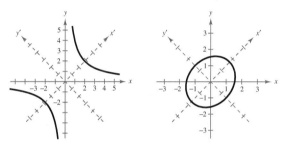

29. 72 miles   **31.** Ellipse

33. $\dfrac{(x')^2}{8} - \dfrac{(y')^2}{8} = 1$ **35.** $\dfrac{(x')^2}{3} + \dfrac{(y')^2}{2} = 1$

37. (a) Parabola

(b) $y = \dfrac{8x - 5 \pm \sqrt{(8x - 5)^2 - 4(16x^2 - 10x)}}{2}$

(c)

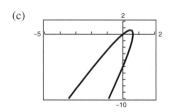

39. (a) Parabola

(b) $y = $

$\dfrac{-\left(2x - 2\sqrt{2}\right) \pm \sqrt{\left(2x - 2\sqrt{2}\right)^2 - 4\left(x^2 + 2\sqrt{2}x + 2\right)}}{2}$

(c)

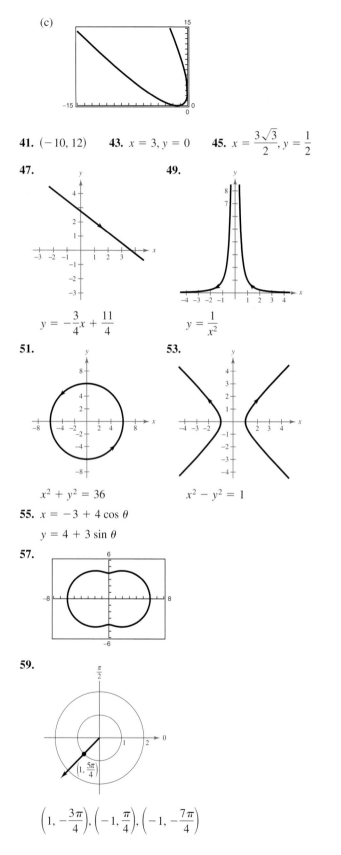

41. $(-10, 12)$ **43.** $x = 3, y = 0$ **45.** $x = \dfrac{3\sqrt{3}}{2}, y = \dfrac{1}{2}$

47.

$y = -\dfrac{3}{4}x + \dfrac{11}{4}$

49.

$y = \dfrac{1}{x^2}$

51.

$x^2 + y^2 = 36$

53.

$x^2 - y^2 = 1$

55. $x = -3 + 4\cos\theta$
  $y = 4 + 3\sin\theta$

57.

59.

$\left(1, \dfrac{5\pi}{4}\right)$

$\left(1, -\dfrac{3\pi}{4}\right), \left(-1, \dfrac{\pi}{4}\right), \left(-1, -\dfrac{7\pi}{4}\right)$

61.

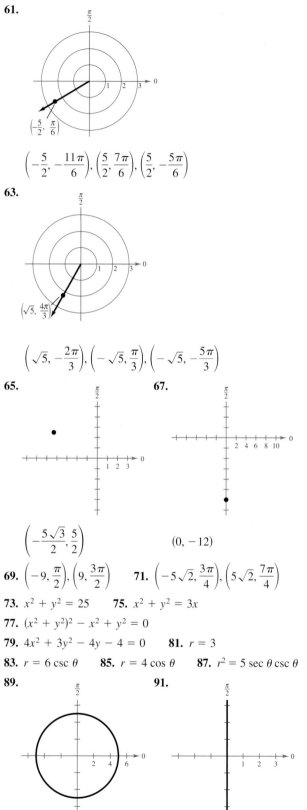

$\left(-\dfrac{5}{2}, \dfrac{\pi}{6}\right)$

$\left(-\dfrac{5}{2}, -\dfrac{11\pi}{6}\right), \left(\dfrac{5}{2}, \dfrac{7\pi}{6}\right), \left(\dfrac{5}{2}, -\dfrac{5\pi}{6}\right)$

63.

$\left(\sqrt{5}, \dfrac{4\pi}{3}\right)$

$\left(\sqrt{5}, -\dfrac{2\pi}{3}\right), \left(-\sqrt{5}, \dfrac{\pi}{3}\right), \left(-\sqrt{5}, -\dfrac{5\pi}{3}\right)$

65.   **67.**

$\left(-\dfrac{5\sqrt{3}}{2}, \dfrac{5}{2}\right)$ $(0, -12)$

69. $\left(-9, \dfrac{\pi}{2}\right), \left(9, \dfrac{3\pi}{2}\right)$ **71.** $\left(-5\sqrt{2}, \dfrac{3\pi}{4}\right), \left(5\sqrt{2}, \dfrac{7\pi}{4}\right)$

73. $x^2 + y^2 = 25$ **75.** $x^2 + y^2 = 3x$

77. $(x^2 + y^2)^2 - x^2 + y^2 = 0$

79. $4x^2 + 3y^2 - 4y - 4 = 0$ **81.** $r = 3$

83. $r = 6\csc\theta$ **85.** $r = 4\cos\theta$ **87.** $r^2 = 5\sec\theta\csc\theta$

89.   **91.**

93.

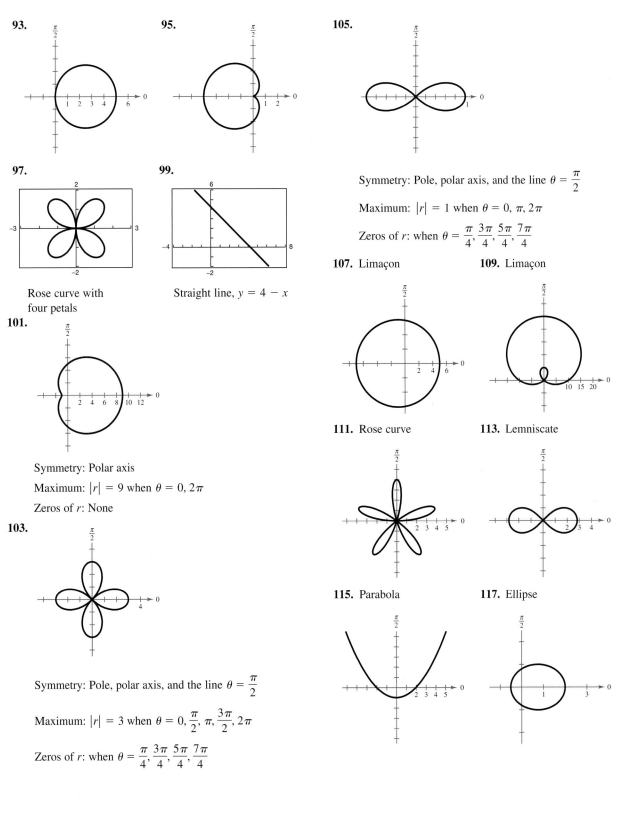

95.

105.

Symmetry: Pole, polar axis, and the line $\theta = \dfrac{\pi}{2}$

Maximum: $|r| = 1$ when $\theta = 0, \pi, 2\pi$

Zeros of r: when $\theta = \dfrac{\pi}{4}, \dfrac{3\pi}{4}, \dfrac{5\pi}{4}, \dfrac{7\pi}{4}$

107. Limaçon

109. Limaçon

97.

99.

Rose curve with four petals

Straight line, $y = 4 - x$

111. Rose curve

113. Lemniscate

101.

Symmetry: Polar axis

Maximum: $|r| = 9$ when $\theta = 0, 2\pi$

Zeros of r: None

115. Parabola

117. Ellipse

103.

Symmetry: Pole, polar axis, and the line $\theta = \dfrac{\pi}{2}$

Maximum: $|r| = 3$ when $\theta = 0, \dfrac{\pi}{2}, \pi, \dfrac{3\pi}{2}, 2\pi$

Zeros of r: when $\theta = \dfrac{\pi}{4}, \dfrac{3\pi}{4}, \dfrac{5\pi}{4}, \dfrac{7\pi}{4}$

119. Ellipse

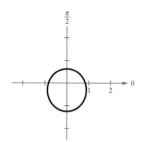

121. $r = 16 \sin \theta$ **123.** $r = \dfrac{5}{3 - 2 \cos \theta}$

125. $r = \dfrac{1.512}{1 - 0.092 \cos \theta}$

Perihelion: 1.3847 astronomical units

Aphelion: 1.6653 astronomical units

127. False. The following are two sets of parametric equations for the line.

$x = t, \quad y = 3 - 2t$

$x = 3t, \quad y = 3 - 6t$

129. (a) Vertical translation

(b) Horizontal translation

(c) Reflection in the y-axis

(d) Parabola opens more slowly.

131. 5; The ellipse becomes more circular and approaches a circle of radius 5.

Chapter Test   *(page 766)*

1.

Vertex: $(0, 0)$

Focus: $(2, 0)$

2.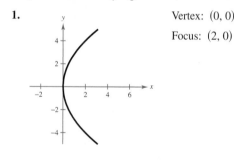

Vertices: $(3, 12), (3, -8)$

Foci: $\left(3, 2 \pm \sqrt{19}\right)$

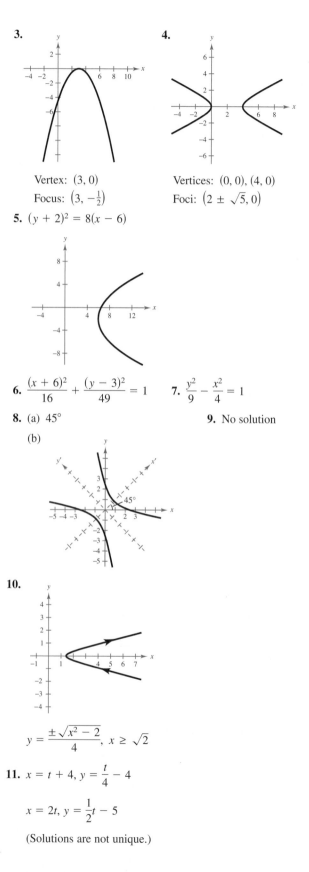

3.

Vertex: $(3, 0)$

Focus: $\left(3, -\frac{1}{2}\right)$

5. $(y + 2)^2 = 8(x - 6)$

4.

Vertices: $(0, 0), (4, 0)$

Foci: $\left(2 \pm \sqrt{5}, 0\right)$

6. $\dfrac{(x + 6)^2}{16} + \dfrac{(y - 3)^2}{49} = 1$ **7.** $\dfrac{y^2}{9} - \dfrac{x^2}{4} = 1$

8. (a) $45°$

(b)

9. No solution

10.

$y = \dfrac{\pm \sqrt{x^2 - 2}}{4}, \; x \geq \sqrt{2}$

11. $x = t + 4, \; y = \dfrac{t}{4} - 4$

$x = 2t, \; y = \dfrac{1}{2}t - 5$

(Solutions are not unique.)

12. $\left(-7, 7\sqrt{3}\right)$

13. $\left(2\sqrt{2}, \frac{5\pi}{4}\right), \left(-2\sqrt{2}, \frac{\pi}{4}\right), \left(2\sqrt{2}, -\frac{3\pi}{4}\right)$

14. $r = 12\sin\theta$

15.

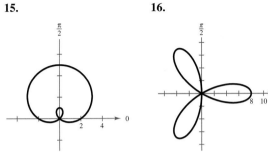

Limaçon

16.

Rose curve

17.

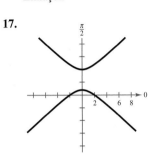

Hyperbola

18. $r = \dfrac{1}{1 + \frac{1}{4}\sin\theta}$

19. $r = \dfrac{34{,}478{,}231}{1 + 0.2056\sin\theta}$

Perihelion: 28,598,400 miles

Aphelion: 43,401,600 miles

Cumulative Test for Chapters 7–10
(page 767)

1. $(4, -3)$ **2.** $\left(\frac{3}{5}, -4, -\frac{1}{5}\right)$

3. $\begin{bmatrix} -7 & -10 & -16 \\ -6 & 18 & 9 \\ -12 & 16 & 7 \end{bmatrix}$ **4.** $\begin{bmatrix} -18 & 15 & -14 \\ 28 & 11 & 34 \\ -20 & 52 & -1 \end{bmatrix}$

5. $\begin{bmatrix} 3 & -31 & 2 \\ 22 & 18 & 6 \\ 52 & -40 & 14 \end{bmatrix}$ **6.** $\begin{bmatrix} 5 & 36 & 31 \\ -36 & 12 & -36 \\ 16 & 0 & 18 \end{bmatrix}$

7. (a) $\begin{bmatrix} -175 & 37 & -13 \\ 95 & -20 & 7 \\ 14 & -3 & 1 \end{bmatrix}$ (b) 1

8. $(4, -2, 1)$ **9.** 22 square units   **10.** 1110

11. 135   **12.** $\frac{47}{52}$ **13.** ≈ 34.480 **14.** 80   **15.** 3685

16. Answers will vary.

17. $x^6 - 12x^5y + 60x^4y^2 - 160x^3y^3 + 240x^2y^4$
$\qquad - 192xy^5 + 64y^6$

18. 151,200

19. Hyperbola

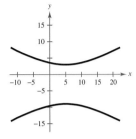

20. Ellipse

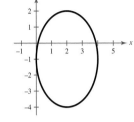

21. Hyperbola

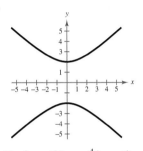

22. Circle

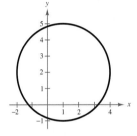

23. $(x - 2)^2 = -\frac{4}{3}(y - 3)$

24. $\dfrac{(x - 1)^2}{25} + \dfrac{(y - 4)^2}{4} = 1$ **25.** $\dfrac{(y + 4)^2}{4} - \dfrac{x^2}{16/3} = 1$

26. $\dfrac{(y - 2)^2}{4/5} - \dfrac{x^2}{16/5} = 1$

27.

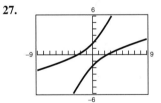

$\theta \approx 38°$

28. $(0, 6), (12, 6)$

29.

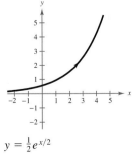

$y = \frac{1}{2}e^{x/2}$

30.

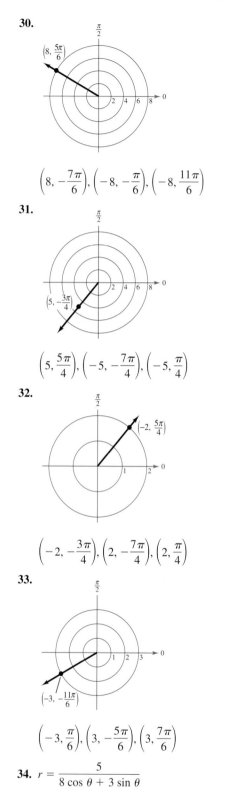

$$\left(8, -\frac{7\pi}{6}\right), \left(-8, -\frac{\pi}{6}\right), \left(-8, \frac{11\pi}{6}\right)$$

31.

$$\left(5, \frac{5\pi}{4}\right), \left(-5, -\frac{7\pi}{4}\right), \left(-5, \frac{\pi}{4}\right)$$

32.

$$\left(-2, -\frac{3\pi}{4}\right), \left(2, -\frac{7\pi}{4}\right), \left(2, \frac{\pi}{4}\right)$$

33.

$$\left(-3, \frac{\pi}{6}\right), \left(3, -\frac{5\pi}{6}\right), \left(3, \frac{7\pi}{6}\right)$$

34. $r = \dfrac{5}{8\cos\theta + 3\sin\theta}$

35. $9x^2 - 16y^2 + 20x + 4 = 0$

36.

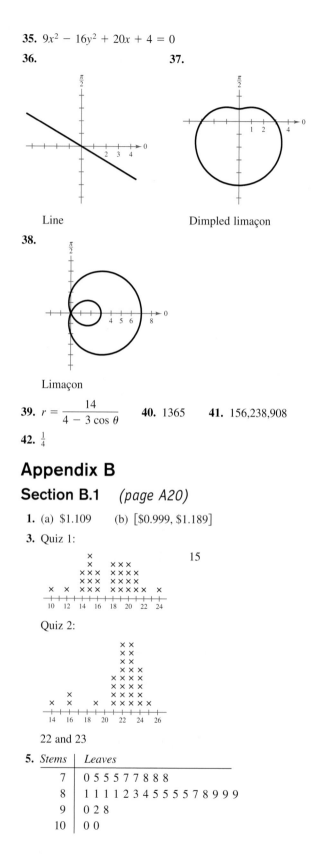

Line

37.

Dimpled limaçon

38.

Limaçon

39. $r = \dfrac{14}{4 - 3\cos\theta}$ **40.** 1365   **41.** 156,238,908

42. $\frac{1}{4}$

Appendix B

Section B.1   *(page A20)*

1. (a) $1.109   (b) [$0.999, $1.189]

3. Quiz 1:

×
×   × × ×
× × ×   × × × ×
× × ×   × × × ×
×   ×   × × ×   × × × × ×   ×

10   12   14   16   18   20   22   24

15

Quiz 2:

× ×
× ×
× ×
× × ×
× × × ×
× × × ×
×   ×   × × × ×
×   ×   ×   × × × × ×

14   16   18   20   22   24   26

22 and 23

5.

Stems	Leaves
7	0 5 5 5 7 7 8 8 8
8	1 1 1 1 2 3 4 5 5 5 5 7 8 9 9 9
9	0 2 8
10	0 0

7.

Stems	Leaves
0	89 66 67 65 80 98 62 93
1	09 01 46 24 96 90
2	92 55 40 61
3	68 35
4	12 96 80 38
5	81 18 50 70 66
6	44 00 34 01
7	61 66 00
8	11 57 41 90
9	
10	
11	60 59 33
12	92
13	19 17 37
27	22
31	32
46	80
65	14

9. (a)

Interval	Tally
90–109	卌
110–129	卌 卌 |
130–149	卌 ||||
150–169	卌 |
170–189	|
190–209	|||

(b)

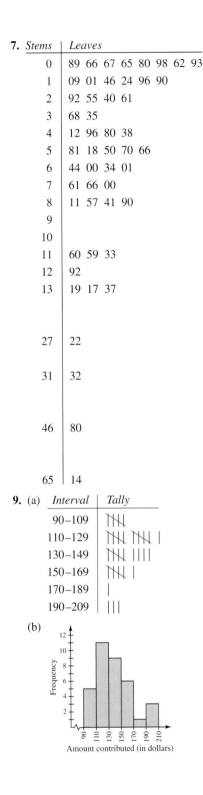

11. (a)

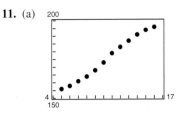

(b) Answers will vary.   (c) $y = 4x + 134$

(d) 2000: 214; 2002: 222; 2005: 234

13. (a) $P = 21.91t + 447$

(b)

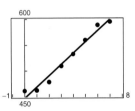

The model fits the data well.

(c)

Year	1990	1991	1992	1993
P (actual)	461.4	462.4	478.6	509.5
P (model)	447	468.9	490.8	512.7

Year	1994	1995	1996	1997
P (actual)	532.9	559.9	589.1	595.6
P (model)	534.6	556.6	578.5	600.4

Section B.2   *(page A28)*

1. Mean: 8.86; median: 8; mode: 7

3. Mean: 10.29; median: 8; mode: 7

5. Mean: 9; median: 8; mode: 7

7. The mean is sensitive to extreme values.

9. Mean: $67.14; median: $65.35

11. Mean: 3.07; median: 3; mode: 3

13. One possibility: $\{4, 4, 10\}$

15. The median gives the most representative description.

17. $\bar{x} = 6$, $v = 10$, $\sigma = 3.16$

19. $\bar{x} = 2$, $v = \frac{4}{3}$, $\sigma = 1.15$ **21.** $\bar{x} = 4$, $v = 4$, $\sigma = 2$

23. $\bar{x} = 47$, $v = 226$, $\sigma = 15.03$ **25.** 3.42

27. 101.55   **29.** 1.65

31. (a) $\bar{x} = 12$; $\sigma = 2.83$ (b) $\bar{x} = 20$; $\sigma = 2.83$

(c) $\bar{x} = 12$; $\sigma = 1.41$ (d) $\bar{x} = 9$; $\sigma = 1.41$

33. $\bar{x} = 12$ and $|x_i - 12| = 8$ for all x_i

35. It will increase the mean by 5, but the standard deviation will not change.

37. First histogram

39. **41.**

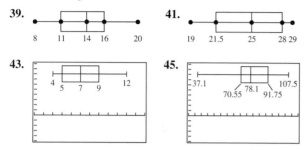

43. **45.**

Appendix C *(page A35)*

1. 3   **3.** 7   **5.** 4   **7.** 20   **9.** 4   **11.** 3

13. $\frac{4}{3}$ **15.** 20   **17.** No solution   **19.** $x < 2$

21. $x < 9$ **23.** $x \leq -14$ **25.** $x > 10$

27. $x < 4$ **29.** $x < 3$ **31.** $x \geq 2$ **33.** $x \leq -5$

35. $x < 6$

Index of Applications

Biology and Life Sciences

Atmospheric pressure, 237, 271
Bacteria growth, 227, 268
Blood pressure, 332
Concentration of medicine, 206
Cost of recycling, 196, 302
Cost of removing pollution, 196
Deer population, 197
Endangered species, 269
Fish population, 212
Forensics, 271
Forest yield, 256
Growth of a red oak tree, 158
Gypsy moth defoliation, 227, 607
Human memory model, 237, 245, 256, 277
Learning curve of factory employees, 268
Lynx population, 87
Moth food consumption, 198
Number of board feet in a log, 146, 497
Predator-prey model, 342
Relative humidity measurement, 65
Reproduction rates of deer, 523
Respiratory cycle, 331
Species population in a new habitat, 280, 302
Sunset times, 365
Weather forecast, 692
Width of a river, 312, 369, 479

Business

Advertising costs, 173
Advertising costs and maximum profit, 145, 209
Advertising costs and sales, 268
Average cost, 86, 206, 212
Average sales, 366
Break-even point, 64, 496, 546, 614
Bristol-Myers Squibb Company, 278
Business loan, 521, 565
Change in annual sales, 35
Change in daily revenue, 35
Coca-Cola Enterprises, Inc., 646
Colgate-Palmolive Company, 35
Compact disc sales, 257
Cost of operating a bulldozer, 36
Cost of Super Bowl advertising, 11
Demand for a product, 53, 226, 256, 277, 507

Depreciation, 23, 36, 228, 268, 270, 275, 280, 645, 689
Estimated sales, 10
Furniture production, 533
Inventory costs, 548
Inventory levels, 533, 580
Kellogg Company, 35
Labor/wage requirements, 580
Lands' End, Inc., 13
Maximum and minimum sales, 331, 371
Maximum profit, 145, 543
Maximum revenue, 543, 548
Minimum cost, 145, 543, 548
Ordering and transportation costs, 206
Partnership costs, 466
Point of diminishing returns, 158
Procter & Gamble Company, 24, 627
Product sales, 268
Production costs, 118
Production levels, 579, 580, 612
Profit, 85, 86, 98, 107, 131, 134, 146, 188
Progressive Corporation, 67
Restaurant sales, 117
Revenue, 465, 580, 702
Sales contacts, 662, 682, 685
Sales of seasonal products, 321, 343, 402
Sbarro, Inc., 67
Starbucks Corp., 13
Supply and demand, 534, 546, 550
Truck scheduling, 507, 522
Wal-Mart Stores, Inc., 11, 627
Whirlpool Corporation, 688

Chemistry and Physics

Boiling and freezing temperatures, 237
Braking force, 465, 482
Buoy oscillating, 365, 372
Cable tension, 455, 481
Current in an electrical network, 301, 522, 565
Design of a guitar, 65
Determining an unknown mass, 197
Engine horsepower, 272
Exhaust temperature of a diesel engine, 129, 523
Explorer 18 Satellite, 759
Fahrenheit and Celsius temperatures, 36

Fluid flow, 99
Force of water against a dam, 86
Forces, 456
Friction, 391
Harmonic motion, 301, 343, 364, 402, 409, 424
Horizontal and vertical components of velocity, 455
Hyperbolic telescope mirror, 721
Impedance of an electrical circuit, 181
Index of refraction, 426
Loaded barge towed by two tug boats, 455
Magnitude of earthquakes, 269, 276, 278
Mixture problems, 181, 205, 506, 522, 533, 547, 614, 647
Motion of oscillating weight, 332, 343
Orbit of comet Encke, 711
Orbit of Halley's comet, 711
Orbit of Mars, 764
Orbit of Mercury, 766
pH of a substance, 269
Planetary motion, 759
Position of the sun, 402
Pulley system, 522
Radioactive decay, 227, 267, 268, 278
Resistance of copper wire, 24
Resultant force, 454, 455, 481, 484, 486
Rope tension, 481
Satellite antenna, 702
Satellite orbit, 703, 712
Sound intensity, 238, 245, 269
Sound location, 721, 762
Springs, 365
Standing waves, 409
Temperature change, 227, 245, 256
Temperature of saturated steam, 53
Tuning fork, 364
Two people lifting a weight, 455
Voltmeter, 293
Wind drag, 107, 146
Work, 237, 465, 482

Construction

Beam deflection, 159
Brick pattern, 636
Bridge design across a lake, 434
Circular saw, 294
Design of a fireplace arch, 711

A169

Time and Distance

U.S. Demographics

Index

COMMON FORMULAS

Temperature

$$F = \frac{9}{5}C + 32$$

F = degrees Fahrenheit

C = degrees Celsius

Distance

$$d = rt$$

d = distance traveled

t = time

r = rate

Simple Interest

$$I = Prt$$

I = interest

P = principal

r = annual interest rate

t = time in years

Compound Interest

$$A = P\left(1 + \frac{r}{n}\right)^{nt}$$

A = balance

P = principal

r = annual interest rate

n = compoundings per year

t = time in years

Coordinate Plane: Midpoint Formula

$$\left(\frac{x_1 + x_2}{2}, \frac{y_1 + y_2}{2}\right)$$

midpoint of line segment joining (x_1, y_1) and (x_2, y_2)

Coordinate Plane: Distance Formula

$$d = \sqrt{(x_2 - x_1)^2 + (y_2 - y_1)^2}$$

d = distance between points (x_1, y_1) and (x_2, y_2)

Quadratic Formula

If $p(x) = ax^2 + bx + c, a \neq 0$ and $b^2 - 4ac \geq 0$, then the real zeros of p are

$$x = \frac{-b \pm \sqrt{b^2 - 4ac}}{2a}.$$

CONVERSIONS

Length and Area

1 foot = 12 inches
1 mile = 5280 feet
1 kilometer = 1000 meters
1 kilometer ≈ 0.621 mile
1 meter ≈ 3.281 feet
1 foot ≈ 0.305 meter

1 yard = 3 feet
1 mile = 1760 yards
1 meter = 100 centimeters
1 mile ≈ 1.609 kilometers
1 meter ≈ 39.370 inches
1 foot ≈ 30.480 centimeters

1 meter = 1000 millimeters
1 centimeter ≈ 0.394 inch
1 inch ≈ 2.540 centimeters
1 acre = 4840 square yards
1 square mile = 640 acres

Volume

1 gallon = 4 quarts
1 gallon = 231 cubic inches
1 liter = 1000 milliliters
1 liter ≈ 1.057 quarts
1 gallon ≈ 3.785 liters

1 quart = 2 pints
1 gallon ≈ 0.134 cubic foot
1 liter = 100 centiliters
1 liter ≈ 0.264 gallon
1 quart ≈ 0.946 liter

1 pint = 16 fluid ounces
1 cubic foot ≈ 7.48 gallons

Weight and Mass on Earth

1 ton = 2000 pounds
1 kilogram ≈ 2.205 pounds

1 pound = 16 ounces
1 pound ≈ 0.454 kilogram

1 kilogram = 1000 grams
1 gram ≈ 0.035 ounce

FORMULAS FROM GEOMETRY

Triangle

$h = a \sin \theta$

$\text{Area} = \dfrac{1}{2}bh$

Laws of Cosines:

$c^2 = a^2 + b^2 - 2ab \cos \theta$

Right Triangle

Pythagorean Theorem:

$c^2 = a^2 + b^2$

Equilateral Triangle

$h = \dfrac{\sqrt{3}s}{2}$

$\text{Area} = \dfrac{\sqrt{3}s^2}{4}$

Parallelogram

$\text{Area} = bh$

Trapezoid

$\text{Area} = \dfrac{h}{2}(a + b)$

Circle

$\text{Area} = \pi r^2$

$\text{Circumference} = 2\pi r$

Sector of Circle

$\text{Area} = \dfrac{\theta r^2}{2}$

$s = r\theta$

(θ in radians)

Circular Ring

$\text{Area} = \pi(R^2 - r^2)$

$\qquad = 2\pi p w$

(p = average radius,

w = width of ring)

Ellipse

$\text{Area} = \pi ab$

$\text{Circumference} \approx 2\pi \sqrt{\dfrac{a^2 + b^2}{2}}$

Cone

(A = area of base)

$\text{Volume} = \dfrac{Ah}{3}$

Right Circular Cone

$\text{Volume} = \dfrac{\pi r^2 h}{3}$

$\text{Lateral Surface Area} = \pi r \sqrt{r^2 + h^2}$

Frustum of Right Circular Cone

$\text{Volume} = \dfrac{\pi(r^2 + rR + R^2)h}{3}$

$\text{Lateral Surface Area} = \pi s(R + r)$

Right Circular Cylinder

$\text{Volume} = \pi r^2 h$

$\text{Lateral Surface Area} = 2\pi rh$

Sphere

$\text{Volume} = \dfrac{4}{3}\pi r^3$

$\text{Surface Area} = 4\pi r^2$

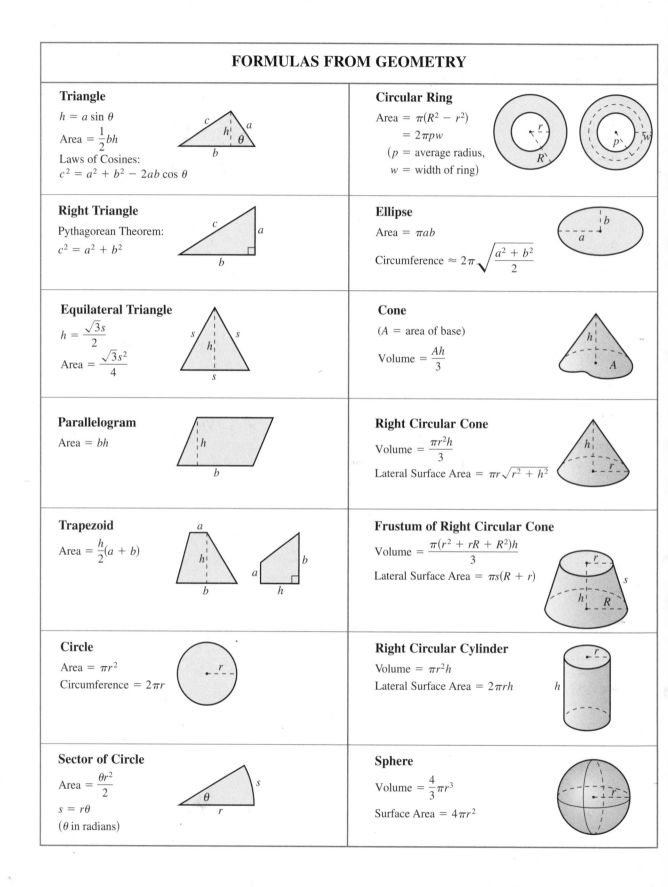